Systems Biology of Cancer

With over 200 types of cancer diagnosed to date, researchers the world over have been forced to rapidly update their understanding of the biology of cancer. In fact, only the study of the basic cellular processes, and how these are altered in cancer cells, can ultimately provide a background for rational therapies.

Bringing together the state-of-the-art contributions of international experts, *Systems Biology of Cancer* proposes an ultimate research goal for the whole scientific community: exploiting systems biology to generate in-depth knowledge based on blueprints that are unique to each type of cancer.

Readers are provided with a realistic view of what is known and what is yet to be uncovered on the aberrations in the fundamental biological processes, deregulation of major signaling networks, alterations in major cancers, and the strategies for using the scientific knowledge for effective diagnosis, prognosis, and drug discovery to improve public health.

Sam Thiagalingam is an Associate Professor of Genetics & Genomics, Medicine, and Pathology & Laboratory Medicine at the Boston University School of Medicine. He played a major role in establishing an association between genomic instability and loss of heterozygosity (LOH) in human cancers. He was the first to show that SMAD4 inactivation is a critical event during the late stages of colon cancer progression, and sustained TGFβ signaling events are required to maintain epigenetic memory during breast cancer progression. Dr. Thiagalingam also proposed a simple minded multi-modular molecular network (MMMN) cancer progression model as a road map to visualize the various gene alterations in modules of networks of pathways. His long-term goal is to identify novel cancer biomarkers and therapeutic targets by contributing to the "big picture" of interconnected networks of events that mediate cancer progression to metastasis using breast and colon cancers as the model systems.

Systems Biology of Cancer

Edited by

Sam Thiagalingam
Boston University School of Medicine

CAMBRIDGE
UNIVERSITY PRESS

University Printing House, Cambridge CB2 8BS, United
Kingdom

Cambridge University Press is part of the University of
Cambridge.

It furthers the University's mission by disseminating
knowledge in the pursuit of education, learning and
research at the highest international levels of excellence.

www.cambridge.org
Information on this title: www.cambridge.org/
9780521493390

© Cambridge University Press 2015

This publication is in copyright. Subject to statutory
exception and to the provisions of relevant collective
licensing agreements, no reproduction of any part may take
place without the written permission of Cambridge
University Press.

First published 2015

Printed in the United Kingdom by T J International Ltd.
Padstow Cornwall

*A catalogue record for this publication is available from the
British Library*

Library of Congress Cataloguing in Publication data

ISBN 978-0-521-49339-0 Hardback

Cambridge University Press has no responsibility for the
persistence or accuracy of URLs for external or third-party
internet websites referred to in this publication, and does
not guarantee that any content on such websites is, or will
remain, accurate or appropriate.

....

Every effort has been made in preparing this book to
provide accurate and up-to-date information which is in
accord with accepted standards and practice at the time of
publication. Although case histories are drawn from actual
cases, every effort has been made to disguise the identities of
the individuals involved. Nevertheless, the authors, editors
and publishers can make no warranties that the information
contained herein is totally free from error, not least because
clinical standards are constantly changing through research
and regulation. The authors, editors and publishers
therefore disclaim all liability for direct or consequential
damages resulting from the use of material contained in this
book. Readers are strongly advised to pay careful attention
to information provided by the manufacturer of any drugs
or equipment that they plan to use.

To my parents

Vanniyan **Sambunathan Seenithamby Sisupalapillai Sambasivamoorthy**
and
Paramsothy Thangaretnaammal Malar Eliyathamby Sambasivamoorthy
for their righteous living and the respect for freedom of expression.

Contents

*The color plate section appears between pages 320
and 321.*

Contributors

Hamid M. Abdolmaleky
Biomedical Genetics
Boston University School of Medicine and
Nutrition/Metabolism Laboratory
Harvard Medical School
Boston, MA
USA

Cory Adamson
Department of Surgery – Neurosurgery
Duke University School of Medicine
Durham, NC
USA

Paola Allavena
Department of Inflammation and Immunology,
Humanitas Clinical and Research Center
Rozzano
Italy

Dimitrios Anastasiou
Division of Physiology and Metabolism
MRC National Institute for Medical Research
The Ridgeway, Mill Hill
London
UK

Johanna Apfel
Molecular Biology
Biotechnology and Biochemistry Group
University of Applied Sciences
Kaiserslautern
Germany

Surinder K. Batra
Department of Biochemistry and Molecular Biology
Eppley Cancer Institute
University of Nebraska Medical Center
Omaha, NE
USA

Mark E. Burkard
Department of Medicine – Hematology/Oncology
University of Wisconsin
Madison, WI
USA

Amancio Carnero
Instituto de Biomedicina de Sevilla
HUVR/Universidad de Sevilla/Consejo
Superior de Investigaciones Cientificas
Sevilla
Spain

Michael J. Clemens
Department of Biochemistry and Molecular Biology
School of Life Sciences
University of Sussex
Brighton
UK

Jeanette Gowen Cook
Department of Biochemistry & Biophysics
University of North Carolina School of Medicine
Chapel Hill, NC
USA

Isabel Dominguez
Hematology-Oncology Section
Department of Medicine
Boston University School of Medicine
Boston, MA
USA

Jeremy S. Edwards
Molecular Genetics and Microbiology
University of New Mexico School of Medicine and
Chemical and Nuclear Engineering
University of New Mexico
Albuquerque, NM
USA

Wafik S. El-Deiry
Fox Chase Cancer Center
Philadelphia, PA
USA

Androulla Elia
Biomedical Sciences
St. George's University of London
London
UK

Mohammad R. Eskandari
Nutrition/Metabolism Laboratory
Harvard Medical School
Boston, MA
USA

Aurora Esquela-Kerscher
Department of Microbiology & Molecular Cell
Biology
Leroy T. Canoles Jr. Cancer Research Center
Eastern Virginia Medical School
Norfolk, VA
USA

Manel Esteller
Cancer Epigenetics and Biology Program (PEBC)
Institut d'Investigació Biomèdica de Bellvitge (IDIBELL)
Barcelona, Catalonia
Spain

Rob M. Ewing
Centre for Biological Sciences
University of Southampton
Southampton
UK

Douglas V. Faller
Cancer Center
Boston University School of Medicine
Boston, MA
USA

Kristopher Frese
Cancer Research UK Cambridge Institute
University of Cambridge
Cambridge
UK

Xijin Ge
Department of Mathematics and Statistics
South Dakota State University
Department of Mathematics and Statistics
Brookings, SD
USA

Giovanni Germano
Department of Inflammation and Immunology
Humanitas Clinical and Research Center
Rozzano
Italy

Daniel A. Haber
Massachusetts General Hospital Cancer Center
Harvard Medical School
Charlestown, MA
USA

William C. Hahn
Department of Medical Oncology
Dana-Farber Cancer Institute
Harvard Medical School
Boston, MA
USA

Antoine Ho
Molecular Genetics and Microbiology
University of New Mexico School of
Medicine
Albuquerque, NM
USA

Christine Iacobuzio-Donahue
Department of Pathology
Johns Hopkins Sidney Kimmel Comprehensive
Cancer Center
Baltimore, MD
USA

Sergii Ivakhno
Cancer Research UK Cambridge Institute
University of Cambridge
Cambridge
UK

Prasad V. Jallepalli
Molecular Biology Program
Memorial Sloan-Kettering Cancer Center
New York, NY
USA

Rosanne Jones
Department of Pathology
Duke University Medical Center
Durham, NC
USA

Sharyn Katz
Department of Radiology
Hospital of the University of Pennsylvania
Philadelphia, PA
USA

Arnaud Krebs
Functional Genomics Department
Institut de Génétique et de Biologie Moléculaire et
Cellulaire (IGBMC)
Université de Strasbourg (UdS)
Strasbourg
France

Karl Krueger
Cancer Biomarkers Research Group
Division of Cancer Prevention
National Cancer Institute
Rockville, MD
USA

Arthur W. Lambert
Whitehead Institute for Biomedical Research
Massachusetts Institute of Technology
Cambridge, MA
USA

Adam Lerner
Hematology/Oncology Section
Departments of Medicine
Boston University School of Medicine
Boston, MA
USA

Holly Lewis
Department of Microbiology & Molecular Cell Biology
Leroy T. Canoles Jr. Cancer Research Center
Eastern Virginia Medical School
Norfolk, VA
USA

Jason W. Locasale
Division of Nutritional Sciences
Cornell University
Ithaca, NJ
USA

Giselle Y. López
Department of Pathology
University of California, San Francisco
San Francisco, CA
USA

Shyamala Maheswaran
Massachusetts General Hospital Cancer Center
Harvard Medical School
Boston, MA
USA

Alberto Mantovani
Department of Inflammation and Immunology
Humanitas Clinical and Research Center
Rozzano
Italy

José Ignacio Martín-Subero
Cancer Epigenetics and Biology Program
(PEBC)
Institut d'Investigació Biomèdica de Bellvitge
(IDIBELL)
Barcelona, Catalonia
Spain

Simon J. Morley
Department of Biochemistry and Molecular Biology
School of Life Sciences
University of Sussex
Brighton
UK

Oliver Müller
Molecular Biology, Biotechnology and Biochemistry
Group
University of Applied Sciences
Kaiserslautern
Germany

Kathleen R. Nevis
Department of Pathology
University of North Carolina School of Medicine
Chapel Hill, NC
USA

Sait Ozturk
Department of Oncological Sciences
Icahn School of Medicine at Mount Sinai
New York, NY
USA

Panagiotis Papageorgis
Departments of Biological Sciences and Mechanical
Engineering
University of Cyprus
Nicosia
Cyprus

Jignesh R. Parikh
Bioinformatics Program
Boston University
Boston, MA
USA

Steven M. Powell
Gastroenterology & Hepatology
University of Virginia School of Medicine
Charlottesville, VA
USA

Kimberly L. Raiford
Department of Biochemistry and Biophysics
University of North Carolina School
of Medicine
Chapel Hill, NC
USA

Andrew M. Rankin
Cancer Center
Boston University School of Medicine
Boston, MA
USA

Patricia Reischmann
Molecular Biology
Biotechnology and Biochemistry Group
University of Applied Sciences
Kaiserslautern
Germany

Simon Rosenfeld
Biometry Research Group
Division of Cancer Prevention
National Cancer Institute
Rockville, MD
USA

Marc Samsky
Department of Pathology
Duke University Medical Center
Durham, NC
USA

Anthony Scott
Department of Genetics
Case Western Reserve University School of
Medicine
Cleveland, OH
USA

Shantibhusan Senapati
Institute of Life Sciences
Bhubaneswar
India

Yashaswi Shrestha
Dana-Farber Cancer Institute
Harvard Medical School and
Broad Institute of Harvard and MIT
Boston, MA
USA

Anurag Singh
Department of Pharmacology and
The Cancer Center
Boston University School of Medicine
Boston, MA
USA

Rakesh K. Singh
Department of Pathology and Microbiology
University of Nebraska Medical Center
Omaha, NE
USA

Gromoslaw A. Smolen
Massachusetts General Hospital Cancer Center
Harvard Medical School
Charlestown, MA
USA

Sudhir Srivastava
Cancer Biomarkers Research Group
Division of Cancer Prevention
National Cancer Institute
Rockville, MD
USA

Simon Tavaré
Cancer Research UK Cambridge Institute
University of Cambridge
Cambridge
UK

Sam Thiagalingam
Biomedical Genetics, Cancer Center
Genetics & Genomics and Pathology & Laboratory
Medicine
Boston University School of Medicine
Boston, MA
USA

László Tora
Functional Genomics Department
Institut de Génétique et de Biologie Moléculaire et
Cellulaire (IGBMC)
Université de Strasbourg (UdS)
Strasbourg
France

David Tuveson
Cancer Research UK Cambridge Research
Institute
University of Cambridge
Cambridge, UK and
Cold Spring Harbor Laboratory
Cold Spring Harbor, NY
USA

Asad Umar
Division of Cancer Prevention
National Cancer Institute
Rockville, MD
USA

Matthew G. Vander Heiden
Koch Institute for Integrative Cancer Research
Massachusetts Institute of Technology
Cambridge, MA
USA

Cyrus Vaziri
Department of Pathology and Laboratory
Medicine
University of North Carolina School of
Medicine
UNC Lineberger Comprehensive Cancer Center
Chapel Hill, NC
USA

Zhenghe John Wang
Department of Genetics
Case Western Reserve University School of
Medicine
Cleveland, OH
USA

Kevin Webster
Cancer Bioscience
AstraZeneca R&D Boston
Waltham, MA
USA

Chen Khuan Wong
Genetics and Genomics
Boston University School of Medicine
Boston, MA
USA

Yu Xia
Department of Bioengineering
McGill University
Montreal, Quebec
Canada

Hai Yan
Department of Pathology
Duke University School of Medicine
Durham, NC
USA

Jian Yu
Department of Pathology
University of Pittsburgh Cancer Institute
University of Pittsburgh School of Medicine
Pittsburgh, PA
USA

Lihua Yu
Cancer Bioscience
AstraZenica R&D Boston
Waltham, MA
USA

Min Yu
Massachusetts General Hospital Cancer Center
Harvard Medical School
Charlestown, MA
USA

Lin Zhang
Department of Pharmacology
University of Pittsburgh Cancer Institute
University of Pittsburgh School of Medicine
Pittsburgh, PA
USA

Jin-Rong Zhou
Nutrition/Metabolism Laboratory
Beth Israel Deaconess Medical Center
Harvard Medical School
Boston, MA
USA

Preface

The heterogeneity in alterations and the failure to detect consistent changes in a unique set of gene(s) or gene products in similar and histologically well defined neoplasms pose a challenge for the accurate diagnosis, prognosis and therapy of cancer. Consequently, there is a need to integrate the individual observations made in tumor cells derived from numerous sources using the systems biology approach to identify a panel of alternate target genes/gene products as biomarkers for diagnosis and/or prognosis and as targets for therapy. This goal could be achieved with efficacy by dissecting alterations in cancer in interconnected modular networks of pathways represented in multi-modular molecular networks (MMMN) specific for progression of individual cancers. This landmark volume consisting of a collection of chapters examines the fundamentals of the molecular basis of the genesis of cancer in parts devoted to the overall big picture, basic biochemical events, manifestation of fingerprints of alterations, units of coordinated events, state of knowledge of the integrated progression of events for specific cancers and the future prospects and implications of the various MMMN cancer progression models in the fight against cancer.

My sincere thanks to the distinguished scientists for graciously contributing chapters on their expertise. My special gratitude to my doctoral thesis advisor, Professor Lawrence Grossman, for his guidance in shaping up my career as a molecular biologist and for stimulating my passion to undertake cancer research as the next step to studying DNA repair mechanisms, and to my post-doctoral advisor, Professor Bert Vogelstein, for being a role model and for sharing his wealth of knowledge and expertise in the field of cancer genetics and biology. My special appreciation to Allan Ross (Former Executive Editor, Medicine and Life Sciences, Cambridge University Press, New York) for inviting me to conceive this volume and for all his assistance at the initial stages of the development of this volume. I am indebted to my students, Arthur Lambert and Chen Wong for critical comments and proofreading and Panagiotis Papageorgis and Sait Ozturk for help with illustrations. I am thankful to Ilaria Tassistro (Assistant Editor, Life Sciences, Cambridge University Press, Cambridge, UK) and Katrina Halliday (Editor and Publisher, Life Sciences, Cambridge University Press, Cambridge, UK) for their patience with the last minute delays and guidance and Kath Pilgrem (Copy Editor) and Jessica Ann Murphy (Production Editor Academic Books, Cambridge University Press, Cambridge, UK) for their help with finalizing this volume for publication. On behalf of the authors, I would also like to thank the American Association for Cancer Research, Nature Publishing Group, Wolters Kluwer Health, and others for allowing partial or full reproduction of their previously published figures.

This project would not have been completed without the encouragement, support and the enduring love of my wife Arunthathi Cumaraswamy Thiagalingam and the unconditional love of my children Natasha Thivya Thiagalingam and Aaron Gajan Thiagalingam.

Systems biology of cancer progression

Sam Thiagalingam

Introduction

Heterogeneity in the genetic and epigenetic alterations of cancers that exhibit similar functional properties during the various stages of cancer progression, including the terminal metastatic stage, has remained as the major challenge to effective diagnosis, prognosis and therapeutic efforts. While there has been significant progress in cataloguing the various genetic and epigenetic alterations with the advent of expanding new high-throughput technologies, streamlining the available and emerging data into a coherent scheme of events depicting drivers, the connectors and the conductors that form multi-modular molecular networks (MMMN) of cancer progression culminating in tumors, requires novel strategies. The ultimate goal of cancer research should be to take advantage of the parallel progress made through both experimental and computational approaches and integrate the data from these fronts using systems biology to generate MMMN cancer progression models. Such models can be cancer specific and can be functionally definable in terms of disease stage to help design biomarker screening tests for effective diagnosis/prognosis and the development of personalized cancer therapies.

Background

Cancer is a genetic and epigenetic disease, which manifests functional properties of target tumor cells at different stages due to the accumulation of specific combinations of alterations. The number of alterations required to assume a given stage of cancer may vary within and between certain types of cancer. While modelling cancer progression has been attempted at various times, the first breakthrough came with the study of the genetics of colon cancer progression, which depicted multiple stages of the disease [1, 2]. Since then, despite an increase in the wealth of knowledge that has emerged on the types of alterations associated with specific cancers as a result of comprehensive profiling and next-generation sequencing (NGS) strategies to decipher genetic and epigenetic alterations, seemingly insurmountable complexity has prevented the streamlining of the various changes into coherent and definable stages, which still awaits the development of novel strategies to make progress.

Multi-modular molecular networks of cancer progression depict heterogeneity in genetic and epigenetic alterations

The lack of consistent and defined genetic and epigenetic alterations affecting a specific set of gene(s) in the majority of sporadic cancers with similar histologic subtypes and stages poses a challenge in understanding the molecular basis for the heterogeneity of molecular aberrations. The inconsistency in these profiles of molecular targets not only imposes a dilemma to gaining a clear understanding of the disease but also complicates efficient early diagnosis, prognosis and strategies for treatment modalities for cancers. To address this challenge faced by the cancer research community, I proposed a strategy for the formulation of a detailed framework known as an MMMN cancer progression model as a road map to dissect the complexity inherent to cancer (Figure 1.1) [3]. This model predicts that cancer initiation and progression are mediated by dysregulation/inactivation of a series of interconnected functional sub-network modules.

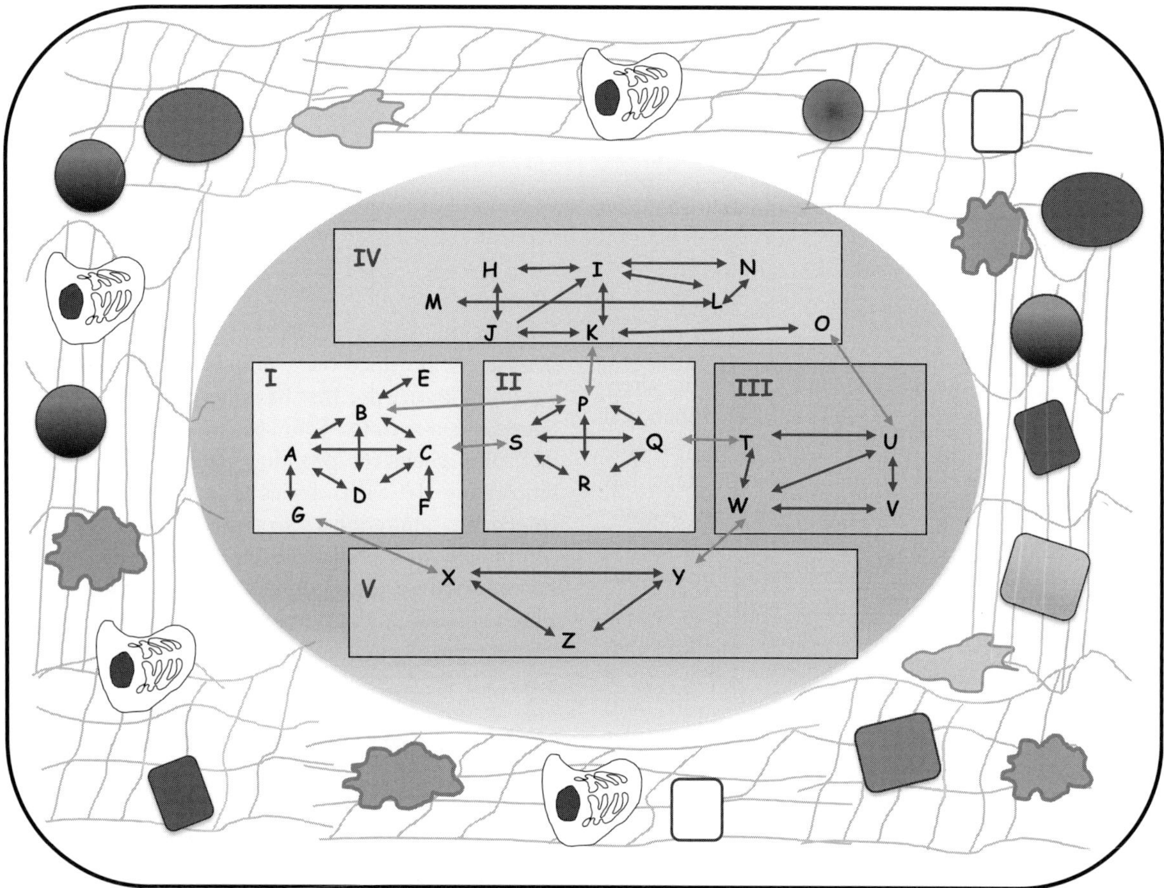

Figure 1.1 A cascade of aberrant network modules defines the multi-modular molecular network (MMMN) model for cancer progression. An MMMN cancer progression model predicts that aberrant activations/inactivations of functional modules of networks in a series of steps would be necessary to elicit properties of metastatic cancer. In this model, cancer initiation is mediated by inactivation of the gatekeeper network module (e.g., module I). We predict that the gatekeeper function is mediated by an interconnecting network of pathways (axes). Dysregulation/inactivation of the gatekeeper module predisposes the cells to become more receptive and susceptible to acquiring additional neoplastic alterations, which occur in a series of modular (modules II, III, IV, etc.) inactivations or hyper-activations leading to intermediate and late carcinoma and finally to the metastatic stage. Modules II and III in this model represent the intermediate stages of tumor progression. The terminal module may represent the metastatic stage (module IV). The fact that there could be alternate target genes in any one of the modules of the network could explain why there is often genetic/epigenetic heterogeneity in multi-step cancer progression resulting in similar histologic subtypes of cancer. In this model, the double-headed light and dark arrows represent intra- and extra-modular connections, respectively. The alphabetical letters represent specific genes or functional protein–protein and protein–DNA interactions that are nodal points/driver alterations in each network. While the modular organization depicted inside the inner oval represents the alterations within the target tumor cells, the web structures that represent the extracellular matrix and the surrounding stromal cells constitute the tumor microenvironment. (A black and white version of this figure will appear in some formats. For the color version, please refer to the plate section.)

The provision in the MMMN cancer progression model, which defines a cascade of events encompassing multiple targets within each module, is that one or more alternate target gene(s) could alter the functionality of each of the specific modules. This provides a molecular basis for the genetic and epigenetic heterogeneity that is observed during the progression of tumors that exhibit similar pathological characteristics (Figure 1.1). Furthermore, the absence of consistent alterations in specific gene(s) in sporadic cancers, and in cancers that are primarily induced by environmental effects to generate neoplastic precursor cells, could be predicted to occur via inactivation/overactivation of multiple alternate gatekeeper gene(s) that act in one or more interconnected axes of events, within a defined sub-network in a module of the global network (Figure 1.1). Thus the first network module that becomes inactivated leading to the initiation of cancer is the gatekeeper functional unit [4]. The cancer precursor cells harboring an inactivated gatekeeper module become receptive to additional genetic and epigenetic alterations that occur in interconnected but defined modules of sub-networks representing multiple stages, leading to the development of advanced and terminal stages. Therefore the functional inactivation or aberrant hyper-activation of network modules occurs in a series of events that advance the tumor from the early to late stages of cancer. It is also noteworthy that overlaps in the functional contributions of the specific gene alterations may be responsible for simultaneous dysregulation of different modules of cancer progression. While any alteration capable of inactivating/ dysregulating a specific sub-network module could occur at any time, its effect will be fully realized to manifest the corresponding cancer stage only when the preceding module(s) have also become inactivated/dysregulated. Thus the rates at which tumor evolution occurs and the time required for the transition from an early to a later stage of cancer will be dependent upon the preexisting genetic and epigenetic alterations (familial or sporadically acquired) and the tumor microenvironment. This notion is also consistent with an accelerated cancer progression when there is a preexisting inherited alteration that corresponds to a specific module as it has been observed with familial cancers. Despite the possibility that the overall phenotypic effects elicited by the target tumor or tumor precursor cells could be influenced by the surrounding cells and/or extracellular matrix (ECM) components, the epigenetic and genetic alterations in the resident target cells are a prerequisite for the effects caused by the microenvironment and surrounding stromal cells [5].

While interdependent interactions of genes and proteins may consist of physical interactions among proteins, representing inter- and intracellular communications and their binding to DNA elements (e.g., transcription factors, histones harboring specific modifications, etc.) and mRNAs/regulatory RNAs (e.g., miRNA, lincRNA, etc.), there could also be metabolic networks of biochemical reactions that involve distinct substrates and products. The modular organization of the various stages of cancer progression consisting of interconnected networks of events also suggests that changes in alternate targets that render similar functional status can lead to the acquisition of drug resistance. Thus developing drugs that target combinations of distinct landscapes of alterations would be necessary for clinical decision making and to select therapies that increase therapeutic efficacy [6].

Driver versus passenger alterations

Cancer phenotypes are driven by gain-of-function alterations as seen with oncogenes such as the *AKT1, ALK, BRAF, CTNNB1, DDR2, EGFR, ERBB2, FGFR1, IDH1, IDH2, KRAS, MDM2, MITF, MYC, MYCN, MYCL1, NKX2.1, PIK3CA, REL* and *SOX2* and/or loss-of-function alterations as frequently observed with specific tumor suppressor genes such as the *APC, BMPR1A, CDH1, CDKN2A, NF1, NF2, MAP2K4, MLH1, MSH2, PIK3R1, PTEN, RB1, SMAD4, SMARCB1* and *TP53*, mediated by either genetic or epigenetic changes [3, 6]. A "20/20 rule," which requires at minimum >20% of the observed missense mutations at recurrent positions in an oncogene, and >20% of inactivating mutations for a tumor suppressor gene has been proposed [7]. The gene alterations that provide a selective advantage during the evolution of a tumor are regarded as the "drivers" while the alterations that are coincidental in their appearance and do not play a role in the cancer progression are termed the "passengers" [6, 8]. While, traditionally, genetic changes are regarded as the drivers and epigenetic alterations as the passengers, there is accumulating evidence for either type of alteration to be passengers or drivers [9]. It is also noteworthy that not all mutations in the same gene are drivers as exemplified by *APC* mutations in colorectal cancer [2, 7]. Furthermore, some driver genes are more frequently mutated and referred to as the "mountains," while others, despite their importance, are less frequently mutated and are known as the "hills," thus shaping the landscape of genetic alterations during cancer progression [10].

In the MMMN model for cancer progression, the driver genes represent the nodal points and activation of a single module could be effected by one, or possibly a few nodal gene alterations [3]. For example, of the more than one hundred pathways aberrantly regulated in breast cancer, several involved phosphatidylinositol 3-kinase (PI3K) signaling, with *PIK3CA* as the most frequent target and others such as *GAB1, IKBKB, IRS4, NFKB1, NFKBIA, NFKBIE, PIK3CB, PIK3CG, PIK3R1, PIK3R4,* and *RPS6KA3* as other potential targets [10, 11]. These observations are consistent with the molecular heterogeneity involving aberrations in alternate target genes in modules of the MMMN model for cancer progression [3].

Emerging MMMN models

Being an autonomous complex genomic disease, cancer presents its characteristics in any group of representative cells and subtypes, based on genetic and epigenetic signatures that are often under the influence of microenvironmental effects. There has been significant progress made in visualizing these signatures through the application of genomic technologies to decipher their functional effects at the level of individual genes, the genome, and the pathways and networks of signaling events.

For example, breast cancer has a well-established genetic component exhibiting a greater than ten-fold risk in individuals harboring familial rare mutations in *BRCA1*, *BRCA2*, *TP53* and *PTEN* but elicit at least 18 morphologically distinct tumor types according to the World Health Organization. Recently, it has been classified into six different intrinsic subtypes, which harbor characteristic gene alterations: luminal A (*CCND1*, *ESR1*, *FOXA1*, *GATA3*, *KRT8*, *KRT18*, *LIV1*, *MAP3K1*, *PIK3CA*, *TFF3* and *XBP1*), luminal B (*ESR1*, *FOXA1*, *GATA3*, *KRT8*, *KRT18*, *LAPTM4B*, *SQLE*, *TFF3* and *XBP1*), HER2-enriched [*ERRB2* (*HER2* or *Neu*) and *GRB7*], basal-like (*CDH3*, *FABP7*, *ID4*, *KRT5*, *KRT17*, *LAMC2* and *TRIM29*), normal breast-like (*AQP7*, *CD36*, *FABP4*, *ITGA7* and *PTN*) and claudin-low (*ALDH1*, *CD29*, *CD44* and *SNAI3*) based on genomic studies [12–14]. Additionally, multiple technology platforms such as mRNA expression profiling, DNA copy number arrays, massively parallel sequencing as well as the high information content assays to probe DNA methylation, miRNA expression and protein expression, have been used to assess the various abnormalities in the cancer state. These efforts identified mutations previously implicated, in breast cancer: *AKT1*, *BRCA1*, *CDH1*, *GATA3*, *PIK3CA*, *PTEN*, *RB1* and *TP53*; and in other cancers: *APC*, *ARID1A*, *ARID2*, *ASXL1*, *BAP1*, *KRAS*, *MAP2K4*, *MLL2*, *MLL3*, *NF1*, *SETD2*, *SF3B1*, *SMAD4* and *STK11*. Interestingly, new lesions were also identified for the first time in breast cancer: *AFF2*, *AKT2*, *ARID1B*, *CASP8*, *CBFB*, *CCND3*, *CDKN1B*, *MAP3K1*, *MAP3K13*, *NCOR1*, *NF1*, *PIK3R1*, *PTPN22*, *PTPRD*, *RUNX1*, *SF3B1*, *SMARCD1* and *TBX3* [15, 16]. Furthermore, while there were cancer subtype specific mutations, only three genes (*GATA3*, *PIK3CA* and *TP53*) exhibited recurrent mutations in >10% of the breast cancers confirming the complexity and heterogeneity in the profiles of alterations that contribute to the formation of each tumor [3, 15, 16].

Similar efforts to catalogue driver genes involved in other cancers are also emerging at this time. The catalogue of genomic alterations in the various cancers are generated using high-throughput technologies at several major institutions such as the Broad Institute and the Johns Hopkins University and through the coordinated efforts of the Cancer Genome Atlas (TCGA) project in the Unites States, the Wellcome Trust Sanger Institute in the United Kingdom and the International Cancer Genome Consortium (ICGC) in Canada. TCGA data can be explored at the gene-based viewing mode using the UCSC Cancer Genomics Browser (https://genome-cancer.ucsc.edu) and the large cohort data also can be analyzed to generate Kaplan–Meier plots [16]. Additionally, pathway-based methods such as the Cytoscape (http://cytoscape.org), Mutual Exclusivity Modules in Cancer (MEMo) (http://cbio.mskcc.org/tools/memo.html), Pathway Recognition Algorithm using Data Integration on Genomic Models (PARADIGM) (https://genome-cancer.ucsc.edu) and cBio Portal (www.cbioportal.org) can be used to elucidate functional connections among the genes of interest [17–19]. While all these efforts are contributing towards building MMMN models for cancer progression of each cancer type, at this time the majority of these alterations are not classifiable to a particular module in the grand scheme of cancer progression. Therefore it will take an improved and organized effort of sampling and profiling strategies of alterations in real time by the shedding of the heavy reliance on snapshots derived from samples corresponding to archived, stationary and/or predetermined randomly fixed time points as is generally the norm at the present time, the use of model systems to infer functional effects, and new bioinformatics tools to achieve what has been predicated by the MMMN hypothesis [3].

Role of the microenvironment in cancer progression

Tumors consist of more than the malignant cells, as the surrounding non-malignant stromal cells such as endothelial cells of blood and lymphatic circulation, fibroblasts, carcinoma-associated fibroblasts (CAFs), myofibroblasts, pericytes, adipocytes, mesenchymal stem cells and immune cells and immunosuppressive cells such as the tumor associated macrophages (TAMs) and myeloid-derived suppressor cells (MDSCs), respectively, embedded in the modified components of the extracellular matrix (ECM) and remodelled vasculature, together form the tumor mass (Figure 1.1) [20, 21]. It is becoming more and more apparent that these diverse components play crucial roles in modulating tumor progression through paracrine/autocrine secretion of cytokines such as transforming growth factor-beta (TGFβ) and interleukin 6 (IL6), growth factors like epidermal growth factor (EGF), fibroblast growth factor (FGF), platelet-derived growth factor (PDGF), insulin-like growth factor-1 (IGF-1) and other factors such as hedgehog (Hh), Notch, periostin (POSTN), vascular endothelial growth factor (VEGF) and Wnts [20–22]. In the context of MMMN models for cancer progression, one can envision that the networks of gene connections and pathways within and between the various modules that constitute the different stages of cancer progression could be influenced by the tumor microenvironment. For example, our previous studies with breast cancer found TGFβ could epigenetically regulate various driver genes involved in epithelial to mesenchymal transition in breast cancer [23]. Thus the functional status of driver genes in the modules of cancer progression could be influenced by TGFβ-like cytokines or other factors and hence impact the functional and phenotypic state of the cancer.

Future perspectives

The success of cancer therapies depends on the fulfillment of two criteria. The first challenge is to offer personalized medicine by treatment with drugs that are tailored to each patient's own tumor(s). The second is the ability to follow up/continue with therapeutic strategies that can prevent therapeutic resistance and the associated relapse to the initial targeted therapy. An optimistic vision for offering the panacea for these major challenges is to develop MMMN models for cancer progression that would provide details of all possible alterations in the tumor and its microenvironment and their contributions, which can be detected in a cancer at the time of diagnosis and used in the future to predict what one could expect to see upon relapse to help with the immediate implementation of effective follow-up therapeutic remedies. While this is not an easy task to achieve at the present time, future research and new technologies may provide the necessary tools to develop combination therapies that achieve the ultimate goal of curing, or at least keeping in check, metastatic disease for the longest term possible.

References

1. Fearon ER and Vogelstein B. 1990. A genetic model for colorectal tumorigenesis. *Cell* 61: 759–767.

2. Kinzler KW and Vogelstein B. 1996. Lessons from hereditary colon cancer. *Cell* 87: 159–170.

3. Thiagalingam S. 2006. A cascade of modules of a network defines cancer progression. *Cancer Res* 66(15): 7379–7385.

4. Kinzler KW and Vogelstein B. 1997. Cancer-susceptibility genes. Gatekeepers and caretakers. *Nature* 386: 761–763.

5. Jacks T and Weinberg RA. 2002. Taking the study of cancer cell survival to a new dimension. *Cell* 111: 923–925.

6. Leary RJ, Kinde I, Diehl F, et al. 2010. Development of personalized tumor biomarkers using massively parallel sequencing. *Sci Transl Med* 2(20): 20ra14.

7. Vogelstein B, Papadopoulos N, Velculescu VE, et al. 2013. Cancer genome landscapes. *Science* 339(6127): 1546–1558.

8. Haber DA and Settleman J. 2007. Cancer: drivers and passengers. *Nature* 446(7132): 145–146.

9. Sawan C, Vaissière T, Murr R, and Herceg Z. 2008. Epigenetic drivers and genetic passengers on the road to cancer. *Mutat Res* 642(1–2): 1–13.

10. Wood LD, Parsons DW, Jones S, et al. 2007. The genomic landscapes of human breast and colorectal cancers. *Science* 318(5853): 1108–1113.

11. Tamborero D, Gonzalez-Perez A, Perez-Llamas C, et al. 2013. Comprehensive identification of mutational cancer driver genes across 12 tumor types. *Sci Rep* 3: 2650.

12. Prat A and Perou CM. 2011. Deconstructing the molecular portraits of breast cancer. *Mol Oncol* 5(1): 5–23.

13. Alizart M, Saunas J, Cummings M, and Lakhani SR. 2012. Molecular classification of breast carcinoma. *Diagnostic Histopathology* 18(3): 97–103.

14. Eroles P, Bosch A, Pérez-Fidalgo JA, and Lluch A. 2012. Molecular

biology in breast cancer: intrinsic subtypes and signaling pathways. *Cancer Treat Rev* 38(6): 698–707.

15. Stephens PJ, Tarpey PS, Davies H, et al. 2012. The landscape of cancer genes and mutational processes in breast cancer. *Nature* 486(7403): 400–404.

16. Cancer Genome Atlas Network. 2012. Comprehensive molecular portraits of human breast tumours. *Nature* 490(7418): 61–70.

17. Goldman M, Craft B, Swatloski T, et al. 2013. The UCSC Cancer Genomics Browser: update 2013. *Nucleic Acids Res* 41(Database issue): D949–954.

18. Cline MS, Craft B, Swatloski T, et al. 2013. Exploring TCGA pan-cancer data at the UCSC Cancer Genomics Browser. *Sci Rep* 3: 2652.

19. Eifert C and Powers RS. 2012. From cancer genomes to oncogenic drivers, tumour dependencies and therapeutic targets. *Nat Rev Cancer* 12(8): 572–578.

20. Joyce JA and Pollard JW. 2009. Microenvironmental regulation of metastasis. *Nat Rev Cancer* 9(4): 239–252.

21. Taddei ML, Giannoni E, Comito G, and Chiarugi P. 2013. Microenvironment and tumor cell plasticity: an easy way out. *Cancer Lett* 341: 80–96.

22. Castaño Z, Fillmore CM, Kim CF, and McAllister SS. 2012. The bed and the bugs: interactions between the tumor microenvironment and cancer stem cells. *Semin Cancer Biol* 22(5–6): 462–470.

23. Papageorgis P, Lambert AW, Ozturk S, et al. 2010. Smad signaling is required to maintain epigenetic silencing during breast cancer progression. *Cancer Res* 70(3): 968–978.

Chapter

2

Lessons from cancer genome sequencing

Antoine Ho and Jeremy S. Edwards

Introduction

The Human Genome Project (HGP) was one of the greatest achievements of the twentieth century, and the publication of the full human genome sequence in 2001 ushered in the new century by starting the post-genome era in human biology. The great success of the HGP has paved the way to many future discoveries. The human genome sequence represents just the beginning of the payoffs for the biomedical community, and many future benefits are promised and expected in the near future. Specifically, the HGP has enabled the rapid sequencing of more genomes, such as cancer genomes, and this holds the potential to transform cancer research and treatment. Therefore it is more appropriate to look at the completion of the human genome as the end of the beginning, rather than the beginning of the end of the era of human genome sequencing. "Next generation" sequencing (NGS) technologies are providing fast, cheap and high-quality sequencing. As these technologies become less expensive and easier to operate, they will become more widely available. However, the bottleneck in the process will quickly shift to the analysis phases. In other words, making sense of the vast amount of sequence data will be a challenging task, and it will require bioinformatics and systems biology. The analysis of sequencing data will likely have a tremendous impact on many areas of medicine and biomedical research.

Background

The sequencing and publication of the human genome was performed simultaneously by two competing groups, one was publicly funded and the other was privately funded. The publicly funded sequencing project was led by Dr. Francis Collins and was performed in the classical clone-by-clone approach using traditional Sanger sequencing. The private sequencing project was based at Celera and was led by Dr. J. Craig Venter. The Celera group sequenced the human genome using the shotgun sequencing approach, which was made possible for three main reasons: (1) they developed novel assembly algorithms, (2) they utilized data from the public project, and (3) they sequenced a very homogeneous sample, as opposed to a sample representative of a large number of individuals [1].

The HGP's impact on future human genome sequencing has two broad implications. First, the HGP has now established a reference human genome sequence, allowing for relatively rapid sequencing of future genomes while using the reference sequence to align reads. Additionally, a major impact of the HGP has been spin-off technologies and bioinformatics tools, which have led to what is now known as "next-generation" sequencing (NGS) technology [2].

Next-generation sequencing technologies

During the HGP, a number of technologies were developed with the goal of increasing sequencing throughput to allow for cheap and rapid human genome sequencing. The first phases of the improvements were essentially advances in instrumentation and miniaturization of the traditional Sanger sequencing approach. However, a number of true next-generation technologies were also developed and have become widely available.

Sequencing template preparation

The first step of the next-generation sequencing pipeline is the construction of the sequencing library. The library preparation step essentially takes a genomic DNA sample, and converts it into DNA molecules

---

Systems Biology of Cancer, ed. S. Thiagalingam. Published by Cambridge University Press. © Cambridge University Press 2015.

that can be sequenced by a given sequencing technology (Figure 2.1). For example, sequencing using the Illumina system fragments the genomic DNA into ~300 bp fragments, amplifies these fragments via PCR and ligates sequencing primer sites to the ends of the fragments [3–5]. These protocols vary in complexity depending on the sequencing platform.

Additionally, genome libraries can be constructed to contain mate-pair sequences. This means that the genome tags will be adjacent in the library molecule, but will have a kilobase or more separation in the genome. The mate-pair approach complicates library preparation, but assists in genome assembly/mapping, especially when dealing with very short read lengths, as is typical in most next-generation sequencing technologies (Figure 2.2) [3–5].

There are many ways to sequence DNA, and because of this there are many ways in which to prepare the DNA libraries for sequencing. First, the template can be clonally amplified unless sequencing can be performed on single molecules without the need for amplification. Methods that do not rely on

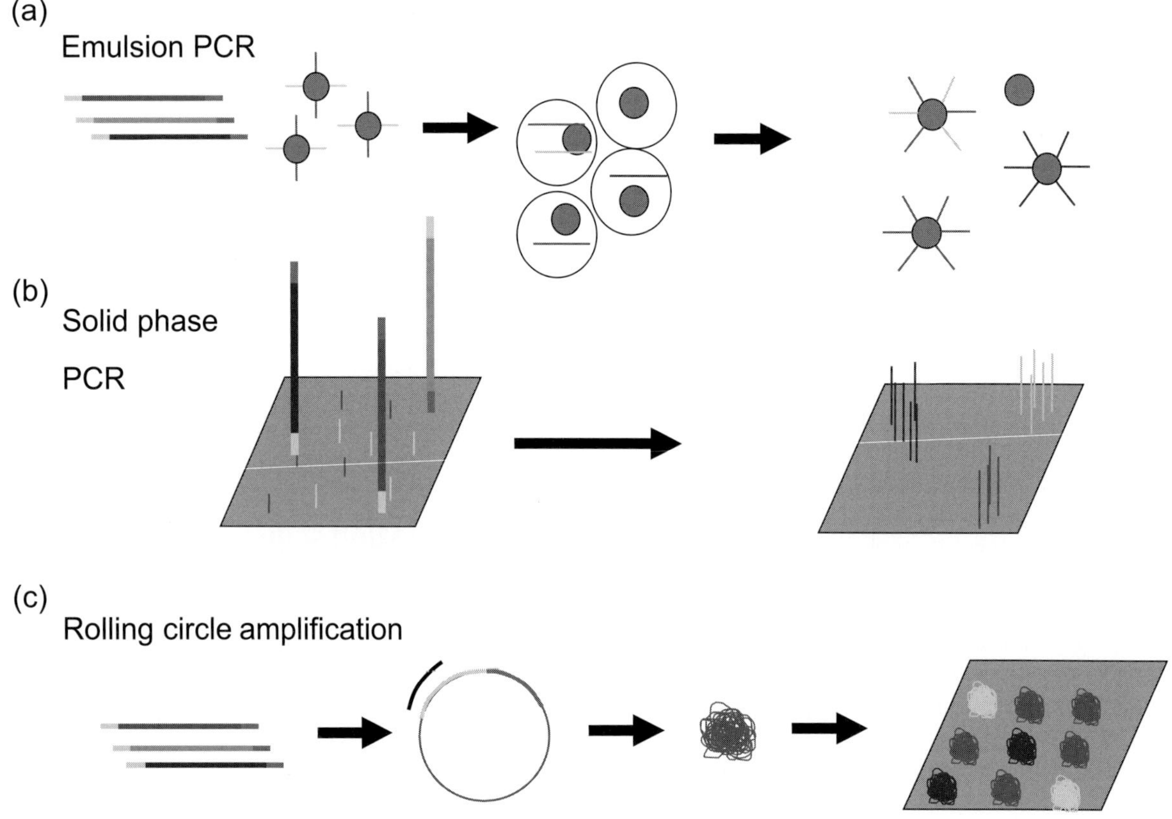

Figure 2.1 (a) Emulsion PCR (ePCR). Template DNA and beads are mixed and then put into an emulsion mixture consisting of an oil phase and an aqueous phase of PCR reagents. These beads have primers complementary to the ends of the template strands coupled to them, allowing the PCR reaction to extend these primers and cover the bead in copies of the template DNA. Template DNA is diluted to maximize the number of emulsions having exactly one template strand and one bead. Proceed with PCR temperature cycling. Sequencing is performed on beads with only clones of a single template DNA, as beads with no DNA and beads with more than one template DNA do not provide usable data. These beads can then be fixed onto an array for sequencing and imaging.
(b) Solid phase PCR. Very similar to ePCR, but without beads. Template DNA is diluted and then added to a slide with primers complementary to end regions of the template DNA coupled to the slide, which allows hybridization and priming. Through a series of PCR temperature cycling, a slide is covered in clonal patches of DNA to be sequenced.
(c) Rolling circle amplification (RCA). A piece of linear DNA is circularized enzymatically. Once circularized, RCA is performed with a polymerase that has displacement activity. This results in a ball of clonal DNA, effectively amplifying the DNA but without the need for emulsions or beads. These balls of DNA are then coupled to an array and sequenced. (A black and white version of this figure will appear in some formats. For the color version, please refer to the plate section.)

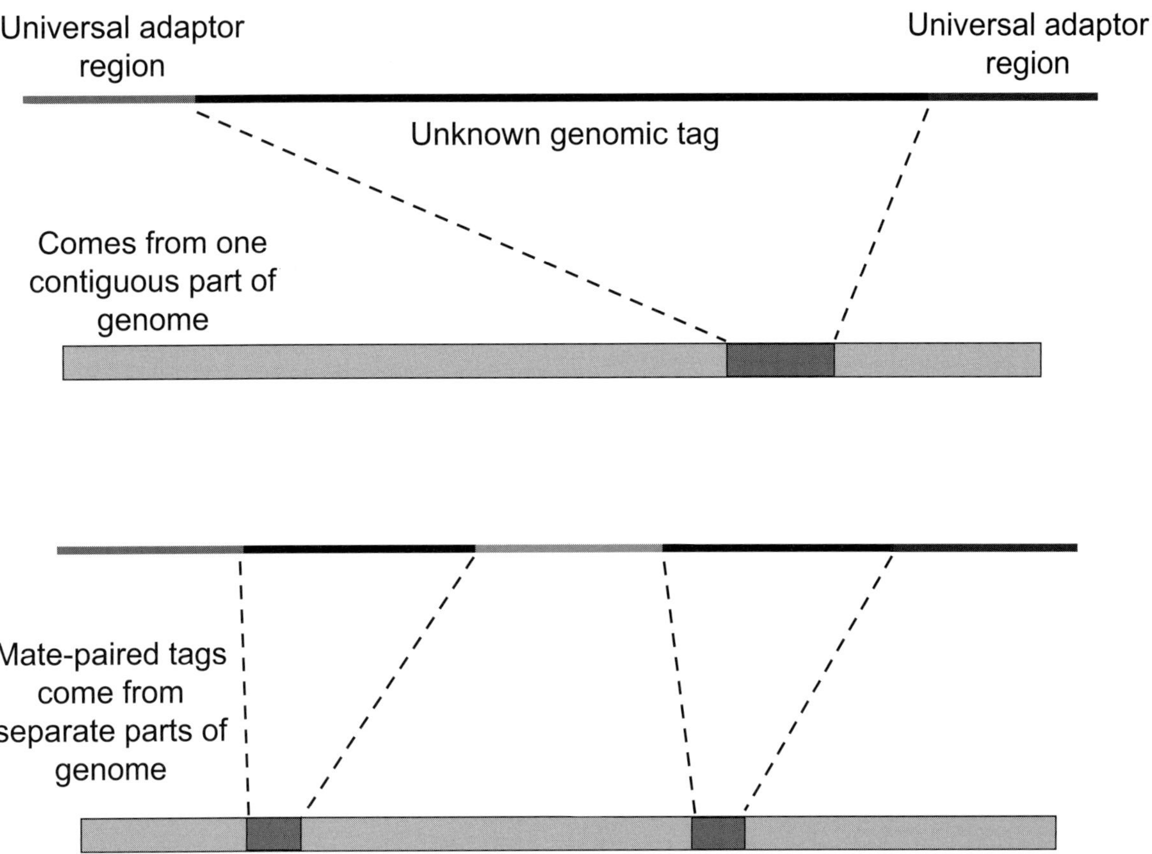

Figure 2.2 Mate-paired libraries. Mate-paired libraries can provide alignment information that is very valuable, especially when trying to sequence large redundant regions with short reads. The most ideal way to sequence a large redundant region is to simply get a single contiguous read of the entire region; however, that may not be technologically possible, which is why this mate-paired strategy is key. Because the mate-paired reads come from two different regions, a set distance apart, it is possible, even with short reads, that one half of the mate-pair will be in a uniquely identifiable region, and even though the other will be in the redundant, difficult to map region, that read will still provide useful alignment data. (A black and white version of this figure will appear in some formats. For the color version, please refer to the plate section.)

an amplification step are known as single-molecule sequencing methods. Amplification is necessary for many sequencing approaches because a signal, whether it is light or electrical, must be amplified or would be too weak to identify otherwise. This amplification can occur through an emulsion PCR (ePCR) step [6] or through solid phase PCR as in the Illumina Inc. system. Additionally, rolling circle amplification (RCA) can be utilized to amplify the DNA into a ball, which may itself be coupled to an array (see Figure 2.1) [7]. Clonal amplification may make certain sequencing approaches possible; however, when clonal amplicons are being sequenced, the issue of phasing arises. For example, when a clonal population of DNA molecules is being sequenced, the initial signals for sequencing each base are near identical for all molecules. However, as sequencing progresses, inefficiencies in biochemistry, enzymatic activity, chemical cleavage steps, or incomplete washing cause the signal to become noisy and may contain an earlier (lag phasing) or later (lead phasing) position.

Single-molecule sequencing template preparation is greatly simplified, as there is no need for amplification, and there are no amplification biases that may occur. Some single-molecule sequencing methods also make real-time sequencing possible, though there are obstacles to single-molecule sequencing that methods must take into account, such as being able to recognize the signal of a single molecule, which requires more expensive and larger sequencing equipment [8].

Sequencing by synthesis

Fluorescent methods

The most popular next-generation sequencing approach is known as sequencing by synthesis (SBS). In SBS, a DNA polymerase is used to extend a primer on the template strand (Figure 2.3) [3–5]. The DNA template to be sequenced must contain a known region at its 3′ end to hybridize a primer. Once hybridized, synthesis is allowed to occur under controlled conditions with specific reagents. The goal is to allow only the incorporation of a single nucleotide onto this growing strand and to visualize the base that was incorporated. The key is to modify (block)

the nucleotides in some fashion that not only allows termination of synthesis once incorporated, but also can be reversible. These can, for example, involve a blocking group on the 3′ OH of the growing DNA strand that can be removed enzymatically or by a chemical cleavage reaction [3–5]. The second element is to attach unique fluorophores onto each of the four different nucleotides to allow visualization. After imaging, and storing this data, the termination must be reversed by removing this blocking group, to allow the addition of another single nucleotide, and then the fluorophores must be cleaved to visualize the signal of the newly incorporated nucleotide. This process is repeated to sequencing up to ~150 bases. Sequencing

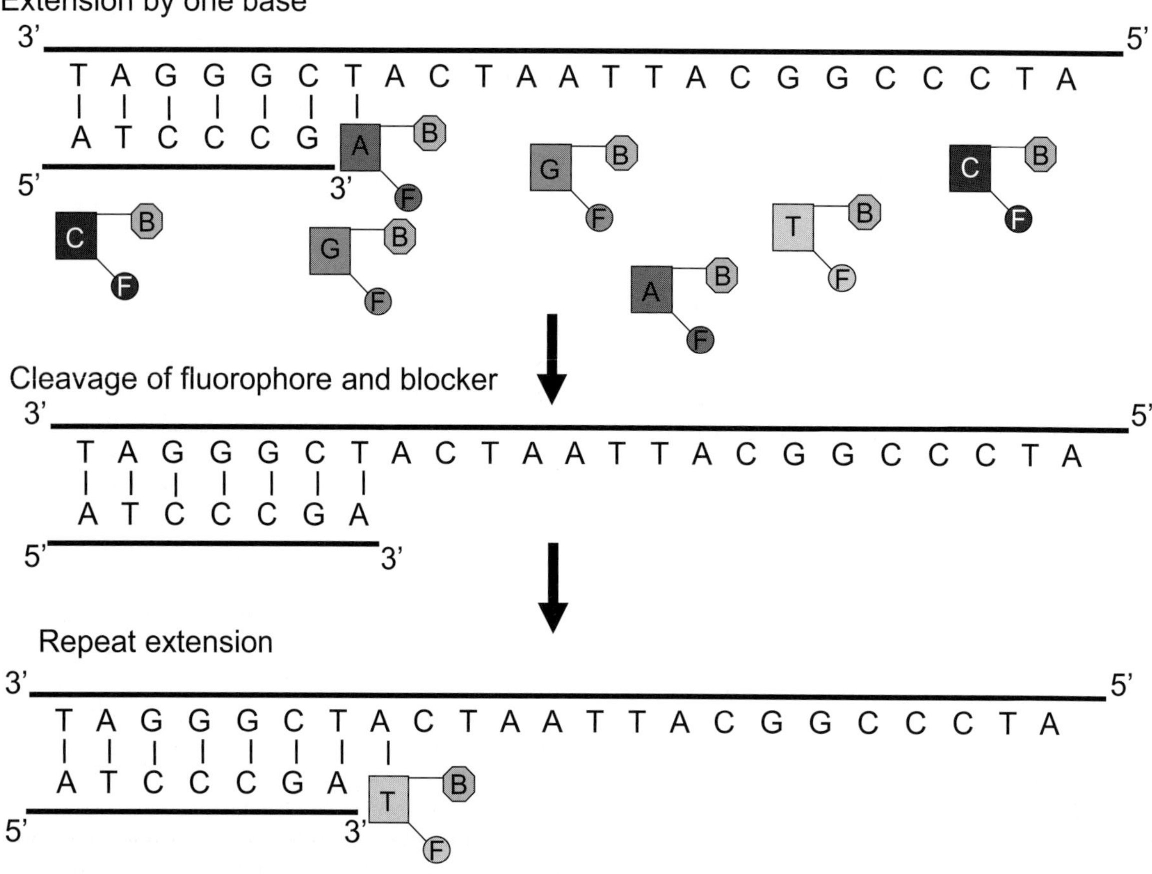

Figure 2.3 Sequencing by synthesis with fluorophores. A primer is hybridized onto the template DNA onto a universal region to allow extension by a polymerase. A single nucleotide will incorporate due to a blocking group on the nucleotides, and the DNA will be able to be visualized by the fluorophores attached to each nucleotide type. If there is a saturation step, as is often the case when dealing with amplified DNA template, it would be performed following the first extension step (not shown). A saturation step is identical to the first step except that there is no fluorophores, though there are still blockers on the nucleotides, and the nucleotides are usually at a very high concentration to saturate. The fluorophores are then cleaved chemically, and then the blocking group is removed so extension can continue to another base. This cycle then repeats. (A black and white version of this figure will appear in some formats. For the color version, please refer to the plate section.)

by synthesis can be performed on clonal amplicons from an amplification step (i.e., sequencing being carried out on beads or a clonal cluster of DNA), or it can be performed on a single molecule [3–5].

Sequencing by synthesis can also be performed in real time with single-molecule visualization. Real-time SBS methods are faster, but constrained to the viewing area limitations of a camera-mounted microscope. Real-time sequencing approaches will likely have a significant impact on cancer systems biology. This is because real-time sequencing has the potential for very long reads, requires a very simple library preparation, and can read out epigenetic markers, such as methylation and hydroxymethylation [9].

Non-fluorescent methods

In addition to using fluorophores to identify incorporated bases, there are other methods to measure and quantify DNA polymerase extension, such as detecting the H^+ or pyrophosphate released during polymerase extension. Since all bases give the same pyrophosphate or H^+ signal this sequencing approach requires cycles of extending with each of the individual nucleotides. This sequencing approach has the advantage of using natural nucleotides; however, this introduces the homopolymer repeat problem. Namely, these types of sequencing methods must record the intensity of such a signal to deduce how many bases of the same type were incorporated in homopolymer repeats. While distinguishing the difference in signal between single or double nucleotide incorporation events is straightforward, it is harder to discern the difference between five or six incorporated nucleotides in a homopolymer repeat [3–5].

The measured signal can be pH changes, as induced by the release of hydrogen atoms when incorporating a nucleotide during synthesis, or there can be other enzymes involved such as luciferase and sulphurylase that create a flash of light when a phosphate is released during the same process. Due to the nature of this method, sequencing is performed in real time, and tends to have lower throughput than fluorescent, sequential array methods.

Sequencing by ligation

Sequencing by ligation (SBL) uses a ligase and a series of query primers to sequence a template strand. The template DNA to be sequenced will contain the unknown genomic tag, flanked by a known region.

The main disadvantage of this sequencing approach is that the read lengths are very short. Therefore to obtain a reasonable read length, a complicated library preparation is required. The sequencing strategy is to hybridize an anchor primer onto a known region, and ligate a query primer to the anchor primer to sequence the unknown genomic tag. Ligation is determined by hybridization of that query primer next to the anchor primer, meaning it must be complementary for the unknown tag region. The query primers are degenerate, a mix that contains all possible combinations for every position except for one. For example, when determining the identity of the first base next to the anchor primer, the query primer set will be degenerate for all positions; however, in this set, all query primers that have an adenine in that first position will have a specific fluorophore attached to the other end of the query primer (Figure 2.4) [3–5]. The clonal features (i.e., beads) will then be imaged, in a manner similarly to fluorescent SBS, and each specific fluorescent signal corresponds with specific bases. This is repeated to obtain the identity of the second base; however, now the fluorophores are specifically linked to bases in the second position of that query primer. This is repeated, generally to a length of seven nucleotides.

The reason for this read-length limitation is that base pairing is specific closer to the site of ligation and less so further out. To get longer reads, cleavage of the ligated query primer is performed, resulting in loss of the fluorophore and effectively extending the anchor primer into unknown regions of the genomic tag (see Figure 2.4). For example, after sequencing the second base of a tag, and imaging the array, the signal and query primer can be cleaved after the fifth base. The anchor primer will be extended by five bases. Now when using the same query primer that sequenced the second base, it will now sequence the seventh base. Repeat the process, extend the anchor primer by another five bases, and sequence the twelfth base with the same query primer. This is repeated to get longer reads. When the signal becomes too weak to continue, the growing ligated template is removed and the sequencing can be repeated to sequencing, for instance, the third, eighth, thirteenth … positions. This is repeated overall to obtain a contiguous sequence for the genomic tag.

The read lengths of SBL are shorter than those obtainable from SBS; however, SBL can be performed in both $5'$ to $3'$ as well as $3'$ to $5'$ directions, whereas SBS must be performed in the direction of DNA

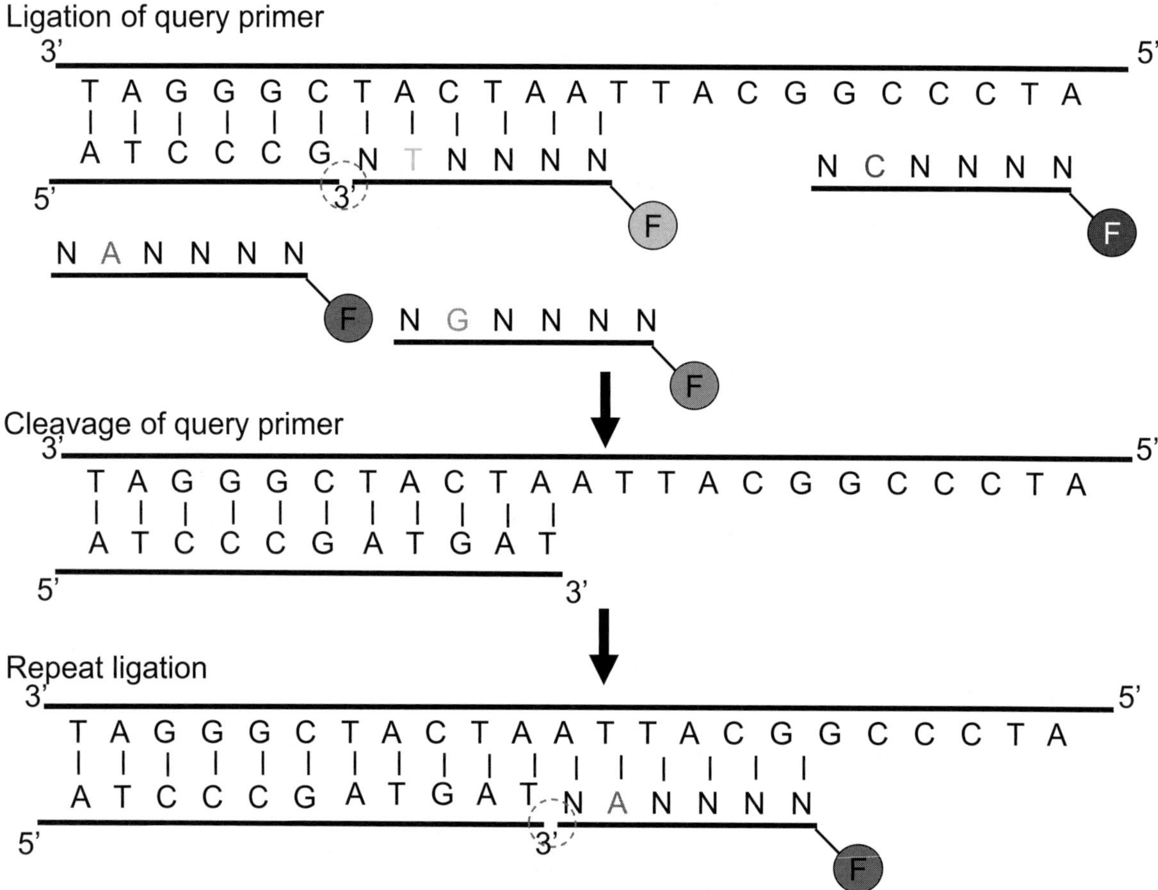

Figure 2.4 Sequencing by ligation. A template strand of DNA is exposed to a population of query primers after hybridizing an anchor primer onto a universal region. These are degenerate for all positions except for the position of interest (2nd shown). The nucleotide in the position of interest will determine what fluorophore is attached to this query primer. The query primer will ligate on, allowing imaging to decode the base at the position of interest. This query primer is then cleaved, either enzymatically or chemically, releasing the fluorophores and exposing a new ligation site. Ligation is repeated to obtain further positions. (A black and white version of this figure will appear in some formats. For the color version, please refer to the plate section.)

synthesis. Similarly to SBS, SBL can be performed on DNA amplified on beads or DNA clusters, and many different types of enzymes or chemicals can be used to cleave the query primer. Sequencing by ligation can suffer from phasing errors as well through repeated ligation and cleavage, but the problem is reduced through changing anchor primers, the entire array is "reset" by washing with a buffer that strips and single strands the DNA.

Sequencing through DNA observation

In addition to methods that involve sequencing a template strand by building a complementary sequencing strand through SBS or SBL, there are emerging methods of sequencing that focus on observing certain traits of the DNA itself. These methods are always single-molecule sequencing methods and, at the time of writing, are not commercially available [10].

A key to observing DNA is to make the DNA single stranded and pulling this single-stranded DNA through a detector or a nanopore. As the DNA passes through the nanopore, a detector must measure the electrical current, which is different for each of the individual nucleotides as they pass through the nanopore [11].

In addition there is also a method to sequence DNA by directly visualizing it using electron microscopy. This involves stretching the DNA on a surface and visualizing the DNA by conjugating metal ions to specific nucleotides, which are read out in the respective order using electron microscopy [12].

Analysis of sequencing information

Sequencing the genome was a monumental task in itself, but deciphering the data is critical and complicated. The human genome is three billion base pairs long, and humans are diploid, and thus each individual carries two sets of homologous chromosomes. Furthermore, the genome is not simply a random arrangement of the four bases. If it were random, sequencing it would be a lot easier. However, when mother nature finds a motif or a protein shape that functions well, she will use it again and again. While this conservation of form and function is elegant and pragmatic, it makes sequencing difficult. These motifs, and regions of similarities may span hundreds of bases and may be located far apart. There are also regions of extreme redundancy called microsatellites, where short patterns, one to six bases in length, will repeat over and over again. These traits make the genome difficult to sequence, but there are sequencing methods to mitigate these obstacles [13].

Various sequencing technologies have varying read lengths and the longer the read length, the easier it is to sequence redundant regions of the genome since many sequencing reads will contain part of the redundant region as well as more uniquely identifying adjacent regions. Long read length assists greatly in allowing one to align, or put together, the sequences obtained from a sequencing run. Another important trait to consider is how many reads one can obtain from the genome. The number of reads multiplied by the average read length gives the total number of bases sequenced, and this product divided by the genome's size (three billion for humans) gives us the coverage. Coverage is important for identifying single nucleotide variations (SNVs), since an altered base pair will not align to a reference genome, it is necessary to resequence that difference to gain confidence [13]. It is estimated that to identify a large percentage of SNVs would require a coverage of 30×, or at least 100 gigabases of sequence. Lastly, the raw accuracy of the sequencing method must be taken into account. Most current next-generation sequencing methods can generate sequence with 98 to 99% raw accuracy.

These factors impact the ability to assemble the sequencing information into a genome. It takes much more information with longer reads to assemble a genome without a reference, or *de novo* sequencing. When a reference sequence is available, shorter reads can be tolerated, since these reads can be aligned to a completed reference genome. This is the most common method for human genome sequencing today; however, information is lost with this approach, namely, information regarding structural variation cannot be resolved from these sequencing studies. Additionally, the phasing of the SNVs is also not determined; in other words, which of the two homologous chromosomes contains which variant cannot be determined.

The goal of genome sequencing is ultimately to use this information to improve medical treatment for various disease states that are influenced by genetic factors, such as heart disease and, of course, cancers [14, 15]. The strategy is to catalogue genetic differences that have led to the development of cancer, as well as use this information to engineer specifically targeted therapeutic measures. The sequence information and what can be inferred varies depending on the nature of the information, how the sequence was obtained, and what it was compared to.

The practice of associating disease states with specific genome information is called genome-wide association studies (GWAS). Such studies were initially performed with microarrays that targeted specific candidate genes and known single nucleotide polymorphisms (SNPs) across the genome. However, in cancer the problem is much more difficult, because it is unlikely that a single SNV (such as the single base difference in a chloride ion channel that leads to cystic fibrosis) may be the direct cause of disease. There are a myriad of genes that contribute and protect against tumor progression, all of which interact in a manifold of ways. Genome-wide association studies therefore require significant sample sizes, and detailed genomic information to determine the nature of the SNVs as they pertain to cancer. Each SNV confers a small percentage of increased or decreased protection to cancer, whether they act in DNA repair pathways, cell growth or metastasis. However, as genome sequencing technology has advanced, it is not simply a matter of categorizing SNVs in patient samples and determining novel cancer genes, but also genome rearrangements or copy number arrangements [16, 17].

Single nucleotide variations

Complete genome sequencing can reveal information about SNVs, which, in turn, can provide information about the resulting protein after translation if the SNV resides in an exon. Even if an SNV is not located within an exon, changes to promoter regions, for example, may impact the transcription of a gene and the subsequent amount of protein product that may then affect cancer development [18].

These SNVs can be substitutions from one base pair to another, which may result in the usual gamut of synonymous, non-synonymous, or non-sense mutations, which may or may not change the amino acid and the protein produced. In addition, there could be insertions or deletions (sometimes collectively referred to as indels), which can also result in a frame shift that completely alters the protein product made.

Cancer sequencing requires a high coverage to accurately detect SNVs. Therefore, high coverage, or repeatedly sequencing the same SNV-containing region many times will allow the SNV to be called with confidence. Without high coverage, the sequence information may simply be thrown out, incorrectly labelled with inaccuracy in the sequence acquisition itself.

Structural variations

Chromosomal rearrangements may be caused by a number of factors, and there is a range in consequences for these events. Even between healthy individuals, genome structure will vary without observable detrimental effects. However, it is also clear that rearrangements can have effects on disease states [19–21].

To obtain information about rearrangements, translocations, insertions, and deletions, genome sequences over a wide range must be obtained, even if that entire range isn't sequenced directly. In other words, mate-pair sequencing is crucial for discerning structural variations. In mate-pair sequencing, two short reads are obtained, but in addition to the sequence, the relative position of these two reads is known. This knowledge of how these two reads are connected is critical for uncovering structural variation. For example, if one cannot map the two short reads to an area in the reference genome, but find that the mate-pairs map too close or too far, it is possible to make inferences about whether a large indel is involved or if that region was rearranged completely. The key is having a library that is constructed with the mate-pair design, as well as having an adequate coverage to increase the confidence of found structural changes.

Copy number variations (CNVs) are another type of structural variation, similar to indels, involving either the deletion or duplication of large parts of the genome, which results in increased or decreased, or even deleted copies of genes [22, 23]. Copy number variations can effectively result in the under-expression of key tumor suppressors or over-expression of oncogenes, resulting in cancer development. Copy number variations are obtainable from genome sequencing, although there are optimized protocols to specifically identify these. Identification of CNVs with genome sequencing can be difficult, and special attention during the sequencing and library preparation must be made if this information is desired.

Somatic mutations and inheritance

A cancer genome will contain more sequence variants than a "normal" germline genome. Specifically, in addition to the natural SNPs in the individual, the tumor will also contain a number of somatic mutations and structural changes [24–26]. Therefore sequencing a genome that comes from a cancer patient's tumor will identify many more alterations in the genome than sequencing a genome from non-tumor tissue. It is assumed that an individual develops cancer due to mutations occurring in cells that results in those cells being positively selected for in terms of growth. There are many key areas in cell growth and regulation that need to be perturbed to allow tumor development: DNA repair pathways, cell growth and division, apoptosis, etc. Therefore the differences between the tumor genome and the germline genome are considered somatic mutations. These somatic mutations are considered important because a subset of these mutations may give rise to tumorigenesis.

This gives researchers options when comparing cancer genomes in order to obtain the information they consider relevant. Comparison of a patient's germline genome with reference genomes will assist in finding inherited genes that may have contributed to or increased a patient's risk for cancer. On the other hand, comparison of a patient's germline and tumor genome will reveal a list of somatic mutations that may have led to the development of cancer. There is a risk, however, in identifying somatic mutations because one of the hallmarks of cancer development is lax DNA repair and reduced apoptosis. Therefore a cancer genome will have many mutations that have nothing to do with cancer development because pathways that would normally stop further mutations have already been disabled.

Drivers and passengers

The difference between a mutation that leads to cancer development and those mutations that are merely the result of a cancerous cell allowing other mutations to randomly arise are the difference between so-called driver and passenger mutations. Drivers are present due to selection during cancer development, whereas passengers have been mutated and have no functional consequence [27, 28]. Consequently, on top of analyzing data and statistically determining which mutations are even real, one must determine which mutations are important. Experimental verification of a potential driver mutation would be time consuming, requiring careful bench science experiments with observations of knockouts and knockdowns of the candidate genes. Depending on the organism used, results may or may not even be relevant. Experimental verification would also run counter to how data from genome sequencing is generated, which is a discovery-based approach to research. There are other potential methods reliant upon pre-existing knowledge about genes and their function, where mutation driver or passenger status can be verified with a literature search. However, this still falls in the same trap of requiring time-consuming experimental verification.

To discover and classify driver and passenger mutations and genes through genome sequencing alone would require a much larger sample size. Only through a large database of high-quality genome sequences will true driver mutations be made evident. Different cancer types most likely have different somatic evolution, creating a need for a large sample size of human genomes, but also for patient genome information for each specific cancer [27, 28].

There are computational approaches that have been developed to look at the complete set of somatic mutations and identify putative cancer genes, or basically separate out the driver and passenger mutations (Figure 2.5). To identify somatic mutations, complete genome sequencing on a patient's germline and tumor tissue must be carried out. Comparisons between the two will yield a list of differences that must be processed thoroughly. Single nucleotide variations that, for example, are in introns or are synonymous, are eliminated and classified as passenger somatic mutations. Once driver and passenger mutations are identified, a validation must ultimately be performed to confirm whether these somatic mutations had an effect on cancer development.

Cancer genome sequencing strategies
Single nucleotide polymorphism profiling

Single nucleotide polymorphism (SNP) profiling is not actually sequencing; however, it is a useful, and relatively low cost, genotyping tool for analyzing a large sample size [29–32]. In fact, it is the ability to perform a study with a large sample size that is SNP profiling's greatest strength. Single nucleotide polymorphism profiling is performed with SNP arrays, which have been following similar trajectories as next-generation sequencing in terms of throughput, increased number of SNPs investigated per array, etc. The SNPs on the array may not have anything to do with cancer, but with large sample sizes, regions of the genome can be identified and these regions can be studied in great detail using targeted resequencing strategies on a very large sample size (a sample size much too large for full genome sequencing). In addition, SNP profiling can provide CNV information that will also be very useful in tracking down cancer-causing genes.

Paired-end mapping

Paired-end mapping is a type of genome sequencing strategy that can more effectively provide information about genome structure and variation [33, 34]. Variations in structure can cause varying expression in cancer developmental pathways by altering expression. Similar to SNPs, even healthy individuals will differ greatly in terms of genome structure [35], but there are obviously variations that can lead to an increased risk of cancer. Genomic structural changes may also impact other factors, for example by disrupting exon and intron organization, leading to altered proteins. Additionally, CNV may be affected as well as gene synteny or order.

Paired-end mapping can be performed on many different sequencing methods, whether it is various SBS or SBL methods. Paired-end mapping requires a library that has been mate-paired, where two reads are separated by a known distance. For the specific applications of pair-end mapping, a larger separation

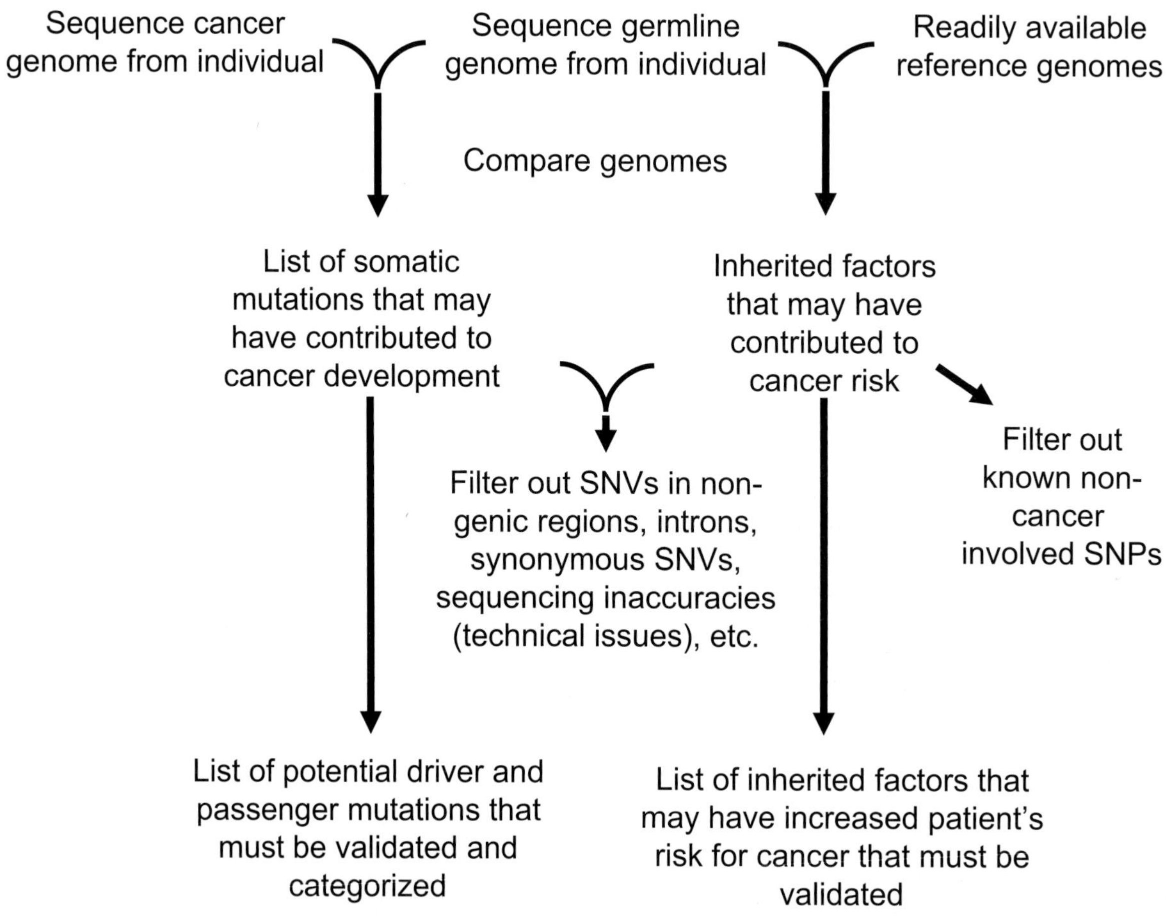

Figure 2.5 Processing SNVs and filtering into somatic mutations. Comparison between germline and tumor genomes provide somatic mutations, whereas comparison between germline and reference genomes can offer information on inherited factors that may have been involved in cancer risk. SNVs that are in non-genic regions, introns, etc. are filtered out, since they either have no effect or don't alter protein function. Validation must follow a highly processed and shortened list of SNVs.

distance is often required, due to the fact that indels may be several kilobases in length. Genome structural variations may be investigated with arrays, but next-generation sequencing methods allow higher resolution mapping.

Targeted resequencing

The human genome is very large, and researchers may be more interested in obtaining more focused information, such as focusing purely on the exome (all exons) [36–38], or even epigenetic changes such as the methylome (all methylated genes) [39, 40], or the kinome (all kinase genes) [38,41]. For example, the exome, in addition to being about 2% of the entire genome, is focused on only the expressed regions of

the genome where many (if not most) of the important somatic driver mutations will lie. Additionally, targeted resequencing could focus on the transcriptome, where sequencing the transcriptome provides information about variation in the expressed exons as well as important information regarding the gene expression level and splice variants. Splice variants as well as expression information on these variants can provide valuable insight into how genomic sequences translate into protein products.

Exome sequencing is the targeted sequencing of all known exons. Exome sequencing has advantages and disadvantages with respect to transcriptome sequencing. First, the advantages of exon sequencing are that all exons are equally represented so the coverage is essentially equal, minus stochastic effects, across all exons, whereas in transcriptome sequencing the highly expressed exons are present in large excess and hence over-sampled, and the lowly expressed exons are often not adequately covered. Also, information is gathered about all exons, not only the expressed exons as in transcriptome sequencing. As in transcriptome sequencing, findings are consolidated into areas of the genome that are translated [42, 43]. The disadvantage to exon sequencing is the complexity associated with isolating all exons from the genome; however, there are currently "kits" available to enrich for all exons and these approaches are becoming easier and cheaper. The most common methods for exon enrichment are PCR and capture-based approaches.

The first complete exon sequencing study was undertaken by Sjoblom et al. [44] in a study focused on colorectal and breast cancer. A total of 11 colorectal cancer samples, along with 11 breast cancer samples, and their corresponding normal tissues were sequenced. The entire exome was sequenced with over 13,000 genes. The identified variations were narrowed down to identify cancer-related mutations by eliminating synonymous mutations, as well as SNVs that were present in the normal germline. This approach has the added benefit of consolidating their findings into exons, which focuses the found changes into actual translated sequence.

Whole genome sequencing

Whole genome sequencing is self-explanatory; sequencing is performed on the entire genome in its entirety with its introns, exons, non-coding regions, repetitive regions, telomeric regions, etc. [45–48]. Everything is obtained and, in effect, will provide all the information that the above methods can give and more, with the exception of expression-based data. In addition to SNVs, no matter whether they reside in introns, exons, and uniquely, non-coding regions will be discovered, as well as structural changes and CNVs. These differences can all be discovered using just whole genome sequencing, as opposed to performing different sequencing methods to get various information [3, 13]. Additionally, chromosomal rearrangements are detectable through whole genome sequencing, as opposed to other methods, which seek to parse down the information. The drawback of whole genome sequencing is the massive and redundant human genome, making whole genome

sequencing expensive and laborious. This often results in greatly reduced sample sizes, making statistically significant observations difficult. Not only is the acquisition of sequence data more stringent in its requirement, but the alignment and assembly of information, even with the aide of a reference, is still problematic. This significant obstacle must be tackled from a computation angle. Regardless, the high-quality information of whole genome sequencing is the most detailed, and therefore has the most potential to be useful.

The first full cancer genome sequencing study was performed by Ley et al. [49]. They used the Illumina sequencing platform, which is SBS based. They were able to identify a complete set of somatic mutations that resulted during tumor progression, and were able to identify ten potential cancer genes with acquired mutations, only two of which were previously described. Large parsing of the data was necessary to find the cancer genes, as the original analysis found 2,647,695 single nucleotide variations (SNVs) after quality control checks; 2,584,418 were also found in the patient's germline, which had to be eliminated. Of the remaining 63,277 genetic variations, 31,645 were previously described in SNP databases, and 20,440 were in the intra-genic regions. This left a total of 11,192 variants. Of these 10,735 were found were in introns, and 216 were in untranslated regions. This left 241 variants, 60 of which were synonymous. The final 181 variants were non-synonymous mutations, which were then actually investigated further using traditional PCR and Sanger methods. Further extension of this vigorous elimination process yielded ten genes with mutations, eight of which were present in nearly all tumor tissue, but whose functions had not previously been described.

Since this first foray into whole cancer genome sequencing, next-generation sequencing methods have continued to be improved and have become even cheaper, allowing more groups to utilize this methodology. The Ley et al. paper, for example, focused primarily on somatic mutations, and categorizing SNVs, insertions, and deletions as passengers and drivers. Cancer genomes sequenced today can also be investigated for chromosomal rearrangements, translocations, and copy number variations. In the near future, we hope to focus on the functional characteristics in the non-coding regions of the genome and the role somatic mutations in these regions have on cancer.

Conclusion

The ability to cheaply generate genome sequences very rapidly will undoubtedly have many medical implications. Ultimately, the value of next-generation sequencing (NGS) technologies will be in the sequencing of large numbers of samples. For example, the ability to sequence hundreds of tumor samples will provide important information toward understanding the microscale evolution that leads to tumor development and will be used to design treatment protocols in the future. Furthermore, sequencing technology is rapidly evolving and will soon allow for large-scale sequencing projects to study thousands of human genomes. Currently, having a personal genome project may be of minimal medical value; however, once many genomes are available, we will have a very powerful tool for uncovering the associations between the genotype and the cancer.

References

1. Collins, F.S., Morgan, M. and Patrinos, A. (2003). The Human Genome Project: lessons from large-scale biology. *Science* **300**, 286–90.

2. Diamandis, E.P. (2009). Next-generation sequencing: a new revolution in molecular diagnostics? *Clin Chem* **55**, 2088–92.

3. Metzker, M.L. (2010). Sequencing technologies – the next generation. *Nat Rev Genet* **11**, 31–46.

4. Reis-Filho, J.S. (2009). Next-generation sequencing. *Breast Cancer Res* **11 Suppl 3**, S12.

5. Shendure, J. and Ji, H. (2008). Next-generation DNA sequencing. *Nat Biotechnol* **26**, 1135–45.

6. Margulies, M. et al. (2005). Genome sequencing in microfabricated high-density picolitre reactors. *Nature* **437**, 376–80.

7. Drmanac, R. et al. (2010). Human genome sequencing using unchained base reads on self-assembling DNA nanoarrays. *Science* **327**, 78–81.

8. Eid, J. et al. (2009). Real-time DNA sequencing from single polymerase molecules. *Science* **323**, 133–8.

9. Flusberg, B.A. et al. (2010). Direct detection of DNA methylation during single-molecule, real-time sequencing. *Nat Methods* **7**, 461–5.

10. Efcavitch, J.W. and Thompson, J.F. (2010). Single-molecule DNA analysis. *Annu Rev Anal Chem (Palo Alto Calif)* **3**, 109–28.

11. Branton, D. et al. (2008). The potential and challenges of nanopore sequencing. *Nat Biotechnol* **26**, 1146–53.

12. Xu, M., Fujita, D. and Hanagata, N. (2009). Perspectives and challenges of emerging single-molecule DNA sequencing technologies. *Small* **5**, 2638–49.

13. Meyerson, M., Gabriel, S. and Getz, G. (2010). Advances in understanding cancer genomes through second-generation sequencing. *Nat Rev Genet* **11**, 685–96.

14. Mardis, E.R. and Wilson, R.K. (2009). Cancer genome sequencing: a review. *Hum Mol Genet* **18**, R163–8.

15. Katsios, C., Zoras, O. and Roukos, D.H. (2010). Cancer genome sequencing and potential application in oncology. *Future Oncol* **6**, 1527–31.

16. Chin, L., Hahn, W.C., Getz, G. and Meyerson, M. (2011). Making sense of cancer genomic data. *Genes Dev* **25**, 534–55.

17. Cui, Q. (2010). A network of cancer genes with co-occurring and anti-co-occurring mutations. *PLoS One* **5**.

18. Goya, R. et al. (2010). SNVMix: predicting single nucleotide variants from next-generation sequencing of tumors. *Bioinformatics* **26**, 730–6.

19. Bignell, G.R. et al. (2007). Architectures of somatic genomic rearrangement in human cancer amplicons at sequence-level resolution. *Genome Res* **17**, 1296–303.

20. Miller, C.A., Hampton, O., Coarfa, C. and Milosavljevic, A. (2011). ReadDepth: a parallel R package for detecting copy number alterations from short sequencing reads. *PLoS One* **6**, e16327.

21. Pang, A.W. et al. (2010). Towards a comprehensive structural variation map of an individual human genome. *Genome Biol* **11**, R52.

22. Ostrovnaya, I., Nanjangud, G. and Olshen, A.B. (2010). A classification model for distinguishing copy number variants from cancer-related alterations. *BMC Bioinformatics* **11**, 297.

23. Rothenberg, S.M. and Settleman, J. (2010). Discovering tumor suppressor genes through genome-wide copy number analysis. *Curr Genomics* **11**, 297–310.

24. Beck, J., Urnovitz, H.B., Mitchell, W.M. and Schutz, E. (2010). Next generation sequencing of serum circulating nucleic acids from patients with invasive ductal breast cancer reveals differences to healthy and nonmalignant controls. *Mol Cancer Res* **8**, 335–42.

25. Bonifaci, N. et al. (2010) Exploring the link between

germline and somatic genetic alterations in breast carcinogenesis. *PLoS One* **5**, e14078.

26. Forbes, S.A. et al. (2010). COSMIC (the Catalogue of Somatic Mutations in Cancer): a resource to investigate acquired mutations in human cancer. *Nucleic Acids Res* **38**, D652–7.

27. Frohling, S. et al. (2007). Identification of driver and passenger mutations of FLT3 by high-throughput DNA sequence analysis and functional assessment of candidate alleles. *Cancer Cell* **12**, 501–13.

28. Youn, A. and Simon, R. (2011). Identifying cancer driver genes in tumor genome sequencing studies. *Bioinformatics* **27**, 175–81.

29. Amato, R. et al. (2009). Genome-wide scan for signatures of human population differentiation and their relationship with natural selection, functional pathways and diseases. *PLoS One* **4**, e7927.

30. Bae, J.S. et al. (2010). Genome-wide profiling of structural genomic variations in Korean HapMap individuals. *PLoS One* **5**, e11417.

31. Easton, D.F. et al. (2007). Genome-wide association study identifies novel breast cancer susceptibility loci. *Nature* **447**, 1087–93.

32. Gamazon, E.R., Zhang, W., Dolan, M.E. and Cox, N.J. (2010). Comprehensive survey of SNPs in the Affymetrix exon array using the 1000 Genomes dataset. *PLoS One* **5**, e9366.

33. Bashir, A., Volik, S., Collins, C., Bafna, V. and Raphael, B.J. (2008). Evaluation of paired-end sequencing strategies for detection of genome rearrangements in cancer. *PLoS Comput Biol* **4**, e1000051.

34. Campbell, P.J. et al. (2008). Identification of somatically acquired rearrangements in cancer using genome-wide massively parallel paired-end sequencing. *Nat Genet* **40**, 722–9.

35. Korbel, J.O. et al. (2007). Paired-end mapping reveals extensive structural variation in the human genome. *Science* **318**, 420–6.

36. Hedges, D.J. et al. (2009). Exome sequencing of a multigenerational human pedigree. *PLoS One* **4**, e8232.

37. Ng, S.B. et al. (2009). Targeted capture and massively parallel sequencing of 12 human exomes. *Nature* **461**, 272–6.

38. Timmermann, B. et al. (2010). Somatic mutation profiles of MSI and MSS colorectal cancer identified by whole exome next generation sequencing and bioinformatics analysis. *PLoS One* **5**, e15661.

39. Cheung, H.H., Lee, T.L., Rennert, O.M. and Chan, W.Y. (2009). DNA methylation of cancer genome. *Birth Defects Res C Embryo Today* **87**, 335–50.

40. Sun, Z. et al. (2011). Integrated analysis of gene expression, CpG island methylation, and gene copy number in breast cancer cells by deep sequencing. *PLoS One* **6**, e17490.

41. Greenman, C. et al. (2007). Patterns of somatic mutation in human cancer genomes. *Nature* **446**, 153–8.

42. Forrest, A.R. and Carninci, P. (2009). Whole genome transcriptome analysis. *RNA Biol* **6**, 107–12.

43. Levin, J.Z. et al. (2009). Targeted next-generation sequencing of a cancer transcriptome enhances detection of sequence variants and novel fusion transcripts. *Genome Biol* **10**, R115.

44. Sjoblom, T. et al. (2006). The consensus coding sequences of human breast and colorectal cancers. *Science* **314**, 268–74.

45. Ahn, S.M. et al. (2009). The first Korean genome sequence and analysis: full genome sequencing for a socio-ethnic group. *Genome Res* **19**, 1622–9.

46. Li, R. et al. (2010). De novo assembly of human genomes with massively parallel short read sequencing. *Genome Res* **20**, 265–72.

47. Pelak, K. et al. (2010). The characterization of twenty sequenced human genomes. *PLoS Genet* **6**.

48. Pleasance, E.D. et al. (2010). A small-cell lung cancer genome with complex signatures of tobacco exposure. *Nature* **463**, 184–90.

49. Ley, T.J. et al. (2008). DNA sequencing of a cytogenetically normal acute myeloid leukaemia genome. *Nature* **456**, 66–72.

Chapter

3

Application of bioinformatics to analyze the expression of tissue-specific and housekeeping genes in cancer

Xijin Ge

Introduction

During the tumorigenesis process, cells can become less differentiated while accumulating genetic mutations that favor proliferative growth. Dedifferentiation is one of the few features (Gabbert et al., 1985) shared by many types of cancer, an extremely heterogeneous disease. During the process, cells can partly lose their committed tissue identity. In fact, the rationale in the pathological tumor grading systems is based on the observation that poorly differentiated tumors are often associated with higher invasiveness and worse prognosis than differentiated tumors of the same tissue of origin (Dukes and Bussey, 1958; Friedell et al., 1976; Thynne et al., 1980). One could speculate that the transformation into a stem cell-like state would be more favorable for the growth and metastasis of cancer cells.

At the molecular level, the dedifferentiation process is associated with changes in the transcriptome. Loss of expression of tissue-specific genes is often observed. Surprisingly, there are few systematic studies on the expression of tissue-specific genes in cancers. This chapter reviews work by the author and others.

The study of tissue-specific gene expression signature is also important clinically. It is estimated that 3 to 5% of all diagnosed cancers are carcinoma of unknown primary (CUP) (Varadhachary et al., 2004). Identification of the primary sites of these tumors is a challenge. Conventional approaches often involve imaging, laboratory tests (liver and kidney function tests, etc.), serum tumor markers (such as alpha-fetoprotein or AFP, and prostate-specific antigen or PSA), as well as immunohistochemistry staining (Carlson, 2009). Better understanding of the gene expression program will facilitate new approaches such as serial analysis of gene expression (Buckhaults et al., 2003), DNA microarray and real-time quantitative polymerase chain reaction (RT-PCR) based approaches. Many techniques have been developed. Some have been approved by the US Food and Drug Administration (FDA) and are available commercially (see next section for details).

Figure 3.1 outlines the flow of work to be presented in this chapter. Systematic gene expression profiling of a variety of normal human tissues was first performed. The data revealed genes whose expression is restricted to specific tissues (tissue-specific genes) as well as others that are expressed ubiquitously (housekeeping or maintenance genes). Taking advantage of available gene expression data for different cancers, my group examined the expression of both types of genes in different types of cancers. The tissue-specific gene expression in cancer was previously reported in *Genomics* (Ge et al., 2005).

Genomic technology for identifying tissue of origin for metastatic cancer

Identification of the origin of metastatic cancer is beneficial to cancer patients because targeted treatments can sometimes be used for particular cancer types. A variety of genomics techniques have been developed to overcome the limitations of conventional diagnostic procedures.

Table 3.1 lists some of these techniques and their accuracy (see Monzon and Koen, 2010, for a more comprehensive review). In a meta-analysis of five published immunohistochemical (IHC) performance studies to identify the primary site of metastatic tumors, on average 65.6% were correctly identified (Anderson and Weiss, 2010). An RT-PCR assay

---

Systems Biology of Cancer, ed. S. Thiagalingam. Published by Cambridge University Press. © Cambridge University Press 2015.

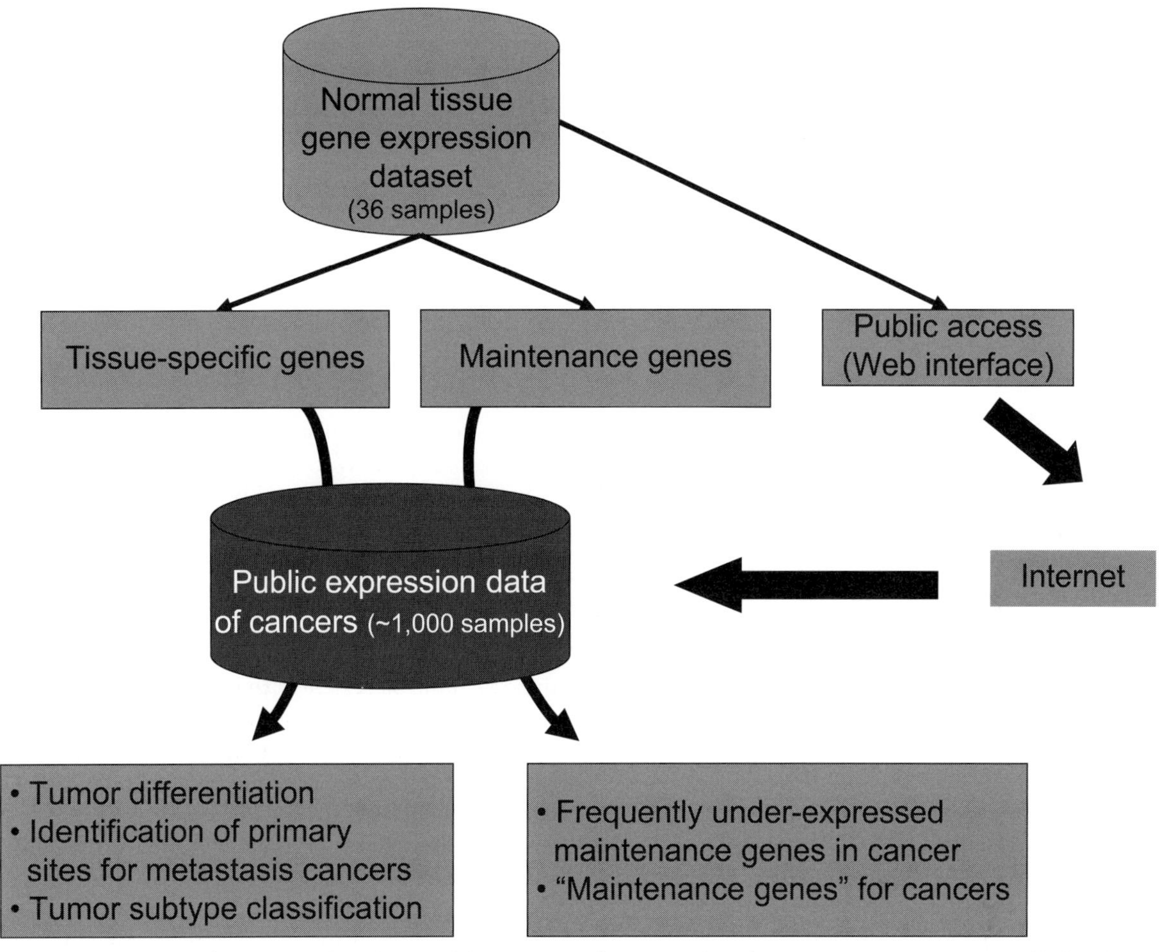

Figure 3.1 Flow chart of a systems biological study of the expression of tissue-specific and maintenance genes in cancer. (A black and white version of this figure will appear in some formats. For the color version, please refer to the plate section.)

targeting gene expression of ten genes showed 76% accuracy when tested in 48 formalin-fixed, paraffin-embedded (FFPE) samples derived from six types of tissue. The accuracy dropped to 61% when a larger set of samples was tested (Varadhachary et al., 2008). A linear discriminant analyses (LDA) method was used for classification. A similar assay with 92 genes accurately (87%) classified 36 types of archival tumors (Ma et al., 2006) with the basic classification algorithm of k-nearest neighbor (kNN).

Based on oligonucleotide microarrays, a 1,550-gene customized chip for "tissue of origin test" was developed by Pathwork Diagnostics, CA (Monzon and Dumur, 2010; Monzon et al., 2009; Wu et al., 2010). This commercially available test can detect 15 tumor types with high accuracy (84.5%). This technology has been tested thoroughly by the developer and associated scientists and has been approved by the FDA. Based on similarity score of expression patterns, the algorithm is also relatively simple.

MicroRNAs (miRNAs) play important roles in developmental processes. Their pattern of expression therefore could be used to predict tissue of origin of metastatic tumors. Rosenfeld et al. used RT-PCR to evaluate the expression profile of 48 miRNAs and used a decision tree and kNN to predict origins of tumors (Rosenfeld et al., 2008). Their method was 85% accurate in a validation study of 25 distinct tissue types (Rosenwald et al., 2010). The technology is commercially available as "miRview[TM] mets" through Rosetta Genomics (Rehovot, Israel) as a part of the miRview[TM] products using miRNAs for diagnosis.

Table 3.1 Existing methods for identification of primary sites of metastatic cancer.

Technology	Algorithm	Number of features	Accuracy	Samples tested	Reference
Immunohistochemical (IHC) staining		One or more markers	82.3% blended; 65.6% metastatic		(Reviewed by Anderson and Weiss, 2010)
Gene expression RT-PCR assay "CUP assay"	Linear discriminant analysis (LDA)	10 genes	76% metastatic	48 FFPE	(Talantov et al., 2006)
			61% metastatic	104 FFPE	(Varadhachary et al., 2008)
microRNA expression by qRT-PCR	Decision tree and k-nearest neighbor (kNN)	48 miRNAs	85%	204 FFPE	(Rosenfeld et al., 2008; Rosenwald et al., 2010)
Gene expression via RT-PCR	k-nearest neighbor (kNN)	92	87%	119 FFPE	(Ma et al., 2006)
Pathwork tissue-of-origin test (DNA microarray)	Similarity score of expression pattern	1,550 genes	84.5% metastatic; 87.8% blended	547 Fresh Frozen	(Monzon et al., 2009)
		2,000 genes	89%	352 FFPE	(Pillai et al., 2011)

Many of the techniques, however, have not been independently tested or validated. Large-scale, independent validation studies are expected to be performed in the next few years. At the same time, other emerging technologies, such as large-scale epigenomics profiling, are likely to introduce new methods for the identification of tissue origin. DNA methylation is a major regulatory mechanism in the differentiation process and studies have confirmed that different tissue types show marked differences in methylation status of CpG islands (De Bustos et al., 2009).

Expression of tissue-specific genes in tumors

Identification of tissue-specific genes

As a starting point, Affymetrix U133A microarrays are used to analyze 36 types of normal human tissues represented by pooled RNA samples (Ge et al., 2005). These tissues include seven different sections of the brain as well as several fetal tissues (brain, liver and lung). Based on the expression level of ~20,000 transcripts, we identified 1,687 *tissue-specific* genes that are exclusively expressed in one particular tissue. Using a combination of empirical criteria and statistical measures (Z scores), we identified 401 testis-specific genes, 329 brain-specific genes and 175 genes

that are only expressed in the liver. These genes are related to the physiological function of the particular organs. For example, many of the brain-specific genes are related to function of neuron or glia cells. In addition, we identified 817 *tissue-selective* genes that are highly expressed in a set of related tissue types, such as colon and small intestine.

Hepatocyte-specific expression signature and dedifferentiation of liver cancer

To study the expression of liver-specific genes in cancer, we re-analyzed a DNA microarray dataset of hepatocellular carcinoma (HCC) (Midorikawa et al., 2002). This dataset includes 25 HCC samples with different degrees of differentiation: well, moderately or poorly differentiated. Through hierarchical clustering analysis of the expression data of liver-specific genes, instead of the whole-genome expression data, we observed clearly that these samples form three groups according to their clinically diagnosed degree of differentiation. We observed that some liver-specific genes are significantly down-regulated in poorly differentiated tumors compared to well differentiated ones. Also, we noticed a general tendency of increased levels of expression of a subset of 64 liver-specific genes in the order of poorly, moderately and well differentiated tumors. Using the kNN method, we can predict the degree of differentiation more accurately based on the 64 liver-specific genes than using all transcripts. Focusing on tissue-specific genes enables us to filter out genes that might be affected by other clinical–pathological parameters, and gain information about tumor differentiation with a higher signal-to-noise ratio.

Interestingly, the genes seem to be divided into several groups. One group is down-regulated in both poorly and moderately differentiated HCC, while another group of genes shows increased expression level in the order of poorly, moderately and well differentiated tumors. The former group includes genes such as SLC22A1 (solute carrier family 22, member 1), CYP2A6 (cytochrome P450, subfamily IIA, polypeptide 6), and ALB (albumin), etc. The latter group includes ADH1A (alcohol dehydrogenase 1A) and AFM (afamin), etc. Another small group of genes, which includes AFP (alpha-fetoprotein) and MKI67 (antigen identified by monoclonal antibody Ki-67), is highly expressed in poorly differentiated HCC. These genes are tissue-specific genes associated with fetal liver. Some of these genes (AFP and Ki-67) are known markers for poor prognosis.

It is important to note that only a small subset of liver-specific genes lost their expression in poorly differentiated tumors. In fact, the majority of liver-specific genes show higher expression even in poorly differentiated HCC when compared to other types of tissues (Figure 3.2).

Expression of neuron and glia-specific genes in brain tumors

Next the expression of brain-specific genes in a microarray dataset of tumors of the central nervous system (CNS) was examined (Pomeroy et al., 2002). The expression data of 329 brain-specific genes were retrieved from this dataset and unsupervised clustering was performed. Our results show brain tumors form two big groups. The first group shows much higher expression of brain-specific genes than the second group. The first group consists of non-embryonal malignant glioma (MG) and medulloblastoma (MD), while the second group includes supratenorial primitive neuroectodermal tumor (PNET) and CNS atypical teratoid/rhabdoid tumor (AT/RT). Within the first group, MD and MG tumors were clearly differentiated from each other, with MD

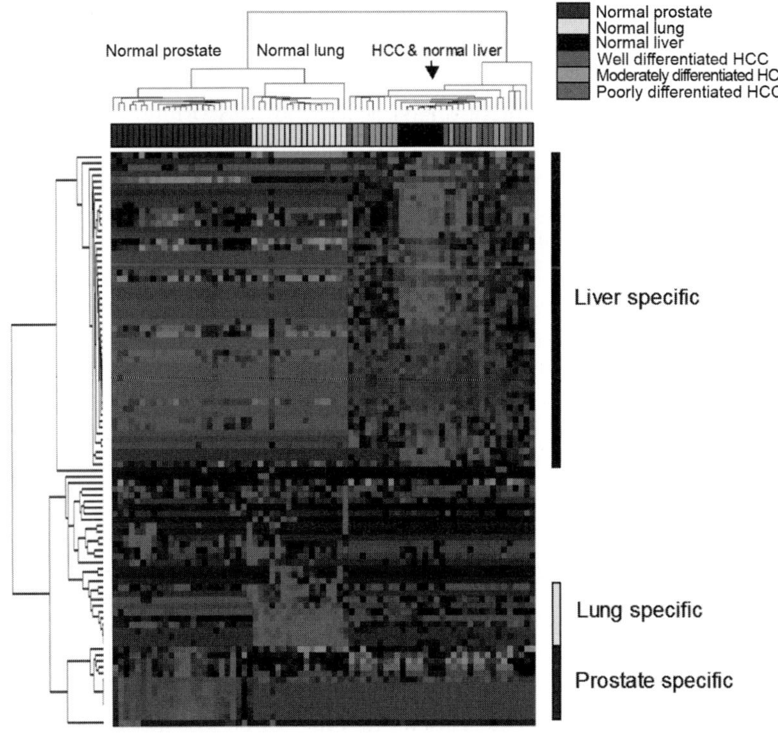

Figure 3.2 Expression of tissue-specific genes in cancer. Only a small portion of tissue-specific genes are under-expressed in poorly differentiated HCC. (A black and white version of this figure will appear in some formats. For the color version, please refer to the plate section.)

showing higher expression in a cluster of genes functionally related to neurons and MG showing elevated expression of glia-related genes. These findings are consistent with our knowledge of the neuronal origin of MD and the glial origin of MG. Even though we started with a list of brain-specific genes, which was a mixture of neuron- and glia-specific genes, expression analyses of these genes in tumors separates them by cell type, as different tumors are the result of expansion of particular cell types. Therefore systematic analyses of the expression of tissue-specific genes in tumors are helpful for studying the cell lineage of tumors.

Breast tumors with two distinct types of differentiation

Starting from a list of genes that are expressed higher in breast tissues than other tissue types, we re-analyzed several breast cancer gene expression datasets. Interestingly, when their expression patterns in breast tumors are analyzed, these genes are further divided into two mutually exclusive groups. This pattern is consistently observed across three independent DNA microarray datasets (Perou et al., 2000; Su et al., 2001; van 't Veer et al., 2002), even though these datasets are based on different patient populations and based on different microarray technology. These two groups of genes define two distinct types of differentiation of breast tumors.

When the expression of these two groups of genes in cell lines is examined, it is apparent that one group of genes is highly expressed in basal epithelial cells while the second group resembles luminal epithelial cells (Ge et al., 2005). In fact, the first group of genes contains conventional markers for breast basal epithelial cells (keratins 5/6 and 17), while the second group includes keratins 8 and 18, which are markers of luminal epithelial cells. This is similar to the phenomenon observed in brain tumors where the clustering of tissue-specific genes pinpoints the cell lineage of tumors.

Of note, some tumors display neither of these two characteristic expression signatures. Also, estrogen receptor (ER) is included in the second group, and ER status is correlated with these two types of differentiation. And, finally, we also found evidence that metastatic tumors in the lung from the breast still show two types of gene expression signatures.

The remarkable heterogeneity of lung cancers

Because of the heterogeneity of lung tumors, all 2,503 tissue-specific genes are used in analyzing the differentiation of a lung tumor dataset. The expression of these genes in a large lung tumor dataset (of Bhattacharjee et al., 2001) was re-analyzed; the dataset includes different lung tumor types: lung adenocarcinoma (AD), squamous cell lung carcinomas (SQ), pulmonary carcinoids (COID), and small-cell lung carcinoma (SCLC). When hierarchical clustering was performed, many different groups of genes defined different tumor subtypes. First, normal lung tissues and some AD tumors showed higher expression of lung-specific genes. Similar to observations for the liver tumors, this expression signature indicates well differentiated tumors.

The second feature is the high-level expression of skin-specific genes in SQ tumors. One of the genes is keratin 16 (KRT16), which is a marker for squamous tumors. Squamous cell tumors are believed to originate from bronchial epithelium and are usually found on central airways (Xing et al., 2010). Therefore the expression of skin-specific genes in SQ is expected.

Consistent with the fact that SCLC and COID are neuroendocrine tumors, these tumors show higher expression of brain-specific genes. Therefore clustering analyses of tissue-specific genes helped identify the original cell types of different tumors.

Interestingly, groups of colon/intestine-specific genes were highly expressed in about a dozen lung tumors. Only some of these tumors were suspected in the original study to be secondary tumors (Bhattacharjee et al., 2001). This expression signature strongly confirms these suspicions, and also provides evidence that some other tumors might be colon metastasis. Similarly, one AD showed higher expression of a set of liver-specific genes, and another tumor showed expression of breast-specific genes. These observations indicate the clustering analyses of tissue-specific genes can help identify the primary origins of metastatic tumor.

Expression of maintenance genes in tumors

During tumorigenesis, cells undergo many genetic and epigenetic changes. The transcription of some genes may be selectively inactivated, whereby significant portions of the chromosome could be lost. It is therefore necessary to study how such events affect the expression of maintenance genes, which are always expressed in all types of tissues under normal circumstances. In this section, maintenance genes that are frequently under-expressed in cancers are identified, followed by others that are always expressed in cancerous tissues.

Maintenance genes frequently under-expressed in cancers: tumor suppressors?

Since maintenance genes are ubiquitously expressed in so many organs, they might play essential roles for normal cellular physiology. Even cancer cells require most maintenance genes. However, a small number of maintenance genes are frequently down-regulated in cancers (illustrated in Figure 3.3).

Several publicly available gene expression datasets constructed for the study of cancer were analyzed. Four datasets of oligonucleotide arrays that contain both normal and cancerous tissues were analyzed. For example, the dataset of Welsh et al. contains the expression of five samples of normal ovary and 29

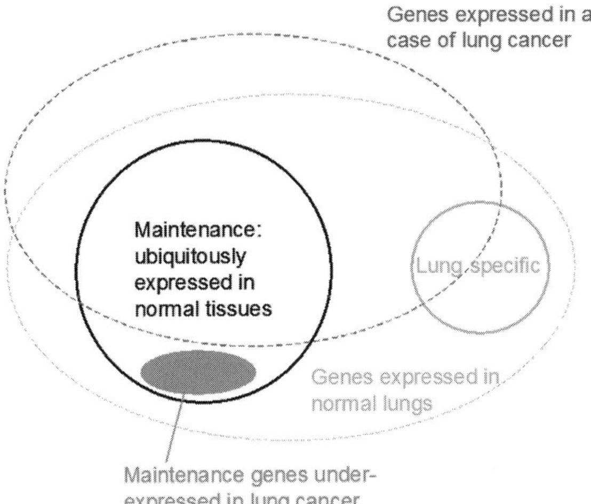

Figure 3.3 Selection of a group of genes ubiquitously expressed in normal tissues, but frequently under-expressed in tumors. Dashed gray circle indicates the genes highly expressed in normal lung samples. These genes include maintenance genes and lung-specific genes. When the total set of genes highly expressed in a case of lung cancer (upper dashed oval) is compared with these categories, some genes are under-expressed. We are interested in the under-expressed maintenance genes (solid gray oval). (A black and white version of this figure will appear in some formats. For the color version, please refer to the plate section.)

samples of ovarian cancer (Welsh et al., 2001). A gene is considered under-expressed if its expression in a cancer sample is reduced by more than five-fold compared to the average expression in normal tissues. A set of 351 genes is under-expressed in more than 20% of cancer samples. Among these genes, only 11 are found on our list of maintenance genes. Table 3.2 lists the top ten frequently under-expressed maintenance genes.

Interestingly, some of the genes are tumor suppressors, while others are related to apoptosis, cell-cycle proliferation and anti-oxidant responses. For example, *BTG2* (BTG family, member 2) is under-expressed in 76% of ovarian cancer samples. As one of the target genes of p53, this gene is an anti-proliferative component of the DNA damage cellular response pathway (Rouault et al., 1996). Montagnoli et al. (1996) found that over-expression of *BTG2* is associated with growth inhibition. Another such gene is *PRSS11* (protease, serine, 11), which is down-regulated during human melanoma progression, and its stable over-expression can strongly inhibit proliferation (Baldi et al., 2002). In addition, under-expression of *DUSP1* (dual specificity phosphatase 1) in ovarian cancer (Manzano et al., 2002; Unoki and Nakamura, 2001) and *CCND2* (cyclin D2) in sporadic breast cancer (Fischer et al., 2002) has been reported. By focusing on down-regulated maintenance genes, the identified set of genes might be important for understanding cancer.

Because these genes are ubiquitously expressed in a variety of normal tissues, but are frequently down-regulated in tumors, they should be investigated further. As in most expression-profiling studies of cancer, it is not known whether the down-regulation of these genes is the major factor responsible for tumorigenesis or is the effect of mutations or deletion of upstream genes. The list of genes should be interpreted in light of the limited available data and of the limitations of microarray technology. By combining expression data of a spectrum of normal tissues and of ovarian cancer, the list of genes may provide candidates for studying cancer with a higher signal-to-noise ratio than a list derived only from ovarian cancer.

Following this strategy, several other datasets were systematically characterized, including lung cancer (Bhattacharjee et al., 2001), leukemia (Yeoh et al., 2002), and a mixed dataset of 14 common tumor types (Ramaswamy et al., 2001). In the dataset of lung cancer, a possible tumor suppressor gene was identified: *TGFBR2* (transforming growth factor, beta receptor II) is located on chromosome 3p22, a region

Table 3.2 List of maintenance genes frequently under-expressed in cancer cells. These genes are ubiquitously expressed in most normal tissues, but show at least a five-fold decrease in their expression levels in some cancer samples (percentage given in the first column). This list includes many tumor suppressors such as BTG2, PRSS11, TGFBR2, and MLH1.

% Under-expression	ProbeSet	Symbol	Title	Location	Sequence ID
Ovarian cancer (5 normal, 27 tumor)					
74%	U72649_at	BTG2	BTG family, member 2	1q32	U72649
63%	M92843_s_at	ZFP36	zinc finger protein 36, C3H type, homolog (mouse)	19q13	M92843
59%	D31884_at	KIAA0063	KIAA0063 gene product	22q13	D31884
52%	X51345_at	JUNB	jun B proto-oncogene	19p13	X51345
52%	X68277_at	DUSP1	dual specificity phosphatase 1	5q34	X68277
37%	D31883_at	ABLIM1	actin binding LIM protein 1	10q25	D31883
30%	U60115_at	FHL1	four and a half LIM domains 1	Xq26	U60115
26%	D78134_at	CIRBP	cold inducible RNA binding protein	19p13	D78134
26%	D87258_at	PRSS11	protease, serine, 11 (IGF binding)	10q26	D87258
15%	D13639_at	CCND2	cyclin D2	12p13	D13639
15%	D84110_at	RBPMS	RNA binding protein gene with multiple splicing	8p12	D84110
15%	J04456_at	LGALS1	lectin, galactoside binding, soluble, 1 (galectin 1)	22q13	J04456
Lung cancer (17 normal, 186 tumor)					
83%	32542_at	FHL1	four and a half LIM domains 1	Xq26	AF063002
52%	39760_at	OKI	homolog of mouse quaking QKI (KH domain RNA binding protein)	6q26-27	AL031781
43%	1814_at	TGFBR2	transforming growth factor, beta receptor II (70/80 kDa)	3p22	D50683
40%	37718_at	SNRK	SNF-1-related kinase	3p21.31	D43636
38%	33862_at	PPAP2B	phosphatidic acid phosphatase type 2B	1pter-p22.1	AF017786
34%	32035_at	HLA-DRS4	major histocompatibility complex, class II, DR beta 4	6p21.3	M16942
30%	1005_at	DUSP1	dual specificity phosphatase 1	5q34	X68277
28%	36915_at	CTSO	cathepsin O	4q31-q32	AI810485
26%	33413_at	PTP4A1	protein tyrosine phosphatase type IVA, member 1	6q12	AF051160
26%	41549_s_at	AP1S2	adaptor-related protein complex 1, sigma 2 subunit	Xp21.3	AF091077
Acute lymphoblastic leukemia (18 normal, 327 cancer)					
17.7%	41136_s_at	APP	amyloid beta (A4) precursor protein (protease nexin-II, Alzheimer disease)	21q21	Y00264
12.6%	36937_s_at	PDLIM1	PDZ and LIM domain 1 (elfin)	10q22	U90878

Table 3.2 (cont.)

% Under-expression	ProbeSet	Symbol	Title	Location	Sequence ID
10.1%	41153_f_at	MAGED1	melanoma antigen, family D, 1	Xpl 1	AF102803
9.5%	40780_at	CTBP2	C-terminal binding protein 2	10q26	AF016507
8.8%	1814_at	TGFBR2	transforming growth factor, beta receptor II (70/80 kDa)	3p22	D50683
8.5%	39839_at	CSDA	cold shock domain protein A	12p13	M24069
8.2%	1102_s_at	NR3C1	nuclear receptor subfamily 3, group C, member 1 (glucocorticoid receptor)	5q31	M10901
8.2%	34685_at	FLJ22028	hypothetical protein FLJ22028	12p12	AI685944
7.6%	1005_at	DUSP1	dual specificity phosphatase 1	5q34	X68277
6%	33261_at	HLA-DRB4	major histocompatibility complex, class II, DR beta 4	6p21	M16941
6%	35259_s_at	SFRS2IP	splicing factor, arginine/serine-rich 2, interacting protein	12q12	Y11251

14 common tumor types (90 normal, 218 tumor)

% Under-expression	ProbeSet	Symbol	Title	Location	Sequence ID
28%	U07418_at	MLH1	mutL homolog 1, colon cancer, nonpolyposis type 2	3p21	U07418
27%	U81556_at	OS4	conserved gene amplified. in osteosarcoma	12q13	U81556
27%	D87258_at	PRSS11	protease, seine, 11 (IGF binding)	10q26	D87258
27%	U34343_at	DAP13	13 kDa differentiation-associated protein (NADH:ubiquinone oxidoreductase)	12q21	U34343
27%	M57710_at	LGALS3	lectin, galactoside binding, soluble, 3 (galectin 3)	14q21	M57710
26%	M28713_at	DIA1	diaphorase (NADH) (cytochrome b-5 reductase)	22q13	M28713
26%	D14662_at	AOP2	anti-oxidant protein 2	1q23	D14662
24%	X(63753_at	SON	SON DNA binding protein	21q22	X63753
23%	U47742_at	ZNF220	zinc finger protein 220	8p11	U47742
22%	U62317 rna3_at				U62317

frequently deleted in solid tumors. This gene might be a potential tumor suppressor, as it reduces tumorigenicity upon transfection (Kok et al., 1997). A gene involved in control of cell growth was also identified: *PTP4A1* (protein tyrosine phosphatase type IVA, member 1) (Diamond et al., 1994).

A dataset consisting of 327 samples of pediatric acute lymphoblastic leukemia (ALL) (Yeoh et al., 2002) was analyzed. Two genes closely related to apoptosis, *APP* (amyloid beta (A4) precursor protein) and *MAGED1* (melanoma antigene, family D, 1), were identified. Note that *MAGED1* is a p75 neurotrophin receptor-interacting protein that can induce caspase activation and cell death through a JNK-dependent mitochondrial pathway (Salehi et al., 2002). *CTBP2* (C-terminal binding protein 2), which is a negative regulator of cell proliferation, was also identified. Another frequently under-expressed maintenance gene is *HLA-DRB4* (major histocompatibility complex, class II, DR beta 4). Polymorphisms of this gene can greatly increase susceptibility to childhood ALL in males (Dorak et al., 1999). This gene is associated with the development of virus-induced or spontaneous leukemia in the mouse, perhaps through deficiency in immune response (Dorak et al., 2002). Again, *TGFBR2* and *DUSP1* are also frequently under-expressed in this leukemia dataset.

A large dataset of 218 tumor samples, spanning 14 common tumor types (Ramaswamy et al., 2001), was analyzed. Because the expression level of maintenance genes varies in different tissues, the detection P values were analyzed instead of the average difference score. In normal tissues, maintenance genes should be confidently detectable, which is characterized by a "present" call based Affymetrix MAS5 algorithm. Thus genes frequently "absent" in tumors were selected. Tumor suppressor gene *MHL1* (mutL homolog 1, colon cancer, non-polyposis type 2), which functions as a mismatch repair gene, was identified. Loss of the wild-type of this gene causes hereditary colon cancer. *PRSS11* gene, a potent tumor suppressor, is again included. *LGALS3*, or galection 3, is believed to be involved in control of cell proliferation and apoptosis. Other genes on the list include *AOP2* (anti-oxidant protein 2), *DIA1* (NADH diaphorase), cytochrome b-5 reductase, and *DAP13* (13 kDa differentiation-associated protein).

Notably, some genes are under-expressed in different types of cancers. These include *DUSP1*, *PRSS11*, *FHL1* and *TGFBR2*. Some of these genes may play important roles in tumorigenesis. For example, *DUSP1* encodes a protein that has intrinsic phosphatase activity and can specifically inactivate mitogen-activated protein (MAP) kinase. In addition to its role in response to oxidative stress, it may be involved in negative regulation of cell proliferation. Additionally, the closely related *DUSP2* was recently shown to be a transcriptional target of p53 in signaling apoptosis and growth suppression (Yin et al., 2003). Therefore further studies are needed to elucidate the function of these genes in normal and tumor cells.

Genes always expressed in tumors define a minimum set of survival genes

As discussed above, genes ubiquitously expressed in normal tissues could be significantly under-expressed in cancer cells. It is easy to imagine, however, that there are other genes that play fundamentally important roles such that a certain level of expression is required for the survival of any kinds of cancer cells. In other words, there should be a minimum set of "critical" maintenance genes whose expression is always required for basic cellular processes of the cell cycle. Cancer cells are the result of multiple iterations of colonial expansion and selection, optimized for aggressive growth, but they should always express these key genes. Analysis of the expression profile of a large ensemble of cancer cells/tissues should always, in theory, define a minimum set of critical maintenance genes.

To identify such genes, the maintenance genes that are commonly expressed in various forms of cancers in different tissues (Figure 3.4) were searched. Starting with the list of maintenance genes, genes whose expression is confidently detectable ("present" calls) in all kinds of tumors were identified. Expression profiles of several types of cancers, namely prostate cancer (Singh et al., 2002), lung cancer (Bhattacharjee et al., 2001), breast cancer (Huang et al., 2003) and ALL (Yeoh et al., 2002), for which reliable data based on the same platform (Affymetrix, Hu-U95) are publicly available were used. "Present" calls were returned by the Affymetrix algorithm in at least 90% of tumors or normal tissues.

Transcripts that represent 331 UniGene clusters were identified. As shown in Figure 3.5, these genes were divided into several functional categories according to gene ontology (GO) annotations. Among the 189 UniGene clusters for which the gene's function is known, 38% are involved in protein synthesis. Many of them are ribosomal proteins. In addition, 12% of the genes are responsible for RNA processing, 10% for energy generation and 10% for transcription. Together, these four categories of genes constitute about 70% of the 189 genes. There are small numbers of genes for cell structure, protein degradation, cell stress, and DNA repair, etc. Therefore most of these genes play fundamentally important cellular roles.

The process of carcinogenesis is often accompanied by damage and deletions to DNA, sometimes even loss of significant portions of a chromosome. For example, loss of 3p is frequently observed in lung cancer. But the inactivation of critical maintenance genes is lethal to cells. With such cells constantly removed from a colony, cancer cells in the final stage must contain all these critical genes for cell survival and division. Thanks to the extensive work done in the area of expression profiling of cancer tissues, a

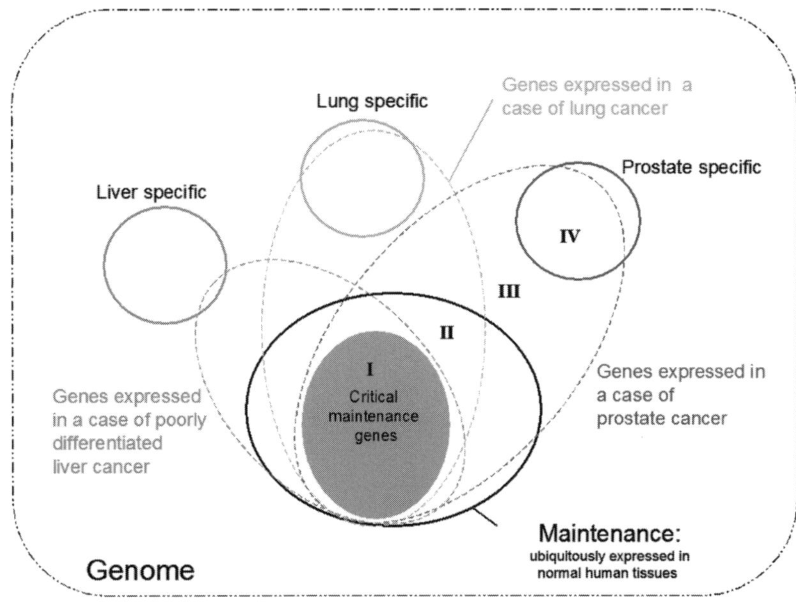

Figure 3.4 "Critical maintenance genes" defined by combining normal and cancer gene expression data. These genes are always expressed in all kinds of cancer cells and may define a minimum set of survival genes. (A black and white version of this figure will appear in some formats. For the color version, please refer to the plate section.)

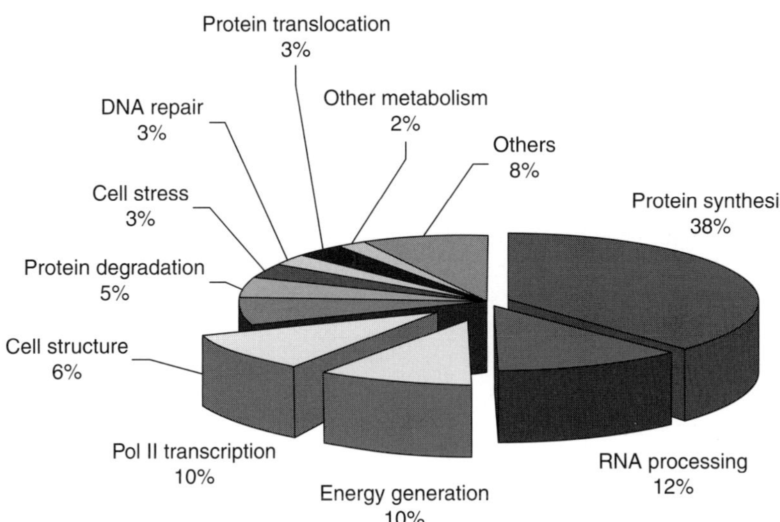

Figure 3.5 Functional categorization of critical maintenance genes according to gene ontology (GO). (A black and white version of this figure will appear in some formats. For the color version, please refer to the plate section.)

"minimal set" of genes required for the survival of cells can be defined.

Since these genes are essential for the survival of single cells, whether cancerous or normal, most of these genes are expected to have orthologs in unicellular species such as yeast. Knockout of such genes should have serious effects in multiple organs of model animals.

Several genes on the list are not well studied, and have unknown functions. These include, for example, genes for protein KIAA0494, KIAA1049, and KIAA0447. If the ubiquitous expression of these genes is not an experimental flaw due to poorly designed probes in the oligonucleotide microarray, which is possible, then determining their cellular functions would be beneficial.

The gene list, including the theoretically "minimum" set of maintenance genes, may be useful for developing cancer treatments. Switching off the expression of critical genes should kill cancer cells. Of course, cancer cells should be targeted initially using other technologies. Finally, 30 genes with relatively constant expression levels in various normal and cancer tissues were selected. The coefficient of variance

Table 3.3 Housekeeping genes that are expressed at constant levels in a variety of normal and cancer cells. Note that ribosomal proteins have been removed from the list. The relative expression level in the right-most column is calculated in average normal tissues by taking the expression level of ACTB as 100.

Rank	Acc#	Symbol	Title	Relative expression level
1	X00351	ACTB	actin, beta	100
2	AB016492	JTB	jumping translocation breakpoint	19
3	X52851	PPIA	peptidylprolyl isomerase A (cyclophilin A)	51
4	A1525652	SRP14	signal recognition particle 14 kDa (homologous Alu RNA binding protein)	18
5	L09159	ARHA	ras homolog gene family, member A	28
6	U73824	EIF4G2	eukaryotic translation initiation factor 4 gamma, 2	32
7	X95404	CFL1	cofilin 1 (non-muscle)	42
8	U41635	OS-9	amplified in osteosarcoma	19
9	AF054187	NACA	nascent-polypeptide-associated complex alpha polypeptide	22
10	AF048977	SRRM1	serine/arginine repetitive matrix 1	5
11	M24194	GNB2L1	guanine nucleotide binding protein (G protein), beta polypeptide 2-like 1	54
12	M74491	ARF3	ADP-ribosylation factor 3	10
13	U17999	BECN1	beclin 1 (coiled-coil, myosin-like BCL2 interacting protein)	9
14	AB007510	PRPF8	PRP8 pre-mRNA processing factor 8 homolog (yeast)	9
15	AB019409	ANAPC5	anaphase promoting complex subunit 5	7
16	U50523	ARPC2	actin-related protein 2/3 complex, subunit 2, 34 kDa	26
17	A1345944	NDUFB1	NADH dehydrogenase (ubiquinone) 1 beta subcomplex, 1, 7 kDa	8
18	D25274	RAC1	ras-related C3 botulinum toxin substrate 1 (rho family, small GTP binding protein Racl)	14
19	M11353	H3F3A	H3 histone, family 3A	28
20	M88108	KHDRBS1	KB domain containing, RNA binding, signal transduction associated 1	4
21	U89322	NPM1	nucleophosrnin (nucleolar phosphoprotein B23, nurnatrin)	23
22	X72727	HNRPK	heterogeneous nuclear ribonucleoprotein K	20
23	M33197	GAPD	glyceraldehyde-3-phosphate dehydrogenase	76
24	A1033692	BANF1	barrier to autointegyation factor 1	10
25	X04347	HNRPA1	heterogeneous nuclear ribonucleoprotein Al	12
26	UO3271	CAPZB	capping protein (actin filament) muscle Z-line, beta	11
27	M10119	FTL	ferritin, light polypeptide	106
28	M17733	TMSB4X	thymosin, beta 4, X-linked	79
29	M16247	ACTG1	actin, gamma 1	50
30	AF044671	GABARAP	GABA(A) receptor-associated protein	17

(CV) of these 331 genes was calculated using each of the four cancer datasets, the dataset of Su et al. (2001) and our normal tissue dataset. Then genes were ranked according to their CV in each dataset. The top 30 genes with smallest average rank were selected (Table 3.3). Notably, the widely used housekeeping gene *ACTB* (actin, beta) appears as number one in this list. Another such gene, *GAPDH* (glyceraldehyde-3-phosphate dehydrogenase), also has a rank of 23. These genes are ubiquitously expressed with relatively constant levels and could serve as positive controls in interpretation of gene expression profiles.

Housekeeping/maintenance genes were previously defined as those genes that are always expressed in cells of a multicellular organism (Watson et al., 1965). The study of gene expression profiles of cancers revealed that some of these genes are significantly under-expressed in cancers. Therefore only a subset of these genes is critical to the survival of single cells. Such genes, referred to as "critical maintenance genes" here, are essential for the activities that must be carried out for successful completion of the cell cycle. Although it is difficult to unequivocally define this critical set of genes, an initial set of about 331 genes is provided based on their constitutive expression in various cancers.

Concluding remarks

Expression profiling of a spectrum of normal human tissues enabled the identification of sets of tissue-specific genes and maintenance genes. Then the expression of these genes in cancers was investigated using a wealth of previously published datasets. The results show that their expression in cancer conveys important information that is often overlooked when datasets are analyzed independently in a stand-alone manner. This systems biological approach to the study of cancer provides insights into the complex disease.

Gene expression profiling of 36 types of normal tissues was performed. Lists of tissue-specific genes and maintenance genes were compiled. A set of liver-specific genes whose expression in HCC reflects the degree of tumor differentiation was identified. By clustering analysis of the expression of brain-specific genes in various brain tumors, neuron-specific expression signatures in medulloblastoma and glia-specific expression signatures in glioma were identified. A small set of 26 genes highly expressed in the normal breast was divided into two groups with mutually exclusive expression in breast tumors and two types of differentiation: basal subtype and luminal subtype. Different subtypes of lung cancers show different patterns of tissue specificity in their gene expression, e.g., high expression of skin-specific genes in SQ and high expression of brain-specific genes in SCLC and COID. Expression signatures of primary sites are detectable in metastatic lung tumors originated from colon, liver and breast. Notably, lung tumors with expression profiles resembling two types of breast cancers were detected. Tissue-specific genes differentially expressed in tumors reveal the cell lineage of different kinds of tumors. Conversely, these genes define the expression signatures of different cell types. Hence, the expression signatures of hepatocytes, luminal and basal cells in mammary gland, and glial and neuron cells in the brain were observed.

Through the analysis of maintenance genes in cancers, a small set of genes frequently under-expressed in cancers was identified, some of which have tumor-suppressive functions. Critical maintenance genes consistently expressed in all kinds of cancer and normal tissues were identified, and their fundamental roles in basic cellular processes were discussed. Theoretically, genes always expressed in all kinds of tumors define a minimum set of maintenance genes for cell survival. Some of these genes with the smallest variances in gene expression are listed in Table 3.3. These genes can serve as a control for measuring expression levels of other genes.

Acknowledgment

This material is based on work supported by the National Science Foundation/EPSCoR Award No. 11A-1355423 and by the State of South Dakota.

References

Anderson, G.G., and Weiss, L.M. (2010). Determining tissue of origin for metastatic cancers: meta-analysis and literature review of immunohistochemistry performance. *Appl Immunohistochem Mol Morphol* 18, 3–8.

Baldi, A., De Luca, A., Morini, M., et al. (2002). The HtrA1 serine protease is down-regulated during human melanoma progression and represses growth of metastatic melanoma cells. *Oncogene 21*, 6684–6688.

Bhattacharjee, A., Richards, W.G., Staunton, J., et al. (2001).

Classification of human lung carcinomas by mRNA expression profiling reveals distinct adenocarcinoma subclasses. *Proc Natl Acad Sci USA 98*, 13790–13795.

Buckhaults, P., Zhang, Z., Chen, Y.C., et al. (2003). Identifying tumor origin using a gene expression-based classification map. *Cancer Res 63*, 4144–4149.

Carlson, H.R. (2009). Carcinoma of unknown primary: searching for the origin of metastases. *JAAPA 22*, 18–21.

De Bustos, C., Ramos, E., Young, J.M., et al. (2009). Tissue-specific variation in DNA methylation levels along human chromosome 1. *Epigenetics Chromatin 2*, 7.

Diamond, R.H., Cressman, D.E., Laz, T.M., Abrams, C.S., and Taub, R. (1994). PRL-1, a unique nuclear protein tyrosine phosphatase, affects cell growth. *Mol Cell Biol 14*, 3752–3762.

Dorak, M.T., Lawson, T., Machulla, H.K., et al. (1999). Unravelling an HLA-DR association in childhood acute lymphoblastic leukemia. *Blood 94*, 694–700.

Dorak, M.T., Oguz, F.S., Yalman, N., et al. (2002). A male-specific increase in the HLA-DRB4 (DR53) frequency in high-risk and relapsed childhood ALL. *Leuk Res 26*, 651–656.

Dukes, C.E., and Bussey, H.J. (1958). The spread of rectal cancer and its effect on prognosis. *Br J Cancer 12*, 309–320.

Fischer, H., Chen, J., Skoog, L., and Lindblom, A. (2002). Cyclin D2 expression in familial and sporadic breast cancer. *Oncol Rep 9*, 1157–1161.

Friedell, G.H., Bell, J.R., Burney, S.W., Soto, E.A., and Tiltman, A.J. (1976). Histopathology and classification of urinary bladder carcinoma. *Urol Clin North Am 3*, 53–70.

Gabbert, H., Wagner, R., Moll, R., and Gerharz, C.D. (1985). Tumor dedifferentiation: an important step in tumor invasion. *Clin Exp Metastasis 3*, 257–279.

Ge, X., Yamamoto, S., Tsutsumi, S., et al. (2005). Interpreting expression profiles of cancers by genome-wide survey of breadth of expression in normal tissues. *Genomics 86*, 127–141.

Huang, E., Cheng, S.H., Dressman, H., et al. (2003). Gene expression predictors of breast cancer outcomes. *Lancet 361*, 1590–1596.

Kok, K., Naylor, S.L., and Buys, C.H. (1997). Deletions of the short arm of chromosome 3 in solid tumors and the search for suppressor genes. *Adv Cancer Res 71*, 27–92.

Ma, X.J., Patel, R., Wang, X., et al. (2006). Molecular classification of human cancers using a 92-gene real-time quantitative polymerase chain reaction assay. *Arch Pathol Lab Med 130*, 465–473.

Manzano, R.G., Montuenga, L.M., Dayton, M., et al. (2002). CL100 expression is down-regulated in advanced epithelial ovarian cancer and its re-expression decreases its malignant potential. *Oncogene 21*, 4435–4447.

Midorikawa, Y., Tsutsumi, S., Taniguchi, H., et al. (2002). Identification of genes associated with dedifferentiation of hepatocellular carcinoma with expression profiling analysis. *Jpn J Cancer Res 93*, 636–643.

Montagnoli, A., Guardavaccaro, D., Starace, G., and Tirone, F. (1996). Overexpression of the nerve growth factor-inducible PC3 immediate early gene is associated with growth inhibition. *Cell Growth Differ 7*, 1327–1336.

Monzon, F.A., and Dumur, C.I. (2010). Diagnosis of uncertain primary tumors with the Pathwork tissue-of-origin test. *Expert Rev Mol Diagn 10*, 17–25.

Monzon, F.A., and Koen, T.J. (2010). Diagnosis of metastatic neoplasms: molecular approaches for identification of tissue of origin. *Arch Pathol Lab Med 134*, 216–224.

Monzon, F.A., Lyons-Weiler, M., Buturovic, L.J., et al. (2009). Multicenter validation of a 1,550-gene expression profile for identification of tumor tissue of origin. *J Clin Oncol 27*, 2503–2508.

Perou, C.M., Sorlie, T., Eisen, M.B., et al. (2000). Molecular portraits of human breast tumours. *Nature 406*, 747–752.

Pillai, R., Deeter, R., Rigl, C.T., et al. (2011). Validation and reproducibility of a microarray-based gene expression test for tumor identification in formalin-fixed, paraffin-embedded specimens. *J Mol Diagn 13*, 48–56.

Pomeroy, S.L., Tamayo, P., Gaasenbeek, M., et al. (2002). Prediction of central nervous system embryonal tumour outcome based on gene expression. *Nature 415*, 436–442.

Ramaswamy, S., Tamayo, P., Rifkin, R., et al. (2001). Multiclass cancer diagnosis using tumor gene expression signatures. *Proc Natl Acad Sci USA 98*, 15149–15154.

Rosenfeld, N., Aharonov, R., Meiri, E., et al. (2008). MicroRNAs accurately identify cancer tissue origin. *Nat Biotechnol 26*, 462–469.

Rosenwald, S., Gilad, S., Benjamin, S., et al. (2010). Validation of a microRNA-based qRT-PCR test for accurate identification of tumor tissue origin. *Mod Pathol 23*, 814–823.

Rouault, J.P., Falette, N., Guehenneux, F., et al. (1996). Identification of BTG2, an antiproliferative p53-dependent component of the DNA damage cellular response pathway. *Nat Genet 14*, 482–486.

Salehi, A.H., Xanthoudakis, S., and Barker, P.A. (2002). NRAGE, a p75 neurotrophin receptor-interacting protein, induces caspase activation and cell death through a JNK-dependent mitochondrial pathway. *J Biol Chem 277*, 48043–48050.

Singh, D., Febbo, P.G., Ross, K., et al. (2002). Gene expression correlates of clinical prostate cancer behavior. *Cancer Cell 1*, 203–209.

Su, A.I., Welsh, J.B., Sapinoso, L.M., et al. (2001). Molecular classification of human carcinomas by use of gene expression signatures. *Cancer Res 61*, 7388–7393.

Talantov, D., Baden, J., Jatkoe, T., et al. (2006). A quantitative reverse transcriptase-polymerase chain reaction assay to identify metastatic carcinoma tissue of origin. *J Mol Diagn 8*, 320–329.

Thynne, G.S., Weiland, L.H., Moertel, C.G., and Silvers, A. (1980). Correlation of histopathologic characteristics of primary tumor and uninvolved regional lymph nodes in Dukes' class C colonic carcinoma with prognosis. *Mayo Clin Proc 55*, 243–245.

Unoki, M., and Nakamura, Y. (2001). Growth-suppressive effects of

BPOZ and EGR2, two genes involved in the PTEN signaling pathway. *Oncogene 20*, 4457–4465.

van 't Veer, L.J., Dai, H., van de Vijver, M.J., et al. (2002). Gene expression profiling predicts clinical outcome of breast cancer. *Nature 415*, 530–536.

Varadhachary, G.R., Abbruzzese, J.L., and Lenzi, R. (2004). Diagnostic strategies for unknown primary cancer. *Cancer 100*, 1776–1785.

Varadhachary, G.R., Talantov, D., Raber, M.N., et al. (2008). Molecular profiling of carcinoma of unknown primary and correlation with clinical evaluation. *J Clin Oncol 26*, 4442–4448.

Watson, J.D., Hopkins, N.H., Roberts, J.W., Steitz, J.A., and Weiner, A.M. (1965). The functioning of higher eukaryotic genes. In Watson, J.D. (ed.) Molecular Biology of the Gene. Vol.1. p. 704. New York: Benjamin.

Welsh, J.B., Sapinoso, L.M., Su, A.I., et al. (2001). Analysis of gene

expression identifies candidate markers and pharmacological targets in prostate cancer. *Cancer Res 61*, 5974–5978.

Wu, A.H., Drees, J.C., Wang, H., et al. (2010). Gene expression profiles help identify the tissue of origin for metastatic brain cancers. *Diagn Pathol 5*, 26.

Xing, L., Todd, N.W., Yu, L., Fang, H., and Jiang, F. (2010). Early detection of squamous cell lung cancer in sputum by a panel of microRNA markers. *Mod Pathol 23*, 1157–1164.

Yeoh, E.J., Ross, M.E., Shurtleff, S.A., et al. (2002). Classification, subtype discovery, and prediction of outcome in pediatric acute lymphoblastic leukemia by gene expression profiling. *Cancer Cell 1*, 133–143.

Yin, Y., Liu, Y.X., Jin, Y.J., Hall, E.J., and Barrett, J.C. (2003). PAC1 phosphatase is a transcription target of p53 in signalling apoptosis and growth suppression. *Nature 422*, 527–531.

Chapter 4

Events at DNA replication origins and genome stability

Kathleen R. Nevis, Kimberly L. Raiford, Cyrus Vaziri and Jeanette Gowen Cook

Introduction

The cell cycle and genome stability

The differences between cancer cells and normal cells can be largely attributed to constellations of genetic alterations in the cancer cells that disrupt growth regulatory pathways. These alterations arise from errors during the cell division cycle that lead to permanent genetic changes in the daughter cells. When mutations occur in the genes for key regulators of cell cycle control, subsequent cell divisions are more likely to generate additional mutations. This *genome instability* provides the opportunity for a single cellular descendant to acquire a combination of mutations that confers the transformed phenotype. For this reason, understanding the mechanisms of genome instability is important for understanding the process of oncogenesis and developing novel strategies for the diagnosis and treatment of cancer.

The primary goal of the cell division cycle is the faithful replication and segregation of chromosomal DNA (Figure 4.1). Human cells must accurately reproduce exactly one copy of the more than 6 billion bases of DNA (3 billion base pairs) that constitute the nuclear genome in a surprisingly short amount of time, which is on average eight hours in culture. This enormous amount of DNA is packaged by histone proteins into chromatin fibers of varying degrees of accessibility. Moreover, DNA replication occurs simultaneously with gene transcription, DNA repair, and chromosome condensation as cells prepare for mitotic chromosome segregation. These and other challenges make it surprising that normal cells manage to accurately replicate their genomes most of the time. Nevertheless, errors do occur that can lead to genome instability. Such errors include

mistakes in DNA polymerase fidelity, incomplete DNA repair, and improper chromosome segregation during mitosis. Another important source of genome instability is defects in the use of DNA replication origins, and it is this particular phenomenon that will be the focus of this chapter.

Origin licensing

Because human cells have such large genomes, DNA replication must initiate at many thousands of chromosomal locations so that the entire genome can be duplicated in a reasonable amount of time. These origins of DNA replication are still, for the most part, unmapped in the human genome largely because the DNA sequence appears to play only a small role in what defines an origin (Cvetic and Walter, 2005; Costa and Blow, 2007; Hamlin et al., 2008). All eukaryotes initiate replication from multiple origins, and pioneering work utilizing the budding yeast *Saccharomyces cerevisiae* has not only mapped the first eukaryotic origins of replication (Williamson, 1985; Wyrick et al., 2001), but also identified the essential proteins that recognize them. Though origin sequences are not conserved (even in yeast the origin sequences differ from one another), the proteins that bind origins are highly conserved in all eukaryotes and to some extent in archaeal species. These proteins assemble into complexes at the origins to generate pre-replication complexes (pre-RCs) during the G1 phase of the cell cycle. A pre-RC must be assembled in order for replication to initiate at any given locus, and for this reason, chromosomal DNA containing pre-RCs is said to be "licensed" for DNA replication. In the S phase, pre-RCs are converted to initiation complexes as each origin "fires" (reviewed in Bell and Dutta, 2002; Machida et al., 2005a;

Systems Biology of Cancer, ed. S. Thiagalingam. Published by Cambridge University Press. © Cambridge University Press 2015.

DePamphilis et al., 2006; Sclafani and Holzen, 2007). It should be noted that most studies of the mammalian origin binding proteins have monitored their association with bulk chromatin rather than the handful of known origins, and it is assumed that proteins bound to chromatin are associated with origins even when sequence-specific DNA binding is not measured.

Pre-replication complexes are assembled onto origin DNA by a series of sequential protein recruitments (Figure 4.2) beginning with the binding of the heterohexameric origin recognition complex (ORC). The ORC is an ATPase composed of six distinct subunits, which (at least in yeast) constitutively binds origin DNA (Bell et al., 1993; Santocanale and Diffley, 1996), though in human cells the ORC may occupy origins preferentially in G1. The ORC associates independently with two monomeric proteins, CDC6 (cell division cycle 6) and CDT1 (Cdc10-dependent transcript 1). Both CDC6 and CDT1 are required for the recruitment of a second heterohexameric complex, the MCM (mini-chromosome maintenance) helicase. The MCM complexes are loaded in multiple copies relative to the ORC through the enzymatic activity of the CDC6 and ORC ATPases (Weinreich et al., 1999; Edwards et al., 2002; Randell et al., 2006). Studies with purified yeast components suggest that CDT1 shuttles between DNA bound and soluble states consistent with a role in delivering MCM complexes from the nucleoplasm to the ORC/CDC6 complex (Randell et al., 2006; Wei et al., 2010). In support of that model, human CDT1 (unlike CDC6 and ORC) is a highly mobile protein throughout the G1 phase during the period of MCM loading (Xouri et al., 2007). CDT1 is not an enzyme and though its precise molecular function remains somewhat unclear, a small amount of free CDT1 supports the licensing of many origins. It is perhaps for this reason that CDT1 abundance and activity are subject to multiple forms of regulation.

Once MCM complexes are loaded at an origin, they do not require the continued presence of ORC, CDC6, or CDT1 to maintain their chromatin-bound state (Donovan et al., 1997; Hua et al., 1997; Hua and Newport, 1998). During the S phase, origins initiate DNA synthesis – or "fire" – at different times with some origins consistently firing early in the S phase and others firing at various times later in the S phase. Nevertheless, all origins are licensed during G1 regardless of when they will fire. Once an origin

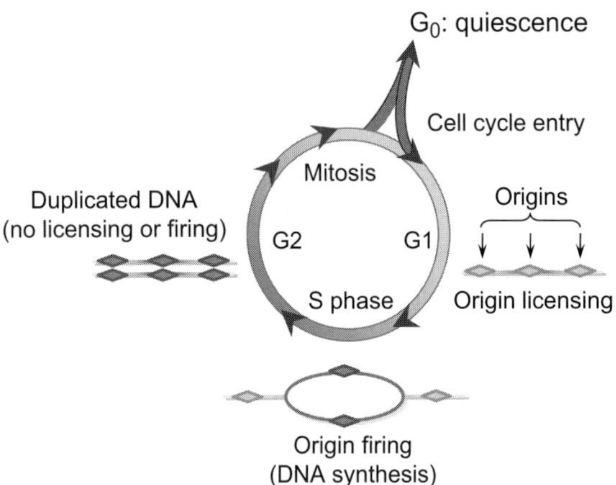

Figure 4.1 DNA replication in the human cell division cycle. Cells enter G1 either from a quiescent state (G_0) or from the preceding mitosis. DNA replication origins (light gray diamonds) are licensed for replication in G1. As origins fire in the S phase, the licenses are inactivated (dark gray diamonds), and throughout the S phase and G2, no new origin licensing is allowed until the next G1. (A black and white version of this figure will appear in some formats. For the color version, please refer to the plate section.)

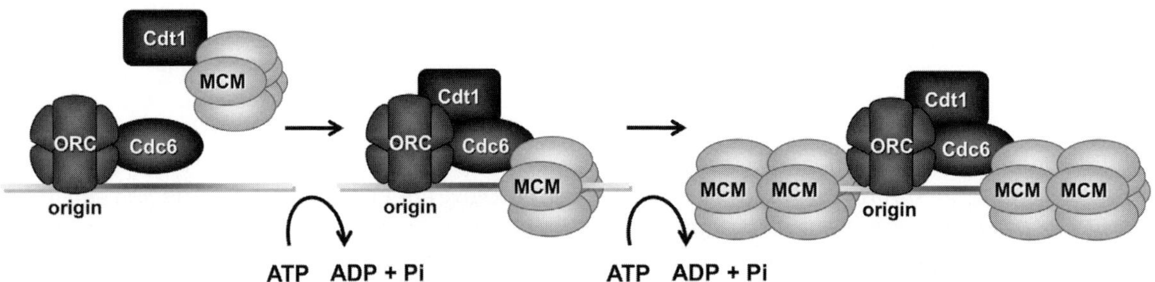

Figure 4.2 Origin licensing by pre-replication complex assembly. Origins are bound by the origin recognition complex (ORC), which then recruits both CDC6 and CDT1. CDT1 moves between the soluble and chromatin-bound states as the MCM helicase complex is brought to the origins. The ATPase activity of both ORC and CDC6 loads multiple copies of the MCM complex at each origin. (A black and white version of this figure will appear in some formats. For the color version, please refer to the plate section.)

fires, the MCM helicases are activated, and they travel with DNA polymerase at replication forks. The MCMs are only unloaded from the DNA as the S phase progresses and replication forks converge.

Positive regulation of origin licensing factors

Transcription

Non-dividing cells, such as terminally differentiated or quiescent cells, are characterized in part by unlicensed chromatin (Stoeber et al., 2001; Cook et al., 2002). As cells enter G1 from quiescence, growth factor signal transduction cascades stimulate the transcription of *cyclin D*, which activates the CDK4 and CDK6 cyclin-dependent kinases. These kinases phosphorylate and inactivate the Rb family of transcriptional repressors that are resident at the promoters of each of the genes required for replication licensing, namely the genes for all of the subunits of ORC, MCM, plus the *CDC6* and *CDT1* genes. Rb family proteins are bound to these promoters through their association with the E2F family of DNA binding proteins. Release of Rb from E2F promotes transcription of these genes and allows their protein products to accumulate (reviewed in Nevins, 2001; Bracken et al., 2004). Mutations that promote aberrant Rb phosphorylation or disrupt the Rb gene itself are frequently found in human cancers. The consequences of loss of Rb control include hyper-accumulation of the proteins in the pre-RC. The high levels of these proteins may make it more difficult to prevent inappropriate origin licensing, such as licensing origins that have already fired.

CDC6 stabilization

During G1, the protein concentration of each component of the pre-RC rises in direct proportion to its corresponding mRNA level with the notable exception of CDC6. The CDC6 protein is subject to ubiquitin-mediated degradation through interaction with the anaphase promoting complex/cyclosome (APC/C) (Petersen et al., 2000). This ubiquitin ligase complex is highly active in G1 cells so the CDC6 protein that is translated is also rapidly degraded throughout most of G1. The binding of CDC6 to APC/C is blocked, however, when CDC6 becomes phosphorylated by the cyclin E/CDK2 kinase (Duursma and Agami, 2005; Mailand and Diffley, 2005). Cyclin E/CDK2 activity rises in late G1, and the subsequent phosphorylation of CDC6 allows the protein to accumulate and promote origin licensing. CDC6 protein levels remain high throughout the S phase and G2 because CDK2 remains active through association with cyclin A, and APC/C is inactive until mitosis. At the metaphase to anaphase transition, APC/C is activated and cyclin A is destroyed leading to dephosphorylation of CDC6 and ultimately its ubiquitination and degradation.

Origin firing

Licensed origins harbor multiple MCM complexes that are poised, but inactive as DNA helicases. MCM helicase activity requires association with additional components, the CDC45 protein and the GINS complex (Figure 4.3), which is composed of four subunits: SLD5, PSF1, PSF2, and PSF3 (Takayama et al., 2003; Moyer et al., 2006). The association of these proteins with the MCM complex requires phosphorylation of MCM subunits by the CDC7/Dbf4 protein kinase (Sheu and Stillman, 2006) and

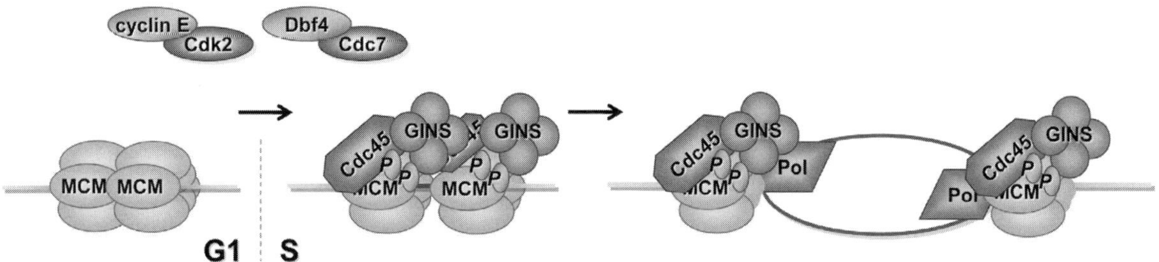

Figure 4.3 Origin firing. At the G1-to-S phase transition, the protein kinase activities of CDC7 and CDK2 stimulate the association of both CDC45 and the GINS complex with the origin-bound MCM complex. These events activate MCM DNA helicase activity leading to DNA unwinding and subsequent association of DNA polymerase. (Many other proteins not shown are also associated with the new replication fork.) (A black and white version of this figure will appear in some formats. For the color version, please refer to the plate section.)

additional phosphorylation events carried out by CDK2 in association with cyclin E (in early S phase) and cyclin A (in late S phase).

In yeast, the association of GINS with MCM is mediated by additional proteins, Sld2 and Sld3. Phosphorylation of Sld2 and Sld3 by CDK are required for origin firing in yeast (Tanaka et al., 2007; Zegerman and Diffley, 2007). RECQL4 is the functional ortholog of Sld2 (Sangrithi et al., 2005), and Treslin/Ticrr is the likely mammalian ortholog of Sld3 (Kumagai et al., 2010; Sanchez-Pulido et al., 2010; Sansam et al., 2010). The rise in both CDK2 and CDC7 activity marks the transition from G1 to S phase and converts the cell from a state that is competent for origin licensing to one that is now competent for origin firing. DNA unwinding by the CDC45/MCM/GINS complex (also known as the CMG complex) (Moyer et al., 2006) provides a single-stranded DNA substrate that is required for the recruitment of DNA polymerase α/primase and the establishment of an active replication fork that includes many other protein factors that will not be described here.

Negative regulation of origin licensing factors

To maintain genome stability it is important that chromosomal DNA is replicated only once during each cell division cycle. This rule of "once and only once" requires that each origin that fires in the S phase is strictly prohibited from firing a second time within the same cell cycle (reviewed in Blow and Dutta, 2005; Machida et al., 2005a; DePamphilis et al., 2006). If origins were to re-fire, the resulting re-replication represents a form of genome instability that promotes tumorigenesis (discussed in more detail later). To avoid re-replication, origin licensing is blocked in all phases of the cell cycle until after chromosomes have segregated. Origin licensing first begins in late telophase as the daughter nuclei form and continues throughout G1 (Dimitrova et al., 2002) (Figure 4.4), but once S phase begins, all new origin licensing is blocked by a variety of overlapping mechanisms that target pre-replication complex components.

CDK-mediated phosphorylation

The activation of origin firing and inhibition of origin re-licensing during the S phase are both regulated by the same protein kinase, CDK2. This dual function of CDK2 makes it a central player in the transition from G1 to S phase and helps ensure that once a cell is competent to fire origins, those origins cannot be re-licensed. CDK2 has multiple substrates central to DNA replication, with some CDK-mediated phosphorylation events stimulating origin firing (see above), and other CDK-mediated phosphorylation events preventing origin licensing (Bates et al., 1998;

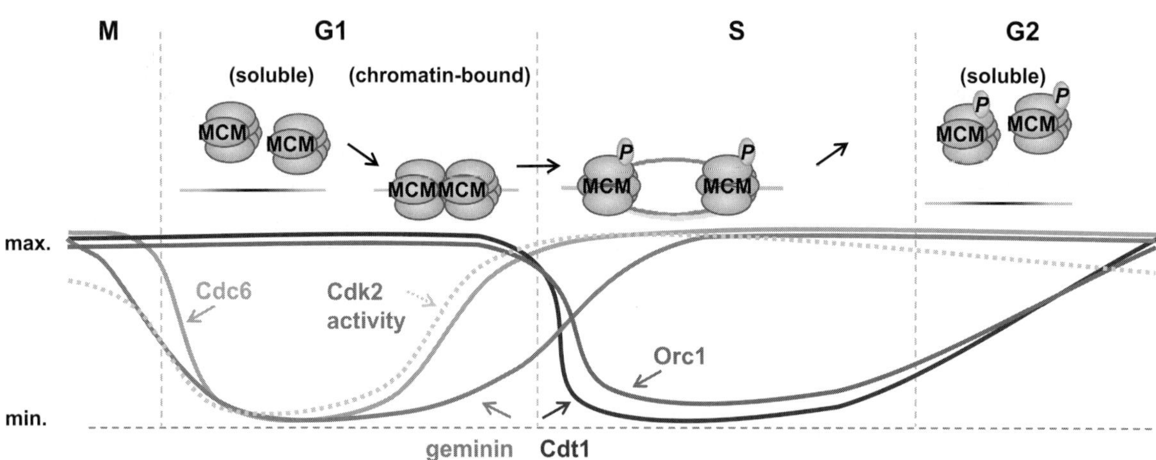

Figure 4.4 Regulation of origin licensing. The regulation of individual activities important for origin licensing control during the human cell cycle is represented. The abundance of each factor relative to its own maximum and minimum is plotted with the exception that CDK activity is plotted rather than CDK abundance. In addition, the distribution of MCM complexes to either the nucleoplasm (soluble) or DNA (chromatin bound) is shown. Note that MCM complexes are loaded in peak amounts in late G1, which is the only time when all of the other pre-RC components are abundant and the two major inhibitors, geminin and CDK activity, are low. MCM complexes travel with replication forks during the S phase and are progressively unloaded as replication completes. (A black and white version of this figure will appear in some formats. For the color version, please refer to the plate section.)

Nguyen et al., 2001). In the absence of CDK2 protein, CDK1 can substitute for both of these functions although normally CDK1 is not active until the late S phase and G2 (Hochegger et al., 2007).

In many cultured human cell lines the largest subunit of ORC, ORC1, is degraded during the S phase as a result of ubiquitination by the S phase kinase-associated protein 1 (Skp1)/Cullin/F-box protein/S phase kinase-associated protein 2 (SCF^{SKP2}, also known as CRL1^{SKP2}) complex (Mendez et al., 2002; Tatsumi et al., 2003). ORC1 interaction with the SCF adaptor, SKP2, is stimulated by CDK2-dependent phosphorylation, most likely carried out by the cyclin A/CDK2 complex (Mendez et al., 2002). Studies with purified yeast and *Drosophila* ORC have shown that ORC1 is essential for origin DNA binding (Bell et al., 1993). By extrapolation, in the absence of human ORC1, the remaining human ORC subunits are also likely to be unable to bind DNA. As a result, the first step in origin licensing is inhibited during the S phase due to low levels of ORC1. In addition, phosphorylation of yeast ORC2 and ORC6 by CDK may inhibit the enzymatic activity of the ORC holoenzyme (Nguyen et al., 2001; Makise et al., 2008; Chen et al., 2011). Furthermore, the CDK-mediated phosphorylation of ORC6 inhibits its interaction with CDT1 preventing MCM loading (Chen et al., 2011); similar regulations could also apply to mammalian ORC subunits.

Even though ORC may become depleted from origins due to ORC1 degradation, the MCM complexes that were loaded at those origins during G1 would remain since MCM does not require ORC, CDC6, or CDT1 to remain chromatin bound once it has been loaded. Several MCM subunits are also substrates for CDKs, and the phosphorylated forms of these proteins are poorly bound to chromatin (Fujita et al., 1998; Zhu et al., 2005). These observations suggest that CDK-dependent MCM phosphorylation may contribute to preventing origin re-licensing, though a mechanistic understanding of how MCM phosphorylation affects chromatin binding is still lacking.

Phosphorylation of CDC6 by CDK2 not only stabilizes CDC6 (Mailand and Diffley, 2005), but also plays a role in CDC6 nuclear localization. Early reports that relied on ectopically expressed CDC6 mutants convincingly demonstrated that unphosphorylated CDC6 was retained in the nucleus whereas phosphorylated CDC6 or a mutant form of CDC6 in which the phosphorylation sites have been

replaced by aspartic acid are largely cytoplasmic (Jiang et al., 1999; Petersen et al., 1999). It seemed likely that export of CDC6 to the cytoplasm during the S phase and G2 would play a role in preventing origin re-licensing after S phase onset. The importance of this mechanism was called into question by subsequent reports that a pool of endogenous CDC6 remains nuclear and chromatin bound throughout the S phase and G2 (Pelizon et al., 2000; Alexandrow and Hamlin, 2004). Assays to determine the consequences of CDC6 phosphorylation on its function in promoting DNA replication also varied in their outcomes (Herbig et al., 2000; Pelizon et al., 2000). More recently, a role for CDK-mediated nuclear export/inactivation of endogenous CDC6 in restricting re-replication was shown for the *C. elegans* CDC6 (Kim et al., 2008). It may be that the relative contribution of CDC6 localization to restricting inappropriate origin licensing depends on cell type, the amount of endogenous CDC6 that is expressed, and the aggregate activity of other mechanisms to restrict origin licensing.

CDT1 degradation

Following S phase entry CDT1 is degraded via at least two independent mechanisms. One mechanism is accomplished by the same CRL1^{SKP2} complex implicated in regulating ORC1. CRL1^{SKP2}-dependent degradation requires cyclin A-CDK2-mediated CDT1 phosphorylation by CDK2. Cyclin A/CDK2 activity rises at the onset of the S phase and coincides with substantial loss of CDT1 protein (Li et al., 2003; Sugimoto et al., 2004). Though CRL1^{SKP2}-mediated degradation of CDT1 contributes to CDT1 downregulation in the S phase, a second mechanism for CDT1 degradation via the CRL4^{CDT2} ubiquitin E3 ligase may play the larger role. During the S phase CDT1 associates with CUL4 independently of CDT1 phosphorylation by CDK2 (Takeda et al., 2005; Nishitani et al., 2006). CRL4^{CDT2}-mediated degradation of CDT1 is dependent on direct association of CDT1 with PCNA via a PCNA-interacting peptide (PIP) box (Arias and Walter, 2006; Hu and Xiong, 2006; Senga et al., 2006). Loading of PCNA onto DNA occurs during the elongation step of DNA synthesis, and CDT1 is only targeted by the CDT2 adaptor subunit of CRL4^{CDT2} when CDT1 is bound to DNA-loaded PCNA. Thus the PCNA dependence of CRL4^{CDT2}-mediated CDT1 proteolysis helps ensure that CDT1 destruction is coupled to DNA synthesis.

The CRL4$^{\text{CDT2}}$ pathway also mediates degradation of CDT1 (and other proteins) in response to DNA damage because PCNA is loaded during the DNA synthesis steps of DNA repair (Higa et al., 2003; Hu et al., 2004).

Geminin

In multicellular eukaryotes (but not yeast), an additional layer of regulation inhibits CDT1 activity to avoid origin re-licensing and re-replication. An origin licensing inhibitor unique to metazoan species, geminin, accumulates from the beginning of the S phase until anaphase onset (McGarry and Kirschner, 1998; Wohlschlegel et al., 2000). Geminin binds tightly to CDT1 and can block its interaction with both the MCM complex and with CDC6 (Yanagi et al., 2002; Cook et al., 2004; Lutzmann et al., 2006; Xouri et al., 2007). Thus throughout the S phase, CDT1 levels are kept low by ubiquitin-mediated degradation, and the remaining CDT1 that has not been degraded is bound by geminin to prevent it from participating in origin licensing. Geminin has additional and separable roles in developmentally regulated gene expression that may influence or be influenced by the status of CDT1 and DNA replication (reviewed in Kroll, 2007).

Like the components of the pre-RC, the *geminin* gene is regulated at the transcriptional level by the E2F/RB pathway; geminin is highly expressed in proliferating cells and undetectable in quiescent cells (Xouri et al., 2004; Yoshida and Inoue, 2004). The geminin protein is ubiquitinated by the APC/C ubiquitin ligase at the beginning of anaphase leading to rapid destruction of geminin (McGarry and Kirschner, 1998). This degradation keeps geminin protein levels low throughout G1 to permit origin licensing. CDC6 is also a substrate for APC/C, but the kinetics of CDC6 degradation in the M phase are delayed relative to geminin allowing for a small window of time in telophase when CDC6 levels are still high but geminin levels are low. In actively dividing cells, MCM loading (or at the least, chromatin recruitment) begins during this window (Dimitrova et al., 2002).

Inappropriate origin licensing

As outlined in the introduction, replication licensing factors are stringently regulated to restrict their expression levels and activities to (telophase and) the G1 phase of the cell cycle. Perturbation of regulatory controls over licensing factor expression and activity may lead to aberrant re-firing of origins multiple times during a single cell cycle leading to replication stress, a source of genomic instability. Recent studies suggest that licensing factors have oncogenic potential, and that aberrant re-replication is a hallmark of pre-malignant cells. The balance between the licensing factor CDT1 and its inhibitory binding partner geminin is particularly important for preventing re-replication. The consequences of CDT1-geminin imbalance and the mechanisms that contribute to maintaining appropriate CDT1 and geminin levels in normal cells are considered below.

Consequences of CDT1 over-expression

From studies with budding and fission yeasts it has long been appreciated that inappropriate expression of replication licensing factors may induce re-replication. For example, in *Schizosaccharomyces pombe*, over-production of Cdc18 (CDC6) during G2 leads to recruitment of MCM (Cdc21) to chromatin and re-initiation of DNA synthesis (Yanow et al., 2001). Moreover, co-expression of Cdc18 with CDT1 leads to uncontrolled DNA synthesis and accumulation of cells with up to >64 N DNA content (Yanow et al., 2001). Therefore CDC6 and CDT1 levels are rate limiting for re-replication of the genome in *S. pombe*.

In many cultured mammalian cancer cell lines over-expression of CDT1 also leads to re-replication and accumulation of large populations of nuclei with >4 N DNA content (Vaziri et al., 2003). Similar to results described in *S. pombe*, the combined expression of CDT1 and CDC6 results in a considerably enhanced re-replication response, although CDC6 expression alone does not significantly promote accumulation of >4 N nuclei in most human cells (Vaziri et al., 2003). In contrast with many cancer cells, which undergo dramatic re-replication, primary untransformed human cells do not typically accumulate large numbers of nuclei with >4 N DNA content following licensing factor over-expression and instead succumb to growth arrest (Vaziri et al., 2003). Therefore it appears that untransformed cells have more robust mechanisms to prevent re-replication in response to aberrant licensing factor activity (Sugimoto et al., 2009). Most probably, normal cellular restraints to re-replication are lost during the course of neoplastic transformation and tumor progression.

TP53 (the gene encoding p53) is deleted or functionally inactivated in ~50% of human tumors, and it has been suggested that p53 deficiency might account for increased propensity for re-replication in cancer cells. p53 is typically activated in response to replication stresses induced by CDT1 and CDC6 over-expression (Vaziri et al., 2003) (discussed below). Moreover, analysis of a limited number of lung carcinoma cell lines demonstrated a correlation between p53 deficiency and the extent of re-replication induced by CDT1 and CDC6 over-expression (Vaziri et al., 2003). Importantly, transient expression of the *Mdm2* oncogene (which antagonizes p53) was found to confer CDT1+CDC6-induced re-replication in a p53-expressing cancer cell line. This result is potentially consistent with a role for p53 in preventing re-replication in response to high-level licensing factor expression. However, MDM2 also has other targets (most notably the tumor suppressor protein Rb), and further work is necessary to elucidate the putative mechanism by which p53 prevents CDT1-induced over-replication. Nevertheless, these studies reveal potential links between responses to aberrant replication licensing and p53 and Mdm2, products of genes whose involvement in human cancer is firmly established.

Consequences of geminin depletion

In vertebrates, CDT1 activity is regulated via interactions with a small binding partner geminin (see introduction). The balance of CDT1 and geminin expression is very important for ensuring appropriate levels of replication licensing. Thus similar to CDT1 expression, depletion of geminin induces re-replication (Melixetian et al., 2004; Zhu et al., 2004). In contrast with CDT1 over-expression, which appears to induce re-replication preferentially in *p53–/–* cells (Vaziri et al., 2003; Aggarwal et al., 2007), re-replication induced due to geminin deficiency can be p53 independent. It is important to note, however, the experiments describing the effect of p53 status on the replication responses to geminin depletion were performed in HCT116 cell lines in which CDT1 induces re-replication regardless of p53 status (unpublished observations; Hall et al., 2008). Moreover, the flow cytometric methods typically used to detect re-replication require extensive increases in DNA mass, meaning that p53+ cells may re-replicate but not as extensively as p53-deficient cells. In support of this

notion, geminin depletion in normal cells can induce re-replication detected by single-molecule analysis (Dorn et al., 2009). Therefore further work is necessary to determine whether p53 selectively restricts CDT1-induced re-replication. Moreover given the numerous effectors identified for p53 it is important to determine whether a specific p53-mediated signal inhibits re-replication, or whether the re-replication seen in many p53-deficient cells occurs secondarily to global changes in the cell cycle and DNA repair-related processes.

Cell cycle-dependent degradation of CDT1 and geminin prevents re-replication

DNA replication factors are generally subject to negative regulation via cell cycle-specific post-translational modifications and protein–protein interactions (e.g., CDC6, CDT1, ORC2), which often lead to re-distribution away from chromatin (e.g., CDC6, MCM) and proteolysis (e.g., ORC, CDT1, geminin). Cell cycle-coupled pathways for proteolysis of CDT1 and geminin have emerged as important mechanisms that restrict replication licensing to specific stages of the cell cycle and prevent re-replication. Proteolytic pathways regulating CDT1 and geminin are described below.

CDT1 degradation

Although CRL1^{SKP2}-mediated ubiquitination contributes to the downregulation of CDT1 during S phase (see above), the CRL1^{SKP2} pathway does not contribute significantly to preventing re-replication. In *C. elegans* inactivation of CUL4 causes massive re-replication and accumulation of cells with up to >100 N DNA content (Zhong et al., 2003). Removing one genomic copy of CDT1 suppresses the re-replication resulting from Cul4 deficiency. Therefore, CUL4 prevents re-initiation in part by destabilizing CDT1 (Zhong et al., 2003). In cultured human cells DDB1 depletion stabilizes CDT1 and induces re-replication (Lovejoy et al., 2006), further consistent with a key role for CRL4^{CDT2} in regulating CDT1 levels and preventing re-replication.

CRL4^{CDT2}-mediated degradation of CDT1 requires both the DDB1 and CUL4-associated factor (DCAF) designated CDT2 (Jin et al., 2006; Sansam et al., 2006). CDT2 is one of a large family of DCAF adaptors and serves as a specificity module that targets CDT1 to CUL4-DDB1. Similar to DDB1 deficiency, CDT2 depletion induces re-replication in zebrafish and human cells (Jin et al., 2006; Sansam et al., 2006). CDT2 depletion in human cells can cause extensive re-replication not only because CDT1 accumulates, but also because a lysine methyltransferase that stimulates origin licensing, the PR-SET7/SET8 enzyme, also aberrantly accumulates (Tardat et al., 2010; Jørgensen et al., 2011).

Geminin degradation

Following anaphase, geminin is degraded by the anaphase-promoting complex (APC/C), thereby relieving the inhibition of CDT1 activity and conferring a state that is permissive for replication licensing. EMI1 is an APC regulator that inhibits APC during the S phase and G2. EMI1 depletion prematurely activates APC and destabilizes geminin, thereby promoting re-replication (Machida and Dutta, 2007).

Similar to CDT1 and geminin, other replication factors (including ORC, CDC6, MCM2-7) are also post-translationally modified, re-distributed, or otherwise functionally inactivated in a manner that is coupled to S phase progression (Fujita et al., 1996; Jiang et al., 1999; Petersen et al., 1999; Ishimi et al., 2000; Mendez et al., 2002; Kim et al., 2007). Potentially, failure to inactivate pre-RC components or other replication factors concomitantly with S phase progression might also confer increased re-replication. Limited evidence exists to support a role for CDC6 degradation in preventing re-replication in some cell types (Vaziri et al., 2003; Hall et al., 2008). The possible significance of cell cycle-dependent ORC downregulation remains to be tested.

Consequences of re-replication
Acute consequences of re-replication: DNA damage and checkpoint activation

Little is known regarding the specific DNA structures generated during re-replication, or the eventual fate of re-replicated DNA. The nature of DNA structures generated during re-replication may depend on many factors such as DNA repair capacity, chromatin context, and the number of re-initiation events. It is widely hypothesized that multiple rounds of initiation from the same origin of replication generate putative DNA structures termed "onion skins." Studies with *Xenopus* nuclei undergoing CDT1-induced re-replication suggest that re-replication results in the formation of DNA fragments due to head-to-tail collisions when replication forks "chase" one another on the same template (Davidson et al., 2006). It appears that collisions between replication forks and other DNA processing events lead to the formation of single-stranded DNA (ssDNA) and DNA double-strand breaks (DSB) (Figure 4.5b "re-replication" compared to Figure 4.5a "normal replication" or Figure 4.5c "incomplete replication"). Consistent with the formation of ssDNA and DSB, many studies have shown that checkpoint pathways are activated following induction of re-replication (Mihaylov et al., 2002; Melixetian et al., 2004; Archambault et al., 2005; Tatsumi et al., 2006).

Cell cycle checkpoints are signal transduction pathways that negatively regulate the cell cycle in response to DNA damage, thereby providing additional time for DNA repair prior to resumption of cell cycle progression. These checkpoints integrate DNA repair with the cell division cycle and are thought to constitute important tumor-suppressive mechanisms. Indeed individuals with inherited defects in checkpoint genes (ATM, ATR, CHK2, TP53) are predisposed to cancer (reviewed in Kastan and Bartek, 2004). Components of checkpoint signaling pathways are broadly categorized as DNA damage sensors, mediators, transducers, and effectors (Sancar et al., 2004; Cimprich and Cortez, 2008).

The DNA damage signaling pathway is often considered to comprise two main branches mediated by 9-1-1/ATR/Chk1 and ATM/Chk2, which respond to replication blocks and DSB respectively (Figure 4.6) (Sancar et al., 2004). The 9-1-1/ATR/Chk1 pathway is triggered by the formation of RPA-coated ssDNA generated by uncoupling of leading/lagging strand synthesis at replication forks (e.g., in response to nucleotide depletion or polymerase-stalling DNA lesions) or by nucleolytic resection of DSB. In response to replication fork stalling, activation of 9-1-1 and ATR/Chk1 serves to inhibit late origin firing (the "S phase checkpoint"), slows rates of elongation, delays entry into mitosis (the "replication checkpoint") and stabilizes DNA replication forks (Cimprich and Cortez, 2008). In contrast with 9-1-1 and ATR activation, the DSB-induced activation of ATM/Chk2 is not replication dependent. DSB-

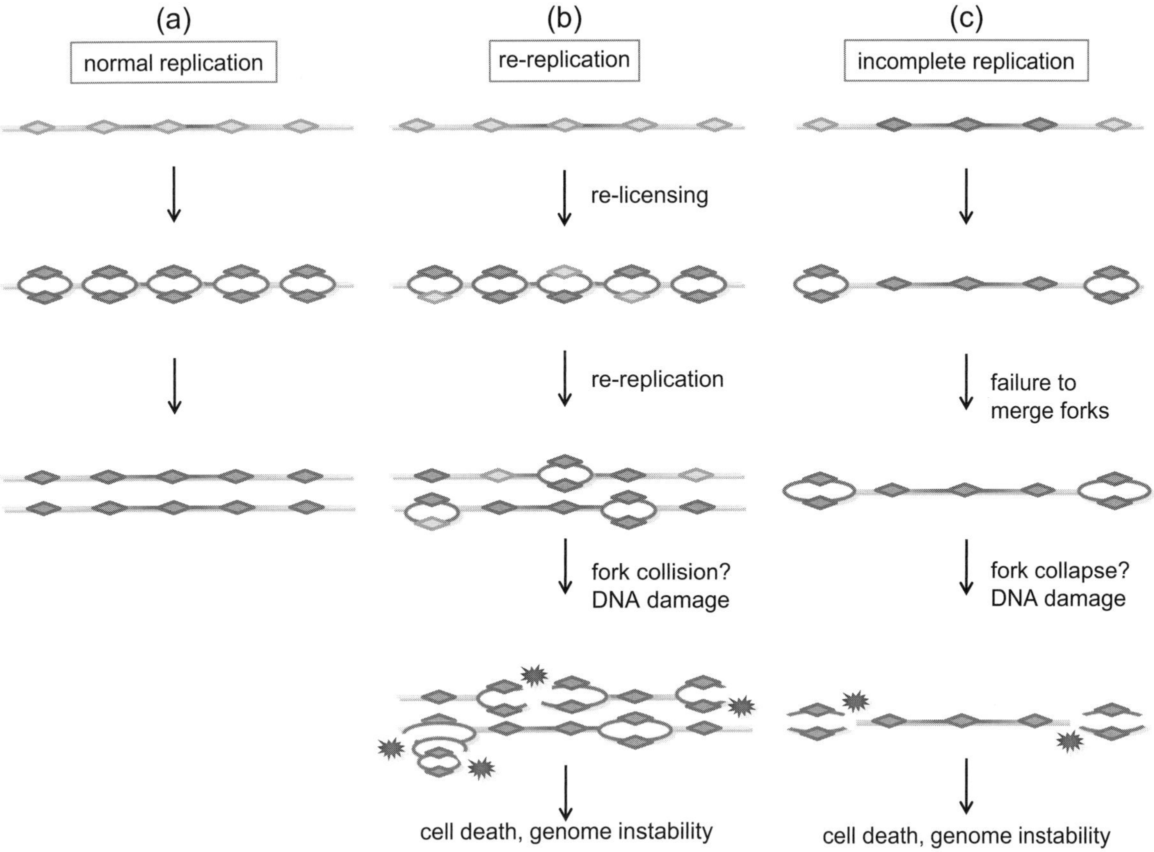

Figure 4.5 Deregulated origin licensing results in DNA damage and genome instability. (a) Normal replication is characterized by origin licensing (light gray diamonds) in G1, and each origin that fires (dark gray diamonds), fires just once in the S phase. (b) Failure to block origin licensing after S phase onset leads to re-licensing of origins that have already fired. The presence of multiple replication forks on the same strand is ultimately resolved as chromosome fragmentation. (c) Failure to complete the origin licensing step in G1 coupled with S phase entry (rather than G1 arrest) leads to incomplete replication and DNA damage. (A black and white version of this figure will appear in some formats. For the color version, please refer to the plate section.)

induced ATM signaling mediates checkpoints in G1, S, and G2/M and may also lead to apoptosis following acquisition of high levels of DNA damage.

It appears that both the 9-1-1/ATR/Chk1 and ATM/Chk2 pathways are activated in response to re replication, but with distinct kinetics. In geminin-depleted colon carcinoma cells, the 9-1-1 and ATR pathways are activated early following the onset of re-replication whereas ATM/Chk2 signaling is induced at later times (Lin and Dutta, 2007). Therefore it may be inferred that ssDNA structures are generated early during re-replication leading to ATR/Chk1 activation. However, the ssDNA-containing replication intermediates (onion skins) are likely to be unstable and eventually collapse (or are otherwise

processed via nucleases) to generate DSB which activate ATM/Chk2 (Figure 4.5b). Potentially, both ATM and ATR pathways can contribute to p53 activation, thereby explaining the p53 response observed in response to CDT1 over-expression or geminin depletion (Vaziri et al., 2003; Melixetian et al., 2004; Zhu et al., 2004).

The significance of checkpoint activation in re-replicating cells is incompletely understood and may differ between cell types. When geminin is depleted from cultured *Drosophila* cells, Chk1 is required for the accumulation of >4 N nuclei. On the other hand, 9-1-1 is necessary for this re-replication phenotype in geminin-depleted human colon carcinoma cells (Mihaylov et al., 2002; Lin and Dutta, 2007).

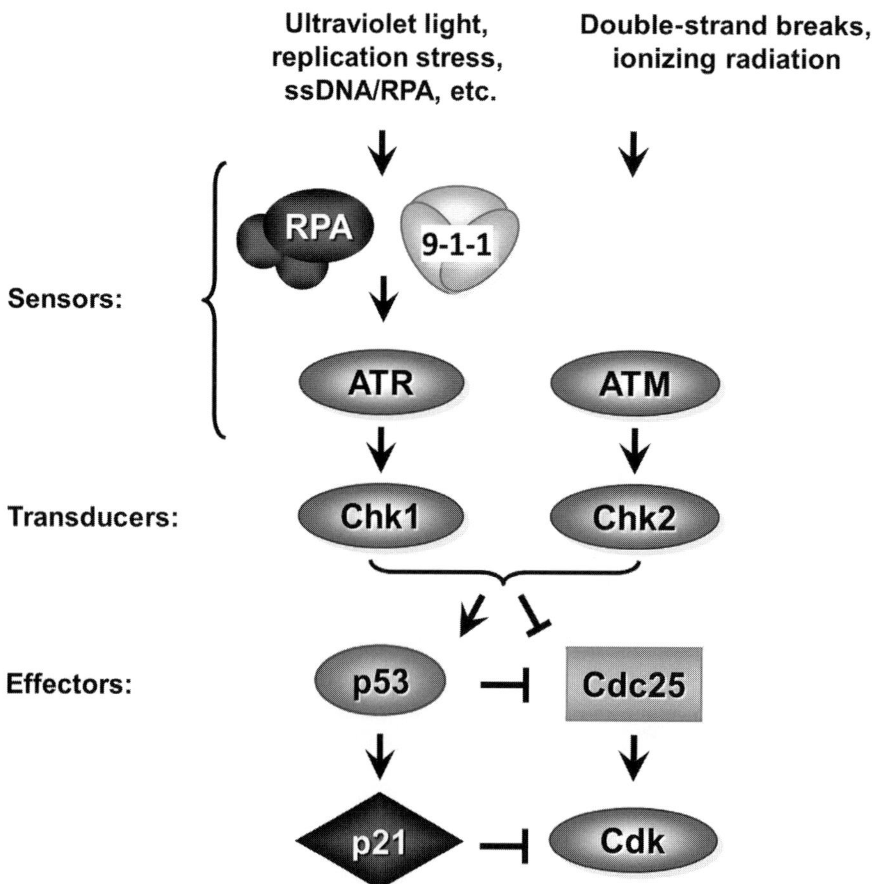

Figure 4.6 DNA damage checkpoint signaling. DNA damage and replication stress are detected in multiple ways depending on the nature of the damage. Detection stimulates the protein kinase activities of ATR or ATM, which, in turn, phosphorylate and activate the CHK1 and CHK2 kinases. These transducer kinases target multiple cellular factors that lead to CDK inhibition and cell cycle arrest, most notably the p53 transcription factor and the CDC25 protein phosphatase. (A black and white version of this figure will appear in some formats. For the color version, please refer to the plate section.)

Moreover, components of the ATR signaling pathway and the MRN checkpoint mediator complex have been implicated in restricting the appearance of cells with grossly re-replicated DNA (Lee et al., 2007; Lin and Dutta, 2007; Liu et al., 2007).

Potentially, ATR/Chk1 signaling might maintain re-replicated cells in a G2 state (via the replication checkpoint), facilitating the detection of >4 N populations by FACS. Alternatively, ATR/Chk1 might stabilize replication forks within re-replicating regions of chromatin, thereby facilitating elongation of daughter strands and accumulation of cells with >4 N DNA content. In geminin-depleted human cells, ATR may also activate the Fanconi anemia (FA) DNA repair pathway, which is necessary for preventing apoptosis and allowing accumulation of >4 N cells (Zhu and Dutta, 2006). In general, it appears that intact ATR-Chk1 signaling is required for accumulation of >4 N nuclei following induction of re-replication.

In contrast with ATR/Chk1 signaling, the ATM/Chk2 pathway does not contribute significantly to the accumulation of >4 N populations following CDT1 over-expression or geminin depletion (Zhu et al., 2004; Liu et al., 2007). However, it is likely that re-replication-associated p53 activation is mediated by ATM/Chk2 (Zhu et al., 2004). CDT1 over-expression and geminin depletion both induce p53 phosphorylation at Ser 20 (Vaziri et al., 2003; Melixetian et al., 2004; Zhu et al., 2004), a phosphorylation event that is thought to be ATM/Chk2 mediated. Moreover, recent results suggest that ATM/Chk2-mediated senescence might play an important role in preventing propagation of cells that have undergone re-replication.

Long-term consequences of re-replication: induction of genome instability and cancer

Abnormal expression/activity of licensing factors likely plays a causal role in tumorigenesis. CDT1 and CDC6 often accumulate abnormally in early dysplastic lesions including lung, larynx, and colon, frequently as a result of gene amplification (Karakaidos et al., 2004; Liontos et al., 2007). Indeed, re-replication from loss-of-origin licensing control can directly lead to gene amplification and copy number expansion (Green et al., 2010). Furthermore, ectopic expression of CDT1 and CDC6 increases the tumorigenicity of cultured cell lines when injected into nude mice (Arentson et al., 2002; Liontos et al., 2007), and CDT1 transgenic mice develop lymphoblastic lymphoma (Seo et al., 2005).

The precise mechanism by which excess replication licensing may induce tumorigenesis is unclear. One possibility is that inaccurate repair of re-replication-induced DNA damage results in genomic instability and contributes to tumorigenesis. Recent work indicates that hyper-replication (re-replication) and activation of the ATM/Chk2-mediated DNA damage response are hallmarks of pre-malignant cells, which may be induced by active oncogenes in untransformed cells (Bartkova et al., 2006; Di Micco et al., 2006; Halazonetis et al., 2008). Such studies have led to a model whereby hyper-replication and replication stress lead to ATM/Chk2-mediated senescence, which, in normal cells, guards against tumorigenesis. However, in the absence of an intact ATM/Chk2 response, oncogene activation leads to continued proliferation of aberrant cells harboring DNA damage, eventually leading to malignancy. The putative mechanism(s) by which oncogenic stimuli affect the DNA replication apparatus and promote re-replication are unknown. One possibility is that oncogenic signaling affects expression levels of licensing factors. Indeed, a recent study showed that constitutively nuclear cyclin D-CDK4 complexes induce CDT1 stabilization due to downregulation of *cul4* mRNA. The resulting CDT1-induced re-replication induces p53-mediated apoptosis, but leads to neoplastic growth in a p53–/– background (Aggarwal et al., 2007). Clearly, further work is necessary to elucidate the links between oncogene signaling, replication licensing factors, re-replication, and cancer.

Future questions

It is clear that perturbation of the balance between replication licensing factors and their inhibitors (geminin) leads to re-replication in many cell types. Potentially, any factors or processes that influence licensing and origin selection during a normal cell cycle could also influence re-replication. For instance, it is increasingly apparent that control of DNA synthesis occurs via epigenetic mechanisms, which affect both origin selection and timing of firing during the S phase (McNairn and Gilbert, 2003; Weinreich et al., 2004; Dorn et al., 2009; Cayrou et al., 2010; Méchali, 2010). Histone acetyltransferases (HATs) and histone deacetylases (HDACs) are emerging as potentially important regulators of DNA replication. HBO1 is a MYST-family HAT originally identified as an ORC- and MCM-interacting protein (Iizuka and Stillman, 1999; Burke et al., 2001). HBO1 promotes replication licensing, most likely via acetylation of histone H4 (and possibly other proteins including replication factors) in the vicinity of replication origins (Iizuka et al., 2006). A recent study demonstrated that HBO1 enhances CDT1-dependent re-replication (Miotto and Struhl, 2008). PR-SET7/SET8-mediated lysine methylation at origins also promotes origin licensing, and the relevant substrate of PR-SET7 is likely histone H4, though non-histone substrates may also be important (Tardat et al., 2010). Further studies are needed to determine the epigenetic mechanisms that regulate replication licensing and re-replication. It will be interesting to determine whether there are differences in which origins re-fire, and how many times they re-fire in response to aberrant licensing factor activity.

A caveat of most re-replication studies is that the experiments described have generally been conducted in cancer cell lines. Thus little information is available regarding responses of normal cells to aberrant replication factor activity. Further studies using untransformed and primary human cells are necessary to define normal cellular responses to deregulated expression of CDT1, geminin, and other replication factors. Such studies will eventually enable elucidation of mechanisms that repress re-replication, preserve genomic stability, and protect normal cells against cancer.

Another limitation of past work is that re-replication assays have relied mainly on measuring numbers of >4 N nuclei by flow cytometry or detecting BrdU-substituted heavy-heavy DNA. Both assays are inherently insensitive and require a substantial accumulation of >4 N DNA content to enable detection of re-replication. Based on failure to detect nuclei containing >4 N DNA some studies have concluded that primary untransformed cells do not re-replicate in response to CDT1–geminin imbalance (Vaziri et al., 2003; Liu et al., 2007; Hall et al., 2008). It is possible, however, that aberrant licensing factor expression does lead to re-licensing and re-initiation events in primary cells, but that the replication forks arising from re-fired origins fail to elongate substantially and generate insufficient re-replicated DNA for detection of >4 N nuclei by flow cytometry. It is thus important to determine whether differences in initiation, elongation, or both phases of DNA synthesis lead to the different re-replication phenotypes observed in different cell types. It is very likely that the capacity of a given cell for elongating aberrant forks generated during re-replication determines the accumulation of nuclei with >4 N DNA. Many cancer cells have elevated recombinational repair activity (Klein, 2008). Thus it is also possible that more efficient processing of ssDNA and DSB-containing re-replication products via recombinational mechanisms accounts for the observed propensity of re-replicating cancer cells to acquire >4 N DNA content. More sensitive methods are required to detect low-level re-replication events that may contribute to genome instability (but not induce significant numbers of >4 N cells).

Immunofluorescence microscopy-based analysis of arrayed DNA fibers (derived from nucleotide analogue-labelled cells) provides a potentially useful technique for analyzing individual initiation events and elongation rates. DNA fiber assays detected low basal levels of re-replication in human cancer cells grown in culture (Dorn et al., 2009). It is probable therefore that low-level re-replication (below the sensitivity limits of other detection methods) occurs in many cancer cells and that the resulting replication stress and DNA damage contribute to genomic instability.

In summary, this section describes potential mechanisms by which excess replication licensing and the ensuing re-replication might contribute to genomic instability and cancer. The next section discusses the detrimental consequences of insufficient replication licensing. Studies reviewed therein demonstrate the existence of mechanisms that ensure S phase entry only when sufficient licensing has occurred, thereby maintaining replication fidelity and preserving genomic stability.

Insufficient origin licensing
Origin licensing checkpoint

The preceding section described the consequences of unscheduled origin licensing; this section will focus on the opposite situation, the consequences of insufficient origin licensing. As outlined in the introduction, origins are normally only licensed in G1 because multiple levels of regulation prevent origin licensing once the S phase has begun. However, if the origin licensing step were to fail in G1, but the S phase were to begin on schedule, DNA replication would be incomplete because sparse replication forks would be unable to merge. Replication forks encounter natural pause sites (reviewed in Labib and Hodgson, 2007), and under normal conditions these sites pose few problems since replication forks from a nearby origin will merge with the fork stalled at the pause site. However, stalling at natural pause sites can also contribute to genome instability (Lemoine et al., 2005; Raveendranathan et al., 2006). In a situation where too few origins have been licensed, forks that stall at pause sites for too long could become unstable and collapse, generating double-strand breaks.

Since the cell cycle is a unidirectional process, cells with partially replicated DNA cannot return to G1 to license more origins. For this reason cells must license enough origins in G1 before the major activities that block origin licensing (geminin and CDK2 activity) accumulate. This principle suggests that origin licensing should be a pre-requisite for the major events of the G1-to-S phase transition, including activation of CDKs. In other words, there should be an *origin licensing cell cycle checkpoint* (Lau and Jiang, 2006). Thus far evidence for such a checkpoint is largely indirect, though recent advances strengthen the case for its existence.

What physiological circumstances would cause origin licensing to fail? Given that the G1 phase in mammalian cells is usually many hours long and origin licensing can begin in telophase (Dimitrova et al., 2002), there should be plenty of time to license enough origins to completely replicate the genome without the need to invoke a licensing checkpoint. Not all cells have long G1 phases, however; for example some embryonic cells have G1 phases of only 30 minutes (Mac Auley et al., 1993). In addition, DNA damage induces the degradation of both CDC6 and CDT1, which are essential for origin licensing (Blanchard et al., 2002; Higa et al., 2003; Hu et al., 2004). Recovery from a DNA damage response involves both activation of CDKs and (presumably) re-synthesis of CDC6 and CDT1. If CDK activation were not delayed relative to the completion of origin licensing, the cells risk incomplete replication. An origin licensing checkpoint that operates to delay activation of CDK2 until sufficient origins have been licensed would ensure complete replication.

Origin licensing defects in yeast

In yeast strains with compromised Cdc6 expression or function, plasmids containing a single origin are lost from the population at a high frequency. The inability to maintain these "minichromosomes" was developed as an indicator of replication origin activity. Similarly, Tah11/CDT1 and MCM2-7 mutants have an elevated minichromosome loss rate (Maine et al., 1984; Devault et al., 2002), and the loss rate is suppressed by increasing the number of origin sequences on the minichromosome (Hogan and Koshland, 1992). The fact that these minichromosomes are lost when origin licensing is perturbed suggests that failure to fire sufficient origins on the endogenous chromosomes may also lead to loss of genetic material in the population. By that argument, origin licensing defects are a potential source of genome instability.

Origin licensing is completely abolished in budding or fission yeast harboring null mutants of pre-RC components. Deletion of the *ORC2*, *CDT1*, or *CDC6* genes results not only in defective origin licensing and replication failure, but also premature entry into mitosis (Hofmann and Beach, 1994; Piatti et al., 1995; Muzi Falconi et al., 1996; Kiely et al., 2000). These mutants then execute a reductional anaphase in which unreplicated chromosomes segregate to one or

the other spindle pole, generating aneuploid and inviable daughter cells. The fact that these mutants do not arrest in G1 suggests that there is no origin licensing checkpoint in yeast.

In contrast to yeast strains that express no Cdc6 protein at all, strains in which Cdc6 is expressed but catalytically inactive arrest with G1 DNA content and do not skip ahead to mitosis (Weinreich et al., 1999). In both cases, the MCM complexes fail to be loaded onto origins (Santocanale and Diffley, 1996; Weinreich et al., 1999). A potential explanation for these apparently contradictory results comes from the discovery that the amino-terminus of yeast Cdc6 is a potent inhibitor of mitotic CDK activity (Mimura et al., 2004). Thus in the absence of any Cdc6 protein, mitotic CDKs are hyperactive. Cdc6 protein may be elevated in yeast strains deleted of other components also, though this relationship remains to be tested. Alternatively, the *cdc6* mutant may retain residual activity that permits a very small number of origins to fire. In such circumstances, the sparsely distributed replication forks could trigger a replication stress response that then triggers an intra-S phase checkpoint arrest. By that argument, the *cdc6* mutant alleles may actually be arresting in the very early S phase rather than in G1. The early yeast studies led to the conclusion that eukaryotic cells do not monitor the status of origin licensing in G1. However, the question of an origin licensing checkpoint that operates in human cells remained an open question until subsequent studies.

Origin licensing defects in higher eukaryotes

The advent of RNAi technology allowed investigators to probe the consequences of origin licensing inhibition in human cells. Cultured human cells transfected with siRNAs to deplete essential licensing factors are not the equivalent of a null allele, but more closely resemble hypomorphic mutants. In that regard, these siRNA-treated cultures are predicted to more closely represent physiological circumstances where origin licensing is still incomplete, either due to early G1 status or to DNA damage-dependent degradation of CDC6 and CDT1.

The outcome of depleting essential origin licensing components, such as CDC6, ORC2, or MCM5, depends on whether the cells are normal or transformed. All cancer cell lines tested thus far die by apoptosis whereas normal cells survive and arrest in an apparent G1 state with 2C DNA content (Feng et al., 2003; Machida et al., 2005b). Furthermore, overproduction of geminin to block CDT1 function has a similar differential effect on cancer cells compared to normal cells (Shreeram et al., 2002). These findings raised two possibilities: (1) robust checkpoint pathways in the normal cells trigger an arrest so early in the S phase that they only appear to be in G1, or (2) normal cells have an active origin licensing checkpoint that arrests them *before* the G1/S transition, and this checkpoint is deficient in cancer cells. In this scenario, the attempted S phase by the cancer cells ultimately results in catastrophic DNA damage, which then triggers the observed apoptosis.

G1 arrest in normal cells

To distinguish a true G1 arrest from arrest in the very early S phase, multiple molecular markers of G1 progression have been examined. Depletion of MCM7, MCM2, CDT1, or CDC6 by transfection of normal human fibroblasts with siRNA induced Rb hypophosphorylation, an indicator of G1 cells rather than S phase cells (Machida et al., 2005b; Liu et al., 2009; Nevis et al., 2009). This Rb hypophosphorylation could be accounted for by a marked decrease in the kinase activity of both CDK4 and CDK2. This inhibition of CDK2 activity was ascribed to an associated delay in nuclear accumulation and the reduction of an essential phosphorylation of T160 (Nevis et al., 2009). The decrease in CDK4 activity was at least partly due to reduced occupancy of RNA PolII on the *CYCD1* promoter and a corresponding decrease in *cyclin D1* mRNA and cyclin D1 protein. However, ectopic expression of cyclin D1 alone was not sufficient to induce S phase in MCM7-deficient cells, suggesting that CDK4-independent events also contribute to the arrest (Liu et al., 2009).

Insufficient origin licensing caused by CDC6 depletion arrested normal human fibroblasts cells in the G1 phase, whereas co-depletion of p53 restored CDK2 and RB phosphorylation allowing cells to prematurely enter the S phase, but then acquire markers of DNA damage (Nevis et al., 2009). It is likely that the inappropriate S phase entry with insufficient licensing led to incomplete replication and subsequent fork collapse, which was recognized as DNA damage. Moreover, CDC6 depletion from cancer cells with documented p53 deficiencies resulted in failure to regulate Rb and CDK2 phosphorylation, and

ultimately these cells died by apoptosis (Nevis et al., 2009). These findings suggest relationships among origin licensing, CDK activity, and tumor suppressors that protect normal cells from inappropriate S phase entry (Nevis et al., 2009).

In support of multiple links between origin licensing and G1 progression, depletion of ORC2 in MCF10A cells (an untransformed, but not fully normal breast epithelial cell line) also led to reduced CDK2 activity. This reduction was attributed to increases in both p27 and p21 CDK inhibitors by stabilization of p27 protein and an increase in p21 mRNA levels (Machida et al., 2005b). MCM7-deficient normal human fibroblasts also accumulated p27, but depletion of p27 was not sufficient to promote S phase entry in those cells (Liu et al., 2009). It may be that the increase in p27 is simply a symptom of G1 arrest rather than a cause, since p27 levels are naturally highest in G1. It may also be that different normal cell types have somewhat different ultimate phenotypes of licensing inhibition. Despite some observations that licensing inhibition can induce p21 expression, a consistent finding in all of these studies has been that licensing inhibition in normal cells does not trigger other markers of a canonical DNA damage response.

One confounding issue in determining the mechanism of the arrest in normal cells is that the most commonly employed experimental approach involves depletion of origin licensing components in an asynchronous population of cells followed by examination of the terminal phenotype of the arrested population. When cells are arrested from an asynchronous population, some cells have been held in the arrest for significantly longer periods of time than others allowing for secondary molecular events to occur. Since many of the markers of cell cycle progression are potent effectors of cell cycle transitions, it is difficult to determine if loss or induction of a factor (such as p27) is the cause or a consequence of the arrest. More recent studies have attempted to minimize this consideration by examining the consequences of insufficient origin licensing in normal cells synchronized by serum deprivation/re-stimulation. By this method, inhibition of both CDK4 (Liu et al., 2009) and CDK2 (Nevis et al., 2009) occurs within the first G1 phase after serum stimulation, and induction of CDK inhibitors plays an apparently minor role if any.

Reduced CDK activity may not be the only effect of origin licensing inhibition, however. Two studies of proteins associated with licensed chromatin compared to unlicensed chromatin prepared in *Xenopus laevis* egg extracts identified a surprisingly wide variety of nuclear activities that are dependent on origin licensing. Addition of geminin to block licensing prevented not only MCM chromatin loading (as predicted) but also the association of cohesin subunits (SMC1 and SMC3), which are utilized later in the cell cycle to maintain sister chromatin cohesion (Takahashi et al., 2004). Licensing inhibition also prevented the chromatin association of nuclear structure and nuclear pore components, DNA repair factors, and multiple chromatin remodelling enzymes such as FACT and RUVBL1 (Khoudoli et al., 2008). These results suggest that, at least in an *in vitro* system, origin licensing promotes extensive changes in chromatin and nuclear structure that may have downstream effects on gene expression in conjunction with effects on core cell cycle factors such as CDKs. The temporal relationship between these global effects on nuclear processes and CDK activation has not yet been determined; thus it remains possible that CDK4 and CDK2 activation are downstream of the link between origin licensing and transcription and/or nuclear structure.

Apoptosis in cancer cells

A wide variety of tumor-derived cell lines die by apoptosis when origin licensing is blocked. Though the threshold for apoptosis is heavily influenced by the status of the p53 tumor suppressor, both p53+ and p53– deficient cell lines are sensitive to killing by origin licensing inhibition (Feng et al., 2003). Furthermore the status of Rb, the other major tumor suppressor implicated in oncogenesis, did not affect the ability of licensing inhibition to induce apoptosis (Shreeram et al., 2002). Sufficient DNA damage triggers apoptosis in virtually all cell types regardless of p53 and Rb status. The appearance of markers of DNA damage in licensing-deficient cancer cell lines (Shreeram et al., 2002; Teer et al., 2006; Nevis et al., 2009) is consistent with a model in which licensing inhibition in transformed cells ultimately results in intolerable levels of DNA damage.

An attractive model to explain why cancer cells do not survive treatments that block origin licensing is incomplete DNA replication from a failed origin licensing checkpoint. Thus far direct evidence to determine the proximate cause of the cancer cell death is not available. It is possible that the cancer cells that enter the S phase with too few licensed origins suffer replication fork collapse in the S phase, and that the resulting double-strand breaks trigger an apoptotic response. Indeed, markers of double-strand breaks such as phosphorylated Chk2 appear in licensing-deficient cancer cell lines (Teer et al., 2006; Nevis et al., 2009). It is also possible, however, that many cancer cells have a weak mitotic checkpoint and initiate mitosis before replication is complete when the S phase has been dramatically slowed due to an origin licensing defect. The attempt to segregate partially replicated chromosomes may be the direct cause of the DNA damage that leads to apoptosis. The particular genetic perturbations associated with a given cell line likely determine which of these models is most relevant. More detailed analysis of the early responses of these cells to licensing inhibition – particularly in synchronized cultures – will provide additional insight into the differential sensitivity of normal and cancer cells.

Although the presence of normal p53 and Rb genes does not protect licensing-deficient cancer cells, Rb status may be an important determinant of which cell cycle phase generates the DNA damage. Cancer cells carrying wild-type Rb (irrespective of their p53 status) underwent apoptosis following an apparent S phase arrest in response to geminin overproduction. On the other hand, cells lacking both Rb and p53 traversed the S phase and then initiated apoptosis upon entry into mitosis (Shreeram et al., 2002).

Future questions

Though it has become increasingly clear that licensing-deficient normal cells arrest in G1 rather than the S phase, the mechanism linking the origin loading of MCM complexes to G1 progression are still virtually unknown. Is there an active signal transduction mechanism initiated by insufficiently licensed origins that inhibits S phase entry? Does loading of the MCM complex at origins recruit or release activators of G1 progression from chromatin? The recent proteomics analysis of licensed *Xenopus* chromatin indicates that many unanticipated nuclear processes are dependent on origin licensing. For example, the RAD50 protein, which is most closely associated with DNA repair, is chromatin bound when origin licensing is blocked, even before DNA replication would normally begin (Khoudoli et al., 2008). Ideally, direct examination of the proteins associated with licensed or unlicensed origins (as opposed to the entire chromatin fraction) would provide some direction to this line of investigation. Ongoing efforts to map mammalian origins (e.g., Mesner et al., 2006; Lucas et al., 2007) will be helpful in this regard.

Finally, the role of chromatin structure in promoting the proper identification and utilization of origins is still unclear. Histone modifications and nucleosome positioning almost certainly influence the accessibility of origin DNA to origin licensing factors. Anecdotal evidence in *Drosophila* cells suggests that histone acetylation may promote ORC binding (Aggarwal and Calvi, 2004), and histone hyper-acetylation accelerates the time in S phase when origins fire (Vogelauer et al., 2002; Weinreich et al., 2004). Acetylation is but one of many histone modifications implicated in replication control (reviewed in Weinreich et al., 2004; Dorn et al., 2009; Cayrou et al., 2010; Méchali, 2010). Much remains to be learned about potential roles for epigenetic regulation to promote origin identification, origin licensing, origin firing, and inhibition of re-replication.

The highly orchestrated sequence of events that occurs at thousands of human origins to permit the complete and precise duplication of the genome is a relatively delicate biological system. Imbalances in the expression or regulation of origin binding factors can lead to genome instability either because replication is incomplete or because replication occurs more than once from even a subset of origins. The genetic lesions associated with oncogenesis consistently include many such imbalances such as overproduction of origin licensing components, hyperactive CDKs, and missing cell cycle checkpoints. These imbalances not only support the aberrant proliferation that defines cancer but also provide the basis for potential novel therapies. Cancer cells are poised on the brink of catastrophic S phases, where even a small additional perturbation can prompt them to self-destruct, either by attempting replication without complete origin licensing or by massively re-replicating their genomes. Continued exploration of the regulation of events at DNA replication origins will allow exploitation of this therapeutic window.

Acknowledgments

The authors are grateful to our colleagues for stimulating discussions and critical reading of this work.

References

Aggarwal, B.D. and Calvi, B.R. 2004. Chromatin regulates origin activity in *Drosophila* follicle cells. *Nature* **430**(6997): 372–376.

Aggarwal, P., Lessie, M.D., Lin, D.I., et al. 2007. Nuclear accumulation of cyclin D1 during S phase inhibits Cul4-dependent Cdt1 proteolysis and triggers p53-dependent DNA rereplication. *Genes Dev* **21**(22): 2908–2922.

Alexandrow, M.G. and Hamlin, J.L. 2004. Cdc6 chromatin affinity is unaffected by serine-54 phosphorylation, S-phase progression, and overexpression of cyclin A. *Mol Cell Biol* **24**(4): 1614–1627.

Archambault, V., Ikui, A.E., Drapkin, B.J., and Cross, F.R. 2005. Disruption of mechanisms that prevent rereplication triggers a DNA damage response. *Mol Cell Biol* **25**(15): 6707–6721.

Arentson, E., Faloon, P., Seo, J., et al. 2002. Oncogenic potential of the DNA replication licensing protein CDT1. *Oncogene* **21**(8): 1150–1158.

Arias, E.E. and Walter, J.C. 2006. PCNA functions as a molecular platform to trigger Cdt1 destruction and prevent re-replication. *Nat Cell Biol* **8**(1): 84–90.

Bartkova, J., Rezaei, N., Liontos, M., et al. 2006. Oncogene-induced senescence is part of the tumorigenesis barrier imposed by DNA damage checkpoints. *Nature* **444**(7119): 633–637.

Bates, S., Ryan, K.M., Phillips, A.C., and Vousden, K.H. 1998. Cell cycle arrest and DNA endoreduplication following p21Waf1/Cip1 expression. *Oncogene* **17**(13): 1691–1703.

Bell, S.P. and Dutta, A. 2002. DNA replication in eukaryotic cells. *Annu Rev Biochem* **71**: 333–374.

Bell, S.P., Kobayashi, R., and Stillman, B. 1993. Yeast origin recognition complex functions in transcription silencing and DNA replication. *Science* **262**(5141): 1844–1849.

Blanchard, F., Rusiniak, M.E., Sharma, K., et al. 2002. Targeted destruction of DNA replication protein Cdc6 by cell death pathways in mammals and yeast. *Mol Biol Cell* **13**(5): 1536–1549.

Blow, J.J. and Dutta, A. 2005. Preventing re-replication of chromosomal DNA. *Nat Rev Mol Cell Biol* **6**(6): 476–486.

Bracken, A.P., Ciro, M., Cocito, A., and Helin, K. 2004. E2F target genes: unraveling the biology. *Trends Biochem Sci* **29**(8): 409–417.

Burke, T.W., Cook, J.G., Asano, M., and Nevins, J.R. 2001. Replication factors MCM2 and ORC1 interact with the histone acetyltransferase HBO1. *J Biol Chem* **276**(18): 15397–15408.

Cayrou, C., Coulombe, P., and Mechali, M. 2010. Programming DNA replication origins and chromosome organization. *Chromosome Res* **18**(1): 137–145.

Chen, S. and Bell, S.P. 2011. CDK prevents Mcm2-7 helicase loading by inhibiting Cdt1 interaction with Orc6. *Genes Dev* **25**(4): 363–372.

Cimprich, K.A. and Cortez, D. 2008. ATR: an essential regulator of genome integrity. *Nat Rev Mol Cell Biol* **9**(8): 616–627.

Cook, J.G., Park, C.H., Burke, T.W., et al. 2002. Analysis of Cdc6 function in the assembly of mammalian prereplication complexes. *Proc Natl Acad Sci USA* **99**(3): 1347–1352.

Cook, J.G., Chasse, D.A., and Nevins, J.R. 2004. The regulated association of Cdt1 with minichromosome maintenance proteins and Cdc6 in mammalian cells. *J Biol Chem* **279**(10): 9625–9633.

Costa, S. and Blow, J.J. 2007. The elusive determinants of replication origins. *EMBO Rep* **8**(4): 332–334.

Cvetic, C. and Walter, J.C. 2005. Eukaryotic origins of DNA replication: could you please be more specific? *Semin Cell Dev Biol* **16**(3): 343–353.

Davidson, I.F., Li, A., and Blow, J.J. 2006. Deregulated replication licensing causes DNA fragmentation consistent with head-to-tail fork collision. *Mol Cell* **24**(3): 433–443.

DePamphilis, M.L., Blow, J.J., Ghosh, S., et al. 2006. Regulating the licensing of DNA replication origins in metazoa. *Curr Opin Cell Biol* **18**(3): 231–239.

Devault, A., Vallen, E.A., Yuan, T., et al. 2002. Identification of Tah11/Sid2 as the ortholog of the replication licensing factor Cdt1 in *Saccharomyces cerevisiae*. *Curr Biol* **12**(8): 689–694.

Di Micco, R., Fumagalli, M., Cicalese, A., et al. 2006. Oncogene-induced senescence is a DNA damage response triggered by DNA hyper-replication. *Nature* **444**(7119): 638–642.

Dimitrova, D.S., Prokhorova, T.A., Blow, J.J., Todorov, I.T., and Gilbert, D.M. 2002. Mammalian nuclei become licensed for DNA replication during late telophase. *J Cell Sci* **115**(Pt 1): 51–59.

Donovan, S., Harwood, J., Drury, L.S., and Diffley, J.F. 1997. Cdc6p-dependent loading of Mcm proteins onto pre-replicative chromatin in budding yeast. *Proc Natl Acad Sci USA* **94**(11): 5611–5616.

Dorn, E.S., Chastain, P.D., 2nd, Hall, J.R., and Cook, J.G. 2009. Analysis of re-replication from deregulated origin licensing by DNA fiber spreading. *Nucleic Acids Res* **37**(1): 60–69.

Duursma, A. and Agami, R. 2005. p53-dependent regulation of Cdc6 protein stability controls cellular proliferation. *Mol Cell Biol* **25**(16): 6937–6947.

Edwards, M.C., Tutter, A.V., Cvetic, C., et al. 2002. MCM2-7 complexes bind chromatin in a distributed pattern surrounding ORC in *Xenopus* egg extracts. *J Biol Chem* **26**: 26.

Feng, D., Tu, Z., Wu, W., and Liang, C. 2003. Inhibiting the expression of DNA replication-initiation proteins induces apoptosis in human cancer cells. *Cancer Res* **63**(21): 7356–7364.

Fujita, M., Kiyono, T., Hayashi, Y., and Ishibashi, M. 1996. hCDC47, a human member of the MCM family. Dissociation of the nucleus-bound form during S phase. *J Biol Chem* **271**(8): 4349–4354.

Fujita, M., Yamada, C., Tsurumi, T., et al. 1998. Cell cycle- and chromatin binding state-dependent phosphorylation of human MCM heterohexameric complexes. A role for cdc2 kinase. *J Biol Chem* **273**(27): 17095–17101.

Green, B.M., Finn, K.J., and Li, J.J. 2010. Loss of DNA replication control is a potent inducer of gene amplification. *Science* **329**(5994): 943–946.

Halazonetis, T.D., Gorgoulis, V.G., and Bartek, J. 2008. An oncogene-induced DNA damage model for cancer development. *Science* **319**(5868): 1352–1355.

Hall, J.R., Lee, H.O., Bunker, B.D., et al. 2008. CDT1 and CDC6 are destablized by rereplication-induced DNA damage. *J Biol Chem* **283**(37): 25356–25363.

Hamlin, J.L., Mesner, L.D., Lar, O., et al. 2008. A revisionist replicon model for higher eukaryotic genomes. *J Cell Biochem* **105**(2): 321–329.

Herbig, U., Griffith, J.W., and Fanning, E. 2000. Mutation of cyclin/cdk phosphorylation sites in HsCdc6 disrupts a late step in initiation of DNA replication in human cells. *Mol Biol Cell* **11**(12): 4117–4130.

Higa, L.A., Mihaylov, I.S., Banks, D.P., Zheng, J., and Zhang, H. 2003. Radiation-mediated proteolysis of CDT1 by CUL4-ROC1 and CSN complexes constitutes a new checkpoint. *Nat Cell Biol* **5**(11): 1008–1015.

Hochegger, H., Dejsuphong, D., Sonoda, E., et al. 2007. An essential role for Cdk1 in S phase control is revealed via chemical genetics in vertebrate cells. *J Cell Biol* **178**(2): 257–268.

Hofmann, J.F. and Beach, D. 1994. cdt1 is an essential target of the Cdc10/Sct1 transcription factor: requirement for DNA replication and inhibition of mitosis. *EMBO J* **13**(2): 425–434.

Hogan, E. and Koshland, D. 1992. Addition of extra origins of replication to a minichromosome suppresses its mitotic loss in cdc6 and cdc14 mutants of *Saccharomyces cerevisiae*. *Proc Natl Acad Sci USA* **89**(7): 3098–3102.

Hu, J. and Xiong, Y. 2006. An evolutionarily-conserved function of PCNA for Cdt1 degradation by the Cul4-Ddb1 ubiquitin ligase in response to DNA damage. *J Biol Chem* **281**(7): 3753–3756.

Hu, J., McCall, C.M., Ohta, T., and Xiong, Y. 2004. Targeted ubiquitination of CDT1 by the DDB1-CUL4A-ROC1 ligase in response to DNA damage. *Nat Cell Biol* **6**(10): 1003–1009.

Hua, X.H. and Newport, J. 1998. Identification of a preinitiation step in DNA replication that is independent of origin recognition complex and cdc6, but dependent on cdk2. *J Cell Biol* **140**(2): 271–281.

Hua, X.H., Yan, H., and Newport, J. 1997. A role for Cdk2 kinase in negatively regulating DNA replication during S phase of the cell cycle. *J Cell Biol* **137**(1): 183–192.

Iizuka, M. and Stillman, B. 1999. Histone acetyltransferase HBO1 interacts with the ORC1 subunit of the human initiator protein. *J Biol Chem* **274**(33): 23027–23034.

Iizuka, M., Matsui, T., Takisawa, H., and Smith, M.M. 2006. Regulation of replication licensing by acetyltransferase Hbo1. *Mol Cell Biol* **26**(3): 1098–1108.

Ishimi, Y., Komamura-Kohno, Y., You, Z., Omori, A., and Kitagawa, M. 2000. Inhibition of Mcm4,6,7 helicase activity by phosphorylation with cyclin A/Cdk2. *J Biol Chem* **275**(21): 16235–16241.

Jiang, W., Wells, N.J., and Hunter, T. 1999. Multistep regulation of DNA replication by Cdk phosphorylation of HsCdc6. *Proc Natl Acad Sci USA* **96**(11): 6193–6198.

Jin, J., Arias, E.E., Chen, J., Harper, J.W., and Walter, J.C. 2006. A family of diverse Cul4-Ddb1-interacting proteins includes Cdt2, which is required for S phase destruction of the replication factor Cdt1. *Mol Cell* **23**(5): 709–721.

Jørgensen, S., Eskildsen, M., Fugger, K., et al. 2011. SET8 is degraded via PCNA-coupled CRL4(Cdt2) ubiquitylation in S phase and after UV irradiation. *J Cell Biol* **192**(1): 43–54.

Karakaidos, P., Taraviras, S., Vassiliou, L.V., et al. 2004. Overexpression of the replication licensing regulators hCdt1 and hCdc6 characterizes a subset of non-small-cell lung carcinomas: synergistic effect with mutant p53 on tumor growth and chromosomal instability – evidence of E2F-1 transcriptional control over hCdt1. *Am J Pathol* **165**(4): 1351–1365.

Kastan, M.B. and Bartek, J. 2004. Cell-cycle checkpoints and cancer. *Nature* **432**(7015): 316–323.

Khoudoli, G.A., Gillespie, P.J., Stewart, G., et al. 2008. Temporal profiling of the chromatin proteome reveals system-wide responses to replication inhibition. *Curr Biol* **18**(11): 838–843.

Kiely, J., Haase, S.B., Russell, P., and Leatherwood, J. 2000. Functions of fission yeast Orp2 in DNA replication and checkpoint control. *Genetics* **154**(2): 599–607.

Kim, J., Feng, H., and Kipreos, E.T. 2007. *C. elegans* CUL-4 prevents rereplication by promoting the nuclear export of CDC-6 via a CKI-1-dependent pathway. *Curr Biol* **17**(11): 966–972.

Kim, Y., Starostina, N.G., and Kipreos, E.T. 2008. The CRL4Cdt2 ubiquitin ligase targets the degradation of p21Cip1 to control replication licensing. *Genes Dev* **22**(18): 2507–2519.

Klein, H.L. 2008. The consequences of Rad51 overexpression for normal and tumor cells. *DNA Repair* **7**(5): 686–693.

Kroll, K.L. 2007. Geminin in embryonic development: coordinating transcription and the cell cycle during differentiation. *Front Biosci* **12**: 1395–1409.

Kumagai, A., Shevchenko, A., Shevchenko, A., and Dunphy, W.G. 2010. Treslin collaborates with TopBP1 in triggering the initiation of DNA replication. *Cell* **140**(3): 349–359.

Labib, K. and Hodgson, B. 2007. Replication fork barriers: pausing for a break or stalling for time? *EMBO Rep* **8**(4): 346–353.

Lau, E. and Jiang, W. 2006. Is there a pre-RC checkpoint that cancer cells lack? *Cell Cycle* **5**(15): 1602–1606.

Lee, A.Y., Liu, E., and Wu, X. 2007. The Mre11/Rad50/Nbs1 complex plays an important role in the prevention of DNA rereplication in mammalian cells. *J Biol Chem* **282**(44): 32243–32255.

Lemoine, F.J., Degtyareva, N.P., Lobachev, K., and Petes, T.D. 2005. Chromosomal translocations in yeast induced by low levels of DNA polymerase: a model for chromosome fragile sites. *Cell* **120**(5): 587–598.

Li, X., Zhao, Q., Liao, R., Sun, P., and Wu, X. 2003. The SCF(Skp2) ubiquitin ligase complex interacts with the human replication licensing factor Cdt1 and regulates Cdt1 degradation. *J Biol Chem* **278**(33): 30854–30858.

Lin, J.J. and Dutta, A. 2007. ATR pathway is the primary pathway for activating G2/M checkpoint induction after re-replication. *J Biol Chem* **282**(42): 30357–30362.

Liontos, M., Koutsami, M., Sideridou, M., et al. 2007. Deregulated overexpression of hCdt1 and hCdc6 promotes malignant behavior. *Cancer Res* **67**(22): 10899–10909.

Liu, E., Lee, A.Y., Chiba, T., et al. 2007. The ATR-mediated S phase checkpoint prevents rereplication in mammalian cells when licensing control is disrupted. *J Cell Biol* **179**(4): 643–657.

Liu, P., Slater, D.M., Lenburg, M., et al. 2009. Replication licensing promotes cyclin D1 expression and G1 progression in untransformed human cells. *Cell Cycle* **8**(1): 125–136.

Lovejoy, C.A., Lock, K., Yenamandra, A., and Cortez, D. 2006. DDB1 maintains genome integrity through regulation of Cdt1. *Mol Cell Biol* **26**(21): 7977–7990.

Lucas, I., Palakodeti, A., Jiang, Y., et al. 2007. High-throughput mapping of origins of replication in human cells. *EMBO Rep* **8**(8): 770–777.

Lutzmann, M., Maiorano, D., and Mechali, M. 2006. A Cdt1-geminin complex licenses chromatin for DNA replication and prevents rereplication during S phase in *Xenopus*. *EMBO J* **25**(24): 5764–5774.

Mac Auley, A., Werb, Z., and Mirkes, P. 1993. Characterization of the unusually rapid cell cycles during rat gastrulation. *Development* **117**(3): 873–883.

Machida, Y.J. and Dutta, A. 2007. The APC/C inhibitor, Emi1, is essential for prevention of rereplication. *Genes Dev* **21**(2): 184–194.

Machida, Y.J., Hamlin, J.L., and Dutta, A. 2005a. Right place, right time, and only once: replication initiation in Metazoans. *Cell* **123**(1): 13–24.

Machida, Y.J., Teer, J.K., and Dutta, A. 2005b. Acute reduction of an origin recognition complex (ORC) subunit in human cells reveals a requirement of ORC for Cdk2 activation. *J Biol Chem* **280**(30): 27624–27630.

Mailand, N. and Diffley, J.F. 2005. CDKs promote DNA replication origin licensing in human cells by protecting Cdc6 from APC/C-dependent proteolysis. *Cell* **122**(6): 915–926.

Maine, G.T., Sinha, P., and Tye, B.K. 1984. Mutants of *S. cerevisiae* defective in the maintenance of minichromosomes. *Genetics* **106**(3): 365–385.

Makise, M., Takehara, M., Kuniyasu, A., et al. 2008. Linkage between phosphorylation of the origin recognition complex and its ATP-binding activity in *Saccharomyces cerevisiae*. *J Biol Chem*: M804293200.

McGarry, T.J. and Kirschner, M.W. 1998. Geminin, an inhibitor of DNA replication, is degraded during mitosis. *Cell* **93**(6): 1043–1053.

McNairn, A.J. and Gilbert, D.M. 2003. Epigenomic replication: linking epigenetics to DNA replication. *BioEssays* **25**(7): 647–656.

Mechali, M. 2010. Eukaryotic DNA replication origins: many choices for appropriate answers. *Nat Rev Mol Cell Biol* **11**(10): 728–738.

Melixetian, M., Ballabeni, A., Masiero, L., et al. 2004. Loss of geminin induces rereplication in the presence of functional p53. *J Cell Biol* **165**(4): 473–482.

Mendez, J., Zou-Yang, X.H., Kim, S.Y., et al. 2002. Human origin recognition complex large subunit is degraded by ubiquitin-mediated proteolysis after initiation of DNA replication. *Mol Cell* **9**(3): 481–491.

Mesner, L.D., Crawford, E.L., and Hamlin, J.L. 2006. Isolating apparently pure libraries of replication origins from complex genomes. *Mol Cell* **21**(5): 719–726.

Mihaylov, I.S., Kondo, T., Jones, L., et al. 2002. Control of DNA replication and chromosome ploidy by geminin and cyclin A. *Mol Cell Biol* **22**(6): 1868–1880.

Mimura, S., Seki, T., Tanaka, S., and Diffley, J.F.X. 2004. Phosphorylation-dependent binding of mitotic cyclins to Cdc6 contributes to DNA replication control. *Nature* **431**(7012): 1118–1123.

Miotto, B. and Struhl, K. 2008. HBO1 histone acetylase is a coactivator of the replication licensing factor Cdt1. *Genes Dev* **22**(19): 2633–2638.

Moyer, S.E., Lewis, P.W., and Botchan, M.R. 2006. Isolation of the Cdc45/Mcm2-7/GINS (CMG) complex, a candidate for the eukaryotic DNA replication fork helicase. *Proc Natl Acad Sci USA* **103**(27): 10236–10241.

Muzi Falconi, M., Brown, G.W., and Kelly, T.J. 1996. cdc18+ regulates initiation of DNA replication in *Schizosaccharomyces pombe*. *Proc Natl Acad Sci USA* **93**(4): 1566–1570.

Nevins, J.R. 2001. The Rb/E2F pathway and cancer. *Hum Mol Genet* **10**(7): 699–703.

Nevis, K.R., Cordeiro-Stone, M., and Cook, J.G. 2009. Origin licensing and p53 status regulate Cdk2 activity during G(1). *Cell Cycle* **8**(12): 1952–1963.

Nguyen, V.Q., Co, C., and Li, J.J. 2001. Cyclin-dependent kinases prevent DNA re-replication through multiple mechanisms. *Nature* **411**(6841): 1068–1073.

Nishitani, H., Sugimoto, N., Roukos, V., et al. 2006. Two E3 ubiquitin ligases, SCF-Skp2 and DDB1-Cul4, target human Cdt1 for proteolysis. *EMBO J* **25**(5): 1126–1136.

Pelizon, C., Madine, M.A., Romanowski, P., and Laskey, R.A. 2000. Unphosphorylatable mutants of Cdc6 disrupt its nuclear export but still support DNA replication once per cell cycle. *Genes Dev* **14**(19): 2526–2533.

Petersen, B.O., Lukas, J., Sorensen, C.S., Bartek, J., and Helin, K. 1999. Phosphorylation of mammalian CDC6 by cyclin A/CDK2 regulates its subcellular localization. *EMBO J* **18**(2): 396–410.

Petersen, B.O., Wagener, C., Marinoni, F., et al. 2000. Cell cycle- and cell growth-regulated proteolysis of mammalian CDC6 is dependent on APC-CDH1. *Genes Dev* **14**(18): 2330–2343.

Piatti, S., Lengauer, C., and Nasmyth, K. 1995. Cdc6 is an unstable protein whose de novo synthesis in G1 is important for the onset of S phase and for preventing a "reductional" anaphase in the budding yeast *Saccharomyces cerevisiae*. *EMBO J* **14**(15): 3788–3799.

Randell, J.C., Bowers, J.L., Rodriguez, H.K., and Bell, S.P. 2006. Sequential ATP hydrolysis by Cdc6 and ORC directs loading of the Mcm2-7 helicase. *Mol Cell* **21**(1): 29–39.

Raveendranathan, M., Chattopadhyay, S., Bolon, Y.T., et al. 2006. Genome-wide replication profiles of S-phase checkpoint mutants reveal fragile sites in yeast. *EMBO J* **25**(15): 3627–3639.

Sancar, A., Lindsey-Boltz, L.A., Unsal-Kacmaz, K., and Linn, S.

2004. Molecular mechanisms of mammalian DNA repair and the DNA damage checkpoints. *Annu Rev Biochem* **73**: 39–85.

Sanchez-Pulido, L., Diffley, J.F., and Ponting, C.P. 2010. Homology explains the functional similarities of Treslin/Ticrr and Sld3. *Curr Biol* **20**(12): R509–510.

Sangrithi, M.N., Bernal, J.A., Madine, M., et al. 2005. Initiation of DNA replication requires the RECQL4 protein mutated in Rothmund–Thomson syndrome. *Cell* **121**(6): 887–898.

Sansam, C.L., Shepard, J.L., Lai, K., et al. 2006. DTL/CDT2 is essential for both CDT1 regulation and the early G2/M checkpoint. *Genes Dev* **20**(22): 3117–3129.

Sansam, C.L., Cruz, N.M., Danielian, P.S., et al. 2010. A vertebrate gene, ticrr, is an essential checkpoint and replication regulator. *Genes Dev* **24**(2): 183–194.

Santocanale, C. and Diffley, J.F. 1996. ORC- and Cdc6-dependent complexes at active and inactive chromosomal replication origins in *Saccharomyces cerevisiae*. *EMBO J* **15**(23): 6671–6679.

Sclafani, R.A. and Holzen, T.M. 2007. Cell cycle regulation of DNA replication. *Annu Rev Genet* **41**: 237–280.

Senga, T., Sivaprasad, U., Zhu, W., et al. 2006. PCNA is a co-factor for Cdt1 degradation by CUL4/DDB1 mediated N-terminal ubiquitination. *J Biol Chem* **281**(10): 6246–6252.

Seo, J., Chung, Y.S., Sharma, G.G., et al. 2005. Cdt1 transgenic mice develop lymphoblastic lymphoma in the absence of p53. *Oncogene* **24**(55): 8176–8186.

Sheu, Y.J. and Stillman, B. 2006. Cdc7-Dbf4 phosphorylates MCM proteins via a docking site-mediated mechanism to promote S phase progression. *Mol Cell* **24**(1): 101–113.

Shreeram, S., Sparks, A., Lane, D.P., and Blow, J.J. 2002. Cell type-specific responses of human cells to inhibition of replication licensing. *Oncogene* **21**(43): 6624–6632.

Stoeber, K., Tlsty, T.D., Happerfield, L., et al. 2001. DNA replication licensing and human cell proliferation. *J Cell Sci* **114**(11): 2027–2041.

Sugimoto, N., Tatsumi, Y., Tsurumi, T., et al. 2004. Cdt1 phosphorylation by cyclin A-dependent kinases negatively regulates its function without affecting geminin binding. *J Biol Chem* **279**(19): 19691–19697.

Sugimoto, N., Yoshida, K., Tatsumi, Y., et al. 2009. Redundant and differential regulation of multiple licensing factors ensures prevention of re-replication in normal human cells. *J Cell Sci* **122**(8): 1184–1191.

Takahashi, T.S., Yiu, P., Chou, M.F., Gygi, S., and Walter, J.C. 2004. Recruitment of *Xenopus* Scc2 and cohesin to chromatin requires the pre-replication complex. *Nat Cell Biol* **6**(10): 991–996.

Takayama, Y., Kamimura, Y., Okawa, M., et al. 2003. GINS, a novel multiprotein complex required for chromosomal DNA replication in budding yeast. *Genes Dev* **17**(9): 1153–1165.

Takeda, D.Y., Parvin, J.D., and Dutta, A. 2005. Degradation of Cdt1 during S phase is Skp2-independent and is required for efficient progression of mammalian cells through S phase. *J Biol Chem* **280**(24): 23416–23423.

Tanaka, S., Umemori, T., Hirai, K., et al. 2007. CDK-dependent phosphorylation of Sld2 and Sld3 initiates DNA replication in budding yeast. *Nature* **445**(7125): 328–332.

Tardat, M., Brustel, J., Dirsh, O., et al. 2010. The histone H4 Lys 20 methyltransferase PR-Set7 regulates replication origins in mammalian cells. *Nat Cell Biol* **12**(11): 1086–1093.

Tatsumi, Y., Ohta, S., Kimura, H., Tsurimoto, T., and Obuse, C. 2003. The ORC1 cycle in human cells: I. Cell cycle-regulated oscillation of human ORC1. *J Biol Chem* **278**(42): 41528–41534.

Tatsumi, Y., Sugimoto, N., Yugawa, T., et al. 2006. Deregulation of Cdt1 induces chromosomal damage without rereplication and leads to chromosomal instability. *J Cell Sci* **119**(Pt 15): 3128–3140.

Teer, J.K., Machida, Y.J., Labit, H., et al. 2006. Proliferating human cells hypomorphic for Origin Recognition Complex 2 and pre-replicative complex formation have a defect in p53 activation and Cdk2 kinase activation. *J Biol Chem* **281**(10): 6253–6260.

Vaziri, C., Saxena, S., Jeon, Y., et al. 2003. A p53-dependent checkpoint pathway prevents rereplication. *Mol Cell* **11**(4): 997–1008.

Vogelauer, M., Rubbi, L., Lucas, I., Brewer, B.J., and Grunstein, M. 2002. Histone acetylation regulates the time of replication origin firing. *Mol Cell* **10**(5): 1223–1233.

Wei, Z., Liu, C., Wu, X., et al. 2010. Characterization and structure determination of the Cdt1 binding domain of human minichromosome maintenance (Mcm) 6. *J Biol Chem* **285**(17): 12469–12473.

Weinreich, M., Liang, C., and Stillman, B. 1999. The Cdc6p nucleotide-binding motif is required for loading mcm proteins onto chromatin. *Proc Natl Acad Sci USA* **96**(2): 441–446.

Weinreich, M., Palacios DeBeer, M.A., and Fox, C.A. 2004. The activities of eukaryotic replication origins in chromatin. *Biochim Biophys Acta* **1677**(1–3): 142–157.

Williamson, D.H. 1985. The yeast ARS element, six years on: a progress report. *Yeast* **1**(1): 1–14.

Wohlschlegel, J.A., Dwyer, B.T., Dhar, S.K., et al. 2000. Inhibition of eukaryotic DNA replication by geminin binding to Cdt1. *Science* **290**(5500): 2309–2312.

Wyrick, J.J., Aparicio, J.G., Chen, T., et al. 2001. Genome-wide distribution of ORC and MCM proteins in *S. cerevisiae*: high-resolution mapping of replication origins. *Science* **294**(5550): 2357–2360.

Xouri, G., Lygerou, Z., Nishitani, H., et al. 2004. Cdt1 and geminin are down-regulated upon cell cycle exit and are over-expressed in cancer-derived cell lines. *Eur J Biochem* **271**(16): 3368–3378.

Xouri, G., Squire, A., Dimaki, M., et al. 2007. Cdt1 associates dynamically with chromatin throughout G1 and recruits geminin onto chromatin. *EMBO J* **26**(5): 1303–1314.

Yanagi, K.I., Mizuno, T., You, Z., and Hanaoka, F. 2002. Mouse geminin inhibits not only Cdt1-MCM6 interactions but also a novel intrinsic Cdt1 DNA binding activity. *J Biol Chem* **277**(43): 40871–40880.

Yanow, S.K., Lygerou, Z., and Nurse, P. 2001. Expression of Cdc18/Cdc6 and Cdt1 during G2 phase induces initiation of DNA replication. *EMBO J* **20**(17): 4648–4656.

Yoshida, K. and Inoue, I. 2004. Regulation of geminin and Cdt1 expression by E2F transcription factors. *Oncogene* **23**(21): 3802–3812.

Zegerman, P. and Diffley, J.F. 2007. Phosphorylation of Sld2 and Sld3 by cyclin-dependent kinases promotes DNA replication in budding yeast. *Nature* **445**(7125): 281–285.

Zhong, W., Feng, H., Santiago, F.E., and Kipreos, E.T. 2003. CUL-4 ubiquitin ligase maintains genome stability by restraining DNA-replication licensing. *Nature* **423**(6942): 885–889.

Zhu, W. and Dutta, A. 2006. An ATR- and BRCA1-mediated Fanconi anemia pathway is required for activating the G2/M checkpoint and DNA damage repair upon rereplication. *Mol Cell Biol* **26**(12): 4601–4611.

Zhu, W., Chen, Y., and Dutta, A. 2004. Rereplication by depletion of geminin is seen regardless of p53 status and activates a G2/M checkpoint. *Mol Cell Biol* **24**(16): 7140–7150.

Zhu, Y., Ishimi, Y., Tanudji, M., and Lees, E. 2005. Human CDK2 inhibition modifies the dynamics of chromatin-bound minichromosome maintenance complex and replication protein A. *Cell Cycle* **4**(9): 1254–1263.

Chapter

5 Systems biology approaches bring new insights in the understanding of global gene regulatory mechanisms and their deregulation in cancer

Arnaud Krebs and László Tora

Introduction

Transcription is a tightly regulated mechanism allowing the expression of particular subsets of genes within the genome. This transcription program is defining the identity, the function, and the fate of every cell of an organism. As a consequence, the alteration of transcription programs can change the cell fate and lead to cancer initiation and progression. Over the past 15 years, several major technical advances have allowed the research community to reconsider the way to address questions concerning the analysis of transcription regulation. The completion of the Human Genome Project in 2003 has provided a road map for large-scale interrogation of gene functions and expression regulation. The single-gene scale approach has progressively been replaced and/or complemented by systematic studies at the scale of the whole genome. These studies are progressively building the global rules defining transcription as a complex regulatory system. In this review, which reflects the state of the art in 2010 when the manuscript was submitted, we will try to illustrate through several typical examples how the latest technical improvements have modified our comprehension of the transcriptional mechanisms and their importance in the development of cancer. First, we will describe the technical improvements in the last ten years, which have allowed studying the transcription at the genome-wide scale. Second, we will present the new insights gained in the field of transcription regulation due to those technical revolutions. Finally, we will describe how this progress helped our understanding on the involvement of the transcription machinery and the transcription regulatory networks in the development of cancer.

The high-throughput revolution

In the past ten years several technical breakthroughs have dramatically modified the possibilities to investigate major questions in the transcription field. In the present chapter, we will describe those technical advances, which have had a major impact on our understanding of transcription regulation. We will first summarize how the developments of novel technologies in the monitoring of gene expression levels allow now the routine analysis of the whole-gene expression profile of a given cell type or tissue sample. Then we will describe the different methodologies that have been developed to allow the mapping of the binding sites of DNA binding factors at the scale of the entire genome. Finally, we will present some of the most innovative *in silico* approaches that are used to investigate transcription-regulation mechanisms in a global way.

Monitoring global gene expression levels: from cDNA arrays to RNA deep sequencing

The *in vivo* study of transcription-regulation mechanisms requires the efficient monitoring of gene expression levels in defined conditions. By the traditional PCR-based methods, the measurement of RNA levels was limited to a small set of genes. The first high-throughput tools developed to determine the mRNA steady-state levels at a genome-wide scale were through microarray analysis.

Systems Biology of Cancer, ed. S. Thiagalingam. Published by Cambridge University Press. © Cambridge University Press 2015.

Box 5.1 High-throughput sequencing

DNA sequencing is clearly driving much of the high-throughput revolution in genomics today. For the past 30 years, the denaturing polyacrylamid gel-based Sanger method has been the dominant approach and gold standard for DNA sequencing. The commercial launch of the first massively parallel pyrosequencing platform in 2005 opened a new era of high-throughput genomic analysis, referred to as high-throughput sequencing. Although the sequencing platforms differ in their engineering configurations and sequencing chemistries, they share a technical paradigm in that sequencing of spatially separated, clonally amplified DNA templates or single DNA molecules is performed in a flow cell in a massively parallel manner. Through iterative cycles of polymerase-mediated nucleotide extensions or through successive oligonucleotide ligations, different types of sequencers produce a billion bases of raw data per run, the equivalent of one-third of the human genome. In the relatively short time frame since 2005, high-throughput sequencing has fundamentally altered genomics research and allowed investigators to conduct experiments that were previously not technically feasible or affordable. The various technologies that constitute this new paradigm continue to evolve, and further improvements in technology robustness and process streamlining will pave the path for translation into clinical diagnostics.

Expression microarrays-based gene expression profiling relies on nucleic acid hybridization with complementary nucleic acid probes immobilized on a solid surface. Expression profiling techniques have been used to simultaneously monitor the expression of thousands of genes from various biological samples. They are relatively easy to use and can be applied to a large number of samples in parallel. Comparative expression microarray analyses are based on the concept of competitively hybridizing reference and sample RNA, which have been fluorescently labelled with different dyes, to a glass slide with immobilized single-stranded DNA targets representing sequences of transcripts of interest. By imaging the slides in each fluorescent channel and computing the ratios with which each sample binds to the complementary (c) DNA targets on the slides, a gene expression ratio can be determined. The microarray technology allowed rapid quantification of gene expression without the difficulty of multiplexing PCR reactions, at a significantly

reduced cost per target compared to conventional methods. The technology has rapidly improved by moving to large-scale analyses of thousands of human transcripts and has since evolved toward oligonucleotide-based microarray technologies, which allow consistent and affordable array production with highly specific probes (Coe et al., 2008).

The high demand for low-cost sequencing has driven the development of high-throughput sequencing technologies that have revolutionized the sequencing process (Box 5.1), producing thousands or millions of sequences at once. These advances in sequencing technologies allow today the direct sequencing of the whole transcriptome of a given cell population in a single experiment (Nagalakshmi et al., 2008). One can anticipate that this novel approach, which was shown to be more sensitive (higher detection of low-expressed transcripts) and unbiased when compared to the oligonucleotide microarray-based methods (Sultan et al., 2008), will become the standard gene expression profiling method in the future.

Genome-wide mapping of binding sites: from PCR microarrays to high-throughput sequencing

One of the major issues in transcription regulation studies is the genome-wide mapping of transcription factor binding sites in order to study their function at the scale of the genome. The chromatin immunoprecipitation (ChIP) technique uses antibodies that are specific for a transcription factor or for a given histone in its unmodified or post-translationally modified form, incorporated into the chromatin, to isolate the DNA to which the factor or histone is bound in a cell at a given time. The bound DNA can then be analyzed by quantitative PCR using primers specific for a region of interest. In order to be able to analyze all of the possible target DNA sequences where the analyzed factor can be detected in the genome of a given cell, ChIP has been combined with different high-throughput DNA analysis techniques.

The first of those techniques used was called ChIP-on-chip, where the immunoprecipitated DNA was fluorescently labelled and hybridized to microscopic slides harboring several defined short DNA fragments (Ren et al., 2000). In these earlier studies, arrays were spotted with a set of pre-amplified PCR products allowing the DNA quantification on a restricted set of loci of interest. The technology was then scaled up to build a human genome-wide array with PCR products covering >90% of human non-repetitive DNA sequences (Kim et al., 2005a). Similarly to gene expression arrays, the appearance of oligonucleotide-based technologies helped to build a human tiling DNA array representing all of the human non-repetitive regions with a resolution of 100 bp (Kim et al., 2005b). However, the use of ChIP-on-chip relies on the development of DNA microarray technology, especially the availability of arrayed slides for the model organism used.

An alternative way of identifying the ChIP-ed DNA pulled down by the antibody bound to the factor of interest is to sequence it. Over the years, several methodologies, based on the construction of DNA libraries using the ChIP-ed material followed by traditional Sanger sequencing, were shown to be efficient in identifying transcription factor binding sites genome wide (reviewed in Wu et al., 2006). These methods were particularly interesting for organisms or genomic regions that were not yet covered by the microarray technology. However, the construction and sequencing of the DNA libraries is a time-consuming process, which can be limiting for studies involving a large number of factors of interest. More recently, improvements in the high-throughput DNA sequencing methodologies (see Box 5.1) allowed the direct sequencing of the ChIP-ed DNA (Johnson et al., 2007). The advantage of this method, named ChIP-Seq, is that it is rapid, unbiased, and presents no limitation on the genomic regions and species analyzed. Thus very likely this technology will progressively replace the microarray approaches and allow the unbiased mapping of the transcription factors genome wide.

Bioinformatics approaches: *in silico* deciphering of transcriptional networks

The delivery of the complete sequence of the human genome, by the Human Genome Project, has set the ground for the possibility of investigating the transcriptional regulation networks by searching *in silico* the known transcription factor (TF) recognition motifs over the genome. However, most of the methodologies inherited from the analysis of prokaryotic genomes (reviewed in Zhou and Yang, 2006) were shown to be weakly adaptable for complex eukaryotic genomes. Two major hurdles needed to be overcome

for complex genomes. First, the consensus recognition motif(s) of a given transcription factor needs to be well characterized, as the knowledge of a single binding site is not sufficient. The preferred sequence motifs of DNA-binding factors generally have a "loose" consensus that can only be determined by comparison of a large collection of binding sites. The second major hurdle is the sizes of the genomes to be analyzed. Mammalian genomes are enormous compared to the six to thirteen base-pair DNA motifs generally recognized by transcription factors. Also, because consensus binding motifs are loose, the number of putative binding sites vastly exceeds the number of functional sites (Holstege and Clevers, 2006). One way to circumvent the "problem" of the genome size is to restrict the analyzed regions to the close vicinity of the transcription start sites (TSSs) of the genes. This approach allows the identification of DNA regulatory sequences in the promoters of the genes genome wide. However, this strategy excludes the discoveries of TF binding sites in enhancer regions, often located several kilobases away from the TSSs. An innovative high-throughput approach named enhancer element locator (ELL) combines additional properties of regulatory DNA sequences in order to make *de novo* discovery of enhancer elements in large and complex genomes possible. The robust prediction of enhancer regions by ELL relies on a new algorithm that is based on finding the conserved regions, which contain binding sites of several different TFs clustering together (Hallikas et al., 2006). Interestingly, the conservation properties of regulatory elements could also be combined with histone modification patterns (e.g., histone H3K4me1 for enhancer or histone H3K4me3 for promoters) in order to predict with high confidence different types of regulatory elements and their activity of the factors binding to these elements (Heintzman et al., 2007; Visel et al., 2009). Although still under development, the combination of these different strategies will have a major impact on the building and the comprehension of transcription regulatory networks.

Transcription as a complex gene expression regulatory system

The completion of the full human genome sequence in 2003 was the starting point for new ambitious projects in the so-called "post-genomic era." The overall goal of these projects is to understand why differential gene expression patterns are observed in different cells containing the very same primary DNA sequence; in other words, to define the regulatory code that mediates the controlled expression of a specific subset of genes, which vary according to cell type, developmental stage, or disease state.

The first objectives of the post-genomic research programs were to properly annotate the 3.4 billion base pairs of the human genome by defining the transcription units over the genome. Then by crossing this information with multiple heterogeneous datasets (e.g., expression profiles, TF binding site locations, etc.) these programs were aimed to define the networks regulating gene expression at the scale of the whole genome, in different cellular systems. Finally, by combining the genome-wide observations with previous mechanistic observations at the molecular level, these large genomic programs try to define global transcription regulation and thus gene expression "rules" that would define the fate of a cell.

The next section will describe the different achievements of those genome-wide initiatives over the past years and their perspectives. We will summarize the emerging views in genome organization and their possible functional consequences. Then, through several examples, we will show how, by using genome-wide approaches, new global transcription regulatory rules and networks were discovered.

Systematic annotation of the genome, or find the common marks for regulation

Annotation of functional elements

In order to have a more comprehensive view of the human genome, several steps of annotation of the genomic landscape were initiated. To define where transcription starts exactly from the transcription units several genome-wide sequencing-based high-throughput methods that require reliable isolation of full-length cDNAs, sequencing of their 5′ ends, and mapping of the sequence to a completed genomic DNA sequence were used. The sequencing stage can use the 5′ ends of cloned full-length cDNA libraries (so-called 5′ ESTs), short tags derived from 5′ ends of capped RNAs (cap analysis of gene expression, CAGE) and 5′-SAGE30 (serial analysis of gene expression), or tags derived from 5′–3′ ends (so-called paired-end tags) (reviewed in Sandelin et al., 2007). This intensive mapping allowed the prediction of the

location of the TSSs and defined the promoter regions of most of the expressed genes (Bajic et al., 2006).

In addition to these sequencing-based methods, another, albeit less precise, approach to localize promoters is to use a genome-wide location analysis (GWLA) identifying the genomic binding sites for the general transcription machinery associated with TSSs. By using the ChIP-on-chip technique with specific antibodies against subunits of different components of the basal transcription machinery (e.g., TFIID or RNA polymerase II (Pol II)) that bind to the core promoters of different genes, many active promoters were mapped genome wide from different cell lines (Kim et al., 2005b) demonstrating the validity of the method.

This technical breakthrough was part of a larger ambitious consortium project named encyclopedia of DNA elements (ENCODE). The goal of ENCODE is to identify all functional DNA elements in the human genome using a large set of high-throughput techniques. The pilot phase (from 2003 to 2007) was focused on 1% of randomly and non-randomly selected regions of the genome in order to develop the methodology for the analysis at a reasonable scale. During the first phase of this analysis, the project could nevertheless map previously known as well as novel functional elements using heterogeneous datasets, but also give new insights on the epigenetic signature of those elements (Birney et al., 2007). For example, the study by Birney et al. (2007) shows that chromatin accessibility (presence of DNase I hypersensitivity sites (DHS)) and specific histone modification patterns are highly predictive for both the presence and the activity of a TSS. By contrast, combination of DHS with another set of histone marks at regions distal from detected TSSs could also define regulatory regions (i.e., insulator or enhancer binding sites). One can anticipate that the completion of the production phase of the ENCODE project (launched in 2007), targeting the whole genome, will contribute to transform the human genomic landscape from a plain base-pair sequence to a fully annotated regulatory catalogue.

Pol II promoter characterization

In order to be able to understand Pol II transcription regulation, a genome-wide characterization of core promoter composition in higher eukaryotes is a prerequisite. Using the genome-wide promoter mapping data recently generated (discussed earlier), some general composition and regulation rules have emerged (Figure 5.1). In the generally accepted model, a promoter used to be defined by a TSS and several regulatory elements located in its close vicinity. Among those elements, the canonical TATA box was thought to be the main marker defining a promoter.

Interestingly, it rapidly became clear from the first genome-wide studies that a vast majority of mammalian promoters were lacking the TATA box. Furthermore, the CAGE-based approaches indicated that most of the mouse and human promoters lack a distinctive sharp TSS, but rather harbor a broad array of closely located TSSs over 50 to 100 bp (Carninci et al., 2005; Bajic et al., 2006; Carninci et al., 2006). These observations provide the basis for a new system of promoter classification based on the TSS distribution ("sharp" or "broad"). It has been observed that the presence of a TATA box is more often associated with promoters that have a single, sharply defined TSS. On the contrary, CpG islands have been shown to be overrepresented in the "broad" promoter categories (Sandelin et al., 2007). Recently, some functional relevance of this new classification has emerged from a broad tissue GWLA of Pol II. This study demonstrates that sharp, TATA-containing promoters are primarily used for tissue-specific expression, whereas broad, CpG island-containing promoters are generally associated with ubiquitously expressed genes (Barrera et al., 2008).

Altogether, the genome-wide approaches generated novel datasets that helped to redefine the promoter

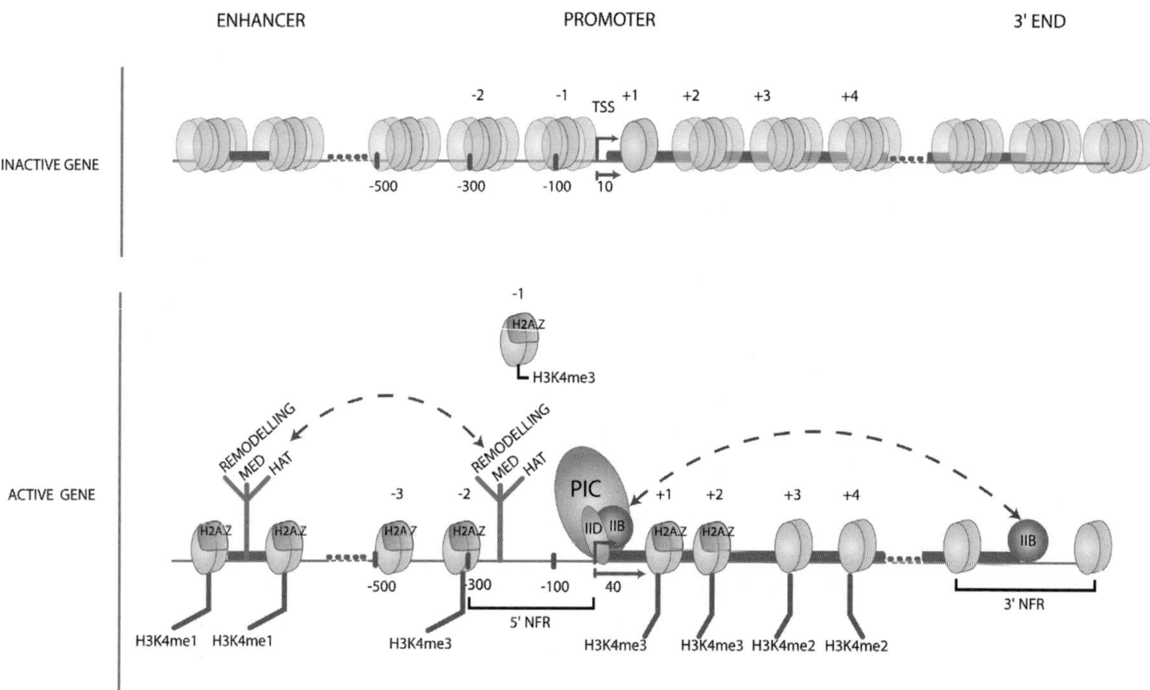

Figure 5.1 Structural and mechanistic changes in the chromatin organization of the Pol II regulatory elements upon gene activation. The figure gives a projection of the structural organization of enhancer, promoter, and transcription termination (3′ end) regions in transcriptionally silent (top panel) and active mode (bottom panel). Upon gene activation, the position of the fuzzy histones around the transcription start sites (TSS) get fixed, the −1 nucleosome is evicted, to form the 5′ nucleosome free region (NFR). A similar positioning of the histones occurs also for histones flanking the enhancer and the 3′ NFR regions. The nucleosomes located outside of the represented regions remain fuzzy after activation (not represented). Several histone modification and histone variants characterize each region upon activation. A looping is believed to happen between enhancer and promoter regions allowing the activity of co-activator complexes (e.g., Mediator, HATs, ATP-dependent remodelling complexes). The cylinders are representing nucleosomes with fixed (single non-transparent cylinder) and fuzzy positioning (three transparent cylinder array). Several histone modifications and variants characteristic of a particular localization or activation state are symbolized (e.g., H2A.Z, methylation marks). Members of the general transcription machinery are symbolized either by circles (PIC: pre-initiation complex; IID: TFIID; IIB: TFIIB) or by their acronyms (REMODELLING: chromatin remodelling complexes; MED: mediator complex; HAT: histone acety-transferase complexes). The dotted arrows show potential long-distance physical interactions within the different regulatory DNA elements (looping). (A black and white version of this figure will appear in some formats. For the color version, please refer to the plate section.)

nomenclature established by single-gene scale studies and opened new perspectives in the understanding of the transcription regulation mechanisms at a larger scale.

Structural and mechanistic insights from genome-wide studies

Besides the functional identification and understanding of regulatory sequences of the mammalian genomes, the use of genome-wide analyses recently allowed the investigation of several transcription regulation mechanisms. The classical approach, studying molecular mechanisms at a single-gene scale and extrapolating the results to the whole genome, has been inverted. Now the observations are directly made at the scale of the genome to try to deduce mechanistic rules applicable at the single-gene level. Furthermore, these new approaches have allowed the construction of more complex regulatory rules by constructing interconnected networks and hubs.

Genome-wide chromatin architecture

When compared to prokaryotes, the presence of chromatin in eukaryotes adds a layer of complexity to the transcriptional regulation mechanisms. Before being able to get access to a particular region of DNA to transcribe it, the transcription machinery has to overcome the nucleosome barrier "problem." Two models have been proposed to explain the functional relationship between the transcription machinery and the chromatin. One possibility is that transcription factor binding at promoters acts through nucleosomes to activate or repress gene expression (Li et al., 2007). However, studies of the *Saccharomyces cerevisiae* *PHO5* promoter suggested that nucleosomes are simply evicted from promoters, and the resulting naked DNA would allow transcription factors to gain access to their binding sites and for the basal transcriptional machinery to assemble (Boeger et al., 2003; Reinke and Horz, 2003). These somewhat contradictory models illustrate the importance of associating the study of transcription regulation processes with the study of chromatin structure dynamics at regulatory regions.

Since the discovery of DHSs in chromatin in the early eighties, the prevalent hypothesis was that those regions would correspond to nucleosome free regions. However, until improvements in the investigation methodology came about, it could not be demonstrated that the DHSs were due to the absence of nucleosomes in these regions. Recently, two different studies in *S. cerevisiae* addressed the question at different scales, reaching different conclusions.

The first strategy, focused on 480 kbp of the *S. cerevisiae* genome at a very high resolution (20 bp tiled DNA microarrays) (Yuan et al., 2005) whereas the other study covered the whole genome at a low resolution (1 kb) (Bernstein et al., 2004). The use of very high-resolution arrays led to the concept of a constitutive nucleosome free region (NFR) spreading on an ~150 bp region immediately upstream of the TSS, flanked on both sides by well positioned nucleosomes, that are deacetylated and enriched in H2A.Z (a histone variant of H2A) (Yuan et al., 2005). Such NFRs were observed for most of the genes studied, with no correlation with their relative expression status. Nucleosome free region were associated in this study with multiple stretches of poly-A or poly-T known to be associated with nucleosome instability. From this data, the authors concluded that NFRs are intrinsic components of promoter regions and thus do not appear to be due to active removal of nucleosomes during transcription initiation. On the other hand, when studied at the whole-genome scale and compared to transcription levels, the presence of nucleosome-depleted regions seemed to correlate mainly with active regulatory elements (Bernstein et al., 2004). This apparent contradiction could be reconciled, however, if upon gene induction regions flanking the fixed constitutive NFRs would lose additional nucleosomes. This model was further supported by a study where nucleosome positioning was investigated at a single DNA molecule level (Gal-Yam et al., 2006). The authors could show for a specific gene that constitutive NFRs are observed on the promoter region and that upon induction, additional nucleosomes are lost downstream, but not upstream, of the constitutive NFRs. The recent extension of nucleosome mapping to regions distant from the TSS by the use of genome-wide ChIP-seq revealed that NFRs are not restricted to TSS, but are also present at the 3′ end of most *S. cerevisiae* genes (Mavrich et al., 2008; Shivaswamy et al., 2008). The location of those NFRs coincides with the polyadenylation sites, suggesting that the 3′ NFRs might be involved in transcription termination.

The conservation of NFRs in higher eukaryote genomes has been addressed by several studies. The first observations were made over limited regions of the human genome, but at a high resolution, assessing the presence of NFRs at promoters of human cells (Heintzman et al., 2007; Ozsolak et al., 2007). These studies associated the presence of NFRs with the presence of a pre-initiation complex and could not detect NFRs at promoters of silenced genes. More recently, a study covering the whole genome using high-throughput sequencing generated genome-wide maps of nucleosome position in both resting and activated human T cells (Schones et al., 2008). The authors found, similar to the above-described *S. cerevisiae* data, that in human cells nucleosomes are highly phased and NFRs are formed around the TSS of expressed genes. However, the nucleosome phasing disappears for silent genes. Furthermore, by comparing the high-resolution nucleosome position map with previously generated histone-modification mark maps, these studies attributed particular modification patterns (mainly histone H3K4me3 and H2A.Z) to the highly positioned NFR flanking nucleosomes (see Figure 5.1). Altogether, these observations show an evolutionarily constrained role for this phenomenon in transcriptional regulation. Furthermore, these results support the model in which the formation and the maintenance of NFR in higher eukaryotes is dependent upon downstream gene activation.

Histone marks and histone variants: transcription in the context of chromatin

As mentioned above, chromatin can be considered as a barrier preventing the transcription machinery from accessing the DNA. The deciphering of the entire regulation of different chromatin states and their functional consequences will constitute a major step in the understanding of the transcription regulation processes.

The presence of NFR is not the only chromatin feature of regulatory regions. The systematic mapping of histone modifications by the ChIP–on-chip technique helped in defining the chromatin signatures associated with DNA regulatory regions. Later these studies allowed the creation of combinatorial maps of these histone modifications, which defined special regulatory regions and thus could also be used to predict such regions. Most of the efforts were dedicated to study acetylation on histones H3, H4 and methylation on the N-terminal tail of histone H3, trying to understand the different combinational

pattern of marks associated with a particular type of regulatory region. The early studies, focused on a limited number of marks, allowed the definition of simple signatures sufficient to characterize higher eukaryotic promoter and enhancer regions (e.g., Heintzman et al., 2007, 2009). According to these studies, active promoters were associated with H3K4 tri-methylation, histone H3–H4 global acetylation, and H2A.Z enrichment. Similarly, enhancer regions were associated with histone H3K4 mono-methylation and H3K27 acetylation, but not histone H3K4 tri-methylation (Heintzman et al., 2009). It was also observed that NFR was a marker shared by both promoter and enhancer regions (Heintzman et al., 2007). Despite the fact that those correlations could robustly be used for predicting promoter or enhancer regions, the limited set of markers considered in these studies did not reflect the entire complexity of all the chromatin marks that have been implicated in potential transcriptional regulatory networks. More recently, two systematic ChIP-seq approaches have mapped 39 acetylation, 20 histone methylation marks, and the enrichment of H2A.Z variant genome wide in human lymphocytes (Barski et al., 2007; Wang et al., 2008). Out of the 4,339 combinatorial patterns possible, only a small fraction was indeed observed at promoter regions. The prevalent patterns associated with promoters and enhancer regions were defined and classified in three expression level categories (Wang et al., 2008). Despite the complexity of the combinations observed, some modification patterns correlate very well with a type of regulatory region. Table 5.1 displays the marks most frequently observed at regulatory regions according to gene expressions.

Interestingly, these combinations of histone marks define flexible transcription regulatory rules in a complex cellular system that can likely be partially extrapolated to other tissues and organisms. However, it becomes clear from several other studies that additional specific rules may govern each cellular system. For example, a ChIP-on-chip study against histone methylation marks H3K4me3 and H3K27me3 in mouse embryonic stem cells (mESCs) observed a specific pattern of modifications in this cell type (Bernstein et al., 2006). H3K4me3 is generally associated with active promoters (Santos-Rosa et al., 2003; Pray-Grant et al., 2005) and H3K27me3 with repressed transcription (Francis et al., 2004; Ringrose et al., 2004). Studies carried out in differentiated cells suggested that the distribution of these marks is

Table 5.1 Histone modification patterns associated with transcription regulatory elements. The table summarizes the most frequently observed histone modifications detected on or around regulatory elements (promoters and enhancers), classified according to the expression levels of the given genes localized in the close vicinity of the regulatory regions.

	Methylation	Acetylation	Histone variant	Expression level
Promoter I	H3K27me3, H3K4me1/2/3, H3K9me1	excluded	H2A.Z	low
Promoter II	H3K36me3	H4K16Ac	H2A.Z	intermediate
Promoter III	H3K4me3, H2BK5me1, H4K20me1, H3K79me1/2/3	H4K16Ac	H2A.Z	high
Enhancer	H3K4me1/2, H3K9me1	H3K18Ac, H3K27Ac	H2A.Z	–

(Based on Heintzman et al., 2007, 2009; Wang et al., 2008)

non-overlapping. However, in the undifferentiated mESCs, the combination of H3K4me3 and H3K27me3 marks is present simultaneously at some promoters. It appeared that this bivalent marking is present at the promoters of genes coding for transcription factors, which will be involved at a subsequent stage in cellular differentiation, but have to be repressed to keep the stem cell state. The authors then hypothesized that this bivalent mark is restricted to mESCs and to those genes that will change their expression during differentiation. However, this hypothesis was contradicted by a study analyzing the change of histone methylation marks in a model where mESCs were differentiated to neurons (Mohn et al., 2008). Similarly to what was observed previously, the authors observed that the promoters harboring bivalent chromatin domains in undifferentiated mESCs lose this particular bivalent mark during the differentiation process. However, they also observed that the bivalent marks were not restricted to mESCs, but during differentiation novel sets of promoters appeared harboring the bivalent chromatin marks. This example illustrates the complexity of generalizing regulation rules deduced from a study in a particular cellular system.

The use of systematic approaches from different systems will help to define the epigenetic landscape that interacts with the transcription machinery and better understand the gene expression regulation processes.

Mechanistic insight of transcription regulation processes

Transcription regulation processes are known to be complex multi-step processes. Most of the mechanistic knowledge accumulated by classical approaches is issued from *in vitro* or *in vivo* studies on inducible model genes. Most of the studies performed in lower or higher eukaryotes supported a model in which transcription factors and cofactors are sequentially recruited, finally leading to the pre-initiation complex (PIC) formation, the loading of Pol II on the DNA, and the production of a functional pre-mRNA transcript. Furthermore, the canonical model assumes that transcription initiation is the rate-limiting step in efficient transcript production. However, the systematic characterization of Pol II location sites on the genome of a given cell type and the mapping of the other actors of the transcription initiation process has not only improved our view of transcription regulation processes, but at the same time also indicated that the transcription initiation process, when studied at a genome-wide level, is much more complex than originally thought. In the following subsections, through several striking examples, we will try to illustrate how high-throughput approaches have brought new insights to our understanding of transcription regulation.

Mechanisms of action of the transcriptional machinery

To date, most of the GWLA studies performed against the transcription initiation machinery in higher eukaryotes were done either on selected regions at high resolution (e.g., ENCODE regions) or at relatively low resolution (~0.5 kb) due to the size of the studied genomes. At low resolution, the precise localization of the position of the investigated factors relative to the promoter of the genes and the surrounding chromatin structure remains limited. However, already at this low resolution a global view emerged that started to give information about the general behavior of those factors. For example, the GWLA of p300 (a general co-activator harboring a histone acetyltransferase activity (HAT)), together with TAF1 (a TFIID subunit) and Pol II (Heintzman et al., 2007), suggest a physical interaction between enhancer and active promoter regions distant from several kilobases (Szutorisz et al., 2005; Heintzman et al., 2007). This observation would confirm the model of action in which enhancer regions are thought to be bridged to promoters by the action of activators and chromatin remodelling complexes resulting in the increase of the promoter activity (Figure 5.1) (Blackwood and Kadonaga, 1998). More recently, the enhancer function has been further investigated at the genome-wide scale in different cell types (Heintzman et al., 2009). Interestingly, this study shows that p300-bound enhancers are not shared in common among different cell types. This observation supports the concept in which tissue-specific activity is mainly controlled at enhancer level rather than at the promoter level.

In contrast to higher eukaryotes, studies in yeast were performed at high resolution, allowing the definition of promoter chromatin architecture (see above). More recently, a majority of the transcription actors (activators, co-activators, PIC subunits, and Pol II) could be mapped around the promoters in *S. cerevisiae* (Venters and Pugh, 2009). It appears in this study that, at least for lower eukaryotes, all the transcription actors are bound within the NFR. The PIC is bound immediately upstream of the TSS at the location of the TATA box. In the middle of the NFR region (–100 bp from the NFR border), co-activators, such as the mediator complex, and the Pol II are bound. Finally, other co-activators such as the histone acetylase complex (SAGA) (Timmers and Tora, 2005) or ATP-dependent remodelling complexes are bound at the 5′ limit of the NFR. This mapping gives some new organizational insights of promoter function and potential role of each member of the machinery with regard to its location. For example, it shows that the active removal of the 5′ NFR nucleosome is necessary prior to Pol II docking to the promoter. These results further explain the observed location of remodelling complexes at this precise location.

Non-productive PIC formation

Expression of genes was long thought to be regulated primarily at the level of RNA polymerase II recruitment to gene promoter regions, and the few genes that did not fit this paradigm were regarded as exceptions. The comparison of the genome-wide mapping by ChIP-on-chip of the pre-initiation complex subunit TAF1 in human fibroblast with the rate of gene expression surprisingly showed that for ~20% of the human genes a PIC is observed, but no transcript can be detected (Kim et al., 2005b). This result may indicate that another mechanism of regulation is responsible for the expression of these genes at a post-transcription initiation step. This concept was further developed by a genome-wide study on hESC and differentiated lymphocytes, showing that most of the silent genes in both of these cell types indeed show PIC formation in a non-productive manner (Guenther et al., 2007). Moreover, recent genome-wide analyses of Pol II distribution in *Drosophila* (Zeitlinger et al., 2007) and mammalian systems have indicated that a large number of genes might be regulated at a step subsequent to Pol II recruitment, during early transcription elongation. At these genes, Pol II begins transcription but stalls after synthesizing a short RNA. The release of this engaged Pol II from the promoter-proximal region is the rate-limiting step for transcription. From these genome-wide studies it seems that Pol II stalling at promoter-proximal is prevalent on genes involved in development and response to stimuli, suggesting that Pol II stalling during early elongation plays important roles in the rapid and precise control of gene expression (reviewed in Nechaev and Adelman, 2008). These results further confirm that regulation at post-initiation steps is a general mechanism widespread in metazoan organisms. However, the mechanistic aspects leading to non-productive PIC formation (e.g., transcriptional pausing, poor processivity, degradation of the transcription machinery) remain to be further investigated.

Transient and permanent regulation of expression

Several processes can be involved in relieving the repression that enables the formation of an active PIC, thus resulting in active transcription. Among those, the transition from a silenced chromatin barrier state to an activated one, the active recruitment of the components of the PIC, and the ability to initiate a productive elongation are thought to be major regulatory mechanisms. The relative contributions of those three phenomena to achieve gene silencing have been recently investigated by comparing cell lines at different phases of the cell cycle and during differentiation (from mESC to embryonic bodies) (Komashko et al., 2008). The authors first isolated subsets of genes that are silent or have a very low expression level in the different conditions tested. Then, using ChIP-on-chip, they could investigate the presence of chromatin repressive marks (H3K9me3, H3K27me3), Pol II, and DNA methylation at the corresponding promoters. They could distinguish between at least two different mechanisms explaining the lack of gene expression. Interestingly, it seems that repressive mechanisms used by the cell during the cell cycle and in differentiation are different. Along the cell cycle, cells achieve transient gene regulation by changes in the recruitment of the PIC, presumably through the action of different transcription factors. On the contrary, during differentiation, changes in different chromatin states seem to establish permanent changes in the expression of genes. This result illustrates how different transcription regulation strategies can be used by cells to achieve transient or permanent changes in their expression patterns.

TFIIB new function: gene looping

TFIIB is a member of the general transcription machinery binding to promoter regions cooperatively with TBP. Recently, TFIIB binding to terminator sequences located downstream of protein-coding regions was reported in *S. cerevisiae* for a limited set of promoters (Singh and Hampsey, 2007). This initial observation suggests an unexpected role for TFIIB in post-initiation processes. It was suggested that TFIIB was involved in looping of the DNA to create physical interaction between the 5′ and 3′ of a gene (gene looping model, see Figure 5.1). This interaction could potentially play a role in the recycling of the Pol II after transcription termination. This result was further extended by a GWLA showing that TFIIB is recruited at the 3′ NFR of a subset of genes in *S. cerevisae* (Mavrich et al., 2008). The selective enrichment of TFIIB at 3′ NFRs in the absence of TBP supports the gene looping model. GWLA of TFIIB across the rat genome showed that only 21% of TFIIB binding sites mapped near the TSSs of known protein-coding genes, while the majority of the binding sites were found at intragenic or 3′ positions of the studied genes suggesting a possible conservation of TFIIB-mediated gene looping mechanism to higher eukaryotes (Yochum et al., 2008). However, due to the complexification of promoter organization with evolution (see above), this model needs to be further investigated in higher eukaryotes.

New insights brought by systems biology in the understanding of transcription deregulation in cancer

Until now we have illustrated how the change of scale from single-gene approaches to genome-wide analyses in the last years has modified the perception of transcription regulation mechanisms. A number of cancer studies showed a major involvement of different transcription factor family members in the promotion and progression of multiple types of cancer. For example, a large number of known oncogenic proteins are TFs (i.e., proteins belonging to the Myc family, p53 …) regulating expression of particular transcription programs. As a consequence, genome-wide studies are going to have a major impact on the understanding of the transcription pathways and networks involved in cancer. In the following section, we are going to illustrate through several examples how the use of GWLA of known oncogenic TFs leads to the discovery of oncogenic regulation networks. Then we will explain how the analysis of these networks in the established genomic context will improve the understanding of the pathways involved in carcinogenesis. Finally, we will try to anticipate how the recent achievements in transcription regulation analysis may lead to new insights in cancer research.

Decipher the oncogenic TF network

The first attempts to understand regulatory networks involved in cancer were based on gene expression analyses using microarrays. By comparing RNA expression profiles from "normal" cells with those of "cancer" cells, differentially expressed genes were identified. This information could then be used to search TF binding sites at the promoter of these genes. However, it rapidly appeared that among those genes that were mis-regulated in cancer cells, only a minor subset showed direct TFs binding evidence, strongly limiting the understanding of cancer pathways in a global way. Later, the adaptation of the GWLA techniques to cancer-related factors provided new resources to investigate efficiently the oncogenic regulatory networks.

p53 network

The transcription factor p53 regulates the expression of genes involved in a variety of cellular functions including cell cycle arrest, DNA repair, and apoptosis. Through those functions, the mis-regulation of p53 pathways are involved in many types of cancer (reviewed in Vousden and Prives, 2005). Several studies have mapped p53 binding sites at different scales and under different physiological conditions (Cawley et al., 2004; Wei et al., 2006; Smeenk et al., 2008). The combination of those results with previous knowledge on p53 allowed the building of a global p53 transcriptional regulation network, which brings new insights in the understanding of the mechanism of p53 action. For example, the p53 family members, p63 and p73, were shown *in vitro* to contribute to p53 recruitment to specific target genes (Flores et al., 2002). The comparison of genome-wide p53 binding sites with p63 and p73 binding profiles shows a strong overlap (Smeenk et al., 2008). This observation reinforced the model in which p63 and p73 are involved in the regulation of p53 binding. Besides molecular mechanism of action of p53, some discoveries having clinical application could be made using these new resources. For example, one of the studies highlighted the importance of p53 in the regulation of genes known for their anti-apoptotic function (*BCL2A1* and *TTNFAIP8*) in p53-dependent breast tumors. The observation that the expression of a set of genes in breast tumors highly correlates with p53 status (mutated or not) and the clinical behavior of those tumors suggests that they could be used as powerful new biomarkers for patient prognosis (Wei et al., 2006). These two examples illustrated the impact of systematic GWLA studies on the understanding of p53-mediated cancer mechanisms.

The network of the Myc family proteins

With a similar strategy as the studies performed for p53, several TFs were in the focus of GWLA studies. Myc proteins have long been modelled to operate strictly as classic gene-specific transcription factors. When the proto-oncogene c-Myc regulation network was first established (Zeller et al., 2006) by ChIP-PET, this study showed that c-Myc is a master regulator of 48 transcription factors for which a function in cell growth and cell cycle regulation was already known (e.g., MAX, NFB, STAT3, JUN, etc.). Furthermore, it shows that c-Myc is also involved in the direct regulation of the mir-17 micro RNA (miRNA) cluster. Besides the identification of target loci, the mechanistic regulation of c-Myc action was investigated by *in*

vivo GWLA (Guccione et al., 2006). According to *in vitro* studies, the c-Myc DNA binding motifs (E-boxes) were thought to be major signals for c-Myc recruitment. However, it was found that c-Myc binding is strongly dependent on the chromatin features present around the target binding site. Euchromatic islands, which are defined as clusters of specific methyl- (H3K4me2, H3K4me3, H3K79me2) and acetyl-histone (H3K9Ac, H3K14Ac, H3K18Ac, H3K27Ac) marks at the 5$'$ end of target genes, are essential for Myc to bind DNA *in vivo*. E-boxes outside of the chromatin islands are not significantly bound by Myc, whereas all E-boxes within euchromatic islands are bound. This implies that the recognition of the E-boxes by c-Myc is an event arising downstream of prior chromatin structure markings and opening.

Another member of the Myc family, N-Myc, has been suggested to have a robust role in regulating global cellular euchromatin, including that of intergenic regions (Cotterman et al., 2008). Strikingly, it was found that 90 to 95% of the total genomic euchromatic marks on histone H3, either acetylated at lysine 9 or methylated at lysine 4, are N-Myc-dependent. However, Myc regulation of transcription of genes, where it directly binds and at which it is required for the maintenance of active chromatin, is generally weak. Thus Myc has a much more potent ability to regulate large domains of euchromatin than to influence the transcription of individual genes. Overall, Myc regulation of chromatin in the human genome includes both specific genes, but also large genomic domains that invoke functions independent of a classic transcription factor. These novel findings including systems biology approaches support a new dual model for Myc chromatin function with important implications for the role of Myc in cancer and stem-cell biology.

These results illustrate the key role of transcription factors on the regulation of the chromatin environment. Moreover, it brings a new layer of complexity to the regulation systems that are known to be mis-regulated in the cancer states. It suggests, for example, that the over-expression of a proto-oncogenic TF as c-Myc will be able to activate different subsets of transcription programs depending on the epigenetic state of the cell. This epigenetic modulation of TF response can partially explain the linkages observed between the origin of the cell and the cause of the cancer.

Tissues do not have the same sensitivity to the action of one or the other TF, depending on their chromatin landscape. Thus building TF regulation networks is necessary, but not sufficient, and will have to be completed by an understanding of the molecular mechanisms governing gene expression and their contribution to cancer development.

Expected application of the high-throughput technology in cancer research

Systematic profiling of tumors using microarray technology allowed the definition of biological markers of tumor types and the refining of patient diagnosis. The role of transcription machinery and epigenetic alteration in modulating gene expression changes leading to cancer development has been the focus of increasing interest in recent years. However, the mechanisms underlying epigenetic and transcriptional alteration in cancer are still poorly characterized. The recent technologic developments will likely allow the quantifying of the contribution of the different mechanistic deregulations responsible for a particular cancer development. One can expect that the use of such information at the diagnosis stage will allow a better clinical targeting of the tumor according to its origin.

For example, a recent approach focusing on hepatocellular carcinoma (HCC) investigated the relative contribution of four molecular mechanisms known to alter RNA levels: the recruitment of the transcription machinery, the chromatin modification, the DNA methylation states, and the changes in gene copy number (Acevedo et al., 2008). It appears that a minority of the genes showing changes in their expression between tumor and normal tissues did not display changes in either Pol II recruitment or chromatin marks at their promoters (~5% of the studied promoters). In contrast, ~30 to 50% of the genes studied displayed alteration in their copy number. Finally, for a majority of the genes studied, none of the tested mechanisms could explain the changes in their expression patterns. This shows that in the case of HCC tumors, the mechanisms of amplification and deletion of chromosomal regions is most often used to confer changes in gene expression. This observation was limited to a small number of liver tumors and cannot be considered as a global rule in cancer development. However, the methodology developed by this systematic approach is very promising since it will allow us, in a relatively cost-effective manner using the ChIP-on-chip technology, to diagnose the mechanistic cause of cancer in a particular tumor. For example, several clinical treatments targeting the epigenetic machinery are currently in development and were shown to be particularly efficient on cancers from a hematologic origin (Sigalotti et al., 2007). The use of such a treatment may not be efficient in those particular HCC cases where the alteration in the transcription and the epigenetic machinery seems to weakly contribute to the changes in expression profiles. The use of such technology initially developed for understanding transcription regulation on a global scale will likely improve the diagnosis quality in the future.

Cancer is, after decades of research, still a devastating disease, responsible for roughly one quarter of the deaths in developed countries. Cancer is clearly one of the most urgent problems we are facing, and will therefore have to have a very high priority, due to the large number of deaths it is responsible for, the enormous human suffering caused by this disease, but also by the enormous health care and other costs associated with it. While progress has been made in the treatment of rare childhood cancers, little progress has been made in the treatment of the common forms of cancer, responsible for most of the death toll. Even highly successful new anti-cancer drugs are successfully used for only a fraction of patients with individual characteristics. Thus in the future coordinated action will be needed that will bring together groups with strong clinical focus, with experience in high-throughput functional genomics as well as with computational and systems-biology resources. These interaction points between the different areas of expertise, including systems-biology approaches, will be crucial in the future to combat cancer.

Acknowledgments

We are grateful to D. Devys, M.E. Torres-Padilla, K. Karmodiya, M. Orpinell, and E. Fournier for critically reading the manuscript. We thank S. Bour for help with preparing the figure. A.K. is supported by a fellowship from INSERM and Alsace Region. This work was supported by funds from CNRS, INSERM, Universite de Strasbourg, INCA (Ubican), AICR and European Community (EUTRACC LSHG-CT-2007-037445) grants.

References

Acevedo, L.G., Bieda, M., Green, R., and Farnham, P.J. (2008). Analysis of the mechanisms mediating tumor-specific changes in gene expression in human liver tumors. *Cancer Res 68*, 2641–2651.

Bajic, V.B., Tan, S.L., Christoffels, A., et al. (2006). Mice and men: their promoter properties. *PLoS Genet 2*, e54.

Barrera, L.O., Li, Z., Smith, A.D., et al. (2008). Genome-wide mapping and analysis of active promoters in mouse embryonic stem cells and adult organs. *Genome Res 18*, 46–59.

Barski, A., Cuddapah, S., Cui, K., et al. (2007). High-resolution profiling of histone methylations in the human genome. *Cell 129*, 823–837.

Bernstein, B.E., Liu, C.L., Humphrey, E.L., Perlstein, E.O., and Schreiber, S.L. (2004). Global nucleosome occupancy in yeast. *Genome Biol 5*, R62.

Bernstein, B.E., Mikkelsen, T.S., Xie, X., et al. (2006). A bivalent chromatin structure marks key developmental genes in embryonic stem cells. *Cell 125*, 315–326.

Birney, E., Stamatoyannopoulos, J.A., Dutta, A., et al. (2007). Identification and analysis of functional elements in 1% of the human genome by the ENCODE pilot project. *Nature 447*, 799–816.

Blackwood, E.M., and Kadonaga, J.T. (1998). Going the distance: a current view of enhancer action. *Science 281*, 60–63.

Boeger, H., Griesenbeck, J., Strattan, J.S., and Kornberg, R.D. (2003). Nucleosomes unfold completely at a transcriptionally active promoter. *Mol Cell 11*, 1587–1598.

Carninci, P., Kasukawa, T., Katayama, S., et al. (2005). The transcriptional landscape of the mammalian genome. *Science 309*, 1559–1563.

Carninci, P., Sandelin, A., Lenhard, B., et al. (2006). Genome-wide analysis of mammalian promoter architecture and evolution. *Nat Genet 38*, 626–635.

Cawley, S., Bekiranov, S., Ng, H.H., et al. (2004). Unbiased mapping of transcription factor binding sites along human chromosomes 21 and 22 points to widespread regulation of noncoding RNAs. *Cell 116*, 499–509.

Coe, B.P., Chari, R., Lockwood, W.W., and Lam, W.L. (2008). Evolving strategies for global gene expression analysis of cancer. *J Cell Physiol 217*, 590–597.

Cotterman, R., Jin, V.X., Krig, S.R., et al. (2008). N-Myc regulates a widespread euchromatic program in the human genome partially independent of its role as a classical transcription factor. *Cancer Res 68*, 9654–9662.

Flores, E.R., Tsai, K.Y., Crowley, D., et al. (2002). p63 and p73 are required for p53-dependent apoptosis in response to DNA damage. *Nature 416*, 560–564.

Francis, N.J., Kingston, R.E., and Woodcock, C.L. (2004). Chromatin compaction by a polycomb group protein complex. *Science 306*, 1574–1577.

Gal-Yam, E.N., Jeong, S., Tanay, A., et al. (2006). Constitutive nucleosome depletion and ordered factor assembly at the GRP78 promoter revealed by single molecule footprinting. *PLoS Genet 2*, e160.

Guccione, E., Martinato, F., Finocchiaro, G., et al. (2006). Myc-binding-site recognition in the human genome is determined by chromatin context. *Nat Cell Biol 8*, 764–770.

Guenther, M.G., Levine, S.S., Boyer, L.A., Jaenisch, R., and Young, R.A. (2007). A chromatin landmark and transcription initiation at most promoters in human cells. *Cell 130*, 77–88.

Hallikas, O., Palin, K., Sinjushina, N., et al. (2006). Genome-wide prediction of mammalian enhancers based on analysis of transcription-factor binding affinity. *Cell 124*, 47–59.

Heintzman, N.D., Hon, G.C., Hawkins, R.D., et al. (2009). Histone modifications at human enhancers reflect global cell-type-specific gene expression. *Nature 459*, 108–112.

Heintzman, N.D., Stuart, R.K., Hon, G., et al. (2007). Distinct and predictive chromatin signatures of transcriptional promoters and enhancers in the human genome. *Nat Genet 39*, 311–318.

Holstege, F.C., and Clevers, H. (2006). Transcription factor target practice. *Cell 124*, 21–23.

Johnson, D.S., Mortazavi, A., Myers, R.M., and Wold, B. (2007). Genome-wide mapping of *in vivo* protein-DNA interactions. *Science 316*, 1497–1502.

Kim, T.H., Barrera, L.O., Qu, C., et al. (2005a). Direct isolation and identification of promoters in the human genome. *Genome Res 15*, 830–839.

Kim, T.H., Barrera, L.O., Zheng, M., et al. (2005b). A high-resolution map of active promoters in the human genome. *Nature 436*, 876–880.

Komashko, V.M., Acevedo, L.G., Squazzo, S.L., et al. (2008). Using ChIP-chip technology to reveal common principles of transcriptional repression in normal and cancer cells. *Genome Res 18*, 521–532.

Li, B., Carey, M., and Workman, J.L. (2007). The role of chromatin during transcription. *Cell 128*, 707–719.

Mavrich, T.N., Ioshikhes, I.P., Venters, B.J., et al. (2008). A barrier nucleosome model for statistical positioning of nucleosomes throughout the yeast genome. *Genome Res 18*, 1073–1083.

Mohn, F., Weber, M., Rebhan, M., et al. (2008). Lineage-specific polycomb targets and de novo DNA methylation define restriction and potential of neuronal progenitors. *Mol Cell 30*, 755–766.

Nagalakshmi, U., Wang, Z., Waern, K., et al. (2008). The transcriptional landscape of the yeast genome defined by RNA sequencing. *Science 320*, 1344–1349.

Nechaev, S., and Adelman, K. (2008). Promoter-proximal Pol II: when stalling speeds things up. *Cell Cycle 7*, 1539–1544.

Ozsolak, F., Song, J.S., Liu, X.S., and Fisher, D.E. (2007). High-throughput mapping of the chromatin structure of human promoters. *Nat Biotechnol 25*, 244–248.

Pray-Grant, M.G., Daniel, J.A., Schieltz, D., Yates, J.R., 3rd, and Grant, P.A. (2005). Chd1 chromodomain links histone H3 methylation with SAGA- and SLIK-dependent acetylation. *Nature 433*, 434–438.

Reinke, H., and Horz, W. (2003). Histones are first hyperacetylated and then lose contact with the activated PHO5 promoter. *Mol Cell 11*, 1599–1607.

Ren, B., Robert, F., Wyrick, J.J., et al. (2000). Genome-wide location and function of DNA binding proteins. *Science 290*, 2306–2309.

Ringrose, L., Ehret, H., and Paro, R. (2004). Distinct contributions of histone H3 lysine 9 and 27 methylation to locus-specific stability of polycomb complexes. *Mol Cell 16*, 641–653.

Sandelin, A., Carninci, P., Lenhard, B., et al. (2007). Mammalian RNA polymerase II core promoters: insights from genome-wide studies. *Nat Rev Genet 8*, 424–436.

Santos-Rosa, H., Schneider, R., Bernstein, B.E., et al. (2003). Methylation of histone H3 K4 mediates association of the Isw1p

ATPase with chromatin. *Mol Cell 12*, 1325–1332.

Schones, D.E., Cui, K., Cuddapah, S., et al. (2008). Dynamic regulation of nucleosome positioning in the human genome. *Cell 132*, 887–898.

Shivaswamy, S., Bhinge, A., Zhao, Y., et al. (2008). Dynamic remodeling of individual nucleosomes across a eukaryotic genome in response to transcriptional perturbation. *PLoS Biol 6*, e65.

Sigalotti, L., Fratta, E., Coral, S., et al. (2007). Epigenetic drugs as pleiotropic agents in cancer treatment: biomolecular aspects and clinical applications. *J Cell Physiol 212*, 330–344.

Singh, B.N., and Hampsey, M. (2007). A transcription-independent role for TFIIB in gene looping. *Mol Cell 27*, 806–816.

Smeenk, L., van Heeringen, S.J., Koeppel, M., et al. (2008). Characterization of genome-wide p53-binding sites upon stress response. *Nucleic Acids Res 36*, 3639–3654.

Sultan, M., Schulz, M.H., Richard, H., et al. (2008). A global view of gene activity and alternative splicing by deep sequencing of the human transcriptome. *Science 321*, 956–960.

Szutorisz, H., Dillon, N., and Tora, L. (2005). The role of enhancers as centres for general transcription factor recruitment. *Trends Biochem Sci 30*, 593–599.

Timmers, H.T., and Tora, L. (2005). SAGA unveiled. *Trends Biochem Sci 30*, 7–10.

Venters, B.J., and Pugh, B.F. (2009). A canonical promoter organization of the transcription machinery and its regulators in the *Saccharomyces* genome. *Genome Res 19*, 360–371.

Visel, A., Blow, M.J., Li, Z., et al. (2009). ChIP-seq accurately predicts tissue-specific activity of enhancers. *Nature 457*, 854–858.

Vousden, K.H., and Prives, C. (2005). P53 and prognosis: new insights and further complexity. *Cell 120*, 7–10.

Wang, Z., Zang, C., Rosenfeld, J.A., et al. (2008). Combinatorial patterns of histone acetylations and methylations in the human genome. *Nat Genet 40*, 897–903.

Wei, C.L., Wu, Q., Vega, V.B., et al. (2006). A global map of p53 transcription-factor binding sites in the human genome. *Cell 124*, 207–219.

Wu, J., Smith, L.T., Plass, C., and Huang, T.H. (2006). ChIP-chip comes of age for genome-wide functional analysis. *Cancer Res 66*, 6899–6902.

Yochum, G.S., Cleland, R., and Goodman, R.H. (2008). A genome-wide screen for beta-catenin binding sites identifies a downstream enhancer element that controls c-Myc gene expression. *Mol Cell Biol 28*, 7368–7379.

Yuan, G.C., Liu, Y.J., Dion, M.F., et al. (2005). Genome-scale identification of nucleosome positions in *S. cerevisiae*. *Science 309*, 626–630.

Zeitlinger, J., Stark, A., Kellis, M., et al. (2007). RNA polymerase stalling at developmental control genes in the *Drosophila melanogaster* embryo. *Nat Genet 39*, 1512–1516.

Zeller, K.I., Zhao, X., Lee, C.W., et al. (2006). Global mapping of c-Myc binding sites and target gene networks in human B cells. *Proc Natl Acad Sci USA 103*, 17834–17839.

Zhou, D., and Yang, R. (2006). Global analysis of gene transcription regulation in prokaryotes. *Cell Mol Life Sci 63*, 2260–2290.

Regulation and dysregulation of protein synthesis in cancer cells

Michael J. Clemens, Androulla Elia and Simon J. Morley

Introduction

The transformation of cells from a normal to a fully malignant phenotype requires both qualitative and quantitative changes in gene expression. Such multiple changes can result from inherited or somatic mutations of cellular genes, but may also be consequent upon infection with tumor viruses. Alterations in gene expression can arise as a result of changes in both transcriptional and post-transcriptional events and, among the latter, there are several mechanisms that act at the level of translation. These include the influence of mRNA structure on translation and the control of the function of various polypeptide chain initiation factors by processes such as reversible protein phosphorylation and irreversible proteolytic cleavage.

There has been a huge expansion of knowledge concerning the signaling mechanisms by which these events are regulated. In this chapter we summarize the nature of the dysregulation of translation in malignant cells and relate this to the mechanisms and pathways that can be involved. The review also presents a brief consideration of the clinical implications of this knowledge. Many of these topics have been described in earlier comprehensive reviews (e.g., Watkins and Norbury, 2002; Bjornsti and Houghton, 2004a; Clemens, 2004; Holland, 2004; Holland et al., 2004; Rosenwald, 2004; Robert and Pelletier, 2009; Sonenberg and Hinnebusch, 2009; Silvera et al., 2010; Spriggs et al., 2010; Yin et al., 2010; Blagden and Willis, 2011). The present chapter concentrates largely on recent developments in the field, up to the date of April 2011.

Mechanisms of regulation of protein synthesis

The process of translation is highly complex and requires the participation of a large number of components (Sonenberg and Dever, 2003; Abbott and Proud, 2004; Merrick, 2004). The rate-limiting step(s) for protein synthesis in most physiological situations occur at the level of polypeptide chain initiation. This process involves several distinct events, which are summarized in Figure 6.1. The steps that commonly constitute points of physiological regulation are the formation of the ternary complex (comprising initiation factor eIF2, GTP and the initiator Met-tRNA$_f$) and the assembly of the eIF4F heterotrimeric complex (comprising initiation factors eIF4E, eIF4G and eIF4A).

When associated with a molecule of GTP, initiation factor eIF2 brings Met-tRNA$_f$ to the smaller (40S) ribosomal subunit (Figure 6.1). At a later stage of the initiation pathway the GTP is hydrolyzed to GDP, which remains bound to the eIF2 after release of the latter from the ribosome. Reactivation of the eIF2 requires subsequent GTP–GDP exchange, which is catalyzed by eIF2B (Proud, 2005a) (Figure 6.1). The α subunit of eIF2 can be phosphorylated at amino acid Ser[51] by the protein kinases PKR, PERK, mGCN2 and HRI and these enzymes can be activated by a range of physiological stresses. Such phosphorylation of eIF2α inhibits overall protein synthesis as a result of impairment of the GTP–GDP exchange activity of eIF2B. Loss of regulation of protein synthesis can result from failure of the stress-regulated kinases to respond to the appropriate stimulus. Similar consequences may follow from changes in expression of eIF2 or eIF2B, potentially affecting the ability of cells to increase their mass and/or to progress through the cell cycle, as well as changing the susceptibility of cells to undergo apoptosis.

Another important regulatory step used to control the rate of translation initiation in eukaryotes is recognition of the mRNA 5′ cap by the eIF4F complex. The formation of the 48S pre-initiation complex

---

Systems Biology of Cancer, ed. S. Thiagalingam. Published by Cambridge University Press. © Cambridge University Press 2015.

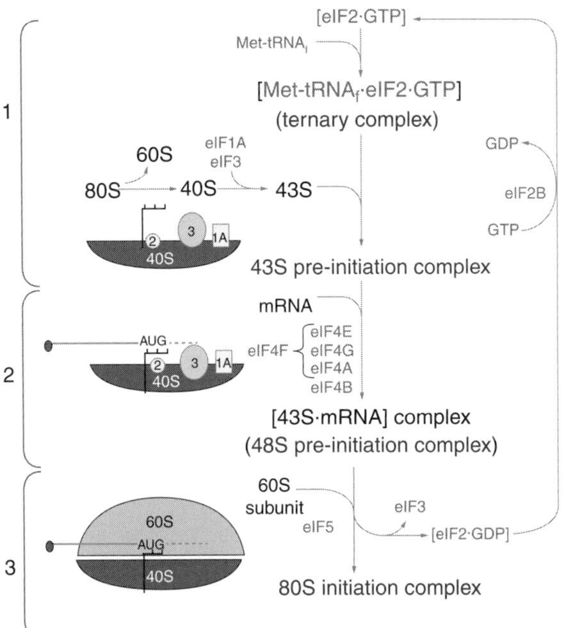

Figure 6.1 The pathway of cap-dependent polypeptide chain initiation. Phase 1: The initiation factor eIF2 forms a ternary complex with GTP and the initiator methionyl-tRNA_f. The 80S ribosome dissociates into its constituent 40S and 60S subunits and the ternary complex, together with eIF1A, eIF3 and other factors (not shown), then associates with the 40S subunit to form a 43S pre-initiation complex. Phase 2: Messenger RNA (mRNA) is bound by the 43S pre-initiation complex in a process that requires the eIF4F complex (consisting of the initiation factors eIF4E, eIF4G and eIF4A) plus eIF4B. The role of eIF4E is to recognize the 5′ mRNA cap structure; eIF4G interacts with the eIF3 already present on the ribosomal subunit (thus forming the link between ribosome and mRNA); eIF4A has RNA helicase activity, which is stimulated by eIF4B. The resulting complex is sometimes referred to as the 48S pre-initiation complex although the actual sedimentation rate depends on the size of the mRNA. Phase 3: The 60S ribosomal subunit then binds, giving rise to the 80S initiation complex. This involves the hydrolysis of the GTP molecule to GDP and phosphate, in a reaction catalyzed by eIF5, and the release of [eIF2.GDP], eIF3 and other factors. The 80S complex is then competent to scan the mRNA from the 5′ end until it reaches the initiator AUG codon, which is recognized by the Met-tRNA_f. The GDP molecule associated with the released eIF2 must be exchanged for a molecule of GTP before the initiation factor can participate in another round of protein synthesis. This is brought about by the guanine nucleotide exchange factor eIF2B. Further details of the initiation pathway are described in reviews (Sonenberg and Dever, 2003; Abbott and Proud, 2004; Merrick, 2004). (A black and white version of this figure will appear in some formats. For the color version, please refer to the plate section.)

(Figure 6.1) relies on the ability of eIF4E to bind to the cap structure and to interact with the large scaffold protein eIF4G, which in turn is associated with eIF4A and eIF3 (reviewed in Gingras et al., 1999; Hershey and Merrick, 2000; Morley, 2001; Morley et al., 2005; Sonenberg and Hinnebusch, 2007,

2009). Binding of eIF4E to eIF4G is critical for the production of all cellular proteins that are translated by the mRNA 5′ cap-dependent mechanism. Because eIF4G also interacts with the 40S ribosomal subunit (via its binding to the multi-subunit factor eIF3) (Korneeva et al., 2000) the association with eIF4E brings the mRNA to the ribosome (Figure 6.1). The availability of eIF4E to associate with eIF4G is regulated by the eIF4E binding proteins, 4E-BP1, -BP2 and -BP3, which bind to eIF4E in competition with eIF4G (Figure 6.2) (Gingras et al., 1999; Marcotrigiano et al., 1999; Raught and Gingras, 2007). The ability of 4E-BP1 and 4E-BP2 to interact with eIF4E is controlled by multi-site phosphorylation. The 4E-BPs are substrates for the rapamycin-sensitive protein kinase mTOR, as well as other enzymes (Figure 6.3). The less highly phosphorylated forms of the 4E-BPs bind tightly to eIF4E, resulting in inhibition of cap-dependent initiation. Activation of mTOR reverses this, permitting higher rates of translation. One of the most striking aspects of the relationship between protein synthesis and cancer that has emerged in recent years is that over-expression of eIF4E is associated with the malignant phenotype. This may arise from the anti-apoptotic effects of high eIF4E activity. At face value it should follow from this that increased levels and/or decreased phosphorylation of the 4E-BPs should bring about the opposite effect, promoting apoptosis and having a tumor suppressive effect. However, there may be complications to this simplistic interpretation, as described later in this chapter.

It is often difficult to determine the relative contributions of the phosphorylation of eIF2α and the dephosphorylation of the 4E-BPs to the inhibition of protein synthesis in response to various cellular stresses, since both commonly occur in parallel. The first of these mechanisms may be able to affect the translation of the majority of mRNAs in the cell, albeit to varying extents. In contrast, regulation of the availability of eIF4E is thought to be more important in influencing the relative selection of different mRNAs for recruitment to ribosomes, since an mRNA that is bound to an inactive [eIF4E.4E-BP] complex cannot be translated (Figure 6.2).

A critical feature of relevance to translational control is that different mRNA species have very different structures. The properties of the 5′ and 3′ untranslated regions (UTRs) of different mRNAs are known to influence the efficiencies with which such

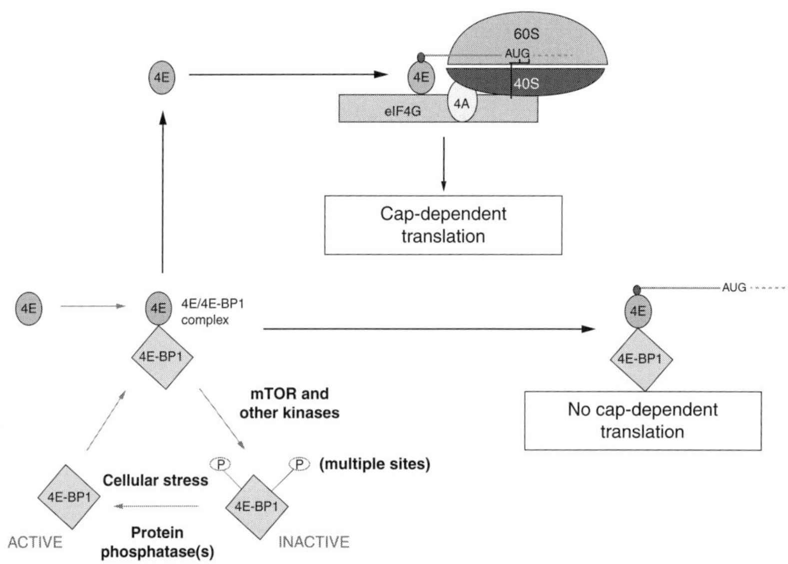

Figure 6.2 Role of the eIF4E binding proteins in the control of cap-dependent translation. During the initiation of protein synthesis eIF4E binds to the scaffold protein eIF4G. These factors, plus eIF4A, constitute the eIF4F complex, as shown in Figure 6.1. The availability of eIF4E for binding to eIF4G is limited by the 4E binding proteins (4E-BP1 is shown here but it is likely that 4E-BP2 has a similar role). These proteins compete with eIF4G for binding to eIF4E in a manner that is regulated by multi-site phosphorylation of the 4E-BPs. Phosphorylation of the 4E-BPs is catalyzed by the mammalian target of rapamycin (mTOR), as well as other protein kinases. In their less phosphorylated state the 4E-BPs bind to eIF4E; more highly phosphorylated forms dissociate from the cap-binding protein, thus permitting association of the eIF4E with eIF4G. Phosphorylation of the 4E-BPs is reversed by protein phosphatases such as PP2A. Messenger RNAs that are bound to eIF4E when the latter is associated with eIF4G (together with eIF4A in the eIF4F complex – see Figure 6.1) can be translated, whereas mRNAs that are bound to eIF4E, that is, associated with the 4E-BPs, are unavailable for cap-dependent translation. (A black and white version of this figure will appear in some formats. For the color version, please refer to the plate section.)

mRNAs can be translated. Thus the presence of a long and/or highly structured 5′ UTR in an mRNA may mean that higher concentrations of the eIF4F complex (i.e., eIF4E + eIF4G + eIF4A) are needed to promote the translation of this mRNA. Of relevance here is the fact that the eIF4A component has RNA helicase activity capable of unwinding mRNA secondary structure. However, some mRNAs can apparently escape the requirement for eIF4E if they contain an internal ribosome entry site (IRES) (Stoneley and Willis, 2004). The presence of IRESs permits mRNAs to be translated in a cap-independent manner as a result of direct internal binding of ribosomes near the start of the coding region, thus abrogating the need for scanning from the 5′ end. It is possible that the translation of these mRNAs is independent of changes in the availability of the eIF4E in the cell and is not impaired (or may even be stimulated) by conditions that favor the dephosphorylation of the 4E-BPs.

Since malignant transformation must involve changes in patterns of gene expression, rather than just being a consequence of quantitative alterations in rates of protein synthesis, we therefore need to understand how alterations of initiation factor phosphorylation or abundance might interact with elements of mRNA structure to implement such effects. Such modulations of gene expression might, for example, result in the over-expression of proteins promoting cell proliferation or inhibiting apoptosis. Conversely, there may be down-regulation of tumor suppressor proteins or pro-apoptotic factors as a result of changes at the level of translation.

Update on protein synthesis factors and cancer

Because of the multiple steps involved in the process of protein synthesis, alterations in the levels or activities of numerous factors could in theory result in aberrant regulation of gene expression at the translational level. When such changes affect cell proliferation or cell survival, malignancy may ensue. Since there are rate-limiting steps at the points where eIF2 and eIF4E mediate the binding of Met-tRNA$_f$ and mRNA respectively to the ribosome, loss of control

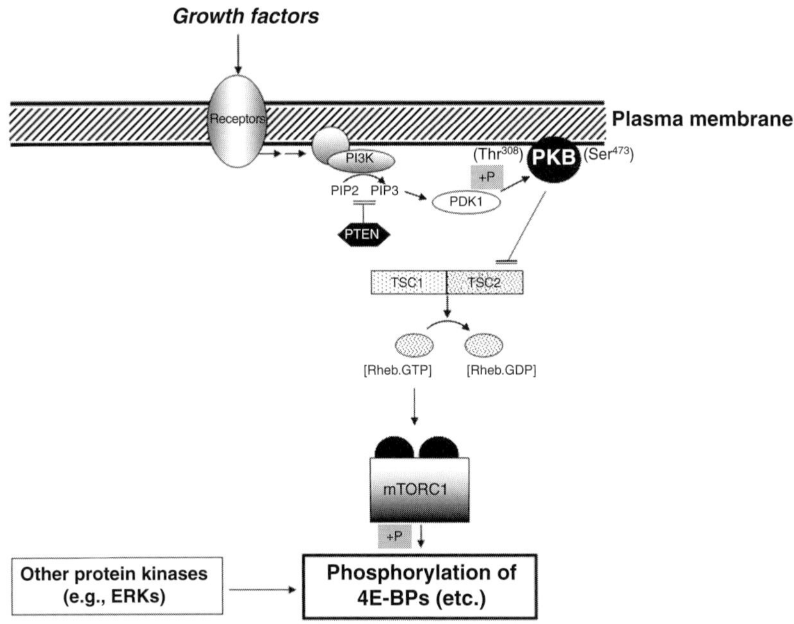

Figure 6.3 Regulation of the mTORC1 protein kinase complex and the phosphorylation of the 4E-BPs. Growth factors and cytokines interact with specific receptors and activate phosphatidylinositol-3-kinase (PI3K) at the plasma membrane. The product of this enzyme, phosphatidylinositol 3,4,5 trisphosphate (PIP3), recruits the protein kinases PDK1 and PKB (Akt) to the membrane and PKB is phosphorylated on Thr^{308} by PDK1. Breakdown of PIP3 is brought about by PTEN, which therefore antagonizes the effect of PI3K. Among the many substrates of PKB is the mTOR inhibitory protein TSC2, which is inactivated by phosphorylation. The TSC1-TSC2 complex impairs the activity of the G protein Rheb by converting [Rheb.GTP] to [Rheb.GDP]. [Rheb.GTP] is an activator of the mTORC1 complex, which also contains the proteins Raptor and mLST8. mTORC1 is also sensitive to regulation by amino acid availability and the oxygen and energy status of the cell (not shown). The complex phosphorylates the 4E-BPs (and other substrates such as p70S6 kinase). The 4E-BPs have several phosphorylation sites, some of which may also be targets for additional protein kinases such as the ERK family of mitogen-activated kinases, the DNA damage-regulated kinase ATM and the cyclin-dependent kinase cdc2.

of these points in the initiation pathway can contribute to tumor development. However, other factors may also play a role. eIF4E merits a complete section to itself, while the contributions of some of the other initiation factors are also described below.

Initiation factor eIF4E and the 4E binding proteins

Polypeptide chain initiation factor eIF4E is undoubtedly the most extensively studied translation component in relation to malignancy and it is now considered to be a *bone fide* oncogene product (Zimmer et al., 2000; De Benedetti and Graff, 2004; Mamane et al., 2004). The evidence for this conclusion comes from transgenic mouse studies (Ruggero et al., 2004) and from numerous observations that many cancers have enhanced levels and/or activity of this mRNA cap-binding factor (Gu et al., 2005; Lee et al., 2005; McClusky et al., 2005; Graff et al., 2009). Progressively enhanced eIF4E expression is associated with

the development of the malignant phenotype (Kim et al., 2009; Wang et al., 2009; Choi et al., 2011) and high levels of the factor correlate with increased risks of tumor recurrence and patient mortality (Coleman et al., 2009; Wang et al., 2009). The conclusion now seems certain that eIF4E can influence tumor progression, and a likely mechanism is the ability of this protein to inhibit apoptosis (see below).

The mechanisms that regulate eIF4E levels may vary in different cell or tumor types. Transcription of the eIF4E gene is positively controlled by the proto-oncogene c-Myc (Lin et al., 2008) and negatively regulated by p53 (Zhu et al., 2005). In addition, however, the stability of eIF4E may be influenced by association of the protein with binding partners such as the chaperone protein Hsp27 (Andrieu et al., 2010).

It is possible that phosphorylation of eIF4E on Ser^{209}, which is catalyzed by the MNK kinases (Buxade et al., 2008), plays a role in promoting the transformed phenotype (Topisirovic et al., 2004; Bianchini et al., 2008). One study has shown that a higher proportion of tumors than of normal tissues exhibits immunostaining for phosphorylated eIF4E (Fan et al., 2009). Although such phosphorylation is unlikely to be required for the function of eIF4E in overall protein synthesis (McKendrick et al., 2001; Morley and Naegele, 2002), it could regulate the eIF4E-dependent nucleocytoplasmic transport of certain key mRNA species (Rousseau et al., 1996b; Topisirovic et al., 2004, 2009a, b) as well as the translation of specific mRNAs such as those encoding proteins required for ribosome biogenesis, cell proliferation and the regulation of apoptosis (Bianchini et al., 2008). A recent report (Konicek et al., 2011) has shown that pharmacological inhibition of MNK activity reverses aspects of the malignant phenotype and induces apoptosis in cultured cancer cell lines. Importantly, inhibition of eIF4E phosphorylation *in vivo* also suppresses the growth of metastases of xenografted human tumor tissues in mice. The oncogenic activity of eIF4E requires the mRNA cap-binding ability of the factor as it is inhibited by the guanosine analogue ribavirin (Kentsis et al., 2004).

As intimated above, high levels of eIF4E are able to confer resistance to apoptosis in cells exposed to a variety of death stimuli. Importantly, the cell death-inducing activity of the proto-oncogene c-Myc is inhibited, by a mechanism requiring the eIF4E-dependent expression of the cell cycle regulatory protein cyclin D1 (Tan et al., 2000; Li et al., 2003). Conversely, c-Myc impairs a senescence-inducing effect of eIF4E. This explains why the two proteins can cooperate in inducing B cell lymphomagenesis (Ruggero et al., 2004). eIF4E also inhibits apoptosis caused by growth factor withdrawal or by disruption of endoplasmic reticulum function following exposure of cells to tunicamycin or the Ca^{2+} ionophore A23187 (Li et al., 2004). Additionally, high levels of eIF4E render cells resistant to the effects of DNA-damaging agents. Under normal conditions expression of the eIF4E gene is repressed by the tumor suppressor protein p53, which is activated by DNA damage (Zhu et al., 2005). However, in p53-negative tumors hyper-production of eIF4E might lead to resistance to the effects of DNA damage, such as that caused by chemotherapy or radiotherapy. eIF4E activity is also regulated by the anti-apoptotic protein kinase PKB (Akt), an enzyme that is implicated in tumor cell survival and resistance to therapy (Wendel et al., 2004). Consistent with these findings, exposure of cells to sodium arsenite, which promotes

the proteasome-mediated degradation of eIF4E, or down-regulation of eIF4E levels using short interfering RNAs (siRNAs), can decrease tumor cell growth (Soni et al., 2008) and restore cellular responsiveness to the DNA damaging agent cisplatin (Oridate et al., 2005; Othumpangat et al., 2005; Dong et al., 2009). Such observations raise the possibility of development of novel therapeutic approaches based on the inhibition of eIF4E expression or activity (Ko et al., 2009).

As indicated earlier in this review, the eIF4E binding proteins (4E-BPs) inhibit the function of eIF4E by competing for the binding of the essential initiation factor eIF4G (Figure 6.2). As a result, these small proteins often have opposite effects to those of eIF4E on the cellular phenotype. Earlier evidence has shown that the 4E-BPs can negatively regulate cell growth and can revert the transformed phenotype in cells over-expressing eIF4E (Rousseau et al., 1996a). Moreover, cell cycle progression is blocked by 4E-BP1 in MCF-7 breast cancer cells (Jiang et al., 2003). This may be due to changes in the expression of proteins that promote or inhibit passage through the cell cycle (e.g., cyclin D1 or the cyclin-dependent kinase inhibitor $p27^{KIP1}$ respectively), without affecting overall protein synthesis. This would explain why 4E-BP1 can prevent the progression of cells from the G1 phase into the S phase of the cell cycle without affecting the increases in cell mass or protein content characteristic of passage of cells through G1 (Lynch et al., 2004). Inhibition of cell cycle progression would be sufficient to block transformation by the c-Myc oncogene. Consistent with these findings, short hairpin RNA (shRNA)-mediated knockdown of 4E-BP1 relieves the inhibition of cell cycle progression induced by cellular stresses such as hypoxia (Barnhart et al., 2008). Since the phosphorylation of 4E-BP1 inactivates the protein, which may be expected to mimic the effects of reduced expression, it is of interest that the presence of phosphorylated 4E-BP1 (in the cell nucleus) has been correlated with an aggressive phenotype and poor prognosis in endometrial cancers (Castellvi et al., 2009). However, cells with decreased 4E-BP1 expression are less able to survive physiological stresses such as exposure to hypoxia *in vitro* (Dubois et al., 2009). Such cells are also more sensitive to ionizing radiation. This is highly relevant to cancer since many tumors have hypoxic regions, due to restricted blood supply, and this can lead to radio resistance. A role for the 4E-BPs as factors that protect cells (and thus favor cell survival) under conditions of physiological stress has been suggested in earlier studies (Teleman et al., 2005). Interestingly, 4E-BP1 knockdown also leads to aberrant mitosis *in vivo* (Barnhart et al., 2008), a phenomenon that could increase genetic instability in tumors.

A very important recent study (She et al., 2010) has shown that 4E-BP1 plays a key role in the oncogenic effects of activation of the PKB and ERK signaling pathways, which converge at the level of 4E-BP1 activity. Thus, in tumors that contain mutations in components of both pathways, inhibition of either pathway alone is not sufficient to cause anti-tumor effects but down-regulation of 4E-BP1 expression can achieve this.

If the function of eIF4E in translation is responsible for its oncogenic effects, one might predict that increased eIF4F complex formation is necessary and/or sufficient for the expression of a malignant phenotype. On the other hand, enhanced nucleocytoplasmic transport of newly synthesized mRNA may not require elevated eIF4F levels. In fact it has been shown that, in breast carcinoma, the transformed state is often characterized by elevated expression of the eIF4F component eIF4GI, as well as by hyperphosphorylation of 4E-BP1 (which inactivates the ability of the latter to block the binding of eIF4E to eIF4G – Figure 6.2). In agreement with these concepts, non-phosphorylatable forms of 4E-BP1 are able to suppress the tumorigenicity of some breast cancer cell lines (Avdulov et al., 2004).

It is likely that the mechanism of action of 4E-BP1 as an anti-oncogenic factor is via the induction of apoptosis, providing a counter-balance to the cell survival-promoting effects of eIF4E (Proud, 2005b). The state of phosphorylation of 4E-BP1 is critical for this effect (Li et al., 2002). Indeed in studies that measured the levels and phosphorylation status of several key signaling molecules in breast, ovary and prostate tumors, the phosphorylated 4E-BP1 signal correlated strongly with poor prognosis and patient survival statistics (Castellvi et al., 2006; Armengol et al., 2007). Recent analysis of head and neck squamous cell carcinomas has also shown significant elevation of phosphorylated 4E-BP1 (Frederick et al., 2011). Somewhat paradoxically, the expression of total 4E-BP1 is also elevated in a variety of tumors showing malignant progression, although the state of phosphorylation of the protein was not always ascertained (Martin et al., 2000; Nathan et al., 2004;

Kremer et al., 2006; Kodali et al., 2011). One report has suggested that 4E-BP2 may have a stronger influence than 4E-BP1 on tumor progression in breast cancer patients (Coleman et al., 2009).

The cell death-inducing effects of the 4E-BPs are unlikely to be a result simply of the repression of total protein synthesis; in fact, too much inhibition can impede apoptosis if the balance between the levels of pro- and anti-apoptotic proteins is changed as a result of translational inhibition. Nevertheless, it has been suggested that tumor cells may have a greater requirement for cap-dependent translation than normal cells (Li et al., 2002), a feature that might be exploitable for the preferential induction of cell death in the former.

Other protein synthesis factors and cancer

Prominent among other initiation factors that have also been implicated in the control of the malignant phenotype are eIF2 and its nucleotide exchange factor eIF2B, as well as eIF3 and members of the eIF4 group. In addition, eIF6, a protein involved in maintaining the dissociation of the large and small ribosomal subunits (and which may also be important in ribosome biogenesis) has recently been suggested to influence oncogene-induced cell transformation (Gandin et al., 2008). These factors are briefly considered below.

eIF2

Levels of initiation factor eIF2 are elevated in some cells transformed by oncogenes (Rosenwald et al., 2003), and over-expression of the protein can result in malignancy (Donze et al., 1995; Perkins and Barber, 2004). Nevertheless, malignancy is not an inevitable consequence of higher than normal eIF2 expression (Rosenwald et al., 2003), suggesting that this change is not sufficient for cell transformation. In fact the state of phosphorylation of the α subunit of eIF2, and the extent to which this can affect eIF2B regulation, may be more important factors for malignant progression. Thus, for example, a non-phosphorylatable mutant of eIF2α (S51A) can cooperate with SV40 virus T antigen and the hTERT component of telomerase to transform human kidney cells (Perkins and Barber, 2004). The protein kinase with specificity for Ser[51] of eIF2α that is most important in relation to carcinogenesis is PKR. Lack of activity of this enzyme has been associated with resistance of cells to a variety of experimental agents that induce apoptosis (Chawla-Sarkar et al., 2003; Garcia et al., 2006). Thus PKR could be considered in some respects to be a tumor-suppressing agent. However, recent studies have reported higher than normal levels or activity of PKR in cancer cells (Pataer et al., 2009; Blalock et al., 2010). In these situations inhibition or knockdown of PKR (by use of a dominant negative mutant or by RNA interference respectively) can induce cell death and sensitize cells to DNA damage. These effects can be rationalized on the basis that the phosphorylation of eIF2α can be cytoprotective against the effects of cell stress (Lu et al., 2004; McEwen et al., 2005). In contrast to this, however, a very thorough examination of the role of PKR in DNA damage-mediated, p53-dependent inhibition of protein synthesis and induction of apoptosis has revealed an important role for the protein kinase in sensitizing cells to the effects of DNA-damaging agents such as etoposide and doxorubicin (Yoon et al., 2009). This may be a consequence of the ability of PKR to activate the pro-apoptotic JNK kinase pathway (Peidis et al., 2011). If the dephosphorylation of eIF2α is blocked this prevents the survival of quiescent cells, which can otherwise result in recurrent disease after treatment of multiple myeloma with the proteasome inhibitor bortezomib (Velcade®) (Schewe and Aguirre-Ghiso, 2009). Thus the exact roles of the phosphorylation of eIF2α, and the kinases that regulate it, in the development of malignant phenotypes or in the control of cellular sensitivity to cytotoxic drugs require further clarification.

eIF2B

In view of the potential importance of eIF2 in the regulation of cellular activities that impinge on malignancy, as described above, it is not surprising that the factor responsible for guanine nucleotide exchange on eIF2 has also been implicated. eIF2B consists of five subunits, one of which (eIF2Bε) mediates the catalytic activity of the factor. This subunit is also a substrate for phosphorylation by at least four protein kinases with important roles in signal transduction (casein kinase-I, casein kinase-II, glycogen synthase kinase-3 and the DYRK kinase). Expression of the eIF2Bε gene is often higher in tumors than in normal cells (Balachandran and Barber, 2004). This may be expected to counteract the effect of eIF2α phosphorylation, since

the latter impairs protein synthesis by inhibiting the exchange activity of eIF2B. Consistent with this, it has recently been shown that shRNA- or siRNA-mediated down-regulation of eIF2Bε can impair several features of the transformed phenotype, including tumor progression in nude mice (Gallagher et al., 2008).

eIF3

The large initiation factor eIF3, which has 13 subunits, has also been linked to malignant transformation of mammalian cells. As is the case with other protein synthesis factors, at least five of the subunits of eIF3 (-3a, -3b, -3c, -3h and -3i) are up-regulated in many tumors, and individual over-expression of these same subunits experimentally can induce oncogenic characteristics in fibroblasts (Zhang et al., 2007). A truncated form of the eIF3e subunit can also cause many phenotypic changes characteristic of tumor cells, including inhibition of apoptosis (Rasmussen et al., 2001; Mayeur and Hershey, 2002). However, expression of full-length eIF3e does not bring about such changes (Mayeur and Hershey, 2002). Inhibition of the expression of the -3a and -3h subunits of eIF3 has been shown to reverse the malignant properties of lung and mammary tumor cells (Dong et al., 2004) and of breast and prostate tumor cells (Zhang et al., 2008a) respectively. eIF3h is also a substrate for a phosphorylation event that can influence the oncogenic effects of this protein (Zhang et al., 2008a).

It is possible that individual eIF3 subunits could have modes of action that are independent of the complete eIF3 complex. However, all of the transforming subunits stimulate both overall translation and the synthesis of critical growth-related proteins such as cyclin D1 and c-Myc, and no alternative mechanisms for their oncogenic effects have yet been reported.

eIF4A, 4B and 4G

As well as eIF4E (see above) a number of other initiation factors involved in the process of mRNA recruitment to the ribosome may potentially play a role in tumorigenesis. The RNA helicase eIF4A is over-expressed in some cancers (Eberle et al., 1997, 2002; Shuda et al., 2000) and can be regulated by the Pdcd4 tumor suppressor protein (Yang et al., 2003). eIF4A is a prototypic member of the DEAD-box family of proteins, a subgroup of the large superfamily of helicases (Linder et al., 1989). The ATPase and helicase activity of free eIF4A is low but it is greatly stimulated in the presence of eIF4B and eIF4H, or when eIF4A is part of the eIF4F complex (Grifo et al., 1984; Pause et al., 1994; Richter-Cook et al., 1998; Feng et al., 2005). eIF4B has recently been shown to be a target for several signaling pathways and protein kinases with growth regulatory activity, including PKB (van Gorp et al., 2009), and activity of the factor can control cell survival (Shahbazian et al., 2010). This establishes a further potential link between signal transduction pathways with relevance to malignancy and the regulation of translation (see below). Pdcd4 inhibits translation through its interaction with eIF4A (Suzuki et al., 2008; Chang et al., 2009; Loh et al., 2009), resulting in the suppression of neoplastic transformation and tumor invasion (Chen et al., 2008; Lu et al., 2008; Wang et al., 2008). The binding of Pdcd4 to eIF4A traps the latter in an inactive conformation and blocks its incorporation into the eIF4F complex. It has also been reported that inhibition of eIF4A activity by the small molecule agent silvestrol synergizes with the effects of standard cytotoxic DNA-damaging drugs in causing apoptosis in acute myelogenous leukemia cell lines (Cencic et al., 2010). Down-regulation of eIF4B expression can also inhibit cell proliferation and promote apoptosis (Shahbazian et al., 2010).

eIF4G, which is expressed as multiple splice variants from two isoforms in mammalian cells (eIF4GI/II), is a critical scaffold protein that functions to recruit the small ribosomal subunit to mRNA (Figure 6.1) (Khan and Goss, 2004; Sonenberg and Hinnebusch, 2007; Sonenberg and Hinnebusch, 2009). It also promotes the functional circularization of mRNA through interactions with the poly(A) binding protein, PABP, which binds to 3′ poly(A) tracts. eIF4G may be capable of cell transformation (Fukuchi-Shimogori et al., 1997; Bauer et al., 2002) and its over-expression is linked to prostate tumor cell growth (Renner et al., 2007) and the pathogenesis of inflammatory breast cancer (Silvera et al., 2009). Since both eIF4A and eIF4G associate with eIF4E in the mRNA cap-binding complex eIF4F (Figure 6.1), an elevated concentration of this complex, however caused, could lead to the selective enhancement of translation of mRNAs encoding growth-promoting, anti-apoptotic or other malignancy-inducing factors (Ramirez-Valle et al., 2008) (see below). Indeed proteomic analysis has implicated the above proteins, as well as other polypeptide initiation and elongation

factors, in the development of malignant phenotypes (Harris et al., 2004). Nevertheless, although the current data indicate that levels of expression of translation factors often correlate with the tumorigenic potential of cells, it is not always possible to determine whether these represent causes or effects of cell transformation. Interference with the interaction between eIF4G and eIF4E using the small molecule inhibitor 4EGI-1 has growth-inhibitory and apoptosis-inducing effects, and can augment the effects of the pro-apoptotic cytokine TRAIL (Fan et al., 2010). However, the data suggest that the latter activity is independent of the inhibition of cap-dependent translation.

eIF6

Although ribosome biogenesis and translation are linked, they are largely independently controlled processes. However, eIF6, an essential protein which is highly conserved, has a role to play in many aspects of cell homeostasis (reviewed in Miluzio et al., 2009), influencing translation, growth and transformation (Gandin et al., 2008). eIF6 has anti-association activity, preventing the interaction of 40S ribosomal subunits with 60S subunits through its binding to the latter. In the nucleolus, eIF6 is a component of pre-ribosomal particles and is required for the biogenesis of 60S subunits, while the cytoplasmic population modulates insulin-stimulated translation rates and can be recruited by the microRNA (miRNA) machinery (Chendrimada et al., 2007). eIF6 is abundant in colon cancer (Sanvito et al., 2000) and some forms of leukemias (Harris et al., 2004) and has a role to play in transformation by activated forms of the oncogenes Ras and Myc (Gandin et al., 2008). However, it is still unclear whether eIF6 is regulating the translation of specific mRNAs involved in tumorigenesis or whether its activity is downstream of oncogenic signaling.

Signal transduction pathways for control of protein synthesis

A number of distinct and overlapping signaling pathways are involved in the control of translation initiation, and loss of correct control of one or more of these may contribute to the dysregulation of initiation factor function that characterizes malignant cells.

Phosphorylation of the α subunit of initiation factor eIF2 occurs in response to many physiological stresses and can cause both down-regulation of overall protein synthesis and mRNA-specific effects. These events may be important for cell survival under adverse conditions (Lu et al., 2004; McEwen et al., 2005) and in the case of tumor cells may play a role in malignant progression. The enzymes responsible are one or more of the protein kinases PKR, PERK, mGCN2 and HRI. In hypoxic tumors endoplasmic reticulum stress activates PERK as part of the unfolded protein response (UPR) (Feldman et al., 2005). The UPR can promote apoptosis (Jiang and Wek, 2005) but can also have a protective effect on tumor cells (Feldman et al., 2005). PKR has also been implicated in apoptotic regulation (Chawla-Sarkar et al., 2003; Garcia et al., 2006; Pataer et al., 2009) but more work is needed to clarify the roles of the eIF2α-specific protein kinases in the control of cell survival and cell death. It is also possible that abnormal PKR activity may contribute to malignant progression, at least in acute leukemia, through mechanisms that are independent of its role as an eIF2α kinase (Blalock et al., 2011).

mTOR is a critical protein kinase for the regulation of translation (Zoncu et al., 2011). When it is associated with a number of other proteins in mTORC1, a complex which senses and integrates signals from extracellular stimuli (Figure 6.3), mTOR phosphorylates the 4E-BPs (Guertin and Sabatini, 2007). The mTORC1 complex also monitors and responds to changes in amino acid availability and the oxygen and energy status of cells, and it is sensitive to inhibition as a result of radiation-induced DNA damage (Paglin et al., 2005). Interestingly, inhibition of mTORC1 by rapamycin itself impairs the repair of double-strand breaks in damaged DNA (Chen et al., 2011). mTORC1 regulates, either directly or indirectly, the phosphorylation not only of the 4E-BPs but of several additional substrates that are relevant to translation. These include eIF4G, the ribosomal protein S6 kinases (S6Ks), eIF4B and eukaryotic elongation factor-2 kinase (eEF2K). The S6Ks also phosphorylate Pdcd4, resulting in the ubiquitination and subsequent degradation of the latter by the proteasome (Yang et al., 2003; Dorrello et al., 2006) (see below).

The phosphorylation of 4E-BP1 is a highly regulated process (reviewed in Gingras et al., 1999; Sonenberg and Hinnebusch, 2007, 2009), and the association of 4E-BP1 with eIF4E is modulated by a series of ordered phosphorylation events that are controlled via the mTORC1 signaling pathway (Gingras et al., 1999; Sonenberg and Dever, 2003; Morley et al., 2005; Mamane et al., 2006; Sonenberg and Hinnebusch, 2009). Current evidence indicates that hyper-phosphorylated 4E-BP1 is released from eIF4E, exposing a surface region on the latter factor that allows it to bind to eIF4G and thus promotes the initiation of translation.

mTORC1 is regulated by the phosphatidylinositol-3-kinase (PI3K), protein kinase B (PKB) and tuberous sclerosis 1/2 (TSC1/2) pathway (Figure 6.3) (Schalm et al., 2003; Ruggero and Sonenberg, 2005; Guertin and Sabatini, 2007; Thoreen et al., 2009). Loss of regulation of PI3K or PKB activity is often observed in tumor cells, in many cases as a result of mutation of the PTEN tumor suppressor gene, and these enzymes are important in promoting cell survival (Guertin and Sabatini, 2007; Kawauchi et al., 2009; Thoreen et al., 2009). There is considerable evidence that PI3K and PKB have important functions in controlling translation and that many of their effects are mediated by mTORC1 activation (Aoki and Vogt, 2004; Beuvink et al., 2005; Dutton et al., 2005; Ruggero and Sonenberg, 2005; Guertin and Sabatini, 2007). Inhibition of the activity of mTOR can be benefical against PKB-driven tumorigenesis (Hsieh et al., 2010). PI3K activity can influence both 4E-BP and eIF4G phosphorylation via its effects on mTOR activity (Raught et al., 2000; Renner et al., 2007). Conversely, mTOR activity is inhibited when the energy-sensing protein kinase AMPK is activated (Hadad et al., 2008) and thus regulation of the latter can also influence translation in cancer cells via effects on mTOR targets (Pradelli et al., 2010).

mTORC1 signaling can regulate the level of expression of the hypoxia-sensitive protein HIF-1α (Majumder et al., 2004), over-production of which is linked to a loss of growth control (Land and Tee, 2007; Wouters and Koritzinsky, 2008; Garcia-Maceira and Mateo, 2009; Harada et al., 2009). A switch from cap-dependent to cap-independent mRNA translation is mediated by hypoxia in large advanced breast cancers and this is required for promoting angiogenesis and tumor survival (Braunstein et al., 2007). The switch is caused by over-expression of 4E-BP1, resulting in the inhibition of cap-dependent translation but enhanced translation of mRNAs which can utilize cap-independent translation pathways. The latter include not only the mRNA encoding HIF-1α but also those for vascular endothelial growth factor (VEGF) and the anti-apoptotic protein Bcl-2, which are additionally required for growth under hypoxic conditions.

Many studies indeed suggest that the balance between cap-dependent and cap-independent translation in the cell plays an important role in tumorigenesis, with inhibition of cap-dependent translation *in vitro* resulting in increased cap-independent translation (Schneider and Sonenberg, 2007; Spriggs et al., 2008). Over-expression of eIF4E or enhanced phosphorylation of the 4E-BPs leads to increased eIF4F levels, which favors the translation of mRNAs with extensive secondary structure in their 5′ UTRs (Figure 6.4) (Koromilas et al., 1992; Morley, 2001; Sonenberg and Hinnebusch, 2009). These include mRNAs encoding ornithine decarboxylase (Rousseau et al., 1996b) and c-Myc (Carter et al., 1999), as well as VEGF and Bcl-2 (Graff and Zimmer, 2003). A recent

study has also suggested that the eIF4E–4E–BP axis has important effects on the translation of the anti-apoptotic protein Mcl-1 (Hsieh et al., 2010). Such proteins contribute not only to primary transformation but also to the metastatic ability of tumors. The Myc protein regulates translation by enhancing the synthesis of ribosomes and other components of the protein synthesis machinery (Ruggero, 2009; van Riggelen et al., 2010) and a raised level of Myc also causes an increase in cap-dependent translation (at the expense of cap-independent translation). One consequence of the latter is reduced expression of the cyclin-dependent protein kinase Cdk11, leading to impaired cytokinesis, aneuploidy and tumorigenesis (Wilker et al., 2007; Barna et al., 2008). Interestingly, the mRNAs encoding VEGF, Bcl-2 and c-Myc (and no doubt several other proteins of importance in oncogenesis) can be translated by both cap-dependent and cap-independent mechanisms. Thus the translation of such messages, although promoted by enhanced eIF4E expression, may also be resistant to

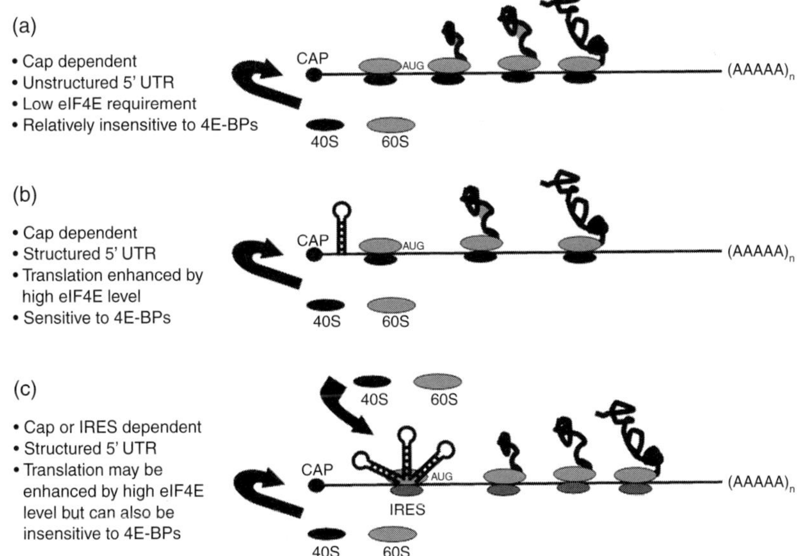

Figure 6.4 Mechanisms by which mRNA structure may influence the sensitivity of translation to variations in eIF4E and 4E-BP levels and activity. (a) Messenger RNAs that are translated exclusively by the cap-dependent pathway and have relatively unstructured 5′ untranslated regions (UTRs) are likely to require only a basal level of eIF4E and are predicted to be relatively insensitive to the effects of the 4E-BPs. (b) Messenger RNAs that are translated exclusively by the cap-dependent pathway and have more highly structured 5′ UTRs are likely to require higher levels of eIF4E for optimal translation and are predicted to be more sensitive to the effects of the 4E-BPs. Such mRNAs may include those encoding critical growth-regulatory and/or anti-apoptotic proteins, expression of which would thus be enhanced in tumor cells that over-express eIF4E. (c) Some mRNAs can be translated by either cap-dependent or IRES-driven internal initiation (these will have structured 5′ UTRs due to the presence of the IRES element). These mRNAs may show enhanced activity in the presence of elevated eIF4E when translated by the first mechanism but can also be independent of eIF4E when being translated by the second mechanism. In the latter situation translation will be insensitive to the effects of the 4E-BPs. These alternative mechanisms can result in efficient expression of the encoded protein under a variety of physiological conditions. Proteins with a role in malignant transformation or progression that can be translated in this way include HIF-1α, VEGF, c-Myc and Bcl-2.

the consequences of elevated 4E-BP expression (Figure 6.4). The latter phenomenon may explain why some tumors are able to show high levels of 4E-BP expression without undergoing apoptosis. Interestingly, Myc has been shown to increase the expression of 4E-BP1 while still exerting oncogenic activity (Balakumaran et al., 2009).

In addition, high levels of eIF4E activity can exert an mRNA-selective, pre-translational effect, enhancing the rate of nucleocytoplasmic transport of the mRNA encoding the critical cell cycle-regulatory protein cyclin D1 (Rousseau et al., 1996b; Topisirovic et al., 2009a, b). It has been reported that an element in the 3′ UTR of cyclin D1 mRNA is critical for its recognition by eIF4E within the nucleus (Culjkovic et al., 2008). The ability of eIF4E to promote cell cycle progression by this mechanism may account for the inhibition of cell differentiation in the granulocyte, monocyte and erythroid lineages (Topisirovic et al., 2003, 2009a, b; Blazquez-Domingo et al., 2005; Culjkovic et al., 2008), with a consequently increased risk of leukemogenesis.

The stability of Pdcd4, the tumor suppressor protein with inhibitory effects on eIF4A (Suzuki et al., 2008; Chang et al., 2009; Loh et al., 2009; Wei et al., 2009), is regulated by phosphorylation driven by S6K (Carayol et al., 2008). Inhibition of S6K activity in renal carcinoma cells with fluvastatin results in the up-regulated expression of Pdcd4 and inhibition of eIF4A activity (Woodard et al., 2008), preventing cell proliferation. As Pdcd4 inhibits transformation and invasion it is not surprising that it is often down-regulated in cancers. In addition, recent studies have shown that the mRNA encoding Pdcd4 has a conserved target site for the miR-21 microRNA within its 3′-UTR and a study with ten colorectal cell lines showed an inverse correlation between the levels of miR-21 and Pdcd4 protein (Asangani et al., 2008). These findings suggest that increased levels of miR-21, and consequent down-regulation of Pdcd4, may have a role in tumor cell invasion and metastasis (Asangani et al., 2008; Woodard et al., 2008).

eIF4B activity can also be regulated by protein phosphorylation, with multiple kinase-mediated modification at different sites being important for its translational activity (Shahbazian et al., 2006; van Gorp et al., 2009). eIF4B is directly targeted by PKB, which can phosphorylate the protein on Ser^{422}. This results in enhanced interaction of eIF4B with the p170

subunit of eIF3 (eIF3a) and plays an important role in mRNA translation (Holz et al., 2005). eIF4B phosphorylation on Ser^{422} following amino acid refeeding of starved cells was found also to be mTORC1 dependent. There is also phosphorylation of eIF4B on Ser^{406}, which is dependent on both MEK and mTORC1 activity.

Significance of mRNA structure

The structure of an mRNA, particularly in its 5′ UTR, is likely to be the main determinant of the way in which the translation of that mRNA is influenced by the regulation of polypeptide chain initiation. For the majority of mRNAs, which are translated via cap-dependent initiation, the presence of extensive secondary structure is likely to render translation sensitive to changes in eIF4E expression or availability (Polunovsky et al., 2000; Schneider and Mohr, 2003; Abbott and Proud, 2004; Parsa and Holland, 2004; Shao et al., 2004; Tee and Blenis, 2005; Mamane et al., 2006; Sonenberg and Hinnebusch, 2009). However, a substantial fraction of cellular mRNAs – perhaps 10 to 15% of the total – have internal ribosome entry sites (IRESs) and can be translated as a result of direct access of the ribosome to the start codon (Stoneley and Willis, 2004; Komar and Hatzoglou, 2005; Pickering and Willis, 2005); the translation of these mRNAs may therefore be independent of eIF4E. The possibility of competition between mRNAs for the available translational machinery means that changes in eIF4E levels could alter the balance in utilization of mRNAs by cap-dependent versus IRES-driven pathways, as indicated above (Svitkin et al., 2005; Braunstein et al., 2007). Good examples are seen in the case of the cdk inhibitor p27^{KIP1} and cyclin D1. Inactivation of cap-dependent initiation by a non-phosphorylatable form of 4E-BP1 or by treatment of cells with the mTOR inhibitor rapamycin enhances the translation of p27^{KIP1} and cyclin D1 because their mRNAs contain IRESs (Jiang et al., 2003; Cho et al., 2005; Shi et al., 2005). Recent data also indicate that over-expression of eIF4GI can re-program cells to increase IRES-mediated translation. This may be of particular importance in the pathogenesis of inflammatory breast cancer, for example by facilitating the synthesis of proteins that promote angiogenesis and/ or resistance to hypoxia (Silvera and Schneider, 2009; Silvera et al., 2009). Mutations in the sequences of IRES elements may also play a role in tumorigenesis, as in some forms of multiple myeloma (Chappell et al., 2000).

In addition to these mechanisms, elements in the 3′ UTR sequences of many mRNAs can affect translation as a result of the binding of miRNAs at such sites and there is likely to be a dynamic balance between events at the 5′ and 3′ ends of mRNA molecules that is important for translational regulation (reviewed in Gregory et al., 2008; Ma and Weinberg, 2008; Hermeking, 2009; Peter, 2009; Santhanam et al., 2009).

Clinical implications

Increased understanding of the role of translational regulation and dysregulation in malignant transformation provides potential new avenues for diagnostic and therapeutic applications. The level of eIF4E in a tumor can be a useful clinical marker for prognosis, both before and after treatment (Hiller et al., 2009; Silvera and Schneider, 2009).

Since the loss of ability of 4E-BP1 to sequester eIF4E as a consequence of dysregulated mTORC1 signaling has an important role to play in cell transformation (Polunovsky et al., 2000; Abbott and Proud, 2004; Parsa and Holland, 2004; Shao et al., 2004; Tee and Blenis, 2005; Mamane et al., 2006; Sonenberg and Hinnebusch, 2009), and mTORC1 is regarded as a central regulator of cell growth and survival in health and disease (Inoki et al., 2005; Tee and Blenis, 2005; Mamane et al., 2006; Peterson et al., 2009; Polak and Hall, 2009; Sonenberg and Hinnebusch, 2009; Efeyan and Sabatini, 2010; Ji and Zheng, 2010), it is not surprising that there is a lot of interest in targeting mTOR for cancer therapy (Bjornsti and Houghton, 2004b; Carraway and Hidalgo, 2004; Tee and Blenis, 2005; Georgakis and Younes, 2006; Mamane et al., 2006; Choo and Blenis, 2009; Lane and Breuleux, 2009; Sonenberg and Hinnebusch, 2009; Dancey, 2010; O'Reilly et al., 2011). Rapamycin treatment of tumors may not only inhibit the translation of undesirable proteins but may also enhance the activity of beneficial ones such as p53 (Kao et al., 2009). However, current agents available for inhibition of mTORC1 (including rapamycin, RAD001 and others) do have drawbacks (such as the feedback activation of cell survival pathways; Wang and Sun, 2009) and future drugs will have to overcome these problems (Choo and Blenis, 2009; Yu et al., 2009). Recent developments have shown that other inhibitors

of mTOR such as PP242, which can block the activity of both mTORC1 and another mTOR complex, mTORC2, may have good therapeutic effects against tumors that are resistant to rapamycin itself (Hsieh et al., 2010; Sparks and Guertin, 2010). However, there may also be additional pathways that act on mTOR substrates such as 4E-BP1 (Tamburini et al., 2009) and these would need to be targeted with other agents.

The use of rapamycin and its derivatives to inhibit 4E-BP1 phosphorylation and other activities of mTOR may, in some cases, sensitize tumor cells to treatments to which they were previously refractory (Dancey, 2003; Chan, 2004; Dutcher, 2004; Easton and Houghton, 2004; Rowinsky, 2004; Gomez-Martin et al., 2005; Dowling et al., 2009; Furic et al., 2009). Inhibition of mTOR can block some of the effects of anti-apoptotic enzymes such as PI3K and PKB (Figure 6.3) and this may be of particular advantage in tumors with activating mutations of these enzymes or in which the PI3K antagonist PTEN is deficient (Panwalkar et al., 2004; Jiang and Liu, 2008; Andreopoulou, 2011). For example, in glioblastoma multiforme with mutant PTEN and an activated PI3K-PKB pathway, deregulation of this signaling axis enhances translation of the caspase inhibitor protein c-FLIP$_s$. This results in inhibition of caspase-8 activation at the receptor-mediated death-inducing signaling complex. This can be counteracted by inhibition of mTOR, which results in the suppression of the translation of c-FLIP$_s$ and renders the cells sensitive to receptor-mediated apoptosis (Panner et al., 2005, 2006). Similarly, the resistance of lung cancer cell lines to the DNA-damaging agent cisplatin can be overcome using the rapamycin analogue CCI-779 (Wu et al., 2005). CCI-779 has also been effective against primary human acute lymphoblastic leukemia, a disease that is often resistant to other types of therapy (Teachey et al., 2006; Guertin and Sabatini, 2007; Choo and Blenis, 2009; Sonenberg and Hinnebusch, 2009). Such findings suggest that therapeutic combinations of rapamycin derivatives with other forms of chemotherapy, or with radiotherapy or cytokine-based treatments, may be of benefit.

In some cases, enzymes that are upstream activators of mTOR, such as the G protein Rheb (Figure 6.3), may themselves be appropriate targets. For example, farnesyltransferase inhibitors can block Rheb activity and show therapeutic effects against some lymphomas (Mavrakis and Wendel, 2008; Mavrakis et al., 2008). Another drug that acts against mTOR, with particular effects on the induction of hypoxia-regulated proteins at the translational level, is the flavinoid silibinin (Garcia-Maceira and Mateo, 2009).

In contrast to the above, in many cases it may be necessary to target eIF4E or the eIF4F complex more directly (Graff et al., 2008; Konicek et al., 2008). For example, high levels of eIF4E expression may render cells insensitive to changes in the phosphorylation state of endogenous 4E-BPs, and hence to the effects of inhibitors of mTOR or its upstream regulators (Wendel et al., 2004). Under these circumstances the balance may be shifted by the ectopic expression of additional 4E-BPs. A gene therapy approach involving adenovirus-mediated expression of 4E-BP1, in combination with the use of rapamycin, has been demonstrated to inhibit the growth of pancreatic ductal adenocarcinoma *in vivo* (Mishra et al., 2009). In this situation the mTOR inhibitor prevented the phosphorylation (and hence inactivation) of the additional 4E-BP1. This could be particularly important in tumors where mTOR is constitutively active. Other agents with selective effects against transformed cells, and/or anti-tumor activity, and which act by inhibiting eIF4F complex assembly include the small molecule 4EGI-1 (Moerke et al., 2007; Tamburini et al., 2009) and the tyrosine kinase inhibitor imatinib (when used in combination with CGP57380 – an inhibitor of eIF4E phosphorylation) (Zhang et al., 2008b).

Other means of preventing eIF4E function may also be beneficial. One example could be the use of ribavirin, a guanosine analogue that interferes with recognition of the 7-methylguanosine mRNA cap structure (Kentsis et al., 2004). This approach might be effective if, as seems to be the case, oncogenic and anti-apoptotic proteins are encoded by highly structured mRNAs with particularly strong dependencies on high levels of the eIF4F complex. A recent report indicates some success in a clinical trial of ribavirin treatment of patients with acute myeloid leukemia (Assouline et al., 2009). Another approach could be the specific down-regulation of eIF4E by RNA interference, but this would require both the effective delivery of vectors expressing siRNAs or miRNAs to a high proportion of the tumor cells and the selective targeting of these cells in order to reduce toxic side effects. However, the use of appropriate viral vectors may ultimately allow this to be achieved (Lundstrom, 2004; Yu et al., 2006). An alternative approach could involve the use of stable antisense oligonucleotides directed against the eIF4E mRNA sequence. Such agents have indeed been shown to be effective in suppressing the growth of human tumor xenografts in mice, without showing significant toxicity toward normal tissues (Graff et al., 2007). The lack of side effects may reflect the increased requirement for eIF4E activity shown by tumor cells compared to normal cells.

It may also be possible to take advantage of high eIF4E activity in tumor cells to promote the expression of suicide genes. For example, expression of a toxic enzyme encoded by an mRNA with a highly structured 5′ UTR may be enhanced in cells overproducing eIF4E. Such an approach has been used to synthesize thymidine kinase, which can then utilize therapeutic substrates such as gancyclovir (DeFatta et al., 2002a, b).

The ability of viruses to manipulate the translational machinery of their hosts (Clemens, 2005) may also be exploitable to bring about the death of tumor cells. Selective replication of such "oncolytic" viruses in cancer cells could be developed into a powerful tool for the treatment of tumors that are resistant to more conventional therapies (Mohr, 2005).

Concluding remarks

This review has attempted to highlight the many facets of translational regulation that impinge on various forms of cancer. It is clear that dysregulation of gene expression at the level of protein synthesis has a major role to play in the development of the malignant phenotype, but it is equally evident that much remains to be learned about the mechanisms by which such effects are mediated. The near future should bring further exciting revelations that will both facilitate our understanding of the biochemistry involved and result in the development of new therapies that can be applied in the clinic.

Acknowledgments

The authors' research has been funded by the BBSRC, the Wellcome Trust and the Ralph Bates Pancreatic Cancer Research Fund.

References

Abbott, C.M. and Proud, C.G. (2004). Translation factors: in sickness and in health. *Trends Biochem Sci 29*, 25–31.

Andreopoulou, E. (2011). The PI3K/AKT/mTOR signaling pathway: implications in the treatment of breast cancer. *Current Breast Cancer Rep 3*, 63–74.

Andrieu, C., Taieb, D., Baylot, V., et al. (2010). Heat shock protein 27 confers resistance to androgen ablation and chemotherapy in prostate cancer cells through eIF4E *Oncogene 29*, 1883–1896.

Aoki, M. and Vogt, P.K. (2004). Retroviral oncogenes and TOR. *Curr Top Microbiol Immunol 279*, 321–338.

Armengol, G., Rojo, F., Castellvi, J., et al. (2007). 4E-binding protein 1: a key molecular "funnel factor" in human cancer with clinical implications. *Cancer Res 67*, 7551–7555.

Asangani, I.A., Rasheed, S.A., Nikolova, D.A., et al. (2008). MicroRNA-21 (miR-21) post-transcriptionally downregulates tumor suppressor Pdcd4 and stimulates invasion, intravasation and metastasis in colorectal cancer. *Oncogene 27*, 2128–2136.

Assouline, S., Culjkovic, B., Cocolakis, E., et al. (2009). Molecular targeting of the oncogene eIF4E in AML: a proof-of-principle clinical trial with ribavirin. *Blood 114*, 257–260.

Avdulov, S., Li, S., Michalek, V., et al. (2004). Activation of translation complex eIF4F is essential for the genesis and maintenance of the malignant phenotype in human mammary epithelial cells. *Cancer Cell 5*, 553–563.

Balachandran, S. and Barber, G.N. (2004). Defective translational control facilitates vesicular stomatitis virus oncolysis. *Cancer Cell 5*, 51–65.

Balakumaran, B.S., Porrello, A., Hsu, D.S., et al. (2009). MYC activity mitigates response to rapamycin in prostate cancer through eukaryotic initiation factor 4E-binding protein 1-mediated inhibition of autophagy. *Cancer Res 69*, 7803–7810.

Barna, M., Pusic, A., Zollo, O., et al. (2008). Suppression of Myc oncogenic activity by ribosomal protein haploinsufficiency. *Nature 456*, 971–975.

Barnhart, B.C., Lam, J.C., Young, R.M., et al. (2008). Effects of 4E-BP1 expression on hypoxic cell cycle inhibition and tumor cell proliferation and survival. *Cancer Biol Ther 7*, 1441–1449.

Bauer, C., Brass, N., Diesinger, I., et al. (2002). Overexpression of the eukaryotic translation initiation factor 4G (eIF4G-1) in squamous cell lung carcinoma. *Int J Cancer 98*, 181–185.

Beuvink, I., Boulay, A., Fumagalli, S., et al. (2005). The mTOR inhibitor RAD001 sensitizes tumor cells to DNA-damaged induced apoptosis through inhibition of p21 translation. *Cell 120*, 747–759.

Bianchini, A., Loiarro, M., Bielli, P., et al. (2008). Phosphorylation of eIF4E by MNKs supports protein synthesis, cell cycle progression and proliferation in prostate cancer cells. *Carcinogenesis 29*, 2279–2288.

Bjornsti, M.A. and Houghton, P.J. (2004a). Lost in translation: dysregulation of cap-dependent translation and cancer. *Cancer Cell 5*, 519–523.

Bjornsti, M.A. and Houghton, P.J. (2004b). The TOR pathway: a target for cancer therapy. *Nat Rev Cancer 4*, 335–348.

Blagden, S.P. and Willis, A.E. (2011). The biological and therapeutic relevance of mRNA translation in cancer. *Nat Rev Clin Oncol 8*, 280–591.

Blalock, W.L., Bavelloni, A., Piazzi, M., Faenza, I., and Cocco, L. (2010). A role for PKR in hematologic malignancies. *J Cell Physiol 223*, 572–591.

Blalock, W.L., Bavelloni, A., Piazzi, M., et al. (2011). Multiple forms of PKR present in the nuclei of acute leukemia cells represent an active kinase that is responsive to stress. *Leukemia 25*, 236–245.

Blazquez-Domingo, M., Grech, G., and von Lindern, M. (2005). Translation initiation factor 4E inhibits differentiation of erythroid progenitors. *Mol Cell Biol 25*, 8496–8506.

Braunstein, S., Karpisheva, K., Pola, C., et al. (2007). A hypoxia-controlled cap-dependent to cap-independent translation switch in breast cancer. *Mol Cell 28*, 501–512.

Buxade, M., Parra-Palau, J.L., and Proud, C.G. (2008). The Mnks: MAP kinase-interacting kinases (MAP kinase signal-integrating kinases). *Front Biosci 13*, 5359–5373.

Carayol, N., Katsoulidis, E., Sassano, A., et al. (2008). Suppression of programmed cell death 4 (PDCD4) protein expression by BCR-ABL-regulated engagement of the mTOR/p70 S6 kinase pathway. *J Biol Chem 283*, 8601–8610.

Carraway, H. and Hidalgo, M. (2004). New targets for therapy in breast cancer: mammalian target of rapamycin (mTOR) antagonists. *Breast Cancer Res 6*, 219–224.

Carter, P.S., Jarquin-Pardo, M., and De Benedetti, A. (1999). Differential expression of Myc1 and Myc2 isoforms in cells transformed by eIF4E: evidence for internal ribosome repositioning in the human c-*myc* 5′UTR. *Oncogene 18*, 4326–4335.

Castellvi, J., Garcia, A., Rojo, F., et al. (2006). Phosphorylated 4E binding protein 1: a hallmark of cell signaling that correlates with survival in ovarian cancer. *Cancer 107*, 1801–1811.

Castellvi, J., Garcia, A., Ruiz-Marcellan, C., et al. (2009). Cell signaling in endometrial carcinoma: phosphorylated 4E-binding protein-1 expression in endometrial cancer correlates with aggressive tumors and prognosis. *Hum Pathol* 40, 1418–1426.

Cencic, R., Carrier, M., Trnkus, A., et al. (2010). Synergistic effect of inhibiting translation initiation in combination with cytotoxic agents in acute myelogenous leukemia cells. *Leuk Res 34*, 535–541.

Chan, S. (2004). Targeting the mammalian target of rapamycin (mTOR): a new approach to treating cancer. *Br J Cancer 91*, 1420–1424.

Chang, J.H., Cho, Y.H., Sohn, S.Y., et al. (2009). Crystal structure of the eIF4A-PDCD4 complex. *Proc Natl Acad Sci USA 106*, 3148–3153.

Chappell, S.A., LeQuesne, J.P.C., Paulin, F.E.M., et al. (2000). A mutation in the c-*myc*-IRES leads to enhanced internal ribosome entry in multiple myeloma: a novel mechanism of oncogene de-regulation. *Oncogene 19*, 4437–4440.

Chawla-Sarkar, M., Lindner, D.J., Liu, Y.F., et al. (2003). Apoptosis and interferons: role of interferon-stimulated genes as mediators of apoptosis. *Apoptosis 8*, 237–249.

Chen, H., Ma, Z., Vanderwaal, R.P., et al. (2011). The mTOR inhibitor rapamycin suppresses DNA double-strand break repair. *Radiat Res 175*, 214–224.

Chen, Y., Liu, W., Chao, T., et al. (2008). MicroRNA-21 down-regulates the expression of tumor suppressor PDCD4 in human glioblastoma cell T98G. *Cancer Lett 272*, 197–205.

Chendrimada, T.P., Finn, K.J., Ji, X., et al. (2007). MicroRNA silencing through RISC recruitment of eIF6. *Nature 447*, 823–828.

Cho, S., Kim, J.H., Back, S.H., and Jang, S.K. (2005). Polypyrimidine tract-binding protein enhances the internal ribosomal entry site-dependent translation of p27Kip1 mRNA and modulates transition from G1 to S phase. *Mol Cell Biol 25*, 1283–1297.

Choi, C.H., Lee, J.S., Kim, S.R., et al. (2011). Direct inhibition of eIF4E reduced cell growth in endometrial adenocarcinoma. *J Cancer Res Clin Oncol 137*, 463–469.

Choo, A.Y. and Blenis, J. (2009). Not all substrates are treated equally: implications for mTOR, rapamycin-resistance and cancer therapy. *Cell Cycle 8*, 567–572.

Clemens, M.J. (2004). Targets and mechanisms for the regulation of translation in malignant transformation. *Oncogene 23*, 3180–3188.

Clemens, M.J. (2005). Translational control in virus-infected cells: models for cellular stress responses. *Semin Cell Dev Biol 16*, 13–20.

Coleman, L.J., Peter, M.B., Teall, T.J., et al. (2009). Combined analysis of eIF4E and 4E-binding protein expression predicts breast cancer survival and estimates eIF4E activity. *Br J Cancer 100*, 1393–1399.

Culjkovic, B., Tan, K., Orolicki, S., et al. (2008). The eIF4E RNA regulon promotes the Akt signaling pathway. *J Cell Biol 181*, 51–63.

Dancey, J. (2010). mTOR signaling and drug development in cancer. *Nat Rev Clin Oncol 7*, 209–219.

Dancey, J.E. (2003). mTOR inhibitors in hematologic malignancies. *Clin Adv Hematol Oncol 1*, 419–423.

De Benedetti, A. and Graff, J.R. (2004). eIF-4E expression and its role in malignancies and metastases. *Oncogene 23*, 3189–3199.

DeFatta, R.J., Chervenak, R.P., and De Benedetti, A. (2002a). A cancer gene therapy approach through translational control of a suicide gene. *Cancer Gene Ther 9*, 505–512.

DeFatta, R.J., Li, Y., and De Benedetti, A. (2002b). Selective killing of cancer cells based on translational control of a suicide gene. *Cancer Gene Ther 9*, 573–578.

Dong, K., Wang, R., Wang, X., et al. (2009). Tumor-specific RNAi targeting eIF4E suppresses tumor growth, induces apoptosis and enhances cisplatin cytotoxicity in human breast carcinoma cells. *Breast Cancer Res Treat 113*, 443–456.

Dong, Z., Liu, L.H., Han, B., Pincheira, R., and Zhang, J.T. (2004). Role of eIF3 p170 in controlling synthesis of ribonucleotide reductase M2 and cell growth. *Oncogene 23*, 3790–3801.

Donze, O., Jagus, R., Koromilas, A.E., Hershey, J.W., and Sonenberg, N. (1995). Abrogation of translation initiation factor eIF-2 phosphorylation causes malignant transformation of NIH 3T3 cells. *EMBO J 14*, 3828–3834.

Dorrello, N.V., Peschiaroli, A., Guardavaccaro, D., et al. (2006). S6K1- and betaTRCP-mediated degradation of PDCD4 promotes protein translation and cell growth. *Science 314*, 467–471.

Dowling, R.J., Pollak, M., and Sonenberg, N. (2009). Current status and challenges associated with targeting mTOR for cancer therapy. *BioDrugs 23*, 77–91.

Dubois, L., Magagnin, M.G., Cleven, A.H., et al. (2009). Inhibition of 4E-BP1 sensitizes U87 glioblastoma xenograft tumors to irradiation by decreasing hypoxia tolerance. *Int J Radiat Oncol Biol Phys 73*, 1219–1227.

Dutcher, J.P. (2004). Mammalian target of rapamycin inhibition. *Clin Cancer Res 10*, 6382S–6387S.

Dutton, A., Reynolds, G.M., Dawson, C.W., Young, L.S., and Murray, P.G. (2005). Constitutive activation of phosphatidyl-inositide 3 kinase contributes to the survival of Hodgkin's lymphoma cells through a mechanism involving Akt kinase and mTOR. *J Pathol 205*, 498–506.

Easton, J.B. and Houghton, P.J. (2004). Therapeutic potential of target of rapamycin inhibitors. *Expert Opin Ther Targets 8*, 551–564.

Eberle, J., Krasagakis, K., and Orfanos, C.E. (1997). Translation initiation factor eIF-4A1 mRNA is consistently overexpressed in human melanoma cells *in vitro*. *Int J Cancer* 71, 396–401.

Eberle, J., Fecker, L.F., Bittner, J.U., Orfanos, C.E., and Geilen, C.C. (2002). Decreased proliferation of human melanoma cell lines caused by antisense RNA against translation factor eIF-4AI. *Br J Cancer* 86, 1957–1962.

Efeyan, A. and Sabatini, D.M. (2010). mTOR and cancer: many loops in one pathway. *Curr Opin Cell Biol* 22, 169–176.

Fan, S., Li, Y., Yue, P., Khuri, F.R., and Sun, S.Y. (2010). The eIF4E/eIF4G interaction inhibitor 4EGI-1 augments TRAIL-mediated apoptosis through c-FLIP down-regulation and DR5 induction independent of inhibition of cap-dependent protein translation. *Neoplasia* 12, 346–356.

Fan, S., Ramalingam, S.S., Kauh, J., et al. (2009). Phosphorylated eukaryotic translation initiation factor 4 (eIF4E) is elevated in human cancer tissues. *Cancer Biol Ther* 8, 1463–1469.

Feldman, D.E., Chauhan, V., and Koong, A.C. (2005). The unfolded protein response: a novel component of the hypoxic stress response in tumors. *Mol Cancer Res* 3, 597–605.

Feng, P., Everly, D.N., Jr., and Read, G.S. (2005). mRNA decay during herpes simplex virus (HSV) infections: protein–protein interactions involving the HSV virion host shutoff protein and translation factors eIF4H and eIF4A. *J Virol* 79, 9651–9664.

Frederick, M.J., VanMeter, A.J., Gadhikar, M.A., et al. (2011). Phosphoproteomic analysis of signaling pathways in head and neck squamous cell carcinoma patient samples. *Am J Pathol* 178, 548–571.

Fukuchi-Shimogori, T., Ishii, I., Kashiwagi, K., et al. (1997). Malignant transformation by overproduction of translation initiation factor eIF4G. *Cancer Res* 57, 5041–5044.

Furic, L., Livingstone, M., Dowling, R.J., and Sonenberg, N. (2009). Targeting mtor-dependent tumours with specific inhibitors: a model for personalized medicine based on molecular diagnoses. *Curr Oncol* 16, 59–61.

Gallagher, J.W., Kubica, N., Kimball, S.R., and Jefferson, L.S. (2008). Reduced eukaryotic initiation factor 2B epsilon-subunit expression suppresses the transformed phenotype of cells overexpressing the protein. *Cancer Res* 68, 8752–8760.

Gandin, V., Miluzio, A., Barbieri, A.M., et al. (2008). Eukaryotic initiation factor 6 is rate-limiting in translation, growth and transformation. *Nature* 455, 684–688.

Garcia, M.A., Gil, J., Ventoso, I., et al. (2006). Impact of protein kinase PKR in cell biology: from antiviral to antiproliferative action. *Microbiol Mol Biol Rev* 70, 1032–1060.

Garcia-Maceira, P. and Mateo, J. (2009). Silibinin inhibits hypoxia-inducible factor-1alpha and mTOR/p70S6K/4E-BP1 signalling pathway in human cervical and hepatoma cancer cells: implications for anticancer therapy. *Oncogene* 28, 313–324.

Georgakis, G.V. and Younes, A. (2006). From Rapa Nui to rapamycin: targeting PI3K/Akt/mTOR for cancer therapy. *Expert Rev Anticancer Ther* 6, 131–140.

Gingras, A.C., Raught, B., and Sonenberg, N. (1999). eIF4 initiation factors: effectors of mRNA recruitment to ribosomes and regulators of translation. *Annu Rev Biochem* 68, 913–963.

Gomez-Martin, C., Rubio-Viqueira, B., and Hidalgo, M. (2005). Current status of mammalian target of rapamycin inhibitors in lung cancer. *Clin Lung Cancer* 7 Suppl 1, S13–S18.

Graff, J.R., Konicek, B.W., Carter, J.H., and Marcusson, E.G. (2008). Targeting the eukaryotic translation initiation factor 4E for cancer therapy. *Cancer Res* 68, 631–634.

Graff, J.R., Konicek, B.W., Lynch, R.L., et al. (2009). eIF4E activation is commonly elevated in advanced human prostate cancers and significantly related to reduced patient survival. *Cancer Res* 69, 3866–3873.

Graff, J.R., Konicek, B.W., Vincent, T.M., et al. (2007). Therapeutic suppression of translation initiation factor eIF4E expression reduces tumor growth without toxicity. *J Clin Invest* 117, 2638–2648.

Graff, J.R. and Zimmer, S.G. (2003). Translational control and metastatic progression: enhanced activity of the mRNA cap-binding protein eIF-4E selectively enhances translation of metastasis-related mRNAs. *Clin Exp Metastasis* 20, 265–273.

Gregory, P.A., Bracken, C.P., Bert, A.G., and Goodall, G.J. (2008). MicroRNAs as regulators of epithelial-mesenchymal transition. *Cell Cycle* 7, 3112–3118.

Grifo, J.A., Abramson, R.D., Satler, C.A., and Merrick, W.C. (1984). RNA-stimulated ATPase activity of eukaryotic initiation factors. *J Biol Chem* 259, 8648–8654.

Gu, X., Jones, L., Lowery-Norberg, M., and Fowler, M. (2005). Expression of eukaryotic initiation factor 4E in astrocytic tumors. *Appl Immunohistochem Mol Morphol* 13, 178–183.

Guertin, D.A. and Sabatini, D.M. (2007). Defining the role of mTOR in cancer. *Cancer Cell* 12, 9–22.

Hadad, S.M., Fleming, S., and Thompson, A.M. (2008). Targeting AMPK: a new therapeutic opportunity in breast cancer. *Crit Rev Oncol Hematol* 67, 1–7.

Harada, H., Itasaka, S., Zhu, Y., et al. (2009). Treatment regimen determines whether an HIF-1 inhibitor enhances or inhibits the effect of radiation therapy. *Br J Cancer* 100, 747–757.

Harris, M.N., Ozpolat, B., Abdi, F., et al. (2004). Comparative proteomic analysis of all-trans-retinoic acid treatment reveals systematic posttranscriptional control mechanisms in acute promyelocytic leukemia. *Blood 104*, 1314–1323.

Hermeking, H. (2009). MiR-34a and p53. *Cell Cycle 8*, 1308.

Hershey, J.W.B. and Merrick, W.C. (2000). The pathway and mechanism of initiation of protein synthesis. In Sonenberg, N., Hershey, J.W.B., and Mathews, M.B. (eds.) *Translational Control of Gene Expression*. New York: Cold Spring Harbor Laboratory Press, pp. 33–88.

Hiller, D.J., Chu, Q., Meschonat, C., et al. (2009). Predictive value of eIF4E reduction after neoadjuvant therapy in breast cancer. *J Surg Res 156*, 265–269.

Holland, E.C. (2004). Regulation of translation and cancer. *Cell Cycle 3*, 452–455.

Holland, E.C., Sonenberg, N., Pandolfi, P.P., and Thomas, G. (2004). Signaling control of mRNA translation in cancer pathogenesis. *Oncogene 23*, 3138–3144.

Holz, M.K., Ballif, B.A., Gygi, S.P., and Blenis, J. (2005). mTOR and S6K1 mediate assembly of the translation preinitiation complex through dynamic protein interchange and ordered phosphorylation events. *Cell 123*, 569–580.

Hsieh, A.C., Costa, M., Zollo, O., et al. (2010). Genetic dissection of the oncogenic mTOR pathway reveals druggable addiction to translational control via 4EBP-eIF4E. *Cancer Cell 17*, 249–261.

Inoki, K., Ouyang, H., Li, Y., and Guan, K.L. (2005). Signaling by target of rapamycin proteins in cell growth control. *Microbiol Mol Biol Rev 69*, 79–100.

Ji, J. and Zheng, P.S. (2010). Activation of mTOR signaling pathway contributes to survival of cervical cancer cells. *Gynecol Oncol 117*, 103–108.

Jiang, B.H. and Liu, L.Z. (2008). Role of mTOR in anticancer drug resistance: perspectives for improved drug treatment. *Drug Resist Updat 11*, 63–76.

Jiang, H., Coleman, J., Miskimins, R., and Miskimins, W.K. (2003). Expression of constitutively active 4EBP-1 enhances p27Kip1 expression and inhibits proliferation of MCF7 breast cancer cells. *Cancer Cell Int 3*, 2.

Jiang, H.Y. and Wek, R.C. (2005). Phosphorylation of eIF2alpha reduces protein synthesis and enhances apoptosis in response to proteasome inhibition. *J Biol Chem 280*, 14189–14202.

Kao, C.L., Hsu, H.S., Chen, H.W., and Cheng, T.H. (2009). Rapamycin increases the p53/MDM2 protein ratio and p53-dependent apoptosis by translational inhibition of mdm2 in cancer cells. *Cancer Lett 286*, 250–259.

Kawauchi, K., Ogasawara, T., Yasuyama, M., Otsuka, K., and Yamada, O. (2009). Regulation and importance of the PI3K/Akt/mTOR signaling pathway in hematologic malignancies. *Anticancer Agents Med Chem 9*, 1024–1038.

Kentsis, A., Topisirovic, I., Culjkovic, B., Shao, L., and Borden, K.L.B. (2004). Ribavirin suppresses eIF4E-mediated oncogenic transformation by physical mimicry of the 7-methyl guanosine mRNA cap. *Proc Natl Acad Sci USA 101*, 18105–18110.

Khan, M.A. and Goss, D.J. (2004). Phosphorylation states of translational initiation factors affect mRNA cap binding in wheat. *Biochemistry 43*, 9092–9097.

Kim, S.H., Miller, F.R., Tait, L., Zheng, J., and Novak, R.F. (2009). Proteomic and phosphoproteomic alterations in benign, premalignant and tumor human breast epithelial cells and xenograft lesions: biomarkers of progression. *Int J Cancer 124*, 2813–2828.

Ko, S.Y., Guo, H., Barengo, N., and Naora, H. (2009). Inhibition of ovarian cancer growth by a tumor-targeting peptide that binds eukaryotic translation initiation factor 4E. *Clin Cancer Res 15*, 4336–4347.

Kodali, D., Rawal, A., Ninan, M.J., et al. (2011). Expression and phosphorylation of eukaryotic translation initiation factor 4E binding protein 1 in B-cell lymphomas and reactive lymphoid tissues. *Arch Pathol Lab Med 135*, 365–371.

Komar, A.A. and Hatzoglou, M. (2005). Internal ribosome entry sites in cellular mRNAs: mystery of their existence. *J Biol Chem 280*, 23425–23428.

Konicek, B.W., Dumstorf, C.A., and Graff, J.R. (2008). Targeting the eIF4F translation initiation complex for cancer therapy. *Cell Cycle 7*, 2466–2471.

Konicek, B.W., Stephens, J.R., McNulty, A.M., et al. (2011). Therapeutic inhibition of MAP kinase interacting kinase blocks eukaryotic initiation factor 4E phosphorylation and suppresses outgrowth of experimental lung metastases. *Cancer Res 71*, 1849–1857.

Korneeva, N.L., Lamphear, B.J., Hennigan, F.L.C., and Rhoads, R.E. (2000). Mutually cooperative binding of eukaryotic translation initiation factor (eIF) 3 and eIF4A to human eIF4G-1. *J Biol Chem 275*, 41369–41376.

Koromilas, A.E., Lazaris-Karatzas, A., and Sonenberg, N. (1992). mRNAs containing extensive secondary structure in their 5′ non-coding region translate efficiently in cells overexpressing initiation factor eIF-4E. *EMBO J 11*, 4153–4158.

Kremer, C.L., Klein, R.R., Mendelson, J., et al. (2006). Expression of mTOR signaling pathway markers in prostate cancer progression. *Prostate 66*, 1203–1212.

Land, S.C. and Tee, A.R. (2007). Hypoxia-inducible factor 1alpha is regulated by the mammalian target of rapamycin (mTOR) via an mTOR signaling motif. *J Biol Chem* 282, 20534–20543.

Lane, H.A. and Breuleux, M. (2009). Optimal targeting of the mTORC1 kinase in human cancer. *Curr Opin Cell Biol* 21, 219–229.

Lee, J.W., Choi, J.J., Lee, K.M., et al. (2005). eIF-4E expression is associated with histopathologic grades in cervical neoplasia. *Hum Pathol* 36, 1197–1203.

Li, S., Perlman, D.M., Peterson, M.S., et al. (2004). Translation initiation factor 4E blocks endoplasmic reticulum-mediated apoptosis. *J Biol Chem* 279, 21312–21317.

Li, S., Sonenberg, N., Gingras, A.C., et al. (2002). Translational control of cell fate: availability of phosphorylation sites on translational repressor 4E-BP1 governs its proapoptotic potency. *Mol Cell Biol* 22, 2853–2861.

Li, S., Takasu, T., Perlman, D.M., et al. (2003). Translation factor eIF4E rescues cells from Myc-dependent apoptosis by inhibiting cytochrome c release. *J Biol Chem* 278, 3015–3022.

Lin, C.J., Cencic, R., Mills, J.R., Robert, F., and Pelletier, J. (2008). c-Myc and eIF4F are components of a feedforward loop that links transcription and translation. *Cancer Res* 68, 5326–5334.

Linder, P., Lasko, P.F., Ashburner, M., et al. (1989). Birth of the D-E-A-D box. *Nature* 337, 121–122.

Loh, P.G., Yang, H.S., Walsh, M.A., et al. (2009). Structural basis for translational inhibition by the tumour suppressor *Pdcd4*. *EMBO J* 28, 274–285.

Lu, P.D., Jousse, C., Marciniak, S.J., et al. (2004). Cytoprotection by pre-emptive conditional phosphorylation of translation initiation factor 2. *EMBO J* 23, 169–179.

Lu, Z., Liu, M., Stribinskis, V., et al. (2008). MicroRNA-21 promotes cell transformation by targeting the programmed cell death 4 gene. *Oncogene* 27, 4373–4379.

Lundstrom, K. (2004). Gene therapy applications of viral vectors. *Technol Cancer Res Treat* 3, 467–477.

Lynch, M., Fitzgerald, C., Johnston, K.A., Wang, S.P., and Schmidt, E.V. (2004). Activated eIF4E-binding protein slows G_1 progression and blocks transformation by c-*myc* without inhibiting cell growth. *J Biol Chem* 279, 3327–3339.

Ma, L. and Weinberg, R.A. (2008). Micromanagers of malignancy: role of microRNAs in regulating metastasis. *Trends Genet* 24, 448–456.

Majumder, P.K., Febbo, P.G., Bikoff, R., et al. (2004). mTOR inhibition reverses Akt-dependent prostate intraepithelial neoplasia through regulation of apoptotic and HIF-1-dependent pathways. *Nat Med* 10, 594–601.

Mamane, Y., Petroulakis, E., LeBacquer, O., and Sonenberg, N. (2006). mTOR, translation initiation and cancer. *Oncogene* 25, 6416–6422.

Mamane, Y., Petroulakis, E., Rong, L.W., et al. (2004). eIF4E – from translation to transformation. *Oncogene* 23, 3172–3179.

Marcotrigiano, J., Gingras, A.C., Sonenberg, N., and Burley, S.K. (1999). Cap-dependent translation initiation in eukaryotes is regulated by a molecular mimic of eIF4G. *Mol Cell* 3, 707–716.

Martin, M.E., Perez, M.I., Redondo, C., et al. (2000). 4E binding protein 1 expression is inversely correlated to the progression of gastrointestinal cancers. *Int J Biochem Cell Biol* 32, 633–642.

Mavrakis, K.J. and Wendel, H.G. (2008). Translational control and cancer therapy. *Cell Cycle* 7, 2791–2794.

Mavrakis, K.J., Zhu, H., Silva, R.L., et al. (2008). Tumorigenic activity and therapeutic inhibition of Rheb GTPase. *Genes Dev* 22, 2178–2188.

Mayeur, G.L. and Hershey, J.W.B. (2002). Malignant transformation by the eukaryotic translation initiation factor 3 subunit p48 (eIF3e). *FEBS Lett* 514, 49–54.

McClusky, D.R., Chu, Q., Yu, H., et al. (2005). A prospective trial on initiation factor 4E (eIF4E) overexpression and cancer recurrence in node-positive breast cancer. *Ann Surg* 242, 584–590.

McEwen, E., Kedersha, N., Song, B., et al. (2005). Heme-regulated inhibitor kinase-mediated phosphorylation of eukaryotic translation initiation factor 2 inhibits translation, induces stress granule formation, and mediates survival upon arsenite exposure. *J Biol Chem* 280, 16925–16933.

McKendrick, L., Morley, S.J., Pain, V.M., Jagus, R., and Joshi, B. (2001). Phosphorylation of eukaryotic initiation factor 4E (eIF4E) at Ser209 is not required for protein synthesis *in vitro* and *in vivo*. *Eur J Biochem* 268, 5375–5385.

Merrick, W.C. (2004). Cap-dependent and cap-independent translation in eukaryotic systems. *Gene* 332, 1–11.

Miluzio, A., Beugnet, A., Volta, V., and Biffo, S. (2009). Eukaryotic initiation factor 6 mediates a continuum between 60S ribosome biogenesis and translation. *EMBO Rep* 10, 459–465.

Mishra, R., Miyamoto, M., Yoshioka, T., et al. (2009). Adenovirus-mediated eukaryotic initiation factor 4E binding protein-1 in combination with rapamycin inhibits tumor growth of pancreatic ductal adenocarcinoma *in vivo*. *Int J Oncol* 34, 1231–1240.

Moerke, N.J., Aktas, H., Chen, H., et al. (2007). Small-molecule inhibition of the interaction between the translation initiation factors eIF4E and eIF4G. *Cell* 128, 257–267.

Mohr, I. (2005). To replicate or not to replicate: achieving selective oncolytic virus replication in cancer cells through translational control. *Oncogene* 24, 7697–7709.

Morley, S.J. (2001). The regulation of eIF4F during cell growth and cell death. *Prog Mol Subcell Biol 27*, 1–37.

Morley, S.J., Coldwell, M.J., and Clemens, M.J. (2005). Initiation factor modifications in the preapoptotic phase. *Cell Death Differ 12*, 571–584.

Morley, S.J. and Naegele, S. (2002). Phosphorylation of eukaryotic initiation factor (eIF) 4E is not required for de novo protein synthesis following recovery from hypertonic stress in human kidney cells. *J Biol Chem 277*, 32855–32859.

Nathan, C.O., Amirghahari, N., Abreo, F., et al. (2004). Overexpressed eIF4E is functionally active in surgical margins of head and neck cancer patients via activation of the Akt/mammalian target of rapamycin pathway. *Clin Cancer Res 10*, 5820–5827.

O'Reilly, T., McSheehy, P.M., Wartmann, M., et al. (2011). Evaluation of the mTOR inhibitor, everolimus, in combination with cytotoxic antitumor agents using human tumor models *in vitro* and *in vivo*. *Anticancer Drugs 22*, 58–78.

Oridate, N., Kim, H.J., Xu, X., and Lotan, R. (2005). Growth inhibition of head and neck squamous carcinoma cells by small interfering RNAs targeting eIF4E or cyclin D1 alone or combined with cisplatin. *Cancer Biol Ther 4*, 318–323.

Othumpangat, S., Kashon, M., and Joseph, P. (2005). Sodium arsenite-induced inhibition of eukaryotic translation initiation factor 4E (eIF4E) results in cytotoxicity and cell death. *Mol Cell Biochem 279*, 123–131.

Paglin, S., Lee, N.Y., Nakar, C., et al. (2005). Rapamycin-sensitive pathway regulates mitochondrial membrane potential, autophagy, and survival in irradiated MCF-7 cells. *Cancer Res 65*, 11061–11070.

Panner, A., James, C.D., Berger, M.S., and Pieper, R.O. (2005). mTOR controls FLIPS translation and

TRAIL sensitivity in glioblastoma multiforme cells. *Mol Cell Biol 25*, 8809–8823.

Panner, A., Parsa, A.T., and Pieper, R.O. (2006). Translational regulation of TRAIL sensitivity. *Cell Cycle 5*, 147–150.

Panwalkar, A., Verstovsek, S., and Giles, F.J. (2004). Mammalian target of rapamycin inhibition as therapy for hematologic malignancies. *Cancer 100*, 657–666.

Parsa, A.T. and Holland, E.C. (2004). Cooperative translational control of gene expression by Ras and Akt in cancer. *Trends Mol Med 10*, 607–613.

Pataer, A., Swisher, S.G., Roth, J.A., Logothetis, C.J., and Corn, P. (2009). Inhibition of RNA-dependent protein kinase (PKR) leads to cancer cell death and increases chemosensitivity. *Cancer Biol Ther 8*.

Pause, A., Methot, N., Svitkin, Y., Merrick, W.C., and Sonenberg, N. (1994). Dominant negative mutants of mammalian translation initiation factor eIF-4A define a critical role for eIF-4F in cap-dependent and cap-independent initiation of translation. *EMBO J 13*, 1205–1215.

Peidis, P., Papadakis, A.I., Muaddi, H., Richard, S., and Koromilas, A.E. (2011). Doxorubicin bypasses the cytoprotective effects of eIF2alpha phosphorylation and promotes PKR-mediated cell death. *Cell Death Differ 18*, 145–154.

Perkins, D.J. and Barber, G.N. (2004). Defects in translational regulation mediated by the α subunit of eukaryotic initiation factor 2 inhibit antiviral activity and facilitate the malignant transformation of human fibroblasts. *Mol Cell Biol 24*, 2025–2040.

Peter, M.E. (2009). Let-7 and miR-200 microRNAs: guardians against pluripotency and cancer progression. *Cell Cycle 8*, 843–852.

Peterson, T.R., Laplante, M., Thoreen, C.C., et al. (2009). DEPTOR is an mTOR inhibitor frequently

overexpressed in multiple myeloma cells and required for their survival. *Cell 137*, 873–886.

Pickering, B.M. and Willis, A.E. (2005). The implications of structured 5′ untranslated regions on translation and disease. *Semin Cell Dev Biol 16*, 39–47.

Polak, P. and Hall, M.N. (2009). mTOR and the control of whole body metabolism. *Curr Opin Cell Biol 21*, 209–218.

Polunovsky, V.A., Gingras, A.C., Sonenberg, N., et al. (2000). Translational control of the antiapoptotic function of Ras. *J Biol Chem 275*, 24776–24780.

Pradelli, L.A., Beneteau, M., Chauvin, C., et al. (2010). Glycolysis inhibition sensitizes tumor cells to death receptors-induced apoptosis by AMP kinase activation leading to Mcl-1 block in translation. *Oncogene 29*, 1641–1652.

Proud, C.G. (2005a). eIF2 and the control of cell physiology. *Semin Cell Dev Biol 16*, 3–12.

Proud, C.G. (2005b). The eukaryotic initiation factor 4E-binding proteins and apoptosis. *Cell Death Differ 12*, 541–546.

Ramirez-Valle, F., Braunstein, S., Zavadil, J., Formenti, S.C., and Schneider, R.J. (2008). eIF4GI links nutrient sensing by mTOR to cell proliferation and inhibition of autophagy. *J Cell Biol 181*, 293–307.

Rasmussen, S.B., Kordon, E., Callahan, R., and Smith, G.H. (2001). Evidence for the transforming activity of a truncated Int6 gene, in vitro. *Oncogene 20*, 5291–5301.

Raught, B. and Gingras A.-C. (2007). Signaling to translation initiation. In Mathews, M.B., Sonenberg, N., and Hershey, J.W.B. (eds.) *Translational Control in Biology and Medicine*. New York: Cold Spring Harbor Laboratory Press, pp. 369–400.

Raught, B., Gingras, A.C., Gygi, S.P., et al. (2000). Serum-stimulated, rapamycin-sensitive phosphorylation sites in the eukaryotic translation initiation factor 4GI. *EMBO J 19*, 434–444.

Renner, O., Fominaya, J., Alonso, S., et al. (2007). Mst1, RanBP2 and eIF4G are new markers for *in vivo* PI3K activation in murine and human prostate. *Carcinogenesis 28*, 1418–1425.

Richter-Cook, N.J., Dever, T.E., Hensold, J.O., and Merrick, W.C. (1998). Purification and characterization of a new eukaryotic protein translation factor – eukaryotic initiation factor 4H. *J Biol Chem 273*, 7579–7587.

Robert, F. and Pelletier, J. (2009). Translation initiation: a critical signalling node in cancer. *Expert Opin Ther Targets 13*, 1279–1293.

Rosenwald, I.B. (2004). The role of translation in neoplastic transformation from a pathologist's point of view. *Oncogene 23*, 3230–3247.

Rosenwald, I.B., Wang, S.T., Savas, L., Woda, B., and Pullman, J. (2003). Expression of translation initiation factor eIF-2α is increased in benign and malignant melanocytic and colonic epithelial neoplasms. *Cancer 98*, 1080–1088.

Rousseau, D., Gingras, A.C., Pause, A., and Sonenberg, N. (1996a). The eIF4E-binding proteins 1 and 2 are negative regulators of cell growth. *Oncogene 13*, 2415–2420.

Rousseau, D., Kaspar, R., Rosenwald, I., Gehrke, L., and Sonenberg, N. (1996b). Translation initiation of ornithine decarboxylase and nucleocytoplasmic transport of cyclin D1 mRNA are increased in cells overexpressing eukaryotic initiation factor 4E. *Proc Natl Acad Sci USA 93*, 1065–1070.

Rowinsky, E.K. (2004). Targeting the molecular target of rapamycin (mTOR). *Curr Opin Oncol 16*, 564–575.

Ruggero, D. (2009). The role of Myc-induced protein synthesis in cancer. *Cancer Res 69*, 8839–8843.

Ruggero, D., Montanaro, L., Ma, L., et al. (2004). The translation factor eIF-4E promotes tumor formation and cooperates with c-Myc in lymphomagenesis. *Nat Med 10*, 484–486.

Ruggero, D. and Sonenberg, N. (2005). The Akt of translational control. *Oncogene 24*, 7426–7434.

Santhanam, A.N., Bindewald, E., Rajasekhar, V.K., et al. (2009). Role of 3′UTRs in the translation of mRNAs regulated by oncogenic eIF4E: a computational inference. *PLoS One 4*, e4868.

Sanvito, F., Vivoli, F., Gambini, S., et al. (2000). Expression of a highly conserved protein, p27BBP, during the progression of human colorectal cancer. *Cancer Res 60*, 510–516.

Schalm, S.S., Fingar, D.C., Sabatini, D.M., and Blenis, J. (2003). TOS motif-mediated raptor binding regulates 4E-BP1 multisite phosphorylation and function. *Curr Biol 13*, 797–806.

Schewe, D.M. and Aguirre-Ghiso, J.A. (2009). Inhibition of eIF2alpha dephosphorylation maximizes bortezomib efficiency and eliminates quiescent multiple myeloma cells surviving proteasome inhibitor therapy. *Cancer Res 69*, 1545–1552.

Schneider, R. and Sonenberg N. (2007). Translational control in cancer development and progression. In Mathews, M.B., Sonenberg, N., and Hershey, J.W.B. (eds.) *Translational Control in Biology and Medicine*. New York: Cold Spring Harbor Laboratory Press, pp. 401–431.

Schneider, R.J. and Mohr, I. (2003). Translation initiation and viral tricks. *Trends Biochem Sci 28*, 130–136.

Shahbazian, D., Parsyan, A., Petroulakis, E., Hershey, J., and Sonenberg, N. (2010a). eIF4B controls survival and proliferation and is regulated by proto-oncogenic signaling pathways. *Cell Cycle 9*, 4106–4109.

Shahbazian, D., Parsyan, A., Petroulakis, E., et al. (2010b). Control of cell survival and proliferation by mammalian eukaryotic initiation factor 4B. *Mol Cell Biol 30*, 1478–1485.

Shahbazian, D., Roux, P.P., Mieulet, V., et al. (2006). The mTOR/PI3K and MAPK pathways converge on eIF4B to control its phosphorylation and activity. *EMBO J 25*, 2781–2791.

Shao, J.Y., Evers, B.M., and Sheng, H.M. (2004). Roles of phosphatidylinositol 3′-kinase and mammalian target of rapamycin/ p70 ribosomal protein S6 kinase in K-Ras-mediated transformation of intestinal epithelial cells. *Cancer Res 64*, 229–235.

She, Q.B., Halilovic, E., Ye, Q., et al. (2010). 4E-BP1 is a key effector of the oncogenic activation of the AKT and ERK signaling pathways that integrates their function in tumors. *Cancer Cell 18*, 39–51.

Shi, Y., Sharma, A., Wu, H., Lichtenstein, A., and Gera, J. (2005). Cyclin D1 and c-myc internal ribosome entry site (IRES)-dependent translation is regulated by AKT activity and enhanced by rapamycin through a p38 MAPK-and ERK-dependent pathway. *J Biol Chem 280*, 10964–10973.

Shuda, M., Kondoh, N., Tanka, K., Ryo, A., et al. (2000). Enhanced expression of translation factor mRNAs in hepatocellular carcinoma. *Anticancer Res 20*, 2489–2494.

Silvera, D., Arju, R., Darvishian, F., Levine, P.H., et al. (2009). Essential role for eIF4GI overexpression in the pathogenesis of inflammatory breast cancer. *Nat Cell Biol 11*, 903–908.

Silvera, D., Formenti, S.C., and Schneider, R.J. (2010). Translational control in cancer. *Nat Rev Cancer 10*, 254–266.

Silvera, D. and Schneider, R.J. (2009). Inflammatory breast cancer cells are constitutively adapted to hypoxia. *Cell Cycle 8*, 3091–3096.

Sonenberg, N. and Dever, T.E. (2003). Eukaryotic translation initiation factors and regulators. *Curr Opin Struct Biol 13*, 56–63.

Sonenberg, N. and Hinnebusch, A.G. (2007). New modes of translational control in development, behavior, and disease. *Mol Cell 28*, 721–729.

Sonenberg, N. and Hinnebusch, A.G. (2009). Regulation of translation initiation in eukaryotes: mechanisms and biological targets. *Cell 136*, 731–745.

Soni, A., Akcakanat, A., Singh, G., et al. (2008). eIF4E knockdown decreases breast cancer cell growth without activating Akt signaling. *Mol Cancer Ther 7*, 1782–1788.

Sparks, C.A. and Guertin, D.A. (2010). Targeting mTOR: prospects for mTOR complex 2 inhibitors in cancer therapy. *Oncogene 29*, 3733–3744.

Spriggs, K.A., Bushell, M., and Willis, A.E. (2010). Translational regulation of gene expression during conditions of cell stress. *Mol Cell 40*, 228–237.

Spriggs, K.A., Stoneley, M., Bushell, M., and Willis, A.E. (2008). Re-programming of translation following cell stress allows IRES-mediated translation to predominate. *Biol Cell 100*, 27–38.

Stoneley, M. and Willis, A.E. (2004). Cellular internal ribosome entry segments: structures, trans-acting factors and regulation of gene expression. *Oncogene 23*, 3200–3207.

Suzuki, C., Garces, R.G., Edmonds, K.A., et al. (2008). PDCD4 inhibits translation initiation by binding to eIF4A using both its MA3 domains. *Proc Natl Acad Sci USA 105*, 3274–3279.

Svitkin, Y.V., Herdy, B., Costa-Mattioli, M., et al. (2005). Eukaryotic translation initiation factor 4E availability controls the switch between cap-dependent and internal ribosomal entry site-mediated translation. *Mol Cell Biol 25*, 10556–10565.

Tamburini, J., Green, A.S., Bardet, V., et al. (2009). Protein synthesis is resistant to rapamycin and constitutes a promising therapeutic target in acute myeloid leukemia. *Blood 114*, 1618–1627.

Tan, A., Bitterman, P., Sonenberg, N., Peterson, M., and Polunovsky, V. (2000). Inhibition of Myc-dependent apoptosis by eukaryotic translation initiation factor 4E requires cyclin D1. *Oncogene 19*, 1437–1447.

Teachey, D.T., Obzut, D.A., Cooperman, J., et al. (2006). The mTOR inhibitor CCI-779 induces apoptosis and inhibits growth in preclinical models of primary adult human ALL. *Blood 107*, 1149–1155.

Tee, A.R. and Blenis, J. (2005). mTOR, translational control and human disease. *Semin Cell Dev Biol 16*, 29–37.

Teleman, A.A., Chen, Y.W., and Cohen, S.M. (2005). 4E-BP functions as a metabolic brake used under stress conditions but not during normal growth. *Genes Dev 19*, 1844–1848.

Thoreen, C.C., Kang, S.A., Chang, J.W., et al. (2009b). An ATP-competitive mammalian target of rapamycin inhibitor reveals rapamycin-resistant functions of mTORC1. *J Biol Chem 284*, 8023–8032.

Topisirovic, I., Guzman, M.L., McConnell, M.J., et al. (2003). Aberrant eukaryotic translation initiation factor 4E-dependent mRNA transport impedes hematopoietic differentiation and contributes to leukemogenesis. *Mol Cell Biol 23*, 8992–9002.

Topisirovic, I., Ruiz-Gutierrez, M., and Borden, K.L.B. (2004). Phosphorylation of the eukaryotic translation initiation factor eIF4E contributes to its transformation and mRNA transport activities. *Cancer Res 64*, 8639–8642.

Topisirovic, I., Siddiqui, N., and Borden, K.L. (2009a). The eukaryotic translation initiation factor 4E (eIF4E) and HuR RNA operons collaboratively regulate the expression of survival and proliferative genes. *Cell Cycle 8*, 960–961.

Topisirovic, I., Siddiqui, N., Lapointe, V.L., et al. (2009b). Molecular dissection of the eukaryotic initiation factor 4E (eIF4E) export-competent RNP. *EMBO J 28*, 1087–1098.

van Gorp, A.G., van der Vos, K.E., Brenkman, A.B., et al. (2009). AGC kinases regulate phosphorylation and activation of eukaryotic translation initiation factor 4B. *Oncogene 28*, 95–106.

van Riggelen, J., Yetil, A., and Felsher, D.W. (2010). MYC as a regulator of ribosome biogenesis and protein synthesis. *Nat Rev Cancer 10*, 301–309.

Wang, R., Geng, J., Wang, J.H., et al. (2009). Overexpression of eukaryotic initiation factor 4E (eIF4E) and its clinical significance in lung adenocarcinoma. *Lung Cancer 66*, 237–244.

Wang, X. and Sun, S.Y. (2009). Enhancing mTOR-targeted cancer therapy. *Expert Opin Ther Targets 13*, 1193–1203.

Wang, X., Wei, Z., Gao, F., et al. (2008). Expression and prognostic significance of PDCD4 in human epithelial ovarian carcinoma. *Anticancer Res 28*, 2991–2996.

Watkins, S.J. and Norbury, C.J. (2002). Translation initiation and its deregulation during tumorigenesis. *Br J Cancer 86*, 1023–1027.

Wei, Z.T., Zhang, X., Wang, X.Y., et al. (2009). PDCD4 inhibits the malignant phenotype of ovarian cancer cells. *Cancer Sci 100*, 1408–1413.

Wendel, H.G., De Stanchina, E., Fridman, J.S., et al. (2004). Survival signalling by Akt and eIF4E in oncogenesis and cancer therapy. *Nature 428*, 332–337.

Wilker, E.W., van Vugt, M.A., Artim, S.A., et al. (2007). 14-3-3sigma controls mitotic translation to facilitate cytokinesis. *Nature 446*, 329–332.

Woodard, J., Sassano, A., Hay, N., and Platanias, L.C. (2008). Statin-dependent suppression of the Akt/mammalian target of rapamycin signaling cascade and programmed cell death 4 up-regulation in renal cell carcinoma. *Clin Cancer Res 14*, 4640–4649.

Wouters, B.G. and Koritzinsky, M. (2008). Hypoxia signalling through mTOR and the unfolded protein response in cancer. *Nat Rev Cancer 8*, 851–864.

Wu, C., Wangpaichitr, M., Feun, L., et al. (2005). Overcoming cisplatin resistance by mTOR inhibitor in lung cancer. *Mol Cancer 4*, 25.

Yang, H.S., Jansen, A.P., Komar, A.A., et al. (2003). The transformation suppressor Pdcd4 is a novel eukaryotic translation initiation factor 4A binding protein that inhibits translation. *Mol Cell Biol 23*, 26–37.

Yin, J.Y., Dong, Z., Liu, Z.Q., and Zhang, J.T. (2010). Translational control gone awry: a new mechanism of tumorigenesis and novel targets of cancer treatments. *Biosci Rep 31*, 1–15.

Yoon, C.H., Lee, E.S., Lim, D.S., and Bae, Y.S. (2009). PKR, a p53 target gene, plays a crucial role in the tumor-suppressor function of p53. *Proc Natl Acad Sci USA 106*, 7852–7857.

Yu, D., Scott, C., Jia, W.W., et al. (2006). Targeting and killing of prostate cancer cells using lentiviral constructs containing a sequence recognized by translation factor eIF4E and a prostate-specific promoter. *Cancer Gene Ther 13*, 32–43.

Yu, K., Toral-Barza, L., Shi, C., et al. (2009). Biochemical, cellular, and *in vivo* activity of novel ATP-competitive and selective inhibitors of the mammalian target of rapamycin. *Cancer Res 69*, 6232–6240.

Zhang, L., Pan, X., and Hershey, J.W. (2007). Individual overexpression of five subunits of human translation initiation factor eIF3 promotes malignant transformation of immortal fibroblast cells. *J Biol Chem 282*, 5790–5800.

Zhang, L., Smit-McBride, Z., Pan, X., Rheinhardt, J., and Hershey, J.W. (2008a). An oncogenic role for the phosphorylated h-subunit of human translation initiation factor eIF3. *J Biol Chem 283*, 24047–24060.

Zhang, M., Fu, W., Prabhu, S., et al. (2008b). Inhibition of polysome assembly enhances imatinib activity against chronic myelogenous leukemia and overcomes imatinib resistance. *Mol Cell Biol 28*, 6496–6509.

Zhu, N., Gu, L., Findley, H.W., and Zhou, M. (2005). Transcriptional repression of the eukaryotic initiation factor 4E gene by wild type p53. *Biochem Biophys Res Commun 335*, 1272–1279.

Zimmer, S.G., DeBenedetti, A., and Graff, J.R. (2000). Translational control of malignancy: the mRNA cap-binding protein, eIF-4E, as a central regulator of tumor formation, growth, invasion and metastasis. *Anticancer Res 20*, 1343–1351.

Zoncu, R., Efeyan, A., and Sabatini, D.M. (2011). mTOR: from growth signal integration to cancer, diabetes and ageing. *Nat Rev Mol Cell Biol 12*, 21–35.

Chapter 7

Genomic instability and carcinogenesis

Mark E. Burkard and Prasad V. Jallepalli

Introduction

The evolution of multicellular organisms requires symbiosis among cells, which cooperate to support organismal development and tissue homeostasis. These tasks require intercellular signaling and regulated cell proliferation and migration. Nevertheless, the program for these functions is encoded within the genome of each individual cell. Thus the maintenance of genome integrity requires surveillance, repair, and if necessary cell death to ensure that this program is not rewired to give rise to malignancy. Such rewiring requires multiple genetic changes, each of which happens at a low frequency under normal conditions. However, environmental factors (e.g., mutagenic chemicals and radiation) or intrinsic defects in genome integrity mechanisms (a "mutator" phenotype) can increase the probability of acquiring the necessary set of mutations. Indeed, considerable evidence indicates that most if not all tumors are much more genetically unstable than normal cells. While this instability provides an evolutionary advantage to cancer cells, it may also create novel liabilities that can be exploited therapeutically. Here we review the current state of knowledge about genetic alterations in cancer, including types and mechanisms of genomic instability and how they elicit oncogenesis.

Genetic alterations in cancer

Historical perspective

A revolution in chemistry emerged in nineteenth-century Germany as organic chemistry emerged in concert with increased availability of coal tar, a ready source of organic heterocycles. As a result, chemists were able to produce synthetic dyes for clothing that could be manufactured rapidly, in great quantity, and without the usual requirements for agricultural raw materials. Astute observers such as Walther Flemming noted that certain such dyes could stain vital stuctures in animal cells, visible by light microscopy. In particular, basophilic dyes labelled a chromatin. This revealed that cells, in an elaborate process, partitioned chromatin accurately into daughter cells. Flemming published his findings in 1882 and termed the division process mitosis (Figure 7.1) [1].

This set the stage for the work of D. Von Hansemann who utilized vital dyes to study cell division in human cancer. This seminal work published in the late nineteenth century demonstrated that cancer cells engage in aberrant mitoses, including multipolar divisions [2]. Such divisions result in unequal segregation of chromosomes, but the cause or effect of this remained obscure, since chromatin was not known to be the heritable genetic material (Figure 7.2a).

In the early twentieth century, Theodor Boveri and Walter Sutton independently came to the conclusion that chromatin held the heritable units – Mendel's genes [3]. Moreover, Boveri observed the effects of a random distribution of chromosomes in double-fertilized sea urchin eggs, resulting in distinct fates of daughter cells (Figure 7.2b) [4]. This led to his idea that the mis-segregation of chromosomes could lead to aberrant cell phenotypes and actually cause cancer [3]. The weight of evidence accumulated over the past century has reinforced this view [3, 5]. The accumulated evidence includes vast databases of genetic alterations, their oncogenic functions, and examples of oncogenesis via carcinogens or mutant genes involved in maintaining genomic integrity. Thus whereas many diseases are influenced by genetics, cancer is fundamentally a disease of altered genetics.

The scales of instability

To describe the genetics of cancer, we first survey the alterations harbored by cancer cells. These are

---

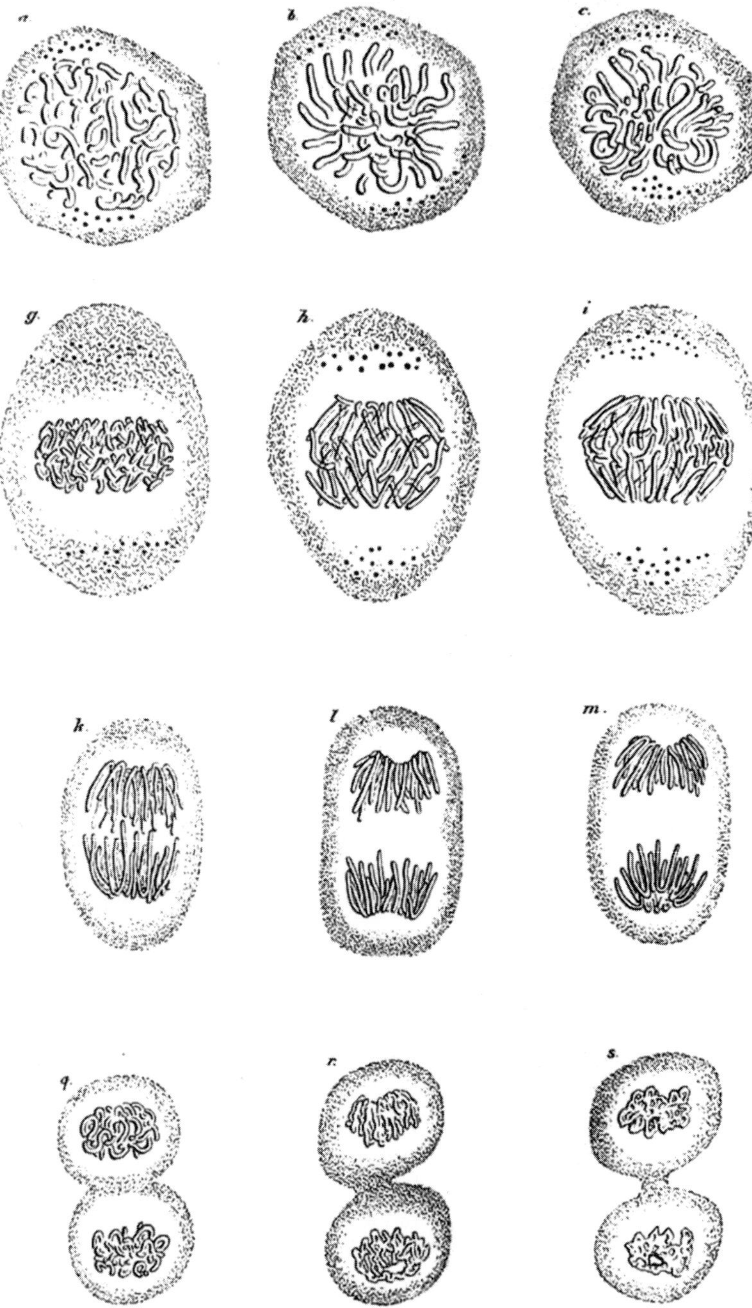

Figure 7.1 Original drawings by Walther Flemming of condensing chromatin, followed by bipolar alignment and equal partitioning into incipient daughter cells. This is a process he termed mitosis (from Greek *mitos*, "thread").

identified by comparing genomes of cancer and normal human cells. When possible, the normal genome is obtained from the same individual with a matched sample. Cells obtained from blood or normal tissue are assumed to have essentially the inherited genetic material, and are termed *germline*. Differences in genetic material in cancer cells are said to be *somatic* – alterations that are acquired in individual cells over the life of the organism, but are not transmitted to offspring.

The human genome is vast, making it especially challenging to develop a complete catalog of somatic alterations in cancer. A complete diploid human genome is comprised of two sets of 23 chromosomes

(a)

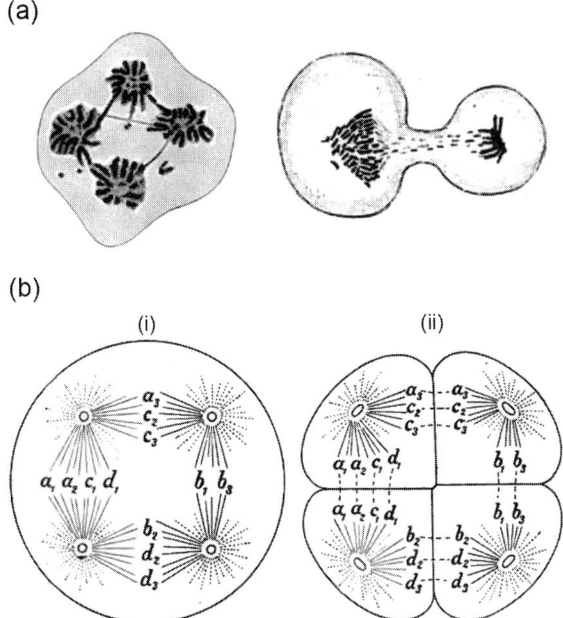

(b)

Figure 7.2 Original drawings of Von Hanseman and Boveri. (a) Multipolar and unequal segregation of chromosomes during mitosis of human cancer cells [2]. (b) An analysis of unequal chromosomal distribution due to multipolar mitosis, resulting in random distribution of chromosomes into multiple daughters [4]. Boveri observed that unequal distributions resulted in distinct cell fates in embryos and hypothesized that such unequal partitioning can lead to genesis of cancer (oncogenesis).

with over 3 billion base pairs of DNA and encodes over 20,000 proteins in addition to multiple RNAs, and non-coding DNA with regulatory or unknown functions. Cancer cells harbor abnormalities on all scales from mutation of a single base pair, to gross alterations in chromosome number, and changes on all scales can be important in oncogenesis (Figure 7.3).

Making the problem of cataloging genetic alterations in cancer even more challenging, there is heterogeneity among individual cells of a given cancer. Because of the scale of the problem, multiple techniques have been used to characterize cancers – each with differences in resolution, scale, throughput, and ability to detect differences among individual cells or an ensemble average of a population of tumor cells.

Mutation, insertion, and deletion

With the increasing availability and higher throughput of DNA sequencing technology, the role of point mutations in cancer is becoming increasingly recognized. An early example is the point mutation in H-Ras resulting in G12V, which is oncogenic [14]. Subsequent analyses have identified oncogenic point mutations in multiple genes including KRAS, EGFR, p53, MYC, PI3KCA, and others [5].

Box 7.1 Methods to detect instability

The genomic analysis of somatic alterations in cancer requires appropriate tools for characterization. Advances in technology have improved the understanding of cancer genomics, but also have revealed that the complexity is tremendous, obscuring the critical genes involved in oncogenesis. Here we describe current technologies available for identifying genetic alterations in cancer.

Karyotyping, banding, and SKY
The oldest tools to detect genetic instability in cancer involved staining condensed chromosomes in metaphase cells, and analysis via light microscopy. Although cytogenetics in humans lagged behind that for other organisms, by 1956 the human karyotype was finally confirmed as diploid, containing 46 chromosomes (Figure 7.4a) [6]. This established the reference genome for comparative analyses of cancer. Additional advances provided higher resolution maps of the genome including banding techniques devised throughout the 1970s [7, 8]. In cancers, this karyotype is frequently aberrant and a large number of such somatic alterations have been observed and cataloged in the Mitelman database [9].

More recently, molecular probes that bind to specific chromosome regions have been linked to fluorophores to allow detection of specific DNA sequences in the genome [10]. Fluorescent *in situ* hybridization (FISH) allows specific regions of chromosomes to be identified and enumerated in a reliable manner. The use of multiple probes with different fluorophores allows rapid identification of locus-specific translocations, amplifications, and deletions. For example, chronic myelogenous leukemia (CML) is defined by translocation between BCR and Abl kinase. Although this usually can be identified by karyotyping of metaphase cells, some cases of CML do not harbor the classic translocation of chromosomes 9 and 22 (abbreviated t(9;22) and known as the Philadelphia chromosome), but nevertheless generate Bcr-Abl fusion gene. Additionally, FISH is useful for identifying amplifications important for

Box 7.1 (*cont.*)

identifying cancer subtypes and is in routine use for Her2 (erbB2) amplifications in breast cancer. Typically, for this application, the ratio of Her2 signals to centrosome signals is averaged for a population of cells and Her2 positive disease is defined as that with a ratio 2.0 or greater.

Multicolor spectral karyotyping (SKY) is a genome-wide extension of FISH that uses numerous probes to all regions of the chromosomes, with those for any given chromosome linked to a unique combination of fluorophores (Figure 7.4b). After image acquisition, signals are deconvoluted and chromosome-specific signals can be readily identified. In cancer cells, duplications and translocations are readily apparent and the donors of altered regions of chromosomes are readily identifiable. Thus this method allows rapid detection of the varied chromosome abnormalities in cancer cells. For karyotyping and SKY, live cells must be obtained, to allow cells to be arrested in metaphase when chromosomes are condensed; FISH can be performed on interphase cells in fixed samples.

Sequencing

Ironically, small-scale alterations were discovered and linked to genetically transmitted diseases well before the discovery of DNA's double-helical structure. As early as 1949, Linus Pauling established sickle cell anemia as a disease of mutant hemoglobin by electrophoresis. In cancer, the development of DNA cloning and sequencing methods has allowed detection of oncogenic point mutations and small gene deletions and insertions. It had long been known that certain chemicals and radiation could impact the fidelity of DNA replication in bacteria, and moreover these same insults were oncogenic in animal models and humans. A logical extension of this idea is that errors in DNA replication can be oncogenic, as put forth by Loeb [11]. With the advent of terminator sequencing of DNA in the 1970s [12, 13] it became feasible to indentify oncogenic mutations and abnormalities detected or suspected through screening or candidate approaches. Screens identified carcinogenic mutations, such as that of Ras [14].

Subsequent refinements in sequencing technology have improved throughput. This has allowed complete sequencing of the human genome, completed over the better part of a decade in the Human Genome Project. Emerging technologies (collectively known as next-generation sequencing) are making a quantum leap in the sequencing throughput, allowing an entire human genome to be sequenced in less than a day. Multiple research groups are poised to use this technology to catalog small-scale somatic mutations in cancer, a molecular companion to the Mitelman chromosome database. This approach will nonetheless be challenging because of difficulty assembling sequences from genomes with altered chromosomes, differences among individual cancer cells in a tumor, and difficulty distinguishing a small number of oncogenic mutations (*driver mutations*) among a large number of adventitious mutations without functional significance (*passenger mutations*).

CGH and digital karyotypes

Molecular analysis of the entire genome has been increasing in importance. One important tool is comparative genomic hybridization (CGH). Originally, this method was conceived to map the DNA of cancer cells onto a reference set of chromosomes from a standard diploid cell [15]. More recent advances of technology incorporate arrayed probes on a DNA, which are hybridized to fluorochrome-labelled DNA fragments from cancer cells and reference DNA [16]. The ratio of intensity from the experimental cells to reference cells is indicative of the relative amplification of regions of DNA. Although this provides information about the copy number of specific DNA fragments, this does not provide structural information about the translocations and duplications that alter copy number of particular regions of DNA. For example, Figure 7.5 compares a karyotype pictogram of Hela cells (human cervical cancer) derived from SKY (Figure 7.5a) with copy number alterations detected by array CGH (Figure 7.5b). Although the latter detects amplifications in chromosomes 1 and 3, for example, it cannot identify the translocation of chromosome 3 to 1 that is, in part, responsible for this amplification. Although CGH is limited in structural detail, it is readily scaled to a large number of samples and provides a higher resolution than SKY, and thus remains of significant utility in describing amplified and deleted portions of the genome in cancer.

Digital karyotyping is an unbiased approach that requires knowledge of a reference genome rather than a reference set of DNA probes. Using this approach, genomic DNA is fragmented with a restriction enzyme and 21-base pair tags are isolated and sequenced. Tags are sequenced and mapped to the genome and the density of observed tags correlates with copy number [17]. Despite the power of CGH and digital karyotyping to describe global alterations in DNA copy number, they are limited in resolving intermediate and small regions of copy number of alteration. However, recent increases in density array have allowed CGH resolution to reach nearly one kilobase. Thus although they do not provide structural information about chromosomal rearrangements, CGH and digital

Box 7.1 (*cont.*)

karyotypes can nevertheless provide important information about recurrent amplifications and deletions, important mechanisms of oncogene upregulation and deletion of tumor suppressors.

Other approaches

A number of other genomic approaches are available to determine genomic alterations in addition to karyotyping, sequencing, and CGH. These include restriction landmark genome scanning [18], restriction optical mapping [19], representation difference analysis [20], and newer methods such as paired-end sequence mapping. The last of these can detect large structural variations by mapping tags at either end of genomic fragments of size 1 to 10 kb and map these onto the genome. When fragments are identified from different chromosomes or from regions far apart on chromosomes, this is indicative of large-scale structural rearrangements.

Ultimately, a complete genome map will require assembly of information acquired from mapping (karyotype, SKY, restriction mapping), sequencing, and high-resolution copy number approaches. This will allow accurate assembly of the genomic structure down to base pair resolution. With recent advances in technology, this is potentially achievable within the next several years. To accomplish this, the Cancer Genome Atlas and the Sanger Institute have undertaken large-scale genomic efforts directed at using these approaches to catalog the manifold genetic alterations in thousands of independent cancers. This catalog will provide a wealth of data, but additional studies will be needed to understand the causes and consequences of the identified genetic alterations with detailed mechanistic studies.

Recent advances in technology are now making it possible to rapidly and inexpensively sequence entire human genomes [21–23]. These tools, when applied to large-scale sequencing of somatic alterations in human cancers will provide a tremendous amount of data about potential oncogenic mutations. However, challenges remain. (1) **SNPs**. Many alterations may be single-nucleotide polymorphisms (SNPs), differences among individuals rather than somatic alterations specific to tumors. These can largely be identified by concurrent sequencing of matched normal tissue. (2) **Passengers**. A majority of observed somatic mutations will be passengers that do not play a role in cancer oncogenesis. One way to eliminate some passenger alterations in protein-coding regions of DNA is to observe whether they are silent, or whether they alter the protein-coding sequence [24]. This approach does not eliminate all passengers, however; nor does it account for the possibility that silent mutations affect gene function on the mRNA level, such as splicing. (3) **Genome structure**. Sequencing information is local; although oversampling reads and computer-based assembly can provide additional information, accurate assembly of the entire genome requires larger scale mapping efforts. For example, it may be difficult to determine whether point mutations in adjacent reads are in *cis* or *trans* across two or more chromosomes in polyploid cells. (4) **Functional data**. Ultimately, the goal is to identify genes that provide cancer-specific phenotypes. There is some debate on whether brute-force sequencing is the optimal method to identify key genes, or whether other approaches, such as unbiased forward genetic screens are more parsimonious [25, 26]. In any case, a significant bottleneck in a genomic pipeline is functional analysis of a large number of potential oncogenes identified by large-scale sequencing efforts. Thus high-throughput functional analysis of gene mutations will be the next challenge for identifying novel driver mutations in cancer.

Large-scale sequencing efforts have provided additional information on point mutations, small insertions and deletions [24, 27–30], and have provided insight into the biology of cancer. Such mutations may have important oncogenic functions, and are termed drivers, or may be present but non-functional bystanders, termed passengers. Large-scale sequencing efforts do not directly assess the functional significance of identified mutations. However, concordance of similar mutants among many independently derived cancers provides indirect evidence that such alterations have functional significance. A recent analysis suggests that the number of genes altered in multiple uniquely derived cancers are surprisingly few [31]. As a result, these efforts have defined a large number of genes that are mutated with low frequency. Although some of these are likely oncogenic, detailed functional analyses will be required to differentiate the oncogenic driver mutations from passengers.

One conclusion drawn from accumulated sequencing studies is that cancers are more genetically heterogeneous than previously thought. Clinicians have long observed differences in biology among cancers of the same tumor site, but genetically, would this represent 10 or 1,000 different types? One sequencing effort analyzed over 13,000 coding regions of 11 breast

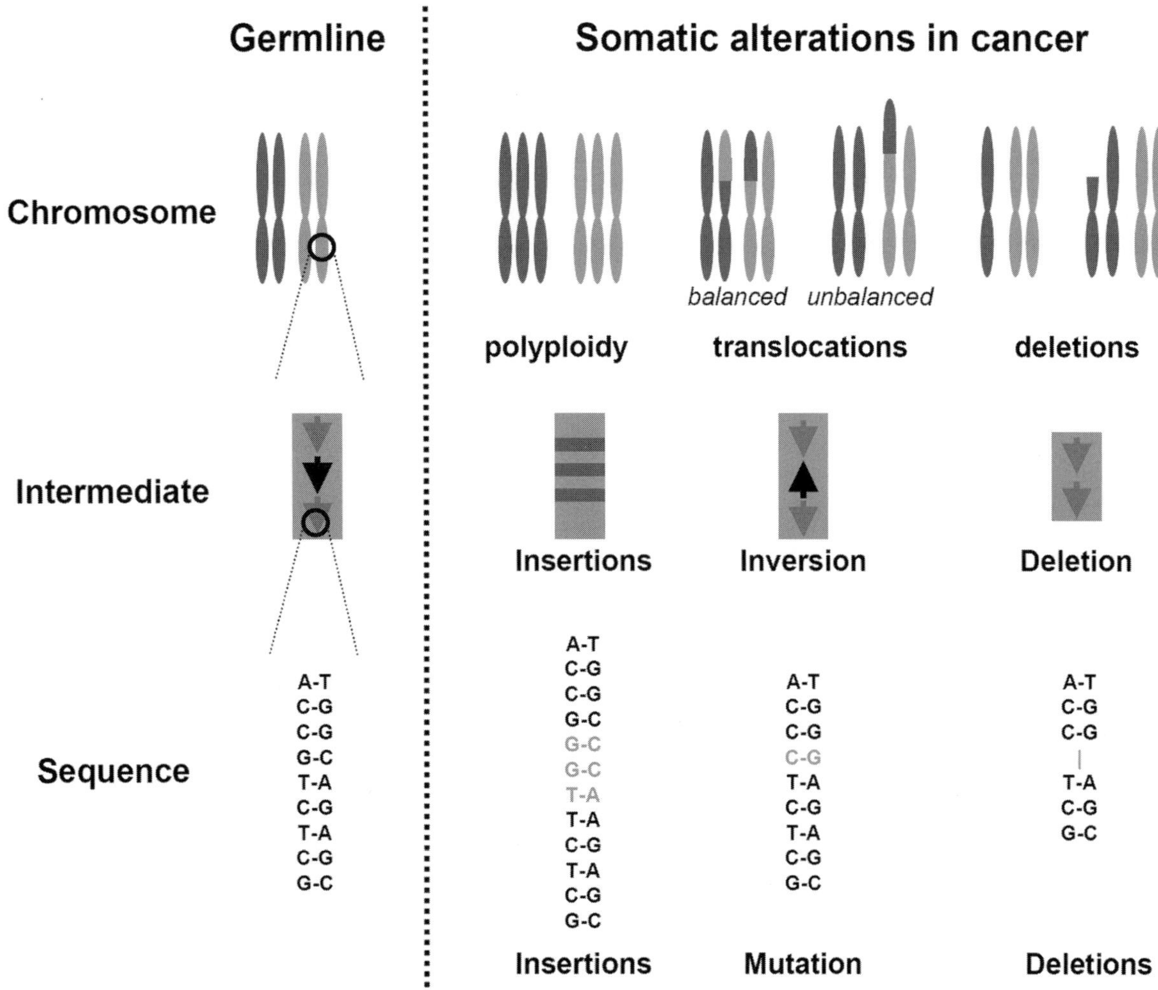

Figure 7.3 Somatic genetic alterations are observed in cancer cells at all scales. Chromosomes (top) are observable by light microscopy and demonstrate alterations in chromosome number and in gain, loss, or exchange of fragments. Intermediate-scale rearrangements (middle) include insertions, deletions, and inversions of chromosomal fragments from 1 kilobase to 1 megabase in length. On the scale of DNA sequence (bottom), insertions, deletions, and point mutations are observed. (A black and white version of this figure will appear in some formats. For the color version, please refer to the plate section.)

and 11 colorectal cancers and identified 189 genes with diverse functions that were mutated at significantly higher frequency than expected from random chance. Each tumor harbored, on average 90 mutant genes and no two cancers were alike [24]. Of these 90 genes, concordance among tumors suggests that perhaps 11 are oncogenic, whereas the remainder may represent bystander mutations. No two cancers had the same set of driver genes, suggesting a high degree of heterogeneity among cancers. Although this suggests the problem of cancer genetics is nearly insurmountable, the key to understanding cancers is to link individual drivers into common pathways [5].

Although some germline alterations, such as p53 in Li–Fraumeni syndrome, herald potential somatic events involved in oncogenesis, other gene alterations are observed only in the germline, such as BRCA1/2, or only in somatic cells (KRAS). It is unclear why some genetic alterations are exclusive to tumor or germline. If, for example, BRCA1 mutation predisposes women to breast cancer, would not somatic mutations in breast tissue also be likely to be an early event in *de novo* cancer development in women who do not carry BRCA1 mutations? Apparently, either somatic alterations in this gene do not occur, or loss of gene function must precede breast or ovarian

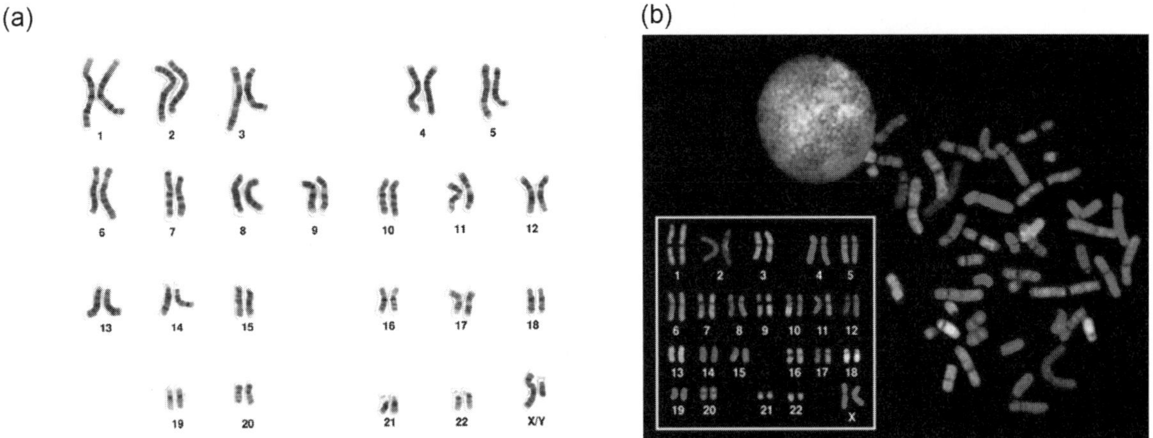

Figure 7.4 Techniques to identify chromosomes. (a) Normal male karyotype. (b) Spectral karyotype of a normal female cell. From NHGRI (www.genome.gov/glossary.cfm). (A black and white version of this figure will appear in some formats. For the color version, please refer to the plate section.)

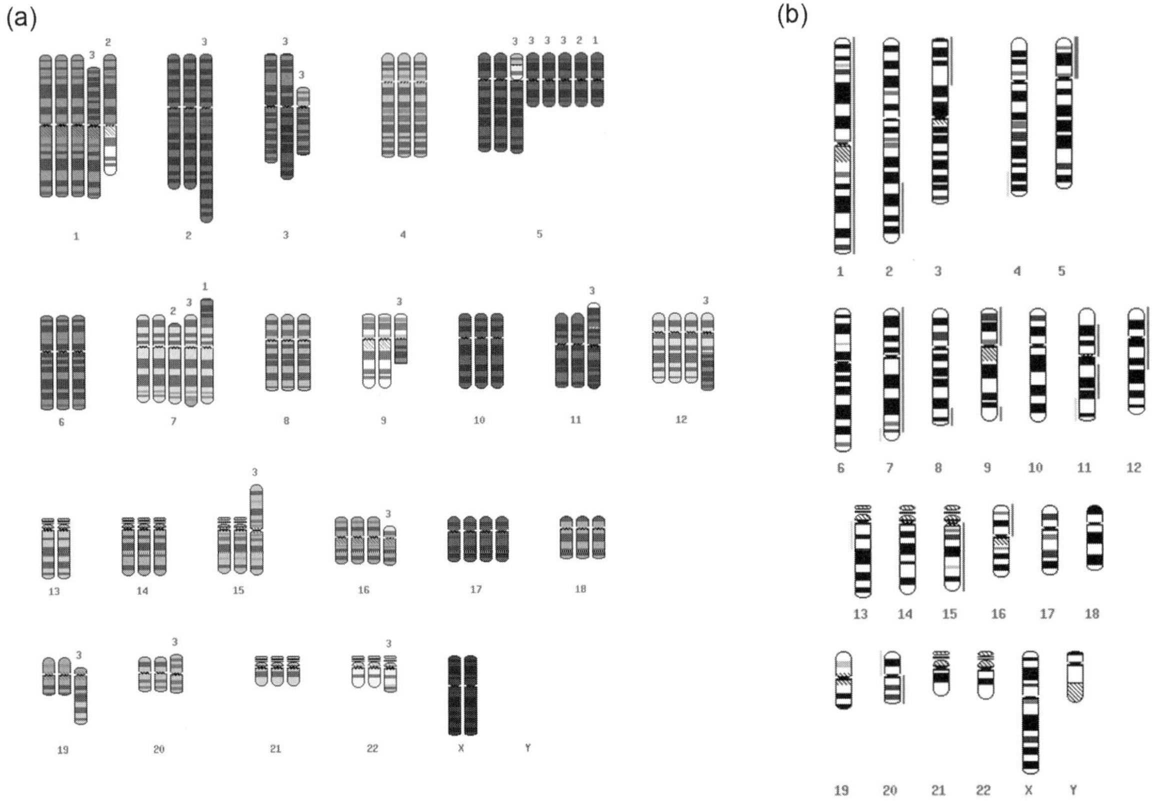

Figure 7.5 (a) Karyogram representing chromosomal aberrations in Hela cells (human cervical cancer) obtained from spectral karyotyping. This cell line is polyploid and harbors multiple translocations and duplicated regions. (b) CGH data on Hela cells reveals regions that have high (green) or low (red) copy number relative to a diploid cell line, but does not provide structural information. Generated by NCBI Cancer Chromosomes (www.ncbi.nlm.nih.gov/sites/entrez?db=cancerchromosomes). (A black and white version of this figure will appear in some formats. For the color version, please refer to the plate section.)

Table 7.1 Selected oncogenic fusions in cancer.

Translocation	Gene fusion	Disease
Sarcomas		
t(5;12)	ETV6-PDGFRB	Myeloproliferative disease
t(11;22)	EWS-FLI1	Ewing's sarcoma
t(12;21)	ETV6-CBFA	Acute lymphocytic leukemia
t(X;18)	SYT-SSX1/2	Synovial sarcoma
t(2;13) or t (1;13)	PAX3-FKHR or PAX7-FKHR	Alveolar rhabdomyosarcoma
t(2;16)	TLS-CHOP	Myxoid liposarcoma
T(9;22)	EWSR1-NR4A3	Soft tissue chondrosarcoma
t(7;17)	JAZF1-SUZ12	Endometrial stromal sarcoma
Blood dyscrasias		
t(9;22)	BCR-ABL	Chronic myelogenous leukemia
t(14;19) t(5;14) t(7;14)	IGH-BCL-3 IGH-IL-3 IGH-CDK6	Chronic lymphocytic leukemia
t(15;17)	RARα-PML	Acute promyelocytic anemia
t(8;21)	AML1-ETO	Acute myelocytic anemia
t(8;14) t(2;8) t(8;22)	IGH-MYC IGK-MYC IGL-MYC	Acute lymphocytic leukemia
t(9;14)	TCRδ-p16/ p19^{ARF}	Acute lymphocytic leukemia
t(7;19) t(1;14) t(7;9) t(11;14) or t(;11) t(7;11) t(14;21) t(10;14)	TCRβ-LYL1 TCRα-TAL1 TCRβ/TAL2 TCR/TCRβ-LMO1/2 TCRβ-HOX11 TCRα-BHLHB1 TCRδ-HOX11	Acute lymphocytic leukemia (T)
t(7;9)	TCRβ-TAL2	Acute lymphocytic leukemia
Lymphomas		
t(14;18)	IGH-BCL2	B-cell lymphoma
t(11;14)	IGH-CCDN1	Mantle cell lymphoma
t(8;14)	IGH-MYC	Burkitt lymphoma
t(11;18)	API2-MLT	MALT lymphoma
t(2;5)	NPM-ALK	Anaplastic large cell T lymphoma

development for tumorigenesis, consistent with findings in murine models [32].

Chromosomal rearrangements and aneuploidy

Large-scale genome rearrangements, observable by DNA stains and light microscopy are among the first alterations recognized in cancer. Over the past century, refinements in the technique, including Giemsa banding and spectral karyotyping (SKY), have allowed detailed mapping of cancer genomes. One of the most comprehensive collections of chromosomal abnormalities is known as the Mitelman database [9, 33]. As of November, 2008 this includes genomic analysis of over 50,000 independently derived cancers.

One common type of alteration evident by these techniques is translocation, where a segment of chromosome is conjoined with another (Figure 7.3). It has long been postulated that these rearrangements harbor oncogenic alterations [34]. By mapping chromosomal breaks, a total of 358 oncogenic gene fusions have been identified, mainly in leukemias and lymphomas (Table 7.1) [35]. In solid tumors, such as cancers of the breast, colon, ovary, and prostate, karyotypes are complex and more commonly vary among individual cancer cells. Moreover, cancer specimens are more difficult to obtain and are admixed with adjacent normal cells. Possibly for these reasons, the data on large-scale chromosome structures has yielded comparatively little information about oncogenic somatic alterations in cancer.

Abnormalities in chromosome number, known as aneuploidy, are also common in cancers. Cell lines derived from solid tumor malignancies frequently harbor ~69 chromosomes (near-triploid) and some have 96 chromosomes or more (tetraploid) [9, 36]. The large number of chromosomes in many cancers suggests that a major event occurred during oncogenic transformation such as failure of cell division, or cell fusion. Alternatively, cell division could be complete, but with gross mis-segregation. With the latter mechanism, both hyperdiploid and hypodiploid daughter cells would be generated, but hypodiploid cells could have a survival disadvantage.

Intermediate-scale rearrangements

Somatic chromosomal aberrations in cancer were observed in the late nineteenth century [2], and alterations in DNA sequence, predicted in 1974 [11] and observed with KRAS G13V in 1982 [14]. In contrast, the technology for rapid global detection of intermediate-scale alterations is just emerging. Emerging data suggests frequent structural alterations too small to visualize with light microscopy, yet too large to be detected in sequencing reactions. Typical approaches are laborious and low throughput, such as CGH coupled with fosmid paired-end mapping. One complication is that many such studies have also identified germline variations in the genome, making it imperative to check matched normal tissue to determine if rearrangements are somatic [37, 38]. For example, a recent study of the genome using a high-throughput paired-end mapping approach revealed numerous deletions, duplications, insertions, inversions, and complex rearrangements on a scale of 10^3 to 10^6 base pairs [39].

There are few studies of intermediate somatic alterations in cancer, and existing studies are rather small due to problems with throughput. One recent effort from the Wellcome Sanger Trust used paired-end sequencing and identified 103 somatic rearrangements mapped from two individuals with lung cancer [40]. Although some of the mapped somatic rearrangements led to the expression of novel transcripts, the functional significance of these alterations is unclear. However, this study establishes the precedent that intermediate-sized rearrangements, not identifiable by sequencing or microscopic (karyotype) methods, are common and generate fusion transcripts. In prostate and lung cancers, such fusion transcripts, not evident from karyotype analyses, were identified by molecular techniques and have been identified as oncogenes [41–43]. Additional fusions generated by intermediate-scale rearrangements may be important for oncogenesis but are difficult to identify by microscopy or direct sequencing. Refined techniques for detecting intermediate-scale rearrangements with high throughput will be important for establishing a complete catalog of molecular oncogenic somatic alterations in cancer.

Copy number changes on the intermediate scale can also be important in cancer development. For example, ~20% of breast cancers have amplification of ErbB2 (Her2) [44, 45]. These can be detected by CGH or FISH, without determining the particular structural rearrangements involved. Such alterations are both prognostic (i.e., correlate with clinical outcome) and predict response to Her2-targeted therapy [46].

Causes of genetic instability
The mutator hypothesis

An enduring hypothesis in oncology is that cancer cells have increased rates of mutation relative to normal cells [11, 47–49]. For our purposes, genetic mutability can be defined broadly to include genetic alterations on all scales such as point mutations and chromosomal instability. There are several lines of evidence supporting this mutator hypothesis. First, development of cancer requires multiple molecular alterations (see multi-hit hypothesis below), and with normal rates of mutation it would likely be difficult to accumulate the necessary genetic alterations. Second, a number of inherited cancer syndromes result from mutations of single genes that result in increased rates of mutation at the nucleotide level [50, 51] or chromosomal instability [52]. Third, most human cancer cells, when cultured, exhibit increased rates of mutation, whether at the nucleotide level or via chromosomal instability [49, 53]. Finally, premalignant lesions frequently harbor genetic alterations [54]. Together, these observations support the idea that an early event in the development of cancer is impaired function of pathways involved in maintaining genomic integrity.

Because it is difficult to precisely enumerate mutation rate in human tissues, it is difficult to prove that the mutator phenotype is responsible for oncogenesis in all cases. Moreover, the current state of technology makes it simpler to observe such mutability on chromosome-wide and sequence level, than intermediate-sized genomic alterations, presenting the possibility that an important class of gene alterations has not been adequately explored. However, the accumulated evidence broadly supports the idea that an increased rate of genetic alterations is common in cancer, although this mutability may be dynamic and the rate of mutability may fall within a broad spectrum.

Aneuploidy, CIN, and MIN

Seminal work in colon cancer has defined two general classes of genetic instability in this disease: chromosomal instability (CIN) and microsatellite instability (MIN). Virtually all colon cancers exhibit CIN or

MIN. Chromosomal instability cancers are characterized by evolving aneuploidy, whereas MIN cancers are generally near-diploid but have impaired mismatch repair. This results in a high mutation rate, most evident in *microsatellite* DNA, highly repetitive sequences of DNA are found throughout the genome. With errors in DNA repair, these regions become unstable and are highly heterogeneous. Of colorectal cancers, 85% are aneuploid (some or all with CIN) and 15% exhibit MIN.

Most MIN cancers occur in the context of hereditary non-polyposis colorectal cancer (HNPCC). This disease is characterized by dominant mutation in one or more mismatch repair genes, most commonly MLH1 and MSH2. Although microsatellites are a marker for this disease, the cancer predisposition in HNPCC is related to mutations in oncogenes and tumor suppressors. For example, HCT116 is a classic MIN colorectal cell line with a stable near-diploid karyotype, and carries a mutation in MLH1 [55]. Consistent with known defects in mismatch repair, this line harbors mutations in several oncogenes and tumor suppressors including PIK3CA, KRAS, CDKN2, and BRCA2 [56]. HNPCC is characterized by multiple solid-tumor malignancies including endometrial, ovarian, biliary cancer, and brain tumors, demonstrating that the oncogenic potential of this inherited mutator phenotype is not limited to colon cancer [57].

Most colorectal and other solid-tumor cancers are aneuploid, and contain supernumerary, often translocated chromosomes. It should be noted that aneuploidy is distinguished from CIN – whereas aneuploidy is defined to be a state of abnormal chromosome number at a point in time, CIN refers to continuing evolution of this aneuploidy in time. Although CIN is thought to be a major cause of aneuploidy – and a number of genes and mechanisms involved in CIN have been identified (Table 7.2 and below) – it is possible that aneuploidy is generated as a result of a single catastrophic event that occurs rarely. In this scenario, a failure in cell division or cell fusion results in a tetraploid cell that may exhibit additional chromosomal losses in divisions that immediately follow (much as Boveri proposed a century ago), leading to a stable aneuploid line.

It should be noted that CIN, by definition, does not require a change of chromosome number with each division, but only an elevated rate of chromosomal abnormalities. Thus it may be expected that each change in chromosome number can be followed by a wave of clonal expansion [58]. Selective pressure may enhance the appearance of stable aneuploidy since many of the derivative cells may harbor deleterious alterations. Only when a subsequent genetic alteration confers an additional selective advantage can a new "plateau" be reached. In support of this, it has been noted that cancer cell lines exhibiting CIN have high cell–cell karyotypic variability but maintain karyotypic stability over many generations [59, 60].

Box 7.2 Can tumors be aneuploid without CIN?

Some studies have identified similar cancer karyotypes among large numbers of tumor cells and in cancer cells sampled at disparate times. These findings have led to the idea that some cancers are aneuploid but do not exhibit CIN. Here we review two studies that have found little variation in CGH or FISH between sampled cells and demonstrate how study methodology and selective pressure can produce a metastable karyotype in the majority of cells.

In one study paired DCIS samples from 18 patients were studied by CGH at diagnosis and at the time of an ipsilateral recurrence [61]. Alterations were scored for entire chromosome arms and smaller genetic abnormalities were not scored. In all cases, aberrations in copy number were seen in the initial DCIS. Chromosome copy numbers from the initial tumor and recurrence were highly concordant and in three cases there was 100% concordance in the CGH abnormalities from the initial tumor and recurrence. Does this demonstrate aneuploidy without CIN? CGH detects the ensemble average of chromosomes from sampled tissue, and thus cannot detect rare chromosomal variations among cells, possibly underestimating CIN [58]. Moreover, some similarities between primary and recurrent DCIS are expected given they are derived from the same clone of origin. The findings do suggest, however, that either CIN is not severe in these DCIS samples or that selection tightly constrains the bounds of the aneuploid karyotype.

In a second study, three pericentromeric probes were used with multicolor FISH to detect aneuploidy of chromosomes 3, 7, and 17 in breast cancer samples [62]. Of 20 tumors, 3 appeared to be near-diploid, 3 had monosomy of chromosome 17 without other significant chromosome alterations, and 14 were aneuploid as defined

Box 7.2 *(cont.)*

by $\geq$15% of cells being polysomic. Chromosomal instability was determined by the fraction of cells that deviated from the modal number of FISH signals. Of the 14 aneuploid samples, nine exhibited high CIN, as evidenced by 40% of cells deviating from modal FISH signals. In five tumors, the fraction of cells deviating from modal FISH signals is less than 20%, and these are characterized as stable aneuploid tumors. In these five tumors, however, FISH signals identified two distinct clones and CIN was defined as the fraction of cells that deviated from both chromosomal values. Although a bimodal copy number mean was used to avoid artificial inflation of CIN values, this technique may have had the opposite effect. Moreover, the use of pericentromeric probes for only three chromosomes cannot detect alterations in the other 18 chromosomes or along chromosomal arms. Thus it is likely that this study underestimates the incidence of CIN. Finally, as we demonstrate below, 20% aneuploidy can be seen in the presence of a significant rate of CIN with moderate selective pressure.

In sum, although it is possible that some tumors are aneuploid but do not exhibit CIN, even studies cited as evidence of this phenomenon demonstrate variability in karyotypes in most aneuploid tumors; in the few exceptions, the nearly stable karyotypes could be explained by experimental technique, selective pressure, and lower rates of CIN. Thus it remains unclear whether aneuploidy can occur without CIN. Two observations support CIN as a major cause of aneuploidy: (1) the degree of aneuploidy is higher in invasive tumors than premalignant tumors [54]; (2) genetic alterations that cause aneuploidy are oncogenic, including mutations in hCDC4 [63] and BubR1 [64].

Recent evidence supports the idea of a conservative selective pressure for a stable aneuploid karyotype, as many chromosomal gains and losses impart a selective disadvantage [65]. To define the role of selective pressure in maintaining a stable karyotype, we established a model simulating division from a single cell, for given rates of CIN and selective pressure. This model demonstrated that when selective pressure dictates that aneuploid cells proliferate at even half the normal rate, the fraction of aneuploid cells remains small, and exceeds 50% only when the rate of CIN is greater than 0.1 chromosomal alterations per division (Figure 7.6a).

Because the rate of CIN is expected to be low, and most chromosomal aberrations provide a selective disadvantage [65], these findings support a model of punctuated equilibrium (Figure 7.6b). In this model, a single clone with a growth advantage proliferates and comes to comprise a large fraction of tumor cells. A subsequent advantageous mutation then occurs within this population that provides an additional selective advantage. Over time, the dual mutant comes to dominate the population. Such mutation followed by clonal expansion of advantaged clones explains the sequence of genetic alterations observed in adenomas to colon cancers and why early steps in oncogenesis require more time than later steps, such as metastases [5, 66, 67].

A recent study supports the view that selection maintains a metastable state in the face of CIN [68]. In this study, the karyotype of newly transformed cells was determined for a sample of the population over multiple generations. At any given time, copy number variation was observed in up to 30% of the cells, but individual clonal karyotypes were stable for up to 30 generations. This supports a model where selective pressure against most chromosomal aberrations creates a metastable aneuploid state.

Oncogenes and tumor suppressors

The precise genetic alterations required for transformation of normal cells to malignant has been of considerable interest since the development of molecular tools. Classically, genes involved in oncogenesis are categorized into oncogenes and tumor suppressors. Oncogenes are defined as genes that are hyperactivated through mutation or increased expression in cancer cells, and therefore exhibit a dominant gain of function relative to normal cells. Initially these were found to be cellular homologs of viral transforming genes, with a capacity to impart oncogenic phenotypes in premalignant cells. Classic examples include HRas, c-Myc, PIK3CA, which, when mutated, constitutively activate mitogenic signaling pathways. Typically oncogenes are dominant and a mutation in a single copy is sufficient to generate an oncogenic phenotype.

Tumor suppressors are genes that tend to check oncogenic transformation and their function is lost in the process of transformation. Loss can be accomplished through deletion of chromosomal segments (loss of heterozygosity) or by inactivating mutations. Because a loss of function is required, inactivation of both copies of the gene is generally required. Classic examples include pRB and p53. Some genes, such as CENP-E, have properties of both oncogenes and tumor suppressors, blurring the distinction [69].

Table 7.2 Genes involved in maintaining genomic integrity.

Genes	Type[a] (G/S/N)	Alteration	Syndrome	Cancers
Replication				
MSH2, MLH1	G	Mutation	Lynch (HNPCC)	Colon, endometrial
DNA damage				
p53	G,S	Mutation, Chr deletion	Li–Fraumeni	
CHK2	G	Mutation		Breast, colorectal
ATM	G	Mutation	Ataxia-teleangiectasia	
FANCA-FANCN	G	Mutation, deletion	Fanconi anemia	Leukemia
BLM	G	Mutation	Bloom syndrome	Leukemia
NBS1	G		Nijmegen breakage syndrome	Leukemia
BRCA1/2	G	Mutation, deletion		Breast, ovarian
Spindle checkpoint				
hCDC4	S	Mutation		Colon [63]
Bub1B	G	Mutation [52]	Mosaic variegated aneuploidy	Leukemia, Wilms
BubR1	S	Expression		Colon [64]
Mad2	S	Expression [91]		Cell line [92]
Securin [93]	N			
CENPE [69]	N			
Telomere maintenance				
TERT	N			
Centrosome maintenance				
Plk4	N			

[a] G = germline; S = somatic; N = alterations not yet observed in human cancer

Genomic integrity during the cell division cycle

A fundamental process for any living organism is cell division. This process is required to develop a complex multicellular organism from a single fertilized egg and, moreover, is required to regenerate tissues with limited lifespan, such as epithelial cells and mature blood cells. In order to maintain genomic integrity, the process of cell division is tightly regulated via a program that is conserved from unicellular organisms. The basic elements of this program consist of (1) duplication of genetic material (i.e., chromosomes) during the S-phase; (2) partitioning of this material and other critical cell components into daughter cells during the M-phase (mitosis). Genomic integrity requires that the S-phase and M-phase programs are unidirectional, executed sequentially (to ensure that chromosomes are duplicated completely prior to partitioning) and in alternating sequence (Figure 7.7), as consecutive S-phases would increase cell ploidy to 4N; consecutive M-phases would reduce ploidy.

Ploidy

The number of chromosome sets can be altered by failure of these regulatory functions of the cell cycle. Three events can lead to polyploid cells, frequently observed in cancer: cytokinesis failure, fusion of diploid cells, and endoreduplication (consecutive

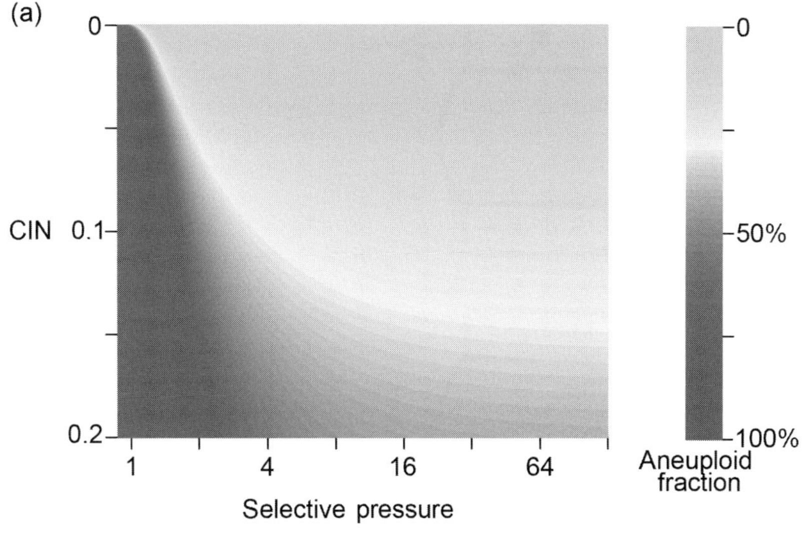

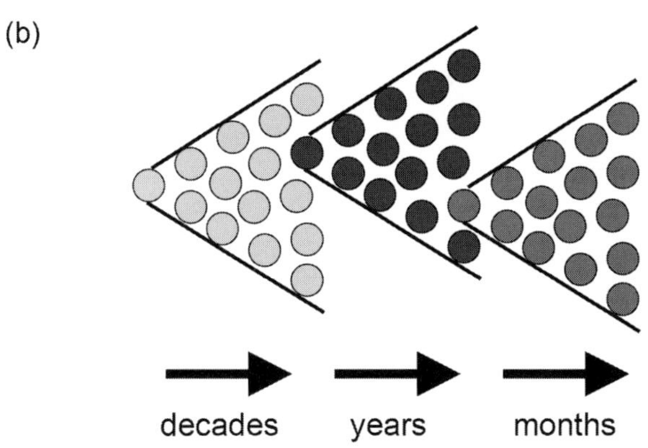

Figure 7.6 A model of CIN, aneuploidy, and selection. (a) Fraction of aneuploid cells among a population of cells that has a given probability of producing aneuploid progeny, given by CIN rate. For each division, a fraction (or probability) of generating aneuploid daughter cells were defined by a rate of CIN. Normal divisions generated two normal daughters; abnormal divisions generate two aneuploid daughters. Aneuploid cells continue to proliferate generating further aneuploid progeny, but at a fraction of the rate of normal cells, as defined by selective pressure (for example, for selection = 2, rate is ½). Modelling was continued for a total of 50 divisions (sufficient to reach steady state), and fraction of cells exhibiting aneuploidy was calculated. (b) A model for clonal expansion and selection of rare favorable aneuploid clones. Because of selection, a metastable karyotype can exist at various steps of oncogenesis, despite CIN. (A black and white version of this figure will appear in some formats. For the color version, please refer to the plate section.)

S-phases without a successful intervening M-phase – Figure 7.7) [70].

In murine models, failure of cytokinesis is oncogenic in p53 null cells [71]. Tetraploid yeast exhibit increased rate of chromosome loss and recombination [72], suggesting that cytokinesis failure is sufficient for inducing CIN [73]. A tetraploid cell has supernumerary centrosomes in addition to chromosomes, leading to multipolar mitoses and irregular segregation of chromosomes. Such multipolar mitoses would further create diversity of daughter cells, potentiating loss of genomic integrity by many processes. It has been proposed that a p53-mediated checkpoint limits proliferation of tetraploid cells [74], although tetraploid cells with intact p53 have been observed to re-enter the M-phase [75]. In sum, these observations suggest that tetraploid cells are frequent intermediates in human oncogenesis.

Translocations, chromosome loss, and mutation

Other cellular functions maintain genomic integrity at the level of individual chromosomes or smaller and these are frequently disrupted in cancer. Figure 7.8 shows the critical cellular functions required to maintain genomic integrity in dividing cells. First, DNA replication inherently is error prone, and adducts and strand breaks can additionally be caused by environmental stress (e.g., oxidative stress, ionizing radiation). As a result, cells are capable of detecting and responding to damage (Figure 7.8a). One type of damage that is particularly difficult to repair is a double-strand break (Figure 7.9a), which generates new DNA ends. Such breaks can be repaired by joining to other chromosomes, resulting in the fused chromosomes frequently observed in cancer cells.

The M-phase, or mitosis, is an exceptionally precarious time for genomic integrity. During this

105

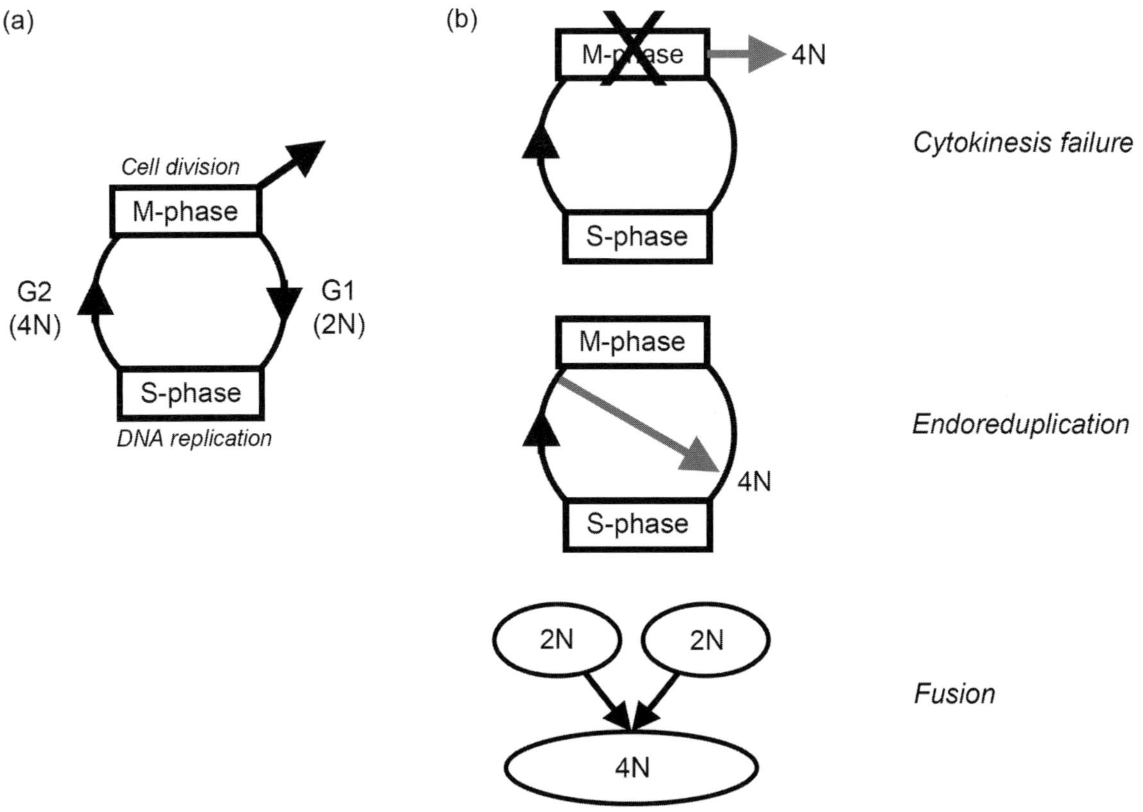

Figure 7.7 Regulation of the cell cycle protects cell ploidy (number of chromosome sets). (a) The normal cell cycle consists of S-phase, where DNA is replicated, and M-phase, where replicated DNA is segregated into daughters. During the G1 phase, ploidy is 2N (two sets of 23 chromosomes in human cells) and in G2 ploidy is 4N. (b) Cell ploidy can be increased by failure of M-phase cell division, endoreduplication, or fusion of G1 cells. (A black and white version of this figure will appear in some formats. For the color version, please refer to the plate section.)

time, chromosomes are condensed, attached to opposite poles of the mitotic spindle via microtubules, and ultimately segregated into incipient daughter cells, divided by a cell membrane. In order to facilitate precise segregation into daughters, cells have evolved a process known as the spindle assembly checkpoint (SAC – Figure 7.8b). The SAC requires assembly of a protein complex, known as a kinetochore, at the centrosome region of each chromosome, and attachment to the spindle poles via microtubules. Moreover, the SAC is not satisfied until tension is generated between adjacent kinetochores of a chromatid pair by attachment of these kinetochores to opposite spindle poles. In the case where both chromatids are attached to the same pole (syntelic attachment), no such force is generated and the SAC is not satisfied until corrected. Only when all chromosome pairs satisfy SAC requirements will cells proceed into anaphase, as accurate spindle formation presages accurate segregation of

chromosomes into daughters in anaphase (Figure 7.8b). Misregulation or abnormal expression of the components of this tightly regulated process can result in mis-segregation of chromosomes or improper resolution of chromosome attachments, resulting in lagging chromosomes and breaks in anaphase (Figure 7.9b). This leads to gain or loss of individual chromosomes or to fused chromosomes.

Telomere function is also thought to play an important role in maintaining chromosomal integrity [76–78]. Telomeres are protein-rich RNA structures that are linked to chromosome ends, and among other functions, protect adjacent coding regions of DNA and prevent non-homologous end joining between different chromosomes (Figure 7.8c). Telomeres are much smaller in cancer cells than in normal cells [79], suggesting that loss of telomere function can be oncogenic. Murine models have supported this view, since loss of a key enzyme, telomerase reverse

(a) DNA damage checkpoint

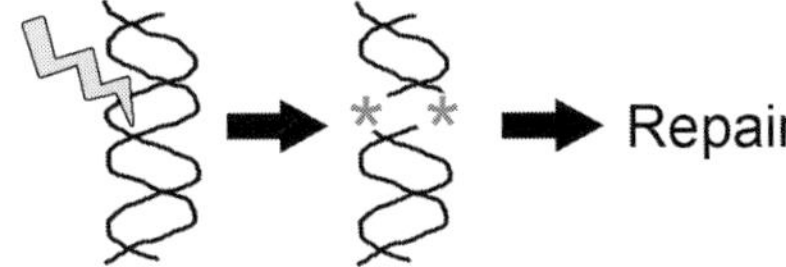

(b) Spindle assembly checkpoint

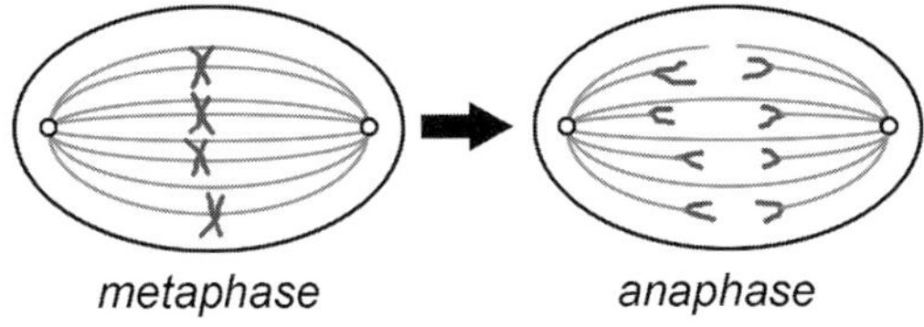

(c) Telomere integrity

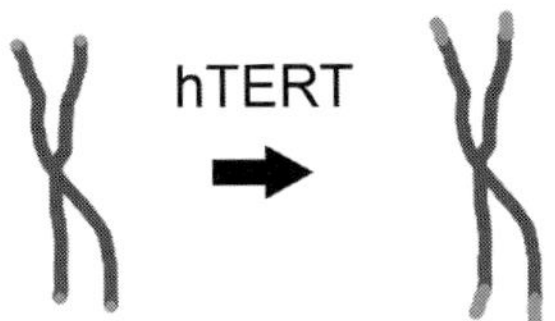

(d) Centrosome number

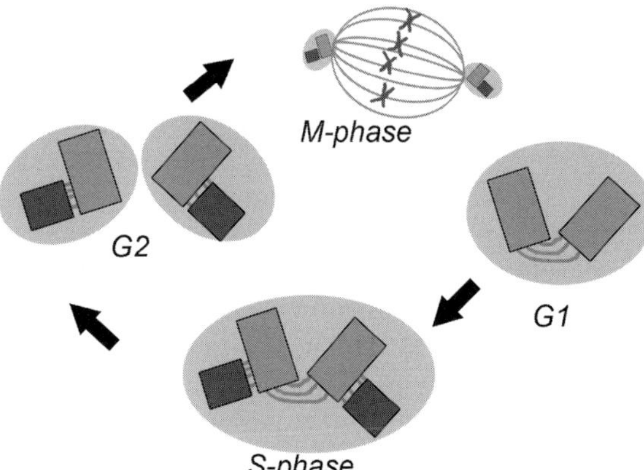

Figure 7.8 Mechanisms to maintain genomic integrity of cells. (a) The DNA damage checkpoint responds to DNA strand breaks and mismatches, delays the cell cycle, and promotes repair. (b) The spindle assembly checkpoint protects cells from mis-segregation of condensed chromosomes in mitosis, but requiring all chromatid pairs to be attached to opposite poles of the spindle prior to severing the links and initiating equal anaphase separation. (c) The telomeres protect DNA ends from erosive loss of genes and from chromosome fusions. Telomerase reverse transcriptase (hTERT) among other genes is required for telomere maintenance and function. (d) The centrosome number is regulated by a cycle, synchronous with the cell cycle, in which they are duplicated and segregated. Each centrosome consists of linked centrioles (rectangles) and pericentriolar material (gray). In the S-phase, centrioles are duplicated and in G2 and M-phases, pairs segregate into incipient daughter cells. (A black and white version of this figure will appear in some formats. For the color version, please refer to the plate section.)

(a) DNA damage repair

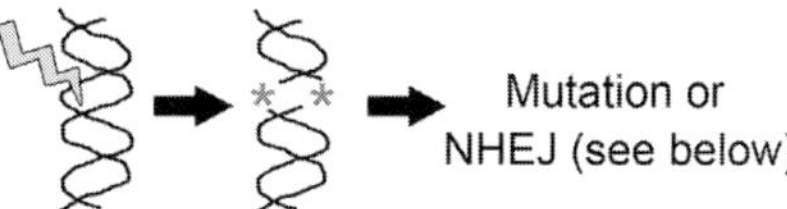

(b) Spindle assembly checkpoint

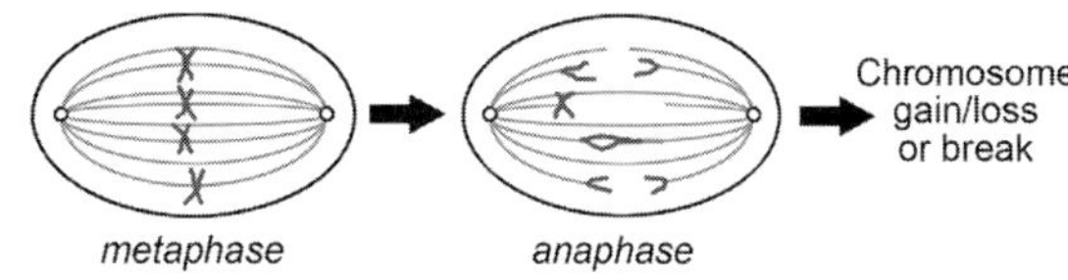

(c) Telomere integrity

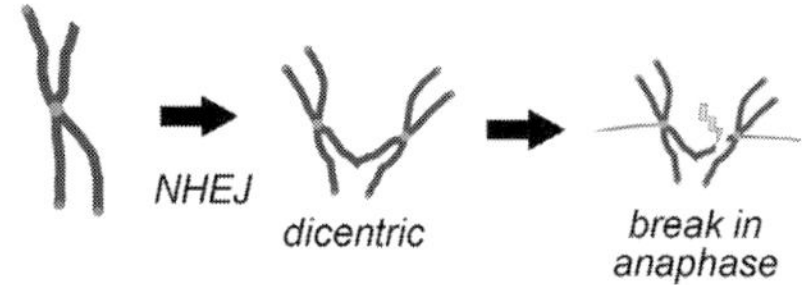

(d) Abnormal centrosome number

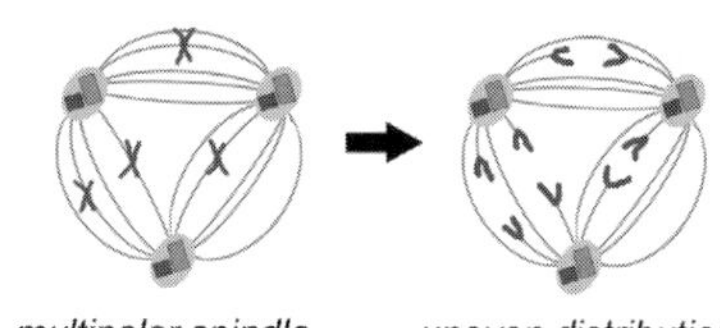

Figure 7.9 Genetic integrity can be lost by interruption of protective mechanisms. (a) If the damage checkpoint is lost, DNA mutants can accumulate in replication. Alternatively, with double-strand breaks, ends can be conjoined with different chromosomes in a process termed non-homologous end joining (NHEJ). (b) If the spindle assembly checkpoint is impaired, chromosome pairs can segregate to the same pole, resulting in gain or loss of this chromosome in daughters. Alternatively chromosome pairs can be incompletely resolved, resulting in a double-strand break and NHEJ. (c) If telomeres are lost, chromosome erosion can occur, as can NHEJ, resulting in dicentric chromosomes – chromosomes with two centromeres. Dicentric chromosomes can attach to opposite poles resulting in chromosome break in anaphase and thus can generate multiple cycles of breakage and fusion. (d) Abnormalities in centrosome number can result in chromosome mis-segregation in multipolar mitoses. (A black and white version of this figure will appear in some formats. For the color version, please refer to the plate section.)

transcriptase (TERT), results in telomere erosion and can cooperate with other genetic alterations to generate cancers [80]. With telomere erosion, genes encoded near telomeres can be lost and, moreover, exposed ends can be conjoined, resulting in dicentric chromosomes (chromosomes with two centrosomes), which are prone to breakage in anaphase when the centromeres are attached to opposite spindle poles (Figure 7.9c). Like DNA double-strand breaks (Figure 7.9a) and chromosome breaks due to entanglement (Figure 7.9b), this can lead to further instability via multiple rounds of break–repair–break until stable, but abnormal unicentric chromosomes are derived (e.g., Figure 7.5a).

Regulation of centrosome number also is important for maintaining genomic integrity, since these generate microtubule organizing centers in mitosis. Because of this, a centrosome cycle maintains appropriate duplication of this structure and is coupled with the cell cycle (Figure 7.8d). Normal cells in G1 have a single centrosome comprised of two centrioles and pericentriolar material. The centrioles are duplicated in the S-phase and separate in G2. During mitosis, these orient at opposite poles of the spindle. When cells contain supernumerary centrosomes, they are prone to multipolar mitosis with chromosomal mis-segregation (Figure 7.9d) [4]. The genesis of centrosome aberrations in cancer is unclear. One possibility is failed cell division or cell fusion; such an alteration would be coupled with increasing cell ploidy, also observed in cancer cells. Other potential mechanisms include decoupling the centrosome cycle from the cell cycle and *de novo* generation of centrosomes [81]. One gene that may be involved in the latter process is polo-like kinase 4 (Plk4), which results in supernumerary centrosomes when overexpressed [82]. However, this has not been demonstrated to be responsible for alterations in centrosome number or aneuploidy in cancer.

Effects of genetic instability

Heretofore, we have surveyed genetic alterations in cancer and the causes of genetic instability. Here we discuss mechanisms by which loss of genomic integrity can lead to oncogenesis.

Multi-hit hypothesis and clonal selection

Oncogenesis can be described as a gradual process starting with normal cells and ultimately ending with the development of cells having properties of cancer: capable of limitless proliferation, invasion, and distant dissemination. It is clear from pathologic analysis of tissues that intermediate states in this process can be identified. For example, *in situ* cancer is frequently identified in ducts of breast and prostate cancer. This has provided an opportunity to examine the various states along the continuum from normal to transformed cells. Based on these observations, researchers have long predicted that the process of tumorigenesis

is multi-step [83] Despite a preponderance of such descriptive analyses, genetic and epidemiologic analyses have more clearly demonstrated that multiple events are required for transformation.

In 1971, Knudson published a statistical analysis of 48 retinoblastoma cases seen at M.D. Anderson from 1944 to 1969 [84]. In the hereditary form, an average of three retinoblastomas is observed per individual whereas in the non-hereditary form, only one is observed per individual. This could be explained by the requirement for two mutational events for transformation, one of which is inherited from the germline in the hereditary form of the disease. Analysis of age at diagnosis and whether retinoblastoma was unilateral or bilateral confirmed that two stochastic events generate retinoblastoma, unless one abnormal gene is inherited.

For retinoblastoma, the two mutational events would eventually be explained by the presence of independent copies of the tumor suppressor, pRB, on homologous chromosomes. However, for other malignancies, the genetic requirements for transformation are considerably more complex. In colorectal tumors, for example, KRAS mutation and loss of the p53 locus are key molecular events in oncogenesis, allowing analysis of these mutations across the premalignant spectrum [67]. In early adenomas, few genetic alterations were present, but alterations in Ras and then p53 frequently accrue in samples obtained in later phases of development of invasive cancer [66]. A detailed analysis of abnormalities in Barrett esophagus is consistent with these observations and define cooperation between p53, CDKN2A, and aneuploidy [85]. This supports an evolutionary model supporting consecutive accumulation of mutations and clonal expansion occurring over years, termed the "multi-hit" model.

The multi-hit hypothesis has been confirmed with forward genetic approaches, which demonstrate that transformation of a normal cell into cancer requires cooperation of multiple genes. In a classic set of experiments, cooperating genes for transformation of normal fibroblasts were established. In cells with limitless replicative potential, a combination of oncogenes (H-Ras and SV40 large T antigen) is required to transform cells to produce malignant phenotypes [86, 87]. Because SV40 components are not found in most cancers, subsequent efforts have been made to identify individual genes involved in transformation. In murine models, cooperation between c-Myc, HRas, and p53 loss is sufficient for transformation. Transforming human fibroblasts, however, required at least six genetic alterations, highlighting the complexity of this process [88].

Although genetic diversity is required *a priori* for oncogenesis, Darwinian selection and clonal expansion play important roles in oncogenic transformation [83, 89]. Because a number of genetic alterations and characteristics need to be acquired to transform from a benign to a malignant phenotype, successive events have to confer these molecular characteristics, including replicative potential/immortalization, evasion of apoptosis, proliferation, and invasion [90]. Consistent with observations, multi-step models for cancer generally predict the establishment of immortalization and proliferation early with later events providing invasiveness and metastatic potential [66, 67, 85, 90]. Cancer therapy can provide additional selective pressures supporting the derivation of resistant clones.

Taken together, these data can, in general terms, explain how cancers are generated. The human genome contains multiple redundant mechanisms to protect against oncogenesis. However, a mutator phenotype can generate the requisite genetic diversity to overcome each mechanism one at a time. When a cell overcomes a mechanism limiting proliferation, it may incur a small advantage over normal cells, although not yet transformed. This premalignant clone can proliferate and generate a population of cells that can acquire additional alterations and develop clonal derivatives with a Darwinian advantage. Thus, over time, multiple somatic alterations can be acquired. In some instances, multiple areas of premalignant cells transform into independent cancers – this phenomenon is known as a field effect. Thus selection and genetic instability allow accumulation of multiple genetic alterations required to transform normal cells into cancer.

Outlook

Cancer is a disease of altered genetics. Cancer develops through enhanced mutagenic propensity and selection of clones that harbor a favorable collection of chromosomes, translocations, deletions, and mutations. Over the next several years, new technologies will allow complete characterization of cancer genomes and facilitate the development of a complete catalog of oncogenes and tumor suppressors. Undoubtedly, some members of this catalog will have weak effects, and others will be present only in a small fraction of tumors. Ultimately, it will remain a major challenge to use these findings for improved detection and treatment of cancer. The presence of a mutator phenotype (including CIN) of many cancers facilitates the development and selection of clones harboring resistance to effective treatment. Overcoming this difficulty will require either simultaneous interruption or multiple oncogenic pathways. One approach may be to target the underlying mutability of cancer cells themselves.

References

1. Flemming W. 1882. *Zell-substanz, Kern und Zelltheilung.* Leipzig: Verlag von F.C.W. Vogel

2. Hansemann D. 1890. Ueber asymmetrische Zelltheilung in Epithelkrebsen und deren biologische Bedeutung. *Virchows Archiv* 119: 299–326.

3. Hardy PA, Zacharias H. 2005. Reappraisal of the Hansemann–Boveri hypothesis on the origin of tumors. *Cell Biol Int* 29: 983–92.

4. Boveri T. 1904. *Ergebnisse uber die Konstitution der chromatischen Substanz des Zellkerns.* Jena: Verlag von Gustav Fischer.

5. Vogelstein B, Kinzler KW. 2004. Cancer genes and the pathways they control. *Nat Med* 10: 789–99.

6. Tjio JH, Levan A. 1956. The chromosome number of man. *Hereditas* 42: 1–6.

7. Caspersson T, Zech L, Johansson C. 1970. Differential binding of alkylating fluorochromes in human chromosomes. *Exp Cell Res* 60: 315–19.

8. Schweizer D, Ambros P, Andrle M. 1978. Modification of DAPI banding on human chromosomes by prestaining with a DNA-binding oligopeptide antibiotic, distamycin A. *Exp Cell Res* 111: 327–32.

9. Mitelman F, Johansson B, Mertens F. 2008 Mitelman Database of Chromosome Aberrations and Gene Fusions in Cancer. http://cgap.nci.nih.gov/Chromosomes/Mitelman.

10. Bauman JG, Wiegant J, Borst P, van Duijn P. 1980. A new method for fluorescence microscopical localization of specific DNA sequences by in situ hybridization of fluorochrome labelled RNA. *Exp Cell Res* 128: 485–90.

11. Loeb LA, Springgate CF, Battula N. 1974. Errors in DNA replication as a basis of malignant changes. *Cancer Res* 34: 2311–21.

12. Sanger F, Coulson AR. 1975. A rapid method for determining sequences in DNA by primed synthesis with DNA polymerase. *J Mol Biol* 94: 441–8.

13. Sanger F, Nicklen S, Coulson AR. 1977. DNA sequencing with chain-terminating inhibitors. *Proc Natl Acad Sci USA* 74: 5463–7.

14. Reddy EP, Reynolds RK, Santos E, Barbacid M. 1982. A point mutation is responsible for the acquisition of transforming properties by the T24 human bladder carcinoma oncogene. *Nature* 300: 149–52.

15. Kallioniemi A, Kallioniemi OP, Sudar D, et al. 1992. Comparative genomic hybridization for molecular cytogenetic analysis of solid tumors. *Science* 258: 818–21.

16. Pinkel D, Segraves R, Sudar D, et al. 1998. High resolution analysis of DNA copy number variation using comparative genomic hybridization to microarrays. *Nat Genet* 20: 207–11.

17. Wang TL, Maierhofer C, Speicher MR, et al. 2002. Digital karyotyping. *Proc Natl Acad Sci USA* 99: 16156–61.

18. Imoto H, Hirotsune S, Muramatsu M, et al. 1994. Direct determination of NotI cleavage sites in the genomic DNA of adult mouse kidney and human trophoblast using whole-range restriction landmark genomic scanning. *DNA Res* 1: 239–43.

19. Aston C, Mishra B, Schwartz DC. 1999. Optical mapping and its potential for large-scale sequencing projects. *Trends Biotechnol* 17: 297–302.

20. Lisitsyn N, Wigler M. 1993. Cloning the differences between two complex genomes. *Science* 259: 946–51.

21. Wadman M. 2008. James Watson's genome sequenced at high speed. *Nature* 452: 788.

22. Wang J, Wang W, Li R, et al. 2008. The diploid genome sequence of an Asian individual. *Nature* 456: 60–5.

23. Wheeler DA, Srinivasan M, Egholm M, et al. 2008. The complete genome of an individual by massively parallel DNA sequencing. *Nature* 452: 872–6.

24. Sjoblom T, Jones S, Wood LD, et al. 2006. The consensus coding sequences of human breast and colorectal cancers. *Science* 314: 268–74.

25. Chng WJ. 2007. Limits to the Human Cancer Genome Project? *Science* 315: 762; author reply 4–5.

26. Strauss BS. 2007. Limits to the Human Cancer Genome Project? *Science* 315: 762–4; author reply 4–5.

27. Greenman C, Stephens P, Smith R, et al. 2007. Patterns of somatic mutation in human cancer genomes. *Nature* 446: 153–8.

28. Cancer Genome Atlas Research Network. 2008. Comprehensive genomic characterization defines human glioblastoma genes and core pathways. *Nature* 455: 1061–8.

29. Parsons DW, Jones S, Zhang X, et al. 2008. An integrated genomic analysis of human glioblastoma multiforme. *Science* 321: 1807–12.

30. Jones S, Zhang X, Parsons DW, et al. 2008. Core signaling pathways in human pancreatic cancers revealed by global genomic analyses. *Science* 321: 1801–6.

31. Wood LD, Parsons DW, Jones S, et al. 2007. The genomic landscapes of human breast and colorectal cancers. *Science* 318: 1108–13.

32. Xu X, Wagner KU, Larson D, et al. 1999. Conditional mutation of Brca1 in mammary epithelial cells results in blunted ductal morphogenesis and tumour formation. *Nat Genet* 22: 37–43.

33. Mitelman F, Mertens F, Johansson B. 1997. A breakpoint map of recurrent chromosomal rearrangements in human neoplasia. *Nat Genet* 15 Spec No: 417–74.

34. Yunis JJ. 1983. The chromosomal basis of human neoplasia. *Science* 221: 227–36.

35. Mitelman F, Johansson B, Mertens F. 2007. The impact of translocations and gene fusions on cancer causation. *Nat Rev Cancer* 7: 233–45.

36. Storchova Z, Kuffer C. 2008. The consequences of tetraploidy and aneuploidy. *J Cell Sci* 121: 3859–66.

37. Redon R, Ishikawa S, Fitch KR, et al. 2006. Global variation in copy number in the human genome. *Nature* 444: 444–54.

38. Iafrate AJ, Feuk L, Rivera MN, et al. 2004. Detection of large-scale variation in the human genome. *Nat Genet* 36: 949–51.

39. Korbel JO, Urban AE, Affourtit JP, et al. 2007. Paired-end mapping reveals extensive structural variation in the human genome. *Science* 318: 420–6.

40. Campbell PJ, Stephens PJ, Pleasance ED, et al. 2008. Identification of somatically acquired rearrangements in cancer using genome-wide massively parallel paired-end sequencing. *Nat Genet* 40: 722–9.

41. Tomlins SA, Rhodes DR, Perner S, et al. 2005. Recurrent fusion of TMPRSS2 and ETS transcription factor genes in prostate cancer. *Science* 310: 644–8.

42. Soda M, Choi YL, Enomoto M, et al. 2007. Identification of the transforming EML4-ALK fusion gene in non-small-cell lung cancer. *Nature* 448: 561–6.

43. Tomlins SA, Laxman B, Dhanasekaran SM, et al. 2007. Distinct classes of chromosomal rearrangements create oncogenic

ETS gene fusions in prostate cancer. *Nature* 448: 595–9.

44. Slamon DJ, Clark GM, Wong SG, Levin WJ, Ullrich A, McGuire WL. 1987. Human breast cancer: correlation of relapse and survival with amplification of the HER-2/neu oncogene. *Science* 235: 177–82.

45. Slamon DJ, Godolphin W, Jones LA, et al. 1989. Studies of the HER-2/neu proto-oncogene in human breast and ovarian cancer. *Science* 244: 707–12.

46. Hudis CA. 2007. Trastuzumab: mechanism of action and use in clinical practice. *N Engl J Med* 357: 39–51.

47. Loeb LA. 1991. Mutator phenotype may be required for multistage carcinogenesis. *Cancer Res* 51: 3075–9.

48. Loeb LA, Loeb KR, Anderson JP. 2003. Multiple mutations and cancer. *Proc Natl Acad Sci USA* 100: 776–81.

49. Lengauer C, Kinzler KW, Vogelstein B. 1998. Genetic instabilities in human cancers. *Nature* 396: 643–9.

50. Ionov Y, Peinado MA, Malkhosyan S, Shibata D, Perucho M. 1993. Ubiquitous somatic mutations in simple repeated sequences reveal a new mechanism for colonic carcinogenesis. *Nature* 363: 558–61.

51. Fishel R, Lescoe MK, Rao MR, et al. 1993. The human mutator gene homolog MSH2 and its association with hereditary nonpolyposis colon cancer. *Cell* 75: 1027–38.

52. Hanks S, Coleman K, Reid S, et al. 2004. Constitutional aneuploidy and cancer predisposition caused by biallelic mutations in BUB1B. *Nat Genet* 36: 1159–61.

53. Lengauer C, Kinzler KW, Vogelstein B. 1997. Genetic instability in colorectal cancers. *Nature* 386: 623–7.

54. Shih IM, Zhou W, Goodman SN, et al. 2001. Evidence that genetic instability occurs at an early stage of colorectal tumorigenesis. *Cancer Res* 61: 818–22.

55. Vilar E, Scaltriti M, Balmana J, et al. 2008. Microsatellite instability due to hMLH1 deficiency is associated with increased cytotoxicity to irinotecan in human colorectal cancer cell lines. *Br J Cancer* 99: 1607–12.

56. Ikediobi ON, Davies H, Bignell G, et al. 2006. Mutation analysis of 24 known cancer genes in the NCI-60 cell line set. *Mol Cancer Ther* 5: 2606–12.

57. Watson P, Vasen HF, Mecklin JP, et al. 2008. The risk of extra-colonic, extra-endometrial cancer in the Lynch syndrome. *Int J Cancer* 123: 444–9.

58. Rajagopalan H, Nowak MA, Vogelstein B, Lengauer C. 2003. The significance of unstable chromosomes in colorectal cancer. *Nat Rev Cancer* 3: 695–701.

59. Yoon DS, Wersto RP, Zhou W, et al. 2002. Variable levels of chromosomal instability and mitotic spindle checkpoint defects in breast cancer. *Am J Pathol* 161: 391–7.

60. Albertson DG, Collins C, McCormick F, Gray JW. 2003. Chromosome aberrations in solid tumors. *Nat Genet* 34: 369–76.

61. Waldman FM, DeVries S, Chew KL, et al. 2000. Chromosomal alterations in ductal carcinomas in situ and their in situ recurrences. *J Natl Cancer Inst* 92: 313–20.

62. Lingle WL, Barrett SL, Negron VC, et al. 2002. Centrosome amplification drives chromosomal instability in breast tumor development. *Proc Natl Acad Sci USA* 99: 1978–83.

63. Rajagopalan H, Jallepalli PV, Rago C, et al. 2004. Inactivation of hCDC4 can cause chromosomal instability. *Nature* 428: 77–81.

64. Shin HJ, Baek KH, Jeon AH, et al. 2003. Dual roles of human BubR1, a mitotic checkpoint kinase, in the monitoring of chromosomal instability. *Cancer Cell* 4: 483–97.

65. Williams BR, Prabhu VR, Hunter KE, et al. 2008. Aneuploidy affects proliferation and spontaneous immortalization in mammalian cells. *Science* 322: 703–9.

66. Fearon ER, Vogelstein B. 1990. A genetic model for colorectal tumorigenesis. *Cell* 61: 759–67.

67. Vogelstein B, Kinzler KW. 1993. The multistep nature of cancer. *Trends Genet* 9: 138–41.

68. Li L, McCormack AA, Nicholson JM, et al. 2009. Cancer-causing karyotypes: chromosomal equilibria between destabilizing aneuploidy and stabilizing selection for oncogenic function. *Cancer Genet Cytogenet* 188: 1–25.

69. Weaver BA, Silk AD, Montagna C, Verdier-Pinard P, Cleveland DW. 2007. Aneuploidy acts both oncogenically and as a tumor suppressor. *Cancer Cell* 11: 25–36.

70. Ganem NJ, Storchova Z, Pellman D. 2007. Tetraploidy, aneuploidy and cancer. *Curr Opin Genet Dev* 17: 157–62.

71. Fujiwara T, Bandi M, Nitta M, et al. 2005. Cytokinesis failure generating tetraploids promotes tumorigenesis in p53-null cells. *Nature* 437: 1043–7.

72. Mayer VW, Aguilera A. 1990. High levels of chromosome instability in polyploids of *Saccharomyces cerevisiae*. *Mutat Res* 231: 177–86.

73. Storchova Z, Pellman D. 2004. From polyploidy to aneuploidy, genome instability and cancer. *Nat Rev Mol Cell Biol* 5: 45–54.

74. Andreassen PR, Lohez OD, Lacroix FB, Margolis RL. 2001. Tetraploid state induces p53-dependent arrest of nontransformed mammalian cells in G1. *Mol Biol Cell* 12: 1315–28.

75. Wong C, Stearns T. 2005. Mammalian cells lack checkpoints for tetraploidy, aberrant centrosome number, and cytokinesis failure. *BMC Cell Biol* 6: 6.

76. Hackett JA, Feldser DM, Greider CW. 2001. Telomere dysfunction increases mutation rate and genomic instability. *Cell* 106: 275–86.

77. Myung K, Chen C, Kolodner RD. 2001. Multiple pathways cooperate in the suppression of genome instability in *Saccharomyces cerevisiae*. *Nature* 411: 1073–6.

78. O'Hagan RC, Chang S, Maser RS, et al. 2002. Telomere dysfunction provokes regional amplification and deletion in cancer genomes. *Cancer Cell* 2: 149–55.

79. Counter CM, Hirte HW, Bacchetti S, Harley CB. 1994. Telomerase activity in human ovarian carcinoma. *Proc Natl Acad Sci USA* 91: 2900–4.

80. Farazi PA, Glickman J, Horner J, Depinho RA. 2006. Cooperative interactions of p53 mutation, telomere dysfunction, and chronic liver damage in hepatocellular carcinoma progression. *Cancer Res* 66: 4766–73.

81. Nigg EA. 2002. Centrosome aberrations: cause or consequence of cancer progression? *Nat Rev Cancer* 2: 815–25.

82. Kleylein-Sohn J, Westendorf J, Le Clech M, et al. 2007. Plk4-induced centriole biogenesis in human cells. *Dev Cell* 13: 190–202.

83. Foulds L. 1958. The natural history of cancer. *J Chronic Dis* 8: 2–37.

84. Knudson AG, Jr. 1971. Mutation and cancer: statistical study of retinoblastoma. *Proc Natl Acad Sci USA* 68: 820–3.

85. Barrett MT, Sanchez CA, Prevo LJ, et al. 1999. Evolution of neoplastic cell lineages in Barrett oesophagus. *Nat Genet* 22: 106–9.

86. Hahn WC, Counter CM, Lundberg AS, et al. 1999. Creation of human tumour cells with defined genetic elements. *Nature* 400: 464–8.

87. Land H, Parada LF, Weinberg RA. 1983. Tumorigenic conversion of primary embryo fibroblasts requires at least two cooperating oncogenes. *Nature* 304: 596–602.

88. Boehm JS, Hession MT, Bulmer SE, Hahn WC. 2005. Transformation of human and murine fibroblasts without viral oncoproteins. *Mol Cell Biol* 25: 6464–74.

89. Nowell PC. 1976. The clonal evolution of tumor cell populations. *Science* 194: 23–8.

90. Hahn WC, Weinberg RA. 2002. Rules for making human tumor cells. *N Engl J Med* 347: 1593–603.

91. Michel LS, Liberal V, Chatterjee A, et al. 2001. MAD2 haplo-insufficiency causes premature anaphase and chromosome instability in mammalian cells. *Nature* 409: 355–9.

92. Li Y, Benezra R. 1996. Identification of a human mitotic checkpoint gene: hsMAD2. *Science* 274: 246–8.

93. Jallepalli PV, Waizenegger IC, Bunz F, et al. 2001. Securin is required for chromosomal stability in human cells. *Cell* 105: 445–57.

Chapter

Epigenomic code

8

José Ignacio Martín-Subero and Manel Esteller

Introduction

Identifying the structure of DNA, deciphering the genetic code, and sequencing the human genome undoubtedly represent some of the greatest achievements in the history of science. However, genetic information alone cannot explain the phenotypic diversity within a population. For instance, monozygotic twins [1, 2] or cloned animals [3] are genetically identical but can show different phenotypes and different susceptibilities to a disease.

The genome encodes for potential information, but the part of the genome that is expressed and translated into function and the part that remains silenced does not directly depend on the sequence itself, but rather on epigenetic mechanisms.

Epigenetics literally means "on top of" or "in addition to" genetics, and the epigenetic code comprises several mechanisms that regulate gene expression and chromatin structure without altering the DNA sequence itself. Furthermore, epigenetics integrates the different chemical languages that genome and environment use to communicate with each other [4–6]. Therefore a more inclusive definition of epigenetics was recently proposed as "the structural adaptation of chromosomal regions so as to register, signal or perpetuate altered activity states" [5].

The most widely studied epigenetic changes are DNA methylation of cytosines within CpG dinucleotides and a growing number of chemical modifications at different amino acid residues of histone tails (to date over 60 have been identified) [7]. Although the present chapter will mostly focus on DNA methylation and histone modifications, other epigenetic factors such as nuclear positioning, non-coding RNAs, and microRNAs are also associated with gene regulation and chromatin structure [8–10].

Epigenetic mechanisms play a key role in multiple physiological processes such as development, establishment of tissue identity, imprinting, X-chromosome inactivation, chromosomal stability, and gene transcription regulation [11]. Additionally, multiple factors, such as aging, nutrition, exposure to metals, or maternal behavior in early childhood are able to induce epigenetic changes [12–14]. These environmentally induced epigenetic modifications are in turn related to susceptibility to malignant and non-malignant diseases in adulthood [15]. Interestingly, there is evidence showing that monozygotic twins acquire epigenetic and phenotypic changes throughout life [2], which supports the concept that life style influences the phenotype through epigenetic modifications.

Given the importance of epigenetic mechanisms, it is not surprising that alterations in the epigenetic pattern are associated with a wide variety of diseases, particularly with cancer [13, 16–20]. As compared to epigenetic patterns in normal cells, cancer cells are characterized by an intense disruption of the epigenomic machinery, which is reflected in multiple aberrations affecting both content and distribution of DNA methylation and histone modifications (Figure 8.1) [16–19, 21–23]. This chapter presents a summary of the current knowledge on the epigenome of normal and cancer cells, along with an overview of techniques to detect epigenomic modifications and a prospect of future directions.

Techniques to study the epigenome

Epigenetics is a rapidly evolving field, and techniques to detect epigenetic changes have undergone a dramatic expansion in the last 30 years [24–28]. The initial efforts in the 1970s were focused on the measurement of global DNA methylation content

---

Systems Biology of Cancer, ed. S. Thiagalingam. Published by Cambridge University Press. © Cambridge University Press 2015.

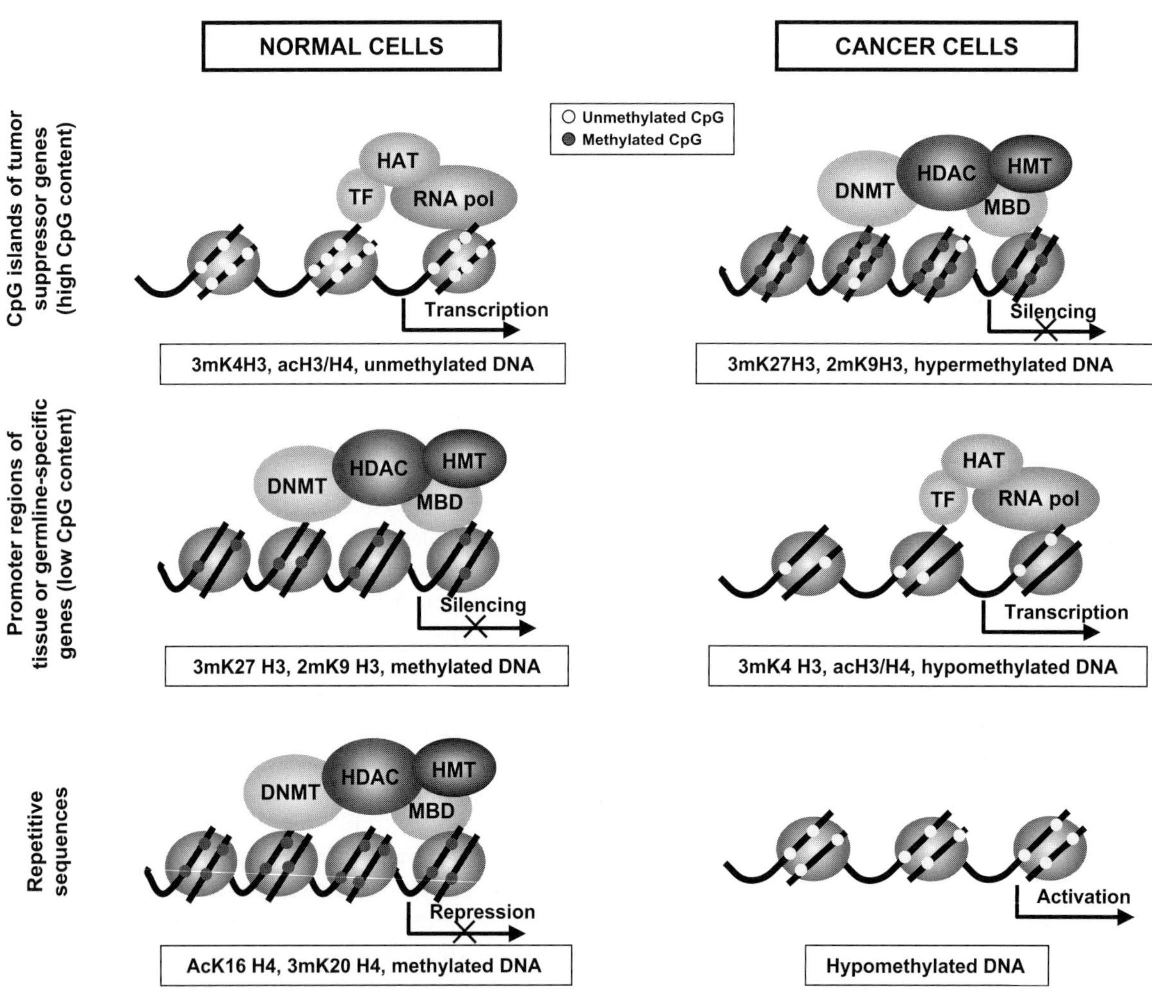

Figure 8.1 Summary of the most common epigenetic changes in normal and cancer cells. Nucleosomes are shown in the context of genomic regions that include promoters of tumor suppressor genes (TSGs), promoters of tissue- and germline-specific genes, and repetitive sequences in heterochromatic regions. DNA (black lines) is wrapped around each nucleosome, being unmethylated and methylated CpGs displayed as white and red spheres, respectively. In normal cells, TSG promoter regions with dense CpG islands show an open chromatin conformation, i.e., unmethylated DNA and enrichment in histone modification marks associated with active transcription, such as acetylation of histones H3 and H4 or trimethylation of histone H3 (at lysine K4). In contrast, genes methylated in a tissue-specific manner show promoters with low CpG content and are frequently silenced by methylation of lysines 9 and 27 of histone 3. These genes can also show an open chromatin if the protein is necessary in a given tissue. DNA repeats also show repressive epigenetic marks, which are associated with chromosomal stability. In cancer cells, TSGs are silenced by the loss of the histone-active marks and gain of promoter hypermethylation. Tissue- and germline-specific genes undergo the opposite process, acquiring an open chromatin conformation leading to gene expression. Repetitive sequences are activated by replacement of the repressive marks, leading to activation of endoparasitic sequences, genomic instability, or loss of imprinting. (A black and white version of this figure will appear in some formats. For the color version, please refer to the plate section.)

and the study of particular sequences by Southern blot analyses using methylation-sensitive restriction endonucleases. The limitations of the latter method (e.g., need for large amounts of high-quality DNA, sequence biases, and problems with incomplete digestion) made the study of specific sequences time consuming and not widely applicable. It was not until 1992, with the introduction of the sodium bisulfite conversion technique, that DNA methylation analyses made a revolutionary step forward [29]. Sodium bisulfite has the property of converting unmethylated cytosine into uracil whereas methylated cytosine remains unmodified. The combination of this chemical modification with genomic sequencing and methylation-specific PCR (MSP) made the study of DNA methylation changes widely available [30] and a large number of studies, especially dealing with CpG island hypermethylation of tumor suppressor genes in cancer, were published in the late 1990s. However, PCR-based approaches are restricted to the study of few candidate genes and are not suitable as screening techniques to identify novel markers. To overcome this, techniques such as amplification of intermethylated sites (AIMS) and restriction landmark genomic scanning (RLGS), which combine the use of methylation-sensitive restriction endonucleases with one-dimensional (1D) or two-dimensional (2D) electrophoresis have been established [31, 32]. These techniques are time consuming, and every new fragment identified as differentially methylated between a control and a test sample has to be cloned and sequenced.

An important step forward has been made in the recent years with the introduction of the microarray technology [33], which allowed the simultaneous study of epigenetic changes of thousands of known sequences. Some of these array-based DNA methylation techniques are based on the enrichment of methylated DNA and sequential hybridization onto a dedicated microarray containing, for instance, thousands of promoters or CpG islands, or even a tiling path array virtually containing the whole genome [34–36]. Enrichment for methylated DNA can be achieved by immunoprecipitation of methylated sequences with an antibody specific for 5-methylcytosine (MeDIP) [37] or by methyl-CpG immunoprecipitation (MCIp) [38, 39]. Methylated DNA can also be isolated by digestion with methylation-specific endonucleases [35, 40, 41]. One of the limitations of these methods is that they only provide a blurry picture of the methylome and it is not possible to determine the methylation status of specific CpGs. This problem can be overcome by combining a bisulfite treatment of the DNA and microarrays able to differentiate methylated and unmethylated CpG dinucleotides [42–45].

The methods mentioned above allow a direct detection of DNA methylation patterns. However, there is an additional, but indirect, way for detecting hypermethylated genes. This method applies gene expression profiling before and after treatment with DNA demethylating agents such as 5-aza-2′-deoxycytosine (5-AZA), so that hypermethylated genes become reactivated after treatment [46, 47]. Although this technique has allowed the detection of novel cancer-related hypermethylated genes, 5-AZA is highly toxic to the cells and can alter the expression levels of many genes regardless of their methylation status, leading to high false positive and false negative rates, and a thorough and time consuming data validation [27].

With regard to histone modifications, global alterations can be detected by isolating histone fractions by high-performance liquid chromatography (HPLC), and then analyzing them by high-performance capillary electrophoresis (HPCE) and liquid chromatography–electrospray mass spectrometry (LC–ES/MS) [48]. Specific modifications at each amino acid residue can also be characterized using antibodies in Western blots, immunostaining [49], or tandem mass spectrometry (MS/MS) [48]. If the goal is to detect histone modifications at specific DNA stretches, a different strategy is necessary. DNA can be crosslinked to the associated histones and in a second step precipitated with antibodies specific to certain histone modifications (a technique called chromatin immunoprecipitation; ChIP) so that a DNA fraction enriched for that specific histone modification can be isolated. Then a PCR using primers for the region of interest can be used. Alternatively, a genome-wide picture of histone modifications can be obtained by the ChIP-on-chip technique. By means of this technique, the immunoprecipitated fraction can be labelled and directly hybridized onto a microarray or compared with the input DNA in two-color hybridization [50, 51].

In spite of the potential of microarrays to characterize DNA methylation and histone modifications across the genome, they are limited either by resolution, type and number of sequences analyzed or by their quantification accuracy. In any case, the complete characterization of the human epigenome of a given sample requires the quantification of the methylation status of each of the ~55 million CpG dinucleotides per diploid cell and the distribution of histone marks of every DNA region, and today this is far outside the possibilities of current microarray platforms. The development of a new generation of sequencers is now revolutionizing both genomics and epigenomics [52–54]. These new sequencing technologies are based, for example, on pyrosequencing using millions of picoliter-scale reactions, sequencing by synthesis, and sequencing by ligation [52], and are able to sequence up to several gigabases of DNA in a single experiment. So far, the DNA methylome of oncogenic viruses [55] and of the plant *Arabidopsis thaliana* have been sequenced [56, 57], but the challenge of sequencing the human methylome has not yet been achieved. In the case of histone modifications, ultrasequencing has been used after ChIP (a technique called ChIP-Seq) and several genome-wide histone modification maps have already been reported [58–60].

The epigenetic machinery

DNA methylation and histone modifications are tightly regulated by DNA methyltransferases (DNMTs) [4] and a large number of histone modifying enzymes [7], respectively (Figure 8.2). As well as this, nucleosome remodelling proteins, which regulate the position and movement of nucleosomes, play an important role in regulating gene transcription

(Figure 8.2). The best characterized DNMTs are DNMT1, DNMT3A, and DNMT3B. Although DNMTs were originally classified as maintenance or de novo DNMTs (depending on their ability to methylate hemimethylated or unmethylated substrates), several lines of evidence indicate that all three DNMTs not only cooperate, but also may show both de novo and maintenance functions *in vivo* [61–63]. The knockout cell lines for DNMT1, DNMT3B, and both enzymes demonstrated that while no effective CpG island demethylation and restoration of gene expression were observed in the single knockouts, the double knockout of DNMT1 and DNMT3B showed complete hypomethylation at the studied CpG islands and corresponding gene activation [62, 64]. The double-knockout cell line has also been shown to be a useful tool for identifying new

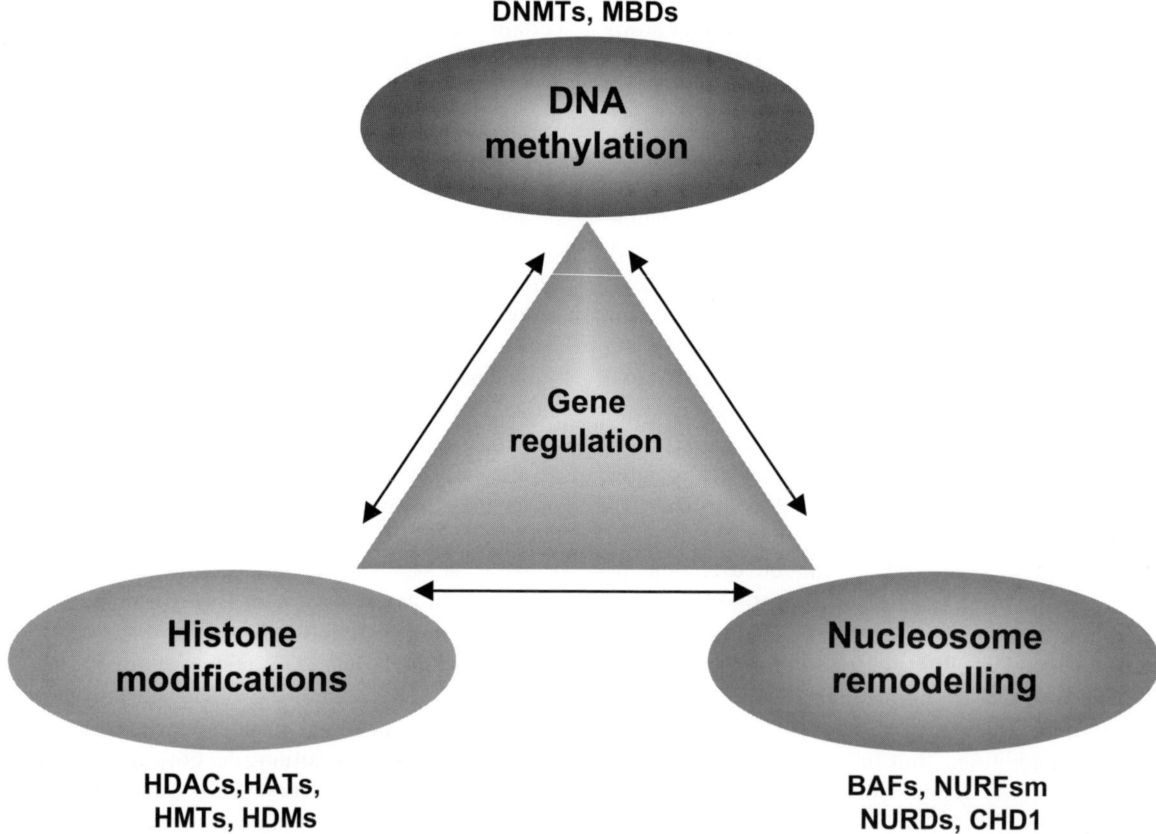

Figure 8.2 Mechanisms of epigenetic regulation. Regulation of gene transcription involves a series of mechanisms that involve DNA methylation, histone modifications, and nucleosome remodelling. (DNMTs: DNA methyltransferases; MBDs: methyl binding domain proteins; HDACs: histone deacetylases; HATs: histone acetyltransferases; HMTs: histone methyltransferases; HDMs: histone demethylases; BAFs: BRG1-associated factors; NURFs: nucleosomal remodelling factors; NURDs: nucleosome remodelling and histone deacelylases; CHD1: chromodomain helicase DNA binding protein 1. (A black and white version of this figure will appear in some formats. For the color version, please refer to the plate section.)

hypermethylated genes in human cancer [64] thanks to the use of global genomic methylation strategies, such as AIMS and CpG island arrays [16, 65]. Taken together, these results strongly suggest that both enzymes, DNMT1 and DNMT3B, are necessary for effective CpG island methylation.

The information for gene silencing contained by methylated CpG islands is in part read by methyl-CpG binding proteins (MBDs). MBDs mechanistically link DNA methylation and histone-modifying enzymes that establish a transcriptionally inactive chromatin environment. This family of proteins consists of five well-characterized members (MeCP2, MBD1, MBD2, MBD3, and MBD4) [66]. MBD proteins are associated with hypermethylated CpG island promoters of tumor suppressor genes and their transcriptional silencing [66], showing remarkable specificity *in vitro* [67] and *in vivo* [68–70]. In fact, most hypermethylated promoters are occupied by MBD proteins, whereas unmethylated promoters generally lack MBDs, with the exception of MBD1 [70]. Several promoters are highly specific in recruiting a particular set of MBDs, while other promoters seem to be less exclusive. Thus it may be speculated that the specific profile of MBD occupancy is gene- and tumor type-specific [70].

The epigenetic machinery associated with DNA methylation is associated to that leading to histone modifications. For instance MeCP2 represses transcription of methylated DNA by recruiting a histone deacetylase (HDAC)-containing complex [71, 72]. Additional connections between DNA methylation and histone modifications have been found: DNMTs are able also to recruit HDACs [73, 74], while, on the other hand, both DNMTs and MBDs recruit histone methyltransferases (HMTs) that modify lysine 9 of histone H3 [75, 76]. Also, it has been reported that EZH2, a member of the polycomb repressor complex leading to methylation of lysine 27 in histone 3, associates with DNMTs and might be involved in establishing DNA methylation in a subset of target genes [77].

Epigenetic marks in normal cells

CpG dinucleotides are not randomly distributed across the genome but are concentrated in the promoter regions of genes (called CpG islands) and in repetitive genomic sequences [16]. In general, it can be said that repetitive sequences are heavily methylated in normal cells (Figure 8.1), which probably prevents chromosomal instability, translocations, and gene disruption caused by the reactivation of transposable DNA sequences [78]. In contrast, most promoter-associated CpG islands contain unmethylated CpG islands (Figure 8.1), which lead to an open chromatin structure and allow gene expression if the appropriate transcriptional activators are present [16]. The methylation of particular subgroups of promoter CpG islands can, however, be detected in normal tissues [79]. In the case of gene promoters lacking the established criteria for CpG islands, which represent approximately 30% of all genes, they are frequently methylated in a tissue-specific manner in normal cells (Figure 8.1).

Histones are not merely DNA-packaging proteins, but molecular structures that participate in the regulation of gene expression. The nucleosome is the fundamental unit of chromatin and it is composed of an octamer of the four core histones (H3, H4, H2A, and H2B) around which 147 base pairs of DNA are wrapped. Protruding from the core histones are the N-terminal "tails," which are subjected to a large number of modified residues [7]. Histone modifications occur in different histone proteins, histone variants (e.g., H3.3), and histone residues such as lysine, arginine, and serine. These modifications also involve different chemical modifications (e.g., acetylation, methylation, phosphorylation, and ubiquitination) and have different degrees of methylation (e.g., monomethylation, dimethylation, and trimethylation). Acetylation and methylation of histones have direct effects on a variety of nuclear processes, including gene transcription, DNA repair, DNA replication, and the organization of chromosomes. It has been proposed that distinct histone modifications form a "histone code," by which post-translational modifications of histones, alone or in combination, determine a given chromatin structure and function [80].

DNA methylation and histone modifications are not independent epigenetic events but together form an "epigenomic code" that functions in an orchestrated manner to achieve a chromatin structure associated with gene silencing or gene expression. In general, active genes are associated with lack of DNA methylation, acetylation of H3 and H4, and methylation of lysine 4 of H3 whereas inactive genes show deacetylation of histones H3 and H4, methylation of lysines 9 and 27 of histone H3, and demethylation of lysine 4 of histone H3 (Figure 8.1).

With the advent of high-throughput technologies based on microarrays or ultrasequencing, a more detailed map of DNA methylation, histone modifications, and their combinations is becoming available.

Although sequencing of the complete human methylome has not been achieved so far, several microarray and sequencing studies have started to provide global patterns of DNA methylation across the genome. These studies clearly demostrated that different normal tissues and cell types display a specific DNA methylation profile (Figure 8.3) and that DNA methylation patterns are mostly bimodal, i.e., most promoter-specific CpG islands are unmethylated whereas promoters with low CpG content are

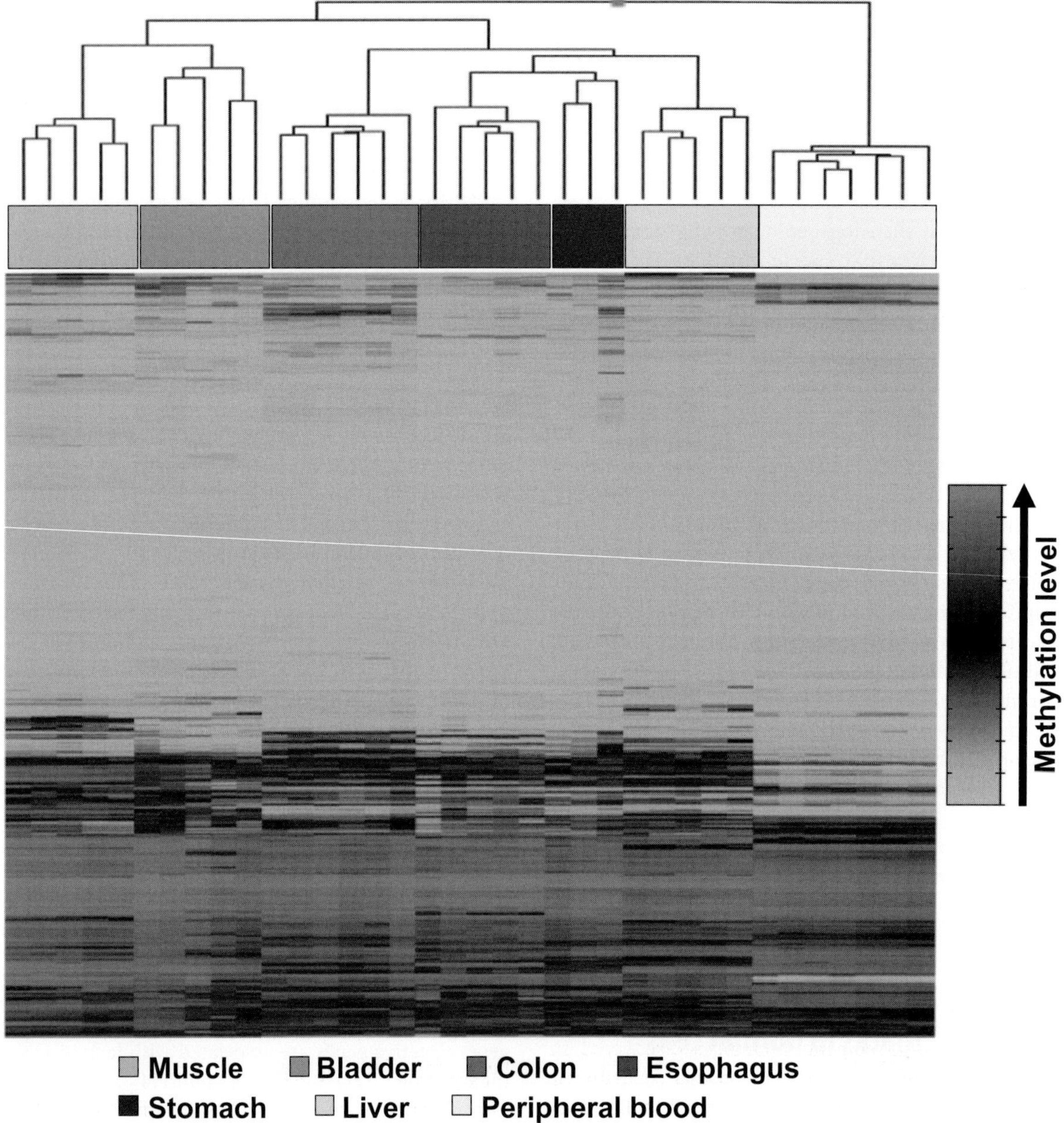

Figure 8.3 Epigenetic profiling of normal tissues. Heat map from a hierarchical cluster analysis of DNA methylation data generated with the bead-array technology (Illumina Inc.) [44] in different normal tissue samples. (A black and white version of this figure will appear in some formats. For the color version, please refer to the plate section.)

methylated [81–85]. An exception to this pattern are the imprinted genes, in which only one allele, either maternal or paternal, is methylated [86]. Furthermore, a recent microarray study has shown that aging and environmental exposures (e.g., to asbestos or smoking) modify the tissue-specific DNA methylation patterns, and several candidate genes were identified [12].

Using the ChIP-Seq technology, Barski and collaborators described common patterns of histone methylations exhibited at promoters, insulators, enhancers, and transcribed regions. The monomethylations of H3K27, H3K9, H4K20, H3K79, and H2BK5 were all linked to gene activation, whereas trimethylations of H3K27, H3K9, and H3K79 are linked to repression [58]. Using the same technology, Wang and collaborators described common patterns of histone methylations and acetylations associated with promoters and enhancers [60].

As epigenetic marks are associated with the establishment of tissue identity, an interesting experimental model to study the epigenome are embryonic stem cells (ESC), which are able to generate any tissue. A recent study using a reduced representation bisulfite ultrasequencing of single CpG dinucleotides in mouse ESCs and differentiated tissues indicated that DNA methylation patterns correlated with histone methylation patterns and methylation of CpGs undergo extensive changes during cellular differentiation, particularly in regulatory regions outside of core promoters [85]. ESCs not only show a specific DNA methylation profile [43] but also a peculiar chromatin state, which has been coined "bivalent" [87]. This bivalent state is characterized by the simultaneous presence of repressive marks such as 3mK27H3, and active marks such as 3mK4H3 [87, 88]. Then, upon differentiation toward a given cell lineage, required genes lose 3mK27H3 and become expressed whereas silencing of unnecessary genes is made permanent by other epigenetic marks such as 3mK9H3 and DNA methylation [89].

DNA hypomethylation in cancer

One of the initial epigenetic changes discovered in cancer is that, in comparison to normal cells, tumor cells are characterized by a global hypomethylation of the DNA [21, 90, 91]. Such global loss of methylation mainly targets repetitive DNA sequences, coding regions, and introns. So far, it has been reported that

DNA hypomethylation leads to the generation of chromosomal instability, reactivation of transposable elements, and loss of imprinting in cancer [17]. Hypomethylation of DNA repeats results in a more open chromatin at those genomic regions and renders the DNA more susceptible toward suffering DNA breaks by mitotic recombination of homologous repetitive sequences in different chromosomes [92–94.]. This leads to chromosomal rearrangements, which represent one of the genetic hallmarks of cancer. Also, germline mutations in the DNMT3B gene in humans are associated with the ICF (immunodeficiency, centromeric region instability, facial abnormalities) syndrome, which displays DNA hypomethylation of repeats, such as satellite 2, and chromosomal instability [95]. Hypomethylation of DNA in malignant cells can reactivate endoparasitic DNA, such as L1 (long interspersed nuclear elements) and Alu (recombinogenic sequence) repeats [78, 96]. These unmethylated transposons can be transcribed or translocated to other genomic regions so that chromosomal instability is further generated. The loss of methyl groups from DNA can also disrupt genomic imprinting. For instance, the hereditary Beckwith–Wiedemann syndrome shows loss of imprinting of IGF2 (the insulin-like growth factor gene) and individuals affected with this syndrome have an increased risk of cancer [97]. Also germline genes, such as the MAGE genes, are methylated and silent in almost all tissues, and become hypomethylated in malignant tumors [98].

Recent analyses with microarrays indicate that promoter-specific hypomethylation is also a frequent event in cancer [45, 99, 100]. Such hypomethylation usually targets tissue-specific genes (Figure 8.1), whose gene promoters show low CpG content and are frequently thought not to be regulated by DNA methylation. However, hypomethylated genes in T-cell prolymphocytic leukemia, a rare type of leukemia, do correlate with increased expression, even though they mostly contain promoters with low CpG content [100].

Gene silencing in cancer by DNA hypermethylation

Tumor suppressor gene silencing by CpG island hypermethylation (Figure 8.1) is perhaps the best studied epigenetic change in cancer development [17–19, 22, 101]. The presence of CpG island promoter hypermethylation affects genes regulating virtually all important cellular functions, such as cell cycle (p16INK4a, p15INK4b, RB1, p14ARF), DNA repair (BRCA1, hMLH1, MGMT, WRN), cell adherence and invasion (CDH1, CDH13, EXT1, SLIT2, EMP3), apoptosis (DAPK, TMS1, SFRP1), carcinogen metabolism (GSTP1), hormonal response (RARB2, ER, PRL, TSH receptors), Ras signaling (RASSF1A, NOREIA), and microRNAs, among others [102]. Table 8.1 shows a summary of the most common hypermethylated genes in human cancer.

Some of the genes hypermethylated in cancer are common to most cancer subtypes whereas others are considered to be cancer subtype-specific [101, 103]. Hierarchical cluster analysis of DNA methylation profiles of tumor suppressor genes in different cancers leads to a classification according to diagnosis and, therefore, each tumor type can be assigned a specific DNA "hypermethylome." Such patterns of epigenetic inactivation occur not only in sporadic tumors but also in inherited cancer syndromes [104], in which hypermethylation can be the second lesion in the classical Knudson's two-hit model of cancer development [104, 105].

So far, most epigenetic studies have targeted known tumor suppressor genes. However, with the advent of the science of epigenomics, a more precise and less biased delineation of the cancer cell epigenome is becoming accessible. This strategy is already providing a complete new generation of epigenetic markers in cancer. An increasing number of microarray-based studies have focused on the detection of differentially methylated biomarkers associated with specific types of solid tumors such as, for example, breast cancer [106–108], colorectal cancer [109–112], prostate cancer [113–115], lung cancer [116–118], head and neck cell carcinoma [34], oligodendroglioma [119], medulloblastoma [120], Wilms' tumors [121], and glioblastomas [99]. In hematological tumors, due to the large number of different cell types of the hematopoietic system, a wide range of different leukemias and lymphomas have been identified by means of morphological, immunohistochemical, and genetic features [122]. Now, several groups are using microarray-based DNA methylation profiling to characterize the epigenome of this heterogeneous group of diseases and to identify diagnostic epigenetic marks. These studies have looked at hematological cancers such as mantle cell lymphoma (MCL), follicular lymphoma (FL),

B-cell chronic lymphocytic leukemia (B-CLL), aggressive B-cell lymphoma, cutaneous T-cell lymphoma, acute lymphoblastic leukemia (ALL), and acute myeloid leukemia (AML) [45, 123–132]. Also, we have recently published the first comprehensive DNA methylation profile of 16 different hematological neoplasms and discovered that lymphoid neoplasias generally show a much higher level of aberrant DNA methylation than myeloid disorders [100].

Taken together, these studies and our recent unpublished experiments indicate that cancer cells, depending on the tumor subtype, might contain up to 2,000 hypermethylated gene promoters. An example of a microarray-based study of cancer cell lines is shown in Figure 8.4.

However, these microarray-based studies focused on gene promoters. Another recent study extended the analysis beyond promoters and discovered that the genomic regions showing the highest hypermethylation in cancer were neither in promoters nor in CpG islands, but in sequences up to 2 kb distant, which were called "CpG island shores," and were correlated with gene repression [133]. This study highlights the importance of analyzing the cancer DNA methylome in an unbiased manner.

MiRNAs and cancer

Epigenetic silencing in cancer not only targets coding sequences but also small non-coding RNAs such as miRNAs. These are made of short stretches of 22 nucleotides that are able to regulate gene transcription of target genes by sequence-specific base pairing in the 3′ UTR regions and subsequent degradation of the target mRNA or inhibition of translation. Target genes of miRNAs are involved in cellular functions such as cell proliferation, differentiation, and apoptosis [134, 135]. Therefore it is not surprising that miRNA expression has been found to be deregulated in cancer development [136, 137]. The role as tumor suppressors of miRNAs has been investigated in more detail for particular cases. For example, the down-regulated let-7 and miR-15/miR-16, and miR-127 are known to target the oncogenic factors RAS and BCL-2, respectively [138, 139]. This may be explained by the failure of these miRNAs during post-transcriptional regulation in cancer cells [140], but additional mechanisms such as CpG island hypermethylation could also be involved. For instance, it has been observed that 5% of human miRNAs are

Table 8.1 Summary of the best characterized genes silenced by CpG island promoter hypermethylation in human cancer.

Gene	Function	Location	Tumor type
APC	Inhibitor of beta-catenin	5q21	Aerodigestive tract
AR	Androgen receptor	Xq11	Prostate
BRCA1	DNA repair, transcription	17q21	Breast, ovary
CDH1	E-cadherin, cell adhesion	16q22.1	Breast, stomach
CDH13	H-cadherin, cell adhesion	16q24	Breast, lung
COX2	Cyclooxygenase-2	1q25	Colon, stomach
CRBP1	Retinol-binding protein	3q23	Colon, stomach, lymphoma
DAPK	Pro-apoptotic	9q34.1	Lymphoma, lung, colon
DKK1	Extracellular Wnt inhibitor	10q11.2	Colon
ER	Estrogen receptor	6q25.1	Breast
EXT1	Heparan sulfate synthesis	8q24	Leukemia, skin
FAT	Cadherin, tumor suppressor	4q35	Colon
GATA4	Transcription factor	8p23	Colon, stomach
GATA5	Transcription factor	20q13	Colon, stomach
GSTP1	Conjugation to glutathione	11q13	Prostate, breast, kidney
HIC1	Transcription factor	17p13.3	Multiple types
HOXA9	Homeobox protein	7p15.2	Neuroblastoma
ID4	Transcription factor	6p22.3	Leukemia
IGFBP3	Growth factor-binding protein	7p13	Lung, skin
Lamin A/C	Nuclear intermediate filament	1q21.2	Lymphoma, leukemia
LKB1/STK11	Serine–threonine kinase	19p13.3	Colon, breast, lung
MGMT	DNA repair of 06–alkyl-guanine	10q26	Multiple types
MLH1	DNA mismatch repair	3p21.3	Colon, endometrium, stomach
NORE1A	Ras effector homolog	1q32	Lung
p14ARF	MDM2 inhibitor	9p21	Colon, stomach, kidney
p15INK4B	Cyclin-dependent kinase inhibitor	9p21	Leukemia
p16INK4A	Cyclin-dependent kinase inhibitor	9p21	Multiple types
p73	p53 homolog	1p36	Lymphoma
PR	Progesterone receptor	11q22	Breast
PRLR	Prolactin receptor	5p13.2	Breast
RARB2	Retinoic acid receptor-beta2	3p24	Colon, lung, head and neck
RASSF1A	Ras effector homolog	3p21.3	Multiple types
RB1	Cell cycle inhibitor	13q14	Retinoblastoma
RIZ1	Histone/protein methyltransferase	1p36	Breast, liver
SFRP1	Secreted frizzled-related protein 1	8p11.21	Colon

Table 8.1 (cont.)

Gene	Function	Location	Tumor type
SLC5A8	Sodium transporter	12q23	Glioma, colon
SOCS1	Inhibitor of JAK–STAT pathway	16p13.13	Liver, myeloma
SOCS3	Inhibitor of JAK–STAT pathway	17q25	Lung
SRBC	BRCA1-binding protein	1p15	Breast, lung
SYK	Tyrosine kinase	9q22	Breast
THBS1	Thrombospondin-1, Anti-angiogenic	15q15	Glioma
TMS1	Pro-apoptotic	16p11	Breast
TPEF/HPP1	Transmembrane protein	2q33	Colon, bladder
TSHR	Thyroid-stimulating hormone receptor	14q31	Thyroid
VHL	Ubiquitin ligase component	3p25	Kidney, hemangioblastoma
WIF1	Wnt inhibitory factor	12q14.3	Colon, lung
WRN	DNA repair	8p12	Colon, stomach, sarcoma

(Adapted from Esteller, 2007) [16]

upregulated by treatment of bladder cancer cells with DNA demethylating agents and HDAC inhibitors [141]. In particular, miR-127 expression was induced by a decrease in DNA methylation levels around the promoter region of the miR-127 gene, and the proto-oncogene BCL6, a potential target of miR-127, was translationally downregulated after treatment [141]. Additionally, using a genetic approach that takes advantage of the genomic disruption of DNMT1 and DNMT3B in cancer cells, we have demonstrated that CpG island hypermethylation is a mechanism that can account for the downregulation of miRNAs in human cancer [142]. The epigenetic silencing of miR-124a, one of the DNA methylation-associated silenced miRNAs isolated using this approach, leads to activation of cyclin D kinase 6 (CDK6), a bona fide oncogenic factor, and phosphorylation of the retinoblastoma (RB1) tumor suppressor protein [142]. Interestingly, hypermethylation of this miRNA is associated with poor prognosis in acute lympho-blastic leukemia [143]. Another study conducted in this type of leukemia identified 13 miRNAs that were silenced by histone modifications (presence of dimethylation of H3K9 and lack of trimethylation of H3 K4) and DNA methylation [144]. Furthermore, epigenetic deregulation of miRNAs is also involved in metastasis, being hypermethylation of miRNAs such as miR-148a, miR34a/b, and miR9 associated with metastasis development and upregulation of onco-genic and metastasis target genes [145].

Mechanisms leading to DNA hypermethylation in cancer

The acquisition of differential methylation in cancer, like hypermethylation of tumor suppressor genes, is thought to provide the tumor clone with a selective (e.g., proliferative) advantage. However, recent reports have proposed that there is an instructive mechanism behind aberrant DNA methylation in cancer [146, 147]. Keshet and colleagues performed a MeDIP-on-chip study in colon and prostate cancer and as well as identifying differentially methylated genes, they studied whether these genes show distinct biological features [146]. They discovered that genes differentially methylated in cancer are enriched for functional categories (e.g., cell adhesion, cell–cell signaling, signal transduction, and ion transport) and that the expression of some of them is already repressed (or expressed at low levels) in cells from matched normal tissues [45, 146]. Furthermore, they detected a significant enrichment of sequence motifs and a significant clustering of such genes in chromo-somal regions [146, 148]. In line with this finding, another study has shown that large stretches of DNA containing several genes can become hypermethylated

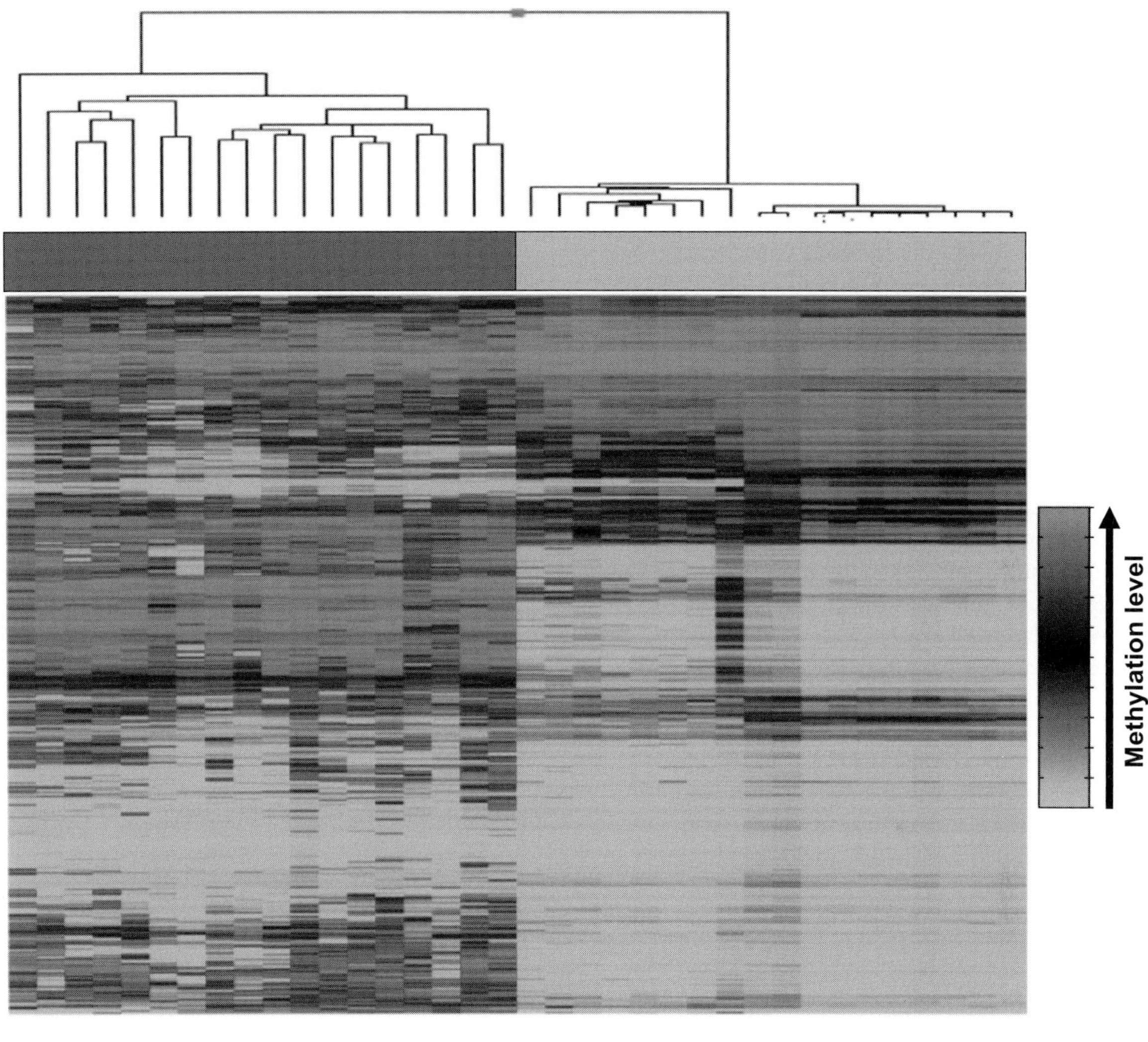

Figure 8.4 Epigenomic profiling of cancer cells. Heat map from a hierarchical cluster analysis of DNA methylation data generated with the bead-array technology (Illumina Inc.) [44] in different B-cell lymphoma cell lines and healthy controls. Lymphoma cell lines are characterized by a large number of hypermethylated genes in comparison with the normal controls. (A black and white version of this figure will appear in some formats. For the color version, please refer to the plate section.)

in cancer [149]. Furthermore, various independent studies have found that a highly significant proportion of genes becoming hypermethylated in cancer were already repressed at the embryonic stem cell stage by members of the polycomb repressive complex 2 [45, 150–153]. These findings might suggest that epigenetic changes of polycomb target genes occurring in a cell with stem cell features might represent the initial event in tumorigenesis, which supports the cancer stem cell theory [147, 154]. However, another study has shown that genes marked with trimethylation H3K27 (a mark established by the polycomb protein EZH2) in normal differentiated colon also tend to acquire DNA hypermethylation in colon cancer, suggesting that cancer can be derived from the clonal expansion of differentiated cells rather than from rare cancer stem cells [155]. Altogether, these studies indicate that there are genes predisposed to acquire aberrant DNA methylation in cancer, and even a recent study proposed a multifactorial signature that was able to predict hypermethylation in cancer with an 81 to 88% accuracy [156].

Histone modifications in cancer cells

Although less studied than DNA methylation, histone modification patterns are also highly disrupted in cancer (Figure 8.1). One of the best studied alterations is the acetylation of histone H4, which is globally hypoacetylated in esophageal squamous cell carcinoma [156–158], gastric cancer [159], testicular cancer [160], and acute promyelocytic leukemia (APL) [161]. Furthermore, monoacetylated lysine 16 of H4 is globally reduced in various types of tumors [48], and lower levels of acetylated lysine 12 of H4 are an indicator of recurrence in prostate cancer [49]. Interestingly, exposure to the carcinogen Ni^{2+} induces a clear decrease in histone acetylation [162]. Another histone H4 modification, the trimethylation of lysine 20 of H4, which is enriched in differentiated cells [163], increases with age [164], is commonly reduced in cancer cells [48, 165]. Global alterations of histone H3 modifications in cancer have been less thoroughly investigated. One study found that low levels of acetylation at lysines 9 and 18 of histone H3 are associated with high recurrence of prostate cancer [49]. In two recent studies, H3 acetylation has been found to be reduced in human colon primary tumors [166] and in several human colon cancer cell lines [167].

Hypermethylation of the CpG islands in the promoter regions of tumor suppressor genes in cancer cells is associated with a particular combination of histone markers: deacetylation of histones H3 and H4, and methylation changes at various amino acid residues of H3, such as loss of trimethylation at lysine 4, and gain of methylation at lysine 9 and trimethylation at lysine 27 (Figure 8.1). The presence of the hypoacetylated and hypermethylated histones H3 and H4 silences certain genes with tumor suppressor-like properties, such as p21WAF1, despite the absence of hypermethylation of the CpG island [168]. On the other hand, silencing of SIRT1, a histone deacetylase, leads to increased H3 and H4 acetylation of cancer genes that become reactivated despite full retention of DNA hypermethylation [169]. Also, a recent publication indicates that gene silencing in cancer can be achieved by trimethylation of lysine 27 of histone 3 independently from the presence of DNA methylation [170].

Cancer as a genetic and epigenetic disease

After decades of cancer research, it is now widely accepted that genetic and epigenetic mechanisms closely interact in carcinogenesis [171]. This interplay between genetic and epigenetic changes can be clearly observed in the case of tumor suppressor gene inactivation, which can be caused by genetic (mutation, deletion) or epigenetic (DNA hypermethylation) means. For instance, the VHL gene is mutated in 60% of the renal carcinomas and hypermethylated in 20% of the remaining cases [172]. In the case of the E-cadherin gene, mutations and methylation are mutually exclusive in breast cancer [173]. Besides genetic alterations (e.g., mutations, translocations, and amplifications) targeting genes directly involved in the regulation of DNA methylation or histone modification processes (e.g., HDAC2, MML, MOZ, MORF, NSD, and GASC1) are frequent in cancer (Table 8.2). Some genes of the polycomb repressor complex such as BMI1, EZH2, and SUZ12 are also targets of chromosomal changes [174, 175]. Interestingly, transgenic mice overexpressing the polycomb member Bmi1 develop lymphomas, and this process of lymphomagenesis is accelerated in double transgenic mice overexpressing Bmi1 and Myc, the latter being frequently translocated in lymphomas [176, 177]. The close interaction between genetic and epigenetic changes is also supported by the discovery of epigenetic changes that affect the stability of the genome. For instance, hypomethylation of DNA repeats leads to chromosomal changes by inducing chromosomal instability. Also, genes targeted by DNA hypermethylation are involved in DNA repair pathways, such as BRCA1, hMLH1, MGMT, or WRN [102]. In these cases, silencing of the DNA repair gene blocks the repair of genetic mistakes, thereby opening the way to neoplastic transformation of the cell.

Overall, these data show that genetic and epigenetic changes represent alternative mechanisms targeting the same genes in cancer, that genetic changes of epigenetic genes can lead to epigenetic modifications, and vice versa, that epigenetic changes of DNA repair genes can lead to genetic alterations.

Future directions

After the completion of the human genome and genomes from many other organisms, epigeneticists worldwide are now calling for an international effort to characterize the epigenome [178–184]. This endeavor is indeed a large-scale project if one considers the presence of inter-individual, tissue-specific,

Table 8.2 Alterations of genes involved in DNA methylation and histone modifications in cancer.

Gene	Alteration	Tumor type
Alterations affecting DNA methylation enzymes (DNMTs)		
DNMT1	Overexpression	Various
DNMT3b	Overexpression	Various
Alterations affecting methyl-CpG-binding proteins (MBPs)		
MeCP2	Overexpression, rare mutations	Various
MBD1	Overexpression, rare mutations	Various
MBD2	Overexpression, rare mutations	Various
MBD3	Overexpression, rare mutations	Various
MBD4	Mutations in microsatellite instable tumors	Colon, stomach, endometrium
Alterations affecting histone acetyltransferases (HATs)		
p300	Mutations in microsatellite instable tumors	Colon, stomach, endometrium
CBP	Mutations, translocations, deletions	Colon, stomach, endometrium, lung, leukemia
pCAF	Rare mutations	Colon
MOZ	Translocations	Hematological malignancies
MORF	Translocations	Hematological malignancies, leiomyomata
Alterations affecting histone deacetylases (HDACs)		
HDAC1	Aberrant expression	Various
HDAC2	Aberrant expression, mutations in microsatellite instable tumors	Various
Alterations affecting histone methyltransferases (HMTs)		
MLL1	Translocation	Hematological malignancies
MLL2	Gene amplification	Glioma, pancreas
MLL3	Deletion	Leukemia
NSD1	Translocation	Leukemia
EZH2	Gene amplification, overexpression	Various
RIZ1	Promoter CpG island hypermethylation	Various
Alterations affecting histone demethylases		
GASC1	Gene amplification	Squamous cell carcinoma

(Adapted from Esteller, 2007) [16]

and disease-specific epigenomes, and that the epigenome is a dynamic system that can be altered throughout life in response to environmental cues [2]. To reach that aim, several initiatives have been already started, such as the Human Epigenome Project (HEP) [81, 183, 185], Alliance for Human Epigenomics and Disease (AHEAD) [178], and the National Methylome Project for Chromosome 21 (NAME 21) [181]. The first complete DNA methylomes have already been obtained for small organisms such as oncogenic viruses [55] and *Arabidopsis thaliana* [56, 57].

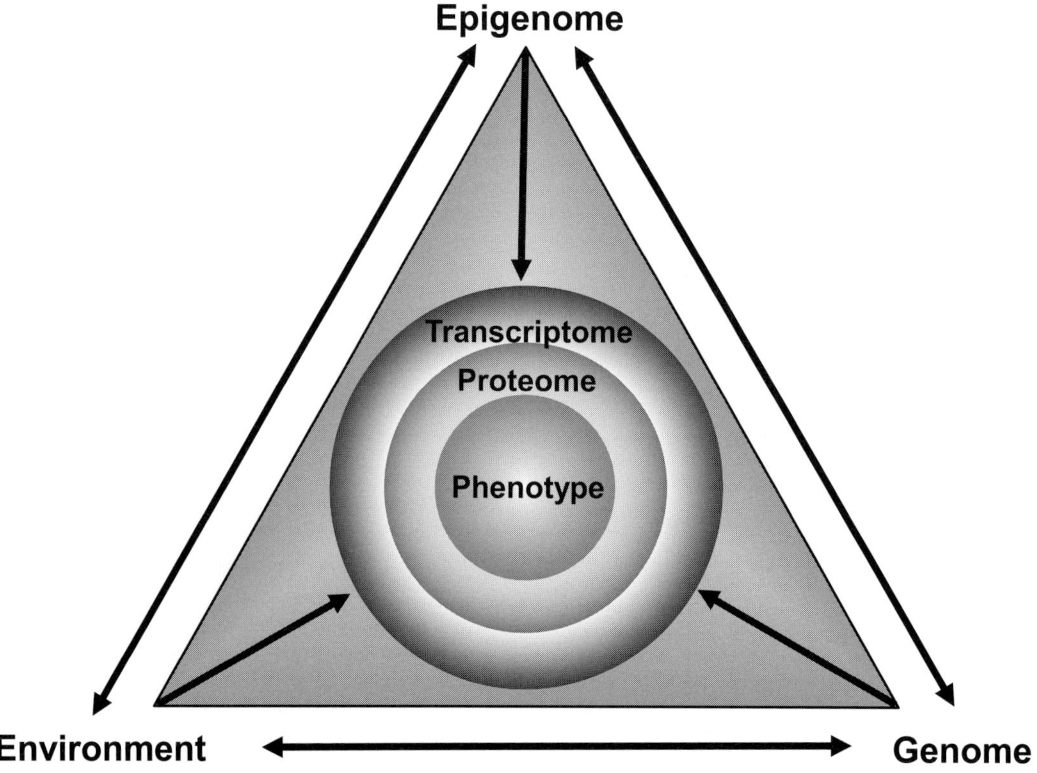

Figure 8.5 Integrative model of environmental, epigenetic, and genetic factors. The genome and epigenome interact with each other and with the enviroment to determine the cell transcriptome and proteome, and finally lead to a certain phenotype. (A black and white version of this figure will appear in some formats. For the color version, please refer to the plate section.)

The precise delineation of the human epigenome, or at least of the DNA methylome, requires sequencing of the methylation status of each of the 55 million CpGs in the genome in different cell types, diseases, and individuals. This is still not possible with current techniques, and alternatives to select part of the epigenome, for example, by reduced representation, array capture, or using padlock probes in combination with ultrasequencing are being applied [85, 186–188]. However, one of the inherent limitations of bisulfite sequencing is the reduced complexity of the genome achieved by converting unmethylated cytosines to uracil (thymine after amplification). Thus the four bases of the genome are mostly reduced to three, which renders the identification and correct mapping of short sequences a hard and sometimes impossible bioinformatic task. This problem might be solved in the near future by the application of the next-generation (i.e., third-generation) sequencers.

These are based on nanopore technology, in which a DNA strand is transported through a nanopore under an electrical potential and individual bases from large fragments can be read. The advantage of this technology for epigenomics is that methylated and unmethylatd cytosines can be differentiated without bisulfite conversion by their different ionic conductivity [189].

However, as long as the new technologies are not affordable for the study of multiple samples and conditions, which is mandatory considering the heterogeneous nature of the epigenome, the application of microarrays in epigenomics will continue to play an important role in research.

Technical limitations and challenges are discussed above. However, at the biological level we have to consider that the epigenetic code is only one layer of information. The complete understanding of a cell and of diseases such as cancer will require the integration of the genome, epigenome, transcriptome, and proteome under the point of view of a systems biology approach (Figure 8.5). This strategy should also include information about life style and environmental exposures because both health and disease are caused by an interplay between genome and environment. The integration of these different layers, or networks, of cell physiology into a unified cellular system will be of great importance to understand the mechanisms underlying normal and altered physiology.

References

1. Kaminsky ZA, Tang T, Wang SC, et al. 2009. DNA methylation profiles in monozygotic and dizygotic twins. *Nat Genet* 41:240–5.

2. Fraga MF, Ballestar E, Paz MF, et al. 2005. Epigenetic differences arise during the lifetime of monozygotic twins. *Proc Natl Acad Sci USA* 102:10604–9.

3. Humpherys D, Eggan K, Akutsu H, et al. 2001. Epigenetic instability in ES cells and cloned mice. *Science* 293:95–7.

4. Bird A. 2002. DNA methylation patterns and epigenetic memory. *Genes Dev* 16:6–21.

5. Bird A. 2007. Perceptions of epigenetics. *Nature* 447:396–8.

6. Jaenisch R, Bird A. 2003. Epigenetic regulation of gene expression: how the genome integrates intrinsic and environmental signals. *Nat Genet* 33 Suppl:245–54.

7. Kouzarides T. 2007. Chromatin modifications and their function. *Cell* 128:693–705.

8. Chen K, Rajewsky N. 2007. The evolution of gene regulation by transcription factors and microRNAs. *Nat Rev Genet* 8:93–103.

9. Fraser P, Bickmore W. 2007. Nuclear organization of the genome and the potential for gene regulation. *Nature* 447:413–7.

10. Zaratiegui M, Irvine DV, Martienssen RA. 2007. Noncoding RNAs and gene silencing. *Cell* 128:763–76.

11. Bernstein BE, Meissner A, Lander ES. 2007. The mammalian epigenome. *Cell* 128:669–81.

12. Christensen BC, Houseman EA, Marsit CJ, et al. 2009. Aging and environmental exposures alter tissue-specific DNA methylation dependent upon CpG island context. *PLoS Genet* 5:e1000602.

13. Jirtle RL, Skinner MK. 2007. Environmental epigenomics and disease susceptibility. *Nat Rev Genet* 8:253–62.

14. Weaver IC, Cervoni N, Champagne FA, et al. 2004. Epigenetic programming by maternal behavior. *Nat Neurosci* 7:847–54.

15. Weidman JR, Dolinoy DC, Murphy SK, Jirtle RL. 2007. Cancer susceptibility: epigenetic manifestation of environmental exposures. *Cancer J* 13:9–16.

16. Esteller M. 2007. Cancer epigenomics: DNA methylomes and histone-modification maps. *Nat Rev Genet* 8:286–98.

17. Esteller M. 2008. Epigenetics in cancer. *N Engl J Med* 358:1148–59.

18. Jones PA, Baylin SB. 2007. The epigenomics of cancer. *Cell* 128:683–92.

19. Laird PW. 2005. Cancer epigenetics. *Hum Mol Genet* 14 Spec No 1:R65–76.

20. Robertson KD. 2005. DNA methylation and human disease. *Nat Rev Genet* 6:597–610.

21. Ehrlich M. 2002. DNA methylation in cancer: too much, but also too little. *Oncogene* 21:5400–13.

22. Feinberg AP, Tycko B. 2004. The history of cancer epigenetics. *Nat Rev Cancer* 4:143–53.

23. Plass C. 2002. Cancer epigenomics. *Hum Mol Genet* 11:2479–88.

24. Fraga MF, Esteller M. 2002. DNA methylation: a profile of methods and applications. *Biotechniques* 33:632, 634, 636–49.

25. Laird PW. 2003. The power and the promise of DNA methylation markers. *Nat Rev Cancer* 3:253–66.

26. Lyko F. 2005. Novel methods for analysis of genomic DNA methylation. *Anal Bioanal Chem* 381:67–8.

27. Shames DS, Minna JD, Gazdar AF. 2007. Methods for detecting DNA methylation in tumors: from bench to bedside. *Cancer Lett* 251:187–98.

28. Shen L, Waterland RA. 2007. Methods of DNA methylation analysis. *Curr Opin Clin Nutr Metab Care* 10:576–81.

29. Frommer M, McDonald LE, Millar DS, et al. 1992. A genomic sequencing protocol that yields a positive display of 5-methylcytosine residues in individual DNA strands. *Proc Natl Acad Sci USA* 89:1827–31.

30. Herman JG, Graff JR, Myohanen S, Nelkin BD, Baylin SB. 1996. Methylation-specific PCR: a novel PCR assay for methylation status of CpG islands. *Proc Natl Acad Sci USA* 93:9821–6.

31. Frigola J, Ribas M, Risques RA, Peinado MA. 2002. Methylome profiling of cancer cells by amplification of inter-methylated sites (AIMS). *Nucleic Acids Res* 30:e28.

32. Kawai J, Hirotsune S, Hirose K, et al. 1993. Methylation profiles of genomic DNA of mouse developmental brain detected by restriction landmark genomic scanning (RLGS) method. *Nucleic Acids Res* 21:5604–8.

33. Ramsay G. 1998. DNA chips: state-of-the art. *Nat Biotechnol* 16:40–4.

34. Adrien LR, Schlecht NF, Kawachi N, et al. 2006. Classification of DNA methylation patterns in tumor cell genomes using a CpG island microarray. *Cytogenet Genome Res* 114:16–23.

35. Ching TT, Maunakea AK, Jun P, et al. 2005. Microarrays identify evolutionary conservation of tissue-specific methylation of SHANK3. *Nat Genet* 37:645–51.

36. Zhang X, Yazaki J, Sundaresan A, et al. 2006. Genome-wide high-resolution mapping and functional analysis of DNA methylation in *Arabidopsis*. *Cell* 126:1189–201.

37. Weber M, Davies JJ, Wittig D, et al. 2005. Chromosome-wide and promoter-specific analyses identify sites of differential DNA methylation in normal and transformed human cells. *Nat Genet* 37:853–62.

38. Gebhard C, Schwarzfischer L, Pham TH, et al. 2006. Rapid and sensitive detection of CpG-methylation using methyl-binding (MB)-PCR. *Nucleic Acids Res* 34:e82.

39. Gebhard C, Schwarzfischer L, Pham TH, et al. 2006. Genome-wide profiling of CpG methylation identifies novel targets of aberrant hypermethylation in myeloid leukemia. *Cancer Res* 66:6118–28.

40. Huang TH, Perry MR, Laux DE. 1999. Methylation profiling of CpG islands in human breast cancer cells. *Hum Mol Genet* 8:459–70.

41. Khulan B, Thompson RF, Ye K, et al. 2006. Comparative isoschizomer profiling of cytosine methylation: the HELP assay. *Genome Res* 16:1046–55.

42. Adorjan P, Distler J, Lipscher E, et al. 2002. Tumour class prediction and discovery by microarray-based DNA methylation analysis. *Nucleic Acids Res* 30:e21.

43. Bibikova M, Chudin E, Wu B, et al. 2006. Human embryonic stem cells have a unique epigenetic signature. *Genome Res* 16:1075–83.

44. Bibikova M, Lin Z, Zhou L, et al. 2006. High-throughput DNA methylation profiling using universal bead arrays. *Genome Res* 16:383–93.

45. Martin-Subero JI, Kreuz M, Bibikova M, et al. 2009. New insights into the biology and origin of mature aggressive B-cell lymphomas by combined epigenomic, genomic, and transcriptional profiling. *Blood* 113:2488–97.

46. Karpf AR. 2007. Epigenomic reactivation screening to identify genes silenced by DNA hypermethylation in human cancer. *Curr Opin Mol Ther* 9:231–41.

47. Shames DS, Girard L, Gao B, et al. 2006. A genome-wide screen for promoter methylation in lung cancer identifies novel methylation markers for multiple malignancies. *PLoS Med* 3:e486.

48. Fraga MF, Ballestar E, Villar-Garea A, et al. 2005. Loss of acetylation at Lys16 and trimethylation at Lys20 of histone H4 is a common hallmark of human cancer. *Nat Genet* 37:391–400.

49. Seligson DB, Horvath S, Shi T, et al. 2005. Global histone modification patterns predict risk of prostate cancer recurrence. *Nature* 435:1262–6.

50. Bernstein BE, Humphrey EL, Liu CL, Schreiber SL. 2004. The use of chromatin immunoprecipitation assays in genome-wide analyses of histone modifications. *Methods Enzymol* 376:349–60.

51. Huebert DJ, Kamal M, O'Donovan A, Bernstein BE. 2006. Genome-wide analysis of histone modifications by ChIP-on-chip. *Methods* 40:365–9.

52. Fan JB, Chee MS, Gunderson KL. 2006. Highly parallel genomic assays. *Nat Rev Genet* 7:632–44.

53. Margulies M, Egholm M, Altman WE, et al. 2005. Genome sequencing in microfabricated high-density picolitre reactors. *Nature* 437:376–80.

54. Shendure J, Porreca GJ, Reppas NB, et al. 2005. Accurate multiplex polony sequencing of an evolved bacterial genome. *Science* 309:1728–32.

55. Fernandez AF, Rosales C, Lopez-Nieva P, et al. 2009. The dynamic DNA methylomes of double-stranded DNA viruses associated with human cancer. *Genome Res* 19:438–51.

56. Lister R, O'Malley RC, Tonti-Filippini J, et al. 2008. Highly integrated single-base resolution maps of the epigenome in *Arabidopsis*. *Cell* 133:523–36.

57. Cokus SJ, Feng S, Zhang X, et al. 2008. Shotgun bisulphite sequencing of the *Arabidopsis* genome reveals DNA methylation patterning. *Nature* 452:215–9.

58. Barski A, Cuddapah S, Cui K, et al. 2007. High-resolution profiling of histone methylations in the human genome. *Cell* 129:823–37.

59. Robertson AG, Bilenky M, Tam A, et al. 2008. Genome-wide relationship between histone H3 lysine 4 mono- and tri-methylation and transcription factor binding. *Genome Res* 18:1906–17.

60. Wang Z, Zang C, Rosenfeld JA, et al. 2008. Combinatorial patterns of histone acetylations and methylations in the human genome. *Nat Genet* 40:897–903.

61. Kim GD, Ni J, Kelesoglu N, Roberts RJ, Pradhan S. 2002. Co-operation and communication between the human maintenance

and de novo DNA (cytosine-5) methyltransferases. *EMBO J* 21:4183–95.

62. Rhee I, Bachman KE, Park BH, et al. 2002. DNMT1 and DNMT3b cooperate to silence genes in human cancer cells. *Nature* 416:552–6.

63. Rhee I, Jair KW, Yen RW, et al. 2000. CpG methylation is maintained in human cancer cells lacking DNMT1. *Nature* 404:1003–7.

64. Paz MF, Wei S, Cigudosa JC, et al. 2003. Genetic unmasking of epigenetically silenced tumor suppressor genes in colon cancer cells deficient in DNA methyltransferases. *Hum Mol Genet* 12:2209–19.

65. Jacinto FV, Ballestar E, Ropero S, Esteller M. 2007. Discovery of epigenetically silenced genes by methylated DNA immunoprecipitation in colon cancer cells. *Cancer Res* 67:11481–6.

66. Ballestar E, Esteller M. 2005. Methyl-CpG-binding proteins in cancer: blaming the DNA methylation messenger. *Biochem Cell Biol* 83:374–84.

67. Fraga MF, Ballestar E, Montoya G, et al. 2003. The affinity of different MBD proteins for a specific methylated locus depends on their intrinsic binding properties. *Nucleic Acids Res* 31:1765–74.

68. Ballestar E, Paz MF, Valle L, et al. 2003. Methyl-CpG binding proteins identify novel sites of epigenetic inactivation in human cancer. *EMBO J* 22:6335–45.

69. Klose RJ, Sarraf SA, Schmiedeberg L, et al. 2005. DNA binding selectivity of MeCP2 due to a requirement for A/T sequences adjacent to methyl-CpG. *Mol Cell* 19:667–78.

70. Lopez-Serra L, Ballestar E, Fraga MF, Alaminos M, Setien F, Esteller M. 2006. A profile of methyl-CpG binding domain protein occupancy of hypermethylated promoter CpG islands of tumor suppressor genes in human cancer. *Cancer Res* 66:8342–6.

71. Jones PL, Veenstra GJ, Wade PA, et al. 1998. Methylated DNA and MeCP2 recruit histone deacetylase to repress transcription. *Nat Genet* 19:187–91.

72. Nan X, Ng HH, Johnson CA, et al. 1998. Transcriptional repression by the methyl-CpG-binding protein MeCP2 involves a histone deacetylase complex. *Nature* 393:386–9.

73. Fuks F, Burgers WA, Godin N, Kasai M, Kouzarides T. 2001. Dnmt3a binds deacetylases and is recruited by a sequence-specific repressor to silence transcription. *EMBO J* 20:2536–44.

74. Robertson KD, Ait-Si-Ali S, Yokochi T, et al. 2000. DNMT1 forms a complex with Rb, E2F1 and HDAC1 and represses transcription from E2F-responsive promoters. *Nat Genet* 25:338–42.

75. Fuks F, Hurd PJ, Deplus R, Kouzarides T. 2003. The DNA methyltransferases associate with HP1 and the SUV39H1 histone methyltransferase. *Nucleic Acids Res* 31:2305–12.

76. Fuks F, Hurd PJ, Wolf D, et al. 2003. The methyl-CpG-binding protein MeCP2 links DNA methylation to histone methylation. *J Biol Chem* 278:4035–40.

77. Vire E, Brenner C, Deplus R, et al. 2006. The Polycomb group protein EZH2 directly controls DNA methylation. *Nature* 439:871–4.

78. Bestor TH. 2005. Transposons reanimated in mice. *Cell* 122:322–5.

79. Shen L, Kondo Y, Guo Y, et al. 2007. Genome-wide profiling of DNA methylation reveals a class of normally methylated CpG island promoters. *PLoS Genet* 3:2023–36.

80. Jenuwein T, Allis CD. 2001. Translating the histone code. *Science* 293:1074–80.

81. Eckhardt F, Lewin J, Cortese R, et al. 2006. DNA methylation profiling of human chromosomes 6, 20 and 22. *Nat Genet* 38:1378–85.

82. Grunau C, Hindermann W, Rosenthal A. 2000. Large-scale methylation analysis of human genomic DNA reveals tissue-specific differences between the methylation profiles of genes and pseudogenes. *Hum Mol Genet* 9:2651–63.

83. Strichman-Almashanu LZ, Lee RS, Onyango PO, et al. 2002. A genome-wide screen for normally methylated human CpG islands that can identify novel imprinted genes. *Genome Res* 12:543–54.

84. Weber M, Hellmann I, Stadler MB, et al. 2007. Distribution, silencing potential and evolutionary impact of promoter DNA methylation in the human genome. *Nat Genet* 39:457–66.

85. Meissner A, Mikkelsen TS, Gu H, et al. 2008. Genome-scale DNA methylation maps of pluripotent and differentiated cells. *Nature* 454:766–70.

86. Reik W, Walter J. 2001. Genomic imprinting: parental influence on the genome. *Nat Rev Genet* 2:21–32.

87. Bernstein BE, Mikkelsen TS, Xie X, et al. 2006. A bivalent chromatin structure marks key developmental genes in embryonic stem cells. *Cell* 125:315–26.

88. Azuara V, Perry P, Sauer S, et al. 2006. Chromatin signatures of pluripotent cell lines. *Nat Cell Biol* 8:532–8.

89. Sparmann A, van Lohuizen M. 2006. Polycomb silencers control cell fate, development and cancer. *Nat Rev Cancer* 6:846–56.

90. Esteller M. 2006. Epigenetics provides a new generation of oncogenes and tumour-suppressor genes. *Br J Cancer* 94:179–83.

91. Feinberg AP, Vogelstein B. 1983. Hypomethylation distinguishes genes of some human cancers from their normal counterparts. *Nature* 301:89–92.

92. Eden A, Gaudet F, Waghmare A, Jaenisch R. 2003. Chromosomal instability and tumors promoted by DNA hypomethylation. *Science* 300:455.

93. Gaudet F, Hodgson JG, Eden A, et al. 2003. Induction of tumors in mice by genomic hypomethylation. *Science* 300:489–92.

94. Karpf AR, Matsui S. 2005. Genetic disruption of cytosine DNA methyltransferase enzymes induces chromosomal instability in human cancer cells. *Cancer Res* 65:8635–9.

95. Ehrlich M. 2003. The ICF syndrome, a DNA methyltransferase 3B deficiency and immunodeficiency disease. *Clin Immunol* 109:17–28.

96. Howard G, Eiges R, Gaudet F, Jaenisch R, Eden A. 2008. Activation and transposition of endogenous retroviral elements in hypomethylation induced tumors in mice. *Oncogene* 27:404–8.

97. Feinberg AP. 1999. Imprinting of a genomic domain of 11p15 and loss of imprinting in cancer: an introduction. *Cancer Res* 59:1743s–6s.

98. Bodey B. 2002. Cancer-testis antigens: promising targets for antigen directed antineoplastic immunotherapy. *Expert Opin Biol Ther* 2:577–84.

99. Martinez R, Martin-Subero JI, Rohde V, et al. 2009. A microarray-based DNA methylation study of glioblastoma multiforme. *Epigenetics* 4:255–64.

100. Martin-Subero JI, Ammerpohl O, Bibikova M, et al. 2009. A comprehensive microarray-based DNA methylation study of 367 hematological neoplasms. *PLoS One* 4:e6986.

101. Paz MF, Fraga MF, Avila S, et al. 2003. A systematic profile of DNA methylation in human cancer cell lines. *Cancer Res* 63:1114–21.

102. Esteller M. 2007. Epigenetic gene silencing in cancer: the DNA hypermethylome. *Hum Mol Genet* 16 Spec No 1:R50–R9.

103. Costello JF, Fruhwald MC, Smiraglia DJ, et al. 2000. Aberrant CpG-island methylation has non-random and tumour-type-specific patterns. *Nat Genet* 24:132–8.

104. Esteller M, Fraga MF, Guo M, et al. 2001. DNA methylation patterns in hereditary human cancers mimic sporadic tumorigenesis. *Hum Mol Genet* 10:3001–7.

105. Grady WM, Willis J, Guilford PJ, et al. 2000. Methylation of the CDH1 promoter as the second genetic hit in hereditary diffuse gastric cancer. *Nat Genet* 26:16–17.

106. Piotrowski A, Benetkiewicz M, Menzel U, et al. 2006. Microarray-based survey of CpG islands identifies concurrent hyper- and hypomethylation patterns in tissues derived from patients with breast cancer. *Genes Chromosomes Cancer* 45:656–67.

107. Versmold B, Felsberg J, Mikeska T, et al. 2007. Epigenetic silencing of the candidate tumor suppressor gene PROX1 in sporadic breast cancer. *Int J Cancer* 121:547–54.

108. Yan PS, Chen CM, Shi H, et al. 2001. Dissecting complex epigenetic alterations in breast cancer using CpG island microarrays. *Cancer Res* 61:8375–80.

109. Estecio MR, Yan PS, Ibrahim AE, et al. 2007. High-throughput methylation profiling by MCA coupled to CpG island microarray. *Genome Res* 17:1529–36.

110. Hayashi H, Nagae G, Tsutsumi S, et al. 2007. High-resolution mapping of DNA methylation in human genome using oligonucleotide tiling array. *Hum Genet* 120:701–11.

111. Model F, Osborn N, Ahlquist D, et al. 2007. Identification and validation of colorectal neoplasia-specific methylation markers for accurate classification of disease. *Mol Cancer Res* 5:153–63.

112. Yan PS, Efferth T, Chen HL, et al. 2002. Use of CpG island microarrays to identify colorectal tumors with a high degree of concurrent methylation. *Methods* 27:162–9.

113. Cottrell S, Jung K, Kristiansen G, et al. 2007. Discovery and validation of 3 novel DNA methylation markers of prostate cancer prognosis. *J Urol* 177:1753–8.

114. Wang Y, Yu Q, Cho AH, et al. 2005. Survey of differentially methylated promoters in prostate cancer cell lines. *Neoplasia* 7:748–60.

115. Yu YP, Paranjpe S, Nelson J, et al. 2005. High throughput screening of methylation status of genes in prostate cancer using an oligonucleotide methylation array. *Carcinogenesis* 26:471–9.

116. Fukasawa M, Kimura M, Morita S, et al. 2006. Microarray analysis of promoter methylation in lung cancers. *J Hum Genet* 51:368–74.

117. Rauch T, Li H, Wu X, Pfeifer GP. 2006. MIRA-assisted microarray analysis, a new technology for the determination of DNA methylation patterns, identifies frequent methylation of homeodomain-containing genes in lung cancer cells. *Cancer Res* 66:7939–47.

118. Rauch T, Wang Z, Zhang X, et al. 2007. Homeobox gene methylation in lung cancer studied by genome-wide analysis with a microarray-based methylated CpG island recovery assay. *Proc Natl Acad Sci USA* 104:5527–32.

119. Ordway JM, Bedell JA, Citek RW, et al. 2006. Comprehensive DNA methylation profiling in a human cancer genome identifies novel epigenetic targets. *Carcinogenesis* 27:2409–23.

120. Pfister S, Schlaeger C, Mendrzyk F, et al. 2007. Array-based profiling of reference-independent methylation status (aPRIMES) identifies frequent promoter methylation and consecutive downregulation of ZIC2 in pediatric medulloblastoma. *Nucleic Acids Res* 35:e51.

121. Bjornsson HT, Brown LJ, Fallin MD, et al. 2007. Epigenetic specificity of loss of imprinting of the IGF2 gene in Wilms tumors. *J Natl Cancer Inst* 99:1270–3.

122. Swerdlow SH, Campo E, Harris NL, et al. 2008. *World Health Organization Classification of Tumours of Haematopoietic and Lymphoid Tissues.* Lyon: IARC Press.

123. Guo J, Burger M, Nimmrich I, et al. 2005. Differential DNA methylation of gene promoters in small B-cell lymphomas. *Am J Clin Pathol* 124:430–9.

124. Rahmatpanah FB, Carstens S, Guo J, et al. 2006. Differential DNA methylation patterns of small B-cell lymphoma subclasses with different clinical behavior. *Leukemia* 20:1855–62.

125. Scholz C, Nimmrich I, Burger M, et al. 2005. Distinction of acute lymphoblastic leukemia from acute myeloid leukemia through microarray-based DNA methylation analysis. *Ann Hematol* 84:236–44.

126. Shi H, Guo J, Duff DJ, et al. 2007. Discovery of novel epigenetic markers in non-Hodgkin's lymphoma. *Carcinogenesis* 28:60–70.

127. Taylor KH, Pena-Hernandez KE, Davis JW, et al. 2007. Large-scale CpG methylation analysis identifies novel candidate genes and reveals methylation hotspots in acute lymphoblastic leukemia. *Cancer Res* 67:2617–25.

128. van Doorn R, Zoutman WH, Dijkman R, et al. 2005. Epigenetic profiling of cutaneous T-cell lymphoma: promoter hypermethylation of multiple tumor suppressor genes including BCL7a, PTPRG, and p73. *J Clin Oncol* 23:3886–96.

129. Figueroa ME, Reimers M, Thompson RF, et al. 2008. An integrative genomic and epigenomic approach for the study of transcriptional regulation. *PLoS One* 3:e1882.

130. Kuang SQ, Tong WG, Yang H, et al. 2008. Genome-wide identification of aberrantly methylated promoter associated CpG islands in acute lymphocytic leukemia. *Leukemia* 22:1529–38.

131. Jiang Y, Dunbar A, Gondek LP, et al. 2009. Aberrant DNA methylation is a dominant mechanism in MDS progression to AML. *Blood* 113:1315–25.

132. Kroeger H, Jelinek J, Estecio MR, et al. 2008. Aberrant CpG island methylation in acute myeloid leukemia is accentuated at relapse. *Blood* 112:1366–73.

133. Irizarry RA, Ladd-Acosta C, Wen B, et al. 2009. The human colon cancer methylome shows similar hypo- and hypermethylation at conserved tissue-specific CpG island shores. *Nat Genet* 41:178–86.

134. He L, Hannon GJ. 2004. MicroRNAs: small RNAs with a big role in gene regulation. *Nat Rev Genet* 5:522–31.

135. Miska EA. 2005. How microRNAs control cell division, differentiation and death. *Curr Opin Genet Dev* 15:563–8.

136. Calin GA, Croce CM. 2006. MicroRNA signatures in human cancers. *Nat Rev Cancer* 6:857–66.

137. Lu J, Getz G, Miska EA, et al. 2005. MicroRNA expression profiles classify human cancers. *Nature* 435:834–8.

138. Cimmino A, Calin GA, Fabbri M, et al. 2005. miR-15 and miR-16 induce apoptosis by targeting BCL2. *Proc Natl Acad Sci USA* 102:13944–9.

139. Johnson SM, Grosshans H, Shingara J, et al. 2005. RAS is regulated by the let-7 microRNA family. *Cell* 120:635–47.

140. Thomson JM, Newman M, Parker JS, et al. 2006. Extensive post-transcriptional regulation of microRNAs and its implications for cancer. *Genes Dev* 20: 2202–7.

141. Saito Y, Liang G, Egger G, et al. 2006. Specific activation of microRNA-127 with downregulation of the proto-oncogene BCL6 by chromatin-modifying drugs in human cancer cells. *Cancer Cell* 9:435–43.

142. Lujambio A, Ropero S, Ballestar E, et al. 2007. Genetic unmasking of an epigenetically silenced microRNA in human cancer cells. *Cancer Res* 67:1424–9.

143. Agirre X, Vilas-Zornoza A, Jimenez-Velasco A, et al. 2009. Epigenetic silencing of the tumor suppressor microRNA Hsa-miR-124a regulates CDK6 expression and confers a poor prognosis in acute lymphoblastic leukemia. *Cancer Res* 69:4443–53.

144. Roman-Gomez J, Agirre X, Jimenez-Velasco A, et al. 2009. Epigenetic regulation of microRNAs in acute lymphoblastic leukemia. *J Clin Oncol* 27:1316–22.

145. Lujambio A, Calin GA, Villanueva A, et al. 2008. A microRNA DNA methylation signature for human cancer metastasis. *Proc Natl Acad Sci USA* 105:13556–61.

146. Keshet I, Schlesinger Y, Farkash S, et al. 2006. Evidence for an instructive mechanism of de novo methylation in cancer cells. *Nat Genet* 38:149–53.

147. Ohm JE, Baylin SB. 2007. Stem cell chromatin patterns: an instructive mechanism for DNA hypermethylation? *Cell Cycle* 6:1040–3.

148. Feltus FA, Lee EK, Costello JF, Plass C, Vertino PM. 2003. Predicting aberrant CpG island methylation. *Proc Natl Acad Sci USA* 100:12253–8.

149. Frigola J, Song J, Stirzaker C, et al. 2006. Epigenetic remodeling in colorectal cancer results in coordinate gene suppression across an entire chromosome band. *Nat Genet* 38:540–9.

150. Ehrich M, Turner J, Gibbs P, et al. 2008. Cytosine methylation profiling of cancer cell lines. *Proc Natl Acad Sci USA* 105:4844–9.

151. Ohm JE, McGarvey KM, Yu X, et al. 2007. A stem cell-like chromatin pattern may predispose tumor suppressor genes to DNA hypermethylation and heritable silencing. *Nat Genet* 39:237–42.

152. Schlesinger Y, Straussman R, Keshet I, et al. 2007. Polycomb-mediated methylation on Lys27 of histone H3 pre-marks genes for de novo methylation in cancer. *Nat Genet* 39:232–6.

153. Widschwendter M, Fiegl H, Egle D, et al. 2007. Epigenetic stem cell signature in cancer. *Nat Genet* 39:157–8.

154. Feinberg AP, Ohlsson R, Henikoff S. 2006. The epigenetic progenitor origin of human cancer. *Nat Rev Genet* 7:21–33.

155. Rada-Iglesias A, Enroth S, Andersson R, et al. 2009. Histone H3 lysine 27 trimethylation in adult differentiated colon associated to cancer DNA hypermethylation. *Epigenetics* 4:107–13.

156. McCabe MT, Lee EK, Vertino PM. 2009. A multifactorial signature of DNA sequence and polycomb binding predicts aberrant CpG island methylation. *Cancer Res* 69:282–91.

157. Toh Y, Ohga T, Endo K, et al. 2004. Expression of the metastasis-associated MTA1 protein and its relationship to deacetylation of the histone H4 in esophageal squamous cell carcinomas. *Int J Cancer* 110:362–7.

158. Toh Y, Yamamoto M, Endo K, et al. 2003. Histone H4 acetylation and histone deacetylase 1 expression in esophageal squamous cell carcinoma. *Oncol Rep* 10:333–8.

159. Yasui W, Oue N, Aung PP, et al. 2005. Molecular-pathological prognostic factors of gastric cancer: a review. *Gastric Cancer* 8:86–94.

160. Faure AK, Pivot-Pajot C, Kerjean A, et al. 2003. Misregulation of histone acetylation in Sertoli cell-only syndrome and testicular cancer. *Mol Hum Reprod* 9:757–63.

161. Nouzova M, Holtan N, Oshiro MM, et al. 2004. Epigenomic changes during leukemia cell differentiation: analysis of histone acetylation and cytosine methylation using CpG island microarrays. *J Pharmacol Exp Ther* 311:968–81.

162. Kang J, Zhang D, Chen J, Liu Q, Lin C. 2005. Antioxidants and trichostatin A synergistically protect against in vitro cytotoxicity of Ni2+ in human hepatoma cells. *Toxicol In Vitro* 19:173–82.

163. Biron VL, McManus KJ, Hu N, Hendzel MJ, Underhill DA. 2004. Distinct dynamics and distribution of histone methyl-lysine derivatives in mouse development. *Dev Biol* 276:337–51.

164. Sarg B, Koutzamani E, Helliger W, Rundquist I, Lindner HH. 2002. Postsynthetic trimethylation of histone H4 at lysine 20 in mammalian tissues is associated with aging. *J Biol Chem* 277:39195–201.

165. Tryndyak VP, Kovalchuk O, Pogribny IP. 2006. Loss of DNA methylation and histone H4 lysine 20 trimethylation in human breast cancer cells is associated with aberrant expression of DNA methyltransferase 1, Suv4-20h2 histone methyltransferase and methyl-binding proteins. *Cancer Biol Ther* 5:65–70.

166. Chen YX, Fang JY, Lu R, Qiu DK. 2007. Expression of p21(WAF1) is related to acetylation of histone H3 in total chromatin in human colorectal cancer. *World J Gastroenterol* 13:2209–13.

167. Chen YX, Fang JY, Zhu HY, Lu R, Cheng ZH, Qiu DK. 2004. Histone acetylation regulates p21WAF1 expression in human colon cancer cell lines. *World J Gastroenterol* 10:2643–6.

168. Richon VM, Sandhoff TW, Rifkind RA, Marks PA. 2000. Histone deacetylase inhibitor selectively induces p21WAF1 expression and gene-associated histone acetylation. *Proc Natl Acad Sci USA* 97:10014–19.

169. Pruitt K, Zinn RL, Ohm JE, et al. 2006. Inhibition of SIRT1 reactivates silenced cancer genes without loss of promoter DNA hypermethylation. *PLoS Genet* 2:e40.

170. Kondo Y, Shen L, Cheng AS, et al. 2008. Gene silencing in cancer by histone H3 lysine 27 trimethylation independent of promoter DNA methylation. *Nat Genet* 40:741–50.

171. Brena RM, Costello JF. 2007. Genome–epigenome interactions in cancer. *Hum Mol Genet* 16 Spec No 1:R96–R105.

172. Jones PA, Baylin SB. 2002. The fundamental role of epigenetic events in cancer. *Nat Rev Genet* 3:415–28.

173. Graff JR, Herman JG, Lapidus RG, et al. 1995. E-cadherin expression is silenced by DNA hypermethylation in human

breast and prostate carcinomas. *Cancer Res* 55:5195–9.

174. Li H, Ma X, Wang J, et al. 2007. Effects of rearrangement and allelic exclusion of JJAZ1/SUZ12 on cell proliferation and survival. *Proc Natl Acad Sci USA* 104:20001–6.

175. Rajasekhar VK, Begemann M. 2007. Concise review: roles of polycomb group proteins in development and disease: a stem cell perspective. *Stem Cells* 25:2498–510.

176. Haupt Y, Bath ML, Harris AW, Adams JM. 1993. bmi-1 transgene induces lymphomas and collaborates with myc in tumorigenesis. *Oncogene* 8:3161–4.

177. Jacobs JJ, Scheijen B, Voncken JW, et al. 1999. Bmi-1 collaborates with c-Myc in tumorigenesis by inhibiting c-Myc-induced apoptosis via INK4a/ARF. *Genes Dev* 13:2678–90.

178. Jones PA, Martienssen R. 2005. A blueprint for a Human Epigenome Project: the AACR Human Epigenome Workshop. *Cancer Res* 65:11241–6.

179. Esteller M. 2006. The necessity of a human epigenome project. *Carcinogenesis* 27:1121–5.

180. Rauscher FJ, 3rd. 2005. It is time for a Human Epigenome Project. *Cancer Res* 65:11229.

181. Jeltsch A, Walter J, Reinhardt R, Platzer M. 2006. German human methylome project started. *Cancer Res* 66:7378.

182. Garber K. 2006. Momentum building for human epigenome project. *J Natl Cancer Inst* 98:84–6.

183. Eckhardt F, Beck S, Gut IG, Berlin K. 2004. Future potential of the Human Epigenome Project. *Expert Rev Mol Diagn* 4:609–18.

184. Bradbury J. 2003. Human epigenome project – up and running. *PLoS Biol* 1:e82.

185. Rakyan VK, Hildmann T, Novik KL, et al. 2004. DNA methylation profiling of the human major histocompatibility complex: a pilot study for the human epigenome project. *PLoS Biol* 2: e405.

186. Ball MP, Li JB, Gao Y, et al. 2009. Targeted and genome-scale strategies reveal gene-body methylation signatures in human cells. *Nat Biotechnol* 27: 361–8.

187. Deng J, Shoemaker R, Xie B, et al. 2009. Targeted bisulfite sequencing reveals changes in DNA methylation associated with nuclear reprogramming. *Nat Biotechnol* 27:353–60.

188. Hodges E, Smith AD, Kendall J, et al. 2009. High definition profiling of mammalian DNA methylation by array capture and single molecule bisulfite sequencing. *Genome Res* 19:1593–605.

189. Clarke J, Wu HC, Jayasinghe L, et al. 2009. Continuous base identification for single-molecule nanopore DNA sequencing. *Nat Nanotechnol* 4:265–70.

Chapter 9

MicroRNA epigenetic systems and cancer

Holly Lewis and Aurora Esquela-Kerscher

Introduction

It is most common to refer to genetic mutations when discussing the causes of cancer. However, there are other mechanisms at work in the initiation and progression of this disease. Cancer occurs when cells proliferate uncontrollably and do not undergo apoptosis upon damage. This is due to inappropriate activation or inhibition of signaling pathways that control cell growth, cell differentiation, and apoptosis. These pathways can be altered by mutations or epigenetic mechanisms. Epigenetics refers to changes that occur in gene expression that are not due to alterations in the DNA sequence. These modifications are retained throughout cell division and can be passed on for multiple generations.

The four main types of epigenetic modifications include DNA methylation, covalent histone modifications, non-covalent modifications, such as nucleosome remodelling or incorporation of histone variants, and non-coding RNA-mediated gene regulation, particularly those involving microRNAs (miRNAs). All of these mechanisms alter the accessibility of chromatin to transcription factors and/or the ability of genes to be activated or silenced and can therefore create a unique gene expression profile for each cell.

Epigenetics is involved in a variety of biological processes that includes X-chromosome inactivation, genomic imprinting, and reprogramming genomes during differentiation and development (Veeck and Esteller, 2010). An epigenetic link to cancer was made over twenty years ago with the observation that the tumor suppressor gene, retinoblastoma, is often hypermethylated in the diseased state (Greger et al., 1989). Hypermethylation of promoter regions is the most common epigenetic change to occur in tumors (Jones and Baylin, 2002). DNA methylation and histone modifications that alter gene expression, usually by silencing genes, are now known to occur frequently in cancer. Recently, epigenetic mechanisms associated with miRNAs have gained considerable attention. These small ~22-nucleotide RNA transcripts are commonly dysregulated in human cancers and a growing subset of miRNAs designated as "oncomirs" participate in the initiation and progression of cancer (Esquela-Kerscher and Slack, 2006).

MicroRNAs and cancer

The discovery of the first miRNA, *lin-4*, occurred in the nematode *Caenorhabditis elegans* in 1993 (Lee et al., 1993) and this growing class of small RNAs has quickly emerged as important regulators of many cellular processes. MiRNAs are small non-coding RNA molecules that post-transcriptionally regulate gene expression (Bartel, 2004; Filipowicz et al., 2008). MiRNAs undergo extensive nuclear and cytoplasmic processing in order to generate mature miRNA species that regulate their targets in a sequence-specific manner via imperfect complementary binding to sites found within the messenger RNA (mRNA) transcript. MiRNAs promote translational inhibition and/or degradation of the mRNA transcript, thereby reducing gene expression of its target. To date, thousands of miRNAs have been identified in animal systems (Griffiths-Jones et al., 2008) (~2,500 in the human genome alone) and are found to control essential biological events such as cellular growth and differentiation, apoptosis, and metabolic and immune responses (Bartel, 2004; Harfe, 2005; Pasquinelli et al., 2005; Stefani and Slack, 2008) that are related to pathways involved in cancer progression. MiRNAs can regulate multiple targets and any given target can be regulated by multiple miRNAs. It

Systems Biology of Cancer, ed. S. Thiagalingam. Published by Cambridge University Press. © Cambridge University Press 2015.

is therefore not surprising that an estimated 30% of protein-coding genes in the human genome are controlled by at least one miRNA.

A direct role for miRNAs in cancer progression was first suggested with the observation that two clustered miRNAs, miR-15a and miR-16-1, were commonly deleted in B-cell chronic lymphocytic leukemia and hypothesized to function as tumor suppressor genes (Calin et al., 2002). Since then, hundreds of miRNAs are reported to be dysregulated in virtually every class of human cancers (Calin and Croce, 2006). The aberrant expression of miRNAs in cancerous tissues can be attributed to multiple mechanisms, which include incorrect miRNA processing, genetic defects, and epigenetic alterations of miRNAs. This chapter will touch upon each of these mechanisms and discuss how the misexpression of certain miRNA subsets likely contributes to tumor formation and metastasis.

MiRNA biogenesis and mRNA targeting

The biogenesis of miRNAs is a complex and highly regulated process resulting in the generation of mature biologically active ~22-nucleotide miRNAs (Figure 9.1) (Breving and Esquela-Kerscher, 2010). MiRNA maturation begins with the transcription of a precursor pri-miRNA in the nucleus by RNA polymerase II in most cases (although RNA polymerase III is sometimes used (Borchert et al., 2006)). The pri-miRNA typically ranges between 100 and 1,000 nucleotides in length, and often contains a 5′ methyl cap (^{7}MGpppG), a 3′ poly A tail, and can encode multiple miRNA genes. The pri-miRNA is processed in the nucleus by the RNase III enzyme Drosha and its co-factor, DiGeorge syndrome critical region 8 (DGCR8), to form a 70-nucleotide hairpin pre-miRNA product. Mirtrons, however, circumvent Drosha-mediated processing altogether, and are pre-miRNA species generated during splicing and

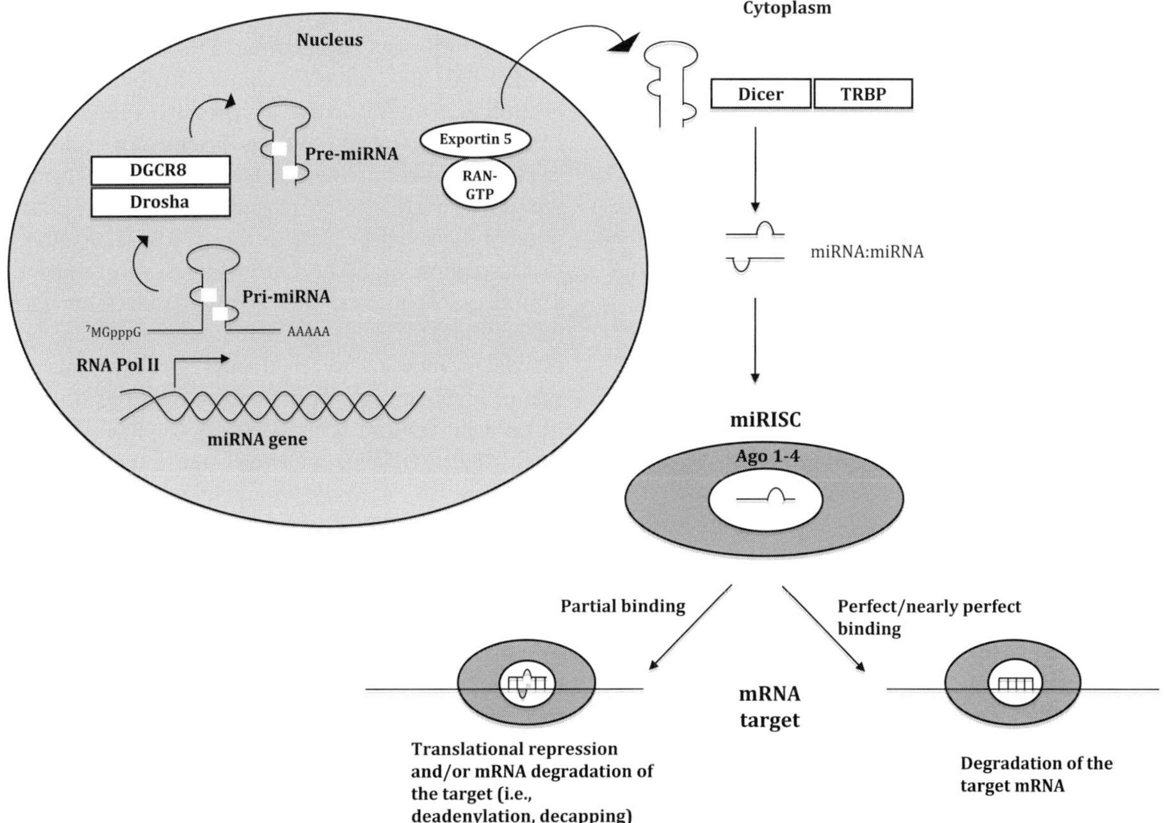

Figure 9.1 MiRNA biogenesis. MiRNA biogenesis is complex and requires separate processing events in the nucleus and the cytoplasm in order to generate the biologically mature ~22 nucleotide miRNA. See text for details.

disbranching events from the intronic regions of mRNA transcripts (Chan and Slack, 2007). The pre-miRNA is subsequently exported into the cytoplasm by RAN-GTP and Exportin 5 and once transported, associates with a second RNase III enzyme, Dicer, and its co-factor, TAR RNA-binding protein (TRBP), which cleave the hairpin structure of the pre-miRNA to release a ~22-nucleotide miRNA:miRNA* double-stranded RNA duplex. One strand of this duplex structure is preferentially loaded into a large multi-protein miRNA-associated RNA-induced silencing complex (miRISC) while the other strand is degraded. The exact composition of the miRISC complex is poorly understood but the Argonaute protein (AGO 1–4) is the key catalytic component of the complex responsible for miRNA strand selection and directing the miRNA toward its target mRNA to alter gene expression. Most Metazoan miRNAs bind with imperfect complementarity to the 3′ UTR of the target mRNA, although miRNA binding within the 5′ UTR and coding regions of the mRNA target have also been observed to modulate target expression levels (Breving and Esquela-Kerscher, 2010). Furthermore, perfect base pairing of a highly conserved "seed sequence" (nucleotides 2–8) in the miRNA is important for proper targeting (Doench and Sharp, 2004). Ultimately, miRNAs mediate negative regulation of their mRNA targets using post-transcriptional mechanisms associated with translational inhibition and/or mRNA degradation (for a more detailed description of miRNA-based regulation please refer to Filipowicz et al., 2008).

Impaired miRNA biogenesis and cancer

Tumors exhibit a global downregulation of miRNAs when compared to normal tissues (Lu et al., 2005). This might occur due to impaired miRNA transcription, miRNA processing, or the ability of the miRNA to load into the miRISC complex and associate with its targets. There are many examples linking defects in miRNA processing and cancer progression. For instance, the oncogene *c-Myc* and the tumor suppressor gene p53 have both been shown to bind to specific miRNA promoter elements and control their expression (Bommer et al., 2007; Chang et al., 2007, 2008; He et al., 2007; Raver-Shapira et al., 2007; Tarasov et al., 2007), suggesting that cells harboring mutations in these proteins could be transformed as a consequence of miRNA misexpression. Furthermore,

approximately 5 to 10% of miRNAs are thought to be epigenetically regulated at CpG islands and miRNA loci such as the promoter regulatory region of miR-124 are found to be hypermethylated in various human cancer cell lines and tumors (e.g., colon, breast, lung, leukemias, and lymphomas) compared to normal tissues (Brueckner et al., 2007; Han et al., 2007; Lujambio et al., 2008; Saito et al., 2006; Toyota et al., 2008).

Altered expression of proteins controlling both pri-miRNA and pre-miRNA processing is also associated with a variety of cellular defects and human cancers. For instance, reduced expression of Drosha (as well as Dicer) correlate with aggressive forms of ovarian cancer and a poor clinical outcome in these patients (Merritt et al., 2008). Defects in the Drosha-DGCR8 complex resulting in reduced pri-miRNA processing of the tumor suppressor miRNA, miR-7, have been reported in glioblastoma tumors (Kefas et al., 2008). Consistent with these findings, decreased production of miRNAs in DGCR8 knockout mice led to defects in cell proliferation and differentiation (Wang et al., 2007). During the progression of cancer, cells become less differentiated and lack of the Drosha-DGCR8 complex may contribute to this phenotype. Interestingly, increased expression of Drosha is also linked to cancer progression. Elevated levels of Drosha are found in late-stage cervical cancer samples and are also correlated with poor prognosis of esophageal cancer patients (Davis and Hata, 2009).

Regulation of Dicer expression is also linked to a number of cancers, with reduced expression seen in non-small cell lung carcinomas that strongly correlates with a shortened post-operative survival rate in lung cancer patients and overexpression seen in prostate tumors (Davis and Hata, 2009; Karube et al., 2005). Deletion of *Dicer1* can increase tumor formation in a *K-RAS*-inducible mouse model (Kumar et al., 2007). Furthermore, an inactivating mutation within TRBP, a protein that helps stabilize Dicer, correlates with hereditary and sporadic colon and gastric carcinomas with microsatellite instability (Melo et al., 2009). Due to this mutation, a 90% reduction of pre-miRNA processing is noted in colorectal cancerous cells and the exogenous delivery of wild type TRBP into these cells leads to decreased tumor formation when injected into immunocompromised mice (Melo et al., 2009). Taken together, mutations in the Drosha and Dicer processing machinery potentially play indirect roles in promoting tumorigenesis due to a global reduction of endogenous miRNAs.

There are also instances in which protein co-factors that aid in the processing of specific subsets of miRNAs are associated with human cancers. The most studied example is *LIN28*, an RNA-binding protein that functions to inhibit the maturation of the *let-7* miRNA family, a group of tumor suppressor miRNAs (described in detail in the next section), at the levels of both Drosha and Dicer processing (Heo et al., 2008; Newman et al., 2008; Piskounova et al., 2008; Rybak et al., 2008; Viswanathan et al., 2008; Wu and Belasco, 2005). Upregulation of *LIN-28* is often observed in human germ-cell tumors, hepatocellular carcinomas, Wilms' tumor, chronic myeloid leukemia, and ovarian cancer (Guo et al., 2006; Viswanathan et al., 2008, 2009; West et al., 2009; Yu et al., 2007b). Exogenous delivery of *LIN28* to human breast cancer cell lines in culture induces their growth and trans-formation, likely due to decreased *let-7* miRNA expression (Viswanathan et al., 2008). Therefore dysregulation of certain classes of "oncomirs" such as the tumor suppressor genes belonging to the *let-7* family could be attributed to defects in processing factors required for miRNA maturation. Thus proteins like *LIN28* are potential therapeutic targets that could be used to control miRNA expression and tumor formation.

Alterations in the Argonaute proteins, essential components of the miRISC complex, have also been associated with cellular transformations. The deletion of the *Ago3*, *Ago1*, and *Ago4* genes (EIF2C1), which are clustered on 1p34–35, is often noted in Wilms tumors of the kidney, in neuroblastoma, and in breast, liver, and colon cancers (Carmell et al., 2002; Koesters et al., 1999; Nelson et al., 2003). Conversely, the overexpression of *Ago2* strongly correlates with estrogen receptor α-negative breast cancers and can induce proliferation and migration of ERα-positive MCF7 cells *in vitro* (Adams et al., 2009). Recent work has indicated that the Argonaute proteins might be useful biomarkers for colon cancer as their expression is found to be signifi-cantly higher in colon tumors and distant metastasis compared to adjacent normal tissue (Li et al., 2010).

in the genome that are more prone to deletions, amplifications, viral insertions, and chromosome fusions – or reside in loci commonly associated with human cancers (Calin et al., 2004b). Altered expression of miRNAs in tumors compared to normal tissue can be attributed to these types of chromosomal abnormalities. For instance, the genomic locus encoding miR-15a and miR-16-1 is commonly deleted in tumors, whereas the miR-17-92 cluster is often amp-lified in diseased tissues. MiRNA expression profiles for tumors of different tissue origins are being exten-sively studied and, for some cancers, unique miRNA signatures have been defined that discriminate between transformed and normal cells (Calin and Croce, 2006; Lu et al., 2005), indicating the future utility of miRNAs as cancer biomarkers.

The altered expression of certain miRNAs in can-cerous tissues has also led to functional studies that indicate that these small RNAs can directly act as tumor suppressor genes and oncogenes (Calin and Croce, 2006; Esquela-Kerscher and Slack, 2006; Kent and Mendell, 2006). A miRNA has a tumor suppressor function if loss of this miRNA leads to the malignant transformation of a cell, whereas a miRNA is thought to have an oncogenic function if overexpression leads to uncontrolled growth and tumor formation. Three criteria must be met in order to distinguish miRNAs as "oncomirs" that function as *bona fide* tumor suppres-sor genes or oncogenes. Specifically, the miRNA must be misexpressed in diseased compared to normal tissues; the miRNA should influence cellular growth and differentiation processes *in vitro* and alter tumor formation in animal models; and the miRNA should regulate the post-transcriptional expression of targets that are pertinent to cancer progression. However, while a growing subset of miRNAs are classified as tumor suppressor genes and/or oncogenes, aberrant expression of additional miRNAs in the diseased state may only be a secondary effect of gross genetic alter-ations and the transformation process. The miRNAs that have experimentally met the criteria to be categor-ized as *bona fide* tumor suppressor genes and onco-genes (as those described in the next sections) hold great promise as therapeutic targets for human cancer.

MiRNAs play a direct role in cancer progression

A large number of miRNAs map within areas previously designated as chromosomal fragile sites – unstable areas

MiRNAs as tumor suppressor genes
The *let-7* family

let-7 was the second miRNA ever discovered and shown in *C. elegans* to play an important role in the timing of cell fate determination during larval development by controlling cell cycle exit and terminal cell differentiation (Reinhart et al., 2000). The *let-7* family, consisting of 13 human homologs, appears to have a conserved role to repress growth and proliferation in mammalian cells (Akao et al., 2006; Calin et al., 2004b; Dong et al., 2010; Esquela-Kerscher et al., 2008; Iorio et al., 2005; Johnson et al., 2005, 2007; Lee et al., 2005; Pasquinelli et al., 2000; Roush and Slack, 2008; Sonoki et al., 2005; Takamizawa et al., 2004; Yu et al., 2007a). For example, *let-7* overexpression in liver, lung, breast, prostate, and colon cancer cell lines results in reduced cellular proliferation via the G1 to S phase cell cycle arrest and inhibition of *let-7* activity leads to increased cellular growth. This family is also found to be downregulated in many different types of cancers, supporting the notion that *let-7* functions as a *bona fide* tumor suppressor gene. Indeed, a number of *let-7* homologs map to fragile sites on human chromosomes preferentially deleted in lung, breast, ovary, cervical, and urothelial cancers (Calin et al., 2004b). In particular, *let-7g* maps to 3p21, a well-known region deleted early in the progression of lung cancer.

Recent work using animal models also supports an *in vivo* role for the *let-7* family as tumor suppressor genes. The overexpression of *let-7* in breast, lung, and prostate cancer cell lines reduces tumor formation and metastases in xenograft experiments using immunodeficient mice (Dong et al., 2010; Esquela-Kerscher et al., 2008; Kumar et al., 2008; Yu et al., 2007a). Furthermore, the exogenous administration of *let-7* dramatically lowers tumor load in the lungs of mice expressing an activating mutation in the *K-ras* oncogene (Esquela-Kerscher et al., 2008; Kumar et al.,

2008; Trang et al., 2010). Evidence also indicates that certain *let-7* members are promising biomarkers for cancer. Patients with non-small cell lung carcinomas that have lower *let-7a* and *let-7f* expression in lung tumors possess a shortened post-operative survival rate compared to patients that exhibit higher *let-7* expression in their lung tumors (Takamizawa et al., 2004; Yanaihara et al., 2006). In addition, reduced levels of *let-7d* correlate with a poor outcome in patients with head and neck squamous cell carcinoma and ovarian cancer (Childs et al., 2009; Shell et al., 2007).

The *let-7* group negatively regulates a large range of targets that are associated with processes related to cellular growth and differentiation (Figure 9.2) (Park et al., 2007). The identification of the *let-7* target, *RAS*, was the first report to show that a miRNA can negatively regulate the expression of an oncogene (Johnson et al., 2005). *RAS* plays a major role in controlling cellular proliferation and apoptosis and activating mutations in *RAS* are found in 15 to 30% of all human tumors. Reciprocal expression of *let-7* and *RAS* is found in lung cancer tumors compared to normal adjacent tissue and overexpression of *let-7* in lung cancer cell lines directly represses *RAS* protein levels (Johnson et al., 2005). Furthermore, a single nucleotide polymorphism (SNP) located within a *let-7* binding site of the KRAS 3′ UTR (LCS6) closely correlates with both non-small cell lung carcinoma and ovarian cancers (Chin et al., 2008; Ratner et al., 2010). The *let-7* family has also been shown to control the expression of other oncogenes that include the transcription factor *c-Myc* and the chromatin remodelling gene HMGA2 (Lee and Dutta, 2007; Mayr et al., 2007; Sampson et al., 2007; Shell et al., 2007). *let-7* also

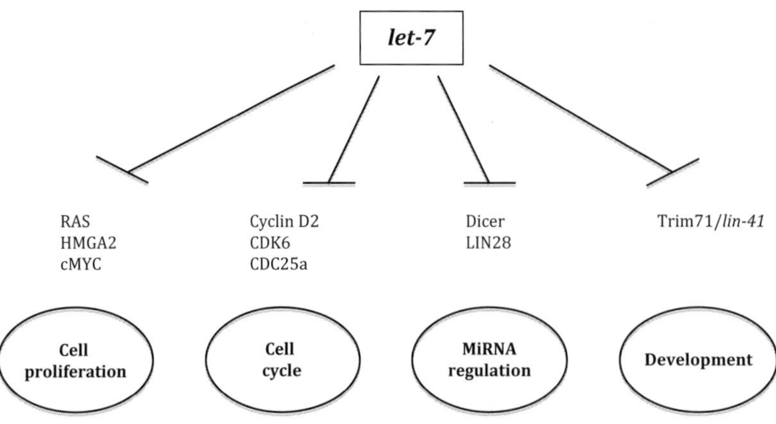

Figure 9.2 Targets of *let-7* reside in multiple signaling pathways to regulate cellular growth. *let-7* functions as a tumor suppressor gene by decreasing the expression of factors required for cellular proliferation, differentiation, cell cycle progression, and miRNA biogenesis (that includes *let-7* autoregulation via Dicer and LIN28).

represses the expression of various cell cycle progression genes, such as Cyclin D2 (CCND2), CDK6, and CDC25a (Johnson et al., 2007), and thus could globally regulate cellular proliferation and cancer progression pathways in a wide range of cell types. Interestingly, *let-7* appears to regulate its own expression (as well as other miRNAs) by targeting the processing factors *Dicer* and *LIN28* (Rybak et al., 2008), all of which are closely associated with human cancers (discussed in the section covering miRNA biogenesis).

Although the *let-7* family has primarily been defined as tumor suppressor genes in many different types of tissues, there have been instances where specific *let-7* members are reported to function as oncogenes in certain cellular and/or disease contexts. For example, elevated levels of *let-7a-3* (due to hypomethylation of miRNA regulatory elements) have been associated with ovarian and lung cancer (Brueckner et al., 2007; Lu et al., 2007) and upregulation of *let-7b* and *let-7i* expression has been linked to diffuse large B-cell lymphoma (Lawrie et al., 2008). Taken together, *let-7* is emerging as a promising therapeutic target for cancer. However, the observation that certain members of this large miRNA family can function both as tumor suppressor genes and oncogenes in tissues, such as the lung, highlights the need to more extensively study the mechanisms of *let-7* function in the diseased state before moving this technology into the clinic.

The miR-15a/miR-16 cluster

The miRNAs miR-15a and miR-16, which are clustered together at chromosome locus 13q14 and located within the intron of the non-coding gene Leu2, are commonly described as tumor suppressor genes. Deletions of 13q4 are found in a number of leukemias, including 65% of chronic lymphocytic leukemias (CLL), 50% of mantle cell lymphomas, and 16 to 40% of multiple myeloma cases (Calin et al., 2002). The miR-15a/miR-16 cluster is also deleted or downregulated in more than 50% of patients with B-cell CLL (Calin et al., 2005, 2008). Aside from having a role in leukemia, this cluster is also downregulated in a large range of human cancers such as non-small cell lung carcinomas, prostate, ovarian, and pancreatic cancer (Bandi et al., 2009; Bhattacharya et al., 2009; Bonci et al., 2008; Spizzo et al., 2009).

Target genes of miR-15a and miR-16 function in cellular processes such as apoptosis, cell cycle progression, and proliferation. The first identified target of miR-15a/miR-16 was the anti-apoptotic gene BCL2 (Cimmino et al., 2005). BCL2 is overexpressed in many cancers, including leukemias and lymphomas, where decreased expression of miR-15a and miR-16 is commonly seen. miR-15a and miR-16 are shown to target BCL2 in various human cancer cell lines, e.g., gastric and CLL cells *in vitro* (Cimmino et al., 2005; Xia et al., 2008). The restored expression of these miRNAs in an acute megakaryocytic leukemia cell line (MEG-01) induces programmed cell death and blocks tumor formation when these cells are implanted into immunocompromised mice (Calin et al., 2008), supporting a role for miR-15a and miR-16 in regulating apoptotic and cancer pathways. Additional targets of miR-15a and miR-16 are genes involved with cell cycle progression and cell division. For example, these miRNAs have both been found to downregulate the polycomb-group oncogene *BMI-1*, a cell cycle regulatory gene, in the context of ovarian cancer (Bhattacharya et al., 2009). In addition, miR-16 induces a block in cellular proliferation at the G0/G1 to S phase transition in the cell cycle in colon, lung, breast, and ovarian cell lines by targeting *CDK6*, *CDC27*, *CARD10*, and *C10orf46* (Calin et al., 2008; Linsley et al., 2007).

miR-15a and miR-16 appear to play an important role as tumor suppressor genes in prostate cancer progression. In a prostate cancer mouse model, miR-15a and miR-16 function to block tumor growth and invasion through the repression of the targets *BCL2*, Cyclin D1, and *WNT3A* (Bonci et al., 2008). Inhibition of these miRNAs using antimir oligonucleotides in non-tumorigenic prostate cell lines leads to the formation of tumors and causes increased tumor cell invasion when injected into immunodeficient mice (Bonci et al., 2008). The opposite also holds true, tumor size is reduced in mice when miR-15a and miR-16 are exogenously overexpressed in human prostate xenografts (Bonci et al., 2008). These studies show that the tumor suppressor miRNAs, miR-15a and miR-16, may be useful in the treatment of a large range of human malignancies.

MiRNAs as oncogenes
miR-17-92 cluster

The miR-17-92 polycistronic cluster is comprised of six miRNA genes, miR-17, miR-18a, miR-19a, miR-20a, miR-19b-1, and miR-92-1, and resides within intron 3 of the non-coding RNA, *c13orf25* that maps to the chromosomal locus 13q31. This locus is frequently amplified in hematopoietic malignancies, such as diffuse large B-cell lymphoma, follicular lymphoma, mantle cell lymphoma, primary cutaneous B-cell lymphoma, as well as in alveolar rhabdomyosarcoma and liposarcoma (Gordon et al., 2000; Ota et al., 2004; Schmidt et al., 2005), and is a common retroviral insertion site in mouse leukemias (Cui et al., 2007a; Tagawa and Seto, 2005; Wang et al., 2006). The miR-17-92 cluster is also upregulated in 65% of B-cell lymphomas and increased in solid cancers including breast, colon, lung, pancreas, prostate, and stomach (Calin et al., 2004a; Hayashita et al., 2005; Volinia et al., 2006).

Experimental evidence first indicated that miR-17-92 functions as an oncogene. Constitutive activation of *MYC* alone can induce tumor formation and in mice transgenic for *MYC* driven by the immunoglobulin heavy chain enhancer (Eμ), B-cell lymphomas form with a long latency of four to six months (He et al., 2005). However, overexpression of a truncated miR-17-19b-1 cluster (lacking miR-92-1) in this same transgenic *MYC* mouse model accelerates the development of lymphoma to 51 days (He et al., 2005). The lymphomas that arise due to *MYC* and miR-17-19b-1 cluster overexpression exhibit increased cellular proliferation, higher invasive potential, and decreased rates of apoptosis compared to malignancies formed due exclusively to increased *MYC* (He et al., 2005). Studies have been performed using this Eμ-*myc* mouse B-cell lymphoma model to determine that out of six miRNAs within the miR-17-92 cluster, miR-19 is the primary oncogenic factor responsible for promoting accelerated malignancies in these animals (Mu et al., 2009; Olive et al., 2009). Evidence also suggests that other miRNAs in the miR-17-92 cluster function in a coordinated fashion to impact growth, adhesion, and migration processes in both negative and positive ways to finely maintain cellular homeostasis (Olive et al., 2010; Shan et al., 2009).

The miR-17-92 cluster regulates a variety of targets that are involved in cell cycle progression, apoptosis, and angiogenesis that are consistent with the role of this cluster as oncogenic factors that promote tumor formation. The miR-17-92 cluster (particularly miR-17 and miR-20) is found to induce cellular proliferation by targeting the cyclin-dependent kinase inhibitor *CDKN1A* (p21) (Ivanovska et al., 2008; Petrocca et al., 2008a), a negative regulator of the G1/S cell cycle checkpoint. The anti-apoptotic activity of the miR-17-92 cluster can be attributed to the cluster's ability to block the function of the phosphatase and tensin homolog (PTEN) via miR-17, miR-20, and miR-19 (Dews et al., 2006; Xiao et al., 2008) and the pro-apoptotic gene *BCL2L11/BIM* by miR-17, miR-20, and miR-92 (Koralov et al., 2008; Petrocca et al., 2008b; Ventura et al., 2008; Xiao et al., 2008). The miR-17-92 cluster is also found to induce tumor cell angiogenesis by targeting the anti-angiogenic proteins thrombospondin-1 (TSP1) via miR-18 and connective tissue growth factor (CTGT) by miR-18 and miR-19 (Dews et al., 2006).

A complex signaling network is observed between the miR-17-92 cluster and the transcription factors *E2F* and *MYC* to regulate cellular proliferation and apoptosis (Figure 9.3). The *MYC* oncogene is responsible for activating the transcription of the miR-17-92 cluster by binding to regulatory sequences found within the first intron of the *c13orf25* gene. The *E2F* family, *E2F1, E2F2,* and *E2F3,* which control cell cycle progression and apoptosis, are positively regulated by *MYC* and repressed by certain members of the miR-17-92 cluster (*E2F1* by miR-17/miR-20 and *E2F2/E2F3* by miR-20) (O'Donnell et al., 2005; Sylvestre et al., 2007; Trimarchi and Lees, 2002; Woods et al., 2007). In turn, *E2F* can activate *MYC* function and miR-17-92 transcription. Therefore these factors reside in reciprocal positive and negative feedback loops with one another in order to keep cell growth and apoptosis in check (Petrocca et al., 2008b). For instance, the regulation of *E2F* protein levels is crucial to determine if this factor will function in a positive or negative fashion. High levels of *E2F* (most notably *E2F1*) induce apoptosis, whereas moderate levels of *E2F* lead to cellular proliferation (Trimarchi and Lees, 2002). In the context of the miR-17-92 cluster functioning as oncogenes, these miRNAs likely negatively regulate *E2F* in order to prevent *E2F*-induced apoptosis and enhance *MYC*-activated cell proliferation.

Due to the intricate regulation between *MYC, E2F* proteins, and miR-17-92, small changes in the cellular concentrations of these molecules could alter the end result of these proliferation signaling pathways. This could determine whether these factors function as oncogenes or tumor suppressor genes and is likely dependent on the cellular context and environmental cues. By targeting *E2F*, the miR-17-92 cluster could interrupt the feedback loop between *MYC* and *E2F*

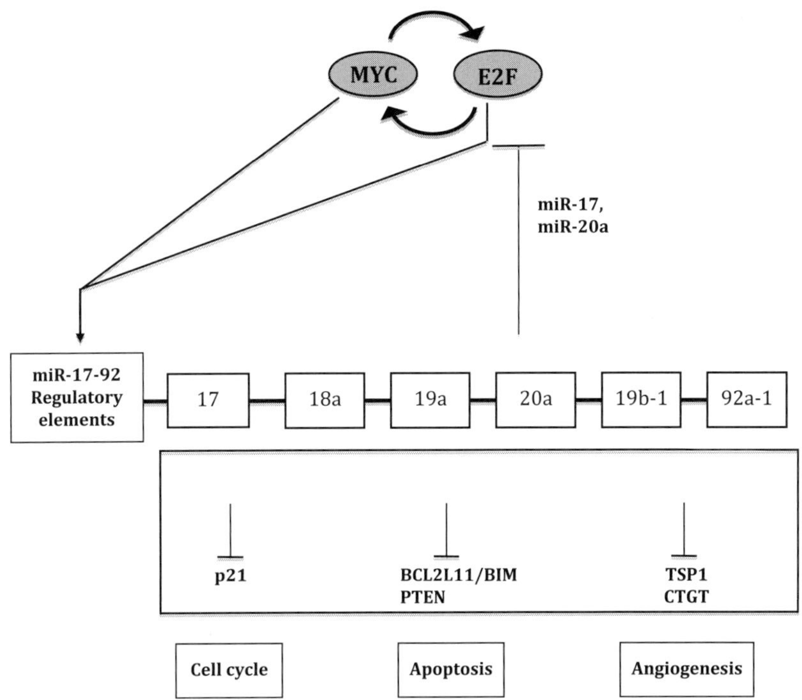

Figure 9.3 Regulatory interactions between *MYC*, *E2F*, and the miR-17-92 cluster. *MYC* and *E2F* can activate transcription of the miR-17-92 cluster, and miR-17 and miR-20a independently function to modulate *E2F* expression. *MYC* and *E2F* exist in a positive feedback loop to control each other's protein levels. Members within the miR-17-92 cluster target various factors that control cancer progression pathways associated with cell cycle progression, proliferation, and angiogenesis.

and lead to decreased *E2F* activity and, in turn, block the cell proliferative effects of *MYC*. Accumulating evidence indicates that miRNAs within the miR-17-92 cluster can indeed act as tumor suppressor genes. Loss of heterozygosity at the 13q31 loci is observed in hepatocellular carcinomas and deletion of the miR-17-92 cluster is reported in 16.5% of ovarian cancers, 21.9% of breast cancers, and 20% of melanomas (Lin et al., 1999; Zhang et al., 2006). Also, miR-17 is reported to inhibit cell proliferation of breast cancer cells *in vitro* (Hossain et al., 2006). The fact that the miR-17-92 cluster possesses both oncogenic and tumor suppressor roles illustrates the complexity of miRNA regulation and tumorigenesis.

miR-21

miR-21 is overexpressed in a large variety of cancers and predicted to function as an oncogene due to its anti-apoptotic and proliferative properties. Upregulation of miR-21 is observed in malignant glioblastomas, breast, colon, lung, pancreas, stomach, and prostate cancer (Chan et al., 2005; Ciafre et al., 2005; Iorio et al., 2005; Landgraf et al., 2007; Volinia et al., 2006). miR-21 is encoded within the tenth intron of the TMEM49 gene (of unknown function) mapping to loci 17q23.2. Altered expression of miR-21 most likely occurs at the transcriptional or

post-transcriptional level as amplification of the 17q23.2 locus is not commonly seen in tumor specimens. Interestingly, many cancer-related factors such as nuclear factor-IB (NFIB), activator protein-1 (AP-1), the androgen receptor (AR), and the estrogen receptor alpha (ERα) have all been shown to bind directly to miR-21 promoter elements and regulate its expression (Jazbutyte and Thum, 2010). Furthermore, miR-21 is a promising prognostic marker for human cancers since this miRNA is often expressed at higher levels in more aggressive and malignant tumors compared to early stages of the disease in a variety of tissues (gliomas, breast, colon, pancreatic, and gastric) and is often correlated with poor patient prognosis (Krichevsky and Gabriely, 2009).

miR-21 has experimentally been shown to induce tumor growth as well as metastasis both *in vitro* in glioblastoma, breast, and prostate cancer cell lines and *in vivo* using breast and prostate cancer orthotopic mouse models (Chan et al., 2005; Li et al., 2009; Si et al., 2007). Transfection of a miR-21 inhibitor into malignant MCF-7 breast cancer cells that are implanted into the mammary pads of immunocompromised mice leads to a 50% decrease in tumor size compared to untreated animals (Si et al., 2007). However, miR-21 activity in cancerous cells appears to be dictated by the cellular context. Highly tumorigenic and metastatic breast (MDA-MB-231) and prostate (PC-3MM) cancer cell lines do not show decreased proliferative rates compared to MCF-7 cells when miR-21 is similarly inhibited. Rather, these cell lines show decreased cell invasion and motility properties *in vitro* and decreased rates of metastasis in mouse xenograft studies (Zhu et al., 2008), potentially due to miR-21 regulation of a distinct set of targets in these cell types.

miR-21 downregulates genes that are associated with a variety of pathways related to cancer progression, including tumor growth, apoptosis, cell migration, and invasion. miR-21 is shown to negatively regulate the tumor suppressor phosphate and tensin homolog (PTEN) resulting in induced cell migration and metastasis (Meng et al., 2007). miR-21 also targets the tumor and metastasis suppressor genes programmed cell death protein 4 (PDC4), and tropomyosin 1 (TPM1) in order to promote survival, invasiveness, and metastatic growth of transformed cells (Asangani et al., 2008; Si et al., 2007; Zhu et al., 2008). Furthermore, miR-21 negatively regulates reversion-inducing cysteine-rich protein with kazal (RECK) and the tissue inhibitor of metalloproteinases (TIMP3) (Gabriely et al., 2008). MiRNA-mediated inhibition of RECK and TIMP3 is shown to increase the migration and invasiveness of cancer cells since these proteins normally function to inhibit matrix metalloproteinases. Regulation of Sprouty 1 and 2 (SPRY1/2) by miR-21 enhances cellular outgrowth, branching, migration, and cell survival and therefore appears to also be associated with inducing cancer progression (Sayed et al., 2008; Thum et al., 2008). It should be mentioned, however, that although miR-21 is generally considered an oncogene in a large range of tissues, validated miRNA targets such as the anti-apoptotic protein BCL2 (Wickramasinghe et al., 2009) indicates that in some contexts miR-21 might function in a tumor suppressor capacity to inhibit cellular growth.

MiRNAs and metastasis

Metastasis is responsible for ~90% of human deaths caused by cancer. Metastasis is a complex multistep process in which cells from the primary tumor must detach from the epithelium, break through the basal lamina, enter the bloodstream, and travel to a distant site. From here, cancer cells must exit the vasculature and proliferate in an unfamiliar milieu to generate a secondary tumor. Less than 1 in 1,000 cells will survive the trip through the bloodstream and initiate new tumor growth. The accumulation of genetic and epigenetic defects including miRNA dysregulation influences the migration and invasive potential of tumor cells. A subset of miRNAs, referred to as "metastamirs" (Hurst et al., 2009), is found to play roles as inducers and suppressors of metastasis that are independent of processes related to primary tumor formation and growth.

MiRNAs as metastatic inducers

The best characterized metastatic inducers, miR-10b, miR-373, and miR-520c, are found to stimulate the migration and invasive behavior of breast cancer cells *in vitro* and *in vivo* (Huang et al., 2008; Ma et al., 2007). When miR-10b, miR-373, or miR-520c are individually overexpressed in non-metastatic breast cancer cells and injected into the mammary fat pads of immunodeficient mice, these cells exhibit increased cell motility and invasion into the surrounding tissues as well as metastasis to distant sites such as the spine and lung (Huang et al., 2008; Ma et al., 2007). Increased expression of all three of these miRNAs is associated with breast cancer metastasis; however, miR-373 but not miR-520c is able to stimulate cell migration in cancer cells of prostate and colon origin (Huang et al., 2008), again indicating that miRNA function is dependent on the cellular context. Interestingly, the expression profiles of miR-10b, miR-373, and miR-520c are also found to closely correlate with metastatic but not primary tumors. For instance, miR-10b is upregulated in metastatic breast cancer cell lines in comparison to primary breast cancer tumors and normal breast tissue (Ma et al., 2007). Breast cancer patients with lymph node metastases exhibit higher expression of miR-373 in the primary tumor than patients without lymph node metastases (Huang et al., 2008). Therefore these miRNAs might function as powerful prognostic biomarkers for aggressive disease in breast cancer patients. It also appears that transcription factors known to activate metastatic events directly control the transcription of certain metastamirs. For example, *TWIST1*, a factor that promotes the epithelial to mesenchymal transition (EMT) of transformed cells, can bind to promoter elements of miR-10b to stimulate this miRNA's transcription and enforce metastasis pathways (Ma et al., 2007).

Direct miRNA targets for miR-10b, miR-373, and miR-520c have all been associated with cell migration and invasive pathways. miR-10b blocks the expression of homeobox D10 (HOXD10), a protein found to inhibit cell migration and invasion *in vitro* via the repression of the pro-metastatic protein RHOC (Ma et al., 2007). miR-373 and miR-520c both negatively regulate the metastatic suppressor gene *CD44* (Huang et al., 2008), a cell surface receptor for the extracellular matrix component hyaluronan that is involved in cell adhesion and cell-matrix interactions (Lou et al., 1999). miR-373 also acts as an oncogene in testicular germ-cell and esophageal tumors by inhibiting the tumor suppressor gene, *LATS2* (Voorhoeve et al., 2006). Identification of additional miR-10b, miR-373, and miR-520c targets will provide further insight into the role miRNAs play in the development of aggressive cancers.

MiRNAs as metastatic suppressors

The miRNAs, miR-335, miR-126, miR-206, and miR-31, function to suppress metastatic progression. These miRNAs are expressed at low levels in metastatic cell lines from lung and/or breast compared to non-aggressive cells and primary tumors (Tavazoie et al., 2008; Valastyan et al., 2009). The individual overexpression of miR-335, miR-126, miR-206, or miR-31 in aggressive human breast cancer cells results in the impaired ability of these cells to seed lung metastases when injected orthotopically into the mammary fat pads of mice (Tavazoie et al., 2008; Valastyan et al., 2009). These miRNAs employ different cellular processes in order to decrease metastasis. For instance, miR-126 inhibits growth and proliferation of metastatic tumors without inducing apoptosis, whereas miR-335 and miR-206 reduce the migration and invasion of cancerous cells but have no effect on cellular proliferation or apoptosis (Tavazoie et al., 2008). Interestingly, miR-31 is found to enhance the proliferation of the primary tumor, implying an additional oncogenic role. However, miR-31 also functions as a metastatic suppressor at multiple levels – and inhibits tumor invasion, migration, and anoikis (i.e., programmed cell death triggered by inadequate/loss of cell adhesion), early post-intravasation events, as well as metastatic colonization (Valastyan et al., 2009). The expression of the miRNAs, miR-335, miR-126, miR-206, and miR-31, also correlates with breast cancer disease prognosis. Patients with a shorter median time to metastatic relapse to the lung, bone, or brain have decreased expression of miR-335, miR-126, and miR-206 in the primary breast tumor. The poor prognosis of a metastasis-free survival also correlates with decreased expression of miR-335 and miR-126 (Tavazoie et al., 2008). Furthermore, low levels of miR-31 are associated with primary tumors that subsequently metastasize compared to normal breast tissue and primary tumors that do not progress (Valastyan et al., 2009). Decreased miR-31 expression in this subset of breast cancer patients is independent of tumor grade and subtype (Valastyan et al., 2009), indicating that miR-31 could be utilized as a powerful prognostic marker for this disease.

Targets associated with controlling metastasis have been identified for miR-335, miR-126, and miR-31. miR-335 negatively regulates at least four targets associated with metastasis, including the cell progenitor transcription factor SOX4, which promotes metastasis and cell migration, tenascin C (TNC), which regulates cell migration, and the extracellular matrix and cytoskeleton genes, receptor-type tyrosine protein phosphatase N2 (PTPRN2) and c-*MER* tyrosine kinase (MERTK) (Tavazoie et al., 2008). The validated targets for miR-126, Sprouty-related *EVH1* domain-containing protein 1 (SPRED1), and the regulatory subunit of *PI3K*, *PIK3R2*, which is responsible for activating *VEGF* signaling (Fish et al., 2008), indicate that this miRNA inhibits metastasis at the level of angiogenesis and vascular integrity. miR-31 has also been shown to inhibit a wide range of targets associated with metastatic events such as frizzled3 (Fzd3), integrin alpha5 (ITGA5), radixin (RDX), and RhoA (Valastyan et al., 2009). Experimental evidence shows that miR-31-mediated suppression of all four of these targets results in suppression of motility and invasion of cancerous cells, whereas miR-31-mediated repression of ITGA5, RDX, and RhoA but not Fzd3 is responsible for a block in anoikis (Valastyan et al., 2009). The wide range of targets currently validated for miR-335, miR-126, and miR-31 indicates that a single miRNA can control metastasis at multiple levels in order to influence disease progression.

MiRNAs and epigenetic machinery

DNA methylation patterns and chromatin modifications can impact gene transcription in both positive and negative ways. For example, methyl groups can be added directly to DNA, typically at CpG-rich sites, by the DNA methyltransferase (DNMT) enzymes and result in transcriptional repression. The mammalian

DNMTs, DNMT1, DNMT3A, and DNMT3B, catalyze the transfer of a methyl group from S-adenosyl methionine (a methyl donor) to the 5′ position on the cytosine ring. All three enzymes have the ability to methylate DNA *de novo*; however, DNMT1 does this inefficiently and is predominantly used to maintain methylation patterns (Pradhan et al., 1999). Histone modifications, e.g., methylation, acetylation, phosphorylation, and ubiquitination, are also employed by cells to finely regulate gene expression. A number of histone modification enzymes exist that are required for the addition and removal of acetyl groups (histone methyltransferases and histone acetyltransferases) and methyl groups (histone demethylases and histone deacetylases). These types of histone modifications alter the compactness of the DNA and thus the ability of DNA-binding factors and cellular machinery to access the chromatin and initiate/maintain gene transcription. In the next section, we will focus on how miRNAs associated with human cancers can directly alter the enzymes responsible for such epigenetic modifications, as well as how miRNAs are themselves regulated by the epigenetic machinery.

MiRNAs as effectors of the epigenetic machinery

MiRNAs are found to target DNA methyltransferase enzymes and subsequently contribute to cellular transformations by repressing the expression of tumor suppressor genes. DNA methylation of promoters is tightly linked to cancer and hypermethylation is found in almost every type of human neoplasm (Jones and Baylin, 2002). The expression of the DNMT enzymes, DNMT1, DMNT3A, and DMNT3B, are elevated in a number of human cancers, such as prostate, breast, colorectal, and lung (Eads et al., 1999; Girault et al., 2003; Patra et al., 2002; Saito et al., 2003). In lung cancer, expression of the miR-29 family (29a, 29b, 29c) inversely correlates with DNMT3A and DNMT3B expression and these two enzymes are shown to be direct targets of the miR-29 family (Fabbri et al., 2007). When miR-29 is ectopically expressed in lung cancer cell lines, normal methylation patterns return and tumorigenesis is inhibited due to the re-expression of the tumor suppressor genes, FHIT and WWOX (Fabbri et al., 2007). In acute myleloid leukemia, overexpression of miR-29b causes global DNA hypomethylation by directly repressing DNMT3A and DNMT3B (Garzon et al., 2009). Interestingly, miR-143 (but not miR-29) is found to target DNMT3A in colorectal tissues (Ng et al., 2009) and emphasizes the fact that miRNAs often function in a tissue- and/or tumor-specific manner.

Aside from DNA methylation, miRNAs can negatively regulate targets that are involved with histone modifications. For example, miR-101 is found to suppress the expression of the polycomb group protein enhancer of zeste homolog 2 (EZH2) (Cao et al., 2010; Varambally et al., 2008). EZH2 is a mammalian histone methyltransferase associated with cellular proliferation, early embryogenesis, and maintaining stem cell pluripotency via methylation of histone H3 lysine 27 on the promoters of target genes. Overexpression of the miR-101 target, EZH2, promotes proliferation and invasion of benign cells and elevated levels of EZH2 is observed in a variety of cancers, including prostate, breast, bladder, and gastric cancer (Varambally et al., 2008). Increased expression of EZH2 inversely correlates with reduced levels of miR-101 (in many cases due to the deletion of the miR-101 genomic loci) during prostate cancer progression (Cao et al., 2010; Varambally et al., 2008).

Histone deacetylase 4 (HDAC4) is a validated target for both miR-1 and miR-140, but exhibits different tissue specificities. miR-1 targets HDAC4 in muscle tissue and is responsible for promoting muscle cell differentiation (Chen et al., 2006). miR-140 functions in a similar manner to miR-1, but is cartilage specific (Tuddenham et al., 2006). miR-1 is expressed at decreased levels in the chordoma cell line, UCH1, and in chordoma tissue (Duan et al., 2010). Transfection of miR-1 into chordoma cell lines leads to a decreased expression of HDAC4 implicating a role for this miRNA and HDAC4 in tumor progression (Duan et al., 2010).

MiRNAs as targets of the epigenetic machinery

The epigenetic machinery can regulate miRNA expression through a variety of mechanisms, including promoter hyper- and hypomethylation and chromosome translocations that can impact cancer progression pathways. Five to ten percent of miRNA loci reside within CpG islands and are estimated to be epigentically regulated (Brueckner et al., 2007; Han et al., 2007; Lujambio et al., 2008; Saito et al., 2006;

Toyota et al., 2008). The proper transcription of miR-148a, miR-34b/c, miR-9, and *let-7a-3* is dependent on their methylation status and the DNMT1 and DNMT3b DNA methyltransferases. In colon cancer cells, promoter CpG hypermethylation is associated with a global loss of miRNA expression (Veeck and Esteller, 2010). This is particularly important in the progression of cancer since miRNAs that act as tumor suppressors are the most common species to be downregulated (Lujambio and Esteller, 2007). For instance, the decreased expression of miR-124a in colon cancer cells due to DNA hypermethylation leads to the increased activation of cyclin-dependent kinase 6 (CDK6) and the phosphorylation of retinoblastoma (Lujambio et al., 2007). Both of these factors are responsible for cell cycle progression and lead to uncontrolled cellular growth.

Conversely, promoter hypomethylation can lead to increased expression of oncogenic miRNAs and promote tumor formation. MiRNA promoter demethylation is attributed to increased expression of miR-126 and miR-128 in acute myeloid leukemia and acute lymphocytic leukemia, respectively (Li et al., 2008). The use of the chromatin-modifying drugs, 5'-aza-2'-deoxycytidine that prevents DNA methylation and 4-phenylbutyric, a histone deacetylase inhibitor, directly modulates miRNA levels. The expression of miR-127 is increased in bladder cancer cells after treatment with both of these drugs and a reciprocal downregulation of the miR-127 target, proto-oncogene BLC6 is also observed (Saito et al., 2006). Thus chromatin-modifying drugs might potentially be developed therapeutically to regulate oncomir expression.

In addition to methylation, chromosomal translocations lead to the epigenetic silencing of miRNAs. A common translocation in acute myeloid leukemia occurs between chromosomes 8 and 21. This results in an AML1-ETO fusion that downregulates the expression of miR-223, a miRNA that controls myeloid cell differentiation (Fazi et al., 2007). Furthermore, increasing the expression of miR-223 through demethylation restores the differentiation of leukemic blasts (Fazi et al., 2007).

Finally, the enzymes responsible for adenosine to inosine (A to I) RNA editing, adenosine deaminases acting on RNA (ADARs) proteins, ADAR1 and ADAR2, are shown to directly impact miRNA activity for approximately 6% of all human miRNAs. Adenosine residues can be edited to inosines at specific nucleotides within miRNA precursors resulting in the block of Drosha microprocessor cleavage (Yang et al., 2006) (i.e., miR-142) as well as Dicer processing (Kawahara et al., 2007a) (i.e., miR-151). A to I RNA editing within the miRNA "seed" region is found to change target recognition for miRNAs. For instance, unedited miR-376a-5p is found to target TTK (threonine and tyrosine kinase) whereas the edited form of miR-376a-5p targets PRPS1 (phosphoribosyl pyrophosphate synthetase 1) (Blow et al., 2006; Kawahara et al., 2007b). Interestingly, A to I miRNA editing appears to be spatially and temporally regulated and the edited forms of miR-151 are present in the central nervous system but absent in tissues such as the lung (Blow et al., 2004, 2006; Kawahara et al., 2007a). Taken together, there are many ways in which the epigenetic machinery can impact the biological activity of miRNAs and in turn have a direct affect on tumor formation.

MiRNAs and systems biology

The targets of miRNAs are diverse and include, but are not limited to, secreted proteins, membrane receptors, downstream signaling molecules, transcription factors, and metabolic enzymes. Therefore miRNAs impact every major cellular process and are integrated into complex gene regulatory networks, protein interaction networks, signaling networks, and metabolic networks. Disruptions in any of these circuits will impact the initiation and progression of diseases such as cancer, and highlights the importance of miRNA-mediated post-transcriptional regulation.

Gene regulatory networks

Gene regulatory networks include all of the information within a cell that determines the rate at which DNA is transcribed into mRNA and are influenced by transcription factors and regulatory RNAs, such as miRNAs. Experimental evidence suggests that genes, which are highly regulated by transcription factors, are also more likely to be regulated by miRNAs (Cui et al., 2007b). For example, the target genes of three transcription factors, OCT4, NANOG, and SOX2, expressed in human embryonic stem cells (determined by ChIP-chip analysis) were initially broken down into groups based on whether they were regulated by one, two, or all three transcription factors. Each of the three groups was then subdivided into two categories: genes that were potential miRNA targets and genes that were not miRNA targets based on computational predictions. Interestingly, the genes found to be regulated by all three transcription factors were also predicted to be direct miRNA targets, revealing a positive correlation between transcription factor control and miRNA-based regulation. This was especially true for genes involved in complex developmental processes that exhibited specific temporal and/or spatial expression patterns.

Complex genetic networks can exist as a result of positive and negative feedback loops between regulatory proteins and miRNAs. Circuits can exist in which an upstream transcription factor simultaneously activates the expression of a protein-coding gene and a miRNA or where the transcription factor activates one factor while inhibiting the other (Drakaki and Iliopoulos, 2009). The *MYC*, *E2F*, and miR-17–92 cluster circuit, previously mentioned, is a good example of a complex regulatory network that functions to keep cellular proliferation in check. To date, 243 transcription factor–miRNA regulatory relationships have been identified and databases, such as TransmiR, are being created to document these interactions (Cui et al., 2007b).

Protein networks

Protein interactions are crucial to many, if not all, cellular functions and are required for processes such as DNA synthesis, DNA transcription, signal transduction, and metabolic pathways. High-throughput methods, e.g., yeast-two hybrid screens and microarrays, have resulted in the characterization of the human protein–protein interactome (PPI) – which classifies all of the protein–protein interactions in a cell (Stelzl et al., 2005; Uetz et al., 2000). The PPI illustrates static interactions between proteins; however, it is the collective expression of individual proteins that controls the functional outcome of the entire network. Protein expression is frequently regulated by miRNAs (Bartel, 2004; Lim et al., 2003), implying that miRNAs play an important role in determining the functional state of the PPI. Comparisons between multiple human PPI datasets and miRNA target prediction datasets indicate a positive correlation between proteins with more interacting partners and the amount of miRNA regulation (Liang and Li, 2007). As the number of interacting partners for a protein increases, the number of miRNA targets within the gene for that protein also increases (Liang and Li, 2007). Furthermore, the genes of two or more interacting proteins are often regulated by the same miRNA. This may be because misexpression of a protein, which generally interacts with a large number of partners, at the incorrect time or place, may produce deleterious effects by interfering with more than one interaction pathway. Also, protein production may need to be more tightly regulated at the post-transcriptional level for genes that are involved in multiple reactions.

MiRNAs are often found in clusters that range in size from two to eight genes. MiRNAs in clusters may functionally overlap because they are composed of miRNAs that share close sequence homology and therefore regulate the same pool of targets. Alternatively, clusters may be composed of distinct miRNA genes that regulate a common target or control individual genes residing in a common biological pathway. MiRNA clusters may also be functionally related by targeting factors that are in proximity with one another in the PPI (Yuan et al., 2009). Indeed, computational studies have shown that miRNA clusters often regulate groups of proteins that control similar functions and the protein products of the regulated genes exist in close proximity to one another in the PPI. As the interactome builds, more trends in miRNA regulation of proteins are likely to emerge.

Cell signaling networks

MiRNAs regulate the output of specific cell signaling by causing an increase or decrease in pathway activity or by regulating crosstalk between pathways (Hagen and Lai, 2008). Cell signaling pathways that are dysregulated in cancer include those that control cellular growth, differentiation, and apoptosis. Alterations in miRNA expression, among other factors, can change the expression of signaling proteins, in turn changing the strength of the transduced signal, and modifying outputs of the pathway. Developmental pathways such as Notch, Hedgehog, TGFβ, and receptor tyrosine kinase (RTK) signaling control many of these processes and all are targeted by miRNAs (Hagen and Lai, 2008). For example, RAS is a component of the RTK signaling pathway and without regulation can contribute to tumor formation by leading to uncontrolled proliferation. However, miRNAs such as *let-7* are able to suppress human RAS and finely regulate cellular growth (Johnson et al., 2005).

MiRNAs also play a role in directing crosstalk between signaling pathways. For example, expression profiling in the zebrafish identified the miRNAs, miR-34a, miR-27b, miR-20a, miR-206, and miR-214, whose expression is modulated by the Notch and Hedgehog signaling pathways (Thatcher et al., 2007). In human glomerular mesangial cells, the TGFß and AKT signaling pathways are interconnected through miR-192, miR-216a, and miR-217 (Inui et al., 2010). TGFß induces the expression of miR-192, which in turn inhibits the transcription factor ZEB2. miR-216a and miR-217 are consequently unrepressed and inhibit PTEN, allowing for increased AKT signaling (Inui et al., 2010). MiRNAs mediate cellular control within linear signaling pathways as well as allowing for regulation between separate signaling pathways.

Global analysis reveals trends in the way miRNAs interact with cell signaling networks. In one study, signaling networks responsible for activating the machinery related to transcription, translation, secretory vesicles, ion channels, and cell motility in mammalian hippocampal CA1 neurons (Ma'ayan et al., 2005) were analyzed computationally for possible miRNA-mediated regulation (Cui et al., 2006). MiRNAs were found to target downstream signaling molecules such as nuclear proteins more frequently than other types of proteins, i.e., ligands, cell surface receptors, and intracellular signaling proteins (Cui et al., 2006). Interestingly, after analyzing 11 different signaling motifs, it was shown that miRNAs preferentially regulate positive signaling networks over negative signaling networks. For instance, this particular study found that within a three-way positive feedback loop comprised of activator protein 1 (AP1), cAMP-responsive element-binding protein (CREB), and CREB-binding protein (CBP), each of these proteins is regulated by miRNAs. This work implies that one role of miRNAs is to provide rapid feedback responses.

Metabolic networks

Metabolic networks consist of the complete set of chemical reactions and regulatory reactions needed for cellular metabolism. Metabolic reactions typically take place in pathways and produce metabolites, small molecules that are intermediates or end products in these pathways. Metabolites are precursors for DNA, RNA, and protein and are involved in cellular processes such as growth, development, reproduction, and cell signaling. Metabolites are crucial to cell function and certain metabolites are transferable between pathways while others are generated from multiple and/or a combination of pathways. The rate of metabolic processes must be amenable to fluctuations in the cellular environment. MiRNAs represent one mechanism of post-transcriptional control that are able to regulate processes such as amino acid catabolism, cholesterol biosynthesis, triglyceride metabolism, insulin secretion, carbohydrate metabolism, and lipid metabolism (Krutzfeldt and Stoffel, 2006). The *Drosophila* miR-14 was the first miRNA found to control metabolic processes. Loss of miR-14 leads to a doubling in the total amount of triacyclglycerides in the body and conversely overexpression of miR-14 decreases triacyclglyceride levels (Xu et al., 2003). There have also been various studies in vertebrates that link miRNA regulation to metabolic processes. miR-29b functions in amino acid catabolism in mammalian cells, specifically, by regulating the amount of branched-chain α-ketoacid dehydrogenase (BCKD). This affects protein synthesis that requires these amino acids as well as nitrogen metabolism (Mersey et al., 2005). Previously in this chapter, miR-29b was described as targeting the DNA methyltransferase enzymes in acute myeloid leukemia, supporting the notion that a single miRNA can control the expression of multiple targets belonging in distinct signaling pathways.

Conclusions

MiRNAs are a class of gene regulators that control essential processes associated with development, apoptosis, cellular proliferation, differentiation, and metabolism. Dysregulation of miRNAs closely correlates with human diseases such as cancer. There is a significant amount of evidence that shows that miRNAs play a direct role in cancer initiation and progression, including metastasis. Several miRNAs are shown to function directly as tumor suppressor genes and oncogenes. Unique miRNA expression profiles have been reported for various human cancers and indicate that miRNA activity is tumor and tissue specific. Essentially, a miRNA "signature" can be created for different disease states. This will make miRNAs very useful as diagnostic and prognostic markers for disease. MiRNAs are an important level of regulation in genetic and cellular networks, signaling pathways, protein interactions, and metabolic reactions. Studies involving cellular networks are becoming increasingly important as medicine shifts toward personalized treatment for cancer based on individualized genomic and proteomic profiles. Recent studies have shown that systemic delivery of miRNA-expressing viruses, synthetic miRNA mimics, and antisense miRNA inhibitors modulate tumor formation, but do not cause toxic effects outside of the targeted organ. Although a more in-depth understanding of miRNAs and delivery systems will be necessary before miRNAs can be used to treat cancer in the clinic, they represent a promising area of therapeutics.

References

Adams, B.D., Claffey, K.P., and White, B.A. (2009). Argonaute-2 expression is regulated by epidermal growth factor receptor and mitogen-activated protein kinase signaling and correlates with a transformed phenotype in breast cancer cells. *Endocrinology 150*, 14–23.

Akao, Y., Nakagawa, Y., and Naoe, T. (2006). let-7 microRNA functions as a potential growth suppressor in human colon cancer cells. *Biol Pharm Bull 29*, 903–906.

Asangani, I.A., Rasheed, S.A., Nikolova, D.A., et al. (2008). MicroRNA-21 (miR-21) post-transcriptionally downregulates tumor suppressor Pdcd4 and stimulates invasion, intravasation and metastasis in colorectal cancer. *Oncogene 27*, 2128–2136.

Bandi, N., Zbinden, S., Gugger, M., et al. (2009). miR-15a and miR-16 are implicated in cell cycle regulation in a Rb-dependent manner and are frequently deleted or down-regulated in non-small cell lung cancer. *Cancer Res 69*, 5553–5559.

Bartel, D.P. (2004). MicroRNAs: genomics, biogenesis, mechanism, and function. *Cell 116*, 281–297.

Bhattacharya, R., Nicoloso, M., Arvizo, R., et al. (2009). MiR-15a and MiR-16 control Bmi-1 expression in ovarian cancer. *Cancer Res 69*, 9090–9095.

Blow, M., Futreal, P.A., Wooster, R., and Stratton, M.R. (2004). A survey of RNA editing in human brain. *Genome Res 14*, 2379–2387.

Blow, M.J., Grocock, R.J., van Dongen, S., et al. (2006). RNA editing of human microRNAs. *Genome Biol 7*, R27.

Bommer, G.T., Gerin, I., Feng, Y., et al. (2007). p53-mediated activation of miRNA34 candidate tumor-suppressor genes. *Curr Biol 17*, 1298–1307.

Bonci, D., Coppola, V., Musumeci, M., et al. (2008). The miR-15a-miR-16-1 cluster controls prostate cancer by targeting multiple oncogenic activities. *Nat Med 14*, 1271–1277.

Borchert, G.M., Lanier, W., and Davidson, B.L. (2006). RNA polymerase III transcribes human microRNAs. *Nat Struct Mol Biol 13*, 1097–1101.

Breving, K., and Esquela-Kerscher, A. (2010). The complexities of microRNA regulation: mirandering around the rules. *Int J Biochem Cell Biol 42*, 1316–1329.

Brueckner, B., Stresemann, C., Kuner, R., et al. (2007). The human let-7a-3 locus contains an epigenetically regulated microRNA gene with oncogenic function. *Cancer Res 67*, 1419–1423.

Calin, G.A., Cimmino, A., Fabbri, M., et al. (2008). MiR-15a and miR-16-1 cluster functions in human leukemia. *Proc Natl Acad Sci USA 105*, 5166–5171.

Calin, G.A., and Croce, C.M. (2006). MicroRNA signatures in human cancers. *Nat Rev Cancer 6*, 857–866.

Calin, G.A., Dumitru, C.D., Shimizu, M., et al. (2002). Frequent deletions and down-regulation of micro-RNA genes miR15 and miR16 at 13q14 in chronic lymphocytic leukemia. *Proc Natl Acad Sci USA 99*, 15524–15529.

Calin, G.A., Ferracin, M., Cimmino, A., et al. (2005). A microRNA signature associated with prognosis and progression in chronic lymphocytic leukemia. *N Engl J Med 353*, 1793–1801.

Calin, G.A., Liu, C.G., Sevignani, C., et al. (2004a). MicroRNA profiling reveals distinct signatures in B cell chronic lymphocytic leukemias. *Proc Natl Acad Sci USA 101*, 11755–11760.

Calin, G.A., Sevignani, C., Dumitru, C.D., et al. (2004b). Human microRNA genes are frequently located at fragile sites and genomic regions involved in cancers. *Proc Natl Acad Sci USA 101*, 2999–3004.

Cao, P., Deng, Z., Wan, M., et al. (2010). MicroRNA-101 negatively regulates Ezh2 and its expression is modulated by androgen receptor and HIF-1alpha/HIF-1beta. *Mol Cancer 9*.

Carmell, M.A., Xuan, Z., Zhang, M.Q., and Hannon, G.J. (2002). The Argonaute family: tentacles that reach into RNAi, developmental control, stem cell maintenance, and tumorigenesis. *Genes Dev 16*, 2733–2742.

Chan, J.A., Krichevsky, A.M., and Kosik, K.S. (2005). MicroRNA-21 is an antiapoptotic factor in human glioblastoma cells. *Cancer Res 65*, 6029–6033.

Chan, S.P., and Slack, F.J. (2007). And now introducing mammalian mirtrons. *Dev Cell 13*, 605–607.

Chang, T.C., Wentzel, E.A., Kent, O.A., et al. (2007). Transactivation of miR-34a by p53 broadly influences gene expression and promotes apoptosis. *Mol Cell 26*, 745–752.

Chang, T.C., Yu, D., Lee, Y.S., et al. (2008). Widespread microRNA repression by Myc contributes to tumorigenesis. *Nat Genet 40*, 43–50.

Chen, J., Mandel, E., Thomson, J., et al. (2006). The role of microRNA-1 and microRNA-133 in skeletal muscle proliferation and differentiation. *Nat Genet 38*, 228–233.

Childs, G., Fazzari, M., Kung, G., et al. (2009). Low-level expression of microRNAs let-7d and miR-205 are prognostic markers of head and neck squamous cell carcinoma. *Am J Pathol 174*, 736–745.

Chin, L.J., Ratner, E., Leng, S., et al. (2008). A SNP in a let-7 microRNA complementary site in the KRAS 3′ untranslated region increases non-small cell lung cancer risk. *Cancer Res 68*, 8535–8540.

Ciafre, S.A., Galardi, S., Mangiola, A., et al. (2005). Extensive modulation of a set of microRNAs in primary glioblastoma. *Biochem Biophys Res Commun 334*, 1351–1358.

Cimmino, A., Calin, G.A., Fabbri, M., et al. (2005). miR-15 and miR-16 induce apoptosis by targeting BCL2. *Proc Natl Acad Sci USA 102*, 13944–13949.

Cui, J.W., Li, Y.J., Sarkar, A., et al. (2007a). Retroviral insertional activation of the Fli-3 locus in erythroleukemias encoding a cluster of microRNAs that convert Epo-induced differentiation to proliferation. *Blood 110*, 2631–2640.

Cui, Q., Yu, Z., Pan, Y., Purisima, E., and Wang, E. (2007b). MicroRNAs preferentially target the genes with transcriptional regulation complexity. *Biochem Biophys Res Commun 352*, 733–738.

Cui, Q., Yu, Z., Purisima, E., and Wang, E. (2006). Principles of microRNA regulation of a human cellular signaling network. *Mol Syst Biol 2*, 46.

Davis, B., and Hata, A. (2009). Regulation of microRNA biogenesis: a miRiad of mechanisms. *Cell Commun Signal 7*, 18.

Dews, M., Homayouni, A., Yu, D., et al. (2006). Augmentation of tumor angiogenesis by a Myc-activated microRNA cluster. *Nat Genet 38*, 1060–1065.

Doench, J.G., and Sharp, P.A. (2004). Specificity of microRNA target selection in translational repression. *Genes Dev 18*, 504–511.

Dong, Q., Meng, P., Wang, T., et al. (2010). MicroRNA let-7a inhibits proliferation of human prostate cancer cells *in vitro* and *in vivo* by targeting E2F2 and CCND2. *PLoS One 5*, e10147.

Drakaki, A., and Iliopoulos, D. (2009). MicroRNA gene networks in oncogenesis. *Curr Genomics 10*, 35–41.

Duan, Z., Choy, E., Nielsen, G.P., et al. (2010). Differential expression of microRNA (miRNA) in chordoma reveals a role for miRNA-1 in Met expression. *J Orthop Res 28*, 746–752.

Eads, C.A., Danenberg, K.D., Kawakami, K., et al. (1999). CpG island hypermethylation in human colorectal tumors is not associated with DNA methyltransferase overexpression. *Cancer Res 59*, 2302–2306.

Esquela-Kerscher, A., and Slack, F.J. (2006). Oncomirs – microRNAs with a role in cancer. *Nat Rev Cancer 6*, 259–269.

Esquela-Kerscher, A., Trang, P., Wiggins, J.F., et al. (2008). The let-7 microRNA reduces tumor growth in mouse models of lung cancer. *Cell Cycle 7*, 759–764.

Fabbri, M., Garzon, R., Cimmino, A., and Liu, Z. (2007). MicroRNA-29 family reverts aberrant methylation in lung cancer by targeting DNA methyltransferases 3A and 3B. *Proc Natl Acad Sci USA 104*, 15805–15810.

Fazi, F., Racanicchi, S., Zardo, G., Starnes, L.M., and Mancini, M. (2007). Epigenetic silencing of the myelopoiesis regulator microRNA-223 by the AML1/ETO oncoprotein. *Cancer Cell 12*, 457–466.

Filipowicz, W., Bhattacharyya, S.N., and Sonenberg, N. (2008). Mechanisms of post-transcriptional regulation by microRNAs: are the answers in sight? *Nat Rev Genet 9*, 102–114.

Fish, J.E., Santoro, M.M., Morton, S.U., et al. (2008). miR-126 regulates angiogenic signaling and vascular integrity. *Dev Cell 15*, 272–284.

Gabriely, G., Wurdinger, T., Kesari, S., et al. (2008). MicroRNA 21 promotes glioma invasion by targeting matrix metalloproteinase regulators. *Mol Cell Biol 28*, 5369–5380.

Garzon, R., Liu, S., Fabbri, M., and Liu, Z. (2009). MicroRNA-29b induces global DNA hypomethylation and tumor suppressor gene reexpression in acute myeloid leukemia by targeting directly DNMT3A and 3B and indirectly DNMT1. *Blood 113*, 6411–6418.

Girault, I., Tozlu, S., Lidereau, R., and Bieche, I. (2003). Expression analysis of DNA methyltransferases 1, 3A, and 3B in sporadic breast carcinomas. *Clin Cancer Res 9*, 4415–4422.

Gordon, A.T., Brinkschmidt, C., Anderson, J., et al. (2000). A novel and consistent amplicon at 13q31 associated with alveolar rhabdomyosarcoma. *Genes Chromosomes Cancer 28*, 220–226.

Greger, V., Passarge, E., Hopping, W., and Messmer, E. (1989). Epigenetic changes may contribute to the formation and spontaneous regression of retinoblastoma. *Hum Genet 83*, 155–158.

Griffiths-Jones, S., Saini, H.K., van Dongen, S., and Enright, A.J. (2008). miRBase: tools for microRNA genomics. *Nucleic Acids Res 36*, D154–158.

Guo, Y., Chen, Y., Ito, H., et al. (2006). Identification and characterization of lin-28 homolog B (LIN28B) in human hepatocellular carcinoma. *Gene 384*, 51–61.

Hagen, J., and Lai, E. (2008). microRNA control of cell-cell signaling during development and disease. *Cell Cycle 7*, 2327–2332.

Han, L., Witmer, P.D., Casey, E., Valle, D., and Sukumar, S. (2007).

DNA methylation regulates microRNA expression. *Cancer Biol Ther 6*, 1284–1288.

Harfe, B.D. (2005). MicroRNAs in vertebrate development. *Curr Opin Genet Dev 15*, 410–415.

Hayashita, Y., Osada, H., Tatematsu, Y., et al. (2005). A polycistronic microRNA cluster, miR-17-92, is overexpressed in human lung cancers and enhances cell proliferation. *Cancer Res 65*, 9628–9632.

He, L., He, X., Lim, L.P., et al. (2007). A microRNA component of the p53 tumour suppressor network. *Nature 447*, 1130–1134.

He, L., Thomson, J.M., Hemann, M.T., et al. (2005). A microRNA polycistron as a potential human oncogene. *Nature 435*, 828–833.

Heo, I., Joo, C., Cho, J., et al. (2008). Lin28 mediates the terminal uridylation of let-7 precursor microRNA. *Mol Cell 32*, 276–284.

Hossain, A., Kuo, M.T., and Saunders, G.F. (2006). Mir-17-5p regulates breast cancer cell proliferation by inhibiting translation of AIB1 mRNA. *Mol Cell Biol 26*, 8191–8201.

Huang, Q., Gumireddy, K., Schrier, M., et al. (2008). The microRNAs miR-373 and miR-520c promote tumour invasion and metastasis. *Nat Cell Biol 10*, 202–210.

Hurst, D.R., Edmonds, M.D., and Welch, D.R. (2009). Metastamir: the field of metastasis-regulatory microRNA is spreading. *Cancer Res 69*, 7495–7498.

Inui, M., Martello, G., and Piccolo, S. (2010). MicroRNA control of signal transduction. *Nat Rev Mol Cell Biol 11*, 252–263.

Iorio, M.V., Ferracin, M., Liu, C.G., et al. (2005). MicroRNA gene expression deregulation in human breast cancer. *Cancer Res 65*, 7065–7070.

Ivanovska, I., Ball, A.S., Diaz, R.L., et al. (2008). MicroRNAs in the miR-106b family regulate p21/CDKN1A and promote cell cycle progression. *Mol Cell Biol 28*, 2167–2174.

Jazbutyte, V., and Thum, T. (2010). MicroRNA-21: from cancer to cardiovascular disease. *Curr Drug Targets 11*, 926–935.

Johnson, C.D., Esquela-Kerscher, A., Stefani, G., et al. (2007). The let-7 microRNA represses cell proliferation pathways in human cells. *Cancer Res 67*, 7713–7722.

Johnson, S.M., Grosshans, H., Shingara, J., et al. (2005). RAS is regulated by the let-7 microRNA family. *Cell 120*, 635–647.

Jones, P., and Baylin, S. (2002). The fundamental role of epigentic events in cancer. *Nat Rev Genet 3*, 415–428.

Karube, Y., Tanaka, H., Osada, H., et al. (2005). Reduced expression of Dicer associated with poor prognosis in lung cancer patients. *Cancer Sci 96*, 111–115.

Kawahara, Y., Zinshteyn, B., Chendrimada, T.P., Shiekhattar, R., and Nishikura, K. (2007a). RNA editing of the microRNA-151 precursor blocks cleavage by the Dicer-TRBP complex. *EMBO Rep 8*, 763–769.

Kawahara, Y., Zinshteyn, B., Sethupathy, P., et al. 2007b). Redirection of silencing targets by adenosine-to-inosine editing of miRNAs. *Science 315*, 1137–1140.

Kefas, B., Godlewski, J., Comeau, L., et al. (2008). microRNA-7 inhibits the epidermal growth factor receptor and the Akt pathway and is down-regulated in glioblastoma. *Cancer Res 68*, 3566–3572.

Kent, O.A., and Mendell, J.T. (2006). A small piece in the cancer puzzle: microRNAs as tumor suppressors and oncogenes. *Oncogene 25*, 6188–6196.

Koesters, R., Adams, V., Betts, D., et al. (1999). Human eukaryotic initiation factor EIF2C1 gene: cDNA sequence, genomic organization, localization to chromosomal bands 1p34–35, and expression. *Genomics 62*, 210–218.

Koralov, S.B., Muljo, S.A., Galler, G.R., et al. (2008). Dicer ablation affects antibody diversity and cell survival in the B lymphocyte lineage. *Cell 132*, 860–874.

Krichevsky, A.M., and Gabriely, G. (2009). miR-21: a small multi-faceted RNA. *J Cell Mol Med 13*, 39–53.

Krutzfeldt, J., and Stoffel, M. (2006). MicroRNAs: a new class of regulatory genes affecting metabolism. *Cell Metab 4*, 9–12.

Kumar, M.S., Erkeland, S.J., Pester, R.E., et al. (2008). Suppression of non-small cell lung tumor development by the let-7 microRNA family. *Proc Natl Acad Sci USA 105*, 3903–3908.

Kumar, M.S., Lu, J., Mercer, K.L., Golub, T.R., and Jacks, T. (2007). Impaired microRNA processing enhances cellular transformation and tumorigenesis. *Nat Genet 39*, 673–677.

Landgraf, P., Rusu, M., Sheridan, R., et al. (2007). A mammalian microRNA expression atlas based on small RNA library sequencing. *Cell 129*, 1401–1414.

Lawrie, C.H., Gal, S., Dunlop, H.M., et al. (2008). Detection of elevated levels of tumour-associated microRNAs in serum of patients with diffuse large B-cell lymphoma. *Br J Haematol 141*, 672–675.

Lee, R.C., Feinbaum, R.L., and Ambros, V. (1993). The *C. elegans* heterochronic gene *lin-4* encodes small RNAs with antisense complementarity to *lin-14*. *Cell 75*, 843–854.

Lee, Y.S., and Dutta, A. (2007). The tumor suppressor microRNA let-7 represses the HMGA2 oncogene. *Genes Dev 21*, 1025–1030.

Lee, Y.S., Kim, H.K., Chung, S., Kim, K.S., and Dutta, A. (2005). Depletion of human micro-RNA miR-125b reveals that it is critical for the proliferation of differentiated cells but not for the down-regulation of putative targets during differentiation. *J Biol Chem 280*, 16635–16641.

Li, L., Yu, C., Gao, H., and Li, Y. (2010). Argonaute proteins: potential biomarkers for human colon cancer. *BMC Cancer 10*, 38.

Li, T., Li, D., Sha, J., Sun, P., and Huang, Y. (2009). MicroRNA-21 directly targets MARCKS and promotes apoptosis resistance and invasion in prostate cancer cells. *Biochem Biophys Res Commun 383*, 280–285.

Li, Z., Lu, J., Sun, M., Mi, S., and Zhang, H. (2008). Distinct microRNA expression profiles in acute myeloid leukemia with common translocations. *Proc Natl Acad Sci USA 105*, 15535–15540.

Liang, H., and Li, W. (2007). MicroRNA regulation of human protein–protein interaction network. *RNA 13*, 1402–1408.

Lim, L.P., Glasner, M.E., Yekta, S., Burge, C.B., and Bartel, D.P. (2003). Vertebrate microRNA genes. *Science 299*, 1540.

Lin, Y.W., Sheu, J.C., Liu, L.Y., et al. (1999). Loss of heterozygosity at chromosome 13q in hepatocellular carcinoma: identification of three independent regions. *Eur J Cancer 35*, 1730–1734.

Linsley, P.S., Schelter, J., Burchard, J., et al. (2007). Transcripts targeted by the microRNA-16 family cooperatively regulate cell cycle progression. *Mol Cell Biol 27*, 2240–2252.

Lou, W., Krill, D., Dhir, R., et al. (1999). Methylation of the CD44 metastasis suppressor gene in human prostate cancer. *Cancer Res 59*, 2329–2331.

Lu, J., Getz, G., Miska, E.A., et al. (2005). MicroRNA expression profiles classify human cancers. *Nature 435*, 834–838.

Lu, L., Katsaros, D., de la Longrais, I.A., Sochirca, O., and Yu, H. (2007). Hypermethylation of let-7a-3 in epithelial ovarian cancer is associated with low insulin-like growth factor-II expression and favorable prognosis. *Cancer Res 67*, 10117–10122.

Lujambio, A., Calin, G.A., Villanueva, A., et al. (2008). A microRNA DNA methylation signature for human cancer metastasis. *Proc Natl Acad Sci USA 105*, 13556–13561.

Lujambio, A., and Esteller, M. (2007). CpG island hypermethylation of tumor suppressor microRNAs in human cancer. *Cell Cycle 6*, 1455–1459.

Lujambio, A., Ropero, S., Ballestar, E., et al. (2007). Genetic unmasking of an epigenetically silenced microRNA in human cancer cells. *Cancer Res 67*, 1424–1429.

Ma, L., Teruya-Feldstein, J., and Weinberg, R.A. (2007). Tumour invasion and metastasis initiated by microRNA-10b in breast cancer. *Nature 449*, 682–688.

Ma'ayan, A., Jenkins, S.L., Neves, S., et al. (2005). Formation of regulatory patterns during signal propagation in a mammalian cellular network. *Science 309*, 1078–1083.

Mayr, C., Hemann, M.T., and Bartel, D.P. (2007). Disrupting the pairing between let-7 and Hmga2 enhances oncogenic transformation. *Science 315*, 1576–1579.

Melo, S.A., Ropero, S., Moutinho, C., et al. (2009). A TARBP2 mutation in human cancer impairs microRNA processing and DICER1 function. *Nat Genet 41*, 365–370.

Meng, F., Henson, R., Wehbe-Janek, H., et al. (2007). MicroRNA-21 regulates expression of the PTEN tumor suppressor gene in human hepatocellular cancer. *Gastroenterology 133*, 647–658.

Merritt, W.M., Lin, Y.G., Han, L.Y., et al. (2008). Dicer, Drosha, and outcomes in patients with ovarian cancer. *N Engl J Med 359*, 2641–2650.

Mersey, B.D., Jin, P., and Danner, D.J. (2005). Human microRNA (miR29b) expression controls the amount of branched chain alpha-ketoacid dehydrogenase complex in a cell. *Hum Mol Genet 14*, 3371–3377.

Mu, P., Han, Y.C., Betel, D., et al. (2009). Genetic dissection of the miR-17∼92 cluster of microRNAs in Myc-induced B-cell lymphomas. *Genes Dev 23*, 2806–2811.

Nelson, P., Kiriakidou, M., Sharma, A., Maniataki, E., and Mourelatos, Z. (2003). The microRNA world: small is mighty. *Trends Biochem Sci 28*, 534–540.

Newman, M.A., Thomson, J.M., and Hammond, S.M. (2008). Lin-28 interaction with the Let-7 precursor loop mediates regulated microRNA processing. *RNA 14*, 1539–1549.

Ng, E., Tsang, W., Ng, S., Jin, H., and Yu, J. (2009). MicroRNA-143 targets DNA methyltransferases 3A in colorectal cancer. *Br J Cancer 101*, 699–706.

O'Donnell, K.A., Wentzel, E.A., Zeller, K.I., Dang, C.V., and Mendell, J.T. (2005). c-Myc-regulated microRNAs modulate E2F1 expression. *Nature 435*, 839–843.

Olive, V., Bennett, M.J., Walker, J.C., et al. (2009). miR-19 is a key oncogenic component of mir-17-92. *Genes Dev 23*, 2839–2849.

Olive, V., Jiang, I., and He, L. (2010). mir-17-92, a cluster of miRNAs in the midst of the cancer network. *Int J Biochem Cell Biol 42*, 1348–1354.

Ota, A., Tagawa, H., Karnan, S., et al. (2004). Identification and characterization of a novel gene, C13orf25, as a target for 13q31-q32 amplification in malignant lymphoma. *Cancer Res 64*, 3087–3095.

Park, S., Shell, S., Radjabi, A.R., and Schickel, R. (2007). Let-7 prevents early cancer progression by suppressing expression of the embryonic gene HMGA2. *Cell Cycle 6*, 2585–2590.

Pasquinelli, A., Reinhart, B., Slack, F., Maller, B., and Ruvkun, G. (2000). Conservation across animal phylogeny of the sequence and temporal regulation of the 21 nucleotide *C. elegans let-7* heterochronic regulatory RNA. *Nature 408*, 86–89.

Pasquinelli, A.E., Hunter, S., and Bracht, J. (2005). MicroRNAs: a developing story. *Curr Opin Genet Dev 15*, 200–205.

Patra, S.K., Patra, A., Zhao, H., and Dahiya, R. (2002). DNA methyltransferase and demethylase in human prostate cancer. *Mol Carcinog 33*, 163–171.

Petrocca, F., Vecchione, A., and Croce, C.M. (2008a). Emerging role of miR-106b-25/miR-17-92 clusters in the control of transforming growth factor beta signaling. *Cancer Res 68*, 8191–8194.

Petrocca, F., Visone, R., Onelli, M.R., et al. (2008b). E2F1-regulated microRNAs impair TGFbeta-dependent cell-cycle arrest and apoptosis in gastric cancer. *Cancer Cell 13*, 272–286.

Piskounova, E., Viswanathan, S.R., Janas, M., et al. (2008). Determinants of microRNA processing inhibition by the developmentally regulated RNA-binding protein Lin28. *J Biol Chem 283*, 21310–21314.

Pradhan, S., Bacolla, A., Wells, R.D., and Roberts, R.J. (1999). Recombinant human DNA (cytosine-5) methyltransferase. I. Expression, purification, and comparison of de novo and maintenance methylation. *J Biol Chem 274*, 33002–33010.

Ratner, E., Lu, L., Boeke, M., et al. (2010). A KRAS-variant in ovarian cancer acts as a genetic marker of cancer risk. *Cancer Res 70*, 6509–6515.

Raver-Shapira, N., Marciano, E., Meiri, E., et al. (2007). Transcriptional activation of miR-34a contributes to p53-mediated apoptosis. *Mol Cell 26*, 731–743.

Reinhart, B., Slack, F., Basson, M., et al. (2000). The 21 nucleotide *let-7* RNA regulates *C. elegans* developmental timing. *Nature 403*, 901–906.

Roush, S., and Slack, F.J. (2008). The let-7 family of microRNAs. *Trends Cell Biol 18*, 505–516.

Rybak, A., Fuchs, H., Smirnova, L., et al. (2008). A feedback loop comprising lin-28 and let-7 controls pre-let-7 maturation during neural stem-cell commitment. *Nat Cell Biol 10*, 987–993.

Saito, Y., Kanai, Y., Nakagawa, T., et al. (2003). Increaed protein expression of DNA methyltransferase (DNMT) 1 is significantly correlated with the malignant potential and poor prognosis of human hepatocellular carcinomas. *Int J Cancer 105*, 527–532.

Saito, Y., Liang, G., Egger, G., et al. (2006). Specific activation of microRNA-127 with downregulation of the proto-oncogene BCL6 by chromatin-modifying drugs in human cancer cells. *Cancer Cell 9*, 435–443.

Sampson, V.B., Rong, N.H., Han, J., et al. (2007). MicroRNA let-7a down-regulates MYC and reverts MYC-induced growth in Burkitt lymphoma cells. *Cancer Res 67*, 9762–9770.

Sayed, D., Rane, S., Lypowy, J., et al. (2008). MicroRNA-21 targets Sprouty2 and promotes cellular outgrowths. *Mol Biol Cell 19*, 3272–3282.

Schmidt, H., Bartel, F., Kappler, M., et al. (2005). Gains of 13q are correlated with a poor prognosis in liposarcoma. *Mod Pathol 18*, 638–644.

Shan, S.W., Lee, D.Y., Deng, Z., et al. (2009). MicroRNA MiR-17 retards tissue growth and represses fibronectin expression. *Nat Cell Biol 11*, 1031–1038.

Shell, S., Park, S.M., Radjabi, A.R., et al. (2007). Let-7 expression defines two differentiation stages of cancer. *Proc Natl Acad Sci USA 104*, 11400–11405.

Si, M.L., Zhu, S., Wu, H., et al. (2007). miR-21-mediated tumor growth. *Oncogene 26*, 2799–2803.

Sonoki, T., Iwanaga, E., Mitsuya, H., and Asou, N. (2005). Insertion of microRNA-125b-1, a human homologue of lin-4, into a rearranged immunoglobulin heavy chain gene locus in a patient with precursor B-cell acute lymphoblastic leukemia. *Leukemia 19*, 2009–2010.

Spizzo, R., Nicoloso, M., Croce, C., and Calin, G. (2009). SnapShot: microRNAs in cancer. *Cell 137*, 586.

Stefani, G., and Slack, F.J. (2008). Small non-coding RNAs in animal development. *Nat Rev Mol Cell Biol 9*, 219–230.

Stelzl, U., Worm, U., Lalowski, M., Haenig, C., and Brembeck, F.H. (2005). A human protein–protein interaction network: a resource for annotating the proteome. *Cell 122*, 957–968.

Sylvestre, Y., De Guire, V., Querido, E., et al. (2007). An E2F/miR-20a autoregulatory feedback loop. *J Biol Chem 282*, 2135–2143.

Tagawa, H., and Seto, M. (2005). A microRNA cluster as a target of genomic amplification in malignant lymphoma. *Leukemia 19*, 2013–2016.

Takamizawa, J., Konishi, H., Yanagisawa, K., et al. (2004). Reduced expression of the *let-7* microRNAs in human lung cancers in association with shortened postoperative survival. *Cancer Res 64*, 3753–3756.

Tarasov, V., Jung, P., Verdoodt, B., et al. (2007). Differential regulation of microRNAs by p53 revealed by massively parallel sequencing: miR-34a is a p53 target that induces apoptosis and G1-arrest. *Cell Cycle 6*, 1586–1593.

Tavazoie, S.F., Alarcon, C., Oskarsson, T., et al. (2008). Endogenous human microRNAs that suppress breast cancer metastasis. *Nature 451*, 147–152.

Thatcher, E.J., Flynt, A.S., Li, N., Patton, J.R., and Patton, J.G. (2007). MiRNA expression analysis during normal zebrafish development and following inhibition of the Hedgehog and Notch signaling pathways. *Dev Dyn 6*, 2172–2180.

Thum, T., Gross, C., Fiedler, J., et al. (2008). MicroRNA-21 contributes to myocardial disease by stimulating MAP kinase signalling in fibroblasts. *Nature 456*, 980–984.

Toyota, M., Suzuki, H., Sasaki, Y., et al. (2008). Epigenetic silencing of microRNA-34b/c and B-cell translocation gene 4 is associated with CpG island methylation in colorectal cancer. *Cancer Res 68*, 4123–4132.

Trang, P., Medina, P.P., Wiggins, J.F., et al. (2010). Regression of murine lung tumors by the let-7 microRNA. *Oncogene 29*, 1580–1587.

Trimarchi, J.M., and Lees, J.A. (2002). Sibling rivalry in the E2F family. *Nat Rev Mol Cell Biol 3*, 11–20.

Tuddenham, L., Wheeler, G., Ntounia-Fousara, S., and Waters, J. (2006). The cartilage specific microRNA-140 targets histone deacetylase 4 in mouse cells. *FEBS Lett 580*, 4214–4217.

Uetz, P., Giot, L., Cagney, G., et al. (2000). A comprehensive analysis of protein-protein interactions in *Saccharomyces cerevisiae*. *Nature 403*, 623–627.

Valastyan, S., Reinhardt, F., Benaich, N., et al. (2009). A pleiotropically acting microRNA, miR-31, inhibits breast cancer metastasis. *Cell 137*, 1032–1046.

Varambally, S., Cao, Q., Mani, R., and Shankar, S. (2008). Genomic loss of microRNA-101 leads to overexpression of histone methyltransferase EZH2 in cancer. *Science 322*, 1695–1699.

Veeck, J., and Esteller, M. (2010). Breast cancer epigenetics: from DNA methylation to microRNAs. *J Mammary Gland Biol Neoplasia 15*, 5–17.

Ventura, A., Young, A.G., Winslow, M.M., et al. (2008). Targeted deletion reveals essential and overlapping functions of the miR-17 through 92 family of miRNA clusters. *Cell 132*, 875–886.

Viswanathan, S.R., Daley, G.Q., and Gregory, R.I. (2008). Selective blockade of microRNA processing by Lin28. *Science 320*, 97–100.

Viswanathan, S.R., Powers, J.T., Einhorn, W., et al. (2009). Lin28 promotes transformation and is associated with advanced human malignancies. *Nat Genet 41*, 843–848.

Volinia, S., Calin, G.A., Liu, C.G., et al. (2006). A microRNA expression signature of human solid tumors defines cancer gene targets. *Proc Natl Acad Sci USA 103*, 2257–2261.

Voorhoeve, P.M., le Sage, C., Schrier, M., et al. (2006). A genetic screen implicates miRNA-372 and miRNA-373 as oncogenes in testicular germ cell tumors. *Cell 124*, 1169–1181.

Wang, C.L., Wang, B.B., Bartha, G., et al. (2006). Activation of an oncogenic microRNA cistron by provirus integration. *Proc Natl Acad Sci USA 103*, 18680–18684.

Wang, Y., Medvid, R., Melton, C., Jaenisch, R., and Blelloch, R. (2007). DGCR8 is essential for microRNA biogenesis and silencing of embryonic stem cell self-renewal. *Nat Genet 39*, 380–385.

West, J.A., Viswanathan, S.R., Yabuuchi, A., et al. (2009). A role for Lin28 in primordial germ-cell development and germ-cell malignancy. *Nature 460*, 909–913.

Wickramasinghe, N.S., Manavalan, T.T., Dougherty, S.M., et al. (2009). Estradiol downregulates miR-21 expression and increases miR-21 target gene expression in MCF-7 breast cancer cells. *Nucleic Acids Res 37*, 2584–2595.

Woods, K., Thomson, J.M., and Hammond, S.M. (2007). Direct regulation of an oncogenic micro-RNA cluster by E2F transcription factors. *J Biol Chem 282*, 2130–2134.

Wu, L., and Belasco, J.G. (2005). Micro-RNA regulation of the mammalian lin-28 gene during neuronal differentiation of embryonal carcinoma cells. *Mol Cell Biol 25*, 9198–9208.

Xia, L., Zhang, D., Du, R., et al. (2008). miR-15b and miR-16 modulate multidrug resistance by targeting BCL2 in human gastric cancer cells. *Int J Cancer 123*, 372–379.

Xiao, C., Srinivasan, L., Calado, D.P., et al. (2008). Lymphoproliferative disease and autoimmunity in mice with increased miR-17-92 expression in lymphocytes. *Nat Immunol 9*, 405–414.

Xu, P., Vernooy, S.Y., Guo, M., and Hay, B.A. (2003). The *Drosophila* microRNA *Mir-14* suppresses cell death and is required for normal fat metabolism. *Curr Biol 13*, 790–795.

Yanaihara, N., Caplen, N., Bowman, E., et al. (2006). Unique microRNA molecular profiles in lung cancer diagnosis and prognosis. *Cancer Cell 9*, 189–198.

Yang, W., Chendrimada, T.P., Wang, Q., et al. (2006). Modulation of microRNA processing and expression through RNA editing by ADAR deaminases. *Nat Struct Mol Biol 13*, 13–21.

Yu, F., Yao, H., Zhu, P., et al. (2007a). let-7 regulates self renewal and tumorigenicity of breast cancer cells. *Cell 131*, 1109–1123.

Yu, J., Vodyanik, M.A., Smuga-Otto, K., et al. (2007b). Induced pluripotent stem cell lines derived from human somatic cells. *Science 318*, 1917–1920.

Yuan, X., Liu, C., Yang, P., et al. (2009). Clustered MicroRNAs' coordination in regulating protein-protein interaction network. *BMC Syst Biol 3*, 65.

Zhang, L., Huang, J., Yang, N., et al. (2006). MicroRNAs exhibit high frequency genomic alterations in human cancer. *Proc Natl Acad Sci USA 103*, 9136–9141.

Zhu, S., Wu, H., Wu, F., et al. (2008). MicroRNA-21 targets tumor suppressor genes in invasion and metastasis. *Cell Res 18*, 350–359.

Chapter

10

Dietary and environmental influences on the genomic and epigenomic codes in cancer

Hamid M. Abdolmaleky, Mohammad R. Eskandari and Jin-Rong Zhou

Introduction

Although genetic and epigenetic codes instruct cellular regeneration that is required to maintain structural integrity and to secure normal functionality of living organisms, food is the main source of materials to construct the building units and the whole structure of an organism. Naturally, any shortage or surplus of these materials may lead to structural and/or functional defects and disease phenotype. While DNA synthesis is dependent on the nutritional state, metabolism of nutrients as well as contaminants and toxins present in food or generated during metabolism of nutritional elements/components (e.g., reactive oxygen species) are reliant on the genetic landscape. Nutritional imbalance can also result in epigenetic aberrations leading to development and/or progression of cancer and other complex diseases. In fact, carcinogenesis is characterized by the contribution of different factors including the inheritance of mutated genes, and the exposure to endogenous and/or exogenous agents during the life span.

Background

It is now well documented that gene–environment interactions are among the most important functions implemented by living organisms to fine tune gene expression levels to adapt with the changing environment. Nutritional elements and components are among immediate environmental factors that modulate gene expression levels on a daily basis and many developmental diseases originate from nutritional imbalance. The half life of most human cells is so short that we require successive and continuous regeneration to maintain the structural and functional integrity of the organism. For instance in adults, the half life of epithelial cells of the elementary tract, the

skin, and blood cells are only a few days, weeks, and months, respectively. Although a limited number of human cells such as neurons are not extensively regenerated, overall we lose and regenerate the equivalent of our total body weight during a year as the result of a balanced apoptosis and reproduction (Davies and Morris, 1997). While genetic codes instruct this cellular regeneration from differentiated and undifferentiated stem cells to make 100 trillion human cells (Trosko, 2003), a vast majority of required materials are provided by food containing amino acids and other essential elements such as vitamins, minerals, and other nutrients in a range of concentrations extended from deficient to toxic levels. Additionally, food may be contaminated by harmful elements and poisons that could impact the quality and quantity of cell regeneration in different stages. Hence, nutrition is among the first few major players in developmental as well as health care medicine, albeit in a close interaction with genetic and epigenetic codes.

Although food is the main source of materials for cell regeneration, the genetic landscape determines different responses of individuals to food and even to specific nutrients. The personalized response to specific nutrients, the topics of nutritional genomics (Stover and Caudill, 2008), and nutritional epigenomics underlie many complex diseases intensified by stereotypic eating behavior and habits, particularly in industrialized communities.

Nutritional genomics investigates gene–diet interactions and nutritional variability of the food considering genomes of humans, plants, and microorganisms to develop a comprehensive food science to prevent diseases through the optimized use of nutrients based on individuals' genetic properties as well as behaviors. The primary step in nutritional genomics is the determination of single nucleotide

---

Systems Biology of Cancer, ed. S. Thiagalingam. Published by Cambridge University Press. © Cambridge University Press 2015.

polymorphisms in genes that interact with or are involved in absorption, metabolism, and functional response to specific nutrients and/or bioactive food components. However, the ultimate goal is to establish "gene–diet–lifestyle–diseases association" to design preventive and therapeutic remedies for human diseases (Jenab et al., 2009). The concept of nutritional genomics has long been a dominant approach for disease prevention and management as well as healthy aging in traditional medicine. Today, the science of nutritional genomics is enriched by the implementation of advanced techniques to define nutrient biomarkers (Prentice et al., 2002) and analysis of the whole genome polymorphisms to modernize clinical and public health nutritional practice.

Nutrients and bioactive food components may also modulate epigenetic codes and change gene expression. For example, the age-related differences in DNA methylation of leukocytes is linked to the daily intake of vitamin B6, magnesium, carbohydrates, lipids, and serum protein levels (Gomes et al., 2012). Thus a comprehensive view for disease prevention and treatment in nutritional medicine requires taking into account the science of nutritional epigenomics as well. Nutritional epigenomics investigates the effects of nutrients and dietary bioactive components on regulating heritable, but reversible, epigenetic modifications that alter gene expression leading to phenotypic variation in health outcomes. The extent to which specific nutritional components may provoke epigenetic responses represents an exciting area of current research that will become fruitful in the near future.

The development and progression of cancer are caused by altered regulation of functional genes. Gene regulation is under the control of multiple factors, ranging from those inherited in each generation, to those responsive to environmental stimuli. Nutritional components, as important environmental factors, may play crucial roles in carcinogenesis and tumorigenesis processes by both genetic and epigenetic alterations of gene functions.

The emerging science of nutriepigenomics has promoted a worldwide interest in exploring the potential roles of nutrition on epigenetic codes in the current decade. Now there is strong evidence that a comprehensive view for understanding the role of diet in the genesis and management of complex diseases will require integrative approaches that consider both nutrigenetics and nutriepigenetics. The ongoing extensive research in this field may revolutionize the current strategies for prevention and treatment of complex diseases in the immediate future. Here we review advances in the understanding of the roles of dietary/nutritional and environmental factors in genomic and epigenomic codes and in biological processes that are significantly related to carcinogenesis and tumorigenesis processes.

Nutritional genomics and cancer

Nutritional genomics includes nutrigenomics and nutrigenetics. Nutrigenomics investigates the effects of nutrients and nutritional deficiency on "genome evolution, mutation rate, in-utero viability," genome programming, and gene expression. Nutrigenetics studies the effects of genetic variations on nutrient requirement, utilization, metabolism, and food tolerances/intolerance (Stover and Caudill, 2008).

The association between nutrients and cancer is well recognized; however, the underlying mechanism of this relation is not well understood and remains controversial. Previous efforts have been focused on understanding the roles that nutrients and environmental factors may play in DNA synthesis and genome stability. Folate is among the most commonly studied nutrients. Since folate deficiency is related to the appearance of the genome fragile sites, frequently observed in various types of cancer (Butterworth, 1992; Chary-Reddy et al., 1994; Kolialexi et al., 1998; Smith et al., 1998), several studies investigated potential roles of folate deficiency in cancer and found confirmatory evidence for chromosomal breaks that may occur in cancer suppressor genes (Beetstra et al., 2005; Gumus et al., 2002; Wang et al., 2004).

Folate is required for DNA synthesis and repair to maintain the genome stability. Some folate-related events, such as defect in methionine metabolism due to methionine synthase (MS or MTR) or methylenetetrahydrofolate reductase (MTHFR) dysfunctional polymorphisms resulting in lower availability of active folate derivatives, are associated with higher cancer risk in susceptible individuals, as in carriers of BRCA genetic variants (Beetstra et al., 2008) or sporadic cases (Van Den Donk et al., 2007). Thus from the nutrigenetics point of view, individuals carrying hypoactive alleles of MTHFR may require more folate intake compared to other people. The same is true in individuals with alcohol abuse, leading to folate deficiency associated with chromosomal break (Teo and Fenech, 2008) and cancer development (Duthie et al., 2010; Seitz et al., 1998; Teo and Fenech, 2008). Although folate appears to prevent cancer development in normal cells, in hypomethylated pre-neoplastic cells containing DNA lesions may "promote DNA synthesis and cancer progression" (Melnyk et al., 1999; Sauer et al., 2009).

In addition to folate, several other dietary and nutritional factors such as fruits, vegetables, low-fat dietary products, fish, vitamin D, calcium, and phytoestrogens are linked to a decrease in cancer risk (Bissonauth et al., 2008; Forte et al., 2008). Selenium and vitamins C and E are other micronutrients that may reduce cancer risk by preventing DNA damage and mutations through trapping reactive oxygen species (Ferguson et al., 2004). On the other hand, high intake of meat, poultry, total energy, total fat, and saturated fatty acids may increase cancer risks in sporadic cases as well as gene mutation carriers of breast or colorectal cancer (Bissonauth et al., 2008; Forte et al., 2008).

Previous research has suggested DNA damage resulting from micronutrient deficiency is likely to be one of the major causes of cancer (Ames, 2001). However, in some specific situations, such as oxidative stress and exposure to contaminants/toxins and radiation, even an adequate intake of nutrients may not be sufficient to protect individuals against cancer and other complex diseases. Hence, the beneficial effects of nutrients in cancer prevention may be more significant in a polluted environment, particularly in individuals carrying dysfunctional polymorphisms involved in cancer prevention (e.g., p53) and/or metabolism of contaminants (e.g., arsenic), toxins (e.g., aflatoxin), and other hazardous materials (Figure 10.1). Although the effects of vitamins and other micronutrients may not be great enough to overcome the impact of potent and/or highly concentrated carcinogens, a large part of the population exposed to low level or mild carcinogens and carrying dysfunctional polymorphisms may benefit from the anticancer effects of micronutrients (Ross, 2007; Sutandyo, 2010). Therefore to overcome inconsistency in research findings investigating the beneficial effects of nutrients and vitamins in cancer prevention may require taking into account the genetic heterogeneity of the population and the potency of hazardous environmental factors.

Nutritional epigenomics and cancer

Epigenetics, which was first defined by Waddington (1939), is one of the main mechanisms of gene–gene and gene–environment interactions and is considered to be "a bridge between [the] genotype and [the] phenotype" (Ross and Milner, 2007), modulated by environmental factors, including food and contaminants. Since all cells of a multicellular organism are genetically homogeneous, epigenetic modifications are essential for cell differentiation to become structurally and functionally heterogeneous in diverse tissues and adapt to different micro- and macro-environments (Bird, 2007; Russo et al., 1996). The current vision of epigenetics refers to heritable but flexible alterations in gene expression controlled by DNA methylation and/or chromatin structure, RNA editing, and RNA interference, which are not associated with DNA sequence variations (Hawkins and Morris, 2008; Lavrov and Kibanov, 2007; Shilatifard, 2008; Thiagalingam et al., 2003; Vaissiere et al., 2008).

Since the concept of epigenetics has been introduced to the field of medicine, investigators have proposed and examined its potential roles in the pathogeneses of cancer, autoimmune, and mental diseases (Abdolmaleky et al., 2004; Issa et al., 2001; Shen et al., 2005). From the etiological point of view, epigenetic dysregulation is proven to be involved in cancer development and several lines of evidence support its role in cardiovascular disease, type-2 diabetes, obesity, and major psychiatric diseases (Abdolmaleky and Thiagalingam, 2011; Waterland, 2009). Environmental influences on epigenetic codes in the effect of diet are believed to be involved in cancer and other complex diseases (Singh et al., 2003).

In the next section, the roles of nutritional/dietary bioactive components and environmental factors in epigenetic modifications and the impacts of these epigenetic alterations, especially at critical stages of life, on cancer risk are discussed.

Nutrition and epigenetic modifications

DNA methylation is one of the important epigenetic events involved in gene–gene and gene–environmental interactions. It is mediated by DNA methyltransferase (DNMT) enzymes (Kim et al., 2002) that catalyze the addition of a methyl group (CH_3) to cytosines, typically followed by guanine (Costello and Plass, 2001; Russo et al., 1996). Choline is the main resource of methyl group (Zeisel, 2007)

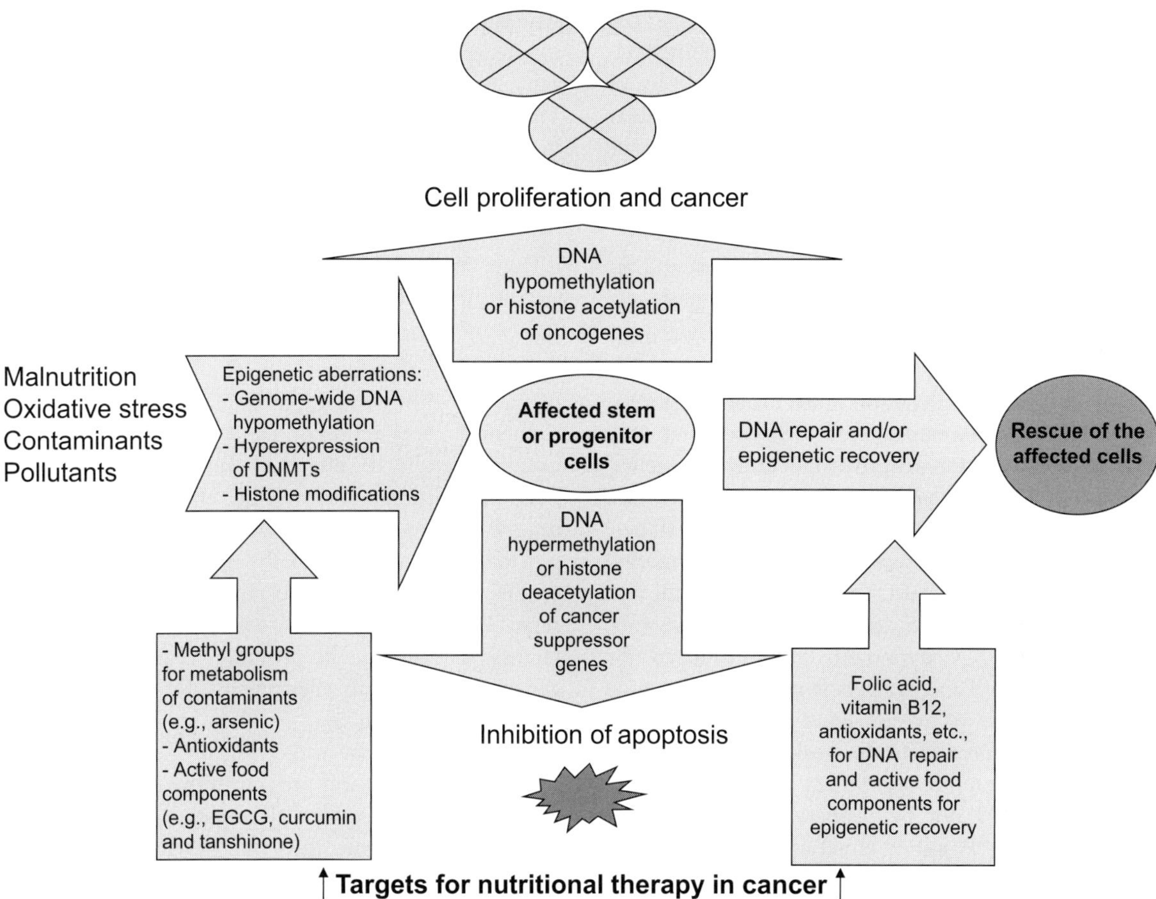

Figure 10.1 Proposed effects of various dietary and environmental factors on epigenetic aberrations in stem and progenitor cells and development and progression of cancer. Nutrients may be helpful for DNA repair and epigenetic recovery, rescuing the affected cells. Promoter DNA hypomethylation of oncogenes and DNA hypermethylation of cancer suppressor genes as well as histone modifications of these genes are among known epigenetic dysregulations that may cause an imbalanced cell apoptosis and regeneration leading to tumor formation. Nutritional intervention can be used in early stages to prevent epigenetic aberrations resulting from malnutrition, dysfunctional polymorphisms involved in the functionality of methylation machinery, and methyl deficiency due to contaminants overconsuming methyl groups. The impact of oxidative stress on methylation machinery as well as the immune system can be inhibited by dietary antioxidants. (A black and white version of this figure will appear in some formats. For the color version, please refer to the plate section.)

and S-adenosyl methionine (SAM) is the major methyl donor. Homocysteine is a by-product of transmethylation reaction that is recycled to methionine catalyzed by methionine synthase (MS or MTR). Folic acid and vitamin B12 also contribute to this remethylation reaction (Fenech, 2001). Diets lacking these required nutrients will impact many cellular processes due to the accumulation of homocysteine, which halts the DNA methylation reaction (Jacob et al., 1998; Jamaluddin et al., 2007). Furthermore, oxidative stress inhibits MS and directs homocysteine toward the transsulfuration pathway to produce glutathione, the main intracellular antioxidant (Mosharov et al., 2000; Sephashvili et al., 2006). Thus

oxidative stress may impact cellular methylation reactions by stalling SAM production.

Since methionine is an essential amino acid (Nimni et al., 2007; Oz et al., 2008), in the case of malnutrition the activity of MS is critical for the cellular methylation process as its functionality is required for homocysteine recycling. Various nutritional/environmental elements such as Cu^+, Pb^{2+}, Hg^{2+}, and Al^{3+} have inhibitory effect on MS activity. Alcohol also inhibits the IGF-1-mediated MS activation. Thus the capability of cellular methylation machinery is influenced by these factors, cellular folate, vitamin B12, and homocysteine levels as well as oxidative stress (Bonsch et al., 2007; Waly et al., 2004).

Studies using various animal models as well as cell culture experiments have provided compelling evidence that both nutritional deficiency and surplus of methyl groups and other contributors in the cellular methylation machinery can have profound impacts on the whole genome DNA methylome that may alter the epigenetic memory, leading to the disease phenotype (Bertino et al., 1996) during exposure and/or later in life, particularly in individuals with genetic susceptibility to specific diseases, including cancer (Beetstra et al., 2008) and metabolic diseases (Sinclair et al., 2007). However, the flexible and dynamic nature of DNA methylation and other epigenetic mechanisms, accompanied with a proper diet may serve the capacity for adaptive fine tuning of genes' expression in erratic environment or in individuals with dysfunctional polymorphisms. For instance, Pogribny et al. (2006) reported that dietary methyl deficiency induces global DNA hypomethylation that is a key step for hepatocarcinogenesis in male rats. Interestingly, DNA hypomethylation induced by short-term dietary methyl deficiency (less than nine weeks) could be recovered by a methyl-rich diet. However, long-term (9 to 36 weeks) dietary methyl deficiency resulted in a permanent DNA hypomethylation that could not be properly recovered by a methyl-rich diet (Pogribny et al., 2006).

Since there are two types of serial (parallel) and sequential cellular differentiation during normal development, the consequence of epigenetic aberrations is not simple. Indeed, any nutritional impact over the functionality of DNA methylation machinery may initially provoke DNA hypo/hypermethylation in a cluster of genes that are under parallel establishment/development of the epigenetic memory. Subsequently, those genes that are under the control of these genes may undergo diverse methylation changes (i.e., hypermethylation, because of the hyperactivation of DNMTs) particularly when the nutritional deficiency was alleviated in the later developmental period. For example, it was shown that folate deficiency in the embryonic stage and early-life period in mice can change epigenetic marks of cancer-related genes in adulthood in a gene-specific manner (McKay et al., 2011).

Experimental evidence indicate that environmentally induced DNA methylation alterations modulated by hormones and other cellular messengers are subject to a kindling-like phenomenon, so that successive stimulations and binding of transcription factors to genes' regulatory regions may change the nucleosome structure, associated with a corresponding decrease in the degree of DNA methylation and thus more capacity for expression in the future (Murray, 2000; Thomassin et al., 2001). Therefore subsequent exposures to those factors or modulators are accompanied by a facilitated DNA methylation and gene expression alterations. While this shows that DNA methylation plays a role in memorizing environmental exposure, the fact that previous exposures facilitate subsequent changes in DNA methylation status would have major implication in clarifying the mechanism of complex diseases with significant clinical applications. Having this along with the fact that human exposure to toxic elements/factors acting on DNA methylation status is usually episodic, the effects of every new exposure will have an additional and stronger effect on the epigenome. In other words, there will be a type of epigenetic sensitization during which the threshold for alteration of DNA methylation level will be decreased upon repeated exposures to environmental and nutritional factors. In contrast, the promoter regions of inactive genes are subject to silencing by DNA methylation over time (Bird, 2002).

In human culture and civilization the daily schedule for food intake and programming for food variety during a week provides a circumstance during which an episodic exposure to active nutritional components is perceived and conditioned by the epigenetic memory. As a result, episodic exposure to epigenetically active nutritional components may have accumulating and prolonged effects on DNA methylation status. In contrast, long-term deprivation of special nutritional components in food may result in intolerance to that component, as a result of promoter DNA methylation of those genes that encode the component-specific digestive enzymes.

Epigenetic impacts of dietary bioactive components on cancer

Some dietary and nutritional components have direct actions in modulating epigenetic events and inhibit cancer development or progression. Among those components, folic acid, vitamins B2, B6, and B12 are important regulators in one-carbon metabolism to provide the methyl group for DNA methylation, and therefore are essential for the maintenance of epigenetic marks (Miller et al., 1994). Anticancer effects of tea polyphenols and bioflavonoids (Fang et al., 2003), soy isoflavones (Fang et al., 2005), and selenium (Das et al., 2003; Davis and Uthus, 2002; Fiala et al., 1998) are shown to be mediated by inhibition of DNMT activity leading to a decrease in DNA methylation levels (Lee et al., 2005; Xiang et al., 2008), as shown in Table 10.1. In addition to modulating DNA methylation, dietary components may effectively modify the epigenetic process by histone modifications and chromatin remodelling. Diallyl disulfide, a component in garlic, functions as a histone deacetylase inhibitor and increases histone acetylation decreasing the risk of cancer (Arunkumar et al., 2007; Druesne et al., 2004; Druesne-Pecollo et al., 2007). Sulforaphane, an isothiocyanate of cruciferous vegetables, such as broccoli and broccoli sprouts, can also inhibit histone deacetylase activity and decrease cancer risk (Dashwood and Ho, 2007). Based on recent reports phenethylisothiocyanate from cruciferous vegetables can also induce site-specific modifications of H3K27me3 and H3K9me2 (trimethylation of lysine 27 and dimethylation of lysine 9 of histone H3,

Table 10.1 Nutrients and dietary active components as cancer preventives via epigenetic modifications.

Active dietary or nutritional components	Epigenetic modification	References
Choline	Main resources of methyl group	Park et al., 2008; Zeisel, 2007
Methionine	Major methyl donor upon conversion to S-adenosylmethionine (SAM)	Ghoshal et al., 2006; Mato et al., 2008; Zeisel, 2007
Folic acid	Carrier and donor of methyl group	Fenech, 2001; Melnyk et al., 1999
Vitamin B12	Cofactor in remethylation of homocysteine	Fenech, 2001; Kune and Watson, 2006
Selenium	Inhibition of DNMT and histone deacetylase activities	Xiang, 2008
Epigallocatechin gallate (EGCG) (tea polyphenols)	DNMT inhibitor Histone acetyltransferase inhibitor	Choi et al., 2009; Lee et al., 2005
Myricetin (tea bioflavonoids)	DNMT inhibitor	Lee et al., 2005
Diallyl disulfide (garlic)	Histone deacetylase inhibitor	Druesne et al., 2004
Sulforaphane (broccoli)	Histone deacetylase inhibitor	Dashwood and Ho, 2007
Genistein (soy isoflavones)	Inhibition of DNMT activity Histone H3-lysine 9 methylation and H3 deacetylation Increase in histone acetyl transferases (HAT) expression	Fang et al., 2005; Kikuno et al., 2008; Majid et al., 2008; Zhang et al., 2013
Curcumin (*Curcuma longa*)	Inhibition of DNMT1, HDAC1,3,8, miRNA186a and miRNA-199a Upregulation of miRNA-22	Reuter et al., 2011 (review)
Fish oil and pectin	Increase in promoter DNA methylation of BCL2 gene	Cho et al., 2012
Tanshinone I (*Salvia miltiorrhiza*)	Decrease of the acetylated histone H3 of Aurora A gene	Gong et al., 2012
Anthocyanins (black raspberry)	Inhibition of DNMT1 and DNMT3B	Wang et al., 2013
Thymus serpyllum	Inhibition of DNMT and HDAC activities	Bozkurt et al., 2012

respectively) at the promoter of cancer-related genes in SW480 epithelial cells derived from a human colon tumor (Liu et al., 2013). In addition, genistein and soy protein isolate could decrease histone H3 acetylation of Sfrp2, Sfrp5, and Wnt5a promoters, associated with reduced binding of RNA polymerase II, an increased HDAC3 nuclear level, and decreased H3K9me3 (Zhang et al., 2013).

The inactivation of tumor suppressor genes and/or activation of oncogenes are considered the major mechanisms responsible for the development and progression of cancer. The dietary components may have cancer preventive and therapeutic activities by epigenetic modifications to reactivate silenced tumor suppressor genes and/or inactivate oncogenes. It has been reported that dietary folate and methyl deficiency could alter (increase) the expression of DNMTs and methyl CpG binding proteins (Ghoshal et al., 2006). In spite of the DNMT1 and DNMT3A increased expression, the dietary methyl deficiency was associated with a genome-wide and/or gene-specific DNA hypomethylation and hepatic carcinogenesis (Steinmetz et al., 1998) likely due to concurrent increased expression of genes involved in cancer development and progression. Additionally, the observed increase in the expression of methyl binding domains 1–3 (MBD1-3), which are involved in epigenetic gene silencing, could also be associated with silencing of tumor suppressor genes inducing tumor genesis in mouse liver (Ghoshal et al., 2006). In other words, it appears that a methyl-deficient diet could induce a genome-wide DNA hypomethylation and hyperexpression of numerous genes, including DNMTs and MBDs. While hyperexpression of DNMTs could not fix the genome-wide DNA hypomethylation resulting from a methyl-deficient diet, hyperexpression of MBDs could silence tumor suppressor genes. As it has been shown that some genes could become hypermethylated during cancer progression, it could be argued that different genes have diverse thresholds for DNMT-mediated methylation based on the structure of the gene/nucleosome, developmental period, the required amount of methyl group for re-methylation, and other unknown mechanisms. In this line, a number of studies have demonstrated that the imprinting status of individual genes could be predicted by DNA sequence characteristics (Luedi et al., 2005, 2007) implying that the gene's structure may have a role in determining the targets for epigenetic modifications.

There is convincing evidence that oxidative stress may induce epigenetic aberrations resulting in many diseases including cancer (Inokuma et al., 2009, Khandrika et al., 2009; Waly et al., 2004). As oxidative stress directs the cellular methylation resources toward glutathione synthesis to avert cellular damage resulting from reactive oxygen species (Waly et al., 2004; reviewed in Abdolmaleky et al., 2008), this halts the cellular methylation capability, which is required for the functionality of hundreds of other pathways. The level of the impact of oxidative stress is controlled by antioxidants. A proper diet is a rich source of nutritional and non-nutritional components that function as antioxidants and/or directly modulate epigenetic processes via DNA methylation and/or histone modifications. Interactions of dietary components with epigenetic machinery may lead to epigenetic alterations that modify the onset and/or outcomes of human diseases. Therefore dietary and nutritional intervention may be an effective strategy to prevent the incidence or to delay/reverse the progress of diseases that result from oxidative stress and epigenetic modification. For instance, polyphenolic components in fruits, vegetables, tea, and other dietary items are shown to have potent antioxidant activities to reduce the impacts of oxidative stress and are known to reduce cancer risk (Gerhauser et al., 2003; Jang and Pezzuto, 1999; Waffo-Teguo et al., 2001). These components include resveratrol (a flavonoid in red wine), limonen (a component of orange), epigallocatechin gallate (EGCG, a major component in green tea), and curcumin (a principal component of turmeric). The distinctive roles of curcumin in cancer prevention and therapy mediated by epigenetic mechanisms has been more evident in recent years (Reuter et al., 2011). Recent *in vitro* and *in vivo* studies have also underscored potent anti-tumor activity of tanshinone-I in diverse malignancies such as prostate, breast, and lung cancer through inhibition of Aurora kinase A gene (*AURKA*), which at least in breast cancer is mediated by histone H3 deacetylation (Gong et al., 2011, 2012; Li et al., 2013).

It is noteworthy that during early gestational life, more than 250,000 neurons and millions of other cells are produced every minute, sequentially differentiated to create 100 trillion human cells (Davies and Morris, 1997; Trosko, 2003). The final destination as well as identity of cells is determined by unknown mechanisms involving DNA methylation and other epigenetic modifications such as histone acetylation and/or methylation and RNA interference. In this way, a disturbed fine tuning of DNA methylation or histone methylation/acetylation status of the affected genes (i.e., hypomethylation/hyperexpression) may lead to dysfunction in the effector genes (i.e., hypermethylation/hypo-expression). For instance, if the hypomethylated genes contain stimulatory effects on the other genes, the promoter DNA of genes under their effects will become hypomethylated as well. However, hypomethylation/hyperexpression of inhibitory genes may lead to the DNA hypermethylation of the promoter region of the affected genes and vice versa. In such a way, the hypomethylation/hyperexpression of genes involved in cell growth and proliferation may lead to DNA hypermethylation of genes contributing to apoptosis and cell cycle arrest, an epigenetic profile that was seen in cancer development and recovered by DNMTs and histone deacetylase inhibitors (Zhu and Otterson, 2003). Note that about 50% of genes lack CpG islands (Costello and Plass, 2001). Thus if these genes are under the stimulatory effects of hypomethylated genes, they will also be overexpressed as long as the stimulatory genes are overactive. Immediate early genes are among the effector genes that may coordinately activate other genes. Nutritional components such as genistein can increases expression of these genes decreasing the risk of cancer (Singletary and Ellington, 2006).

Epigenetic impacts of food contaminants and environmental factors on cancer

In addition to the effects of nutritional methyl group imbalance, there are other factors that may change the cellular methylation capabilities. As an example, contamination of drinking water with arsenic, a known carcinogen (Liu and Waalkes, 2008; Vahter, 2008), burdens the functionality of cellular methylation machinery as arsenic metabolism requires methyl groups to be secreted from the urine (Hopenhayn-Rich et al., 1996). Even low levels of arsenic contamination could result in overconsumption of the methyl group as evidenced by accumulation of homocysteine in the blood of affected individuals (Gamble et al., 2005). Although it has been suggested that exposure to arsenic may have no effect on the mother or the fetus because of the higher production of choline (one of the major resources of methyl group) induced by the higher level of estrogen during pregnancy, such exposure in infants and adult males could reduce the capability of methylation machinery resulting in DNA hypomethylation of several genes. Furthermore, arsenic as an oxidant, like other oxidants, directs the cellular methylation machinery toward glutathione synthesis to evade oxidative stress (Waly et al., 2004).

Thus even in the presence of sufficient methyl groups for normal functions, contaminants could overconsume methyl groups and reduce or paralyze the functionality of cellular methylation machinery required for many other pathways, including the epigenetic modulation of genes' functions. It has to be emphasized that such impacts on epigenetic fine tuning, particularly during the critical developmental period when the genes' methylation pattern is establishing, would have long-term effects on genes' functions and cancer development (Waalkes et al., 2004).

There is also convincing evidence that maternal nutrient supplement could counteract DNA hypomethylation induced by bisphenol A in early development (Dolinoy et al., 2007). Bisphenol A is an endocrine-active component and is used for polycarbonate plastic manufacturing. Fetuses/neonates exposed to bisphenol A show higher body weight and increased incidences of breast and prostate cancer. This component also directs the coat color of agouti (Avy) mouse offspring toward yellow by decreasing CpG methylation of the metastable epiallele located upstream of the agouti gene (Dolinoy et al., 2007). Bisphenol A-induced epigenetic alteration occurred during early stem cell development, and maternal dietary supplementation (methyl donors, such as folic acid or the phytoestrogen genistein) can prevent the DNA hypomethylating effect of bisphenol A. This study uncovered that CpG methylation level at another metastable locus, the CDK5 activator binding protein, was also decreased, indicating that the epigenetic effects of bisphenol A could be widespread (Dolinoy et al., 2007). Considering that a methyl-deficient diet and/or other demethylating elements could also impact methylation level of gonadal cells that are involved in spermatogenesis in adulthood, it is likely that such effects will be transferred to the next generation as well. Indeed, recent human studies uncovered that even paternal obesity is associated with DNA hypomethylation of differentially methylated regions of IGF-2, and imprinted genes, in offspring (Soubry et al., 2013). Other studies with agouti gene in mice also provided strong confirmatory evidence that the induced epigenetic modifications can be transferred to the next generations maternally (Cooney et al., 2002).

Seasonal and diurnal changes in environmental conditions with their episodic and repetitive nature, accompanied by changes in accessibility of nutrient and load of contaminants and pollutant, may result in DNA methylation changes and epigenetic modifications (Baccarelli et al., 2009). Interestingly, cancer is linked to season of birth (Hoffman et al., 2007; Mainio et al., 2006), one of the major determinants of early-life nutrient availability. The circadian and light-induced transcription of clock genes (per1) was also shown to be associated with changes in histone acetylation level (Naruse et al., 2004). Per1 expression, involved in cancer development and progression (Gery and Koeffler, 2007; Lin et al., 2008; Tokunaga et al., 2008), is influenced by early-life nutritional availability of choline (Kovacheva et al., 2009). The early-life effects of nutritional elements and components as well as ecological conditions on genes' promoter DNA methylation and histone acetylation or methylation levels may determine long-life epigenetic marks that could have positive adaptive benefit for those conditions and negative consequences in a different nutritional/ecological environment. In a similar fashion, travel/immigration from a specific ecological condition with its regional specific nutritional components to a different ecological system may demand a re-setting and re-fine tuning of the epigenome to adapt to the new environmental conditions. However, this may not be easy for a compromised epigenome or DNA methylation machinery, particularly in older age, which may result in disease phenotype, including cancer (Kim at al., 2009). For instance, workers exposed to polycyclic aromatic hydrocarbons show promoter DNA hypermethylation of the p16INK4a gene, which is linked to cancer development (Yang et al., 2012).

Critical periods for stable epigenetic aberrations

The epidemiological associations between diet and the risk of human diseases have been studied for as long as the development of human civilization. Accumulating evidence suggests that nutritional imbalance during critical time windows of developmental programming may have persistent effects on the onset and outcomes of diseases (Friso and Choi, 2005; Jiang et al., 2004; Robertson and Wolffe, 2000). This has led to the development/emergence of a new hypothesis supporting the embryonic origin of adult diseases (Delisle, 2002; Junien et al., 2005) that are believed to be mediated via epigenetic mechanisms (Cutfield et al., 2007; Kim et al., 2009; Waterland, 2009).

Extensive studies of agouti gene in mice revealed that feeding pregnant mice with a methyl-supplemented diet (folic acid, vitamin B$_{12}$, choline,

and betaine) alters *agouti* expression in their offspring and directs the agouti/black mottling toward pseudoagouti phenotype and these epigenetic phenotypes are maternally heritable (Wolff et al., 1998). It was shown that CpG methylation of metastable epiallele located in the 5' region of the gene is responsible for variability of the gene expression in animals with identical genotype, and that hypomethylation of this region is associated with yellow fur, obesity, diabetes, and cancer, which could be prevented by a methyl-supplemented diet (Waterland and Jirtle, 2003).

Other studies (Waterland et al., 2006) showed that not only methyl deficiency but also methyl donor supplementation of female mice before and during pregnancy could increase DNA methylation level at Axin fused metastable epiallele and change the related phenotype (tail kinking) in the offspring. This methylation change is not restricted to tail tissue as it was also seen in other tissues such as brain and liver (Waterland et al., 2006). Furthermore, the impacts of nutrition on genes' epigenetic status is not limited to the pregnancy period or a number of genes as even a post-weaning diet could permanently affect genomic imprinting at the IGF-2 locus (Waterland et al., 2006).

The methylation machinery is highly active during early developmental periods and in the uterus (Reik and Dean, 2001). Therefore at this time the epigenome would be more vulnerable for permanent aberrations as a result of inappropriate supply of methyl groups and other elements required for the functionality of epigenetic machinery leading to disease development later in life (Godfrey and Barker, 1995; Sinclair et al., 2007). However, fine tuning and the established pattern of the genes' promoter methylation and histone methylation/acetylation status may also be disturbed in later developmental periods (e.g., childhood, puberty, and adulthood) by the micro/macro (i.e., internal/external) environment such as cells' metabolic states or correlated external conditions (e.g., nutritional, social, seasonal, geographical, etc.).

Compelling evidence indicates that the aberrant DNA methylation pattern established during the nutritional methyl deficiency and/or other environmental conditions may not be fully recovered later on, even though the nutritional defects or interfering factors are alleviated (Pogribny et al., 2006). It appears that the acquired conditional epigenetics pattern will be imprinted to the developmental memory and likely will be transmitted to the next generations of cells during mitosis as well as meiosis. Because the gonads' cells will also be affected by the nutritional defects – at least in the testis – any epigenetic insult can be inherited to the next generation as well. Thus it is likely that the pattern and degree of DNA methylation in each historical period and geographical region would be different from those of other times and locations.

From an evolutionary point of view, the generation-specific epigenetic programming may have beneficial effects for saving the species at the expense of the affected individuals. Indeed, early-life nutritional perturbations lead to special programming for adaptation to those conditions. Although it is useful to save the life of the affected individuals in those conditions in order to reach a reproductive age to save the species, this change in programming may be harmful for longevity, particularly in a nutritionally rich condition (McMillen and Robinson, 2005).

With the current humanistic view that long-term health and longevity of every living individual is important, those elements and mechanisms of this reprogramming procedure that are not compatible with the longevity and health of the affected individuals need to be uncovered for therapeutic and preventive measures. In today's world, in addition to malnutrition, it is also necessary to pay more attention to the consequences of overfeeding and overnutrition that may be correlated with a different programming leading to the current pandemic of obesity and related diseases such as cancer (Delisle, 2002; Park et al., 2010). Since methyl donor supplementation can prevent "transgenerational amplification of obesity" (Waterland et al., 2008) nutritional therapy targeting the epigenome may be an effective preventive strategy in individuals exposed to food deprivation in early life (or even previous generations). These individuals are prone to subsequent obesity, diabetes, and cardiovascular diseases (Delisle, 2002; Ong and Loos, 2006; Ong et al., 2000) and probably higher cancer risk (Jee et al., 2008).

Summary and future directions

The human body contains more than 100 trillion cells with the same genetic makeup but different phenotypes. During early gestational life, millions of cells are produced every minute, sequentially differentiated, and their final destination as well as identity are determined involving epigenetic mechanisms such as DNA methylation, histone acetylation and/or methylation, RNA editing, and RNA interference. With the exception of a number of cells such as neurons, the half-life of most human cells ranges

from a few days to a few months. Thus continuous cell regeneration is required to maintain the structural and functional integrity of the organism. While genetic codes instruct cellular regeneration from stem and progenitor cells, the required materials (e.g., amino acids, vitamins, and minerals) are provided by food in a range of concentrations extending from deficient to toxic levels.

DNA synthesis, epigenetic modifications, gene expression, and cellular structure/function are all influenced by nutrients and bioactive food components, as well as food contaminants. The genetic landscape also determines different responses of individuals to specific food, nutrients, and contaminants. The personalized response to specific nutrients, the topic of nutritional genomics, underlies many complex diseases intensified by stereotypic eating behavior and habits. The flexible and dynamic nature of epigenetic modifications can serve a capacity for adaptive fine tuning of genes' expression in variable environments or in individuals with dysfunctional polymorphisms. Nevertheless, experimental manipulation of the players involved in the establishment of epigenetic memory was shown to provoke disease phenotype during exposure and later in life. DNA damage and epigenetic aberrations resulted from contaminants, micronutrient deficiencies and surplus during critical time windows of the developmental programming may have persistent effects on the onset and outcomes of many complex diseases, including cancer.

In the twentieth century, with the implementation of human knowledge, mostly acquired during the last few decades, human life expectancy has doubled. This was mostly achieved by taking advantage of antibiotics and vaccinations against the living microelements, bacteria and viruses. Now is the time to consider strategies against non living microelements: toxins, contaminants, and nutritional imbalances. Until now, the dynamicity and adaptability of the DNA methylation machinery was the major determinant of the survival and evolution of humans and other species that survived under haphazard conditions. The facts that over 50% of mammals' fertilized eggs are not successfully implanted in the uterus, more than 20% of fetuses are spontaneously aborted during pregnancy, and 10% of successfully delivered children die during the childhood period indicate that genetic and environmentally induced epigenetic defects have profound effects on human health and fate. Altogether, approximately 20% of fertilized eggs successfully pass this natural selection gate to create active reproductive individuals to save the species. However, a majority of these individuals will suffer from subtle genetic and epigenetic aberrations that may pass to the next generations (Junien et al., 2005). Despite these natural dilemmas, the human being has been successful in survival and evolution over the last few million years, while many other species have become extinct.

Now, with the current status of human knowledge of nutritional science and advanced techniques for the detection of genetic and epigenetic vulnerabilities, it is time for this knowledge to be refined and implemented to improve the quality and length of human life. In this direction, extensive experience in the dissection of the effects of malnutrition on the pathogenesis of complex diseases should be a priority to investigate the most complex issues dealing with nutrition and human health. Considering the dynamicity and adaptability of epigenetic regulation during life, and the impacts of nutritional elements and dietary components in determining the functionality of the epigenetic landscape, one of the long-term goals of research in this field could be the identification of bioactive dietary and nutritional components that modulate epigenetic codes to be used for the prevention and treatment of cancer and other complex diseases. In order to test the efficacy of candidate components, a series of clinically relevant animal models for various diseases should be used. Advanced techniques for cellular and molecular biology and genetics/epigenomics/proteomics should be applied to elucidate the mechanisms of actions of active components. By combining efficacy evaluation and mechanistic study, one would be able to identify effective dietary and nutritional regimens for the prevention and treatment of common diseases such as cancer. The research findings need to be directly translated to clinical practice and provide evidence to make dietary guidelines for the prevention of related diseases. Considering the exponential escalation of biological science, it is expected that the current living generation will be the observer of the next revolution in human health and longevity being bridged by the emerging science of nutrigenomics and nutriepigenomics.

References

Abdolmaleky HM, and Thiagalingam S (2011). Can the schizophrenia epigenome provide clues for the molecular basis of pathogenesis? *Epigenomics*, 3(6):679–83.

Abdolmaleky HM, Smith CL, Faraone SV, et al. (2004). Methylomics in psychiatry: modulation of gene–environment interactions may be through DNA methylation. *Am J Med Genet B Neuropsychiatr Genet*, 127B:51–9.

Abdolmaleky HM, Zhou JR, Thiagalingam S, and Smith CL (2008). Epigenetic and pharmacoepigenomic studies of major psychoses and potentials for therapeutics. *Pharmacogenomics*, 9(12):1809–23.

Ames BN (2001). DNA damage from micronutrient deficiencies is likely to be a major cause of cancer. *Mutat Res*, 475:7–20.

Arunkumar A, Vijayababu MR, Gunadharini N, Krishnamoorthy G, and Arunakaran J (2007). Induction of apoptosis and histone hyperacetylation by diallyl disulfide in prostate cancer cell line PC-3. *Cancer Lett*, 251:59–67.

Baccarelli A, Wright RO, Bollati V, et al. (2009). Rapid DNA methylation changes after exposure to traffic particles. *Am J Respir Crit Care Med*, 179(7):572–8.

Beetstra S, Thomas P, Salisbury C, Turner J, and Fenech M (2005). Folic acid deficiency increases chromosomal instability, chromosome 21 aneuploidy and sensitivity to radiation-induced micronuclei. *Mutat Res*, 578(1–2):317–26.

Beetstra S, Suthers G, Dhillon V, et al. (2008). Methionine-dependence phenotype in the de novo pathway in BRCA1 and BRCA2 mutation carriers with and without breast cancer. *Cancer Epidemiol Biomarkers Prev*, 17(10):2565–71.

Bertino L, Ruffini MC, Copani A, et al. (1996). Growth conditions influence DNA methylation in cultured cerebellar granule cells. *Dev Brain Res*, 95:38–43.

Bird A (2002). DNA methylation patterns and epigenetic memory. *Genes Dev*, 16:6–21.

Bird A (2007). Perceptions of epigenetics. *Nature*, 447:396–8.

Bissonauth V, Shatenstein B, and Ghadirian P (2008). Nutrition and breast cancer among sporadic cases and gene mutation carriers: an overview. *Cancer Detect Prev*, 32(1):52–64.

Bonsch D, Hothorn T, Krieglstein C, et al. (2007). Daily variations of homocysteine concentration may influence methylation of DNA in normal healthy individuals. *Chronobiol Int*, 24:315–26.

Bozkurt E, Atmaca H, Kisim A, et al. (2012). Effects of Thymus serpyllum extract on cell proliferation, apoptosis and epigenetic events in human breast cancer cells. *Nutr Cancer*, 64(8):1245–50.

Butterworth CE Jr. (1992). Effect of folate on cervical cancer. Synergism among risk factors. *Ann N Y Acad Sci*, 30(669):293–9.

Chary-Reddy S, Prasad VS, and Ahuja YR (1994). Expression of common fragile sites in untreated non-Hodgkin's lymphoma with aphidicolin and folate deficiency. *Cancer Lett*, 86(1):111–17.

Cho Y, Turner ND, Davidson LA, et al. (2012). A chemoprotective fish oil/pectin diet enhances apoptosis via Bcl-2 promoter methylation in rat azoxymethane-induced carcinomas. *Exp Biol Med*, 237(12):1387–93.

Choi KC, Jung MG, Lee YH, et al. (2009). Epigallocatechin-3-gallate, a histone acetyltransferase inhibitor, inhibits EBV-induced B lymphocyte transformation via suppression of RelA acetylation. *Cancer Res*, 69(2):583–92.

Cooney CA, Dave AA, and Wolff GL (2002). Maternal methyl supplements in mice affect epigenetic variation and DNA methylation of offspring. *J Nutrition*, 132:2393S–400S.

Costello JF, and Plass C (2001). Methylation matters. *J Med Genet*, 38(5):285–303.

Cutfield WS, Hofman PL, Mitchell M, and Morison IM (2007). Could epigenetics play a role in the developmental origins of health and disease? *Pediatr Res*, 61(5 Pt 2):68R–75R.

Das A, Desai D, Pittman B, Amin S, and El-Bayoumy K (2003). Comparison of the chemopreventive efficacies of 1,4-phenylenebis(methylene) selenocyanate and selenium-enriched yeast on 4-(methylnitrosamino)-1-(3-pyridyl)-1-butanone induced lung tumorigenesis in A/J mouse. *Nutr Cancer*, 46:179–85.

Dashwood RH, and Ho E (2007). Dietary histone deacetylase inhibitors: from cells to mice to man. *Semin Cancer Biol*, 17: 363–9.

Davies R, and Morris B (1997). *Molecular Biology of Neuron*. Oxford, UK: Oxford University Press.

Davis CD, and Uthus EO (2002). Dietary selenite and azadeoxycytidine treatments affect dimethylhydrazine-induced aberrant crypt formation in rat colon and DNA methylation in HT-29 cells. *J Nutr*, 132:292–7.

Delisle H (2002). Foetal programming of nutrition-related chronic diseases. *Sante*, 12(1):56–63.

Dolinoy DC, Huang D, and Jirtle RL (2007). Maternal nutrient supplementation counteracts bisphenol A-induced DNA hypomethylation in early development. *Proc Natl Acad Sci USA*, 104(32):13056–61.

Druesne N, Pagniez A, Mayeur C, et al. (2004). Repetitive treatments of colon HT-29 cells with diallyl disulfide induce a prolonged hyperacetylation of histone H3 K14. *Ann NY Acad Sci*, 1030:612–21.

Druesne-Pecollo N, Chaumontet C, Pagniez A, et al. (2007). *In vivo* treatment by diallyl disulfide increases histone acetylation in rat colonocytes. *Biochem Biophys Res Commun*, 354:140–7.

Duthie SJ, Grant G, Pirie LP, Watson, AJ, and Margison GP (2010). Folate deficiency alters hepatic and colon MGMT and OGG-1 DNA repair protein expression in rats but has no impact on genome-wide DNA methylation. *Cancer Prev Res*, 3(1):92–100.

Fang MZ, Wang Y, Ai N, et al. (2003). Tea polyphenol (-)-epigallocatechin-3-gallate inhibits DNA methyltransferase and reactivates methylation-silenced genes in cancer cell lines. *Cancer Res*, 63:7563–70.

Fang MZ, Chen D, Sun Y, Jin Z, Christman JK, and Yang CS (2005). Reversal of hypermethylation and reactivation of p16INK4a, RARbeta, and MGMT genes by genistein and other isoflavones from soy. *Clin Cancer Res*, 11:7033–41.

Fenech M (2001). The role of folic acid and vitamin B12 in genomic stability of human cells. *Mutat Res*, 475:57–67.

Ferguson LR, Philpott M, and Karunasinghe N (2004). Dietary cancer and prevention using antimutagens. *Toxicology*, 198 (1–3):147–59.

Fiala ES, Staretz ME, Pandya GA, El-Bayoumy K, and Hamilton SR (1998). Inhibition of DNA cytosine methyltransferase by chemopreventive selenium compounds, determined by an improved assay for DNA cytosine methyltransferase and DNA cytosine methylation. *Carcinogenesis*, 19:597–604.

Forte A, De Sanctis R, Leonetti G, et al. (2008). Dietary chemoprevention of colorectal cancer. *Ann Ital Chir*, 79(4):261–7.

Friso S, and Choi SW (2005). Gene-nutrient interactions in one-carbon metabolism. *Curr Drug Metab*, 6(1):37–46.

Gamble MV, Liu X, Ahsan H, et al. (2005). Folate, homocysteine, and arsenic metabolism in arsenic-exposed individuals in Bangladesh. *Environ Health Perspect*, 113(12):1683–8.

Gerhauser C, Klimo K, Heiss E, et al. (2003). Mechanism-based in vitro screening of potential cancer chemopreventive agents. *Mutat Res*, 523–4:163–72.

Gery S, and Koeffler HP (2007). The role of circadian regulation in cancer. *Cold Spring Harb Symp Quant Biol*, 72:459–64.

Ghoshal K, Li X, Datta J, et al. (2006). A folate- and methyl-deficient diet alters the expression of DNA methyltransferases and methyl CpG binding proteins involved in epigenetic gene silencing in livers of F344 rats. *J Nutr*, 136(6):1522–7.

Godfrey KM, and Barker DJ (1995). Maternal nutrition in relation to fetal and placental growth. *Eur J Obstet Gynecol Reprod Biol*, 61(1):15–22.

Gomes MV, Toffoli LV, Arruda DW, et al. (2012). Age-related changes in the global DNA methylation profile of leukocytes are linked to nutrition but are not associated with the MTHFR C677T genotype or to functional capacities. *PLoS One*, 7(12):e52570.

Gong Y, Li Y, Lu Y, et al. (2011). Bioactive tanshinones in *Salvia miltiorrhiza* inhibit the growth of prostate cancer cells in vitro and in mice. *Int J Cancer*, 129(5):1042–52.

Gong Y, Li Y, Abdolmaleky HM, Li L, and Zhou JR (2012). Tanshinones inhibit the growth of breast cancer cells through epigenetic modification of Aurora A expression and function. *PLoS One*, 7(4):e33656.

Gumus G, Sunguroglu A, Tükün A, Sayin DB, and Bakesoy I (2002). Common fragile sites associated with the breakpoints of chromosomal aberrations in hematologic neoplasms. *Cancer Genet Cytogenet*, 133(2):168–71.

Hawkins PG, and Morris KV (2008). RNA and transcriptional modulation of gene expression. *Cell Cycle*, 7(5):602–7.

Hoffman S, Schellinger KA, Propp JM, et al. (2007). Seasonal variation in incidence of pediatric medulloblastoma in the United States, 1995–2001. *Neuroepidemiology*, 29(1–2):89–95.

Hopenhayn-Rich C, Biggs ML, Smith AH, Kalman DA, and Moore LE (1996). Methylation study of a population environmentally exposed to arsenic in drinking water. *Environ Health Perspect*, 104(6):620–8.

Inokuma T, Haraguchi M, Fujita F, Tajima Y, and Kanematsu T (2009). Oxidative stress and tumor progression in colorectal cancer. *Hepatogastroenterology*, 56(90):343–7.

Issa JP, Ahuja N, Toyota M, Bronner MP, and Brentnall TA (2001). Accelerated age-related CpG island methylation in ulcerative colitis. *Cancer Res*, 61(9):3573–7.

Jacob RA, Gretz DM, Taylor PC, et al. (1998). Moderate folate depletion increases plasma homocysteine and decreases lymphocyte DNA methylation in postmenopausal women. *J Nutr*, 128(7):1204–12.

Jamaluddin MD, Chen I, Yang F, et al. (2007). Homocysteine inhibits endothelial cell growth via DNA hypomethylation of the cyclin A gene. *Blood*, 110(10):3648–55.

Jang M, and Pezzuto JM (1999). Cancer chemopreventive activity of resveratrol. *Drugs Exp Clin Res*, 25(2–3):65–77.

Jee SH, Yun JE, Park EJ, et al. (2008). Body mass index and cancer risk in Korean men and women. *Int J Cancer*, 123(8):1892–6.

Jenab M, Slimani N, Bictash M, Ferrari P, and Bingham SA (2009). Biomarkers in nutritional epidemiology: applications, needs and new horizons. *Hum Genet*, 125(5–6):507–25.

Jiang YH, Bressler J, and Beaudet AL (2004). Epigenetics and human disease. *Annu Rev Genomics Hum Genet*, 5:479–510.

Junien C, Gallou-Kabani C, Vigé A, and Gross MS (2005). Nutritionnal epigenomics: consequences of unbalanced diets on epigenetics processes of programming during lifespan and between generations. *Ann Endocrinol*, 66(2 Pt 3):2S19–28.

Khandrika L, Kumar B, Koul S, Maroni P, and Koul HK (2009). Oxidative stress in prostate cancer. *Cancer Lett*, 282(2):125–36.

Kikuno N, Shiina H, Urakami S, et al. (2008). Genistein mediated histone acetylation and demethylation activates tumor suppressor genes in prostate cancer cells. *Int J Cancer*, 123(3):552–60.

Kim GD, Ni J, Kelesoglu N, Roberts RJ, and Pradhan S (2002). Co-operation and communication between the human maintenance and de novo DNA (cytosine-5) methyltransferases. *EMBO J*, 21:4183–95.

Kim KC, Friso S, and Choi SW (2009). DNA methylation, an epigenetic mechanism connecting folate to healthy embryonic development and aging. *J Nutr Biochem*, 20(12):917–26.

Kolialexi A, Mavrou A, Tsenghi C, et al. (1998). Chromosome fragility and predisposition to childhood malignancies. *Anticancer Res*, 18(4A):2359–64.

Kovacheva VP, Davison JM, Mellott TJ, et al. (2009). Raising gestational choline intake alters gene expression in DMBA-evoked mammary tumors and prolongs survival. *FASEB J*, 23(4):1054–63.

Kune G, and Watson L (2006). Colorectal cancer protective effects and the dietary micronutrients folate, methionine, vitamins B6, B12, C, E, selenium, and lycopene. *Nutr Cancer*, 56(1):11–21.

Lavrov SA, and Kibanov MV (2007). Noncoding RNAs and chromatin structure. *Biochemistry*, 72(13):1422–38.

Lee WJ, Shim JY, and Zhu BT (2005). Mechanisms for the inhibition of DNA methyltransferases by tea catechins and bioflavonoids. *Mol Pharmacol*, 68(4):1018–30.

Li Y, Gong Y, Li L, Abdolmaleky HM, and Zhou JR (2013). Bioactive tanshinone I inhibits the growth of lung cancer in part via downregulation of Aurora A function. *Mol Carcinog*, 52(7):535–43.

Lin YM, Chang JH, Yeh KT, et al. (2008). Disturbance of circadian gene expression in hepatocellular carcinoma. *Mol Carcinog*, 47(12):925–33.

Liu J, and Waalkes MP (2008). Liver is a target of arsenic carcinogenesis. *Toxicol Sci*, 105(1):24–32.

Liu Y, Chakravarty S, and Dey M (2013). Phenethylisothiocyanate alters site- and promoter-specific histone tail modifications in cancer cells. *PLoS One*, 8(5):e64535.

Luedi PP, Hartemink AJ, and Jirtle RL (2005). Genome-wide prediction of imprinted murine genes. *Genome Res*, 15(6):875–84.

Luedi PP, Dietrich FS, Weidman JR, et al. (2007). Computational and experimental identification of novel human imprinted genes. *Genome Res*, 17(12):1723–30.

Mainio A, Hakko H, Koivukangas J, Niemelä A, and Räsänen P (2006). Winter birth in association with a risk of brain tumor among a Finnish patient population. *Neuroepidemiology*, 27(2):57–60.

Majid S, Kikuno N, Nelles J, et al. (2008). Genistein induces the p21WAF1/CIP1 and p16INK4a tumor suppressor genes in prostate cancer cells by epigenetic mechanisms involving active chromatin modification. *Cancer Res*, 68(8):2736–44.

Mato JM, Martínez-Chantar ML, and Lu SC (2008). Methionine metabolism and liver disease. *Annu Rev Nutr*, 28:273–93.

McKay JA, Williams EA, and Mathers JC (2011). Effect of maternal and post-weaning folate supply on gene-specific DNA methylation in the small intestine of weaning and adult apc and wild type mice. *Front Genet*, 2:23.

McMillen IC, and Robinson JS (2005). Developmental origins of the metabolic syndrome: prediction, plasticity, and programming. *Physiol Rev*, 85(2):571–633.

Melnyk S, Pogribna M, Miller BJ, et al. (1999). Uracil misincorporation, DNA strand breaks, and gene amplification are associated with tumorigenic cell transformation in folate deficient/repleted Chinese hamster ovary cells. *Cancer Lett*, 146(1):35–44.

Miller JW, Nadeau MR, Smith D, and Selhub J (1994). Vitamin B-6 deficiency vs folate deficiency: comparison of responses to methionine loading in rats. *Am J Clin Nutr*, 59(5):1033–9.

Mosharov E, Cranford MR, and Banerjee R (2000). The quantitatively important relationship between homocysteine metabolism and glutathione synthesis by the transsulfuration pathway and its regulation by redox changes. *Biochemistry*, 39(42):13005–11.

Murray RK (2000). Principles of medical biochemistry. In *Harper's Biochemistry*. St. Louis: Mosby Elsevier.

Naruse Y, Oh-hashi K, Iijima N, et al. (2004). Circadian and light-induced transcription of clock gene Per1 depends on histone acetylation and deacetylation. *Mol Cell Biol*, 24(14):6278–87.

Nimni ME, Han B, and Cordoba F (2007). Are we getting enough sulfur in our diet? *Nutr Metab (Lond)*, 4:24.

Ong KK, and Loos RJ (2006). Rapid infancy weight gain and subsequent obesity: systematic reviews and hopeful suggestions. *Acta Paediatr*, 95(8):904–8.

Ong KK, Ahmed ML, Emmett PM, Preece MA, and Dunger DB (2000). Association between postnatal catch-up growth and obesity in childhood: prospective cohort study. *BMJ*, 320(7240):967–71.

Oz HS, Chen TS, and Neuman M (2008). Methionine deficiency and hepatic injury in a dietary steatohepatitis model. *Dig Dis Sci*, 53(3):767–76.

Park CS, Cho K, Bae DR, et al. (2008). Methyl-donor nutrients inhibit breast cancer cell growth. *In Vitro Cell Dev Biol Anim*, 44(7):268–72.

Park SL, Goodman MT, Zhang ZF, et al. (2010). Body size, adult BMI gain and endometrial cancer risk: the multiethnic cohort. *Int J Cancer*, 126:490–9.

Pogribny IP, Ross SA, Wise C, et al. (2006). Irreversible global DNA hypomethylation as a key step in hepatoc arcinogenesis induced by dietary methyl deficiency. *Mutat Res*, 593(1–2):80–7.

Prentice RL, Sugar E, Wang CY, Neuhouser M, and Patterson R (2002). Research strategies and the use of nutrient biomarkers in studies of diet and chronic disease. *Public Health Nutr*, 5(6A):977–84.

Reik W, and Dean W (2001). DNA methylation and mammalian epigenetics. *Electrophoresis*, 22 (14):2838–43.

Reuter S, Gupta SC, Park B, Goel A, and Aggarwal BB (2011). Epigenetic changes induced by curcumin and other natural compounds. *Genes Nutr*, 6(2):93–108.

Robertson KD, and Wolffe AP (2000). DNA methylation in health and disease. *Nat Rev Genet*, 1(1):11–19.

Ross SA (2007). Nutritional genomic approaches to cancer prevention research. *Exp Oncol*, 29(4):250–6.

Ross SA, and Milner JA (2007). Epigenetic modulation and cancer:

effect of metabolic syndrome? *Am J Clin Nutr*, 86(3):S872–7.

Russo V, Martienssen R, and Riggs A (1996). *Epigenetic Mechanisms of Gene Regulation*. Plainview, NY: Cold Spring Harbor Laboratory Press.

Sauer J, Mason JB, and Choi SW (2009). Too much folate: a risk factor for cancer and cardiovascular disease? *Curr Opin Clin Nutr Metab Care*, 12(1): 30–6.

Seitz HK, Pöschl G, and Simanowski UA (1998). Alcohol and cancer. *Recent Dev Alcohol*, 14:67–95.

Sephashvili M, Zhuravliova E, Barbakadze T, et al. (2006). L-NAME has opposite effects on the productions of S-adenosylhomocysteine and S-adenosylmethionine in V12-H-Ras and M-CR3B-Ras pheochromocytoma cells. *Neurochem Res*, 31(10):1205–10.

Shen L, Kondo Y, Rosner GL, et al. (2005). MGMT promoter methylation and field defect in sporadic colorectal cancer. *J Natl Cancer Inst*, 97(18):1317–19.

Shilatifard A (2008). Molecular implementation and physiological roles for histone H3 lysine 4 (H3K4) methylation. *Curr Opin Cell Biol*, 20(3):341–8.

Sinclair KD, Allegrucci C, Singh R, et al. (2007). DNA methylation, insulin resistance, and blood pressure in offspring determined by maternal periconceptional B vitamin and methionine status. *Proc Natl Acad Sci USA*, 104(49):19351–6.

Singh SM, Murphy B, and O'Reilly RL (2003). Involvement of gene-diet/drug interaction in DNA methylation and its contribution to complex diseases: from cancer to schizophrenia. *Clin Genet*, 64(6):451–60.

Singletary K, and Ellington A (2006). Genistein suppresses proliferation and MET oncogene expression and induces EGR-1 tumor suppressor expression in immortalized human breast epithelial cells. *Anticancer Res*, 26(2A):1039–48.

Smith DI, Huang H, and Wang L (1998). Common fragile sites and cancer. *Int J Oncol*, 12(1):187–96.

Soubry A, Schildkraut JM, Murtha A, et al. (2013). Paternal obesity is associated with IGF2 hypomethylation in newborns: results from a Newborn Epigenetics Study (NEST) cohort. *BMC Med*, 11:29.

Stover PJ, and Caudill MA (2008). Genetic and epigenetic contributions to human nutrition and health: managing genome-diet interactions. *J Am Diet Assoc*, 108(9):1480–7.

Steinmetz KL, Pogribny IP, James SJ, and Pitot HC (1998). Hypomethylation of the rat glutathione S-transferase p (GSTP) promoter region isolated from methyl-deficient livers and GSTP-positive liver neoplasms. *Carcinogenesis*, 19(8):1487–94.

Sutandyo N (2010). Nutritional carcinogenesis. *Acta Med Indones*, 42(1):36–42.

Teo T, and Fenech M (2008). The interactive effect of alcohol and folic acid on genome stability in human WIL2-NS cells measured using the cytokinesis-block micronucleus cytome assay. *Mutat Res*, 657(1):32–8.

Thiagalingam S, Cheng KH, Lee HJ, et al. (2003). Histone deacetylases: unique players in shaping the epigenetic histone code. *Ann NY Acad Sci*, 983:84–100.

Thomassin H, Flavin M, Espinás ML, and Grange T (2001). Glucocorticoid-induced DNA demethylation and gene memory during development. *EMBO J*, 20(8):1974–83.

Tokunaga H, Takebayashi Y, Utsunomiya H, et al. (2008). Clinicopathological significance of circadian rhythm-related gene expression levels in patients with epithelial ovarian cancer. *Acta Obstet Gynecol Scand*, 87(10):1060–70.

Trosko JE (2003). Human stem cells as targets for the aging and diseases of aging processes. *Med Hypothesis*, 60(3):439–47.

Vahter M (2008). Health effects of early life exposure to arsenic. *Basic Clin Pharmacol Toxicol*, 102(2):204–11.

Vaissiere T, Sawan C, and Herceg Z (2008). Epigenetic interplay between histone modifications and DNA methylation in gene silencing. *Mutat Res*, 659(1–2):40–8.

Van den Donk M, van Engeland M, Pellis L, et al. (2007). Dietary folate intake in combination with MTHFR C677T genotype and promoter methylation of tumor suppressor and DNA repair genes in sporadic colorectal adenomas. *Cancer Epidemiol Biomarkers Prev*, 16(2):327–33.

Waalkes MP, Liu J, Ward JM, and Diwan BA (2004). Mechanisms underlying arsenic carcinogenesis: hypersensitivity of mice exposed to inorganic arsenic during gestation. *Toxicology*, 198(1–3):31–8.

Waddington CH (1939). *Introduction to Modern Genetics*. London: Allen and Unwin.

Waffo-Téguo P, Hawthorne ME, Cuendet M, et al. (2001). Potential cancer-chemopreventive activities of wine stilbenoids and flavans extracted from grape (*Vitis vinifera*) cell cultures. *Nutr Cancer*, 40(2):173–9.

Waly M, Olteanu H, Banerjee R, et al. (2004). Activation of methionine synthase by insulin-like growth factor-1 and dopamine: a target for neurodevelopmental toxins and thimerosal. *Mol Psychiatry*, 9(4):358–70.

Wang LS, Kuo CT, Cho SJ, et al. (2013). Black raspberry-derived anthocyanins demethylate tumor suppressor genes through the inhibition of DNMT1 and DNMT3B in colon cancer cells. *Nutr Cancer*, 65(1):118–25.

Wang X, Thomas P, Xue J, and Fenech M (2004). Folate deficiency induces aneuploidy in human lymphocytes in vitro – evidence using cytokinesis-blocked cells and probes specific for chromosomes 17 and 21. *Mutat Res*, 551(1–2):167–80.

Waterland RA (2009). Is epigenetics an important link between early life events and adult disease? *Horm Res*, 71(Suppl 1):13–16.

Waterland RA, and Jirtle RL (2003). Transposable elements: targets for early nutritional effects on epigenetic gene regulation. *Mol Cell Biol*, 23(15):5293–300.

Waterland RA, Lin JR, Smith CA, and Jirtle RL (2006). Post-weaning diet affects genomic imprinting at the insulin-like growth factor 2 (Igf2) locus. *Hum Mol Genet*, 15:705–16.

Waterland RA, Travisano M, Tahiliani KG, Rached MT, and Mirza S (2008). Methyl donor supplementation prevents transgenerational amplification of obesity. *Int J Obes*, 32(9):1373–9.

Wolff GL, Kodell RL, Moore SR, and Cooney CA (1998). Maternal epigenetics and methyl supplements affect agouti gene expression in Avy/a mice. *FASEB J*, 12(11):949–57.

Xiang N, Zhao R, Song G, and Zhong W (2008). Selenite reactivates silenced genes by modifying DNA methylation and histones in prostate cancer cells. *Carcinogenesis*, 29(11):2175–81.

Yang P, Ma J, Zhang B, et al. (2012). CpG site-specific hypermethylation of p16INK4α in peripheral blood lymphocytes of PAH-exposed workers. *Cancer Epidemiol Biomarkers Prev*, 21(1):182–90.

Zeisel SH (2007). Gene response elements, genetic polymorphisms and epigenetics influence the human dietary requirement for choline. *IUBMB Life*, 59(6): 380–7.

Zhang Y, Li Q, and Chen H (2013). DNA methylation and histone modifications of Wnt genes by genistein during colon cancer development. *Carcinogenesis*, 34(8):1756–63.

Zhu WG, and Otterson GA (2003). The interaction of histone deacetylase inhibitors and DNA methyltransferase inhibitors in the treatment of human cancer cells. *Curr Med Chem Anticancer Agents*, 3(3):187–99.

Chapter 11

Regulatory signaling networks in cell transformation and cancer

Yashaswi Shrestha and William C. Hahn

Introduction

An intricate balance of signaling networks regulates homeostasis and protects the integrity of cells. Signals that promote cell growth and proliferation are kept in check by anti-proliferative signals and pro-cell death pathways. Disruption of signaling pathways may compromise the viability of the cell and/or alter cellular function, leading to disease. Cancer is the end product of multiple genetic and epigenetic alterations that accumulate over time, and the resulting signaling imbalances cooperate to promote tumorigenesis (Hanahan and Weinberg, 2000).

Although cancers are heterogeneous, all malignant tumors share the hallmark of cancer cells – the fundamental ability to proliferate. To attain this ability, cancer cells develop resistance toward anti-growth signals and exhibit activated growth and proliferation signaling pathways that act independent of growth factors. The cells also acquire the ability to overcome cell death signals, to achieve immortalization and often exhibit arrested differentiation. Ability to migrate enables cancer cells to invade tissues and metastasize. Finally, the source of nourishment to maintain all cancer cell activities is established by gaining neoangiogenic properties.

A comprehensive understanding of signaling pathways, which confer neoplastic capabilities when perturbed, is essential to gain insight into the biology of this disease and establish a foundation for rational drug development. Although the study of tumor-derived cells is an important approach in understanding the events that lead to cancer, these cells exhibit widespread genomic instability that complicates the identification of driver among passenger alterations. Nevertheless, the presence of these autochthonous alterations in tumor-derived cell lines make them invaluable in the analyses and verification of oncogenic alterations. A complementary approach to determine cancer-promoting alterations is through the use of cell and animal models of cancer. Cancer models contain specific alterations that disrupt the minimum signaling pathways required to execute a malignant program.

Cancer models have revolutionized the process of understanding cancer initiation and progression through defined alterations. Experimental models of several cancer types have been developed by transgenic or endogenous expression of oncogenes and dominant negative tumor suppressors. For example, several transgenic mouse models of lung cancer have been created through the engineered expression of the oncogenic allele of *KRAS*, while germline deletion of the p53 tumor suppressor gene not only confers tumor susceptibility but cooperates with other genetic alterations to create models that recapitulate particular types of tumors (Cichowski et al., 1999; Donehower et al., 1995; Fisher et al., 2001; Jackson et al., 2001). Cell-based experimental models of transformation for several cancer types including lung, prostate, and breast also have been created (Zhao et al., 2004). Together, these human and murine cancer models have been invaluable in the discovery and characterization of novel oncogenes and tumor suppressors.

This chapter will discuss the roles of several signaling pathways using the context of experimental transformation models. We will also discuss technological advances that have made unbiased genome-scale studies possible, and more importantly also promise to transform cancer therapeutics.

Signaling pathways involved in cancer

Transformation models

The earliest studies of transformed cells involved immortal murine fibroblasts that expressed retroviral proto-oncogenes (Bishop, 1985). Although single oncogenes sufficed to transform immortal fibroblasts, the same oncogenes when introduced into murine primary cells were unable to transform them (Balmain and Pragnell, 1983; Eva and Aaronson, 1983; Krontiris and Cooper, 1981; Perucho et al., 1981; Pulchiani et al., 1982; Shih et al., 1979, 1981; Sukumar et al., 1983). Ultimately it was shown that a combination of oncogenes, such as HRAS and MYC or HRAS and E1A, could transform murine cells (Land et al., 1983; Ruley, 1983). The same combinations of oncogenes, however, failed to transform human cells (Sager et al., 1983; Stevenson and Volsky, 1986). Prolonged culture of murine embryonic fibroblasts *in vitro* resulted in spontaneous immortalization while similar treatment of human cells experienced growth arrest (Wright and Shay, 2000). Such observations identified a fundamental difference between murine and human cells. Subsequent studies recognized the differences in telomere length and telomerase expression between human and murine cells, suggesting the possible role of telomerase in immortalization (Prowse and Grieder, 1995). Since human tumors exhibited stable telomere length with passage in culture, expression of telomerase was predicted to play a role in cell immortalization (Counter et al., 1992).

Expression of an oncogene, such as HRAS^{V12}, in primary cells led to growth arrest, also known as oncogene-induced senescence. Although inhibition of p53 with telomerase expression was found to be sufficient for murine cells to bypass oncogene-induced senescence, human cells also required inhibition of the RB pathway for immortalization and transformation (Hara et al., 1991; Rangarajan et al., 2004; Serrano et al., 1997; Shay et al., 1991). Human cells immortalized with telomerase activity and perturbation of the p53 and RB pathways were then receptive to one or a combination of oncogenes. These calculated sets of perturbations led to transformation of human cells and permitted the development of cancer models of defined genetic composition (Boehm et al., 2005; Hahn et al., 1999, 2002). Various human cell types have now been successfully transformed in such a manner (Berger et al., 2004; Elenbaas et al., 2001; Lundberg et al., 2002; MacKenzie et al., 2002; Rich et al., 2001).

These experimental models substantiate that altered function of a limited number of pathways can lead to cancer. Using one such transformation model, we and others have found that cooperative disruption of five pathways – telomerase, p53, RB, PP2A, and RAS – leads to the transformation of a wide range of human cell types (Hahn et al., 1999, 2002; Zhao et al., 2004). A summary of the pathways and their function in transformation is summarized in Table 11.1.

Table 11.1 Perturbation of five pathways successfully transforms primary human cells.

Transforming alteration		Target	Substitutes	Activation	Inhibition
hTERT/ALT		Maintain telomere length		Immortalization	Crisis
SV40 ER	LT	p53 inhibition	HPV E6, MDM2 overexpression, p14ARF loss	Proliferation, neo-angiogenesis	Apoptosis, senescence, cell cycle checkpoint
		RB inhibition	Overexpression of cyclin D1 or CDK4, loss of p16INK4A	Cell cycle inhibition, transcription progresssion	
	ST	PP2A inhibition	AKT + RAC1	Proliferation and survival pathways	
HRAS G12V			Combination of RAS effector pathways depending upon cell type	Proliferation pathways, cell cycle progression, survival pathways, neo-angiogenesis, metastasis/migration	

Telomerase

Telomerase is a ribonucleoprotein enzyme complex that consists of a reverse transcriptase subunit (TERT), an RNA template (TR/TERC), and dyskerin in its active form (Cohen et al., 2007). The complex adds six-nucleotide repeats (TTAGGG) to the ends of chromosomes. Maintenance of these chromosome end telomeres, protects chromosomal integrity by preventing end-to-end fusions and subsequent genomic instability (Blasco et al., 1997). Each cycle of DNA replication results in subsequently shorter telomeres due to the inability of DNA polymerase to replicate the complete chromosomal length (Harley, 1991). Functional telomeres facilitate complete replication of chromosomes during cell division. Without telomerase activity, the number of divisions a cell can undergo is finite.

Somatic human cells suppress *hTERT* expression (Mantell and Greider, 1994). After a finite number of cell cycles, human cells undergo senescence – a state of irreversible proliferative arrest with metabolic activity (Allsopp et al., 1992, 1995; Harley et al., 1990; Hastie et al., 1990; Hayflick and Morehead, 1961; Levy et al., 1992; Yu et al., 1990). Inhibition of two tumor suppressor pathways, p53 and RB, allows cells to bypass senescence and cell division continues. Further shortening of telomeres force cells to enter another proliferative hurdle – crisis (Ishikawa, 1997). The state of crisis occurs when cells undergo chromosomal rearrangements and an eventual cell death. Rarely, one in 10 million, does a colony of cells overcome crisis and is henceforth termed immortal (Wright et al., 1989).

Interestingly, the spontaneously immortalized cells exhibit stable telomere lengths, suggesting that during the process of immortalization, cells acquire the ability to maintain telomeres (Counter et al., 1992, 1994, 1998; Kim et al., 1994). Almost all human cancers have activated mechanisms to stabilize telomeres, the majority of which are due to reactivation of telomerase (Shay and Bacchetti, 1997). A small fraction of cancers have elongated telomeres without detectable telomerase activity – a mechanism termed alternative lengthening of telomeres or ALT (Bryan et al., 1997). ALT involves inter-telomeric recombination that results in maintenance of telomeres and continued cell division (Dunham et al., 2000). An *in vivo* model that establishes the importance of telomere length in cancer shows that tumor-prone *INK4A*-null mice that also lack telomerase have reduced tumor incidence (Greenberg et al., 1999). Ectopic expression of hTERT in combination with SV40 ER in several types of normal human cells facilitates bypass of senescence and crisis, conferring the ability to proliferate incessantly.

p53

The tumor suppressor gene *TP53* is inactivated in most human tumors (Vogelstein, 2000). p53-null mice develop spontaneous tumors of different cells of origin in as early as ten weeks of age (Donehower et al., 1995). Cellular stress due to DNA damage or oncogene activation triggers the activity of p53, a transcription factor that activates signaling pathways regulating DNA repair, cell cycle arrest, angiogenesis, senescence, and programmed cell death.

Double-stranded breaks in the DNA activate cell cycle checkpoint proteins that activate the DNA repair machinery before the cell resumes replication. Cell cycle checkpoint activation phosphorylates p53 at the amino-terminus (Berger et al., 2005). Expression of oncogenes activates a tumor suppressor ARF, which sequesters MDM2 (Zhang et al., 1998). In both of these cases, MDM2 is no longer able to interact with p53, resulting in p53 stabilization. p53 directly regulates transcription of *CDNK1A* (p21$^{\text{WAF1/CIP}}$), a cyclin-dependent kinase (CDK) inhibitor that facilitates DNA repair by blocking cell cycle progression from G1 to S as well as G2 to M phase. Therefore p53-dependent cell cycle arrest and DNA repair allows the cell to maintain its genetic integrity.

Another important cancer repressive function of p53 is to curb angiogenesis by interfering with regulators of hypoxia. p53 inhibits pro-angiogenic factors and increases expression of anti-angiogenic factors. p53 binds to HIF1α, which is a subunit of hypoxia-inducing factor that senses oxygen deprivation and initiates angiogenesis, and targets it for degradation (Pugh and Ratcliffe, 2003). Secondly, pro-angiogenic factors, such as vascular endothelial growth factor (*VEGF*) and cyclooxygenase-2 (*COX2*), are transcriptionally repressed by p53 (Pal, 2001; Subbaramaiah et al., 1999). Finally, p53 activates anti-angiogenic factors such as brain-specific angiogenesis inhibitor 1 (*BAI1*) and thrombospondin 1 (*TSP1*) for a complete inhibition of neo-angiogenesis (Teodoro et al., 2007).

During oncogene activation, p53 induces replicative senescence to limit cell proliferation. As cells replicate and telomere length decreases to 4 to 7 kb, p53 arrests cells in G1. Senescent fibroblasts show change in morphology, increase in acidic beta-galactosidase staining, resistance to apoptosis, and inability to be stimulated by growth factors (Dimri et al., 1995; Linskens et al., 1995; Seshadri and Campisi, 1989). Senescence prevents further telomere shortening and protects cells from chromosomal rearrangements. Inhibition of p53 activity enables cells to proliferate beyond the point of senescence, resulting in extremely short telomere repeats and increased karyotypic disarray.

p53 activates programmed cell death via apoptosis in cells that are damaged beyond repair. A cell may die by apoptosis due to activation of the DNA damage pathway triggered by critically short telomeres or cellular stress induced by hypoxia and growth factor deprivation (Artandi and Attardi, 2005). p53 also promotes apoptosis by transcriptional repression of surviving (BIRC5) – a member of the inhibitor of apoptosis proteins (IAPs) – that is involved in chemoresistance and angiogenesis (Hoffman et al., 2002). Apoptosis is induced by p53 through the extrinsic or intrinsic pathways by activating death receptors FAS, DR4, and DR5 or through expression of pro-apoptotic genes (*BAX)* and other BH3-only proteins (NOXA, PUMA, and BID), respectively (Chipuk et al., 2006; Yu et al., 2005). Finally, p53 regulates several other members of the apoptotic pathway such as *APAF-1*, caspase-6, and REDOX genes (Yu et al., 2005).

Therefore p53 inactivation disables one of the key mechanisms that regulates cell death. Disruption of p53 function leads to accumulation of DNA damage, uncontrolled proliferation, neo-angiogenesis, and apoptosis evasion – generating an ideal condition for tumor initiation and growth. In transformation models, p53 function can be inhibited by SV40 large T antigen (LT), human papillomavirus gene E6, or *MDM2* overexpression. p53 function is also positively regulated by E2F-1, a transcription factor that activates ARF (or p14 in humans, p19 in mice). ARF is a tumor suppressor that inhibits MDM2, consequently stabilizing p53 (Quelle et al., 1995). ARF is one of two proteins encoded by the *CDKN2A* locus and connects the p53 and RB tumor suppressor pathways (Bates et al., 1998; Sherr, 2001). The alternatively spliced variant of the *CDKN2A* transcript gives rise to INK4A (or p16), a crucial regulator of the RB pathway.

RB

Retinoblastoma protein (RB) is a tumor suppressor that negatively regulates entry into the G1 phase of the cell cycle. INK4A is a CDK inhibitor that prevents the association of CDK4 or CDK6 with D-type cyclins. CDK4/6-cyclin D complexes determine the phosphorylation status of RB and consequently its function (Serrano et al., 1993). Hypo-phosphorylated RB sequesters the E2F transcription factors and inhibits their function by concealing their transactivation domain (Phillips and Vousden, 2001). Additionally, association of RB with E2F allows recruitment of histone deacetylases to the complex, which actively repress the expression of E2F target genes, such as ARF (Brehm et al., 1998; Ferreira et al., 1998; Luo et al., 1998; Magnaghi-Jaulin et al., 1998).

When post-mitotic cells in G0 receive mitogenic stimulation, CDK4/6-cyclin D complexes phosphorylate RB. Hyperphosphorylation of RB inhibits its interaction with E2F, which then activates the expression of genes required for G1/S transition and cell cycle progression (Mittnacht, 1998). Constitutive hyperphosphorylation of RB inactivates the RB pathway. SV40 LT antigen, cyclin D1 or CDK4 amplification, or INK4A loss inhibits RB function and renders the cells insensitive to anti-proliferative signals.

PP2A

PP2A is a heterotrimeric complex that functions as a serine/threonine phosphatase. PP2A is a tumor suppressor that is found in inactivated forms in human cancers (Calin et al., 2000; Ruediger et al., 2001a, b; Takagi et al., 2000; Tamaki et al., 2004; Wang et al., 1998). The protein complex comprises a structural A subunit, a catalytic C subunit, and a variable B subunit that is responsible for substrate specificity, localization, and function. Two A subunit isoforms, Aα and Aβ, two C subunit isoforms, Cα and Cβ, and over twenty B subunits cause PP2A to form more than a hundred unique complexes, providing each PP2A complex with the specificity and flexibility of unique substrate and function (Sablina and Hahn, 2008).

SV40 ST antigen structurally resembles the B subunit and forms a heterotrimeric complex with PP2A AC complexes. ST interaction with PP2A inhibits PP2A tumor suppressive function by activating survival and proliferative pathways and inhibiting anti-growth, cell death signals. Deregulation of PP2A function by ST leads to activation of PI3K, MAPK, and JNK pathways (Howe et al., 1998; Sontag et al., 1993). ST inhibits the function of PP2A B56γ subunit that dephosphorylates and activates p53 and APC tumor suppressors (Li et al., 2007; Okamoto et al., 2002). PP2A Aβ complex dephosphorylates and deactivates RAL-A GTPase (Sablina et al., 2007). ST inhibition of PP2A promotes RAL-A-dependent tumorigenic functions such as cell mobility, transformation, and tumor metastasis (Feig, 2003; Tchevkina et al., 2005; Ward et al., 2001).

RAS

The RAS family of proteins was the first to indicate the ability to regulate cell growth (Downward, 2006). *HRAS* was the first proto-oncogene to be discovered by several research groups simultaneously. Human bladder cancer cell lines expressed oncogenic HRAS that contained a single mutation compared to the normal human HRAS gene (McCoy et al., 1984; Pulciani et al., 1982; Taparowsky et al., 1982). These observations led to intensive study of the RAS genes and to the pioneering realization that three RAS genes, *HRAS*, *NRAS*, and *KRAS*, were most frequently altered in human cancers.

RAS proteins are small GTPases, active in the GTP-bound state and inactive when bound to GDP. RAS GTPase activating proteins (GAPs) hydrolyze GTP while guanine nucleotide exchange factors (GEFs) are required for RAS activity in the membrane. GTPases process extracellular stimuli through receptors and activate intracellular signaling pathways to regulate various cellular functions. Growth factor binding to cognate receptors induces dimerization and autophosphorylation of the receptors. The phosphotyrosine residues of growth factor receptors bind to the SH2 domain of GRB2, which in turn recruits SOS to the membrane through interaction with its SH3 domain (Downward, 2003). Farnesylation of RAS relocates it to the membrane, where its proximity to SOS, a RAS-GEF, results in its activation by nucleotide exchange of GDP to GTP (Aronheim et al., 1994).

Genetic alterations in RAS, RAS regulators such as NF1, and RAS effectors such as RAF and PI3K are common in human cancers. Despite tremendous efforts, no clear strategy to block RAS activity exists and emerging clinical data suggest that tumors that harbor mutant forms of RAS fail to respond to existing therapies (Engelman et al., 2008). Hence the need for an in-depth understanding of RAS signaling pathways has gained importance to identify efficient targets for cancer drug development.

RAS, in its active GTP-bound state, interacts with several effectors to regulate cell growth, proliferation, survival, and transformation. RAS effectors include the RAF, PI3K, RALGDS, and PLCε pathways (Downward, 2003). RAS interacts with RAF to activate the MAPK-ERK pathway that regulates cell growth and proliferation. Activation of the transcription factors such as ELK1, JUN, and FOS regulates expression of cyclin D that promotes cell cycle progression (Pruitt and Der, 2001). Ectopic activation of such signaling pathways renders cancer cells independent of growth factor signaling.

RAS activation of the PI3K pathway in turn activates the survival signaling kinase, AKT/PKB in a PDK-dependent manner (Bader et al., 2005; Stephens et al., 1998). AKT activation in cancer cells enables them to evade cell death signals by phosphorylation and inactivation of pro-apoptotic proteins, such as BAD and forkhead (FKHR) transcription factors (Hennessy et al., 2005 ; Vivanco and Sawyers, 2002). PI3K also activates RAC, a RHO-GTPase, which functions in cytoskeleton reorganization. Direct interaction of RAS with TIAM1, a RAC-GEF, in a PI3K-dependent manner leads to activation of RAC, which may enhance invasive and metastatic properties of cancer cells (Lambert et al., 2002; Malliri et al., 2002). In endothelial cells, hypoxia induces expression of the RHO family of GTPases including RHO-A, RAC1, and CDC42 that lead to HIF1α accumulation as well as change in cell polarity and migration, promoting neo-angiogenesis (Bryan and D'Amore, 2007; Turcotte et al., 2003). Therefore oncogenic activation of RAS signals effector pathways to carry out functions that enable survival of cancer cells.

Contribution of individual RAS effector pathways in transformation has been studied using specific point mutant alleles of oncogenic H-RAS^{V12} – T35S, Y40C, or E37G, which exclusively bind and activate RAF, PI3K, or RAL-GDS, respectively. Several studies have used mutant alleles of oncogenic HRAS to examine the importance of each pathway in transformation of various rodent and human cell types. Hamad et al. (2002) showed that rodent fibroblasts require activation of the RAF-MAPK pathway for transformation, while RAL-GDS play a more important role in RAS-dependent transformation of immortalized human embryonic kidney epithelial cells. In contrast to this observation, Rangarajan et al. (2004) found that PI3K pathway activation was required in addition to RAL-GDS for transformation of human fibroblasts. Similarly, human mammary epithelial cells require activation of all three major RAS effector pathways, PI3K, RAF, and RALGDS. Boehm et al. (2007) replaced the oncogenic activity of HRAS with the combined activation of the PI3K and RAF pathways by using the activated allele of MEK1 and myristoylated AKT, respectively.

Transformation models have highlighted a handful of signaling pathways that play critical roles in tumorigenesis. Targeting these pathways for cancer drug development is promising; however, the determination of exact molecular targets within the pathway for efficient therapy may prove to be a major hurdle in the process. Moreover, discovery of novel pathways that may transform human cells alone or in combination with other signaling pathways is crucial to address the diverse alterations seen in human tumors. Such findings of novel pathways and targets are possible via unbiased approaches of discovery.

Technical tools for the discovery of oncogenes and tumor suppressors

Technological advances have greatly modified the process of understanding cancer. It is no longer necessary to limit the study of tumor initiation and growth to a handful of signaling pathways or genes implicated in cancer. Tools have now been developed to manipulate each gene in the genome. These tools provide the means to assign novel tumorigenic functions to known genes and discover cancer vulnerabilities. Manipulation of every gene has been possible due to the development of genome-scale libraries of open reading frames (ORFs) and short-hairpin RNAs (shRNAs) which facilitate gain-of-function and loss-of-function studies, respectively. Such libraries permit whole-genome studies as well as candidate-based approaches to study cancer. These genetic devices pledge to revolutionize the approach to understanding cancer and lead the field toward personalized therapy.

Genetic tools

Loss-of-function tools

RNA interference (RNAi) is an unparalleled tool for loss-of-function studies. Small interfering RNA (siRNA) and short-hairpin RNA (shRNA) libraries have been developed and used effectively in the study

of cancer to determine vulnerabilities that can be exploited as targets for drug development.

Organisms, such as fruit flies and worms, which are easier to manipulate, have been efficiently used in whole-genome RNAi screens. *Drosophila melanogaster* has been used to screen RNAi libraries for novel regulators of pathways implicated in cancer, such as JAK/STAT and WNT, in order to reveal promising targets for cancer drug development (DasGupta et al., 2005; Müller et al., 2005). Similarly, Gort et al. (2008) performed an RNAi screen in *Caenorhabditis elegans* to determine hypoxia-dependent genes and found *TWIST1* to be a direct target of HIF2α, forming an association between hypoxic tumors and metastasis.

The discovery that RNAi is functional in mammalian cells has made similar studies in human cells feasible (Meister and Tuschl, 2004). Genome-scale RNAi screens in pooled and arrayed formats have successfully identified oncogenes and tumor suppressors. For example, Firestein et al. (2008) used an shRNA library for two parallel screens: first, an unbiased genome-scale phenotypic screen for cell proliferation and second, a screen directed to measure the activity of the Wnt/β-catenin pathway. Among the genes that scored in both screens, CDK8 was found to be necessary for the proliferation of colon cancer cells as well as β-catenin transcriptional activity in colon cancer cells. Another study showed that multiple myeloma cells are dependent on the expression of IRF-4, a transcription factor that may thus be an effective target for myeloma therapy (Shaffer et al., 2008).

In addition, others have used RNAi libraries to dissect the signaling pathways that promote cancer. Berns et al. (2004) used an shRNA library against 8,000 genes and found five genes with functions crucial for p53-dependent proliferation arrest of immortalized fibroblasts. Similarly, Tang et al. (2008) conferred a novel role to TCF7L2 in the WNT/β-catenin pathway response. Previously known to promote cancer, TCF7L2 in this study was found to repress WNT/β-catenin-dependent transcription, resulting in restriction of colorectal cancer cell growth. Moerover, a genome-scale, pooled, loss-of-function screen used immortalized human mammary epithelial cells, which depend on PI3K activity for transformation, and found that *REST/NSRF* behaved as a tumor suppressor in the PI3K pathway (Westbrook et al., 2005). In a very similar manner

Kolfschoten et al. (2005) found another novel tumor suppressor gene, *PITX1*, that represses RAS activity.

More recently, investigators have used directed RNAi screens to characterize genes involved in cancer *in situ*. An *in vivo* RNAi screen for tumor suppressors in liver cancer conducted by Zender et al. (2008) focused only on genes deleted in human hepatocellular carcinomas and identified 13 tumor suppressor genes, including *XPO4* and *SET*. Other functional screens such as suppression of metastasis used an RNAi library in an *ex vivo* three-dimensional cell culture system to look for genes whose loss of function promoted satellite colony formation and enhanced metastasis in mice (Gobeil et al., 2008).

Using a loss-of-function approach with RNAi libraries, several groups have searched for genes that contribute to cancer drug resistance. An siRNA library focused on kinases and phosphatases was used to determine those that promoted survival of HeLa cells (MacKeigan et al., 2005). The study identified 73 survival kinases including CDK6, NLK, and ROR1, which when knocked down increased apoptosis as well as sensitivity of the cells to chemotherapeutic agents such as cisplatin, Taxol®, and etoposide. Additionally, suppression of phosphatases such as MK-STYX, previously implicated in the regulation of cell death, increased resistance to the same chemotherapeutic agents, indicating their tumor suppressive function. Likewise, Iorns et al. (2008, 2009) carried out a kinase RNAi screen and identified the importance of the PDK1 pathway and CDK10 for tamoxifen sensitivity. A parallel chemical screen identified PDK1 inhibitors that may be used to sensitize the cells to tamoxifen. A similar study involving paclitaxel-dependent synthetic lethal screen found that loss of function of certain genes sensitized NSCLC cells to 1,000-times lower concentrations of paclitaxel than otherwise required for a significant response, thus improving possibility of paclitaxel as a potent anti-cancer drug in combination therapy (Whitehurst et al., 2007).

Gain-of-function tools

ORFs and cDNAs

A powerful complementary approach to study gene function is to overexpress the wild type or mutant versions of a gene. Transient expression of cDNAs is adequate for short-term assays while long-term assays, such as proliferation, anchorage-independent growth, tumor formation, and metastasis that test oncogenic function, require stable expression of genes. Human retroviral expression libraries have been created to facilitate such gain-of-function studies (Lamesch et al., 2007). The advantage of using such ORF libraries compared with cDNA expression libraries is that each gene in the collection is represented at the same level as every other gene. Boehm et al. (2007) developed a human kinase retroviral expression library that contained a myristoylation tag. Myristoylated kinases are tethered to membranes, which facilitates their activation. From the activated kinase library of 354 kinases, Boehm et al. (2007) found *IKBKE* to cooperate with the activated MAPK pathway to replace the transforming ability of oncogenic HRAS.

A candidate approach taken by Mavrakis et al. (2008) included only the cDNAs of regulators and effectors of mTOR to screen for the ability to promote lymphomagenesis. They found Rheb GTPases to promote tumor formation *in vivo*, which was suppressed by rapamycin treatment. This finding indicates the possibility of using rapamycin in tumors that are driven by Rheb GTPase activity. Another cDNA library screen successfully identified *DRIL1* as the oncogene that neutralizes the ARF/p53 block and allows human fibroblasts to bypass oncogene-induced senescence (Peeper et al., 2002).

MicroRNAs

MicroRNAs are a recently discovered family of non-coding RNAs that regulate gene expression in a post-transcriptional manner (Ambros, 2001). Acting on a mature mRNA, microRNAs downregulate a single or a family of genes by inhibiting the expression of mRNAs to fulfill cellular function.

MicroRNA libraries have also been used in unbiased studies to identify oncogenes and tumor suppressors. Immortalized human fibroblasts with functional p53 were used in a study to show that *miR-372* and *miR-373* cooperated with oncogenic HRAS to transform these cells (Voorhoeve et al., 2006). *miR-372* and *miR-373* function as oncogenes because they inhibit expression of a tumor suppressor *LATS2*. Another screen conducted by the same group discovered that *miR-372* and *miR-520c* promoted invasion and metastasis of non-invasive breast cancer cells (Huang et al., 2008).

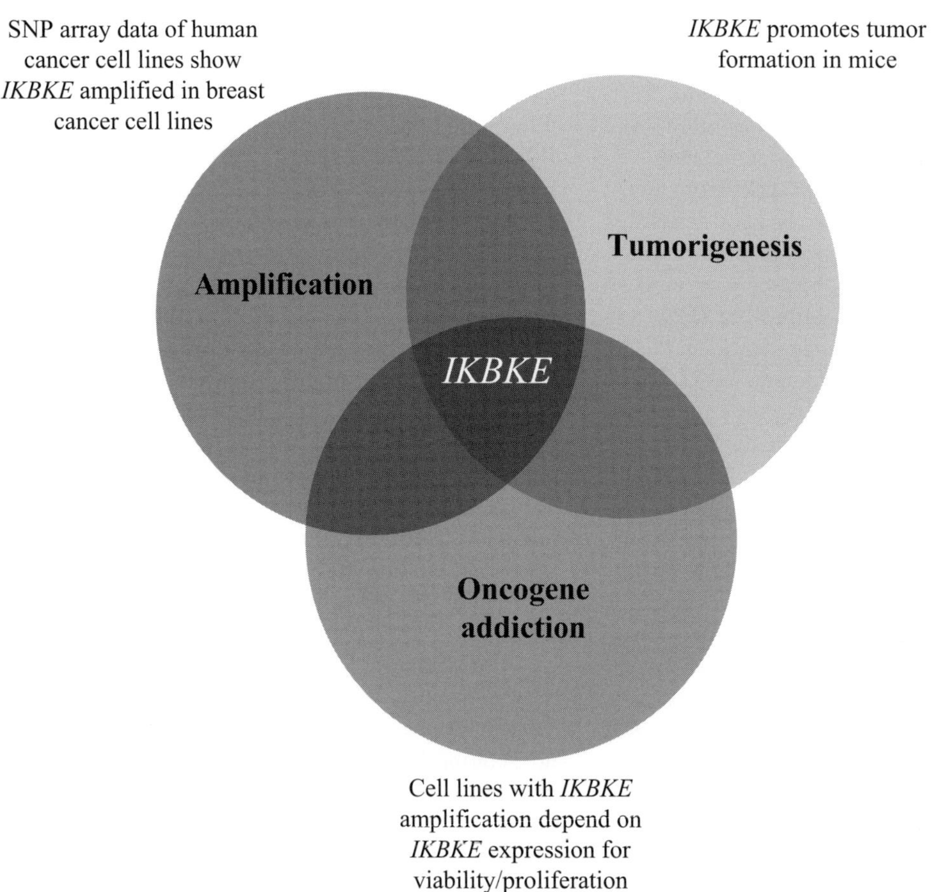

Figure 11.1 Collaboration of unbiased studies and genomic data facilitates the identification of novel cancer promoting genes. The study conducted by Boehm et al. (2007) involved a gain-of-function screen of the human kinase ORF library to look for kinases that promoted soft-agar growth. Genomic SNP array data from human cancer cell lines showed *IKBKE* was amplified in breast samples, and comprehensive RNAi screens revealed that *IKBKE* was essential for the survival of breast cancer cell lines. Further validations established *IKBKE* as a breast cancer oncogene. (A black and white version of this figure will appear in some formats. For the color version, please refer to the plate section.)

The availability of such libraries has made studies such as those conducted by Tavazoie et al. (2008) possible. In this study, miRNAs that lost their expression during the development of metastatic ability were reintroduced to the metastatic human breast cancer cells. This study led to the identification of *miR-335* and *miR-126* as metastatic suppressors. A study that focused on the loss of expression of the *APC* gene in human colorectal cancers found that *miR-35a* and *b* negatively regulated *APC* expression

(Nagel et al., 2008). The same study also found that *miR-35a* and *b* were highly expressed in human color-ectal tumor samples, corroborating the novel finding. In a different study, tumor suppressor *p27^{Kip1}* was suppressed, whereas its negative regulators *miR-221* and *222* were highly expressed in glioblastoma (le Sage et al., 2007).

MicroRNA libraries have also been useful to identify tumor suppressors. Cole et al. (2008) used a directed approach to study miRNAs that were located in frequently amplified or deleted regions in neuroblastoma. When *miR-34a* and *miR-34c*, that lie in the 1p36 hemizygous deletion locus, were re-introduced to corresponding cells, dramatic growth arrest was observed. They concluded that *BCL2* and *MYCN* were the possible targets of *miR-34*, through which the microRNA enforced its tumor suppressive functions.

The cases discussed in this chapter are only a few examples that have used genetic tools for unbiased studies. The application of these libraries in more sophisticated assays will permit the discovery of other genes in oncogenesis.

Integration of genomic approaches

The aforementioned technical tools are invaluable for the discovery of novel functions of genes and path-ways. The relevance of these discoveries in human cancer may only be realized when corroborating aber-rations are observed in human tumors. For this pur-pose, tumor samples and their matched normal counterparts are carefully collected from patients. Samples are analyzed for genomic alterations that may provide a comprehensive understanding of cancer progression and treatment.

Sophisticated technologies have been developed to analyze tumor samples for various aberrations (Cancer Genome Atlas Research Network, 2008; Chin and Gray, 2008). For instance, DNA sequencing technology can identify transforming mutations, deletions, and translocations in tumor samples (Weir et al., 2004). Other techniques such as single nucleotide polymorphism (SNP) analysis, compara-tive genomic hybridization (CGH), and representa-tional oligonucleotide microarray analysis (ROMA) recognize copy number alterations of chromosomal regions and may aid in the identification of amplified oncogenes and deleted tumor suppressors. The

Cancer Genome Atlas (TCGA) Research Network has pioneered the generation and assembly of geno-mic aberrations in human tumor samples. A pilot database of mutations, copy number alterations, gene expression, and DNA methylation of 206 glioblastoma samples has been created.

Integration of such genomic findings is critical in establishing oncogenes and tumor suppressors. Human breast cancer copy number alteration infor-mation led to the realization that *IKBKE* is a breast cancer oncogene and may prove to be an excellent target for cancer drug development (Figure 11.1; Boehm et al., 2007). *CDK8*, a credible member of the WNT/β-catenin that was required for colon cancer cell viability, was amplified in colorectal cancer samples (Firestein et al., 2008). *PITX1* and *REST/ NRSF* were regarded as tumor suppressors with cor-roborating evidence from human tumor samples that had deletions of the gene (Kolfschoten et al., 2005; Westbrook et al., 2005).

Concluding remarks

The field of cancer research has evolved from a single-candidate or single-pathway approach to unbiased whole-genome studies. From the discovery of the first oncogene and tumor suppressor to subsequent under-standing of their regulators and effectors, the funda-mentals of cancer have been revealed. Cancer is a genetic disease that can now be created by introdu-cing known alterations in normal cells. Transform-ation models have proved invaluable in the discovery of novel cancer-promoting functions of signaling pathways. Currently available technical tools employ as well as contribute to the present knowledge of cancer initiation and development. While gain-of-function studies bring novel molecules and novel functions of known signaling pathways into attention, loss-of-function studies are designed to reveal cancer vulnerabilities. Integration of information from human tumors substantiates the findings of such unbiased studies. High-throughput resources are responsible for revealing molecules that can be trans-lated from the bench to be applied to the bedside for targeted, and hence more efficient, cancer therapy. These promising targets for therapy can then replace more general and invasive methods of cancer therapy with much more personalized and effective treatment options.

References

Allsopp RC, Vaziri H, Patterson C, et al. 1992. Telomere length predicts replicative capacity of human fibroblasts. *Proc Natl Acad Sci USA* 89(21):10114–8.

Allsopp RC, Chang E, Kashefi-Aazam M, et al. 1995. Telomere shortening is associated with cell division *in vitro* and *in vivo*. *Exp Cell Res* 220(1):194–200.

Ambros V. 2001. MicroRNAs: tiny regulators with great potential. *Cell* 107(7):823–6.

Aronheim A, Engelberg D, Li N, et al. 1994. Membrane targeting of the nucleotide exchange factor Sos is sufficient for activating the Ras signaling pathway. *Cell* 78(6): 949–61.

Artandi SE, Attardi LD. 2005. Pathways connecting telomeres and p53 in senescence, apoptosis, and cancer. *Biochem Biophys Res Commun* 331(3):881–90.

Bader AG, Kang S, Zhao L, Vogt PK. 2005. Oncogenic PI3K deregulates transcription and translation. *Nat Rev Cancer* 5(12):921–9.

Balmain A, Pragnell IB. 1983. Mouse skin carcinomas induced *in vivo* by chemical carcinogens have a transforming Harvey-ras oncogene. *Nature* 303(5912):72–4.

Bates S, Phillips AC, Clark PA, et al. 1998. p14ARF links the tumour suppressors RB and p53. *Nature* 395(6698):124–5.

Berger M, Stahl N, Del Sal G, Haupt Y. 2005. Mutations in proline 82 of p53 impair its activation by Pin1 and Chk2 in response to DNA damage. *Mol Cell Biol* 25(13): 5380–8.

Berger R, Febbo PG, Majumder PK, et al. 2004. Androgen-induced differentiation and tumorigenicity of human prostate epithelial cells. *Cancer Res* 64(24):8867–75.

Berns K, Hijmans EM, Mullenders J, et al. 2004. A large-scale RNAi screen in human cells identifies new components of the p53 pathway. *Nature* 428(6981):431–7.

Bishop JM. 1985. Viral oncogenes. *Cell* 42(1):23–38.

Blasco MA, Lee HW, Hande MP, et al. 1997. Telomere shortening and tumor formation by mouse cells lacking telomerase RNA. *Cell* 91(1):25–34.

Boehm JS, Hession MT, Bulmer SE, Hahn WC. 2005. Transformation of human and murine fibroblasts without viral oncoproteins. *Mol Cell Biol* 25(15):6464–74.

Boehm JS, Zhao JJ, Yao J, et al. 2007. Integrative genomic approaches identify IKBKE as a breast cancer oncogene. *Cell* 129(6): 1065–79.

Brehm A, Miska EA, McCance DJ, et al. 1998. Retinoblastoma protein recruits histone deacetylase to repress transcription. *Nature* 391(6667):597–601.

Bryan BA, D'Amore PA. 2007. What tangled webs they weave: Rho-GTPase control of angiogenesis. *Cell Mol Life Sci* 64(16):2053–65.

Bryan TM, Englezou A, Dalla-Pozza L, Dunham MA, Reddel RR. 1997. Evidence for an alternative mechanism for maintaining telomere length in human tumors and tumor-derived cell lines. *Nat Med* 3(11):1271–4.

Calin GA, di Iasio MG, Caprini E, et al. 2000. Low frequency of alterations of the alpha (PPP2R1A) and beta (PPP2R1B) isoforms of the subunit A of the serine-threonine phosphatase 2A in human neoplasms. *Oncogene* 19(9): 1191–5.

Cancer Genome Atlas Research Network. 2008. Comprehensive genomic characterization defines human glioblastoma genes and core pathways. *Nature* 455(7216): 1061–8.

Chin L, Gray JW. 2008. Translating insights from the cancer genome into clinical practice. *Nature* 452(7187):553–63.

Chipuk JE, Green DR. 2006. Dissecting p53-dependent apoptosis. *Cell Death Differ* 13(6):994–1002.

Cichowski K, Shih TS, Schmitt E, et al. 1999. Mouse models of tumor development in neurofibromatosis type 1. *Science* 286(5447):2172–6.

Cohen SB, Graham ME, Lovrecz GO, et al. 2007. Protein composition of catalytically active human telomerase from immortal cells. *Science* 315(5820):1850–3.

Cole KA, Attiyeh EF, Mosse YP, et al. 2008. A functional screen identifies miR-34a as a candidate neuroblastoma tumor suppressor gene. *Mol Cancer Res* 6(5): 735–42.

Counter CM, Avilion AA, LeFeuvre CE, et al. 1992. Telomere shortening associated with chromosome instability is arrested in immortal cells which express telomerase activity. *EMBO J* 11(5):1921–9.

Counter CM, Botelho FM, Wang P, Harley CB, Bacchetti S. 1994. Stabilization of short telomeres and telomerase activity accompany immortalization of Epstein–Barr virus-transformed human B lymphocytes. *J Virol* 68(5): 3410–4.

Counter CM, Hahn WC, Wei W, et al. 1998. Dissociation among in vitro telomerase activity, telomere maintenance, and cellular immortalization. *Proc Natl Acad Sci USA* 95(25):14723–8.

DasGupta R, Kaykas A, Moon RT, Perrimon N. 2005. Functional genomic analysis of the Wnt-wingless signaling pathway. *Science* 308(5723):826–33.

Derksen PW, Liu X, Saridin F, et al. 2006. Somatic inactivation of E-cadherin and p53 in mice leads to metastatic lobular mammary carcinoma through induction of anoikis resistance and angiogenesis. *Cancer Cell* 10(5):437–49.

Dimri GP, Lee X, Basile G, et al. 1995. A biomarker that identifies senescent human cells in culture and in aging skin *in vivo*. *Proc Natl Acad Sci USA* 92(20):9363–7.

Donehower LA, Godley LA, Aldaz CM, et al. 1995. Deficiency of p53 accelerates mammary tumorigenesis in Wnt-1 transgenic mice and promotes chromosomal instability. *Genes Dev* 9(7):882–95.

Downward J. 2003. Targeting RAS signalling pathways in cancer therapy. *Nat Rev Cancer* 3(1):11–22.

Downward J. 2006. Signal transduction. Prelude to an anniversary for the RAS oncogene. *Science* 314(5798):433–4.

Dunham MA, Neumann AA, Fasching CL, Reddel RR. 2000. Telomere maintenance by recombination in human cells. *Nat Genet* 26(4): 447–50.

Elenbaas B, Spirio L, Koerner F, et al. 2001. Human breast cancer cells generated by oncogenic transformation of primary mammary epithelial cells. *Genes Dev* 15(1):50–65.

Engelman JA, Chen L, Tan X, et al. 2008. Effective use of PI3K and MEK inhibitors to treat mutant Kras G12D and PIK3CA H1047R murine lung cancers. *Nat Med* 14(12):1351–6.

Eva A, Aaronson SA. 1983. Frequent activation of c-kis as a transforming gene in fibrosarcomas induced by methylcholanthrene. *Science* 220(4600):955–6.

Feig LA. 2003. Ral-GTPases: approaching their 15 minutes of fame. *Trends Cell Biol* 13(8):419–25.

Ferreira R, Magnaghi-Jaulin L, Robin P, Harel-Bellan A, Trouche D. 1998. The three members of the pocket proteins family share the ability to repress E2F activity through recruitment of a histone deacetylase. *Proc Natl Acad Sci USA* 95(18): 10493–8.

Firestein R, Bass AJ, Kim SY, et al. 2008. CDK8 is a colorectal cancer oncogene that regulates beta-catenin activity. *Nature* 455(7212): 547–51.

Fisher GH, Wellen SL, Klimstra D, et al. 2001. Induction and apoptotic regression of lung adenocarcinomas by regulation of a K-Ras transgene in the presence and absence of tumor suppressor genes. *Genes Dev* 15(24):3249–62.

Greenberg RA, Chin L, Femino A, et al. 1999. Short dysfunctional telomeres impair tumorigenesis in the INK4a (delta2/3) cancer-prone mouse. *Cell* 97(4):515–25.

Gobeil S, Zhu X, Doillon CJ, Green MR. 2008. A genome-wide shRNA screen identifies GAS1 as a novel melanoma metastasis suppressor gene. *Genes Dev* 22(21):2932–40.

Gort EH, van Haaften G, Verlaan I, et al. 2008. The TWIST1 oncogene is a direct target of hypoxia-inducible factor-2α. *Oncogene* 27(11):1501–10.

Hahn WC, Counter CM, Lundberg AS, et al. 1999. Creation of human tumour cells with defined genetic elements. *Nature* 400(6743): 464–8.

Hahn WC, Dessain SK, Brooks MW, et al. 2002. Enumeration of the simian virus 40 early region elements necessary for human cell transformation. *Mol Cell Biol* 22(7): 2111–23. (Erratum in *Mol Cell Biol* 2002 22(10):3562)

Hamad NM, Elconin JH, Karnoub AE, et al. 2002. Distinct requirements for Ras oncogenesis in human versus mouse cells. *Genes Dev* 16(16):2045–57.

Hanahan D, Weinberg RA. 2000. The hallmarks of cancer. *Cell* 100(1): 57–70.

Hara E, Tsurui H, Shinozaki A, Nakada S, Oda K. 1991. Cooperative effect of antisense-Rb and antisense-p53 oligomers on the extension of life span in human diploid fibroblasts, TIG-1. *Biochem Biophys Res Commun* 179(1):528–34.

Harley CB. 1991. Telomere loss: mitotic clock or genetic time bomb? *Mutat Res* 256(2–6): 271–82.

Harley CB, Futcher AB, Greider CW. 1990. Telomeres shorten during ageing of human fibroblasts. *Nature* 345(6274):458–60.

Hastie ND, Dempster M, Dunlop MG, et al. 1990. Telomere reduction in human colorectal carcinoma and with ageing. *Nature* 346(6287): 866–8.

Hayflick L, Moorhead PS. 1961. The serial cultivation of human diploid cell strains. *Exp Cell Res* 25: 585–621.

Hennessy BT, Smith DL, Ram PT, Lu Y, Mills GB. 2005. Exploiting the PI3K/AKT pathway for cancer drug discovery. *Nat Rev Drug Discov* 4(12):988–1004.

Hoffman WH, Biade S, Zilfou JT, Chen J, Murphy M. 2002. Transcriptional repression of the anti-apoptotic survivin gene by wild type p53. *J Biol Chem* 277(5):3247–57.

Howe AK, Gaillard S, Bennett JS, Rundell K. 1998. Cell cycle progression in monkey cells expressing simian virus 40 small T antigen from adenovirus vectors. *J Virol* 72(12):9637–44.

Huang Q, Gumireddy K, Schrier M, et al. 2008. The microRNAs miR-373 and miR-520c promote tumour invasion and metastasis. *Nat Cell Biol* 10(2):202–10.

Iorns E, Turner NC, Elliott R, et al. 2008. Identification of CDK10 as an important determinant of resistance to endocrine therapy for breast cancer. *Cancer Cell* 13(2):91–104.

Iorns E, Lord CJ, Ashworth A. 2009. Parallel RNAi and compound screens identify the PDK1 pathway as a target for tamoxifen sensitization. *Biochem J* 417(1): 361–70.

Ishikawa F. 1997. Telomere crisis, the driving force in cancer cell evolution. *Biochem Biophys Res Commun* 230(1):1–6.

Jackson EL, Willis N, Mercer K, et al. 2001. Analysis of lung tumor initiation and progression using conditional expression of oncogenic K-ras. *Genes Dev* 15(24):3243–8.

Kim NW, Piatyszek MA, Prowse KR, et al. 1994. Specific association of human telomerase activity with immortal cells and cancer. *Science* 266(5193):2011–15.

Kolfschoten IG, van Leeuwen B, Berns K, et al. 2005. A genetic screen identifies PITX1 as a suppressor of RAS activity and tumorigenicity. *Cell* 121(6):849–58.

Krontiris TG, Cooper GM. 1981. Transforming activity of human tumor DNAs. *Proc Natl Acad Sci USA* 78(2):1181–4.

Lambert JM, Lambert QT, Reuther GW, et al. 2002. Tiam1 mediates Ras activation of Rac by a PI(3)K-independent mechanism. *Nat Cell Biol* 4(8):621–5.

Lamesch P, Li N, Milstein S, et al. 2007. hORFeome v3.1: a resource of human open reading frames representing over 10,000 human genes. *Genomics* 89(3):307–15.

Land H, Parada LF, Weinberg RA. 1983. Tumorigenic conversion of primary embryo fibroblasts requires at least two cooperating oncogenes. *Nature* 304(5927):596–602.

le Sage C, Nagel R, Egan DA, et al. 2007. Regulation of the p27(Kip1) tumor suppressor by miR-221 and miR-222 promotes cancer cell proliferation. *EMBO J* 26(15): 3699 708.

Levy MZ, Allsopp RC, Futcher AB, Greider CW, Harley CB. 1992. Telomere end-replication problem and cell aging. *J Mol Biol* 225(4): 951–60.

Li HH, Cai X, Shouse GP, Piluso LG, Liu X. 2007. A specific PP2A regulatory subunit, B56gamma, mediates DNA damage-induced dephosphorylation of p53 at Thr55. *EMBO J* 26(2):402–11.

Linskens MH, Harley CB, West MD, Campisi J, Hayflick L. 1995. Replicative senescence and cell death. *Science* 267(5194):17.

Lundberg AS, Randell SH, Stewart SA, et al. 2002. Immortalization and transformation of primary human airway epithelial cells by gene transfer. *Oncogene* 21(29):4577–86.

Luo RX, Postigo AA, Dean DC. 1998. Rb interacts with histone deacetylase to repress transcription. *Cell* 92(4):463–73.

MacKeigan JP, Murphy LO, Blenis J. 2005. Sensitized RNAi screen of human kinases and phosphatases identifies new regulators of apoptosis and chemoresistance. *Nat Cell Biol* 7(6):591–600.

MacKenzie KL, Franco S, Naiyer AJ, et al. 2002. Multiple stages of malignant transformation of human endothelial cells modelled by co-expression of telomerase reverse transcriptase, SV40 T antigen and oncogenic N-ras. *Oncogene* 21(27): 4200–11.

Magnaghi-Jaulin L, Groisman R, Naguibneva I, et al. 1998. Retinoblastoma protein represses transcription by recruiting a histone deacetylase. *Nature* 391(6667): 601–5.

Malliri A, van der Kammen RA, Clark K, et al. 2002. Mice deficient in the Rac activator Tiam1 are resistant to Ras-induced skin tumours. *Nature* 417(6891):867–71.

Mantell LL, Greider CW. 1994. Telomerase activity in germline and embryonic cells of *Xenopus*. *EMBO J* 13(13):3211–17.

Mavrakis KJ, Zhu H, Silva RL, et al. 2008. Tumorigenic activity and therapeutic inhibition of Rheb GTPase. *Genes Dev* 22(16):2178–88.

McCoy MS, Bargmann CI, Weinberg RA. 1984. Human colon carcinoma Ki-ras2 oncogene and its corresponding proto-oncogene. *Mol Cell Biol* 4(8):1577–82.

Meister G, Tuschl T. 2004. Mechanisms of gene silencing by double-stranded RNA. *Nature* 431(7006):343–9.

Mittnacht S. 1998. Control of pRB phosphorylation. *Curr Opin Genet Dev* 8(1):21–7.

Müller P, Kuttenkeuler D, Gesellchen V, Zeidler MP, Boutros M. 2005. Identification of JAK/STAT signalling components by genome-wide RNA interference. *Nature* 436(7052):871–5.

Nagel R, le Sage C, Diosdado B, et al. 2008. Regulation of the adenomatous polyposis coli gene by the miR-135 family in colorectal cancer. *Cancer Res* 68(14):5795–802.

Okamoto K, Li H, Jensen MR, et al. 2002. Cyclin G recruits PP2A to dephosphorylate Mdm2. *Mol Cell* 9(4):761–71.

Pal S, Datta K, Mukhopadhyay D. 2001. Central role of p53 on regulation of vascular permeability factor/ vascular endothelial growth factor (VPF/VEGF) expression in mammary carcinoma. *Cancer Res* 61(18):6952–7.

Peeper DS, Shvarts A, Brummelkamp T, et al. 2002. A functional screen identifies hDRIL1 as an oncogene that rescues RAS-induced senescence. *Nat Cell Biol* 4(2): 148–53.

Perucho M, Goldfarb M, Shimizu K, et al. 1981. Human-tumor-derived cell lines contain common and different transforming genes. *Cell* 27(3 Pt 2):467–76.

Phillips AC, Vousden KH. 2001. E2F-1 induced apoptosis. *Apoptosis* 6(3): 173–82.

Prowse KR, Greider CW. 1995. Developmental and tissue-specific regulation of mouse telomerase and telomere length. *Proc Natl Acad Sci USA* 92(11):4818–22.

Pruitt K, Der CJ. 2001. Ras and Rho regulation of the cell cycle and oncogenesis. *Cancer Lett* 171(1): 1–10.

Pugh CW, Ratcliffe PJ. 2003. Regulation of angiogenesis by hypoxia: role of the HIF system. *Nat Med* 9(6):677–84.

Pulciani S, Santos E, Lauver AV, et al. 1982. Oncogenes in human tumor cell lines: molecular cloning of a transforming gene from human bladder carcinoma cells. *Proc Natl Acad Sci USA* 79(9):2845–9.

Quelle DE, Zindy F, Ashmun RA, Sherr CJ. 1995. Alternative reading frames of the INK4a tumor suppressor gene encode two unrelated proteins capable of inducing cell cycle arrest. *Cell* 83(6):993–1000.

Rangarajan A, Hong SJ, Gifford A, Weinberg RA. 2004. Species- and cell type-specific requirements for cellular transformation. *Cancer Cell* 6(2):171–83.

Rich JN, Guo C, McLendon RE, et al. 2001. A genetically tractable model of human glioma formation. *Cancer Res* 61(9):3556–60.

Ruediger R, Pham HT, Walter G. 2001a. Disruption of protein phosphatase 2A subunit interaction in human cancers with mutations in the A alpha subunit gene. *Oncogene* 20(1):10–15.

Ruediger R, Pham HT, Walter G. 2001b. Alterations in protein phosphatase 2A subunit interaction in human carcinomas of the lung and colon with mutations in the A beta subunit gene. *Oncogene* 20(15):1892–9.

Ruley HE. 1983. Adenovirus early region 1A enables viral and cellular transforming genes to transform primary cells in culture. *Nature* 304(5927):602–6.

Sablina AA, Hahn WC. 2008. SV40 small T antigen and PP2A phosphatase in cell transformation. *Cancer Metastasis Rev* 27(2):137–46.

Sablina AA, Chen W, Arroyo JD, et al. 2007. The tumor suppressor PP2A Aβ regulates the RalA GTPase. *Cell* 129(5):969–82.

Sager R, Tanaka K, Lau CC, Ebina Y, Anisowicz A. 1983. Resistance of human cells to tumorigenesis induced by cloned transforming genes. *Proc Natl Acad Sci USA* 80(24):7601–5.

Serrano M, Hannon GJ, Beach D. 1993. A new regulatory motif in cell-cycle control causing specific inhibition of cyclin D/CDK4. *Nature* 366(6456):704–7.

Serrano M, Lin AW, McCurrach ME, Beach D, Lowe SW. 1997. Oncogenic ras provokes premature cell senescence associated with accumulation of p53 and p16INK4a. *Cell* 88(5):593–602.

Seshadri T, Campisi J. 1989. Growth-factor-inducible gene expression in senescent human fibroblasts. *Exp Gerontol* 24(5–6):515–22.

Shaffer AL, Emre NC, Lamy L, et al. 2008. IRF4 addiction in multiple myeloma. *Nature* 454(7201): 226–31.

Shay JW, Bacchetti S. 1997. A survey of telomerase activity in human cancer. *Eur J Cancer* 33(5):787–91.

Shay JW, Pereira-Smith OM, Wright WE. 1991. A role for both RB and p53 in the regulation of human cellular senescence. *Exp Cell Res* 196(1):33–9.

Sherr CJ. 2001. The INK4a/ARF network in tumour suppression. *Nat Rev Mol Cell Biol* 2(10): 731–7.

Shih C, Shilo BZ, Goldfarb MP, Dannenberg A, Weinberg RA. 1979. Passage of phenotypes of chemically transformed cells via transfection of DNA and chromatin. *Proc Natl Acad Sci USA* 76(11):5714–18.

Shih C, Padhy LC, Murray M, Weinberg RA. 1981. Transforming genes of carcinomas and neuroblastomas introduced into mouse fibroblasts. *Nature* 290(5803):261–4.

Sontag E, Fedorov S, Kamibayashi C, et al. 1993. The interaction of SV40 small tumor antigen with protein phosphatase 2A stimulates the map kinase pathway and induces cell proliferation. *Cell* 75(5):887–97.

Stephens L, Anderson K, Stokoe D, et al. 1998. Protein kinase B kinases that mediate phosphatidylinositol 3,4,5-trisphosphate-dependent activation of protein kinase B. *Science* 279(5351):710–14.

Stevenson M, Volsky DJ. 1986. Activated v-myc and v-ras oncogenes do not transform normal human lymphocytes. *Mol Cell Biol* 6(10):3410–17.

Subbaramaiah K, Altorki N, Chung WJ, et al. 1999. Inhibition of cyclooxygenase-2 gene expression by p53. *J Biol Chem* 274(16): 10911–15.

Sukumar S, Notario V, Martin-Zanca D, Barbacid M. 1983. Induction of mammary carcinomas in rats by nitroso-methylurea involves malignant activation of H-ras-1 locus by single point mutations. *Nature* 306(5944):658–61.

Takagi Y, Futamura M, Yamaguchi K, Aoki S, Takahashi T, Saji S. 2000. Alterations of the PPP2R1B gene located at 11q23 in human colorectal cancers. *Gut* 47(2):268–71.

Tamaki M, Goi T, Hirono Y, Katayama K, Yamaguchi A. 2004. PPP2R1B gene alterations inhibit interaction of PP2A-Abeta and PP2A-C proteins in colorectal cancers. *Oncol Rep* 11(3):655–9.

Tang W, Dodge M, Gundapaneni D, et al. 2008. A genome-wide RNAi screen for Wnt/beta-catenin pathway components identifies unexpected roles for TCF transcription factors in cancer. *Proc Natl Acad Sci USA* 105(28): 9697–702.

Taparowsky E, Suard Y, Fasano O, et al. 1982. Activation of the T24 bladder carcinoma transforming gene is linked to a single amino acid change. *Nature* 300(5894):762–5.

Tavazoie SF, Alarcón C, Oskarsson T, et al. 2008. Endogenous human microRNAs that suppress breast cancer metastasis. *Nature* 451(7175):147–52.

Tchevkina E, Agapova L, Dyakova N, et al. 2005. The small G-protein RalA stimulates metastasis of transformed cells. *Oncogene* 24(3): 329–35.

Teodoro JG, Evans SK, Green MR. 2007. Inhibition of tumor angiogenesis by p53: a new role for the guardian of the genome. *J Mol Med* 85(11):1175–86.

Turcotte S, Desrosiers RR, Béliveau R. 2003. HIF-1α mRNA and protein upregulation involves Rho GTPase expression during hypoxia in renal cell carcinoma. *J Cell Sci* 116(Pt 11): 2247–60.

Vivanco I, Sawyers CL. 2002. The phosphatidylinositol 3-kinase AKT pathway in human cancer. Review. *Nat Rev Cancer* 2(7):489–501.

Vogelstein B, Lane D, Levine AJ. 2000. Surfing the p53 network. *Nature* 16(408):306–10.

Voorhoeve PM, le Sage C, Schrier M, et al. 2006. A genetic screen implicates miRNA-372 and miRNA-373 as oncogenes in testicular germ cell tumors. *Cell* 124(6):1169–81.

Wang SS, Esplin ED, Li JL, et al. 1998. Alterations of the PPP2R1B gene in human lung and colon cancer. *Science* 282(5387):284–7.

Ward Y, Wang W, Woodhouse E, et al. 2001. Signal pathways which promote invasion and metastasis: critical and distinct contributions of extracellular signal-regulated kinase and Ral-specific guanine exchange factor pathways. *Mol Cell Biol* 21(17):5958–69.

Weir B, Zhao X, Meyerson M. 2004. Somatic alterations in the human cancer genome. *Cancer Cell* 6(5): 433–8.

Westbrook TF, Martin ES, Schlabach MR, et al. 2005. A genetic screen for candidate tumor suppressors identifies REST. *Cell* 121(6):837–48.

Whitehurst AW, Bodemann BO, Cardenas J, et al. 2007. Synthetic lethal screen identification of chemosensitizer loci in cancer cells. *Nature* 446(7137):815–19.

Wright WE, Shay JW. 2000. Telomere dynamics in cancer progression and prevention: fundamental differences in human and mouse telomere biology. *Nat Med* 6(8):849–51.

Wright WE, Pereira-Smith OM, Shay JW. 1989. Reversible cellular senescence: implications for immortalization of normal human diploid fibroblasts. *Mol Cell Biol* 9(7):3088–92.

Yu GL, Bradley JD, Attardi LD, Blackburn EH. 1990. *In vivo* alteration of telomere sequences and senescence caused by mutated *Tetrahymena* telomerase RNAs. *Nature* 344(6262):126–32.

Yu J, Zhang L. 2005. The transcriptional targets of p53 in apoptosis control. *Biochem Biophys Res Commun* 331(3):851–8.

Zender L, Xue W, Zuber J, et al. 2008. An oncogenomics-based *in vivo* RNAi screen identifies tumor suppressors in liver cancer. *Cell* 135(5):852–64.

Zhang Y, Xiong Y, Yarbrough WG. 1998. ARF promotes MDM2 degradation and stabilizes p53: ARF-INK4a locus deletion impairs both the Rb and p53 tumor suppression pathways. *Cell* 92(6): 725–34.

Zhao JJ, Roberts TM, Hahn WC. 2004. Functional genetics and experimental models of human cancer. *Trends Mol Med* 10(7): 344–50.

RAS signaling networks

Douglas V. Faller and Andrew M. Rankin

Introduction

The RAS proteins regulate signal transduction underlying diverse cellular activities, including proliferation, survival, growth, migration, differentiation, or cytoskeletal dynamism. GTP-bound ("on-state") RAS proteins convert extracellular stimuli into intracellular signaling cascades, which eventually evoke changes in cellular activities; this signaling ceases when RAS-bound GTP is hydrolyzed to GDP as the result of another signaling cascade. Thus, in normal cells, RAS proteins function as molecular switches for critical changes in cellular activities, such as cell proliferation and survival, and their proper and tight regulation is indispensable to maintain the homeostasis of cells and, ultimately, the entire organism. Conversely, uncontrolled activity of the RAS proteins, or the molecular components of their downstream pathways, can result in serious pathophysiological consequences, including cancers and other diseases. Approximately 30% of human tumors harbor activating mutations in one of the three RAS isoforms: KRAS, NRAS, and HRAS. KRAS is most frequently mutated among three isoforms in malignancies; its mutation rate in all tumors is estimated to be 25 to 30%. KRAS mutation is especially prominent in colorectal carcinoma (40 to 45% mutation rate), non-small cell lung cancer (NSCLC) (16 to 40%), and pancreatic ductal carcinoma (69 to 95%). In contrast, activating mutations of NRAS and HRAS are less common (8 and 3% mutation rate, respectively). Malignant melanomas predominantly harbor NRAS mutations (20 to 30% prevalence). The activating oncogenic mutations most commonly occur in codons 12, 13, and 61, in the GTPase catalytic domains, identically among the three isoforms. Eighty percent of KRAS mutations are observed in codon 12, whereas NRAS mutations preferentially involve codon 61

(60%) compared to codon 12 (35%). HRAS mutations are divided almost equally among codon 12 (50%) and codon 61 (40%). Regardless of isoform type or codon location, all these activating mutations render RAS proteins resistant to GTP hydrolysis (and consequent Ras inactivation) mediated by GTPase-activating proteins (GAPs). These constitutively activated oncogenic Ras mutant proteins therefore initiate intracellular signaling cascades without the input of extracellular stimuli, resulting in uncontrolled cell proliferation and abnormal cell survival. Although Ras proteins therefore represent a potential critical target for therapeutic intervention, and multiple approaches to target aberrant Ras activity have been attempted, direct Ras-targeted therapeutics remain elusive. However, indirect approaches, including antibodies and small molecules that interrupt signaling upstream or downstream of Ras, have recently revolutionized the treatment of a number of diseases.

RAS proteins

RAS proteins comprise a family of small GTPase proteins associated with the inner leaflet of the plasma membrane, which function to transmit signals generated by extracellular stimuli into the cytoplasm, thereby facilitating cell cycle progression, proliferation, and survival. These functions are chiefly regulated by the ability of the RAS protein to bind a guanosine triphosphate (GTP) or diphosphate (GDP) molecule, and thus switch between a functionally active (GTP-bound) or inactive (GDP-bound) state.

RAS isoforms and their homologies

The human genome harbors 36 genes, which code for 39 RAS family proteins, ranging from 20 to 29 kilodaltons in size. In nearly all cases, these multiple RAS GTPases are responsible for regulating a myriad of

---

Systems Biology of Cancer, ed. S. Thiagalingam. Published by Cambridge University Press. © Cambridge University Press 2015.

cellular processes, including adhesion, cytoskeletal rearrangements, golgi trafficking, motility, differentiation, transcription, and translation. Common to the RAS family proteins is a tertiary structure composed of five α-helices and six β-sheets, connected by ten loops, producing a hydrophobic core. Also shared by all RAS proteins are four separate regions of primary structure exhibiting a high degree of sequence homology: the N-terminal phosphate-binding loop; the nucleotide-sensitive switch regions I and II; and the C-terminal membrane-targeting domain. The HRAS, NRAS, and KRAS genes in mammalian cells are composed of five similarly sized exons, the fifth of which is non-coding and designated exon Φ. Due to varying lengths of intronic sequences, the HRAS, NRAS, and KRAS genes extend three, seven, and fifteen kilobases respectively, and are located on chromosomes eleven, three, and six [1, 2].

Post-translational modifications for membrane trafficking

Post-translational modifications of RAS are critical for its association with plasma membranes. There are three basic modifications immediately following cytosolic translation of the nascent RAS protein, each of which are regulated in part by the CAAX motif located on the carboxy terminus of the RAS protein. Initially, farnesyl transferase covalently attaches a 15-carbon farnesyl isoprenoid lipid to the cysteine residue within the CAAX motif, dramatically increasing the hydrophobicity of the carboxy terminus. This is followed by cleavage of the AAX residues by an endopeptidase, the RAS-converting enzyme-1. Finally, to increase the ability of RAS to associate with plasma membranes, isoprenylcysteine carboxymethyltransferase-1 adds a methyl group on the carboxy group of the now C-terminal cysteine residue. After anchoring to the inner leaflet, HRAS, NRAS, and KRAS 4A (a splice variant of the ubiquitously expressed KRAS 4B) specifically undergo further palmitoylation reactions on cysteine residues upstream of cysteine 186, enhancing the stability of the RAS-membrane association [2, 3].

Regulation of RAS protein activation and inactivation

RAS activity regulatory mechanisms

The activation of RAS, and thus its ability to signal, is critically dependent upon whether it is bound to GDP or GTP. Intrinsically, RAS has only a modest ability to self-activate by exchanging bound GDP for GTP from the surrounding environment. Likewise, as an ATPase, isolated RAS protein is also capable of self-inactivation to some extent, by cleaving the terminal phosphate from bound GTP. In the intracellular environment, however, the balance between RAS activation and inactivation relies heavily on the competing activities of two main families of proteins: the guanine nucleotide exchange factors (GEFs) and the GTPase-activating proteins (GAPs).

RAS-guanine nucleotide exchange factors

GEFs are primarily responsible for enzymatically switching the GDP bound to RAS with GTP, resulting in modulation of the tertiary structure of the RAS protein. This conformational change depends on rearrangement of the switch I region, located in loop 2 (encompassing residues 30 to 40), following binding of GTP. Specifically, the side chain containing Thr35 reorients upon binding of GTP to interact non-covalently with the GTP γ phosphate and a magnesium ion, stabilizing the interaction. The switch II region located in loop 4 and helix 2 (encompassing residues 60 to 76) cooperates with this structural change via the interaction of Gly60 with the GTP, thus stabilizing RAS in an active form and allowing for its downstream signaling capabilities [1, 2, 4].

Signaling networks initiating RAS activation

There exist multiple mechanisms for RAS activation, all of which involve extracellular signaling events resulting in the activation of GEFs as a prelude to the activation of RAS (Figure 12.1). Examples include signaling via cytokines, neurotransmitters, hormones, and growth factors. The epidermal growth factor receptor (EGFR), for example, is a receptor tyrosine kinase typically found in close proximity to membrane-bound, inactive RAS proteins. When the EGFR is bound by its ligand, epidermal growth factor (EGF), the EGFR undergoes dimerization and auto-phosphorylation of its cytoplasmic domain. These specific phosphorylated–tyrosine residues recruit Shc proteins and induce their auto-phosphorylation and activation. The Grb2 adaptor protein, which is normally found stably associated with SOS, a GEF for RAS, co-localizes with the activated Shc proteins bound to the EGFR. Consequently, the GEF SOS is approximated to RAS, allowing the exchange of inactivating

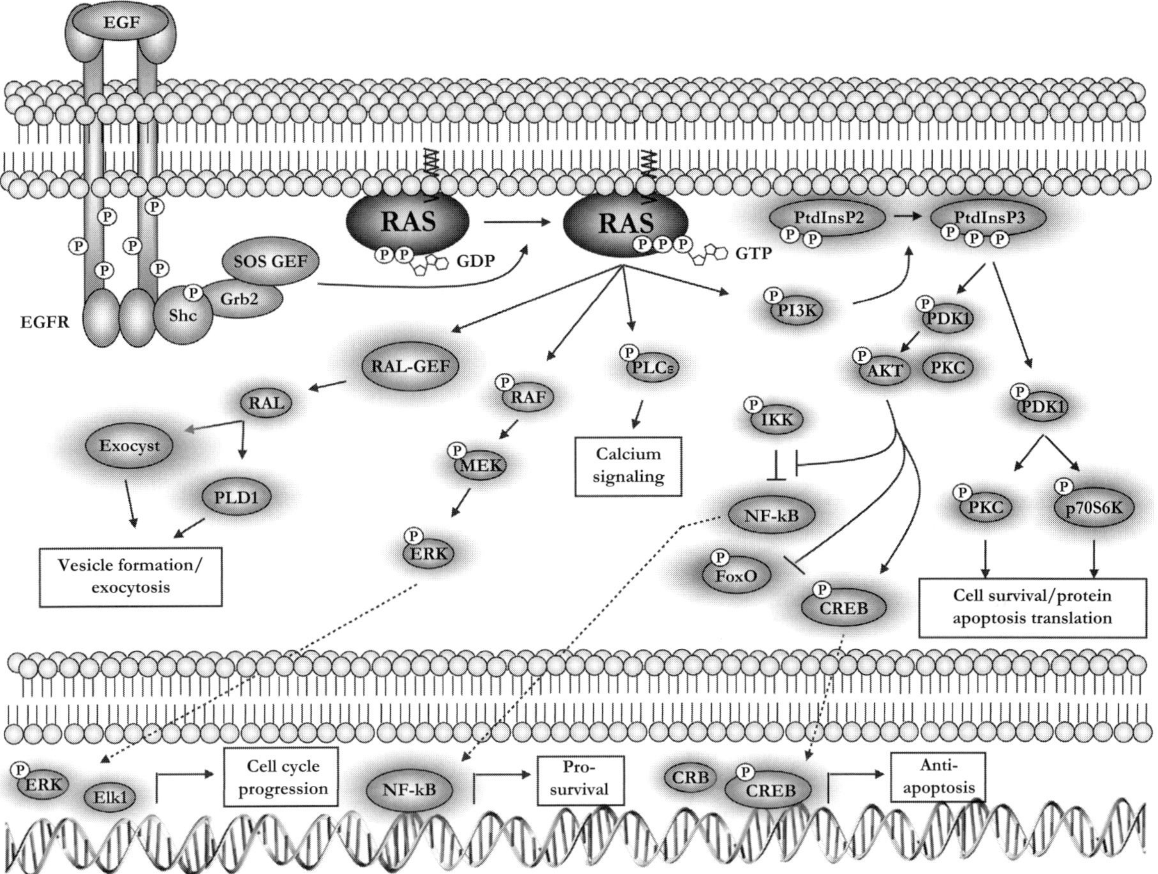

Figure 12.1 RAS-associated signaling events. RAS proteins, small GTPases associated with the plasma membrane inner leaflet, transmit extracellular signals, such as those initiated by EGF when bound to its cognate receptor, through the cytoplasm. RAS function is regulated by the exchange of inactivating-GDP for activating-GTP within the RAS active site, thereby inhibiting or facilitating context-dependent and cell type-dependent downstream signaling pathways. Active RAS may signal through RAF to influence cell cycle progression, through RAL to regulate vesicle synthesis, through PLCε in calcium signaling, or through PI3K to inhibit apoptosis, enhance cell survival, and/or modulate protein synthesis.
(Key: circled P: phosphorylation; solid arrows: post-translational-activating events; solid lines: post-translational-inhibiting events; dashed arrows: nuclear translocation events; EGF: epidermal growth factor; EGFR: epidermal growth factor receptor; Grb2: growth factor receptor bound protein 2; SOS-GEF: son of sevenless-guanine nucleotide exchange factor; RAS: rat sarcoma protein; RAL-GEF: RAL-guanine nucleotide exchange factor; PLD1: phospholipase D1; MEK: mitogen-activated protein kinase/extracellular signal-related kinase; ERK: extracellular signal-related kinase; ELK1: Ets-like gene 1; PI3K: phosphatidylinositol 3- kinase; PtdInsP2: phosphatidylinositol (3,4)-bisphosphate; PtdInsP3: phosphatidylinositol (3,4,5)-trisphosphate; PDK1: phosphoinositide-dependent kinase-1; PKC: protein kinase C; p70S6K: p70 ribosomal protein S6 kinase; IKK: IκB kinase; NF-κB: nuclear factor kappa-light-chain-enhancer of activated B cells; FoxO: forkhead box O; CREB: cAMP response-element binding)
(A black and white version of this figure will appear in some formats. For the color version, please refer to the plate section.)

GDP with activating GTP [5]. The T-cell receptor, when engaged by an antigen in the context of the appropriate class I major histocompatibility complex, activates RAS through the tyrosine kinase lck, via recruitment of Grb2, SOS, RAS, and LAT (linker for activated T cells) into a lipid raft assembly [6]. An example of hormone-induced activation of RAS involves the effects of estrogen signaling. Estrogen-dependent breast cancer cells produce and release growth factors upon exposure to estrogen. These cells also express receptor tyrosine kinases, such as ErbB2/HER, which can be engaged by these growth factors in an autocrine fashion. Upon their activation, the ErbB2/HER receptor tyrosine kinases dimerize and activate RAS, allowing downstream activation of Erk and ultimately resulting in increased rates of proliferation [7].

Interestingly, calcium signaling has been shown to be important in certain contexts for RAS activation.

RASGRF2, a GEF for RAS, is normally found associated with calmodulin, a protein critically regulated by cytosolic calcium levels. Upon an influx of calcium into the cytosol from the endoplasmic reticulum, RASGRF2 translocates to the plasma membrane, bringing it into close proximity with RAS and inducing its activation [2].

RAS-GTPase-activating proteins

As already indicated, the RAS-activating function of GEFs are tightly and reciprocally regulated by the ability of the GAPs to inhibit RAS signaling. Three RAS-specific GAP proteins, p120GAP, Neurofibromin 1 (NF1), and GAP1, have been extensively characterized to date. p120GAP is capable of increasing RAS-associated GTPase activity by a factor of 10^5 compared with the activity of RAS protein alone. It is thought that p120GAP proteins can also functionally participate in downstream RAS signaling, this ability being dependent on the binding of activated p120GAP proteins to phosphotyrosine-specific residues on the cytoplasmic domains of receptor tyrosine kinases as well as non-receptor tyrosine kinases, including Shc and Irs-1. NF1 proteins, named for the disease neurofibromatosis type I, which results from inactivating mutations in the gene that encodes NF1, co-localize to cytoplasmic microtubules and inhibit RAS activity following their growth factor-dependent serine and threonine phosphorylation. GAP1 proteins interact with the heterotrimeric G protein subunit Gα12, which stimulates the ability of GAP1 to hydrolyze RAS-bound GTP [8].

Downstream signaling by activated RAS

Activation of downstream RAS effectors depends on their physical interactions with GTP-bound RAS and, in some cases, recruitment of the effector substrate to the plasma membrane. RAF, for instance, undergoes recruitment and plasma membrane translocation, which is dependent upon the switch between GDP- and GTP-bound RAS. Following recruitment, two specific domains in RAF: the N-terminal RAS binding domain (RBD) and cysteine-rich domain (CRD) physically interact with GTP-bound RAS, inducing conformational changes in RAF tertiary structure and initiating its downstream kinase-dependent signaling [9]. Activated RAS signals through multiple molecular pathways promoting proliferation, cell survival, gene expression, and, in some cases, rearrangement of the actin cytoskeleton. In certain non-physiological settings, unregulated RAS activity elicits a "tumor suppressor" function by initiating cell cycle arrest.

Cell cycle progression through the G_1 phase is largely dependent on the activation of RAS, initiated via ligand-binding, phosphorylation, and dimerization of growth factor-bound receptor tyrosine kinases.

Signaling through RAF, influencing cell cycle progression and transcription

One of the earliest signaling events initiated by GTP-bound RAS is the phosphorylation and activation of the RAF serine/threonine kinases, A-RAF, B-RAF, and C-RAF-1 [10]. This small family of kinases is best known for its critical relationship to a key branch of the mitogen-activated protein kinase (MAPK) signaling cascade. Early in the G_1 phase, activated RAF phosphorylates the dual-specificity kinase, Mek kinase, which in turn activates the Erk kinases by phosphorylation and homodimerization. These undergo nuclear translocation and interact with a host of transcription factors, including ELK1, an ETS family member, to upregulate transcription of genes promoting cell cycle progression through the G_1 phase. One critical transcriptional target of Erk is cyclin D1, a member of the greater cyclin family of proteins, which serve the primary role of cell cycle regulation. Cyclin D1 binds to, and initiates the activation of, cyclin-dependent kinases (CDKs) 4 and 6. Following an additional activating phosphorylation by the CDK-activating kinase (CAK), and the cleavage of inhibitory phosphate groups in the ATP-binding cleft by the CDC25 dual-specificity phosphatases, CDK4/6 can hyperphosphorylate the retinoblastoma (Rb) protein, thereby inactivating it. As a result of conformational changes in Rb tertiary structure, certain E2F family transcription factors such as E2F-1, once held inactive by hypophosphorylated Rb, are released, and facilitate the transcriptional initiation of genes critical to the G_1 to S phase transition, and thus cell cycle progression [11–14].

Signaling through PI₃Kinase, influencing cell survival and protein translation

RAS regulation and facilitation of cell survival relies heavily on the activation of the type I phosphatidylinositol 3-kinases (PI₃Ks). Activated RAS can bind to the PI$_3$K catalytic subunit, facilitating its activation and downstream phosphorylation and modification of phosphatidylinositol-4,5-bisphosphate (PtdIns (4,5)P$_2$) into phosphatidylinositol-3,4,5-triphosphate (PtdIns(3,4,5)P$_3$). This second messenger lipid can then signal through multiple effector proteins, the most important of which, in the context of cell survival, are 3-phosphoinositide-dependent protein kinase 1 (PDK1) and AKT/PKB. PDK1 phosphorylates and activates the p70 ribosomal protein S6 kinase (p70S6K), which in turn serves a critical regulatory role in protein translation. Members of the protein kinase C (PKC) family also serve as downstream targets of PDK1 and are activated by phosphorylation. This diverse group of kinases functions in an isoform- and cell type-specific manner to regulate cell survival and apoptosis [15]. The α, β, ε, and ζ PKC isoforms, for example, have pro-survival functions, while the δ isoform plays a pro-apoptotic role in certain cell types [16]. Conversely, recent findings have demonstrated a critical pro-survival role for PKC δ in cancer cells with mutated, constitutively activated RAS isoforms [17]. PKC δ activity and protein levels are upregulated by activation of KRAS and HRAS, via PDK1, and this upregulation is critical for the survival of the cell, by increasing AKT activity and offsetting the pro-apoptotic signals generated by the coincident dysregulated RAF pathway activation [18].

AKT can phosphorylate a myriad of target proteins, and thus regulates their ability to facilitate or inhibit apoptosis. One target, the BAD protein, when phosphorylated and inactivated by AKT, is rendered unable to bind and inhibit the anti-apoptotic proteins BCL-2 and BCL-xL. The pro-apoptotic forkhead family of transcription factors, when inactivated by AKT-mediated phosphorylation, undergo proteolysis, and are thereby prevented from initiating transcription of pro-apoptotic genes. Likewise, the AKT-dependent inhibition of the IκB kinase complex (IKK), which binds to and inhibits the nuclear factor-κB (NF-κB), results in the transcriptional upregulation of pro-survival genes, including caspase inhibitors, c-Myb and BCL-xL. Activated AKT can also phosphorylate and increase the ability of the CREB transcription factors to bind their co-activators, the CRB proteins, resulting in increased expression of anti-apoptotic genes including BCL-2 and Mcl-1. PtdIns(3,4,5)P$_3$ also has the distinctive function of stimulating the activity of the RAC proteins, members of the RHO protein family. The activation of RAC is accomplished by the PtdIns (3,4,5)P$_3$-mediated activation of RAC guanine nucleotide exchange factors. RAC proteins then facilitate cytoskeletal regulation, as well as the activation of the pro-survival NF-κB transcription factors [1, 11, 19].

Signaling through RAL GDS, influencing vesicle transport, and cell cycle

RAS activation also results in stimulation of effector signaling pathways regulating vesicle transport and cell cycle through the activation of the RAL proteins, a protein family of small GTPases sharing considerable homology with RAS. RAS functions in this pathway by targeting, via phosphorylation, three different guanine nucleotide exchange factors for RAL, including RAL guanine nucleotide dissociation stimulator (RALGDS), RGL2/RLF, and RALGDS-like gene (RGL/RSB2). The activation of RAL has at least three primary sequelae. First, RAL is capable of phosphorylating and inhibiting forkhead transcription factors, in concert with activated AKT. Second, activated RAL can facilitate the catalytic breakdown of phosphatidylcholine (PC) into phosphatidic acid (PA) and choline, via its interaction with phospholipase D1 (PLD1). Third, the binding of RAL to the Exo84 and Sec 5 subunits of the exocyst complex, a large protein complex necessary for exocytosis, enables the RAL-dependent regulation of exocytosis, thus demonstrating a key regulatory function for RAS in membrane trafficking and vesicle formation [20, 21].

Signaling through PLCε, influencing calcium signaling

Regulation of calcium signaling is an important function of activated RAS, and is mediated predominantly by phospholipase Cε (PLCε). PLCε serves two main roles in RAS effector signaling. The first involves its ability to hydrolyze PtdIns(4,5)P$_2$ into inositol 1,4,5 trisphosphate (Ins(1,4,5)P$_3$) and diacylglycerol (DAG), which both serve as secondary messengers in cell signaling. Ins(1,4,5)P$_3$ binds to its specific receptor on the endoplasmic reticulum membrane outer leaflet, thereby inducing the release of calcium into the cytoplasm through calcium channels, and serving as a major signaling event facilitating various dynamics of cell proliferation and movement [1, 2].

Mechanisms of aberrant RAS signaling in cancer

Aberrant signaling through RAS pathways occurs as the result of several different classes of mutational damage in tumor cells, of which activating mutations in the Ras protein(s) itself are but one type [11].

RAS-activating mutations

The extensive roles that RAS has in tightly regulating cell cycle progression and proliferation can be dramatically compromised through the acquisition of genomic mutations in critical locations within RAS genes. Mutations of the RAS genes themselves are common in human tumors. Roughly 20% of human tumors present with specific point mutations resulting in constitutively activated, GTP-bound RAS proteins. These mutations all compromise the GTPase activity of RAS, preventing GAPs from promoting hydrolysis of GTP on RAS, and therefore causing RAS to accumulate in the GTP-bound, active form. Nearly all RAS activation in tumors is accounted for by mutations in codons 12, 13, and 61 [22]. Of the many RAS family members, the vast majority of genomic mutations thought to contribute to tumor formation are found in the KRAS, NRAS, and HRAS genes at frequencies of 85%, 15%, and 1% respectively. Interestingly, specific RAS genes are found to be mutated with varying frequencies depending on tumor type and location. Approximately 90% of pancreatic tumors exhibit point mutations in codon 12 of KRAS. Structurally, this results in amino acid substitutions of glycine, the amino acid normally encoded by codon 12, for aspartic acid (49%), valine (32%), arginine (13%), cysteine (5%), serine (approximately 2%), and alanine (less than 1%) [23]. Colorectal tumors demonstrate an approximately 45% frequency of activating mutations in codons 12, 13, and 61 of KRAS, with glycine to valine substitutions in codon 12 being the most frequently observed mutation [24]. Approximately 15% of melanomas demonstrate mutations in NRAS codons 12 and 61, with the most common amino acid substitutions being glutamine to arginine or lysine [25]. The presence or absence of activating mutations in KRAS in colorectal tumors has prognostic significance, and can now also be employed to direct the use of targeted anti-cancer therapeutic agents.

Activating mutations in RAS genes allow for the propagation of cell signaling in the absence of growth factor receptor activation. This aberrant process serves to engage the cell in unscheduled cell cycle progression, and thereby in aberrant proliferation. Even more frequently in human tumors, however, oncogenic dysregulation in upstream or downstream components of the RAS-dependent signaling pathways, due to genomic mutations, facilitates tumorigenesis independently of mutations in the RAS proteins themselves.

GAP deletion

RAS can also be activated in tumors by loss or inactivating mutations of Ras-GAPs (GTPase-activating proteins). In these cases, GAP proteins are incapable of hydrolyzing RAS-bound GTP, thus allowing for the accumulation of active RAS, and enhanced RAS signaling. The GAP protein NF1, when inactivated by mutation in schwann cells and astrocytes, results in the pathological condition neurofibromatosis, characterized by the formation of neurofibromas, small benign tumors below the skin, as well as optic pathway gliomas, which can evolve over time into malignancies with the acquisition of inactivating genomic mutations in the second NF1 allele [11, 26, 27].

Mutation or amplification of RAS effectors

Aberrant signaling downstream of functional RAS results from mutations or gene duplications in three key proteins: BRAF, AKT, and PI$_3$K.

B-RAF

B-RAF somatic mismatch mutations occur in malignant melanomas at a rate of around 66% and in colon carcinomas at a rate of 15% [28, 29]. Mutations in B-RAF occur in a very limited number of residues in the kinase domain (with a single substitution of valine to glutamic acid at residue 599 accounting for 80%), all of which result in kinase activation [30].

PI$_3$K/AKT

In nearly 40% of ovarian carcinomas, the gene encoding the p110α catalytic subunit of PI$_3$K is amplified, resulting in a significant increase in activity of the kinase. The PI$_3$K pathway is activated by the amplification of its downstream target AKT2 in ovarian, pancreatic, breast, and head and neck tumors [31, 32].

However, the most significant direct activation of this pathway in tumors comes from deletion of the tumor suppressor gene PTEN (phosphatase and tensin homolog). This gene encodes a lipid phosphatase that removes the phosphate from the $3'$ position of $PtdIns(3,4,5)P_3$ and $PtdIns(3,4)P_2$, thus reversing the accumulation of these second messengers generated by PI_3K. PTEN is deleted in approximately 30 to 40% of human tumors [33], making it the second most commonly lost tumor-suppressor gene, after TP53 (which encodes p53 in humans).

Growth factor receptor activation

RAS signaling pathways are also commonly activated in tumors in which growth factor receptor tyrosine kinases have been overexpressed. The most common examples are EGF-R and ErbB2 (also known as HER2/neu), which are frequently activated by their overexpression in many types of cancer, including breast, ovarian, and gastric carcinomas [34]. Greater than 50% of human carcinomas demonstrate overexpression of EGFRs specifically. The overexpression of ERBB2 (HER2/NEU) receptors, enabling their aberrant growth factor-independent dimerization and activation, is seen in approximately 30% of breast cancers [11]. Alternatively, a common mutation in the EGF-R gene results in the expression of a truncated receptor that lacks part of the extracellular domain [35], and this mutated receptor is found to be overactivated in a significant proportion of glioblastomas and some other tumor types. EGF-R family tyrosine kinases are also commonly activated by the autocrine production of EGF-like factors in tumors, such as transforming growth factor-α (TGF-α). The exact frequency of this activation is hard to establish in human tumors, but it seems to be very high in tumors of epithelial origin.

Current anti-cancer therapeutics targeting RAS signaling networks

In light of the enormous tumorigenic potential of aberrant RAS signaling pathways, a great deal of research has focused on ways of interfering with these processes, using a variety of molecular inhibitors specific for RAS and its effectors for therapeutic anti-cancer effects. The central role of RAS and RAS signaling pathways in multiple and diverse aspects of normal physiology, however, invariably complicates these approaches.

Farnesyl transferase inhibitors

The first such effort at a RAS-directed therapeutic focused on inhibiting the ability of RAS proteins to localize to the plasma membrane, a process required for RAS activity. This can be accomplished with farnesyl transferase-specific inhibitors (FTIs), which block addition of the 15-carbon farnesyl isoprenoid lipid to the C-terminal CAAX motif of RAS. Although HRAS depends entirely on farnesylation for activity, KRAS and NRAS can undergo plasma membrane localization following geranylgeranylation with a 20-carbon isoprenoid group by geranyl transferase in the absence of farnesyl transferase activity. Thus KRAS and NRAS activity cannot be quenched, or their tumorigenesis halted, by FTIs alone, and activation of these family members constitute 99% of Ras-induced malignancies. Geranylgeranyl transferase inhibitors (GGTIs) were developed to overcome this limitation, but clinical trials employing a combination of FTIs and GGTIs produced unacceptable non-specific toxicities, as might have been predicted by the critical role of RAS signaling in the functioning of normal cellular processes [11].

Antisense oligonucleotides

Targeting RAS expression, rather than its posttranslational activation, has also received much attention. Using antisense phosphorothioate oligodeoxynucleotides with complementary sequences to that of HRAS-specific messenger RNA, translation of HRAS can be effectively halted. This technology relies on endogenous RNASE H, an enzyme that recognizes and cleaves the RNA component of RNA:DNA complexes it encounters in the cytoplasm. Certain issues involving the feasibility of such a therapeutic technique employing infusion of oligonucleotides have been partially alleviated by the generation of modified oligonucleotides that demonstrate half-lives of up to two weeks in rodent plasma, and accumulation within tissues as early as 24 hours post treatment. Other potential problems with this approach, however, involve the relatively non-tumor-specific nature of the treatment, which could elicit off-target toxicities [36].

Kinase inhibitors of RAS effectors

Other technologies seeking to target RAS-associated carcinogenesis involve the design of small molecules to inhibit downstream RAS effector pathways, or to target upstream activators of RAS. Both constitutively active RAS and its principal downstream effector, B-RAF, signal through the MAPK pathway to promote cellular survival and induce proliferation. To date, several small molecules have been developed to destabilize the dysregulation of this pathway. The ERK inhibitor CI-1040 has been employed in clinical trials and was shown effective at inhibiting ERK signaling in peripheral blood mononuclear cells, while the cRAF1 inhibitor BAY43–9006 has shown the ability to inhibit B-RAF in tumor cells, suggesting its potential as a therapeutic in melanoma.

Inhibitors of upstream activators of RAS (receptor tyrosine kinase inhibitors)

Efforts to develop therapeutics targeting upstream components of RAS signaling pathways have focused on the inhibition of the growth factor receptor tyrosine kinases, such as EGFR and ERBB2, by monoclonal antibodies or small molecules. This strategy is obviously limited to tumors wherein the dysregulated RAS signaling is secondary to mutation in the receptor tyrosine kinase or its proximal effectors. Gefitinib and erlotinib are small molecule tyrosine kinase inhibitors with some selectivity for EGFR with activity in colorectal and lung cancer. Cetuximab is an EGFR-specific monoclonal antibody, which results in internalization and downregulation of expression of the EGFR. Its lack of complete specificity for the EGFR may be an advantage, in that it also appears moderately effective in downregulating the vascular endothelial growth factor, thereby reducing angiogenesis [11, 37]. The development of monoclonal antibodies against ERBB2 (Her2/NEU) for the treatment of certain breast cancers overexpressing ERBB2 is a clear success in the evolution of targeted anti-cancer therapeutics, and has revolutionized the treatment of these breast cancers, but it does not appear that their mode of action primarily involves interruption of the downstream signaling of ERBB2 through RAS.

Interference with survival pathways (AKT, PKC δ)

Another strategy devised to inhibit RAS signaling pathways focuses on the inhibition of the RAS effector, AKT. Cancers bearing either a constitutively activated AKT or AKT gene duplications (amplification) are often resistant to current chemotherapy and radiation treatment modalities. The small-molecule kinase inhibitor perifosine is the most widely studied AKT inhibitor to date. Perifosine is thought to prevent the membrane translocation of AKT, thereby inhibiting its potential activation, and has demonstrated synergy with radiation treatment in the induction of apoptotic and cell cycle arresting pathways in cancer cells. A second class of AKT-specific inhibitors are the phosphatidylinositol ether lipid analogues (PIAs). These are thought to bind AKT within its plextrin homology domain, thus inhibiting the downstream signaling ability of AKT [38].

All of these cancer-directed therapeutic approaches, however, whether inhibiting RAS activity itself, or inhibiting aberrant RAS activation or signaling by constitutively active RAS effectors, are potentially limited by the critical need of normal cells and tissues for these same RAS-associated signaling pathways in their maintenance of homeostasis, survival, and/or growth. Different approaches, currently under development, seek to exploit those signaling pathways unique to cells expressing dysregulated RAS or RAS effector pathways.

The discovery that constitutive activation of RAS results in the efficient suppression of molecular anti-viral defenses in tumor cells formed the basis of the novel viral oncolytic therapeutic approach [39]. A reovirus is a benign virus that can replicate in, and induce death of, tumor cells having an activated RAS pathway, which suppresses the normal translation block on reovirus transcripts. This approach has shown dramatic and specific activity in clinical models of RAS-associated cancers [40].

PKC δ activity and protein levels are upregulated by activation of KRAS or HRAS in tumor cells, via PDK1, and this upregulation is critical for the survival of the cell, by increasing AKT activity, and offsetting the pro-apoptotic signals generated by the coincidently dysregulated RAF pathway activation [17]. Interference with PKC δ activity induces rapid apoptosis in cancer cells containing an activated RAS protein or constitutive activation of RAS effector pathways [18]. PKC δ inhibitors are being developed as RAS pathway-targeted cancer therapeutics and have the potential to selectively target tumor cells with dysregulation of RAS signaling, while sparing normal cells.

The high frequency of dysregulation of RAS pathway activity observed in a wide spectrum of cancers confirms the attractiveness of RAS/RAS pathways as a therapeutic target. Unfortunately, efforts at inhibiting RAS protein function through the use of FTIs and GGTIs have been unsuccessful in clinical trials. Despite these relative failures, however, the opportunity still exists to design therapeutics targeting both upstream and downsteam RAS pathway effectors, with the hope that aberrant RAS signaling can be subverted in malignancies. The central and critical roles of RAS in diverse pathways, in cells from species as diverse as nematodes and humans, however, will complicate efforts to selectively and safely suppress RAS signaling in cancer. As our understanding of the complexities and interactions of the RAS signaling network continues to expand, so should our ability to selectively manipulate components of this network for targeted therapies.

References

1. Karnoub A.E., and Weinberg R.A. (2008) Ras oncogenes: split personalities. *Nat Rev* **9**, 517–531.

2. Malumbres M., and Pellicer A. (1998) Ras pathways to cell cycle control and cell transformation. *Front Biosci* **3**, 887–912.

3. Hancock J.F. (2003) RAS proteins: different signals from different locations. *Nat Rev* **4**, 373–384.

4. Yamamoto T., Taya S., and Kaibuchi K. (1999) Ras-induced transformation and signaling pathway. *J Biochem* **126**, 799–803.

5. Campbell S.L., Khosravi-Far R., Rossman K.L., Clark G.J., and Der C.J. (1998) Increasing complexity of Ras signaling. *Oncogene* **17**, 1395–1413.

6. Nel A.E. (2002) T-cell activation through the antigen receptor. Part 1: signaling components, signaling pathways, and signal integration at the T-cell antigen receptor synapse. *J Allergy Clin Immunol* **109**, 758–770.

7. Keshamouni V.G., Mattingly R.R., and Reddy K.B. (2002) Mechanism of 17-β-estradiol-induced Erk1/2 activation in breast cancer cells. *J Biol Chem* **277**, 22558–22565.

8. Rojas J.M., and Santos E. (2006) RAS-GEFS and RAS GAPS. *RAS Family GTPases*, 15–43.

9. Bondeva T., Balla A., Várnai P., and Balla T. (2002) Structural determinants of Ras-Raf interaction analyzed in live cells. *Mol Biol Cell* **13**, 2323–2333.

10. Pruitt K., and Der C.J. (2001) Ras and Rho regulation of the cell cycle and oncogenesis. *Cancer Lett* **171**, 1–10.

11. Downward J. (2003) Targeting RAS signalling pathways in cancer therapy. *Nat Rev Cancer* **3**, 11–22.

12. Kerkhoff E., and Rapp U.R. (1998) Cell cycle targets of Ras/Raf signalling. *Oncogene* **17**, 1457–1462.

13. Marshall C. (1999) How do small GTPase signal transduction pathways regulate cell cycle entry? *Curr Opin Cell Biol* **11**, 732–736.

14. Downward J. (1997) Cell cycle: routine role for Ras. *Curr Biol* **7**, R258–R260.

15. Parker P.J., and Murray-Rust J. (2004) PKC at a glance. *J Cell Sci* **117**, 131–132.

16. Reyland M.E. (2007) Protein kinase Cδ and apoptosis. *Biochem Soc Trans* **35**, 1001–1004.

17. Xia S., Forman L.W., and Faller D.V. (2007) Protein kinase C δ is required for survival of cells expressing activated p21RAS. *J Biol Chem* **282**, 13199–13210.

18. Xia S., Forman L.W., Chen Z., and Faller D.V. (2009) PKCδ survival signaling in cells containing an activated p21^{Ras} oncoprotein. *Cell Signal* **21**, 502–508.

19. Song G., Ouyang G., and Bao S. (2005) The activation of Akt/PKB signaling pathway and cell survival. *J Cell Mol Med* **9**, 59–71.

20. van Dam E.M., and Robinson P.J. (2006) Ral: mediator of membrane trafficking. *Int J Biochem Cell Biol* **38**, 1841–1847.

21. Malumbres M., and Barbacid M. (2003) RAS oncogenes: the first 30 years. *Nat Rev Cancer*. **3**, 459–465.

22. Bos J.L. (1989) RAS oncogenes in human cancer: a review. *Cancer Res* **4**, 4682–4689.

23. Mu D., Peng Y., and Xu Q. (2004) Values of mutations of K-ras oncogene at codon 12 in detection of pancreatic cancer: 15-year experience. *World J Gastroenterol* **10**, 471–475.

24. Andreyev H.J.N., Norman A.R., Cunningham D., et.al. (2001) Kirsten ras mutations in patients with colorectal cancer: the 'RASCAL II' study. *Br J Cancer* **85**, 692–696.

25. Kumar R., Angelini S., and Hemminki K. (2003) Activating BRAF and N-Ras mutations in sporadic primary melanomas: an inverse association with allelic loss on chromosome 9. *Oncogene* **22**, 9217–9224.

26. Lynch T.M., and Gutmann D.H. (2002) Neurofibromatosis 1. *Neurol Clin* **3**, 841–865.

27. Weiss B., Bollag G., Shannon K. (1999) Hyperactive Ras as a therapeutic target in neurofibromatosis type 1. *Am J Med Genet* **89**, 14–22.

28. Mercer K.E., and Pritchard C.A. (2003) Raf proteins and cancer: B-Raf is identified as a mutational target. *Biochim Biophys Acta* **1653**, 25–40.

29. Brose M.S., Volpe P., Feldman M., et al. (2002) BRAF and RAS mutations in human lung cancer and melanoma. *Cancer Res* **62**, 6997–7000.

30. Davies H., Bignell G.R., Cox C., et al. (2002) Mutations of the BRAF gene in human cancer. *Nature* **417**, 949–954.

31. Yuan T.L., and Cantley L.C. (2008) PI3K pathway alterations in cancer: variations on a theme. *Oncogene* **27**, 5497–5510.

32. Bellacosa A., de Feo D., Godwin A.K., et.al. (1995) Molecular alterations of the AKT2 oncogene in ovarian and breast carcinomas. *Int J Cancer* **64**, 280–285.

33. Simpson L., and Parsons R. (2001) PTEN: life as a tumor suppressor. *Exp Cell Res* **264**, 29–41.

34. Mendelsohn J., and Baselga J. (2000) The EGF receptor family as targets for cancer therapy. *Oncogene* **19**, 6550–6565.

35. Kuan C.T., Wikstrand C.J., and Bigner D.D. (2001) EGF mutant receptor vIII as a molecular target in cancer therapy. *Endocr Relat Cancer* **8**, 83–96.

36. Dean N.M., and Bennett C.F. (2003) Antisense oligonucleotide-based therapeutics for cancer. *Oncogene* **22**, 9087–9096.

37. Harari P.M. (2004) Epidermal growth factor receptor inhibition strategies in oncology. *Endocr Relat Cancer* **11**, 689–708.

38. LoPiccolo J., Blumenthal G.M., Bernstein W.B., and Dennis P.A. (2008) Targeting the PI3K/Akt/mTOR pathway: effective combinations and clinical considerations. *Drug Resist Updat* **11**, 32–50.

39. Mundschau L.J., and Faller D.V. (1992) Oncogenic *ras* induces an inhibitor of double-stranded rna-dependent eukaryotic initiation factor 2α-kinase activation. *J Biol Chem* **267**, 23092–23098.

40. Marcato P., Shmulevitz M., and Lee P.W. (2005) Connecting reovirus oncolysis and Ras signaling. *Cell Cycle* **4**, 556–559.

Chapter

13

PI3K pathway in cancer

Amancio Carnero

Introduction

Phosphatidylinositol 3-kinases (PI3Ks) are a family of lipid kinases that phosphorylate the hydroxyl group at position 3 of phosphatidylinositol (4,5) bisphosphate (PIP2) to produce the second messenger phosphatidylinositol (3,4,5) trisphosphate (PIP3), a lipid second messenger that controls a wide range of cellular responses. PI3K isoforms are divided into three classes and composed of several subunits. Class I PI3Ks are the best characterized and are subdivided into classes IA and IB. The most popular class of PI3Ks (class I) consists of a catalytic subunit (p110s) and a regulatory subunit (p85 or p101). They are activated by receptor tyrosine kinases or by G-protein-coupled receptors. Other tyrosine kinase receptors, such as BCR-ABL and ErbB2, or oncogenes, such as ras, also signal through the PI3K pathway. Phosphatase and tensin homolog (PTEN) negatively regulates PI3K signaling by dephosphorylating PIP3 to PIP2. Akt is activated downstream of PIP3 to mediate physiologic processes. Substantial crosstalk exists with other signaling networks at all levels of the PI3K pathway. The PI3K pathway is one of the most frequently found altered pathways in human tumors. PTEN is lost in a great variety of cancers, especially metastases. PI3K is found point mutated with activated mutations and amplified in several tumors, and activating mutations in AKT have also been recently found in human tumor cells.

The pathway

PTEN (phosphatase and tensin homolog deleted on chromosome 10) is a dual lipid and protein phosphatase. Its primary target is the lipid phosphatidylinositol-3,4,5-triphosphate (PIP3) (Maehama and Dixon, 1998), the product of the phosphatidylinositol 3 kinase (PI3K). Loss of PTEN function, as well as PI3K activation, results in accumulation of PIP3 triggering the activation of its downstream effectors, PDK1, AKT/PKB, and Rac1/cdc42. The PI3K family is divided into four classes, three of them phosphorylate lipids as main targets, while class IV (mTOR, ATM, ATR, and DNA-PK) phosphorylate proteins. Class I, the more broadly studied, and the one we generally refer to in this review, is composed of heterodimers formed by a catalytic subunit (p110α, β, γ, and δ) and a regulatory subunit. Class I can be subdivided into two subclasses: Ia, formed by the combination of p110α, β, and δ and a regulatory subunit (p85, p65, or p55) and Ib, formed by p110γ and p101 regulatory subunits. Activation of PI3K is induced by growth factors and insulin targeting the catalytic subunit to the membrane where it is in close proximity with its substrate, mainly PIP2. PDK1 contains a C-terminal pleckstrin homology (PH) domain, which binds the membrane-bound PIP3 triggering PDK1 activation. Activated PDK1 phosphorylates AKT at T308 activating its ser/thr kinase activity. Once phosphorylated in T308, further activation occurs by PDK2 (the complex rictor/mTOR or DNA-PK) by phosphorylation at S473. AKT activation stimulates cell cycle progression, survival, metabolism, and migration through phosphorylation of many physiological substrates (Dahia, 2000; Downward, 2004; Kandel and Hay, 1999; Stokoe, 2001; Vivanco and Sawyers, 2002). AKT is a serine–threonine kinase downstream of PTEN/PI3K that has three family members: AKT1, AKT2, and AKT3, which are encoded by three different genes (Datta et al., 1999). They are ubiquitously expressed, but their levels are variable depending upon the tissue type. The amino terminus also contains a PH domain that binds phospholipids. AKT activity is regulated by PI3K, which recruits AKT to the cell membrane

---

Systems Biology of Cancer, ed. S. Thiagalingam. Published by Cambridge University Press. © Cambridge University Press 2015.

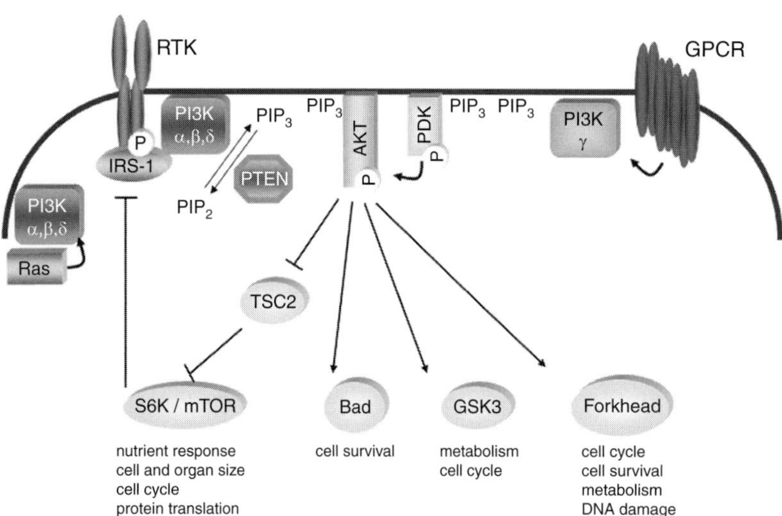

Figure 13.1 General scheme of the PKB/AKT pathway. (A black and white version of this figure will appear in some formats. For the color version, please refer to the plate section.)

through PIP3 binding, permitting its activation by PDK1 (Kandel and Hay, 1999) (Figure 13.1). Activation of AKT results in the suppression of apoptosis induced by a number of stimuli including growth factor withdrawal, detachment of extracellular matrix, UV irradiation, cell cycle discordance, and activation of FAS signaling (Downward, 2004; Kandel and Hay, 1999; Plas and Thompson, 2005). Hyperactivated AKT has also been shown to promote cell proliferation, cell growth and metabolism, resistance to hypoxia, and migration (Blanco-Aparicio et al., 2007; Downward, 2004).

Different genetic approaches have been used to directly assess the role of AKT in PTEN null-induced phenotype. Deletion of AKT1 reversed the survival phenotype in PTEN null cells and abrogated its growth advantage (Stiles et al., 2002). Furthermore, deleting both alleles of AKT1 appears to have additional effects, since mutated cells were more sensitive to serum starvation-induced cell death. AKT1 KO cells lose the ability to compete with wild type or PTEN null cells (Stiles et al., 2002). Similarly, inactivation of AKT by dominant negative mutants inhibits the survival advantage provided by activated class I PI3K (Link et al., 2005). These and other results underline the essential role of AKT in the PTEN/PI3K pathway (Brazil and Hemmings, 2001; Chen et al., 2005; Mayo and Donner, 2002; Samuels and Ericson, 2006; Toker and Yoeli-Lerner, 2006).

PI3K may control multiple pathways (PIP3 dependent or independent), besides the AKT pathway. AKT is one of the proteins recruited to the membrane by PIP3 where it is activated by another

PIP3-activated protein, PDK1. PIP3-dependent functions, not related to AKT, might be PDK1 dependent as suggested by the hypomorphic PDK1 mice (Bayascas et al., 2005). Reduced levels of PDK1 expression in PTEN(+/−) mice markedly protects these animals from developing a wide range of tumors. PDK1 has been shown to phosphorylate the critical residue in the activation loops of all AGC kinase family members including AKTs, SGKs, S6K, PKA, PKC, RSK, and protein kinase N (Blanco-Aparicio et al., 2007; Carnero et al., 2008). Furthermore, other proteins might also be recruited and activated by PIP3 increase (Downward, 2004; Stiles et al., 2004). The PH domain was the first phosphoinositide-binding domain identified. It is present in the largest number of proteins and is associated with the formation of signaling complexes on the plasma membrane. Recent studies identified other novel phosphoinositide-binding domains (Fab1p, YOTB, Vps27p, EEA1, Phox homology, and epsin N-terminal homology (ENTH)), extending the functional versatility of the pathway (Figure 13.2).

RAS

Activating point mutations in the genes encoding the Ras subfamily of small GTP-binding proteins contribute to the formation of a large proportion of human tumors (Bos, 1989). The expression of this active version of Ras promotes tumor initiation by activating at least three different effectors: Raf, PI3K, and RalGEFs (Carnero and Beach, 2004; Carnero and Lacal, 1995; Carnero et al., 1995; Shields et al., 2000;

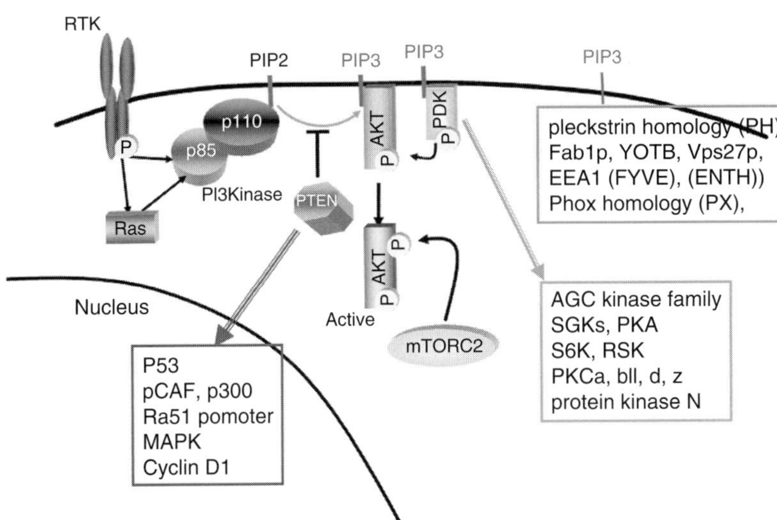

Figure 13.2 General scheme of the alternative signaling to canonical AKT activation. (A black and white version of this figure will appear in some formats. For the color version, please refer to the plate section.)

Ulku and Der, 2003). Raf is a serine/threonine kinase that is localized to the plasma membrane from the cytoplasm and activated by GTP-Ras. Activated Raf proteins then initiate a MAP kinase (MAPK) signal transduction cascade leading to transformed morphologies, anchorage-independent growth, and angiogenesis (Morrison and Cutler, 1997; Shields et al., 2000). Finally, the RalGEFs family of guanine exchange factors are activated via their recruitment to the plasma membrane by GTP-Ras (Feig, 2003). In human cells it has been reported that the Ras effector pathways MAPK, RalGEF, and PI3K are required to initiate tumor growth (Lim and Counter, 2005; McFarlin and Gould, 2003; McFarlin et al., 2003; Rangarajan et al., 2004). Conversely, activation of the PI3K/AKT pathway replaced Ras once tumors formed, although other effectors were still activated independently of Ras, presumably by factors provided upon the establishment of the tumor microenvironment. Thus as tumorigenesis progresses the addiction of cancers to their initiating oncogene is reduced to, at least in the case of *Ras*, the PI3K/AKT pathway (Lim and Counter, 2005).

AKT "master" kinase

Localization to the membrane seems to be critical for AKT activation. Modifications that target AKT to the membrane, such as the gag sequence found in v-Akt or fusion of a myristoylation sequence from Src-like kinases to the amino terminus of AKT results in the constitutive activation of the kinase (Aoki et al., 1998; Bellacosa et al., 1991; Kennedy et al., 1997).

AKT phosphorylates proteins containing the amino acid sequence RXRXXS/T-B (Alessi et al., 1996) (where X represents any amino acid and B is any bulky hydrophobic residue). Curently more than a hundred non-redundant AKT substrates have been reported and many of them do not contain the canonical AKT recognition motif (Manning and Cantley, 2007). This variety of substrates elicits a broad physiological response mediated by multiple downstream effectors (Figure 13.1). The three AKT isoforms are very similar, and it is unclear whether they have distinguishable substrate specificities. However, knockout mice have uncovered distinct physiological functions for the three AKT isoforms (Dummler and Hemmings, 2007; Dummler et al., 2006). Although combined knockouts have illustrated that functional redundancy exists between the three isoforms, the distinct phenotypes resulting from loss of each individually suggests that there are likely to be functional differences at the cellular level.

AKT enhances cell survival by blocking the function of pro-apoptotic proteins. AKT negatively regulates the function or expression of several pro-apoptotic proteins that inactivate pro-survival Bcl-2 family members. Survival factors stimulate AKT-mediated phosphorylation of BAD, and this creates a binding site for 14-3-3 proteins, which sequesters BAD from its target proteins (Datta et al., 1997). AKT also inhibits the expression of BIM through effects on transcription factors, such as FOXO and p53 (Dijkers et al., 2002). AKT phosphorylates the FOXO family members FOXO, FOXO3a, and FOXO4 (Tran et al., 2003) while they are in the nucleus, creating a binding site for 14-3-3 proteins, which triggers their export from the nucleus. Through this mechanism, AKT blocks FOXO-mediated transcription of target genes that promote apoptosis, cell cycle arrest, and metabolic processes. AKT also promotes survival by phosphorylating MDM2, an E3 ubiquitin ligase that triggers p53 degradation. AKT phosphorylates MDM2 promoting translocation of MDM2 to the nucleus, where it negatively regulates p53 function (Mayo and Donner, 2001; Zhou and Hung, 2002; Zhou et al., 2003). Two transcriptional targets of p53 are the proteins Puma and Noxa, which appear to be the essential targets in p53-induced apoptosis (Villunger et al., 2003). AKT has also been found to directly phosphorylate S196 on human procaspase-9, and this phosphorylation correlates with a decrease in the protease activity of caspase-9 *in vitro* (Cardone et al., 1998). AKT also exerts some of its cell survival effects through the modification of nutrient uptake and metabolism (reviewed in Plas and Thompson, 2005; Robey and Hay, 2006).

One of the best conserved functions of AKT is its role in the cell mass increase through activation of the mTOR complex 1 (mTORC1 or the mTOR-raptor complex), which is regulated by both nutrients and growth factor signaling. mTORC1 is a critical regulator of translation initiation and ribosome biogenesis and plays an evolutionarily conserved role in cell growth control (Wullschleger et al., 2006). The enhanced sensitivity to mTORC1 inhibitors of cancer cells with oncogenic activation of the PI3K-Akt pathway illustrates the importance of mTORC1 activation downstream of Akt (Sabatini, 2006). AKT activates mTORC1 indirectly by inhibiting TSC2, thereby allowing Rheb-GTP to activate mTORC1 signaling. Recently, Akt has been shown to directly phosphorylate PRAS40 (Kovacina et al., 2003), and this phosphorylation was shown to be important for 14-3-3 binding. PRAS40 associates and negatively regulates mTORC1 signaling (Sancak et al., 2007).

AKT activation can stimulate proliferation through multiple downstream targets impinging on cell cycle regulation. AKT phosphorylates the cyclin-dependent kinase inhibitors p21Cip1/WAF1 and p27Kip1 promoting its cytosolic localization (Liang et al., 2002; Shin et al., 2002; Viglietto et al., 2002; Zhou and Hung, 2002) and preventing its cell cycle inhibitory effects. AKT-dependent phosphorylation of other targets such as GSK3, TSC2, and PRAS40 is also likely to drive cell proliferation through

regulation of the stability and synthesis of proteins involved in cell cycle entry.

In response to growth factors, AKT signaling regulates nutrient uptake and metabolism. One of the most important physiological functions of AKT is to stimulate glucose uptake in response to insulin. Akt2, the primary isoform in insulin-responsive tissues, has been found to associate with glucose transporter 4 (Glut4)-containing vesicles upon insulin stimulation of adipocytes (Calera et al., 1998), and AKT activation leads to Glut4 translocation to the plasma membrane (Kohn et al., 1996). AKT activation can enhance the rate of glycolysis by promoting its ability to express glycolytic enzymes through HIFα (Lum et al., 2007; Majumder et al., 2004). In a cell context-dependent manner, Akt-mediated phosphorylation and inhibition of FOXO1 also contribute to glucose homeostasis (reviewed in Accili and Arden, 2004). AKT plays important roles in both physiological and pathological angiogenesis through effects in both endothelial cells and cells producing angiogenic signals, such as tumor cells. The PI3K–AKT pathway is robustly activated by vascular endothelial growth factor (VEGF; reviewed in Olsson et al., 2006), promoting endothelial cell survival, growth, and proliferation.

The AKT pathway is highly connected with other pathways. AKT signaling can activate the NF-κB transcription factor by direct phosphorylation of IκB kinase α (IKKα), thereby leading to activation of the kinase upstream of NF-κB (Ozes et al., 1999). Akt can directly phosphorylate c-Raf and this can lead to an inhibitory effect on the Erk pathway (Zimmermann and Moelling, 1999). The stress-activated MAPKs JNK and p38 have also been shown to be inhibited by AKT signaling.

All these physiological implications suggest that the PTEN/PI3K/AKT pathway is a target of genetic alterations causing cancer.

AKT-independent pathways

Loss of PTEN function results in the constitutive activation of AKT, which plays a key role in PTEN-mediated tumorigenesis via multiple mechanisms (Downward, 2004; Gottlieb et al., 2002; Mayo and Donner, 2001; Oren et al., 2002; Plas and Thompson, 2005; Radu et al., 2003; Stiles et al., 2004; Weng et al., 2001b). There are, though, alternative mechanisms of PTEN-mediated tumorigenesis, possibly independent of AKT (Chang et al., 2004; Freeman et al., 2003; Li et al., 2006; Weng et al., 2001a). It has been shown that PTEN directly associates with p53 increasing its stability, protein levels, and transcriptional activity. $Pten^{+/-};p53^{+/-}$ double heterozygous mice show an onset of lymphoma development similar to that seen in $p53^{-/-}$ animals. p53 protein levels are dramatically reduced in $Pten^{-/-}$ cells due to PTEN-mediated stabilization of p53, which increases its half-life. Furthermore, ectopic expression of PTEN phosphatase-dead mutants also leads to a significant increase in p53 protein levels, and PTEN can stabilize p53 even in the absence of MDM2, suggesting that PTEN can regulate p53 levels in a phosphatase-independent and MDM2-independent manner (Figure 13.3). Moreover, PTEN physically associates with endogenous p53, and regulates the transcriptional activity of p53 by modulating its DNA binding capability (Freeman et al., 2003).

PTEN contains a sequence motif highly conserved among the protein tyrosine phosphatase family members. PTEN has been shown to possess phosphatase activity on phosphotyrosyl and phosphothreonyl-containing substrates (Li et al., 1997; Myers et al., 1997, 1998) *in vitro*, and on phosphatidylinositol (3,4,5) trisphosphate, the product of phosphatidylinositol 3-kinase (PI3K), both *in vitro* and *in vivo* (Maehama and Dixon, 1998; Myers et al., 1998; Stambolic et al., 1998; Sun et al., 1999). The fact that naturally occurring mutations in the PTEN phosphatase domain (such as PTEN-C124S and PTEN-G129E mutants) are tumorigenic indicates that the effects of PTEN's phosphatase-independent activity in tumorigenesis may be tissue specific or associated with a

more aggressive phenotype upon loss of PTEN function. PTEN regulates cell cycle arrest via protein phosphatase-dependent interaction with cyclin D. The oncogenic potential of PTEN is further highlighted by roles in integrin signaling and the ability to dephosphorylate focal adhesion kinase that reduce cell adhesion and enhance migration (Gu et al., 1998, 1999; Tamura et al., 1999). PTEN appears to inhibit cell cycle progression through the cooperation of its protein phosphatase activity, which leads to the downregulation of cyclin D1, while its lipid phosphatase activity leads to upregulation of p27 (Weng et al., 2001b). PTEN also abrogates insulin-stimulated ETS2 activation independently of PI3K, probably through the protein phosphatase activity (Weng et al., 2001a). ETS transforms NIH3T3 allowing tumor formation in mice. Inhibition of ETS2 can inhibit ras-dependent transformation and abolish breast carcinoma cells' anchorage-independent growth and migration. ETS2 can also activate the cyclin D1 promoter.

PTEN is also found in the nucleus (Baker, 2007; Gil et al., 2007). Nuclear localization of PTEN may contribute to its tumor suppressor activity in several ways. It is possible that PTEN is a nuclear PIP3 phosphatase, although it has been suggested that nuclear PTEN does not dephosphorylate the nuclear pool of PIP3 (Lindsay et al., 2006). The protein phosphatase functions of nuclear PTEN downregulating the MAPK pathway and cyclin D1 have been implicated in G1 cell cycle arrest. The regulation of p53 activity and stability by direct protein–protein interaction also occurs within the nucleus (Li et al., 2006; Lian and Di Cristofano, 2005) (Figure 13.2).

Recently, another nuclear tumor suppressor activity of PTEN has been reported. PTEN loss increased genomic instability by causing increased AKT-mediated sequestration of Chk1 and compromising DNA damage response (Puc and Parsons, 2005). The nuclear PTEN co-localized with centromeres, binding to Cenp-C, which is required for proper kinetochore assembly and for the metaphase to anaphase transition. The C-terminus but not the phosphatase domain of PTEN was required for this function (Shen et al., 2007). Therefore PTEN null MEFs contain an increased number or chromosomal abnormalities (Shen et al., 2007). These chromosomal aberrations suggest a potential defect in the DNA damage checkpoint. Furthermore, PTEN was found to be bound to the Rad51 promoter suggesting a more direct impact of PTEN on transcriptional regulation

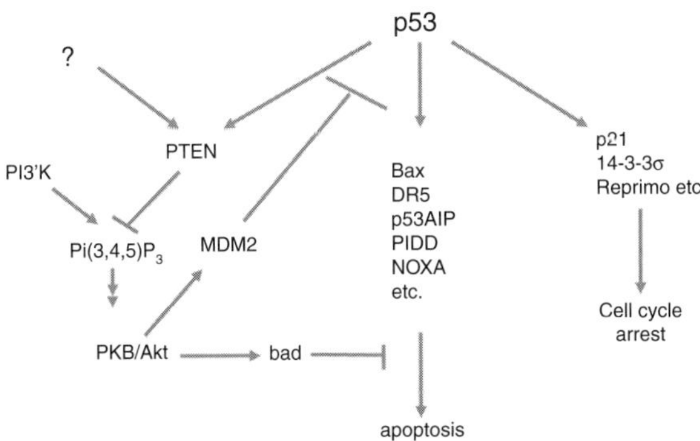

Figure 13.3 Functional relationship between PTEN and p53 tumor suppressor pathways. (A black and white version of this figure will appear in some formats. For the color version, please refer to the plate section.)

(Shen et al., 2007). Although it does not directly activate the transcription of Rad51, PTEN enhanced E2F-mediated transactivation. Other studies have also shown interactions of PTEN with PCAF and p300 transcriptional co-activators that function as histone acetyltransferases (Li et al., 2006; Okumura et al., 2005).

PI3K pathway in tumors

The PTEN/PI3K pathway is highly involved in cancer. PTEN activity is lost by mutations, deletions, or promoter methylation silencing at high frequency in many primary and metastatic human cancers (Parsons et al., 2005; Vivanco and Sawyers, 2002; Figure 13.3). Germline mutations of PTEN are found in Cowden, Bannayan–Riley–Ruvalcaba, and Proteus-like syndromes, all of which are familial cancer predisposition syndromes (Liaw et al., 1997; Marsh et al., 1999; Nelen et al., 1997; Schrager et al., 1998). Recently, many activating mutations have been described in the PI3KCA gene (coding for the p110α catalytic subunit of PI3K) to be present in human tumors (Parsons et al., 2005; Samuels et al., 2004). The three most frequently observed PI3-kinase mutations: E542K, E545K, and H1047R showed enhanced catalytic activity (Kang et al., 2005), comparable to membrane-bound myr-p110. They are able to constitutively activate AKT and produce transcriptional activation. These enhanced biochemical capabilities translate to enhanced oncogenic activity of the PI3K mutants (Bader et al., 2006; Zhao et al., 2005). The oncogenic mutations have only been detected in the PI3KCA gene (p110α isoform), despite the observations that the activation, by membrane tagging, of all the class I PI3K isoforms have oncogenic potential (Bader et al., 2006; Link et al., 2005; Zhao et al., 2005).

Recently (Carpten et al., 2007), it has been reported that a somatic mutation in human breast, colorectal, and ovarian cancers results in a glutamic acid to lysine substitution at amino acid 17 (E17K) in the lipid-binding pocket of AKT1. Lys 17 alters the electrostatic interactions of the pocket and forms new hydrogen bonds with a phosphoinositide ligand. Furthermore, this mutation activates AKT1 by means of pathological localization to the plasma membrane, stimulates downstream signaling, transforms cells, and induces leukemia in mice. This mechanism indicates a direct role of AKT1 in human cancer, and adds to the known genetic alterations that promote oncogenesis through the phosphatidylinositol-3-OH kinase/AKT pathway (Carpten et al., 2007).

Activation without mutations of PI3K and AKT are reported to occur in breast (Bachman et al., 2004; Campbell et al., 2004; Kirkegaard et al., 2005), ovarian (Campbell et al., 2004; Nakayama et al., 2006; Xing and Orsulic, 2005), pancreatic (Asano et al., 2004), esophageal (Okano et al., 2003), thyroid (Garcia-Rostan et al., 2005), and other cancers (Broderick et al., 2004; Parsons et al., 2005).

This pathway is unique in that every major node is frequently mutated or amplified in a wide variety of solid tumors. Receptor tyrosine kinases upstream of PI3K, the P110alpha, AKT, and the negative regulator PTEN are all frequently altered in cancer. However, a closer analysis of these mutations indicated that they do not exist mutually exclusively but in many cases with co-existing mutations (Table 13.1; Yuan and Cantley, 2008). Co-existence of two or more PI3K pathway mutations in a single tumor would suggest differences in oncogenic mechanisms, given that there would be no selective advantage for cells bearing redundant mutations. Overall, PI3KCA mutations and PTEN loss co-exist in breast, endometrial, and colon cancers (Table 13.1). In the breast, the observed frequency of tumors with co-existing PTEN and PI3KCA mutations is 8.7%. Ras and PI3KCA mutations are mutually exclusive in endometrial cancers but co-existed in colorectal cancers. In this scenario, it is possible that an early RAS mutation committed or

Table 13.1 Summary of co-existence of PI3K pathway genetic alterations.

Genetic variants	Tumor type	Found
PTEN null + PI3KCA mut	Breast	Co-existent
	Endometrium	Co-existent
	Colon	Co-existent
Ras mut + PI3KCA mut	Breast	Exclusive
	Endometrium	Mutually exclusive
	Colon	Co-existent
Ras mut + PTEN null	Breast	Exclusive
	Endometrium	Co-existent
	Colon	Co-existent
Her2$^+$ + PI3KCA mut	Breast	Co-existent
Her2$^+$ + PTEN null	Breast	Co-existent

(Table adapted from Yuan and Cantley, 2008)

addicted the cell to a network independent of the PI3K network, as appears to be the case for most K-RAS mutant pancreatic and lung cancers. This would render a secondary PIK3CA mutation ineffectual and potentially disadvantageous if overactivation of mitogenic signaling pathways leads to oncogene-induced senescence (Sarkisian et al., 2007). During tumor initiation, PI3K signaling could be activated by autocrine factors or chemokines secreted by the tumor stroma. Unlike endometrial cancers, the observed frequency of RAS and PIK3CA mutations in the colon is 7.3%, which is slightly higher than the expected frequency of 5.4% (Table 13.1 and reference therein). This suggests that constitutively active RAS and PIK3CA may function synergistically in the colorectal epithelium to confer an important selective advantage. Previous studies have indicated that the ability of PIK3CA to bind to activated RAS is critical for initiation of lung tumors by mutant K-RAS in a mouse model (Ramjaun and Downward, 2007).

The differences we see between tissue types with regard to mutations in PIK3CA and PTEN may reflect the possibility that in certain tissues PTEN loss alone or PIK3CA mutations alone are insufficient to enhance cell growth or survival. This could be due to higher expression of redundant negative regulators of this pathway such as the lipid phosphatases, SHIP and INPP4B, or due to the fact that mutations in this pathway without prior loss of p53 result in senescence (Chen et al., 2005).

In the breast, HER2 is amplified in 30% of tumors (Downward, 2003) and appears to co-exist with both PIK3CA mutations and PTEN loss (Table 13.1). In fact, the co-existence of PTEN loss and HER2 amplification was critical to understanding trastuzumab resistance in HER2-positive breast cancers. Three studies showed that loss of PTEN expression occurred at high frequencies in HER2-overexpressing breast tumors (11/55, 21.7%; 17/47, 36.2%; and 8/17, 47.1%) and led to resistance due to incomplete shut off of PI3K signaling (Fujita et al., 2006; Ma et al., 2005; Nagata et al., 2004). This suggests that PTEN loss and HER2 overexpression have redundant abilities to activate PI3K. HER2 overexpression may also provide PIK3CA mutants and PTEN−/− cells with sufficient growth factor stimulation to achieve the high levels of PIP3 needed for AKT activation and transformation.

The co-existence of PIK3CA mutations and PTEN loss in all the three tissues examined provides striking evidence that the two genetic alterations have both redundant and non-overlapping mechanisms of oncogenicity. HER2 overexpression and PI3K pathway alterations likely also have divergent but additive mechanisms, whereas RAS mutations and PI3K pathway alterations may drive tumorigenesis by completely different mechanisms in certain tissues.

Given that we rarely see mutual exclusivity between PI3K pathway alterations, it is unlikely that PI3K pathway alterations function with absolute redundancy. A more likely explanation for the data described above is that co-existing mutations can arise if each targeted gene activates non-overlapping

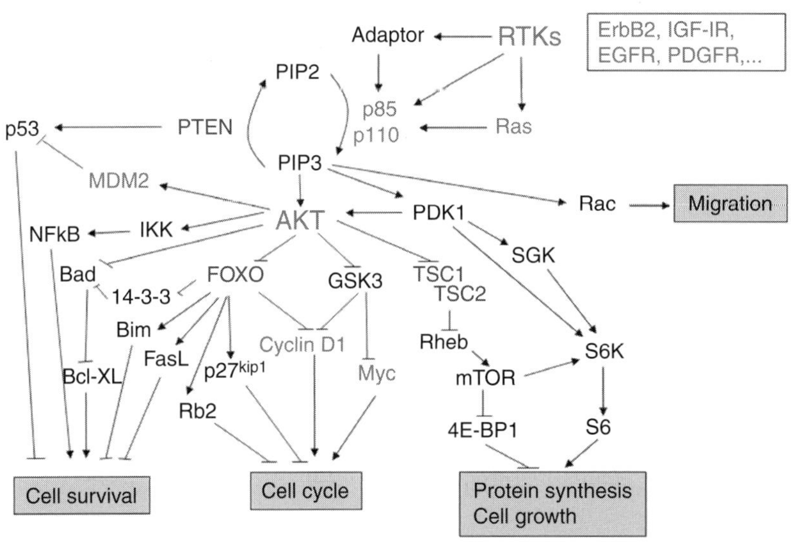

Figure 13.4 General overview of PI3K pathway alterations found in human tumors. (A black and white version of this figure will appear in some formats. For the color version, please refer to the plate section.)

pathways that (1) also induce tumorigenesis (growth/proliferation/motility) or (2) relieve negative feedback on the PI3K pathway. These additional pathways, along with redundant activation of PI3K signaling, would justify retention of multiple PI3K alterations within the same tumor.

Several other genes of the pathway can act as tumor suppressors such as TSC1, TSC2, or LKB1, which carry germline familial mutations, FOXO proteins, and, probably, the phosphatases PHLPP and SHIP (Figure 13.4).

Acknowledgments

This work has been funded by grants from the Spanish Ministry of Economy and Competitivity, ISCIII (Fis: PI12/00137, RTICC: RD12/0036/0028) co-funded by FEDER from Regional Development European Funds (European Union), Consejeria de Ciencia e Innovacion (CTS-6844 and CTS-1848) and Consejeria de Salud of the Junta de Andalucia (PI-0135-2010 and PI-0306-2012). This work has been also possible thanks to the Grant PIE13/0004 co-funded by the ISCIII and FEDER funds..

References

Accili, D. and Arden, K.C. (2004) FoxOs at the crossroads of cellular metabolism, differentiation, and transformation. *Cell*, 117, 421–426.

Alessi, D.R., Caudwell, F.B., Andjelkovic, M., Hemmings, B.A. and Cohen, P. (1996) Molecular basis for the substrate specificity of protein kinase B; comparison with MAPKAP kinase-1 and p70 S6 kinase. *FEBS Lett*, 399, 333–338.

Aoki, M., Batista, O., Bellacosa, A., Tsichlis, P. and Vogt, P.K. (1998) The akt kinase: molecular determinants of oncogenicity. *Proc Natl Acad Sci USA*, 95, 14950–14955.

Asano, T., Yao, Y., Zhu, J., Li, D., Abbruzzese, J.L. and Reddy, S.A. (2004) The PI 3-kinase/Akt signaling pathway is activated due to aberrant Pten expression and targets transcription factors NF-kappaB and c-Myc in pancreatic cancer cells. *Oncogene*, 23, 8571–8580.

Bachman, K.E., Argani, P., Samuels, Y., et al. (2004) The PIK3CA gene is mutated with high frequency in human breast cancers. *Cancer Biol Ther*, 3, 772–775.

Bader, A.G., Kang, S. and Vogt, P.K. (2006) Cancer-specific mutations in PIK3CA are oncogenic *in vivo. Proc Natl Acad Sci USA*, 103, 1475–1479.

Baker, S.J. (2007) PTEN enters the nuclear age. *Cell*, 128, 25–28.

Bayascas, J.R., Leslie, N.R., Parsons, R., Fleming, S. and Alessi, D.R. (2005) Hypomorphic mutation of PDK1 suppresses tumorigenesis in PTEN(+/-) mice. *Curr Biol*, 15, 1839–1846.

Bellacosa, A., Testa, J.R., Staal, S.P. and Tsichlis, P.N. (1991) A retroviral oncogene, akt, encoding a serine-threonine kinase containing an SH2-like region. *Science*, 254, 274–277.

Blanco-Aparicio, C., Renner, O., Leal, J.F. and Carnero, A. (2007) PTEN, more than the AKT pathway. *Carcinogenesis*, 28, 1379–1386.

Bos, J.L. (1989) ras oncogenes in human cancer: a review. *Cancer Res*, 49, 4682–4689.

Brazil, D.P. and Hemmings, B.A. (2001) Ten years of protein kinase B signalling: a hard Akt to follow. *Trends Biochem Sci*, 26, 657–664.

Broderick, D.K., Di, C., Parrett, T.J., et al. (2004) Mutations of PIK3CA in anaplastic oligodendrogliomas, high-grade astrocytomas, and medulloblastomas. *Cancer Res*, 64, 5048–5050.

Calera, M.R., Martinez, C., Liu, H., et al. (1998) Insulin increases the association of Akt-2 with Glut4-containing vesicles. *J Biol Chem*, 273, 7201–7204.

Campbell, I.G., Russell, S.E., Choong, D.Y., et al. (2004) Mutation of the PIK3CA gene in ovarian and breast cancer. *Cancer Res*, 64, 7678–7681.

Cardone, M.H., Roy, N., Stennicke, H.R., et al. (1998) Regulation of cell death protease caspase-9 by phosphorylation. *Science*, 282, 1318–1321.

Carnero, A. and Beach, D.H. (2004) Absence of p21WAF1 cooperates with c-myc in bypassing Ras-induced senescence and enhances oncogenic cooperation. *Oncogene*, 23, 6006–6011.

Carnero, A., Blanco-Aparicio, C., Renner, O., Link, W. and Leal, J.F. (2008) The PTEN/PI3K/AKT signalling pathway in cancer, therapeutic implications. *Curr Cancer Drug Targets*, 8, 187–198.

Carnero, A. and Lacal, J.C. (1995) Activation of intracellular kinases in *Xenopus* oocytes by p21ras and phospholipases: a comparative study. *Mol Cell Biol*, 15, 1094–1101.

Carnero, A., Liyanage, M., Stabel, S. and Lacal, J.C. (1995) Evidence for different signalling pathways of PKC zeta and ras-p21 in *Xenopus* oocytes. *Oncogene*, 11, 1541–1547.

Carpten, J.D., Faber, A.L., Horn, C., et al. (2007) A transforming mutation in the pleckstrin homology domain of AKT1 in cancer. *Nature*, 448, 439–444.

Chang, C.J., Freeman, D.J. and Wu, H. (2004) PTEN regulates Mdm2 expression through the P1 promoter. *J Biol Chem*, 279, 29841–29848.

Chen, Z., Trotman, L.C., Shaffer, D., et al. (2005) Crucial role of p53-dependent cellular senescence in suppression of Pten-deficient tumorigenesis. *Nature*, 436, 725–730.

Dahia, P.L. (2000) PTEN, a unique tumor suppressor gene. *Endocr Relat Cancer*, 7, 115–129.

Datta, S.R., Brunet, A. and Greenberg, M.E. (1999) Cellular survival: a play in three Akts. *Genes Dev*, 13, 2905–2927.

Datta, S.R., Dudek, H., Tao, X., et al. (1997) Akt phosphorylation of BAD couples survival signals to the cell-intrinsic death machinery. *Cell*, 91, 231–241.

Dijkers, P.F., Birkenkamp, K.U., Lam, E.W., et al. (2002) FKHR-L1 can act as a critical effector of cell death induced by cytokine withdrawal: protein kinase B-enhanced cell survival through maintenance of mitochondrial integrity. *J Cell Biol*, 156, 531–542.

Downward, J. (2003) Targeting RAS signalling pathways in cancer therapy. *Nat Rev Cancer*, 3, 11–22.

Downward, J. (2004) PI 3-kinase, Akt and cell survival. *Semin Cell Dev Biol*, 15, 177–182.

Dummler, B. and Hemmings, B.A. (2007) Physiological roles of PKB/Akt isoforms in development and disease. *Biochem Soc Trans*, 35, 231–235.

Dummler, B., Tschopp, O., Hynx, D., et al. (2006) Life with a single isoform of Akt: mice lacking Akt2 and Akt3 are viable but display impaired glucose homeostasis and growth deficiencies. *Mol Cell Biol*, 26, 8042–8051.

Feig, L.A. (2003) Ral-GTPases: approaching their 15 minutes of fame. *Trends Cell Biol*, 13, 419–425.

Freeman, D.J., Li, A.G., Wei, G., et al. (2003) PTEN tumor suppressor regulates p53 protein levels and activity through phosphatase-dependent and -independent mechanisms. *Cancer Cell*, 3, 117–130.

Fujita, T., Doihara, H., Kawasaki, K., et al. (2006) PTEN activity could be a predictive marker of trastuzumab efficacy in the treatment of ErbB2-overexpressing breast cancer. *Br J Cancer*, 94, 247–252.

Garcia-Rostan, G., Costa, A.M., Pereira-Castro, I., et al. (2005) Mutation of the PIK3CA gene in anaplastic thyroid cancer. *Cancer Res*, 65, 10199–10207.

Gil, A., Andres-Pons, A. and Pulido, R. (2007) Nuclear PTEN: a tale of many tails. *Cell Death Differ*, 14, 395–399.

Gottlieb, T.M., Leal, J.F., Seger, R., Taya, Y. and Oren, M. (2002) Cross-talk between Akt, p53 and Mdm2: possible implications for the regulation of apoptosis. *Oncogene*, 21, 1299–1303.

Gu, J., Tamura, M., Pankov, R., et al. (1999) Shc and FAK differentially regulate cell motility and directionality modulated by PTEN. *J Cell Biol*, 146, 389–403.

Gu, J., Tamura, M. and Yamada, K.M. (1998) Tumor suppressor PTEN inhibits integrin- and growth factor-mediated mitogen-activated protein (MAP) kinase signaling pathways. *J Cell Biol*, 143, 1375–1383.

Kandel, E.S. and Hay, N. (1999) The regulation and activities of the multifunctional serine/threonine kinase Akt/PKB. *Exp Cell Res*, 253, 210–229.

Kang, S., Bader, A.G. and Vogt, P.K. (2005) Phosphatidylinositol 3-kinase mutations identified in human cancer are oncogenic. *Proc Natl Acad Sci USA*, 102, 802–807.

Kennedy, S.G., Wagner, A.J., Conzen, S.D., et al. (1997) The PI 3-kinase/Akt signaling pathway delivers an anti-apoptotic signal. *Genes Dev*, 11, 701–713.

Kirkegaard, T., Witton, C.J., McGlynn, L.M., et al. (2005) AKT activation predicts outcome in breast cancer patients treated with tamoxifen. *J Pathol*, 207, 139–146.

Kohn, A.D., Summers, S.A., Birnbaum, M.J. and Roth, R.A. (1996) Expression of a constitutively active Akt Ser/Thr kinase in 3T3-L1 adipocytes stimulates glucose uptake and glucose transporter 4 translocation. *J Biol Chem*, 271, 31372–31378.

Kovacina, K.S., Park, G.Y., Bae, S.S., et al. (2003) Identification of a proline-rich Akt substrate as a 14-3-3 binding partner. *J Biol Chem*, 278, 10189–10194.

Li, A.G., Piluso, L.G., Cai, X., et al. (2006) Mechanistic insights into maintenance of high p53 acetylation by PTEN. *Mol Cell*, 23, 575–587.

Li, J., Yen, C., Liaw, D., et al. (1997) PTEN, a putative protein tyrosine phosphatase gene mutated in human brain, breast, and prostate cancer. *Science*, 275, 1943–1947.

Lian, Z. and Di Cristofano, A. (2005) Class reunion: PTEN joins the nuclear crew. *Oncogene*, 24, 7394–7400.

Liang, J., Zubovitz, J., Petrocelli, T., et al. (2002) PKB/Akt phosphorylates p27, impairs nuclear import of p27 and opposes p27-mediated G1 arrest. *Nat Med*, 8, 1153–1160.

Liaw, D., Marsh, D.J., Li, J., et al. (1997) Germline mutations of the PTEN gene in Cowden disease, an inherited breast and thyroid cancer syndrome. *Nat Genet*, 16, 64–67.

Lim, K.H. and Counter, C.M. (2005) Reduction in the requirement of oncogenic Ras signaling to activation of PI3K/AKT pathway during tumor maintenance. *Cancer Cell*, 8, 381–392.

Lindsay, Y., McCoull, D., Davidson, L., et al. (2006) Localization of agonist-sensitive PtdIns(3,4,5)P3 reveals a nuclear pool that is insensitive to PTEN expression. *J Cell Sci*, 119, 5160–5168.

Link, W., Rosado, A., Fominaya, J., Thomas, J.E. and Carnero, A. (2005) Membrane localization of all class I PI 3-kinase isoforms suppresses c-Myc-induced apoptosis in Rat1 fibroblasts via Akt. *J Cell Biochem*, 95, 979–989.

Lum, J.J., Bui, T., Gruber, M., et al. (2007) The transcription factor HIF-1alpha plays a critical role in the growth factor-dependent regulation of both aerobic and anaerobic glycolysis. *Genes Dev*, 21, 1037–1049.

Ma, X., Ziel-van der Made, A.C., Autar, B., et al. (2005) Targeted biallelic inactivation of Pten in the mouse prostate leads to prostate cancer accompanied by increased epithelial cell proliferation but not by reduced apoptosis. *Cancer Res*, 65, 5730–5739.

Maehama, T. and Dixon, J.E. (1998) The tumor suppressor, PTEN/MMAC1, dephosphorylates the lipid second messenger, phosphatidylinositol 3,4,5-trisphosphate. *J Biol Chem*, 273, 13375–13378.

Majumder, P.K., Febbo, P.G., Bikoff, R., et al. (2004) mTOR inhibition reverses Akt-dependent prostate intraepithelial neoplasia through regulation of apoptotic and HIF-1-dependent pathways. *Nat Med*, 10, 594–601.

Manning, B.D. and Cantley, L.C. (2007) AKT/PKB signaling: navigating downstream. *Cell*, 129, 1261–1274.

Marsh, D.J., Kum, J.B., Lunetta, K.L., et al. (1999) PTEN mutation spectrum and genotype-phenotype correlations in Bannayan–Riley–Ruvalcaba syndrome suggest a single entity with Cowden syndrome. *Hum Mol Genet*, 8, 1461–1472.

Mayo, L.D. and Donner, D.B. (2001) A phosphatidylinositol 3-kinase/Akt pathway promotes translocation of Mdm2 from the cytoplasm to the nucleus. *Proc Natl Acad Sci USA*, 98, 11598–11603.

Mayo, L.D. and Donner, D.B. (2002) The PTEN, Mdm2, p53 tumor suppressor-oncoprotein network. *Trends Biochem Sci*, 27, 462–467.

McFarlin, D.R. and Gould, M.N. (2003) Rat mammary carcinogenesis induced by in situ expression of constitutive Raf kinase activity is prevented by tethering Raf to the plasma membrane. *Carcinogenesis*, 24, 1149–1153.

McFarlin, D.R., Lindstrom, M.J. and Gould, M.N. (2003) Affinity with Raf is sufficient for Ras to efficiently induce rat mammary carcinomas. *Carcinogenesis*, 24, 99–105.

Morrison, D.K. and Cutler, R.E. (1997) The complexity of Raf-1 regulation. *Curr Opin Cell Biol*, 9, 174–179.

Myers, M.P., Pass, I., Batty, I.H., et al. (1998) The lipid phosphatase activity of PTEN is critical for its tumor supressor function. *Proc Natl Acad Sci USA*, 95, 13513–13518.

Myers, M.P., Stolarov, J.P., Eng, C., et al. (1997) P-TEN, the tumor suppressor from human chromosome 10q23, is a dual-specificity phosphatase. *Proc Natl Acad Sci USA*, 94, 9052–9057.

Nagata, Y., Lan, K.H., Zhou, X., et al. (2004) PTEN activation contributes to tumor inhibition by trastuzumab, and loss of PTEN predicts trastuzumab resistance in patients. *Cancer Cell*, 6, 117–127.

Nakayama, K., Nakayama, N., Kurman, R.J., et al. (2006) Sequence mutations and amplification of PIK3CA and AKT2 genes in purified ovarian serous neoplasms. *Cancer Biol Ther*, 5, 779–785.

Nelen, M.R., van Staveren, W.C., Peeters, E.A., et al. (1997) Germline mutations in the PTEN/MMAC1 gene in patients with Cowden disease. *Hum Mol Genet*, 6, 1383–1387.

Okano, J., Snyder, L. and Rustgi, A.K. (2003) Genetic alterations in esophageal cancer. *Methods Mol Biol*, 222, 131–145.

Okumura, K., Zhao, M., DePinho, R.A., Furnari, F.B. and Cavenee, W.K. (2005) PTEN: a novel anti-oncogenic function independent of phosphatase activity. *Cell Cycle*, 4, 540–542.

Olsson, A.K., Dimberg, A., Kreuger, J. and Claesson-Welsh, L. (2006) VEGF receptor signalling – in control of vascular function. *Nat Rev Mol Cell Biol*, 7, 359–371.

Oren, M., Damalas, A., Gottlieb, T., et al. (2002) Regulation of p53: intricate loops and delicate balances. *Biochem Pharmacol*, 64, 865–871.

Ozes, O.N., Mayo, L.D., Gustin, J.A., et al. (1999) NF-kappaB activation by tumour necrosis factor requires the Akt serine-threonine kinase. *Nature*, 401, 82–85.

Parsons, D.W., Wang, T.L., Samuels, Y., et al. (2005) Colorectal cancer: mutations in a signalling pathway. *Nature*, 436, 792.

Plas, D.R. and Thompson, C.B. (2005) Akt-dependent transformation: there is more to growth than just surviving. *Oncogene*, 24, 7435–7442.

Puc, J. and Parsons, R. (2005) PTEN loss inhibits CHK1 to cause double stranded-DNA breaks in cells. *Cell Cycle*, 4, 927–929.

Radu, A., Neubauer, V., Akagi, T., Hanafusa, H. and Georgescu, M.M. (2003) PTEN induces cell cycle arrest by decreasing the level and nuclear localization of cyclin D1. *Mol Cell Biol*, 23, 6139–6149.

Ramjaun, A.R. and Downward, J. (2007) Ras and phosphoinositide 3-kinase: partners in development and tumorigenesis. *Cell Cycle*, 6, 2902–2905.

Rangarajan, A., Hong, S.J., Gifford, A. and Weinberg, R.A. (2004) Species- and cell type-specific requirements for cellular transformation. *Cancer Cell*, 6, 171–183.

Robey, R.B. and Hay, N. (2006) Mitochondrial hexokinases, novel mediators of the antiapoptotic effects of growth factors and Akt. *Oncogene*, 25, 4683–4696.

Sabatini, D.M. (2006) mTOR and cancer: insights into a complex relationship. *Nat Rev Cancer*, 6, 729–734.

Samuels, Y. and Ericson, K. (2006) Oncogenic PI3K and its role in cancer. *Curr Opin Oncol*, 18, 77–82.

Samuels, Y., Wang, Z., Bardelli, A., et al. (2004) High frequency of mutations of the PIK3CA gene in human cancers. *Science*, 304, 554.

Sancak, Y., Thoreen, C.C., Peterson, T.R., et al. (2007) PRAS40 is an insulin-regulated inhibitor of the mTORC1 protein kinase. *Mol Cell*, 25, 903–915.

Sarkisian, C.J., Keister, B.A., Stairs, D.B., et al. (2007) Dose-dependent oncogene-induced senescence *in vivo* and its evasion during mammary tumorigenesis. *Nat Cell Biol*, 9, 493–505.

Schrager, C.A., Schneider, D., Gruener, A.C., Tsou, H.C. and Peacocke, M. (1998) Clinical and pathological features of breast disease in Cowden's syndrome: an underrecognized syndrome with an increased risk of breast cancer. *Hum Pathol*, 29, 47–53.

Shen, W.H., Balajee, A.S., Wang, J., et al. (2007) Essential role for nuclear PTEN in maintaining chromosomal integrity. *Cell*, 128, 157–170.

Shields, J.M., Pruitt, K., McFall, A., Shaub, A. and Der, C.J. (2000) Understanding Ras: "it ain't over 'til it's over". *Trends Cell Biol*, 10, 147–154.

Shin, I., Yakes, F.M., Rojo, F., et al. (2002) PKB/Akt mediates cell-cycle progression by phosphorylation of p27(Kip1) at threonine 157 and modulation of its cellular localization. *Nat Med*, 8, 1145–1152.

Stambolic, V., Suzuki, A., de la Pompa, J.L., et al. (1998) Negative regulation of PKB/Akt-dependent cell survival by the tumor suppressor PTEN. *Cell*, 95, 29–39.

Stiles, B., Gilman, V., Khanzenzon, N., et al. (2002) Essential role of AKT-1/protein kinase B alpha in PTEN-controlled tumorigenesis. *Mol Cell Biol*, 22, 3842–3851.

Stiles, B., Groszer, M., Wang, S., Jiao, J. and Wu, H. (2004) PTENless means more. *Dev Biol*, 273, 175–184.

Stokoe, D. (2001) Pten. *Curr Biol*, 11, R502.

Sun, H., Lesche, R., Li, D.M., et al. (1999) PTEN modulates cell cycle progression and cell survival by regulating phosphatidylinositol 3,4,5,-trisphosphate and Akt/protein kinase B signaling pathway. *Proc Natl Acad Sci USA*, 96, 6199–6204.

Tamura, M., Gu, J., Takino, T. and Yamada, K.M. (1999) Tumor suppressor PTEN inhibition of cell invasion, migration, and growth: differential involvement of focal adhesion kinase and p130Cas. *Cancer Res*, 59, 442–449.

Toker, A. and Yoeli-Lerner, M. (2006) Akt signaling and cancer: surviving but not moving on. *Cancer Res*, 66, 3963–3966.

Tran, H., Brunet, A., Griffith, E.C. and Greenberg, M.E. (2003) The many forks in FOXO's road. *Sci STKE*, 2003, RE5.

Ulku, A.S. and Der, C.J. (2003) Ras signaling, deregulation of gene expression and oncogenesis. *Cancer Treat Res*, 115, 189–208.

Viglietto, G., Motti, M.L., Bruni, P., et al. (2002) Cytoplasmic relocalization and inhibition of the cyclin-dependent kinase inhibitor p27(Kip1) by PKB/Akt-mediated phosphorylation in breast cancer. *Nat Med*, 8, 1136–1144.

Villunger, A., Michalak, E.M., Coultas, L., et al. (2003) p53- and drug-induced apoptotic responses mediated by BH3-only proteins puma and noxa. *Science*, 302, 1036–1038.

Vivanco, I. and Sawyers, C.L. (2002) The phosphatidylinositol 3-Kinase AKT pathway in human cancer. *Nat Rev Cancer*, 2, 489–501.

Weng, L., Brown, J. and Eng, C. (2001a) PTEN induces apoptosis and cell cycle arrest through phosphoinositol-3-kinase/Akt-dependent and -independent pathways. *Hum Mol Genet*, 10, 237–242.

Weng, L.P., Brown, J.L. and Eng, C. (2001b) PTEN coordinates G(1) arrest by down-regulating cyclin D1 via its protein phosphatase activity and up-regulating p27 via its lipid phosphatase activity in a breast cancer model. *Hum Mol Genet*, 10, 599–604.

Wullschleger, S., Loewith, R. and Hall, M.N. (2006) TOR signaling in growth and metabolism. *Cell*, 124, 471–484.

Xing, D. and Orsulic, S. (2005) A genetically defined mouse ovarian carcinoma model for the molecular characterization of pathway-targeted therapy and tumor resistance. *Proc Natl Acad Sci USA*, 102, 6936–6941.

Yuan, T.L. and Cantley, L.C. (2008) PI3K pathway alterations in cancer: variations on a theme. *Oncogene*, 27, 5497–5510.

Zhao, J.J., Liu, Z., Wang, L., et al. (2005) The oncogenic properties of mutant p110alpha and p110beta phosphatidylinositol 3-kinases in human mammary epithelial cells. *Proc Natl Acad Sci USA*, 102, 18443–18448.

Zhou, B.P. and Hung, M.C. (2002) Novel targets of Akt, p21(Cipl/WAF1), and MDM2. *Semin Oncol*, 29, 62–70.

Zhou, M., Gu, L., Findley, H.W., Jiang, R. and Woods, W.G. (2003) PTEN reverses MDM2-mediated chemotherapy resistance by interacting with p53 in acute lymphoblastic leukemia cells. *Cancer Res*, 63, 6357–6362.

Zimmermann, S. and Moelling, K. (1999) Phosphorylation and regulation of Raf by Akt (protein kinase B). *Science*, 286, 1741–1744.

14

TGFβ and BMP signaling in cancer

Panagiotis Papageorgis, Arthur W. Lambert, Sait Ozturk
and Sam Thiagalingam

Introduction

Cancer progression is modulated by aberrant expression and secretion of cytokines at various stages. TGFβ and BMP belong to a superfamily of around 40 secreted cytokines that regulate a plethora of biological responses in normal as well as in cancer cells. Studies have shown that these molecules can regulate a large number of processes such as cell proliferation, apoptosis, senescence, differentiation, angiogenesis, immunosuppression, cell migration, and cancer metastasis. Recent studies have shed light into the molecular mechanisms and signaling networks that govern the effects of these pivotal pathways during cancer progression. Therefore it is becoming increasingly clear that unraveling the mechanistic complexity and clinical relevance of these pathways will greatly enhance our therapeutic efforts against tumor development and evolution of malignant cells.

TGFβ synthesis and activation

Most of the knowledge pertaining to the TGFβ superfamily of cytokines has been elucidated from studies regarding the TGFβ isoforms. There are three TGFβ isoforms, TGFβ1, TGFβ2, and TGFβ3, which are initially synthesized as inactive 75-kDa homodimeric pro-proteins, known as pro-TGFβ. These propeptides, referred to as the latency-associated proteins (LAPs), are part of the TGFβ large latent complex (LLC) which consists of LAPs and latent TGFβ binding proteins (LTBPs) assembled together by the formation of disulfide bonds between cysteine residues [1–3]. LTBPs are members of the LTBP/fibrillin protein family, which consists of fibrillin-1, 2, and 3 as well as LTBP-1, 2, 3, and 4. Out of these proteins, LTBP-1, 3, and 4 have the unique ability to bind LAP through the third of their four 8-cystein domains [4].

The remaining cysteine domains are likely to localize LTBPs to the extracellular matrix (ECM) [5]. As a part of the LLC, TGFβ remains in an inactive form. In this state, LAPs form a non-covalent, high-affinity association with TGFβ preventing the receptor–ligand interaction [6]. LLC is primarily localized at the matrix via covalent association of the N-terminal region of LTBPs with ECM proteins [7]. During the activation step, LAPs undergo conformational changes induced by thrombospondin-1 (TSP-1) [8, 9] and cleavage by furins and other convertases leading to the release of the mature 24-kDa TGFβ dimer [10, 11], which can bind to and activate TGFβ receptors resulting in the propagation of downstream signaling events.

Smads mediate TGFβ signaling

Smads comprise a small family of structurally related proteins that play a pivotal role in mediating intracellular TGFβ signaling [12–14]. The role of vertebrate Smad proteins in TGFβ signaling was predicted from their high degree of homology to Mad (Mothers against decapentaplegic) from *Drosophila melanogaster* and the Sma2, Sma3, and Sma4 proteins from *Caenorhabditis elegans* in analogous signaling pathways [15–17]. In fact, the name Smad was derived from the combination of the homologs' names in these organisms (Sma and mad) [18]. The discovery of the human *SMAD* genes, more than ten years ago, as downstream effectors of the TGFβ signaling pathway was a major breakthrough in the understanding of the molecular basis of TGFβ-mediated effects [19–26]. Within the human genome, eight homologs of the *SMAD* genes are known to date and have been shown to function downstream of the TGFβ serine/threonine kinase receptors. In addition to the eight human *SMAD* family members, homologous genes

---

Systems Biology of Cancer, ed. S. Thiagalingam. Published by Cambridge University Press. © Cambridge University Press 2015.

are also known in mouse, rat, *Xenopus*, zebrafish, *Drosophila*, and *Caenorhabditis elegans*.

Smad proteins can be divided into three functional groups: the receptor-activated Smads (R-Smads), which includes Smad1, Smad2, Smad3, Smad5, Smad8; the common mediator Smad (Co-Smad), Smad4; and the inhibitory Smads (I-Smads), Smad6, and Smad7 [13, 27]. They have a relative mass of 42 to 60 kDa and are composed of two regions of homology (Mad Homology (MH) domains) at the amino- and carboxy-terminals of the protein, named the MH1 and MH2 domains, respectively. These domains are separated by a proline-rich acidic linker region of variable length and sequence [28–30]. The MH2 domain is involved in homo- and heteromeric complex formation, as well as in transcriptional activation and repression. On the other hand, the MH1 domain exhibits DNA binding activity [28, 31, 32]. Under basal conditions, MH1 and MH2 domains interact with each other resulting in auto-inhibition of their respective functions. TGFβ stimulation induces conformational changes to relieve this inhibition allowing the MH2 domain of R-Smads to interact with the TGFβ receptors [29, 33].

TGFβ signal transduction pathway

There are three types of TGFβ receptors (TGFβRI, TGFβRII, and TGFβRIII (also known as endoglin or betaglycan)) that transduce signals from the three different TGFβ isoforms. Seven TGFβRIs and five TGFβRIIs have been characterized to date. While the TGFβRIs include activin-like receptors 1–7 (ALK1-7), the TGFβRIIs include TGFβRII, BMPRII, ACTRII, ACTRIIB, and anti-Mullerian hormone receptor II-AMHRII [34]. In most cell types, the TGFβRII–ALK5 complex transduces the signal from all three TGFβ isoforms, whereas TGFβRII associates with ALK1 in endothelial cells and ALK2 in cell types that are related to cardiovascular development. An important difference between the alternative heteromeric TGFβRII signaling complexes that are initiated by TGFβ is the activation of different Smad family members. ALK5 activates Smad2 and Smad3 (canonical TGFβ signaling pathway) whereas ALK2, ALK3, and ALK6 activate receptor-associated Smad proteins Smad1, Smad5, and Smad8 (BMP signaling pathway) [35–41].

The active TGFβ1 ligand initiates intracellular signaling by binding to type II receptor Ser/Thr kinase (TGFβRII) on the cell membrane. This recruits type I receptor (TGFβRI also known as ALK5) in order to form a heterotetrameric receptor–ligand complex where the type II phosphorylates the type I receptor in a conserved Glycine–Serine (GS)-rich domain [42]. Activated type I receptor interacts with R-Smads (Smad2 and 3) through the MH2 domain [43] resulting in phosphorylation at the conserved SSXS C-terminal motif [33, 44]. Subsequently, a heterotrimeric complex between R-Smads (Smads2/3) and Co-Smad (Smad4) is formed and translocated into the nucleus to mediate differential regulation of gene expression by acting as transcription factors in cooperation with coactivators, such as CBP/p300, P/CAF, SMIF, FoxO, Sp1, and c-Jun/c-Fos or corepressors, such as E2F4/5-p107, ATF3, TGIF, Ski, SnoN, and FoxG1 [31, 44–57]. Expression of the inhibitory Smad (I-Smad), Smad7, is also induced by TGFβ signaling and acts as a negative feedback regulator of the pathway. Smad7 normally resides in the nucleus of unstimulated cells and translocates to the plasma membrane upon TGFβ-mediated receptor activation (Figure 14.1) [58]. Smad7 has been shown to inhibit TGFβ signaling by at least three different mechanisms: (1) by interfering and blocking interactions between the R-Smads and the activated receptors [22, 59]; (2) by interacting with the E3-ubiquitin ligases Smurf1 or Smurf2 in the nucleus – upon TGFβ stimulation, the Smad7–Smurf complex translocates to the plasma membrane, where Smurf induces ubiquitination and degradation of the TGFβ receptors; (3) binding to DNA via its MH2 domain and therefore blocking the TGFβ pathway by antagonizing the formation of a functional Smad–DNA complex formation [60–62].

SARA (Smad anchor for receptor activation) is an FYVE domain-containing protein, which only interacts with the MH2 domain of Smad2 and Smad3. It preferentially associates with unphosphorylated Smad2 and is released upon Smad3 phosphorylation by TGFβRI. The FYVE domain of SARA is involved in the localization of the R-Smads to the plasma membrane to facilitate receptor-mediated phosphorylation [63].

Recent studies have also indicated that TGFβ-mediated effects could be exerted through non-canonical pathways. Contrary to the *in vivo* ablation of the TGFβ receptors [64–66] studies have shown that disruption of Smad4 in the mammary gland, heart, or pancreas of mice does not impair the development of these organs. TIF1γ (transcription

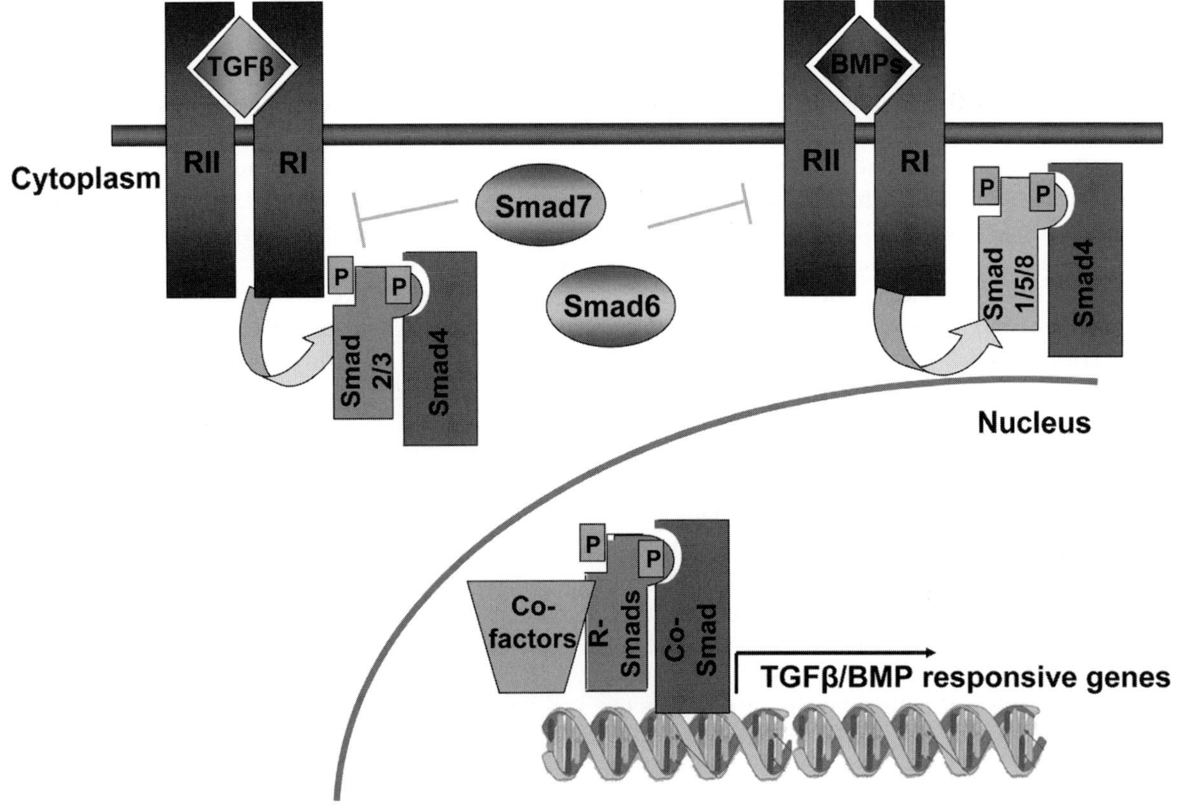

Figure 14.1 TGFβ and BMP signal transduction pathways. (A black and white version of this figure will appear in some formats. For the color version, please refer to the plate section.)

intermediate factor 1γ), which participates in erythroid differentiation, was identified as a Smad2/3-interaction partner competing against Smad4 [67]. These observations support the notion that Smad4 is essential for many but not all TGFβ-regulated transcriptional responses. Moreover, Smads2/3 were also found to interact with IκB kinase α (IKKα), in a Smad4-independent manner, to regulate the expression of Mad1, a Myc oncogene antagonist, and control keratinocyte differentiation [68].

The bone morphogenetic protein (BMP) signaling pathway

Members of the TGFβ superfamily, bone morphogenetic proteins (BMPs) were originally identified as factors capable of inducing bone and cartilage formation [69]. Subsequently, more than 20 BMPs have been characterized and clearly demonstrated to be involved in various processes including skeletal development and fracture repair [70].

Similar to TGFβ, BMPs signal through type II and type I serine/threonine kinase receptors; however, they bind to a distinct subset of these receptors and activate different downstream effectors. BMPs can bind to three different type II receptors (BMPR-II, ActR-II, and ActR-IIB) and a similar number of alternate type I receptors (ALK-2, ALK-3, and ALK-6) to initiate signal transduction [41]. Binding of the BMP ligand facilitates the interaction of the two kinase receptors and activation of the type I receptor but in contrast to TGFβ, BMPs have an increased affinity for the type I receptors and initially form a complex with them before binding to the type II receptor [71]. Activation of the type I receptor kinase by BMP results in the recruitment and phosphorylation of R-Smads distinct from those activated by TGFβ, which include Smad1, Smad5, and Smad8 (Figure 14.1) [70]. Phosphorylated Smad1/5/8 can interact with Smad4 and translocate to the nucleus to positively or negatively regulate the transcription of specific downstream genes [72]. Targets of particular interest include the Id (inhibitors of differentiation) proteins [1–4, 41], which act to maintain an undifferentiated state and have been implicated in tumorigenesis [73]. Furthermore, mutations in the BMP signaling pathway have been described in patients with juvenile polyposis [74], a condition that leads to the development of hamartomas, although their relevance to sporadic colon cancer has been debated [75]. Intriguingly, epigenetic silencing of Smad8 by DNA methylation has been reported in breast and colon tumors [76] suggesting a tumor suppressive role.

TGFβ in development – mouse models

The emergence of TGFβ expression and activation of the Smad pathway begins early in development. This cytokine is critically important for a number of biological functions during development, including gastrulation, neural crest and anterior–posterior axis formation, vasculogenesis, hematopoiesis, and cardiovascular system development. Because the TGFβ pathway is so vital for the normal development of an organism, its activity is tightly regulated by elaborate mechanisms in order to produce meaningful signals that the cells can properly respond to [77]. Often these signals superimpose or merge with others from auxiliary pathways, such as Wnt, in a timely and spatially defined manner. On the other hand, BMP signaling has been implicated in gastrulation, organ patterning [78], neural crest formation [79], and is also critically important for bone and cartilage differentiation [80].

The fundamental role of these signaling pathways during embryonic development has been extensively studied *in vivo* by several gene-targeting studies in mice. In summary, the phenotypes of the knockout mice that are currently available for the TGFβ/Smad signaling components are listed in Table 14.1.

Overexpression of the inhibitory Smad (I-Smad), Smad7, in transgenic mice results in several pathological abnormalities in a number of tissues, including corneal defects, delayed and aberrant hair follicle morphogenesis, and hyperproliferation of the epidermis, followed by death ten days after birth [102].

TGFβ in cell proliferation and apoptosis

It has long been noted that TGFβ has a cytostatic effect on normal epithelial tissues. TGFβ can induce the expression of genes involved in anti-proliferative

Table 14.1 TGFβ and SMAD knockout mice studies.

Gene	–/– Phenotype	References
TGFβ1	50% die due to defects in yolk sac vasculogenesis and hematopoiesis while the rest develop multifocal inflammatory disease.	[81, 82]
TGFβ2	Death before or after birth due to congenital cyanosis and multiple developmental defects.	[83]
TGFβ3	Death 24 h after birth – cleft palate and abnormal pulmonary histology.	[84]
TGFβRI	Embryonic lethality at E10.5. Defects in vascular development of the yolk sac but intact hematopoiesis.	[85]
TGFβRII	Embryonic lethality at E10.5. Defects in yolk sac hematopoiesis and vasculogenesis.	[86]
SMAD1	Impaired primordial germ cell development and short allantois.	[87]
SMAD2	Embryonic lethality by E7.5–E8.5. Defects in egg cylinder elongation, mesoderm formation, and gastrulation. Defective anterior–posterior axis formation.	[88–91]
SMAD3	Death within 1 to 10 months. Metastatic intestinal cancer at 4 to 6 months of age. Immune dysregulation, severe mucosal infection, accelerated wound healing, osteoporosis, and skeletal defects.	[92–95]
SMAD4	Embryonic lethality by E7.5–E8.5. Growth retardation, abnormal endoderm formation, abnormalities in anterior–posterior axis formation.	[96, 97]
SMAD5	Embryonic lethality by E9.5–E10.5. Defects in angiogenesis and in gut, heart, and craniofacial development.	[98–100]
SMAD6	Cardiovascular abnormalities.	[101]

cell responses in all phases of the cell cycle, but primarily targets G1 to S phase events [103]. This is achieved predominantly by two major mechanisms: (1) induction of the cyclin-dependent kinase inhibitors *CDKN2B* (encoding INK4B/p15) [104], *CDKN1A* (encoding Waf1/Cip/p21) [105], and p27/Kip1 [106]; and (2) the repression of proliferation-inducing transcription factors c-Myc [107], Id1, Id2, and Id3 [50]. TGFβ signaling was shown to induce the expression and protein stability of p15, enhancing the formation of p15–cdk4 complexes and therefore inhibiting cyclin D1–cdk4 association in mammary epithelial cells [108]. Under normal conditions, during the early G1 phase, cyclin D1–cdk4 complex formation is required for mitogen sensing and progression through the S phase. However, upon p15 induction by TGFβ, p15 binds cdk4 inhibiting its catalytic activity and preventing association with cyclin D1, resulting in cell cycle arrest. TGFβ can also inhibit G1/S phase progression by inhibiting the formation of cyclin E–cdk2 and cyclin A–cdk2 via induction of p21 and p27, which bind to these cyclin–CDK complexes causing their functional inactivation [109, 110]. Furthermore, recent genome-wide transcriptional profiling studies using normal human epithelial cell lines from mammary gland, skin, and lung have identified a common set of genes that are transcriptionally regulated by TGFβ in order to mediate its cytostatic effects. This transcriptional program includes upregulation of p15 and p21 and downregulation of c-Myc, Id1, Id2, and Id3 [50]. However, the effect of TGFβ in proliferation may vary in different cell types. For example, while TGFβ can inhibit proliferation of epithelial, endothelial, neuronal, hematopoietic, and T cells, it has also been shown to enhance the proliferation of fibroblasts [111].

Under physiological conditions, TGFβ is also known to trigger apoptosis through molecular mechanisms that are not entirely clear. Even though the induction of TGFβ-mediated apoptosis remains to be established *in vivo*, studies using cell lines have identified several candidate mediators linked to this effect. Proposed mechanisms include the induction of death-associated protein kinase Dapk in a hepatoma cell line, the signaling factor Gadd45b, the death receptor Fas, and the proapoptotic molecule Bim [112].

TGFβ in immunosuppression

The maintenance of a delicate balance in the immune system is critically important in order to sustain an effective defense against foreign pathogens as well as to prevent autoimmune disease. This is achieved through a complex network interplay between several cytokines, including TGFβ [113]. The fact that TGFβ1–/– mice die shortly after birth due to severe inflammation in several organs provided the first evidence for the important role of TGFβ1 in maintaining immune system homeostasis [81, 82]. Similarly, the blockade of TGFβ signaling in T cells by transfection of a dominant-negative TGFβRII or in bone marrow by conditional knockout of the TGFβRII results in similar multifocal inflammatory responses [114, 115].

It is now widely accepted that TGFβ plays a critical role in suppressing the immune system and therefore preventing an autoimmune response. This is achieved by affecting the properties of various types of cells of the immune system (see below). On the other hand, overexpression of TGFβ, which is commonly observed in most cancers, can lead to immunosuppression and attenuation of anti-tumor immune system responses. As a result, cancer cells can escape immune surveillance and initiate malignant progression [116].

TGFβ exerts its effects on the immune system predominantly by regulating the functions of T lymphocytes – which can differentiate into effector cytotoxic T lymphocytes (CD8+; CTLs) or helper (CD4+; Th1 and Th2) T cells during an immune response – and natural killer (NK) cells, which are mainly responsible for the prevention, killing, and clearance of the tumor cells. More than 20 years ago, *in vitro* studies gave the initial clues for the anti-proliferative effects of TGFβ on T cells [117]. Studies using transgenic mice expressing a dominant-negative TGFβRII specifically in either CD4+ or CD8+ T cells have shown that both T-lymphocyte populations are important targets for TGFβ-mediated immune suppression [118]. The inhibition of proliferation is mediated through suppression of IL-2 production, a lymphokine capable of activating T and NK cells [119, 120], as well as through upregulation of cyclin-dependent kinase inhibitors p15, p21, and p27 and suppression of cell cycle-promoting transcription factors c-Myc, cyclin D2, cyclin E, and cyclin-dependent kinase 2 (CDK2) [104–107]. Besides proliferation, TGFβ also regulates T-cell function by inhibiting the expression of IFNγ and perforin, which are CTL effector molecules [121, 122]. TGFβ is also able to inhibit T-cell activation through negative regulation on antigen-presenting cells (APCs), such as dendritic cells, which have the ability to mature and effectively stimulate T cells during an immune response [123].

TGFβ further contributes to immune evasion by promoting the generation of regulatory T cells (Tregs), which are found in large numbers in the peripheral blood, lymph nodes, and tumor sites of cancer patients [124]. Moreover, TGFβ attenuates the activity of the natural killer (NK) cells, which participate in the innate and early immune defense by recognizing and eliminating infected or neoplastic cells. This is mediated by blocking the production of IFNγ by NK cells, which is considered essential for the NK-mediated responses [117]. Collectively, this evidence highlights a crucial role for TGFβ in facilitating cancer progression through host immunosuppression.

TGFβ signaling in cancer progression and metastasis

It is well established that TGFβ plays dual roles during carcinogenesis: an early tumor suppressive role, by promoting growth inhibition in normal epithelial cells and incipient tumors, and a late-stage pro-oncogenic/pro-metastatic role that accompanies a progressive increase in the locally secreted TGFβ levels [125–127]. Whereas in normal cells the TGFβ-mediated anti-proliferative effects are often dominant over opposing mitogenic signals, in cancer cells the potent mitogenic action of certain oncogenes can overwhelm the anti-mitogenic capacity of TGFβ. Therefore one of the hallmarks of all cancers is that the vast majority of cases exhibit insensitivity to TGFβ-mediated growth inhibition. Recent findings provide molecular evidence to explain the reasons for the switch in TGFβ functions and suggest that mutant p53 can form a complex with Smads under the influence of oncogenic Ras to empower TGFβ-induced metastasis by opposing p63 activity [128]. This elaborate model is a great example of how interconnected signaling pathways can often coordinately modulate the function of key signaling molecules, such as TGFβ, to regulate cancer progression and metastasis.

Moreover, TGFβ overexpression is often detected in several metastatic human tumor types, including colon, mammary, prostate, and renal cell cancers [129–133]. Increased levels of TGFβ are usually detected locally in the microenvironment surrounding the tumor and in the tumor stroma [134]. The excessive amounts of TGFβ are believed to enhance tumor progression by promoting local tissue invasion and by inducing tumor angiogenesis and metastasis [135, 136]. Recent evidence suggests that increased LRG1 expression in endothelial cells could modulate TGFβ signaling to further promote angiogenesis [137].

Under physiological conditions of tissue injury, TGFβ is released by blood platelets and stromal components in order to prevent uncontrolled regenerative cell proliferation and inflammation. Similarly, in premalignant tumors TGFβ is secreted in the microenvironment initially to control proliferation and cancer progression but ultimately it is utilized by malignant cells to promote their invasive and metastatic properties. The detection of high levels of this cytokine by immunohistochemical analysis in several tumor samples from patients with poor outcome [138] clearly indicates the prominent association of TGFβ with cancer progression.

TGFβ in tumors could be derived from several different sources: (1) epithelial cancer cells themselves secrete this cytokine affecting their properties within the tumor mass in an autocrine or paracrine fashion [139]; (2) the presence of various cell types of the stroma infiltrating the tumor, including leukocytes, macrophages, bone marrow-derived endothelial, mesenchymal and myeloid precursor cells, coincides with TGFβ secretion in the leading front of the tumor and is therefore a suspected source of this cytokine [130]; and (3) the bone matrix has the ability to store TGFβ, which can be mobilized during development of osteolytic metastatic lesions [140].

In addition to tumor cells, TGFβ can have profound effects on the surrounding microenvironment. Stromal cells can play an active role during cancer progression as transformed stroma has been shown to promote epithelial tumorigenesis [141]. Interestingly, TGFβ signaling in fibroblasts seems to have an important role in maintaining proper stromal–epithelial interaction. This was demonstrated in mice with fibroblast-specific conditional inactivation of the type II receptor (TβRII). The mice developed prostatic intraepithelial neoplasia and invasive squamous cell carcinoma of the forestomach demonstrating that loss of TGFβ signaling in the stroma can result in epithelial tumors [142].

TGFβ in epithelial to mesenchymal transition (EMT) and cancer stem-like cells (CSCs)

One of the major mechanisms by which TGFβ has been shown to promote cell motility, invasiveness, and metastasis of cancer cells [143] is through induction of epithelial to mesenchymal transition (EMT). Studies have shown that TGFβ stimulation of both non-transformed and carcinoma-derived cell populations in culture can lead to the activation of this reversible process [40, 135, 136]. *In vivo* studies have further shown that expression of TGFβ1 in the skin of transgenic mice enhanced the conversion of benign skin tumors to carcinomas and highly invasive spindle-cell carcinomas [144]. Moreover, expression of a dominant-negative TGFβRII prevented squamous carcinoma cells from undergoing EMT in response to TGFβ *in vivo* [145].

Epithelial to mesenchymal transition is a vital process for morphogenesis during embryonic development, which occurs via activation of a highly orchestrated program whereby epithelial cells lose polarity and cell–cell contacts and undergo dramatic remodelling of their cytoskeleton. During this process, the expression of epithelial marker genes, such as E-cadherin, P-cadherin, γ-catenin, β-catenin, claudin1, claudin4, and claudin7, is suppressed with concurrent expression of mesenchymal components, such as N-cadherin, vimentin, fibronectin, and alpha smooth-muscle actin [146]. Several pleiotropically acting transcription factors have been characterized as "master regulators" of EMT and have been shown to activate this program. Snail [147], Slug [148], ZEB1/deltaEF1 [149], ZEB2/SIP1 [150], twist [146], HMGA2 [151], and FOXC2 [152] are the best examples of such genes reported to date. In addition to these mechanisms, recent studies indicate that an overactive TGFβ–TGFβR–Smad2 signaling axis could further contribute to the manifestation of an EMT phenotype by maintaining the epigenetic silencing of epithelial genes during this process. This appears to be mediated via Smad2-dependent regulation of DNMT1 binding activity and DNA methylation of the corresponding gene promoter regions [153]. Importantly, EMT has been correlated with induction of intravasation of carcinoma *in situ* cells through the basement membrane, survival in the circulation, extravasation at the distal tissues, and formation of micrometastases [146, 154, 155].

Besides Smads, several other signaling pathways have been implicated in TGFβ-induced EMT, including PI3K–Akt, RhoA, p38-MAPK, and cofilin [139, 156, 157]. Although EMT might involve Smad-independent TGFβ signals, studies using mutant TGFβRI constructs that are defective in binding Smads, but that can still signal via MAPKs, show that Smads are required for the EMT process [158, 159]. These results are consistent with reports demonstrating cooperation between the TGFβ and Ras–Raf–MAPK pathways in promoting EMT [135, 136, 160, 161]. Thus it is becoming increasingly clear that TGFβ signaling synergizes with a number of auxiliary signaling pathways to mediate EMT, which promotes the migratory, invasive, and metastatic properties of cancer cells (Figure 14.2).

Recent evidence strongly suggests that a subset of undifferentiated cancer cells exhibit stem cell-like properties and have the ability to initiate tumor formation, even in very low numbers [162]. Importantly, these cancer stem-like cells exhibit resistance to drug treatment and are often enriched upon chemotherapy [163, 164]. This subpopulation of tumor cells exhibit a CD44^{high}/CD24^{low} cell surface marker phenotype and have been shown to emerge, at least in part, as a result of TGFβ-induced EMT [165]. Therefore aberrant activity of the TGFβ signaling pathway, observed in almost 90% of solid tumors, is likely to be functionally linked to cancer stem-like cell development and maintenance. This evidence further supports the notion that targeting the TGFβ–Smad signaling cascade is an attractive strategy against this axis of evil in the war against cancer [166].

Context and tissue-specific Smad signaling alterations and defective cytostatic responses

Several studies have revealed genetic and epigenetic alterations affecting the components of the TGFβ signaling pathway in gastrointestinal cancers, especially pancreatic and colon cancers to explain, at least partially, the observed loss of its growth inhibitory effect (Table 14.2) [167–171]. Subsequent reports have further elucidated functional consequences of Smad signaling inactivation in pancreatic and colon cancers. For example, restoration of Smad4 expression in pancreatic cancer cell lines can suppress tumor growth and angiogenesis by inhibiting VEGF levels [172]. Similarly, concurrent Smad4 inactivation and

Table 14.2 Genetic alterations of TGFβ signaling pathway components in cancer.

Gene	Map position	Alterations in cancers	References
TGFβRI	9q22	Mutations (prostate)	[178]
TGFβRII	3p22	Mutations (colon, gastric, head and neck)	[168, 170, 171]
SMAD1	4q28	Not detected	[25]
SMAD2	18q21	Mutations (colon)	[26, 179]
SMAD3	15q21	Not detected	[25, 180]
SMAD4	18q21	Mutations (colon, pancreas, head and neck, biliary tract), 18q loss (colon, pancreas)	[19, 20, 181–185]
SMAD5	5q31	Not detected	[25]
SMAD6	15q21	Not detected	[25]
SMAD7	18q21	Overexpression (endometrial)	[186]
SMAD8	13q12	DNA promoter methylation (breast, colon, lung)	[76]

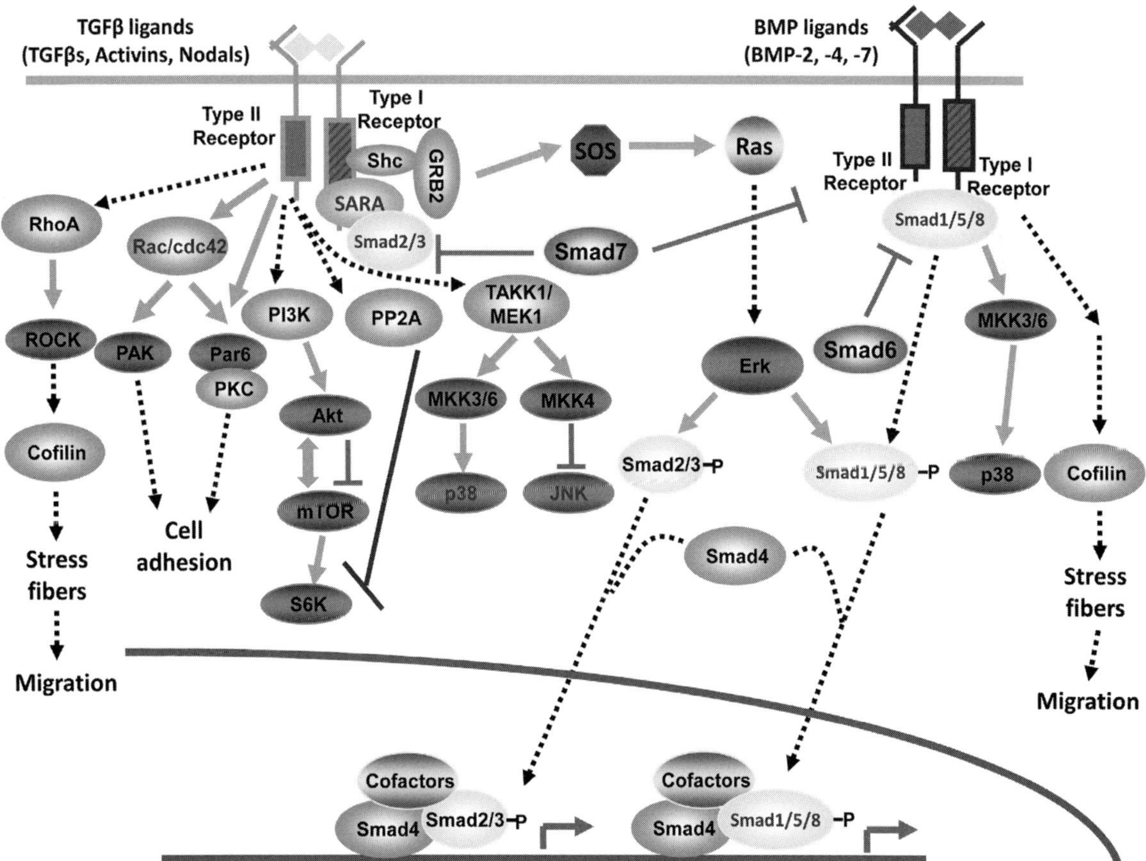

Figure 14.2 Crosstalk of TGFβ/BMP components with auxiliary signaling molecules. (A black and white version of this figure will appear in some formats. For the color version, please refer to the plate section.)

TGFβ overexpression can induce VEGF expression via MEK-Erk signaling, thus facilitating colon cancer progression and drug resistance [173]. However, in breast cancer the frequency of alterations in the Smad signaling components is rare [138] and it is likely that loss of the TGFβ-mediated growth inhibitory effects is mediated through alternate mechanisms. One of the most prevalent alternate pathways affected in breast cancer is the Ras signaling pathway. Activating mutations in the Ras oncogene have been shown to inhibit the TGFβ-dependent anti-proliferative events [174]. In a different example, breast cancer cells from pleural fluids of patients, despite normal expression of all components of the Smad pathway, were found to be unresponsive to TGFβ-mediated growth inhibition. Interestingly, this phenomenon appeared to be due to lack of p15 induction and c-Myc repression, which was associated with overexpression of the dominant-negative form of C/EBPβ isoform LIP, which binds and inhibits the transcriptionally active isoform LAP [175]. Furthermore, expression of Id1 in patient-derived metastatic breast cancer cells was found to be induced rather than repressed in response to TGFβ [176]. It is noteworthy that Id1 is part of a lung-metastasis gene expression signature, which is correlated with relapse in patients with estrogen receptor negative (ER-) breast tumors [177].

The TGFβ signaling pathway as a therapeutic target in cancer

Because TGFβ can act as both a tumor suppressor and a tumor promoter molecule, inhibiting TGFβ signaling as a therapeutic strategy against cancer must be approached with caution. In normal epithelial tissues and early in tumorigenesis, TGFβ can induce growth arrest in many epithelial cells and occasionally apoptosis [111]. Consistent with a growth inhibitory role, loss-of-function mutations of the type II TGFβ receptor (TGFβRII) [170] as well as mutations and deletions of the Smad proteins have been reported in colon and pancreatic cancer [19, 187, 188], suggesting that the TGFβ signaling pathway can be a target for inactivation during tumorigenesis. On the other hand, elevated TGFβ levels have been observed in advanced stages of various cancer types, including colon [189] and breast [130], and are correlated with a poor prognosis [190] reflecting the positive effect of TGFβ in malignant progression.

Given the well established pro-oncogenic functions of TGFβ, such as the induction of epithelial–mesenchymal transition (EMT) [155], suppression of immune surveillance [191], and promoting of invasion and metastasis [126, 145], considerable effort has been directed to developing inhibitors of the TGFβ pathway to treat cancer. The results from *in vivo* studies targeting TGFβ signaling have supported the development of these inhibitors and generated promise for this strategy. For example, MMTV-polyomavirus middle T antigen transgenic mice develop metastatic mammary tumors and serve as a model for metastatic disease [192]. Inhibition of TGFβ receptor binding and downstream signaling in these mice using an injectable soluble fusion protein (SR2F), consisting of the immunoglobulin G constant fragment (Fc) and the extracellular TGFβRII domain, resulted in increased apoptosis of tumor cells and reduced metastasis [193]. This evidence demonstrated *in vivo* that inhibiting TGFβ may be a viable option to treat or prevent metastatic tumors. Furthermore, recent evidence indicates that targeting TGFβ receptor signaling in drug-resistant tumors lacking MED12 could restore sensitivity to gefitinib, MEK, and BRAF inhibitors [194].

T-cell specific blocking of TGFβ signaling has also yielded encouraging results in mouse models [118]. In these experiments TGFβ signaling in transgenic mice was blocked using a dominant-negative TβRII specifically expressed in CD4$^+$ and CD8$^+$ T cells. When injected with thymoma and melanoma cells, wild-type animals developed tumors while transgenic animals did not. CD8$^+$ cells with disrupted TGFβ signaling were found to be responsible for the tumor-specific cytotoxicity but a CD4$^+$ population was also required to eradicate the tumor. This study highlighted the importance of the immune system during tumorigenesis as well as the central role of TGFβ in regulating these responses and provided additional evidence supporting the validity of targeting this signaling pathway for tumor therapy.

However, the pleiotropic nature of TGFβ has raised concerns about potential harmful side effects that may result from inhibiting this signaling cascade. For example, TGFβ1 null mice die from a severe inflammatory response [81, 82]. However, transgenic animals expressing the soluble TGFβ antagonist SR2F throughout their lives displayed limited metastasis similar to the systemic administration mentioned previously, with no major adverse side effects [195], indicating that inhibition of TGFβ signaling is not associated with the severe toxicities observed upon complete TGFβ ablation in knockout mice.

Inhibitors

The majority of TGFβ inhibitors being developed in preclinical studies and clinical trials can be broadly classified into three categories: (1) antibodies, (2) oligonucleotides, and (3) small-molecule inhibitors. In addition, a soluble fusion protein SR2F [193], a TGFβ2 antisense tumor vaccine [196], and various interacting peptides [195] are also being tested as potential cancer therapeutic molecules.

Antibodies

Two monoclonal antibodies, Lerdelimumab and Metelimumab, have been developed to specifically neutralize the TGFβ2 and TGFβ1 ligands, respectively. Lerdelimumab has shown promise in preventing postoperative scarring following glaucoma surgery and is currently being tested in Phase III trials [197], while Metelimumab is being investigated as a treatment for scleroderma [198]. In addition, GC-1008, a pan-TGFβ neutralizing antibody that targets all three TGFβ isoforms, has entered Phase I trials for patients with malignant melanoma or renal cell carcinoma [199]. Recently, the structure of GC-1008 in complex with the TGFβ3 ligand has been resolved and appears to be similar to the structure of the ligand when bound to both type II and type I TGFβ receptors, indicating that it can functionally mimic the binding of TGFβ3 [200].

While the efficacy of monoclonal TGFβ antibodies is yet to be fully determined in cancer patients, results from other clinical trials indicate that they are relatively well tolerated [201]. In combination with the preclinical *in vitro* and mouse model data [202], this evidence argues for the continued development of TGFβ-specific antibodies and their evaluation in clinical trials for the treatment of various malignancies.

Antisense oligonucleotides

Antisense oligonucleotides are used to target specific mRNA transcripts for degradation based on complimentary base pairing. In a mouse mesothelioma model, liposome-mediated transfection of a TGFβ2 antisense oligonucleotide effectively reduced mRNA levels and protein secretion, resulting in reduced proliferation of the malignant cells *in vitro* and reduced tumor growth *in vivo* [203]. Similarly, AP-12009 is a synthetic oligonucleotide (a phosphorothioate derivate to enhance stability) that targets TGFβ2 transcripts, preventing the translation and secretion of this isoform. AP-12009 is currently being explored as a therapy for tumors that have been found to overexpress TGFβ2, such as malignant glioma, pancreatic and colorectal carcinoma, and melanoma [204].

In patients with high-grade gliomas, AP-12009 was found to be safe and well tolerated when administered by convection-enhanced delivery and no maximum tolerated dose was reached during dose-escalation studies [19]. Median overall survival time was increased in both anaplastic astrocytoma and glioblastoma patients compared to previously reported survival data. Furthermore, two patients were reported to have complete tumor remission [205]. AP-12009 can mitigate the immunosuppressant action of TGFβ2 resulting in an enhanced antitumor cytotoxic response *in vitro*, indicating that this may be a major mechanism of action [113, 204]. Phase I/II trials for AP-12009 are ongoing in patients with pancreatic and colorectal carcinoma and melanoma. In addition, AP-11014, another synthetic oligonucleotide, specific for TGFβ1 is being evaluated in advanced preclinical studies for non-small cell lung carcinoma and prostate cancer [206].

A similar approach involves the use of TGFβ2 antisense-modified tumor vaccine. Efficacy was first demonstrated in the rat 9L gliosarcoma model where allogenic tumor cells were transfected with TGFβ2 antisense oligonucleotide and injected into rats with established gliomas [207]. A recent Phase II study concluded that Belagenpumatucel-L, a TGFβ2 antisense-modified tumor vaccine, injected subcutaneously was safe, well tolerated, and provided a potential survival advantage in non-small cell lung cancer patients [196]. Subsequent Phase III trials are being conducted and antisense-modified vaccines are also being investigated in patients with advanced glioma [208].

Small-molecule inhibitors

Current small-molecule inhibitors are under development to target the kinase domain of the type I TGFβ receptor (TGFβRI) through competitive inhibition of ATP [206] thereby preventing Smad2/Smad3 phosphorylation and downstream signaling events.

Numerous inhibitors are in preclinical development [113, 206] while one compound, LY573636, has been evaluated in patients with advanced solid tumors in a Phase I trial, which found bone marrow suppression to be the dose-limiting toxicity [206]. Several Phase II trials currently underway will determine the efficacy of this compound in several malignancies, such as melanoma, non-small cell lung cancer, and ovarian cancer [113]. It should be noted that in contrast to antisense oligonucleotides and monoclonal antibodies, small-molecule inhibitors are specific only for TGFβRI while the other inhibitors aim to inhibit the initial signaling event – ligand binding to the TGFβRII. Functional differences and potential diverse therapeutic outcomes in these two approaches are yet to be determined.

Acknowledgments

Some of the information and concepts included in this chapter were presented in the PhD thesis work of Panagiotis Papageorgis.

References

1. Gleizes PE, Beavis RC, Mazzieri R, Shen B, Rifkin DB. 1996. Identification and characterization of an eight-cysteine repeat of the latent transforming growth factor-beta binding protein-1 that mediates bonding to the latent transforming growth factor-beta1. *J Biol Chem*. 271:29891–6.

2. Saharinen J, Taipale J, Keski-Oja J. 1996. Association of the small latent transforming growth factor-beta with an eight cysteine repeat of its binding protein LTBP-1. *EMBO J*. 15:245–53.

3. Miyazono K, Olofsson A, Colosetti P, Heldin CH. 1991. A role of the latent TGF-beta 1-binding protein in the assembly and secretion of TGF-beta 1. *EMBO J*. 10:1091–101.

4. Saharinen J, Keski-Oja J. 2000. Specific sequence motif of 8-Cys repeats of TGF-beta binding proteins, LTBPs, creates a hydrophobic interaction surface for binding of small latent TGF-beta. *Mol Biol Cell*. 11:2691–704.

5. Unsold C, Hyytiainen M, Bruckner-Tuderman L, Keski-Oja J. 2001. Latent TGF-beta binding protein LTBP-1 contains three potential extracellular matrix interacting domains. *J Cell Sci*. 114:187–97.

6. Lawrence DA, Pircher R, Kryceve-Martinerie C, Jullien P. 1984. Normal embryo fibroblasts release transforming growth factors in a latent form. *J Cell Physiol*. 121:184–8.

7. Nunes I, Gleizes PE, Metz CN, Rifkin DB. 1997. Latent transforming growth factor-beta binding protein domains involved in activation and transglutaminase-dependent cross-linking of latent transforming growth factor-beta. *J Cell Biol*. 136:1151–63.

8. Crawford SE, Stellmach V, Murphy-Ullrich JE, et al. 1998. Thrombospondin-1 is a major activator of TGF-beta1 *in vivo*. *Cell*. 93:1159–70.

9. Ribeiro SM, Poczatek M, Schultz-Cherry S, Villain M, Murphy-Ullrich JE. 1999. The activation sequence of thrombospondin-1 interacts with the latency-associated peptide to regulate activation of latent transforming growth factor-beta. *J Biol Chem*. 274:13586–93.

10. Annes JP, Munger JS, Rifkin DB. 2003. Making sense of latent TGFbeta activation. *J Cell Sci*. 116:217–24.

11. Dubois CM, Laprise MH, Blanchette F, Gentry LE, Leduc R. 1995. Processing of transforming growth factor beta 1 precursor by human furin convertase. *J Biol Chem*. 270:10618–24.

12. Derynck R, Zhang Y, Feng XH. 1998. Smads: transcriptional activators of TGF-beta responses. *Cell*. 95:737–40.

13. Massague J. 1998. TGF-beta signal transduction. *Annu Rev Biochem*. 67:753–91.

14. Massague J, Blain SW, Lo RS. 2000. TGFbeta signaling in growth control, cancer, and heritable disorders. *Cell*. 103: 295–309.

15. Sekelsky JJ, Newfeld SJ, Raftery LA, Chartoff EH, Gelbart WM. 1995. Genetic characterization and cloning of mothers against dpp, a gene required for decapentaplegic function in *Drosophila melanogaster*. *Genetics*. 139:1347–58.

16. Raftery LA, Twombly V, Wharton K, Gelbart WM. 1995. Genetic screens to identify elements of the decapentaplegic signaling pathway in *Drosophila*. *Genetics*. 139: 241–54.

17. Savage C, Das P, Finelli AL, et al. 1996. *Caenorhabditis elegans* genes sma-2, sma-3, and sma-4 define a conserved family of transforming growth factor beta pathway components. *Proc Natl Acad Sci USA*. 93:790–4.

18. Derynck R, Gelbart WM, Harland RM, et al. 1996. Nomenclature: vertebrate mediators of TGFbeta family signals. *Cell*. 87:173.

19. Hahn SA, Schutte M, Hoque AT, et al. 1996. DPC4, a candidate tumor suppressor gene at human chromosome 18q21.1. *Science*. 271:350–3.

20. Thiagalingam S, Lengauer C, Leach FS, et al. 1996. Evaluation of

candidate tumour suppressor genes on chromosome 18 in colorectal cancers. *Nat Genet.* 13:343–6.

21. Lagna G, Hata A, Hemmati-Brivanlou A, Massague J. 1996. Partnership between DPC4 and SMAD proteins in TGF-beta signalling pathways. *Nature.* 383:832–6.

22. Nakao A, Afrakhte M, Moren A, et al. 1997. Identification of Smad7, a TGFbeta-inducible antagonist of TGF-beta signalling. *Nature.* 389:631–5.

23. Nakao A, Imamura T, Souchelnytskyi S, et al. 1997. TGF-beta receptor-mediated signalling through Smad2, Smad3 and Smad4. *EMBO J.* 16:5353–62.

24. Nakao A, Roijer E, Imamura T, et al. 1997. Identification of Smad2, a human Mad-related protein in the transforming growth factor beta signaling pathway. *J Biol Chem.* 272:2896–900.

25. Riggins GJ, Kinzler KW, Vogelstein B, Thiagalingam S. 1997. Frequency of Smad gene mutations in human cancers. *Cancer Res.* 57:2578–80.

26. Eppert K, Scherer SW, Ozcelik H, et al. 1996. MADR2 maps to 18q21 and encodes a TGFbeta-regulated MAD-related protein that is functionally mutated in colorectal carcinoma. *Cell.* 86:543–52.

27. Heldin CH, Miyazono K, ten Dijke P. 1997. TGF-beta signalling from cell membrane to nucleus through SMAD proteins. *Nature.* 390:465–71.

28. de Caestecker MP, Hemmati P, Larisch-Bloch S, et al. 1997. Characterization of functional domains within Smad4/DPC4. *J Biol Chem.* 272:13690–6.

29. Hata A, Shi Y, Massague J. 1998. TGF-beta signaling and cancer: structural and functional consequences of mutations in Smads. *Mol Med Today.* 4:257–62.

30. Shi Y, Wang YF, Jayaraman L, et al. 1998. Crystal structure of a Smad MH1 domain bound to DNA: insights on DNA binding in TGF-beta signaling. *Cell.* 94:585–94.

31. Luo K, Stroschein SL, Wang W, et al. 1999. The Ski oncoprotein interacts with the Smad proteins to repress TGFbeta signaling. *Genes Dev.* 13:2196–206.

32. Wotton D, Lo RS, Lee S, Massague J. 1999. A Smad transcriptional corepressor. *Cell.* 97:29–39.

33. Zhang Y, Feng X, We R, Derynck R. 1996. Receptor-associated Mad homologues synergize as effectors of the TGF-beta response. *Nature.* 383:168–72.

34. Bierie B, Moses HL. 2006. Tumour microenvironment: TGFbeta: the molecular Jekyll and Hyde of cancer. *Nat Rev Cancer.* 6:506–20.

35. Derynck R, Zhang YE. 2003. Smad-dependent and Smad-independent pathways in TGF-beta family signalling. *Nature.* 425:577–84.

36. Olivey HE, Mundell NA, Austin AF, Barnett JV. 2006. Transforming growth factor-beta stimulates epithelial-mesenchymal transformation in the proepicardium. *Dev Dyn.* 235:50–9.

37. Lebrin F, Deckers M, Bertolino P, ten Dijke P. 2005. TGF-beta receptor function in the endothelium. *Cardiovasc Res.* 65:599–608.

38. Desgrosellier JS, Mundell NA, McDonnell MA, Moses HL, Barnett JV. 2005. Activin receptor-like kinase 2 and Smad6 regulate epithelial-mesenchymal transformation during cardiac valve formation. *Dev Biol.* 280:201–10.

39. Lai YT, Beason KB, Brames GP, et al. 2000. Activin receptor-like kinase 2 can mediate atrioventricular cushion transformation. *Dev Biol.* 222:1–11.

40. Miettinen PJ, Ebner R, Lopez AR, Derynck R. 1994. TGF-beta induced transdifferentiation of mammary epithelial cells to mesenchymal cells: involvement of type I receptors. *J Cell Biol.* 127:2021–36.

41. Miyazono K, Maeda S, Imamura T. 2005. BMP receptor signaling: transcriptional targets, regulation of signals, and signaling cross-talk. *Cytokine Growth Factor Rev.* 16:251–63.

42. Souchelnytskyi S, ten Dijke P, Miyazono K, Heldin CH. 1996. Phosphorylation of Ser165 in TGF-beta type I receptor modulates TGF-beta1-induced cellular responses. *EMBO J.* 15:6231–40.

43. Lo RS, Chen YG, Shi Y, Pavletich NP, Massague J. 1998. The L3 loop: a structural motif determining specific interactions between SMAD proteins and TGF-beta receptors. *EMBO J.* 17:996–1005.

44. Abdollah S, Macias-Silva M, Tsukazaki T, et al. 1997. TbetaRI phosphorylation of Smad2 on Ser465 and Ser467 is required for Smad2-Smad4 complex formation and signaling. *J Biol Chem.* 272:27678–85.

45. Feng XH, Zhang Y, Wu RY, Derynck R. 1998. The tumor suppressor Smad4/DPC4 and transcriptional adaptor CBP/p300 are coactivators for smad3 in TGF-beta-induced transcriptional activation. *Genes Dev.* 12:2153–63.

46. Janknecht R, Wells NJ, Hunter T. 1998. TGF-beta-stimulated cooperation of smad proteins with the coactivators CBP/p300. *Genes Dev.* 12:2114–19.

47. Itoh S, Ericsson J, Nishikawa J, Heldin CH, ten Dijke P. 2000. The transcriptional co-activator P/CAF potentiates TGF-beta/

Smad signaling. *Nucleic Acids Res.* 28:4291–8.

48. Bai RY, Koester C, Ouyang T, et al. 2002. SMIF, a Smad4-interacting protein that functions as a co-activator in TGFbeta signalling. *Nat Cell Biol.* 4:181–90.

49. Chen CR, Kang Y, Siegel PM, Massague J. 2002. E2F4/5 and p107 as Smad cofactors linking the TGFbeta receptor to c-myc repression. *Cell.* 110:19–32.

50. Kang Y, Chen CR, Massague J.A 2003. Self-enabling TGFbeta response coupled to stress signaling: Smad engages stress response factor ATF3 for Id1 repression in epithelial cells. *Mol Cell.* 11:915–26.

51. Wotton D, Knoepfler PS, Laherty CD, Eisenman RN, Massague J. 2001. The Smad transcriptional corepressor TGIF recruits mSin3. *Cell Growth Differ.* 12:457–63.

52. Akiyoshi S, Inoue H, Hanai J, et al. 1999. c-Ski acts as a transcriptional co-repressor in transforming growth factor-beta signaling through interaction with smads. *J Biol Chem.* 274: 35269–77.

53. Stroschein SL, Wang W, Zhou S, Zhou Q, Luo K. 1999. Negative feedback regulation of TGF-beta signaling by the SnoN oncoprotein. *Science.* 286:771–4.

54. Sun Y, Liu X, Eaton EN, et al. 1999. Interaction of the Ski oncoprotein with Smad3 regulates TGF-beta signaling. *Mol Cell.* 4:499–509.

55. Seoane J, Le HV, Shen L, Anderson SA, Massague J. 2004. Integration of Smad and forkhead pathways in the control of neuroepithelial and glioblastoma cell proliferation. *Cell.* 117: 211–23.

56. Pardali K, Kurisaki A, Moren A, et al. 2000. Role of Smad proteins and transcription factor Sp1 in p21(Waf1/Cip1) regulation by transforming growth factor-beta. *J Biol Chem.* 275:29244–56.

57. Zhang Y, Feng XH, Derynck R. 1998. Smad3 and Smad4 cooperate with c-Jun/c-Fos to mediate TGF-beta-induced transcription. *Nature.* 394:909–13.

58. Itoh S, Landstrom M, Hermansson A, et al. 1998. Transforming growth factor beta1 induces nuclear export of inhibitory Smad7. *J Biol Chem.* 273:29195–201.

59. Hayashi H, Abdollah S, Qiu Y, et al. 1997. The MAD-related protein Smad7 associates with the TGFbeta receptor and functions as an antagonist of TGFbeta signaling. *Cell.* 89:1165–73.

60. Ebisawa T, Fukuchi M, Murakami G, et al. 2001. Smurf1 interacts with transforming growth factor-beta type I receptor through Smad7 and induces receptor degradation. *J Biol Chem.* 276:12477–80.

61. Kavsak P, Rasmussen RK, Causing CG, et al. 2000. Smad7 binds to Smurf2 to form an E3 ubiquitin ligase that targets the TGF beta receptor for degradation. *Mol Cell.* 6:1365–75.

62. Zhang S, Fei T, Zhang L, et al. 2007. Smad7 antagonizes transforming growth factor beta signaling in the nucleus by interfering with functional Smad-DNA complex formation. *Mol Cell Biol.* 27:4488–99.

63. Tsukazaki T, Chiang TA, Davison AF, Attisano L, Wrana JL. 1998. SARA, a FYVE domain protein that recruits Smad2 to the TGFbeta receptor. *Cell.* 95: 779–91.

64. Bardeesy N, Cheng KH, Berger JH, et al. 2006. Smad4 is dispensable for normal pancreas development yet critical in progression and tumor biology of pancreas cancer. *Genes Dev.* 20:3130–46.

65. Li W, Qiao W, Chen L, et al. 2003. Squamous cell carcinoma and mammary abscess formation through squamous metaplasia in

Smad4/Dpc4 conditional knockout mice. *Development.* 130:6143–53.

66. Wang J, Xu N, Feng X, et al. 2005. Targeted disruption of Smad4 in cardiomyocytes results in cardiac hypertrophy and heart failure. *Circulation Research.* 97:821–8.

67. He W, Dorn DC, Erdjument-Bromage H, et al. 2006. Hematopoiesis controlled by distinct TIF1gamma and Smad4 branches of the TGFbeta pathway. *Cell.* 125:929–41.

68. Descargues P, Sil AK, Sano Y, et al. 2008. IKKalpha is a critical coregulator of a Smad4-independent TGFbeta-Smad2/3 signaling pathway that controls keratinocyte differentiation. *Proc Natl Acad Sci USA.* 105:2487–92.

69. Reddi AH. 1997. Bone morphogenetic proteins: an unconventional approach to isolation of first mammalian morphogens. *Cytokine Growth Factor Rev.* 8:11–20.

70. Xiao YT, Xiang LX, Shao JZ. 2007. Bone morphogenetic protein. *Biochem Biophys Res Commun.* 362:550–3.

71. Shi Y, Massague J. 2003. Mechanisms of TGF-beta signaling from cell membrane to the nucleus. *Cell.* 113:685–700.

72. Kawabata M, Imamura T, Miyazono K. 1998. Signal transduction by bone morphogenetic proteins. *Cytokine Growth Factor Rev.* 9:49–61.

73. Iavarone A, Lasorella A. 2006. ID proteins as targets in cancer and tools in neurobiology. *Trends Mol Med.* 12:588–94.

74. Howe JR, Bair JL, Sayed MG, et al. 2001. Germline mutations of the gene encoding bone morphogenetic protein receptor 1A in juvenile polyposis. *Nat Genet.* 28:184–7.

75. Hardwick JC, Kodach LL, Offerhaus GJ, van den Brink GR. 2008. Bone morphogenetic

protein signalling in colorectal cancer. *Nat Rev Cancer.* 8: 806–12.

76. Cheng KH, Ponte JF, Thiagalingam S. 2004. Elucidation of epigenetic inactivation of SMAD8 in cancer using targeted expressed gene display. *Cancer Res.* 64:1639–46.

77. Massague J. 2000. How cells read TGF-beta signals. *Nat Rev Mol Cell Biol.* 1:169–78.

78. Hogan BL. 1996. Bone morphogenetic proteins in development. *Curr Opin Genet Dev.* 6:432–8.

79. Harland R. 2000. Neural induction. *Curr Opin Genet Dev.* 10:357–62.

80. Reddi AH. 1994. Bone and cartilage differentiation. *Curr Opin Genet Dev.* 4:737–44.

81. Kulkarni AB, Huh CG, Becker D, et al. 1993. Transforming growth factor beta 1 null mutation in mice causes excessive inflammatory response and early death. *Proc Natl Acad Sci USA.* 90:770–4.

82. Shull MM, Ormsby I, Kier AB, et al. 1992. Targeted disruption of the mouse transforming growth factor-beta 1 gene results in multifocal inflammatory disease. *Nature.* 359:693–9.

83. Sanford LP, Ormsby I, Gittenberger-de Groot AC, et al. 1997. TGFbeta2 knockout mice have multiple developmental defects that are non-overlapping with other TGFbeta knockout phenotypes. *Development.* 124:2659–70.

84. Proetzel G, Pawlowski SA, Wiles MV, et al. 1995. Transforming growth factor-beta 3 is required for secondary palate fusion. *Nat Genet.* 11:409–14.

85. Larsson J, Goumans MJ, Sjostrand LJ, van Rooijen MA, Ward D, Leveen P, et al. 2001. Abnormal angiogenesis but intact hematopoietic potential in TGF-beta type I receptor-deficient mice. *EMBO J.* 20:1663–73.

86. Oshima M, Oshima H, Taketo MM. 1996. TGF-beta receptor type II deficiency results in defects of yolk sac hematopoiesis and vasculogenesis. *Dev Biol.* 179: 297–302.

87. Hayashi K, Kobayashi T, Umino T, et al. 2002. SMAD1 signaling is critical for initial commitment of germ cell lineage from mouse epiblast. *Mechanisms of Development.* 118:99–109.

88. Waldrip WR, Bikoff EK, Hoodless PA, Wrana JL, Robertson EJ. 1998. Smad2 signaling in extraembryonic tissues determines anterior-posterior polarity of the early mouse embryo. *Cell.* 92:797–808.

89. Nomura M, Li E. 1998. Smad2 role in mesoderm formation, left-right patterning and craniofacial development. *Nature.* 393: 786–90.

90. Weinstein M, Yang X, Li C, et al. 1998. Failure of egg cylinder elongation and mesoderm induction in mouse embryos lacking the tumor suppressor smad2. *Proc Natl Acad Sci USA.* 95:9378–83.

91. Heyer J, Escalante-Alcalde D, Lia M, et al. 1999. Postgastrulation Smad2-deficient embryos show defects in embryo turning and anterior morphogenesis. *Proc Natl Acad Sci USA.* 96:12595–600.

92. Datto MB, Frederick JP, Pan L, et al. 1999. Targeted disruption of Smad3 reveals an essential role in transforming growth factor beta-mediated signal transduction. *Mol Cell Biol.* 19:2495–504.

93. Zhu Y, Richardson JA, Parada LF, Graff JM. 1998. Smad3 mutant mice develop metastatic colorectal cancer. *Cell.* 94:703–14.

94. Ashcroft GS, Yang X, Glick AB, et al. 1999. Mice lacking Smad3 show accelerated wound healing and an impaired local inflammatory response. *Nat Cell Biol.* 1:260–6.

95. Yang X, Letterio JJ, Lechleider RJ, et al. 1999. Targeted disruption of SMAD3 results in impaired mucosal immunity and diminished T cell responsiveness to TGF-beta. *EMBO J.* 18: 1280–91.

96. Sirard C, de la Pompa JL, Elia A, et al. 1998. The tumor suppressor gene Smad4/Dpc4 is required for gastrulation and later for anterior development of the mouse embryo. *Genes Dev.* 12:107–19.

97. Yang X, Li C, Xu X, Deng C. 1998. The tumor suppressor SMAD4/ DPC4 is essential for epiblast proliferation and mesoderm induction in mice. *Proc Natl Acad Sci USA.* 95:3667–72.

98. Yang X, Castilla LH, Xu X, et al. 1999. Angiogenesis defects and mesenchymal apoptosis in mice lacking SMAD5. *Development.* 126:1571–80.

99. Chang H, Zwijsen A, Vogel H, Huylebroeck D, Matzuk MM. 2000. Smad5 is essential for left-right asymmetry in mice. *Dev Biol.* 219:71–8.

100. Chang H, Huylebroeck D, Verschueren K, et al. 1999. Smad5 knockout mice die at mid-gestation due to multiple embryonic and extraembryonic defects. *Development.* 126: 1631–42.

101. Galvin KM, Donovan MJ, Lynch CA, et al. 2000. A role for smad6 in development and homeostasis of the cardiovascular system. *Nat Genet* 24:171–4.

102. He W, Li AG, Wang D, et al. 2002. Overexpression of Smad7 results in severe pathological alterations in multiple epithelial tissues. *EMBO J.* 21:2580–90.

103. Laiho M, DeCaprio JA, Ludlow JW, Livingston DM, Massague J. 1990. Growth inhibition by TGF-beta linked to suppression of retinoblastoma protein phosphorylation. *Cell.* 62:175–85.

104. Hannon GJ, Beach D. 1994. p15INK4B is a potential effector of TGF-beta-induced cell cycle arrest. *Nature.* 371:257–61.

105. Datto MB, Li Y, Panus JF, et al. 1995. Transforming growth factor beta induces the cyclin-dependent kinase inhibitor p21 through a p53-independent mechanism. *Proc Natl Acad Sci USA.* 92:5545–9.

106. Polyak K, Kato JY, Solomon MJ, et al. 1994. p27Kip1, a cyclin-Cdk inhibitor, links transforming growth factor-beta and contact inhibition to cell cycle arrest. *Genes Dev.* 8:9–22.

107. Pietenpol JA, Stein RW, Moran E, et al. 1990. TGF-beta 1 inhibition of c-myc transcription and growth in keratinocytes is abrogated by viral transforming proteins with pRB binding domains. *Cell.* 61:777–85.

108. Sandhu C, Garbe J, Bhattacharya N, et al. 1997. Transforming growth factor beta stabilizes p15INK4B protein, increases p15INK4B-cdk4 complexes, and inhibits cyclin D1-cdk4 association in human mammary epithelial cells. *Mol Cell Biol.* 17:2458–67.

109. Reynisdottir I, Polyak K, Iavarone A, Massague J. 1995. Kip/Cip and Ink4 Cdk inhibitors cooperate to induce cell cycle arrest in response to TGF-beta. *Genes Dev.* 9: 1831–45.

110. Reynisdottir I, Massague J. 1997. The subcellular locations of p15 (Ink4b) and p27(Kip1) coordinate their inhibitory interactions with cdk4 and cdk2. *Genes Dev.* 11:492–503.

111. Siegel PM, Massague J. 2003. Cytostatic and apoptotic actions of TGF-beta in homeostasis and cancer. *Nat Rev Cancer.* 3:807–21.

112. Pardali K, Moustakas A. 2007. Actions of TGF-beta as tumor suppressor and pro-metastatic factor in human cancer. *Biochim Biophys Acta.* 1775:21–62.

113. Wrzesinski SH, Wan YY, Flavell RA. 2007. Transforming growth factor-beta and the immune response: implications for anticancer therapy. *Clin Cancer Res.* 13:5262–70.

114. Gorelik L, Flavell RA. 2000. Abrogation of TGFbeta signaling in T cells leads to spontaneous T cell differentiation and autoimmune disease. *Immunity.* 12:171–81.

115. Leveen P, Larsson J, Ehinger M, et al. 2002. Induced disruption of the transforming growth factor beta type II receptor gene in mice causes a lethal inflammatory disorder that is transplantable. *Blood.* 100:560–8.

116. Li MO, Wan YY, Sanjabi S, Robertson AK, Flavell RA. 2006. Transforming growth factor-beta regulation of immune responses. *Ann Rev Immunol.* 24:99–146.

117. Rook AH, Kehrl JH, Wakefield LM, et al. 1986. Effects of transforming growth factor beta on the functions of natural killer cells: depressed cytolytic activity and blunting of interferon responsiveness. *J Immunol.* 136:3916–20.

118. Gorelik L, Flavell RA. 2001. Immune-mediated eradication of tumors through the blockade of transforming growth factor-beta signaling in T cells. *Nat Med.* 7:1118–22.

119. Ma A, Koka R, Burkett P. 2006. Diverse functions of IL-2, IL-15, and IL-7 in lymphoid homeostasis. *Ann Rev Immunol.* 24:657–79.

120. Becknell B, Caligiuri MA. 2005. Interleukin-2, interleukin-15, and their roles in human natural killer cells. *Adv Immunol.* 86:209–39.

121. Bonig H, Banning U, Hannen M, et al. 1999. Transforming growth factor-beta1 suppresses interleukin-15-mediated interferon-gamma production in human T lymphocytes. *Scand J Immunol.* 50:612–18.

122. Ahmadzadeh M, Rosenberg SA. 2005. TGF-beta 1 attenuates the acquisition and expression of effector function by tumor antigen-specific human memory CD8 T cells. *J Immunol.* 174:5215–23.

123. Geissmann F, Revy P, Regnault A, et al. 1999. TGF-beta 1 prevents the noncognate maturation of human dendritic Langerhans cells. *J Immunol.* 162:4567–75.

124. Lopez M, Aguilera R, Perez C, et al. 2006. The role of regulatory T lymphocytes in the induced immune response mediated by biological vaccines. *Immunobiology.* 211:127–36.

125. Roberts AB, Wakefield LM. 2003. The two faces of transforming growth factor beta in carcinogenesis. *Proc Natl Acad Sci USA.* 100:8621–3.

126. Tang B, Vu M, Booker T, et al. 2003. TGF-beta switches from tumor suppressor to prometastatic factor in a model of breast cancer progression. *J Clin Invest.* 112:1116–24.

127. Wakefield LM, Roberts AB. 2002. TGF-beta signaling: positive and negative effects on tumorigenesis. *Curr Opin Genet Dev.* 12:22–9.

128. Adorno M, Cordenonsi M, Montagner M, et al. 2009. A mutant-p53/Smad complex opposes p63 to empower TGFbeta-induced metastasis. *Cell.* 137:87–98.

129. Steiner MS, Zhou ZZ, Tonb DC, Barrack ER. 1994. Expression of transforming growth factor-beta 1 in prostate cancer. *Endocrinology.* 135:2240–7.

130. Dalal BI, Keown PA, Greenberg AH. 1993. Immunocytochemical localization of secreted transforming growth factor-beta 1 to the advancing edges of primary tumors and to lymph node metastases of human mammary carcinoma. *Am J Pathol.* 143:381–9.

131. Walker RA, Dearing SJ, Gallacher B. 1994. Relationship of transforming growth factor beta 1 to extracellular matrix and stromal infiltrates in invasive breast carcinoma. *Br J Cancer.* 69:1160–5.

132. Sargent ER, Gomella LG, Wade TP, et al. 1989. Expression of mRNA for transforming growth factors-alpha and -beta and secretion of transforming growth factor-beta by renal cell carcinoma cell lines. *Cancer Commun.* 1:317–22.

133. Tsushima H, Kawata S, Tamura S, et al. 1996. High levels of transforming growth factor beta 1 in patients with colorectal cancer: association with disease progression. *Gastroenterology.* 110:375–82.

134. Sieweke MH, Bissell MJ. 1994. The tumor-promoting effect of wounding: a possible role for TGF-beta-induced stromal alterations. *Crit Rev Oncog.* 5:297–311.

135. Oft M, Peli J, Rudaz C, et al. 1996. TGF-beta1 and Ha-Ras collaborate in modulating the phenotypic plasticity and invasiveness of epithelial tumor cells. *Genes Dev.* 10:2462–77.

136. Oft M, Heider KH, Beug H. 1998. TGFbeta signaling is necessary for carcinoma cell invasiveness and metastasis. *Curr Biol.* 8:1243–52.

137. Wang X, Abraham S, McKenzie JA, et al. 2013. LRG1 promotes angiogenesis by modulating endothelial TGF-beta signalling. *Nature.* 499:306–11.

138. Xie W, Mertens JC, Reiss DJ, et al. 2002. Alterations of Smad signaling in human breast carcinoma are associated with poor outcome: a tissue microarray study. *Cancer Res.* 62:497–505.

139. Derynck R, Akhurst RJ, Balmain A. 2001. TGF-beta signaling in tumor suppression and cancer progression. *Nat Genet.* 29: 117–29.

140. Kingsley LA, Fournier PG, Chirgwin JM, Guise TA. 2007. Molecular biology of bone metastasis. *Mol Cancer Ther.* 6:2609–17.

141. Barcellos-Hoff MH, Ravani SA. 2000. Irradiated mammary gland stroma promotes the expression of tumorigenic potential by unirradiated epithelial cells. *Cancer Res.* 60:1254–60.

142. Bhowmick NA, Chytil A, Plieth D, et al. 2004. TGF-beta signaling in fibroblasts modulates the oncogenic potential of adjacent epithelia. *Science.* 303:848–51.

143. Deckers M, van Dinther M, Buijs J, et al. 2006. The tumor suppressor Smad4 is required for transforming growth factor beta-induced epithelial to mesenchymal transition and bone metastasis of breast cancer cells. *Cancer Res.* 66:2202–9.

144. Cui W, Fowlis DJ, Bryson S, et al. 1996. TGFbeta1 inhibits the formation of benign skin tumors, but enhances progression to invasive spindle carcinomas in transgenic mice. *Cell.* 86:531–42.

145. Portella G, Cumming SA, Liddell J, et al. 1998. Transforming growth factor beta is essential for spindle cell conversion of mouse skin carcinoma *in vivo*: implications for tumor invasion. *Cell Growth Differ.* 9:393–404.

146. Yang J, Mani SA, Donaher JL, et al. 2004. Twist, a master regulator of morphogenesis, plays an essential role in tumor metastasis. *Cell.* 117:927–39.

147. Cano A, Perez-Moreno MA, Rodrigo I, et al. 2000. The transcription factor snail controls epithelial-mesenchymal transitions by repressing E-cadherin expression. *Nat Cell Biol.* 2:76–83.

148. Savagner P, Yamada KM, Thiery JP. 1997. The zinc-finger protein slug causes desmosome dissociation, an initial and necessary step for growth factor-induced epithelial-mesenchymal transition. *J Cell Biol.* 137:1403–19.

149. Eger A, Aigner K, Sonderegger S, et al. 2005. DeltaEF1 is a transcriptional repressor of E-cadherin and regulates epithelial plasticity in breast cancer cells. *Oncogene.* 24:2375–85.

150. Comijn J, Berx G, Vermassen P, et al. 2001. The two-handed E box binding zinc finger protein SIP1 downregulates E-cadherin and induces invasion. *Mol Cell.* 7:1267–78.

151. Thuault S, Valcourt U, Petersen M, et al. 2006. Transforming growth factor-beta employs HMGA2 to elicit epithelial-mesenchymal transition. *J Cell Biol.* 174:175–83.

152. Mani SA, Yang J, Brooks M, et al. 2007. Mesenchyme Forkhead 1 (FOXC2) plays a key role in metastasis and is associated with aggressive basal-like breast cancers. *Proc Natl Acad Sci USA.* 104:10069–74.

153. Papageorgis P, Lambert AW, Ozturk S, et al. 2010. Smad signaling is required to maintain epigenetic silencing during breast cancer progression. *Cancer Res.* 70:968–78.

154. Kang Y, Massague J. 2004. Epithelial–mesenchymal transitions: twist in development and metastasis. *Cell,* 118:277–9.

155. Thiery JP. 2002. Epithelial–mesenchymal transitions in tumour progression. *Nat Rev Cancer.* 2:442–54.

156. Yang J, Weinberg RA. 2008. Epithelial–mesenchymal transition: at the crossroads of development and tumor metastasis. *Dev Cell.* 14:818–29.

157. Zhu B, Fukada K, Zhu H, Kyprianou N. 2006. Prohibitin and cofilin are intracellular effectors of transforming growth factor beta signaling in human prostate cancer cells. *Cancer Res.* 66:8640–7.

158. Yu L, Hebert MC, Zhang YE. 2002. TGF-beta receptor-activated p38 MAP kinase mediates Smad-independent TGF-beta responses. *EMBO J.* 21:3749–59.

159. Itoh S, Thorikay M, Kowanetz M, et al. 2003. Elucidation of Smad requirement in transforming growth factor-beta type I receptor-induced responses. *J Biol Chem.* 278:3751–61.

160. Janda E, Lehmann K, Killisch I, et al. 2002. Ras and TGF(beta) cooperatively regulate epithelial cell plasticity and metastasis: dissection of Ras signaling pathways. *J Cell Biol.* 156:299–313.

161. Oft M, Akhurst RJ, Balmain A. 2002. Metastasis is driven by sequential elevation of H-ras and Smad2 levels. *Nat Cell Biol.* 4:487–94.

162. Al-Hajj M, Wicha MS, Benito-Hernandez A, Morrison SJ, Clarke MF. 2003. Prospective identification of tumorigenic breast cancer cells. *Proc Natl Acad Sci USA.* 100:3983–8.

163. Sharma SV, Lee DY, Li B, et al. 2010. A chromatin-mediated reversible drug-tolerant state in cancer cell subpopulations. *Cell.* 141:69–80.

164. Witta SE, Gemmill RM, Hirsch FR, et al. 2006. Restoring E-cadherin expression increases sensitivity to epidermal growth factor receptor inhibitors in lung cancer cell lines. *Cancer Res.* 66:944–50.

165. Mani SA, Guo W, Liao MJ, et al. 2008. The epithelial–mesenchymal transition generates cells with properties of stem cells. *Cell.* 133:704–15.

166. Singh A, Settleman J. 2010. EMT, cancer stem cells and drug resistance: an emerging axis of evil in the war on cancer. *Oncogene.* 29:4741–51.

167. Carcamo J, Zentella A, Massague J. 1995. Disruption of transforming growth factor beta signaling by a mutation that prevents transphosphorylation within the receptor complex. *Mol Cell Biol.* 15:1573–81.

168. Garrigue-Antar L, Munoz-Antonia T, Antonia SJ, et al. 1995. Missense mutations of the transforming growth factor beta type II receptor in human head and neck squamous carcinoma cells. *Cancer Res.* 55:3982–7.

169. Kim DH, Kim SJ. 1996. Transforming growth factor-beta receptors: role in physiology and disease. *J Biomed Sci.* 3:143–58.

170. Markowitz S, Wang J, Myeroff L, et al. 1995. Inactivation of the type II TGF-beta receptor in colon cancer cells with microsatellite instability. *Science.* 268:1336–8.

171. Park K, Kim SJ, Bang YJ, et al. 1994. Genetic changes in the transforming growth factor beta (TGF-beta) type II receptor gene in human gastric cancer cells: correlation with sensitivity to growth inhibition by TGF-beta. *Proc Natl Acad Sci USA.* 91:8772–6.

172. Schwarte-Waldhoff I, Volpert OV, Bouck NP, et al. 2000. Smad4/DPC4-mediated tumor suppression through suppression of angiogenesis. *Proc Natl Acad Sci USA.* 97:9624–9.

173. Papageorgis P, Cheng K, Ozturk S, et al. 2011. Smad4 inactivation promotes malignancy and drug resistance of colon cancer. *Cancer Res.* 71:998–1008.

174. Kretzschmar M, Doody J, Timokhina I, Massague J. 1999. A mechanism of repression of TGFbeta/Smad signaling by oncogenic Ras. *Genes Dev.* 13:804–16.

175. Gomis RR, Alarcon C, Nadal C, Van Poznak C, Massague J. 2006. C/EBPbeta at the core of the TGFbeta cytostatic response and its evasion in metastatic breast cancer cells. *Cancer Cell.* 10:203–14.

176. Padua D, Zhang XH, Wang Q, et al. 2008. TGFbeta primes breast tumors for lung metastasis seeding through angiopoietin-like 4. *Cell.* 133:66–77.

177. Minn AJ, Gupta GP, Siegel PM, et al. 2005. Genes that mediate breast cancer metastasis to lung. *Nature.* 436:518–24.

178. Kim IY, Ahn HJ, Zelner DJ, et al. 1996. Genetic change in transforming growth factor beta (TGF-beta) receptor type I gene correlates with insensitivity to TGF-beta 1 in human prostate cancer cells. *Cancer Res.* 56:44–8.

179. Riggins GJ, Thiagalingam S, Rozenblum E, et al. 1996. Mad-related genes in the human. *Nat Genet.* 13:347–9.

180. Arai T, Akiyama Y, Okabe S, et al. 1998. Genomic structure of the human Smad3 gene and its infrequent alterations in colorectal cancers. *Cancer Lett.* 122:157–63.

181. Hahn SA, Bartsch D, Schroers A, et al. 1998. Mutations of the DPC4/Smad4 gene in biliary tract carcinoma. *Cancer Res.* 58:1124–6.

182. Miyaki M, Iijima T, Konishi M, et al. 1999. Higher frequency of Smad4 gene mutation in human colorectal cancer with distant metastasis. *Oncogene.* 18:3098–103.

183. Xie W, Bharathy S, Kim D, et al. 2003. Frequent alterations of Smad signaling in human head and neck squamous cell carcinomas: a tissue microarray analysis. *Oncol Res.* 14:61–73.

184. Howe JR, Roth S, Ringold JC, et al. 1998. Mutations in the SMAD4/DPC4 gene in juvenile polyposis. *Science.* 280:1086–8.

185. Takagi Y, Kohmura H, Futamura M, et al. 1996. Somatic alterations of the DPC4 gene in human colorectal cancers *in vivo*. *Gastroenterology.* 111:1369–72.

186. Dowdy SC, Mariani A, Reinholz MM, et al. 2005. Overexpression

of the TGF-beta antagonist Smad7 in endometrial cancer. *Gynecol Oncol.* 96:368–73.

187. Grady WM, Myeroff LL, Swinler SE, et al. 1999. Mutational inactivation of transforming growth factor beta receptor type II in microsatellite stable colon cancers. *Cancer Res.* 59:320–4.

188. Hahn SA, Hoque AT, Moskaluk CA, et al. 1996. Homozygous deletion map at 18q21.1 in pancreatic cancer. *Cancer Res.* 56:490–4.

189. Picon A, Gold LI, Wang J, Cohen A, Friedman E. 1998. A subset of metastatic human colon cancers expresses elevated levels of transforming growth factor beta1. *Cancer Epidemiol Biomarkers Prev.* 7:497–504.

190. Friedman E, Gold LI, Klimstra D, et al. 1995. High levels of transforming growth factor beta 1 correlate with disease progression in human colon cancer. *Cancer Epidemiol Biomarkers Prev.* 4:549–54.

191. Thomas DA, Massague J. 2005. TGF-beta directly targets cytotoxic T cell functions during tumor evasion of immune surveillance. *Cancer Cell.* 8:369–80.

192. Guy CT, Cardiff RD, Muller WJ. 1992. Induction of mammary tumors by expression of polyomavirus middle T oncogene: a transgenic mouse model for metastatic disease. *Mol Cell Biol.* 12:954–61.

193. Muraoka RS, Dumont N, Ritter CA, et al. 2002. Blockade of TGF-beta inhibits mammary tumor cell viability, migration, and metastases. *J Clin Invest.* 109:1551–9.

194. Huang S, Holzel M, Knijnenburg T, et al. 2012. MED12 controls the response to multiple cancer drugs through regulation of TGF-beta receptor signaling. *Cell.* 151:937–50.

195. Yang YA, Dukhanina O, Tang B, et al. 2002. Lifetime exposure to a soluble TGF-beta antagonist protects mice against metastasis without adverse side effects. *J Clin Invest.* 109:1607–15.

196. Nemunaitis J, Dillman RO, Schwarzenberger PO, et al. 2006. Phase II study of belagenpumatucel-L, a transforming growth factor beta-2 antisense gene-modified allogeneic tumor cell vaccine in non-small-cell lung cancer. *J Clin Oncol.* 24:4721–30.

197. Mead AL, Wong TT, Cordeiro MF, Anderson IK, Khaw PT. 2003. Evaluation of anti-TGF-beta2 antibody as a new postoperative anti-scarring agent in glaucoma surgery. *Invest Ophthalmol Vis Sci.* 44:3394–401.

198. Denton CP, Merkel PA, Furst DE, et al. 2007. Recombinant human anti-transforming growth factor beta1 antibody therapy in systemic sclerosis: a multicenter, randomized, placebo-controlled phase I/II trial of CAT-192. *Arthritis Rheum.* 56:323–33.

199. Morris JC, Tan AR, Lawrence DP, et al. 2008. Phase I/II study of GC1008: a human anti-transforming growth factor-beta (TGFβ) monoclonal antibody (MAb) in patients with advanced malignant melanoma (MM) or renal cell carcinoma (RCC). *ASCO Annual Meeting Proceeding Part I J Clin Oncol.* May 20, Abstract # 9028.

200. Grutter C, Wilkinson T, Turner R, et al. 2008. A cytokine-neutralizing antibody as a structural mimetic of 2 receptor interactions. *Proc Natl Acad Sci USA.* 105:20251–6.

201. Akhurst RJ. 2006. Large- and small-molecule inhibitors of transforming growth factor-beta signaling. *Curr Opin Investig Drugs.* 7:513–21.

202. Saunier EF, Akhurst RJ. 2006. TGF beta inhibition for cancer therapy. *Curr Cancer Drug Targets.* 6:565–78.

203. Marzo AL, Fitzpatrick DR, Robinson BW, Scott B. 1997. Antisense oligonucleotides specific for transforming growth factor beta2 inhibit the growth of malignant mesothelioma both *in vitro* and *in vivo*. *Cancer Res.* 57:3200–7.

204. Schlingensiepen KH, Schlingensiepen R, Steinbrecher A, et al. 2006. Targeted tumor therapy with the TGF-beta 2 antisense compound AP 12009. *Cytokine Growth Factor Rev.* 17: 129–39.

205. Hau P, Jachimczak P, Schlingensiepen R, et al. 2007. Inhibition of TGF-beta2 with AP 12009 in recurrent malignant gliomas: from preclinical to phase I/II studies. *Oligonucleotides.* 17:201–12.

206. Yingling JM, Blanchard KL, Sawyer JS. 2004. Development of TGF-beta signalling inhibitors for cancer therapy. *Nat Rev Drug Discov.* 3:1011–22.

207. Fakhrai H, Dorigo O, Shawler DL, et al. 1996. Eradication of established intracranial rat gliomas by transforming growth factor beta antisense gene therapy. *Proc Natl Acad Sci USA.* 93:2909–14.

208. Fakhrai H, Mantil JC, Liu L, et al. 2006. Phase I clinical trial of a TGF-beta antisense-modified tumor cell vaccine in patients with advanced glioma. *Cancer Gene Ther.* 13:1052–60.

15

The Wnt signaling network in cancer

Johanna Apfel, Jignesh R. Parikh, Patricia Reischmann, Rob M. Ewing, Oliver Müller, Yu Xia and Isabel Dominguez

Introduction

Wnt genes are a family of evolutionary conserved genes with specific expression patterns and a species-specific number of genes expressed (e.g., 19 Wnts expressed in mammals) (van Amerongen and Nusse, 2009). *Wnt* genes code for secreted lipid-modified cysteine-rich glycoprotein ligands that act in autocrine and paracrine fashion (Hausmann et al., 2007; Mikels and Nusse, 2006; Port and Basler, 2010).

Wnt signaling is involved in embryonic development by regulating processes such as progenitor cell expansion, proliferation, cell fate decisions, migration, cell polarity, and differentiation among others (Aoki and Taketo, 2008; Grigoryan et al., 2008; Komiya and Habas, 2008; Logan and Nusse, 2004; van Amerongen and Berns, 2006; van Amerongen and Nusse, 2009). Research in the past decades supports the idea that signaling pathways involved in embryonic development can be co-opted by cells during tumorigenesis; the Wnt pathway is not an exception. It was the first cloned Wnt gene, *Wnt1*, which linked embryonic development and cancer. Indeed, the name Wnt is a fusion between the names of two homologous genes, the fruit fly *Drosophila melanogaster* wingless (*wg*) gene, whose mutation causes segmentation defects, and the mouse *Int1* gene, a gene identified as a preferential integration site-1 of the mouse mammary tumor virus (MMTV), a retrovirus capable of inducing mammary gland tumors (Nusse and Varmus, 1982; Nusse et al., 1991). During tumorigenesis, the Wnt pathway has been involved in the regulation of proliferation, survival, differentiation, and migration (Giles et al., 2003; Jessen, 2009; MacDonald et al., 2009). Since the homology between *wg* and *Int-1* was discovered, parallel research *in vitro* and in embryonic and tumor animal models has led to significant advances in our knowledge of the biological and oncogenic potential of

Wnts and Wnt pathway components. In this chapter, we will describe the Wnt signaling pathways and their involvement in tumorigenesis, and we will review several systems biology approaches and the impact that they are having in characterizing, defining, and integrating the complex Wnt signaling network.

Wnt signaling components and pathways

Before discussing the involvement of Wnts in cancer, we will first describe the Wnt signaling pathways. After the cloning of *wg/Int-1* a major challenge was to identify Wnt signaling components and determine the Wnt signaling mechanism. Based on the concept that genes with similar phenotypes may participate in the same signaling pathway, many Wnt signaling components have been identified in *Drosophila* in forward genetic screens as segmentation mutants similar to *wg*. These genes were tested in genetic epistasis experiments in *Drosophila* to establish a hierarchy among them, leading to the first model for the Wnt signaling pathway (Noordermeer et al., 1994). In addition, a number of other assays have been used to test the biological role of Wnts and to characterize Wnt signaling components. Ectopic expression of Wnts in mammary epithelial cell lines such as C57MG and RAC311 can cause malignant transformation characterized by elongated refractile morphology, increased proliferation, and increased plate saturation density (Brown et al., 1986; Rijsewijk et al., 1987). Ectopic expression of Wnts in ventral blastomeres of early *Xenopus laevis* frog embryos can induce an ectopic body axis leading to tadpoles looking like conjoined Siamese twins (Kuhl and Pandur, 2008a; McMahon and Moon, 1989; Smith and Harland, 1991; Sokol et al., 1991). This biological assay in *Xenopus laevis*

Systems Biology of Cancer, ed. S. Thiagalingam. Published by Cambridge University Press. © Cambridge University Press 2015.

embryos has also been used for expression cloning experiments to identify novel Wnt signaling components (Kuhl and Pandur, 2008a). The differential ability of Wnt ligands to transform epithelial cells and to induce ectopic axis formation in *Xenopus* led to their historical assignment into two classes: the Wnt1-like Wnt ligands that behave positively in these biological assays and the Wnt5a-like Wnt ligands that have no effect in these assays but inhibit Wnt1-induced transformation and can affect convergent extension movements in *Xenopus* (Chien et al., 2009). *In vitro* and biochemical studies defined different intracellular signaling mechanisms that these two classes of Wnts elicit, historically called the canonical or Wnt/β-catenin pathway and the non-canonical or β-catenin-independent Wnt pathways, respectively (James et al., 2008; Kestler and Kuhl, 2008; Komiya and Habas, 2008; van Amerongen and Berns, 2006; van Amerongen and Nusse, 2009). We succinctly describe these Wnt pathways below and provide references where a more detailed description of the molecular mechanisms of the Wnt signaling pathways can be found.

Canonical or Wnt/β-catenin pathway

The Wnt/β-catenin pathway, the most widely studied Wnt signaling pathway, is involved in cellular processes such as cell fate determination, progenitor expansion, and differentiation (Angers and Moon, 2009; Cadigan and Peifer, 2009; Dominguez et al., 2009; Gao and Chen, 2010; Kikuchi et al., 2007; MacDonald et al., 2009; Mosimann et al., 2009; Verheyen and Gottardi, 2010). Activation of Wnt/β-catenin signaling results in the induction of Wnt/β-catenin-responsive target genes (Figure 15.1a). The key step in the Wnt/β-catenin pathway is the stabilization and nuclear translocation of cytoplasmic β-catenin, a protein also involved in E-cadherin cell adhesion complexes when localized to the membrane. In the absence of Wnts, free cytoplasmic β-catenin is constitutively ubiquitinated by the E3 ligase receptor βTrcp and subsequently degraded via the ubiquitin–proteasome pathway. The targeting of β-catenin for degradation is achieved by N-terminus phosphorylation of β-catenin at S33/S37/T41/S45 by casein kinase 1 (CK1) and glycogen synthase kinase-3β (GSK-3β), facilitated by the scaffold proteins axin and adenomatous polyposis coli (Apc). In addition, basal Wnt/β-catenin-responsive target gene transcription is repressed by high mobility group (HMG box) DNA-binding factors of the T-cell factor/lymphoid enhancer-binding factor (TCF/LEF) family through recruitment of transcriptional co-repressors such as Groucho (Gro)/TLE. Wnt/β-catenin signaling is

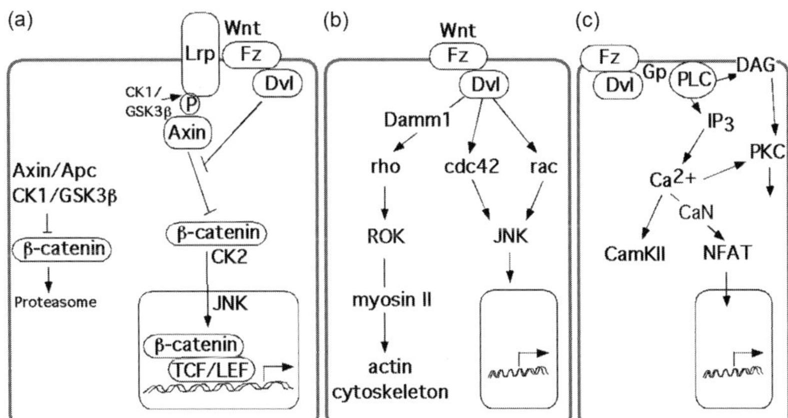

Figure 15.1 Wnt/β-catenin and β-catenin-independent Wnt signaling pathways.
(a) The β-catenin multiprotein destruction complex, active in the absence of Wnts, functions to target free cytoplasmic β-catenin for proteasome-dependent degradation through phosphorylation by CK1 and GSK-3β facilitated by the scaffold proteins Axin and Apc. Wnts bind to co-receptors Fz and Lrp leading to membrane recruitment of Dsh and Axin, respectively. This results in the inhibition of the β-catenin destruction complex and stabilization of N-terminus unphosphorylated β-catenin that may be further stabilized by CK2 phosphorylation. β-catenin then translocates into the nucleus, partners with a TCF/LEF family member to induce the expression of Wnt/β-catenin-target genes. (b) Wnt/PCP signaling is initiated upon binding of Wnt to Fz receptors, leading to activation of Rac and Rho through Dsh. Activation of Rac stimulates JNK and activation of Rho stimulates the Rho-associated kinase (ROCK) and myosin leading to remodelling of the actin cytoskeleton. (c) Wnt/Ca²⁺ signaling is initiated by the binding of Wnt to Fz receptors and mediated by Dvl and heterotrimeric G-proteins. PLC activation by G-proteins generates Ca²⁺ and DAG that activate PKC. Ca²⁺, in addition, activates CamKll and NAFT.

initiated by the binding of Wnt to two co-receptors: a member of the Frizzled G-protein-coupled receptor family (Fz) and a member of the low-density lipoprotein-related protein receptor family (Lrp). Wnt binding leads to the recruitment of the cytoplasmic protein Dishevelled (Dvl, Dsh) to the membrane through interaction with Fz or membrane phospholipids, and also to the recruitment of axin to Lrp mediated by the phosphorylation of Lrp by CK1/GSK3β. These events appear to result in axin degradation and Dsh-mediated inhibition of the β-catenin destruction complex. As a consequence, β-catenin is no longer phosphorylated at S33/S37/T41/S45 and is more stable. β-catenin can be further stabilized by phosphorylation at T393 by the kinase CK2, a positive Wnt/β-catenin regulator in *Xenopus* bioassays (Dominguez et al., 2004; Song et al., 2003). Subsequently, β-catenin translocates into the nucleus through a mechanism that may involve phosphorylation at S191 by the c-jun NH2-terminal kinase (JNK). In the nucleus, β-catenin partners with TCF/LEF to induce the expression of Wnt/β-catenin target genes together with co-activators such as BCL9, pygopus (pygo), and p300/CBP. A number of putative Wnt/β-catenin target genes have been identified; however, only a small subset seem to be direct targets (MacDonald et al., 2009; Vlad et al., 2008).

The study of Wnt/β-catenin function has been greatly facilitated by the development of assays to measure its activation *in vivo* and *in vitro*. In addition to the *Xenopus* bioassays described above, these assays include detection of elevated nuclear levels of N-terminus unphosphorylated β-catenin, expression of target genes such as *c-myc*, and activation of Wnt/β-catenin reporters based on multimerized TCF/LEF binding sites (Barolo, 2006; Chien et al., 2009).

Non-canonical or β-catenin-independent Wnt pathways

β-catenin-independent Wnt signaling pathways have been historically divided into two main branches: the Wnt/JNK or Wnt/planar cell polarity (PCP) and the Wnt/Ca^{2+} (Croce and McClay, 2008; James et al., 2008; Jenny and Mlodzik, 2006; Kohn and Moon, 2005; Komiya and Habas, 2008; Roszko et al., 2009; Simons and Mlodzik, 2008). The Wnt/JNK or Wnt/PCP pathway promotes changes in cellular cytoskeletal organization, cell polarity, and migration, and is involved in wing hair orientation, convergent extension movements during gastrulation, and migration of neural crest cells. The Wnt/Ca^{2+} pathway affects migration, cell fate specification, and survival, and is involved in dorsal–ventral patterning and convergent extension movements during gastrulation. The activation of both pathways is initiated upon binding of Wnt to Fz receptors. In Wnt/JNK signaling, Fz activation of Dvl leads to the activation of the small GTPases Rac, Cdc42, and Rho (Figure 15.1b). Activated Rac stimulates JNK, potentially leading to transcriptional responses. Activation of Rho, mediated by the formin homology protein Damm1, leads to the activation of the Rho-associated kinase (ROCK) and phopshorylation of myosin, leading to remodelling of the actin cytoskeleton required for tissue polarity and migration processes. It should be noted that, in *Drosophila*, the Wnt/PCP pathway does not seem to require Wnts and may be activated by asymmetric localization of two Dsh-antagonists, the LIM domain protein prickle (pk) and the transmembrane protein van gogh/strabismus (Vang/stbm), while the ankyrin repeat protein, Diego, promotes Dsh activation and inhibits pk.

In Wnt/Ca^{2+} signaling, Wnt binding to Fz activates heterotrimeric G-proteins through Dsh (Figure 15.1c). Phospholipase C (PLC) activation by G-proteins generates Ca^{2+} and DAG that activate protein kinase C (PKC). Ca^{2+}, in addition, activates calcium/calmodulin-dependent kinase II (CamKII) and causes nuclear translocation of the trancription factor, NAFT, through the Ca^{2+}/calmodulin-dependent phosphatase calcineurin.

Activation of the Wnt/PCP and Wnt/Ca^{2+} pathways is measured by analysis of biological and cellular readouts, such as cell polarity and cell migration, and biochemically by measuring the activity of intracellular components such as JNK, PKC, and CamKII and the release of Ca^{2+} (Kuhl, 2004; Kuhl and Pandur, 2008b). However, these biochemical analyses are not specific for β-catenin-independent Wnt pathways.

Extracellular and membrane-bound modulators of Wnt signaling

In addition to all these intracellular Wnt components, a number of extracellular and membrane-bound proteins modulate Wnt signaling (MacDonald et al., 2009; van Amerongen and Nusse, 2009). These include (1) Wnt-binding extracellular Wnt-antagonists such as Cerberus (Cer), soluble frizzled-related proteins

(SFRP), Frzb, and Wnt-inhibitory factor (WIF-1); (2) the Wnt-binding dual Wnt-inhibitor/activator Wise; and (3) Lrp-binding extracellular proteins such as the Wnt-antagonist sclerostin (SOST) and the Wnt-modulator Dickkopf1 (DKK1) and its transmembrane co-receptor Kremen (Krm).

Wnt pathway specificity and Wnt pathway network

The paradigm that canonical and non-canonical ligands and receptors are distinct has recently been challenged (Chien et al., 2009; Kestler and Kuhl, 2008; Komiya and Habas, 2008; MacDonald et al., 2009; van Amerongen and Nusse, 2009). For example, the finding that historical non-canonical Wnts can also signal through Wnt/β-catenin provided that the appropriate receptor is expressed, that Fz receptors mediate Wnt/β-catenin and β-catenin-independent Wnt pathways, and that the Lrp receptor can also mediate non-canonical biological functions, suggest that the specificity of the pathway may not lie at the Wnt/receptor level. The specificity may lie at the intracellular level, particularly at the level of Dsh, where the Wnt pathways bifurcate, and may be mediated by differences in Dsh localization and/or in the interaction with different sets of proteins that modify Dsh function (Gao and Chen, 2010; Wallingford and Habas, 2005; Wharton, 2003). An exception to the paradigm that Wnt responses can only be elicited by Wnts was first found in the Wnt-independent *Drosophila* Wnt/PCP pathway. Nowadays, adding to the complexity of Wnt signaling, ligands other than Wnt such as Norrin and R-spondin (Rspo) are able to activate Wnt signaling through Wnt receptors in contexts such as vasculogenesis (Kikuchi et al., 2007), and receptors other than Fz, such as tyrosine kinase receptors Ror2 and Ryk, can bind Wnt and activate the Wnt/JNK pathway (MacDonald et al., 2009; van Amerongen and Nusse, 2009). The paradigm that canonical and non-canonical pathways act separately has also shifted, as recent studies show that in different cellular and biological contexts there is crosstalk between the Wnt/β-catenin and β-catenin-independent Wnt signaling pathways and other pathways. For example, the Wnt/Ca^{2+} pathway can antagonize the Wnt/PCP pathway during convergent extension and the Wnt/β-catenin pathway during dorsal axis formation (Kestler and Kuhl, 2008; Komiya and Habas, 2008; Roszko et al., 2009). In addition, components of β-catenin-independent Wnt signaling pathways, such as Rac and Rho, may positively regulate Wnt/β-catenin signaling (Schlessinger et al., 2009). All these findings led to the proposal that the specific biochemical changes and cellular processes activated in response to a ligand depends on the Wnt components expressed and on the differential activation of the network of intracellular Wnt components (Kestler and Kuhl, 2008; van Amerongen and Nusse, 2009).

Wnts also cooperate with other signaling pathways to regulate embryological and homeostasis processes such as the Notch pathway, the bone morphogenetic protein (BMP) pathway, and the insulin/IGF-1 pathway (Hayward et al., 2008; Itasaki and Hoppler, 2010; Jin et al., 2008). Pathways such as the Hippo pathway and nuclear hormone receptors, such as peroxisome proliferator-activated receptor gamma (PPARγ) and estrogen receptor (ER), interact with Wnt signaling to regulate gene expression (Beildeck et al., 2010; McNeill and Woodgett, 2010; van Amerongen and Berns, 2006). Furthermore, other signaling pathways are able to stimulate β-catenin/TCF transcriptional activation such as the insulin pathway (Jin et al., 2008).

In order to analyze and understand the complexity of the cellular and biological responses to Wnts and cooperating pathways at the single and organism levels, we require the novel high-throughput approaches and quantitative analysis that systems biology provides. In this chapter, we will review several high-throughput approaches that are facilitating the identification of Wnt network components and their interactions in different cellular and activation contexts, and the identification of Wnt-specific responsive genes. We will also describe methods for high-throughput data analysis and integration. In addition, we will review computer simulation models of Wnt circuits that facilitate the understanding of Wnt signaling pathways.

Wnt signaling in cancer

A number of studies have implicated activated Wnt/β-catenin and β-catenin-independent Wnt signaling pathways in the initiation and progression of cancer, suggesting that targeting Wnt signaling components is a potential cancer treatment (Barker and Clevers, 2006; Camilli and Weeraratna, 2010; Giles et al., 2003; Polakis, 2000, 2007). Deregulated Wnt signaling leads to increased stem cell renewal, proliferation, migration, invasiveness, and survival (Giles et al., 2003; Jessen, 2009; MacDonald et al., 2009), underlying the important role that Wnt signaling plays in cell commitment, proliferation, metabolism, growth, and stem cell maintenance during normal adult tissue homeostasis (Chen et al., 2008; Matushansky et al., 2008; Reya and Clevers, 2005; Sethi and Vidal-Puig, 2010). Importantly, a number of Wnt components at all levels within the pathways have been found deregulated in human tumors suggesting Wnt signaling dysregulation. If should be noted that Wnt signaling is also linked to other human diseases and disorders such as sex reversal, osteoporosis, obesity, and coronary disease (Clevers, 2006; MacDonald et al., 2009; Moon et al., 2004).

Tumorigenic potential of Wnts in animal models

In addition to *Wnt1*, MMTV insertions also occur in the regions of the *Wnt3* (Roelink et al., 1990, 1992) and *Wnt10a* genes (Nusse and Varmus, 1982), suggesting that Wnts have tumorigenic potential. A number of assays have been used to test the tumorigenic potential of Wnt-transformed cells *in vivo*, including subcutaneous injection or transplantation into mammary fat pads (Edwards et al., 1992; Rijsewijk et al., 1987). Tissue-specific transgenic mice have confirmed the tumorigenic potential of several Wnt components such as Wnt-1, activated forms of β-catenin lacking the N-terminus, and positive regulators such as CK2α (Aoki and Taketo, 2008; Giles et al., 2003; Grigoryan et al., 2008; Landesman-Bollag et al., 2001). In addition, *APC* and *axin* mutant mice and transgenic mice for kinase-inactive GSK3β also display tumors, supporting their role as tumor suppressors (Aoki and Taketo, 2008; Grigoryan et al., 2008). Furthermore, mutant mice with activation of Wnt/β-catenin signaling due to misexpression of diverse Wnt/β-catenin components showed similar tumor characteristics (Miyoshi et al., 2002).

Upregulated Wnt/β-catenin signaling in human tumors

Forty types of solid human tumors, including colorectal, breast, lung, ovarian, liver, prostate, cervical, and melanoma, have increased levels of cytoplasmic and nuclear β-catenin, suggesting upregulation of Wnt/β-catenin signaling (Barker and Clevers, 2006; Giles et al., 2003; Ilyas, 2005; Polakis, 2000, 2007). Constitutive Wnt/β-catenin signaling activation in human tumors occurs through genetic and epigenetic mechanisms that may be tissue specific.

Germline nonsense mutations in *APC* are found in patients with familial adenomatous polyposis. These nonsense mutations are also associated with 80% of sporadic colorectal adenomas and carcinomas. Nonsense mutations in *APC* result in the loss of function of *APC* as they generate a truncated protein lacking the region of APC that is involved in β-catenin degradation, i.e., removal of β-catenin and axin binding sites and putative phosphorylation site for GSK3β (*GSK3B*). APC mutations are mostly confined to colorectal cancers but are also rarely found in sporadic lung, ovarian, and breast cancer (Polakis, 2007). Gain-of-function mutations in β-catenin (*CTNNB1*) are found in a large number of primary human cancers, such as endometrioid ovarian cancer, hepatoblastoma, medulloblastoma, Wilms' kidney tumors, liver, skin, prostate, and rarely in sporadic colorectal cancer and melanoma. *CTNNB1* gain of function is caused by in-frame deletions in the CK1/GSK3β phosphorylation sites or missense mutations in exon 3 of *CTNNB1*. These mutations eliminate the CK1/GSK3β phosphorylation sites and make β-catenin refractory to CK1/GSK3β phosphorylation-dependent degradation, leading to inappropriate accumulation of β-catenin. Interestingly, mutations of *APC* and *CTNNB1* are rarely found in the same tumor, suggesting that there are alternatives for Wnt/β-catenin activation (Polakis, 2007). Mutations in *AXIN* are found in medulloblastoma, endometrioid ovarian cancer, hepatomas, and hepatocellular carcinomas. These missense mutations in the *AXIN1* and *AXIN2* genes should be inactivating but the significance of the mutations in tumorigenesis has not yet been determined (Barker and Clevers, 2006; Polakis, 2007). In addition, loss of *AXIN* is found in hepatocellular carcinomas (Satoh et al., 2000). Frameshift mutations in *TCF4* (*TCF7L2*) in colorectal cancers affect the C-terminus of the protein and may influence its binding to the transcriptional repressor carboxy-terminal binding protein (CtBP) (Barker and Clevers, 2006; Polakis, 2007). Truncating mutations in *sFRP1* exon 1, predicted to lead to loss of inhibitory activity, are rarely found in sporadic colorectal cancer (Barker and Clevers, 2006; Polakis, 2007).

In contrast, in other human cancers with upregulated β-catenin, such as breast cancer and leukemia, mutations in *APC*, *CTNNB1*, or *AXIN* are rare (Polakis, 2007). Two other distinct mechanisms have been described to explain upregulated β-catenin in these tumors: overexpression of *WNT* ligands, *FZ* receptors, and intracellular components such as *DVL*, found in cancers such as breast, prostate, and colorectal; and downregulation of extracellular Wnt modifiers (e.g., *SFRPs*, *WIF1*, *DKKs*) and intracellular negative regulators such as *APC* and *AXIN*. Downregulation of Wnt negative regulators may be due to epigenetic silencing through promoter-gene hypermethylation. Silencing of Wnt negative regulators is also found in tumors with mutated Wnt components, such as colon tumors with *APC* mutations, providing an added advantage for cell survival and proliferation (MacDonald et al., 2009).

Upregulated β-catenin-independent Wnt signaling in human tumors

Traditionally, based on the biological assays described above (transformation and ectopic axis formation), historical non-canonical Wnts did not transform cells but had the ability to inhibit canonical Wnt-induced transformation (Jessen, 2009). In support of this, non-canonical Wnts such as *Wnt5A* were found decreased in human tumors. However, recent studies show that *Wnt5A* is overexpressed in highly aggressive melanoma, lung, breast, prostate, gastric, and other human cancers (Jessen, 2009; Kikuchi and Yamamoto, 2008; McDonald and Silver, 2009; O'Connell and Weeraratna, 2009; Pukrop and Binder, 2008; Wang, 2009). In addition, *Wnt5A* affects melanoma breast and gastric cancer cell motility and invasion *in vitro* (McDonald and Silver, 2009, Pukrop and Binder, 2008). Based on these data and on the biological role of β-catenin-independent Wnt signaling in embryonic development, the idea that β-catenin-independent Wnt signaling regulates cytoskeletal remodelling, cell migration, and invasiveness during tumorigenesis has been put forward (Jessen, 2009; Lai et al., 2009). This idea is further supported by studies finding other β-catenin-independent Wnt signaling components such as *WNT5A*, *WNT11*, *VANGL*, and *PRICKLE* overexpressed in tumor cell lines (Jessen, 2009). For example, *VANGL1* is highly expressed in colon cancer cells with high metastatic potential, and in metastatic gastric cancer cells (Jessen, 2009).

Wnt signaling activation and Wnt network in cancer

Since the Wnt signaling pathway is frequently dysregulated in human cancer, it has a great potential as a target for therapeutics. From the standpoint of diagnosis, nuclear β-catenin immunostaining has been used as a specific diagnostic marker of elevated Wnt/β-catenin signaling (Barker and van den Born, 2008; Chien et al., 2009); however, there is a variability between reports in the ability to detect nuclear β-catenin, which may be due to technical reasons or lack of segregation by tumor characteristics such as molecular subtyping (Fodde and Tomlinson, 2010; Khramtsov et al., 2010). In addition, recent biological and biochemical studies show that increased β-catenin levels and nuclear β-catenin presence do not always result in a corresponding activation of Wnt/β-catenin signaling and may be tissue specific (Fagotto et al., 1998; Geng et al., 2003; Goentoro and Kirschner, 2009; Guger and Gumbiner, 2000; Hagen et al., 2004; Kim and Hay, 2001; MacDonald et al., 2009; Shimizu et al., 2008; Wu et al., 2008). This may explain the heterogeneity of β-catenin staining intensity between tumors and suggests that additional/alternative diagnostic markers/methods need to be developed, such as quantitation of *bona fide* Wnt/β-catenin target genes. In addition, specific assays are needed to pinpoint Wnt/PCP and Wnt/Ca^{2+} pathway activation. In this regard, recently, a specific Wnt/PCP/JNK reporter has been described that can be utilized in *Xenopus* studies (Ohkawara and Niehrs, 2011). Future studies may identify characteristic gene expression signatures or other markers of Wnt/PCP and Wnt/Ca^{2+} signaling activation. The identification of unequivocal markers for Wnt pathway activation will be extremely useful for prognosis. In this regard, elevated levels of β-catenin, assessed by immunohistochemical analysis, correlate with poor prognosis in cancers such as breast and hepatocellular carcinoma (Camilli and Weeraratna, 2010; Lin et al., 2000).

From the standpoint of therapeutics, a number of different strategies are being developed to inhibit overactive Wnt signaling (see later in this chapter). As with any other cancer drugs, anti-Wnt signaling drugs will have to have the least side effects, so as not to affect normal tissue homeostasis (Barker and Clevers, 2006; Takahashi-Yanaga and Kahn, 2010; Takemaru et al., 2008; Verkaar and Zaman, 2011). However, the findings that in several biological contexts Wnt signaling may operate as a network and that there is crosstalk with other signaling pathways to regulate biochemical and cellular processes has profound implications (Beachy et al., 2004; Hu and Li, 2010; Katoh, 2007; Saadeddin et al., 2009). The high-throughput analyses and bioinformatic tools that systems biology provides may need to be applied in developing cancer therapeutics, in order to find the optimal targets for cancer treatment that may actually affect more than one signaling pathway.

Delineating the Wnt signaling network through large-scale genetic screens

Signal transduction pathways can be viewed as networks of interacting proteins that communicate information within the cell. Accordingly, deconstructing these networks into their constituent proteins and delineating the interactions between proteins is a major goal of signal transduction research. Genome-wide approaches to identification of regulatory components in signaling pathways typically lead to the implication of diverse (and frequently unexpected) sets of additional genes and proteins. Indeed, comparison of the lists of hits from genetic or proteomic screens often shows little or no overlap between studies. The approach that many studies take therefore is the "candidate gene" approach whereby the initial large-scale screen is gradually reduced to a smaller list of confident hits that can be validated using lower-throughput, but more laborious, methods. Despite this, RNAi and associated techniques that perturb the expression of specific genes are proving to be powerful means of identifying Wnt pathway components that might not be revealed using more classical genetics techniques. This section reviews how two complementary techniques, large-scale functional interaction screens using RNAi and large-scale interaction proteomics screens, are being used to delineate the Wnt signaling network. Rather than attempt to comprehensively review all studies in this area, we have selected several representative examples that illustrate the approaches and highlight some of the challenges.

RNAi techniques

The ability to selectively and specifically silence gene expression in mammalian cells has revolutionized the world of functional genomics. RNA interference (RNAi), in particular, whereby double-stranded RNAs mediate gene-specific, post-transcriptional silencing is an extremely powerful tool for gene function analysis. There are multiple techniques for achieving RNAi in mammalian cells, including the use of small interfering RNAs (siRNAs) that are short (19 to 22 bp) double-stranded RNAs. These can be of synthetic origin or processed within the cell from short hairpin RNAs encoded on plasmids (Moffat and Sabatini, 2006). In typical genome-wide screens in mammalian cells, siRNAs are pooled and introduced into cultured cells with an appropriate cell-based readout to record the effects of the gene silencing in an automated fashion. In spite of the power of genome-wide genetic screens, however, technologies such as RNAi are not without limitations and issues. In particular, the false positive and false negative rates of RNAi screens in mammalian cells are not well characterized. In addition, off-target effects of siR-NAs, whereby siRNAs perturb the expression of non-targeted genes, have been observed in multiple systems (Jackson et al., 2006). To mitigate these challenges, therefore, most successful large-scale screens take the form of primary screens followed by secondary screens with subsequent low-throughput validation of selected hits (Moffat and Sabatini, 2006). Successful screens are also highly dependent upon robust and reproducible readouts to assess perturbation. Many of the published RNAi screens in Wnt biology have made use of transcriptional reporters, described in the next section.

Wnt/β-catenin assays and readouts

Several classic readouts of perturbations to Wnt signaling have been used to identify components of the Wnt signaling pathway, such as axis duplication in *Xenopus laevis* (McMahon and Moon, 1989). In addition, as the mechanisms by which the transcriptional responses to Wnt signaling have been defined, robust transcriptional reporters have been developed that are now enabling genome-wide RNAi screens for Wnt modulators to be performed. Various transcription reporter systems have been developed that allow quantitative measurement of Wnt/β-catenin pathway activity with good sensitivity and dynamic range (Biechele and Moon, 2008). The first of these to be developed, TOPFlash, is a luciferase reporter construct with a promoter that either contains repeated wild-type TCF response elements or repeated mutant

TCF response elements (FOPFlash) as a control (Korinek et al., 1997). This reporter system, and subsequent more sensitive variants of it (Biechele and Moon, 2008), offers robust, convenient, and sensitive detection of Wnt/β-catenin signaling with high dynamic range. Coupled with large-scale RNAi, transcriptional reporters offer a powerful means of identifying genes that regulate the Wnt transcriptional response.

Applications to Wnt/β-catenin signaling

Several studies have used cDNA or RNAi screens to identify novel Wnt/β-catenin components and further delineate the Wnt signaling network. The TOPFlash reporter was used in conjunction with an arrayed cDNA library of approximately 20,000 clones to identify novel Wnt regulatory genes in human HEK 293T cells (Liu et al., 2005). This screen identified "leucine-rich repeat in Flightless interaction protein 2" (LRRFIP2) as a Wnt regulator. LRRFIP2 was shown to interact with Dishevelled (DVL) and to modulate Wnt/β-catenin signaling by increasing the cellular levels of β-catenin. The function of LRRFIP2 in embryogenesis was explored and it was shown that LRRFIP2 induces axis duplication and suppression of Wnt/β-catenin target genes in *Xenopus* embryos.

Model organisms such as *Drosophila* are tractable systems for performing genome-wide genetic screens and identification of potentially conserved Wnt regulators. RNAi screens have been used to uncover Wnt regulatory components using the TOPFlash reporter in *Drosophila* imaginal cell cultures (DasGupta et al., 2005). In the latter study, over 200 potential Wnt/β-catenin regulatory genes were found, and several of these validated and positioned in the Wnt/β-catenin pathway. The evolutionary conservation of Wnt signaling as well as the multiple tractable models for studying the Wnt pathway at the cellular and whole organism levels means that candidate genes from large-scale screens can typically be validated using multiple systems. Thus in the large-scale *Drosophila* screen (DasGupta et al., 2005), several of the candidate genes were also found to function as Wnt/β-catenin regulators in both mammalian cells and in zebrafish.

Aside from identification of new Wnt regulatory genes, high-throughput RNAi has been used to identify new *functions* for transcriptional regulators of the Wnt/β-catenin pathway such as the TCF/LEF genes TCF7L2 and TCF7 (Tang et al., 2008). Using the SuperTOPFlash reporter (Veeman et al., 2003), which contains eight TCF response elements, and screening approximately 21,000 pools of siRNAs in human cells, 530 siRNAs were identified from the primary screen. After several additional secondary screens, including cross-testing of significant genes using independent siRNAs from another library, approximately 100 positive and negative regulators were identified. Interestingly, additional characterization of TCF7L2, previously categorized as a positive regulator of Wnt signaling, suggested that TCF7L2 is potentially a repressor of Wnt/β-catenin activity and a tumor suppressor.

RNAi approaches may also be coupled with small molecule screens to increase the sensitivity and specificity of the screens. In a screen combining data from independent RNAi and bioactive compound screens, BTK (Bruton's tyrosine kinase) was identified as a negative regulator of Wnt/β-catenin signaling (James et al., 2009). To gain mechanistic insight into this observation, affinity-purification mass-spectrometry experiments were used to identify that a component of the PAF complex (CDC73), a known nuclear component of Wnt signaling, interacts with BTK.

More focused RNAi screens have targeted specific classes of proteins important for Wnt/β-catenin pathway regulation. Although CK1 and GSK3β are known to phosphorylate LRP6 under conditions of Wnt activation (Niehrs and Shen, 2010), it has been hypothesized that additional kinases regulate LRP6 phosphorylation (Davidson et al., 2005). Using a *Drosophila* kinome RNAi screen, a cyclin-dependent kinase (CDK) was shown to phosphorylate LRP6 (Davidson et al., 2009). Additional experiments showed that phosphorylation of LRP6 is indeed cell cycle regulated through the CDK-associated Cyclin Y.

These examples illustrate the important role that RNAi and associated techniques can play in delineating the larger Wnt signaling network. RNAi screens complement the more classical genetics and biochemical approaches that have played such an important part in understanding Wnt signaling. Although RNAi screens are a powerful means of identifying functional regulators of Wnt signaling, they do not provide mechanistic information on the specific interactions between proteins that make up the Wnt signaling network. The next section focuses on functional proteomics techniques that may be used to determine interactions between pathway components or mechanisms of regulation such as phosphorylation.

Defining the Wnt signaling network through interaction proteomics

With the capability to comprehensively identify and quantify peptides and proteins from biological samples, proteomics is a powerful suite of techniques for signal transduction research. Three principal layers of perturbation are key to signal transduction pathways and amenable to analysis through proteomics techniques. Changes in protein abundance (proteome dynamics), post-translational modifications (PTMs) such as phosphorylation, and physical interactions between proteins or re-configuration of protein complexes are all key mediators of signal transduction, including Wnt signaling (Choudhary and Mann, 2010). The aim of this section is to provide an overview of proteomics techniques and to highlight representative examples of the application of functional proteomics to delineation of the Wnt signaling network. We focus on large-scale techniques and applications of protein–protein interaction mapping to signaling networks.

Proteomics: fundamentals

There is an extensive literature on mass-spectrometry-based proteomics (Choudhary and Mann, 2010), and the intention here is to outline those techniques most relevant to the delineation of signal transduction networks. Proteomics is a suite of many different techniques, employing different methods of protein or peptide enrichment and separation, and different techniques for peptide identification and quantification. Enrichment techniques particularly useful for studying signal transduction networks include affinity-purification, using an antibody with affinity for a protein of interest or a class of post-transcriptional modification such as phosphorylation. Separation techniques can be broadly divided into gel-based (e.g., 2D-DIGE) techniques that separate proteins prior to protease digestion and peptide identification and those techniques that separate peptides prior to identification by mass-spectrometry. In the former technique, gel-based quantification is used to quantify proteins by using fluorescent tags, whereas in the latter, peptides are quantified in the mass-spectrometer.

We outline widely used techniques for shotgun peptide quantification since this is the method of choice for many large-scale proteomics studies of signal transduction networks. Large-scale shotgun proteomics experiments typically enable the identification and quantification of thousands of proteins across samples. Two of the most commonly used methods for quantification are "isotopic labelling" and "label-free" quantification of peptides. In isotopic labelling, heavy and light isotopes are separately incorporated (via growth of cells in labelled media or via chemical linkage) into proteins from the samples to be compared. The samples are then mixed and analyzed using mass-spectrometry. Peptides from the samples to be compared can then be distinguished by a mass shift corresponding to the label. In contrast, label-free proteomics compares samples by running each sample separately and comparing ion signals between mass-spectrometry runs. Both of these methods enable quantification of thousands of peptides per sample and have been successfully used in signal transduction studies.

Protein–protein interactions and complexes

Physical interactions between proteins and the higher-order organization of protein complexes and networks are the fundamental building blocks of signal transduction pathways. Traditionally, protein–protein interactions within signal transduction pathways have been defined using relatively low-throughput methods. Although these methods have contributed enormously to our in-depth understanding of pathways such as Wnt signaling, the focus here is on high-throughput technologies that provide the necessary data for network-based views of Wnt signaling (see later in this chapter). The two principal high-throughput interaction proteomics techniques that have contributed thus far to systems-wide protein–protein interaction mapping are affinity-purification mass-spectrometry (AP-MS) and yeast two-hybrid (Y2H). These two approaches to identifying protein interactions and complexes provide different but complementary views of protein interaction networks; AP-MS detects co-membership in protein complexes while Y2H detects binary protein interactions (Yu et al., 2008). In this section, we review these technologies and highlight several examples of their application to delineating the Wnt signaling network.

Affinity-purification mass-spectrometry (AP-MS)

AP-MS uses biochemical purification of protein complexes with subsequent identification of complex members using mass-spectrometry and thus combines the specificity of antibody-based protein purification with the sensitivity of mass-spectrometry (Collins and Choudhary, 2008). Current AP-MS strategies typically make use of epitope-tagged bait proteins that are expressed in cells and then purified with an antibody against the tag (Köcher and Superti-Furga, 2007). Several variants of this approach have been developed such as the tandem affinity purification (TAP) method, that includes two N-terminus or C-terminus tags, and can be used to purify complexes in mammalian cells (Gingras et al., 2005). Although truly genome-wide applications of AP-MS have thus far only been performed in yeast (Gavin et al., 2002; Ho et al., 2002; Krogan et al., 2006), medium-scale AP-MS studies in mammalian cells are beginning to delineate selected mammalian protein complexes and networks (Ewing et al., 2007; Glatter et al., 2009; Sowa et al., 2009).

Application of AP-MS to the Wnt signaling network has yielded both detailed mechanistic understanding of core Wnt components and has been used to identify novel interactors and regulators. By using epitope-tagged CTNNB1 (β-catenin), APC (Adenomatous Polyposis Coli), and BTRC (beta-transducin repeat containing) proteins as baits, FAM123B protein (also known as WTX, Wilms' Tumor on the X) was identified in associated protein complexes (Major et al., 2007). WTX was previously identified as being commonly mutated in Wilms' tumors (Rivera et al., 2007). Therefore the significance of finding WTX in association with CTNNB1 and APC, and showing that WTX negatively regulates Wnt/β-catenin signaling in mammalian cells, *Xenopus*, and zebrafish, is that it provides a possible mechanistic basis for the function of WTX as a tumor suppressor (Major et al., 2007).

Components that directly mediate the transcriptional response to Wnt/β-catenin signaling in colorectal tissues have also been identified using AP-MS. For example, by separating mouse intestinal crypt epithelia from villus epithelia, and using AP-MS, the Traf2 and Nck-interacting kinase (Tnik) was identified as a partner of mouse Tcf4 and β-catenin (Mahmoudi et al., 2009). Furthermore, Tnik is required for β-catenin-driven transcriptional activation in mouse cells, suggesting that it functions as a co-activator of Wnt signaling (Mahmoudi et al., 2009). Consistent with this, Tnik is essential for ectopic axis formation in *Xenopus* and for Wnt signaling in mammalian cells (Satow et al., 2010; Shitashige

et al., 2010); however, Tnik can also activate Jnk activity and regulate the cytoskeleton (Fu et al.,1999).

One of the principal challenges of interpreting AP-MS experiments is distinguishing specific interactors of the proteins of interest from promiscuously binding contaminants. Typically, control experiments are performed in parallel and quantitative proteomics techniques are used to identify those proteins that are more abundant in the "bait" samples than in the control. An ingenious extension of this approach was developed and used to identify β-catenin interactors (Selbach and Mann, 2006). Using RNAi in control samples to reduce the expression level of the bait protein allows true interactors to be distinguished from contaminants, since the contaminants are expressed at similar levels in bait and control samples while true interactors are found more abundantly in the untreated sample as compared to the RNAi-treated sample. This approach identified several of the known β-catenin interacting proteins important for Wnt/β-catenin signaling to be identified, such as TCF4 and CTNNBIP1, and will be a useful addition to the suite of interaction proteomics techniques for mapping signal transduction networks.

Yeast two-hybrid (Y2H)

In contrast to AP-MS, Y2H techniques specifically test whether two individual proteins physically interact in the cell, thus providing a pairwise binary readout of protein–protein interactions. High-throughput Y2H techniques developed from the initial observation that transcription factor activity can be reconstituted from physically separate activation (AD) and binding (BD) domains brought into close physical proximity (Fields and Song, 1989). By fusing the BD and AD to bait and prey proteins, respectively, the interaction of the bait and prey may be tested through reconstitution of transcription factor activity. Large-scale application of this technique is achieved by mating thousands of different yeast strains each expressing a different bait or prey fusion (Chien et al., 1991), and several related two-hybrid related techniques, such as mammalian two-hybrid, have been developed (Lievens et al., 2009).

Large-scale, unbiased Y2H screens have the potential to reveal novel and functional interactions pertaining to Wnt signaling. For example, several novel interactions in the Wnt signaling network were identified in a large-scale human Y2H screen in which a total of 3,186 interactions among 1,705 distinct proteins were identified (Stelzl et al., 2005). By intersecting the Y2H screen with known pathways, this study identified two novel AXIN1 interactors, ANP32A and CRMP1. Both of these were shown to negatively regulate Wnt/β-catenin signaling in cell-based assays, showing the utility of large-scale screens to identify both physically interacting and functional components of the Wnt signaling pathway.

A limitation of the original Y2H technique is the requirement for both bait and prey protein to interact in the nucleus of yeast cells, making it unsuitable for detecting protein–protein interactions that occur in other cellular compartments, such as at the cell membrane. Since membrane protein interactions are critical for receptor signaling, several techniques have been developed that specifically target membrane protein–protein interactions. The split-ubiquitin membrane two-hybrid system works by separately fusing two halves of the ubiquitin protein to bait and prey and tethering one of these to the cell membrane, thus enabling the detection of interactions that occur at the cell membrane (Stagljar et al., 2003). Application of this technique in Wnt signaling has led for example to the identification of novel Frizzled1 interactors that can mediate either Wnt/β-catenin or β-catenin-independent Wnt pathways (Dirnberger et al., 2008). Another variant two-hybrid assay, the Ras recruitment system (RRS) (Kohler and Muller, 2003), was used to identify Chibby (CBY1 in human), a negative regulator of Wnt/β-catenin signaling, in a screen for interactions with the β-catenin armadillo domain (Takemaru et al., 2003).

In summary, interaction proteomics techniques and RNAi-based techniques provide complementary approaches to identifying functional and/or physically interacting proteins in the Wnt signaling network. Furthermore, integration of data from different types of large-scale screen (RNAi, proteomic, small-molecule, etc.) is anticipated to play an increasingly important role in this area. An early example of this comes from a study in which data from an initial RNAi screen was integrated with protein interaction data to reveal AGGF1 as a novel chromatin remodelling-associated protein that regulates Wnt signaling in multiple systems (Major et al., 2008). As these techniques are used more widely, and as both specificity and sensitivity of the datasets improves, we can expect to reveal a more complete picture of the wider cellular network that regulates Wnt signaling.

The targetome of the Wnt/β-catenin pathway

Introduction

The Wnt/β-catenin pathway unfolds its various effects in cell differentiation, proliferation, and apoptosis by modifying the transcriptome (Vlad et al. 2008). Based on this awareness, it is significant to find out how exactly the Wnt/β-catenin pathway controls such various genes and so many different cellular functions. The so-called targetome, the totality of Wnt/β-catenin target genes, represents a key to understanding the manifold roles of the Wnt pathway. Analyzing and understanding the Wnt/β-catenin targetome and its multiple effects is the area of interest of many scientists in nearly all fields of biological research, e.g., embryonic development, cell biology, bioinformatics, cancer research, and therapeutic screening.

Over the last decades several techniques have been adopted to identify Wnt/β-catenin target genes. These include classical low-throughput methods as well as novel techniques that analyze entire genomes in a parallel, high-throughput fashion. The former include those "tortoise" methods that are invented to study gene by gene, for example RT-PCR, EMSA (Electrophoretic Mobility Shift Assay), Western blotting, and Sanger sequencing, the latter includes ChIP-chip (Chromatin ImmunePrecipitation), SAGE (Serial Analysis of Gene Expression), differential display RT-PCR, next-generation sequencing, and DNA microarrays. The last three exemplified methods will be briefly introduced in the following section of this chapter. The section after that then summarizes the most important aspects of the Wnt targetome and its functions, based on data mainly taken from the Wnt homepage (www.stanford.edu/group/nusselab/cgi-bin/wnt).

Methods: how to identify Wnt/β-catenin target genes

Differential display PCR (DDRT-PCR)

Differential display PCR is a technical variant of the conventional RT-PCR. Messenger RNA from eukaryotic cells is reverse transcribed and amplified using oligo-(dT)-primers that bind to the 3′ poly(A)-tail of all cDNA copies and a mix of random primers, which are supposed to bind to the 5′ ends. With this method, it is possible to separate different DNA segments corresponding to each mRNA molecule in the original mixture in a gel electrophoresis. DDRT-PCR allows the simultaneous analysis of the expression of many more genes than conventional RT-PCR (Liang et al., 1992). The method was used by Haertel-Wiesmann et al. to identify new Wnt3 target genes in two mammary epithelial cell lines (C57MG) that were transduced with either Wnt3 or control vector. For this, PCR reactions were performed with 240 primer pairs. The analysis revealed 26 differentially expressed genes. Nineteen of these were upregulated, e.g., *cyclooxygenase-2*, and seven were downregulated, e.g., *periostin*, in Wnt3 cells versus control cells (Haertel-Wiesmann et al., 2000).

DNA microarray

Expression levels of thousands of genes can be analyzed in a single experiment with a microarray chip, on which the entire transcriptome is represented by short immobilized DNA oligonucleotides (Schena et al., 1995). These single-stranded oligonucleotides can be spotted, printed, or directly synthesized onto the solid support, usually a glass slide, but can also be a nylon membrane or a silicon chip. After extraction the transcriptome is labelled with a radioactive isotope, or more often with a fluorescent dye, and hybridized to the microarray. A scanner detects the qualitative and quantitative binding intensity of the fluorescence and thus the amount of hybridized mRNA for every spot on the microarray. Huang et al. identified new direct or indirect Wnt targets playing a role in apoptosis by designing a microarray covering 1,384 apoptosis-related genes coupled with an RNAi knockdown of β-catenin in HeLaT-β-catenin-RNAi cells (Huang et al., 2006). Some of these 130 detected target genes may represent new possibilities for pharmacological intervention in tumor cells, because it has already been proved that an inhibition of the activity of the Wnt/β-catenin pathway induces apoptosis in some cancer cell lines, e.g., the blockade of Wnt-1 signaling, which induces apoptosis in human colorectal cancer cells (He et al., 2005).

Next-generation sequencing

Pyrosequencing is an accelerated sequencing method based on sequencing by synthesis taking advantage of the released pyrophosphate whenever a nucleotide is incorporated in an open $3'$ DNA strand. The ATP sulfurylase converts adenosine phosphosulfate (APS) plus pyrophosphate to ATP and this in turn is needed for the firefly luciferase to oxidize luciferin and emit a light pulse, which is recognized by a photo detector. If several identical nucleotides are incorporated in a row, the signal intensity is proportional to the number of nucleotides. Before the next dNTP is added, unused dNTP and ATP are removed by the enzyme apyrase (Ronaghi et al. 1998, 2001). In the near future, more genomes will be sequenced by high-throughput methods (up to 3 Gbp per run), like pyrosequencing, e.g., Illumina or 454. With the use of quantitative bisulfite pyrosequencing analyses, RT-PCR, and Western blotting, Svedlund et al. figured out that the Wnt/β-catenin signaling pathway is aberrantly activated in parathyroid carcinomas. The activation might be due to lost expression of *APC* caused by promoter DNA methylation (Svedlund et al., 2010).

Results of exemplified Wnt/β-catenin targetome studies

The transcriptomic side of Wnt

Microarray studies resulted in 1,886 target candidates that are differentially expressed in cells with activated Wnt/β-catenin pathway (Hallikas et al., 2006; Huang et al., 2005; van de Wetering et al., 2002). The identified target genes affect different kinds of metabolic and signaling pathways, regulational cascades, or cellular and biochemical functions, e.g., cell cycle, differentiation, metabolism, immune response, and cell adhesion. Overall, the identified genes play a role in at least 36 different pathways. Table 15.1 summarizes the various target pathways sorted according to the number of genes targeted by the Wnt/β-catenin pathway. The data is taken from the Wnt homepage.

The Wnt/β-catenin pathway has a multi-level targetome

The genes of the Wnt/β-catenin targetome can be categorized as direct or indirect Wnt/β-catenin target genes (Vlad et al., 2008; Ziegler et al., 2005), where direct target genes are regulated by a TCF/LEF promoter motif. Many direct targets operate as transcription factors, autocrine or paracrine factors, which themselves regulate the expression of secondary target genes or pathways. This results in a multi-level regulation, or a multi-level targetome. On the "first level," TCF/LEF regulates primary targets as an amplifier of the original Wnt signal. These targets are effectors, transcriptional regulators, and pathway regulators.

Table 15.1 Target pathways of Wnt signaling and the number of targets, which are involved in the corresponding pathway, sorted according to their biological effects.

Function	Target pathway	Number of targets involved in the pathways
Proliferation	CREB signaling	1
	VEGF signaling	2
	G-protein-MAPK activation	4
	erythropoietin signaling, NGF signaling	5
	MAPK signaling	7
	JAK-STAT signaling, EGF signaling	8
	ERK-PI3K (collagen) signaling	9
	hedgehog signaling, PDGF signaling, Wnt signaling (calcium)	10
Apoptosis	ACH-R apoptosis signaling, mitochondrial apoptosis control	1
	CD40L signaling, PTEN signaling	2
	anti-apoptotic pathway	3
	death receptor signaling, NF-κB signaling	4
	FAS signaling	5
Cell cycle	Rb signaling	1
	AKT signaling,	3
	p38 signaling, p53 signaling,	9
Differentiation	Notch pathway	1
	Regulation of myogenesis	2
	Embryonic stem cell development	3
Metabolism	Insulin signaling	6
	Lipid signaling	2
Inflammation, immune response	IFNα signaling, T-cell response	1
	IL-1 and IL-6 signaling, IL-2 signaling, IL-3 signaling, TGFβ signaling, toll-like receptor signaling	5
Cell adhesion	β-catenin/E-cadherin	2
	Integrin signaling, SAPK-JNK signaling	7
Regulation	WNT signaling	16
Diseases	Parkinson's disease	1
	Alzheimer's disease, Huntington's disease, prion disease	2
	Asthma	4
	Atherosclerosis	5
	HIV	6
	Breast cancer	13

In the "next level," the direct targeted transcriptional and pathway regulators activate further effectors and pathways. Hence the Wnt/β-catenin pathway can be considered as a highly ranked "master pathway" or network, which regulates genes with direct biological effects, transcription regulators, and other signaling cascades.

The regulational side of Wnt

By evaluating these identified target genes, it becomes clear that only a few genes are directly regulated via the TCF/LEF-transcription factor and the Wnt/β-catenin pathway. Most target genes are themselves regulated in a multi-level manner by other target genes or target pathways.

Surprisingly the pathway with the highest number of target genes is the Wnt/β-catenin pathway itself. This fact indicates the importance of a feedback regulation. The high investment in autoregulation indicates the importance of a tight control of the Wnt/β-catenin pathway. In addition, various temporal and functional levels of regulation assure that the cellular metabolism of a Wnt-activated cell does not get out of control. For example, *Dkk1* modulates the formation of the ternary complex LRP5/6, Frizzled, and Wnt ligand (Semënov et al., 2001). *Dkk1* is a target and a negative modulator of the canonical Wnt pathway at the same time, which results in a negative feedback loop. Another regulational step occurs through binding of sFRPs and WIF to the extracellular Wnt ligands and modulating its binding to Frizzled (Bovolenta et al., 2008).

Target genes differ in time point after activation

Not all Wnt/β-catenin target genes are induced simultaneously and immediately after activation of TCF/LEF. For example, the targetome of the colon cancer cell line named LS174T comprises 2,411 genes after transfection with dominant negative TCF.

Twenty-three hours after transfection it includes 1,199 genes, whereas the expression of 1,971 genes terminates and 759 additional genes are induced (van de Wetering et al., 2002). Another study analyzes the chronological expression of exemplary Wnt/β-catenin target genes in murine mammary epithelial cells (C57MG cells). In this study, the immediate expression (within one hour after activation) of *c-Myc* was identified, ensuing activation of cell proliferation (Röhrs et al., 2009).

Wnt signaling results in a waterfall modification of the transcriptome

Following this exemplified pathway, c-Myc activates a row of next-level genes resulting in regulation of cell cycle-dependent genes and preparing the cell for proliferation (for a list of c-Myc target genes see also www.myccancergene.org). In Wnt-activated cells c-Myc target genes can be considered as indirect targets of the Wnt/β-catenin pathway. Thus every next level of targeting results in more and more pathways affected by the "drop of Wnt activation" by β-catenin. The biological effects are immense: allocating the 36 identified Wnt/β-catenin target pathways regarding their regulation, each pathway is regulated by up to 16 target genes. Remembering the differentially expressed genes in different cell types, the impact of effective Wnt/β-catenin regulation and the consequences of malfunction become obvious.

The biological side of Wnt

Wnt signaling has manifold biological significance. It affects nearly all stages and aspects of cellular homeostasis, proliferation, and development. The biochemical functions of the Wnt targets, such as cell cycle kinase regulation, cell adhesion, hormone signaling, or transcription regulation, affect a wide variety of biological processes, e.g., proliferation, apoptosis, cell cycle, differentiation, metabolism, inflammation, immune response, and cell adhesion. Thus the targetome as a whole is responsible for a biological effect (Vlad et al., 2008).

Wnt targetomes are specific for cell type

Bioinformatical comparison of the different targetomes identified by microarray analysis found on the Wnt homepage, using an affymetrix data mining tool, revealed that less than 5% of all target candidates were found (Ziegler et al., 2005). Depending on species, organ, and cell type, most genes are differentially expressed. Interestingly, few genes reported to be in relation with Wnt signaling in one cell type are not affected in a second cell type (e.g., *Cyclin D1*; see Sansom et al., 2005; Shtutman et al., 1999; van de Wetering et al., 2002; Willert et al., 2002). This finding often comes along with differentially expressed Frizzled receptors and Lrp co-receptors on the cell surface. Nineteen different members of the WNT family, ten FRIZZLED (Fz) receptors, and two LRP co-receptors are known in human cells so far (Rao et al., 2010). Depending on the combinations of Wnts, Fz, and Lrp receptor, specific cellular responses are regulated: Wnt3a in combination with Dvl and Fz is known to activate the canonical pathway, whereas Wnt5a is known to activate the β-catenin-independent pathway (Cadigan et al., 2006).

Conclusions

Because of its various effects and its high biological impact, the Wnt/β-catenin pathway is one of the best studied pathways. Various methods have been developed and refined in order to better understand the connectivity and the context of Wnt/β-catenin signaling. The more we know, the higher is the probability to find specific pharmaceuticals against diseases such as cancer resulting from a malfunction of the Wnt/β-catenin pathway.

In order to analyze this information it is necessary to handle masses of data revealed from experimental and bioinformatical approaches. Conventional gene databases neither include details of the function nor data about the expression of genes. For example, an interactive platform might connect multiple available databases for a favored pathway and create targetome-maps (e.g., "Pathway Studio" by Ariadne Genomics). With the help of innovative experimental and bioinformatical tools we will be able to dissect the complex network of Wnt extracellular receptor protein interactions, intracellular protein modifications, and transcriptional regulation that give rise to specific multi-level biological responses (van Amerongen et al., 2009).

Wnt signaling as target for small interfering molecules

Introduction

Dysregulation of the Wnt signaling pathway leads to different diseases. The hyperactivation of Wnt/

β-catenin signaling induces cancers, whereas hypoactivation is linked to bone malformations and neurodegenerative disorders (Clevers, 2006; Verkaar and Zaman, 2011). For this reason there is a high medical interest in inhibitors and activators of the pathway. Unfortunately, the Wnt/β-catenin pathway lacks a specific enzymatic reaction that can be easily and specifically targeted: targeting the Wnt pathway means targeting a complex multidimensional signaling network. For a number of years, the focus has been on finding new small molecules that specifically target the Wnt/β-catenin pathway, particularly the regulation of β-catenin activation and its interactions with intracellular proteins. Hence several promising small molecules for drug intervention have been detected so far. This section introduces several molecules with a major focus on small molecules with inhibitory effects on the Wnt/β-catenin pathway and it gives a forecast in current drug discovery and clinical studies.

Strategies to find new modulators of the Wnt/β-catenin pathway

Hyperactivation of the Wnt/β-catenin pathway results in constitutive β-catenin-mediated transactivation of the TCF/LEF-dependent genes (Dihlmann and von Knebel Doeberitz, 2005). In order to interfere with this deregulated signal, a hypothetical drug would be optimal if it interferes with one well-defined and clear-cut biochemical activity of one single protein in the Wnt/β-catenin pathway while not affecting or affecting minimally the functions of the target protein in other pathways (Röhrs, 2007). A common strategy for drug screening is the cell-based reporter gene assay. In the case of Wnt/β-catenin signaling, TCF/LEF reporters, such as the TOPFlash described above, can be used to quantify relative transcriptional activity by assaying for luciferase activity (Korinek et al., 1997). In addition, bioinformatical approaches, such as molecular modelling and virtual screening, can be useful for detecting inhibitors, for example, of the β-catenin/TCF interaction based on pivotal amino acids of the β-catenin sequence (Trosset et al., 2006). For any drug candidates that influence the Wnt/β-catenin reporter activity, the exact target has yet to be identified in further biochemical experiments followed by biological experiments to explore cross-reactivity and toxicity. For example, the *APC–/–* mouse model that displays intestinal carcinomas

(Moser et al., 1992; Su et al., 1992) is suitable for different further experiments with interfering drug compounds. Upon biochemical and biological drug evaluation, subsequent directed chemical modifications will improve the specificity or the chemical attributes *in vivo*.

Exemplified molecules interfering with the Wnt pathway

Potential drug target candidates are found throughout the entire Wnt/β-catenin signaling cascade, including β-catenin and its interactors because of its central role in the cascade. Figure 15.2 shows the different levels of interfering classified into five groups. Remarkably, none of the detected inhibitors have a significant common chemical structure, although several molecules are poly-hydroxylated planar polyrings (Table 15.2).

Among the approved drugs for cancer treatment that could target Wnt/β-catenin activation is Gleevec. Gleevec, a small molecule inhibiting protein tyrosine kinases, is used for treatment of specific cancers such as gastrointestinal stromal tumors and has been an important achievement of drug development over the last few years (Capdeville et al., 2002). Gleevec® was connected with the Wnt/β-catenin pathway based on the suggestion that β-catenin is degraded after growth factor-mediated tyrosine phosphorylation (Danilkovitch-Miagkova, 2003; Papkoff, 1997; Playford et al., 2000). Indeed, Gleevec® is able to downregulate Wnt/β-catenin signaling activation in human colorectal cancer cells with mutation-activated Wnt/β-catenin signaling and in Wnt-1-induced cancer cells (Zhou et al., 2003). In addition, there are also some compounds with various or uncharacterized effects on the Wnt/β-catenin pathway. These include non-steroidal anti-inflammatory drugs (NSAIDs), such as acetylsalicylic acid, sulindac, and indomethacin, that inhibit cyclo-oxygenases and can lower the risk of colorectal cancer development (Herendeen and Lindley, 2003). Particularly, sulindac inhibits nuclear β-catenin localization (Boon et al., 2004) and acetylsalicylic acid activates N-terminal CK1/GSK3β phoshorylation of β-catenin promoting its degradation (Dihlmann et al., 2003). The NSAID celecoxib is approved for the treatment of familial adenomatous polyposis by the US Food and Drug Admistration (FDA).

A number of novel compounds have been identified. These include two low molecular weight

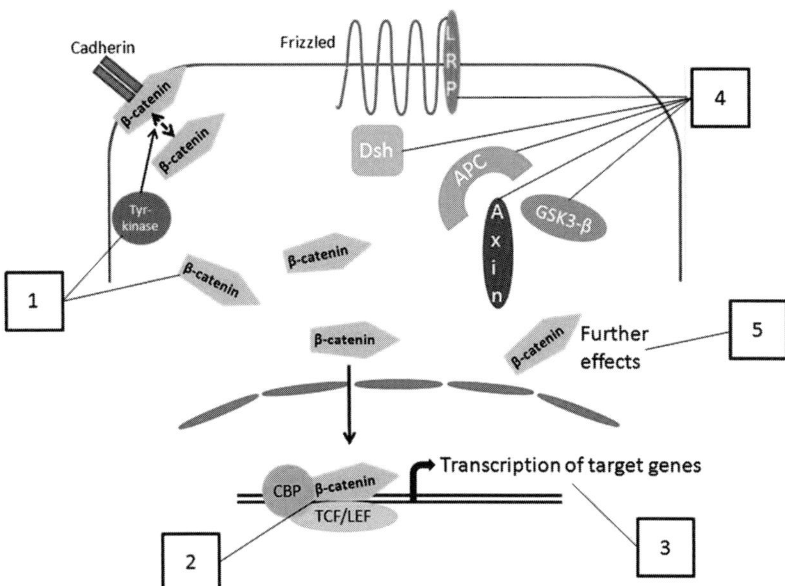

Figure 15.2 β-catenin/Wnt signaling activity in cancer cells and possible points of interfering: (1) modulators of β-catenin activation; (2) inhibitors of Tcf/β-catenin transcription complex; (3) inhibitors of target genes and target gene expression; (4) modulators of β-catenin regulating proteins (destruction complex and receptors); (5) molecules with further or unspecific effects on the Wnt pathway. (A black and white version of this figure will appear in some formats. For the color version, please refer to the plate section.)

compounds that were identified in a high-throughput ELISA screening study, PKF118–310 and PKF118–744, that were able to inhibit the interaction between β-catenin and the transcriptional activator TCF-4 (Lepourcelet et al., 2004). These two compounds are rare examples of molecules with influence on a specific protein–protein interaction. Recently, Chen et al. published two new possibilities to block Wnt/β-catenin signaling. First, small molecules (IWP) were identified that specifically target the membrane-bound acyltransferase porcupine, a protein important for Wnt secretion (Chen et al., 2009). Second, a compound class (IWR) that was able to stabilize the Axin degradation complex leading to β-catenin upregulation (Chen et al., 2009; Lu et al., 2009). The effects of IWR were tested on re-sected caudal fins of adult zebrafish, a process dependent on the Wnt/β-catenin pathway. Fish treated with IWR-1 failed to regenerate fin tissue; however, this effect is reversible as, after removal of IWR-1, treated fin tissue re-grows (Chen et al., 2009). In addition, IWR compounds were able to stabilize Axin proteins and induce β-catenin destruction even in the presence of mutated Apc protein. On the other hand, interfering with the activities of Wnt/β-catenin target gene products and other downstream modulators of the Wnt/β-catenin pathway has the potential for unspecific side effects. For example, flavopiridol inhibits cyclin D1, a Wnt-β-catenin target gene, which activates the

cyclin-dependent kinase 1 (CDK1) and consequently the progression of the cell cycle (Carlson et al., 1999). Thus cyclin D1 inhibition might lead to general inhibition of cellular proliferation.

Small molecules that activate Wnt/β-catenin signaling may be needed to treat diseases other than cancer. In this regard, Zhang et al. discovered a purine derivative, QS11, in a cell-based assay that is able to bind and inhibit the GTPase activating protein factor 1 (ARFGAP1) and synergizes with Wnt3a in the activation of Wnt/β-catenin signaling (Zhang et al., 2007).

The Wnt pathway and clinical trials

Some new promising small molecules targeting Wnt/β-catenin signaling have been identified, though none of them have entered the clinical phase of drug discovery until now (Verkaar and Zaman, 2011). Furthermore, the Wnt/β-catenin pathway is assumed to be linked with so-called cancer stem cells (CSCs). While CSCs behave similarly to normal adult stem cells, it is important to target CSCs in particular. First screenings of molecules interfering with the CBP/β-catenin interaction showed first approaches to specific target CSCs. Since 2007, several small and large pharmaceutical companies have concentrated their research on inhibiting Wnt signaling and announced their collaboration to find and to develop modulators

Table 15.2 Small compounds with influence on the Wnt pathway.

Compound	Experiment for identification	Molecular target	Effects
Quercetin Park et al., 2005; van Erk et al., 2005	Various	Various	Anti-oxidant Inhibits inflammation via affecting several initial processes of inflammation Inhibits TCF/β-catenin signaling Changes gene expression patterns of Wnt targets
Pyrvinium Saraswati et al., 2010	TOPFlash reporter gene assay	Activates CK1α	Stabilizes β-catenin, inhibits Axin degradation Inhibition of Myc, Dkk-1, Axin2 gene expression
QS11 Zhang et al., 2007	TOPFlash reporter gene assay	ARFGAPs, e.g., GSK3β	Synergist of Wnt31A Induces Axin2 and DKK1 expression
SKI-606 Coluccia et al., 2006	Yeast screen for Src inhibitors	ATP competitor for Src kinase	Inhibits β-catenin phosphorylation
ICG-001 Emami et al., 2004	TOPFlash reporter gene assay	CBP/β-catenin interaction	Inhibits CBP/β-catenin interaction
Sulindac sulfone (Exisulinid, Aptosyn) Boon et al., 2004	Selective test for anti-tumorigenic activity	cGMP phosphodiesterase	Induces and activates protein kinase G Induce apoptosis Increases β-catenin phosphorylation promoting its degradation
10058-F4 Yin et al., 2003	Yeast two-hybrid screen for inhibitors of c-Myc/Max interaction	c-Myc/Max interaction	Inhibits cell cycle progression Inhibits tumor growth
NO-ASA Nath et al., 2003	Molecular design of anti-inflammatory and anti-thrombic drugs	COX TCF/β-catenin interaction	Inhibits COX pathway Activates guanylate cyclase Inhibits TCF/β-catenin interaction
SMAF-1 Karaguni et al., 2004	TOPFlash reporter gene assay	COX, APC	Inhibits COX pathway Stimulates APC function
NSAIDS Dihlmann et al., 2003	Various	COX, various	Various effects on the Wnt pathway, e.g., increase in β-catenin phosphorylation
Flavopiridol Carlson et al., 1999; Shapiro, 2006	Cell proliferation screen	Cyclin-dependent kinases	Inhibits cyclin D1 expression
NSC668036 Shan et al., 2005	Structure-based virtual ligand screen	Dsh PDZ domain	Inhibits interaction with Fz
6-bromoindirubin-3′-oxime (BIO) Meijer and Raymond, 2003	Indirubins testing and screening for activity	GSK3β inhibitor	Activates Wnt signaling
Piperidinyl diphenylsulfonyl sulfonamides Moore et al., 2010	Competitive binding assay TOPFlash reporter gene assay Computational optimization	Inhibitor of sFRP-1	Disruption of sFRP/Wnt interaction

Table 15.2 *(cont.)*

Compound	Experiment for identification	Molecular target	Effects
Hexachlorophene Park et al., 2006	TOPFlash reporter gene assay	Siah-1	Promotes β-catenin degradation Inhibits tumor growth
PNU-74654 Trosset et al., 2006	Molecular modelling and virtual docking	TCF/β-catenin interaction	Inhibits TCF/β-catenin interaction
PKF118–310 and PKF118–744 Lepourcelet et al., 2004	HTS ELISA of TCF/ β-catenin interaction	TCF-4/β-catenin interaction	Inhibits TCF-4/β-catenin interaction
Gleevec Zhou et al., 2003	Biochemical screen for PDGFR inhibitor	Tyrosine kinase PDGFR, abl-kinase, c-kit	Inhibits Wnt-induced activation of β-catenin
Differentation-inducing factors (DIFs) Takahashi-Yanaga and Sasaguri, 2009	Testings for Wnt activity	unknown	Activation of GSK-3β Suppression of cyclinD1 expression Inhibit proliferation
SU5416 Ye et al., 2006	Cellular tyrosine kinase assay	VEGF receptor	Inhibits angiogenesis Inhibits VEGF-dependent mitogenesis of endothelial cells, inhibits tumor growth
Sorafenib (BAY 43–9006) Keating and Santoro, 2009	Raf-interaction screens Clinical trials (phase III studies)	VEGF-2, VEGF-3, PDGFR-β, c-KIT, FLT3, Raf-1	Inhibition of tumor growth
IWPs and IWRs Chen et al., 2009; Lu et al., 2009	TOPFlash reporter gene assay Zebrafish tail regeneration studies	Wnt production, Porcupine acyltransferase Wnt pathway response	Inhibit aberrant cell growth Stabilize APC Decrease Axin2 expression Promote β-catenin degradation
Murrayafoline A Choi et al., 2010	TOPFlash reporter gene assay	β-catenin response transcription	Represses expression of cyclin D1 and c-myc Promoting degradation of intracellular β-catenin

(Modified from Röhrs, 2007)

of the Wnt signaling as new therapeuticals. Principal molecules include small molecules as well as antibodies and small therapeutic proteins, while the latter actually are to enter phase I clinical tests (Bayer Schering Pharma, 2010). Clinical tests for new small-molecules compounds were due to start in 2011.

An indirect Wnt target drug is sorafenib, an anticancer drug (marketed as Nexavar®) that targets the VEGF-receptors in renal cell carcinoma and primary liver carcinoma. This drug finally entered phase III clinical tests (Keating and Santoro, 2009). VEGFR is a target of the β-catenin response transcription (CRT).

Conclusions

The identification of compounds targeting Wnt signaling is a main task in therapeutical cancer research. Over the last few years several substances have been identified that affect Wnt signaling at different levels. Thus it is a question of time until the first compound enters clinical trials.

Omics data analysis and integration

The landscape of data analysis methods has evolved concurrently with the advancement of high-throughput technologies such as microarray, mass-spectrometry, robotics, and next-generation sequencing. The modern fields of bioinformatics and systems biology analyze and integrate high-throughput data to gain biological insights, adapting principles and techniques from classical fields such as mathematics, statistics, computer science, and physics. This section discusses popular methods for analyzing and integrating heterogeneous high-throughput data to further understand the Wnt signaling pathway.

Processing omics data

High-throughput proteomic and transcriptomic technologies can analyze thousands of proteins or tens of thousands of genes, respectively. However, with increased sample complexity and throughput, it becomes vital to account for biological and technical noise, the effect of which (and other confounding variables) is removed via normalization. For example, Robust Multichip Average (RMA), a popular normalization method for processing microarray data (1) subtracts the background intensity, (2) integrates multiple probe values into a single value for each probe set, and (3) sets all arrays to the same scale using quantile normalization (Irizarry et al., 2003). After normalization, the next step typically is to identify the subset of detected genes/proteins with good signal that are meaningful in a given condition. The simplest strategy is to threshold the data, e.g., select genes with at least two-fold differential expression. However, it is important to take into account the reliability of the fold-change calculation, which is dependent on the overall signal in the data (Zhang et al., 2010). Furthermore, identifying groups of meaningful genes or proteins is hindered when analyzing multidimensional omics data where thousands of biomolecules are measured across several potentially correlated conditions.

Clustering multidimensional data

Threshold-based approaches are often replaced or combined with pattern recognition strategies for identifying meaningful groups of genes in multidimensional high-throughput data. A popular pattern recognition strategy is unsupervised clustering where biomolecules (genes or proteins) are typically grouped based on common patterns of expression across all samples. Similarly, clustering can also be applied to group samples (or conditions) based on common patterns of expression across all genes. Two-dimensional hierarchical clustering is often applied to gene expression microarray data to identify groups of related genes or samples with similar gene expression. Kool et al. clustered gene expression profiles of 1,300 genes across 62 medulloblastoma samples and identified five distinct clusters of medulloblastoma subtypes. Pathway enrichment analysis (see below) revealed that the smallest but most distinctive cluster was characterized by overexpression of several genes in the Wnt signaling pathway including *LEF1*, *AXIN2*, and *WNT11* (Kool et al., 2008). The nine tumors in this cluster were associated with a medulloblastoma subtype involving activation of the Wnt/β-catenin pathway and β-catenin mutations. Clustering of other omic datasets, besides gene expression data, has also been noted in Wnt-related literature. Tang et al. identified protein groups from quantitative phosphoproteomic data (Tang et al., 2007) using another popular clustering algorithm called k-means. HEK293 cells were stimulated for 0, 1, or 30 minutes with Wnt3a and the relative phosphorylation levels of 1,057 proteins were quantified, out of which 287 proteins showed ≥ 1.5-fold changes. These 287 proteins included new potential Wnt/β-catenin intermediates and were clustered into four groups based on their quantitative phosphorylation profile pattern. Clustering is one of several methods borrowed from the field of machine learning, which has also contributed classification and other statistical modelling algorithms for biological data analysis.

Enrichment analyses

While clustering facilitates organizing multidimensional omics data to extract groups of meaningful genes, gaining biological insight requires further analyses. Since manual literature exploration for each gene is tedious and subject to human error, several databases have been developed in recent years for automating biological analysis of high-throughput data. These databases organize gene sets based on common biological themes. For example, literature-curated pathway databases such as KEGG (Kanehisa et al., 2006), BioCarta (www.biocarta.com), WikiPathways (Pico et al., 2008), and Panther (Mi and

Thomas, 2009) provide pathway maps and the ability to search genes within them. The Wnt pathway definition across each of the databases varies and it is therefore important to analyze omics data using multiple databases (Figure 15.3). Aggregate database and web servers such as DAVID (Dennis et al., 2003) and GATHER (Chang and Nevins, 2006) give scientists the ability to search genes across multiple databases at once. Furthermore, they provide statistical tools for evaluating the overlap between an input gene list and the pathway definitions. The most common approach is to conduct a Fisher's exact test, which utilizes the hypergeometric distribution to calculate a p-value representing significant overlap while accounting for the size of the gene list, pathway definition, and background. More advanced strategies such as gene set enrichment analysis (GSEA) (Subramanian et al., 2005) incorporate relative quantitation of the genes to compute overlap significance. Such strategies where the enrichment of a particular biological theme of interest is enriched within experimentally derived genes are collectively referred to as enrichment

analyses. Enrichment analyses provide the advantage of being robust to unidentified genes in high-throughput data, as experimental data are evaluated on a whole system or theme level as opposed to at the specific gene level. Pathway enrichment analysis, as exemplified by the aforementioned medulloblastoma study by Kool et al., is used to reveal (dys)regulated pathways within omics datasets (Kool et al., 2008). It is important to note that there is often overlap between biological themes, which is most apparent in mammalian signaling pathways where crosstalk is evident (Dumont et al., 2001). KEGG pathway enrichment analysis of the Wnt pathway members identifies several signaling pathways that crosstalk with the Wnt pathway (Figure 15.4).

The Gene Ontology (GO) (Ashburner et al., 2000) database is another resource for exploring biological themes within data. GO is divided into three categories: (1) molecular function, (2) cellular component, and (3) biological processes. Molecular function distinguishes proteins based on their functional role in the cell and includes terms such as "MAP kinase activity." The cellular component category groups proteins based on cellular localization and, at a higher resolution, complexes such as "IkappaB kinase complex." The biological processes category is similar to the idea of pathways and includes terms such as "MAPKKK cascade" that comprise proteins participating in a particular biological event. In the same manner as pathway enrichment, GO term enrichment can also reveal functional groups within omics data. For example, Rossol-Allison et al. identified GO terms related to cell growth and size, development, and morphogenesis to be enriched in genes derived from microarray data upon Wnt3 stimulation and Rho activation (Rossol-Allison et al., 2009).

Networks

Meta analysis provides a quick way for finding biological themes present in omics data. However, what defines the specific context of an experiment is usually the specific alterations of relative activity or expression of individual pathway members. To understand how these specific alterations lead to downstream, genome-wide changes in cellular behavior, a higher resolution map of biomolecular connectivity is required. Protein–protein interaction (PPI) networks, where each node represents a single protein and an edge connects two nodes if the protein pair physically

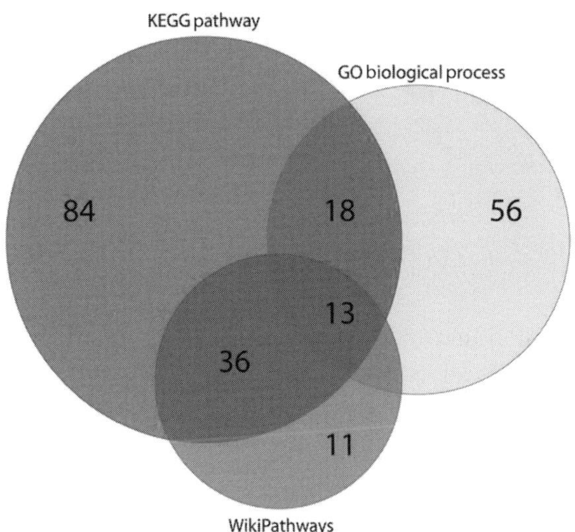

Figure 15.3 Disparity in Wnt signaling pathway definitions across annotation databases. Biological annotation databases such as KEGG and GO are widely used in omics data analysis. However, independent curation processes lead to different pathway definitions. The Venn diagram highlights partial overlap across KEGG pathway, WikiPathways, and GO definitions of the Wnt signaling pathway that contain 151, 60, and 87 protein members, respectively. Although all three definitions claim to include the canonical Wnt/β-catenin pathway as well as non-canonical planar cell polarity (PCP) and Wnt/Ca^{2+} pathways, there is an overlap of only 13 proteins out of a total 218 unique proteins across all three databases. (A black and white version of this figure will appear in some formats. For the color version, please refer to the plate section.)

interacts, have been widely used to visualize high-throughput data. PPI databases such as BioGRID (Stark et al., 2006) and HPRD (Keshava Prasad et al., 2009) store hundreds of thousands of physical interactions from yeast two-hybrid and affinity-purification experiments described earlier. Context-specific sub-networks, where the nodes are genes or proteins experimentally determined to be of interest, can be constructed from high-throughput data with edges extracted from PPI databases. It is often informative to also examine the direct physical inter-actors that may be relevant but were not identified in the experiment. Figure 15.5, constructed using Path-way Palette (Askenazi et al., 2010), shows physical interactors of Wnt pathway members. The advantage of generating context-specific networks is that they can be further analyzed using graph theoretic algo-rithms. For example, Kestler and Kuhl (2008) gener-ated an integrated PPI network of Wnt pathway components and showed that it exhibited small-world and scale-free properties present in known cellular networks; the authors computed the mean shortest path length to be 3.5 and mean node degree to be 2.5. The network contained several hubs such as

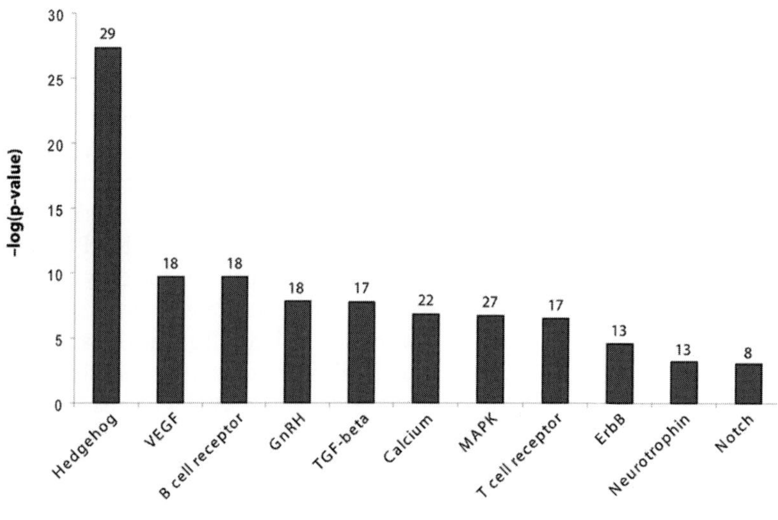

Figure 15.4 Wnt pathway crosstalk with other signaling pathways determined using enrichment analysis. Several proteins in the Wnt pathway participate in other signaling pathways as well, indicating crosstalk. Enrichment analysis via Fisher's Exact Test was used to compute the degree of overlap between Wnt pathway and other signaling pathways in KEGG. Only significantly overlapping, i.e., enriched, signaling pathways (FDR corrected p-value <0.05) are shown. The height of the bar indicates degree of significant enrichment and the number above each bar indicates the number of proteins in both Wnt and the respective signaling pathway. (A black and white version of this figure will appear in some formats. For the color version, please refer to the plate section.)

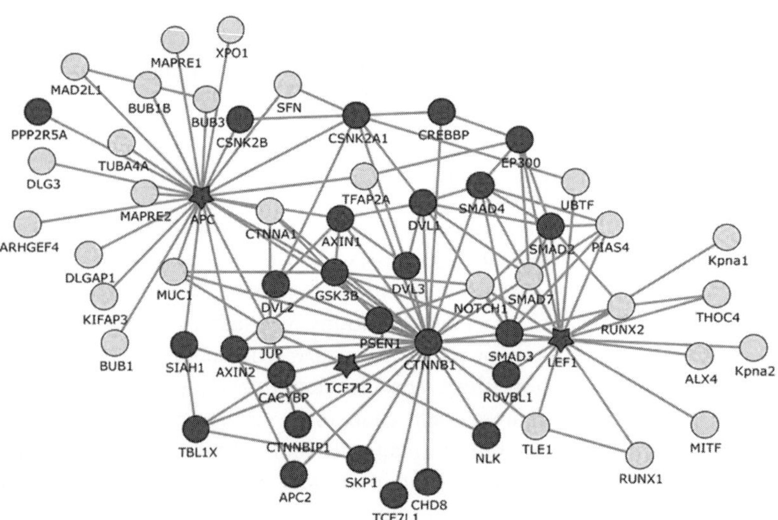

Figure 15.5 Partial Wnt pathway protein–protein interaction network. Protein–protein interactions with APC, TCF, and LEF with Wnt pathway members (dark nodes) are displayed as a network using the Pathway Palette web tool. Physical interactions are retrieved from the BioGrid database. Interacting partners outside the Wnt pathway are also shown as light nodes. Note that the neighborhood of the Wnt pathway members resembles a complex network as opposed to a linear pathway. (A black and white version of this figure will appear in some formats. For the color version, please refer to the plate section.)

Dishevelled, Axin, β-catenin, and Lef with more than 15 interactions each. The authors also applied a network motif discovery algorithm to identify recurring patterns such as triads and tetrads within the network.

In silico prediction of Wnt targets

Pathway enrichment analysis and PPI networks provide insight into if and how the Wnt pathway is regulated in an experiment. However, in order to determine whether the Wnt pathway was causally responsible for regulating a set of genes, it is first necessary to identify targets of the Wnt pathway. One strategy for determining gene targets of pathways is to create transcriptional signatures derived from gene expression data following perturbation of the pathway. Segditsas et al. identified a signature consisting of 211 genes that were consistently differentially expressed across 16 human and 63 mouse adenomas, each with germline *APC* mutations (Segditsas et al., 2008). Since loss of *APC* immediately perturbs Wnt signaling (Sansom et al., 2004), genes differentially expressed in such a condition are potential targets of the Wnt pathway. However, some differentially expressed genes may be indirect Wnt targets given crosstalk between Wnt and other pathways or secondary target genes as described earlier. Furthermore, signatures may differ in different cell types. The Signaling Pathway Enrichment using Experimental Datasets (SPEED) database contains curated perturbation gene expression datasets for several pathways (Parikh et al., 2010), including six Wnt/β-catenin pathway perturbation experiments from different studies in heterogeneous cell types with 916 genes differentially expressed in at least one study. However, there are no genes that are differentially expressed in all six studies, and only three genes are differentially expressed in five out of six studies. Notably, one of the three genes is DKK1, a negative regulator of the Wnt/β-catenin pathway (Niida et al., 2004). The authors clustered all signaling pathways in SPEED according to the perturbation signatures of differentially expressed genes, and revealed that the major clusters separated immune response pathways (JAK/STAT, TLR, IL1, and TNFα) from pathways controlling cell cycle and cell growth (VEGF, MAPK, PI3K, TGFβ, H_2O_2, and Wnt/β-catenin). Although the Wnt/β-catenin signaling pathway belonged to the cell cycle and growth cluster, it had the least overlap with other pathways in the cluster. The varying targets across different conditions as well as the low similarity with targets of other related pathways demonstrates the challenges in elucidating downstream effects of alterations in the Wnt/β-catenin pathway.

A more direct approach for target identification is to use chromatin immunoprecipitation followed by massively parallel sequencing (CHIP-seq), where the downstream transcription factors crosslinked with chromatin are immunoprecipitated and the DNA is sequenced. Bottomly et al. used CHIP-seq to identify gene targets of the β-catenin/TCF4 complex (Bottomly et al., 2010). Binding sites of transcription factors have variation at some nucleotide positions and are therefore represented with weights at each position. Such weighted representations called position weight matrices (PWMs), often visualized as sequence logos, are subsequently used for *in silico* discovery of new targets. For example, Hodar et al. identified 89 novel Wnt pathway targets using PWMs for several transcription factors with TCF/LEF1 being most relevant for target identification (Hodar et al., 2010).

Conclusion

The use of computational methods for integrating new experimental data with previous knowledge has become an indispensable part of modern biological research. Existing tools from computational sciences have been successfully adapted and applied, such as probabilistic and statistical procedures for cleaning messy data, machine learning methods for finding patterns in multidimensional data, hypothesis testing protocols for identifying enriched biological themes, and graph theoretic algorithms for making sense of complex context-dependent cellular networks. New algorithms and applications need to be developed that specifically address issues such as crosstalk in analyzing signaling pathways. For example, graph theory methods such as the Steiner Tree have recently been used to combine proteomic, transcriptional, and interactome data for integrative pathway analysis in *S. cerevisiae* (Huang and Fraenkel, 2009). Mammalian signaling systems are far more complex than single-celled yeast, and continued efforts will be required to develop novel analytic and integrative methods to maximize the utility of available omics data, and to gain further insights into signaling systems.

Computer simulation of Wnt circuits

Introduction

Is it possible to map signal transduction networks in sufficient detail to permit computer simulations that can be qualitatively and quantitatively compared to experiments? While whole-genome computer simulations of metabolic networks are becoming more widespread, computer simulations of signal transduction networks face several significant challenges. In metabolic networks, metabolites are related to one another via well-defined stoichiometric chemical transformations catalyzed by metabolic enzymes or enzyme complexes. Although it is still not possible to construct and simulate a full chemical kinetic model of the entire metabolic network due to the many missing reaction rate constants and binding constants, the stoichiometry of the metabolic reactions poses powerful constraints on metabolic fluxes. This enables a wide variety of whole-genome simulation methods such as the flux balance analysis. In addition, the functional interactions between metabolic enzymes usually follow simple rules of logic: two proteins are connected to a metabolic reaction via an "OR" gate if they are isoenzymes that catalyze the same reaction; and via an "AND" gate if they are essential subunits of an enzymatic complex. In contrast, in signal transduction networks, what flows though the network is not metabolites but information. Unlike metabolic transformations, information "transformations" are quantified via dose-response curves relating input signals to output signals, which is difficult to measure in practice. Information flow does not obey stoichiometric flux balance, as one activated receptor can in turn activate many downstream targets. Integration of multiple upstream signals may be guided by highly complex rules of logic that go beyond simple "OR" and "AND" gates. As a result of these inherent complexities, whole-genome computer simulations of signal transduction networks are currently not feasible. However, it is becoming more widespread to construct and simulate pathway-specific circuits. Several models have already been built on different aspects of Wnt circuits, and the resulting simulations have revealed significant insights into the dynamics, evolution, and design principles of Wnt circuits. These simulation studies are summarized below.

Simulation of the *Drosophila* segment polarity network

An early simulation work on Wnt circuits is the analysis of the segment polarity gene network in *Drosophila* embryonic development by von Dassow et al. (von Dassow and Odell, 2002; von Dassow et al., 2000). This gene network involves interactions among products of five genes, including the *Drosophila* Wnt homolog *wg* and a homeodomain transcription factor *engrailed* (*en*), and it plays a key role in the establishment and maintenance of a highly conserved periodic expression pattern in spatially adjacent cells. A set of non-linear ordinary differential equations (ODEs) was used to describe how concentrations of component molecules change over time. The geometric patterns of cell–cell interactions were fixed, and cell–cell communications were modelled as simple transport reactions. There are nearly 50 free parameters in the model, including various rate constants, binding constants, and cooperativity coefficients (Hill coefficients) that describe chemical reactions and dose-response curves. Most of these free parameters are unknown, and they are randomly sampled within biologically realistic ranges. It was found that the model maintains the desired segment polarity pattern at steady state for a large fraction of these random parameter sets. This suggests that the *Drosophila* segment polarity gene network is a robust developmental module that is resistant to environmental or genetic perturbations to kinetic parameters, and this robustness is dictated by the topology of the network. This segment polarity network model was subsequently improved to account for cell division (Ingolia, 2004). A key feature of these polarity network models is the two positive feedback loops for *wg* and *en*. These positive feedback loops act as bi-stable switches under certain parameter settings, which are necessary for the formation of the correct segment polarity pattern (Ingolia, 2004).

Because the robustness of the segment polarity gene network is dictated by network topology, the function of the network should be largely independent of the detailed parameter values as well as the detailed rules for interactions among components. Albert and colleagues tested this hypothesis by proposing a Boolean model of the *Drosophila* segment polarity network, where component levels are either ON or OFF, and interactions among components are formulated as logical functions (Albert and Othmer,

2003; Chaves and Albert, 2008). It was found that the Boolean network model can indeed reproduce the correct segment polarity patterns. Finally, hybrid models that combine the advantages of both ODE and Boolean models have also been applied to the segment polarity gene network (Dayarian et al., 2009).

Is it possible to design a segment polarity gene network that is ever more robust than the *Drosophila* network? To answer this question, Ma et al. exhaustively enumerated alternative network topologies, and ranked them according to their robustness against parameter variation (Ma et al., 2006). The *Drosophila* network is ranked among the top, suggesting that it is near optimal in terms of robustness. In the most robust model, *wg* and *en* are mutually exclusive within a cell but activate each other between cells, consistent with a previously proposed minimalist theoretical model (Meinhardt and Gierer, 1980). As a robust and stable scaffold, the *Drosophila* segment polarity network could be easily modified or extended in different temporal and spatial contexts (Ma et al., 2006), consistent with the observed high level of pleiotropy for this network (Galis et al., 2002). The principle that some biological scaffolds are more robust than others against environmental and mutational perturbations seems to be broadly applicable to a wide range of biological systems, from biological networks to protein structures (Li et al., 1996; Xia and Levitt, 2004). Finally, although the segment polarity gene network is robust against many types of parameter perturbations, there is enormous variation ($\sim 10^6$-fold) in robustness when the parameters are collectively perturbed in different ways (Gutenkunst et al., 2007).

Simulation of the Wnt/β-catenin pathway

Lee et al. developed a mathematical model for the core components of the Wnt/β-catenin pathway including Wnt and β-catenin, and five additional components: Dsh, GSK3β, Apc, Axin, and TCF (Lee et al., 2003). All reactions in the pathway were assumed to be elementary reactions, and the law of mass action was used to construct a set of ODEs that describes how concentrations of component molecules change over time. The free parameters in the model are mostly rate constants, and they were calibrated against a variety of thermodynamic, kinetic, and flux data from experimental data in *Xenopus* egg extracts. Many

predictions from the resulting model were experimentally validated in the *Xenopus* egg extract system. Metabolic control analysis was used to characterize the effect of a particular reaction or species on the function of this protein network (for example, in terms of the level of β-catenin) (Kruger and Heinrich, 2004; Lee et al., 2003). Computer simulations revealed that although the level of β-catenin protein is sensitive to small perturbations in the Wnt/β-catenin pathway, the fold-change in β-catenin protein levels (before vs. after Wnt stimulation) is robust to most perturbations in the Wnt/β-catenin pathway (Goentoro and Kirschner, 2009). This prediction was subsequently verified experimentally (Goentoro and Kirschner, 2009). Moreover, the fold-change in β-catenin level, rather than its absolute level, determines the ultimate phenotypic effects of Wnt/β-catenin pathway perturbations in experiments (Goentoro and Kirschner, 2009). This means that the downstream transcriptional system is responding to fold-changes in β-catenin level rather than β-catenin absolute level. The fold-change reading task can be accomplished by an incoherent feed-forward loop, a network motif previously known to be capable of biological adaptation (Goentoro et al., 2009). By responding to fold-changes in signals rather than absolute levels, the Wnt/β-catenin pathway and the downstream transcriptional system become more robust against environmental or genetic perturbations.

The core Wnt/β-catenin pathway model developed by Lee et al. (2003) provided the basis for subsequent analysis in tumorigenesis. Cho et al. used the Lee model to study the effects of tumorigenic mutations in Wnt/β-catenin pathway core components *APC*, *CTNNB1*, and *AXIN* (Cho et al., 2006). It was found that while in colorectal cancer the majority of the mutations occur in *APC*, *APC* mutations are not selected to maximize the level of β-catenin protein, but to optimize the balance between the level of *CTNNB1* and its target and negative regulator, *AXIN2*. Cho and co-workers have further coupled the core Wnt/β-catenin pathway model with the ERK pathway via positive feedback loops between the two pathways that generates stability in the activation of both pathways during carcinogenesis. In contrast, Wnt/ERK modelling suggests that EGF and Wnt pathways may play different roles in epithelial-to-mesenchymal transition (EMT) processes associated with metastasis (Kim et al., 2007; Shin et al., 2010).

Simulation of Wnt pathway oscillatory dynamics and pattern formation

The described Wnt circuit models are largely analyzed in terms of the steady-state behavior. However, expression of proteins in the Wnt pathway is known to change upon Wnt signaling and to oscillate during embryo development (Aulehla et al., 2003; Dequeant et al., 2006). To model such oscillatory dynamics, Wawra et al. introduced time-delayed negative feedback loops into the core Wnt/β-catenin pathway model, and showed that the resulting model can display robust oscillatory behavior (Wawra et al., 2007). Time-delayed negative feedback loops are essential features of other recently proposed Wnt oscillator models as well (Goldbeter and Pourquie, 2008; Jensen et al., 2010; Rodriguez-Gonzalez et al., 2007).

The above Wnt circuit models also assume that the components are well mixed within the cell. Cell–cell communications are usually modelled as simple transport reactions across a fixed cell–cell interaction boundary. However, diffusion plays a central role in many developmental pattern formation and morphogenesis phenomena, and the reaction-diffusion model of Turing (1952) is often necessary to capture the essential features of the spatial pattern establishment and maintenance. To model reaction-diffusion systems, a set of partial differential equations (PDEs) is used to describe how concentrations of component molecules vary over time as well as space. These pattern formation models have produced predictions that are in quantitative agreement with experiments for Wnt-DKK interactions (Sick et al., 2006) as well as the Wnt/PCP pathway (Amonlirdviman et al., 2005; Ma et al., 2008).

Conclusions

Quantitative experimental data on signal transduction networks is rapidly accumulating. As a result, computer modelling and simulation of signaling circuits such as Wnt are becoming increasingly important. However, significant challenges lie ahead. First, there are a plethora of modelling and simulation methods. For a specific biological problem under investigation, models can be too simple to be relevant, or too complex to be accurately parameterized. It remains a challenge to formulate the right model, possibly a hybrid one, with the maximal explanatory and predictive power. Second, most of the model parameters are likely unknown, and they need to be calibrated against a heterogeneous collection of thermodynamic, kinetic, concentration, and flux measurements *in vitro*, *in vivo*, at steady state, and in response to perturbations. Currently this process is done largely through trial and error. It remains a challenge to completely automate this process. Third, current simulations focus on small parts of the Wnt circuit. However, the Wnt circuit as a whole is highly pleiotropic. The Wnt circuit interacts heavily and dynamically with other signaling and regulatory modules. It affects a multitude of downstream events, and is associated with a large number of phenotypes. It remains a significant challenge to simulate the Wnt circuit while taking full account of its interdependencies with other functional modules in the cell. Finally, and perhaps most importantly, in addition to these small-scale circuit models, Wnt-mediated signal transduction is also being quantified by large-scale functional genomic, proteomic, and network mapping experiments (reviewed in earlier sections). It remains a central challenge to integrate these models and datasets across multiple scales to capture the full complexity of Wnt-mediated signal transduction networks within and between cells.

Outlook

Wnts control different cellular processes that ultimately regulate embryonic development, homeostasis, and diseases such as cancer. The challenge is to utilize all the tools that biochemistry, cellular biology, and systems biology provide to help understand how the Wnt network functions normally and during tumorigenesis, and to use the information for better tumor diagnosis, prognosis, and cancer therapeutics. The above sections describe these specific challenges in detail. First, to identify new Wnt components, *bona fide* Wnt targets, and cross-regulators from other signaling pathways and their importance *in vivo* to regulate Wnt controlled processes. Second, to examine cell-specific tissue responses to Wnts based on their tissue- and cell-specific expression pattern and the cellular concentration of Wnt components and cross-interactors. Third, to characterize meaningful molecular interactions between these molecules and changes in expression levels and localization upon Wnt signaling in real time, which will allow us to design molecules that can prevent these interactions or biochemical changes specifically. Fourth, to develop sophisticated bioinformatic models that can integrate all this biochemical and biological data to identify new targets. And fifth, to improve computer modelling and simulation of Wnt circuits utilizing the rapid accumulation of experimental data to help identify contradictory experimental information, improve existing Wnt circuit diagrams, make qualitative and quantitative predictions on Wnt signaling dynamics and function that are subsequently validated experimentally, and analyze and predict mutations with tumorigenic and metastatic potentials. It is expected that such circuit simulations, when tightly coupled with experiments, will play an increasingly important role in our quest to understand the function of the highly dynamic and pleiotropic Wnt signaling network, as well as its role in the highly complex and heterogeneous disease of cancer.

Acknowledgments

The support by the Bundesministerium fuer Bildung und Forschung (BMBF) (grants to J.A. and P.R.), and by the National Institutes of Health (NIGMS) and the American Heart Association (AHA) (grants to I.D.) is acknowledged.

References

Albert, R., and Othmer, H.G. (2003). The topology of the regulatory interactions predicts the expression pattern of the segment polarity genes in *Drosophila melanogaster*. *J Theoret Biol 223*, 1–18.

Amonlirdviman, K., Khare, N.A., Tree, D.R., et al. (2005). Mathematical modeling of planar cell polarity to understand domineering nonautonomy. *Science 307*, 423–426.

Angers, S., and Moon, R.T. (2009). Proximal events in Wnt signal transduction. *Nat Rev Mol Cell Biol 10*, 468–477.

Aoki, K., and Taketo, M.M. (2008). Tissue-specific transgenic, conditional knockout and knock-in mice of genes in the canonical Wnt signaling pathway. *Methods Mol Biol 468*, 307–331.

Ashburner, M., Ball, C.A., Blake, J.A., et al. (2000). Gene ontology: tool for the unification of biology. The Gene Ontology Consortium. *Nat Genet 25*, 25–29.

Askenazi, M., Li, S., Singh, S., and Marto, J.A. (2010). Pathway Palette: a rich internet application for peptide-, protein- and network-oriented analysis of MS data. *Proteomics 10*, 1880–1885.

Aulehla, A., Wehrle, C., Brand-Saberi, B., et al. (2003). Wnt3a plays a major role in the segmentation clock controlling somitogenesis. *Dev Cell 4*, 395–406.

Barker, N., and Clevers, H. (2006). Mining the Wnt pathway for cancer therapeutics. *Nat Rev Drug Discov 5*, 997–1014.

Barker, N., and van den Born, M. (2008). Detection of beta-catenin localization by immunohistochemistry. *Methods Mol Biol 468*, 91–98.

Barolo, S. (2006). Transgenic Wnt/TCF pathway reporters: all you need is Lef? *Oncogene 25*, 7505–7511.

Bayer Schering Pharma (2010). Bayer Schering Pharma and OncoMed pharmaceuticals enter strategic alliance to develop anti-cancer stem cell therapeutics. June 17, 2010. www.investor.bayer.com/en/nc/news/archive/investor-news-2010/investor-news-2010/bayer-schering-pharma-and-oncomed-pharmaceuticals-enter-strategic-alliance-to-develop-anti-cancer-st.

Beachy, P.A., Karhadkar, S.S., and Berman, D.M. (2004). Tissue repair and stem cell renewal in carcinogenesis. *Nature 432*, 324–331.

Beildeck, M.E., Gelmann, E.P., and Byers, S.W. (2010). Cross-regulation of signaling pathways: an example of nuclear hormone receptors and the canonical Wnt pathway. *Exp Cell Res 316*, 1763–1772.

Biechele, T.L., and Moon, R.T. (2008). Assaying beta-catenin/TCF transcription with beta-catenin/TCF transcription-based reporter constructs. *Methods Mol Biol 468*, 99–110.

Boon, E.M., Keller, J.J., Wormhoudt, T.A., et al. 2004). Sulindac targets nuclear beta-catenin accumulation and Wnt signalling in adenomas of patients with familial adenomatous polyposis and in human colorectal cancer cell lines. *Br J Cancer 90*, 224–229.

Bottomly, D., Kyler, S.L., McWeeney, S.K., and Yochum, G.S. (2010). Identification of {beta}-catenin binding regions in colon cancer cells using ChIP-Seq. *Nucleic Acids Res 38*, 5735–5745.

Bovolenta, P., Esteve, P., Ruiz, J.M., Cisneros, E., and Lopez-Rios, J. (2008). Beyond Wnt inhibition: new functions of secreted Frizzled-related proteins in development and disease. *J Cell Sci 121*, 737–746.

Brown, A.M., Wildin, R.S., Prendergast, T.J., and Varmus, H.E. (1986). A retrovirus vector expressing the putative mammary oncogene int-1 causes partial transformation of a mammary epithelial cell line. *Cell 46*, 1001–1009.

Cadigan, K.M., and Liu, Y.I. (2006). Wnt signaling: complexity at the surface. *J Cell Sci 119*, 395–402.

Cadigan, K.M., and Peifer, M. (2009). Wnt signaling from development to disease: insights from model systems. *Cold Spring Harb Perspect Biol 1*, a002881.

Camilli, T.C., and Weeraratna, A.T. (2010). Striking the target in Wnt-y conditions: intervening in Wnt signaling during cancer progression. *Biochem Pharmacol 80*, 702–711.

Capdeville, R., Buchdunger, E., Zimmermann, J., and Matter, A. (2002). Glivec (STI571, imatinib), a rationally developed, targeted anticancer drug. *Nat Rev Drug Discov 1*, 493–502.

Carlson, B., Lahusen, T., Singh, S., et al. (1999). Down-regulation of cyclin D1 by transcriptional repression in MCF-7 human breast carcinoma cells induced by flavopiridol. *Cancer Res 59*, 4634–4641.

Chang, J.T., and Nevins, J.R. (2006). GATHER: a systems approach to interpreting genomic signatures. *Bioinformatics 22*, 2926–2933.

Chaves, M., and Albert, R. (2008). Studying the effect of cell division on expression patterns of the segment polarity genes. *J R Soc Interface 5* Suppl 1, S71–84.

Chen, B., Dodge, M.E., Tang, W., et al. (2009). Small molecule-mediated disruption of Wnt-dependent signaling in tissue regeneration and cancer. *Nat Chem Biol 5*, 100–107.

Chen, X., Yang, J., Evans, P.M., and Liu, C. (2008). Wnt signaling: the good and the bad. *Acta Biochim Biophys Sin 40*, 577–594.

Chien, A.J., Conrad, W.H., and Moon, R.T. (2009). A Wnt survival guide: from flies to human disease. *J Invest Dermatol 129*, 1614–1627.

Chien, C.T., Bartel, P.L., Sternglanz, R., and Fields, S. (1991). The two-hybrid system: a method to identify and clone genes for proteins that interact with a protein of interest. *Proc Natl Acad Sci USA 88*, 9578–9582.

Cho, K.H., Baek, S., and Sung, M.H. (2006). Wnt pathway mutations selected by optimal beta-catenin signaling for tumorigenesis. *FEBS Lett 580*, 3665–3670.

Choi, H.Y., Lim, J.E., and Hong, J.H. (2010). Curcumin interrupts the interaction between the androgen receptor and Wnt/beta-catenin signaling pathway in LNCaP prostate cancer cells. *Prostate Cancer Prostatic Dis 13*, 343–349.

Choudhary, C., and Mann, M. (2010). Decoding signalling networks by mass spectrometry-based proteomics. *Nat Rev Mol Cell Biol 11*, 427–439.

Clevers, H. (2006). Wnt/beta-catenin signaling in development and disease. *Cell 127*, 469–480.

Collins, M.O., and Choudhary, J.S. (2008). Mapping multiprotein complexes by affinity purification and mass spectrometry. *Curr Opin Biotechnol 19*, 324–330.

Coluccia, A.M., Benati, D., Dekhil, H., et al. (2006). SKI-606 decreases growth and motility of colorectal cancer cells by preventing pp60(c-Src)-dependent tyrosine phosphorylation of beta-catenin and its nuclear signaling. *Cancer Res 66*, 2279–2286.

Croce, J.C., and McClay, D.R. (2008). Evolution of the Wnt pathways. *Methods Mol Biol 469*, 3–18.

Danilkovitch-Miagkova, A. (2003). Oncogenic signaling pathways activated by RON receptor tyrosine kinase. *Curr Cancer Drug Targets 3*, 31–40.

DasGupta, R., Kaykas, A., Moon, R.T., and Perrimon, N. (2005). Functional genomic analysis of the Wnt-wingless signaling pathway. *Science 308*, 826–833.

Davidson, G., Shen, J., Huang, Y.-L., et al. (2009). Cell cycle control of wnt receptor activation. *Dev Cell 17*, 788–799.

Davidson, G., Wu, W., Shen, J., et al. (2005). Casein kinase 1 gamma couples Wnt receptor activation to cytoplasmic signal transduction. *Nature 438*, 867–872.

Dayarian, A., Chaves, M., Sontag, E.D., and Sengupta, A.M. (2009). Shape, size, and robustness: feasible regions in the parameter space of biochemical networks. *PLoS Comp Biol 5*, e1000256.

Dennis, G., Jr., Sherman, B.T., Hosack, D.A., et al. (2003). DAVID: Database for Annotation, Visualization, and Integrated Discovery. *Genome Biol 4*, P3.

Dequeant, M.L., Glynn, E., Gaudenz, K., et al. (2006). A complex oscillating network of signaling genes underlies the mouse segmentation clock. *Science 314*, 1595–1598.

Dihlmann, S., Klein, S., and Doeberitz Mv, M.K. (2003). Reduction of beta-catenin/T-cell transcription factor signaling by aspirin and indomethacin is caused by an increased stabilization of phosphorylated beta-catenin. *Mol Cancer Ther 2*, 509–516.

Dihlmann, S., and von Knebel Doeberitz, M. (2005). Wnt/beta-catenin-pathway as a molecular target for future anti-cancer therapeutics. *Int J Cancer 113*, 515–524.

Dirnberger, D., Messerschmid, M., and Baumeister, R. (2008). An optimized split-ubiquitin cDNA-library screening system to identify novel interactors of the human Frizzled 1 receptor. *Nucleic Acids Res 36*, e37.

Dominguez, I., Mizuno, J., Wu, H., et al. (2004). Protein kinase CK2 is required for dorsal axis formation in *Xenopus* embryos. *Dev Biol 274*, 110–124.

Dominguez, I., Sonenshein, G.E., and Seldin, D.C. (2009). Protein kinase CK2 in health and disease: CK2 and its role in Wnt and NF-kappaB signaling: linking development and cancer. *Cell Mol Life Sci 66*, 1850–1857.

Dumont, J.E., Pecasse, F., and Maenhaut, C. (2001). Crosstalk and specificity in signalling. Are we crosstalking ourselves into general confusion? *Cell Signal 13*, 457–463.

Edwards, P.A., Hiby, S.E., Papkoff, J., and Bradbury, J.M. (1992).

Hyperplasia of mouse mammary epithelium induced by expression of the Wnt-1 (int-1) oncogene in reconstituted mammary gland. *Oncogene 7*, 2041–2051.

Emami, K.H., Nguyen, C., Ma, H., et al. (2004). A small molecule inhibitor of beta-catenin/CREB-binding protein transcription (corrected). *Proc Natl Acad Sci USA 101*, 12682–12687.

Ewing, R.M., Chu, P., Li, H., et al. (2007). Large-scale mapping of human protein-protein interactions by mass spectrometry. *Mol Syst Biol 3*, 89.

Fagotto, F., Gluck, U., and Gumbiner, B.M. (1998). Nuclear localization signal-independent and importin/karyopherin-independent nuclear import of beta-catenin. *Curr Biol 8*, 181–190.

Fields, S., and Song, O. (1989). A novel genetic system to detect protein-protein interactions. *Nature 340*, 245–246.

Fodde, R., and Tomlinson, I. (2010). Nuclear beta-catenin expression and Wnt signalling: in defence of the dogma. *J Pathol 221*, 239–241.

Fu, C.A., Shen, M., Huang, B.C., et al. (1999). TNIK, a novel member of the germinal center kinase family that activates the c-Jun N-terminal kinase pathway and regulates the cytoskeleton. *J Biol Chem 274*, 30729–30737.

Galis, F., van Dooren, T.J., and Metz, J.A. (2002). Conservation of the segmented germband stage: robustness or pleiotropy? *Trends Genet 18*, 504–509.

Gao, C., and Chen, Y.G. (2010). Dishevelled: the hub of Wnt signaling. *Cell Signal 22*, 717–727.

Gavin, A.C., Bosche, M., Krause, R., et al. (2002). Functional organization of the yeast proteome by systematic analysis of protein complexes. *Nature 415*, 141–147.

Geng, X., Xiao, L., Lin, G.F., et al. (2003). Lef/Tcf-dependent Wnt/beta-catenin signaling during

Xenopus axis specification. *FEBS Lett 547*, 1–6.

Giles, R.H., van Es, J.H., and Clevers, H. (2003). Caught up in a Wnt storm: Wnt signaling in cancer. *Biochim Biophys Acta 1653*, 1–24.

Gingras, A.-C., Aebersold, R., and Raught, B. (2005). Advances in protein complex analysis using mass spectrometry. *J Physiol 563*, 11–21.

Glatter, T., Wepf, A., Aebersold, R., and Gstaiger, M. (2009). An integrated workflow for charting the human interaction proteome: insights into the PP2A system. *Mol Syst Biol 5*, 237.

Goentoro, L., and Kirschner, M.W. (2009). Evidence that fold-change, and not absolute level, of beta-catenin dictates Wnt signaling. *Mol Cell 36*, 872–884.

Goentoro, L., Shoval, O., Kirschner, M.W., and Alon, U. (2009). The incoherent feedforward loop can provide fold-change detection in gene regulation. *Mol Cell 36*, 894–899.

Goldbeter, A., and Pourquie, O. (2008). Modeling the segmentation clock as a network of coupled oscillations in the Notch, Wnt and FGF signaling pathways. *J Theoret Biol 252*, 574–585.

Grigoryan, T., Wend, P., Klaus, A., and Birchmeier, W. (2008). Deciphering the function of canonical Wnt signals in development and disease: conditional loss- and gain-of-function mutations of beta-catenin in mice. *Genes Dev 22*, 2308–2341.

Guger, K.A., and Gumbiner, B.M. (2000). A mode of regulation of beta-catenin signaling activity in *Xenopus* embryos independent of its levels. *Dev Biol 223*, 441–448.

Gutenkunst, R.N., Waterfall, J.J., Casey, F.P., et al. (2007). Universally sloppy parameter sensitivities in systems biology models. *PLoS Comp Biol 3*, 1871–1878.

Haertel-Wiesmann, M., Liang, Y., Fantl, W.J., and Williams, L.T. (2000). Regulation of

cyclooxygenase-2 and periostin by Wnt-3 in mouse mammary epithelial cells. *J Biol Chem 275*, 32046–32051.

Hagen, T., Sethi, J.K., Foxwell, N., and Vidal-Puig, A. (2004). Signalling activity of beta-catenin targeted to different subcellular compartments. *Biochem J 379*, 471–477.

Hallikas, O., Palin, K., Sinjushina, N., et al. (2006). Genome-wide prediction of mammalian enhancers based on analysis of transcription-factor binding affinity. *Cell 124*, 47–59.

Hausmann, G., Banziger, C., and Basler, K. (2007). Helping Wingless take flight: how WNT proteins are secreted. *Nat Rev Mol Cell Biol 8*, 331–336.

Hayward, P., Kalmar, T., and Arias, A.M. (2008). Wnt/Notch signalling and information processing during development. *Development 135*, 411–424.

He, B., Reguart, N., You, L., et al. (2005). Blockade of Wnt-1 signaling induces apoptosis in human colorectal cancer cells containing downstream mutations. *Oncogene 24*, 3054–3058.

Herendeen, J.M., and Lindley, C. (2003). Use of NSAIDs for the chemoprevention of colorectal cancer. *Ann Pharmacother 37*, 1664–1674.

Ho, Y., Gruhler, A., Heilbut, A., et al. (2002). Systematic identification of protein complexes in *Saccharomyces cerevisiae* by mass spectrometry. *Nature 415*, 180–183.

Hodar, C., Assar, R., Colombres, M., et al. (2010). Genome-wide identification of new Wnt/beta-catenin target genes in the human genome using CART method. *BMC Genomics 11*, 348.

Hu, T., and Li, C. (2010). Convergence between Wnt-beta-catenin and EGFR signaling in cancer. *Mol Cancer 9*, 236.

Huang, M., Wang, Y., Sun, D., et al. (2006). Identification of genes regulated by Wnt/beta-catenin pathway and involved in apoptosis via microarray analysis. *BMC Cancer 6*, 221.

Huang, S., Li, Y., Chen, Y., et al. (2005). Changes in gene expression during the development of mammary tumors in MMTV-Wnt-1 transgenic mice. *Genome Biol 6*, R84.

Huang, S.S., and Fraenkel, E. (2009). Integrating proteomic, transcriptional, and interactome data reveals hidden components of signaling and regulatory networks. *Sci Signal 2*, ra40.

Ilyas, M. (2005). Wnt signalling and the mechanistic basis of tumour development. *J Pathol 205*, 130–144.

Ingolia, N.T. (2004). Topology and robustness in the *Drosophila* segment polarity network. *PLoS Biol 2*, e123.

Irizarry, R.A., Hobbs, B., Collin, F., et al. (2003). Exploration, normalization, and summaries of high density oligonucleotide array probe level data. *Biostatistics 4*, 249–264.

Itasaki, N., and Hoppler, S. (2010). Crosstalk between Wnt and bone morphogenic protein signaling: a turbulent relationship. *Dev Dyn 239*, 16–33.

Jackson, A.L., Burchard, J., Schelter, J., et al. (2006). Widespread siRNA "off-target" transcript silencing mediated by seed region sequence complementarity. *RNA 12*, 1179–1187.

James, R.G., Biechele, T.L., Conrad, W.H., et al. (2009). Bruton's tyrosine kinase revealed as a negative regulator of Wnt-beta-catenin signaling. *Sci Signal 2*, ra25.

James, R.G., Conrad, W.H., and Moon, R.T. (2008). Beta-catenin-independent Wnt pathways: signals, core proteins, and effectors. *Methods Mol Biol 468*, 131–144.

Jenny, A., and Mlodzik, M. (2006). Planar cell polarity signaling: a common mechanism for cellular polarization. *Mt Sinai J Med 73*, 738–750.

Jensen, P.B., Pedersen, L., Krishna, S., and Jensen, M.H. (2010). A Wnt oscillator model for somitogenesis. *Biophysical J 98*, 943–950.

Jessen, J.R. (2009). Noncanonical Wnt signaling in tumor progression and metastasis. *Zebrafish 6*, 21–28.

Jin, T., George Fantus, I., and Sun, J. (2008). Wnt and beyond Wnt: multiple mechanisms control the transcriptional property of beta-catenin. *Cell Signal 20*, 1697–1704.

Kanehisa, M., Goto, S., Hattori, M., et al. (2006). From genomics to chemical genomics: new developments in KEGG. *Nucleic Acids Res 34*, D354–357.

Karaguni, I.M., Gourzoulidou, E., Carpintero, M., et al. (2004). SMAF-1 inhibits the APC/beta-catenin pathway and shows properties similar to those of the tumor suppressor protein APC. *Chembiochem 5*, 1267–1270.

Katoh, M. (2007). Networking of WNT, FGF, Notch, BMP, and Hedgehog signaling pathways during carcinogenesis. *Stem Cell Rev 3*, 30–38.

Keating, G.M., and Santoro, A. (2009). Sorafenib: a review of its use in advanced hepatocellular carcinoma. *Drugs 69*, 223–240.

Keshava Prasad, T.S., Goel, R., Kandasamy, K., et al. (2009). Human Protein Reference Database – 2009 update. *Nucleic Acids Res 37*, D767–772.

Kestler, H.A., and Kuhl, M. (2008). From individual Wnt pathways towards a Wnt signalling network. *Philos Trans R Soc Lond B Biol Sci 363*, 1333–1347.

Khramtsov, A.I., Khramtsova, G.F., Tretiakova, M., et al. (2010). Wnt/beta-catenin pathway activation is enriched in basal-like breast cancers and predicts poor outcome. *Am J Pathol 176*, 2911–2920.

Kikuchi, A., and Yamamoto, H. (2008). Tumor formation due to abnormalities in the beta-catenin-independent pathway of Wnt signaling. *Cancer Sci 99*, 202–208.

Kikuchi, A., Yamamoto, H., and Kishida, S. (2007). Multiplicity of the interactions of Wnt proteins and their receptors. *Cell Signal 19*, 659–671.

Kim, D., Rath, O., Kolch, W., and Cho, K.H. (2007). A hidden oncogenic positive feedback loop caused by crosstalk between Wnt and ERK pathways. *Oncogene 26*, 4571–4579.

Kim, K., and Hay, E.D. (2001). New evidence that nuclear import of endogenous beta-catenin is LEF-1 dependent, while LEF-1 independent import of exogenous beta-catenin leads to nuclear abnormalities. *Cell Biol Int 25*, 1149–1161.

Köcher, T., and Superti-Furga, G. (2007). Mass spectrometry-based functional proteomics: from molecular machines to protein networks. *Nat Methods 4*, 807–815.

Kohler, F., and Muller, K.M. (2003). Adaptation of the Ras-recruitment system to the analysis of interactions between membrane-associated proteins. *Nucleic Acids Res 31*, e28.

Kohn, A.D., and Moon, R.T. (2005). Wnt and calcium signaling: beta-catenin-independent pathways. *Cell Calcium 38*, 439–446.

Komiya, Y., and Habas, R. (2008). Wnt signal transduction pathways. *Organogenesis 4*, 68–75.

Kool, M., Koster, J., Bunt, J., et al. (2008). Integrated genomics identifies five medulloblastoma subtypes with distinct genetic profiles, pathway signatures and clinicopathological features. *PLoS One 3*, e3088.

Korinek, V., Barker, N., Morin, P.J., et al. (1997a). Constitutive transcriptional activation by a beta-catenin-Tcf complex in APC-/- colon carcinoma. *Science 275*, 1784–1787.

Korinek, V., Barker, N., Morin, P.J., et al. (1997b). Constitutive transcriptional activation by a beta-catenin-Tcf complex in APC–/– colon carcinoma. *Science 275*, 1784–1787.

Krogan, N.J., Cagney, G., Yu, H., et al. (2006). Global landscape of protein complexes in the yeast *Saccharomyces cerevisiae*. *Nature 440*, 637–643.

Kruger, R., and Heinrich, R. (2004). Model reduction and analysis of robustness for the Wnt/beta-catenin signal transduction pathway. *Genome Informatics 15*, 138–148.

Kuhl, M. (2004). The WNT/calcium pathway: biochemical mediators, tools and future requirements. *Front Biosci 9*, 967–974.

Kuhl, M., and Pandur, P. (2008a). Dorsal axis duplication as a functional readout for Wnt activity. *Methods Mol Biol 469*, 467–476.

Kuhl, M., and Pandur, P. (2008b). Measuring CamKII activity in *Xenopus* embryos as a read-out for non-canonical Wnt signaling. *Methods Mol Biol 468*, 173–186.

Lai, S.L., Chien, A.J., and Moon, R.T. (2009). Wnt/Fz signaling and the cytoskeleton: potential roles in tumorigenesis. *Cell Res 19*, 532–545.

Landesman-Bollag, E., Romieu-Mourez, R., Song, D.H., et al. (2001). Protein kinase CK2 in mammary gland tumorigenesis. *Oncogene 20*, 3247–3257.

Lee, E., Salic, A., Kruger, R., Heinrich, R., and Kirschner, M.W. (2003). The roles of APC and Axin derived from experimental and theoretical analysis of the Wnt pathway. *PLoS Biol 1*, e10.

Lepourcelet, M., Chen, Y.N., France, D.S., et al. (2004). Small-molecule antagonists of the oncogenic Tcf/beta-catenin protein complex. *Cancer Cell 5*, 91–102.

Li, H., Helling, R., Tang, C., and Wingreen, N. (1996). Emergence of preferred structures in a simple model of protein folding. *Science 273*, 666–669.

Liang, P., and Pardee, A.B. (1992). Differential display of eukaryotic messenger RNA by means of the polymerase chain reaction. *Science 257*, 967–971.

Lievens, S., Lemmens, I., and Tavernier, J. (2009). Mammalian two-hybrids come of age. *Trends Biochem Sci 34*, 579–588.

Lin, S.Y., Xia, W., Wang, J.C., et al. (2000). Beta-catenin, a novel prognostic marker for breast cancer: its roles in cyclin D1 expression and cancer progression. *Proc Natl Acad Sci USA 97*, 4262–4266.

Liu, J., Bang, A.G., Kintner, C., et al. (2005). Identification of the Wnt signaling activator leucine-rich repeat in Flightless interaction protein 2 by a genome-wide functional analysis. *Proc Natl Acad Sci USA 102*, 1927–1932.

Logan, C.Y., and Nusse, R. (2004). The Wnt signaling pathway in development and disease. *Annu Rev Cell Dev Biol 20*, 781–810.

Lu, J., Ma, Z., Hsieh, J.C., et al. (2009). Structure-activity relationship studies of small-molecule inhibitors of Wnt response. *Bioorg Med Chem Lett 19*, 3825–3827.

Ma, D., Amonlirdviman, K., Raffard, R.L., et al. (2008). Cell packing influences planar cell polarity signaling. *Proc Natl Acad Sci USA 105*, 18800–18805.

Ma, W., Lai, L., Ouyang, Q., and Tang, C. (2006). Robustness and modular design of the *Drosophila* segment polarity network. *Mol Syst Biol 2*, 70.

MacDonald, B.T., Tamai, K., and He, X. (2009). Wnt/beta-catenin signaling: components, mechanisms, and diseases. *Dev Cell 17*, 9–26.

Mahmoudi, T., Li, V.S.W., Ng, S.S., et al. (2009). The kinase TNIK is an essential activator of Wnt target genes. *EMBO J 28*, 3329–3340.

Major, M.B., Camp, N.D., Berndt, J.D., et al. (2007). Wilms tumor suppressor WTX negatively regulates WNT/beta-catenin signaling. *Science 316*, 1043–1046.

Major, M.B., Roberts, B.S., Berndt, J.D., et al. (2008). New regulators of Wnt/beta-catenin signaling revealed by integrative molecular screening. *Sci Signal 1*, ra12.

Matushansky, I., Maki, R.G., and Cordon-Cardo, C. (2008). A context dependent role for Wnt signaling in tumorigenesis and stem cells. *Cell Cycle 7*, 720–724.

McDonald, S.L., and Silver, A. (2009). The opposing roles of Wnt-5a in cancer. *Br J Cancer 101*, 209–214.

McMahon, A.P., and Moon, R.T. (1989). Ectopic expression of the proto-oncogene int-1 in *Xenopus* embryos leads to duplication of the embryonic axis. *Cell 58*, 1075–1084.

McNeill, H., and Woodgett, J.R. (2010). When pathways collide: collaboration and connivance among signalling proteins in development. *Nat Rev Mol Cell Biol 11*, 404–413.

Meijer, L., and Raymond, E. (2003). Roscovitine and other purines as kinase inhibitors. From starfish oocytes to clinical trials. *Acc Chem Res 36*, 417–425.

Meinhardt, H., and Gierer, A. (1980). Generation and regeneration of sequence of structures during morphogenesis. *J Theoret Biol 85*, 429–450.

Mi, H., and Thomas, P. (2009). PANTHER pathway: an ontology-based pathway database coupled with data analysis tools. *Methods Mol Biol 563*, 123–140.

Mikels, A.J., and Nusse, R. (2006). Wnts as ligands: processing, secretion and reception. *Oncogene 25*, 7461–7468.

Miyoshi, K., Rosner, A., Nozawa, M., et al. (2002). Activation of different Wnt/beta-catenin signaling components in mammary epithelium induces transdifferentiation and the formation of pilar tumors. *Oncogene 21*, 5548–5556.

Moffat, J., and Sabatini, D.M. (2006). Building mammalian signalling pathways with RNAi screens. *Nat Rev Mol Cell Biol 7*, 177–187.

Moon, R.T., Kohn, A.D., De Ferrari, G.V., and Kaykas, A. (2004). WNT and beta-catenin signalling: diseases and therapies. *Nat Rev Genet 5*, 691–701.

Moore, W.J., Kern, J.C., Bhat, R., et al. (2010). Modulation of Wnt signaling through inhibition of secreted frizzled-related protein I (sFRP-1) with N-substituted piperidinyl diphenylsulfonyl sulfonamides: part II. *Bioorg Med Chem 18*, 190–201.

Moser, A.R., Dove, W.F., Roth, K.A., and Gordon, J.I. (1992). The Min (multiple intestinal neoplasia) mutation: its effect on gut epithelial cell differentiation and interaction with a modifier system. *J Cell Biol 116*, 1517–1526.

Mosimann, C., Hausmann, G., and Basler, K. (2009). Beta-catenin hits chromatin: regulation of Wnt target gene activation. *Nat Rev Mol Cell Biol 10*, 276–286.

Nath, N., Kashfi, K., Chen, J., and Rigas, B. (2003). Nitric oxide-donating aspirin inhibits beta-catenin/T cell factor (TCF) signaling in SW480 colon cancer cells by disrupting the nuclear beta-catenin-TCF association. *Proc Natl Acad Sci USA 100*, 12584–12589.

Niehrs, C., and Shen, J. (2010). Regulation of Lrp6 phosphorylation. *Cell Mol Life Sci 67*, 2551–2562.

Niida, A., Hiroko, T., Kasai, M., et al. (2004). DKK1, a negative regulator of Wnt signaling, is a target of the beta-catenin/TCF pathway. *Oncogene 23*, 8520–8526.

Noordermeer, J., Klingensmith, J., Perrimon, N., and Nusse, R. (1994). Dishevelled and armadillo act in the wingless signalling pathway in *Drosophila*. *Nature 367*, 80–83.

Nusse, R., Brown, A., Papkoff, J., et al. (1991). A new nomenclature for int-1 and related genes: the Wnt gene family. *Cell 64*, 231.

Nusse, R., and Varmus, H.E. (1982). Many tumors induced by the mouse mammary tumor virus contain a provirus integrated in the same region of the host genome. *Cell 31*, 99–109.

O'Connell, M.P., and Weeraratna, A.T. (2009). Hear the Wnt Ror: how melanoma cells adjust to changes in Wnt. *Pigment Cell Melanoma Res 22*, 724–739.

Ohkawara, B., and Niehrs, C. (2011). An ATF2-based luciferase reporter to monitor non-canonical Wnt signaling in *Xenopus* embryos. *Dev Dyn 240*, 188–194.

Papkoff, J. (1997). Regulation of complexed and free catenin pools by distinct mechanisms. Differential effects of Wnt-1 and v-Src. *J Biol Chem 272*, 4536–4543.

Parikh, J.R., Klinger, B., Xia, Y., Marto, J.A., and Bluthgen, N. (2010). Discovering causal signaling pathways through gene-expression patterns. *Nucleic Acids Res 38*, W109–117.

Park, C.H., Chang, J.Y., Hahm, E.R., et al. (2005). Quercetin, a potent inhibitor against beta-catenin/Tcf signaling in SW480 colon cancer cells. *Biochem Biophys Res Commun 328*, 227–234.

Park, S., Gwak, J., Cho, M., et al. (2006). Hexachlorophene inhibits Wnt/{beta}-catenin pathway by promoting Siah-mediated {beta}-catenin degradation. *Mol Pharmacol 70*, 960–966.

Pico, A.R., Kelder, T., van Iersel, M.P., et al. (2008). WikiPathways: pathway editing for the people. *PLoS Biol 6*, e184.

Playford, M.P., Bicknell, D., Bodmer, W.F., and Macaulay, V.M. (2000). Insulin-like growth factor 1 regulates the location, stability, and transcriptional activity of beta-catenin. *Proc Natl Acad Sci USA 97*, 12103–12108.

Polakis, P. (2000). Wnt signaling and cancer. *Genes Dev 14*, 1837–1851.

Polakis, P. (2007). The many ways of Wnt in cancer. *Curr Opin Genet Dev 17*, 45–51.

Port, F., and Basler, K. (2010). Wnt trafficking: new insights into Wnt maturation, secretion and spreading. *Traffic 11*, 1265–1271.

Pukrop, T., and Binder, C. (2008). The complex pathways of Wnt 5a in cancer progression. *J Mol Med 86*, 259–266.

Rao, T.P., and Kuhl, M. (2010). An updated overview on Wnt signaling pathways: a prelude for more. *Circ Res 106*, 1798–1806.

Reya, T., and Clevers, H. (2005). Wnt signalling in stem cells and cancer. *Nature 434*, 843–850.

Rijsewijk, F., van Deemter, L., Wagenaar, E., Sonnenberg, A., and Nusse, R. (1987). Transfection of the int-1 mammary oncogene in cuboidal RAC mammary cell line results in morphological transformation and tumorigenicity. *EMBO J 6*, 127–131.

Rivera, M.N., Kim, W.J., Wells, J., et al. (2007). An X chromosome gene, WTX, is commonly inactivated in Wilms tumor. *Science 315*, 642–645.

Rodriguez-Gonzalez, J.G., Santillan, M., Fowler, A.C., and Mackey, M.C. (2007). The segmentation clock in mice: interaction between the Wnt and Notch signalling pathways. *J Theoret Biol 248*, 37–47.

Roelink, H., Wagenaar, E., Lopes da Silva, S., and Nusse, R. (1990). Wnt-3, a gene activated by proviral insertion in mouse mammary tumors, is homologous to int-1/ Wnt-1 and is normally expressed in mouse embryos and adult brain. *Proc Natl Acad Sci USA 87*, 4519–4523.

Roelink, H., Wagenaar, E., and Nusse, R. (1992). Amplification and proviral activation of several Wnt genes during progression and clonal variation of mouse mammary tumors. *Oncogene 7*, 487–492.

Röhrs, S., Kutzner, N., Vlad, A., et al. (2009). Chronological expression of Wnt target genes Ccnd1, Myc, Cdkn1a, Tfrc, Plf1 and Ramp3. *Cell Biol Int 33*, 501–508.

Röhrs, S.a.M., O. (2007). The Wnt pathway as a target for small synthetic anti-cancer molecules. *Curr Chem Biol 1*, 227–239.

Ronaghi, M. (2001). Pyrosequencing sheds light on DNA sequencing. *Genome Res 11*, 3–11.

Ronaghi, M., Uhlen, M., and Nyren, P. (1998). A sequencing method based on real-time pyrophosphate. *Science 281*, 363–365.

Rossol-Allison, J., Stemmle, L.N., Swenson-Fields, K.I., et al. (2009). Rho GTPase activity modulates Wnt3a/beta-catenin signaling. *Cell Signal 21*, 1559–1568.

Roszko, I., Sawada, A., and Solnica-Krezel, L. (2009). Regulation of convergence and extension movements during vertebrate gastrulation by the Wnt/PCP pathway. *Semin Cell Dev Biol 20*, 986–997.

Saadeddin, A., Babaei-Jadidi, R., Spencer-Dene, B., and Nateri, A.S. (2009). The links between transcription, beta-catenin/JNK signaling, and carcinogenesis. *Mol Cancer Res 7*, 1189–1196.

Sansom, O.J., Reed, K.R., Hayes, A.J., et al. (2004). Loss of Apc *in vivo* immediately perturbs Wnt signaling, differentiation, and migration. *Genes Dev 18*, 1385–1390.

Sansom, O.J., Reed, K.R., van de Wetering, M., et al. (2005). Cyclin D1 is not an immediate target of beta-catenin following Apc loss in the intestine. *J Biol Chem 280*, 28463–28467.

Saraswati, S., Alfaro, M.P., Thorne, C.A., et al. (2010). Pyrvinium, a potent small molecule Wnt inhibitor, promotes wound repair and post-MI cardiac remodeling. *PLoS One 5*, e15521.

Satoh, S., Daigo, Y., Furukawa, Y., et al. (2000). AXIN1 mutations in hepatocellular carcinomas, and growth suppression in cancer cells by virus-mediated transfer of AXIN1. *Nat Genet 24*, 245–250.

Satow, R., Shitashige, M., Jigami, T., et al. (2010). Traf2- and Nck-interacting kinase is essential for canonical Wnt signaling in *Xenopus* axis formation. *J Biol Chem 285*, 26289–26294.

Schena, M., Shalon, D., Davis, R.W., and Brown, P.O. (1995). Quantitative monitoring of gene expression patterns with a complementary DNA microarray. *Science 270*, 467–470.

Schlessinger, K., Hall, A., and Tolwinski, N. (2009). Wnt signaling pathways meet Rho GTPases. *Genes Dev 23*, 265–277.

Segditsas, S., Sieber, O., Deheragoda, M., et al. (2008). Putative direct and indirect Wnt targets identified through consistent gene expression changes in APC-mutant intestinal adenomas from humans and mice. *Hum Mol Genet 17*, 3864–3875.

Selbach, M., and Mann, M. (2006). Protein interaction screening by quantitative immunoprecipitation combined with knockdown (QUICK). *Nat Methods 3*, 981–983.

Semenov, M.V., Tamai, K., Brott, B.K., et al. (2001). Head inducer Dickkopf-1 is a ligand for Wnt coreceptor LRP6. *Curr Biol 11*, 951–961.

Sethi, J.K., and Vidal-Puig, A. (2010). Wnt signalling and the control of cellular metabolism. *Biochem J 427*, 1–17.

Shan, J., Shi, D.L., Wang, J., and Zheng, J. (2005). Identification of a specific inhibitor of the dishevelled PDZ domain. *Biochemistry 44*, 15495–15503.

Shapiro, G.I. (2006). Cyclin-dependent kinase pathways as targets for cancer treatment. *J Clin Oncol 24*, 1770–1783.

Shimizu, M., Fukunaga, Y., Ikenouchi, J., and Nagafuchi, A. (2008). Defining the roles of beta-catenin and plakoglobin in LEF/T-cell factor-dependent transcription using beta-catenin/plakoglobin-null F9 cells. *Mol Cell Biol 28*, 825–835.

Shin, S.Y., Rath, O., Zebisch, A., et al. (2010). Functional roles of multiple feedback loops in extracellular signal-regulated kinase and Wnt signaling pathways that regulate epithelial-mesenchymal transition. *Cancer Res 70*, 6715–6724.

Shitashige, M., Satow, R., Jigami, T., et al. (2010). Traf2- and Nck-interacting kinase is essential for Wnt signaling and colorectal cancer growth. *Cancer Res 70*, 5024–5033.

Shtutman, M., Zhurinsky, J., Simcha, I., et al. (1999). The cyclin D1 gene is a target of the beta-catenin/LEF-1 pathway. *Proc Natl Acad Sci USA 96*, 5522–5527.

Sick, S., Reinker, S., Timmer, J., and Schlake, T. (2006). WNT and DKK determine hair follicle spacing through a reaction-diffusion mechanism. *Science 314*, 1447–1450.

Simons, M., and Mlodzik, M. (2008). Planar cell polarity signaling: from fly development to human disease. *Annu Rev Genet 42*, 517–540.

Smith, W.C., and Harland, R.M. (1991). Injected Xwnt-8 RNA acts early in *Xenopus* embryos to promote formation of a vegetal dorsalizing center. *Cell 67*, 753–765.

Sokol, S., Christian, J.L., Moon, R.T., and Melton, D.A. (1991). Injected Wnt RNA induces a complete body axis in *Xenopus* embryos. *Cell 67*, 741–752.

Song, D.H., Dominguez, I., Mizuno, J., et al. (2003). CK2 phosphorylation of the armadillo repeat region of beta-catenin potentiates Wnt signaling. *J Biol Chem 278*, 24018–24025.

Sowa, M.E., Bennett, E.J., Gygi, S.P., and Harper, J.W. (2009). Defining the human deubiquitinating enzyme interaction landscape. *Cell 138*, 389–403.

Stagljar, I. (2003). Finding partners: emerging protein interaction technologies applied to signaling networks. *Sci STKE 2003*, pe56.

Stark, C., Breitkreutz, B.J., Reguly, T., et al. (2006). BioGRID: a general repository for interaction datasets. *Nucleic Acids Res 34*, D535–539.

Stelzl, U., Worm, U., Lalowski, M., et al. (2005). A human protein-protein interaction network: a resource for annotating the proteome. *Cell 122*, 957–968.

Su, L., Kinzler, K., Vogelstein, B., et al. (1992). Multiple intestinal neoplasia caused by a mutation in the murine homolog of the APC gene. *Science 256*, 668–670.

Subramanian, A., Tamayo, P., Mootha, V.K., et al. (2005). Gene set enrichment analysis: a knowledge-based approach for interpreting genome-wide expression profiles. *Proc Natl Acad Sci USA 102*, 15545–15550.

Svedlund, J., Auren, M., Sundstrom, M., et al. (2010). Aberrant WNT/beta-catenin signaling in parathyroid carcinoma. *Mol Cancer 9*, 294.

Takahashi-Yanaga, F., and Kahn, M. (2010). Targeting Wnt signaling: can we safely eradicate cancer stem cells? *Clin Cancer Res 16*, 3153–3162.

Takahashi-Yanaga, F., and Sasaguri, T. (2009). Drug development targeting the glycogen synthase kinase-3beta (GSK-3beta)-mediated signal transduction pathway: inhibitors of the Wnt/beta-catenin signaling pathway as novel anticancer drugs. *J Pharmacol Sci 109*, 179–183.

Takemaru, K.-I., Yamaguchi, S., Lee, Y.S., et al. (2003). Chibby, a nuclear beta-catenin-associated antagonist of the Wnt/Wingless pathway. *Nature 422*, 905–909.

Takemaru, K.I., Ohmitsu, M., and Li, F.Q. (2008). An oncogenic hub: beta-catenin as a molecular target

for cancer therapeutics. *Handb Exp Pharmacol*, 261–284.

Tang, L.Y., Deng, N., Wang, L.S., et al. (2007). Quantitative phosphoproteome profiling of Wnt3a-mediated signaling network: indicating the involvement of ribonucleoside-diphosphate reductase M2 subunit phosphorylation at residue serine 20 in canonical Wnt signal transduction. *Mol Cell Proteomics 6*, 1952–1967.

Tang, W., Dodge, M., Gundapaneni, D., et al. (2008). A genome-wide RNAi screen for Wnt/beta-catenin pathway components identifies unexpected roles for TCF transcription factors in cancer. *Proc Nat Acad Sci USA 105*, 9697–9702.

Trosset, J.Y., Dalvit, C., Knapp, S., et al. (2006). Inhibition of protein-protein interactions: the discovery of druglike beta-catenin inhibitors by combining virtual and biophysical screening. *Proteins 64*, 60–67.

Turing, A.M. (1952). The chemical basis of morphogenesis. *Philos Trans R Soc Lond B Biol Sci 237*, 37–72.

van Amerongen, R., and Berns, A. (2006). Knockout mouse models to study Wnt signal transduction. *Trends Genet 22*, 678–689.

van Amerongen, R., and Nusse, R. (2009). Towards an integrated view of Wnt signaling in development. *Development 136*, 3205–3214.

van de Wetering, M., Sancho, E., Verweij, C., et al. (2002). The beta-catenin/TCF-4 complex imposes a crypt progenitor phenotype on colorectal cancer cells. *Cell 111*, 241–250.

van Erk, M.J., Roepman, P., van der Lende, T.R., et al. (2005). Integrated assessment by multiple gene expression analysis of quercetin bioactivity on anticancer-related mechanisms in colon cancer cells in vitro. *Eur J Nutr 44*, 143–156.

Veeman, M.T., Slusarski, D.C., Kaykas, A., Louie, S.H., and Moon, R.T.

(2003). Zebrafish prickle, a modulator of noncanonical Wnt/Fz signaling, regulates gastrulation movements. *Curr Biol 13*, 680–685.

Verheyen, E.M., and Gottardi, C.J. (2010). Regulation of Wnt/beta-catenin signaling by protein kinases. *Dev Dyn 239*, 34–44.

Verkaar, F., and Zaman, G.J. (2011). New avenues to target Wnt/beta-catenin signaling. *Drug Discov Today 16*, 35–41.

Vlad, A., Röhrs, S., Klein-Hitpass, L., and Müller, O. (2008). The first five years of the Wnt targetome. *Cell Signal 20*, 795–802.

von Dassow, G., Meir, E., Munro, E.M., and Odell, G.M. (2000). The segment polarity network is a robust developmental module. *Nature 406*, 188–192.

von Dassow, G., and Odell, G.M. (2002). Design and constraints of the *Drosophila* segment polarity module: robust spatial patterning emerges from intertwined cell state switches. *J Exp Zoology 294*, 179–215.

Wallingford, J.B., and Habas, R. (2005). The developmental biology of Dishevelled: an enigmatic protein governing cell fate and cell polarity. *Development 132*, 4421–4436.

Wang, Y. (2009). Wnt/Planar cell polarity signaling: a new paradigm for cancer therapy. *Mol Cancer Ther 8*, 2103–2109.

Wawra, C., Kuhl, M., and Kestler, H.A. (2007). Extended analyses of the Wnt/beta-catenin pathway: robustness and oscillatory behaviour. *FEBS Lett 581*, 4043–4048.

Wharton, K.A., Jr. (2003). Runnin' with the Dvl: proteins that associate with Dsh/Dvl and their significance to Wnt signal transduction. *Dev Biol 253*, 1–17.

Willert, J., Epping, M., Pollack, J.R., Brown, P.O., and Nusse, R. (2002). A transcriptional response to Wnt protein in human embryonic carcinoma cells. *BMC Dev Biol 2*, 8.

Wu, X., Tu, X., Joeng, K.S., et al. (2008). Rac1 activation controls nuclear localization of beta-catenin during canonical Wnt signaling. *Cell 133*, 340–353.

Xia, Y., and Levitt, M. (2004). Simulating protein evolution in sequence and structure space. *Curr Opin Struct Biol 14*, 202–207.

Ye, C., Sweeny, D., Sukbuntherng, J., et al. (2006). Distribution, metabolism, and excretion of the anti-angiogenic compound SU5416. *Toxicol In Vitro 20*, 154–162.

Yin, X., Giap, C., Lazo, J.S., and Prochownik, E.V.(2003). Low molecular weight inhibitors of Myc-Max interaction and function. *Oncogene 22*, 6151–6159.

Yu, H., Braun, P., Yildirim, M.A., et al. (2008). High-quality binary protein interaction map of the yeast interactome network. *Science 322*, 104–110.

Zhang, Q., Major, M.B., Takanashi, S., et al. (2007). Small-molecule synergist of the Wnt/beta-catenin signaling pathway. *Proc Natl Acad Sci USA 104*, 7444–7448.

Zhang, Y., Askenazi, M., Jiang, J., et al. (2010). A robust error model for iTRAQ quantification reveals divergent signaling between oncogenic FLT3 mutants in acute myeloid leukemia. *Mol Cell Proteomics 9*, 780–790.

Zhou, L., An, N., Haydon, R.C., et al. (2003). Tyrosine kinase inhibitor STI-571/Gleevec down-regulates the beta-catenin signaling activity. *Cancer Lett 193*, 161–170.

Ziegler, S., Rohrs, S., Tickenbrock, L., et al. (2005). Novel target genes of the Wnt pathway and statistical insights into Wnt target promoter regulation. *FEBS J 272*, 1600–1615.

Chapter

16

Apoptotic pathways and cancer

Jian Yu and Lin Zhang

Introduction

Apoptosis, also known as programmed cell death, is an evolutionarily conserved cell suicidal process for dealing with stress conditions and maintenance of tissue homeostasis. It consists of a series of well-ordered biochemical events, which are regulated by a network of proteins containing distinctive functional domains. Apoptosis also serves as a safeguard mechanism against tumorigenesis. During oncogenic transformation, neoplastic cells become resistant to apoptosis as a result of genetic and epigenetic alterations. Defective apoptosis regulation drives other tumorigenic events, such as extension of lifespan, accumulation of further genetic mutations, growth under stress conditions, and tumor angiogenesis (Hanahan and Weinberg, 2011). Defective apoptosis regulation also plays an important role in therapeutic resistance (Yu and Zhang, 2004). Elucidation of the pathways involved in apoptosis signaling has stimulated intensive efforts to restore apoptosis in cancer cells for therapeutic purposes, and has led to the discoveries of several promising anticancer agents (Reed, 2006). This chapter provides an overview of the current understanding of the apoptotic machinery and its regulation, alterations of the apoptotic pathways in cancer cells, and the role of apoptosis in cancer therapies.

Apoptosis proteins

Overview

Apoptotic events are carried out by a network of proteins that contain distinctive structural and functional domains. These domains are evolutionarily conserved and can be identified using sequence-based approaches. The best characterized protein domains involved in apoptosis regulation include Bcl-2 homology (BH) domains, caspase protease domains, caspase recruitment domains (CARDs), death effector domains (DEDs), death domains (DDs), baculovirus inhibitor of apoptosis protein (IAP) repeat (BIR) domains, and cell death-inducing DFF45-like effector (CIDE) domains (Table 16.1) (Reed et al., 2004). In lower organisms such as *Caenorhabditis elegans* and *Drosophila*, approximately 10 to 20 genes are known to contain these domains (Reed et al., 2004). Genetic screenings identified several of these genes, such as *ced–3*, *ced–4*, and *ced–9* in *C. elegans* (Horvitz, 1999), as well as *ark, dronc*, and *DriCE* in *Drosophila*, as key regulators of cell death (Steller, 2008). Compared with these lower organisms, humans and mice have much more complex apoptotic machineries, with over a hundred genes encoding proteins that contain one or more of these signature domains. These proteins make up protein families (Table 16.1), thereby allowing apoptotic events triggered by diverse stimuli to be regulated in a highly specialized fashion in different tissues and cell types. The list and description of the genes encoding apoptosis proteins have been complied and organized into a searchable database (www.deathbase.org) (Diez et al., 2010). This database is likely to evolve as the functional domains are not always identified using sequence-based approaches.

Bcl-2 family of proteins

The Bcl-2 family members are key regulators of apoptosis. These proteins are structurally and functionally related to Bcl-2, the founding member of this family, which is altered in B cell lymphomas (Adams and Cory, 2007; Danial and Korsmeyer, 2004). In humans, 25 Bcl-2 family proteins have been identified (Table 16.1), whereas *Drosophila* and *C. elegans* each have only two (Reed et al., 2004). Bcl-2 family members all contain at least one characteristic BH

Systems Biology of Cancer, ed. S. Thiagalingam. Published by Cambridge University Press. © Cambridge University Press 2015.

Table 16.1 Human apoptotic machinery.

Protein families	Signature domains	Interacting proteins	Family members	
			Number	**Name**
Bcl-2 family	Bcl-2 homology (BH) domains	Other Bcl-2 family members	25	Bcl-2, Bcl-X, Mcl-1, Bcl-w, Bcl-B, A1, Bax, Bak, Bok, Mil-1, Bcl-G, Bik, Hrk, Bim, Bad, Bid, Puma, Bmf, Noxa, Bnip3, Nix, Bcl-L12, MAP-1, P193, Spike
Caspases	Caspase protease domains	CARD proteins, DD proteins, IAPs	11	Capsase-1, Capsase-2, Capsase-3, Capsase-4, Capsase-5, Capsase-6, Capsase-7, Capsase-8, Capsase-9, Capsase-10, Capsase-14
CARD family	Caspase recruitment domains (CARDs)	Caspases, other CARD proteins	22 (excluding caspases)	Apaf1, Arc, ASC, Bcl-10, Bimp1 (CARD10), Bimp2 (CARD14), Bimp3 (CARD11), Cardiak, CARD6, CARD9, CARP, c-IAP1, c-IAP2, CLAN (CARD12), COP, Helicard, Iceberg, NAC, Nod1 (CARD4), Nod2 (CARD15), RAIDD, TUCAN
DED family	Death effector domains (DEDs)	Caspases, other DED proteins	7	Caspase-8, Caspase-10, DEDD, DEDD2 (FLAME-3), FADD, c-FLIP, PEA-15
DD family	Death domains (DDs)	Other DD proteins	32	Ankyrin-2, Ankyrin-3, DAP-kinase, DR-3, DR-4, DR-5, DR-6, EDAR, EDARADD, FADD, Fas, IRAK-1, IRAK-2, IRAK-4, IRAK-3 (IRAK-M), MALT-1, MyD88, NF-κB-1, NF-κB-2, NGFR, NMP-84, Pidd, RAIDD, RIP, TNFR1, Tradd, UNC5H1, UNC5H2, Unc5H3 (Unc5H-C), Unc5H4 (Unc5H-D), UNC5H5
IAPs	Baculovirus IAP repeats (BIR) domains	Caspases, IAP antagonists such as SMAC	8	Apollon, cIAP1, cIAP2, ILP2, ML-IAP, NAIP, XIAP, Survivin
CIDE family	Cell death-inducing DFF45-like effector (CIDE) domains	Other CIDE proteins	5	CPAN, DFF45, FSP27 (CIDE-3), DFFA-likeA, DFFA-likeB

domain, and are divided into three groups (Figure 16.1). The antiapoptotic (pro-survival) group, including Bcl-2, Bcl-X$_L$, Mcl-1, Bcl-w, and A1/Bfl1, contain four BH domains. They protect cells from diverse cytotoxic conditions. The second group, including Bax and Bak, are proapoptotic and contain multiple BH domains. The third group is also proapoptotic and termed "BH3-only proteins"; with sequence homology with other Bcl-2 family members only within the amphipathic and α-helical BH3 segments (Figure 16.1) (Lomonosova and Chinnadurai, 2008). While there is only a single BH3-only protein, EGL-1, in *C. elegans* (Horvitz, 1999), at least nine BH3-only proteins have been identified in mammals, including Bad, Bid, Bik, Bim, Bmf, Bnip3, Hrk, Noxa, and PUMA (Bouillet and Strasser, 2002). The BH domains of the Bcl-2 family mediate interactions among the family members. Some Bcl-2 family members contain a trans-membrane (TM) domain in the carboxyl (C)-terminal of the proteins (Figure 16.1), which directs their localization to the mitochondria or other membrane-bound organelles.

Caspase proteases

Caspases are a family of intracellular cysteine proteases that are responsible for the execution of apoptosis (Li and Yuan, 2008). To date, 11 human caspase encoding genes have been identified, including *caspase-1* to *-10* and *caspase-14*. Caspase-11 and -12, which were found in mice, are functional orthologs of human caspase-4 and -5. Caspases are synthesized as inactive zymogens consisting of a prodomain and a conserved C-terminal catalytic domain including a p20 large subunit and a p10 small subunit (Figure 16.1). Proteolytic removal of the prodomain

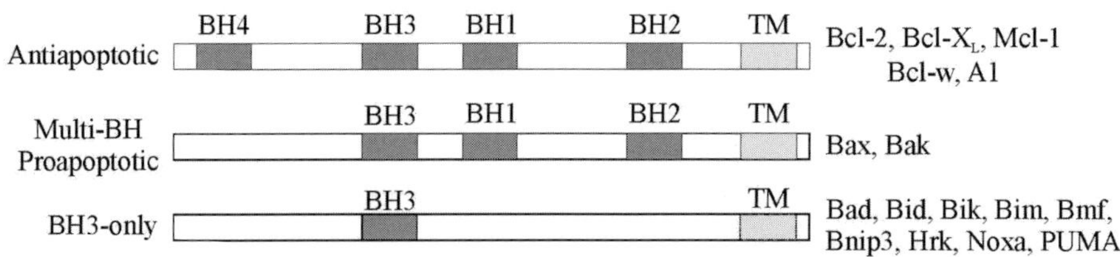

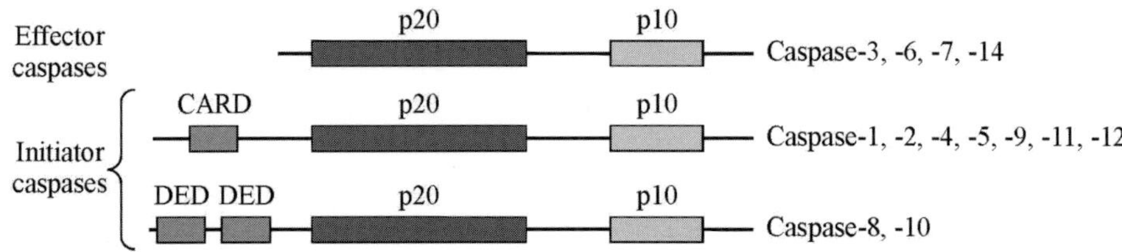

Figure 16.1 Major groups of apoptotic regulators and their signature domains. (BH, Bcl-2 homology domain; TM, trans-membrane domain; CARD, caspase recruitment domain; DED, death effector domain; EC, extracellular domain; DD, death domain; BIR, baculovirus IAP repeat domain)

and separation of the p20 and p10 subunits activate caspases (Li and Yuan, 2008). These two subunits then assemble into a symmetrical tetramer with two active sites formed at their junction. Caspases can be grouped into effector (executioner) caspases (caspase-3, -6, -7, -14) and initiator (activator) caspases (caspase-1, –2, -4, -5, -8, -9, -10, -11, and -12). The initiator caspases possess a larger prodomain containing one or more DED or CARD protein–protein interaction motif (Figure 16.1), which is responsible for binding to the adapter molecules required for their activation. The effector caspases contain a short prodomain and are generally activated by initiator caspases to cleave multiple substrates. Caspases recognize at least four contiguous amino acid residues in their substrates, and cleave after the C-terminal of the fourth residue, which is usually an aspartic acid (ASP) (Li and Yuan, 2008). To date, nearly 400 mammalian caspase substrates have been described (http://bioinf.gen.tcd.ie/casbah). In addition to caspases, several other proteases such as granzymes and calpain are also involved in apoptosis regulation (Reed et al., 2004).

Death domain (DD) superfamily

The death domain (DD) superfamily consists of proteins with one or more CARD, DED, or DD domain. CARD is a protein interaction domain which is composed of a bundle of six α-helices. In addition to CARD-containing caspases, 22 human CARD proteins have been identified (Table 16.1) (Reed et al., 2004). The CARD domains are commonly involved in the regulation of CARD-containing caspases, including human caspase-1, -2, -4, -5, -9, and -11, and that of nuclear factor κB (NF-κB) (Burstein and Duckett, 2003). Activation of caspases or NF-κB by CARD proteins is often associated with assembly of multiprotein complexes. For example, the adapter protein Apaf-1 can form oligomers and interact with pro-caspase-9 through their respective CARD domains, leading to caspase-9 activation (Wang, 2001). The IAP family members cIAP1 and cIAP2 inhibit caspase activation through CARD-mediated protein–protein interaction (Reed et al., 2004).

Similar to the CARD, the DED and DD domains are also α-helical protein interaction motifs. DED- and DD-containing proteins are major components of the extrinsic (death receptor) pathway. Seven DED proteins have been identified (Table 16.1), such as the adaptor protein Fadd, caspase-8, and caspase-10. The DED domains mediate interactions among these proteins to launch caspase activation cascade. DD is the signature domain of the tumor necrosis factor (TNF) receptor family of death receptors. As many as 32 DD proteins have been recognized (Table 16.1), such as TNF receptor 1 (TNFR1), Fas, death receptor-4 (DR-4), and DR-5 (Figure 16.1). They mediate the interactions among the death receptors, as well as the interactions between death receptors and adaptor proteins such as Tradd. DD domains are commonly involved in activation of caspases or NF-κB (Yu and Shi, 2008).

IAPs and their antagonists

All IAP family members contain at least one BIR domain (Figure 16.1), which is capable of interacting with caspases and IAP antagonists. There are eight known human IAP family members (Table 16.1), several of which, including XIAP, cIAP1, cIAP2, survivin, and ILP2, directly bind to and suppress caspases through their BIR domains (Deveraux and Reed, 1999). Several IAPs such as cIAP1 and cIAP2 also contain a RING domain (Figure 16.1), which is the signature domain of E3 ligases of the ubiquitin-proteasome machinery (Reed et al., 2004). The RING-containing IAPs promote degradation of proapoptotic proteins to suppress cell death. The antiapoptotic activities of the IAP family are relieved by IAP antagonists, including SMAC/Diablo and Omi/HtrA2. SMAC binds to IAPs through its amino (N)-terminal APVI tetrapeptide motif to facilitate caspase activation (Wu et al., 2000). Similar IAP antagonists have also been identified in *Drosophila*, including Hid, Grim, and Reaper (Steller, 2008).

Endonucleases

In the late stage of apoptosis, DNA is cleaved by several endonucleases, all of which contain a CIDE domain consisting of a β sheet and two α-helices. Five human CIDE family members have been identified (Table 16.1). The CIDE-containing endonuclease CAD (DFF40) is bound and inhibited by another CIDE family member ICAD via CIDE–CIDE interactions. Cleavage of ICAD by caspases during apoptosis leads to the activation of its endonuclease activity. Two other proteins, EndoG and AIF, also contribute to DNA degradation upon their release into the cytosol from the mitochondria (Danial and Korsmeyer, 2004).

Apoptotic pathways
The intrinsic pathway

Apoptosis triggered by stress conditions, such as DNA damage, oncogene expression, or nutrient/growth factor deprivation, is largely regulated by the Bcl-2 family of proteins through the mitochondria. Bcl-2 family members sense, transmit, and integrate a variety of inter- and intracellular signals (Danial and Korsmeyer, 2004). A linear apoptotic pathway was first delineated in *C. elegans* through genetic and biochemical analyses (Horvitz, 1999). EGL-1, a BH3-only protein, is transcriptionally activated and required for all developmental cell death in *C. elegans*. EGL-1 binds to the Bcl-2 homolog CED-9, thereby releasing CED-4, an ATPase and Apaf 1 homolog normally sequestered by CED-9. The freed CED-4 assembles into a tetrameric apoptosome and activates the protease activity of the caspase CED-3.

This apoptotic pathway is evolutionarily conserved, and becomes much more complex in mammals. There are multiple BH3-only and Bcl-2-like proteins, as well as the Bax/Bak group not found in the worm (Figure 16.1). The existence of multiple proteins in each group allows for fine tuning of apoptotic responses. Among the Bcl-2 members, the BH3-only proteins are apical sensors of apoptotic stimuli and are responsible for transmitting apoptotic signals to other Bcl-2 family members (Adams and Cory, 2007). Each BH3-only protein has specialized functions and can be activated by a different set of

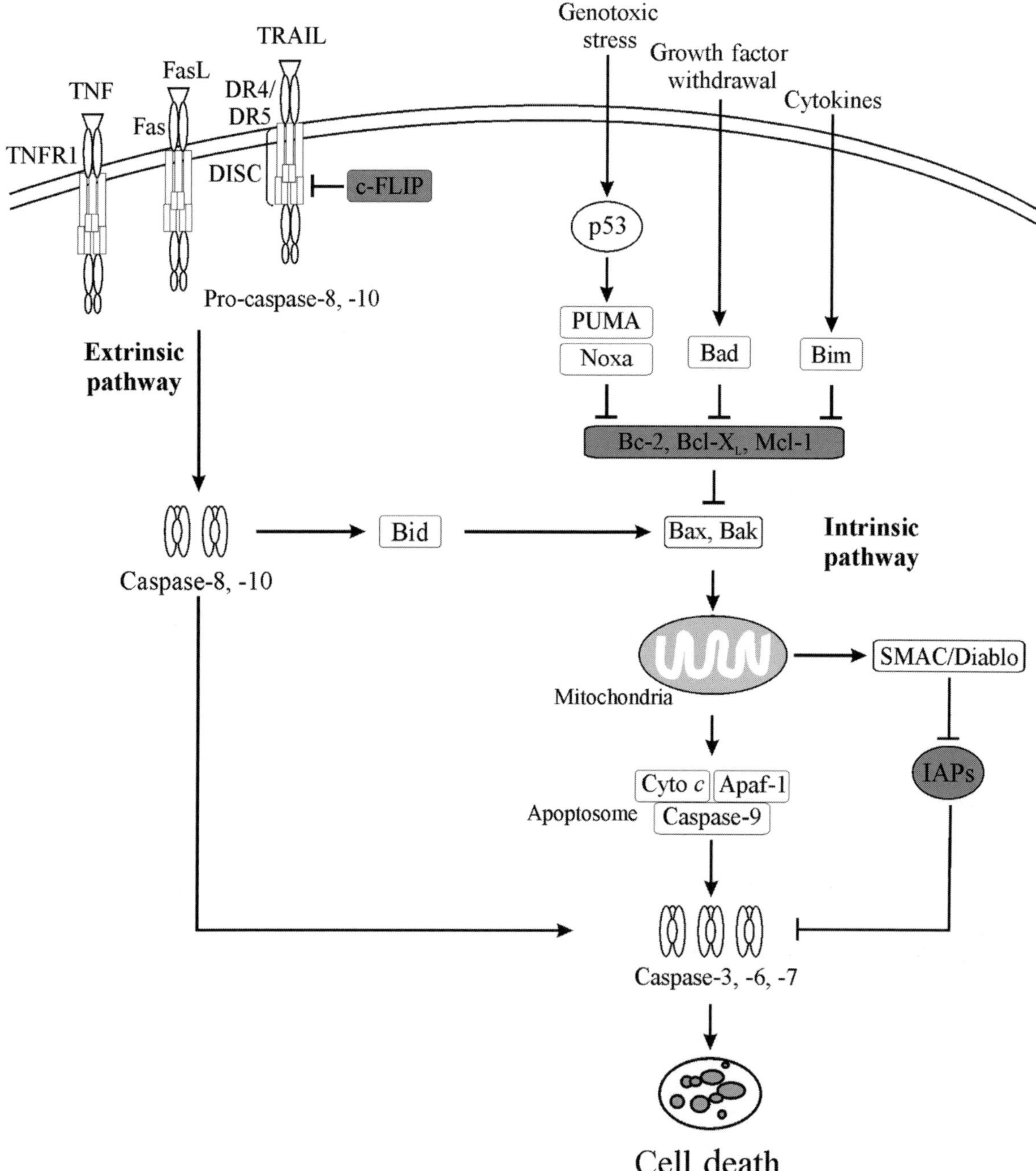

Figure 16.2 The intrinsic and extrinsic apoptotic pathways. Apoptosis in mammalian cells is initiated and executed through the intrinsic or extrinsic pathway. The intrinsic pathway is activated by stress conditions such as DNA damage, growth factor withdrawal, or cytokine exposure. It is regulated by the Bcl-2 family through the mitochondria. The extrinsic pathway is engaged upon binding of proapoptotic ligands, such as TNF, FasL, and TRAIL, to their respective cell surface receptors. Both pathways trigger a caspase activation cascade. The two pathways can crosstalk through the proapoptotic Bcl-2 family member Bid. The antiapoptotic proteins are shaded in gray. (Cyto c, cytochrome c)

conditions in a tissue- and cell type-specific manner (Figure 16.2) (Puthalakath and Strasser, 2002). For example, Bid mediates apoptosis induced by anti-Fas antibody in hepatocytes (Yin et al., 1999), while Bim is required for cytokine-mediated homeostasis of hematopoietic cells (Bouillet et al., 1999). PUMA and Noxa are transcriptionally activated by the tumor suppressor p53 in response to DNA damage or oncogene activation, and are required for stress-induced apoptosis in epithelial and several other types of cells (Ploner et al., 2008; Yu and Zhang, 2008). Bad regulates apoptosis induced by growth factor or nutrient deprivation (Danial, 2008).

Death signals received by the BH3-only proteins are transmitted to the Bcl-2 family members containing multiple BH domains (Figure 16.2). Bax and Bak are essential for apoptosis initiated by various stimuli (Wei et al., 2001). In healthy cells, the proapoptotic activities of Bax and Bak are kept in check by antiapoptotic Bcl-2 family members, such as Bcl-2, Bcl-X$_L$, and Mcl-1. Structural studies revealed that the BH1, BH2, and BH3 regions in these antiapoptotic proteins fold into a globular domain containing a surface hydrophobic groove (Sattler et al., 1997). Upon activation or induction, BH3-only proteins competitively bind to the same hydrophobic grooves of the antiapoptotic proteins to displace Bax or Bak (Cheng et al., 2001). Through yet unclear mechanisms, Bax then forms multimeric complexes, undergoes a conformational change, and translocates to the mitochondria. Some studies indicated that Bax and Bak can also be activated by the BH3-only proteins Bid, Bim, and perhaps PUMA through direct binding (Kim et al., 2009).

The cell death mediated by the Bcl-2 family members is characterized by mitochondrial outer membrane permeabilization (MOMP) (Chipuk and Green, 2008), and escape of several mitochondrial apoptogenic proteins including cytochrome c, SMAC/Diablo, Omi/HtrA2, AIF, and EndoG (Wang, 2001). The release of these proteins triggers execution of apoptosis through caspase-dependent and -independent mechanisms. In the cytosol, cytochrome c, Apaf-1, and pro-caspase-9 form an apoptosome complex, leading to the activation of caspase-9 (Wang, 2001). SMAC/Diablo and Omi/HtrA2 bind to IAPs to activate caspases (Du et al., 2000; Suzuki et al., 2001a). On the other hand, AIF and EndoG promote DNA degradation (Joza et al., 2001; Wang et al., 2002).

Cell death is culminated by the activation of caspases and destruction of cellular machinery including structural proteins, kinases, transcription/translation regulators, and other signaling molecules, resulting in eventual cell death (Li and Yuan, 2008). Although it was widely believed that caspases are mainly responsible for execution of apoptosis, several studies suggest

that they may play a role in initiating apoptosis prior to the mitochondrial events in some cells. For example, deficiencies in both *caspase-3* and *caspase-7* surprisingly affected mitochondrial dysfunction that was previously thought to be upstream of caspase activation (Lakhani et al., 2006). However, the general role of caspases in cell death initiation remains unsolved.

Regulation of the intrinsic pathway

The intrinsic apoptotic pathway is regulated at the protein level through several different mechanisms. Most notably, the stimuli- and cell type-specific regulation is achieved through differential interactions among the Bcl-2 family members. For example, BH3-only proteins have different capabilities of inducing apoptosis in different tissues and cell types, likely owing to their differential binding specificities for antiapoptotic proteins (Chen et al., 2005; Kuwana et al., 2005). The BH3 domains of PUMA and Bim bind to all five antiapoptotic Bcl-2 family members, and are potent inducers of apoptosis in diverse tissues and cell types. In contrast, those of other BH3-only proteins bind to a subset of antiapoptotic proteins and have limited capabilities of apoptosis induction. For example, the BH3 domain of Bad preferentially interacts with Bcl-2, Bcl-X$_L$, and Bcl-w, while that of Noxa binds to Mcl-1 and A1/Bfl1 (Chen et al., 2005). The molecular basis of this difference remains poorly understood. Protein–protein interactions also play a key role in other stages of apoptosis, such as the formation of apoptosome and assembly of active caspases (Adams and Cory, 2007).

A variety of post-translational modifications play a key role in the regulation of apoptosis (Puthalakath and Strasser, 2002). For example, the proapoptotic activity of the BH3-only protein Bad is suppressed by phosphorylation, which mediates its binding to 14-3-3 proteins and cytoplasmic sequestration (Zha et al., 1996). Phosphorylation of Bcl-2 enhances its antiapoptotic function (Deng et al., 2004). Targeting of the truncated form of Bid (tBid) to the mitochondria is facilitated by N-myristoylation of a residue that is exposed after caspase-mediated cleavage (Zha et al., 2000). The antiapoptotic activity of Bcl-X$_L$ can be modulated by the removal of amino groups from two ASP residues in response to DNA damage (Deverman et al., 2002). In addition, nitrosylation of the active sites of caspases by nitric oxide was shown to be an important mechanism for inhibiting apoptosis (Brune, 2003). Furthermore, ubiquitination also regulates apoptosis. For example, IAPs can promote ubiquitination and proteasomal degradation of caspases in normal healthy cells (Suzuki et al., 2001b). Following treatment with apoptotic stimuli, IAPs can promote self-ubiquitination (Yang et al., 2000). Degradation of Mcl-1, which is mediated by the E3 ligase Mule, is necessary for UV-induced apoptosis (Zhong et al., 2005).

The extrinsic pathway

The extrinsic apoptotic pathway is predominantly utilized by specialized immune cells, such as natural killer cells or cytotoxic T cells, to induce apoptosis in infected or damaged cells. This pathway is engaged upon binding of proapoptotic ligands to cell surface death receptors. The death receptors of the TNFR family all contain a cysteine-rich extracellular domain in addition to a DD domain (Figure 16.1). Over 20 TNFR family members have been identified (Ashkenazi and Dixit, 1998). Their extracellular domains are responsible for ligand binding, while the intracellular death domains play a crucial role in transmitting death signals from cell surface to intracellular signaling molecules. The proapoptotic ligands belong to the greater cytokine TNF superfamily, and are either presented on the cell surface or secreted into the extracellular space (Ashkenazi, 2008). Due to the complexities of the death receptor and ligand families, only few of them are relatively well characterized (Ashkenazi, 2008; Fulda and Debatin, 2006). For example, apoptosis ligand 2 (Apo2L)/TNF-related apoptosis-inducing ligand (TRAIL) signals through two related proapoptotic receptors, DR4 (TRAIL-R1) and DR5 (TRAIL-R2). Fas/Apo1/CD95 ligand (FasL) signals apoptosis through the Fas/Apo1/CD95 receptor (Ashkenazi and Dixit, 1998; Nagata, 1997; Peter and Krammer, 2003). In limited cases, TNF-α, the founding member of the TNF family, also acts as a proapoptotic ligand by signaling through TNFR1 when NF-κB activation is blocked (Karin and Lin, 2002).

Upon binding of proapoptotic ligands to their respective receptors, each receptor can independently form a death-inducing signaling complex (DISC) by recruiting the adapter protein Fadd, along with pro-caspase-8 and pro-caspase-10 (Figure 16.2) (Kischkel et al., 2000, 2001). The formation of DISC is mediated by the interactions between the DED domains in these proteins. Fadd recruitment and DISC formation

lead to proximity-induced activation of caspase-8 and caspase-10. These caspases then undergo self-processing, which releases active caspase molecules into the cytoplasm (Figure 16.2) (Ashkenazi and Dixit, 1998). The active caspase-8 and caspase-10 in turn cleave and activate caspase-3, -6, and -7 to execute apoptosis (LeBlanc et al., 2002). Among the apoptotic mechanisms regulated by the extrinsic pathway, TRAIL-induced apoptosis mediated by DR4 or DR5 is best exploited in relation to cancer (Ashkenazi, 2008; Johnstone et al., 2008).

Regulation of the extrinsic pathway

The extrinsic pathway is regulated through the death receptors on the cell surface and DISC formation in the cytosol. The DISC can be negatively regulated by c-FLIP or decoy death receptors. c-FLIP has sequence homology to caspase-8 and caspase-10, but lacks caspase enzymatic activity. The recruitment of c-FLIP to the DISC blocks caspase activation (Krueger et al., 2001). The decoy receptors can be bound by ligands but lack the ability to transmit apoptotic signals. The decoy receptors DcR1 (TRAIL-R3), DcR2 (TRAIL-R4), and osteoprotegerin (OPG) bind to TRAIL (Ashkenazi, 2008; Ashkenazi and Dixit, 1998), whereas DcR3 binds to FasL26.

The DISC can be positively regulated by proapoptotic death receptors through post-translational modifications. O-linked glycosylation of the extracellular domains of DR4 and DR5 promotes apoptosis by facilitating ligand-induced receptor clustering and DISC stimulation (Wagner et al., 2007). Similarly, membrane proximal palmitoylation of Fas leads to Fas clustering and activation of DISC (Feig et al., 2007; Muppidi and Siegel, 2004). Moreover, vesicle trafficking can differentially affect DISC activation in a cell type-dependent manner. In some cell types, Fas internalization through clathrin-mediated endocytosis enhances DISC recruitment (Lee et al., 2006), while that of DR4 and DR5 can inhibit caspase activation (Austin et al., 2006; Kohlhaas et al., 2007).

Crosstalk between the intrinsic and extrinsic pathways

Under certain conditions, crosstalk can occur between the intrinsic and extrinsic pathways at the level of initiator caspases (Ashkenazi, 2008). For example, caspase-8-dependent cleavage of the BH3-only protein Bid generates the active tBid. tBid serves in an amplification loop of apoptotic signaling in some cells with low levels of DISC formation to recruit the intrinsic pathway and the mitochondria for efficient apoptosis induction (Figure 16.2). Recent studies indicated that degradation of IAPs induced by the N-terminal residues of SMAC/Diablo can result in stimulation of NF-κB-mediated TNF-α production and subsequent apoptosis in some cancer cells (Petersen et al., 2007), suggesting multiple crosstalk mechanisms between the intrinsic and extrinsic apoptotic pathways.

Integration of proximal apoptotic signals

The complexity of apoptosis regulation is reflected not only by the sheer number of proteins in the two major pathways, but also by their dynamic inter-actions and post-translational modifications. Adding another level of complexity, the expression of these proteins is coordinately controlled by several tran-scription factors. In response to stress conditions, these transcription factors are activated through post-translational modifications as a result of a chain of signaling events. For example, the ATM kinase becomes phosphorylated within minutes of exposure to γ-irradiation (Bakkenist and Kastan, 2003). This triggers a protein phosphorylation cascade leading to the activation of p53 and other transcription factors. These transcription factors in turn modulate the expression of the proteins in the apoptotic machinery, and those mediating rescuing attempts such as cell cycle arrest and DNA repair, before a decision of irreversible cell demise is reached. This section sum-marizes three major transcription factor families involved in apoptosis regulation.

p53 and genotoxic stress-induced apoptosis

The p53 tumor suppressor is essential for preventing inappropriate cell proliferation and for maintaining genome integrity following genotoxic stress (Vogel-stein et al., 2000). Transcriptional activation by p53 appears to be essential for its proapoptotic function *in vitro* and *in vivo* (Yu and Zhang, 2005). In response to genotoxic stress, p53 undergoes extensive post-translational modifications, including phosphoryl-ation and acetylation, which stabilize p53 and enhance its transcriptional activities (Vogelstein

et al., 2000). Genes encoding proteins involved in virtually every step of apoptosis can be regulated by p53 (Yu and Zhang, 2005).

p53 mediates DNA damage-induced apoptosis primarily via the intrinsic pathway, in particular the BH3-only proteins PUMA and Noxa. PUMA and Noxa are transcriptionally activated by p53 in response to DNA damage (Nakano and Vousden, 2001; Oda et al., 2000; Yu et al., 2001). Much like p53, PUMA is required for DNA damage-induced apoptosis in several tissues of mice, including thymocytes, embryonic fibroblasts, and developing neurons (Jeffers et al., 2003; Villunger et al., 2003). Noxa-deficient mice exhibited reduced apoptosis following DNA damage in some cells, albeit at a much lesser extent compared to PUMA-deficient mice (Villunger et al., 2003). Other proapoptotic Bcl-2 family members, such as Bax and Bid, can also be activated by p53 (Yu and Zhang, 2005). Several p53 targets exert their effects on the mitochondria or Bcl-2 family proteins, including the developmentally regulated gene *Pw1/Peg3*, the mitochondrial protein p53AIP1, and p53DINP1. p53 can also facilitate the execution step of apoptosis by enhancing the expres-sion of *Apaf-1*, *caspase-1*, *-6*, *-8*, and *Omi/HtrA2*. In some cells, p53 suppresses the expression of Bcl-2 and Bcl-X$_L$, though the mechanism of p53-mediated tran-scriptional repression is not well understood (Yu and Zhang, 2005).

The extrinsic pathway can also be modulated by p53 (Yu and Zhang, 2005). p53 engages this pathway by activating the expression of *Fas*, *DR4*, and *DR5*. The DD family protein Pidd and the PMP-22/gas family protein PERP contribute modestly to p53-dependent apoptosis induced by γ-irradiation. More-over, genotoxic stress can activate the p53 family proteins p73 and p63, which are capable of activating many proapoptotic p53 targets (Melino et al., 2003; Moll and Slade, 2004). In addition to the p53 family, other transcription factors implicated in genotoxic stress-induced apoptosis include AP-1 (c-Jun/c-Fos) and E2F-1 transcription factors (Eferl and Wagner, 2003; Stevens and La Thangue, 2004).

NF-κB and cytokine-induced apoptosis

NF-κB is a dimeric transcription factor complex formed by members of a highly conserved protein family that shares the Rel homology domain (RHD) (Burstein and Duckett, 2003). In unstimulated cells,

NF-κB complexes are sequestered in the cytoplasm through their interactions with members of the inhibitor of the κB (IκB) family. Upon stimulation by pro-inflammatory cytokines, such as TNF-α and interleukin-1β, IκB is phosphorylated by IKK kinases, which results in ubiquitination and proteasomal degradation of IκB, allowing the nuclear translocation of NF-κB complexes. Nuclear NF-κB dimers in turn bind to gene promoters to activate or repress the expression of its downstream targets (Burstein and Duckett, 2003).

A large body of evidence has demonstrated an antiapoptotic role of NF-κB (Deveraux and Reed, 1999). For example, NF-κB is the major cellular factor responsible for resistance to TNF-α-induced cell death *in vitro* and *in vivo* (Burstein and Duckett, 2003). NF-κB regulates the expression of several antiapoptotic proteins involved in the intrinsic pathway, such as Bcl-2, Bcl-X$_L$, A1/Bfl1, and IAPs (Karin and Lin, 2002). Furthermore, NF-κB can also exert its antiapoptotic function via the extrinsic pathway through the upregulation of DcR1 or c-FLIP (Bernard et al., 2001; Micheau et al., 2001). However, the functions of NF-κB in apoptosis appear to be paradoxical, as it can promote apoptosis under some conditions through the induction of DR4 and DR5 (Ravi et al., 2001).

FoxO and growth factor deprivation-induced apoptosis

Growth factor withdrawal triggers apoptosis in multiple tissues via irreversible damage to the mitochondria and a crisis in energy metabolism. One of the hallmarks of cancer is inappropriate activation of growth-promoting kinase pathways in the absence of growth factor stimulation (Hanahan and Weinberg, 2000). In addition to the BH3-only protein Bad, the forkhead transcription factor O (FoxO) family of transcription factors play an important role in apoptosis induced by cytokine/growth factor withdrawal. This family includes FoxO1 (FKHR), FoxO3A (FKHRL1), FoxO4 (AFX), and FoxO6 (Fu and Tindall, 2008).

In the presence of growth factors, FoxO proteins can be phosphorylated by numerous kinases such as Akt and JNK. Phosphorylation of FoxO proteins masks their nuclear localization signals and interferes with their ability to transactivate genes (Myatt and Lam, 2007). FoxO proteins promote apoptosis via both the intrinsic and extrinsic pathways. Their proapoptotic targets in the extrinsic pathway include FasL, TRAIL, and Tradd (Fu and Tindall, 2008). For the intrinsic pathway, FoxO3A mediates the induction of the BH3-only proteins PUMA and Bim following cytokine/serum withdrawal (Dijkers et al., 2000; You et al., 2006). FoxO4 can downregulate Bcl-X$_L$ indirectly by inducing the expression of Bcl-6, which represses the *Bcl-X$_L$* promoter (Tang et al., 2002). FoxO proteins can also activate the expression of the p19/Arf tumor suppressor, and inactivation of FoxO proteins leads to inhibition of apoptosis and accelerated lymphomagenesis in the Eμ-myc lymphoma model (Bouchard et al., 2007).

Deregulation of apoptosis in cancer
Alterations in the proximal regulators

Maintenance of tumor phenotypes is highly dependent on suppression of apoptosis by certain pro-survival proteins, because so-called "oncogenic stress" imposed by neoplastic transformation would normally trigger an apoptotic response (Hanahan and Weinberg, 2011). Several key pathways controlling apoptosis are commonly altered in cancer (Vogelstein and Kinzler, 2004). More than half of human tumors contain mutations in the *p53* tumor suppressor gene. *p53* is also the target gene inactivated in the germline of Li–Fraumeni cancer syndrome patients (Vogelstein et al., 2000). The great majority of tumor-derived p53 mutants are defective in apoptosis induction due to their inability to transactivate the proapoptotic downstream targets such as PUMA and Noxa (Yu and Zhang, 2005). Other major transcription factors involved in integrating apoptosis signals are also common targets for deregulation in cancer. For example, NF-κB is overexpressed in many types of cancer, and has been linked to chemoresistance through the induction of various antiapoptotic proteins (Wang et al., 1999). Loss of the PTEN tumor suppressor or constitutive PI3K/Akt signaling leads to phosphorylation of FoxO3A and inhibition of apoptosis (You et al., 2006).

Deregulation of the intrinsic pathway in cancer

Antiapoptotic Bcl-2 family members are frequently overexpressed in tumors, and some of them are oncogenes. For instance, *bcl-2* is located at the breaking point of a frequent chromosomal translocation in human lymphomas, leading to Bcl-2 overexpression at the transcriptional level (Tsujimoto et al., 1985). Upregulation of Bcl-2 also can be caused by other mechanisms including gene amplification, promoter hypomethylation, and loss of endogenous micro RNA expression (Yip and Reed, 2008). Overexpression of other antiapoptotic proteins, such as Bcl-X$_L$, Mcl-1, and IAPs, are common in cancer (Adams and Cory, 2007). Human transcriptome analysis revealed that one of the most universal changes in human tumors is the elevated expression of the IAP family member survivin (Velculescu et al., 1999).

On the other hand, proapoptotic Bcl-2 family members have been found to be mutated, downregulated, or compromised in tumor cells. Homozygous deletions or inactivating mutations of *BAX* have been identified in a variety of tumors, in particular DNA mismatch repair-deficient colorectal tumors (Rampino et al., 1997). The expression of PUMA is downregulated in malignant cutaneous melanoma (Karst et al., 2005). *PUMA* is also inactivated in a subset of human Burkitt lymphomas in part through DNA methylation (Garrison et al., 2008). The activities of proapoptotic Bcl-2 family members can also be suppressed in tumors through post-translational modifications. For example, BAD can be phosphorylated by Akt and other protein kinases that are hyperactive in cancer (Zha et al., 1996). Virtually all tumor suppressor genes and oncogenes have been directly or indirectly linked to apoptosis regulation (Green and Evan, 2002). However, most genetic alterations are found in the more proximal components rather than the core apoptotic machinery, suggesting that shutting off parallel and redundant apoptotic pathways at nodal points such as p53 is preferred during tumor development.

Deregulation of the extrinsic pathway in cancer

Cancer cells are able to evade the extrinsic apoptotic pathway by blocking death receptor signaling. For example, Fas is downregulated or absent in a variety of hematological malignancies and solid tumors due to mutations or promoter methylation (Fulda and Debatin, 2006). Loss of surface expression of DR4 and DR5 in cancer cells can be caused by loss of heterozygosity (LOH), mutations, or homozygous deletions (Dechant et al., 2004; LeBlanc and Ashkenazi, 2003; Pai et al., 1998). On the other hand,

alterations in the antagonistic decoy receptors can also inhibit death receptor-mediated apoptosis. Overexpression of DcR3 or DcR1 interferes with FasL- or TRAIL-triggered apoptosis in cancer cells, respectively (Fulda and Debatin, 2006). Alterations in the proteins that modulate the cytoplasmic function of the death receptors are also found in tumors. For example, c-FLIP overexpression occurs in many tumor cells (Fulda and Debatin, 2006; Krueger et al., 2001).

Changes in several downstream mediators of the extrinsic pathway, such as caspase-8 and caspase-10, have been implicated in cancer. Polymorphisms of *caspase-8* (*CASP8*) are associated with altered risks of cancer (Sun et al., 2007). Deletions of *CASP8* were found in advanced childhood neuroblastoma (Teitz et al., 2000). Promoter methylation of either *DCR2* or *CASP8* in primary neuroblastoma was associated with high-risk disease and poor clinical outcome (Ullenhag et al., 2007; Yang et al., 2007). In general, direct impairment of the extrinsic pathway by mutations is rare in human cancer. The evidence for a definitive involvement of this pathway in cancer initiation and/or progression is relatively weak in comparison to that of the intrinsic pathway (Ashkenazi, 2008; Johnstone et al., 2008).

Apoptosis modulation for cancer therapy

Induction of apoptosis is a major cytotoxic mechanism of anticancer therapies including chemotherapy and radiotherapy (Johnstone et al., 2002). Cancer cells can develop various mechanisms to evade apoptosis by blocking death receptor signaling or, more often, by disabling the intrinsic pathway. Deficiency in p53, Bax, PUMA, or caspase activation can cause resistance to chemotherapeutic agents or γ-irradiation (Bunz et al., 1999; Schimmer et al., 2003; Yu et al., 2003; Zhang et al., 2000). On the other hand, overexpression of antiapoptotic proteins such as Bcl-X$_L$ and c-FLIP is frequently associated with therapeutic resistance (Amundson et al., 2000; Fulda and Debatin, 2006). Unfolding of the complex apoptotic pathways has stimulated intensive efforts to restore apoptosis in cancer cells for therapeutic purposes (Reed, 2006). Several strategies have been devised to manipulate the apoptotic pathways at various regulatory points to facilitate the killing of cancer cells. The protein-based agents trigger apoptosis from the cell surface by stimulating proapoptotic signaling. The small-molecule or antisense inhibitors of pro-survival Bcl-2 family members or the IAPs promote apoptosis from within cells by inhibiting antiapoptotic signaling. Several major classes of apoptosis-based anticancer agents are discussed in this section.

Exploiting the intrinsic pathway for cancer therapy

Deregulation of the intrinsic apoptotic pathway in cancer cells appears to primarily affect the signaling pathways upstream of Bax/Bak and the mitochondria, leaving the downstream core apoptotic machinery mostly intact (Adams and Cory, 2007; Danial and Korsmeyer, 2004). It is therefore attractive to target the antiapoptotic proteins that are highly expressed in cancer cells. An initial approach is to use antisense DNA to induce degradation of the mRNA of anti-apoptotic Bcl-2 family members. Antisense oligonucleotides for various antiapoptotic members, including Bcl-2, Bcl-X$_L$, and Mcl-1, have been described in preclinical studies (Reed and Pellecchia, 2005). The Bcl-2 antisense oligonucleotide oblimersen sodium has advanced into phase III clinical trials (O'Brien et al., 2007). Another method is to use peptides containing the proapoptotic BH3 domains to neutralize antiapoptotic Bcl-2 family proteins (Shangary et al., 2004). However, this method is limited by the unfavorable pharmacological properties of peptides (Denicourt and Dowdy, 2004).

The most promising approach thus far is to use small-molecule inhibitors of antiapoptotic proteins. The relatively small and deep hydrophobic grooves on the surface of antiapoptotic Bcl-2 family members such as Bcl-X$_L$ allow for design of highly specific inhibitors (Petros et al., 2000). A number of small-molecule inhibitors of antiapoptotic Bcl-2 family members have been described, such as HA14-1, Antimycin A and analogs, BH3Is, gossypol and analogs, and GX015-070 (Zhang et al., 2007). The most potent and specific Bcl-2/Bcl-X$_L$ inhibitor discovered to date is a synthetic compound called ABT-737. ABT-737 mimics the BH3 domain of Bad in binding to Bcl-2 and Bcl-X$_L$ (Oltersdorf et al., 2005), and can induce Bax/Bak- and caspase-dependent apoptosis (van Delft et al., 2006). As a single agent, ABT-737 effectively kills small-cell lung carcinoma (SCLC) and several types of leukemia and lymphoma cells (Oltersdorf et al., 2005). It exhibited striking synergy when combined with a variety of anticancer agents in preclinical models (Oltersdorf et al., 2005). Several studies showed that Mcl-1 expression is associated with ABT-737 resistance (Zhang et al., 2007). ABT-263, an orally available derivative of ABT-737, is currently being evaluated in clinical trials (Tse et al., 2008).

In addition to Bcl-2 inhibition, targeting the IAPs using SMAC mimetics has emerged as a promising approach to induce apoptosis in cancer cells. SMAC expression enhances the proapoptotic effects of a variety of stimuli (Arnt and Kaufmann, 2003). The N-terminal AVPI motif of SMAC linked to carrier peptides enhanced apoptosis induced by several conventional and experimental anticancer agents (Fulda and Debatin, 2006). Small molecules mimicking the AVPI motif of SMAC have also been developed, and exhibited *in vitro* antitumor activities mostly when combined with other classes of anticancer agents (Bank et al., 2008; Li et al., 2004). Several of these compounds are anticipated to enter clinical trials. Recent studies revealed that SMAC mimetics alone can induce apoptosis in some cancer cells by promoting degradation of IAPs and stimulating NF-κB-mediated TNF-α production (Petersen et al., 2007).

Exploring the extrinsic pathway for cancer therapy

Targeting the extrinsic apoptotic pathway is attractive for several reasons. First, the alterations in this pathway in cancer are much more limited compared with those in the intrinsic pathway, with most of the death receptors expressed in cancer cells (Ashkenazi, 2008). Second, apoptosis triggered by death receptors occurs largely independent of p53 (Ashkenazi, 2008; El-Deiry, 2001). Third, overexpression of oncogenes such as *KRAS* and *c-Myc* can sensitize cells to death receptor-mediated apoptosis (Nieminen et al., 2007; Wang et al., 2005). Two major classes of protein-based agents have been developed to specifically target the extrinsic pathway to trigger apoptosis. They are collectively called pro-apoptotic receptor agonists (PARAs) (Ashkenazi, 2008).

The clinical applications of death receptor ligands have been focused on TRAIL and FasL, as other ligands such as TNF-α are associated with severe side effects (Walczak and Krammer, 2000). Manipulating the TRAIL pathway has shown promise in both preclinical and clinical studies, perhaps reflecting the more direct involvement of its receptors DR4 and DR5 in cancer compared with other TNFR family members (Ashkenazi, 2008; Finnberg and El-Deiry, 2008; Johnstone et al., 2008). Recombinant human TRAIL (rhTRAIL) triggers p53-independent apoptosis in cancer cells, and has little effect on most non-malignant cells. TRAIL and FasL are both being tested in clinical trials (Call et al., 2008). The clinical development of the agonistic death receptor antibodies has been focused on the TRAIL receptors DR4 and DR5 (Ashkenazi, 2008; Finnberg and El-Deiry, 2008; Johnstone et al., 2008). Mapatumumab (HGS-ETR1) is the only human agonistic antibody for DR4 that has advanced into clinical trials, showing some efficacy in combination regimens (Call et al., 2008). Currently, at least four DR5 agonistic antibodies are in early phase clinical development, but the data on their efficacies is very limited (Ashkenazi, 2008; Call et al., 2008).

Although it is unclear what determines the response to PARAs in patients, a number of potential biomarkers of TRAIL sensitivity have been identified. For example, high expression of certain *O*-glycosylation enzymes is associated with TRAIL sensitivity (Wagner et al., 2007), while mutations in *CASP8* or *BAX* are associated with TRAIL resistance (Ashkenazi, 2008; Johnstone et al., 2008).

Other proapoptotic cancer therapies

Several classes of targeted therapeutic agents used clinically have been shown to directly induce apoptosis in cancer cells with certain genetic alterations. The selective tyrosine kinase inhibitors (TKIs) of epidermal growth factor receptor (EGFR) are particularly useful in treating a subset of lung cancer patients whose tumors harbor oncogenic mutations in the EGFR kinase domain. These agents induce apoptosis via the intrinsic pathway, in part, through the BH3-only protein Bim or Bad (Costa et al., 2007; Gong et al., 2007; She et al., 2005). Interestingly, secondary mutations in EGFR result in resistance to EGFR TKIs (Sharma et al., 2007). Moreover, the inhibitors of c-MET, a receptor kinase that is deregulated in various cancers, preferentially induce apoptosis in the cancer cells with *c-MET* amplification (Comoglio et al., 2008). These examples elegantly illustrate how oncogenic addiction can be targeted by selective kinase inhibitors to kill cancer cells.

Inhibition of proteasome-mediated protein degradation is another approach to promote apoptosis in cancer cells. Small-molecule proteasome inhibitors such as bortezomib can trigger apoptosis in many cancer cells, or sensitize cancer cells to chemotherapy through apoptosis induction. Bortezomib induces apoptosis by modulating the expression of proapoptotic and antiapoptotic proteins, or by eliciting stress responses (Milano et al., 2007). For example, bortezomib treatment results in reduced levels of Bcl-2 and Bcl-X$_L$, and increased levels of Fas, p53, Bax, or TRAIL receptors (Call et al., 2008). It has been approved for the treatment of multiple myeloma or mantle-cell lymphoma (Milano et al., 2007).

Conclusion

Apoptosis in mammalian cells is regulated through two major pathways by several protein families each containing distinctive functional domains. Protein–protein interactions mediated by these domains determine the balance between proapoptotic and antiapoptotic activities, thus the decision between life and death. Integration of apoptotic signals is achieved through the crosstalk among many upstream signals and downstream effectors. Alterations in the apoptotic pathways appear to be the Achilles' heel of cancer, and serve as potential targets for drug development. The hope is to be able to develop targeted therapies to greatly enhance the sensitivity of cancer cells to commonly used cytotoxic treatments. A great challenge ahead is to understand in detail the integration modules and nodes of each apoptotic pathway, which will help drug development and delineation of therapeutic resistance mechanisms. Such information will ultimately allow for tailored therapies based on the genetic makeup of individual patients.

Acknowledgments

Research in the authors' laboratories is supported by NIH grants CA106348, CA121105, CA172136, American Cancer Society grant RSG-07-156-01-CNE, the Career Investigator Award from the American Lung Association/CHEST Foundation, the V Foundation for Cancer Research (L.Z.), the Flight Attendant Medical Research Institute (FAMRI), the Alliance for Cancer Gene Therapy (ACGT), American Cancer Society grant RSG-10-124-10-CCE, and NIH grants CA129829 and U01DK085570 (J.Y.).

References

Adams, J.M. and Cory, S. (2007). The Bcl-2 apoptotic switch in cancer development and therapy. *Oncogene*, 26, 1324–37.

Amundson, S.A., Myers, T.G., Scudiero, D., et al. (2000). An informatics approach identifying markers of chemosensitivity in human cancer cell lines. *Cancer Res*, 60, 6101–10.

Arnt, C.R. and Kaufmann, S.H. (2003). The saintly side of Smac/DIABLO: giving anticancer drug-induced apoptosis a boost. *Cell Death Differ*, 10, 1118–20.

Ashkenazi, A. (2008). Directing cancer cells to self-destruct with pro-apoptotic receptor agonists. *Nat Rev Drug Discov*, 7, 1001–12.

Ashkenazi, A. and Dixit, V.M. (1998). Death receptors: signaling and modulation. *Science*, 281, 1305–8.

Austin, C.D., Lawrence, D.A., Peden, A.A., et al. (2006). Death-receptor activation halts clathrin-dependent endocytosis. *Proc Natl Acad Sci USA*, 103, 10283–8.

Bakkenist, C.J. and Kastan, M.B. (2003). DNA damage activates ATM through intermolecular autophosphorylation and dimer dissociation. *Nature*, 421, 499–506.

Bank, A., Wang, P., Du, C., Yu, J. and Zhang, L. (2008). SMAC mimetics sensitize nonsteroidal anti-inflammatory drug-induced apoptosis by promoting caspase-3-mediated cytochrome c release. *Cancer Res*, 68, 276–84.

Bernard, D., Quatannens, B., Vandenbunder, B. and Abbadie, C. (2001). Rel/NF-kappaB transcription factors protect against tumor necrosis factor (TNF)-related apoptosis-inducing ligand (TRAIL)-induced apoptosis by up-regulating the TRAIL decoy receptor DcR1. *J Biol Chem*, 276, 27322–8.

Bouchard, C., Lee, S., Paulus-Hock, V., et al. (2007). FoxO transcription factors suppress Myc-driven lymphomagenesis via direct activation of Arf. *Genes Dev*, 21, 2775–87.

Bouillet, P., Metcalf, D., Huang, D.C., et al. (1999). Proapoptotic Bcl-2 relative Bim required for certain apoptotic responses, leukocyte homeostasis, and to preclude autoimmunity. *Science*, 286, 1735–8.

Bouillet, P. and Strasser, A. (2002). BH3-only proteins – evolutionarily conserved proapoptotic Bcl-2 family members essential for initiating programmed cell death. *J Cell Sci*, 115, 1567–74.

Brune, B. (2003). Nitric oxide: NO apoptosis or turning it ON? *Cell Death Differ*, 10, 864–9.

Bunz, F., Hwang, P.M., Torrance, C., et al. (1999). Disruption of p53 in human cancer cells alters the responses to therapeutic agents. *J Clin Invest*, 104, 263–9.

Burstein, E. and Duckett, C.S. (2003). Dying for NF-kappaB? Control of cell death by transcriptional regulation of the apoptotic machinery. *Curr Opin Cell Biol*, 15, 732–7.

Call, J.A., Eckhardt, S.G. and Camidge, D.R. (2008). Targeted manipulation of apoptosis in cancer treatment. *Lancet Oncol*, 9, 1002–11.

Chen, L., Willis, S.N., Wei, A., et al. (2005). Differential targeting of prosurvival Bcl-2 proteins by their BH3-only ligands allows complementary apoptotic function. *Mol Cell*, 17, 393–403.

Cheng, E.H., Wei, M.C., Weiler, S., et al. (2001). BCL-2, BCL-X(L) sequester BH3 domain-only molecules preventing BAX- and BAK-mediated mitochondrial apoptosis. *Mol Cell*, 8, 705–11.

Chipuk, J.E. and Green, D.R. (2008). How do BCL-2 proteins induce mitochondrial outer membrane permeabilization? *Trends Cell Biol*, 18, 157–64.

Comoglio, P.M., Giordano, S. and Trusolino, L. (2008). Drug development of MET inhibitors: targeting oncogene addiction and expedience. *Nat Rev Drug Discov*, 7, 504–16.

Costa, D.B., Halmos, B., Kumar, A., et al. (2007). BIM mediates EGFR tyrosine kinase inhibitor-induced apoptosis in lung cancers with oncogenic EGFR mutations. *PLoS Med*, 4, 1669–79; discussion 1680.

Danial, N.N. (2008). BAD: undertaker by night, candyman by day. *Oncogene*, 27 Suppl 1, S53–70.

Danial, N.N. and Korsmeyer, S.J. (2004). Cell death. Critical control points. *Cell*, 116, 205–19.

Dechant, M.J., Fellenberg, J., Scheuerpflug, C.G., Ewerbeck, V. and Debatin, K.M. (2004). Mutation analysis of the apoptotic "death-receptors" and the adaptors TRADD and FADD/MORT-1 in osteosarcoma tumor samples and osteosarcoma cell lines. *Int J Cancer*, 109, 661–7.

Deng, X., Gao, F., Flagg, T. and May, W.S., Jr. (2004). Mono- and multisite phosphorylation enhances Bcl2's antiapoptotic function and inhibition of cell cycle entry functions. *Proc Natl Acad Sci USA*, 101, 153–8.

Denicourt, C. and Dowdy, S.F. (2004). Medicine. Targeting apoptotic pathways in cancer cells. *Science*, 305, 1411–13.

Deveraux, Q.L. and Reed, J.C. (1999). IAP family proteins – suppressors of apoptosis. *Genes Dev*, 13, 239–52.

Deverman, B.E., Cook, B.L., Manson, S.R., et al. (2002). Bcl-xL deamidation is a critical switch in the regulation of the response to DNA damage. *Cell*, 111, 51–62.

Diez, J., Walter, D., Munoz-Pinedo, C. and Gabaldon, T. (2010). DeathBase: a database on structure, evolution and function of proteins involved in apoptosis and other forms of cell death. *Cell Death Differ*, 17, 735–6.

Dijkers, P.F., Medema, R.H., Lammers, J.W., Koenderman, L. and Coffer, P.J. (2000). Expression of the pro-apoptotic Bcl-2 family member Bim is regulated by the forkhead transcription factor FKHR-L1. *Curr Biol*, 10, 1201–4.

Du, C., Fang, M., Li, Y., Li, L. and Wang, X. (2000). Smac, a mitochondrial protein that promotes cytochrome c-dependent caspase activation by eliminating IAP inhibition. *Cell*, 102, 33–42.

Eferl, R. and Wagner, E.F. (2003). AP-1: a double-edged sword in tumorigenesis. *Nat Rev Cancer*, 3, 859–68.

El-Deiry, W.S. (2001). Insights into cancer therapeutic design based on p53 and TRAIL receptor signaling. *Cell Death Differ*, 8, 1066–75.

Feig, C., Tchikov, V., Schutze, S. and Peter, M.E. (2007). Palmitoylation of CD95 facilitates formation of SDS-stable receptor aggregates that initiate apoptosis signaling. *EMBO J*, 26, 221–31.

Finnberg, N. and El-Deiry, W.S. (2008). TRAIL death receptors as tumor suppressors and drug targets. *Cell Cycle*, 7, 1525–8.

Fu, Z. and Tindall, D.J. (2008). FOXOs, cancer and regulation of apoptosis. *Oncogene*, 27, 2312–19.

Fulda, S. and Debatin, K.M. (2006). Extrinsic versus intrinsic apoptosis pathways in anticancer chemotherapy. *Oncogene*, 25, 4798–811.

Garrison, S.P., Jeffers, J.R., Yang, C., et al. (2008). Selection against PUMA gene expression in Myc-driven B-cell lymphomagenesis. *Mol Cell Biol*, 28, 5391–402.

Gong, Y., Somwar, R., Politi, K., et al. (2007). Induction of BIM is essential for apoptosis triggered by EGFR kinase inhibitors in mutant EGFR-dependent lung adenocarcinomas. *PLoS Med*, 4, e294.

Green, D.R. and Evan, G.I. (2002). A matter of life and death. *Cancer Cell*, 1, 19–30.

Hanahan, D. and Weinberg, R.A. (2000). The hallmarks of cancer. *Cell*, 100, 57–70.

Hanahan, D. and Weinberg, R.A. (2011). Hallmarks of cancer: the next generation. *Cell*, 144, 646–74.

Horvitz, H.R. (1999). Genetic control of programmed cell death in the nematode *Caenorhabditis elegans*. *Cancer Res*, 59, 1701s–6s.

Jeffers, J.R., Parganas, E., Lee, Y., et al. (2003). Puma is an essential mediator of p53-dependent and -independent apoptotic pathways. *Cancer Cell*, 4, 321–8.

Johnstone, R.W., Frew, A.J. and Smyth, M.J. (2008). The TRAIL apoptotic pathway in cancer onset, progression and therapy. *Nat Rev Cancer*, 8, 782–98.

Johnstone, R.W., Ruefli, A.A. and Lowe, S.W. (2002). Apoptosis: a link between cancer genetics and chemotherapy. *Cell*, 108, 153–64.

Joza, N., Susin, S.A., Daugas, E., et al. (2001). Essential role of the mitochondrial apoptosis-inducing factor in programmed cell death. *Nature*, 410, 549–54.

Karin, M. and Lin, A. (2002). NF-kappaB at the crossroads of life and death. *Nat Immunol*, 3, 221–7.

Karst, A.M., Dai, D.L., Martinka, M. and Li, G. (2005). PUMA expression is significantly reduced in human cutaneous melanomas. *Oncogene*, 24, 1111–16.

Kim, H., Tu, H.C., Ren, D., et al. (2009). Stepwise activation of BAX and BAK by tBID, BIM, and PUMA initiates mitochondrial apoptosis. *Mol Cell*, 36, 487–99.

Kischkel, F.C., Lawrence, D.A., Chuntharapai, A., et al. (2000). Apo2L/TRAIL-dependent recruitment of endogenous FADD and caspase-8 to death receptors 4 and 5. *Immunity*, 12, 611–20.

Kischkel, F.C., Lawrence, D.A., Tinel, A., et al. (2001). Death receptor recruitment of endogenous caspase-10 and apoptosis initiation in the absence of caspase-8. *J Biol Chem*, 276, 46639–46.

Kohlhaas, S.L., Craxton, A., Sun, X.M., Pinkoski, M.J. and Cohen, G.M.

(2007). Receptor-mediated endocytosis is not required for tumor necrosis factor-related apoptosis-inducing ligand (TRAIL)-induced apoptosis. *J Biol Chem*, 282, 12831–41.

Krueger, A., Baumann, S., Krammer, P.H. and Kirchhoff, S. (2001). FLICE-inhibitory proteins: regulators of death receptor-mediated apoptosis. *Mol Cell Biol*, 21, 8247–54.

Kuwana, T., Bouchier-Hayes, L., Chipuk, J.E., et al. (2005). BH3 domains of BH3-only proteins differentially regulate Bax-mediated mitochondrial membrane permeabilization both directly and indirectly. *Mol Cell*, 17, 525–35.

Lakhani, S.A., Masud, A., Kuida, K., et al. (2006). Caspases 3 and 7: key mediators of mitochondrial events of apoptosis. *Science*, 311, 847–51.

LeBlanc, H., Lawrence, D., Varfolomeev, E., et al. (2002). Tumor-cell resistance to death receptor-induced apoptosis through mutational inactivation of the proapoptotic Bcl-2 homolog Bax. *Nat Med*, 8, 274–81.

LeBlanc, H.N. and Ashkenazi, A. (2003). Apo2L/TRAIL and its death and decoy receptors. *Cell Death Differ*, 10, 66–75.

Lee, K.H., Feig, C., Tchikov, V., et al. (2006). The role of receptor internalization in CD95 signaling. *EMBO J*, 25, 1009–23.

Li, J. and Yuan, J. (2008). Caspases in apoptosis and beyond. *Oncogene*, 27, 6194–206.

Li, L., Thomas, R.M., Suzuki, H., et al. (2004). A small molecule Smac mimic potentiates TRAIL- and TNFalpha-mediated cell death. *Science*, 305, 1471–4.

Lomonosova, E. and Chinnadurai, G. (2008). BH3-only proteins in apoptosis and beyond: an overview. *Oncogene*, 27 Suppl 1, S2–19.

Melino, G., Lu, X., Gasco, M., Crook, T. and Knight, R.A. (2003). Functional regulation of p73 and p63: development and cancer. *Trends Biochem Sci*, 28, 663–70.

Micheau, O., Lens, S., Gaide, O., Alevizopoulos, K. and Tschopp, J. (2001). NF-kappaB signals induce the expression of c-FLIP. *Mol Cell Biol*, 21, 5299–305.

Milano, A., Iaffaioli, R.V. and Caponigro, F. (2007). The proteasome: a worthwhile target for the treatment of solid tumours? *Eur J Cancer*, 43, 1125–33.

Moll, U.M. and Slade, N. (2004). p63 and p73: roles in development and tumor formation. *Mol Cancer Res*, 2, 371–86.

Muppidi, J.R. and Siegel, R.M. (2004). Ligand-independent redistribution of Fas (CD95) into lipid rafts mediates clonotypic T cell death. *Nat Immunol*, 5, 182–9.

Myatt, S.S. and Lam, E.W. (2007). The emerging roles of forkhead box (Fox) proteins in cancer. *Nat Rev Cancer*, 7, 847–59.

Nagata, S. (1997). Apoptosis by death factor. *Cell*, 88, 355–65.

Nakano, K. and Vousden, K.H. (2001). PUMA, a novel proapoptotic gene, is induced by p53. *Mol Cell*, 7, 683–94.

Nieminen, A.I., Partanen, J.I., Hau, A. and Klefstrom, J. (2007). c-Myc primed mitochondria determine cellular sensitivity to TRAIL-induced apoptosis. *EMBO J*, 26, 1055–67.

O'Brien, S., Moore, J.O., Boyd, T.E., et al. (2007). Randomized phase III trial of fludarabine plus cyclophosphamide with or without oblimersen sodium (Bcl-2 antisense) in patients with relapsed or refractory chronic lymphocytic leukemia. *J Clin Oncol*, 25, 1114–20.

Oda, E., Ohki, R., Murasawa, H., et al. (2000). Noxa, a BH3-only member of the Bcl-2 family and candidate mediator of p53-induced apoptosis. *Science*, 288, 1053–8.

Oltersdorf, T., Elmore, S.W., Shoemaker, A.R., et al. (2005). An inhibitor of Bcl-2 family proteins induces regression of solid tumours. *Nature*, 435, 677–81.

Pai, S.I., Wu, G.S., Ozoren, N., et al. (1998). Rare loss-of-function mutation of a death receptor gene in head and neck cancer. *Cancer Res*, 58, 3513–18.

Peter, M.E. and Krammer, P.H. (2003). The CD95(APO-1/Fas) DISC and beyond. *Cell Death Differ*, 10, 26–35.

Petersen, S.L., Wang, L., Yalcin-Chin, A., et al. (2007). Autocrine TNFalpha signaling renders human cancer cells susceptible to Smac-mimetic-induced apoptosis. *Cancer Cell*, 12, 445–56.

Petros, A.M., Nettesheim, D.G., Wang, Y., et al. (2000). Rationale for Bcl-xL/Bad peptide complex formation from structure, mutagenesis, and biophysical studies. *Protein Sci*, 9, 2528–34.

Ploner, C., Kofler, R. and Villunger, A. (2008). Noxa: at the tip of the balance between life and death. *Oncogene*, 27 Suppl 1, S84–92.

Puthalakath, H. and Strasser, A. (2002). Keeping killers on a tight leash: transcriptional and post-translational control of the pro-apoptotic activity of BH3-only proteins. *Cell Death Differ*, 9, 505–12.

Rampino, N., Yamamoto, H., Ionov, Y., et al. (1997). Somatic frameshift mutations in the BAX gene in colon cancers of the microsatellite mutator phenotype. *Science*, 275, 967–9.

Ravi, R., Bedi, G.C., Engstrom, L.W., et al. (2001). Regulation of death receptor expression and TRAIL/Apo2L-induced apoptosis by NF-kappaB. *Nat Cell Biol*, 3, 409–16.

Reed, J.C. (2006). Drug insight: cancer therapy strategies based on restoration of endogenous cell death mechanisms. *Nat Clin Pract Oncol*, 3, 388–98.

Reed, J.C., Doctor, K.S. and Godzik, A. (2004). The domains of apoptosis: a genomics perspective. *Sci STKE*, 2004, re9.

Reed, J.C. and Pellecchia, M. (2005). Apoptosis-based therapies for hematologic malignancies. *Blood*, 106, 408–18.

Sattler, M., Liang, H., Nettesheim, D., et al. (1997). Structure of Bcl-xL-Bak peptide complex: recognition between regulators of apoptosis. *Science*, 275, 983–6.

Schimmer, A.D., Pedersen, I.M., Kitada, S., et al. (2003). Functional blocks in caspase activation pathways are common in leukemia and predict patient response to induction chemotherapy. *Cancer Res*, 63, 1242–8.

Shangary, S., Oliver, C.L., Tillman, T.S., Cascio, M. and Johnson, D.E. (2004). Sequence and helicity requirements for the proapoptotic activity of Bax BH3 peptides. *Mol Cancer Ther*, 3, 1343–54.

Sharma, S.V., Bell, D.W., Settleman, J. and Haber, D.A. (2007). Epidermal growth factor receptor mutations in lung cancer. *Nat Rev Cancer*, 7, 169–81.

She, Q.B., Solit, D.B., Ye, Q., O'Reilly, K.E., Lobo, J. and Rosen, N. (2005). The BAD protein integrates survival signaling by EGFR/MAPK and PI3K/Akt kinase pathways in PTEN-deficient tumor cells. *Cancer Cell*, 8, 287–97.

Steller, H. (2008). Regulation of apoptosis in *Drosophila*. *Cell Death Differ*, 15, 1132–8.

Stevens, C. and La Thangue, N.B. (2004). The emerging role of E2F-1 in the DNA damage response and checkpoint control. *DNA Repair*, 3, 1071–9.

Sun, T., Gao, Y., Tan, W., et al. (2007). A six-nucleotide insertion-deletion polymorphism in the CASP8 promoter is associated with susceptibility to multiple cancers. *Nat Genet*, 39, 605–13.

Suzuki, Y., Imai, Y., Nakayama, H., et al. (2001a). A serine protease, HtrA2, is released from the mitochondria and interacts with XIAP, inducing cell death. *Mol Cell*, 8, 613–21.

Suzuki, Y., Nakabayashi, Y. and Takahashi, R. (2001b). Ubiquitin-protein ligase activity of X-linked inhibitor of apoptosis protein promotes proteasomal degradation of caspase-3 and enhances its anti-apoptotic effect in Fas-induced cell death. *Proc Natl Acad Sci USA*, 98, 8662–7.

Tang, T.T., Dowbenko, D., Jackson, A., et al. (2002). The forkhead transcription factor AFX activates apoptosis by induction of the BCL-6 transcriptional repressor. *J Biol Chem*, 277, 14255–65.

Teitz, T., Wei, T., Valentine, M.B., et al. (2000). Caspase 8 is deleted or silenced preferentially in childhood neuroblastomas with amplification of MYCN. *Nat Med*, 6, 529–35.

Tse, C., Shoemaker, A.R., Adickes, J., et al. (2008). ABT-263: a potent and orally bioavailable Bcl-2 family inhibitor. *Cancer Res*, 68, 3421–8.

Tsujimoto, Y., Cossman, J., Jaffe, E. and Croce, C.M. (1985). Involvement of the bcl-2 gene in human follicular lymphoma. *Science*, 228, 1440–3.

Ullenhag, G.J., Mukherjee, A., Watson, N.F., et al. (2007). Overexpression of FLIPL is an independent marker of poor prognosis in colorectal cancer patients. *Clin Cancer Res*, 13, 5070–5.

van Delft, M.F., Wei, A.H., Mason, K.D., et al. (2006). The BH3 mimetic ABT-737 targets selective Bcl-2 proteins and efficiently induces apoptosis via Bak/Bax if Mcl-1 is neutralized. *Cancer Cell*, 10, 389–99.

Velculescu, V.E., Madden, S.L., Zhang, L., et al. (1999). Analysis of human transcriptomes. *Nat Genet*, 23, 387–8.

Villunger, A., Michalak, E.M., Coultas, L., et al. (2003). p53- and drug-induced apoptotic responses mediated by BH3-only proteins Puma and Noxa. *Science*, 302, 1036–8.

Vogelstein, B. and Kinzler, K.W. (2004). Cancer genes and the pathways they control. *Nat Med*, 10, 789–99.

Vogelstein, B., Lane, D. and Levine, A.J. (2000). Surfing the p53 network. *Nature*, 408, 307–10.

Wagner, K.W., Punnoose, E.A., Januario, T., et al. (2007). Death-receptor O-glycosylation controls tumor-cell sensitivity to the proapoptotic ligand Apo2L/TRAIL. *Nat Med*, 13, 1070–7.

Walczak, H. and Krammer, P.H. (2000). The CD95 (APO-1/Fas) and the TRAIL (APO-2L) apoptosis systems. *Exp Cell Res*, 256, 58–66.

Wang, C.Y., Cusack, J.C., Jr., Liu, R. and Baldwin, A.S., Jr. (1999). Control of inducible chemoresistance: enhanced anti-tumor therapy through increased apoptosis by inhibition of NF-kappaB. *Nat Med*, 5, 412–17.

Wang, X. (2001). The expanding role of mitochondria in apoptosis. *Genes Dev*, 15, 2922–33.

Wang, X., Yang, C., Chai, J., Shi, Y. and Xue, D. (2002). Mechanisms of AIF-mediated apoptotic DNA degradation in *Caenorhabditis elegans*. *Science*, 298, 1587–92.

Wang, Y., Quon, K.C., Knee, D.A., Nesterov, A. and Kraft, A.S. (2005). RAS, MYC, and sensitivity to tumor necrosis factor-alpha-related apoptosis-inducing ligand-induced apoptosis. *Cancer Res*, 65, 1615–16; author reply 1616–17.

Wei, M.C., Zong, W.X., Cheng, E.H., et al. (2001). Proapoptotic BAX and BAK: a requisite gateway to mitochondrial dysfunction and death. *Science*, 292, 727–30.

Wu, G., Chai, J., Suber, T.L., et al. (2000). Structural basis of IAP recognition by Smac/DIABLO. *Nature*, 408, 1008–12.

Yang, Q., Kiernan, C.M., Tian, Y., et al. (2007). Methylation of CASP8, DCR2, and HIN-1 in neuroblastoma is associated with poor outcome. *Clin Cancer Res*, 13, 3191–7.

Yang, Y., Fang, S., Jensen, J.P., Weissman, A.M. and Ashwell, J.D. (2000). Ubiquitin protein ligase activity of IAPs and their degradation in proteasomes in response to apoptotic stimuli. *Science*, 288, 874–7.

Yin, X.M., Wang, K., Gross, A., et al. (1999). Bid-deficient mice are resistant to Fas-induced hepatocellular apoptosis. *Nature*, 400, 886–91.

Yip, K.W. and Reed, J.C. (2008). Bcl-2 family proteins and cancer. *Oncogene*, 27, 6398–406.

You, H., Pellegrini, M., Tsuchihara, K., et al. (2006). FOXO3a-dependent regulation of Puma in response to cytokine/growth factor withdrawal. *J Exp Med*, 203, 1657–63.

Yu, J., Wang, Z., Kinzler, K.W., Vogelstein, B. and Zhang, L. (2003). PUMA mediates the apoptotic response to p53 in colorectal cancer cells. *Proc Natl Acad Sci USA*, 100, 1931–6.

Yu, J. and Zhang, L. (2004). Apoptosis in human cancer cells. *Curr Opin Oncol*, 16, 19–24.

Yu, J. and Zhang, L. (2005). The transcriptional targets of p53 in apoptosis control. *Biochem Biophys Res Commun*, 331, 851–8.

Yu, J. and Zhang, L. (2008). PUMA, a potent killer with or without p53. *Oncogene*, 27 Suppl 1, S71–83.

Yu, J., Zhang, L., Hwang, P.M., Kinzler, K.W. and Vogelstein, B. (2001). PUMA induces the rapid apoptosis of colorectal cancer cells. *Mol Cell*, 7, 673–82.

Yu, J.W. and Shi, Y. (2008). FLIP and the death effector domain family. *Oncogene*, 27, 6216–27.

Zha, J., Harada, H., Yang, E., Jockel, J. and Korsmeyer, S.J. (1996). Serine phosphorylation of death agonist BAD in response to survival factor results in binding to 14-3-3 not BCL-X(L). *Cell*, 87, 619–28.

Zha, J., Weiler, S., Oh, K.J., Wei, M.C. and Korsmeyer, S.J. (2000). Posttranslational N-myristoylation of BID as a molecular switch for targeting mitochondria and apoptosis. *Science*, 290, 1761–5.

Zhang, L., Ming, L. and Yu, J. (2007). BH3 mimetics to improve cancer therapy; mechanisms and examples. *Drug Resist Updat*, 10, 207–17.

Zhang, L., Yu, J., Park, B.H., Kinzler, K.W. and Vogelstein, B. (2000). Role of BAX in the apoptotic response to anticancer agents. *Science*, 290, 989–92.

Zhong, Q., Gao, W., Du, F. and Wang, X. (2005). Mule/ARF-BP1, a BH3-only E3 ubiquitin ligase, catalyzes the polyubiquitination of Mcl-1 and regulates apoptosis. *Cell*, 121, 1085–95.

17

Molecular links between inflammation and cancer

Paola Allavena, Giovanni Germano and Alberto Mantovani

Introduction

Already in the nineteenth century it was realized that cancer was linked to inflammation. This perception has essentially disappeared for a long time. The inflammation–cancer connection is now part of an accepted paradigm based on different lines of work and leading to a generally accepted paradigm (Balkwill and Mantovani, 2001; Coussens and Werb, 2002; Balkwill et al., 2005; Mantovani et al., 2008) (Figure 17.1).

Chronic inflammation predisposes to different forms of cancer. The triggers of chronic inflammation, which increase cancer risk, include microbial infections (e.g., *H.pilori* for gastric cancer and mucosal lymphoma), autoimmune diseases (e.g., inflammatory bowel disease for colon cancer), and inflammatory conditions of uncertain origin (e.g., prostatitis for prostate cancer). Usage of non-steroidal anti-inflammatory agents is associated with protection against various tumors, a finding that to a large extent mirrors that of inflammation as a risk factor for certain cancers. The "inflammation–cancer" connection is not restricted to increased risk for a subset of tumors. An inflammatory component is present in the microenvironment of most neoplastic tissues, including those not causally related to an obvious inflammatory process. Key features of cancer-related inflammation include the infiltration of white blood cells, prominently tumor associated macrophages (TAM); the presence of polipeptide messengers of inflammation (cytokines such as tumor necrosis factor (TNF) or interleukin-1 (IL-1) and chemokines such as CCL2); the occurrence of tissue remodelling and angiogenesis.

Recent efforts have shed new light on the molecular and cellular pathways linking inflammation and cancer (Mantovani et al., 2008). Two pathways link inflammation and cancer. In the intrinsic pathway, activation of different classes of oncogenes drives the expression of inflammation-related programs, which guide the construction of an inflammatory microenvironment. In the extrinsic pathway inflammatory conditions promote cancer development (e.g., colitis-associated cancer of the intestine). Key orchestrators at the intersection of the intrinsic and extrinsic pathway include transcription factors (e.g., NF-kb) (Rius et al., 2008), cytokines (e.g., TNF), and chemokines. Thus cancer-related inflammation (CRI) is a key component of the tumor microenvironment and a target for pharmacologic intervention (e.g., Harrison et al., 2007).

Orchestrating cancer-related inflammation: transcription factors

In the panoply of molecular players involved in CRI one can identify prime movers (endogenous promoters). These include transcription factors such as nuclear factor-κB (NF-κB) and signal transducer activator of transcription-3 (Stat3), and primary inflammatory cytokines such as IL-1b, IL-6, IL-23, and TNF-a (Voronov et al., 2003; Karin, 2006; Langowski et al., 2006; Yu et al., 2007).

NF-κB is a key orchestrator of innate immunity and inflammation and has emerged as a major endogenous tumor promoter (Karin, 2006). NF-κB is key both in the context of tumor or would-be tumor cells and in the context of inflammatory cells. In both cellular contexts, NF-κB can lie downstream of the sensing of microbes or tissue damage by the toll-like receptor (TLR)-MyD88 pathway, the inflammatory cytokines TNF-a and IL-1b, and downstream of tissue damage that results in release of alarm signals. In addition, NF-κB activation can be the result of cell-autonomous genetic alterations (amplification, mutations, or deletions) in cancer cells.

Systems Biology of Cancer, ed. S. Thiagalingam. Published by Cambridge University Press. © Cambridge University Press 2015.

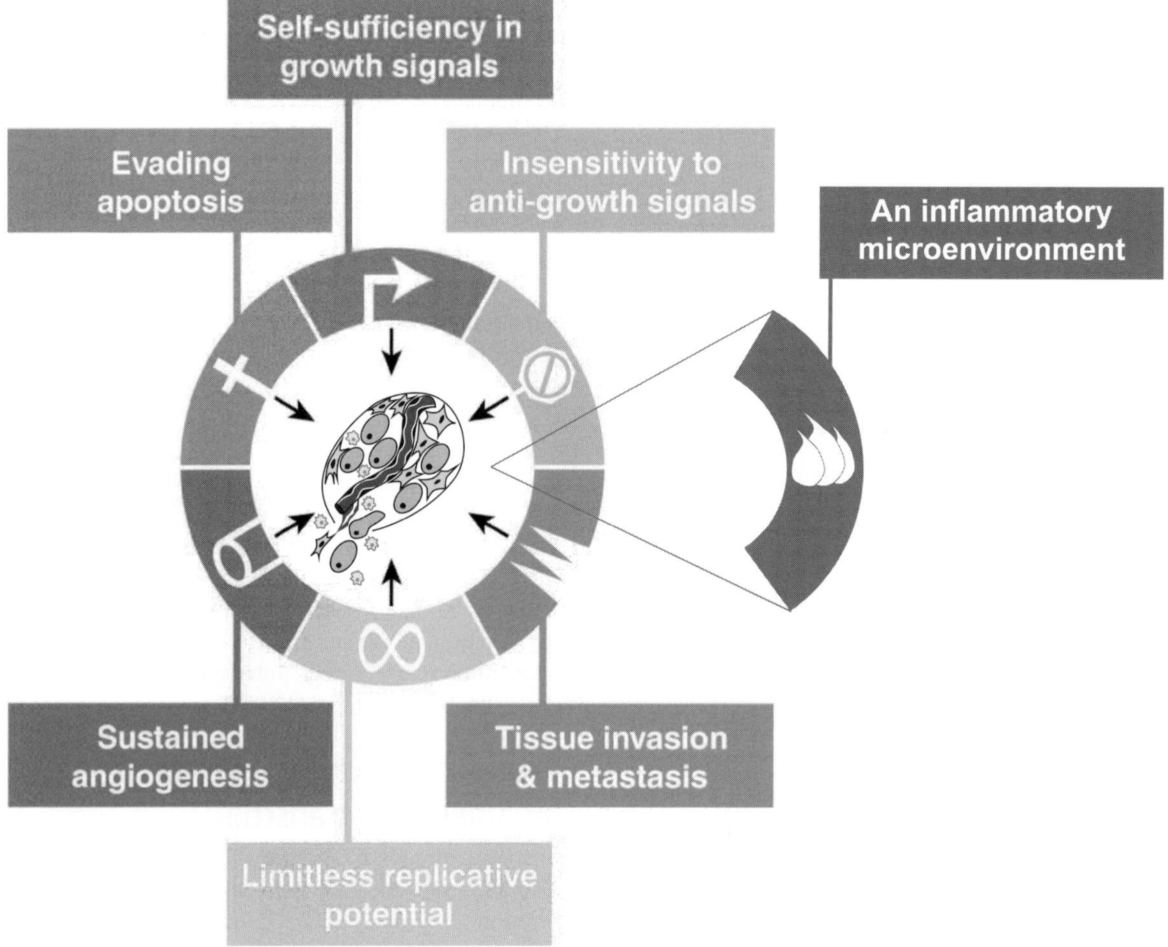

Figure 17.1 Inflammation as the seventh hallmark of cancer. An integration to the six hallmarks of cancer. (Image reproduced from Mantovani, A. 2009. *Nature* 457: 36–7. © 2009, Nature Publishing Group. Reprinted with permission of Macmillan Publishers Ltd) (A black and white version of this figure will appear in some formats. For the color version, please refer to the plate section.)

In both tumor cells and epithelial cells at risk of transformation by carcinogens, as well as in inflammatory cells, NF-κB activates the expression of inflammatory cytokines, adhesion molecules, key enzymes in the prostaglandin synthase pathway (COX-2), nitric oxide (NO) synthase, and angiogenic factors. NF-κB also promotes cell survival by inducing anti-apoptotic genes (e.g., Bcl2), a major function in tumor cells or cells targeted by carcinogenic agents. Accumulating evidence suggests that intersections and compensatory pathways may exist between the NF-κB and hypoxia inducible factor (HIF-1) systems (Carbia-Nagashima et al., 2007; Rius et al., 2008), linking innate immunity to the hypoxic response.

Strong genetic evidence, including tissue-specific gene targeting of components of the Ikk complex, such as IkappaB-kinase beta (IKKβ), have provided unequivocal evidence for a role of NF-κB in tumor promotion in prototypic tissues of CRI (e.g., gastrointestinal tract and liver) (Pikarsky et al., 2004). Gene targeting in liver epithelial cells can have divergent effects in different models of carcinogenesis possibly depending on the balance between promotion of initiated cell apoptosis and triggering of compensatory cell proliferation (Pikarsky et al., 2004).

The NF-κB pathway is tightly controlled by inhibitors acting at different levels. Toll-interleukin receptor 8 (TIR8), also known as single immunoglobulin interleukin-1 receptor-related protein (SIGIRR), is a fringe member of the IL-1 receptor family with a single Ig domain, a long cytoplasmic tail, and an altered TIR domain. It inhibits TLR and

IL-1R signaling and is highly expressed in intestinal mucosa. TIR8 gene deficiency is associated with increased susceptibility to intestinal inflammation and carcinogenesis (Garlanda et al., 2007; Xiao et al., 2007) as well as to B cell lymphoproliferation and autoimmunity (Lech et al., 2008). Thus balancing inhibitors and activators tunes the action of the NF-κB pathway as an endogenous tumor promoter.

NF-κB can act in infiltrating leukocytes as well as in cells targeted by carcinogens. Myeloid-specific gene inactivation of IKKβ inhibited CRI and colitis-associated cancer thus providing an unequivocal genetic evidence for the role of inflammatory cells in carcinogenesis in this tissue. In established advanced tumors, where inflammation is typically smouldering (Balkwill et al., 2005), TAM have defective and delayed NF-κB activation (Biswas et al., 2006). Evidence suggests that p50 homodimers negative regulator is responsible for the sluggish NF-κB activation in TAM and for their protumor phenotype. Thus NF-κB appears to act as a rheostat tuned at different levels in florid inflammatory conditions predisposing to cancer (e.g., inflammatory bowel disease, IBD) and in TAM sustaining the smouldering inflammatory milieu of established metastatic neoplasia. NF-kB has also been recently involved in driving the M2 polarization of TAM (Hagemann et al., 2008).

Along with NF-kB, STAT3 is a point of convergence for numerous oncogenic signaling pathways (Yu et al., 2007). This transcription factor is constitutively activated both in tumor cells and in immune cells and plays a role both in oncogenesis and inhibition of apoptosis. STAT3 activation in cancer also enhances tumor immune evasion by inhibiting DC maturation and promoting immunosuppression (Kortylewski et al., 2005).

Orchestrating cancer-related inflammation: cytokines and chemokines

Studies on TNF demonstrating enhancement of cancer cell invasion ability provided early proof for a protumor function of inflammatory cytokines. In addition, the finding of protection of TNF-deficient mice against skin carcinogenesis offered genetic evidence linking inflammation and cancer (Balkwill, 2004, 2009).

Tumor necrosis factor (TNF)-mediated tumor promotion can involve different pathways (Popivanova et al., 2008), a direct effect on tumor cells of low concentration of this cytokine; an interplay with the chemokine system with induction of CXCR4; stimulation of epithelial to mesenchymal transition (Kulbe et al., 2007). TAM promote wnt signaling through TNF in gastric cancer (Oguma et al., 2008). These findings provide a rationale for the development of clinical protocols employing TNF antagonists in cancer therapy (Harrison et al., 2007). Decoy receptor 3 (DcR3) is a member of the TNF receptor superfamily and has been involved in the control of MHC class II in TAM (Chang et al., 2008). IL-1, IL-6, TNF, and the related receptor activator of NFkB ligand (RANKL) have long been known to augment the capacity to metastatize by affecting multiple steps in the dissemination and inflammation cascade (Giavazzi et al., 1990; Luo et al., 2007; Mantovani et al., 2008). A major source of inflammatory cytokines in the tumor microenvironment are tumor-associated macrophages (TAM) (Mantovani et al., 2002). TAM assist tumor cell malignant behavior in many ways, including producing cytokines, growth factors, and matrix degrading enzymes (Condeelis and Pollard, 2006; Mantovani et al., 2006; Wyckoff et al., 2007; Yang et al., 2008) (Figure 17.1). Kim et al. have recently explored the molecular link(s) between tumor cells, macrophages, and metastasis (Kim et al., 2009). They found that tumor cell supernatants contained an inducer of cytokine production in macrophages. Unexpectedly, the tumor-derived macrophage activation was identified as a component of the extracellular matrix, versican, which is frequently upregulated in human tumors, and was found to be recognized by toll-like receptor 2 and 6 (TLR2/TLR6), which belong to a family of sensors of microbial moieties and tissue damage. Thus in the Lewis lung carcinoma model this study identifies a cascade of amplification of metastasis initiated by tumor-derived versican acting on myeloid cells via TLR2/6, leading to inflammatory cytokine production (Kim et al., 2009).

Early on, IL-1 was shown to augment metastasis, a finding that at the time was related to induction of adhesion molecules in target organs. IL-1R1 gene targeted mice have provided unequivocal evidence for the protumor potential of IL-1 (Voronov et al., 2003; Dinarello, 2006). In addition strong genetic evidence has linked a proinflammatory haplotype at the IL-1 gene locus, comprising agonist and antagonist, to the pathogenesis of gastric carcinoma (El-Omar et al., 2000).

In a pancreatic islet tumor model, a first wave of myc-driven angiogenesis is induced by the inflammatory cytokine IL-1. In an unexpected twist, IL-1α was also shown recently to play a pivotal role in the pathogenesis of liver cancer (Sakurai, 2008). Recent studies have uncovered a novel relationship between sex steroid hormones and cancer. In carcinoma of the prostate, an androgen-dependent tumor, sensitivity to hormonal stimulation is regulated by selective androgen receptor modulators (SARM). The inflammatory cytokine IL-1 produced by macrophages in the tumor microenvironment converts SARMs from inhibitors to stimulators.

The signaling by IL-1 receptor and TLRs is under control by negative pathways of regulation (Mantovani et al., 2008) There is now evidence that TIR8, an orphan member of the IL-1R family (also known as SIGIRR), inhibits signaling from the IL-1R/TLR complexes, possibly by trapping IRAK-1 and TRAF-6. In a mouse model of intestinal carcinogenesis in response to Dextran sulfate sodium salt (DSS) and azoxymethane administration, Tir8-deficient mice exhibited a dramatic susceptibility to inflammation in terms of weight loss, bleeding, and mortality, and showed increased susceptibility to colon carcinogenesis, associated with increased permeability and local production of prostaglandin E_2, pro-inflammatory cytokines (IL-1, IL-6), and chemokines (KC/CXC, JE/CCL2 and CCL3) (Garlanda et al., 2007; Xiao et al., 2007). These mediators are downstream of NF-kB and have been shown to promote inflammation-propelled neoplasia (Karin, 2006). Thus the lack of a checkpoint (TIR8) of NF-kB activation leads to increased carcinogenesis in the gastrointestinal tract, underlying once more the connection between chronic inflammation and cancer promotion.

IL-6 is a key growth-promoting and anti-apoptotic inflammatory cytokine (Lin and Karin, 2007; Naugler and Karin, 2008) and is also one of the effector signals of activated NF-kB in the promotion of neoplasia. Recent studies suggest an alternative pathway of connection between IL-6 and cancer, having NF-kB as a linker (Annunziata et al., 2007; Keats et al., 2007).

Another example of a role for IL-6 is in hepatocarcinogenesis. Hepatocellular carcinoma (HCC), the most common type of liver cancer, is a frequent outcome after years of chronic inflammation, induced either by chronic infection (HBV, HCV) or sustained alcohol consumption.

Naugler et al. have explored the mechanisms underlying gender difference in HCC. They started from the observation that, in response to injections of the carcinogen diethylnitrosamine, male mice had higher levels of IL-6. The sex difference in susceptibility to cancer was absent in mice deficient in IL-6 and in the adaptor protein MyD88. The latter operates downstream of the toll-like receptors that sense microbial invasion and tissue damage, or (in the form of the IL-1 receptor) act as amplifiers of inflammation. From *in vivo* studies in mice (Naugler et al., 2007; Rakoff-Nahoum and Medzhitov, 2007) carcinogen-induced tissue damage results in the release of debris, which in turn causes MyD88-dependent activation of Kupffer cells in the liver. The macrophage-like Kupffer cells produce IL-6, which promotes liver injury, inflammation, compensatory cell proliferation, and carcinogenesis. In females, however, estrogen steroid hormones act through gene transcription factors (such as NF-$_k$B) to inhibit IL-6 production in Kupffer cells, and so to protect female mice from cancer.

Chemokines have long been associated with the recruitment of leukocytes in tumors. Recent results with gene-targeted mice have provided unequivocal evidence for a role of CC chemokines in carcinogenesis. Mice deficient in D6, a decoy and scavenger receptor for inflammatory CC chemokines, show increased susceptibility to skin carcinogenesis and colitis-associated cancer (Nibbs et al., 2007 and our unpublished data).

The contribution of chemokines to angiogenesis and tumor promotion has been the object of intensive investigation. A variety of chemokines, including CCL2, CXCL12, CXCL8, CXCL1, CXCL13, CCL5 (Garlanda et al., 2007), CCL17, and CCL22, have been detected in neoplastic tissues as products of either tumor cells or stromal elements. CXCL1 and related molecules (CXCL2, CXCL3, CXCL8, or interleukin-8 (IL-8)) have an important role in melanoma progression by stimulating neoplastic growth, promoting inflammation, and inducing angiogenesis. Strong evidence demonstrates that levels of CCL2 are associated with TAM accumulation (Balkwill and Mantovani, 2001) and that CCL2 may play an important role in the regulation of angiogenesis. Chemokines are also involved in the rapid turnover of myeloid-derived suppressor cells (Sawanobori et al., 2008).

Expression of chemokine receptors plays an important role in guiding metastasis. Upregulation of CXCR4 is downstream to the Von Hippel Lindau activation of tyrosine kinase oncogenes and TNF (Mantovani et al., 2008). CXCR4 is the most frequently upregulated chemokine receptor in cancer cells and it is associated with advanced stages and metastasis (Mantovani et al., 2008).

An intrinsic and an extrinsic pathway link inflammation and cancer

The occurrence of an inflammatory component in the microenvironment of tumors that are not epidemiologically related to inflammation raised the question of whether genetic events causing neoplasia are responsible for the construction of an inflammatory milieu. To address this question in a clinically relevant setting, advantage was taken of papillary carcinoma of the thyroid (PTC), a tumor characterized by the presence of chemokine-guided macrophage and dendritic cell infiltration (Russell et al., 2003, 2004; Borrello et al., 2005). Rearrangements of tyrosine kinase genes are involved in the pathogenesis of this tumor. In particular, the RET/PTC rearrangement represents a frequent, early, causative and sufficient genetic event in the pathogenesis of PTC. In an appropriate cellular context provided by primary human thyrocytes, RET/PTC activates a genetic program related to inflammation (Borrello et al., 2005). In particular, the RET/PTC-activated transcriptome profile includes: colony-stimulating factors (CSFs), which promote leukocyte recruitment and survival; interleukin-1 (IL-1), a primary inflammatory cytokine; cyclooxygenase 2 (COX2), frequently expressed in cancer; chemokines attracting monocytes and dendritic cells (CCL2, CCL20); angiogenic chemokines; coordinate induction and inhibition of matrix degrading enzymes and inhibitors; L-selectin and expression of the chemokine receptor CXCR4 on the initiated cells (Cerutti et al., 2007). Key elements of the RET/PTC-activated inflammatory program were found in biopsy specimens and patients with lymph node metastasis showed higher levels of the inflammatory molecules in their primary tumors (Borrello et al., 2005; Cerutti et al., 2007; De Falco et al., 2007). These results show that an early, causative, and sufficient genetic event (RET/PTC) involved in the pathogenesis of a human tumor directly promotes the buildup of an inflammatory

microenvironment to its direct advantage (Borrello et al., 2005).

Members of the epidermal growth factor receptor (EGFR) family have tyrosine kinase activity and are frequently involved in human cancer. EGFR activation in glioma induces COX2 expression via p38-MAPK activation of Sp1/Sp3 (Xu and Shu, 2007); COX2 is an independent prognostic factor in glioma.

Activated oncogenic components of the ras–raf pathway also induce expression and production of inflammatory mediators. In HeLa cells, ras induces the production of IL-8 (or CXCL8) (Masih-Khan et al., 2006; Wang et al., 2006), a chemokine that promotes angiogenesis and progression. Adult mice are resistant to K-ras pancreatic carcinogenesis. However, mild chronic pancreatitis, possibly mirroring clinical epidemiology, acts in concert with K-ras mutation to induce pancreatic intraepithelial neoplasia and invasive ductal carcinoma (Guerra et al., 2007). Along the same line Braf, frequently activated in malignant melanoma, induces cytokines that contribute to a protumor milieu.

Myc is a transcription factor overexpressed in many human tumors – its activation initiates and maintains key aspects of the tumor phenotype. In addition to promoting cell autonomous proliferation, myc instructs remodelling of the extracellular microenvironment with inflammatory cells and mediators playing key roles. In a pancreatic islet tumor system, a first wave of myc-driven angiogenesis is induced by IL-1. The myc-activated genetic program also includes several CC chemokines, which recruit mast cells. Mast cells have long been known to drive angiogenesis and here, after IL-1, they sustain new vessel formation and tumor growth (Soucek et al., 2007).

Tumor suppressor genes have the capacity to regulate production of inflammatory mediators. The chemokine receptor CXCR4, which is frequently expressed on malignant cells and implicated in cell survival and metastasis, lies downstream of the von Hippel–Lindau HIF axis, as does the inflammatory cytokine TNFα (reviewed in Balkwill et al., 2005). In non-small cell lung cancer (NSCLC) mutation of the tumor suppressor gene phosphatase and tensin homolog (PTEN) results in upregulation of HIF-1 activity and in HIF-1-dependent transcription of the CXCR4 gene, which promotes metastasis formation. Alpha catenin is more than a tumor suppressor sequestering beta-catenin. Its ablation results in NF-κB activation, induction of genes involved in inflammation, cell proliferation, wound healing, and, ultimately, squamous cell carcinoma (Kobielak and Fuchs, 2006). The putative tumor suppressor semaphorin 3B triggers an IL-8-mediated prometastatic program (Rolny et al., 2008). p53 has been connected to CXCR4 ligand expression (Moskovits et al., 2006). In prostate cancer, the inflammatory cytokines TNF and IL-1beta induce phosphorylation, ubiquitination, and degradation of the tumor suppressor NKX3.1 (Markowski et al., 2008).

Transforming growth factor beta (TGFβ) is a tumor suppressor frequently involved in human tumors and associated with metastasis – though it can also act under certain conditions as a tumor promoter. In a mammary carcinoma model, deficiency in the type II TGFβ receptor is associated with metastasis and recruitment of myeloid-derived suppressor cells (MDSC), defined as Gr1+ Mac-1+ (Yang et al., 2008). The chemokines CXCL5 and CXCL12, acting on the chemokine receptors CXCR2 and CXCR4, respectively, mediate MDSC recruitment. MDSC facilitate metastasis via metalloproteinase activity. Interestingly, chemokine-mediated recruitment in TGFβRII-deficient tumors was associated with increased levels of TGFβ1.

Thus oncogenes representative of different molecular classes and mode of action (tyrosine kinases; ras–raf; nuclear oncogenes; tumor suppressors) share the capacity to orchestrate pro-inflammatory programs (Figure 17.1). While these may share common elements (e.g., a link to angiogenesis; recruitment of cells of myelo-monocytic origin) the diversity of inflammatory components calls for detailed analysis of essential, common versus tumor/tissue-specific inflammatory players. In the extrinsic pathway, it remains uncertain whether chronic inflammation per se is sufficient for carcinogenesis (Mantovani et al., 2008). Reactive oxygen and nitrogen intermediates are obvious inflammation-generated candidate mediators for DNA damage, and evidence obtained *in vitro* and *in vivo* is consistent with this view (Mantovani et al., 2008). In addition, aberrant expression of activation-induced cytidine deaminase (AID) is induced in gastric epithelium by *Helicobacter pilori* (Meira et al., 2008) and by TNF in bile duct cells (Campregher et al., 2008). Indeed, cholangiocarcinoma is associated with liver fluke-induced inflammation. AID is an initiator of somatic hypermutation in B cells and may link chronic inflammation to DNA damage in these tumors.

Overview

Recent efforts have shed new light on molecular and cellular pathways linking inflammation and cancer (Mantovani et al., 2008). Two pathways link inflammation and cancer. In the intrinsic pathway, activation of different classes of oncogenes drives the expression of inflammation-related programs, which guide the construction of an inflammatory microenvironment. In the extrinsic pathway, inflammatory conditions promote cancer development. Key orchestrators at the intersection of the intrinsic and extrinsic pathway include transcription factors (e.g., NF-kB) (Karin, 2006), cytokines (e.g., TNF), and chemokines. Thus inflammation is a key component of the tumor microenvironment. It should be emphasized that inflammatory reaction can also have anti-tumor activity (the bright side of the force) (Apetoh et al., 2007; Allavena et al., 2008; Mantovani et al., 2008). This dual function of inflammatory cells and mediators is reflected by studies on correlations between parameters of CRI and clinical behavior in different contexts (Liu et al., 2007; Seike et al., 2007; Taskinen et al., 2007; Byers et al., 2008; Shabo et al., 2008; Tamimi et al., 2008).

Cytokines are a key component and orchestrator of the inflammatory microenvironment of tumors. As such they represent a prime target in therapeutic efforts aimed at taming tumor-promoting CRI. For many years all efforts to treat cancer have concentrated on the destruction/inhibition of tumor cells. Strategies to modulate the host microenvironment offer a complementary perspective. Initial results in this direction justify continuing efforts (e.g., Harrison et al., 2007). The complexity and ambivalent potential of CRI calls for a systems biology approach.

Acknowledgments

The authors are supported by the Italian Association for Cancer Research, Italian Ministry of Health, University and Research, European Commission (INNOCHEM project), Fondazione Cariplo (NOBEL project). GG is the recipient of a fellowship by Fondazione De Andrè.

References

Allavena, P., Sica, A., Garlanda, C. and Mantovani, A. (2008). The Yin-Yang of tumor-associated macrophages in neoplastic progression and immune surveillance. *Immunol Rev 222*: 155–61.

Annunziata, C.M., Davis, R.E., Demchenko, Y., et al. (2007). Frequent engagement of the classical and alternative NF-kappaB pathways by diverse genetic abnormalities in multiple myeloma. *Cancer Cell 12*: 115–30.

Apetoh, L., Ghiringhelli, F., Tesniere, A., et al. (2007). Toll-like receptor 4-dependent contribution of the immune system to anticancer chemotherapy and radiotherapy. *Nat Med 13*: 1050–9.

Balkwill, F. (2004). Cancer and the chemokine network. *Nat Rev Cancer 4*: 540–50.

Balkwill, F. (2009). Tumor Necrosis Factor and cancer. *Nat Rev Cancer 9*: 361–71.

Balkwill, F., Charles, K.A. and Mantovani, A. (2005). Smoldering and polarized inflammation in the initiation and promotion of malignant disease. *Cancer Cell 7*: 211–17.

Balkwill, F. and Mantovani, A. (2001). Inflammation and cancer: back to Virchow? *Lancet 357*: 539–45.

Biswas, S.K., Gangi, L., Paul, S., et al. (2006). A distinct and unique transcriptional programme expressed by tumor-associated macrophages: defective NF-kB and enhanced IRF-3/STAT1 activation. *Blood 107*: 2112–22.

Borrello, M.G., Alberti, L., Fischer, A., et al. (2005). Induction of a proinflammatory program in normal human thyrocytes by the RET/PTC1 oncogene. *Proc Natl Acad Sci USA 102*: 14825–30.

Byers, R.J., Sakhinia, E., Joseph, P., et al. (2008). Clinical quantitation of immune signature in follicular lymphoma by RT-PCR-based gene expression profiling. *Blood 111*: 4764–70.

Campregher, C., Luciani, M.G. and Gasche, C. (2008). Activated neutrophils induce an hMSH2-dependent G2/M checkpoint arrest and replication errors at a (CA) 13-repeat in colon epithelial cells. *Gut 57*: 780–7.

Carbia-Nagashima, A., Gerez, J., Perez-Castro, C., et al. (2007). RSUME, a small RWD-containing protein, enhances SUMO conjugation and stabilizes HIF-1alpha during hypoxia. *Cell 131*: 309–23.

Cerutti, J.M., Oler, G., Michaluart, P., Jr., et al. (2007). Molecular profiling of matched samples identifies biomarkers of papillary thyroid carcinoma lymph node metastasis. *Cancer Res 67*: 7885–92.

Chang, Y.C., Chen, T.C., Lee, C.T., et al. (2008). Epigenetic control of MHC class II expression in tumor-associated macrophages by decoy receptor 3. *Blood 111*: 5054–63.

Condeelis, J. and Pollard, J.W. (2006). Macrophages: obligate partners for tumor cell migration, invasion, and metastasis. *Cell 124*: 263–6.

Coussens, L.M. and Werb, Z. (2002). Inflammation and cancer. *Nature 420*: 860–7.

De Falco, V., Guarino, V., Avilla, E., et al. (2007). Biological role and potential therapeutic targeting of the chemokine receptor CXCR4 in undifferentiated thyroid cancer. *Cancer Res 67*: 11821–9.

Dinarello, C.A. (2006). The paradox of pro-inflammatory cytokines in cancer. *Cancer Metastasis Rev 25*: 307–13.

El-Omar, E.M., Carrington, M., Chow, W.H., et al. (2000). Interleukin-1 polymorphisms associated with increased risk of gastric cancer. *Nature 404*: 398–402.

Garlanda, C., Riva, F., Veliz, T., et al. (2007). Increased susceptibility to colitis-associated cancer of mice lacking TIR8, an inhibitory member of the IL-1 receptor family. *Cancer Res 67*: 6017–21.

Giavazzi, R., Garofalo, A., Bani, M.R., et al. (1990). Interleukin 1-induced augmentation of experimental metastases from a human melanoma in nude mice. *Cancer Res 50*: 4771–5.

Guerra, C., Schuhmacher, A.J., Canamero, M., et al. (2007). Chronic pancreatitis is essential for induction of pancreatic ductal adenocarcinoma by K-Ras oncogenes in adult mice. *Cancer Cell 11*: 291–302.

Hagemann, T., Lawrence, T., McNeish, I., et al. (2008). "Re-educating" tumor-associated macrophages by targeting NF-kappaB. *J Exp Med 205*: 1261–8.

Harrison, M.L., Obermueller, E., Maisey, N.R., et al. (2007). Tumor necrosis factor alpha as a new target for renal cell carcinoma: two sequential phase II trials of infliximab at standard and high dose. *J Clin Oncol 25*: 4542–9.

Karin, M. (2006). Nuclear factor-kappaB in cancer development and progression. *Nature 441*: 431–6.

Keats, J.J., Fonseca, R., Chesi, M., et al. (2007). Promiscuous mutations activate the noncanonical NF-kappaB pathway in multiple myeloma. *Cancer Cell 12*: 131–44.

Kim, S., Takahashi, H., Lin, W.-W., et al. (2009). Carcinoma produced factors activate myeloid cells via TLR2 to stimulate metastasis. *Nature 457*: 102–6.

Kobielak, A. and Fuchs, E. (2006). Links between alpha-catenin, NF-kappaB, and squamous cell carcinoma in skin. *Proc Natl Acad Sci USA 103*: 2322–7.

Kortylewski, M., Kujawski, M., Wang, T., et al. (2005). Inhibiting Stat3 signaling in the hematopoietic system elicits multicomponent antitumor immunity. *Nat Med 11*: 1314–21.

Kulbe, H., Thompson, R., Wilson, J.L., et al. (2007). The inflammatory cytokine tumor necrosis factor-alpha generates an autocrine tumor-promoting network in epithelial ovarian cancer cells. *Cancer Res 67*: 585–92.

Langowski, J.L., Zhang, X., Wu, L., et al. (2006). IL-23 promotes tumour incidence and growth. *Nature 442*: 461–5.

Lech, M., Kulkarni, O.P., Pfeiffer, S., et al. (2008). Tir8/Sigirr prevents murine lupus by suppressing the immunostimulatory effects of lupus autoantigens. *J Exp Med 205*: 1879–88.

Lin, W.W. and Karin, M. (2007). A cytokine-mediated link between innate immunity, inflammation, and cancer. *J Clin Invest 117*: 1175–83.

Liu, R., Wang, X., Chen, G.Y., et al. (2007). The prognostic role of a gene signature from tumorigenic breast-cancer cells. *N Engl J Med 356*: 217–26.

Luo, J.L., Tan, W., Ricono, J.M., et al. (2007). Nuclear cytokine-activated IKKalpha controls prostate cancer metastasis by repressing Maspin. *Nature 446*: 690–4.

Mantovani, A. (2009). Cancer: inflaming metastasis. *Nature 457*: 36–7.

Mantovani, A., Allavena, P., Sica, A. and Balkwill, F. (2008). Cancer-related inflammation. *Nature 454*: 436–44.

Mantovani, A., Schioppa, T., Porta, C., Allavena, P. and Sica, A. (2006). Role of tumor-associated macrophages in tumor progression and invasion. *Cancer Metastasis Rev 25*: 315–22.

Mantovani, A., Sozzani, S., Locati, M., Allavena, P. and Sica, A. (2002). Macrophage polarization: tumor-associated macrophages as a paradigm for polarized M2 mononuclear phagocytes. *Trends Immunol 23*: 549–55.

Markowski, M.C., Bowen, C. and Gelmann, E.P. (2008). Inflammatory cytokines induce phosphorilation and ubiquitination of prostate suppressor protein NKX3.1. *Cancer Res 68*: 6896–901.

Masih-Khan, E., Trudel, S., Heise, C., et al. (2006). MIP-1alpha (CCL3) is a downstream target of FGFR3 and RAS-MAPK signaling in multiple myeloma. *Blood 108*: 3465–71.

Meira, L.B., Bugni, J.M., Green, S.L., et al. (2008). DNA damage induced by chronic inflammation contributes to colon carcinogenesis in mice. *J Clin Invest 118*: 2516–25.

Moskovits, N., Kalinkovich, A., Bar, J., Lapidot, T. and Oren, M. (2006). p53 attenuates cancer cell migration and invasion through repression of SDF-1/CXCL12 expression in stromal fibroblasts. *Cancer Res 66*: 10671–6.

Naugler, W.E. and Karin, M. (2008). The wolf in sheep's clothing: the role of interleukin-6 in immunity, inflammation and cancer. *Trends Mol Med 14*: 109–19.

Naugler, W.E., Sakurai, T., Kim, S., et al. (2007). Gender disparity in liver cancer due to sex differences in MyD88-dependent IL-6 production. *Science 317*: 121–4.

Nibbs, R.J., Gilchrist, D.S., King, V., et al. (2007). The atypical chemokine receptor D6 suppresses the development of chemically induced skin tumors. *J Clin Invest 117*: 1884–92.

Oguma, K., Oshima, H., Aoki, M., et al. (2008). Activated macrophages promote Wnt signalling through tumour necrosis factor-alpha in gastric tumour cells. *EMBO J 27*: 1671–81.

Pikarsky, E., Porat, R.M., Stein, I., et al. (2004). NF-kappaB functions as a tumour promoter in inflammation-associated cancer. *Nature 431*: 461–6.

Popivanova, B.K., Kitamura, K., Wu, Y., et al. (2008). Blocking TNF-alpha in mice reduces colorectal carcinogenesis associated with chronic colitis. *J Clin Invest 118*: 560–70.

Rakoff-Nahoum, S. and Medzhitov, R. (2007). Regulation of spontaneous intestinal tumorigenesis through the adaptor protein MyD88. *Science 317*: 124–7.

Rius, J., Guma, M., Schachtrup, C., et al. (2008). NF-kappaB links innate immunity to the hypoxic response through transcriptional regulation of HIF-1alpha. *Nature 453*: 807–11.

Rolny, C., Capparuccia, L., Casazza, A., et al. (2008). The tumor suppressor semaphorin 3B triggers a prometastatic program mediated by interleukin 8 and the tumor microenvironment. *J Exp Med 205*: 1155–71.

Russell, J.P., Engiles, J.B. and Rothstein, J.L. (2004). Proinflammatory mediators and genetic background in oncogene mediated tumor progression. *J Immunol 172*: 4059–67.

Russell, J.P., Shinohara, S., Melillo, R.M., et al. (2003). Tyrosine kinase oncoprotein, RET/PTC3, induces the secretion of myeloid growth and chemotactic factors. *Oncogene 22*: 4569–77.

Sakurai, T., He, G., Matsuzawa, A., Yu, G., et al. (2008). Hepatocyte necrosis induced by oxidative stress and IL-1α release mediate carcinogen-induced compensatory proliferation and liver tumorigenesis. *Cancer Cell 14*:156–65.

Sawanobori, Y., Ueha, S., Kurachi, M., et al. (2008). Chemokine-mediated rapid turnover of myeloid-derived suppressor cells in tumor-bearing mice. *Blood 111*: 5457–66.

Seike, M., Yanaihara, N., Bowman, E.D., et al. (2007). Use of a cytokine gene expression signature in lung adenocarcinoma and the surrounding tissue as a prognostic classifier. *J Natl Cancer Inst 99*: 1257–69.

Shabo, I., Stal, O., Olsson, H., Dore, S. and Svanvik, J. (2008). Breast cancer expression of CD163, a macrophage scavenger receptor, is related to early distant recurrence and reduced patient survival. *Int J Cancer 123*: 780–6.

Soucek, L., Lawlor, E.R., Soto, D., et al. (2007). Mast cells are required for angiogenesis and macroscopic expansion of Myc-induced pancreatic islet tumors. *Nat Med 13*: 1211–18.

Tamimi, R.M., Brugge, J.S., Freedman, M.L., et al. (2008). Circulating colony stimulating factor-1 and breast cancer risk. *Cancer Res 68*: 18–21.

Taskinen, M., Karjalainen-Lindsberg, M.L., Nyman, H., Eerola, L.M. and Leppa, S. (2007). A high tumor-associated macrophage content predicts favorable outcome in follicular lymphoma patients treated with rituximab and cyclophosphamide-doxorubicin-vincristine-prednisone. *Clin Cancer Res 13*: 5784–9.

Voronov, E., Shouval, D.S., Krelin, Y., et al. (2003). IL-1 is required for tumor invasiveness and angiogenesis. *Proc Natl Acad Sci USA 100*: 2645–50.

Wang, D., Wang, H., Brown, J., et al. (2006). CXCL1 induced by prostaglandin E2 promotes angiogenesis in colorectal cancer. *J Exp Med 203*: 941–51.

Wyckoff, J.B., Wang, Y., Lin, E.Y., et al. (2007). Direct visualization of macrophage-assisted tumor cell intravasation in mammary tumors. *Cancer Res 67*: 2649–56.

Xiao, H., Gulen, M.F., Qin, J., et al. (2007). The Toll-interleukin-1 receptor member SIGIRR regulates colonic epithelial homeostasis, inflammation, and tumorigenesis. *Immunity 26*: 461–75.

Xu, K. and Shu, H.K. (2007). EGFR activation results in enhanced cyclooxygenase-2 expression through p38 mitogen-activated protein kinase-dependent activation of the Sp1/Sp3 transcription factors in human gliomas. *Cancer Res 67*: 6121–9.

Yang, L., Huang, J., Ren, X., et al. (2008). Abrogation of TGFbeta signaling in mammary carcinomas recruits Gr-1+CD11b+ myeloid cells that promote metastasis. *Cancer Cell 13*: 23–35.

Yu, H., Kortylewski, M. and Pardoll, D. (2007). Crosstalk between cancer and immune cells: role of STAT3 in the tumour microenvironment. *Nat Rev Immunol 7*: 41–51.

Chapter

18

Cancer metastasis

Sait Ozturk, Arthur W. Lambert, Chen Khuan Wong, Panagiotis Papageorgis and Sam Thiagalingam

Introduction

Metastasis is the cause of almost all cancer-related deaths. It is an extremely complex, multistep process defined as the spreading of cancer cells from their primary site to distant tissues. Once metastasis occurs it causes catastrophic damage to the critical organs, which is ultimately detrimental to patients. Collective efforts of many scientists have revealed the underlying molecular mechanisms of metastasis by a considerable extent, but there is still a colossal job to be undertaken by researchers to solve this life-threatening health problem.

In this chapter, metastasis is explained by focusing on underlying molecular pathways. We define the steps that a cancer cell needs to climb in order to metastasize and discuss the significant molecular actors aberrantly regulated during this process. First, we outline how these molecules are deregulated in cancer cells in order to circumvent natural barriers against metastasis. Then, we give a molecular explanation on why some cancer types metastasize to certain organs. Lastly, we look into recent therapeutic trials, which involve targeting of pathways in the metastatic cascade (Figure 18.1).

Metastasis as a complex, multistep event

Metastasis begins with the formation of an invasive tumor. As its name implies, an invasive tumor has the ability to invade neighboring tissues. During this process many cancer cells can enter the blood stream or lymphatic vessels by degrading the basement membrane and migrating through endothelial cells [3, 4]. This first step in the metastatic cascade, entrance of cancer cells into blood or lymphatic vessels, is referred to as intravasation. Many of the cells that successfully intravasate die due to a specific type of apoptosis called anoikis [2, 4–6]. Anoikis is initiated by the loss of cells' supportive contacts to the extracellular matrix and to their neighboring cells. Yet another group of intravasated cells die because of the sheer mechanical forces they encounter in the circulation [2–4]. Still, some survive and stick to capillaries in distant tissues. These stuck cancer cells may enter quiescence and live in a dormant phase or may migrate through endothelial cells and exit from capillaries into the tissue. This latter step in metastasis is called extravasation. An extravasated cancer cell is in a completely "alien world" compared to its tissue of origin. It is exposed to different sets of extracellular apoptotic signals and growth factors. Here, it needs to "learn" how to cope with a new set of death signals and how to benefit from the new set of growth factors [7]. Once a cancer cell adapts to the new microenvironment, it can start to divide and form a colony. This final step of the metastatic cascade is called colonization and it is the rate-limiting step of metastasis. Compared to colonization, intravasation and extravasation steps are "easier" for a cancer cell to be successful at. A cell in an invasive tumor may enter and exit circulatory system because of already accumulated mutations at the primary site. However, colonization of a distant tissue requires another distinct set of mutations and epigenetic changes that equip or prime the cell to be successful in a completely different tissue.

Only a tiny fraction of cells in the primary tumor can successfully climb all these steps of the metastatic cascade and gain the ability to colonize [2]. In the next sections, we look into the molecular details of these three main steps of metastasis.

Systems Biology of Cancer, ed. S. Thiagalingam. Published by Cambridge University Press. © Cambridge University Press 2015.

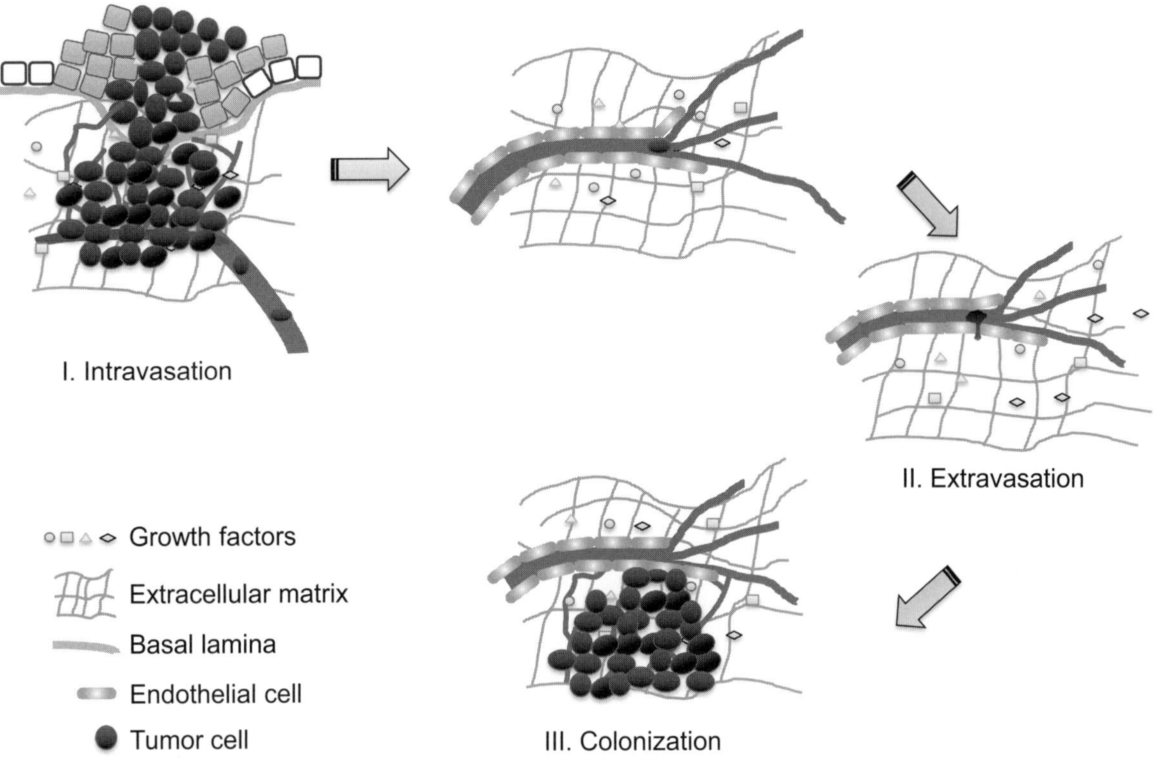

Figure 18.1 The metastatic cascade. Metastasis is a multistep process, which consists of intravasation, extravasation, and colonization. The spreading of cancer cells to distant tissues from their primary site is called metastasis. Although it is an extremely complex event, metastasis (or the metastatic cascade, named due to its multistep nature) can be analyzed in three main steps: intravasation, extravasation, and colonization [1, 2]. (A black and white version of this figure will appear in some formats. For the color version, please refer to the plate section.)

Intravasation

When a benign *in situ* carcinoma turns into an invasive carcinoma, the cancer cells undergo a transformation highlighted by their ability to alter their microenvironment (Figure 18.2). They recruit stromal cells to "nurse" themselves, initiate formation of blood vessels by attracting endothelial cells, and start to degrade the basement membrane and extracellular matrix that may otherwise block their growth [4]. Elucidating the underlying molecular mechanisms of these changes has been crucial to understanding how an invasive tumor enters the metastasic cascade.

One of the major changes occurs on the surface of cancer cells. Adhesion to neighboring cells is very critical in defining the architecture of epithelial tissues. Interactions with neighboring cells give clues (signals) to a cell about when to divide, when to migrate, and how to function. Cadherins, a family of transmembrane proteins, play important roles in these cell to cell interactions and downstream signaling. Cancer cells alter their expression of cadherins according to their advantage during metastasis [8]. E-cadherin (CDH1) is mainly expressed in epithelial cells but is often downregulated in invasive cancer cells. Loss of E-cadherin leads to loss of cell to cell adhesions, resulting in a more migratory phenotype, which facilitates intravasation [3, 9].

Members of the transforming growth factor-beta (TGFβ) signaling pathway are among commonly altered proteins by cancer cells [10]. TGFβ receptors recognize extracellular ligands and activate SMAD proteins to transcriptionally regulate a large set of genes, which have roles in a wide array of cellular events including differentiation and migration [9–11]. Activation of the TGFβ pathway results in transcriptional activation of master regulators such as Snail, Slug, and Twist [9, 12, 13]. This results in loss of epithelial cell markers such as E-cadherin, γ-catenin and gain of mesenchymal markers such as N-cadherin (CDH2), vimentin, and alpha smooth muscle actin.

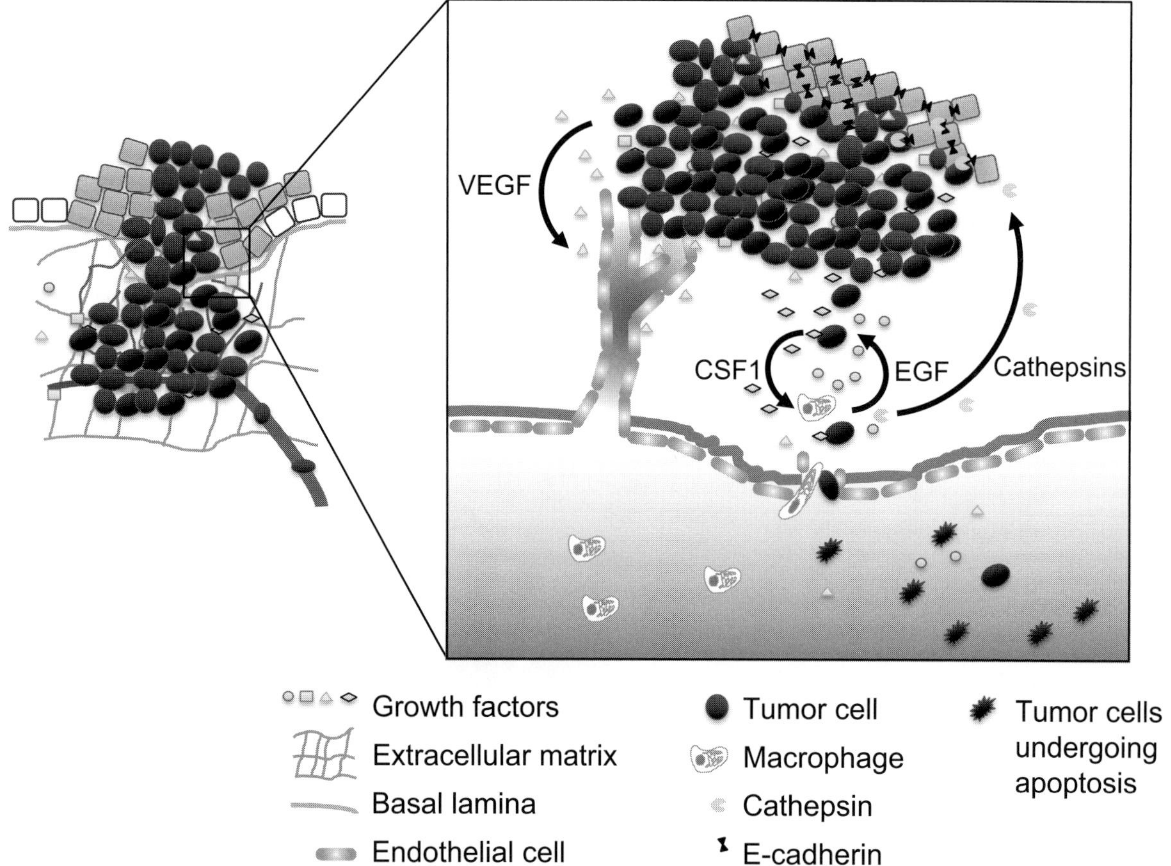

Figure 18.2 Intravasation. During the intravasation stage, a malignant cell can manipulate the microenvironment around it. VEGF secreted from a cancer cell can recruit and activate endothelial cells and induce angiogenesis. Moreover, abnormal VEGF activity makes the endothelial cell layer "leaky." Similarly, CSF1 secreted from cancer cells recruits macrophages, which are critical for extracellular matrix remodelling. Recruited macrophages secrete EGF and cathepsins. Cathepsins can cleave membrane proteins such as E-cadherin that in turn disturb cell to cell interactions. This results in liberation of malignant cells from neighboring cells. EGF can act as a chemoattractant and can also initiate migration of cancer cells, facilitating intravasation. (A black and white version of this figure will appear in some formats. For the color version, please refer to the plate section.)

This de-differentiation is called epithelial to mesenchymal transition (EMT). Cancer cells that have undergone EMT acquire a migratory phenotype [3, 12] and this enhances tumor cell invasion and thereby intravasation.

Tumor cells require the formation of blood vessels in order to sustain their growth. Cancer cells can meet this requirement by secreting vascular endothelial growth factor (VEGF) protein at very high levels [2, 14]. VEGF act as chemoattractant for endothelial cells and causes their migration, proliferation, and differentiation [4, 14]. This leads to the formation of blood vessels that can support the continued growth of the tumor. Since the level of VEGF is chronically high and lacks a directive gradient, it results in formation of a chaotic network of vessels. In this environment, endothelial cells cannot form proper vessels, where each of them attaches to one another in an orderly structure. Instead, they form "leaky" blood vessels providing cancer cells with more opportunities to enter the circulatory system [14].

One of the hallmarks of invasive tumors is their ability to degrade the basement membrane and surrounding extracellular matrix. The basement membrane defines and supports epithelial cell layers in normal tissues and it blocks the unchecked growth of tumor cells. The extracellular matrix creates a three-dimensional environment for cells to function and regulates their growth. Cancer cells circumvent the unfavorable effects of the basement membrane and extracellular matrix by expressing metalloproteases (MMPs) such as MMP2 and MMP9 [4, 13, 15, 16]. Degradation of these structures by MMPs removes the physical barriers that confront tumor cells and, moreover, liberates sequestered growth factors. This combination acts to fuel cancer cell growth and aids in intravasation.

Another significant family of proteins on the surface of cancer cells is the integrin receptor family. Integrins are transmembrane proteins that function in cell to matrix interactions. Overexpression of certain integrins by tumor cells facilitates their migration throught the extracellular matrix [17, 18]. Integrins αvβ3 and α3β1 bind to ligands in the extracellular matrix such as collogen, fibronectin, and laminin, which provides "grip" for cancer cell migration [3, 18]. By creating a complex with ErbB2 receptor (also known as HER2), integrin β4 can also increase the invasiveness of cancer cells [18, 19]. The ability of invasive cancer cells to overexpress integrins can, in many cases, foster migration, which in turn enables intravasation.

In a tumor, cancer cells are far from being the only residents. Cancer cells recruit many other cell types (called "tumor-associated" cell types) to create the optimum settings for their growth. Tumor-associated macrophages (TAMs) are among these cells. They are recruited to the tumor by the expression of cytokines such as interleukins-8 (IL-8), -6 (IL-6), and colony stimulating factor-1 (CSF1) [4, 16, 20, 21]. TAMs can then recruit endothelial cells to form blood vessels. Blood vessel formation is required to sustain tumor growth and provides more alleyways for cancer cells to escape from the primary site [21]. Additionally, TAMs secrete epidermal growth factor (EGF), which initiates tumor cell migration and CSF1 production in tumor cells [4]. CSF1 production, in turn, recruits more TAMs to complete the "cycle." During this process, cancer cells migrate toward the source of the EGF and TAMs migrate toward the source of CSF1. The directional migration of cancer cells (cell streaming), along with the aid of TAMs may end with the entrance of cancer cells to the blood stream. TAMs also secrete cysteine cathepsins, proteases whose targets include surface proteins [4, 22, 23]. One of the most interesting targets of capthepsins is E-cadherin. Cleavage of E-cadherin on the cancer cell surface disrupts cell to cell junctions and releases cancer cells that are free to migrate [23].

The beginning of the metastatic cascade is marked by entrance of cancer cells into blood or lymphatic vessels. This first step, intravasation, requires many alterations at the molecular level involving diverse cellular functions. To intravasate, a cancer cell must detach from neighboring cells (loss of E-cadherin expression, cathepsin activity), recruit endothelial cells to form the escape routes (VEGF expression and TAMs recruitment), degrade the basement membrane and extracellular matrix to migrate through them (secretion of MMPs and expression of integrins), and attract TAMs to guide them during migration (secretion of cytokines).

Extravasation

Extravasation is defined as the exit of circulating cancer cells from a blood or lymphatic vessel into a distant tissue. A successfully extravasated cell must have survived in circulation, docked to the vessel wall, migrated through endothelial cells, and entered into the distant tissue by degrading the basement membrane supporting endothelial cells (Figure 18.3). In this section, we analyze the molecular actors and their respective functions during these events.

Once a cancer cell has intravasated, it is in a suspension. This is a completely different environment for a carcinoma cell, compared to the solid tumor where it used to reside. In a solid tumor, a cancer cell can attach to the extracellular matrix, neighboring cancer cells, or tumor-associated cells. These interactions present survival signals and a structural framework that supports growth of the tumor. Intravasated cancer cells lack all of these support elements. Furthermore, they are challenged by an internal pre-programmed mechanism that is triggered upon loss of solid connections: anoikis [5].

Anoikis is a form of apoptosis. A cancer cell adheres to its neighbors and the extracellular matrix through membrane proteins such as cadherins and integrins. These transmembrane proteins relay survival and anti-apoptotic signals. Withdrawal of these signals upon loss of attachment results in automatic induction of the apoptotic cascade [2, 24]. Bid and Bim are Bcl-2 family proteins that act as activators of the apoptotic cascade [5]. They bind to Bax and Bak proteins and facilitate their insertion to the outer mitochondrial membrane. This causes permeabilization of the membrane and release of cytochrome-c protein. In the cytoplasm, cytochrome-c binds to

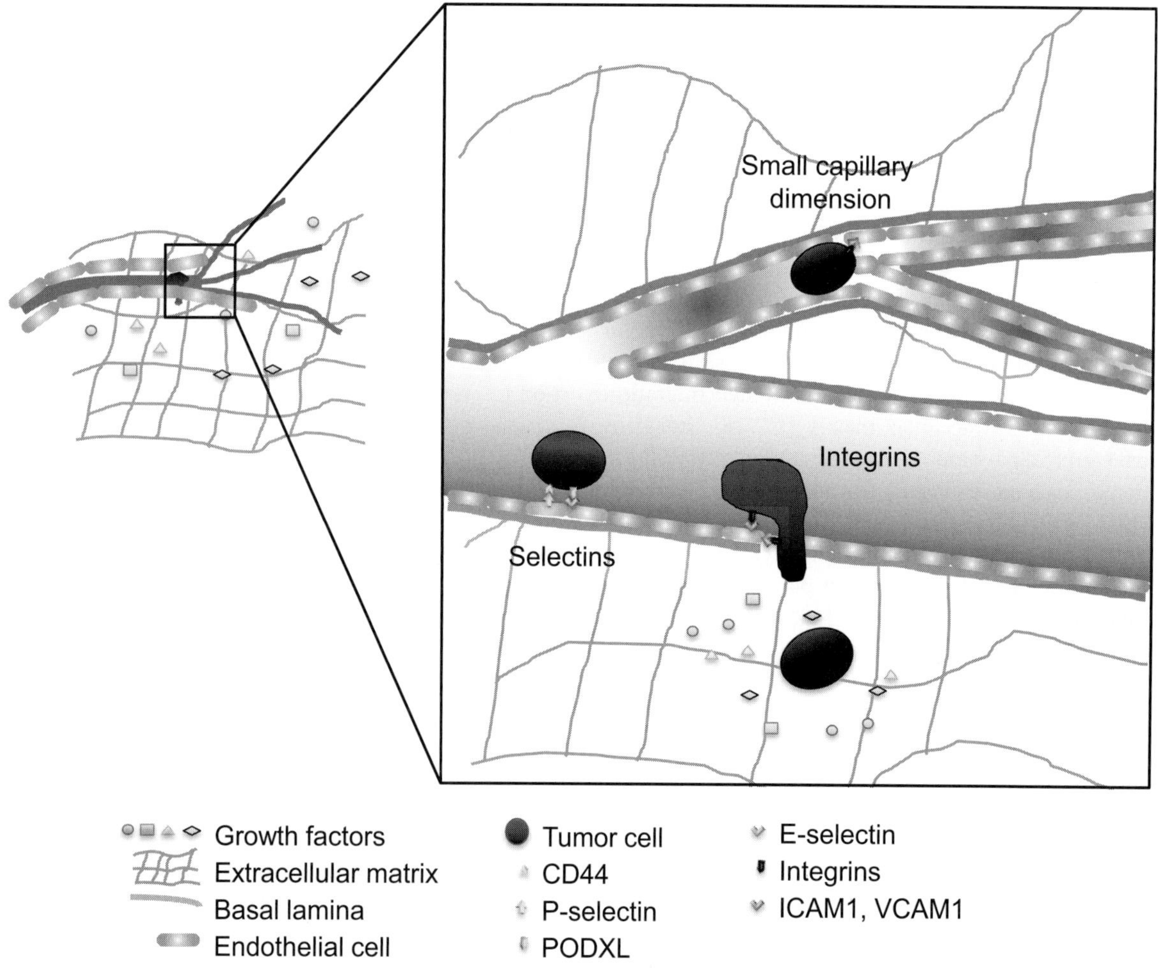

Figure 18.3 Extravasation. Cancer cells, which can escape from anoikis following intravasation, may dock to the vessel walls via weak interaction with selectins on endothelial cells. Integrins on the cancer cell membrane can sustain stronger interactions to the endothelial cells. Moreover, molecules such as VEGF and ANGPTL4 disturb the endothelial cell layer. Together these molecular players facilitate entry and migration of the cancer cell through the endothelial cell layer into distant tissue. (A black and white version of this figure will appear in some formats. For the color version, please refer to the plate section.)

Apaf-1. The Apaf-1/cytochrome-c complex then activates caspase-9, which initiates the caspase cascade and results in mass protein degradation and eventual cell death [5]. The main advantage of this cascade of interactions for apoptosis is that it presents many checkpoints before committing an extremely critical decision: suicide. Cells have inhibitors and activators for each of the mentioned steps. As a whole, this gives cells a very fine-tuned decision mechanism but it also gives a cancer cell many possible targets to alter in order to survive. Indeed cancer cells have many mutated genes that function to inhibit apoptosis.

One of the most effective mechanisms for cancer cells to block apoptosis is activation of the RAS pathway. RAS proteins are GTPases that play pivotal roles in many vital signaling pathways and cellular events. Transmembrane receptors we have mentioned above, such as integrins and growth factor receptors, converge on the RAS pathway through downstream signaling events. Cancer cells often have constitutively active RAS pathways that relay anti-apoptotic/ survival signals even when they lose their attachment to a solid surrounding [25].

Active TGFβ signaling and resulting initiation of EMT in cancer cells increases the population of cells equipped with stem cell-like features [12, 26]. Cancer stem cells are marked by CD44 overexpression and CD24 repression. One of the main advantages CD44$^+$/CD24$^-$ cells have is anchorage independent growth. This allows them to survive in circulation without a need for solid attachments after intravasation, and represents another pathway that cancer cells can manipulate to their benefit during metastasis.

To continue with their distal growth program, surviving cancer cells need to exit the circulatory system. A cancer cell can extravasate upon becoming trapped in small capillaries or by actively docking to endothelial cells of larger diameter veins. Glycoproteins such as CD44, podocalyxin-like (PODXL), carcinoembryonic antigen (CEA), and integrins expressed on the surface of the cancer cells are critical for this process [3]. CD44 and PODXL are recognized by P- and E-selectins on endothelial cells [27–29]. These interactions are temporary, because physical forces between selectins and glycoproteins are weak, but they gain precious time for stronger bonds to occur. More stable attachments take place between integrins and their ligands on endothelial cells, intercellular adhesion molecule 1 (ICAM1) and vascular cell adhesion molecule 1 (VCAM1) [3, 17, 30, 31]. Binding of integrins to ICAM and VCAM on the endothelial surface provides an opportunity for cancer cells to extravasate.

Tumor-associated platelets are loyal allies of cancer cells during extravasation [18, 31]. Following intravasation, tumor-associated platelets and cancer cells can be joined together through "fibrinogen bridges" where both cells attach to the same fibrinogen molecule by the integrins on their surface. This allows cancer cells to receive survival signals even when they are in circulation and helps them to cope with anoikis. Moreover, platelets assist cancer cells in sticking to the endothelial cells and this enhances their chance to extravasate.

Once cancer cells bind to the vessel walls, they start to "claw" their way into the new tissue. Similar sets of proteins used during invasion at the primary site aid in their infiltration. Secretion of VEGF makes endothelial cells more permeable; integrins let cancer cells attach to the basement membrane supporting endothelial cells and this initiates migration; MMPs digest basement membrane and supporting extracellular matrix to "dig" a tunnel where cancer cells migrate further into the tissue. As a result of the collective action of these molecules and tumor-associated cells, a cancer cell can extravasate successfully to a new nest.

Colonization

Even when a cancer cell is successful at extravasation, this does not guarantee it will grow in the foreign "soil." The vast majority of extravasated cells cannot sustain their growth and either die or enter into a quiescence stage. At best, it is only around 1 out of 4,000 extravasated cells that eventually colonize a distant tissue [6]. In order to colonize, cancer cells not only have to adapt to their new microenvironment but also redesign it. This may take decades for a cancer cell to achieve. That is why colonization is the rate-limiting step of the metastatic cascade. A cancer cell needs to find creative solutions to survive and thrive in distant tissues. Next, we will briefly look into the molecular mechanism underlying colonization.

Cancer cells undergo EMT in order to facilitate their intravasation and extravasation, but this has a major drawback: inhibition of proliferation. Activators of EMT, such as TGFβ, slow down cell division. Therefore extravasated cells revert back to a more epithelial-like phenotype through mesenchymal to epithelial transition (MET). This is marked by upregulation of epithelial proteins such as E-cadherin and CD24 [2, 32]. miR-200 is a key regulator of MET and downregulates the transcriptional repressors zinc finger E-box-binding homeobox 1 (Zeb1) and 2 (Zeb2) [2, 12, 32–35]. These proteins are responsible for E-cadherin suppression during EMT. Once E-cadherin levels reach a certain threshold it can further augment MET by inhibiting Snail through sequestering β-catenin and p65 [36]. Also, extracellular signals received upon arrival to distant tissue such as BMP7 can induce MET [12, 37]. After a cancer cell has gone through MET and been released from growth arrest, it may have an opportunity to colonize.

As stated previously, a cancer cell must learn how to benefit from the growth factors and surrounding host cells in the distant tissue in order to colonize. Cancer cells that extravasate in bone tissue may sustain their growth by expressing C-X-C chemokine receptor 4 (CXCR4) to recognize stromal derived factor 1 (SDF1) [38]. Moreover, these cancer cells secrete IL-6, IL-11, tumor necrosis factor-α (TNF-α), and parathyroid hormone-related protein (PTHRP), which activate resident osteoblast cells to secrete receptor activator of nuclear factor kappa-B ligand

(RANKL) [4, 39]. RANKL promotes the differentiation of mature osteoclasts. Osteoclasts are responsible for the bone matrix degradation during bone remodelling (osteolysis). Degradation of bone matrix releases a plethora of growth factors including TGFβ, bone morphogenetic proteins (BMPs), and insulin-like growth factor 1 (IGF1). These growth factors boost cancer cell growth further and complete a vicious positive feedback loop that may eventually result in colonization.

A very intriguing hypothesis is that tumor cells may cause remodelling of a possible site of colonization even before their arrival [2, 7]. Cancer cells secrete VEGF and lysyl oxidase (LOX), which can be transported to possible distant metastatic sites through systemic circulation and initiate remodelling of the extracellular matrix [40]. Transported VEGF recruits VEGF receptor 1 (VEGFR1) expressing bone marrow derived cells (BMDCs) and LOX can directly transform the composition of the extracellular matrix by crosslinking collagens and elastins. Recruited VEGFR1[+] BMDCs express MMP9, which further changes the architecture of the extracellular matrix and by doing so releases growth factors and chemokines sequestered in it [2]. The effects of LOX on the extracellular matrix activate existing fibroblasts and initiate expression of fibronectin. Secreted fibronectin can be recognized by integrins on cancer cells to support their growth. Distant tissue sites affected by this remodelling are referred to as metastatic niches. Metastatic niches may support the survival of extravasated cells and possibly assist in colonization.

Cancer cells that extravasate successfully, and are able to survive initially, have to induce angiogenesis in order to colonize the distant tissue. Angiogenic factors such as VEGF and FGF are expressed by cancer cells to initiate blood vessel formation. VEGF, again, plays a crucial role by recruiting endothelial progenitor cells (EPCs) [41]. EPCs are required for colonization because they support angiogenesis.

In order to persist at a distant site, cancer cells have to cope with the extracellular death signals that may lead to apoptosis. Encountered death signals includes Fas ligand (FASL), TNF-α, and tumor necrosis factor (ligand) superfamily, member 10 (TRAIL) [42]. Once recognized by receptors expressed on the cell surface, these death signals activate death-inducing signaling complex (DISC). Activation of DISC liberates caspase-8, which induces the caspase cascade and subsequent apoptosis. Cancer cells can circumvent apoptosis by repressing expression of caspase-8, a metastasis suppressor, at the transcriptional level through methylation of DNA in the promoter region of the gene [43, 44]. This makes cancer cells more resistant to extracellular death signals and gives them an opportunity to colonize.

Cancer cells also have to cope with internal blocks that confront metastasis – metastasis suppressor genes. Metastasis suppressors are defined as genes that inhibit metastasis but do not have any apparent effects on the primary tumor [45, 46]. These genes probably affect colonization since this is the rate-limiting step where most cancer cells fail. Metastasis suppressors include non-metastatic cells 1, 2 (NME1, NME 2 – also known as NM23, NM23B respectively), KISS1, KAI1 (also known as CD82), mitogen activated protein kinase 4 (MAP2K4) [47–51]. Surprisingly, some of these metastasis suppressor genes converge on the RAS pathway [45]. As mentioned earlier, the RAS pathway has critical roles in apoptosis and cell proliferation. A subtle push toward apoptosis or a hiccup in proliferation provoked by the action of metastasis suppressors may make cancer cells incompetent at colonization. MicroRNA-31 is another significant metastasis suppressor [2]. It represses a set of pro-metastatic genes including integrin α5, MMP16, and ras homolog gene family, member A (RhoA) [52]. As a result miR-31 function effectively impairs colonization of cancer cells. A cancer cell has to repress these internal blockades, metastasis suppressors, in order to colonize a distant tissue.

Colonization is the stage at which cancer cells are least efficient. The traits they have acquired during invasion and intravasation, such as activation of EMT program, may work against them at the distant site. Only the cells that have the plasticity to transform into an adequate phenotype may have a chance to prosper. They have to be equipped with the receptors to benefit from new growth factors that are presented to them in the new site. A cancer cell also has to express growth factors, chemokines and proteases that will remodel the microenvironment around them into a more hospitable host (expression of molecules such as interleukins, VEGF, FGF, MMPs). Moreover, cancer cells need to be resistant to both extracellular death signals and the intracellular apoptotic program in order to survive and colonize. Having all these favorable traits ready even before extravasation is equal to winning the lottery for a cancer cell and explains why colonization is the most inefficient and rate-limiting step during metastasis.

Fortunately, the many steps a cancer cell has to climb in order to successfully metastasize at a distant tissue are very steep. As demonstrated by the complexity of the underlying molecular mechanisms, metastasis is a multidimensional, multivariable problem for a cancer cell to solve. The formation of metastasis requires a cancer cell to be extremely plastic in order to adapt to different microenvironments. They need to cope with intrinsic and extrinsic death signals and, moreover, they need to educate other cell types to assist them during metastasis.

Organ-specific metastasis

From very early on, it has been realized that certain cancers prefer certain target organs to metastasize. Prostate cancer almost exclusively metastasizes to the bones; colorectal and pancreatic cancers metastasize to the liver and the lungs; whereas breast cancer can metastasize to a wider range of organs including the bones, lungs, liver, and brain [2, 16]. A straightforward explanation for this organ-specific metastasis may lie in the anatomy of the circulatory system. Cancer cells shed from the primary site after intravasation follow the course of the blood stream and stick in the capillaries of the organ next in line. Therefore most of the intravasated colorectal cancer cells accumulate in liver tissue while most breast cancer cells arrive in the lungs. Still, the anatomy of the circulatory system does not give the full explanation of organ-specific metastasis and the underlying molecular biology seems to play a significant role. In this section, we look at the molecular actors and their roles underlying organ-specific metastasis.

Chemokines are among the most intriguing actors for organ-specific metastasis. They are relatively small, secreted proteins, which can induce migration of target cells [21]. They function in organogenesis during development and during immune responses in adults. Cancer cells expressing chemokine receptors can take advantage of these pathways during metastasis. The most commonly expressed chemokine receptor in cancer cells is CXCR4 [21, 53]. Its ligand, SDF1, is expressed in bone marrow, liver, lungs, and brain. As a result, CXCR4$^+$ cancer cells can actively extravasate to the SDF1-expressing organs and cause metastasis. A striking example of chemokine involvement in organ-specific metastasis comes from

different subsets of melanomas [54]. Chemokine (C-C motif) receptor 9 (CCR9) positive melanoma cells specifically metastasize to the small intestine because CCR9 recognizes chemokine (C-C motif) ligand 25 (CCL25) expressed in this tissue [55]. However, expression of CCR10 by melanoma cells leads to skin metastasis where the CCR10 ligand, CCL27, is highly expressed [54, 56]. Yet another subset of melanoma cells, expressing CCR7, metastasizes to the lymph nodes [57]. CCR7's interaction with its ligand CCL21 also facilitates lymph node-specific metastasis of lung and gastric cancers [58, 59].

As stated earlier, breast cancer can metastasize to a rather wider range of organs including bones, lungs, brain, and liver. Each of these metastases has a particular set of genes that facilitate organ specificity. A lung metastasis-specific gene signature of breast cancer includes secreted protein, acidic, cysteine-rich (SPARC), interleukin 13 receptor, alpha 2 (IL13RA2), VCAM1, and MMP2 [60]. These genes have roles in extracellular functions, which suggest that they allow breast cancer cells to alter their microenvironment in the lungs. Active TGFβ signaling can also facilitate breast cancer metastasis to the lungs [61]. TGFβ signaling induces angiopoietin-like 4 (ANGPTL4) expression by SMAD transcriptional factors. In turn, ANGPTL4 assists transendothelial migration of cancer cells by disturbing cell to cell junctions on the blood vessel walls.

ST6(alpha-N-acetyl-neuraminyl-2,3-beta-galactosyl-1,3)-N-acetylgalactosaminide alpha-2,6-sialyltransferase 5 (ST6GALNAC5) was discovered as a determinant for brain-specific metastasis of breast cancer [62]. ST6GALNAC5 in cooperation with epiregulin (EREG) and prostaglandin-endoperoxide synthase 2 (COX2) assists in breast cancer cell adhesion to endothelial cells and facilitates migration through the blood–brain barrier.

Liver-specific metastasis of breast cancer cells is marked by expression of claudin 2 (CLDN2) [63]. CLDN2 is necessary for breast cancer cells to metastasize to the liver. It increases detectable levels of α2β1 and α5β1 integrin complexes on the cell surface, which in turn enhances breast cancer cell adhesion to the extracellular matrix components.

IL-11 expressing breast cancer cells specifically metastasize to bones where they initiate a vicious osteolytic cycle (see previous section).

Bone seems to present an attractive setting for metastasis since many different types of cancers can metastasize to this tissue [64]. Interestingly, though, they do not use a similar molecular machinery to do so. As stated above, breast cancer cells express IL-11 and CXCR4 to colonize bone. On the other hand, neuroblastoma cells express RANKL or stimulate mesenchymal stem cells to synthesize IL-6, which, similar to breast cancer, initiates a detrimental osteolytic cycle. Prostate cancer uses yet another strategy to metastasize to bone. Instead of activating an osteolytic cycle, they predominantly cause aberrant osteoblast proliferation by producing FGF8, TGFβ2, and endothelin 1. This leads to excess bone depositions, which serve as rich sources of sequestered growth factors. Meanwhile prostate cancer cells also secrete proteases such as urpkinase and prostate-specific antigen to harness these sequestered growth factors.

Organ-specific metastasis depends not only on the anatomy of the circulatory system but also on the molecular signatures of the cancer cells. The expression of certain genes primes cancer cells to metastasize to distinct organs. Membrane receptors such as chemokines and integrins play critical roles as they allow cancer cells to interact with the surrounding microenvironment. Autocrine signaling molecules such as interleukins and growth factors also play significant roles in organ specificity because particular ligands expressed by cancer cells permit them to influence particular cell types in specific tissues. Therefore aberrantly regulated molecules by cancer cells play a dominant role in organ-specific metastasis.

Cancer therapies directed at metastasis

It is a striking fact that surgery is still the most effective treatment for cancer. Unfortunately, since metastasis can affect inaccessible parts of critical organs or reach a scale where numerous body parts are affected, surgery becomes a less feasible option. That is why the development of cancer therapies and strategies directed at metastasis bears utmost importance. Understanding the molecular mechanisms underlying metastasis will let the development of new therapeutic reagents specifically target these metastatic pathways.

A group of anti-metastatic drugs target surrounding stromal cells instead of cancer cells. Endothelial cells are responsible for new blood vessel formation and are required for metastatic growth. As a result, the main actor driving endothelial cell recruitment

and activation during metastasis, VEGF, has been a major target for anti-metastatic drug development. Bevacizumab, an inhibitory antibody against VEGF, was designed to block angiogenesis [65]. Unfortunately, it failed to show significant success in clinical trials to stop metastasis indicating that further research is required to understand the angiogenesis–metastasis axis [2, 66]. Osteoclasts are prime targets for anti-metastatic drugs due to their role in bone metastasis [64]. Bisphosphonates are chemicals that inhibit enzymes catalyzing post-translational protein modifications. Uptake of bisphosphonates results in accumulation of these chemicals in the bone matrix. Therefore, during the osteolytic cycle, osteoclasts are exposed to bisphosphonates. This causes the apoptosis of osteoclasts, and breaks the vicious osteolytic cycle depriving cancer cells of the benefit from the released growth factors [67]. Another method to target osteoclasts is inhibiting RANK signaling. The inhibitory RANKL antibody denosumab was shown to be effective in preventing skeletal complications related to bone metastasis [68, 69]. Another way to deprive cancer cells of growth factors, such as TGFβ ligands released during bone metastasis, is targeting the corresponding pathways activated in cancer cells. An inhibitor of the TGFβ receptor I kinase, SD-208, and small molecule TGFβ inhibitor, LY2157299, are potential candidate drugs to use in clinical trials [2, 70, 71].

A very attractive target protein family for anti-metastasis therapy is integrins. An inhibitory antibody, etaracizumab, against αvβ3 and αvβ5 integrins has been the first of similar antibodies used in clinical trials [18, 72]. It prevents angiogenesis and blocks bone metastasis by inhibiting integrin–extracellular matrix interactions. CNTO 95 is also an antibody targeting the same set of integrin complexes and was under clinical trials [73, 74]. Cilengitide, a small peptide molecule, is also used to inhibit αvβ3 and αvβ5 integrins and was in clinical trials against glioblastoma. Promising results in phase I and II trials resulted in initiation of a phase III trial, which unfortunately failed [18, 75].

MMPs are used by cancer cells to remodel their microenvironment during metastasis and hence MMP inhibitors have the potential to be anti-metastatic therapeutic agents. Unfortunately, however, clinical trials including MMP inhibitors failed to show a benefit of these drugs [2]. This may be due to involvement of MMPs during the early stages of metastasis, and once colonization takes place the requirement for MMP by cancer cells may be diminished. That is why MMP inhibitors may have roles in the prevention of metastasis rather than treatment of clinically relevant metastases.

NME1, suppressed by cancer cells during metastasis, is the target for an anti-metastatic therapeutic drug, medroxy-progesterone acetate (MPA) [76]. MPA induces expression of NME1 in cancer cells at the transcriptional level, which in turn impairs formation of metastasis. Another target metastasis suppressor for cancer therapy may be KAI1. It has been shown that etoposide can increase expression of KAI1 by a p53 response element dependent manner [77].

So far, most anti-metastatic drugs could not reach their desired efficacy. This should motivate further research in order to understand the underlying molecular pathways of metastasis, the main cause of cancer deaths. Metastasis is a complex process and revealing the relevant molecular players and their function will require considerable further research.

Conclusion

Metastasis is the major cause of cancer-related deaths. That is why understanding of the whole metastatic cascade in detail is vital for our efforts against the disease. In this chapter, we tried to demonstrate the complex nature of metastasis mainly by focusing on the underlying molecular events. Each step of the metastatic cascade displays the remarkable interplay between signaling pathways in a tumor cell. Moreover, interactions between tumor cells and surrounding stromal cells in both the primary tumor and distant tissue microenvironment play critical roles in disease progression.

We are still far from reaching a desired level of understanding of metastasis. However, advances made regarding understanding of the metastatic cascade have been remarkable, and some have resulted in the development of new targeted therapies that have been in clinical trials with various degrees of success. Therefore understanding of the metastatic cascade and underlying molecular players should be one of the main targets in cancer research.

References

1. Gupta, G.P., and Massague, J. 2006. Cancer metastasis: building a framework. *Cell* 127:679–695.

2. Valastyan, S., and Weinberg, R.A. 2011. Tumor metastasis: molecular insights and evolving paradigms. *Cell* 147:275–292.

3. Wirtz, D., Konstantopoulos, K., and Searson, P.C. 2011. The physics of cancer: the role of physical interactions and mechanical forces in metastasis. *Nat Rev Cancer* 11:512–522.

4. Joyce, J.A., and Pollard, J.W. 2009. Microenvironmental regulation of metastasis. *Nat Rev Cancer* 9: 239–252.

5. Taddei, M.L., Giannoni, E., Fiaschi, T., and Chiarugi, P. 2012. Anoikis: an emerging hallmark in health and diseases. *J Pathol* 226: 380–393.

6. Luzzi, K.J., MacDonald, I.C., Schmidt, E.E., et al. 1998. Multistep nature of metastatic inefficiency: dormancy of solitary cells after successful extravasation and limited survival of early micrometastases. *Am J Pathol* 153: 865–873.

7. Psaila, B., and Lyden, D. 2009. The metastatic niche: adapting the foreign soil. *Nat Rev Cancer* 9:285–293.

8. Cavallaro, U., and Christofori, G. 2004. Cell adhesion and signalling by cadherins and Ig-CAMs in cancer. *Nat Rev Cancer* 4: 118–132.

9. Polyak, K., and Weinberg, R.A. 2009. Transitions between epithelial and mesenchymal states: acquisition of malignant and stem cell traits. *Nat Rev Cancer* 9: 265–273.

10. Massague, J., Blain, S.W., and Lo, R.S. 2000. TGFbeta signaling in growth control, cancer, and heritable disorders. *Cell* 103: 295–309.

11. Massague, J., Seoane, J., and Wotton, D. 2005. Smad transcription factors. *Genes Dev* 19:2783–2810.

12. Brabletz, T. 2012. To differentiate or not – routes towards metastasis. *Nat Rev Cancer* 12:425–436.

13. Sethi, N., and Kang, Y. 2011. Unravelling the complexity of metastasis – molecular understanding and targeted therapies. *Nat Rev Cancer* 11: 735–748.

14. Carmeliet, P., and Jain, R.K. 2011. Principles and mechanisms of vessel normalization for cancer and other angiogenic diseases. *Nat Rev Drug Discov* 10:417–427.

15. Egeblad, M., and Werb, Z. 2002. New functions for the matrix metalloproteinases in cancer progression. *Nat Rev Cancer* 2:161–174.

16. Nguyen, D.X., Bos, P.D., and Massague, J. 2009. Metastasis: from dissemination to organ-specific colonization. *Nat Rev Cancer* 9:274–284.

17. Avraamides, C.J., Garmy-Susini, B., and Varner, J.A. 2008. Integrins in angiogenesis and lymphangiogenesis. *Nat Rev Cancer* 8:604–617.

18. Desgrosellier, J.S., and Cheresh, D.A. 2010. Integrins in cancer: biological implications and therapeutic opportunities. *Nat Rev Cancer* 10:9–22.

19. Guo, W., Pylayeva, Y., Pepe, A., et al. 2006. Beta 4 integrin amplifies ErbB2 signaling to promote mammary tumorigenesis. *Cell* 126:489–502.

20. Bingle, L., Brown, N.J., and Lewis, C.E. 2002. The role of tumour-associated macrophages in tumour progression: implications for new anticancer therapies. *J Pathol* 196:254–265.

21. Roussos, E.T., Condeelis, J.S., and Patsialou, A. 2011. Chemotaxis in cancer. *Nat Rev Cancer* 11: 573–587.

22. Gocheva, V., Wang, H.W., Gadea, B.B., et al. 2010. IL-4 induces cathepsin protease activity in tumor-associated macrophages to promote cancer growth and invasion. *Genes Dev* 24:241–255.

23. Gocheva, V., Zeng, W., Ke, D., et al. 2006. Distinct roles for cysteine cathepsin genes in multistage tumorigenesis. *Genes Dev* 20:543–556.

24. Guo, W., and Giancotti, F.G. 2004. Integrin signalling during tumour progression. *Nat Rev Mol Cell Biol* 5:816–826.

25. Pylayeva-Gupta, Y., Grabocka, E., and Bar-Sagi, D. 2011. RAS oncogenes: weaving a tumorigenic web. *Nat Rev Cancer* 11: 761–774.

26. Scheel, C., and Weinberg, R.A. 2012. Cancer stem cells and epithelial–mesenchymal transition: concepts and molecular links. *Semin Cancer Biol* 22:396–403.

27. Napier, S.L., Healy, Z.R., Schnaar, R.L., and Konstantopoulos, K. 2007. Selectin ligand expression regulates the initial vascular interactions of colon carcinoma cells: the roles of CD44v and alternative sialofucosylated selectin ligands. *J Biol Chem* 282:3433–3441.

28. Thomas, S.N., Schnaar, R.L., and Konstantopoulos, K. 2009. Podocalyxin-like protein is an E-/L-selectin ligand on colon carcinoma cells: comparative biochemical properties of selectin ligands in host and tumor cells. *Am J Physiol Cell Physiol* 296: C505–513.

29. Thomas, S.N., Zhu, F., Schnaar, R.L., Alves, C.S., and Konstantopoulos, K. 2008. Carcinoembryonic antigen and CD44 variant isoforms cooperate to mediate colon carcinoma cell adhesion to E- and L-selectin in shear flow. *J Biol Chem* 283:15647–15655.

30. Konstantopoulos, K., and Thomas, S.N. 2009. Cancer cells in transit: the vascular interactions of tumor cells. *Annu Rev Biomed Eng* 11:177–202.

31. Gay, L.J., and Felding-Habermann, B. 2011. Contribution of platelets to tumour metastasis. *Nat Rev Cancer* 11:123–134.

32. Thiery, J.P., Acloque, H., Huang, R.Y., and Nieto, M.A. 2009. Epithelial-mesenchymal transitions in development and disease. *Cell* 139:871–890.

33. Cieply, B., Farris, J., Denvir, J., Ford, H.L., and Frisch, S.M. 2013. Epithelial–mesenchymal transition and tumor suppression are controlled by a reciprocal feedback loop between ZEB1 and Grainyhead-like-2. *Cancer Res* 73:6299–6309.

34. Burk, U., Schubert, J., Wellner, U., et al. 2008. A reciprocal repression between ZEB1 and members of the miR-200 family promotes EMT and invasion in cancer cells. *EMBO Rep* 9:582–589.

35. Park, S.M., Gaur, A.B., Lengyel, E., and Peter, M.E. 2008. The miR-200 family determines the epithelial phenotype of cancer cells by targeting the E-cadherin repressors ZEB1 and ZEB2. *Genes Dev* 22:894–907.

36. Solanas, G., Porta-de-la-Riva, M., Agusti, C., et al. 2008. E-cadherin controls beta-catenin and NF-kappaB transcriptional activity in mesenchymal gene expression. *J Cell Sci* 121:2224–2234.

37. Buijs, J.T., Henriquez, N.V., van Overveld, P.G., et al. 2007. Bone morphogenetic protein 7 in the development and treatment of bone metastases from breast cancer. *Cancer Res* 67: 8742–8751.

38. Sohara, Y., Shimada, H., and DeClerck, Y.A. 2005. Mechanisms of bone invasion and metastasis in human neuroblastoma. *Cancer Lett* 228:203–209.

39. Mundy, G.R. 2002. Metastasis to bone: causes, consequences and therapeutic opportunities. *Nat Rev Cancer* 2:584–593.

40. Erler, J.T., Bennewith, K.L., Cox, T.R., et al. 2009. Hypoxia-induced lysyl oxidase is a critical mediator of bone marrow cell recruitment to form the premetastatic niche. *Cancer Cell* 15:35–44.

41. Weis, S.M., and Cheresh, D.A. 2011. Tumor angiogenesis: molecular pathways and therapeutic targets. *Nat Med* 17:1359–1370.

42. Mehlen, P., and Puisieux, A. 2006. Metastasis: a question of life or death. *Nat Rev Cancer* 6:449–458.

43. Stupack, D.G., Teitz, T., Potter, M.D., et al. 2006. Potentiation of neuroblastoma metastasis by loss of caspase-8. *Nature* 439:95–99.

44. Martinez, R., Setien, F., Voelter, C., et al. 2007. CpG island promoter hypermethylation of the pro-apoptotic gene caspase-8 is a common hallmark of relapsed

glioblastoma multiforme. *Carcinogenesis* 28:1264–1268.

45. Stafford, L.J., Vaidya, K.S., and Welch, D.R. 2008. Metastasis suppressor genes in cancer. *Int J Biochem Cell Biol* 40:874–891.

46. Smith, S.C., and Theodorescu, D. 2009. Learning therapeutic lessons from metastasis suppressor proteins. *Nat Rev Cancer* 9: 253–264.

47. Steeg, P.S., Bevilacqua, G., Kopper, L., et al. 1988. Evidence for a novel gene associated with low tumor metastatic potential. *J Natl Cancer Inst* 80:200–204.

48. Baba, H., Urano, T., Okada, K., et al. 1995. Two isotypes of murine nm23/nucleoside diphosphate kinase, nm23-M1 and nm23-M2, are involved in metastatic suppression of a murine melanoma line. *Cancer Res* 55:1977–1981.

49. Welch, D.R., Chen, P., Miele, M.E., et al. 1994. Microcell-mediated transfer of chromosome 6 into metastatic human C8161 melanoma cells suppresses metastasis but does not inhibit tumorigenicity. *Oncogene* 9: 255–262.

50. Dong, J.T., Lamb, P.W., Rinker-Schaeffer, C.W., et al. 1995. KAI1, a metastasis suppressor gene for prostate cancer on human chromosome 11p11.2. *Science* 268:884–886.

51. Yoshida, B.A., Dubauskas, Z., Chekmareva, M.A., et al. 1999. Mitogen-activated protein kinase kinase 4/stress-activated protein/ Erk kinase 1 (MKK4/SEK1), a prostate cancer metastasis suppressor gene encoded by human chromosome 17. *Cancer Res* 59:5483–5487.

52. Valastyan, S., Reinhardt, F., Benaich, N., et al. 2009. A pleiotropically acting microRNA, miR-31, inhibits breast cancer metastasis. *Cell* 137:1032–1046.

53. Balkwill, F. 2004. The significance of cancer cell expression of the chemokine receptor CXCR4. *Semin Cancer Biol* 14:171–179.

54. Murakami, T., Cardones, A.R., and Hwang, S.T. 2004. Chemokine receptors and melanoma metastasis. *J Dermatol Sci* 36:71–78.

55. Amersi, F.F., Terando, A.M., Goto, Y., et al. 2008. Activation of CCR9/CCL25 in cutaneous melanoma mediates preferential metastasis to the small intestine. *Clin Cancer Res* 14:638–645.

56. Murakami, T., Cardones, A.R., Finkelstein, S.E., et al. 2003. Immune evasion by murine melanoma mediated through CC chemokine receptor-10. *J Exp Med* 198:1337–1347.

57. Wiley, H.E., Gonzalez, E.B., Maki, W., Wu, M.T., and Hwang, S.T. 2001. Expression of CC chemokine receptor-7 and regional lymph node metastasis of B16 murine melanoma. *J Natl Cancer Inst* 93:1638–1643.

58. Takanami, I. 2003. Overexpression of CCR7 mRNA in nonsmall cell lung cancer: correlation with lymph node metastasis. *Int J Cancer* 105: 186–189.

59. Mashino, K., Sadanaga, N., Yamaguchi, H., et al. 2002. Expression of chemokine receptor CCR7 is associated with lymph node metastasis of gastric carcinoma. *Cancer Res* 62: 2937–2941.

60. Minn, A.J., Gupta, G.P., Siegel, P.M., et al. 2005. Genes that mediate breast cancer metastasis to lung. *Nature* 436:518–524.

61. Padua, D., Zhang, X.H., Wang, Q., et al. 2008. TGFbeta primes breast tumors for lung metastasis seeding through angiopoietin-like 4. *Cell* 133:66–77.

62. Bos, P.D., Zhang, X.H., Nadal, C., et al. 2009. Genes that mediate breast cancer metastasis to the brain. *Nature* 459:1005–1009.

63. Tabaries, S., Dong, Z., Annis, M.G., et al. 2011. Claudin-2 is selectively enriched in and promotes the formation of breast cancer liver metastases through engagement of integrin complexes. *Oncogene* 30: 1318–1328.

64. Weilbaecher, K.N., Guise, T.A., and McCauley, L.K. 2011. Cancer to bone: a fatal attraction. *Nat Rev Cancer* 11:411–425.

65. Ferrara, N., Hillan, K.J., Gerber, H.P., and Novotny, W. 2004. Discovery and development of bevacizumab, an anti-VEGF antibody for treating cancer. *Nat Rev Drug Discov* 3:391–400.

66. Los, M., Roodhart, J.M., and Voest, E.E. 2007. Target practice: lessons from phase III trials with bevacizumab and vatalanib in the treatment of advanced colorectal cancer. *Oncologist* 12:443–450.

67. Lipton, A., Cook, R., Saad, F., et al. 2008. Normalization of bone markers is associated with improved survival in patients with bone metastases from solid tumors and elevated bone resorption receiving zoledronic acid. *Cancer* 113:193–201.

68. Rizzoli, R., Yasothan, U., and Kirkpatrick, P. 2010. Denosumab. *Nat Rev Drug Discov* 9:591–592.

69. Stopeck, A.T., Lipton, A., Body, J.J., et al. 2010. Denosumab compared with zoledronic acid for the treatment of bone metastases in patients with advanced breast cancer: a randomized, double-blind study. *J Clin Oncol* 28: 5132–5139.

70. Akhurst, R.J., and Hata, A. 2012. Targeting the TGFbeta signalling pathway in disease. *Nat Rev Drug Discov* 11:790–811.

71. Yingling, J.M., Blanchard, K.L., and Sawyer, J.S. 2004.

Development of TGF-beta signalling inhibitors for cancer therapy. *Nat Rev Drug Discov* 3:1011–1022.

72. Cox, D., Brennan, M., and Moran, N. 2010. Integrins as therapeutic targets: lessons and opportunities. *Nat Rev Drug Discov* 9:804–820.

73. Martin, P.L., Jiao, Q., Cornacoff, J., et al. 2005. Absence of adverse effects in cynomolgus macaques treated with CNTO 95, a fully human anti-alphav integrin monoclonal antibody, despite widespread tissue binding.

Clin Cancer Res 11: 6959–6965.

74. Mullamitha, S.A., Ton, N.C., Parker, G.J., et al. 2007. Phase I evaluation of a fully human anti-alphav integrin monoclonal antibody (CNTO 95) in patients with advanced solid tumors. *Clin Cancer Res* 13:2128–2135.

75. Kurozumi, K., Ichikawa, T., Onishi, M., Fujii, K., and Date, I. 2012. Cilengitide treatment for malignant glioma: current status and future direction. *Neurol Med Chir (Tokyo)* 52:539–547.

76. Palmieri, D., Halverson, D.O., Ouatas, T., et al. 2005. Medroxyprogesterone acetate elevation of Nm23-H1 metastasis suppressor expression in hormone receptor-negative breast cancer. *J Natl Cancer Inst* 97:632–642.

77. Mashimo, T., Bandyopadhyay, S., Goodarzi, G., et al. 2000. Activation of the tumor metastasis suppressor gene, KAI1, by etoposide is mediated by p53 and c-Jun genes. *Biochem Biophys Res Commun* 274:370–376.

Cancer metabolism

Dimitrios Anastasiou, Jason W. Locasale and Matthew G. Vander Heiden

Introduction

Recent technological developments in combination with novel computational approaches for analysis of large-scale datasets have provided novel insights into genetic changes that are associated with tumorigenesis. Better genetic understanding of this process forms the basis for personalized approaches to pharmacologic intervention that promises more successful therapeutics. Yet the vast diversity of cancer-promoting genetic changes along with the economic burden associated with the development of individualized therapies and corresponding diagnostic tools constitute a great challenge: to devise drugs that target a broad spectrum of cancers while minimizing unwanted effects. Thus better understanding the common characteristics that distinguish cancer cells from normal tissues is imperative for devising novel therapeutic strategies to complement "targeted" treatments that focus on specific genetic events.

Aberrations in cell cycle regulatory circuits underlie one such common feature of cancer cells, which is uncontrolled proliferation. During proliferation, accumulation of biomass for cell growth is the first discernable process preceding cell division. This occurs primarily during the G1 phase of the eukaryotic cell cycle before committing to a new round of DNA replication and ensures that sufficient nucleotides, lipids, and amino acids are available to build a functioning daughter cell. Thus highly proliferative cancer cells have an increased demand for biosynthetic processes to provide these building blocks.

The studies of Louis Pasteur on the effects of oxygen availability and glucose metabolism in yeast first hinted at a link between nutrient consumption and growth rate (Racker, 1974). Pasteur observed that, in the presence of oxygen, glucose was oxidized to CO_2, while under limiting oxygen conditions glucose was consumed at higher rates and metabolized to ethanol. Interestingly, the latter metabolic state is also implemented during proliferation of yeast cells (DeRisi et al., 1997). The observation that tumor cells exhibit these same metabolic characteristics, and that these characteristics are distinct from those of normal cells, was formally proposed by Otto Warburg in the early twentieth century (Warburg, 1956b). By extending Pasteur's studies to mammalian cells, Warburg discovered that in contrast to normal, differentiated tissues, the rapidly proliferating cancer and embryonic tissues metabolize glucose into lactate irrespective of the available oxygen tension (a phenomenon known now as the "Warburg effect"). This led Warburg to postulate that mitochondrial defects underlie the characteristic features of tumor metabolism (Warburg, 1956a), a theory widely contested even to date.

It has been recently appreciated that, despite their multifaceted nature, many known oncogenic events result in common modifications to cellular metabolism that recapitulate Warburg's observations (Figure 19.1) (Locasale et al., 2009; Vander Heiden et al., 2009). In multicellular organisms, nutrient uptake is dictated by growth factors, a coupling mechanism that prevents aberrant proliferation of individual cells (Vander Heiden et al., 2001). Cancer cells often exhibit deregulated mitogenic signaling, which, among other effects, contributes to enhanced nutrient uptake. As the goal of cellular metabolic function shifts away from that of a differentiated tissue to that of accommodating cell growth, tumors need to rewire existing pathways in order to funnel these nutrients into biosynthetic routes. This is achieved by changing the expression levels, isoform expression, and/or regulation of relevant enzymes, especially those that catalyze rate-limiting steps.

Systems Biology of Cancer, ed. S. Thiagalingam. Published by Cambridge University Press. © Cambridge University Press 2015.

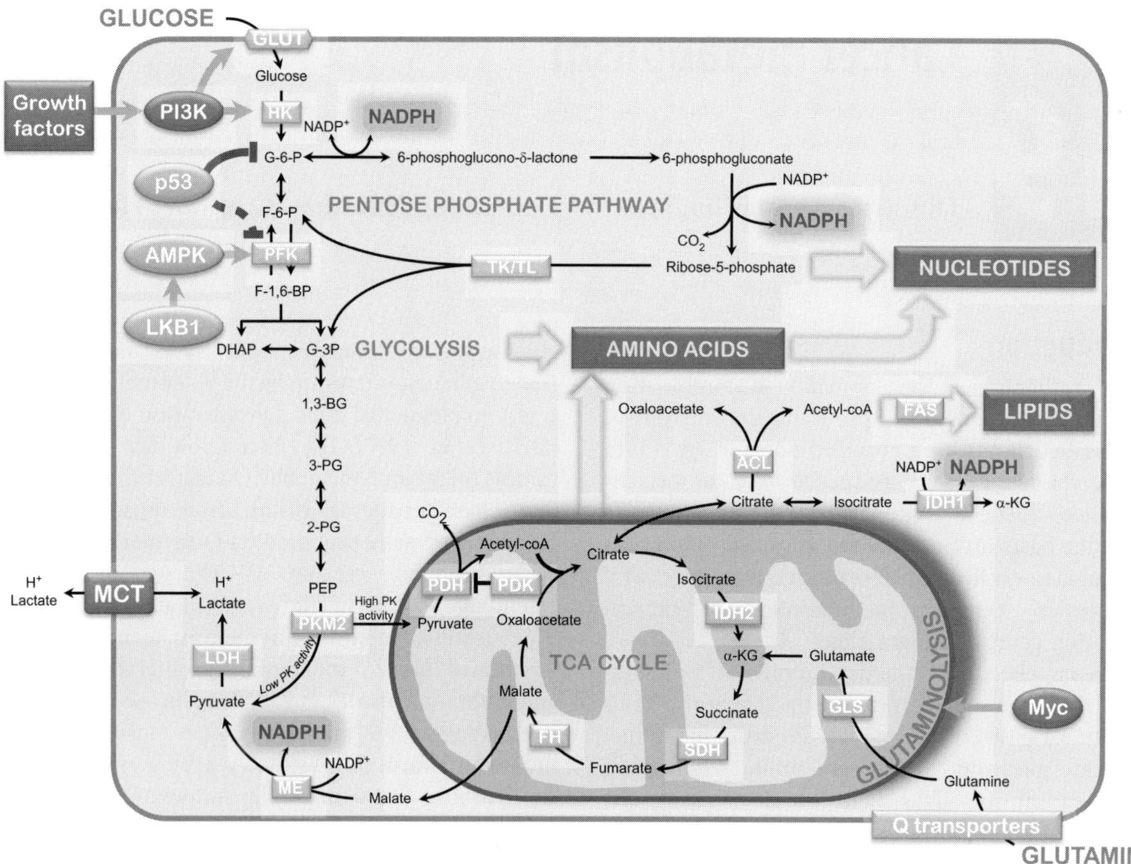

Figure 19.1 Overview of central metabolism. Major oncogenic (PI3K and Myc) and tumor suppressor (p53, AMPK, and LKB1) events impose changes in the wiring of metabolic networks (block arrows) to promote utilization of nutrients (mainly glucose and glutamine) for the synthesis of key cellular building blocks (amino acids, nucleotides, and lipids). See text for details. (A black and white version of this figure will appear in some formats. For the color version, please refer to the plate section.)

Understanding the intricate nature of metabolic reprogramming during cell transformation, how this is implemented by oncogenic events, and the influence of mitogenic signaling on this process provides a basis for selecting novel drug targets and therapeutic strategies for cancer. In the first part of this chapter, we will review recent advances in our understanding of how metabolic pathways are utilized in cancer cells to meet their biosynthetic demands for proliferation and sustained survival. In the second part we will review the technologies currently employed that allow us to obtain a systems-level vista of cancer metabolism.

Enhanced glucose metabolism is a central feature of cancer cell metabolism

Glucose plays a central role in nutrient homeostasis both at the cellular and the organismal level. This is best exemplified by the metabolic syndrome where defective glucose homeostasis caused by disruption of the insulin signaling axis results in a complex multi-organ disease phenotype (Moller and Kaufman, 2005). In addition, there are tissues such as the brain and red blood cells that utilize glucose as the primary source for energy whereas other cell types predominantly utilize other catabolic sources such as lipids. Similarly, most tumors exhibit increased glucose uptake, a characteristic exploited for diagnostic purposes in the form of positron emission tomography (PET) scanning using the radiolabelled glucose analogue ^{18}F-deoxy-glucose (^{18}FDG), which accumulates in cancer tissues allowing visualization of tumors in the clinic (Ak et al., 2000).

Glycolysis is a primary route of glucose metabolism involving a series of enzymatic steps that convert glucose into pyruvate. Pyruvate can subsequently enter mitochondria where it is decarboxylated and converted to acetyl-coA by the action of pyruvate dehydrogenase (PDH). Acetyl-coA then combines with oxaloacetate to form citrate, which enters the TCA cycle where it fuels the reduction of mitochondrial FAD and NAD^+ to $FADH_2$ and NADH, respectively. This provides electrons to the electron transport chain to produce a proton gradient across the mitochondrial inner membrane that drives the generation of ATP in the process of oxidative phosphorylation.

While this pathway is the predominant route for glucose metabolism in most differentiated tissues, cancer cells exhibit a reduced rate of pyruvate entry into mitochondria. Rather, pyruvate is reduced to lactate by lactate dehydrogenase (LDH). This reaction is coupled to the oxidation of NADH to NAD^+, which is critical for maintaining high glycolytic rates as NAD^+ is required for glyceraldehyde-3-phosphate dehydrogenase (GAPDH) to catalyze an intermediate step in glycolysis involving the oxidation of glyceraldehyde-3-phosphate to 1,3-bisphosphoglycerate.

Glucose enters cells via transmembrane glucose transporters (GLUTs). There are more than a dozen such proteins, most of which have been found to be upregulated in cancer (Calvo et al., 2010). GLUT1 is responsible for basal glucose uptake in most cells and is the target of hypoxia inducible factor-1 (HIF-1), a transcription factor frequently upregulated in cancer (Semenza, 2003). Another transcription factor, the oncogene myc, also drives glucose transporter expression, in particular GLUT2 and GLUT4 (Osthus et al., 2000). In contrast, GLUT1 and GLUT4 are transcriptionally repressed by the tumor suppressor p53 (Schwartzenberg-Bar-Yoseph et al., 2004). Glucose uptake can also be stimulated by post-translational modifications elicited downstream of growth factor receptors. Much of this effect is mediated by the phosphoinositide-3-kinase (PI3K)/Akt pathway, which comprises proteins encoded by genes frequently mutated in cancer (Engelman et al., 2006). PI3K/Akt signaling stimulates the transport of GLUT-containing intracellular vesicles to the plasma membrane and leads to increased activity of the proximal enzymes in glycolysis (Elstrom et al., 2004; Plas et al., 2001). In particular, Akt promotes the retention of glucose in cells by inducing the association of hexokinase to the outer mitochondrial membrane (Gottlob et al., 2001). Hexokinase, the enzyme that

catalyzes the first committed step in glycolysis, phosphorylates glucose in the proximity to mitochondria-produced ATP. Enhanced glucose metabolism is further achieved by HIF- and myc-mediated transcriptional upregulation of all glycolytic enzymes except for phosphoglycerate mutase (PGAM) (Gordan et al., 2007; Hu et al., 2003; Osthus et al., 2000). Interestingly, PGAM is suppressed by the p53 tumor suppressor, suggesting a coherent strategy during oncogenesis to increase glucose metabolism.

Metabolic reprogramming mediates flux of glucose carbons to anabolic pathways for proliferation and survival

While the events described above can account for the enhanced glycolytic rates observed in tumors, they do not explain the qualitative differences in the way glucose is metabolized compared to normal tissues. For this, glucose-derived carbons need to be re-routed into biosynthetic pathways (see Figure 19.1). Furthermore, apart from its contribution to the cellular carbon pool, glucose metabolism also plays an important role in establishing redox homeostasis in cells. Redox balance is primarily determined by the relative abundance of the nicotinamide adenine dinucleotides NAD^+ and $NADP^+$ and their respective reduced forms NADH and NADPH. Maintenance of an appropriate redox balance is critical for ensuring cell viability and proliferation because the directionalities of several metabolic pathways, including lipid and nucleic acid biosynthesis, depend on $NAD(P)^+$/$NAD(P)H$ ratios.

The fate of glycolytic metabolites

While TCA cycle-derived metabolites contribute to the production of pyrimidine and some amino acids required for protein synthesis, pathways that emanate from glycolysis are central for the generation of nucleotides, other carbohydrates, and lipids. Thus a key event in cancer metabolism is limiting the entry of glycolytic metabolites into mitochondria to allow their diversion into these upstream biosynthetic pathways. Multiple mechanisms are employed by cancer cells to achieve this.

Limiting the activity of pyruvate kinase, the enzyme that catalyzes the last step in glycolysis, is one such mechanism (Eigenbrodt et al., 1992). While most differentiated tissues express the PKM1 isoform, cancer cells express its alternative splice variant PKM2 (Christofk et al., 2008a). PKM1 and PKM2 share high sequence similarity (Noguchi et al., 1986), yet they are regulated in critically different ways. In contrast to PKM1, PKM2 is allosterically activated by fructose-1,6-bisphosphate (FBP), an upstream glycolytic metabolite, as part of a feed-forward mechanism to activate glycolysis (Imamura and Tanaka, 1982). PKM2 has been shown to bind phosphotyrosine-containing peptides through a moiety proximal to the FBP binding pocket that results in the release of FBP and enzyme inactivation (Christofk et al., 2008b). PKM2 is itself a substrate of FGFR1 tyrosine kinase, an event that negatively regulates its activity by the aforementioned mechanism in a transmolecular manner (Hitosugi et al., 2009). This renders the protein susceptible to inhibition by tyrosine kinase-elicited signaling, which is frequently enhanced in tumors. Because tyrosine kinases are associated with mitogenic signaling, this provides an additional regulatory input for growth factors to regulate central metabolism in cancer.

Splicing of the PKM pre-mRNA is also under the control of oncogenes and is regulated by members of the heterogeneous nuclear ribonucleoprotein (hnRNP) family of proteins, which bind to and actively repress inclusion of the PKM1-specific exon (Clower et al., 2010). Cancer cells express high levels of these hnRNPs, an event driven at least in part by the myc oncoprotein (David et al., 2010). It is unclear whether an actual switch from PKM1 to PKM2 expression occurs during tumorigenesis; however, many proliferative tissues and all cancer tissues to our knowledge express PKM2. Therefore it is also possible that the exclusive expression of PKM2 reflects the genetic profile of a PKM2-expressing tumor-initiating cell. Other glycolysis enzymes are also known to be alternatively spliced, raising the possibility that in addition to PKM2, a broader, coordinated splicing switch may underlie the complex metabolic regulation of cancer cells.

Replacement of PKM2 with PKM1 in cultured cancer cells impedes tumor cell proliferation under metabolic stress conditions and in mouse xenograft studies (Christofk et al., 2008a). Replacement of PKM2 with PKM1 also results in reduced lactate production and enhanced oxygen consumption, i.e., a partial reversal of the Warburg phenotype. A PKM2 mutant defective in phosphotyrosine-mediated inhibition is also incompatible with cancer cell

proliferation and this correlates with decreased incorporation of glucose carbons into lipids (Christofk et al., 2008b). These observations are consistent with a model where inhibition of PKM2 results in the accumulation of upstream glycolytic metabolites for biomass production. It is noteworthy that another metabolite upstream of PKM2, 3-phosphoglycerate (3-PG), is a precursor for the serine and glycine biosynthesis pathway. The enzymes participating in this pathway have been shown to have increased activity in certain cancer cells (Snell et al., 1988). Both serine and glycine act as methylene group donors to folic acid derivatives that are essential for amino acid and nucleotide biosynthesis. It is conceivable that PKM2 inhibition may also cause increased entry of glycolytic metabolites into this pathway.

The pentose phosphate pathway

In addition to its role as an intermediate in glycolysis, glucose-6-phosphate can enter the pentose phosphate pathway (PPP), to generate both ribose-5-phosphate (R-5-P), a precursor of nucleotides (Reitzer et al., 1980) as well as metabolic cofactors such as NAD(P)H. When R-5-P is not limiting, carbon from the PPP can re-enter glycolysis via the non-oxidative branch of the pathway through the action of transketolases-transaldolases (TK-TA). One TK enzyme, TKTL1, is upregulated in tumors and appears to be required for proliferation of some cancer cells (Langbein et al., 2006; Zhang et al., 2007). In the mouse, genetic ablation of glucose-6-phosphate dehydrogenase (G-6-PDH), which catalyzes the first committed step into the oxidative branch of the PPP, results in embryonic lethality (Pandolfi et al., 1995). Notably embryonic stem (ES) cells from these mice exhibit enhanced sensitivity to oxidative stress. This reflects another important role for the PPP, in mediating NADPH production to facilitate cellular antioxidant responses.

As a result of oxidative metabolism and growth factor signaling, oxygen-containing molecules with unpaired electrons called reactive oxygen species (ROS) are generated. ROS are reactive toward all classes of biomolecules and can be detrimental to cellular functions. Cells employ elaborate mechanisms to neutralize ROS and repair oxidatively damaged molecules. The reducing equivalents for these defenses are mainly provided by reduced glutathione and peroxiredoxin molecules that contain reactive cysteine residues to act as electron acceptors. The uninterrupted operation of these defenses depends on mechanisms that ensure the continuous regeneration of reduced glutathione and peroxiredoxin when the electron-carrying cysteines are re-oxidized. The reducing power for this process is provided by NADPH.

There is abundant evidence that most tumors have high levels of ROS and, in some cases, ROS appear to be an integral part of the oncogenic process. Loss of the p53 tumor suppressor results in increased oxidative stress, and mouse xenograft models suggest that antioxidants can limit the tumorigenic potential of p53-deficient cell lines (Sablina et al., 2005). This is in part attributed to reduced expression of the p53 target, TP53-induced glycolysis and apoptosis regulator (TIGAR). TIGAR has been proposed to harbor a phosphatase activity that degrades fructose-2,6-bisphosphate (F-2,6-BP) (Bensaad et al., 2006). F-2,6-BP is an allosteric activator of phosphofructokinase 1 (PFK1), the major enzyme controlling commitment of glucose into glycolysis as opposed to flux into the oxidative branch of the PPP that generates NADPH. Thus PFK1 inhibition allows the accumulation of G-6-P to increase NADPH and fuel cellular antioxidant defenses. Tumorigenic events can also lead to increases in ROS levels by inducing ROS production. The Ras oncogene induces ROS production via activation of NADPH oxidase (Hancock et al., 2001; Irani et al., 1997) although other mechanisms involving ROS generated by mitochondrial complex III have also been proposed (Weinberg et al., 2010).

By reacting with DNA, increased ROS can lead to the introduction of mutations in the cancer genome and contribute to the emergence of more aggressive cancer phenotypes (Busuttil et al., 2003). However, this damage also renders tumors more vulnerable to acute oxidative stress as it effectively lowers the threshold for ROS-mediated cell death (Nogueira et al., 2008; Trachootham et al., 2006, 2009). Indeed, oncogenic events have also been shown to require PPP-mediated antioxidant capacity to establish anchorage-independent growth (Schafer et al., 2009; Weinberg et al., 2010). Matrix detachment elicits a cell death process called anoikis due to inhibition of glucose uptake and oxidative stress, which inhibits fatty acid oxidation, an ATP-generating process. Increased glucose uptake by oncogenes allows efficient operation of the PPP and enhanced ROS neutralization, which aids in the survival of cells during the energy crisis induced by matrix detachment. Thus because of the observed high levels of ROS in tumors, these data would predict that cancer cells have enhanced antioxidant capacity.

Mechanisms that allow cancer cells to derive benefit from ROS as well as remain viable despite increased oxidative stress are likely at play. In tissue culture, acute oxidative stress elicits a rapid upregulation of glucose flux through the oxidative branch of the PPP (Tuttle et al., 2007). A similar phenomenon in yeast has been attributed to the inhibition by ROS of two downstream glycolytic enzymes: glyceraldehyde-3-phosphate dehydrogenase (GAPDH) and triose-phosphate isomerase (Ralser et al., 2007). Slowing glycolysis at this step makes more G-6-P available to enter the PPP. However, whether this mechanism also applies to cancer cells remains to be determined.

The impact of mitochondrial metabolism changes in cancer

For most of the twentieth century, much of the conceptual progress in the field of cancer metabolism was shaped by Warburg's proposed explanation for his namesake phenomenon (Warburg, 1956b). Based on meticulous respiratory quotient measurements, Warburg suggested that cancer cells must harbor respiratory defects that he hypothesized were due to mitochondrial damage (Warburg, 1956a). Evidence that this may be true was subsequently obtained only for a limited number of tumors despite the near universal occurrence of aerobic glycolysis. This dissuaded further pursuit of understanding the phenomenon (Burk and Schade, 1956; Weinhouse, 1956, 1976).

To date, in the context of cancer, only a handful of genes that encode mitochondrial proteins have been found to be mutated, namely succinate dehydrogenase (SDH) (Baysal et al., 2000), fumarate hydratase (FH) (Tomlinson et al., 2002), and isocitrate dehydrogenase (IDH) (Parsons et al., 2008). Furthermore, it is not clear that in all cases these mutations prevent oxidative phosphorylation while all can influence cancer development by other mechanisms. Loss-of-function mutations in SDH and FH result in accumulation of their substrates succinate and fumarate, respectively (King et al., 2006). These metabolites inhibit the action of a class of enzymes called prolyl-hydroxylases (PHDs). PHDs modify HIF1 so that it is recognized by the VHL E3-ligase complex for degradation. Inhibition of PHDs by succinate and fumarate leads to HIF stabilization, an event thought to mediate the tumorigenic effects of SDH and FH mutation (King et al., 2006).

IDH catalyzes the oxidative decarboxylation of isocitrate to α-ketoglutarate (α-KG). There are three known mammalian isoforms, IDH1, which is cytoplasmic, and IDH2 and IDH3, which are both mitochondrial. IDH1 and IDH2 are NADP$^+$-dependent enzymes whereas IDH3 is NAD$^+$-dependent. IDH1 and IDH2 have been found to be frequently mutated in gliomas and leukemias, and infrequently in a variety of other cancers (Parsons et al., 2008; Yan et al., 2009). Notably, the identified mutations involve an equivalent arginine residue in both IDH1 and IDH2 that resides in the active site of the enzyme. However, instead of impairing catalytic activity, this mutation was surprisingly shown to favor the reductive reaction and produce an alternative metabolic product, 2-hydroxyglutarate (2-HG) (Dang et al., 2009). Consistent with this, 2-HG levels were shown to be elevated in tumors harboring IDH mutations and 2-HG may therefore serve as a useful cancer biomarker (Yan et al., 2009).

While such alterations in metabolic enzymes could impede mitochondrial processes, in most tumors mitochondria appear to be functional. Furthermore, mitochondrial metabolism has been shown to play an important role in the oncogenic process (Weinberg and Chandel, 2009). For example, genes involved in mitochondrial biogenesis are induced by myc overexpression (Gordan et al., 2007). This may ensure the supply of sufficient mitochondrial intermediates required for pyrimidine biosynthesis, another aspect of metabolism upregulated by myc (Coller et al., 2000; O'Connell et al., 2003). Similarly, Ras-driven tumorigenesis depends on STAT3-mediated upregulation of mitochondrial activity and is attenuated when mitochondrial function is disrupted in mice (Gough et al., 2009; Weinberg et al., 2010).

The above considerations raise the question about the mechanism underlying the attenuated utilization of glycolytic pyruvate by mitochondria in the absence of obvious mitochondrial defects. Myc- or HIF-driven overexpression of lactate dehydrogenase (LDH), a ubiquitous event in tumors, could channel pyruvate away from mitochondria. In addition, pyruvate conversion to acetyl-coA is actively repressed by inhibition of the pyruvate dehydrogenase (PDH) complex. A well-documented mechanism for PDH complex inhibition also involves a HIF and myc target gene, PDH kinase (PDK). PDK phosphorylates the E1α subunit of PDH and this results in inhibition of its activity (Holness and Sugden, 2003; Kim et al., 2006). Pharmacologic inhibition of PDK by dichloroacetate (DCA) has been shown to result in enhanced oxidation of glucose through mitochondria, and has been explored as a cancer therapy for glioblastoma (Bonnet et al., 2007; Michelakis et al., 2008, 2010).

Once in mitochondria, the fate of pyruvate is also altered compared to normal tissues. Cancer cells appear to operate a "truncated" TCA cycle characterized by an enhanced efflux of mitochondrial citrate back to the cytoplasm (Parlo and Coleman, 1984). Citrate in the cytoplasm is a substrate for ATP-dependent citrate lyase (ACL), which converts it back to acetyl-coA and oxaloacetate. Cytoplasmic acetyl-coA is required for many biosynthetic reactions including those involved in the *de novo* biosynthesis of lipids and sterols. This mechanism is corroborated by glucose tracer data that show a high proportion of glucose carbon incorporation into lipids (DeBerardinis et al., 2007; Medes et al., 1953). Upregulation of downstream lipid synthesis pathways further promotes this process. Acetyl-coA carboxylase and fatty acid synthase (FASN) are frequently found overexpressed in tumors (Menendez and Lupu, 2007). In addition, both ACL and FASN are also under the positive control of the PI3K pathway as both substrates and transcriptional targets, respectively, of Akt signaling. This contributes to growth factor-stimulated promotion of lipid biosynthesis. It is also noteworthy that an ACL fraction that resides in the nucleus has been identified as the source of nuclear acetyl-coA for histone acetylation and consequent regulation of gene expression (Wellen et al., 2009). The interplay between cytoplasmic and nuclear pools of acetyl-coA to cancer development remains largely unexplored.

Glutamine requirements for cancer cell proliferation

The enhanced efflux of mitochondrial citrate to the cytoplasm for lipid biosynthesis raises the question of how continuous operation of the TCA cycle is sustained. The main source for replenishing mitochondrial intermediates downstream of citrate is glutamine. Thus glutamine is said to play an "anaplerotic" function for the metabolism of cancer cells. Utilization of glutamine has long been thought to rival that of glucose as a substrate for cancer cell metabolism (Eagle, 1955). HeLa cells were shown to derive a significant amount of their ATP from glutamine-fueled oxidative phosphorylation and in the presence of sugars that are less efficiently metabolized through glycolysis, glutamine may function as the sole energy source (Reitzer et al., 1979). The rationale behind this may lie with the fact that glutamine is the most abundant amino acid in the free cytoplasmic pool as well as in the plasma (Eagle, 1955) and acts as the major nitrogen donor for biosynthetic reactions. Nevertheless, cells have mechanisms to couple glutamine utilization with glucose availability, possibly as a means to support efficient use of organismal resources to support individual cell proliferation.

A surge of recent studies implicate the myc oncogene in the regulation of glutamine utilization by cancer cells. Myc expression sensitizes cells to apoptosis induced by glutamine deprivation. Cells overexpressing myc exhibit enhanced levels of the glutamine transporters ASCT2 and SN2 as well as glutaminase (GLS1), the enzyme that converts glutamine to glutamate (Wise et al., 2008). The latter may be indirect, as glutaminase expression is negatively controlled by two microRNAs, miR-23a and miR-23b, which are transcriptionally repressed by myc (Gao et al., 2009).

Another glutaminase isozyme GLS2 is also a direct transcriptional target of p53 but its expression has opposing effects on tumor metabolism. GLS2 expression enhances respiration and ATP production by oxidative phosphorylation (Hu et al., 2010). In parallel, GLS2 expression suppresses oxidative stress by providing glutamate for the biosynthesis of glutathione (GSH) (Hu et al., 2010; Suzuki et al., 2010). It is unclear whether GLS1 has a similar antioxidant function, although this is possible given that both enzymes catalyze the same reaction. Nevertheless, there are clear differences as expression of GLS1 and GLS2 can have opposite effects on cell growth (Hu et al., 2010; Suzuki et al., 2010; Wise et al., 2008). Remarkably, despite efficient uptake into cells, glutamate, which can be a precursor for mitochondrial α-KG, does not rescue growth inhibition by glutamine deprivation (Eagle, 1955) further underlying the importance of GLS activity in cellular dependence on glutamine for growth.

Up to 60% of glutamine entering the TCA cycle via its products is converted to lactate in some glioblastoma cell lines (DeBerardinis et al., 2007). This involves the transport of mitochondrial malate to the cytosol and the action of malic enzyme (a.k.a. decarboxylating malate dehydrogenase) to generate pyruvate. Malic enzyme concomitantly reduces $NADP^+$ to NADPH to provide reducing power lipid biosynthesis in the same cell lines. This could be a more general mechanism for providing NADPH for anabolic processes as current evidence indicates that glucose flux through the PPP at basal cell growth conditions in culture may not provide sufficient NADPH for anabolism. It is unclear how different NADPH pools contribute differentially to various cellular processes requiring reducing power. As discussed earlier, altering glycolytic flux in response to elevated ROS may participate in diverting glucose to the PPP to acutely increase NADPH production for defenses against oxidative stress. Whether similar adaptive mechanisms coordinate malic enzyme activity with biosynthetic demands for NADPH provision is not known.

In addition to its anaplerotic role for TCA cycle metabolites in mitochondria, glutamine is involved in many metabolic pathways acting as a nitrogen donor. The side-chain nitrogen (referred to here as the γ-nitrogen) of glutamine is incorporated in the pyridine and pyrimidine moiety of nucleotide bases. As a result, following glutamine starvation cancer cells in tissue culture arrest in S-phase due, in part, to depletion of intracellular nucleotide pools. This phenotype can be rescued by providing nucleotides in the extracellular media (Gaglio et al., 2009).

Glutamine also contributes its γ-nitrogen for the synthesis of hexosamines, which are precursors for glycosylation reactions. Glycosylation of protein and lipid has an increasingly appreciated but complex role in various cell processes. Many growth factor receptors are glycosylated. Notably, flux through the hexosamine pathway has recently been shown to influence the quality of growth factor receptor glycosylation. This is part of an elaborate system that modifies cell surface exposure of glycosylated proteins, which ultimately determines proliferation-or-differentiation decisions (Lau et al., 2007). Similarly, because glycosylation is necessary for folding and surface expression of the receptors, growth signaling only occurs when hexosamine precursors are available (Wellen et al., 2010).

General considerations

Based on the evidence described above, it is clear that mitochondria in most cancer cells are functional. The direct comparison of mitochondrial function between cancer and the originating healthy tissue remains a challenge; yet, in most experimental systems, oncogenic modifications correlate with reduced oxygen consumption, a surrogate of mitochondrial activity. Given the up to 16-fold deficit in ATP yield from glycolysis versus oxidative phosphorylation (2 versus 38 ATPs per glucose as the maximum theoretical yield) the field of cancer metabolism has long been faced with an apparent paradox: why is an energetically less efficient metabolic strategy favored by cancer cells? It has been proposed that enhanced glucose uptake and glycolytic rates could compensate for ATP production. However, in the absence of a detailed balance sheet of the bioenergetic demands for proliferation, whether ATP limitation is the strategic determinant of cancer cell metabolism remains debatable. ATP is itself a potent glycolytic inhibitor, which led to the proposal that, for glycolysis to operate at the observed high rates, it is necessary to in fact limit ATP production (Racker, 1976). In fact, glucose uptake into Pten-deficient cancer cells has been shown to be limited by ATP consumption (Fang et al., 2010). In contrast, a stoichiometrically higher amount of glucose needs to be metabolized to provide sufficient reducing power and carbons for biosynthesis compared to the ATP required for the same reactions (Vander Heiden et al., 2009). This suggests that an advantage of aerobic glycolysis to cancer cells may involve balancing biosynthetic demands with ATP production allowing conditions that are optimal for cell proliferation.

Much of our current knowledge of metabolic pathways was derived by heroic experimental efforts throughout the twentieth century from the study of differentiated tissues. As a result of this, while the linear relationships between individual metabolites have been exhaustively charted, the interconnections between metabolic pathways largely reflect conditions where cells are quiescent and poised to perform specific functions in the context of differentiated tissues. Thus the intricate details of how such pathways are altered to meet the biosynthetic demands of proliferating tissues remain to be determined. Given the complexity of these networks, deciphering the functional significance of changes found at the genomic or proteomic level requires a direct look at the dynamic behavior of the metabolome. Powerful and versatile new technologies allow us to obtain such measurements in a high-throughput manner.

Analytical and computational frameworks for studying cancer metabolism

Advances in analytical chemistry originating from technological breakthroughs in mass spectrometry (MS) and nuclear magnetic resonance (NMR) spectroscopy have allowed for the simultaneous, quantitative measurements of metabolite compositions of cells, tissues, and biological fluids (Dettmer et al., 2007; Fan and Lane, 2008). The technologies are rapidly developing for applications involving both the understanding of the mechanisms of metabolic regulation as well as the development of medical diagnostics (Kind et al., 2007). These technologies require first the preparation of a metabolite extract from the biological material of interest. This involves subjecting biological samples to prescribed conditions in controlled settings and undertaking a biochemical separation process that isolates water or organic soluble small molecules, such as sugars, amino acids, lipids, and metabolic pathway intermediates (Teng et al., 2009). There is no single solvent that is suitable to analyze all metabolites in cells, but extraction conditions can be optimized to interrogate specific molecules or classes of metabolites.

NMR spectroscopy involves a quantum mechanical excitation process whereby nuclear spins are excited in applied magnetic fields (Fan and Lane, 2008). The dynamics of their relaxation provides a direct measurement of their local chemical environment that allows for, in many cases, unambiguous determination of chemical structure. Knowledge of the resonance frequencies of a set of nuclei allows for structure assignment. The most common nuclei used are protons, and ^{1}H proton NMR can routinely be used to detect specific metabolites with high accuracy and precise quantitation. In extracts derived from cells, tissue, or biological fluids, ^{1}H proton NMR can routinely distinguish 30 to 50 compounds. Despite its relatively high sensitivity, ^{1}H proton NMR and other one-dimensional NMR spectroscopy methodologies suffer from limited spectral resolution that results in substantial peak overlap when considering metabolite extracts that may contain as many as 5,000 compounds. Greater resolution of compounds in metabolite extracts can be obtained by resolving the NMR spectra across multiple dimensions. Such methods exploit quantum mechanical interactions between multiple covalently bound nuclei. With spectral resolution across multiple dimensions, a larger number of compounds can be distinguished. Multidimensional NMR can allow for separation of more compounds. These techniques, in principle, could distinguish simultaneously hundreds of metabolites. In practice, multidimensional NMR spectroscopy is limited by acquisition times (single samples take several hours to run) and the lack of readily available software to conveniently interpret the spectra for metabolite assignment and quantitation.

Mass spectrometry offers a complementary and arguably more versatile approach to metabolomics (Dettmer et al., 2007). Mass spectrometry requires ionization of each metabolite and detection of ion fragments based on mass and charge. Several different modes of ionization are used, and each mass spectrometer has advantages and disadvantages. For example, some mass spectrometry techniques use compound targeting in which a set of ion fragments are specified for each molecule of interest (Bajad et al., 2006). These techniques usually offer greater quantitative capabilities at the expense of biasing the compound search to those that have known ion fragmentation patterns. Alternatively, unbiased methods require no prior selection of fragment ions and use database matching for compound identification. These methods have traditionally been less quantitative but recent advances have dramatically improved their quantitative abilities (Lu et al., 2010). Nevertheless, because different metabolites in cells can have identical masses, additional experiments are often required to unequivocally assign a given mass to a particular metabolite. Since mass spectrometry is inherently limited by the number of ions that can be sampled over a given time window, some form of chromatography that couples directly to the mass spectrometer is required. Using chromatography to add a temporal component to the analysis can help with assignment of a given mass to an individual metabolite as conditions can be optimized to separate different molecules with identical mass. Traditionally, gas chromatography was used in metabolite analysis. Despite inclusion of a chemical derivatization step for most metabolite analyses, gas chromatography is

limited to those metabolites that can be ionized in the gas phase. This remains an extremely sensitive method to look at some metabolites and has applications today. Electrospray ionization has allowed the use of liquid chromatography in metabolomics. For many, liquid phase chromatography is the method of choice and metabolite extracts are routinely separated using both normal-phase and reverse-phase chromatography. In normal-phase chromatography, compounds bind to a non-polar substrate termed a stationary phase. Compounds are then eluted using a polar organic solvent such as acetonitrile and this is termed the mobile phase. Separation of the compounds is achieved due to differential interactions between the mobile solvent and stationary substrate. In reverse-phase chromatography the same principles apply except that the polarity of the stationary and mobile phases are reversed. Both reverse and normal phase are routinely used and have been shown to achieve good separation of metabolic intermediates such as sugars, phosphates, lipids, amino acids, and other aqueous compounds found in the biological milieu. Reverse-phase chromatography is the method of choice for separation of lipids.

While current metabolomics technologies allow for the relative quantification of metabolite levels, measurements of metabolite conversion rates (or "flux") generally require isotopic tracing. Such tracing involves the introduction of an isotopically labelled version of a metabolic compound of interest. The compound is then metabolized and its incorporation into other compounds within its pathway can be monitored. Historically, metabolite flux experiments involved the use of radioactive tracers, and these techniques still have an important role in cell metabolism research today (Vander Heiden et al., 2010).

One way to monitor flux involves monitoring the kinetics of heavy-label incorporation of the metabolite into the intermediate pathways of interest. A typical experiment involves replacing the media at time zero with a media containing an identical formula but with a nutrient of interest replaced by its corresponding isotopic compound (^{13}C glucose is the most common isotopic reagent). The kinetics of incorporation are measured by extracting metabolites at defined time points (Munger et al., 2008). Fluxes can be extrapolated from the observed kinetics of the label incorporation. If a complete reaction network is known, fluxes can be measured either by simulation of a model of time-dependent ordinary differential equations that consider small deviations from steady-state kinetics or more straightforwardly by simple curve fitting and extraction of relevant parameters.

While powerful, these approaches are technically difficult, expensive, and often time consuming. Alternative steady-state measurements of isotopic incorporation can also provide an estimate of flux. The principle is that the relative amount of label incorporation of a given metabolite gives a direct measure of the flux from the labelled compound to the metabolite of interest. A more detailed analysis involves the relative amount of label incorporation as compared to different regions of a metabolic pathway. These methods have been termed isotopomer measurements in that they involve measuring the relative incorporation of elementally identical but isotopically distinct metabolic intermediates (Schmidt et al., 1997). While there are limitations with either method, both a steady-state measurement of label incorporation and a steady-state kinetic time course provide distinct and often complementary information about metabolic flux.

An additional approach toward measuring flux that bypasses the use of tracers involves treating cells with specific stimuli and measuring the dynamics of metabolite changes. Correlating the dynamics of changes in the relative levels of metabolites at different points in metabolic pathways provides a powerful measurement of the flux changes in metabolic pathways upon a given stimulation. Nevertheless, each of these approaches requires some degree of computational modelling and data analysis. Computational and theoretical strategies for understanding metabolism are widely used and have proven invaluable. These approaches involve algorithms for finding patterns in diverse sets of metabolomics data. Principle components analysis and unsupervised hierarchical clustering are both commonly used for clustering data and can help to identify specific biomarkers or biochemical pathways that coordinate overall changes in metabolism. Other machine learning and general pattern-finding algorithms are used as well.

Kinetic models are invaluable to study the regulation of biochemical pathways (Aldridge et al., 2006). These models involve a set of coupled differential equations that are derived from phenomenological enzyme kinetics of the different steps of interest. Such models are often useful in understanding feedback and allosteric regulation. Additionally, kinetic models of metabolism have the advantage of being constrained to mass conservation. As a result, the number of equations in a metabolic system is dramatically reduced. Also, the equations can be solved in terms of their fluxes instead of relying on kinetic parameters that are required to solve these equations for metabolite concentrations. Solving steady-state kinetic equations in terms of their fluxes provides the basis of useful computational modelling strategies such as flux balance analysis (FBA). FBA is a theoretical framework by which a metabolic network is constructed that defines a stoichiometry matrix (Segre et al., 2002). A set of fluxes is then chosen to be optimized (usually a heuristic model involving a set of fluxes that defines growth) subject to the constraints of the stoichiometry matrix. Additional constraints can also be imposed. Efforts have focused on compilations of entire organism-level metabolic reaction networks. Recent work has recreated a full network of human metabolism (Duarte et al., 2007). Many predictions of lethality and synthetic-lethality have been obtained from these models (Imielinski and Belta, 2010). Other methods of deconstructing fluxes are also useful (Stelling et al., 2002).

Conclusion

Most oncogenic pathways have been shown to impact some aspect of metabolism. Metabolic changes found in cancer reflect the need of cells to modify the way extracellular nutrients are managed so that they are efficiently utilized for the accumulation of biomass necessary to support proliferation. Understanding what these changes are and why they occur could allow the design of novel therapeutic strategies.

As discussed above, a single oncogenic alteration may mediate the coordinate expression of multiple genes or implement novel regulatory mechanisms. Arguably targeting these tumor-initiating events directly could have a broader therapeutic impact than individual downstream pathways. However, many of these events are driven by transcription factors (such as HIF, myc, or p53), which have proven elusive to productive pharmacologic targeting. In contrast, metabolic enzymes are considered more easily "druggable" targets.

Further investigation into the metabolic basis of cancer and disease in general will likely yield new insights into less understood roles of small molecule metabolites in cellular processes. Metabolic intermediates may directly affect gene expression either by modulating the activity of transcription factors or by contributing material for other types of protein or DNA modifications such as acetylation and glycosylation. The impact of metabolic processes beyond the classical pathways remains a largely uncharted territory.

References

Ak, I., Stokkel, M.P., and Pauwels, E.K. (2000). Positron emission tomography with 2-[18F]fluoro-2-deoxy-D-glucose in oncology. Part II. The clinical value in detecting and staging primary tumours. *J Cancer Res Clin Oncol 126*, 560–574.

Aldridge, B.B., Burke, J.M., Lauffenburger, D.A., and Sorger, P.K. (2006). Physicochemical modelling of cell signalling pathways. *Nature Cell Biol 8*, 1195–1203.

Bajad, S.U., Lu, W.Y., Kimball, E.H., et al. (2006). Separation and quantitation of water soluble cellular metabolites by hydrophilic interaction chromatography-tandem mass spectrometry. *J Chromatography A 1125*, 76–88.

Baysal, B.E., Ferrell, R.E., Willett-Brozick, J.E., et al. (2000). Mutations in SDHD, a mitochondrial complex II gene, in hereditary paraganglioma. *Science 287*, 848–851.

Bensaad, K., Tsuruta, A., Selak, M.A., et al. (2006). TIGAR, a p53-inducible regulator of glycolysis and apoptosis. *Cell 126*, 107–120.

Bonnet, S., Archer, S.L., Allalunis-Turner, J., et al. (2007). A mitochondria-K+ channel axis is suppressed in cancer and its normalization promotes apoptosis and inhibits cancer growth. *Cancer Cell 11*, 37–51.

Burk, D., and Schade, A.L. (1956). On respiratory impairment in cancer cells. *Science 124*, 270–272.

Busuttil, R.A., Rubio, M., Dolle, M.E., Campisi, J., and Vijg, J. (2003). Oxygen accelerates the accumulation of mutations during the senescence and immortalization of murine cells in culture. *Aging Cell 2*, 287–294.

Calvo, M.B., Figueroa, A., Pulido, E.G., Campelo, R.G., and Aparicio, L.A. (2010). Potential role of sugar transporters in cancer and their relationship with anticancer therapy. *Int J Endocrinol.*

Christofk, H.R., Vander Heiden, M.G., Harris, M.H., et al. (2008a). The M2 splice isoform of pyruvate kinase is important for cancer metabolism and tumour growth. *Nature 452*, 230–233.

Christofk, H.R., Vander Heiden, M.G., Wu, N., Asara, J.M., and Cantley, L.C. (2008b). Pyruvate kinase M2 is a phosphotyrosine-binding protein. *Nature 452*, 181–186.

Clower, C.V., Chatterjee, D., Wang, Z., et al. (2010). The alternative splicing repressors hnRNP A1/A2 and PTB influence pyruvate kinase isoform expression and cell metabolism. *Proc Natl Acad Sci USA 107*, 1894–1899.

Coller, H.A., Grandori, C., Tamayo, P., et al. (2000). Expression analysis with oligonucleotide microarrays reveals that MYC regulates genes involved in growth, cell cycle, signaling, and adhesion. *Proc Natl Acad Sci USA 97*, 3260–3265.

Dang, L., White, D.W., Gross, S., et al. (2009). Cancer-associated IDII1 mutations produce 2-hydroxyglutarate. *Nature 462*, 739–744.

David, C.J., Chen, M., Assanah, M., Canoll, P., and Manley, J.L. (2010). HnRNP proteins controlled by c-Myc deregulate pyruvate kinase mRNA splicing in cancer. *Nature 463*, 364–368.

DeBerardinis, R.J., Mancuso, A., Daikhin, E., et al. (2007). Beyond aerobic glycolysis: transformed cells can engage in glutamine metabolism that exceeds the requirement for protein and nucleotide synthesis. *Proc Natl Acad Sci USA 104*, 19345–19350.

DeRisi, J.L., Iyer, V.R., and Brown, P.O. (1997). Exploring the metabolic and genetic control of gene expression on a genomic scale. *Science 278*, 680–686.

Dettmer, K., Aronov, P.A., and Hammock, B.D. (2007). Mass spectrometry-based metabolomics. *Mass Spectrometry Rev 26*, 51–78.

Duarte, N.C., Becker, S.A., Jamshidi, N., et al. (2007). Global reconstruction of the human metabolic network based on genomic and bibliomic data. *Proc Natl Acad Sci USA 104*, 1777–1782.

Eagle, H. (1955). Nutrition needs of mammalian cells in tissue culture. *Science 122*, 501–514.

Eigenbrodt, E., Reinacher, M., Scheefers-Borchel, U., Scheefers, H., and Friis, R. (1992). Double role for pyruvate kinase type M2 in the expansion of phosphometabolite pools found in tumor cells. *Crit Rev Oncog 3*, 91–115.

Elstrom, R.L., Bauer, D.E., Buzzai, M., et al. (2004). Akt stimulates aerobic glycolysis in cancer cells. *Cancer Res 64*, 3892–3899.

Engelman, J.A., Luo, J., and Cantley, L.C. (2006). The evolution of phosphatidylinositol 3-kinases as regulators of growth and metabolism. *Nat Rev Genet 7*, 606–619.

Fan, T.W.M., and Lane, A.N. (2008). Structure-based profiling of metabolites and isotopomers by NMR. *Prog Nucl Magn Res Spectroscopy 52*, 69–117.

Fang, M., Shen, Z., Huang, S., et al. (2010). The ER UDPase ENTPD5 promotes protein N-glycosylation, the Warburg effect, and proliferation in the PTEN pathway. *Cell 143*, 711–724.

Gaglio, D., Soldati, C., Vanoni, M., Alberghina, L., and Chiaradonna, F. (2009). Glutamine deprivation induces abortive S-phase rescued by deoxyribonucleotides in k-ras transformed fibroblasts. *PLoS One 4*, e4715.

Gao, P., Tchernyshyov, I., Chang, T.C., et al. (2009). c-Myc suppression of miR-23a/b enhances mitochondrial glutaminase expression and glutamine metabolism. *Nature 458*, 762–765.

Gordan, J.D., Thompson, C.B., and Simon, M.C. (2007). HIF and c-Myc: sibling rivals for control of cancer cell metabolism and proliferation. *Cancer Cell 12*, 108–113.

Gottlob, K., Majewski, N., Kennedy, S., et al. (2001). Inhibition of early apoptotic events by Akt/PKB is dependent on the first committed step of glycolysis and mitochondrial hexokinase. *Genes Dev 15*, 1406–1418.

Gough, D.J., Corlett, A., Schlessinger, K., et al. (2009). Mitochondrial STAT3 supports Ras-dependent oncogenic transformation. *Science 324*, 1713–1716.

Hancock, J.T., Desikan, R., and Neill, S.J. (2001). Role of reactive oxygen species in cell signalling pathways. *Biochem Soc Trans 29*, 345–350.

Hitosugi, T., Kang, S., Vander Heiden, M.G., et al. (2009). Tyrosine phosphorylation inhibits PKM2 to promote the Warburg effect and tumor growth. *Sci Signal 2*, ra73.

Holness, M.J., and Sugden, M.C. (2003). Regulation of pyruvate dehydrogenase complex activity by reversible phosphorylation. *Biochem Soc Trans 31*, 1143–1151.

Hu, C.J., Wang, L.Y., Chodosh, L.A., Keith, B., and Simon, M.C. (2003). Differential roles of hypoxia-inducible factor 1alpha (HIF-1alpha) and HIF-2alpha in hypoxic

gene regulation. *Mol Cell Biol 23*, 9361–9374.

Hu, W., Zhang, C., Wu, R., et al. (2010). Glutaminase 2, a novel p53 target gene regulating energy metabolism and antioxidant function. *Proc Natl Acad Sci USA 107*, 7455–7460.

Imamura, K., and Tanaka, T. (1982). Pyruvate kinase isozymes from rat. *Methods Enzymol 90Pt E*, 150–165.

Imielinski, M., and Belta, C. (2010). Deep epistasis in human metabolism. *Chaos 20*, 12.

Irani, K., Xia, Y., Zweier, J.L., et al. (1997). Mitogenic signaling mediated by oxidants in Ras-transformed fibroblasts. *Science 275*, 1649–1652.

Kim, J.W., Tchernyshyov, I., Semenza, G.L., and Dang, C.V. (2006). HIF-1-mediated expression of pyruvate dehydrogenase kinase: a metabolic switch required for cellular adaptation to hypoxia. *Cell Metab 3*, 177–185.

Kind, T., Tolstikov, V., Fiehn, O., and Weiss, R.H. (2007). A comprehensive urinary metabolomic approach for identifying kidney cancer. *Anal Biochem 363*, 185–195.

King, A., Selak, M.A., and Gottlieb, E. (2006). Succinate dehydrogenase and fumarate hydratase: linking mitochondrial dysfunction and cancer. *Oncogene 25*, 4675–4682.

Langbein, S., Zerilli, M., Zur Hausen, A., et al. (2006). Expression of transketolase TKTL1 predicts colon and urothelial cancer patient survival: Warburg effect reinterpreted. *Br J Cancer 94*, 578–585.

Lau, K.S., Partridge, E.A., Grigorian, A., et al. (2007). Complex N-glycan number and degree of branching cooperate to regulate cell proliferation and differentiation. *Cell 129*, 123–134.

Locasale, J.W., Cantley, L.C., and Vander Heiden, M.G. (2009).

Cancer's insatiable appetite. *Nat Biotechnol 27*, 916–917.

Lu, W.Y., Clasquin, M.F., Melamud, E., et al. (2010). Metabolomic analysis via reversed-phase ion-pairing liquid chromatography coupled to a stand alone orbitrap mass spectrometer. *Anal Chem 82*, 3212–3221.

Medes, G., Thomas, A., and Weinhouse, S. (1953). Metabolism of neoplastic tissue. IV. A study of lipid synthesis in neoplastic tissue slices in vitro. *Cancer Res 13*, 27–29.

Menendez, J.A., and Lupu, R. (2007). Fatty acid synthase and the lipogenic phenotype in cancer pathogenesis. *Nat Rev Cancer 7*, 763–777.

Michelakis, E.D., Sutendra, G., Dromparis, P., et al. (2010). Metabolic modulation of glioblastoma with dichloroacetate. *Sci Transl Med 2*, 31–34.

Michelakis, E.D., Webster, L., and Mackey, J.R. (2008). Dichloroacetate (DCA) as a potential metabolic-targeting therapy for cancer. *Br J Cancer 99*, 989–994.

Moller, D.E., and Kaufman, K.D. (2005). Metabolic syndrome: a clinical and molecular perspective. *Annu Rev Med 56*, 45–62.

Munger, J., Bennett, B.D., Parikh, A., et al. (2008). Systems-level metabolic flux profiling identifies fatty acid synthesis as a target for antiviral therapy. *Nat Biotechnol 26*, 1179–1186.

Noguchi, T., Inoue, H., and Tanaka, T. (1986). The M1- and M2-type isozymes of rat pyruvate kinase are produced from the same gene by alternative RNA splicing. *J Biol Chem 261*, 13807–13812.

Nogueira, V., Park, Y., Chen, C.C., et al. (2008). Akt determines replicative senescence and oxidative or oncogenic premature senescence and sensitizes cells to oxidative apoptosis. *Cancer Cell 14*, 458–470.

O'Connell, B.C., Cheung, A.F., Simkevich, C.P., et al. (2003). A large scale genetic analysis of c-Myc-regulated gene expression patterns. *J Biol Chem 278*, 12563–12573.

Osthus, R.C., Shim, H., Kim, S., et al. (2000). Deregulation of glucose transporter 1 and glycolytic gene expression by c-Myc. *J Biol Chem 275*, 21797–21800.

Pandolfi, P.P., Sonati, F., Rivi, R., et al. (1995). Targeted disruption of the housekeeping gene encoding glucose 6-phosphate dehydrogenase (G6PD): G6PD is dispensable for pentose synthesis but essential for defense against oxidative stress. *EMBO J 14*, 5209–5215.

Parlo, R.A., and Coleman, P.S. (1984). Enhanced rate of citrate export from cholesterol-rich hepatoma mitochondria. The truncated Krebs cycle and other metabolic ramifications of mitochondrial membrane cholesterol. *J Biol Chem 259*, 9997–10003.

Parsons, D.W., Jones, S., Zhang, X., et al. (2008). An integrated genomic analysis of human glioblastoma multiforme. *Science 321*, 1807–1812.

Plas, D.R., Talapatra, S., Edinger, A.L., Rathmell, J.C., and Thompson, C.B. (2001). Akt and Bcl-xL promote growth factor-independent survival through distinct effects on mitochondrial physiology. *J Biol Chem 276*, 12041–12048.

Racker, E. (1974). History of the Pasteur effect and its pathobiology. *Mol Cell Biochem 5*, 17–23.

Racker, E. (1976). Why do tumor cells have a high aerobic glycolysis? *J Cell Physiol 89*, 697–700.

Ralser, M., Wamelink, M.M., Kowald, A., et al. (2007). Dynamic rerouting of the carbohydrate flux is key to counteracting oxidative stress. *J Biol 6*, 10.

Reitzer, L.J., Wice, B.M., and Kennell, D. (1979). Evidence that glutamine, not sugar, is the major energy source for cultured HeLa cells. *J Biol Chem 254*, 2669–2676.

Reitzer, L.J., Wice, B.M., and Kennell, D. (1980). The pentose cycle. Control and essential function in HeLa cell nucleic acid synthesis. *J Biol Chem 255*, 5616–5626.

Sablina, A.A., Budanov, A.V., Ilyinskaya, G.V., et al. (2005). The antioxidant function of the p53 tumor suppressor. *Nat Med 11*, 1306–1313.

Schafer, Z.T., Grassian, A.R., Song, L., et al. (2009). Antioxidant and oncogene rescue of metabolic defects caused by loss of matrix attachment. *Nature 461*, 109–113.

Schmidt, K., Carlsen, M., Nielsen, J., and Villadsen, J. (1997). Modeling isotopomer distributions in biochemical networks using isotopomer mapping matrices. *Biotechnol Bioengin 55*, 831–840.

Schwartzenberg-Bar-Yoseph, F., Armoni, M., and Karnieli, E. (2004). The tumor suppressor p53 down-regulates glucose transporters GLUT1 and GLUT4 gene expression. *Cancer Res 64*, 2627–2633.

Segre, D., Vitkup, D., and Church, G.M. (2002). Analysis of optimality in natural and perturbed metabolic networks. *Proc Natl Acad Sci USA 99*, 15112–15117.

Semenza, G.L. (2003). Targeting HIF-1 for cancer therapy. *Nat Rev Cancer 3*, 721–732.

Snell, K., Natsumeda, Y., Eble, J.N., Glover, J.L., and Weber, G. (1988). Enzymic imbalance in serine metabolism in human colon carcinoma and rat sarcoma. *Br J Cancer 57*, 87–90.

Stelling, J., Klamt, S., Bettenbrock, K., Schuster, S., and Gilles, E.D. (2002). Metabolic network structure determines key aspects of functionality and regulation. *Nature 420*, 190–193.

Suzuki, S., Tanaka, T., Poyurovsky, M.V., et al. (2010). Phosphate-activated glutaminase (GLS2), a p53-inducible regulator of glutamine metabolism and reactive oxygen species. *Proc Natl Acad Sci USA 107*, 7461–7466.

Teng, Q., Huang, W.L., Collette, T.W., Ekman, D.R., and Tan, C. (2009). A direct cell quenching method for cell-culture based metabolomics. *Metabolomics 5*, 199–208.

Tomlinson, I.P., Alam, N.A., Rowan, A.J., et al. (2002). Germline mutations in FH predispose to dominantly inherited uterine fibroids, skin leiomyomata and papillary renal cell cancer. *Nat Genet 30*, 406–410.

Trachootham, D., Alexandre, J., and Huang, P. (2009). Targeting cancer cells by ROS-mediated mechanisms: a radical therapeutic approach? *Nat Rev Drug Discov 8*, 579–591.

Trachootham, D., Zhou, Y., Zhang, H., et al. (2006). Selective killing of oncogenically transformed cells through a ROS-mediated mechanism by beta-phenylethyl isothiocyanate. *Cancer Cell 10*, 241–252.

Tuttle, S.W., Maity, A., Oprysko, P.R., et al. (2007). Detection of reactive oxygen species via endogenous oxidative pentose phosphate cycle activity in response to oxygen concentration: implications for the mechanism of HIF-1alpha stabilization under moderate hypoxia. *J Biol Chem 282*, 36790–36796.

Vander Heiden, M.G., Cantley, L.C., and Thompson, C.B. (2009). Understanding the Warburg effect: the metabolic requirements of cell proliferation. *Science 324*, 1029–1033.

Vander Heiden, M.G., Locasale, J.W., Swanson, K.D., et al. (2010). Evidence for an alternative glycolytic pathway in rapidly proliferating cells. *Science 329*, 1492–1499.

Vander Heiden, M.G., Plas, D.R., Rathmell, J.C., et al. (2001). Growth factors can influence cell growth and survival through effects on glucose metabolism. *Mol Cell Biol 21*, 5899–5912.

Warburg, O. (1956a). On respiratory impairment in cancer cells. *Science 124*, 269–270.

Warburg, O. (1956b). On the origin of cancer cells. *Science 123*, 309–314.

Weinberg, F., and Chandel, N.S. (2009). Mitochondrial metabolism and cancer. *Ann NY Acad Sci 1177*, 66–73.

Weinberg, F., Hamanaka, R., Wheaton, W.W., et al. (2010). Mitochondrial metabolism and ROS generation are essential for Kras-mediated tumorigenicity. *Proc Natl Acad Sci USA 107*, 8788–8793.

Weinhouse, S. (1956). On respiratory impairment in cancer cells. *Science 124*, 267–269.

Weinhouse, S. (1976). The Warburg hypothesis fifty years later. *Z Krebsforsch Klin Onkol Cancer Res Clin Oncol 87*, 115–126.

Wellen, K.E., Hatzivassiliou, G., Sachdeva, U.M., et al. (2009). ATP-citrate lyase links cellular metabolism to histone acetylation. *Science 324*, 1076–1080.

Wellen, K.E., Lu, C., Mancuso, A., et al. (2010). The hexosamine biosynthetic pathway couples growth factor-induced glutamine uptake to glucose metabolism. *Genes Dev 24*, 2784–2799.

Wise, D.R., DeBerardinis, R.J., Mancuso, A., et al. (2008). Myc regulates a transcriptional program that stimulates mitochondrial glutaminolysis and leads to glutamine addiction. *Proc Natl Acad Sci USA 105*, 18782–18787.

Yan, H., Parsons, D.W., Jin, G., et al. (2009). IDH1 and IDH2 mutations in gliomas. *N Engl J Med 360*, 765–773.

Zhang, S., Yang, J.H., Guo, C.K., and Cai, P.C. (2007). Gene silencing of TKTL1 by RNAi inhibits cell proliferation in human hepatoma cells. *Cancer Lett 253*, 108–114.

Chapter 20

Tumor microenvironment: blood vascular system in cancer metastasis

Shantibhusan Senapati, Rakesh K. Singh and Surinder K. Batra

Introduction

Despite improvements in diagnosis and other therapeutic methods for cancer treatment, mortality has increased only by a few months at best. It is a well-known fact that most cancer patients die due to metastasis. Different factors that contribute to the severity of metastasis include: (a) occurrence of metastasis before diagnosis of the primary tumor; (b) multiple localization of metastatic tumors; (c) biological heterogeneity of the primary neoplasm and metastatic tumors; and (d) the negative influence of the microenvironment on the response of a metastatic tumor cell to systemic therapy.

For the past decade, various studies have shown the role of tumor microenvironments in cancer progression. Now it is well established that metastasis is an outcome of cancer cell adaptation to its microenvironment. Metastatic cancer cells have properties that enable them not only to adapt to an alien environment, but also to modify it in such a way that it becomes conducive for the establishment and continuous growth of the secondary tumor. Importantly, the tumor microenvironment, which is genetically more stable than tumor cells, is an attractive target for therapeutic purposes. Therefore only a better understanding of the tumor microenvironment will lead to improvements in the design of a more effective therapy for cancer metastasis.

Metastasis to distant organs requires a series of coordinated steps. Out of these steps, dissemination of tumor cells from the primary tumor site is a crucial step, which is carried out through different biological routes. From the primary tumor site, tumor cells may disseminate through three major routes: lymphatic vessels (lymphatic metastasis), blood vessels (hematogenous metastasis), and serosal surfaces (transcoelomic metastasis). The vascular systems involved in the process of metastasis are not just channels that transport the circulating tumor cells from one place to other; rather they provide a microenvironment that has significant roles in tumor cell survival, proliferation, and establishment in different organs. Out of the blood and lymphatic vascular systems, the blood vascular system plays a major role in distant organ metastasis. In many cases, tumor cells that disseminated into the lymphatic system channeled into the blood vascular system before producing secondary metastasis. In this chapter, realizing the paramount significance of the blood vascular system/microenvironment in tumor cell metastasis, we discuss the role of the blood vascular microenvironment in tumor cell metastasis and highlight the therapeutic implications.

Blood vascular system in the pathogenesis of cancer metastasis

Blood vascular system in the spread of cancer cells

The metastatic process consists of a cascade of sequential events [1, 2]. The major steps in metastasis are as follows. (a) After the transformation of normal epithelial cells into a carcinoma, initially the continuous growth of the tumor mass depends on the diffused oxygen and nutrients from nearby blood vessels. (b) For further growth of the tumor mass (beyond 1–2 mm diameter) angiogenesis occurs to establish a new capillary network from the existing surrounding vasculature. (c) Tumor cells become more invasive by different molecular mechanisms, such as the loss of cell adhesion junctions followed by basement membrane degradation and local invasion. (d) The invasive tumor cells invade the surrounding stroma, migrate and intravasate into blood or lymph vessels. (e) Intravasation of tumor cells may occur as a single cell or a group of cells (emboli), but the vast majority of tumor cells are destroyed in the circulation. (f) The circulating tumor cells that aggregate with host cells and survive in the circulation get lodged in the capillary beds of the organs, either by the interaction with endothelial cells or by adhesion to exposed sub-endothelial basement membranes. (g) The arrested tumor cells can proliferate intravascularly or extravasate. (h) The extravasated tumor cells initiate and maintain growth to form pre-angiogenic micrometastasis. Further neo-angiogenesis must occur to sustain the growth and formation of macrometastasis. (i) Tumor cells from an established metastatic tumor may enter into the blood vessel and produce other metastases by a process that is known as metastasis of metastases.

Tumors that are <1 to 2 mm in diameter can receive all their nutrients by diffusion, but further growth depends on the development of an adequate blood supply [1]. The prevascular stage of a tumor is associated with local, non-invasive, benign tumors, whereas the vascular stage is associated with aggressive, invasive, and metastatic tumors [3]. The extent of neovascularization in different malignancies, such as melanoma, breast carcinoma, and prostate cancer, is correlated with their potential for invasion and metastasis [4, 5]. Indeed, histopathological analyses suggest that the number of blood vessels in invasive breast carcinoma of patients predicts the capacity of the tumors to produce metastasis. Taken together, the blood vascular system plays a major role in the progression of cancer, from the initiation of cancer at the primary site to metastasis and the establishment of secondary tumors at the distant organs.

Blood vascular system in organ-specific metastasis

Repeated observations by different investigators (Table 20.1) have shown that metastases are not found randomly in the body [6]. To explain this phenomenon, two hypotheses, known as the anatomical and the seed and soil hypothesis of metastatic spread, have been put forward. The anatomical hypothesis argues that dissemination of metastatic tumor cells occurs by mechanical factors, and that "anatomical arrangement of the vascular system at target organ" is the key determinate factor for organ-specific metastasis. But,

Table 20.1 Organ-specific metastasis.

Cancer type	Site of primary tumor	Site of secondary metastasis
Prostate cancer	prostate	bone
Breast cancer	breast	bone, brain, adrenal, lung, liver
Clear-cell carcinoma	kidney	lung, bone, adrenal
Gastrointestinal cancer	stomach/ intestine	liver
Small-cell carcinoma	lung	brain, liver,
Melanoma	skin	liver, brain, bowel
Melanoma	eye	liver
Neoblastoma		liver, adrenal
Follicular carcinoma	thyroid	bone, lung
Ovarian cancer	ovary	peritoneum, liver, lung
Pancreatic cancer	pancreas	liver, lung
Colorectal	colon	liver

the seed and soil hypothesis, on the other hand, argues that the environment surrounding the tumor cells (the organ microenvironment) is the most important factor for the growth of tumor cells. Also, the outcome of metastasis is determined by factors that can be independent of the vascular anatomy, rate of blood flow, physical presence of tumor cells, and the number of tumor cells delivered to each organ [7]. A century ago, Paget questioned whether the distribution of metastases was due to chance. He analyzed the autopsies of a large number of patients with metastatic neoplasms [8]. The non-random pattern of visceral metastases suggested that the process of metastasis is not due to chance but rather that certain tumor cells (the seeds) have a specific affinity for certain organs (the soil) [8]. Metastasis results only when seed and soil are matched. In recent years, Paget's hypothesis has received considerable experimental and clinical support [3, 6]. Site-specific metastasis occurs with many transplantable experimental tumors and has been reported in autochthonous human tumors in patients with peritoneovenous shunts [9]. To date there is no convincing reason to consider the two hypotheses to be mutually exclusive. In fact, the current evidence supports a role for both factors [10]. As far as the role of the blood vascular system in organ-specific metastasis is concerned, it may be a key determinant factor in both of the possible ways.

In support of the anatomical hypothesis, *in vivo* video microscopy and other analysis indicate that the lungs and the liver are the two major organs where circulating tumor cells are mostly entrapped due to size restrictions [11]. Factors such as the relative size of circulating tumor cells and the capillaries, the blood pressure in the organ, and the deformity of the cells can be the determinant factors for this mechanical entrapment of tumor cells. Studies support the fact that a clump of cells would have a greater chance of lodging than a single cell. Tumor cell clumps may also provide a degree of protection to some of the enclosed cells. Mechanical trapping of tumor cells may cause endothelial injury. Additionally, mechanical entrapment by plugging of a blood vessel may also lead to stasis and this could induce clot formation. Some tumor cells may be associated with the adherence of platelets and deposition of fibrin, even if only temporarily. Clot formation may be enhanced by the secretion of fibrinolytic substances such as plasminogen activator. In this form of adhesion, lodgement results from molecular interactions that are not exclusive to the adhering surfaces.

Although lodgement can undoubtedly result from mechanical entrapment, there is now considerable evidence showing that some tumor cells may adhere selectively to blood vessel endothelium in a similar way to the mechanisms employed by leukocytes. Studies have shown that when endothelium was activated with cytokine (IL-1), cancer cells also adhered to pre-capillary vessels (portal venules) of the liver, having larger size lumens than the tumor cell diameter [12]. Therefore it is practical to propose that organ-specific adhesive interactions are organ-specific signaling dependent, rather than a matter of just physical arrest of tumor cells at secondary sites. At the same time, experimental evidence also indicates that cancer cells do not usually arrest by receptor-mediated leukocyte-like adhesive interactions with the endothelium [13]. Therefore the mechanism through which tumor cells lodge at the specific target organs may be cancer specific, and along with the blood vascular systems, many other factors may be involved in the homing and growth of tumor cells at specific organs.

The first close interaction between circulating malignant cells and their target occurs at the level of the microvasculature, where tumor cells will contact the capillary endothelium. Capillary endothelial cells derived from different organs are not alike, and organ-specific patterns of cell surface glycoproteins and ligands have been identified on various organ endothelia and endothelial cells [14–18]. Experimental evidence suggests an active role for vascular endothelial cells in directing the organ preference of metastasis [15, 18]. Auerbach and colleagues [19, 20] have shown that different malignant cell lines preferentially attach to endothelial cells of their *in vivo* target organs. Similar findings have been reported from other laboratories in a detailed study with different murine cell lines selected for their organ-specific homing to the lungs, liver, and brain [16, 21, 22]. Several mechanisms for the adhesion of malignant cells to organ-specific endothelial cells and eventual metastases have been proposed, including the role of local components of the extracellular matrix [15, 18, 23], ICAM-1 [19, 20], and the expression of variant forms of membrane glycoprotein CD44 [24]. In addition to well-characterized anatomical diversity *in situ*, specific differences are increasingly being recognized between surface antigens on endothelial cells from different tissues, including the absence of the classic endothelial marker factor VIII-related antigen (von Willebrand factor) from many endothelial cells [25–27]. Microvascular heterogeneity extends to properties of endothelial cells thought to be involved in tumor angiogenesis and metastasis, such as growth factor responsiveness and the expression of cell adhesion molecules [25, 27]. These findings are not only of relevance to the unambiguous identification and characterization of cultured endothelial cells, but also may explain the phenomenon of preferential organ tumor metastasis and provide novel opportunities for therapy.

Blood vascular system associated factors/events that play a role in cancer metastasis

Vascular pathways

In a human body, deoxygenated blood from most of the organs is carried to the heart by the venous system and passes through the lungs for oxygenation. It then returns to the heart and is circulated to all organs of the body by the systemic arterial system. From the aorta, blood goes through arteries of narrowing diameter, finally emptying into the capillary bed of different organs. The vascular pathways play a significant role in determining organ-specific metastasis. Tumor cells from the primary site gain access to the vena cava either through venules (small vessels) or via the lymphatic system, which drains into the venous blood system. Metastatic tumor cells, downstream to their entry into the vascular system, get trapped in the first capillary network they encounter. For example, breast cancer cells that escape from the primary site of the tumor and intravasate into the blood circulation will be circulated through the heart to the lungs, where many will be trapped in the capillaries. Those cells that manage to escape through the lung capillaries enter into the systemic arterial circulation and are taken to the capillary beds present in other organs of the body. Some breast cancer cells from the primary tumor site may invade the adjacent lymphatics and get carried to the nearby draining lymph node, where these cells might grow. For distant organ metastasis, cells in the lymphatic system enter into the blood circulation and are transported to metastatic sites such as the bone, liver, brain, and lungs.

On the other hand, cancer cells from the splanchnic organs, such as the intestines, first pass through the liver before entering into the venous system. For example, in the case of colon cancer, cancer cells living in the primary tumor would enter the hepatic-portal system and would be taken to the capillaries of the liver. Those cells that escape the liver capillary beds enter into the venous blood circulation and are transported to the heart, followed by lungs. Some tumor cells get trapped in the lung capillaries, and those that escape enter into the systemic blood circulation and get transported to the other organs. Therefore the homing of cancer cells, preferentially to specific organs, depends on the blood-flow pattern.

Cellular and non-cellular components of blood

The circulating tumor cells may interact with host vascular components such as lymphocytes, monocytes, and platelets. This interaction may promote the formation of large tumor cell clumps/emboli and this may enhance the chance of cancer metastasis. Various studies also support a good correlation between the size of the tumor cell clump, the rate of

arrest, and the probability of secondary tumor formation [28]. Furthermore, accumulating evidence suggests the role of the cell-associated tissue factor and circulating homeostatic factors to increase the hematogenous metastatic potential of tumor cells [29]. On the other hand, many studies argue that those host cells that adhere to tumor cells in the circulation may inhibit the incidence of metastasis by promoting tumor cell destruction. These conflicting mechanisms are yet to be explored and the extent to which these two processes contribute to the establishment of metastasis remains to be established.

Cellular components

Among the cellular components of blood, platelets and leukocytes play a major role in hematogenous metastasis.

Platelets

Experimental evidence has shown that tumor cells can selectively lodge in areas of endothelial damage [30]. However, under less damage of the endothelium, lodgement involving indirect interaction via platelet plug or thrombus seem to be favored over direct interaction between the tumor cell and the basement membrane [31]. A series of experiments showed that platelets, a major component of the coagulation process, contribute significantly to the initial arrest of tumor cells in organs. The most convincing evidence is the inhibition of metastasis by experimental platelet depletion and the re-establishment of metastasis after platelet repletion [32, 33]. Studies also indicate that inhibition of platelet function may be beneficial for prophylaxis against metastasis [34]. Platelets may contribute to metastasis by stabilizing tumor cell arrest in the vasculature or by many other mechanisms such as promoting tumor cell proliferation, survival, secondary tethering to endothelium (adhesion of circulating tumor cells to an already attached tumor cell), and extravasations [35–38]. The interaction of P-selectin present on the surface of platelets with its ligands (mucin and non-mucin type) present on the surface of the tumor is mostly responsible for the formation of microemboli that facilitates their arrest in the vasculature [39].

Leukocytes

L-selectin is constitutively expressed on the surface of leukocytes. Its role in leukocyte rolling along the vessel wall is well established. Although the significance of platelets to metastasis has been confirmed in several experimental settings, for a long period of time the role and mechanism of leukocyte involvement in tumor metastasis remained largely unknown. Later on, it was found that L-selectin deficiency attenuates metastasis. Although the exact mechanism is not known, experimental evidence indicates that L-selectin enhances metastasis by recruiting leukocytes to platelet–tumor emboli in the circulation [39]. The increased size of the emboli may facilitate their mechanical trapping in the microvasculature. Additionally, it has been suggested that leukocytes could bridge the emboli to the vascular endothelial cells expressing L-selectin ligands.

Non-cellular components of blood

It has been apparent for many years that the coagulation system can enhance the colonization of circulating tumor cells at distant organs. Coagulation does not only help the adhesion of tumor cells on the surface of the endothelium, it also helps in the spreading of tumor cells on the vascular surface, which is a crucial component of the metastasis cascade [40]. A plethora of evidence has shown that many non-cellular components of the coagulation process also influence metastasis. For example, both experimental and spontaneous metastases are reduced in fibrinogen-deficient mice, with no effect on tumor growth [41, 42]. At many other places, the role of factors such as plasminogen activator, tissue factor, and thrombin have also been addressed as contributing factors for metastasis [43, 44].

Blood vessels

Blood vessels, which consist of endothelial cells, perivascular extracellular matrix (ECM), and pericytes, can contribute to metastasis in different ways. Angiogenesis, the process of blood vessel formation, itself is a key event for the establishment of tumor metastasis. During the process of normal angiogenesis, newly formed blood vessels become stabilized through the recruitment of vascular mural cells (VSMC/pericytes) and by the formation of perivacular ECM, which includes the vascular basement membrane. In solid tumors, the neovasculature is structurally and functionally abnormal. These vessels are leaky, tortuous, dilated, and have a disorganized pattern of interconnections [45]. Additionally, the tumor blood vessels have abnormalities such as endothelial cells with deformed morphology, the absence or loose attachment of pericytes with the endothelial cells, and the abnormal basement membrane (Figure 20.1). These vascular abnormalities create a conducive tumor microenvironment having elevated hydrostatic pressure outside the blood vessels, hypoxia, and acidosis. Furthermore, the endothelial cells, pericytes, and basement membrane also play major roles in the adhesion and extravasation of circulating tumor cells.

Neovascularization/angiogenesis

Angiogenesis – the growth of new vessels from pre-existing, large vessels – plays an important role in development, reproduction, and repair. The rate of blood vessel growth is very high during the normal development of an embryo. However, in the adult, the vasculature is generally quiescent. Physiological angiogenesis, which is distinct from arteriogenesis and lymphangeogenesis, is usually localized/focal and well regulated. But pathological angiogenesis is a de-regulated process and can persist for years. Pathological angiogenesis plays a significant role in the progression of various cancers. An extensive body of evidence generated over a period of several decades has concluded that angiogenesis is the primary compensatory mechanism through which tumor cells compensate for the metabolic pressure on them. Tumor cells need nutrients to grow. Data from cancer metastasis signify that metastatic cells residing beyond 150 μm from a vessel undergo a programmed cell death. These observations are further supported by the fact that the diffusion coefficient of oxygen in tumor tissue is approximately 120 μm. Therefore the expansion of a small tumor mass residing just within the range of availability of oxygen requires an additional vascular density. Additionally, experimental and clinical evidence also suggest that spreading of tumor cells is angiogenesis dependent. The neovasculature of tumors tends to be fenestrated and leaky. Therefore the burgeoning blood vessels can provide an escape route by which metastatic tumor cells can intravasate more efficiently than non-metastatic tumors [46]. Further, angiogenic factors from tumors, such as bFGF and VPF/VEGF, induce an increased production of plasminogen activator and collagenases in proliferating endothelial cells, thus promoting the degradation of the vascular basement membrane.

To exploit the angiogenesis event for anticancer therapy, it is essential to have a clear understanding of

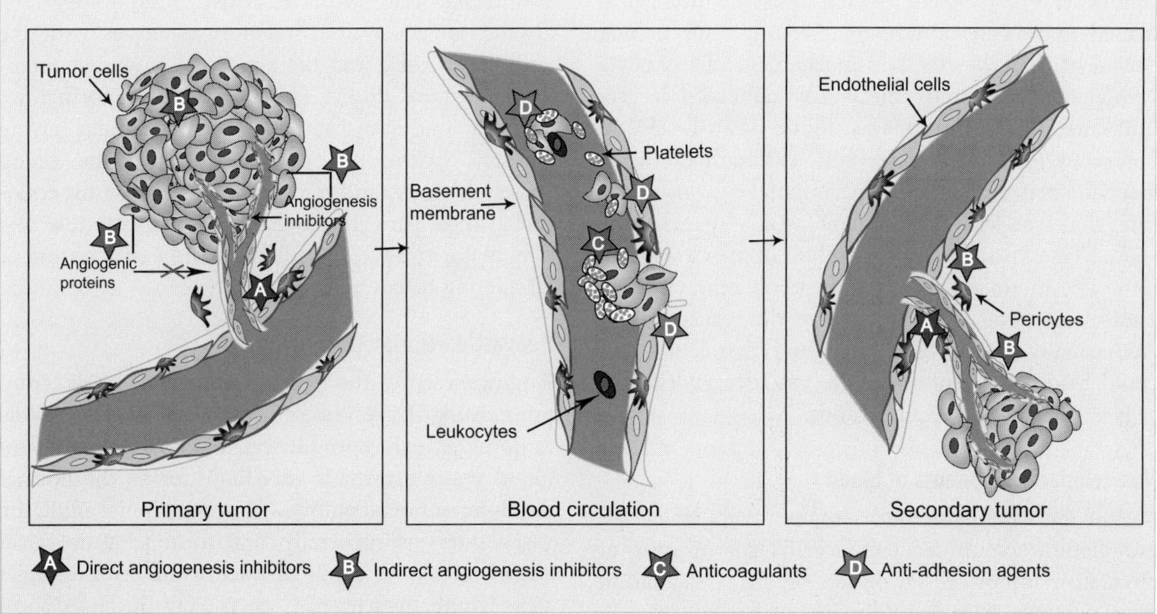

Figure 20.1 Hematogenous metastasis of cancer is a multi-step process. To combat hematogenous metastasis, overwhelming clinical and experimental findings provide a rationale for combination therapy. Anti-angiogenesis drugs (direct and indirect) can help to restrict the growth of primary and secondary tumors, and also can inhibit dissemination of tumor cells to distant organs. Direct anti-angiogenesis inhibitors (A) are used to target endothelial cells and/or pericytes, which are complementary targets to inhibit angiogenesis. On the other hand, indirect angiogenesis inhibitors (B) are used to target tumor cells and/or different angiogenesis-regulating products produced by them. Some indirect angiogenesis agents are used to target the receptors for tumor-originated angiogenesis-regulating proteins, which are present on the endothelial cell surface. In the blood circulation, anticoagulants (C) and anti-adhesion agents (D) can prevent colonization and homing of circulating tumor cells at the secondary metastasis site. (A black and white version of this figure will appear in some formats. For the color version, please refer to the plate section.)

this process. Angiogenesis is a consequence of an imbalance between pro-angiogenic and anti-angiogenic factors and is referred to as the "angiogenic switch" [47]. The angiogenic switch can occur at any stage of tumor progression. It depends on the tumor type and its microenvironment. The molecular mediators involved in tumor-induced vascularization include hypoxia-inducible transcription factors (HIF1-α), which elicit a well-coordinated response of angiogenesis by inducing the expression of growth factors. HIF1-α has been shown to be localized to hypoxic regions of tumors, and its overexpression has been reported in various primary and metastatic tumors. Interestingly, Blouw et al. have demonstrated that the anatomical location of a tumor determines HIF1-α-mediated angiogenesis [48]. The investigators observed that tumors growing in the brain, where the tissues are well vascularized, may be less dependent on angiogenesis than tumors located in subcutaneous space, where fewer vascular structures are present.

Among the different genes targeted by HIF, VEGF plays a major role in the angiogenesis process. The VEGF family consists of placental growth factor (PIGF1 and 2), VEGFA, VEGFB, VEGFC, VEGFD, and VEGFE. These ligands have a specific affinity to their distinct VEGF receptors (VEGFR), VEGFR1 (Flt-1), VEGFR2 (KDR/Flk-1), and VEGFR3 (Flt-4). VEGFA isoforms, commonly known as "VEGF," are the predominant angiogenic molecule expressed by tumor cells and bind to both VEGFR1 and VEGFR2. VEGF induce most of the events such as endothelial cell migration, proliferation, survival, and permeability of blood vessels.

Other angiogenesis-regulating factors that get activated through HIF1-α are platelet-derived growth factors (PDGF). The five identified PDGF isoforms (PDGF-AA, PDGF-AB, PDGF-BB, PDGF-CC, and PDGF-DD) mediate their effect by binding to PDGFRa, PDGFRb, and tyrosine kinase receptors. Studies suggest that the functional role of PDGF in tumor angiogenesis largely depends on the anatomical location of the tumor. For example, in the case of pancreatic cancer, by recruiting pericytes around the developing vasculature, PDGF helps to stabilize the blood vessels (discussed later). In CNS tumors, PDGF stimulate VEGF release, therefore promoting angiogenesis. Likewise, in metastatic prostate cancer, PDGF stimulate the survival signal in tumor endothelial cells. In the case of skin cancer, PDGF signaling helps to regulate the level of tumor interstitial fluid pressure.

Other than angiogenesis, several other mechanisms, such as sprouting, co-option of pre-existing vessels, mosaic vessels, vasculogenic mimicry, and mobilization of latent vessels, also have a role in tumor neovascularization. In the future, further studies about all of these events will essentially strengthen the endeavor to find a potent anticancer therapy.

Endothelial cells

Although lodgement of circulating tumor cells can undoubtedly result from mechanical entrapment, there is now considerable evidence showing that some tumor cells may adhere selectively to blood vessel endothelium [49]. As discussed previously, direct interaction between tumor cells and tumor endothelial cells may occur due to physical entrapment of tumor cells or interaction between some specific adhesion molecules present on the tumor cell and endothelial cell surface. However, the actual molecular nature of this interaction is not yet clear.

Although it has been hypothesized that tumor cells extravasate from the circulation in a similar way as that employed by leukocytes, this remains an open controversial question. Many findings support this argument that the processes of leukocyte and tumor cell adhesion with endothelial cells are very similar. Particularly, selectin–selectin ligand-mediated interaction between tumor cells and endothelial cells seem to be a key factor in regulating metastasis of cancers. Quite a large body of experiments and clinical findings indicate that the interaction of sLe-a and sLe-x-expressing tumor cells with E- or P-selectin-expressing endothelial cells is a crucial event in tumor cell transendothelial migration [50–55]. As discussed later, L-selectin ligands present on the endothelial cell surface may also promote leukocyte-mediated adhesion of tumor cells to the endothelial surface.

Empirical evidence has shown the involvement of many non-selectin adhesion molecules in the attachment of tumor cells to the endothelial surface.

Interestingly, *in vitro* and *ex vivo* experimental evidence have shown that microvascular endothelium activation leads to an increase in galectin-3 cell surface expression promoting metastatic breast and prostate carcinoma cell adhesion to the endothelium. Apart from leukocyte adhesion, this activation occurs in a non-cytokine-mediated manner and this is induced by cancer-associated carbohydrate structures (TF antigen disaccharide) present on the tumor cell surface [56]. Likewise, studies suggest that melanoma cells use their surface very late activation antigen-4 (VLA-4) integrin to adhere to the endothelial VCAM-1 protein [57]. Also, in the spontaneous murine model of melanoma, an organ-selective (brain, heart, and liver) upregulation of VCAM-1 in metastatic melanoma was also reported [58].

Molecular profiles of tumor endothelial cells indicate a significant genetic difference between tumor-associated endothelial cells and endothelial cells present in the adjacent normal blood vessel [59]. Experimental evidence shows that tumor endothelial cells have a tendency to express EGF receptor (EGF-R), and binding of its ligands (TGF-a and/or EGF) produced by the adjacent tumor cells activate these receptors. The activation of EGF-R on endothelial cells of tumor blood vessels may play an important role in tumor progression [7]. Interestingly, pharmacological suppression of this signaling cascade in animal models showed tumor growth inhibition and reduced metastasis frequency [60].

Pericytes

Pericytes, also known as Rouget cells, periendothelial cells, or mural cells, are an integral component of the vasculature. Pericytes are located within the basement membrane of capillaries and post-capillary venules. The pericytes extend long cytoplasmic processes over the endothelial cell surface and help in the stabilization of newly formed blood vessels. Pericytes may also contribute to endothelial cell survival, proliferation, migration, permeability, and maturation [61–64]. Animal studies indicate that β tumor cells metastasize to distant organs in primary pericyte-deficient Pdgf$^{ret/ret}$ mice [65]. These results demonstrate a role of pericytes in limiting tumor cell metastasis. Further studies have shown that tumor cells can perturb pericyte–endothelial cell interaction and promote hematogenous metastasis [65]. Other than cancer, the role of pericytes in many other diseases has also been elucidated [66].

Endothelial basement membrane

Stabilization of the tumor cell endothelium contact after the initial arrest in the blood vessel is an important step in the successful development of extravasation. For stable adhesion, the adhesion of tumor cells to the sub-endothelial matrix appears to be a crucial step [67]. Experimental evidence suggests that highly metastatic tumor cells attach more quickly and avidly to the endothelial basement membrane than to the endothelial cells [68, 69]. The vascular basement membrane is dynamic and consists of a self-assembled layer of proteins, glycoproteins, and proteoglycans. It consists of cell-adhesive molecules such as collagen IV, laminin, entactin/nidogen, fibronectin, and the heparin-sulfate proteoglycan perlecan [70]. These molecules may collectively be responsible for the binding of tumor cells to the basement membrane [69]. Receptors for laminin and fibronectin have been identified in the plasma membrane of malignant cells [71]. Therefore the interaction of these receptors with laminin and fibronectin may promote the adhesion of tumor cells to the endothelial basement membrane. Experimental evidence has shown that laminin receptors on the tumor cell membrane help in the establishment of lung colony formation, which may be essential for extravasation. Furthermore, attachment of tumor cells to the endothelial basement membrane may lead to the proliferation of tumor cells within the vascular space [72]. The intravascular proliferation of tumor cells eventually develops multiple interruptions or degradation in the vascular basement membrane. The degradation of the basement membrane can be carried out by the proteases secreted by the tumor cells or by platelets. In comparison with the apical surface of endothelial cells, the vascular basement membrane constitutes a highly adhesive and thrombogenic surface. Interaction of platelets with components of the basement membrane activates platelets. Activated platelets secrete heparitinase, which degrades heparin sulfate of the basement membrane and with this, tumor cells complete the final step in the extravasation cascade [73].

Therapeutic prospects

A plethora of biological, pharmacological, and genetic evidence has confirmed that tumors are angiogenesis dependent. Anti-angiogenic agents are potential therapeutics not only for the treatment of cancer, but also for the prevention of cancer relapse or metastasis. There are two classes of angiogenesis inhibitors: direct and indirect (Figure 20.1). The indirect angiogenesis inhibitors can mediate their function by decreasing or blocking the expression of angiogenic products produced by the tumor cells, or neutralizing the activity of the tumor product itself, or by blocking its receptor on endothelial cells. Recently, it has been recognized that many oncogenes are pro-angiogenic, which upregulate the expression of angiogenic proteins and/or downregulate inhibitors of angiogenesis. Therefore targeting oncogenes and/or their product and/or their corresponding receptors can not only directly inhibit the growth of tumor cells, but at the same time can disrupt tumor-related angiogenesis and inhibit tumor growth and spread. On the other hand, direct angiogenesis inhibitors block vascular endothelial cell proliferation, survival, or migration in response to various pro-angiogenic proteins. Direct angiogenesis inhibitors are less likely to develop acquired drug resistance, because these drugs target genetically stable endothelial cells, rather than unstable heterogeneous tumor cells. There is additional information on these agents in other articles [74–77].

It is a well-established fact that, for survival, endothelial cells depend on other cells, such as pericytes. Thus the important and dynamic function of pericytes in maintaining homeostasis and carcinogenesis makes them a potential target for pharmacological targeting (Figure 20.1). The pericytes that surround the tumor blood vessels are known to produce VEGF, which is essential for the survival of endothelial cells. Therefore it is logical to expect that in the absence of pericytes, the endothelium will become vulnerable to VEGF blockade [78]. Further, the endothelial cells and pericytes present in tumor vasculature are complementary targets for anticancer drugs (Figure 20.1). Animal studies have shown that a combinatorial targeting of VEGFRs in endothelial cells by VEGFR inhibitor (SU5416) and a kinase inhibitor selectively for PDGFRs in pericytes (SU6668 or Gleevec) has a beneficial effect on late-stage solid tumors [79].

Coagulation has long been known to facilitate metastasis. Many encouraging preliminary results and compelling biochemical rationale indicate the potential use of anticoagulants for the prevention and treatment of cancer (Figure 20.1). At present, only a few studies have qualified for controlled randomized clinical studies. Interpretation of results indicating the effect of anticoagulants or fibrinolytic agents on cancer metastasis is complicated because most studies have tested anticoagulants in combination with cytotoxic drugs. These anticoagulant drugs may also alter immunologic mechanisms or directly inhibit tumor cell growth and motility. It has been proposed that use of anticoagulants during surgical manipulation of a tumor can help to prevent metastasis. This idea seems to be logical and impressive, because it has been suspected that after surgical intervention the chance of metastasis increases. At the same time, anticoagulant with low-molecular-weight (LMW) heparin is mostly used in the perioperative period; therefore use of anticoagulants will not put the patients at additional risk. Interestingly, in a pre-clinical study, one orally absorbable heparin derivative (LDH) dramatically attenuated the metastasis of murine melanoma or human lung carcinoma cells [80]. Therefore these studies indicate a possible development in the anticoagulant therapy for the treatment of cancer metastasis. Additionally, studies have shown that different heparin derivatives can also be used as angiogenesis inhibitors [81]. In this context, it has been reported that LMW glycol-split heparin (ST2184) is a potential non-anticoagulant, angiostatic compound [82].

Targeting glycans in hematogenous metastasis is a promising area for anti-metastasis therapy (anti-adhesion type) (Figure 20.1). Agents that can interfere with selectin–carbohydrate interactions are considered for anti-metastasis treatment. These carbohydrate-based interactions can be targeted by using neutralizing anti-selectin antibodies or small molecules that mimic the SLe^x or SLe^A selectin ligands [83]. In this connection, it is of interest to note that cimetidine, which downregulated the expression of E-selectin on endothelial cells, blocked the *in vitro* adhesion of colorectal tumor cells to endothelial cells and suppressed metastasis by these tumor cells in nude mice [84]. But, along with the role of cimetidine on E-selectin expression, the anticancer effect of cimetidine in colorectal cancer may also be mediated through its immunomodulatory effect [85]. Like E-selectin, P- and L-selectin are also two potential targets for anti-metastasis therapy. Studies have also shown that the glycosaminoglycan heparin can inhibit both P- and L-selectin [39].

Another kind of anti-adhesion agents, which target the adhesion of tumor cells to the sub-endothelial extracellular matrix, could be potentially used in anticancer therapeutics. Adhesion of tumor cells to the sub-endothelium extracellular matrix occurs by the interaction of specific receptors present on the surface of the tumor cells and their corresponding ligands present in the sub-endothelium. Preclinical experiments using non-peptide agonists targeting both the tumor cells avb3 and platelet aIIbb3 integrins together in breast adenocarcinoma cells showed an effective anti-metastatic effect [86]. But further extensive studies are warranted to explore the possible use of the anti-adhesive therapy in cancer treatment.

Epilogue

Tumors are heterogeneous in nature, leading to heterogeneity in the tumor vasculature that is a major problem for an effective blood vessel targeted therapy. Keeping this in mind, combination therapy that targets different key factors might be more effective in the treatment and prevention of cancer progression [87]. During the study design for anti-vascular agents, impact on translational research, effect on multiple end points, and efficacy of the compound should be given priority. To validate the efficacy of the drug response, identification of specific and sensitive biomarkers is also warranted. Imaging studies could also have a potential role in assessing the efficacy of treatment. Further development of target-specific contrast agents will definitely improve functional imaging and aid in better evaluation of drug efficacy. Anti-adhesive therapeutics has a tremendous potential to prevent hematogenous metastasis. Targeting selectins might have undesirable side effects due to the prevention of the normal selectin-dependent physiological process, such as normal leukocyte and platelet functions. Studies by different investigators, including our group (unpublished data), have shown that glycoproteins, such as mucins, can play a major role in the adhesion of circulating tumor cells to endothelial cells, directly and indirectly. Therefore, in the future, dissection of the mechanisms through which these molecules mediate their functions may help to develop new anti-adhesive therapeutics.

Acknowledgments

The authors of this article are supported by grants from the US Department of Defense (PC074289 and BC074639) and National Institutes of Health (RO1 CA78590, TMEN U54 CA 163120 and CA163120). S.S. is supported by the Ramalingaswami Fellowship, Department of Biotechnology, India.

We thank Shonali Deb for the critical reading and editing of the manuscript.

References

1 Langley, R.R. and Fidler, I.J. 2007. Tumor cell-organ microenvironment interactions in the pathogenesis of cancer metastasis, *Endocr Rev, 28*: 297–321.

2 Woodhouse, E.C., Chuaqui, R.F. and Liotta, L.A. 1997. General mechanisms of metastasis, *Cancer, 80*: 1529–1537.

3 Singh, R.K. and Fidler, I.J. 1996. Regulation of tumor angiogenesis by organ-specific cytokines, *Curr Top Microbiol Immunol, 213*: 1–11.

4 Hart, I.R. 1982. "Seed and soil" revisited: mechanisms of site-specific metastasis, *Cancer Metastasis Rev, 1*: 5–16.

5 Tarin, D., Price, J.E., Kettlewell, M.G., et al. 1984. Mechanisms of human tumor metastasis studied in patients with peritoneovenous shunts, *Cancer Res, 44*: 3584–3592.

6 Fidler, I.J. 1990. Critical factors in the biology of human cancer metastasis: twenty-eighth G.H.A. Clowes memorial award lecture, *Cancer Res, 50*: 6130–6138.

7 Baker, C.H., Kedar, D., McCarty, M.F., et al. 2002. Blockade of epidermal growth factor receptor signaling on tumor cells and tumor-associated endothelial cells for therapy of human carcinomas, *Am J Pathol, 161*: 929–938.

8 Paget, S. 1989. The distribution of secondary growths in cancer of the breast. 1889, *Cancer Metastasis Rev, 8*: 98–101.

9 Rusciano, D. and Burger, M.M. 1992. Why do cancer cells metastasize into particular organs? *Bioessays, 14*: 185–194.

10 Weiss, L. 1992. Comments on hematogenous metastatic patterns in humans as revealed by autopsy, *Clin Exp Metastasis, 10*: 191–199.

11 Chambers, A.F., Groom, A.C. and MacDonald, I.C. 2002. Dissemination and growth of cancer cells in metastatic sites, *Nat Rev Cancer, 2*: 563–572.

12 Orr, F.W. and Wang, H.H. 2001. Tumor cell interactions with the microvasculature: a rate-limiting step in metastasis, *Surg Oncol Clin N Am, 10*: 357–381.

13 Thorlacius, H., Prieto, J., Raud, J., et al. 1997. Tumor cell arrest in the microcirculation: lack of evidence for a leukocyte-like rolling adhesive interaction with vascular endothelium *in vivo*, *Clin Immunol Immunopathol, 83*: 68–76.

14 Pauli, B.U. and Knudson, W. 1988. Tumor invasion: a consequence of destructive and compositional matrix alterations, *Hum Pathol, 19*: 628–639.

15 Pauli, B.U., Augustin-Voss, H.G., el-Sabban, M.E., Johnson, R.C. and Hammer, D.A. 1990. Organ preference of metastasis. The role of endothelial cell adhesion molecules, *Cancer Metastasis Rev, 9*: 175–189.

16 Nicolson, G.L. 1991. Molecular mechanisms of cancer metastasis: tumor and host properties and the role of oncogenes and suppressor genes, *Curr Opin Oncol, 3*: 75–92.

17 Rajotte, D., Arap, W., Hagedorn, M., et al. 1998. Molecular heterogeneity of the vascular endothelium revealed by *in vivo* phage display, *J Clin Invest, 102*: 430–437.

18 Pauli, B.U. and Lee, C.L. 1988. Organ preference of metastasis. The role of organ-specifically modulated endothelial cells, *Lab Invest, 58*: 379–387.

19 Auerbach, R., Lu, W.C., Pardon, E., et al. 1987. Specificity of adhesion between murine tumor cells and capillary endothelium: an *in vitro* correlate of preferential metastasis *in vivo*, *Cancer Res, 47*: 1492–1496.

20 Auerbach, R. 1988. Patterns of tumor metastasis: organ selectivity in the spread of cancer cells, *Lab Invest, 58*: 361–364.

21 Nicolson, G.L. 1991. Tumor and host molecules important in the organ preference of metastasis, *Semin Cancer Biol, 2*: 143–154.

22 Nicolson, G.L. 1991. Gene expression, cellular diversification and tumor progression to the metastatic phenotype, *Bioessays, 13*: 337–342.

23 Johnson, R.C., Augustin-Voss, H.G., Zhu, D.Z. and Pauli, B.U. 1991. Endothelial cell membrane vesicles in the study of organ preference of metastasis, *Cancer Res, 51*: 394–399.

24 Gunthert, U., Hofmann, M., Rudy, W., et al. 1991. A new variant of glycoprotein CD44 confers metastatic potential to rat carcinoma cells, *Cell, 65*: 13–24.

25 Zetter, B.R. 1997. On target with tumor blood vessel markers, *Nat Biotechnol, 15*: 1243–1244.

26 McCarthy, J.B., Skubitz, A.P., Iida, J., et al. 1991. Tumor cell adhesive mechanisms and their relationship to metastasis, *Semin Cancer Biol, 2*: 155–167.

27 Zetter, B.R. 1990. The cellular basis of site-specific tumor metastasis, *N Engl J Med, 322*: 605–612.

28 Fidler, I.J. 1973. The relationship of embolic homogeneity, number, size and viability to the incidence of experimental metastasis, *Eur J Cancer, 9*: 223–227.

29 Palumbo, J.S., Talmage, K.E., Massari, J.V., et al. 2007. Tumor cell-associated tissue factor and circulating hemostatic factors cooperate to increase metastatic potential through natural killer cell-dependent and -independent mechanisms, *Blood, 110*: 133–141.

30 Cotmore, S.F. and Carter, R.L. 1973. Mechanisms of enhanced intrahepatic metastasis in surfactant-treated hamsters: an electron microscopy study, *Int J Cancer, 11*: 725–738.

31 Crissman, J.D., Hatfield, J., Schaldenbrand, M., Sloane, B.F. and Honn, K.V. 1985. Arrest and extravasation of B16 amelanotic melanoma in murine lungs. A light and electron microscopic study, *Lab Invest, 53:* 470–478.

32 Gasic, G.J., Gasic, T.B. and Stewart, C.C. 1968. Antimetastatic effects associated with platelet reduction, *Proc Natl Acad Sci USA, 61:* 46–52.

33 Karpatkin, S., Pearlstein, E., Ambrogio, C. and Coller, B.S. 1988. Role of adhesive proteins in platelet tumor interaction *in vitro* and metastasis formation *in vivo*, *J Clin Invest, 81:* 1012–1019.

34 Camerer, E., Qazi, A.A., Duong, D.N., et al. 2004. Platelets, protease-activated receptors, and fibrinogen in hematogenous metastasis, *Blood, 104:* 397–401.

35 Honn, K.V., Tang, D.G. and Crissman, J.D. 1992. Platelets and cancer metastasis: a causal relationship? *Cancer Metastasis Rev, 11:* 325–351.

36 Honn, K.V., Tang, D.G., Grossi, I.M., et al. 1994. Enhanced endothelial cell retraction mediated by 12(S)-HETE: a proposed mechanism for the role of platelets in tumor cell metastasis, *Exp Cell Res, 210:* 1–9.

37 Burdick, M.M. and Konstantopoulos, K. 2004. Platelet-induced enhancement of LS174T colon carcinoma and THP-1 monocytoid cell adhesion to vascular endothelium under flow, *Am J Physiol Cell Physiol, 287:* C539–C547.

38 Palumbo, J.S., Talmage, K.E., Massari, J.V., et al. 2005. Platelets and fibrin(ogen) increase metastatic potential by impeding natural killer cell-mediated elimination of tumor cells, *Blood, 105:* 178–185.

39 Borsig, L., Wong, R., Hynes, R.O., Varki, N.M. and Varki, A. 2002. Synergistic effects of L- and P-selectin in facilitating tumor metastasis can involve non-mucin ligands and implicate leukocytes as enhancers of metastasis, *Proc Natl Acad Sci USA, 99:* 2193–2198.

40 Im, J.H., Fu, W., Wang, H., et al. 2004. Coagulation facilitates tumor cell spreading in the pulmonary vasculature during early metastatic colony formation, *Cancer Res, 64:* 8613–8619.

41 Palumbo, J.S., Potter, J.M., Kaplan, L.S., et al. 2002. Spontaneous hematogenous and lymphatic metastasis, but not primary tumor growth or angiogenesis, is diminished in fibrinogen-deficient mice, *Cancer Res, 62:* 6966–6972.

42 Palumbo, J.S., Kombrinck, K.W., Drew, A.F., et al. 2000. Fibrinogen is an important determinant of the metastatic potential of circulating tumor cells, *Blood, 96:* 3302–3309.

43 Nierodzik, M.L., Chen, K., Takeshita, K., et al. 1998. Protease-activated receptor 1 (PAR-1) is required and rate-limiting for thrombin-enhanced experimental pulmonary metastasis, *Blood, 92:* 3694–3700.

44 Amirkhosravi, A., Meyer, T., Chang, J.Y., et al. 2002. Tissue factor pathway inhibitor reduces experimental lung metastasis of B16 melanoma, *Thromb Haemost, 87:* 930–936.

45 Jain, R.K. 2005. Normalization of tumor vasculature: an emerging concept in antiangiogenic therapy, *Science, 307:* 58–62.

46 Wyckoff, J.B., Jones, J.G., Condeelis, J.S. and Segall, J.E. 2000. A critical step in metastasis: *in vivo* analysis of intravasation at the primary tumor, *Cancer Res, 60:* 2504–2511.

47 Folkman, J. 1992. The role of angiogenesis in tumor growth, *Semin.Cancer Biol, 3:* 65–71.

48 Blouw, B., Song, H., Tihan, T., et al. 2003. The hypoxic response of tumors is dependent on their microenvironment, *Cancer Cell, 4:* 133–146.

49 Alby, L. and Auerbach, R. 1984. Differential adhesion of tumor cells to capillary endothelial cells in vitro, *Proc Natl Acad Sci USA, 81:* 5739–5743.

50 Burdick, M.M., McCaffery, J.M., Kim, Y.S., Bochner, B.S. and Konstantopoulos, K. 2003. Colon carcinoma cell glycolipids, integrins, and other glycoproteins mediate adhesion to HUVECs under flow, *Am J Physiol Cell Physiol, 284:* C977–C987.

51 Dimitroff, C.J., Lechpammer, M., Long-Woodward, D. and Kutok, J.L. 2004. Rolling of human bone-metastatic prostate tumor cells on human bone marrow endothelium under shear flow is mediated by E-selectin, *Cancer Res, 64:* 5261–5269.

52 Kannagi, R., Izawa, M., Koike, T., Miyazaki, K. and Kimura, N. 2004. Carbohydrate-mediated cell adhesion in cancer metastasis and angiogenesis, *Cancer Sci, 95:* 377–384.

53 Renkonen, R., Mattila, P., Majuri, M.L., et al. 1997. In vitro experimental studies of sialyl Lewis x and sialyl Lewis a on endothelial and carcinoma cells: crucial glycans on selectin ligands, *Glycoconj J, 14:* 593–600.

54 Witz, I.P. 2006. Tumor-microenvironment interactions: the selectin-selectin ligand axis in tumor-endothelium cross talk, *Cancer Treat Res, 130:* 125–140.

55 Ito, K., Ye, C.L., Hibi, K., et al. 2001. Paired tumor marker of soluble E-selectin and its ligand sialyl Lewis A in colorectal cancer, *J Gastroenterol, 36:* 823–829.

56 Glinskii, O.V., Turk, J.R., Pienta, K.J., Huxley, V.H. and Glinsky, V.V. 2004. Evidence of porcine and human endothelium activation by cancer-associated

carbohydrates expressed on glycoproteins and tumour cells, *J Physiol*, 554: 89–99.

57 Okahara, H., Yagita, H., Miyake, K. and Okumura, K. 1994. Involvement of very late activation antigen 4 (VLA-4) and vascular cell adhesion molecule 1 (VCAM-1) in tumor necrosis factor alpha enhancement of experimental metastasis, *Cancer Res*, 54: 3233–3236.

58 Langley, R.R., Carlisle, R., Ma, L., et al. 2001. Endothelial expression of vascular cell adhesion molecule-1 correlates with metastatic pattern in spontaneous melanoma, *Microcirculation*, 8: 335–345.

59 St, C.B., Rago, C., Velculescu, V., et al. 2000. Genes expressed in human tumor endothelium, *Science*, 289: 1197–1202.

60 Yokoi, K., Thaker, P.H., Yazici, S., et al. 2005. Dual inhibition of epidermal growth factor receptor and vascular endothelial growth factor receptor phosphorylation by AEE788 reduces growth and metastasis of human colon carcinoma in an orthotopic nude mouse model, *Cancer Res*, 65: 3716–3725.

61 Hirschi, K.K. and D'Amore, P.A. 1996. Pericytes in the microvasculature, *Cardiovasc Res*, 32: 687–698.

62 Jain, R.K. and Booth, M.F. 2003. What brings pericytes to tumor vessels? *J Clin Invest*, 112: 1134–1136.

63 Lindahl, P., Hellstrom, M., Kalen, M. and Betsholtz, C. 1998. Endothelial-perivascular cell signaling in vascular development: lessons from knockout mice, *Curr Opin Lipidol*, 9: 407–411.

64 Wesseling, P., Schlingemann, R.O., Rietveld, F.J., et al. 1995. Early and extensive contribution of pericytes/vascular smooth muscle cells to microvascular proliferation in glioblastoma multiforme: an immuno-light and immuno-electron microscopic study, *J Neuropathol Exp Neurol*, 54: 304–310.

65 Xian, X., Hakansson, J., Stahlberg, A., et al. 2006. Pericytes limit tumor cell metastasis, *J Clin Invest*, 116: 642–651.

66 Allt, G. and Lawrenson, J.G. 2001. Pericytes: cell biology and pathology, *Cells Tissues Organs*, 169: 1–11.

67 Crissman, J.D., Hatfield, J.S., Menter, D.G., Sloane, B. and Honn, K.V. 1988. Morphological study of the interaction of intravascular tumor cells with endothelial cells and subendothelial matrix, *Cancer Res*, 48: 4065–4072.

68 Kramer, R.H., Gonzalez, R. and Nicolson, G.L. 1980. Metastatic tumor cells adhere preferentially to the extracellular matrix underlying vascular endothelial cells, *Int J Cancer*, 26: 639–645.

69 Nicolson, G.L., Irimura, T., Nakajima, M. and Estrada, J. 1983. Metastatic cell attachment to and invasion of vascular endothelium and its underlying basal lamina using endothelial cell monolayers, *Symp Fundam Cancer Res*, 36: 145–167.

70 Baluk, P., Morikawa, S., Haskell, A., Mancuso, M. and McDonald, D.M. 2003. Abnormalities of basement membrane on blood vessels and endothelial sprouts in tumors, *Am J Pathol*, 163: 1801–1815.

71 Liotta, L.A. 1986. Tumor invasion and metastases – role of the extracellular matrix: Rhoads Memorial Award lecture, *Cancer Res*, 46: 1–7.

72 Crissman, J.D., Hatfield, J.S., Menter, D.G., Sloane, B. and Honn, K.V. 1988. Morphological study of the interaction of intravascular tumor cells with endothelial cells and subendothelial matrix, *Cancer Res*, 48: 4065–4072.

73 Yahalom, J., Eldor, A., Biran, S., Fuks, Z. and Vlodavsky, I. 1985. Platelet-tumor cell interaction with the subendothelial extracellular matrix: relationship to cancer metastasis, *Radiother Oncol*, 3: 211–225.

74 Hicklin, D.J. and Ellis, L.M. 2005. Role of the vascular endothelial growth factor pathway in tumor growth and angiogenesis, *J Clin Oncol*, 23: 1011–1027.

75 Spannuth, W.A., Sood, A.K. and Coleman, R.L. 2008. Angiogenesis as a strategic target for ovarian cancer therapy, *Nat Clin Pract Oncol*, 5: 194–204.

76 Folkman, J. 2007. Angiogenesis: an organizing principle for drug discovery? *Nat Rev Drug Discov*, 6: 273–286.

77 Folkman, J. 2006. Angiogenesis, *Annu Rev Med*, 57: 1–18.

78 Jain, R.K. and Booth, M.F. 2003. What brings pericytes to tumor vessels? *J Clin Invest*, 112: 1134–1136.

79 Bergers, G., Song, S., Meyer-Morse, N., Bergsland, E. and Hanahan, D. 2003. Benefits of targeting both pericytes and endothelial cells in the tumor vasculature with kinase inhibitors, *J Clin Invest*, 111: 1287–1295.

80 Lee, D.Y., Park, K., Kim, S.K., et al. 2008. Antimetastatic effect of an orally active heparin derivative on experimentally induced metastasis, *Clin Cancer Res*, 14: 2841–2849.

81 Presta, M., Leali, D., Stabile, H., et al. 2003. Heparin derivatives as angiogenesis inhibitors, *Curr Pharm Des*, 9: 553–566.

82 Pisano, C., Aulicino, C., Vesci, L., et al. 2005. Undersulfated, low-molecular-weight glycol-split heparin as an antiangiogenic VEGF antagonist, *Glycobiology*, 15: 1C–6C.

83 Fuster, M.M. and Esko, J.D. 2005. The sweet and sour of cancer:

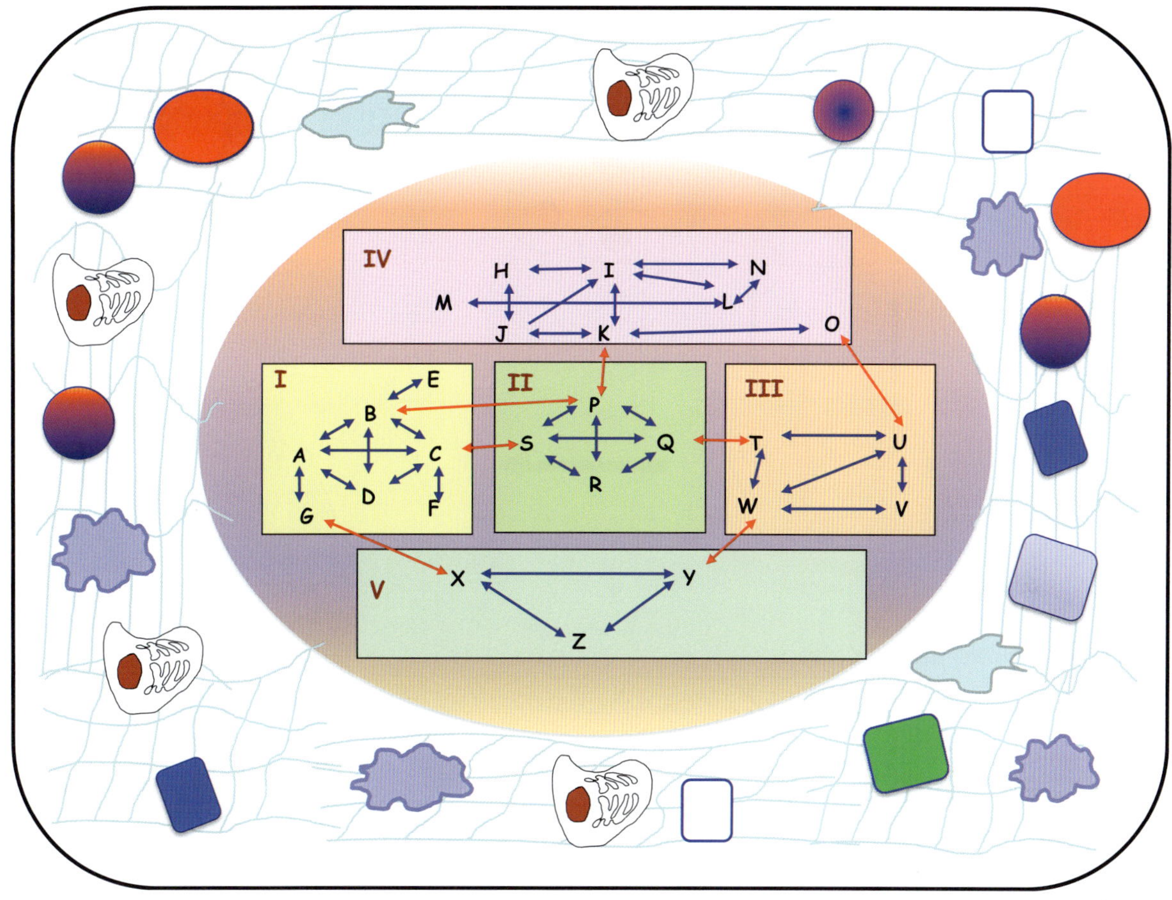

Figure 1.1 A cascade of aberrant network modules defines the multi-modular molecular network (MMMN) model for cancer progression.

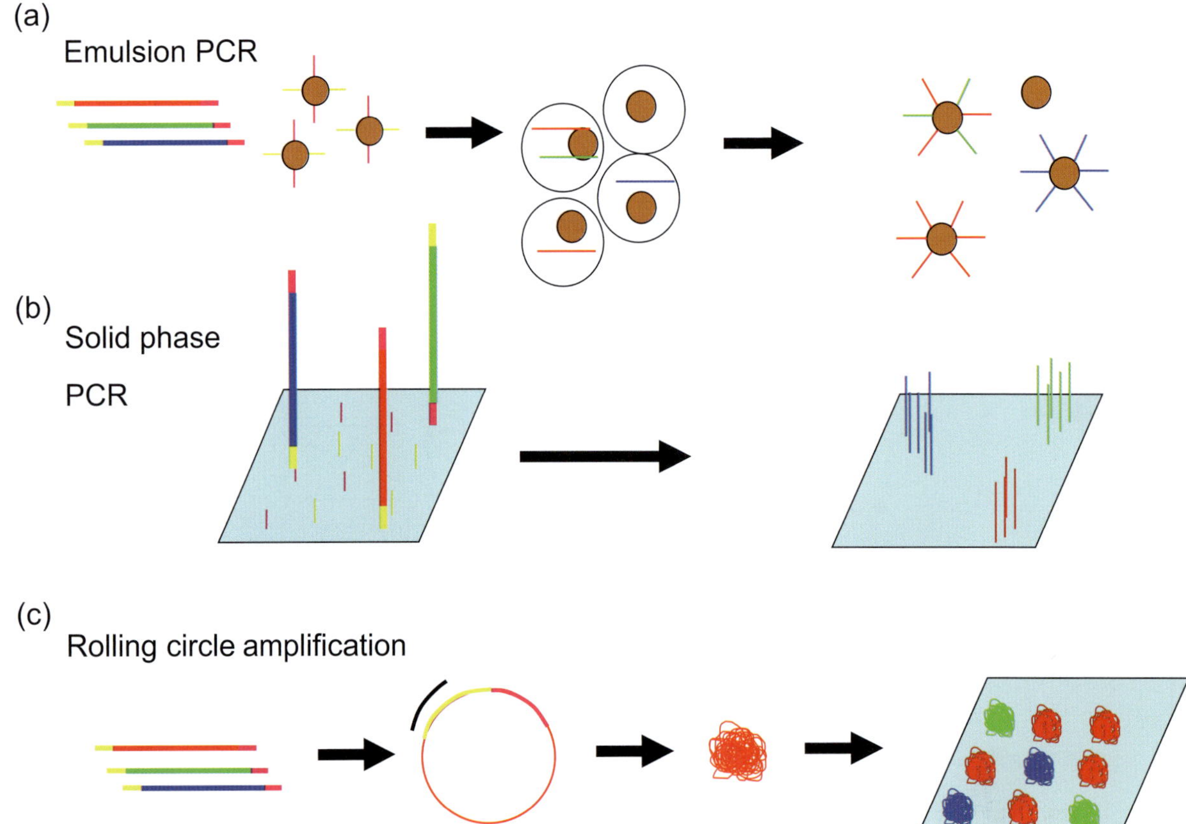

Figure 2.1 (a) Emulsion PCR.
(b) Solid phase PCR.
(c) Rolling circle amplification.

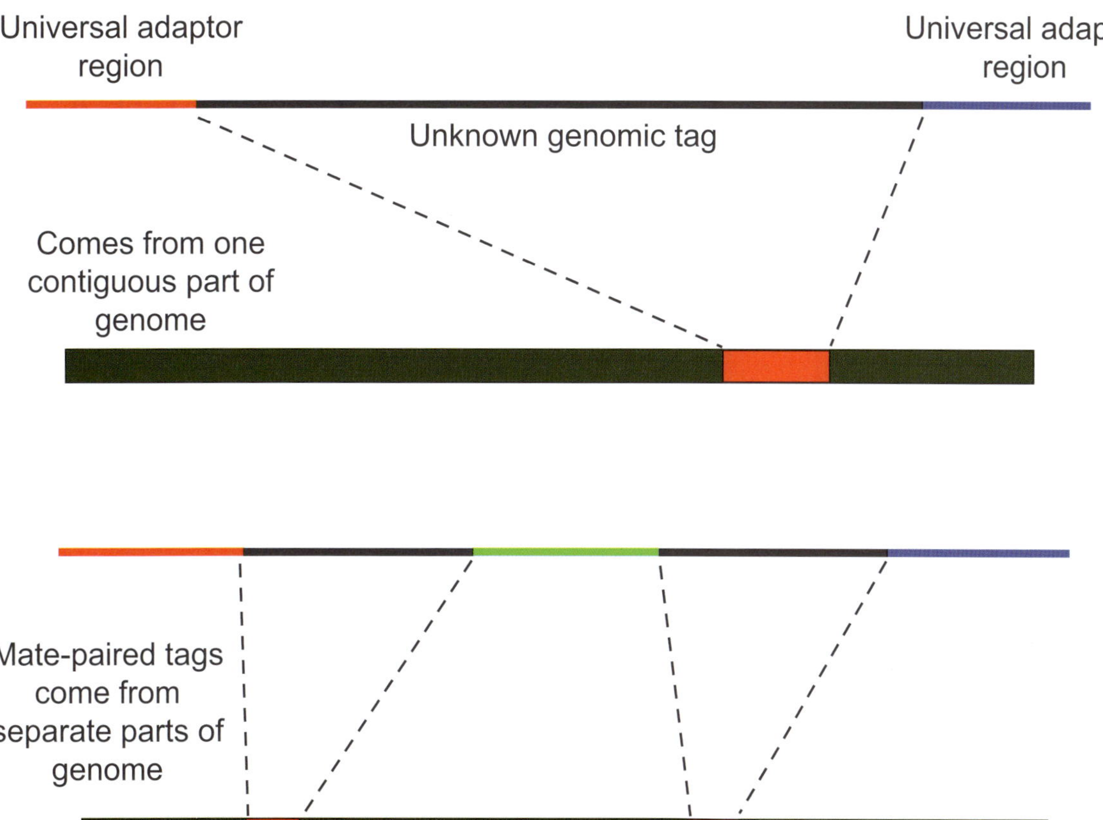

Figure 2.2 Mate-paired libraries.

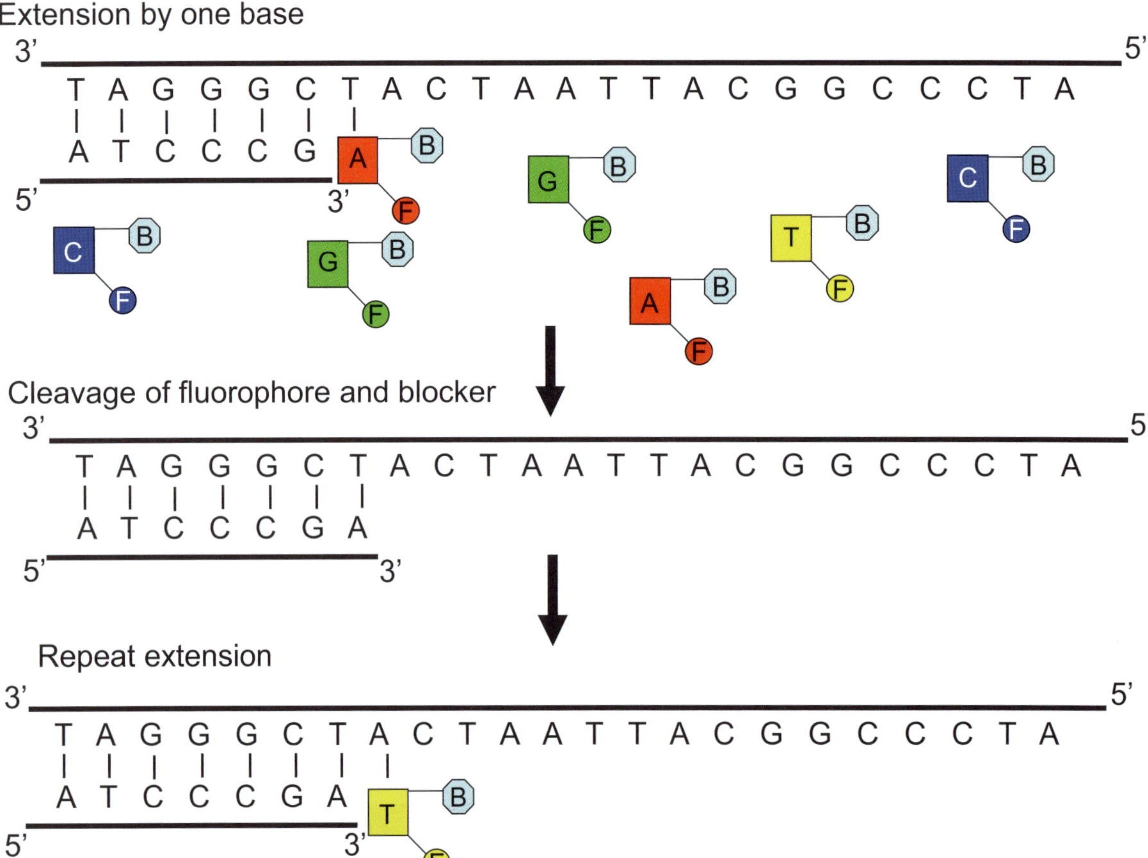

Figure 2.3 Sequencing by synthesis with fluorophores.

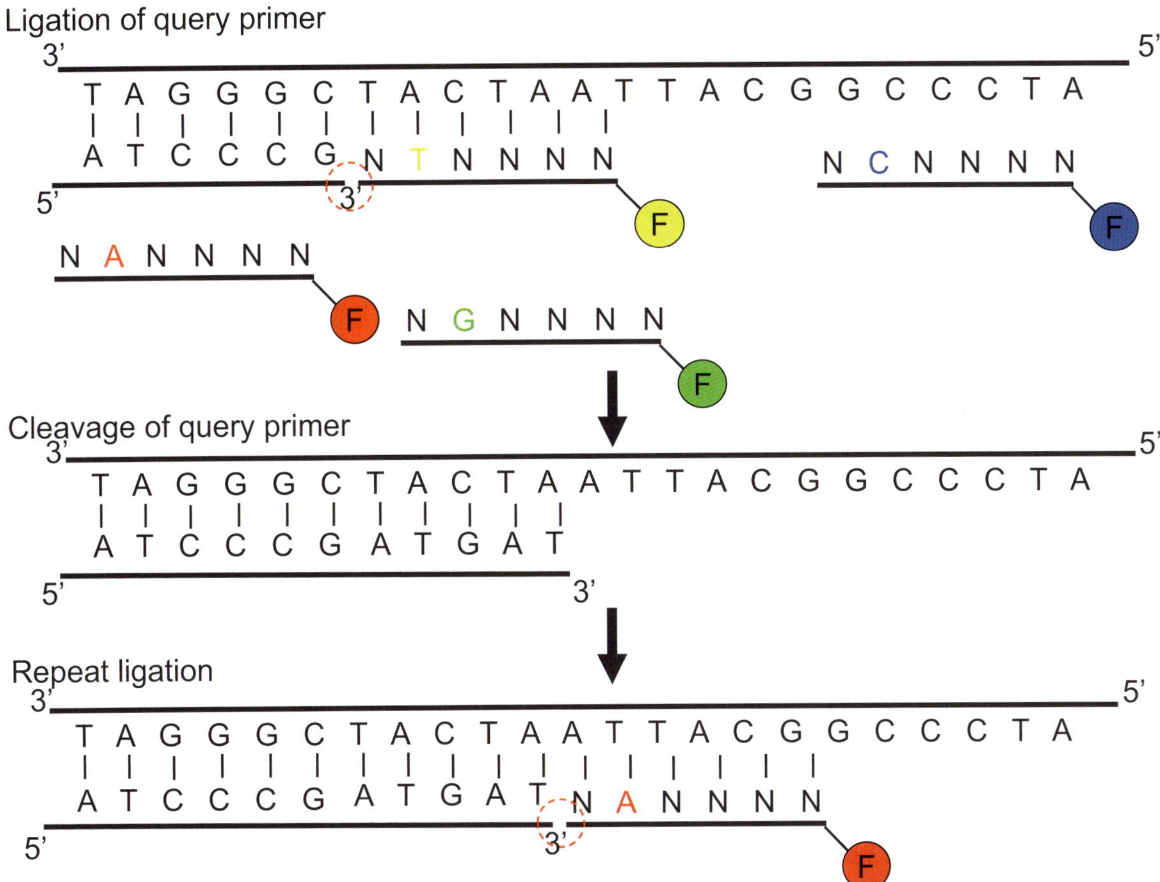

Figure 2.4 Sequencing by ligation.

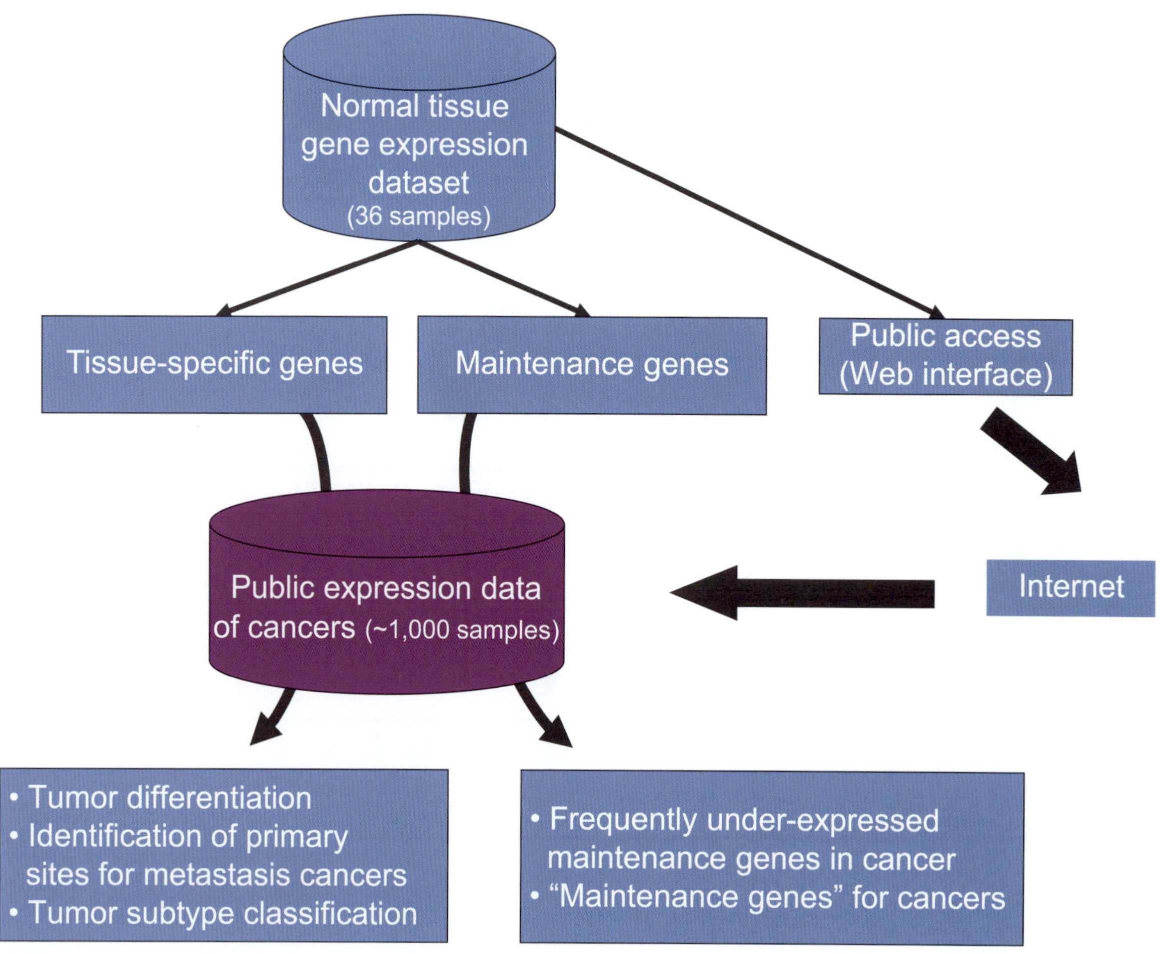

Figure 3.1 Flow chart of a systems biological study of the expression of tissue-specific and maintenance genes in cancer.

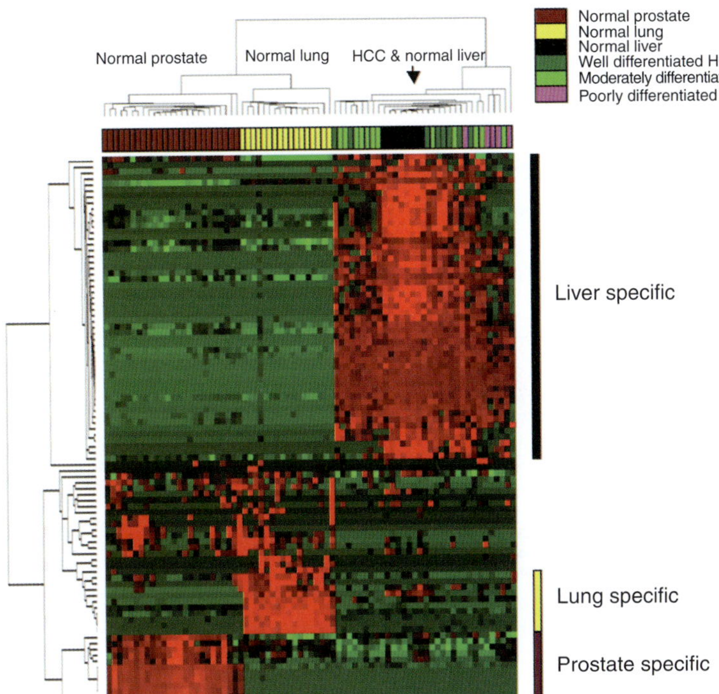

Figure 3.2 Expression of tissue-specific genes in cancer. Only a small portion of tissue-specific genes are under-expressed in poorly differentiated HCC.

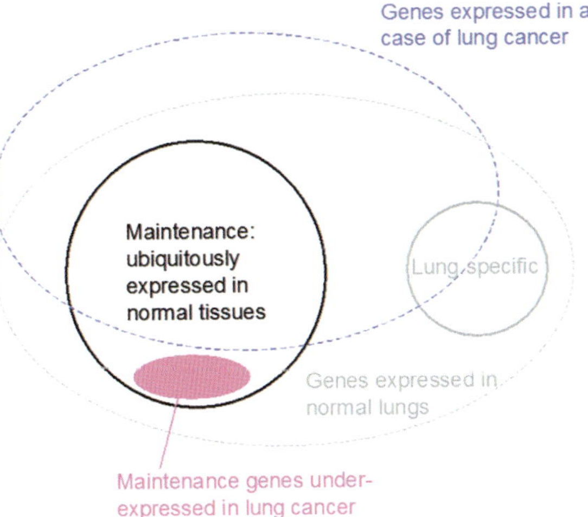

Figure 3.3 Selection of a group of genes ubiquitously expressed in normal tissues, but frequently under-expressed in tumors.

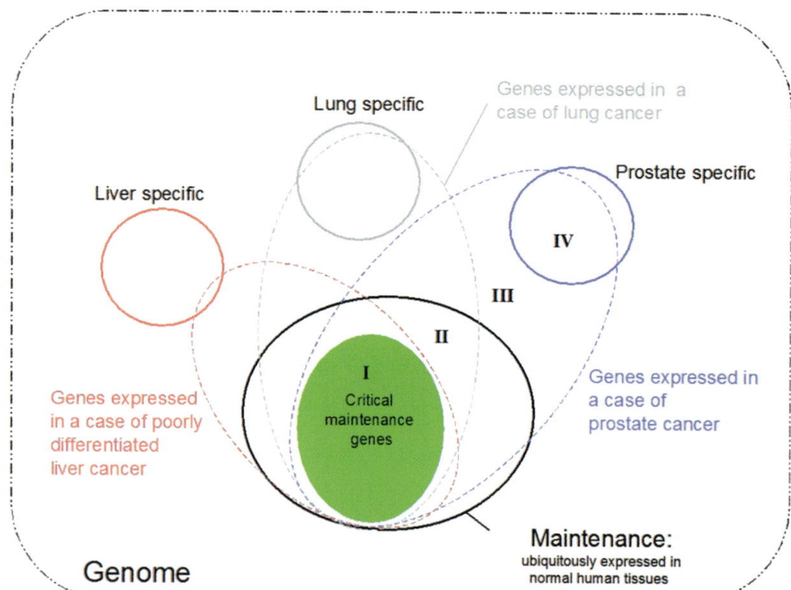

Figure 3.4 "Critical maintenance genes" defined by combining normal and cancer gene expression data.

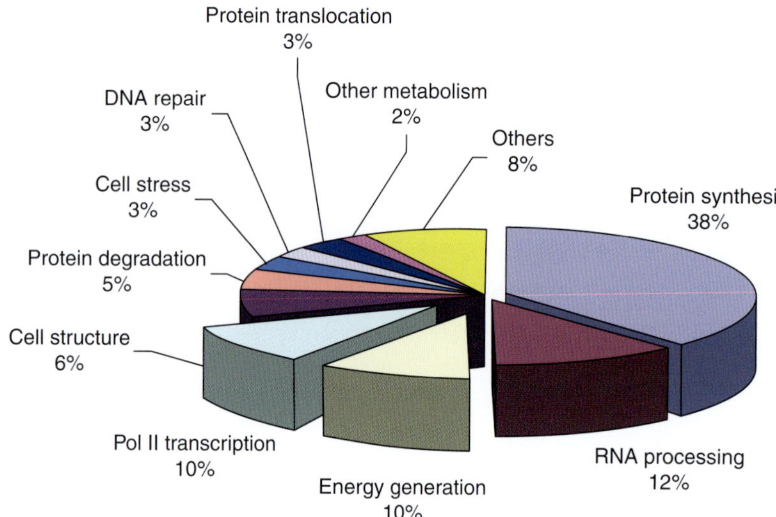

Figure 3.5 Functional categorization of critical maintenance genes according to gene ontology (GO).

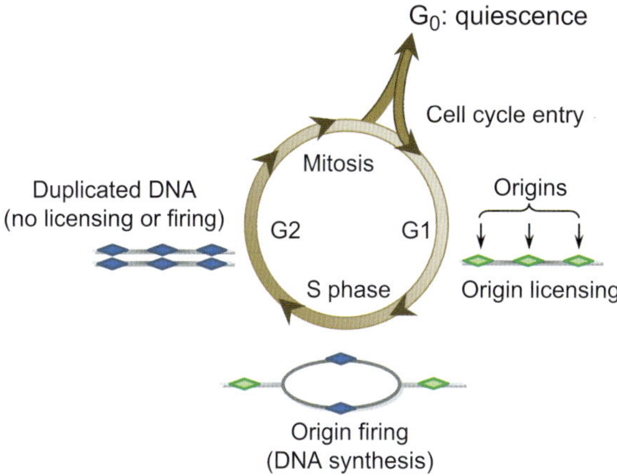

Figure 4.1 DNA replication in the human cell division cycle.

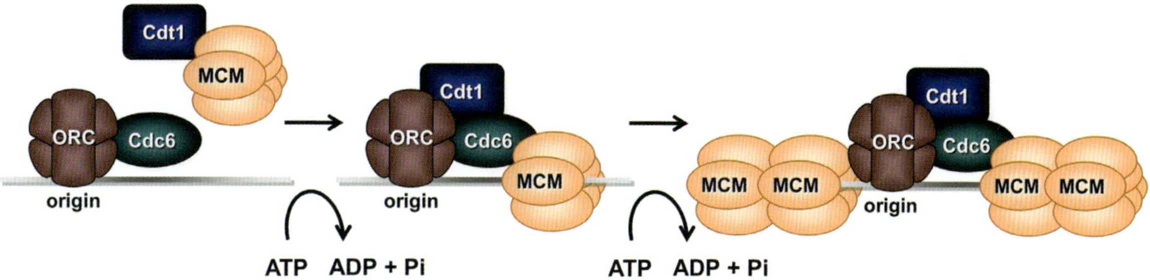

Figure 4.2 Origin licensing by pre-replication complex assembly.

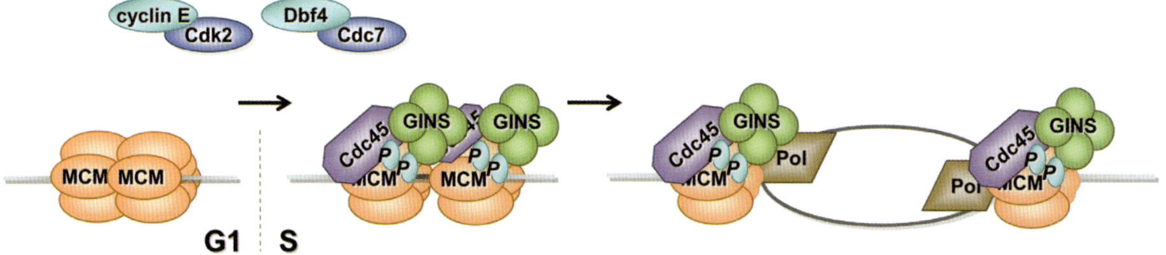

Figure 4.3 Origin firing.

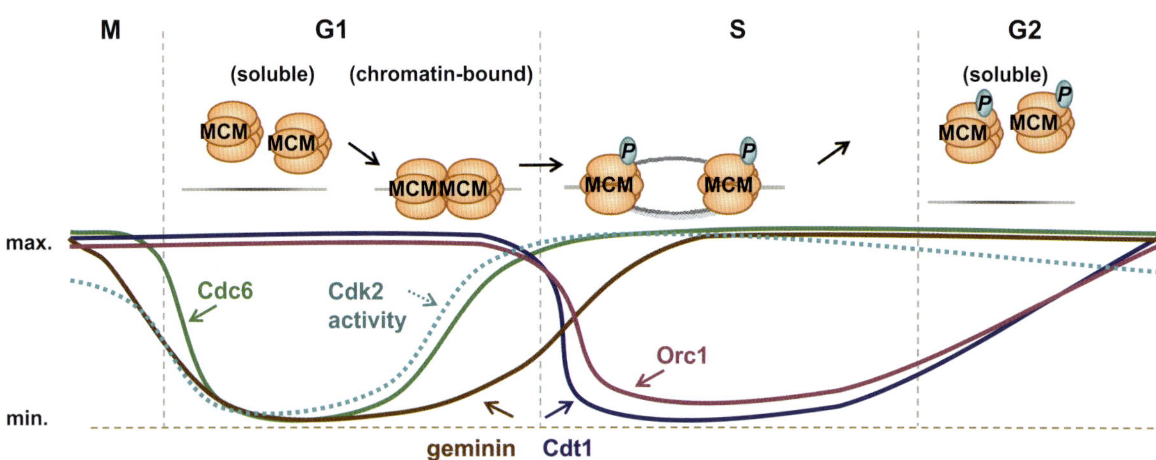

Figure 4.4 Regulation of origin licensing.

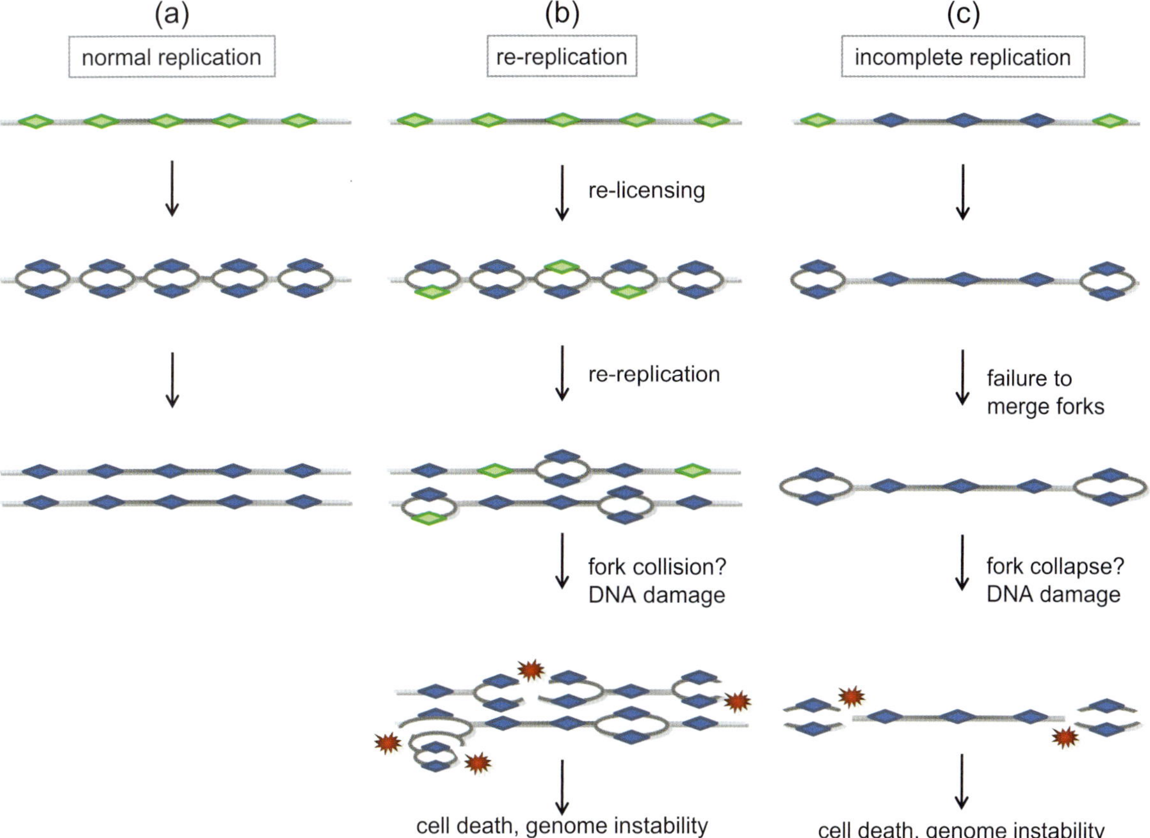

Figure 4.5 Deregulated origin licensing results in DNA damage and genome instability.

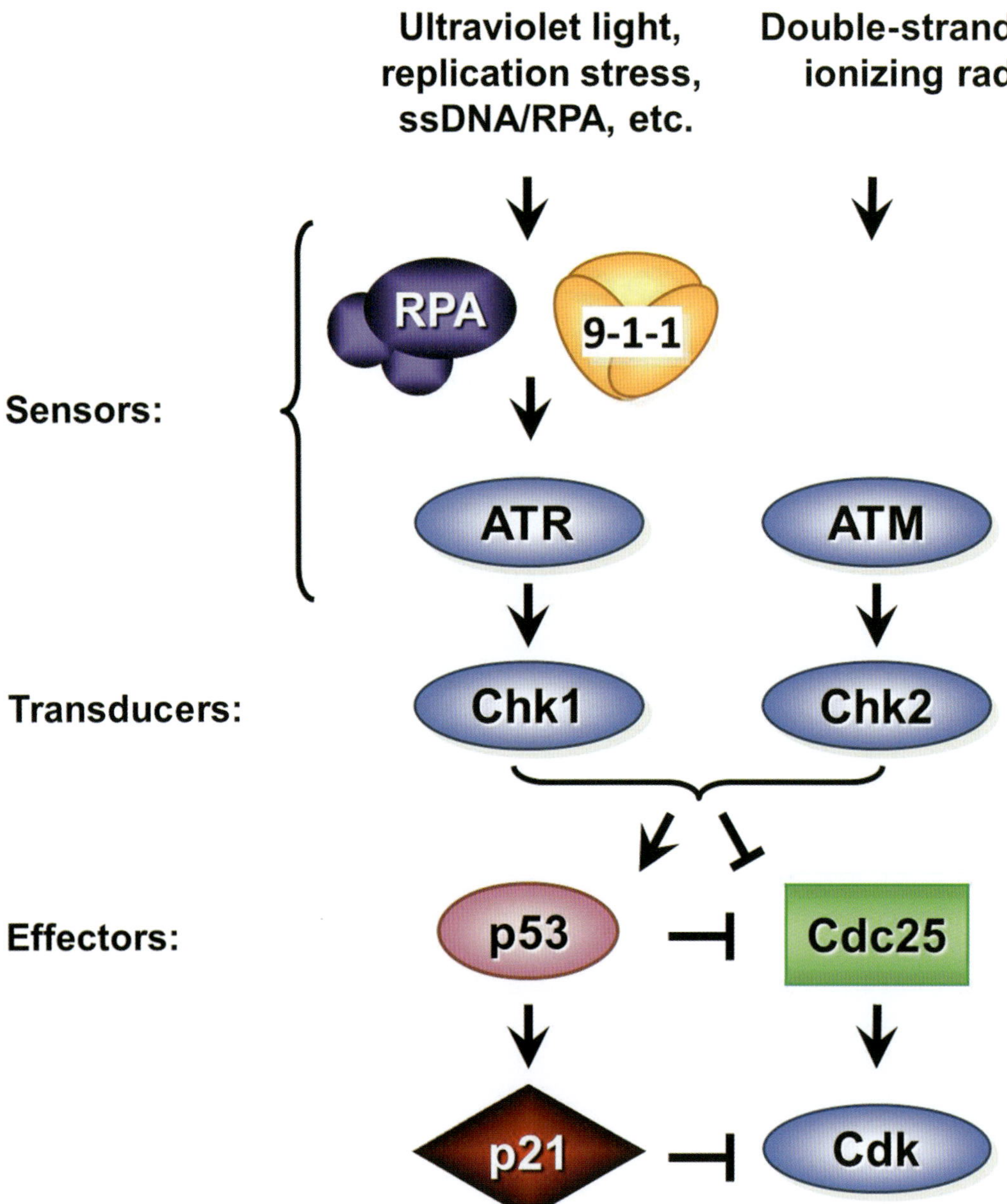

Figure 4.6 DNA damage checkpoint signaling.

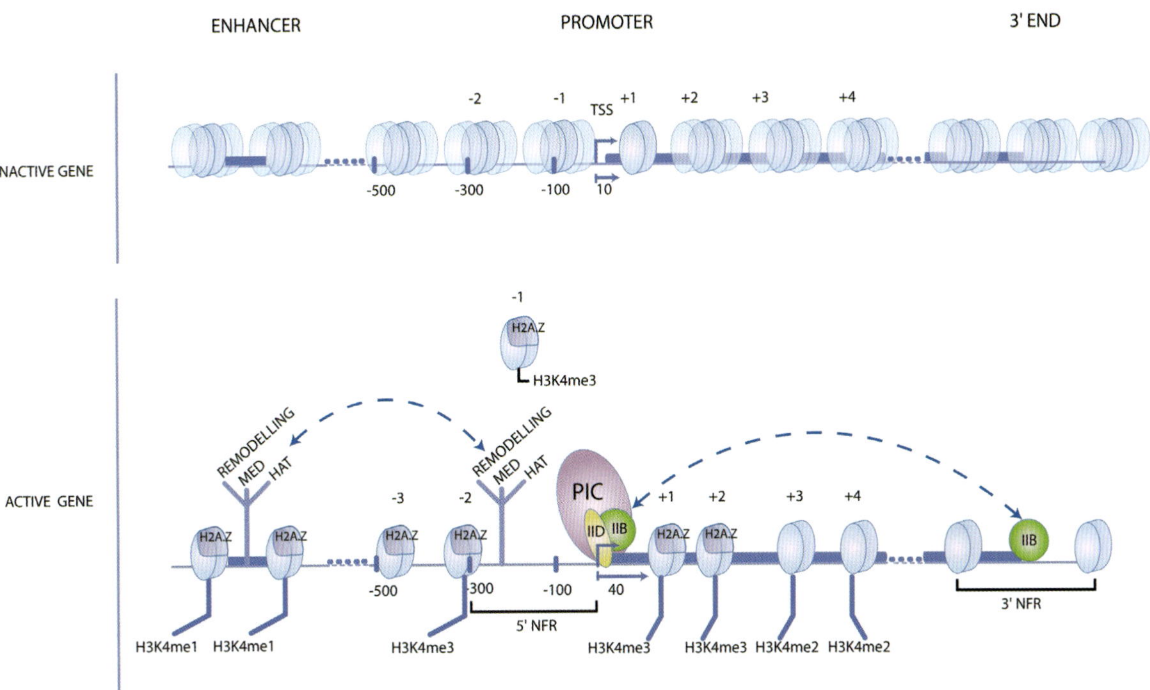

Figure 5.1 Structural and mechanistic changes in the chromatin organization of the Pol II regulatory elements upon gene activation.

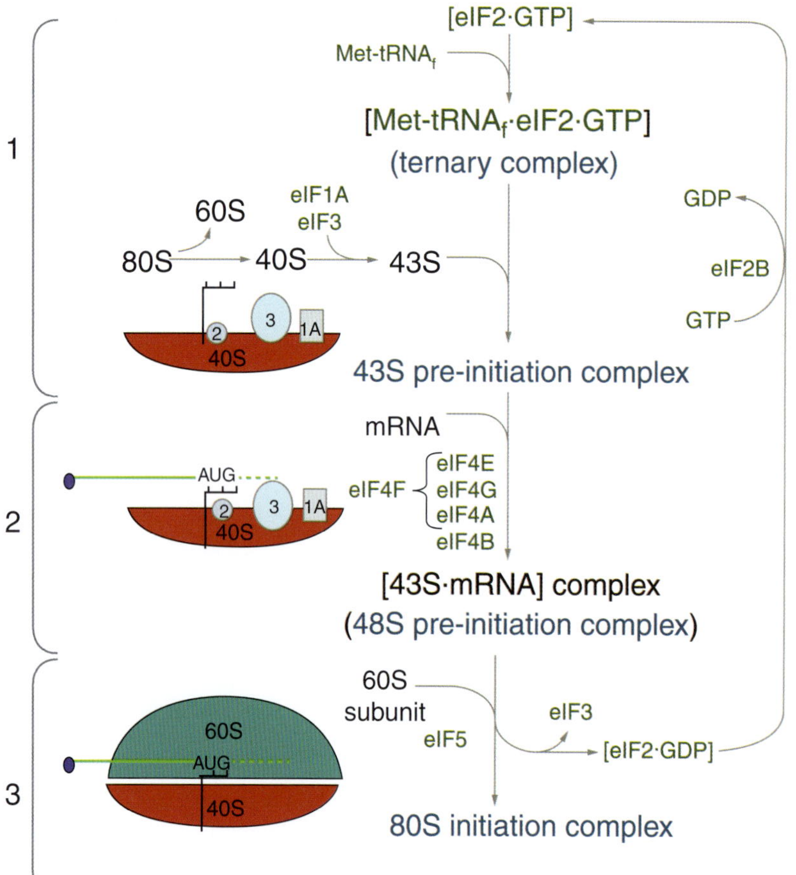

Figure 6.1 The pathway of cap-dependent polypeptide chain initiation.

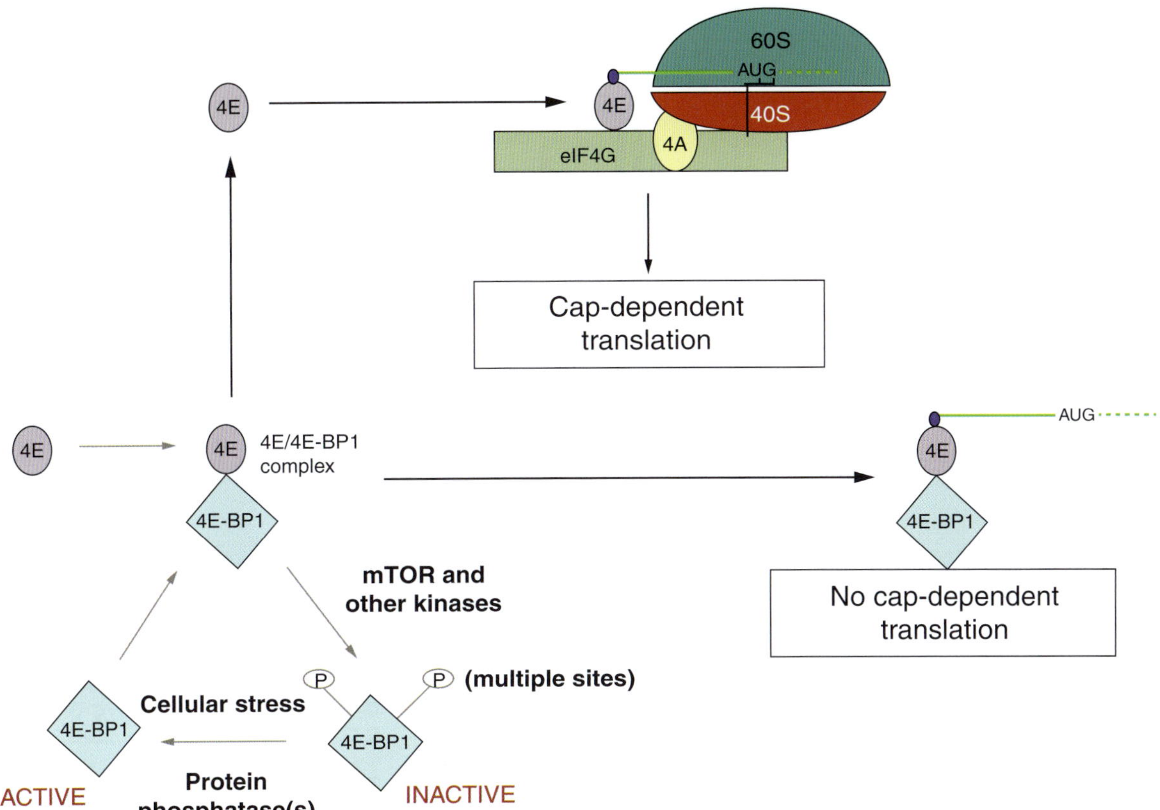

Figure 6.2 Role of the eIF4E binding proteins in the control of cap-dependent translation.

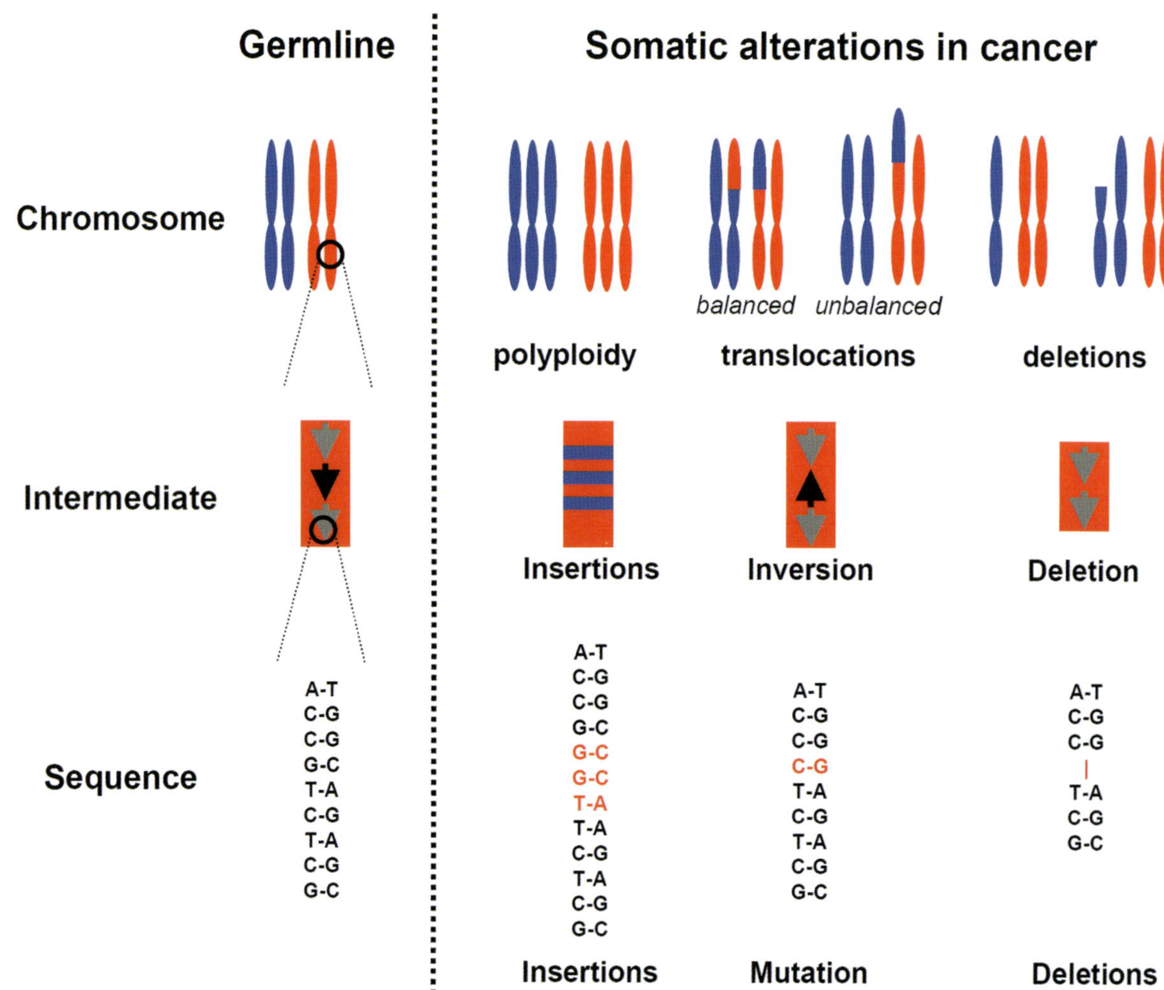

Figure 7.3 Somatic genetic alterations are observed in cancer cells at all scales.

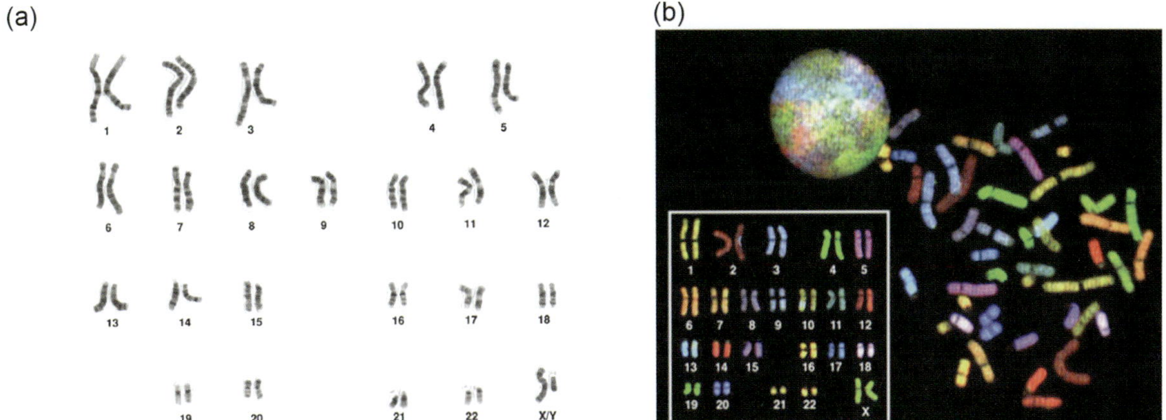

Figure 7.4 Techniques to identify chromosomes. (a) Normal male karyotype. (b) Spectral karyotype of a normal female cell. From NHGRI (www.genome.gov/glossary.cfm).

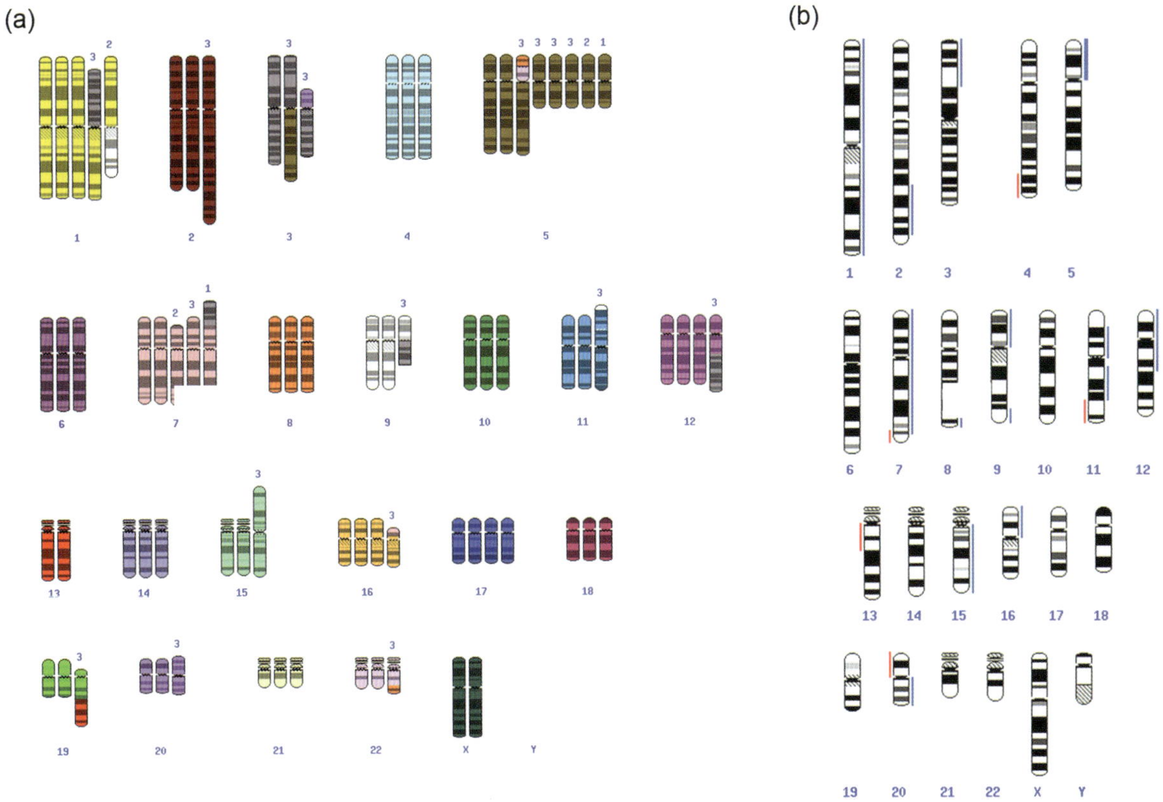

Figure 7.5 (a) Karyogram representing chromosomal aberrations in Hela cells. (b) CGH data on Hela cells reveals regions that have high (green) or low (red) copy number relative to a diploid cell line. Generated by NCBI Cancer Chromosomes (www.ncbi.nlm.nih.gov/sites/entrez?db=cancerchromosomes).

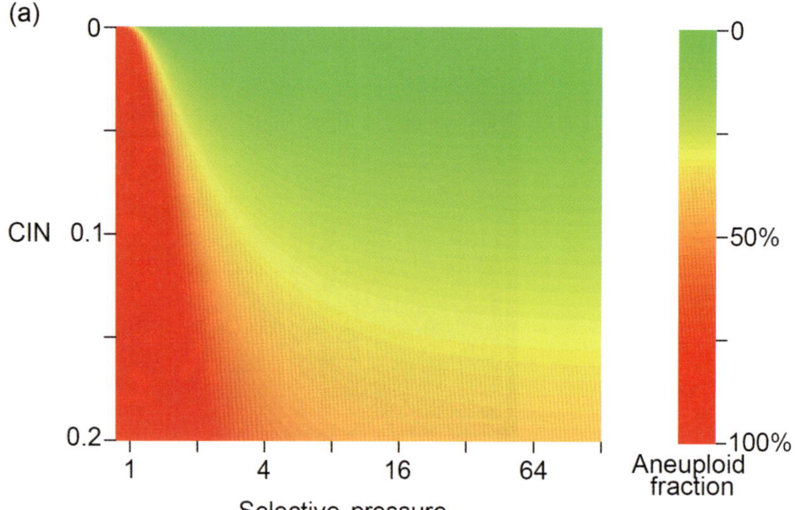
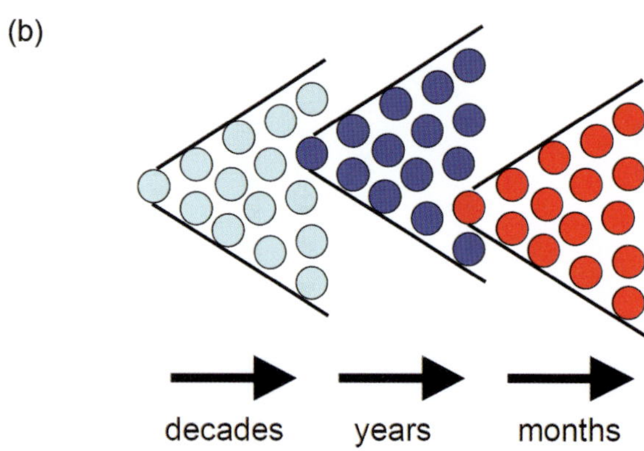

Figure 7.6 A model of CIN, aneuploidy, and selection.

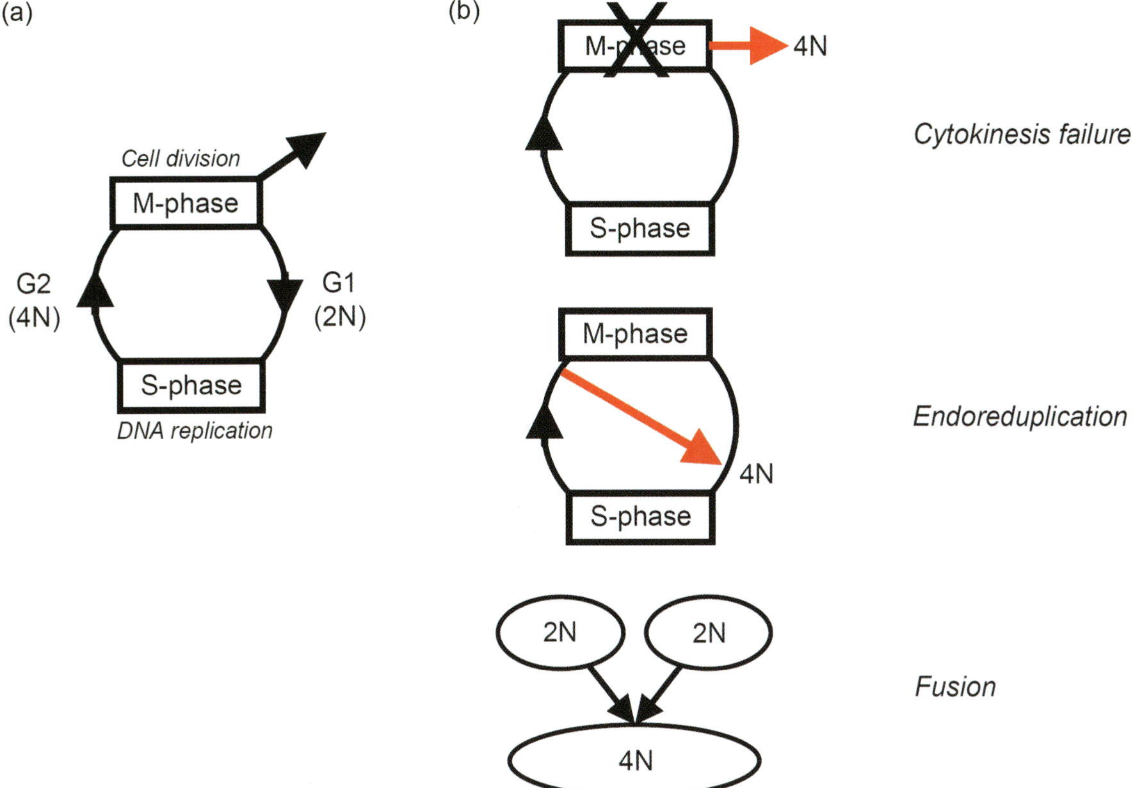

Figure 7.7 Regulation of the cell cycle protects cell ploidy (number of chromosome sets).

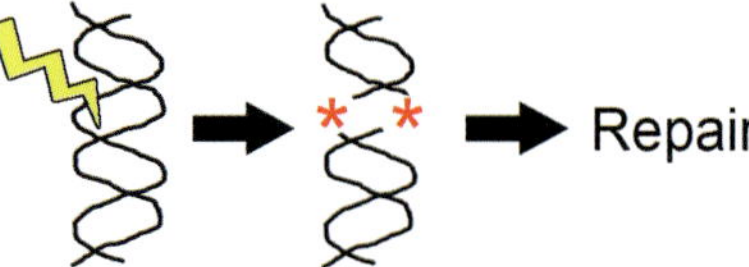

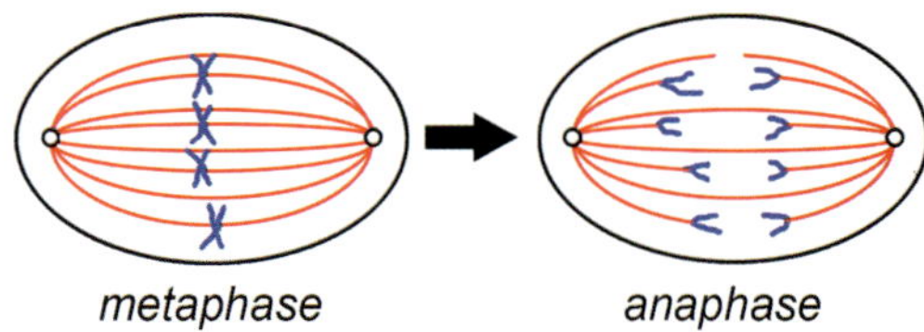

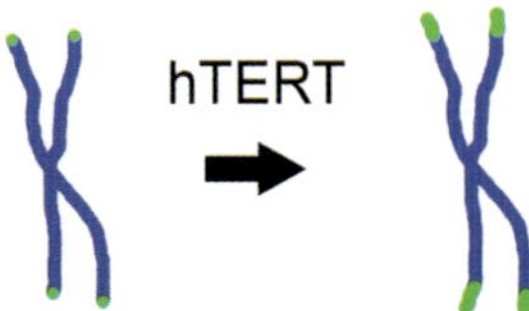

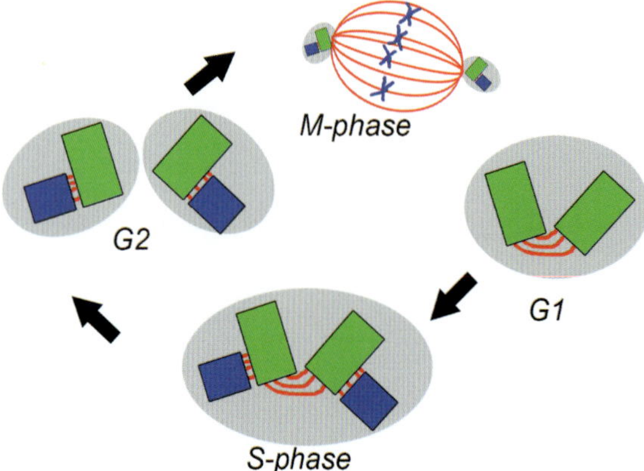

Figure 7.8 Mechanisms to maintain genomic integrity of cells.

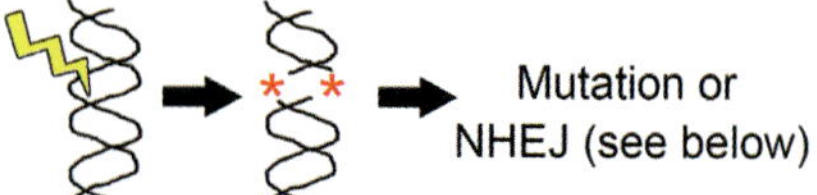

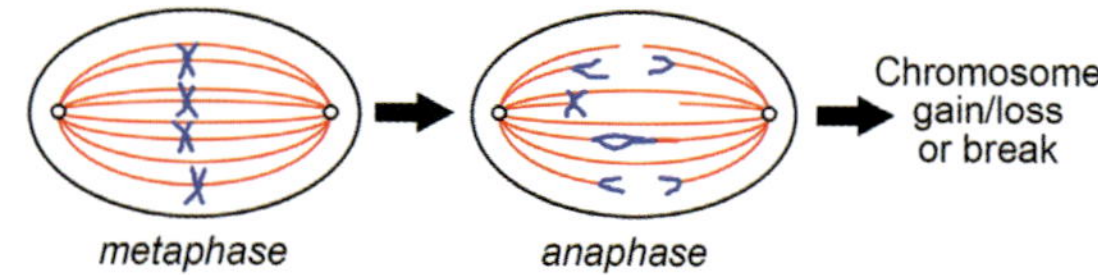

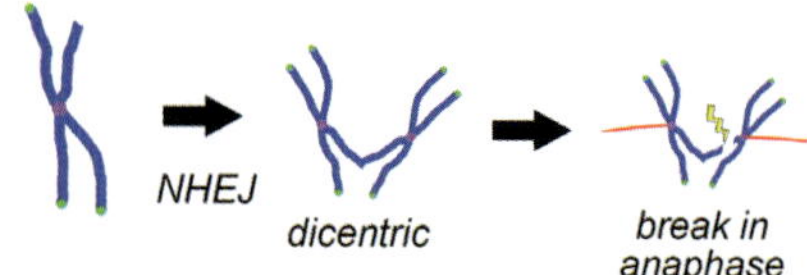

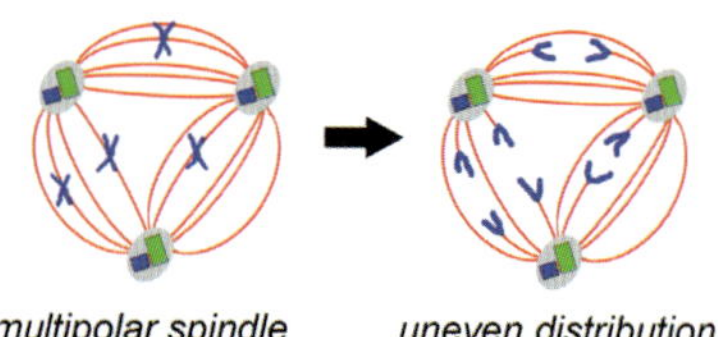

Figure 7.9 Genetic integrity can be lost by interruption of protective mechanisms.

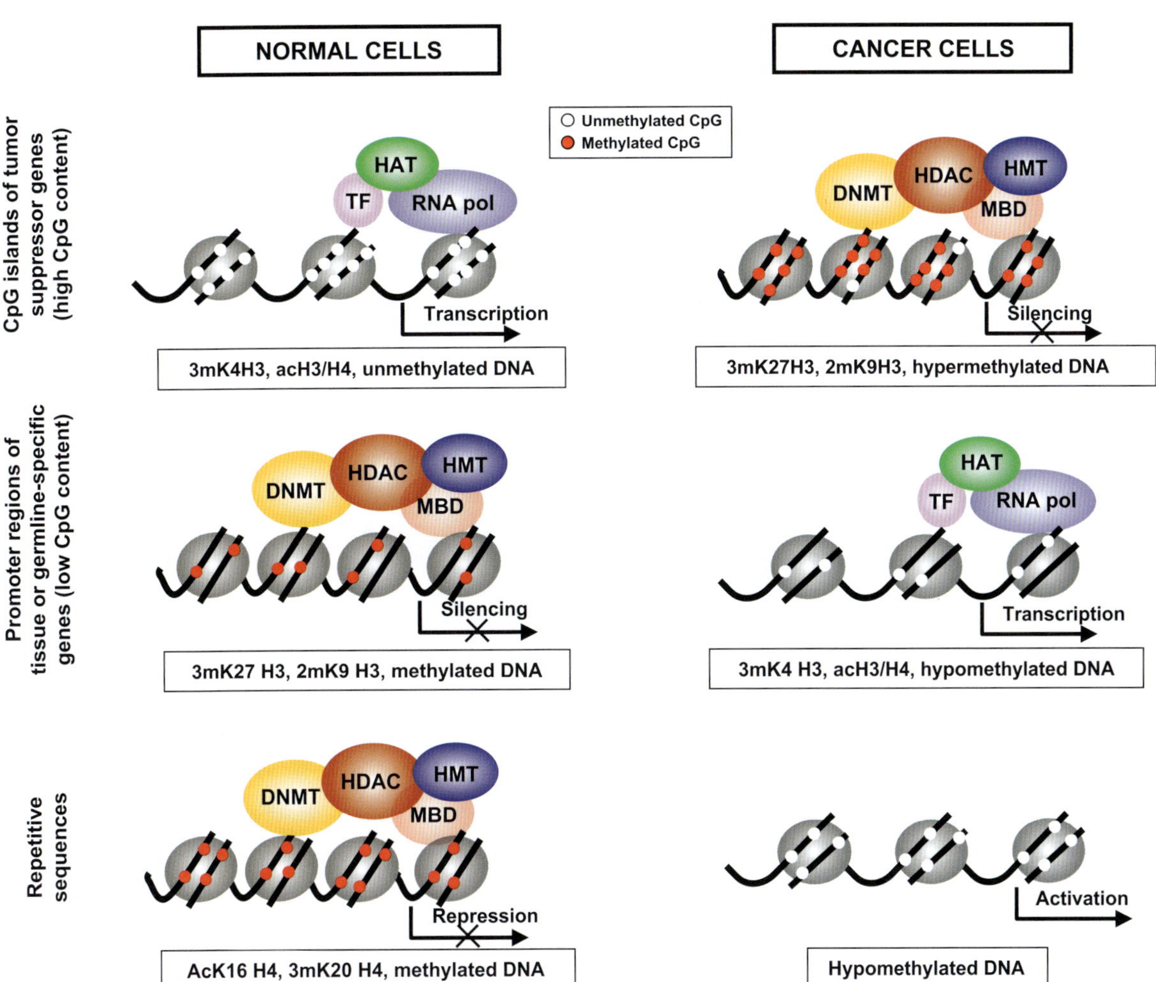

Figure 8.1 Summary of the most common epigenetic changes in normal and cancer cells.

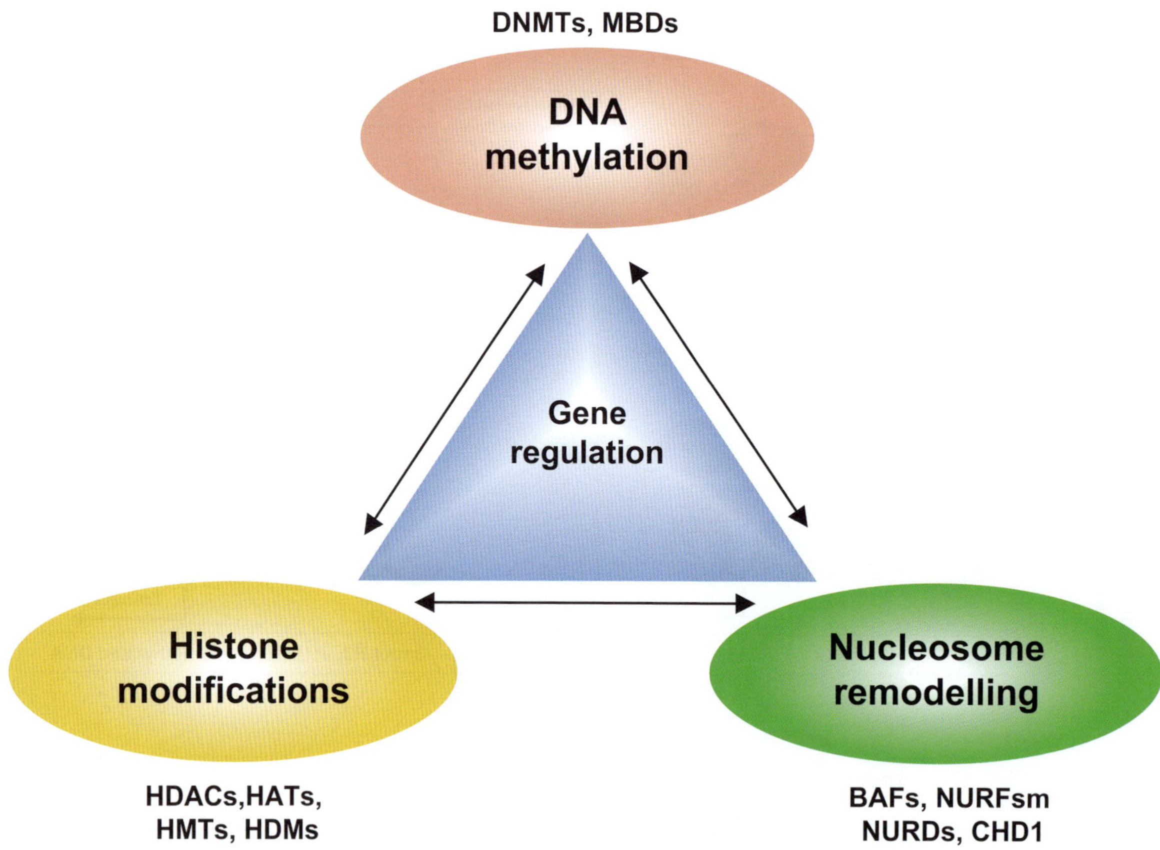

Figure 8.2 Mechanisms of epigenetic regulation.

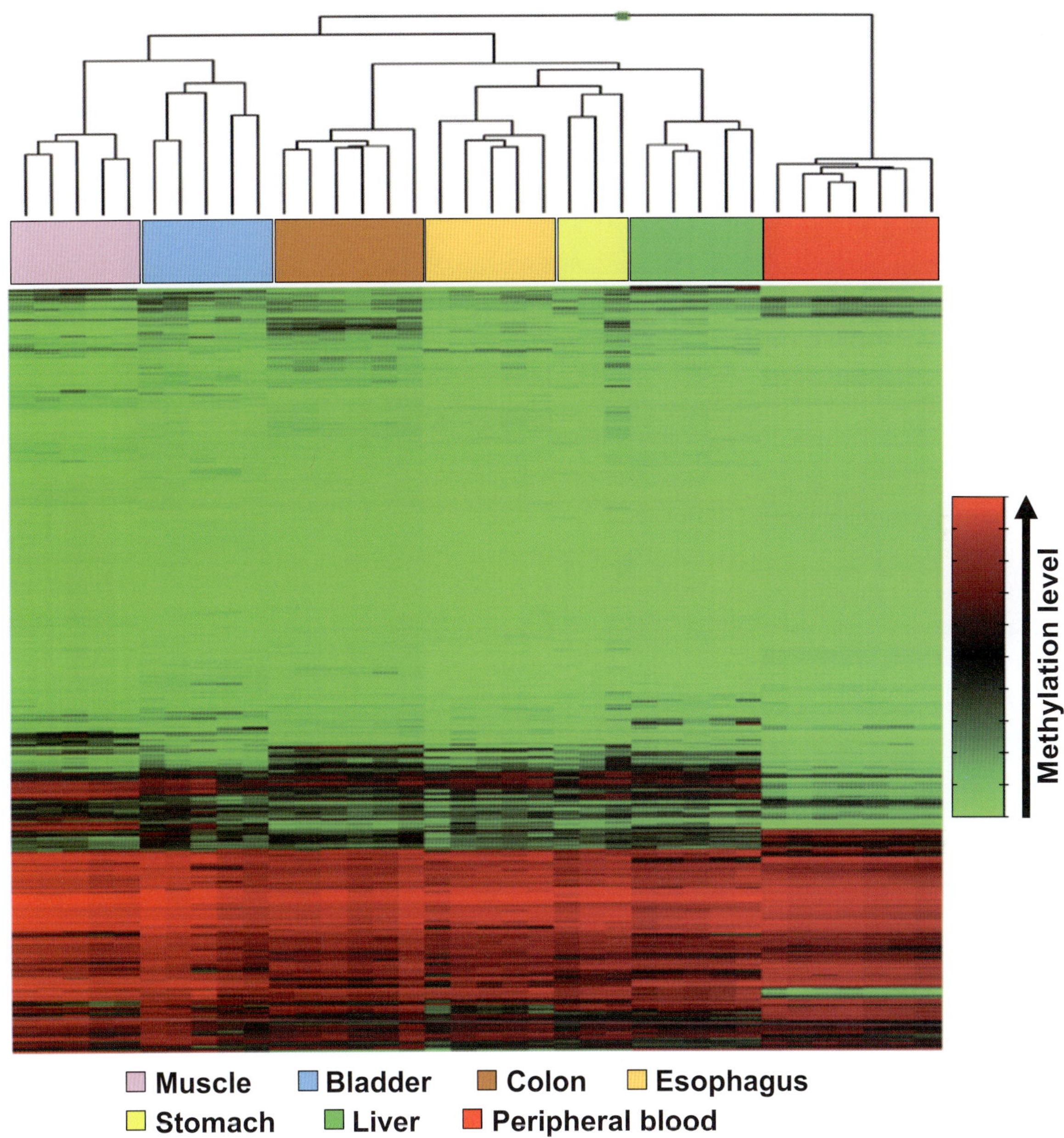

Figure 8.3 Epigenetic profiling of normal tissues.

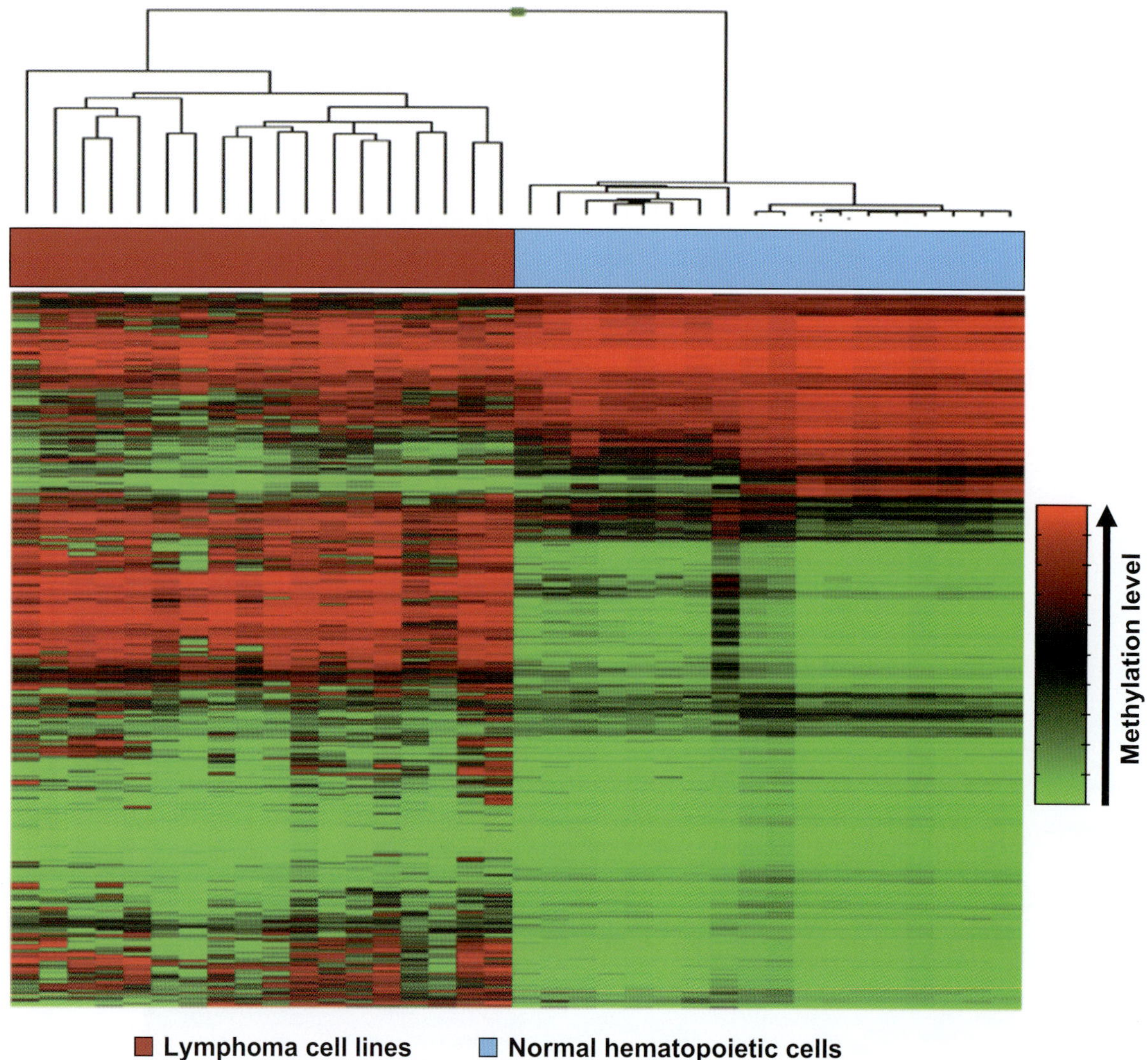

Figure 8.4 Epigenomic profiling of cancer cells.

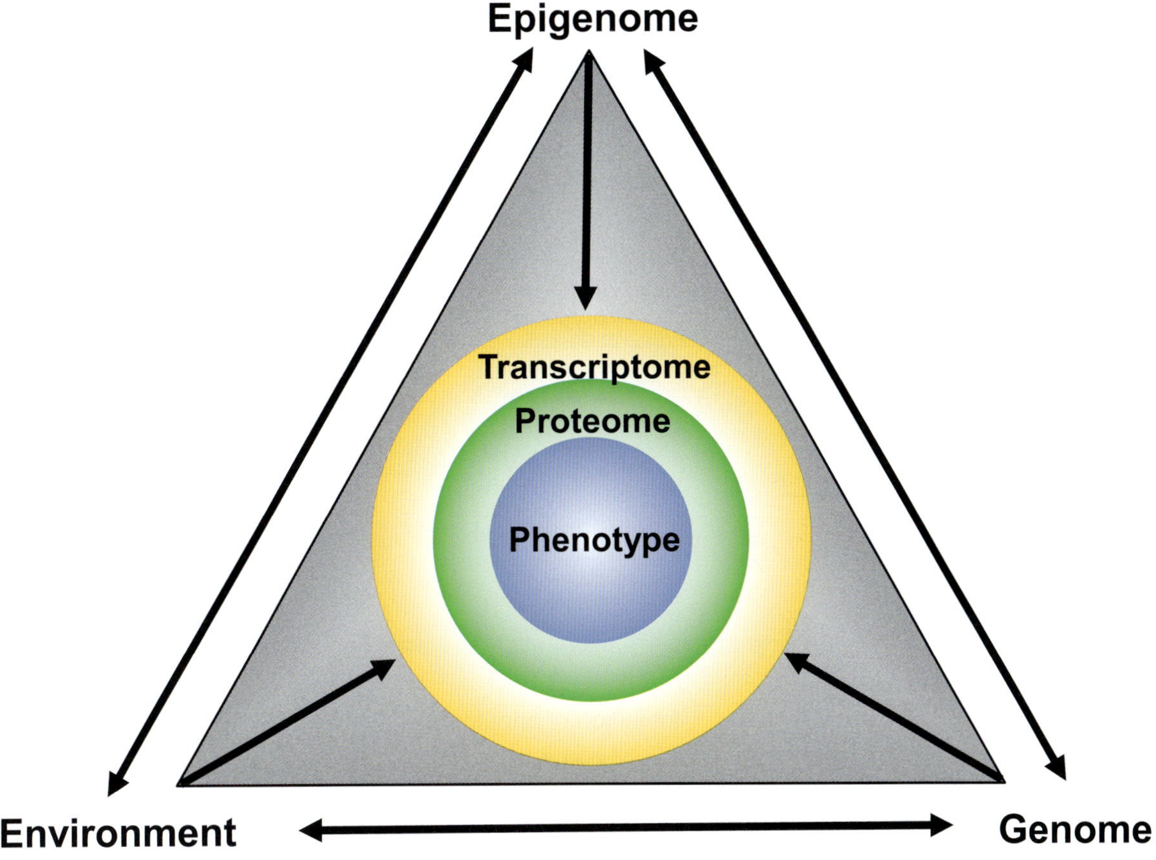

Figure 8.5 Integrative model of environmental, epigenetic, and genetic factors.

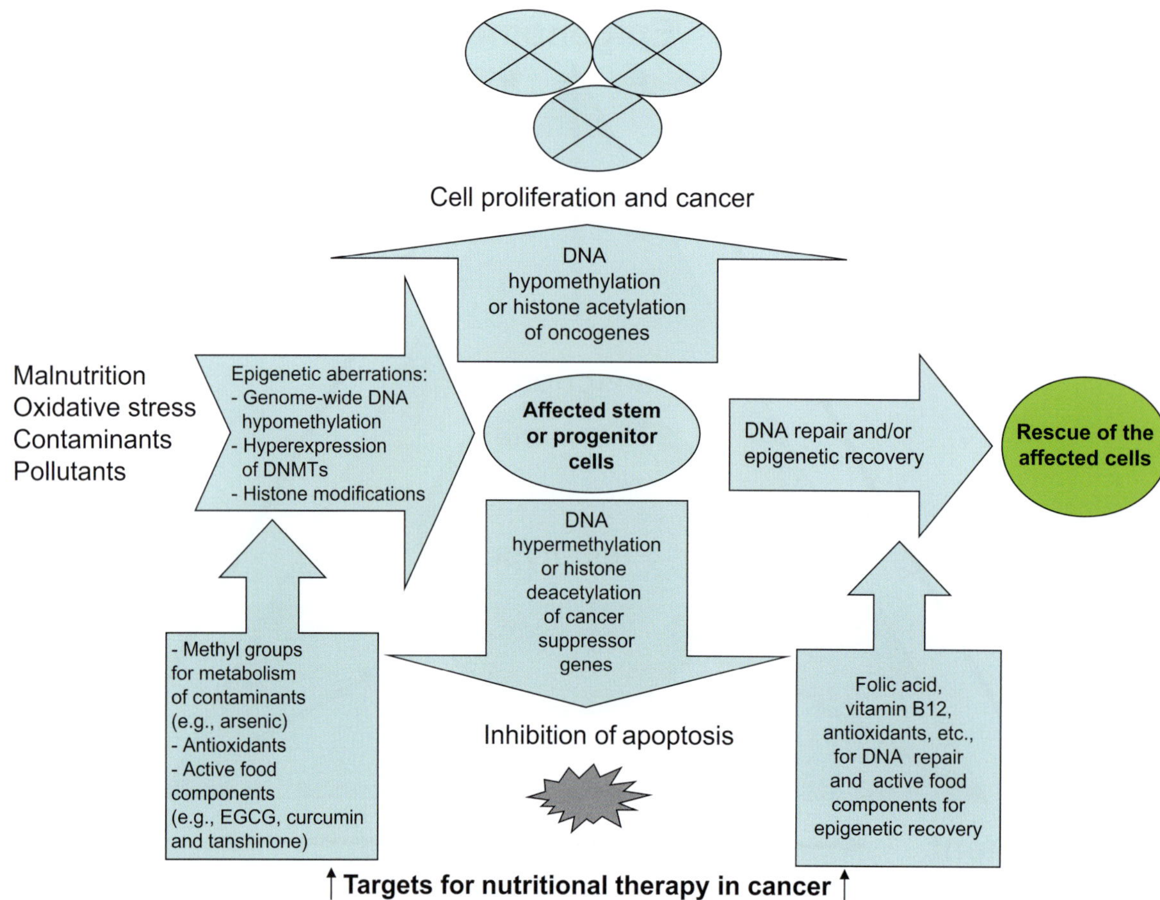

Figure 10.1 Proposed effects of various dietary and environmental factors on epigenetic aberrations in stem and progenitor cells and development and progression of cancer.

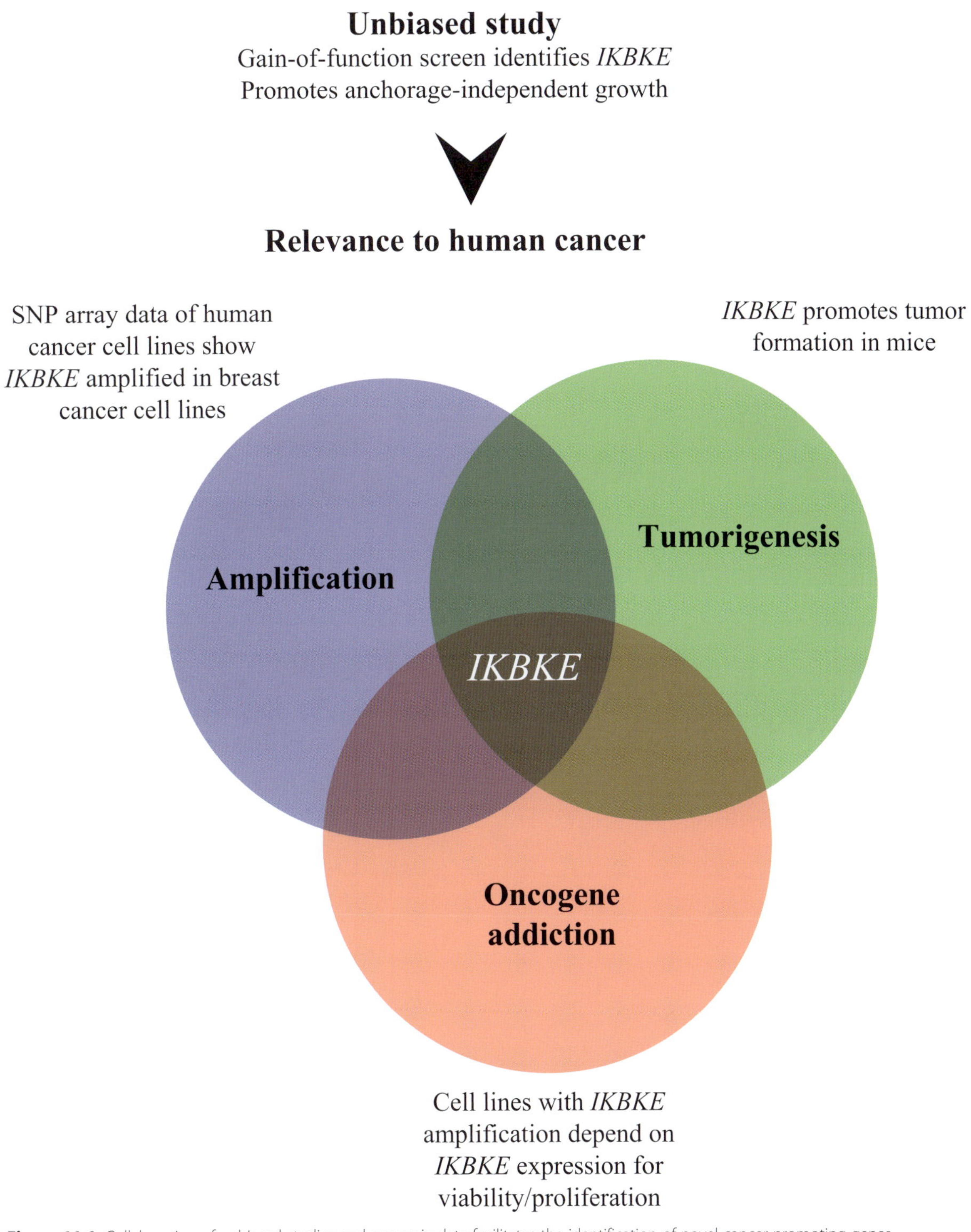

Figure 11.1 Collaboration of unbiased studies and genomic data facilitates the identification of novel cancer promoting genes.

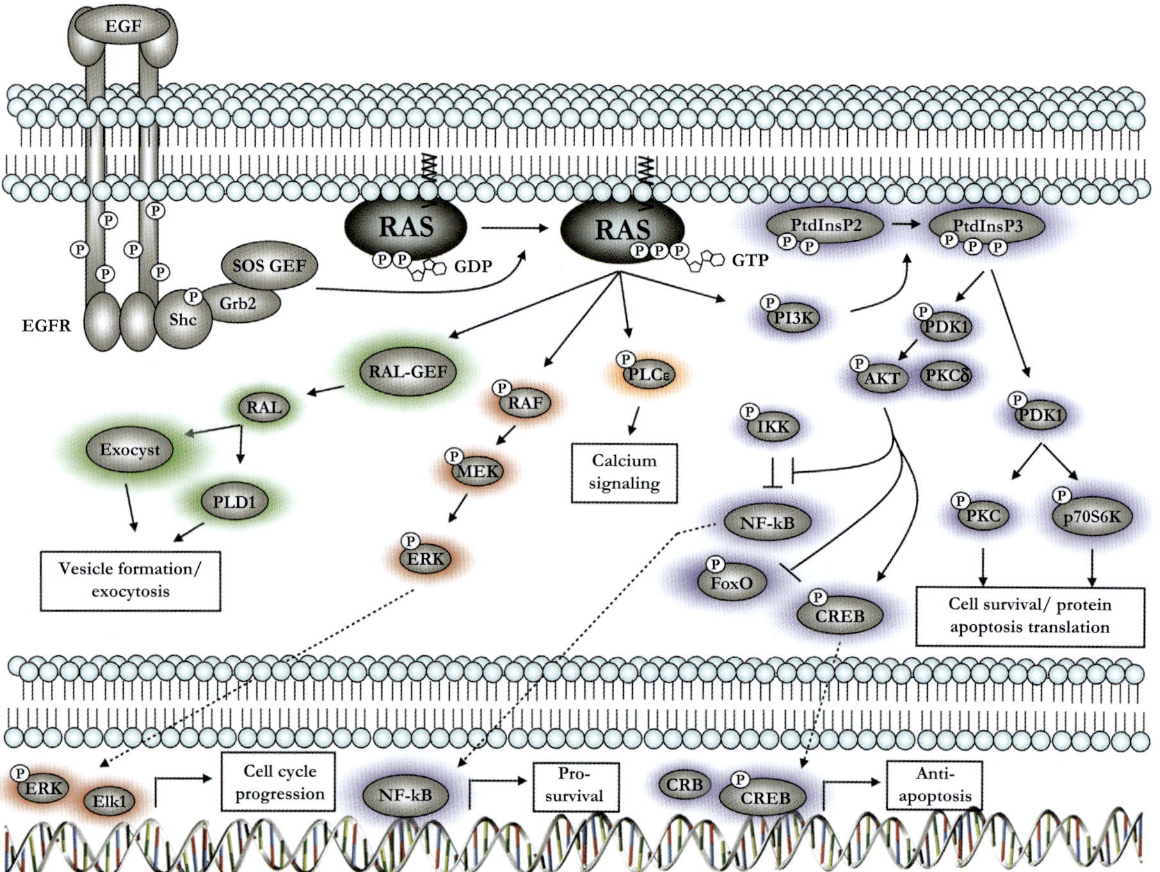

Figure 12.1 RAS-associated signaling events.

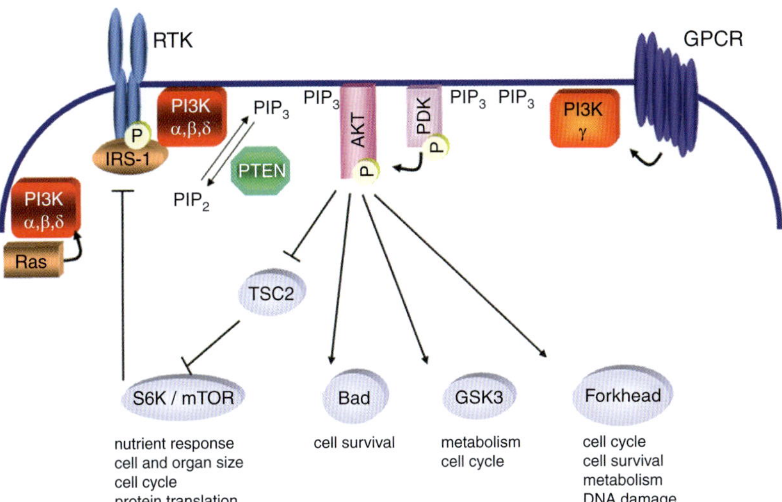

Figure 13.1 General scheme of the PKB/AKT pathway.

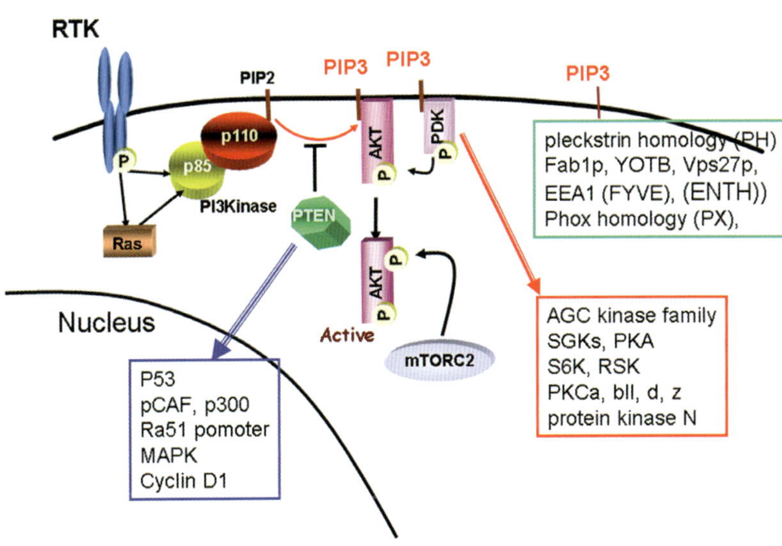

Figure 13.2 General scheme of the alternative signaling to canonical AKT activation.

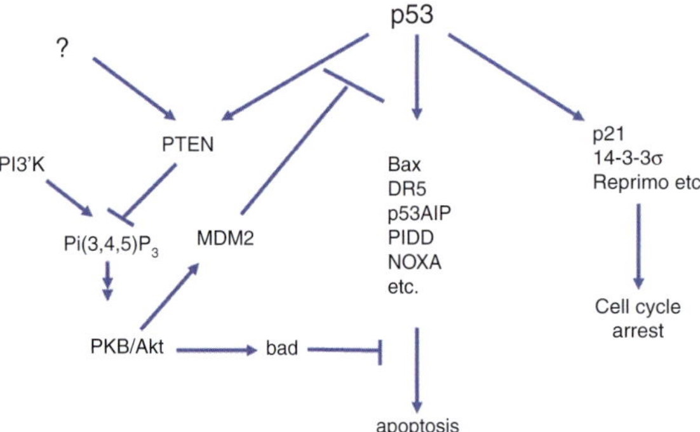

Figure 13.3 Functional relationship between PTEN and p53 tumor suppressor pathways.

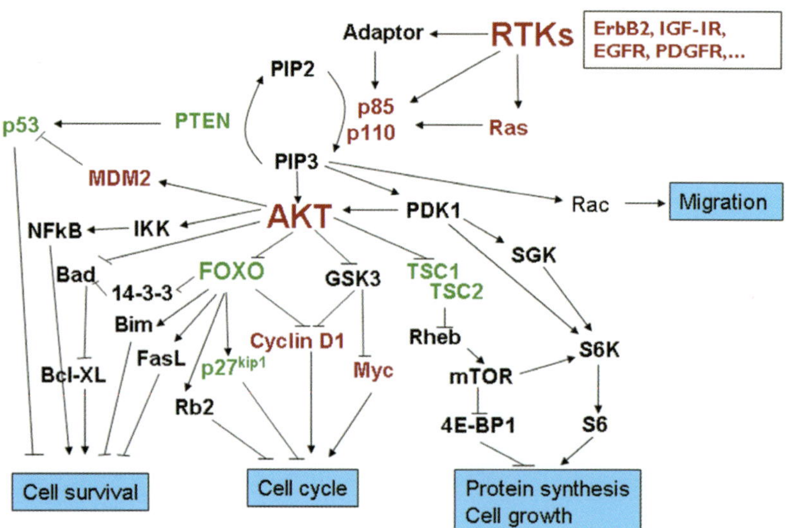

Figure 13.4 General overview of PI3K pathway alterations found in human tumors.

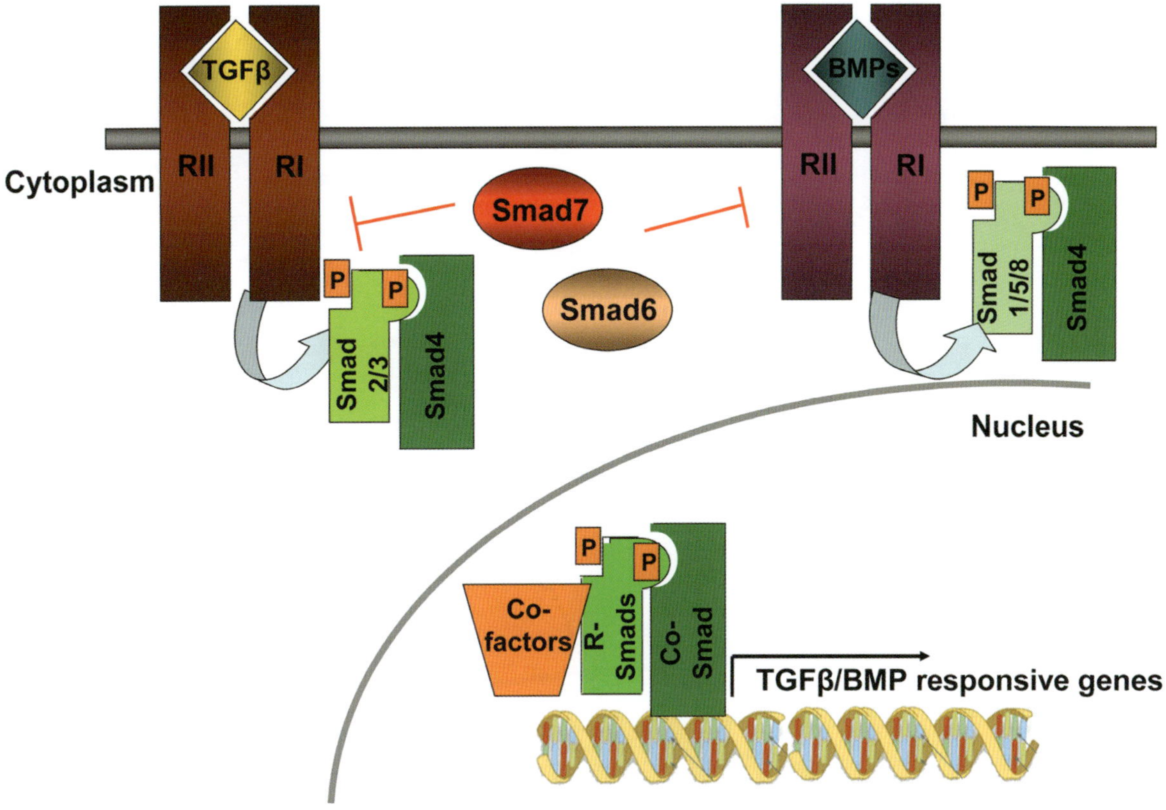

Figure 14.1 TGFβ and BMP signal transduction pathways.

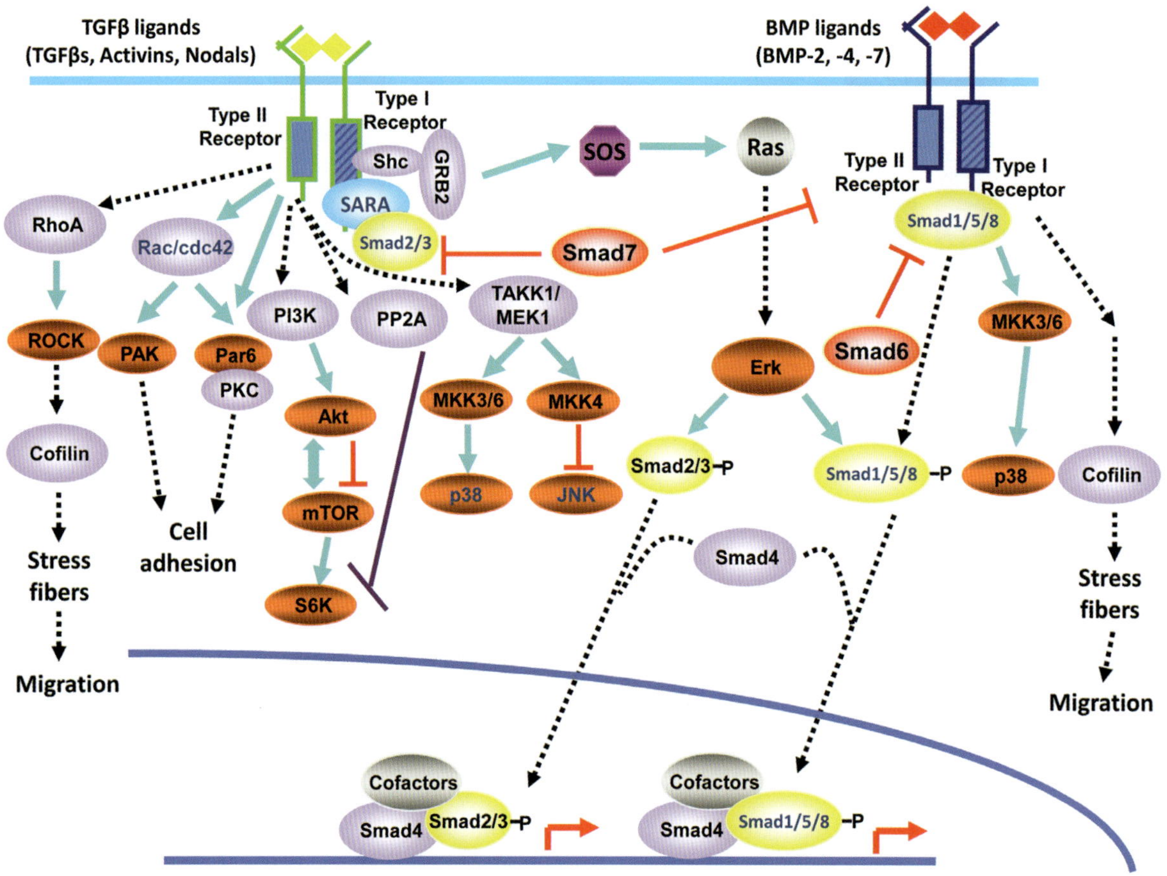

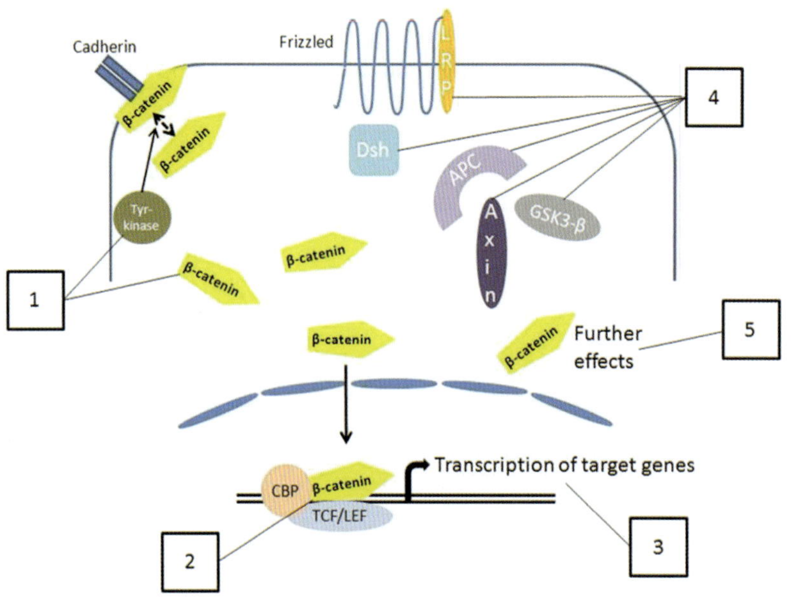

Figure 14.2 Crosstalk of TGFβ/BMP components with auxiliary signaling molecules.

Figure 15.2 β-catenin/Wnt signaling activity in cancer cells and possible points of interfering.

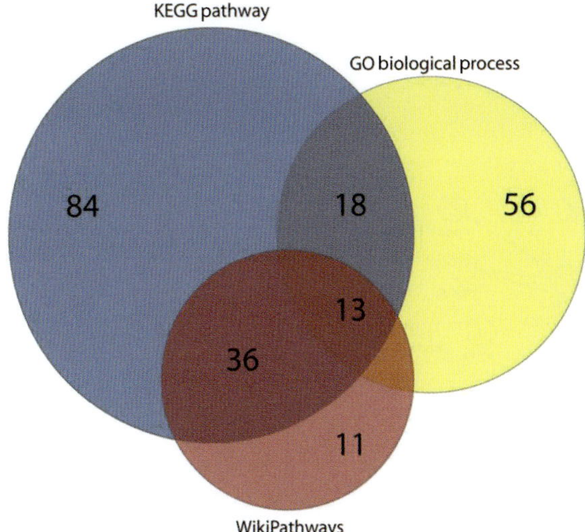

Figure 15.3 Disparity in Wnt signaling pathway definitions across annotation databases.

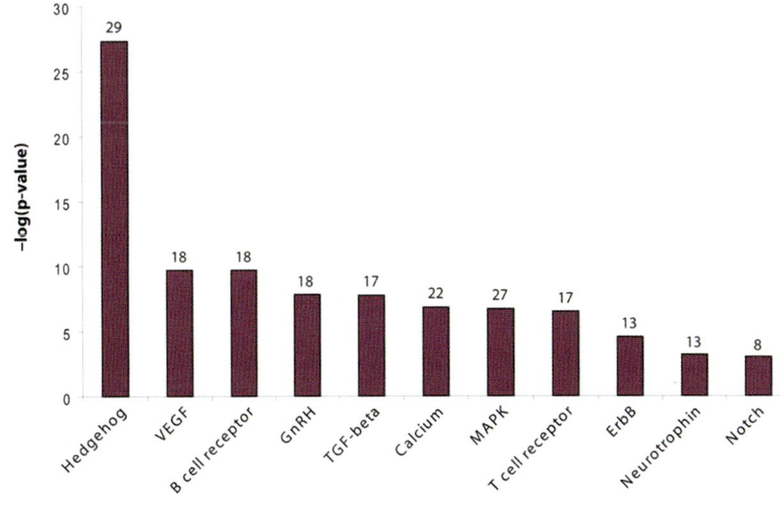

Figure 15.4 Wnt pathway crosstalk with other signaling pathways determined using enrichment analysis.

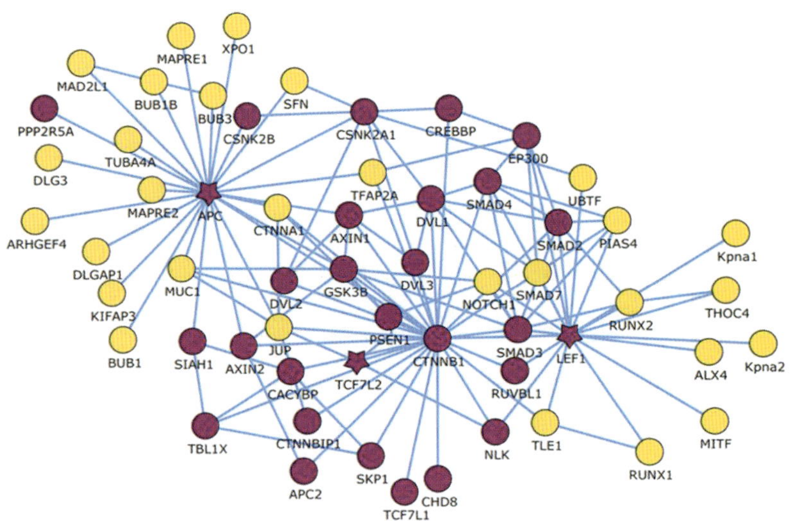

Figure 15.5 Partial Wnt pathway protein–protein interaction network.

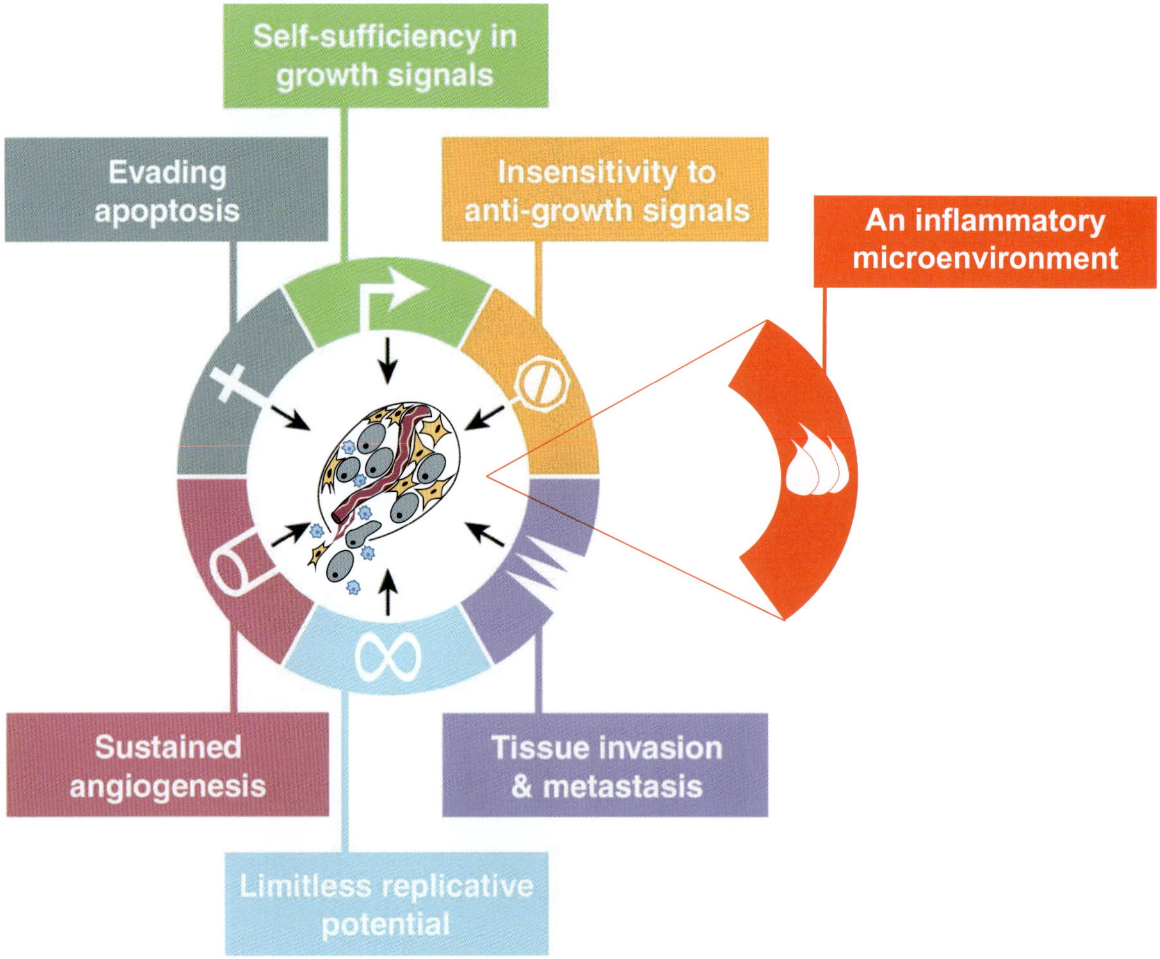

Figure 17.1 Inflammation as the seventh hallmark of cancer. An integration to the six hallmarks of cancer.

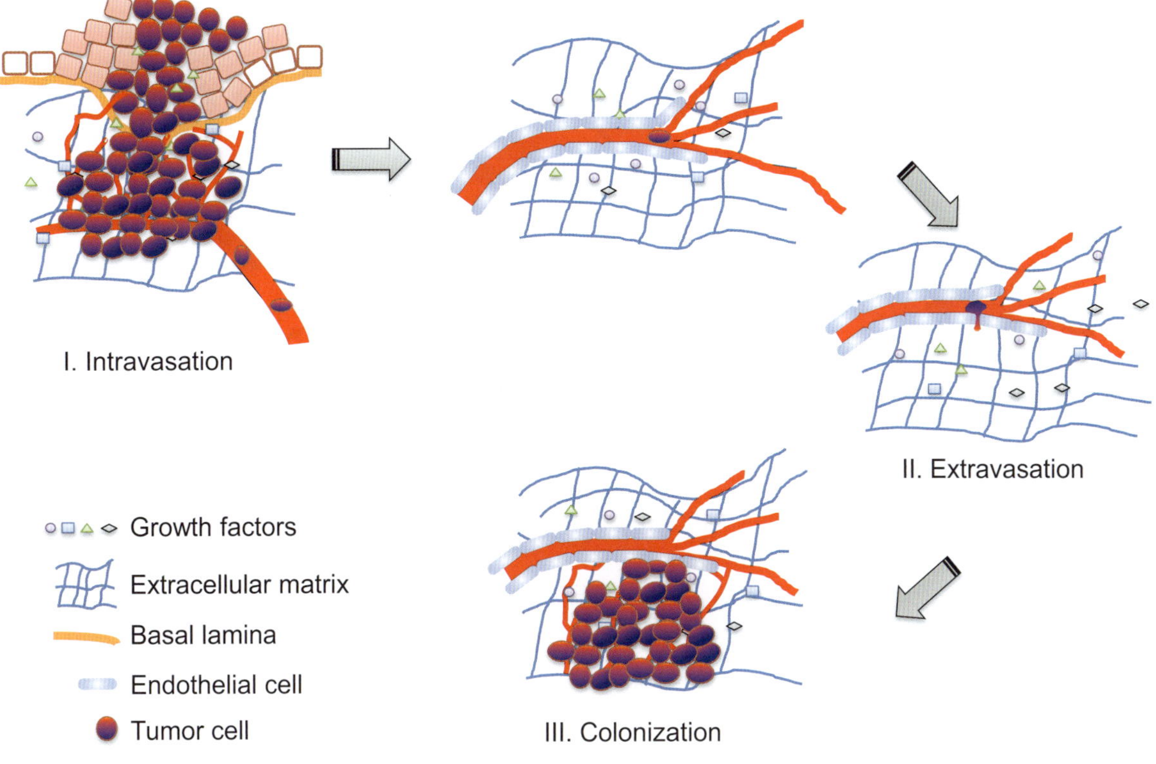

Figure 18.1 The metastatic cascade.

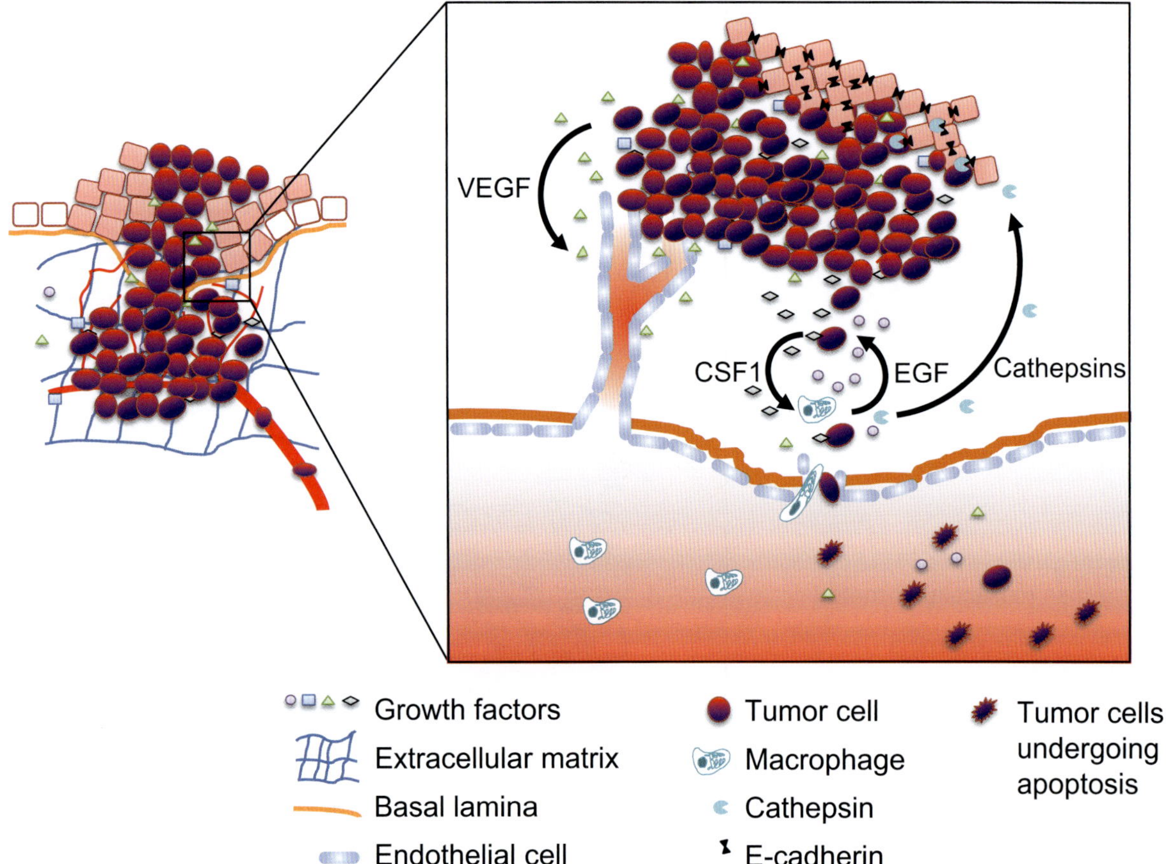

Figure 18.2 Intravasation.

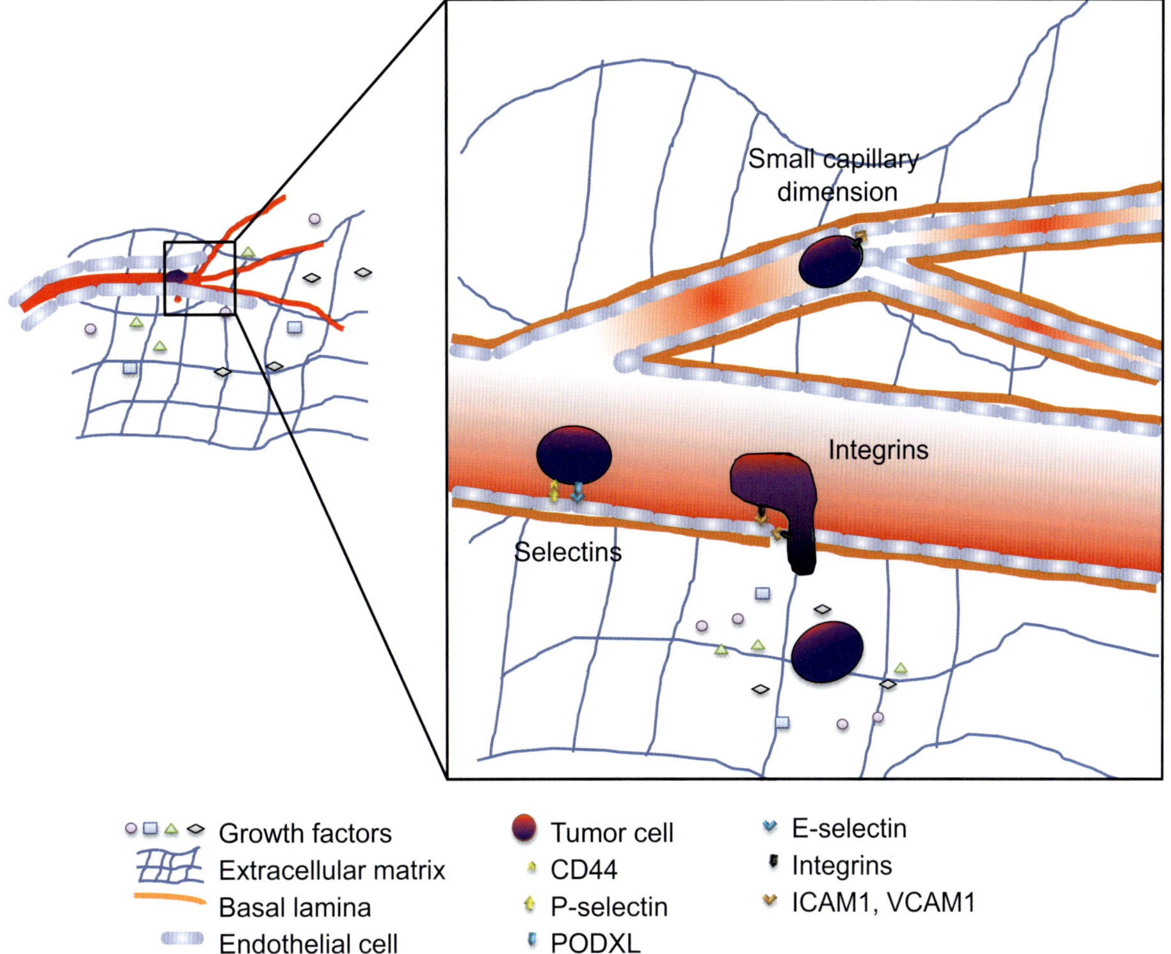

Figure 18.3 Extravasation.

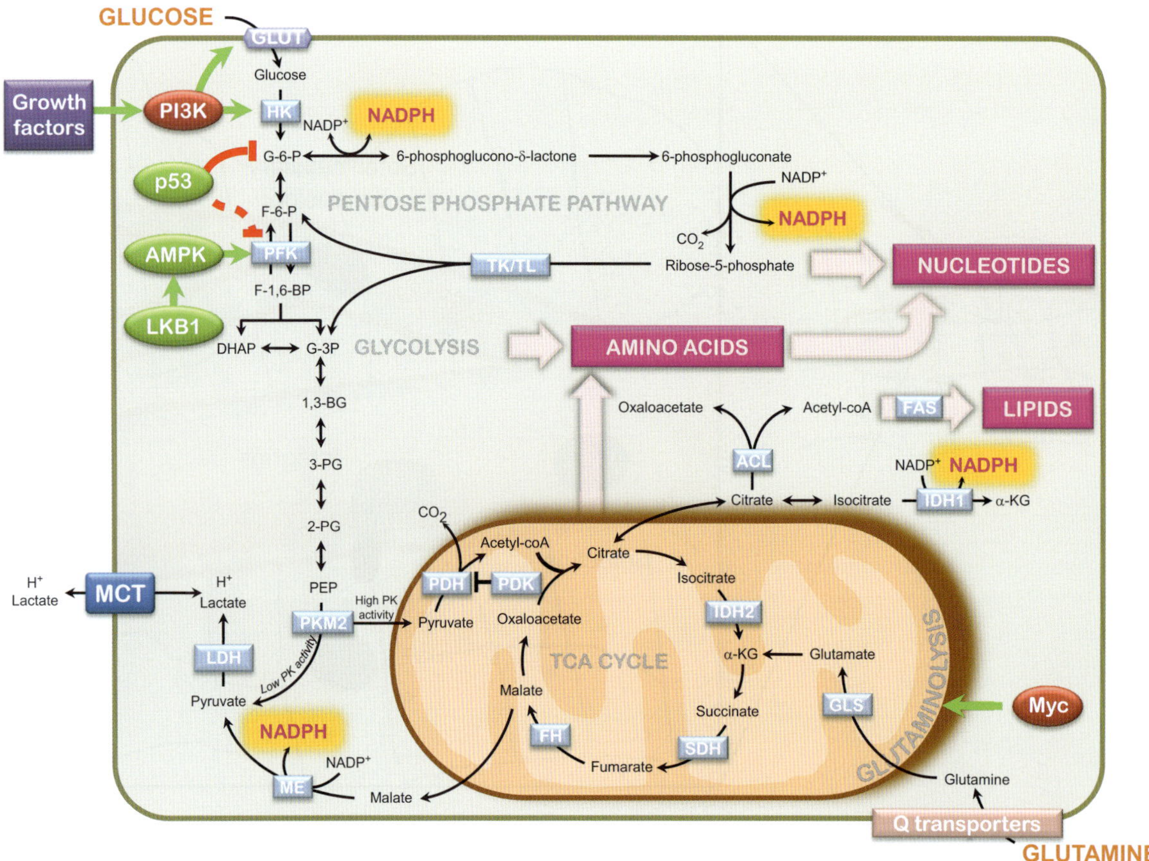

Figure 19.1 Overview of central metabolism.

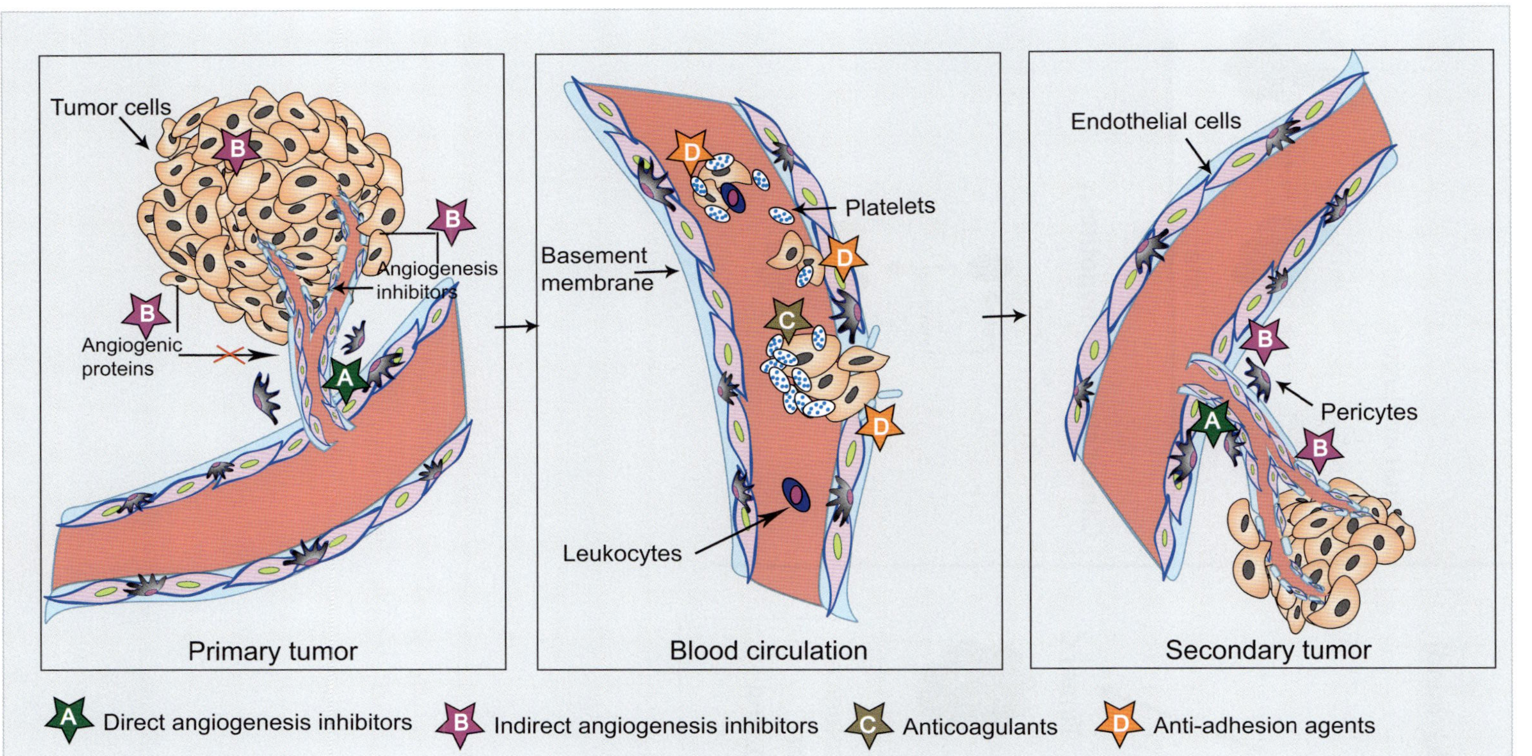

Figure 20.1 Hematogenous metastasis of cancer is a multi-step process. To combat hematogenous metastasis, overwhelming clinical and experimental findings provide a rationale for combination therapy.

(a) INTERTUMOR HETEROGENEITY

- Cell of origin (epigenetic background)
- Mutational profile
- Transcriptional networks
- Microenvironment

Tumor A   Tumor B   Tumor C

Differences could include: histological type, molecular subtype, driver mutations, metastatic tropism, and response to chemotherapy.

(b) INTRATUMOR HETEROGENEITY

Clonal evolution

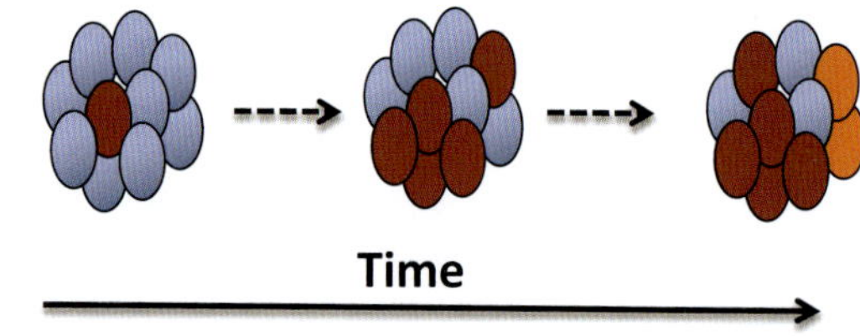

Cancer stem cell hypothesis

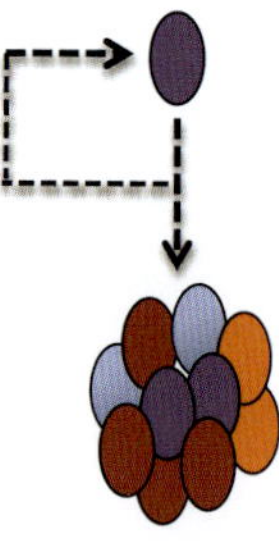

Figure 22.1 Heterogeneity in breast cancer.

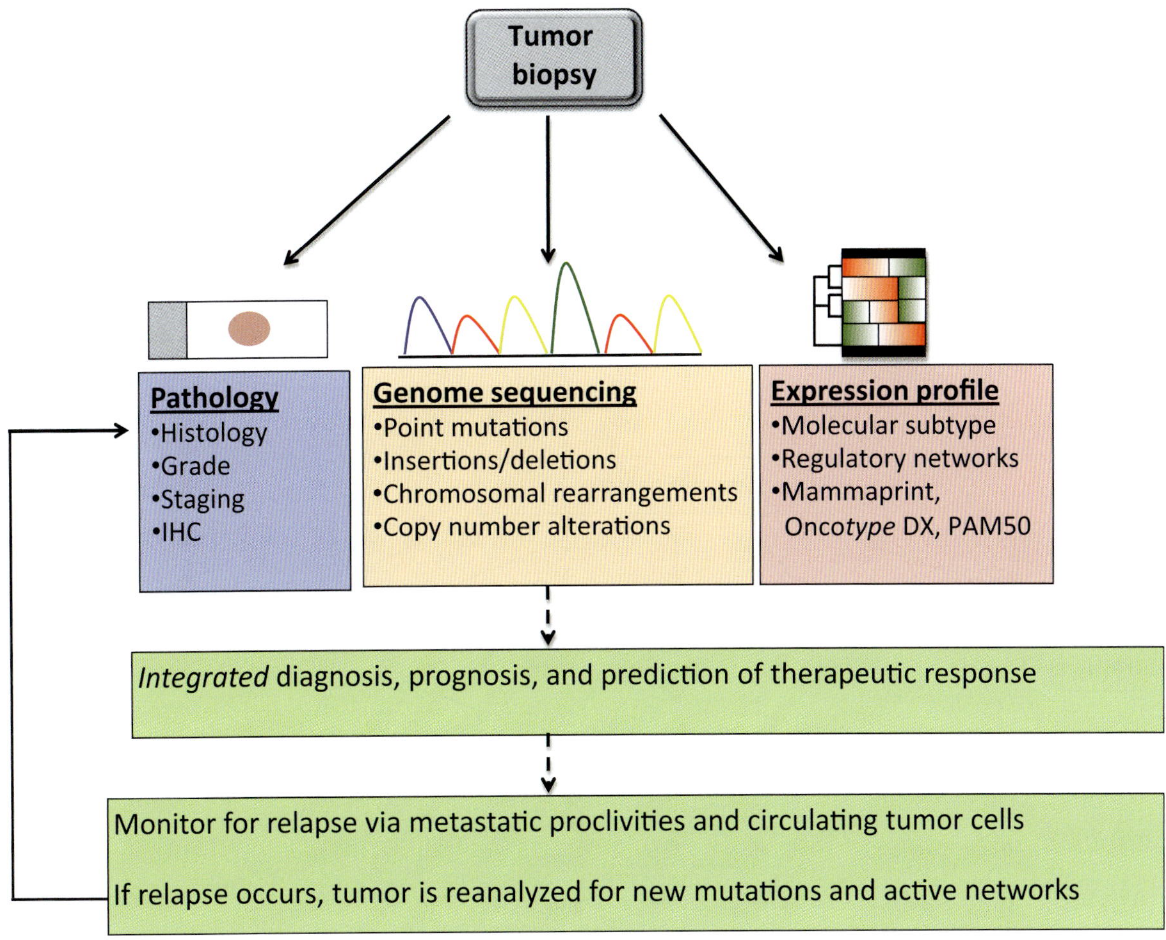

Figure 22.2 Integrated approaches to diagnose and treat breast cancer.

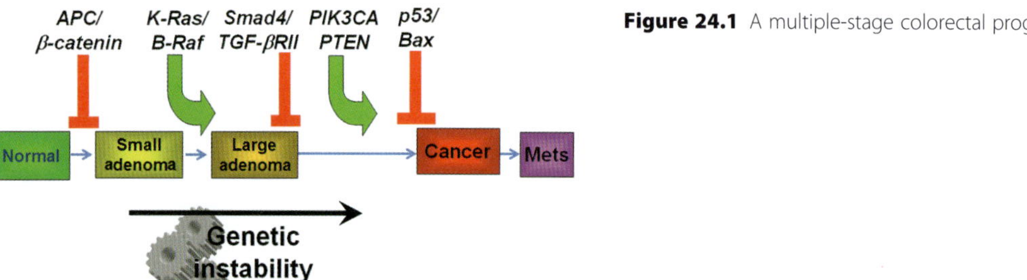

Figure 24.1 A multiple-stage colorectal progression model.

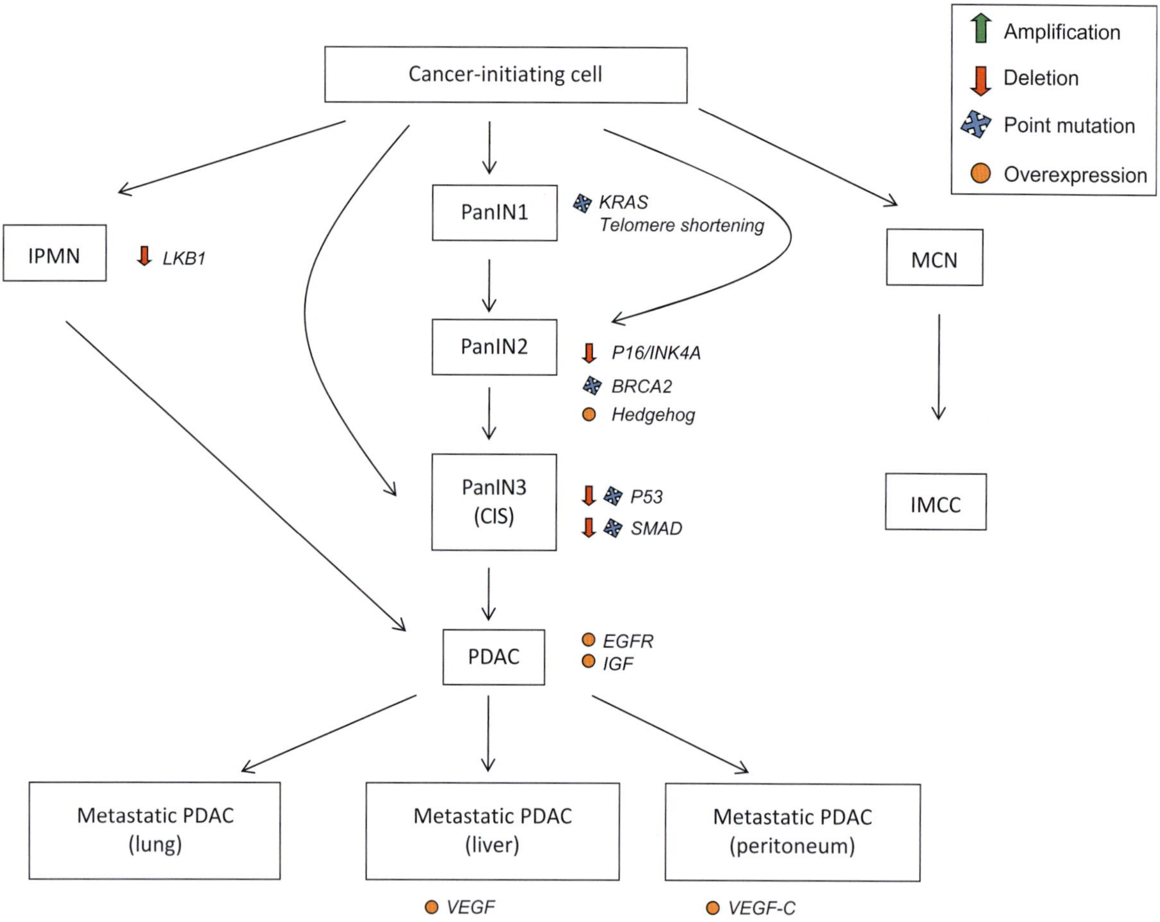

Figure 26.1 Histological and molecular patterns of pancreatic cancer progression.

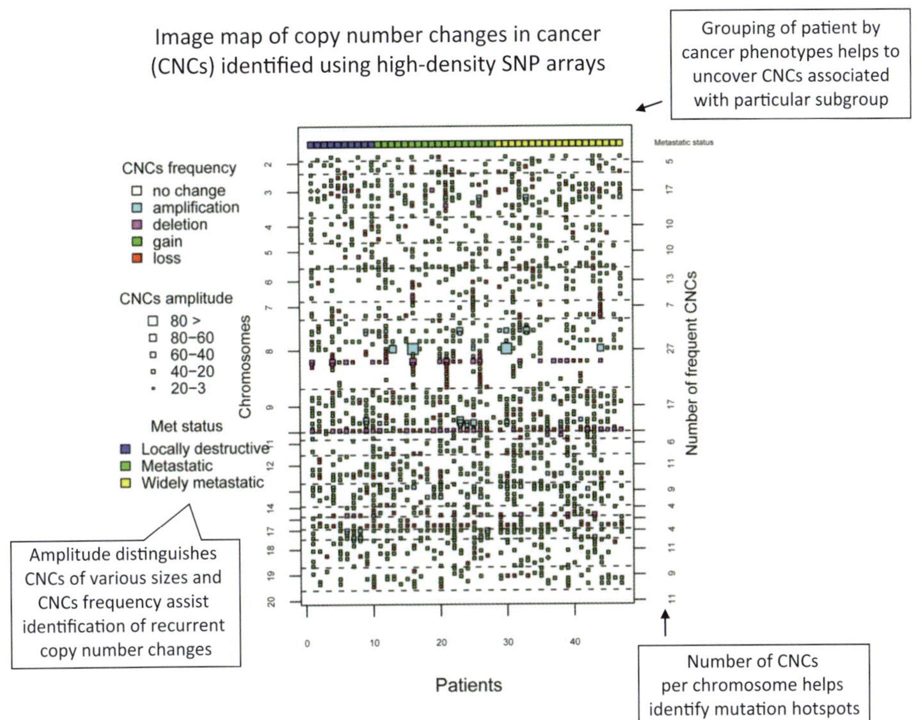

Figure 26.2 Example of an image map of recurrent copy number changes (CNCs) in cancer.

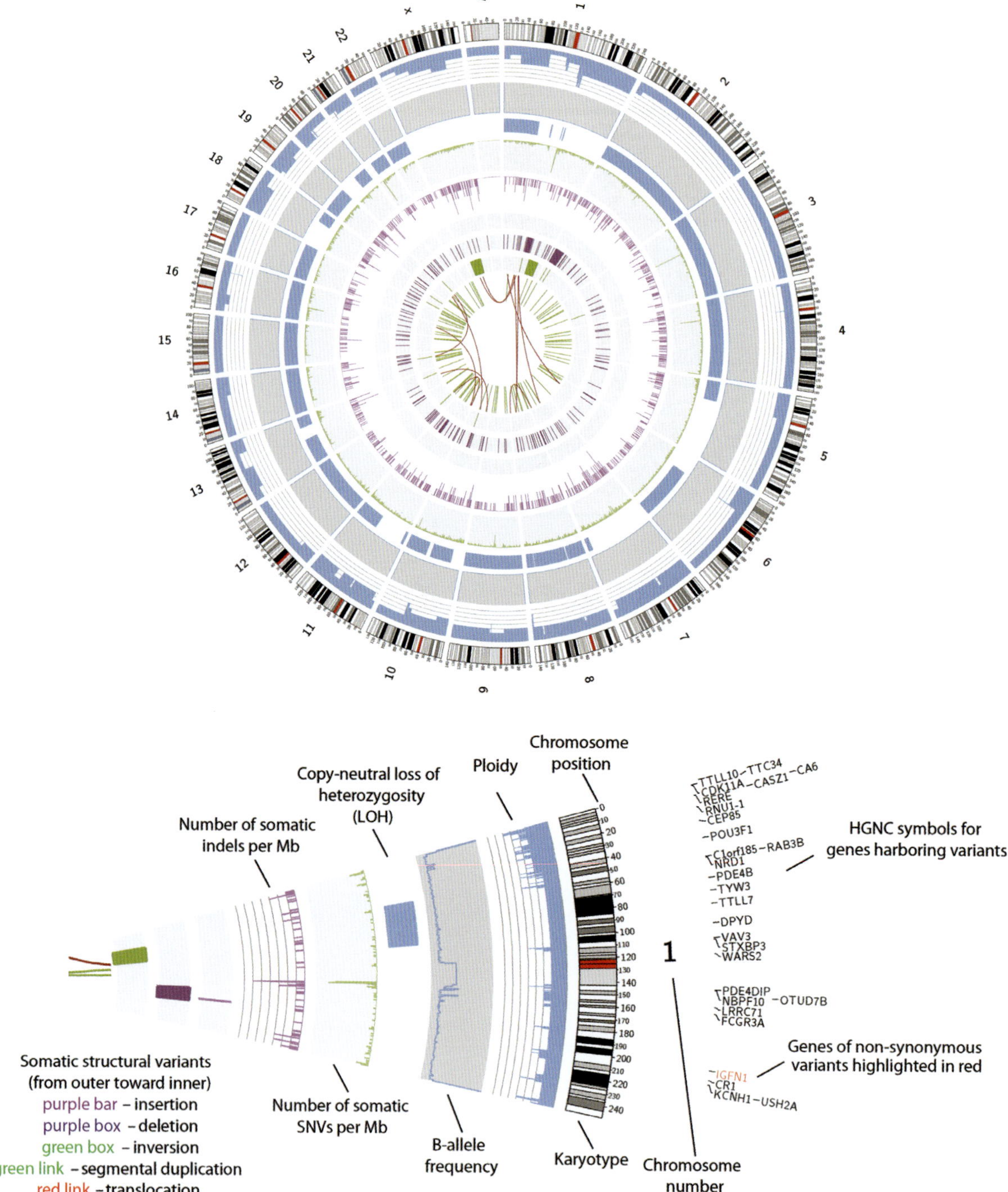

Figure 26.3 Comprehensive categorization of genetic abnormalities in cancers using next-generation sequencing technologies.

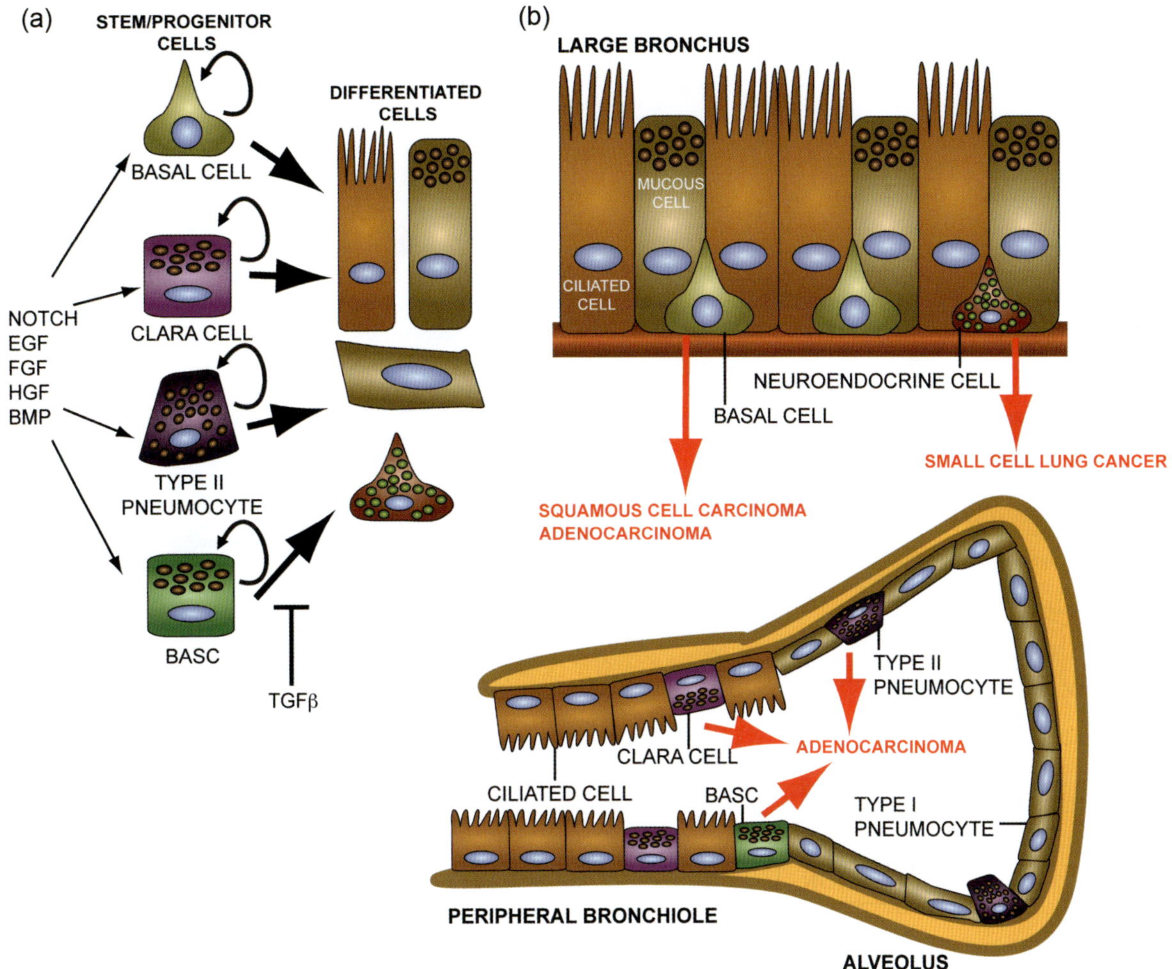

Figure 27.1 Lung morphogenesis and tissue structure.

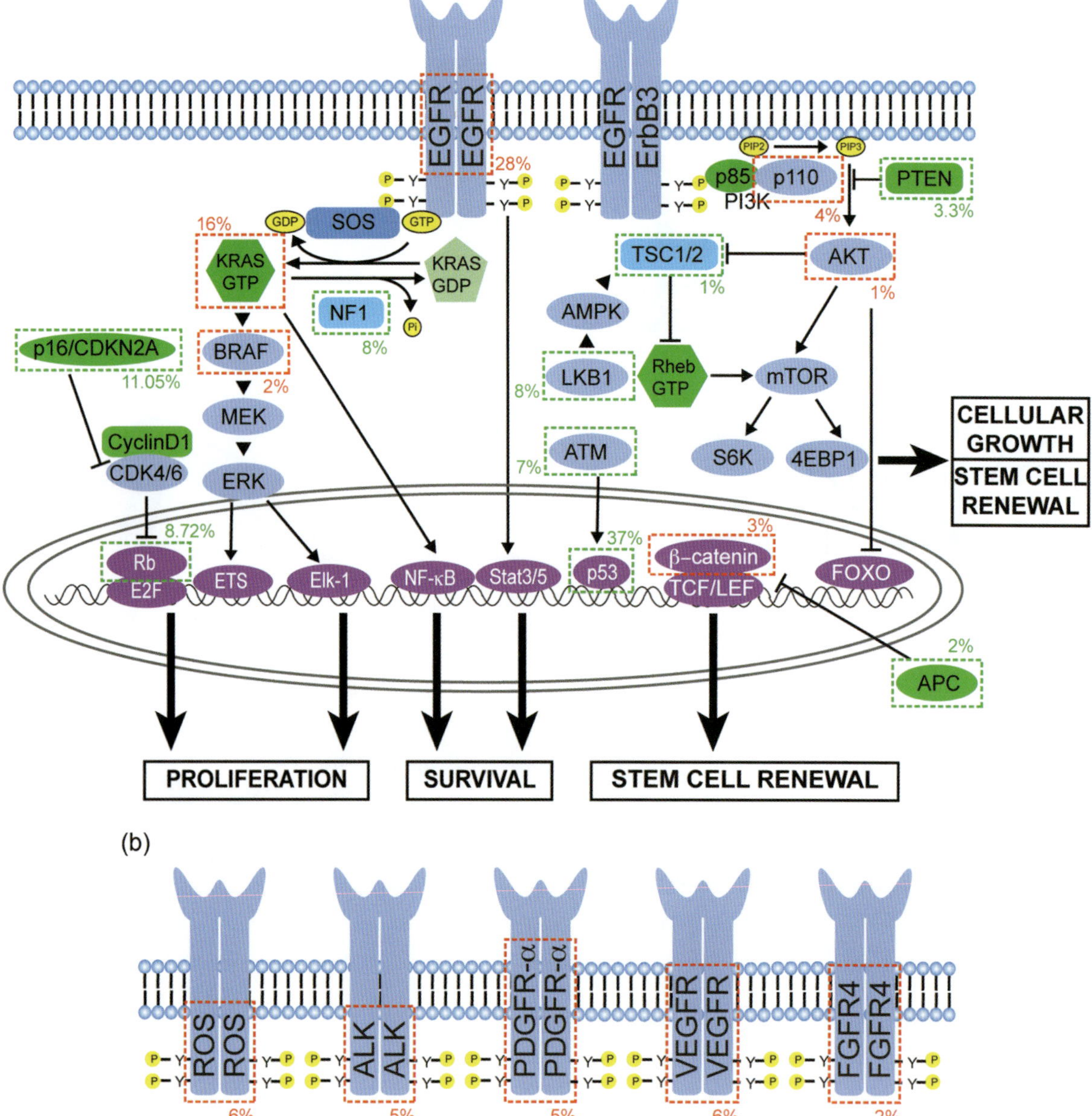

Figure 27.2 Mutationally activated signaling networks in lung cancer.

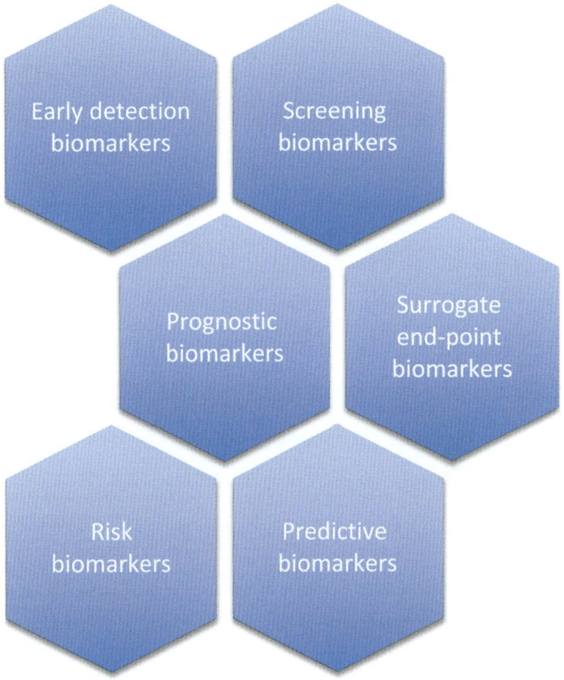

Figure 29.1 Biomarkers in the continuum of carcinogenesis.

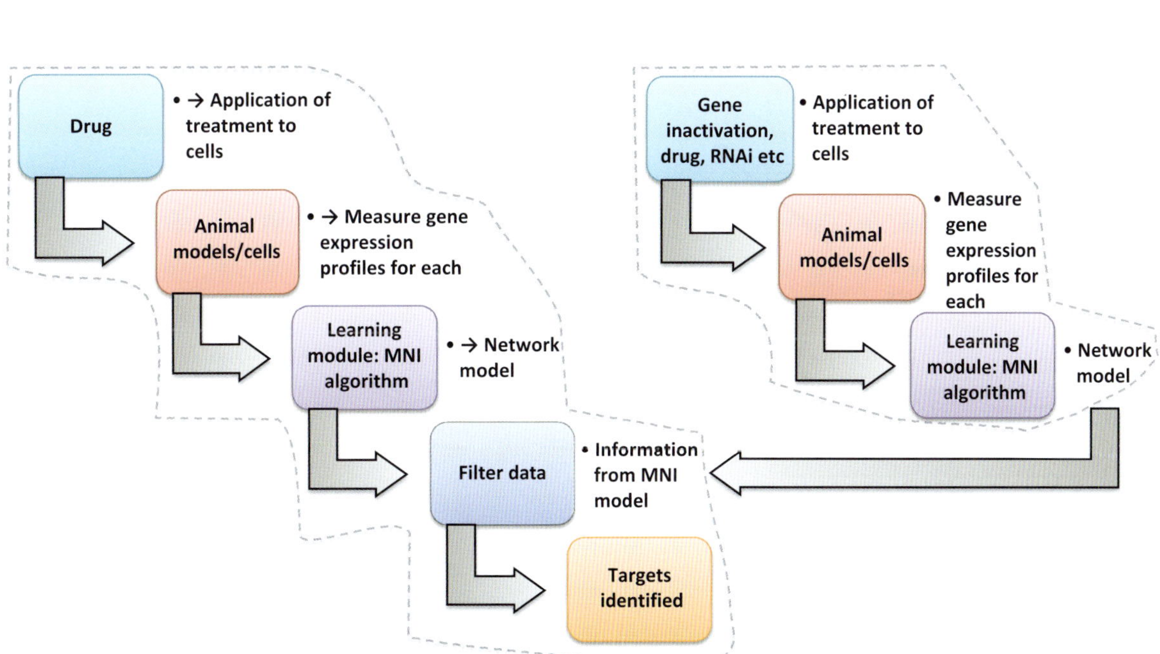

Figure 29.2 A set of treatments, including gene inactivation, interventions, and RNAi, may be used in the initial screen.

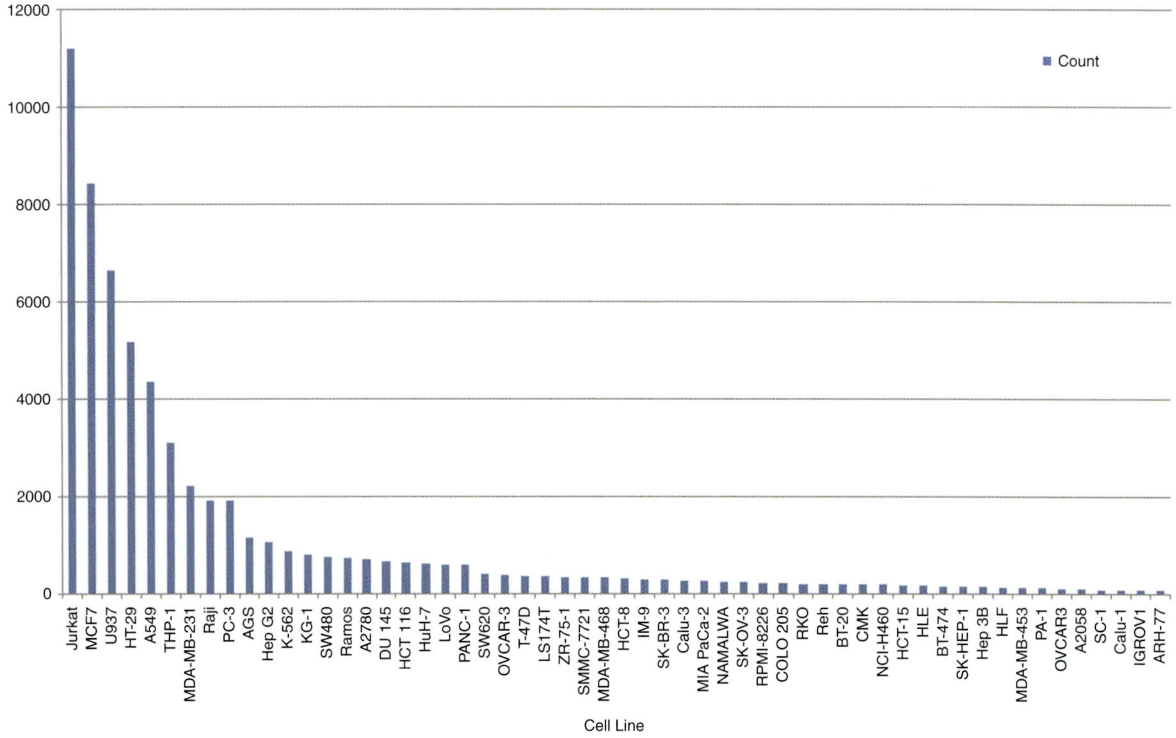

Figure 32.2 Number of published papers citing cancer cell lines. Total of 56 commonly used and public accessible cancer cell lines covering diverse tumor types were included in the search. The numbers in this figure were obtained by using search terms "cell line" and cell line name as it is.

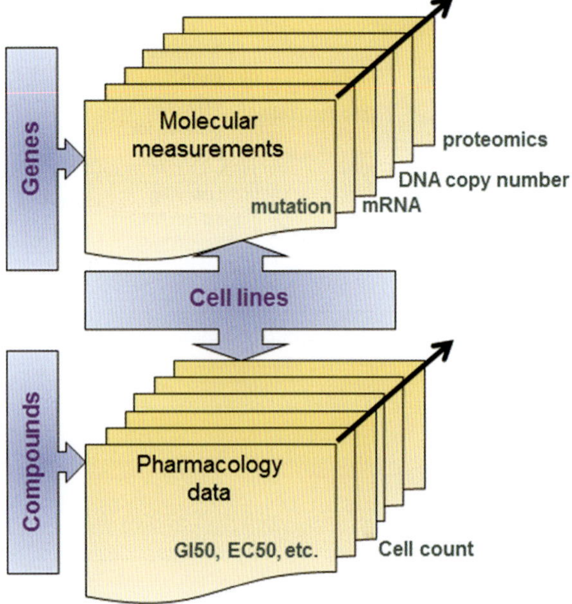

Figure 32.3 Schematic view of a cancer pharmacogenomics data integration system developed by AstraZeneca.

glycans as novel therapeutic targets, *Nat Rev Cancer, 5:* 526–542.

84 Kobayashi, K., Matsumoto, S., Morishima, T., Kawabe, T. and Okamoto, T. 2000. Cimetidine inhibits cancer cell adhesion to endothelial cells and prevents metastasis by blocking E-selectin expression, *Cancer Res, 60:* 3978–3984.

85 Eaton, D. and Hawkins, R.E. 2002. Cimetidine in colorectal cancer: are the effects immunological or adhesion-mediated? *Br J Cancer, 86:* 159–160.

86 Gomes, N., Vassy, J., Lebos, C., et al. 2004. Breast adenocarcinoma cell adhesion to the vascular subendothelium in whole blood and under flow conditions: effects of alphavbeta3 and alphaIIbbeta3 antagonists, *Clin Exp Metastasis, 21:* 553–561.

87 Steeg, P.S. and Theodorescu, D. 2008. Metastasis: a therapeutic target for cancer, *Nat Clin Pract Oncol, 5:* 206–219.

Chapter

21

Genetic alterations in glioblastoma multiforme

Giselle Y. López, Marc Samsky, Rosanne Jones, Cory Adamson and Hai Yan

Introduction

Glioblastoma multiforme (GBM), or Grade IV astrocytoma, is one of the most common and malignant primary tumors of the central nervous system in adults [1]. Despite years of research and advances, the average survival after diagnosis remains about 12 months [1–4]. GBM tumors can be divided into two groups: primary and secondary. Primary GBM presents *de novo* as GBM with no preceding lower grade tumor, representing over 90% of GBM cases [5]. On the other hand, secondary GBM first presents as a lower grade glioma and is part of a progression of increasing grade of astrocytoma: World Health Organization (WHO) Grade II diffuse astrocytoma (often called low-grade glioma), WHO Grade III anaplastic astrocytoma, and WHO Grade IV glioblastoma [5]. WHO Grade I pilocytic astrocytomas are not thought to progress [6]. Despite a similar histopathology, primary and secondary GBM each comprise a distinct combination of genetic changes and tend to occur in distinct populations. Whereas primary GBM generally occurs in older patients, usually over 60 years old [3], secondary GBM tends to occur in middle-aged patients, of an average age of 45 years [3]. Because of these differences between primary and secondary GBM, it is entirely possible that treatments that are effective in one GBM patient population will be ineffective in others, and vice versa.

The importance of understanding the genetic differences in GBM tumors from different patients is becoming increasingly clear, as genetic differences between tumors may more accurately predict prognosis than histological differences alone [7, 8]. Furthermore, because different GBM patients present with distinct combinations of genetic mutations, it is likely that therapies targeted to the specific genetic pathways that are altered in a patient's tumor will be the most efficacious way of treating this disease. However, due to the heterogeneous genome of GBM tumors, with different cells within one tumor having different mutations, tailored treatments targeting the multiple dysregulated oncogenic pathways in cancer cells of differing genotypes may achieve the most effective outcome. Identification of the genetic changes that occur in subsets of cells is critical to optimizing treatment for each patient. In this review, we focus on the genetic pathways identified as commonly altered in GBMs and the implications of changes in these pathways for prognosis and treatment.

Large-scale genetic changes

A large number of chromosomal aberrations have been identified in GBM. The most common is loss of heterozygosity (LOH) of chromosome 10q [3, 9–12], which is seen in 75 to 89% of GBM tumor samples [13, 14]. The presence of LOH 10q is correlated with a shorter survival in patients with GBM [9, 15]. LOH 10p is also common [11], with two thirds of GBM samples in one study also showing loss of 10p [9], and another study finding that almost all samples with LOH 10q also had LOH 10p [10]. Gain of chromosome 7 has also been identified as a frequent copy number alteration [11, 16], particularly in primary GBM [16]. In addition, LOH along chromosome 10 is often correlated with gains on chromosome 7 [12], with both aberrations being seen in approximately 90% of primary GBM samples [16], and this particular combination being associated with a shorter overall survival [10].

Chromosomal alterations also assist in differentiating primary and secondary GBM. For example, gain of chromosome 19 [11, 16] is exclusive to primary GBM, whereas loss of chromosome 19 is exclusive to secondary GBM [16]. As LOH chromosome 10 is seen

Systems Biology of Cancer, ed. S. Thiagalingam. Published by Cambridge University Press. © Cambridge University Press 2015.

in primary GBM and LOH chromosome 19 in secondary GBM, one study found them to be mutually exclusive [12]. In addition, LOH 19q is frequently correlated with LOH 1p [12–14]. Other broad copy number alterations identified include LOH 6q [11], 9p [11, 14, 17], 13 [11, 12, 17], 14 [11], 15q [11], and 22 [11, 12, 17]. In addition, broad gains of copy number have been identified for chromosome 20 [11, 17], in addition to the previously mentioned gains of chromosome 7 [11, 17], and chromosome 19 [11]. As can be seen, many studies in recent years have identified a large number of copy number alterations in human GBM [10–12, 15, 17–21]. These studies are starting to yield information regarding new areas of the genome that harbor tumor suppressor genes or oncogenes.

Mutations in genetic pathways

Several pathways have been identified with a high probability of playing key roles in the development of GBM. In particular, the receptor tyrosine kinase, p53, and RB1 pathways are altered in a majority of GBM tumors [11, 18]. In most cases, more than one pathway is altered in any one tumor sample [11]. While tumors tend to harbor multiple mutations, mutations in any one of the major pathways are mutually exclusive. In other words, a mutation or alteration in one component of a pathway decreases the selective pressure for a second mutation in the same pathway [11, 18]. However, it would appear that there is a selective pressure for an alteration in each of the three pathways; 74% of the tumor samples in one study contained mutations in all three pathways [11].

Receptor tyrosine kinase pathways

Receptor tyrosine kinases, and the growth factors related to them, are frequently overexpressed in GBM tumor samples. Epidermal growth factor receptor (EGFR), vascular endothelial growth factor receptor (VEGFR), insulin-like growth factor 2 (ILGF2), insulin-like growth factor binding proteins 2, 4, and 6 (IGFBP 2, 4, and 6), platelet-derived growth factor receptor (PDGFR), and tumor growth factor β receptor (TβRI) have all been identified as amplified and/or overexpressed in various GBM tumor samples [3, 8–11, 15, 18, 19, 22–32]. Alterations in various elements of receptor tyrosine kinase/PI3K pathways (Figure 21.1) have been identified in 86% of GBM tumor samples [11].

EGFR pathway

EGFR amplification in GBM, described early on by Libermann et al. [23], is one of the most frequent alterations in GBM [reviewed in 33]. In numerous studies, one third or more of GBM tumors have demonstrated amplification and/or overexpression

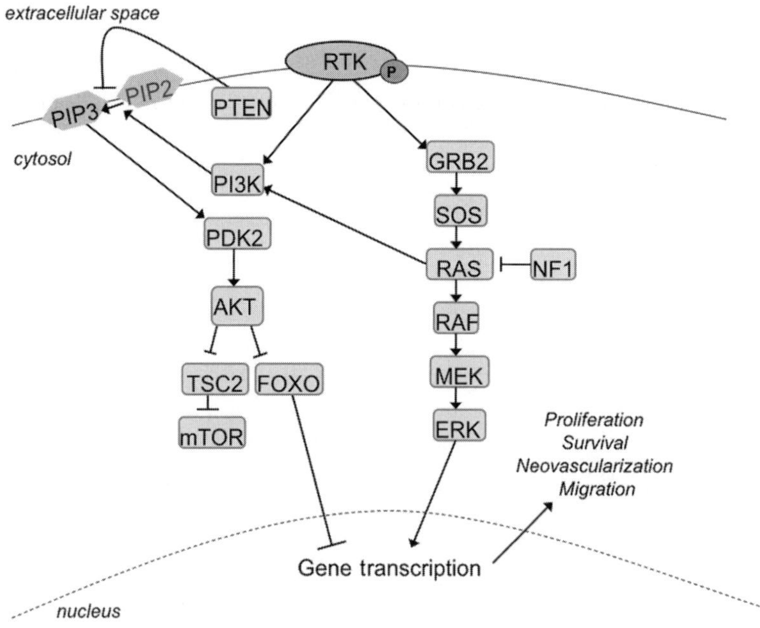

Figure 21.1 Receptor tyrosine kinase pathways.

of *EGFR* [3, 9, 11, 14, 15, 18,19, 24–28, 34], with overexpression being seen most commonly in primary GBM [3, 14, 35, 36] and higher degrees of overexpression being seen toward the invading edge of GBM tumors [37]. *EGFR* gene amplification is not seen in low-grade astrocytomas, and there is only minimal increase in expression of EGFR in low-grade astrocytomas [38]. *EGFR* amplification is highly correlated with LOH 10q [3], and is also correlated with $p16^{INK4A}/p14^{ARF}$ deletion [3, 11]. However, the presence and degree of *EGFR* amplification is inversely correlated with *TP53* mutations [3, 11, 14, 26, 37].

Of particular note, alterations and mutations in *EGFR* occur only in the presence of amplified *EGFR* [27]. Of the mutations seen in *EGFR*, the variant *EGFRvIII* is the most prevalent [39]. As with other mutations in *EGFR*, it is found almost exclusively in GBM with *EGFR* amplification [39–41], with 30 to 50% of GBM tumors with *EGFR* amplification having this variant [41, 42]. The *EGFRvIII* mutation results in the deletion of exons 2 to 7, producing a truncated receptor [43]. When the concentration of EGFRvIII increases above a certain level, autophosphorylation begins [44]. EGFRvIII is considered to be constitutively active [25, 41, 43, 45–49]. However, unlike wild-type EGFR, EGFRvIII is not internalized and downregulated in response to EGF signaling [48]. Studies have shown that the presence of EGFRvIII leads to increased tumor growth rates *in vivo* [47, 50] and a competitive growth advantage compared to cells without EGFRvIII [51]. It also appears to affect cell proliferation and apoptosis through a number of pathways, including via induction of the Ras–Shc–Grb2 [36, 52], Bcl-X$_L$ [51, 53] and PI3K [42, 44, 50, 54] pathways, and through the activation of c-Met, a receptor tyrosine kinase [44]. *MET* itself is amplified in 4% of GBM tumor samples [11], and amplification of *MET* is seen exclusively in secondary GBM [16]. Phosphorylation of EGFRvIII leads to a notable increase in the PI3K pathway [42, 44, 50, 54] and c-Met activation [44]. The increased phosphorylation of c-Met, a receptor tyrosine kinase, in cell lines with EGFRvIII suggests that EGFRvIII may cross-activate other receptor tyrosine kinases [44], indicating that successful treatment of tumors containing EGFRvIII will likely require the inhibition of multiple receptor tyrosine kinase receptors [44]. Concordantly, while the presence of EGF increases the growth rates of GBM tumors *ex vivo*, the presence of multiple growth factors is needed for optimal growth [55].

In addition to GBM tumor samples with *EGFRvIII*, approximately 14 to 20% of GBM tumors contain missense mutations in *EGFR* [11, 18, 56]. They most often occur in the same extracellular region that is found deleted in *EGFRvIII* [56], although rare mutations have also been identified in the 3′ intracellular kinase domain [27]. Of samples showing *EGFR* amplification, about 42% of the samples had point mutations in the amplified *EGFR* [11]. These mutations have been found to lead to increased tumorigenicity [56]. When identified mutants are transfected into NIH3T3 fibroblast cells and injected into nude mice, those with mutant *EGFR* form large tumors, whereas wild-type *EGFR* and empty vector lead to no measurable tumors [56]. In addition, *EGFR* mutations lead to increased basal and post-stimulation EGFR phosphorylation levels, as well as a decrease in receptor downregulation [56], similar results as those seen with *EGFRvIII*.

Numerous studies have been conducted comparing EGFR expression and survival in patients. Although EGFR expression has not been found to correlate with overall survival [57], some studies suggest that its overexpression affects survival following certain treatments. Increased EGFR expression in tumors of patients who are treated with a combination of temozolomide and radiation is correlated with a poorer survival than that experienced by patients whose tumors do not express EGFR [58]. In cell lines, EGFRvIII appears to confer decreased rates of apoptosis following treatment with cisplatin [53]. Resistance to chemotherapy has been correlated with an increase in Bcl-X$_L$ [51, 53] due to increased PI3K/AKT activation [53]. The increase in Bcl-X$_L$ leads to inhibition of caspase 3-like protease, which appears to lead to the resistance to cisplatin-induced apoptosis [53]. EGFRvIII also confers radioresistance [54], making this a mutation that confers resistance to multiple treatment modalities. However, it appears that radioresistance occurs via a different mechanism, as neither Bcl-X$_L$ levels nor rates of apoptosis are altered in astrocyte cell lines transfected with *EGFRvIII* following treatment with radiation [54]. However, radioresistance may occur via the PI3K/AKT pathway [54]. Increased activity of EGFR leads to an increase in activity in both the MAPK and PI3K pathways [55, 59]. The presence of EGFRvIII also leads to an increase in activity in some pathways, including the PI3K pathway [55, 59]. In GBM cell lines, abrogation of only one of these pathways is not sufficient to inhibit growth [59]. However, it is suggested that inhibiting both pathways may be necessary for inhibition of anchorage independence, a common phenotype of tumor cells [59]. It is worth noting that this study also found varying responses to treatment with ERK1/2 inhibitors, PTEN reconstitution, or PI3K inhibitors, depending on the serum status [59]. Therefore treatments targeted specifically against EGFRvIII could potentially prove successful in sensitizing the cells to multiple treatment modalities.

Studies involving treatment with EGFR inhibitors demonstrate the importance of the interactions of multiple pathways. It is unclear if there is a survival advantage associated with treatment with EGFR inhibitors such as erlotinib in patients whose tumors overexpress EGFR. Some studies suggest an advantage [60] while others show no advantage [41, 61]. Studies of cell lines and of brain tumor patients suggest that mutations in *EGFR* or the presence of *EGFRvIII* may sensitize cells to EGFR kinase inhibitors [56], showing that patients whose tumors express EGFRvIII are more likely to respond to EGFR inhibitors than patients whose tumors do not express EGFRvIII [41]. However, only half of the patients expressing EGFRvIII show a response; there are likely to be other factors that also play a role in determining response [41]. One factor may be the presence of intact wild-type *PTEN*, in that patients with intact *PTEN* are more likely to respond to EGFR kinase inhibitors than patients without it [41]. The patients most likely to respond to EGFR kinase inhibitors therefore have both *EGFRvIII* and intact *PTEN* [41], which confirms that multiple genetic alterations must be taken into account when deciding on a GBM treatment for a particular patient. Even these two genes do not fully predict responses: another pathway likely involved is EGF signaling mediated by AKT [42, 62]. Positive EGFR staining in GBM samples shows a strong correlation with AKT activation [42, 62]. Likewise, the presence of constitutively active AKT in GBM cell lines also affects the response to EGFR inhibitors [41], and patients with phosphorylated AKT prior to treatment with erlotinib are less likely to respond to treatment than those with unphosphorylated AKT [60, 63]. Because multiple genetic alterations must be considered, it is likely that treatment will require the use of multiple therapies. Treatment for GBMs expressing EGFRvIII may involve inhibition of both the EGF receptor and other proteins in the pathway,

thus attacking the pathway at various levels. Huang et al. found that c-Met was often activated in cell lines expressing high levels of EGFRvIII [44]. Treatment of these lines with inhibitors of both EGFRvIII and c-Met decreased cell viability more than treatment with either inhibitor alone [44].

Other investigators are attempting to utilize the overexpression of EGFR in GBM tumor cells as a way of targeting treatments specifically to those cells. One approach has been by attaching dsRNA to vectors that allow entry into EGFR-overexpressing cells [64]. This is done through targeting to EGFR via polyethylenimine–polyethylene-glycol–mEGF complexes, leading to receptor-mediated endocytosis via EGFR, thus specifically increasing uptake only into cells that overexpress EGFR. This may prove doubly advantageous via a "bystander" effect, as the cells that take up the dsRNA release cytokines that lead to decreased proliferation and increased apoptosis of surrounding cells, including non-EGFR-expressing cells [64]. Thus although the treatment targets one set of cells, it is able to cause a more dramatic cell death within the larger population of surrounding tumor cells. New treatment modalities include "cancer vaccines," which attempt to stimulate the immune system to target proteins that are either overexpressed or specific to the cancer cells, such as EGFR. Clinical trials are currently underway to study the efficacy of dendritic cell-based vaccine therapies. These cells are removed from the patient and exposed to select antigens prior to being returned to the patient. While some trials are focusing on antigens that are unique to the tumor, such as EGFRvIII [65], others use entire tumor lysates [66, 67]. While EGFR-vIII vaccines have the potential to be very specific to the tumor cells, because of the heterogeneity of tumors, it is possible that not all tumor cells will be targeted. On the other hand, vaccines with tumor lysates have the potential to target all cells in the tumor, but also increase the risk of targeting normal cells as well. Future studies may further elucidate the potential for these "cancer vaccines" in the treatment of GBM.

VEGFR pathway

VEGF is believed to play a role in the development of GBM by leading to increased angiogenesis [68, 69]. VEGF mRNA, while found in low levels in low-grade astrocytoma, has significantly increased expression in GBM [29, 32] and appears to be preferentially produced by glioma stem cells [69]. Although only a subset of GBM tumor cells are CD133+ (a cell surface marker that has been used to enrich for brain tumor stem cells), these tumor cells, through release of VEGF, may promote angiogenesis throughout the entire tumor [69]. In a convergence of different signaling pathways, cells overexpressing EGFR have increased expression of VEGF [28], and it appears that increased EGFR or EGFRvIII signaling works through Ras to lead to the induction of VEGF secretion [70]. This increased VEGF may lead to increased morbidity through increased tumor vascularization and hemorrhage [71].

As mentioned previously, targeting different molecular pathways in the GBM tumor may require multiple pharmacological agents. In a study of mice with xenografts of GBM exposed to both irinotecan, a topoisomerase inhibitor, and bevacizumab, a VEGF inhibitor, the mice had longer survival on combination treatment than with either treatment alone [72]. It has been proposed that the combination treatment leads to an improved outcome because irinotecan targets the more differentiated cancer cells, while bevacizumab targets the angiogenic effects of VEGF secretion by CD133+ cells, thus providing a drug that targets each population of cells [69]. Indeed, a clinical trial has demonstrated improved survival in GBM patients with irinotecan and bevacizumab combination therapy [72].

PDGFR pathway

PDGF and its receptors are often upregulated in GBM tumors. Some studies have demonstrated that all astrocytoma samples, regardless of grade, have increased expression of at least one form of PDGF (A, B, or C) and one receptor (PDGFRα or β) [30, 31]. The level of expression positively correlates with the grade of tumor [31, 73]. However, the increase in expression of PDGF or its receptor is not correlated with an amplification of the corresponding genes. Amplification of the *PDGFA* and *PDGFRα* genes is seen in only 10% and 12% of samples, respectively [11], even though increased expression is seen in a much larger percentage of samples, indicating that overexpression of the genes occurs predominantly through the use of other genetic mechanisms [30, 31, 74]. Of note, overexpression of PDGFRα is correlated with *TP53* mutation, in that the presence of PDGFRα increases the likelihood of a *TP53* mutation [38]. Neither *TP53* null mice, nor mice overexpressing

PDGFB in glial cells, develop GBM [75]. However, mice that both overexpress PDGFB and are null for *TP53* have high rates of tumors with a similar histology to GBM, suggesting an interaction between the two may be needed for the development of some GBM tumors [75].

TβRI pathway

The TβRI pathway also appears to play a role in some GBM tumors, leading to increased proliferation and neovascularization. TGF-β and its receptors have been identified as overexpressed in GBM [76]. However, these proteins are not expressed in lower grade gliomas or normal brain tissue [76]. In addition, expression levels of TGF-β1, TβRI, and TβRII are correlated with grade of tumor [76]. TGF-β also induces expression of PDGF-A, which may serve as the primary mediator of TGF-β's growth stimulatory effects [77]. TGF-β appears to promote tumorigenesis through a number of routes, including serving as a tumor suppressor gene, mitogen, or invasion promoter. TGF-β works as a tumor suppressor gene via inhibition of expression of cdks and downregulation of cdk activity. This downregulation occurs through induction of cdk inhibitors p15, p27, and Cip/WAF1/p21 [78, 79]. The TβRI pathway also contributes to tumorigenesis through downregulation of cell adhesion proteins, leading to an increase in cell migration and invasion [80–82]. TGF-β can alter collagen synthesis, integrin expression, cell adhesion to reconstituted basement membranes, and invasiveness in gliomas [83, 84]. Instead of the standard Smad pathway that results in growth inhibition, TGF-β works upstream of Smad to activate other pathways that may play a role in GBM, including MAPK (Ras-Erk) and SAPK (Rho-JNK, TAK1-p38 kinase). Activation of these pathways results in increased expression of proteins involved in proliferation and transformation. Compared to other commonly studied signaling pathways in GBM, TGF-β is a unique therapeutic target in that it may represent an important cross-link to various intracellular processes. However, its complex and sometimes opposing actions have made it difficult to clearly understand when and how to target it.

Growth factor networks

Studies looking at the increased expression of growth factor receptors in tumors have found that some tumors may overexpress EGFR while others overexpress ILGF2, and that these are mutually exclusive [22]. However, either one can lead to formation of neurospheres and increased proliferative potential in tissue culture lines [22]. Inhibition of PIK3R3, a regulatory subunit of PI3K, blocks proliferation in cell lines overexpressing either receptor, suggesting that both receptors may act upon overlapping pathways [22]. While these two receptors are mutually exclusive in GBM, this is not the case with all tyrosine kinase receptors. The Cancer Genome Atlas study looked at four different receptor tyrosine kinase genes that are altered in GBMs, including *EGFR*, *MET*, *PDGFRα*, and *ERBB2* [11]. Approximately 10% of samples analyzed in this study demonstrated activation or alteration of more than one receptor tyrosine kinase.

As mentioned previously, successful treatment of GBM is likely to require multiple modalities. For example, research has shown that multiple receptor tyrosine kinases are activated in many GBM samples [85]. In one study, 95% of the cell lines showed three or more activated receptor tyrosine kinases, which appear to act redundantly in PI3K signaling [85]. Therefore treatment with inhibitors of only one receptor tyrosine kinase has only a minimal effect on PI3K activity, whereas multiple inhibitors can yield a significant decrease in signaling [85]. These data underscore the need for combination treatment of GBMs on the basis of their activated oncogenic pathways.

PI3K/AKT pathway

PI3K

Through signaling via EGFR and receptor tyrosine kinases, the PI3K pathway appears to play a key role in GBM pathogenesis through its subsequent increased activation, and many studies are evaluating targeting this pathway for treatment [86–89]. EGFR is activated by growth factors, leading to activation of PI3K, which phosphorylates phosphitidylinositol-4,5-bisphosphate (PIP$_2$) to phosphatidylinositol-3,4,5-triphosphate (PIP$_3$) and in turn activates a number of downstream molecules, including AKT. The PI3K pathway activation regulates numerous pathways, including pathways involved in cell cycle regulation, cell growth, metabolism, polarity, survival, proliferation, and migration [reviewed in 90, 91]. PI3K consists of multiple subunits: a regulatory p85α subunit encoded by *PIK3R1*, and a catalytic p110α subunit

encoded by *PIK3CA* [reviewed in 91]. PI3K has been found to be mutated in some GBM tumors [92], with mutations tending to occur in exons 9 and 20 of *PIK3CA* [92, 93], in positions that would confer an increase in kinase activity [94]. In various studies of brain tumor samples, up to 27% of Grade III astrocytomas or GBMs contained mutations in *PIK3CA* [11, 18, 92, 93, 95, 96], although there was no significant amplification noted [92]. *PIK3CA* mutations have been found to occur more often in younger patients with high-grade astrocytomas, a population that often has more secondary glioblastomas and that coincidentally also has a higher rate of *PIK3CA* mutations [92]. It is likely that *PIK3CA* mutations tend to occur later in the development of glioblastoma, rather than as an initiating factor [92]. Mutations in *PIK3R1* have also been identified [11]. Approximately 10% of GBM samples analyzed in a collaborative study had a deletion or mutation in *PIK3R1* [11]. Mutations in *PIK3R1* are mutually exclusive of mutations in *PIK3CA* [11]. Mutations in *PIK3R1* tend to cluster in the SH2 domain, which is responsible for interacting with and inhibiting *PIK3CA* [11]. Likewise, mutations in *PIK3CA* tend to cluster around the C2 domain, which is responsible for its interaction with *PIK3R1* [11]. Taken together, this suggests a selective pressure against the interaction of the inhibitory subunit p85α with p110α, the active subunit. Studies in mice utilizing p110α inhibitors found that a combination treatment of this inhibitor and radiation led to a greater decrease in the growth of the tumor than either treatment alone, again supporting the importance of this pathway in tumorigenesis [86].

PTEN

Phosphatase and tensin homolog (PTEN), found at chromosome 10q23, is an inhibitor of the PI3K pathway [97, 98]. It inhibits PIP$_3$ signaling via dephosphorylation of the lipid products of the PI3K pathway, thus inhibiting the pathway [99]. *PTEN* gene deletion can lead to AKT activation [98]. Mutations in *PTEN* have been found in 25 to 35% of cell lines, xenografts, and GBM patient samples [3, 9, 18, 19, 100, 101]. Although *PTEN* loss is identified in a significant fraction of primary GBMs, it is rarely seen in lower grade astrocytomas [101, 102]. The preferential loss of *PTEN* in primary GBM suggests that *PTEN* mutation may play a key role in the development of primary GBM, and only a minor role if any in the development of secondary GBM. It has been postulated that while the presence of *EGFRvIII* leads to increased PI3K signaling, *PTEN* loss increases the duration of signaling through increased duration of PI3K lipid products, thus working cooperatively with *EGFRvIII* [59].

AKT

Deletion of *PTEN* leads to activation of the kinase AKT, as seen above. AKT works through several pathways. It phosphorylates and inactivates Bad, suppressing apoptosis [103], and it also inhibits AFX/Forkhead transcription factors by phosphorylation to increase survival [104]. It appears that AKT activation is part of the progression from anaplastic astrocytoma to glioblastoma [105]. In one study, addition of AKT to cells already expressing Ras was enough to change the cellular characteristics to those of GBM [105], although it is likely that progression is through more than this one pathway. This result nevertheless suggests that AKT activation may be required for tumor progression. Indeed, AKT activation is a characteristic associated with higher grade tumors [106], as demonstrated by one study that identified amplification of *AKT* in up to one third of tumors [19]. Another mechanism for activating AKT is increased signaling through the EGF pathway [62]. In brain tumors of mice with *NF1* mutations and *TP53* deletions, only tumors of high grade were found to have AKT activation, with no Grade II tumors, 1/5 Grade III tumors, and 5/5 Grade IV tumors (GBM) demonstrating AKT activation [106].

Ras

The *Ras* oncogene activates portions of the PI3K pathway by binding to the p110α domain of PI3K [107]. In mice with a propensity for lung tumors due to increased Ras activity in some cells, mutations in the Ras binding domain of p110α in these mice led to a significant decrease in the number of macroscopic lung tumors formed [107]. However, the presence of both intact Ras and intact p110α is not sufficient for PI3K pathway activation. Receptor tyrosine kinase signaling/coactivators, such as EGF or PDGF, appear to be necessary as well [107]. Guha et al. found an increase in Ras activity, in the absence of mutations, in astrocytoma samples and cell lines, such that the Ras activity in these lines was comparable to that seen in fibroblast lines transfected with a

mutated oncogenic Ras [108]. One study showed that GBM tumors formed in brains in mice only when both Ras and AKT were activated [109], again showing the interplay necessary between different pathways. Clinical trials attempting to increase survival in patients with GBM have begun exploring the use of drugs that may inhibit Ras [110]. However, it appears that Ras may have both oncogenic and tumor suppressor properties, complicating the situation. Some studies have also found that the presence of wild-type Ras led to an increase in cell death through non-apoptotic pathways [111, 112].

STAT

Signal transducer and activator of transcription (STAT) proteins are transcription factors involved in a number of pathways, including cell cycle regulation and apoptosis [113]. In GBM, STAT appears to be activated via signaling through overexpressed EGFR [42, 62], and it has also been correlated with the presence of *EGFRvIII* [42] and with VEGF signaling via VEGFR-2 [114], leading to constitutively active STAT in many GBM tumor samples and cell lines [115, 116]. Other studies have also found decreased levels of STAT inhibitors in GBM compared to normal tissue, providing another mechanism for STAT activation [117]. STAT activation is positively correlated with tumor grade [116]. Inactivation of STAT in cell lines leads to decreased proliferation and increased apoptosis [115]. STAT3 is strongly correlated with AKT activation; immunohistochemistry reveals that 97% of tumors staining positive for active STAT3 also stain positive for active AKT [42], providing a potential explanation for some of the findings. In addition to a potential role in proliferation and apoptosis regulation, STAT is believed to play a role in inhibiting proinflammatory signaling and inhibiting T-cell-mediated immune responses [118]. Indeed, research from Hussain et al. has determined that in some glioma samples, although MHCII is present, the tumor cells lack the co-stimulatory molecules needed for T-cell activation [87]. Addition of STAT3 inhibitors to T cells taken from patients induces activation of T cells that were previously refractory to activation [87]. These data suggest that STAT3 inhibitors may prove promising as drug adjuvants to boost the immune system while other drugs target the tumors; in addition, STAT3 inhibition could also help to increase rates of apoptosis in the tumors.

p53 pathway

TP53

TP53 is one of the genes that is most often mutated in human cancers [3, 119]. Activation of p53 leads to the transcription of both *PTEN* and *Tuberous Sclerosis Complex 2 (TSC2)* [119], both of which lead to PI3K pathway downregulation [119] (Figure 21.2). In addition, p53 activation also leads to transcriptional repression of the p110α subunit of PI3K, further downregulating the pathway [119]. Somatic mutations or alterations in the p53 pathway occur in 60 to 70% of GBM tumor samples [11, 18]. *TP53* in particular is mutated in about 24 to 40% of GBM samples [3, 9, 11, 18, 26, 120, 121], and this rate increases to approximately two thirds when considering just secondary GBM [3, 35]. Mutations in *TP53* tend to cluster around the DNA binding domain [11] and they appear to be an early event in secondary GBM, as a study of low-grade astrocytomas identified *TP53* mutations in 65 to 75% of low-grade astrocytoma (Grade II) samples [38, 121]. In addition, one study found that for each secondary GBM containing *TP53* mutations, the corresponding lower grade astrocytoma also contained the same mutation [32]. Recent evidence has begun to suggest that *TP53* mutation may play a role not only in secondary GBM but also in primary GBM [11, 120], although mutations in *TP53* are less frequent in primary GBM [35, 120]. Accumulation of p53 protein, often seen as a result of mutation or other form of p53 inactivation, was found in one third of primary GBM samples under study [35]. The p53 inactivation may be due to alterations in other genes in the p53 pathway, such as *MDM2* amplification [35].

Again demonstrating the importance of the interplay between pathways in the development of GBM, one study found that neural stem cells from mice that were null for both *TP53* and *PTEN* were unable to differentiate following exposure to differentiation cues, whereas cells null for only one or the other were still able to differentiate [120]. This trait appears to be mediated through activation of myc [120], which is altered in 73% of low-grade astrocytomas [38]. These findings point to one mechanism through which GBM stem cells may be able to maintain their stem cell phenotype [120].

p14^{ARF}/MDM2/MDM4

There are a number of different alterations that can disrupt the p53 pathway besides mutation in *TP53* itself, such as *MDM2* amplification, *MDM4* amplification, or *CDKN2A* deletion [18]. Among the mutations seen most frequently in GBM is *CDKN2A* mutation or deletion, seen in approximately half of GBM tumor samples [11, 18, 26, 121–125]. The *CDKN2A* locus contains the gene *p14^{ARF}*, which normally acts to increase the half-life of p53 [126]. This leads to a selective pressure for deletions or mutations in *p14^{ARF}*, which account for 55% of p53 pathway alterations [11].

The rate of deletion of the *CDKN2A* locus increases as tumor grade increases. No Grade II

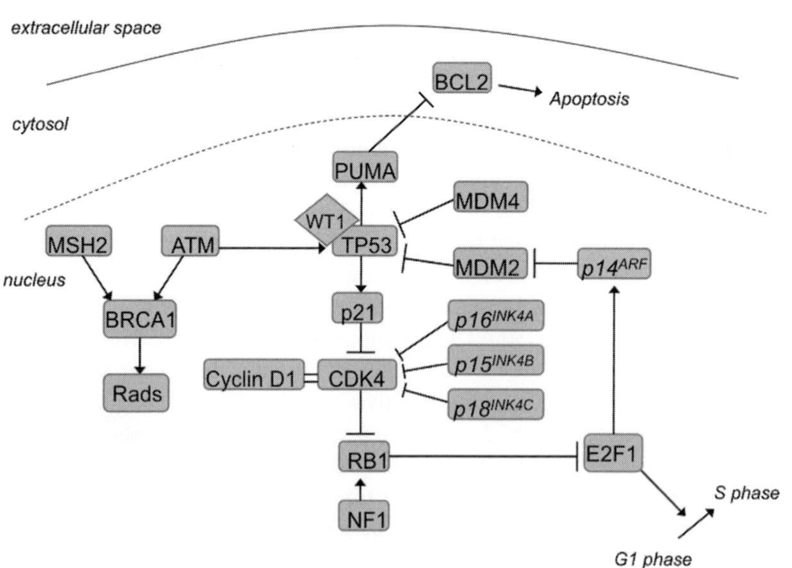

Figure 21.2 RB1 and p53 pathways.

astrocytomas show homozygous deletion [124], 10 to 25% of anaplastic astrocytomas show homozygous deletion [121, 122, 124], and 38 to 52% of GBM tumors have homozygous deletion of this locus [11, 121, 122, 124]. Of the GBM tumors showing heterozygous deletion of this region, a small number additionally show a mutation in the non-deleted copy of *CDKN2A* [11, 121, 122, 124]. In addition, a study looking at *CDKN2A/CDKN2B* deletions present in secondary GBM did not find the deletion in the corresponding lower grade astrocytoma [32]. All these findings suggest that there is a selective pressure for loss of this region and that deletion of this locus may be involved in progression from anaplastic astrocytoma to GBM.

Amplification of *MDM2* is seen in approximately 6 to 19% of tumors [11, 14, 121, 123], and amplification of *MDM4* occurs in approximately 4% of GBM tumors [11, 127]. MDM2 and MDM4 work by binding to and inhibiting p53. MDM2 leads to increased degradation of p53 [126]. In addition, MDM2 works in concert with p14^{ARF}, and it appears to serve as the bridge for formation of a tetramer including p14^{ARF} and p53 [126]. *MDM2* amplification and *p14^{ARF}* deletion are mutually exclusive [14]. In an interesting exception to the mutual exclusivity rule, while *TP53* mutations are mutually exclusive with amplification of *MDM2* and *MDM4* [11, 121, 123, 127], *TP53* mutations are not mutually exclusive with loss of *p14^{ARF}* (the percentage ranging from 22 to 83%) [11, 121, 123]. However, *p14^{ARF}* is located at the same locus as *p16^{INK4A}*, the *CDKN2A* locus, and it is possible that deletion of *p14^{ARF}* may be driven in part by a selective pressure for deletion of *p16^{INK4A}* [11], which would not be relieved by a mutation in *TP53*.

Although mutations along the p53 pathway are common in GBM tumors, it is questionable whether they lead to a change in prognosis. In one study, mutations leading to downregulation of this pathway led to a worse prognosis and increased rate of progression [128]. Patients with low-grade astrocytic tumors with *TP53* mutations generally had a little over four years of progression-free survival, about two years less than patients without the *TP53* mutation [128]. However, another study found that mutations in *TP53* in primary GBM had no effect on the survival of the patient [129], and yet another found that the presence of mutations in *TP53* appears to predict a longer survival time in patients with GBM [9]. These contrasting results make the prognostic value of this

mutation unclear. As it appears to be in many cases, it is probable that a specific combination of mutations in the gene, along with particular alterations in other genes, may have a prognostic value, although mutations in *TP53* alone may not.

RB1 pathway

Mutations and alterations have been frequently identified in the RB1 pathway (Figure 21.2), including *RB1*, *CDK4*, and *CDKN2A/CDKN2B*, with anywhere from 50 to 85% of tumors showing a loss of one component of the pathway [11, 18, 26, 121–125], and 94% of GBM tumors showing at least a heterozygous deletion of *RB1* or *CDKN2A* or amplification of *CDK4* [124]. The rates of alterations of the *RB1* pathway increase as the grade of tumor increases [121, 124, 125]. No Grade II astrocytomas, one third of Grade III astrocytomas, and more than two thirds of Grade IV astrocytomas (GBM) show either homozygous deletion of *CDKN2A*, homozygous deletion of *RB1*, or amplification of *CDK4* [124]. A common alteration is LOH of the *RB1* locus, which also occurs in increasing frequency as tumor grade increases [121, 124]. Deletion of *RB1* is not seen in lower grade (Grade II or III) astrocytomas [121]. LOH is found in up to 45% of samples in GBM tumors [124, 130]. *CDK4* amplification is also seen, with 8% of anaplastic astrocytomas [121] and 6 to 34% of GBM tumors [11, 14, 19, 121, 125] having amplification of *CDK4*. Amplifications of *CDK4* are seen in about half of GBM samples without *CDKN2A* deletion [125]. Mutations of *RB1* or *CDKN2A*, and amplification of *CDK4* appear to be mutually exclusive [14, 121].

CDKN2A encodes both *p14^{ARF}* and *p16^{INK4A}*, hence playing a role in both the p53 pathway and the RB1 pathway. Mutations and alterations at the *CDKN2A/CDKN2B* locus on chromosome 9p represent the most common alterations in the RB1 pathway, and they are seen in approximately half of GBM tumor samples [11, 18, 26, 121–125, 131]. The protein product of *p16^{INK4A}*, which is found at the *CDKN2A* locus, binds to CDK4, inhibiting formation of the CDK4/cyclinD1 complex [132], leading to phosphorylation and inactivation of RB. *p16^{INK4A}* is homozygously deleted in about 31% of tumors [3], and appears to show decreased expression by immunohistochemistry in approximately 60% of tumors [133]. This suggests that even when *p16^{INK4A}* is not deleted, there may be other methods of decreasing its expression.

The *CDKN2B* locus, which is located next to the *CDKN2A* locus, is also frequently deleted in GBM [121, 131, 134] and is often co-deleted with *CDKN2A* due to its proximity [121]. The *CDKN2B* locus encodes p15^{INK4B} which, like p16^{INK4A}, binds to and inhibits CDK4 [reviewed in 135]. Transfection of either *p16^{INK4A}* or *p15^{INK4B}* into cell lines with a homozygous deletion of both is sufficient to suppress cell growth in these cell lines [134], suggesting that deletion of both may be required to prevent CDK4 inhibition.

A related gene, *p18^{INK4C}*, is deleted in approximately 20% of GBM samples [133]. In addition, one study found no detectable expression by immunohistochemistry of p18^{INK4C} in 40 to 50% of GBM samples studied for expression [133]. Of note, in nearly all cases where *p18^{INK4C}* is either homozygously or heterozygously deleted, *p16^{INK4A}* and *p15^{INK4B}* are also deleted [21, 133], with 60% of samples that show no expression of *p18^{INK4C}* also having no expression of *p16^{INK4A}* [133]. Interestingly, the reverse is not necessarily true [21, 133]. E2F1 transcription factor can bind not only the *p16^{INK4A}* promoter but also the *p18^{INK4C}* promoter [21]. Suppression of *p16^{INK4A}* leads to increased expression of *p18^{INK4C}*, leading to the conclusion that in samples where *p16^{INK4A}* is deleted there is selective pressure to also delete *p18^{INK4C}*, which would otherwise be upregulated to compensate for deletion of *p16^{INK4A}* [21].

In considering the interactions of pathways, one study found that the majority of GBM samples with mutations in the RB1 pathway also have mutations in the p53 pathway. In particular, 96% of GBMs and 88% of anaplastic astrocytomas have both mutations [121]. In addition, many tumors have deleted both the *CDKN2A* and *CDKN2B* loci; thus one alteration results in the removal of a protein from each pathway [121]. Ichimura et al. point out that although p53 pathway mutations are found in all grades of GBM, RB1 pathway alterations occur more frequently as grade increases, suggesting that the combination of alterations of both pathways may be necessary for tumor progression [121].

Other mutations

MGMT

Temozolomide is a methylating agent that is used as a chemotherapeutic treatment for GBM. One way it functions is through methylation of the O [6] position on guanines in DNA, which is highly mutagenic when not repaired [reviewed in 136]. O-methylguanine-DNA methyltransferase [6] (MGMT) is able to remove this, and thereby repair the residue, conferring resistance to therapy in tumors in patients in which the tumor cells express MGMT [137]. A majority of somatic mutations in GBM occur as GC-to-AT transitions [11]. The Cancer Genome Atlas study identified that the location of the mutations vary based on the *MGMT* status [11]. In samples from patients treated with alkylating agents, in which the samples had a non-methylated *MGMT* promoter (i.e., samples had MGMT expression), 23% of somatic mutations were GC-to-AT transitions at non-CpG dinucleotides, whereas 29% were the same transitions, but on CpG dinucleotides [11]. However, in tumor samples from patients treated with alkylating agents, in which the samples had a methylated *MGMT* promoter, in addition to there being a higher mutation rate overall, 81% of the mutations were GC-to-AT transitions at non-CpG dinucleotides [^{11}], versus 4% at CpG dinucleotides [11]. This would suggest that *MGMT* promoter methylation leads to an inability to repair the alkylated guanines that result from treatment, therefore leading to an increase in the rate of mutations at non-CpGs [11].

MGMT promoter methylation decreases MGMT activity through decreased transcription [11, 138, 139]. Studies have determined that approximately 30 to 40% of gliomas have methylated *MGMT* promoters [140–142]. Over half of GBM tumor samples in one study had no detectable MGMT activity [143]. Decreased MGMT activity is seen as patients age: 22% of primary GBM patients below age 20 do not show MGMT activity in normal brain tissue; this increases to 75% in patients over 50 years of age [144]. Notably, the lack of MGMT activity is more frequent in normal brain tissue from brain tumor patients than in normal brain tissue from non-tumor patients; in the age group 15 to 44 years of age, 13% of non-tumor patients had no MGMT activity, whereas 55% of tumor patients had no MGMT activity in normal brain tissue. In addition, non-tumor patients do not show this decrease in MGMT activity with age [144]. Silber et al. postulate that the lack of MGMT activity in normal brain tissue might increase the rate of mutagenesis and lead to the development of GBM [143, 144].

Research over the past decade suggests that the presence of a methylated *MGMT* promoter, and the subsequent decrease in MGMT activity, leads to greater sensitivity to temozolomide and other alkylating agents. Some studies have found improved survival in those GBM patients with a methylated *MGMT* promoter on alkylating agents versus unmethylated promoter on alkylating agents [137, 141, 145, 146] and, likewise, increased survival with low MGMT expression/activity versus high expression/activity [147–149]. Temozolomide, compared to other treatments, appears to have less survival benefit in patients with unmethylated *MGMT* promoters [137, 145], which indicates a need to find treatments other than alkylating agents for these patients. A higher percentage of some tumors are positive for MGMT expression following treatment with alkylating agents than before, suggesting that cells with active MGMT are able to survive and preferentially proliferate following treatment with alkylating agents [150]. In one study, 18% were positive for *MGMT* promoter methylation prior to treatment with alkylating agents, and 30% were positive for *MGMT* following treatment [11]. Although some studies find that *MGMT* promoter methylation status does not correlate with a shorter progression-free survival [150] or tumor growth rate [147], other studies dispute this, finding that *MGMT* promoter methylation is correlated with a longer overall survival following surgery, radiation treatment, and treatment with an alkylating agent [58, 151, 152]. In addition, recurrence within the radiation field is more likely to occur in patients with an unmethylated *MGMT* promoter [152]. Although there are conflicting results, it appears that the schedule for administering temozolomide may play a large role in its effect on MGMT activity [153]. Because many studies suggest that decreased MGMT activity leads to a greater response to alkylating agents, studies have also begun looking at MGMT inhibitors in MGMT-expressing tumor cell lines, finding that MGMT inhibitors may increase sensitivity to alkylating agents [154]. Clinical trials utilizing MGMT inhibitors in combination with alkylating agents have begun [155–157].

In addition to MGMT expression, deletion or mutation of *MSH6*, a mismatch repair gene, may play a role in resistance to temozolomide [147]. This gene's protein product normally heterodimerizes with MSH2, and this allows it to recognize single nucleotide mismatches [158]. It has been suggested that *MSH6* becomes mutated after temozolomide treatment and that cells in which *MSH6* is mutated subsequently clonally expand [147, 159]. Further, it has been postulated that inactivation of MSH6, by no longer repairing mutations, could prove to be a mechanism leading to the hypermutation phenotype seen in some patients following temozolomide treatment [159]. In fact, in one study of a large number of tumor samples, tumors that exhibited a hypermutation phenotype were tumors that had a mutation in a mismatch repair gene and, in addition, had been treated with either temozolomide or a combination therapy involving lomustine [11]. In looking more broadly at tumors that were treated versus untreated, the Cancer Genome Atlas study determined that treated tumors exhibited an average of 5.8 silent mutations per sample, compared to 1.4 silent mutations for samples that were untreated [11]. The majority of this difference is accounted for by the tumors exhibiting the hypermutation phenotype [11].

NF1

The *neurofibromatosis 1* (*NF1*) gene has recently been identified as probably playing some role in GBM. NF1 is a GTPase-activating protein that inhibits Ras, an oncogene [reviewed elsewhere 160]. Previous studies have identified cases of GBM in patients with germline *NF1* mutations [161]. They have also shown an increased prevalence of non-optic brain tumors in adolescents and adults with *NF1* mutations [161, 162], with these tumors showing genetic traits very similar to those of sporadic anaplastic astrocytomas, including *TP53* mutations, *CDKN2A* inactivation, and *PTEN* deletion [161]. One study that reviewed the death certificates of patients with NF1 deficiency found a five-fold increase in reports of malignant brain tumors over what would be anticipated in the general population [163]. More recent studies have also identified somatic mutations in *NF1* within GBM tumors, the *NF1* gene being somatically mutated in 14% of GBM samples studied [11]. There were also a number of heterozygous deletions, with 23% of samples demonstrating some sort of alteration in the *NF1* gene [11]. In mouse studies, a combination of both *TP53* and *NF1* mutation leads to the development of low-grade tumors, anaplastic astrocytomas, and GBMs [106, 164]. This suggests this genetic combination as a potential cause of GBM in patients with *NF1* mutations, as well as a mechanism for development of secondary GBM [106, 164]. In a further interplay of pathways, one study using the

NF1/TP53 deficient mouse model found that addition of somatic heterozygosity of *PTEN* leads to faster formation of GBM tumors and decreased survival in these mice compared to *NF1/TP53* deficient mice without LOH *PTEN*, with an average survival of one week instead of eight [165].

In again considering the interplay of multiple pathways, Kwon et al. argue that PTEN deficiency alone is not sufficient to lead to formation of GBM [165]. However, in the context of certain mutations such as *NF1* and *TP53* together, it can drive the formation of a higher grade tumor from a lower grade tumor [165]. The rate of LOH *PTEN* and subsequent AKT activation increases significantly in high-grade tumors, which suggests that it is the subsequent AKT activation that may play a role in progression to high-grade astrocytoma [165], as discussed previously.

Newly identified pathways

Recent research has uncovered a number of pathways not previously well studied in GBM, including some not previously identified in other cancers; these pathways hold the potential to serve as future targets of treatment [18]. They include cation transport channels and pathways specific to the nervous system [18]. For example, aquaporin 1 appears to be overexpressed in some GBMs [10, 166], and its overexpression is correlated with a poorer prognosis [10]. Aquaporin 4 has also been found to be overexpressed in some GBM tumors [167]. Parsons et al. speculate that changes in these pathways in particular may reflect the utilization of pathways already in place to further increase the growth and proliferation of the cancer cells [18]. Serial analysis of gene expression has also identified secreted proteins and cell surface proteins that, due to alterations in their levels, hold potential as future therapeutic targets [18].

Isocitrate dehydrogenases

In addition, there have been some recently identified proteins that do not fall into the categories described previously. One such example is isocitrate dehydrogenase 1 (IDH1), a metabolic protein that increases production of NADPH, which protects against oxidative damage [168, 169]. *IDH1* was found to be altered in about 12% of all GBM tumor samples [18]. Further studies revealed that between 60 and 90% of Grade II astrocytomas and oligodendrogliomas, Grade III astrocytomas and oligodendrogliomas, and secondary

Grade IV astrocytomas (GBM) contained a mutation in either *IDH1* [170, 171] or *IDH2* [171], whereas only 5% of primary GBM contained mutations in these genes [171]. This suggests that *IDH1* and *IDH2* may be involved in early tumorigenesis in secondary GBM and progressive gliomas [170, 171]. These mutations lead to loss of wild-type activity [171–173] and lead to the generation of a novel metabolite, 2-hydroxyglutarate [174, 175]. Nonetheless, the ramifications for the cell and the roles these mutations and the resulting novel metabolite play in the development of tumors are still being elucidated. IDH1/2 were recently identified as frequently mutated in acute myelogenous leukemias [176], and appear to inhibit hematopoietic differentiation and lead to a hypermethylation phenotype [177]. It has also recently been correlated with a hypermethylation phenotype in gliomas [178]. This may be due to inhibition of α-ketoglutarate dependent dioxygenases, such as TET2 [177, 179]. Inhibition of histone demethylases [179] may also affect gene expression patterns. Additionally, signaling pathways such as HIF-1α have the potential to be altered by 2-hydroxyglutarate [173, 179, 180]. The Warburg effect, in which there is an increased rate of aerobic glycolysis, has also been proposed as a potential result of changes in metabolism resulting from mutations in IDH1/2 [181, 182]. Further studies may identify key pathways altered during the development of gliomas.

Stem cells

Current research suggests that astrocytomas may derive from a small set of glioma stem cells [183–186]. These cells have been shown to be identified by CD133+ (a neural stem cell marker) [184]. CD133, or prominin, is a transmembrane protein that often localizes to cellular protrusions [187], and which has also been utilized for isolation of neural stem cells [188]. Higher expression levels of CD133 have been identified in GBM cells than in normal tissue [189, 190]. In addition, the degree of expression of CD133 increases as the grade of astrocytoma increases [189], with the highest levels of expression in GBM tumors [189, 190]. Likewise, the five-year survival decreases as CD133 expression increases [189], and tumor samples showing no CD133 expression come from patients with the longest progression-free survival [191]. In addition, CD133+ cells are present not only in the highest grade astrocytomas, but also in the highest grade oligodendrogliomas, and CD133 is a prognostic marker for poor survival in those patients as well [191].

These CD133+ enriched cells are believed to have unique characteristics that make them important for the creation and maintenance of tumors. CD133+ enriched cells are able to retain proliferative capacity *in vitro* [69, 191] and are able to recapitulate a tumor when injected into NOD-SCID mice [184, 185]. In contrast, injection of a 1,000-fold higher number of CD133– cells is unable to recapitulate the tumor [185]. Tumors formed by CD133+ cells contain both CD133+ and CD133– cells, suggesting that CD133+ cells are able to form CD133– cells in the new tumor [185]. Tumors formed by CD133+ cells are larger, more hemorrhagic, more necrotic, secrete increased levels of VEGF, and demonstrate increased vascularity, compared to those tumors formed by CD133– cells [69]. CD133+ cells are also more resistant to radiation treatment [192] and can repopulate tumors, with a larger percentage of tumor cells being CD133+ following radiation treatment [192].

While CD133 appears to be a potential marker for glioma stem cells, the significance is still being addressed. It has been postulated that CD133+ cells in particular may have increased expression of *HOX* genes, which are often involved in developmental processes, which allow for the increased proliferative potential [191]. The expression pattern of *HOX* genes in normal human astrocytes appears to differ substantially from that in GBM samples [193]. Nevertheless, some recent studies suggest that the identification of stem cells may prove more complex than selection by CD133+ expression alone. One study found that only a fraction of CD133+ cells retain increased proliferative capacity, after only a fraction of CD133+ cells tested positive for Ki67 [190], suggesting that while the CD133+ cell population contains stem cells, there are also non-proliferating cells in the CD133+ population [190], which may or may not be tumor cells. Although CD133+ cells appear to represent a subset of tumor cells with unique characteristics, rarely, some CD133– tumor cells appear to also contain the proliferative capacity to form tumors [194]. Tumors formed by CD133– cells appear to have a decreased proliferative capacity compared to tumors formed by CD133+ cells [69]. While the majority of CD133– tumor cells are not able to proliferate through asymmetric division in a stem cell-like manner, a small subset of CD133– tumors are able to do so,

which suggests that CD133+ represents one type of tumor stem cell, whereas other tumors may be maintained through a specific type of CD133– stem cell [194, 195]. Tumors formed in NOD-SCID mice after injection of CD133+ cells versus those injected with CD133– cells produce different phenotypes [194]. Thus, although both of the cell types appear to have stem cell potential, their different expression profiles and phenotypes suggest that they cause distinct tumors with different characteristics. Alternatively, there may be subsets of glioma stem cells that initiate gliomas versus those that maintain them.

Nestin, an intermediate protein expressed in neuronal precursor cells during development [196], has also been identified as having increased expression in GBM, potentially in stem cells [189, 190]. It has been found that the degree of nestin expression increases as the grade of astrocytoma increases [189], with the highest level being found in GBM [189, 190]. In accordance with this determination, rates of five-year survival decrease as nestin expression increases [189]. Patients with tumors expressing high levels of both prominin and nestin had a worse prognosis than patients expressing high levels of just one of those proteins [189]. Studies have begun to look at ways to target genes preferentially expressed in CD133+ cells. One study identified L1CAM, an adhesion molecule involved in signaling, growth, and migration [reviewed in 197] as a potential target [198]. Treatment targeting this molecule in mouse xenografts led to decreased growth of the xenografts [198]. It is hoped that future testing will identify further molecules for the targeting of glioma stem cells.

Future paths

Genomic fingerprinting may prove to be a useful method for both predicting survival and identifying the ideal treatment modality for different patients. Several studies have begun to analyze brain tumor samples by microarray for gene expression or copy number alterations, arriving at genes and pathways that are often altered in various brain tumors [8, 10, 17, 20, 32, 199, 200]. Following analysis of these pathways, the investigators in these studies were able to use the pathways as predictors of outcome in multiple tumor types, including not just GBM [8, 10, 200], but also anaplastic astrocytoma, anaplastic oligodendrogliomas, and anaplastic mixed oligoastrocytomas [8, 12]. Some genes that have been identified as altered include genes in the following pathways: proliferation [8, 17, 32, 199, 200], extracellular matrix/migration/invasion [8, 32, 199], angiogenesis [17, 32, 200], neurogenesis [8, 32, 200], and signal transduction [17, 32]. Being able to identify the category a patient falls into allows for some predictions into prognosis [8, 57, 200, 201]. In addition, the identification of pathways that are commonly altered provides new opportunities for the development of therapies to target GBM. The identification of the specific genetic alterations in individual tumors may allow for treatment using a combination of drugs and modalities to target the combination of alterations specific to each patient.

References

1. Maher, E.A., et al. (2001) Malignant glioma: genetics and biology of a grave matter. *Genes Development* **15**, 1311–1333.

2. Legler, J.M., et al. (1999) Cancer surveillance series (corrected): brain and other central nervous system cancers: recent trends in incidence and mortality. *Journal of the National Cancer Institute* **91**, 1382–1390.

3. Ohgaki, H., et al. (2004) Genetic pathways to glioblastoma: a population-based study. *Cancer Research* **64**, 6892–6899.

4. Louis, D.N., et al. (2007) The 2007 WHO classification of tumours of the central nervous system. *Acta Neuropathologica* **114**, 97–109.

5. *WHO Classification of Tumours of the Central Nervous System*, (2007) Lyon: IARC.

6. Parsa, C.F. and Givrad, S. (2008) Juvenile pilocytic astrocytomas do not undergo spontaneous malignant transformation: grounds for designation as hamartomas. *British Journal of Ophthalmology* **92**, 40–46.

7. Nutt, C.L., et al. (2003) Gene expression-based classification of malignant gliomas correlates better with survival than histological classification. *Cancer Research* **63**, 1602–1607.

8. Freije, W.A., et al. (2004) Gene expression profiling of gliomas strongly predicts survival. *Cancer Research* **64**, 6503–6510.

9. Schmidt, M.C., et al. (2002) Impact of genotype and morphology on the prognosis of glioblastoma. *Journal of Neuropathology and Experimental Neurology* **61**, 321–328.

10. Nigro, J.M., et al. (2005) Integrated array-comparative genomic hybridization and expression array profiles identify clinically relevant molecular subtypes of glioblastoma. *Cancer Research* **65**, 1678–1686.

11. The Cancer Genome Atlas Research (2008) Comprehensive genomic characterization defines human glioblastoma genes and core pathways. *Nature* **455**, 1061–1068.

12. Bredel, M., et al. (2005) High-resolution genome-wide mapping of genetic alterations in human glial brain tumors. *Cancer Research* **65**, 4088–4096.

13. Gresner, S.M., et al. (2007) Gliomas: association of histology and molecular genetic analysis of chromosomes 1p, 10q, and 19q. *Acta Neurobiologiae Experimentalis* **67**, 103–112.

14. Houillier, C., et al. (2006) Prognostic impact of molecular markers in a series of 220 primary glioblastomas. *Cancer* **106**, 2218–2223.

15. Kotliarov, Y., et al. (2006) High-resolution global genomic survey of 178 gliomas reveals novel regions of copy number alteration and allelic imbalances. *Cancer Research* **66**, 9428–9436.

16. Maher, E.A., et al. (2006) Marked genomic differences characterize primary and secondary glioblastoma subtypes and identify two distinct molecular and clinical secondary glioblastoma entities. *Cancer Research* **66**, 11502–11513.

17. de Tayrac, M., et al. (2009) Integrative genome-wide analysis reveals a robust genomic glioblastoma signature associated with copy number driving changes in gene expression. *Genes, Chromosomes and Cancer* **48**, 55–68.

18. Parsons, D.W., et al. (2008) An integrated genomic analysis of human glioblastoma multiforme. *Science* **321**, 1807–1812.

19. Korshunov, A., Sycheva, R. and Golanov, A. (2006) Genetically distinct and clinically relevant subtypes of glioblastoma defined by array-based comparative genomic hybridization (array-CGH). *Acta Neuropathologica* **111**, 465–474.

20. Liu, F., et al. (2006) A genome-wide screen reveals functional gene clusters in the cancer genome and identifies EphA2 as a mitogen in glioblastoma. *Cancer Research* **66**, 10815–10823.

21. Wiedemeyer, R., et al. (2008) Feedback circuit among INK4 tumor suppressors constrains human glioblastoma development. *Cancer Cell* **13**, 355–364.

22. Soroceanu, L., et al. (2007) Identification of IGF2 signaling through phosphoinositide-3-kinase regulatory subunit 3 as a growth-promoting axis in glioblastoma. *Proceedings of the National Academy of Sciences of the United States of America* **104**, 3466–3471.

23. Libermann, T.A., et al. (1985) Amplification, enhanced expression and possible rearrangement of EGF receptor gene in primary human brain tumours of glial origin. *Nature* **313**, 144–147.

24. Wong, A.J., et al. (1987) Increased expression of the epidermal growth factor receptor gene in malignant gliomas is invariably associated with gene amplification. *Proceedings of the National Academy of Sciences of the United States of America* **84**, 6899–6903.

25. Schlegel, J., et al. (1994) Amplification of the epidermal-growth-factor-receptor gene correlates with different growth behaviour in human glioblastoma. *International Journal of Cancer* **56**, 72–77.

26. Hayashi, Y., et al. (1997) Association of EGFR gene amplification and CDKN2 (p16/MTS1) gene deletion in glioblastoma multiforme. *Brain Pathology* **7**, 871–875.

27. Ekstrand, A.J., Sugawa, N., James, C.D. and Collins, V.P. (1992) Amplified and rearranged epidermal growth factor receptor genes in human glioblastomas reveal deletions of sequences encoding portions of the N- and/or C-terminal tails. *Proceedings of the National Academy of Sciences of the United States of America* **89**, 4309–4313.

28. Mischel, P.S., et al. (2003) Identification of molecular subtypes of glioblastoma by gene expression profiling. *Oncogene* **22**, 2361–2373.

29. Plate, K.H., Breier, G., Weich, H.A. and Risau, W. (1992) Vascular endothelial growth factor is a potential tumour angiogenesis factor in human gliomas *in vivo*. *Nature* **359**, 845–848.

30. Lokker, N.A., Sullivan, C.M., Hollenbach, S.J., Israel, M.A. and Giese, N.A. (2002) Platelet-derived growth factor (PDGF) autocrine signaling regulates survival and mitogenic pathways in glioblastoma cells: evidence that the novel PDGF-C and PDGF-D ligands may play a role in the development of brain tumors. *Cancer Research* **62**, 3729–3735.

31. Guha, A., Dashner, K., Black, P.M., Wagner, J.A. and Stiles, C.D. (1995) Expression of PDGF and PDGF receptors in human astrocytoma operation specimens supports the existence of an autocrine loop. *International Journal of Cancer* **60**, 168–173.

32. van den Boom, J., et al. (2003) Characterization of gene expression profiles associated with glioma progression using oligonucleotide-based microarray analysis and real-time reverse transcription-polymerase chain reaction. *American Journal of Pathology* **163**, 1033–1043.

33. Wikstrand, C.J. and Bigner, D.D. (2009) Constitutive activation of truncated EGF receptors in glioblastoma. In Haley, J.D. and Gullick, W.J. (eds) *EGFR Signaling Networks in Cancer Therapy*. Totowa, New Jersey: Springer-Humana.

34. Ruano, Y., et al. (2009) Worse outcome in primary glioblastoma multiforme with concurrent epidermal growth factor receptor and p53 alteration. *American Journal of Clinical Pathology* **131**, 257–263.

35. Watanabe, K., et al. (1996) Overexpression of the EGF receptor and p53 mutations are mutually exclusive in the evolution of primary and secondary glioblastomas. *Brain Pathology* **6**, 217–223; discussion 223–224.

36. Wen, P.Y. and Kesari, S. (2008) Malignant gliomas in adults. *New England Journal of Medicine* **359**, 492–507.

37. Okada, Y., et al. (2003) Selection pressures of TP53 mutation and microenvironmental location influence epidermal growth factor receptor gene amplification in human glioblastomas. *Cancer Research* **63**, 413–416.

38. Huang, H., et al. (2000) Gene expression profiling of low-grade diffuse astrocytomas by cDNA arrays. *Cancer Research* **60**, 6868–6874.

39. Frederick, L., Wang, X.Y., Eley, G. and James, C.D. (2000) Diversity and frequency of epidermal growth factor receptor mutations in human glioblastomas. *Cancer Research* **60**, 1383–1387.

40. Aldape, K.D., et al. (2004) Immunohistochemical detection of EGFRvIII in high malignancy grade astrocytomas and evaluation of prognostic significance. *Journal of Neuropathology and Experimental Neurology* **63**, 700–707.

41. Mellinghoff, I.K., et al. (2005) Molecular determinants of the response of glioblastomas to EGFR kinase inhibitors. *New England Journal of Medicine* **353**, 2012–2024.

42. Mizoguchi, M., et al. (2006) Activation of STAT3, MAPK, and AKT in malignant astrocytic gliomas: correlation with EGFR status, tumor grade, and survival. *Journal of Neuropathology and Experimental Neurology* **65**, 1181–1188.

43. Sugawa, N., Ekstrand, A.J., James, C.D. and Collins, V.P. (1990) Identical splicing of aberrant epidermal growth factor receptor transcripts from amplified rearranged genes in human glioblastomas. *Proceedings of the National Academy of Sciences of the United States of America* **87**, 8602–8606.

44. Huang, P.H., et al. (2007) Quantitative analysis of EGFRvIII cellular signaling networks reveals a combinatorial therapeutic strategy for glioblastoma. *Proceedings of the National*

Academy of Sciences of the United States of America **104**, 12867–12872.

45. Pelloski, C.E., et al. (2007) Epidermal growth factor receptor variant III status defines clinically distinct subtypes of glioblastoma. *Journal of Clinical Oncology* **25**, 2288–2294.

46. Wong, A.J., et al. (1992) Structural alterations of the epidermal growth factor receptor gene in human gliomas. *Proceedings of the National Academy of Sciences of the United States of America* **89**, 2965–2969.

47. Nishikawa, R., et al. (1994) A mutant epidermal growth factor receptor common in human glioma confers enhanced tumorigenicity. *Proceedings of the National Academy of Sciences of the United States of America* **91**, 7727–7731.

48. Huang, H.S., et al. (1997) The enhanced tumorigenic activity of a mutant epidermal growth factor receptor common in human cancers is mediated by threshold levels of constitutive tyrosine phosphorylation and unattenuated signaling. *Journal of Biological Chemistry* **272**, 2927–2935.

49. Fernandes, H., Cohen, S. and Bishayee, S. (2001) Glycosylation-induced conformational modification positively regulates receptor-receptor association: a study with an aberrant epidermal growth factor receptor (EGFRvIII/DeltaEGFR) expressed in cancer cells. *Journal of Biological Chemistry* **276**, 5375–5383.

50. Narita, Y., et al. (2002) Mutant epidermal growth factor receptor signaling down-regulates p27 through activation of the phosphatidylinositol 3-kinase/Akt pathway in glioblastomas. *Cancer Research* **62**, 6764–6769.

51. Nagane, M., et al. (1996) A common mutant epidermal growth factor receptor confers enhanced tumorigenicity on human glioblastoma cells by increasing proliferation and reducing apoptosis. *Cancer Research* **56**, 5079–5086.

52. Prigent, S.A., et al. (1996) Enhanced tumorigenic behavior of glioblastoma cells expressing a truncated epidermal growth factor receptor is mediated through the Ras-Shc-Grb2 pathway. *Journal of Biological Chemistry* **271**, 25639–25645.

53. Nagane, M., Levitzki, A., Gazit, A., Cavenee, W.K. and Huang, H.J. (1998) Drug resistance of human glioblastoma cells conferred by a tumor-specific mutant epidermal growth factor receptor through modulation of Bcl-XL and caspase-3-like proteases. *Proceedings of the National Academy of Sciences of the United States of America* **95**, 5724–5729.

54. Li, B., et al. (2004) Mutant epidermal growth factor receptor displays increased signaling through the phosphatidylinositol-3 kinase/AKT pathway and promotes radioresistance in cells of astrocytic origin. *Oncogene* **23**, 4594–4602.

55. Zhu, H., et al. (2009) Oncogenic EGFR signaling cooperates with loss of tumor suppressor gene functions in gliomagenesis. *Proceedings of the National Academy of Sciences of the United States of America.*

56. Lee, J.C., et al. (2006) Epidermal growth factor receptor activation in glioblastoma through novel missense mutations in the extracellular domain. *PLoS Medicine* **3**, e485.

57. Rich, J.N., et al. (2005) Gene expression profiling and genetic markers in glioblastoma survival. *Cancer Research* **65**, 4051–4058.

58. Murat, A., et al. (2008) Stem cell-related "self-renewal" signature and high epidermal growth factor receptor expression associated with resistance to concomitant chemoradiotherapy in glioblastoma. *Journal of Clinical Oncology* **26**, 3015–3024.

59. Klingler-Hoffmann, M., Bukczynska, P. and Tiganis, T. (2003) Inhibition of phosphatidylinositol 3-kinase signaling negates the growth advantage imparted by a mutant epidermal growth factor receptor on human glioblastoma cells. *International Journal of Cancer* **105**, 331–339.

60. Haas-Kogan, D.A., et al. (2005) Biomarkers to predict response to epidermal growth factor receptor inhibitors. *Cell Cycle* **4**, 1369–1372.

61. Brown, P.D., et al. (2008) A phase II trial (N0177) of erlotinib and temozolomide (TMZ) combined with radiation therapy (RT) in glioblastoma multiforme (GBM). *Journal of Clinical Oncology* **26**, 2016.

62. Ghosh, M.K., Sharma, P., Harbor, P.C., Rahaman, S.O. and Haque, S.J. (2005) PI3K-AKT pathway negatively controls EGFR-dependent DNA-binding activity of Stat3 in glioblastoma multiforme cells. *Oncogene* **24**, 7290–7300.

63. Haas-Kogan, D.A., et al. (2005) Epidermal growth factor receptor, protein kinase B/Akt, and glioma response to erlotinib. *Journal of the National Cancer Institute* **97**, 880–887.

64. Shir, A., Ogris, M., Wagner, E. and Levitzki, A. (2006) EGF receptor-targeted synthetic double-stranded RNA eliminates glioblastoma, breast cancer, and adenocarcinoma tumors in mice. *PLoS Medicine* **3**, e6.

65. Sampson, J.H., et al. (2009) An epidermal growth factor receptor variant III-targeted vaccine is safe and immunogenic in patients with glioblastoma multiforme. *Molecular Cancer Therapeutics* **8**, 2773–2779.

66. De Vleeschouwer, S., et al. (2008) Postoperative adjuvant dendritic cell-based immunotherapy in patients with relapsed glioblastoma multiforme. *Clinical Cancer Research* **14**, 3098–3104.

67. Wheeler, C.J., et al. (2008) Vaccination elicits correlated immune and clinical responses in glioblastoma multiforme patients. *Cancer Research* **68**, 5955–5964.

68. Jain, R.K., et al. (2007) Angiogenesis in brain tumours. *Nature Reviews* **8**, 610–622.

69. Bao, S., et al. (2006) Stem cell-like glioma cells promote tumor angiogenesis through vascular endothelial growth factor. *Cancer Research* **66**, 7843–7848.

70. Feldkamp, M.M., Lau, N., Rak, J., Kerbel, R.S. and Guha, A. (1999) Normoxic and hypoxic regulation of vascular endothelial growth factor (VEGF) by astrocytoma cells is mediated by Ras. *International Journal of Cancer* **81**, 118–124.

71. Oka, N., et al. (2007) VEGF promotes tumorigenesis and angiogenesis of human glioblastoma stem cells. *Biochemical and Biophysical Research Communications* **360**, 553–559.

72. Vredenburgh, J.J., et al. (2007) Bevacizumab plus irinotecan in recurrent glioblastoma multiforme. *Journal of Clinical Oncology* **25**, 4722–4729.

73. Hermanson, M., et al. (1992) Platelet-derived growth factor and its receptors in human glioma tissue: expression of messenger RNA and protein suggests the presence of autocrine and paracrine loops. *Cancer Research* **52**, 3213–3219.

74. Fleming, T.P., et al. (1992) Amplification and/or overexpression of platelet-derived growth factor receptors and epidermal growth factor receptor in human glial tumors. *Cancer Research* **52**, 4550–4553.

75. Hede, S.M., et al. (2008) GFAP promoter driven transgenic expression of PDGFB in the mouse brain leads to glioblastoma in a Trp53 null background. *Glia* **57**, 1143–1153.

76. Yamada, N., et al. (1995) Enhanced expression of transforming growth factor-beta and its type-I and type-II receptors in human glioblastoma. *International Journal of Cancer* **62**, 386–392.

77. Jennings, M.T., et al. (1997) Transforming growth factor beta as a potential tumor progression factor among hyperdiploid glioblastoma cultures: evidence for the role of platelet-derived growth factor. *Journal of Neuro-oncology* **31**, 233–254.

78. Johnson, M.D., Jennings, M.T., Gold, L.I. and Moses, H.L. (1993) Transforming growth factor-beta in neural embryogenesis and neoplasia. *Human Pathology* **24**, 457–462.

79. Matsuura, I., et al. (2004) Cyclin-dependent kinases regulate the antiproliferative function of Smads. *Nature* **430**, 226–231.

80. Cui, W., et al. (1996) TGFbeta1 inhibits the formation of benign skin tumors, but enhances progression to invasive spindle carcinomas in transgenic mice. *Cell* **86**, 531–542.

81. Oft, M., et al. (1996) TGF-beta1 and Ha-Ras collaborate in modulating the phenotypic plasticity and invasiveness of epithelial tumor cells. *Genes and Development* **10**, 2462–2477.

82. Yin, J.J., et al. (1999) TGF-beta signaling blockade inhibits PTHrP secretion by breast cancer cells and bone metastases development. *Journal of Clinical Investigation* **103**, 197–206.

83. Merzak, A., McCrea, S., Koocheckpour, S. and Pilkington, G.J. (1994) Control of human glioma cell growth, migration and invasion in vitro by transforming growth factor beta 1. *British Journal of Cancer* **70**, 199–203.

84. Paulus, W., et al. (1995) Effects of transforming growth factor-beta 1 on collagen synthesis, integrin expression, adhesion and invasion of glioma cells. *Journal of Neuropathology and Experimental Neurology* **54**, 236–244.

85. Stommel, J.M., et al. (2007) Coactivation of receptor tyrosine kinases affects the response of tumor cells to targeted therapies. *Science* **318**, 287–290.

86. Chen, J.S., et al. (2008) Characterization of structurally distinct, isoform-selective phosphoinositide 3′-kinase inhibitors in combination with radiation in the treatment of glioblastoma. *Molecular Cancer Therapeutics* **7**, 841–850.

87. Hussain, S.F., et al. (2007) A novel small molecule inhibitor of signal transducers and activators of transcription 3 reverses immune tolerance in malignant glioma patients. *Cancer Research* **67**, 9630–9636.

88. Gallia, G.L., et al. (2009) Inhibition of Akt inhibits growth of glioblastoma and glioblastoma stem-like cells. *Molecular Cancer Therapeutics* **8**, 386–393.

89. Cloughesy, T.F., et al. (2008) Antitumor activity of rapamycin in a Phase I trial for patients with recurrent PTEN-deficient glioblastoma. *PLoS Medicine* **5**, e8.

90. Cantley, L.C. (2002) The phosphoinositide 3-kinase pathway. *Science* **296**, 1655–1657.

91. Engelman, J.A., Luo, J. and Cantley, L.C. (2006) The evolution of phosphatidylinositol 3-kinases as regulators of growth and metabolism. *Nature Reviews Genetics* **7**, 606–619.

92. Broderick, D.K., et al. (2004) Mutations of PIK3CA in anaplastic oligodendrogliomas, high-grade astrocytomas, and

medulloblastomas. *Cancer Research* **64**, 5048–5050.

93. Hartmann, C., Bartels, G., Gehlhaar, C., Holtkamp, N. and von Deimling, A. (2005) PIK3CA mutations in glioblastoma multiforme. *Acta Neuropathologica* **109**, 639–642.

94. Samuels, Y., et al. (2004) High frequency of mutations of the PIK3CA gene in human cancers. *Science* **304**, 554.

95. Gallia, G.L., et al. (2006) PIK3CA gene mutations in pediatric and adult glioblastoma multiforme. *Molecular Cancer Research* **4**, 709–714.

96. Mueller, W., et al. (2005) Mutations of the PIK3CA gene are rare in human glioblastoma. *Acta Neuropathologica* **109**, 654–655.

97. Wu, X., Senechal, K., Neshat, M.S., Whang, Y.E. and Sawyers, C.L. (1998) The PTEN/MMAC1 tumor suppressor phosphatase functions as a negative regulator of the phosphoinositide 3-kinase/ Akt pathway. *Proceedings of the National Academy of Sciences of the United States of America* **95**, 15587–15591.

98. Choe, G., et al. (2003) Analysis of the phosphatidylinositol 3′-kinase signaling pathway in glioblastoma patients *in vivo. Cancer Research* **63**, 2742–2746.

99. Maehama, T. and Dixon, J.E. (1998) The tumor suppressor, PTEN/MMAC1, dephosphorylates the lipid second messenger, phosphatidylinositol 3,4,5-trisphosphate. *Journal of Biological Chemistry* **273**, 13375–13378.

100. Li, J., et al. (1997) PTEN, a putative protein tyrosine phosphatase gene mutated in human brain, breast, and prostate cancer. *Science* **275**, 1943–1947.

101. Duerr, E.M., et al. (1998) PTEN mutations in gliomas and glioneuronal tumors. *Oncogene* **16**, 2259–2264.

102. Rasheed, B.K., et al. (1997) PTEN gene mutations are seen in high-grade but not in low-grade gliomas. *Cancer Research* **57**, 4187–4190.

103. Datta, S.R., et al. (1997) Akt phosphorylation of BAD couples survival signals to the cell-intrinsic death machinery. *Cell* **91**, 231–241.

104. Brunet, A., et al. (1999) Akt promotes cell survival by phosphorylating and inhibiting a Forkhead transcription factor. *Cell* **96**, 857–868.

105. Sonoda, Y., et al. (2001) Akt pathway activation converts anaplastic astrocytoma to glioblastoma multiforme in a human astrocyte model of glioma. *Cancer Research* **61**, 6674–6678.

106. Zhu, Y., et al. (2005) Early inactivation of p53 tumor suppressor gene cooperating with NF1 loss induces malignant astrocytoma. *Cancer Cell* **8**, 119–130.

107. Gupta, S., et al. (2007) Binding of ras to phosphoinositide 3-kinase p110alpha is required for ras-driven tumorigenesis in mice. *Cell* **129**, 957–968.

108. Guha, A., Feldkamp, M.M., Lau, N., Boss, G. and Pawson, A. (1997) Proliferation of human malignant astrocytomas is dependent on Ras activation. *Oncogene* **15**, 2755–2765.

109. Holland, E.C., et al. (2000) Combined activation of Ras and Akt in neural progenitors induces glioblastoma formation in mice. *Nature Genetics* **25**, 55–57.

110. da Fonseca, C.O., Linden, R., Futuro, D., Gattass, C.R. and Quirico-Santos, T. (2008) Ras pathway activation in gliomas: a strategic target for intranasal administration of perillyl alcohol. *Archivum Immunologiae et Therapiae Experimentalis* **56**, 267–276.

111. Chi, S., et al. (1999) Oncogenic Ras triggers cell suicide through the activation of a caspase-independent cell death program in human cancer cells. *Oncogene* **18**, 2281–2290.

112. Overmeyer, J.H., Kaul, A., Johnson, E.E. and Maltese, W.A. (2008) Active ras triggers death in glioblastoma cells through hyperstimulation of macropinocytosis. *Molecular Cancer Research* **6**, 965–977.

113. Chin, Y.E., Kitagawa, M., Kuida, K., Flavell, R.A. and Fu, X.Y. (1997) Activation of the STAT signaling pathway can cause expression of caspase 1 and apoptosis. *Molecular and Cellular Biology* **17**, 5328–5337.

114. Schaefer, L.K., Ren, Z., Fuller, G.N. and Schaefer, T.S. (2002) Constitutive activation of Stat3alpha in brain tumors: localization to tumor endothelial cells and activation by the endothelial tyrosine kinase receptor (VEGFR-2). *Oncogene* **21**, 2058–2065.

115. Rahaman, S.O., et al. (2002) Inhibition of constitutively active Stat3 suppresses proliferation and induces apoptosis in glioblastoma multiforme cells. *Oncogene* **21**, 8404–8413.

116. Lo, H.W., Cao, X., Zhu, H. and Ali-Osman, F. (2008) Constitutively activated STAT3 frequently coexpresses with epidermal growth factor receptor in high-grade gliomas and targeting STAT3 sensitizes them to Iressa and alkylators. *Clinical Cancer Research* **14**, 6042–6054.

117. Brantley, E.C., et al. (2008) Loss of protein inhibitors of activated STAT-3 expression in glioblastoma multiforme tumors: implications for STAT-3 activation and gene expression. *Clinical Cancer Research* **14**, 4694–4704.

118. Wang, T., et al. (2004) Regulation of the innate and adaptive immune responses by Stat-3

signaling in tumor cells. *Nature Medicine* **10**, 48–54.

119. Cully, M., You, H., Levine, A.J. and Mak, T.W. (2006) Beyond PTEN mutations: the PI3K pathway as an integrator of multiple inputs during tumorigenesis. *Nature Reviews Cancer* **6**, 184–192.

120. Zheng, H., et al. (2008) p53 and Pten control neural and glioma stem/progenitor cell renewal and differentiation. *Nature* **455**, 1129–1133.

121. Ichimura, K., et al. (2000) Deregulation of the p14ARF/MDM2/p53 pathway is a prerequisite for human astrocytic gliomas with G1-S transition control gene abnormalities. *Cancer Research* **60**, 417–424.

122. Ueki, K., et al. (1996) CDKN2/p16 or RB alterations occur in the majority of glioblastomas and are inversely correlated. *Cancer Research* **56**, 150–153.

123. Ghimenti, C., et al. (2003) Deregulation of the p14ARF/Mdm2/p53 pathway and G1/S transition in two glioblastoma sets. *Journal of Neuro-oncology* **61**, 95–102.

124. Ichimura, K., Schmidt, E.E., Goike, H.M. and Collins, V.P. (1996) Human glioblastomas with no alterations of the CDKN2A (p16INK4A, MTS1) and CDK4 genes have frequent mutations of the retinoblastoma gene. *Oncogene* **13**, 1065–1072.

125. Schmidt, E.E., Ichimura, K., Reifenberger, G. and Collins, V.P. (1994) CDKN2 (p16/MTS1) gene deletion or CDK4 amplification occurs in the majority of glioblastomas. *Cancer Research* **54**, 6321–6324.

126. Kamijo, T., et al. (1998) Functional and physical interactions of the ARF tumor suppressor with p53 and Mdm2. *Proceedings of the National Academy of Sciences of the United States of America* **95**, 8292–8297.

127. Riemenschneider, M.J., et al. (1999) Amplification and overexpression of the MDM4 (MDMX) gene from 1q32 in a subset of malignant gliomas without TP53 mutation or MDM2 amplification. *Cancer Research* **59**, 6091–6096.

128. Ishii, N., et al. (1999) Cells with TP53 mutations in low grade astrocytic tumors evolve clonally to malignancy and are an unfavorable prognostic factor. *Oncogene* **18**, 5870–5878.

129. Kraus, J.A., et al. (2001) TP53 gene mutations, nuclear p53 accumulation, expression of Waf/p21, Bcl-2, and CD95 (APO-1/Fas) proteins are not prognostic factors in de novo glioblastoma multiforme. *Journal of Neuro-oncology* **52**, 263–272.

130. Henson, J.W., et al. (1994) The retinoblastoma gene is involved in malignant progression of astrocytomas. *Annals of Neurology* **36**, 714–721.

131. Jen, J., et al. (1994) Deletion of p16 and p15 genes in brain tumors. *Cancer Research* **54**, 6353–6358.

132. Serrano, M., Hannon, G.J. and Beach, D. (1993) A new regulatory motif in cell-cycle control causing specific inhibition of cyclin D/CDK4. *Nature* **366**, 704–707.

133. Solomon, D.A., et al. (2008) Identification of p18 INK4c as a tumor suppressor gene in glioblastoma multiforme. *Cancer Research* **68**, 2564–2569.

134. Simon, M., Koster, G., Menon, A.G. and Schramm, J. (1999) Functional evidence for a role of combined CDKN2A (p16-p14 (ARF))/CDKN2B (p15) gene inactivation in malignant gliomas. *Acta Neuropathologica* **98**, 444–452.

135. Kim, W.Y. and Sharpless, N.E. (2006) The regulation of INK4/ARF in cancer and aging. *Cell* **127**, 265–275.

136. Drablos, F., et al. (2004) Alkylation damage in DNA and RNA: repair mechanisms and medical significance. *DNA Repair* **3**, 1389–1407.

137. Hegi, M.E., et al. (2005) MGMT gene silencing and benefit from temozolomide in glioblastoma. *New England Journal of Medicine* **352**, 997–1003.

138. Qian, X.C. and Brent, T.P. (1997) Methylation hot spots in the 5′ flanking region denote silencing of the O6-methylguanine-DNA methyltransferase gene. *Cancer Research* **57**, 3672–3677.

139. Watts, G.S., et al. (1997) Methylation of discrete regions of the O6-methylguanine DNA methyltransferase (MGMT) CpG island is associated with heterochromatinization of the MGMT transcription start site and silencing of the gene. *Molecular and Cellular Biology* **17**, 5612–5619.

140. Esteller, M., Hamilton, S.R., Burger, P.C., Baylin, S.B. and Herman, J.G. (1999) Inactivation of the DNA repair gene O6-methylguanine-DNA methyltransferase by promoter hypermethylation is a common event in primary human neoplasia. *Cancer Research* **59**, 793–797.

141. Esteller, M., et al. (2000) Inactivation of the DNA-repair gene MGMT and the clinical response of gliomas to alkylating agents. *New England Journal of Medicine* **343**, 1350–1354.

142. Paz, M.F., et al. (2004) CpG island hypermethylation of the DNA repair enzyme methyltransferase predicts response to temozolomide in primary gliomas. *Clinical Cancer Research* **10**, 4933–4938.

143. Silber, J.R., Mueller, B.A., Ewers, T.G. and Berger, M.S. (1993) Comparison of O6-methylguanine-DNA methyltransferase activity in brain tumors and adjacent

normal brain. *Cancer Research* **53**, 3416–3420.

144. Silber, J.R., et al. (1996) Lack of the DNA repair protein O6-methylguanine-DNA methyltransferase in histologically normal brain adjacent to primary human brain tumors. *Proceedings of the National Academy of Sciences of the United States of America* **93**, 6941–6946.

145. Herrlinger, U., et al. (2006) Phase II trial of lomustine plus temozolomide chemotherapy in addition to radiotherapy in newly diagnosed glioblastoma: UKT-03. *Journal of Clinical Oncology* **24**, 4412–4417.

146. Donson, A.M., Addo-Yobo, S.O., Handler, M.H., Gore, L. and Foreman, N.K. (2007) MGMT promoter methylation correlates with survival benefit and sensitivity to temozolomide in pediatric glioblastoma. *Pediatric Blood and Cancer* **48**, 403–407.

147. Cahill, D.P., et al. (2007) Loss of the mismatch repair protein MSH6 in human glioblastomas is associated with tumor progression during temozolomide treatment. *Clinical Cancer Research* **13**, 2038–2045.

148. Chinot, O.L., et al. (2007) Correlation between O6-methylguanine-DNA methyltransferase and survival in inoperable newly diagnosed glioblastoma patients treated with neoadjuvant temozolomide. *Journal of Clinical Oncology* **25**, 1470–1475.

149. Nagane, M., Kobayashi, K., Ohnishi, A., Shimizu, S. and Shiokawa, Y. (2007) Prognostic significance of O6-methylguanine-DNA methyltransferase protein expression in patients with recurrent glioblastoma treated with temozolomide. *Japanese Journal of Clinical Oncology* **37**, 897–906.

150. Silber, J.R., et al. (1999) O6-methylguanine-DNA methyltransferase-deficient phenotype in human gliomas: frequency and time to tumor progression after alkylating agent-based chemotherapy. *Clinical Cancer Research* **5**, 807–814.

151. Hegi, M.E., et al. (2004) Clinical trial substantiates the predictive value of O-6-methylguanine-DNA methyltransferase promoter methylation in glioblastoma patients treated with temozolomide. *Clinical Cancer Research* **10**, 1871–1874.

152. Brandes, A.A., et al. (2009) Recurrence pattern after temozolomide concomitant with and adjuvant to radiotherapy in newly diagnosed patients with glioblastoma: correlation with MGMT promoter methylation status. *Journal of Clinical Oncology* **27**, 1275–1279.

153. Tolcher, A.W., et al. (2003) Marked inactivation of O6-alkylguanine-DNA alkyltransferase activity with protracted temozolomide schedules. *British Journal of Cancer* **88**, 1004–1011.

154. Kokkinakis, D.M., Bocangel, D.B., Schold, S.C., Moschel, R.C. and Pegg, A.E. (2001) Thresholds of O6-alkylguanine-DNA alkyltransferase which confer significant resistance of human glial tumor xenografts to treatment with 1,3-bis (2-chloroethyl)-1-nitrosourea or temozolomide. *Clinical Cancer Research* **7**, 421–428.

155. Quinn, J.A., et al. (2009) Phase II trial of Gliadel plus O6-benzylguanine in adults with recurrent glioblastoma multiforme. *Clinical Cancer Research* **15**, 1064–1068.

156. Quinn, J.A., et al. (2009) Phase II trial of temozolomide plus O6-benzylguanine in adults with recurrent, temozolomide-resistant malignant glioma. *Journal of Clinical Oncology* **27**, 1262–1267.

157. Schold, S.C., Jr., et al. (2004) O6-benzylguanine suppression of O6-alkylguanine-DNA alkyltransferase in anaplastic gliomas. *Neuro-oncology* **6**, 28–32.

158. Karran, P., Offman, J. and Bignami, M. (2003) Human mismatch repair, drug-induced DNA damage, and secondary cancer. *Biochimie* **85**, 1149–1160.

159. Hunter, C., et al. (2006) A hypermutation phenotype and somatic MSH6 mutations in recurrent human malignant gliomas after alkylator chemotherapy. *Cancer Research* **66**, 3987–3991.

160. Cichowski, K. and Jacks, T. (2001) NF1 tumor suppressor gene function: narrowing the GAP. *Cell* **104**, 593–604.

161. Gutmann, D.H., et al. (2003) Molecular analysis of astrocytomas presenting after age 10 in individuals with NF1. *Neurology* **61**, 1397–1400.

162. Gutmann, D.H., et al. (2002) Gliomas presenting after age 10 in individuals with neurofibromatosis type 1 (NF1). *Neurology* **59**, 759–761.

163. Rasmussen, S.A., Yang, Q. and Friedman, J.M. (2001) Mortality in neurofibromatosis 1: an analysis using U.S. death certificates. *American Journal of Human Genetics* **68**, 1110–1118.

164. Reilly, K.M., Loisel, D.A., Bronson, R.T., McLaughlin, M.E. and Jacks, T. (2000) Nf1;Trp53 mutant mice develop glioblastoma with evidence of strain-specific effects. *Nature Genetics* **26**, 109–113.

165. Kwon, C.H., et al. (2008) Pten haploinsufficiency accelerates formation of high-grade astrocytomas. *Cancer Research* **68**, 3286–3294.

166. Saadoun, S., Papadopoulos, M.C., Davies, D.C., Bell, B.A. and Krishna, S. (2002) Increased aquaporin 1 water channel expression in human brain

tumours. *British Journal of Cancer* **87**, 621–623.

167. Saadoun, S., Papadopoulos, M.C., Davies, D.C., Krishna, S. and Bell, B.A. (2002) Aquaporin-4 expression is increased in oedematous human brain tumours. *Journal of Neurology, Neurosurgery, and Psychiatry* **72**, 262–265.

168. Minard, K.I. and McAlister-Henn, L. (1999) Dependence of peroxisomal beta-oxidation on cytosolic sources of NADPH. *Journal of Biological Chemistry* **274**, 3402–3406.

169. Lee, S.M., et al. (2002) Cytosolic NADP(+)-dependent isocitrate dehydrogenase status modulates oxidative damage to cells. *Free Radical Biology and Medicine* **32**, 1185–1196.

170. Balss, J., et al. (2008) Analysis of the IDH1 codon 132 mutation in brain tumors. *Acta Neuropathologica* **116**, 597–602.

171. Yan, H., et al. (2009) IDH1 and IDH2 mutations in gliomas. *New England Journal of Medicine* **360**, 765–773.

172. Ichimura, K., et al. (2009) IDH1 mutations are present in the majority of common adult gliomas but are rare in primary glioblastomas. *Neuro-oncology* **11**, 341–347.

173. Zhao, S., et al. (2009) Glioma-derived mutations in IDH1 dominantly inhibit IDH1 catalytic activity and induce HIF-1alpha. *Science* **324**, 261–265.

174. Dang, L., et al. (2009) Cancer-associated IDH1 mutations produce 2-hydroxyglutarate. *Nature* **462**, 739–744.

175. Ward, P.S., et al. (2010) The common feature of leukemia-associated IDH1 and IDH2 mutations is a neomorphic enzyme activity converting alpha-ketoglutarate to 2-hydroxyglutarate. *Cancer Cell* **17**, 225–234.

176. Mardis, E.R., et al. (2009) Recurring mutations found by sequencing an acute myeloid leukemia genome. *New England Journal of Medicine* **361**, 1058–1066.

177. Figueroa, M.E., et al. (2010) Leukemic IDH1 and IDH2 mutations result in a hypermethylation phenotype, disrupt TET2 function, and impair hematopoietic differentiation. *Cancer Cell* **18**, 553–567.

178. Noushmehr, H., et al. (2010) Identification of a CpG island methylator phenotype that defines a distinct subgroup of glioma. *Cancer Cell* **17**, 510–522.

179. Xu, W., et al. (2011) Oncometabolite 2-hydroxyglutarate is a competitive inhibitor of alpha-ketoglutarate-dependent dioxygenases. *Cancer Cell* **19**, 17–30.

180. Koivunen, P., et al. (2012) Transformation by the (R)-enantiomer of 2-hydroxyglutarate linked to EGLN activation. *Nature* **483**, 484–488.

181. Vander Heiden, M.G., Cantley, L.C. and Thompson, C.B. (2009) Understanding the Warburg effect: the metabolic requirements of cell proliferation. *Science* **324**, 1029–1033.

182. Thompson, C. (2009) Metabolic enzymes as oncogenes or tumor suppressors. *New England Journal of Medicine* **360**, 813–815.

183. Galli, R., et al. (2004) Isolation and characterization of tumorigenic, stem-like neural precursors from human glioblastoma. *Cancer Research* **64**, 7011–7021.

184. Singh, S.K., et al. (2003) Identification of a cancer stem cell in human brain tumors. *Cancer Research* **63**, 5821–5828.

185. Singh, S.K., et al. (2004) Identification of human brain tumour initiating cells. *Nature* **432**, 396–401.

186. Yuan, X., et al. (2004) Isolation of cancer stem cells from adult glioblastoma multiforme. *Oncogene* **23**, 9392–9400.

187. Corbeil, D., et al. (2000) The human AC133 hematopoietic stem cell antigen is also expressed in epithelial cells and targeted to plasma membrane protrusions. *Journal of Biological Chemistry* **275**, 5512–5520.

188. Uchida, N., et al. (2000) Direct isolation of human central nervous system stem cells. *Proceedings of the National Academy of Sciences of the United States of America* **97**, 14720–14725.

189. Zhang, M., et al. (2008) Nestin and CD133: valuable stem cell-specific markers for determining clinical outcome of glioma patients. *Journal of Experimental and Clinical Cancer Research* **27**, 85.

190. Ma, Y.H., et al. (2008) Expression of stem cell markers in human astrocytomas of different WHO grades. *Journal of Neuro-oncology* **86**, 31–45.

191. Beier, D., et al. (2008) CD133 expression and cancer stem cells predict prognosis in high-grade oligodendroglial tumors. *Brain Pathology* **18**, 370–377.

192. Bao, S., et al. (2006) Glioma stem cells promote radioresistance by preferential activation of the DNA damage response. *Nature* **444**, 756–760.

193. Abdel-Fattah, R., et al. (2006) Differential expression of HOX genes in neoplastic and non-neoplastic human astrocytes. *Journal of Pathology* **209**, 15–24.

194. Joo, K.M., et al. (2008) Clinical and biological implications of CD133-positive and CD133-negative cells in glioblastomas. *Laboratory Investigation* **88**, 808–815.

195. Beier, D., et al. (2007) CD133(+) and CD133(-) glioblastoma-derived cancer stem cells show differential growth characteristics and molecular profiles. *Cancer Research* **67**, 4010–4015.

196. Lendahl, U., Zimmerman, L.B. and McKay, R.D. (1990) CNS stem cells express a new class of intermediate filament protein. *Cell* **60**, 585–595.

197. Maness, P.F. and Schachner, M. (2007) Neural recognition molecules of the immunoglobulin superfamily: signaling transducers of axon guidance and neuronal migration. *Nature Neuroscience* **10**, 19–26.

198. Bao, S., et al. (2008) Targeting cancer stem cells through L1CAM suppresses glioma growth. *Cancer Research* **68**, 6043–6048.

199. Liang, Y., et al. (2005) Gene expression profiling reveals molecularly and clinically distinct subtypes of glioblastoma multiforme. *Proceedings of the National Academy of Sciences of the United States of America* **102**, 5814–5819.

200. Phillips, H.S., et al. (2006) Molecular subclasses of high-grade glioma predict prognosis, delineate a pattern of disease progression, and resemble stages in neurogenesis. *Cancer Cell* **9**, 157–173.

201. Liang, Y., Bollen, A.W., Aldape, K.D. and Gupta, N. (2006) Nuclear FABP7 immunoreactivity is preferentially expressed in infiltrative glioma and is associated with poor prognosis in EGFR-overexpressing glioblastoma. *BMC Cancer* **6**, 97.

Chapter

22

Breast cancer

Arthur W. Lambert, Sait Ozturk, Chen Khuan Wong, Panagiotis Papageorgis and Sam Thiagalingam

Introduction

The notion that breast cancer is not a single disease but many – that there is considerable heterogeneity among different tumors – is not new, having been observed by physicians for decades. But with the advent of powerful analytical and computational tools, systems biology has provided a fascinating view of the underlying mutations, biology, and networks that drive the process of breast tumorigenesis. The information derived from analyzing these networks is also proving to be clinically useful and is beginning to be incorporated into the standard clinical management of breast cancer patients – to determine prognosis and guide the choice of therapeutics, for example. However, these advances are complicated by the heterogeneity of cancer cell phenotypes that exist within a single tumor, which drives metastasis, resistance to treatment, and, eventually, recurrence. Systems biology offers a strong tool for investigating both levels of heterogeneity to comprehensively define, and ultimately attack, the aberrant molecular networks governing breast cancer cells.

Epidemiology

Breast cancer is a staggering public health problem. The American Cancer Society expects that over 232,000 cases of invasive breast cancer will be diagnosed in 2014, accounting for the largest number of cancer cases in women (ACS, 2014). Nearly 40,000 deaths are predicted to occur, which rank it the second leading cause of cancer deaths in women, behind only lung cancer. Fortunately, the number of newly diagnosed cases has decreased since 2000, largely as a result of a reduction in number of post-menopausal women on hormone replacement therapy, which has been strongly linked to the development of breast cancer (Rossouw et al., 2002). Death rates have also decreased in recent years and this has been attributed to improvements in screening and early detection with mammography as well as better treatment options (ACS, 2012). However, racial disparities do exist as African American women have a lower incidence rate but an increased chance of death (ACS, 2011). While this may reflect socioeconomic factors and access to health care, there also seems to be a biological difference since African American women are more likely to be diagnosed with aggressive cancers (Amend et al., 2006; Stead et al., 2009). Additionally, over 62,000 cases of carcinoma *in situ* are expected in 2014 (ACS, 2014). These are non-invasive neoplasms and likely represent a pre-malignant stage in the progression toward invasive breast cancer (Burstein et al., 2004).

Numerous risk factors have been identified for breast carcinomas. The most important factors that increase the risk of breast cancer are gender, age, and family history, and modifiable factors such as obesity, physical inactivity, and alcohol consumption are thought to confer risk as well (ACS, 2011). There is also a strong relationship between exposure to reproductive hormones, such as estrogen and progesterone, and the development of breast cancer in post-menopausal women (Key et al., 2002), and hence, the connection between hormone replacement therapy and cancer incidence mentioned above. There do seem to be inherited factors that predispose one to breast cancer because having a relative with a history of breast cancer increases risk. But this contribution is somewhat limited as most patients diagnosed do not have a close relative with breast cancer, and conversely, the majority of women that do have a relative with breast cancer will never develop the disease (ACS, 2011). The exception here is individuals who

---

Systems Biology of Cancer, ed. S. Thiagalingam. Published by Cambridge University Press. © Cambridge University Press 2015.

inherit a mutant allele of *BRCA1* or *BRCA2*, two well-characterized breast and ovarian cancer susceptibility genes. Women with these inherited mutations have up to an 80% chance of developing breast cancer, but these cases account for only 5 to 10% of all breast cancer diagnoses (Fackenthal and Olopade, 2007).

Pathological classification of breast cancer

Breast cancer arises from malignant transformation of epithelial cells in the mammary gland. The normal epithelium of the breast is organized as a series of ducts and lobules, designed to secrete milk and transport it to the nipple. On the cellular level, luminal cells line the ducts and, along with alveolar cells, are present in the lobules. Myoepithelial cells circumscribe these cells and form a basal layer of epithelium, which rests on the basement membrane separating the epithelial and connective tissue. The vast majority of carcinomas – upwards of 75 to 80% – arise in the ductal region and have the histological designation of invasive ductal carcinoma, not otherwise specified (IDC NOS) (Bertos and Park, 2011; Li et al., 2005; Weigelt et al., 2010). The remaining 20 to 25% of cases are referred to as "special types" and occupy one of 17 different histological subtypes such as invasive lobular, cribriform, medullary, mucinous, aporcine, and metaplastic, to name a few of the more common – but each with less than a 5% overall incidence rate – special types (Weigelt et al., 2010).

In addition to histological type, which categorizes tumors based on tissue morphology, the histological grade of the cancer reflects the degree of cellular differentiation as well as the mitotic rate. Histological grade is routinely used to classify cases of breast cancer and it serves as a good prognostic indicator, where poorly differentiated and highly proliferative cancers generally have worse outcomes (Elston and Ellis, 1991). Breast cancers are also subject to a staging process that evaluates the size of the primary tumor, the dissemination to regional lymph nodes, and the presence or absence of distant metastases. The disease stage is an indicator of the extent to which a cancer has spread and is the most important factor in determining prognosis and treatment options (Singletary et al., 2003).

Finally, there are a number of well-characterized molecules that are also used to classify, prognosticate, and guide treatment decisions. Foremost here is the estrogen receptor (ER), a nuclear hormone receptor that, upon binding estrogen, translocates to the nucleus and regulates the transcription of numerous target genes, many of which are involved in cell proliferation (Pearce and Jordan, 2004). Roughly 70% of breast tumors are classified as ER-positive and the majority of these also express the progesterone receptor (PR), another member of the nuclear receptor superfamily that is commonly assessed by pathologists (Osborne, 1998a). Hormone receptor expression is less of a prognostic factor because it is correlated to histological grade (Elston et al., 1999). However, because the growth of most of these cancers is hormone dependent, ER/PR expression is strongly predictive of a response to endocrine therapies such as selective estrogen receptor modulators (SERMs) and aromatase inhibitors (Baum et al., 2002). It is worth noting, though, that not all patients with hormone receptor-positive disease will respond to endocrine therapy and of those that do many will develop resistance over time (Ali and Coombes, 2002; Massarweh and Schiff, 2006), implying that there is significant heterogeneity within this class of tumors.

The human epidermal receptor-2 (*HER2*, also known as *ERBB2*) is an oncogenic growth factor receptor that is now included in the standard pathological assessment of breast cancers. Up to 30% of cases harbor amplifications of this gene (Slamon et al., 1987) that generally result in an increased abundance of the protein in the plasma membrane, where it functions to stimulate intracellular growth and division signaling pathways. Protein levels are routinely assessed by immunohistochemistry (IHC) and gene amplification can be verified by fluorescent *in situ* hybridization (FISH) to obtain a more accurate measure (Bartlett et al., 2003). Patients with amplified *HER2* have an aggressive form of the disease and a substantially worse prognosis (Seshadri et al., 1993; Slamon et al., 1987) but with the advent of targeted therapies, such as trastuzumab, which specifically counteracts the HER2 protein, these patients have seen remarkable response rates and a corresponding benefit in terms of disease outcome (Slamon et al., 2001). Today, the *HER2* oncogene serves as a prime example of a patient-tailored molecular target, subject to standard pathologic detection and amenable to drug inhibition.

Another collection of breast tumors, referred to as triple-negative cancer, is defined based on the absence of ER, PR, and HER2 expression. About 15% of breast cancers are classified as triple-negative and these patients tend to be younger and can expect poor short-term outcomes, which improve after five years (Foulkes et al., 2010). These patients are not eligible for targeted therapies such as tamoxifen and trastuzumab because they do not express ER and HER2; however, standard chemotherapy seems to be particularly effective in some patients (Foulkes et al., 2010).

Molecular classification of breast cancer

A major milestone in breast cancer research was reached in 2000 when it was reported that breast tumors could be classified into different subgroups based on their gene expression profiles (Perou et al., 2000; Sorlie et al., 2001). This unprecedented molecular view of breast tumors suggested that common transcriptional programs, shared between different tumors, could explain some of the biological variability inherent to breast cancer that had been observed by pathologists and oncologists for years. Five subtypes were originally described: two luminal subtypes (luminal A and luminal B), which are hormone-receptor positive; an HER-2 positive subtype; a basal-like group that shares a gene expression profile with myoepithelial cells of the normal breast and generally lacks expression of HER-2 and the hormone receptors (and thus is often classified as triple-negative tumors); and a heterogeneous, normal-like collection of tumors. These subtypes have been repeatedly observed (Perou and Borresen-Dale, 2011; Sorlie et al., 2003) and, importantly, are associated with distinct clinical outcomes, where the HER-2 and basal-like subtypes have the worst prognosis (Sorlie et al., 2001); however, these two subtypes do seem to be the most responsive to neoadjuvant therapy (Rouzier et al., 2005). Recently, an additional claudin-low subtype that has reduced expression of tight junction and cell–cell adhesion genes has been observed as well (Herschkowitz et al., 2007). These tumors exhibit an epithelial–mesenchymal transition (EMT) and cancer stem-like phenotype, are chemoresistant, and appear to include most cases of metaplastic breast cancer (Hennessy et al., 2009). Recent studies have also found that some of these subtypes can be subdivided even further. For example, triple-negative breast cancer, which often falls in the basal-like group, appears to encompass at least six discrete subtypes that can be classified on the basis of gene expression profiling and are potentially susceptible to different targeted therapies (Lehmann et al., 2011). The

appearance of these intrinsic molecular subtypes may reflect a different cell of origin, varying driver mutations, or some combination of the two, and understanding the factors that generate the different subtypes of breast cancer is a major focus of current research efforts. Regardless, these gene expression signatures provide a significant amount of insight that helps to conceptualize the basic biological differences underlying heterogeneous breast cancers.

Furthermore, molecular signatures are beginning to make their way into the routine clinical management of breast cancer. Mammaprint was developed using a supervised classification strategy to define a gene expression signature consisting of 70 different genes that together could accurately stratify young, lymph node-negative patients, based on whether or not they developed metastases (van 't Veer et al., 2002). This prognostic signature has been validated in larger samples of patients (Buyse et al., 2006; van de Vijver et al., 2002) and was approved by the FDA in 2007 to assess the risk of recurrence five to ten years after diagnosis of the primary tumor. Using previously published genome-wide expression data, another group prospectively identified a 21-gene set (consisting of 16 genes involved in cancer and five reference genes) that could be evaluated by reverse transcription polymerase chain reaction (RT-PCR) to derive a recurrence score that is predictive of relapse in lymph node-negative, ER-positive patients treated with tamoxifen (Paik et al., 2004). This diagnostic test, referred to as Onco*type* DX, has the advantage of being compatible with standard paraffin-embedded pathology samples and also seems to highlight a patient population especially responsive to chemotherapy, as those patients with a high recurrence score benefited the most from the addition of adjuvant chemotherapy (Paik et al., 2006). This finding is notable given that there is considerable heterogeneity within clinically defined ER-positive tumors (Russnes et al., 2011). Finally, as the intrinsic molecular subtypes have prognostic value themselves, recent work has established that a 50-gene quantitative RT-PCR assay, called the PAM50, is able to accurately classify patients in essentially the same manner (Parker et al., 2009), and thus has the potential to bring the molecular subtypes into routine clinical use. Although there is minimal overlap in the genes that define the intrinsic subtypes and those that comprise Mammaprint and Onco*type* DX, they appear to classify tumors in much the same way (Fan et al., 2006).

These diagnostic tests illustrate the power of using genomic technologies to interrogate patient tumors (Sotiriou and Pusztai, 2009). While both measures are primarily predictive, they also provide a new method to select patients who most likely need – and perhaps more importantly those who do not need – chemotherapy that is based on the molecular profile of the individual tumor. Interestingly, the subtype and prognostic gene expression signatures imply that the ability of a cancer to metastasize is, in a sense, pre-determined within the primary tumor, providing a diagnostic, and potentially therapeutic, window of opportunity.

New work is beginning to conflate gene expression data with other molecular information to derive new and informative subgroups of breast cancer. This has largely focused on somatic copy number alterations (CNAs), assessed by array or high-resolution comparative genomic hybridization (CGH), which reflect genomic changes present in the tumor cells. For example, one study analyzed CNAs and found that ER-negative tumors could be subdivided further based on their level of genomic stability (Chin et al., 2007). Other work has shown that recurrent CNAs differ between the intrinsic subtypes (Bergamaschi et al., 2006) and that high-level amplification is associated with a poor clinical outcome (Chin et al., 2006). These studies are also consistent with the idea that basal-like tumors are marked by complex patterns of genomic alterations (Russnes et al., 2010) and tend to harbor more somatic rearrangements than the other subgroups (Stephens et al., 2009). In an effort to uncover the functional drivers of breast cancer subtypes, a recent report described a large-scale analysis of over 2,000 patients that linked copy number and gene expression profiles (Curtis et al., 2012). This "integrated" analysis revealed ten novel subgroups of breast cancer, based on the combined genomic and transcriptional profiling, with varying risk of disease progression. While the application of such a complex classifier in the clinical setting is difficult to imagine at present, the power of this methodology lies in the ability to identify mutations, even at low frequency, that are likely driving tumor growth. Identifying, characterizing, and targeting such alterations are central to the idea of personalized cancer therapy – administered to the correct subgroups of patients, defined by the molecular profile of their tumor (Russnes et al., 2011). The challenge now is to reconcile the numerous molecular classification systems that have been used to subdivide breast cancers, identify the most accurate in terms of their prognostic and predictive value, and find a cost-effective way to deliver these technologies to patients.

Mutational landscape of breast cancer genomes

Cancer growth ultimately results from the presence of mutant cellular genes, which are predicted to impact at least six hallmark characteristics of cancer cells: self-sufficiency in growth factor signaling; insensitivity to anti-growth signals; evasion of apoptosis; replicative immortality; the ability to induce angiogenesis; and, eventually, invasive and metastatic behavior (Hanahan and Weinberg, 2011). Broadly, cellular genes could be perturbed through nucleotide substitutions, insertions and deletions, chromosomal rearrangements, and copy number alterations (Hudson et al., 2010; Stratton, 2011). Recent work has leveraged the complete human genome sequence in an effort to reveal all of the mutations present within a tumor and, hopefully, to specify those few mutations that actually drive cellular proliferation and cancer progression. An early demonstration of this approach was published in 2006 when the sequence of over 13,000 protein-coding genes was reported for 11 colon and 11 breast cancers (Sjoblom et al., 2006). In the breast tumor set, this sequencing study found over 900 distinct somatic mutations affecting 673 genes, with each tumor harboring roughly 67 mutant genes. However, it was estimated that each tumor contained an average of 12 mutations in candidate cancer genes (*CAN* genes) – genes likely to be positively selected for, and thus potential drivers of tumorigenesis. Importantly, many of these genes had not been previously linked to breast cancer, highlighting the value of investigating breast cancer genomes using an unbiased sequencing approach. Building upon this work by extending the sequencing to over 18,000 genes – all of the Reference Sequence (RefSeq) genes at the time – revealed a more comprehensive picture of the mutational landscape present in breast cancer cells (Wood et al., 2007). Most interesting here was the fact that only a small number of genes were recurrently mutated in breast cancer (as well as colon cancer); the majority of mutations appear at a low frequency, giving rise to numerous mutational "hills," amongst a few frequently mutated "mountains" (Wood et al., 2007). The most frequently mutated genes found in breast cancer, represented by two "mountains," include the tumor suppressor *TP53* found on chromosome 17 and the oncogene *PIK3CA* on chromosome 3. With the inclusion of the additional RefSeq genes, it was estimated that approximately 15 mutations might be responsible for tumorigenesis. And although most of these mutations are not shared across different breast cancer cases, it seems likely that they may converge on similar signaling pathways (Thiagalingam, 2006; Vogelstein and Kinzler, 2004), providing hope that the apparent mutational heterogeneity is in fact clinically tractable.

Mutational profiling of tumor genomes has produced other important insights as well. For example, one large-scale sequencing study examined somatic point mutations in 518 protein kinase genes from 210 cancer cases including colorectal, gastric, lung, ovarian, and breast, to name a few (Greenman et al., 2007). This analysis indicated that the prevalence of somatic mutations was highly variable both between and within cancer types, but in general breast cancers displayed a reduced prevalence of mutations compared to other tumors such as gliomas, lung carcinomas, and gastric cancers. The patterns of the somatic mutations were distinct as well, with breast cancers exhibiting a high level of C>G and C>A transversions, possibly reflecting a distinct mutational path or differences in DNA repair activity (Stratton, 2011). Similar to the previously described sequencing studies, a large proportion (over 80%) of the mutations were classified, based on the ratio of non-synonymous to synonymous mutations, as passenger mutations because they were not selected for during the process of tumorigenesis, but the number of potential driver genes – 120 of the 518 kinases – was larger than originally expected and may be another reflection of the numerous, low-frequency mutation "hills."

The extent and nature of copy number alterations and chromosomal rearrangements have also been examined in a genome-wide fashion for breast cancer. One study, which examined 45 breast tumors, reported that each tumor harbored 18 copy number alterations, with 11 amplifications and 7 homozygous deletions (Leary et al., 2008). While the list of candidate cancer genes affected by these changes included some expected genes, such as *ERBB2* and *CCND1* in the amplified regions, and *CDKN2A* in the deleted region, many genes found within these loci were not previously linked to cancer. Similar to the mutations uncovered by sequencing, copy number alterations are infrequently recurrent, although somewhat more so than point mutations (Leary et al., 2008). Further, the data on copy number alterations were integrated with that of point mutations obtained from sequencing (Sjoblom et al., 2006; Wood et al., 2007) to obtain a comprehensive view of the genes altered in breast tumors and the pathways predicted to be perturbed as a result. This analysis indicated that the ERBB2, PI3K, and, surprisingly, the pathway regulating DNA topology were highly enriched for genes subject to point mutations and copy number changes (Leary et al., 2008).

With the advent of second-generation, massively parallel DNA sequencing (Mardis, 2011) it is now possible to examine entire cancer genomes, along with their normal or metastatic counterparts, and early reports of these efforts are beginning to come online. The first report of a whole breast cancer genome was published in late 2009 (Shah et al., 2009). This study sequenced an ER-positive metastatic lobular breast cancer at 40× genomic coverage, as well as the corresponding primary tumor. Thirty-two non-synonymous somatic point mutations were identified in the metastatic sample. When ten of these coding mutations were screened in an additional collection of 192 breast cancers no mutation was found to be recurrent, although a very small number of cases (5 of 192) contained neighboring mutations in the same gene (Shah et al., 2009). This serves as yet another indication of the spectacular diversity of mutations found among different breast tumors. There was also a level of heterogeneity within this individual case, as the mutant allele frequencies found in the metastatic sample were different from the matched primary tumor. Only a small number of mutations detected in the metastasis were also present in the primary tumor, with some present at low frequencies, suggesting a high degree of genomic evolution occurred between resection of the primary tumor and the appearance of metastatic disease nine years later (Shah et al., 2009). A similar whole-genome sequencing approach was used to study the progression of a basal-like breast cancer; this genomic analysis included peripheral blood, primary tumor, a brain metastasis, and a xenograft tumor (Ding et al., 2010). Fifty somatic point mutations were uncovered along with a collection of copy number and structural alterations. A few of the genes were recurrently mutated in a handful of cancers but, once again, the percentage of cases found to harbor these mutations was relatively low. Like the lobular breast cancer, mutation frequencies were highly variable, but in this case the metastasis retained a subset of the mutations that were present in the primary tumor, with only two new mutations and a large deletion acquired, suggesting the metastasis (and xenograft for that matter) represented an outgrowth of a positively selected clone (Ding et al., 2010). A similar clonal expansion has been documented using whole-genome analysis of single tumor cells (Navin et al., 2011).

These studies likely represent the tip of the iceberg for sequencing of whole cancer genomes. As the financial, technical, and computational barriers begin to be lowered the number of tumors sequenced is likely to grow at a rapid rate and groups such as the International Cancer Genome Consortium and the Cancer Genome Atlas (TCGA) (Hudson et al., 2010; Network, 2008) are sure to shepherd these efforts along. Whole-genome sequencing seems particularly well suited to identify low-frequency mutations in individual cancers and would be the ideal means to truly personalize a cancer diagnosis, and thus therapy. And, as these initial reports have demonstrated, this approach should also yield a wealth of biological information, including new cancer genes, novel disease subtypes, and insights into tumor evolution and progression. But, it is reasonable to wonder, with the unexpectedly high number of potential driver genes revealed and the fact that most are non-recurrent, whether this information can be practically applied. To this end, many reports have analyzed global gene expression patterns in an attempt to understand the aberrant biological pathways, rather than the specific genetic mutations, that are operative in breast cancer. Intriguingly, there is evidence to support the idea that the networks active in embryonic stem cells are also at work in high-grade tumors, as they share similar expression of defined gene sets (Ben-Porath et al., 2008). Others have speculated that this overlap can in large part be attributed to activation of an oncogenic Myc network (Kim et al., 2010). With continued analysis of expression data, and integration of this with the corresponding genomic mutations, it should be possible to consolidate the large number of genetic lesions and aberrant signaling pathways into a set of network modules that can be used to better understand the drivers of tumorigenesis (Akavia et al., 2010) and to personalize cancer therapy (Pe'er and Hacohen, 2011; Thiagalingam, 2006).

Groundbreaking work from the Vogelstein laboratory established a model for colorectal tumorigenesis where sequential mutation of defined oncogenes and tumor suppressors could be linked to various pathological intermediates – from hyperplasia to carcinoma – in order to understand the genetic basis of cancer development (Fearon and Vogelstein, 1990). It is worth noting that a similar multi-step mutational progression for human breast cancer has not been identified to date. It is possible that such a progression model remains to be elucidated but, at present, it seems more likely that the mutational path to a breast tumor may simply be more varied and might be less stringent in terms of the inciting genetic mutations.

The cell of origin

While genetic mutations undeniably drive carcinogenesis, recent work has provided evidence to support the idea that the cell initially transformed – the cell of origin – can impart certain characteristics to the cancer that eventually arises. In light of the intrinsic molecular subtypes described above, it seems reasonable to speculate that some of the intertumor heterogeneity is a reflection of a different cell of origin (Polyak, 2007; Visvader, 2011). The extent to which oncogenic mutations or the initial cell type in which they occur contribute to heterogeneity in breast cancer is not completely known at present, and the two ideas are not mutually exclusive, but it is worth noting that the cell of origin hypothesis would seem to place a premium on the pre-existing epigenetic state in the cell, which has been found to be maintained in certain epithelial cells of the breast during transformation (Bloushtain-Qimron et al., 2008).

Until recently, it was not possible to interrogate the cellular origins of breast cancer because the various epithelial cell types of the mammary gland were difficult to purify and the hierarchical relationship between them was not well understood. However, it is now known that the breast contains functional mammary stem cells, which can be enriched for by flow cytometry and are capable of regenerating the complete gland from a single cell (Shackleton et al., 2006; Stingl et al., 2006a). The gland is also likely to contain bipotent, as well as luminal and myoepithelial restricted progenitor cells, in addition to the mature, fully differentiated cells mentioned previously (Stingl et al., 2006b). Many of these distinct cell types were first identified and characterized in the mouse mammary gland but, other than a difference in the cell surface proteins used to isolate each population, there seems to be a strong functional overlap between the various stem and progenitor cells of the mouse and human mammary gland (Visvader, 2009).

With this initial framework in place, a landmark study, published in 2009, began to explore connections between distinct epithelial populations in the breast and different subtypes of breast cancer (Lim et al., 2009). First, using the cell surface markers CD49f (α6 integrin) and EpCAM (epithelial cell adhesion molecule) to fractionate epithelial subsets from normal human mammary tissue, and then by immunohistochemical examination, *in vitro* colony formation and mammary repopulating activity, this group was able to isolate three different cellular populations: a mammary stem cell/basal-enriched, a luminal progenitor, and a mature luminal. These populations were subjected to genome-wide expression profiling and the expression signatures were correlated to previously generated tumor profiles, classified according to the intrinsic molecular subtypes. Interestingly, the luminal progenitor signature was highest in basal-like tumors while the mammary stem cell/basal signature scored highly in claudin-low and normal-like tumors; as expected the mature luminal profile signature was highest in luminal tumors. This finding provided a strong link between the transcriptional programs operating in normal subsets of mammary epithelial cells and biologically distinct breast tumors, and immediately suggested the potential cellular origins of these cancer subtypes (Prat and Perou, 2009). Further, women carrying a *BRCA1* germline mutation had an expanded pool of luminal progenitors and the expression profile of the corresponding pre-neoplastic tissue was highly correlated to the luminal progenitor signature (Lim et al., 2009), which is consistent with the fact that most *BRCA1* mutant cancers are of the basal-like subtype (Sorlie et al., 2003) and that BRCA1 has been linked to the regulation of mammary stem and progenitor cells (Liu et al., 2008).

These correlative findings have been substantiated in genetically engineered mouse models. In one example, *Brca1* was conditionally deleted (on a $p53^{+/-}$ background) in distinct cellular populations in the mammary gland (Molyneux et al., 2010). The beta-lactoglobulin (*Blg*) promoter, driving Cre recombinase, was used to target this disruption to progenitors of the luminal lineage while a keratin 14 (*K14*) promoter targeted this inactivation to basal cells, where *K14* is expressed. When *Brca1* was deleted in the luminal lineage these mice developed tumors

that closely resembled – both histologically and at the gene expression level – human *BRCA1* mutant cancers as well as sporadic basal-like cancers, providing additional evidence that these cancers arise from transformation of luminal cells. Interestingly, *K14* targeted deletion produced metaplastic cancers and malignant adenomyoepitheliomas, relatively rare tumor types that may arise from stem cells within the mammary gland (Molyneux et al., 2010). Similar results have been obtained using a humanized mouse model where human breast tissue was transplanted into a cleared mammary fat pat that had been supplemented with human stromal cells (Proia and Kuperwasser, 2006). Normal breast epithelial cells transformed with a cocktail of oncogenes (mutant p53, cyclin D1, constitutively active PI3K, and oncogenic K-Ras) developed tumors with a mixed histological phenotype (Proia et al., 2011). However, when epithelial cells derived from heterozygous *BRCA1* mutation carriers were implanted following transformation with the identical set of oncogenes, tumors with basal-like features predominated, likely as a result of impaired cellular differentiation caused by deregulated expression of the transcription factor Slug. While all of these studies provide strong evidence that basal-like tumors originate from luminal progenitor cells, it is worth noting that some plasticity appears to be present within the mammary epithelial cell hierarchy (Chaffer et al., 2011). Since BRCA1 is thought to regulate this differentiation at some level (Liu et al., 2008) it seems likely then, at least in this case, that this underlying genetic mutation may actually skew the cell of origin (Tlsty, 2011). Whether sporadic basal-like breast cancer (in the absence of a *BRCA1* mutation) also originates from luminal progenitors remains to be seen.

An increased understanding of the different mammary epithelial cell populations present in the normal mammary gland is also helping to reveal the possible molecular and cellular basis for breast cancers linked to reproductive history, such as the increased risk that is associated with a greater number of menstrual cycles or the transient risk that occurs immediately after a pregnancy (Lydon, 2010). Two studies, published from different groups (Asselin-Labat et al., 2010; Joshi et al., 2010), revealed that the steroid hormones progesterone and estrogen indirectly increase the pool of mammary stem cells via RANKL secretion. This expanded population of stem cells could potentially serve as targets for oncogenic mutations, which would have the potential to accumulate over time in these self-renewing cells. Although such a phenomenon remains to be proven, these studies, as well as those discussed above, clearly demonstrate how insights into normal mammary gland biology can be quickly applied to breast cancer and how they may help to pinpoint the cellular origins of this diverse malignancy.

Breast cancer stem cells

It is abundantly clear, as discussed above, that there are considerable differences among individual breast tumors – in their cellular origins, pathological and molecular classifications, genomic mutations, and, importantly, in their propensity to metastasize and respond to chemotherapy (Figure 22.1a). However,

cancer cells within the same tumor are also heterogeneous and exhibit a range of phenotypes, a phenomenon referred to as intratumoral heterogeneity (Figure 22.1b) (Marusyk et al., 2012). This added level of complexity presents daunting challenges to both cancer research and clinical oncology, so recent efforts have been directed at understanding the basis for this heterogeneity (Marusyk and Polyak, 2010). Broadly, two ideas have been proposed to account for this: clonal evolution and the cancer stem cell (also referred to as tumor-initiating cell) hypothesis (Campbell and Polyak, 2007; Shackleton et al., 2009). The clonal evolution model is essentially Darwinian natural selection and posits that the cancer cells that acquire the most advantageous set of mutations will be selected for over time, and thus contribute to disease progression. On the other hand, the cancer

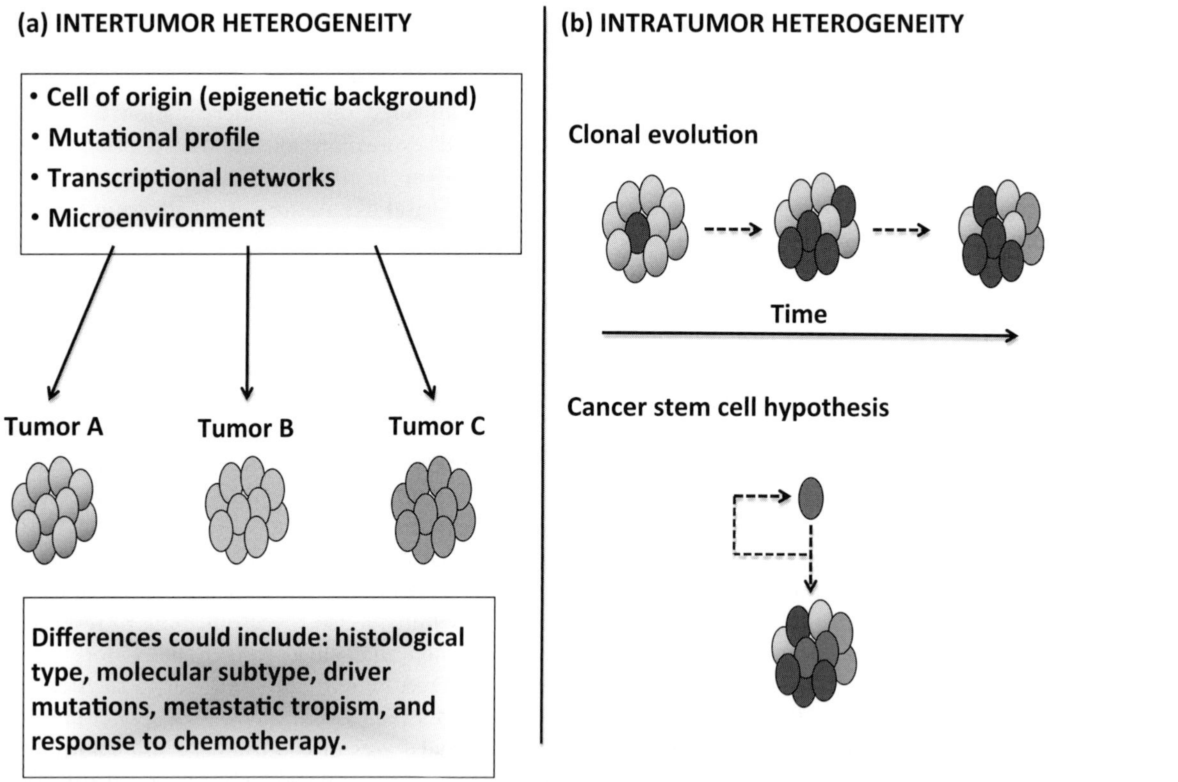

Figure 22.1 Heterogeneity in breast cancer. (a) Intertumor heterogeneity is used to describe the varying characteristics, such as those listed in the box below, that exist between tumors arising in different individuals. Potential sources of this heterogeneity are listed above. (b) Intratumor heterogeneity refers to differences between cancer cells within the same monoclonal tumor. In the clonal evolution model, cancer cells with a growth or survival advantage are selected for over time and, due to the accumulation of new genetic mutations this results in heterogeneous cellular populations. The cancer stem cell model predicts that tumors, similar to tissues, are organized as a hierarchy in which only the cancer stem cells are tumorigenic. Cancer stem cells can self-renew and generate the various other cancer cell types. Note, that although these two models are depicted separately for conceptual purposes, they are not mutually exclusive. (A black and white version of this figure will appear in some formats. For the color version, please refer to the plate section.)

stem cell hypothesis predicts that tumors are organized as a hierarchy in which the cancer stem cells lie at the apex, where they fuel tumor growth, give rise to other types of tumor cells, and serve as the seeds of metastasis and recurrence. Importantly, these two concepts are not mutually exclusive, as even a population of cancer stem cells could be subject to the pressures of clonal evolution, but it does seem logical that genetic differences between tumor cells (see the section on mutational landscape of breast cancer genomes, above) would play a prominent role in the clonal evolution model while the differences between cancer stem cells and non-cancer stem cells are more likely mediated by epigenetic changes (Shackleton et al., 2009). Determining the nature of intratumor heterogeneity has profound implications for cancer therapy and has been the subject of intense study, and controversy, for the past decade.

Breast cancer stem cells were first identified in 2003 using xenotransplantation of primary human tumor cells (Al-Hajj et al., 2003). Here, tumor cells were first separated on the basis of the cell surface expression of CD44, CD24, and ESA (epithelial specific antigen) and subsequently injected subcutaneously into immunocompromised (NOD/SCID) mice at limiting dilution. Only a subset of cells, those that were ESA$^+$/CD44$^+$/CD24$^{-/low}$ (hereafter referred to as CD44$^+$/CD24$^-$), were capable of initiating tumor growth when injected with as few as a couple of hundred cells. Furthermore, these tumors recapitulated the original tumors in terms of the surface expression of CD44 and CD24, implying that the tumorigenic CD44$^+$/CD24$^-$ population was enriched with cancer stem cells that were able to regenerate at least some of the original tumor heterogeneity. Importantly, the proportion of CD44$^+$/CD24$^-$ cells was relatively low, roughly 15% of all tumor cells, indicating that the majority of tumor cells were not in fact tumorigenic. Since then, other markers have also been used to select tumor cell populations enriched with cancer stem cells. One good example is the enzyme aldehyde dehydrogenase (ALDH); normal mammary stem and progenitor cells exhibit ALDH activity and ALDH activity marks a population of highly tumorigenic breast cancer stem cells (Ginestier et al., 2007). In line with this, high ALDH expression is associated with a poor prognosis in breast cancer (Ginestier et al., 2007). Interestingly, there is minimal overlap between the cells marked as CD44$^+$/CD24$^-$ and the cells with high ALDH activity, although the few cells that co-expressed both markers (estimated at fewer than 1% of cells) were capable of generating tumors with as few as 20 cells (Ginestier et al., 2007).

The ability to separate cancer stem cells from the bulk non-cancer stem cells has allowed for detailed comparisons of these different tumor cell populations, with the hope of understanding the signaling networks active specifically in the tumorigenic cancer stem cells. Collectively, this work has implicated numerous developmental regulators such as Wnt, Notch, Hedgehog, and the polycomb transcription factor Bmi1 – pathways activated in normal stem cells – as also being operative in cancer stem cell populations (Liu et al., 2005; Lobo et al., 2007). TGF-β, another important developmental regulator and well-established culprit in breast cancer metastasis, is also highly active in the CD44$^+$ tumor cells (Shipitsin et al., 2007). It is not entirely clear whether this reactivation of dormant stem cell programs reflects the underlying genetic mutations, the cell of origin, or an increased level of cellular plasticity, but nevertheless, identifying such regulators of the cancer stem cell state is clearly of interest, both biologically and clinically.

Consistent with the notion of cancer stem cells, there is evidence to support their role in metastasis and resistance to chemotherapy. In one example, a gene expression signature for breast cancer stem cells was derived by sorting CD44$^+$/CD24$^-$ cells and comparing their expression profile to that of normal mammary epithelial cells (Liu et al., 2007). This analysis led to the identification of a 186-gene set that was termed the invasiveness gene signature (IGS). As expression of this signature in human tumors was associated with reduced overall and metastasis-free survival, it suggests that breast cancer stem cells might play a functional role in disease progression (Liu et al., 2007), not a trivial point considering that the majority of cancer stem cell studies are carried out using xenograft mouse models. A separate study looked at the percentage of CD44$^+$/CD24$^-$ cells and mammosphere formation another indication of cancer stem cell content (Dontu et al., 2003) in patients treated with neoadjuvant chemotherapy (Li et al., 2008). Following chemotherapy, both the percentage of CD44$^+$/CD24$^-$ cells and mammosphere forming efficiency increased, implying that in human tumors, cancer stem cells are more resistant to chemotherapy (Li et al., 2008). There is experimental evidence in support of the idea that cancer stem cells are also resistant to radiotherapy (Diehn et al., 2009).

While the cancer stem cell field has been rife with controversy, ambiguity, and doubt, the data increasingly support the presence of subpopulations of breast cancer cells, with distinct phenotypic properties that closely parallel those predicted for cancer stem cells, which make substantial contributions to disease progression and recurrence.

Metastasis

Metastasis is the systemic form of human cancer, where cancer cells leave the primary tumor and migrate to distant tissues throughout the body. In breast cancer, the most common sites of dissemination are bone, lungs, liver, and brain, and metastasis to any of these organs almost always results in a fatal outcome (Chiang and Massague, 2008). The successful outgrowth of a macroscopic metastasis, referred to as colonization, is an incredibly inefficient process and represents the culmination of numerous steps including local invasion; intravasation into, and survival in, the circulatory system; extravasation at the distant site; and, finally, metastatic colonization (Valastyan and Weinberg, 2011). Incredible progress has been made in making sense of this complex series of events but, relative to other areas of cancer research much remains to be learned and it has proved difficult to translate these findings into meaningful therapeutic options for patients with metastatic disease.

Epithelial–mesenchymal transition (EMT) is, physiologically, a developmental process but tumor cells are capable of reactivating it, thereby gaining the ability to move and invade the surrounding tissue, atypical properties for epithelial cells (Thiery et al., 2009). Numerous signals can induce EMT including TGF-β, Wnt signaling, and hypoxia, but the program is executed by various transcription factors, sometimes referred to as "master regulators," such as Slug, Snail, and Twist (Polyak and Weinberg, 2009). These proteins disrupt cell polarity by coordinately suppressing a subset of epithelial cadherins, most notably E-cadherin (*CDH1*), while at the same time promoting the expression of mesenchymal genes and matrix metalloproteinases capable of degrading the surrounding tissue stroma (Thiery et al., 2009). In recent years an interesting link has formed between EMT, cancer stem cells, and metastasis. In 2008 it was reported that induction of EMT, by either overexpression of EMT-inducing transcription factors or through TGF-β

treatment, generated cells with a cancer stem cell-like phenotype (Mani et al., 2008). This connection is particularly important as it may help to explain how metastases are able to regenerate cellular phenotypes present in the primary tumor (Brabletz et al., 2005) and it underscores the potential clinical relevance of the cancer stem cell hypothesis. New technologies are being developed to capture and characterize circulating tumor cells and here it is important to note that many of these rely on antibodies specific to epithelial proteins (Nagrath et al., 2007). This approach is clearly capable of isolating tumor cells from the circulation but there remains some uncertainty as to whether these are the same cells contributing to disease progression, since the studies above imply that metastatic cancer stem cells might not express certain proteins associated with the epithelial state.

The study of organ-specific metastases is progressing at a rapid rate (Nguyen et al., 2009). Pioneering work from the Massague laboratory has used extensive gene expression profiling to characterize derivatives of the MDA-MB-231 cell line, as well as cell lines established from clinical samples, that exhibit distinct proclivities to metastasize to certain organs when injected into immunocompromised mice. This approach has led to the identification of numerous genes that can assist tumor cells in overcoming tissue-specific obstacles to metastasis. For instance, *IL-11* and *CTGF* – both induced by TGF-β – aid in the activation of osteoclasts and angiogenesis, respectively, to promote the growth of osteolytic bone metastases (Kang et al., 2003). TGF-β, produced in the primary tumor, can also activate *ANGPTL4* (angiopoietin-like 4), which enhances the ability of tumor cells to extravasate through lung capillaries and thereby specifically promotes lung metastasis (Padua et al., 2008). On the other hand, *ST6GALNAC5*, a sialytransferase, is able to help tumor cells pass through the blood–brain barrier in order to colonize the brain (Bos et al., 2009). These studies help to shed light on the molecular and cellular programs activated during metastasis, and provide potentially useful biomarkers and therapeutic targets. But more broadly, they demonstrate that the ability to metastasize can be, as mentioned before, pre-determined in the primary tumor. If this is indeed the case for organ-specific metastasis, then it may be possible to tailor specific therapies and screening based on the gene expression signature in the primary tumor. But it is not yet clear how these expression changes are selected for and maintained in the primary tumor, where these signatures are extracted from, if they confer advantages at distant anatomical sites (Valastyan and Weinberg, 2011).

Treatment

A major focus of molecular oncology has been to identify, characterize, and define the mechanisms by which underlying genetic aberrations are able to drive tumorigenesis, and over the past 30 years researchers have made significant progress in understanding the basis for cancer development and progression. For breast cancer, the textbook example is the *HER2* oncogene discussed above. Gene amplification, and less often truncating mutations (Arribas et al., 2011), leads to excessive activity of this membrane-bound tyrosine kinase receptor, producing an especially aggressive cancer but one that is now susceptible to targeted therapy with the humanized monoclonal antibody trastuzumab (Herceptin®) (Slamon et al., 2001). This paradigm, over 15 years in the making, effectively validated the clinical utility of understanding a breast tumor at the molecular level. The efficacy of tamoxifen, an estrogen receptor antagonist, for ER-positive cancers provides another good example of therapy directed at a specific molecular target (Osborne, 1998b). While the maintenance of some tumors, such as HER2-positive cancer, may be largely dependent on activation or mutation of a single gene – what has been termed oncogene addiction (Weinstein, 2002) – this does not seem to be the norm in the majority of breast cancer cases. Thus a successful attack on many cancers is likely to rely upon a more complete view of the operative signaling networks, and here, a systems biology approach appears especially useful.

PARP inhibitors (olaparib, veliparib, rucaparib) that target PARP1 and PARP2, two enzymes involved in the response to DNA damage, appear to be promising new therapeutic agents and are currently in clinical trials (Lord and Ashworth, 2012). These drugs selectively kill tumor cells with mutant *BRCA1* or *BRCA2* by taking advantage of their inability to make use of homologous DNA repair (Farmer et al., 2005). This type of sensitivity is referred to as synthetic lethality, where neither *BRCA* mutation nor PARP inhibition alone results in cell death, but when the functionality of both is impaired it becomes lethal (Ashworth et al., 2011). A phase 1 clinical trial of olaparib in patients with a wide variety of *BRCA* mutant tumors (including nine breast cancers) was published in 2009 and reported few side effects along with durable anti-tumor activity (Fong et al., 2009). This raises the possibility of whether these agents might be effective against basal-like tumors that, at least by expression profiling, closely resemble *BRCA*-mutant cancers (Turner et al., 2004). The concept of synthetic lethality may become more widely applicable as whole-genome sequencing reveals the extensive genetic defects present in tumor cells, and it offers the prospect of targeting tumor suppressors, albeit indirectly (Ashworth et al., 2011).

There is also substantial interest in developing therapies directed at cancer stem cells. This approach has been demonstrated in principle using high-throughput screening of cells induced by EMT into a cancer stem cell-like state, and salinomycin was identified as a selective inhibitor of breast cancer stem cells (Gupta et al., 2009). Others have found that cancer stem cells rely heavily on cytokines such as IL-8, and that inhibiting its receptor, CXCR1, with the drug repertaxin is a potential means to eliminate cancer stem cells (Ginestier et al., 2010). Two companies, OncoMed Pharmaceuticals and Verastem, are developing agents targeting cancer stem cells and initiating clinical trials.

The outlook for tailored therapy is quite exciting but it is mitigated slightly by two persistent therapeutic challenges. For one, acquired resistance to treatment seems unavoidable so far, and already mutations have been described in cells resistant to PARP inhibitors, which revert *BRCA2* back to a functional state (Edwards et al., 2008). Second, in spite of the progress made in understanding metastasis, there is a dearth of efficacious anti-metastatic agents. The emerging picture of modern oncology is one of personal tumor genotyping – based on standard pathological assessments, expression profiling, and, potentially, whole-genome sequencing – combinatorial therapy targeting the bulk population of tumor cells along with cancer stem cells, and continued monitoring for new oncogenic mutations (Figure 22.2).

Summary and outlook

Through an immense amount of work we are beginning to see breast cancer in all of its complexity – the once occult molecular subtypes, the genomic

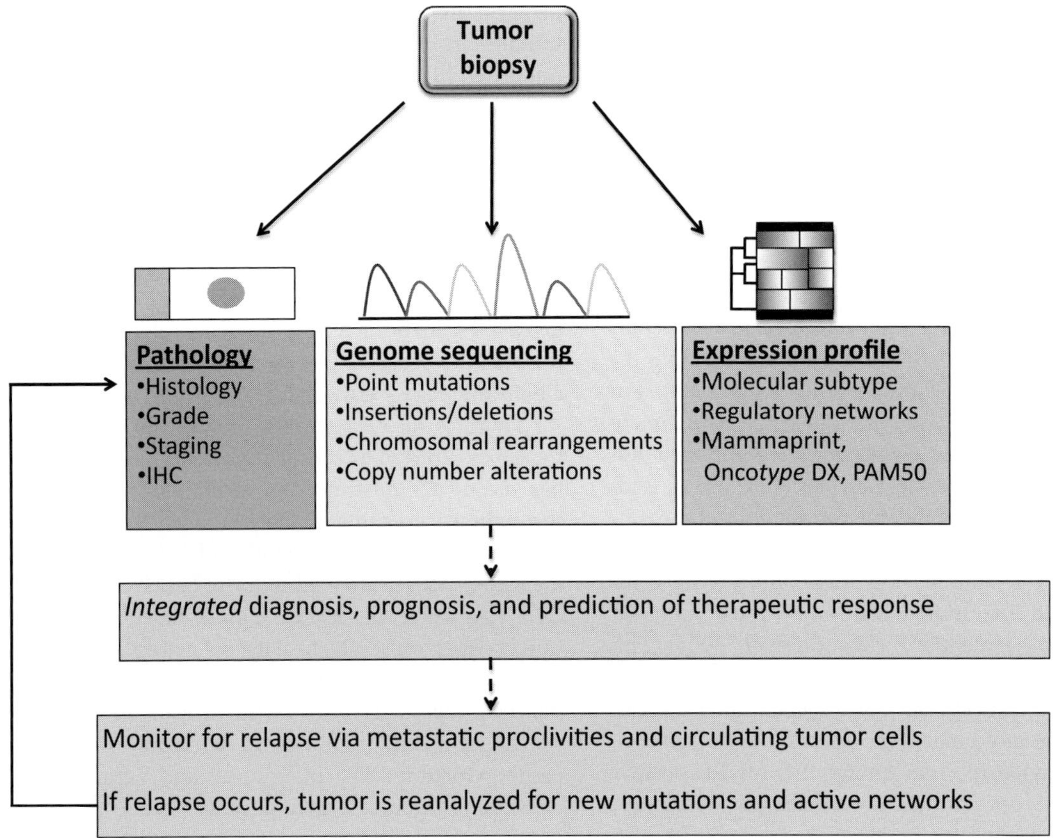

Figure 22.2 Integrated approaches to diagnose and treat breast cancer. Following surgery, the breast tumor is subjected to standard pathological assessment as well as extensive genetic analysis (potentially whole-genome sequencing) and gene expression profiling. This information is integrated in order to determine the molecular subtype, driving oncogenic mutations and prognosis. This can be used to identify those patients most likely to benefit from certain treatments and may help to inform future screening, especially in regard to metastatic tropism. Circulating tumor cells or relapsed tumors can be subjected to similar analyses to determine newly activated mutations or regulatory networks driving relapse. (A black and white version of this figure will appear in some formats. For the color version of this figure, please refer to the plate section.)

landscapes at single base pair resolution, the cellular origins, and the stunning heterogeneity between and within tumors. And yet this represents only the beginning. An explosion of recent research has begun to characterize non-genetic aberrations in cancer such as microRNA dysregulation and epigenetic changes (Croce, 2009; Esteller, 2008) and it is possible that these changes will prove to be just as numerous and consequential as the genetic drivers of tumorigenesis. Moreover, the cancer cell does not exist in isolation and the surrounding microenvironment is known to exert a potent effect on the developing tumor or metastasis (Bissell and Radisky, 2001). Defining a comprehensive picture of tumor-stromal signaling networks is ongoing (Hanahan and Weinberg, 2011)

and will help to complement the analysis of whole cancer genomes. We still know too little about how tumor genomes and their cellular populations evolve during the course of cancer progression but recent deep sequencing efforts and computational advances are beginning to reveal the clonal dynamics and mutational processes that underlie tumor evolution (Nik-Zainal, 2012a, b). Perhaps most important here are the mutations that result from standard cytotoxic chemotherapy (Azim et al., 2011). Collectively, the desired goal of this research should be nothing short of personalized cancer diagnosis and therapy that is safe for patients and without the possibility for additional genetic mutations and the development of secondary malignancies.

References

ACS (2011). *Breast Cancer Facts & Figures 2011–2012* (Atlanta: American Cancer Society, Inc.).

ACS (2012). *Cancer Facts & Figures* (Atlanta: American Cancer Society, Inc.).

ACS (2014). *Cancer Facts & Figures* (Atlanta: American Cancer Society, Inc.).

Akavia, U.D., Litvin, O., Kim, J., et al. (2010). An integrated approach to uncover drivers of cancer. *Cell 143*, 1005–1017.

Al-Hajj, M., Wicha, M.S., Benito-Hernandez, A., Morrison, S.J., and Clarke, M.F. (2003). Prospective identification of tumorigenic breast cancer cells. *Proceedings of the National Academy of Sciences of the United States of America 100*, 3983–3988.

Ali, S., and Coombes, R.C. (2002). Endocrine-responsive breast cancer and strategies for combating resistance. *Nature Reviews Cancer 2*, 101–112.

Amend, K., Hicks, D., and Ambrosone, C.B. (2006). Breast cancer in African-American women: differences in tumor biology from European-American women. *Cancer Research 66*, 8327–8330.

Arribas, J., Baselga, J., Pedersen, K., and Parra-Palau, J.L. (2011). p95HER2 and breast cancer. *Cancer Research 71*, 1515–1519.

Ashworth, A., Lord, C.J., and Reis-Filho, J.S. (2011). Genetic interactions in cancer progression and treatment. *Cell 145*, 30–38.

Asselin-Labat, M.L., Vaillant, F., Sheridan, J.M., et al. (2010). Control of mammary stem cell function by steroid hormone signalling. *Nature 465*, 798–802.

Azim, H.A., Jr., de Azambuja, E., Colozza, M., Bines, J., and Piccart, M.J. (2011). Long-term toxic effects of adjuvant chemotherapy in breast cancer. *Annals of Oncology 22*, 1939–1947.

Bartlett, J., Mallon, E., and Cooke, T. (2003). The clinical evaluation of HER-2 status: which test to use? *Journal of Pathology 199*, 411–417.

Baum, M., Budzar, A.U., Cuzick, J., et al. (2002). Anastrozole alone or in combination with tamoxifen versus tamoxifen alone for adjuvant treatment of postmenopausal women with early breast cancer: first results of the ATAC randomised trial. *Lancet 359*, 2131–2139.

Ben-Porath, I., Thomson, M.W., Carey, V.J., et al. (2008). An embryonic stem cell-like gene expression signature in poorly differentiated aggressive human tumors. *Nature Genetics 40*, 499–507.

Bergamaschi, A., Kim, Y.H., Wang, P., et al. (2006). Distinct patterns of DNA copy number alteration are associated with different clinicopathological features and gene-expression subtypes of breast cancer. *Genes, Chromosomes and Cancer 45*, 1033–1040.

Bertos, N.R., and Park, M. (2011). Breast cancer – one term, many entities? *Journal of Clinical Investigation 121*, 3789–3796.

Bissell, M.J., and Radisky, D. (2001). Putting tumours in context. *Nature Reviews Cancer 1*, 46–54.

Bloushtain-Qimron, N., Yao, J., Snyder, E.L., et al. (2008). Cell type-specific DNA methylation patterns in the human breast. *Proceedings of the National Academy of Sciences of the United States of America 105*, 14076–14081.

Bos, P.D., Zhang, X.H., Nadal, C., et al. (2009). Genes that mediate breast cancer metastasis to the brain. *Nature 459*, 1005–1009.

Brabletz, T., Jung, A., Spaderna, S., Hlubek, F., and Kirchner, T. (2005). Opinion: migrating cancer stem cells – an integrated concept of malignant tumour progression. *Nature Reviews Cancer 5*, 744–749.

Burstein, H.J., Polyak, K., Wong, J.S., Lester, S.C., and Kaelin, C.M. (2004). Ductal carcinoma in situ of the breast. *New England Journal of Medicine 350*, 1430–1441.

Buyse, M., Loi, S., van 't Veer, L., et al. (2006). Validation and clinical utility of a 70-gene prognostic signature for women with node-negative breast cancer. *Journal of the National Cancer Institute 98*, 1183–1192.

Campbell, L.L., and Polyak, K. (2007). Breast tumor heterogeneity: cancer stem cells or clonal evolution? *Cell Cycle 6*, 2332–2338.

Chaffer, C.L., Brueckmann, I., Scheel, C., et al. (2011). Normal and neoplastic nonstem cells can spontaneously convert to a stem-like state. *Proceedings of the National Academy of Sciences of the United States of America 108*, 7950–7955.

Chiang, A.C., and Massague, J. (2008). Molecular basis of metastasis. *New England Journal of Medicine 359*, 2814–2823.

Chin, K., DeVries, S., Fridlyand, J., et al. (2006). Genomic and transcriptional aberrations linked to breast cancer pathophysiologies. *Cancer Cell 10*, 529–541.

Chin, S.F., Teschendorff, A.E., Marioni, J.C., et al. (2007). High-resolution aCGH and expression profiling identifies a novel genomic subtype of ER negative breast cancer. *Genome Biology 8*, R215.

Croce, C.M. (2009). Causes and consequences of microRNA dysregulation in cancer. *Nature Reviews Genetics 10*, 704–714.

Curtis, C., Shah, S.P., Chin, S.F., et al. (2012). The genomic and transcriptomic architecture of 2,000 breast tumours reveals novel subgroups. *Nature 486*, 346–352.

Diehn, M., Cho, R.W., Lobo, N.A., et al. (2009). Association of reactive oxygen species and radioresistance in canceer stem cells. *Nature 458*, 780–783.

Ding, L., Ellis, M.J., Li, S., et al. (2010). Genome remodelling in a basal-like breast cancer metastasis and xenograft. *Nature 464*, 999–1005.

Dontu, G., Abdallah, W.M., Foley, J.M., et al. (2003). In vitro propagation and transcriptional profiling of human mammary stem/progenitor cells. *Genes and Development 17*, 1253–1270.

Edwards, S.L., Brough, R., Lord, C.J., et al. (2008). Resistance to therapy caused by intragenic deletion in BRCA2. *Nature 451*, 1111–1115.

Elston, C.W., and Ellis, I.O. (1991). Pathological prognostic factors in breast cancer. I. The value of histological grade in breast cancer: experience from a large study with long-term follow-up. *Histopathology 19*, 403–410.

Elston, C.W., Ellis, I.O., and Pinder, S.E. (1999). Pathological prognostic factors in breast cancer. *Critical Reviews in Oncology/Hematology 31*, 209–223.

Esteller, M. (2008). Epigenetics in cancer. *New England Journal of Medicine 358*, 1148–1159.

Fackenthal, J.D., and Olopade, O.I. (2007). Breast cancer risk associated with BRCA1 and BRCA2 in diverse populations. *Nature Reviews Cancer 7*, 937–948.

Fan, C., Oh, D.S., Wessels, L., et al. (2006). Concordance among gene-expression-based predictors for breast cancer. *New England Journal of Medicine 355*, 560–569.

Farmer, H., McCabe, N., Lord, C.J., et al. (2005). Targeting the DNA repair defect in BRCA mutant cells as a therapeutic strategy. *Nature 434*, 917–921.

Fearon, E.R., and Vogelstein, B. (1990). A genetic model for colorectal tumorigenesis. *Cell 61*, 759–767.

Fong, P.C., Boss, D.S., Yap, T.A., et al. (2009). Inhibition of poly(ADP-ribose) polymerase in tumors from BRCA mutation carriers. *New England Journal of Medicine 361*, 123–134.

Foulkes, W.D., Smith, I.E., and Reis-Filho, J.S. (2010). Triple-negative breast cancer. *New England Journal of Medicine 363*, 1938–1948.

Ginestier, C., Hur, M.H., Charafe-Jauffret, E., et al. (2007). ALDH1 is a marker of normal and malignant human mammary stem cells and a predictor of poor clinical outcome. *Cell Stem Cell 1*, 555–567.

Ginestier, C., Liu, S., Diebel, M.E., et al. (2010). CXCR1 blockade selectively targets human breast cancer stem cells in vitro and in xenografts. *Journal of Clinical Investigation 120*, 485–497.

Greenman, C., Stephens, P., Smith, R., et al. (2007). Patterns of somatic mutation in human cancer genomes. *Nature 446*, 153–158.

Gupta, P.B., Onder, T.T., Jiang, G., et al. (2009). Identification of selective inhibitors of cancer stem cells by high-throughput screening. *Cell 138*, 645–659.

Hanahan, D., and Weinberg, R.A. (2011). Hallmarks of cancer: the next generation. *Cell 144*, 646–674.

Hennessy, B.T., Gonzalez-Angulo, A.M., Stemke-Hale, K., et al. (2009). Characterization of a naturally occurring breast cancer subset enriched in epithelial-to-mesenchymal transition and stem cell characteristics. *Cancer Research 69*, 4116–4124.

Herschkowitz, J.I., Simin, K., Weigman, V.J., et al. (2007). Identification of conserved gene expression features between murine mammary carcinoma models and human breast tumors. *Genome Biology 8*, R76.

Hudson, T.J., Anderson, W., Artez, A., et al. (2010). International network of cancer genome projects. *Nature 464*, 993–998.

Joshi, P.A., Jackson, H.W., Beristain, A.G., et al. (2010). Progesterone induces adult mammary stem cell expansion. *Nature 465*, 803–807.

Kang, Y., Siegel, P.M., Shu, W., et al. (2003). A multigenic program mediating breast cancer metastasis to bone. *Cancer Cell 3*, 537–549.

Key, T., Appleby, P., Barnes, I., and Reeves, G. (2002). Endogenous sex hormones and breast cancer in postmenopausal women: reanalysis of nine prospective studies. *Journal of the National Cancer Institute 94*, 606–616.

Kim, J., Woo, A.J., Chu, J., et al. (2010). A Myc network accounts for similarities between embryonic stem and cancer cell transcription programs. *Cell 143*, 313–324.

Leary, R.J., Lin, J.C., Cummins, J., et al. (2008). Integrated analysis of homozygous deletions, focal amplifications, and sequence alterations in breast and colorectal cancers. *Proceedings of the National Academy of Sciences of the United States of America 105*, 16224–16229.

Lehmann, B.D., Bauer, J.A., Chen, X., et al. (2011). Identification of human triple-negative breast cancer subtypes and preclinical models for selection of targeted therapies. *Journal of Clinical Investigation 121*, 2750–2767.

Li, C.I., Uribe, D.J., and Daling, J.R. (2005). Clinical characteristics of different histologic types of breast cancer. *British Journal of Cancer 93*, 1046–1052.

Li, X., Lewis, M.T., Huang, J., et al. (2008). Intrinsic resistance of tumorigenic breast cancer cells to chemotherapy. *Journal of the National Cancer Institute 100*, 672–679.

Lim, E., Vaillant, F., Wu, D., et al. (2009). Aberrant luminal progenitors as the candidate target population for basal tumor development in BRCA1 mutation carriers. *Nature Medicine 15*, 907–913.

Liu, R., Wang, X., Chen, G.Y., et al. (2007). The prognostic role of a gene signature from tumorigenic breast-cancer cells. *New England Journal of Medicine 356*, 217–226.

Liu, S., Dontu, G., and Wicha, M.S. (2005). Mammary stem cells, self-renewal pathways, and carcinogenesis. *Breast Cancer Research 7*, 86–95.

Liu, S., Ginestier, C., Charafe-Jauffret, E., et al. (2008). BRCA1 regulates human mammary stem/progenitor cell fate. *Proceedings of the National Academy of Sciences of the United States of America 105*, 1680–1685.

Lobo, N.A., Shimono, Y., Qian, D., and Clarke, M.F. (2007). The biology of cancer stem cells. *Annual Review of Cell and Developmental Biology 23*, 675–699.

Lord, C.J., and Ashworth, A. (2012). The DNA damage response and cancer therapy. *Nature 481*, 287–294.

Lydon, J.P. (2010). Stem cells: cues from steroid hormones. *Nature 465*, 695–696.

Mani, S.A., Guo, W., Liao, M.J., et al. (2008). The epithelial-mesenchymal transition generates cells with properties of stem cells. *Cell 133*, 704–715.

Mardis, E.R. (2011). A decade's perspective on DNA sequencing technology. *Nature 470*, 198–203.

Marusyk, A., Almendro, V., and Polyak, K. (2012). Intra-tumour heterogeneity: a looking glass for cancer? *Nature Reviews Cancer 12*, 323–334.

Marusyk, A., and Polyak, K. (2010). Tumor heterogeneity: causes and consequences. *Biochimica et Biophysica Acta 1805*, 105–117.

Massarweh, S., and Schiff, R. (2006). Resistance to endocrine therapy in breast cancer: exploiting estrogen receptor/growth factor signaling crosstalk. *Endocrine-Related Cancer 13 Suppl 1*, S15–24.

Molyneux, G., Geyer, F.C., Magnay, F.A., et al. (2010). BRCA1 basal-like breast cancers originate from luminal epithelial progenitors and not from basal stem cells. *Cell Stem Cell 7*, 403–417.

Nagrath, S., Sequist, L.V., Maheswaran, S., et al. (2007). Isolation of rare circulating tumour cells in cancer patients by microchip technology. *Nature 450*, 1235–1239.

Navin, N., Kendall, J., Troge, J., et al. (2011). Tumour evolution inferred by single-cell sequencing. *Nature 472*, 90–94.

Network, C.G.A.R. (2008). Comprehensive genomic characterization defines human glioblastoma genes and core pathways. *Nature 455*, 1061–1068.

Nguyen, D.X., Bos, P.D., and Massague, J. (2009). Metastasis: from dissemination to organ-specific colonization. *Nature Reviews Cancer 9*, 274–284.

Nik-Zainal, S., Loo, P.V., Wedge, D.C., et al. (2012a). The life history of 21 breast cancers. *Cell 149*, 994–1007.

Nik-Zainal, S., Alexandrov, L.B., Wedge, D.C., et al. (2012b). Mutational processes molding the genomes of 21 breast cancers. *Cell 149*, 979–993.

Osborne, C.K. (1998a). Steroid hormone receptors in breast cancer management. *Breast Cancer Research and Treatment 51*, 227–238.

Osborne, C.K. (1998b). Tamoxifen in the treatment of breast cancer. *New England Journal of Medicine 339*, 1609–1618.

Padua, D., Zhang, X.H., Wang, Q., et al. (2008). TGFbeta primes breast tumors for lung metastasis seeding through angiopoietin-like 4. *Cell 133*, 66–77.

Paik, C.H., Nagasaka, S., and Hirashita, A. (2006). Class III nonextraction treatment with miniscrew anchorage. *Journal of Clinical Orthodontics : JCO 40*, 480–484.

Paik, S., Shak, S., Tang, G., et al. (2004). A multigene assay to predict recurrence of tamoxifen-treated, node-negative breast cancer. *New England Journal of Medicine 351*, 2817–2826.

Parker, J.S., Mullins, M., Cheang, M.C., et al. (2009). Supervised risk predictor of breast cancer based on intrinsic subtypes. *Journal of Clinical Oncology 27*, 1160–1167.

Pe'er, D., and Hacohen, N. (2011). Principles and strategies for developing network models in cancer. *Cell 144*, 864–873.

Pearce, S.T., and Jordan, V.C. (2004). The biological role of estrogen receptors alpha and beta in cancer. *Critical Reviews in Oncology/Hematology 50*, 3–22.

Perou, C.M., and Borresen-Dale, A.L. (2011). Systems biology and genomics of breast cancer. *Cold Spring Harbor Perspectives in Biology 3*.

Perou, C.M., Sorlie, T., Eisen, M.B., et al. (2000). Molecular portraits of human breast tumours. *Nature 406*, 747–752.

Polyak, K. (2007). Breast cancer: origins and evolution. *Journal of Clinical Investigation 117*, 3155–3163.

Polyak, K., and Weinberg, R.A. (2009). Transitions between epithelial and mesenchymal states: acquisition of malignant and stem cell traits. *Nature Reviews Cancer 9*, 265–273.

Prat, A., and Perou, C.M. (2009). Mammary development meets cancer genomics. *Nature Medicine 15*, 842–844.

Proia, D.A., and Kuperwasser, C. (2006). Reconstruction of human mammary tissues in a mouse model. *Nature Protocols 1*, 206–214.

Proia, T.A., Keller, P.J., Gupta, P.B., et al. (2011). Genetic predisposition directs breast cancer phenotype by dictating progenitor cell fate. *Cell Stem Cell 8*, 149–163.

Rossouw, J.E., Anderson, G.L., Prentice, R.L., et al. (2002). Risks and benefits of estrogen plus progestin in healthy postmenopausal women: principal results from the Women's Health Initiative randomized controlled trial. *Journal of the American Medical Association 288*, 321–333.

Rouzier, R., Perou, C.M., Symmans, W.F., et al. (2005). Breast cancer molecular subtypes respond differently to preoperative chemotherapy. *Clinical Cancer Research 11*, 5678–5685.

Russnes, H.G., Navin, N., Hicks, J., and Borresen-Dale, A.L. (2011). Insight into the heterogeneity of breast cancer through next-generation sequencing. *Journal of Clinical Investigation 121*, 3810–3818.

Russnes, H.G., Vollan, H.K., Lingjaerde, O.C., et al. (2010). Genomic architecture characterizes tumor progression paths and fate in breast cancer patients. *Science Translational Medicine 2*, 38ra47.

Seshadri, R., Firgaira, F.A., Horsfall, D.J., McCaul, K., Setlur, V., and Kitchen, P. (1993). Clinical significance of HER-2/neu oncogene amplification in primary breast cancer. The South Australian Breast Cancer Study Group. *Journal of Clinical Oncology: Official Journal of the American Society of Clinical Oncology 11*, 1936–1942.

Shackleton, M., Quintana, E., Fearon, E.R., and Morrison, S.J. (2009). Heterogeneity in cancer: cancer stem cells versus clonal evolution. *Cell 138*, 822–829.

Shackleton, M., Vaillant, F., Simpson, K.J., et al. (2006). Generation of a functional mammary gland from a single stem cell. *Nature 439*, 84–88.

Shah, S.P., Morin, R.D., Khattra, J., et al. (2009). Mutational evolution in a lobular breast tumour profiled at single nucleotide resolution. *Nature 461*, 809–813.

Shipitsin, M., Campbell, L.L., Argani, P., et al. (2007). Molecular definition of breast tumor heterogeneity. *Cancer Cell 11*, 259–273.

Singletary, S.E., Allred, C., Ashley, P., et al. (2003). Staging system for breast cancer: revisions for the 6th edition of the AJCC Cancer Staging Manual. *Surgical Clinics of North America 83*, 803–819.

Sjoblom, T., Jones, S., Wood, L.D., et al. (2006). The consensus coding sequences of human breast and colorectal cancers. *Science 314*, 268–274.

Slamon, D.J., Clark, G.M., Wong, S.G., et al. (1987). Human breast cancer: correlation of relapse and survival with amplification of the HER-2/neu oncogene. *Science 235*, 177–182.

Slamon, D.J., Leyland-Jones, B., Shak, S., et al. (2001). Use of chemotherapy plus a monoclonal antibody against HER2 for metastatic breast cancer that overexpresses HER2. *New England Journal of Medicine 344*, 783–792.

Sorlie, T., Perou, C.M., Tibshirani, R., et al. (2001). Gene expression patterns of breast carcinomas distinguish tumor subclasses with clinical implications. *Proceedings of the National Academy of Sciences of the United States of America 98*, 10869–10874.

Sorlie, T., Tibshirani, R., Parker, J., et al. (2003). Repeated observation of breast tumor subtypes in independent gene expression data sets. *Proceedings of the National Academy of Sciences of the United States of America 100*, 8418–8423.

Sotiriou, C., and Pusztai, L. (2009). Gene-expression signatures in breast cancer. *New England Journal of Medicine 360*, 790–800.

Stead, L.A., Lash, T.L., Sobieraj, J.E., et al. (2009). Triple-negative breast cancers are increased in black women regardless of age or body mass index. *Breast Cancer Research 11*, R18.

Stephens, P.J., McBride, D.J., Lin, M.L., et al. (2009). Complex landscapes of somatic rearrangement in human breast cancer genomes. *Nature 462*, 1005–1010.

Stingl, J., Eirew, P., Ricketson, I., et al. (2006a). Purification and unique properties of mammary epithelial stem cells. *Nature 439*, 993–997.

Stingl, J., Raouf, A., Eirew, P., and Eaves, C.J. (2006b). Deciphering the mammary epithelial cell hierarchy. *Cell Cycle 5*, 1519–1522.

Stratton, M.R. (2011). Exploring the genomes of cancer cells: progress and promise. *Science 331*, 1553–1558.

Thiagalingam, S. (2006). A cascade of modules of a network defines cancer progression. *Cancer Research 66*, 7379–7385.

Thiery, J.P., Acloque, H., Huang, R.Y., and Nieto, M.A. (2009). Epithelial-mesenchymal transitions in development and disease. *Cell 139*, 871–890.

Tlsty, T.D. (2011). A twist of cell fate. *Cell Stem Cell 8*, 126–127.

Turner, N., Tutt, A., and Ashworth, A. (2004). Hallmarks of "BRCAness" in sporadic cancers. *Nature Reviews Cancer 4*, 814–819.

Valastyan, S., and Weinberg, R.A. (2011). Tumor metastasis: molecular insights and evolving paradigms. *Cell 147*, 275–292.

van 't Veer, L.J., Dai, H., van de Vijver, M.J., et al. (2002). Gene expression profiling predicts clinical outcome of breast cancer. *Nature 415*, 530–536.

van de Vijver, M.J., He, Y.D., van 't Veer, L.J., et al. (2002). A gene-expression signature as a predictor of survival in breast cancer. *New England Journal of Medicine 347*, 1999–2009.

Visvader, J.E. (2009). Keeping abreast of the mammary epithelial hierarchy and breast tumorigenesis. *Genes and Development 23*, 2563–2577.

Visvader, J.E. (2011). Cells of origin in cancer. *Nature 469*, 314–322.

Vogelstein, B., and Kinzler, K.W. (2004). Cancer genes and the pathways they control. *Nature Medicine 10*, 789–799.

Weigelt, B., Geyer, F.C., and Reis-Filho, J.S. (2010). Histological types of breast cancer: how special are they? *Molecular Oncology 4*, 192–208.

Weinstein, I.B. (2002). Cancer. Addiction to oncogenes – the Achilles heel of cancer. *Science 297*, 63–64.

Wood, L.D., Parsons, D.W., Jones, S., et al. (2007). The genomic landscapes of human breast and colorectal cancers. *Science 318*, 1108–1113.

Chapter

23

The role of growth factor-induced changes in cell fate in prostate cancer progression

Min Yu, Gromoslaw A. Smolen, Daniel A. Haber and Shyamala Maheswaran

Introduction

Prostate cancer mortality is primarily due to distal metastasis, preferentially to the bone. A microenvironment enriched with growth factors can lead to inappropriate activation of genetic programs that alter cell fates resulting in the loss of cell adhesion, reorganization of cytoskeleton, and increased cell motility, thus playing a critical role in establishing metastatic lesions at distal sites. In this chapter we review the molecular pathways that induce mesenchymal fates in prostate cancer model systems, and evaluate the evidence of their involvement in clinically annotated prostate cancers with emphasis on their significance in prognosis and therapy.

Prostate gland development and prostate cancer

Prostate development requires a complex interaction between the epithelium and mesenchyme (Cunha, 1972a). The inductive signals emanating from the androgen receptor (AR) expressing urogenital mesenchyme (UGM) are responsible for androgen-induced prostatic epithelial development (Cunha, 1972b) and the prostatic epithelium in turn induces the UGM to differentiate into smooth muscle (Aumuller, 1983, 1989; Cunha, 1994; Cunha et al., 1986; Hayward and Cunha, 2000). The reciprocal interactions between these two tissue compartments are critical in the maintenance of the differentiated state and proliferative control of the adult prostate (Cunha et al., 1985). The UGM, upon androgen activation, releases several paracrine factors including Hedgehogs, Wnts, FGFs, and TGFß (activins and BMPs), which control epithelial cell proliferation and differentiation. Consistent with the hypothesis that cancer recapitulates normal development,

prostate oncogenesis results from disruption of these reciprocal interactions, which in turn lead to dedifferentiation of the prostatic epithelium and smooth muscle cells, culminating in epithelial cell proliferation and tumor progression.

Prostate cancer is the most common male cancer worldwide. Benign prostate hyperplasia (BPH) and prostate carcinoma represent the two most common proliferative lesions in the prostate. BPH generally occurs within the transitional and peripheral zones due to "reawakening of embryonic inductive interactions," whereas carcinomas preferentially originate in the peripheral zone (Kumar and Majumder, 1995). Widespread screening with measurement of serum PSA (prostate-specific antigen) and digital rectal examination enables prostate cancer to be diagnosed at an earlier stage and the cancer is generally assessed using pre-treatment PSA, histological grade, and tumor stage. This together with disease states measured as organ-confined disease, extracapsular extension, seminal vesicle invasion, and lymph node involvement are helpful in guiding treatment.

Since the prostate is an androgen responsive tissue, clinical interventions generally involve the reduction of serum testosterone and/or prevention of testosterone action. Most prostate tumors initially respond to androgen withdrawal, but they recur and the recurrent tumors are androgen independent (Klocker et al., 1994; Wilding, 1992). The survival of malignant cells in the absence of androgen is dependent on the presence of high levels of polypeptide growth factors in the prostate, especially members of the fibroblast growth factor (FGF), epidermal growth factor (EGF), insulin-like growth factor (IGF), and the transforming growth factor (TGF)-beta family (Kim et al., 1999; Thompson, 1990; Wong and Wang, 2000). Many of these ligands, as mentioned earlier,

---

Systems Biology of Cancer, ed. S. Thiagalingam. Published by Cambridge University Press. © Cambridge University Press 2015.

are also involved in prostate development (Hogan, 1999; Prins and Putz, 2008).

Prostate cancer mortality, as with most cancers, is due to metastasis, with bone marrow being the preferred site of metastasis. Metastatic tendency arises from loss of cell–cell, and cell–matrix interactions, increased cell motility, intravasation, extravasation, and site-specific establishment of metastatic lesions. Experimental evidence suggests that genetic programs induced by tumorally secreted growth factors lead to the acquisition of mesenchymal cell fates resulting in metastases of many epithelial cancers, including that of the prostate. In this chapter, we will review the molecular mechanisms involved in mesenchymal transition of prostate epithelial cells, with emphasis on their significance in driving prostate tumor metastasis and their utility as potential diagnostic, prognostic, and therapeutic targets.

Epithelial–mesenchymal transition (EMT)

EMT refers to a process by which stationary epithelial cells acquire a migratory mesenchymal character (Figure 23.1a). The EMT program is used repeatedly during embryogenesis in such diverse settings as gastrulation, neural crest migration, and heart valve and limb muscle formation (Thiery, 2002, 2003). The shift to the mesenchymal state during embryonic development in many cases is reversible, as cells can

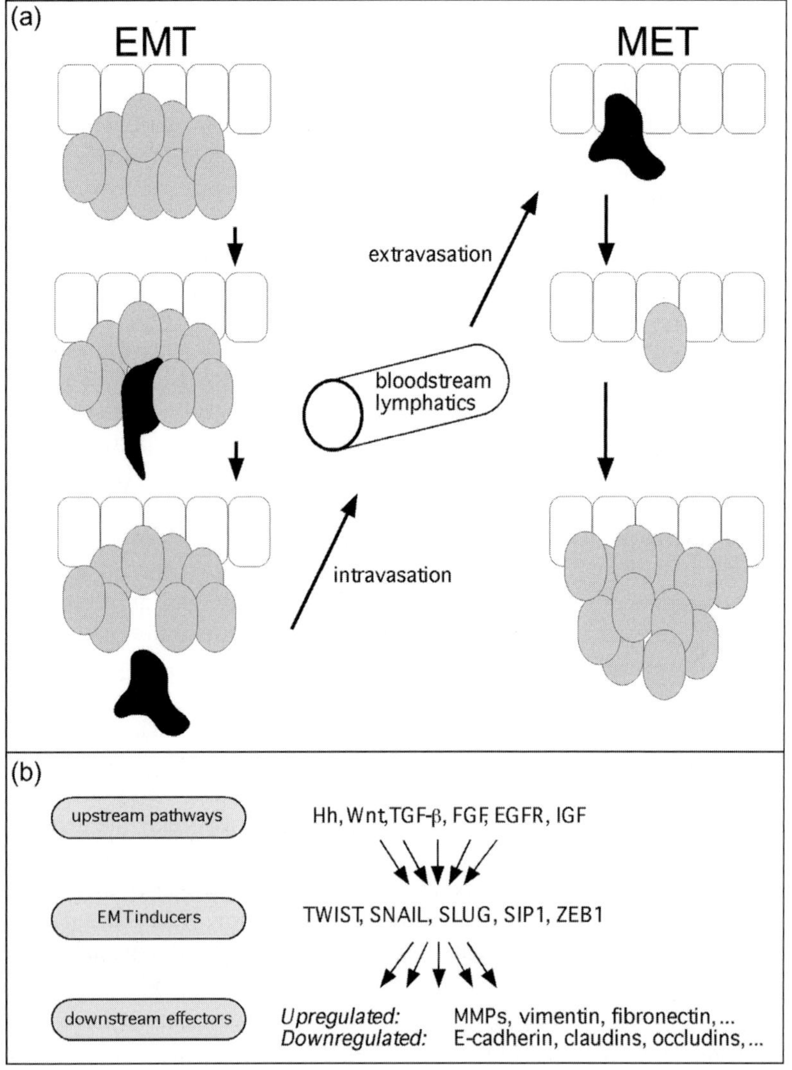

Figure 23.1 An overview of EMT in metastasis.

(a) Model for the phenotypic plasticity of tumor cells. Tumor cells (gray) of epithelial origin (white) display disorganized architecture, but maintain their overall epithelial character. Cells activate the mesenchymal program through increased activity of one or more of the EMT-inducing pathways. These cell(s) progressively lose epithelial polarization and gain invasive, mesenchymal properties. The invasive, mesenchymal cell (black) enters the circulatory system and upon extravasation seeds a suitable distal site, where inductive signals from the underlying normal epithelium convert the mesenchymal tumor cell into a more epithelial cell. Metastatic cells proliferate in the new location and display epithelial features characteristic of the primary tumor.
(b) Signaling pathways involved in the induction of EMT. Paracrine and/or autocrine signaling pathways modulate the expression of EMT-inducing transcription factors, which in turn regulate the expression of effector molecules that ultimately result in the changes in cell fates.

participate in mesenchymal–epithelial transition (MET) and form such epithelial structures as kidneys and somites. Similar cell plasticity occurs during normal urogenital development. Particularly well described has been the regression of Müllerian duct, which relies in part on apoptosis and on EMT (Allard et al., 2000; Zhan et al., 2006).

Morphologically, EMT is associated with the dissolution of the regular "cobblestone" appearance of cells due to loss of cell–cell adhesion and cell polarity, with the resultant gain of a fibroblast-like spindle shape. At the molecular level, these morphological changes are accompanied by changes in expression levels of markers characteristic for epithelial cells (e.g., E-cadherin, claudins, occludins) and mesenchymal cells (e.g., fibronectin, vimentin, FSP1). Functionally, EMT is also associated with increased resistance to apoptosis. This property is quite important during embryonic development because as former epithelial cells migrate to their new location they are deprived of cell–cell contacts usually present in epithelial cells: adherens junctions, desmosomes, tight junctions, and gap junctions. While normal epithelial cells undergo programmed cell death upon detachment, termed anoikis, the EMT process activates anti-apoptotic programs that result in anoikis resistance. Another recently emergent property of cells that have undergone EMT is their acquisition of stem cell-like properties (Mani et al., 2008). While many of the embryonic cell populations that undergo EMT are multipotent, the relationship between "stemness" and EMT is not completely understood. However, these latest results hint that the multipotency of certain early progenitor cells could be in part due to EMT.

Models to study EMT and invasion of prostate cancer

The majority of EMT and invasion studies in prostate cancer have utilized PC-3 and DU145 human prostate cancer cells. These cell lines are androgen independent and exhibit partially altered cytokeratin and cell adhesion profiles. ARCaP, another human prostate cancer-derived cell line, has also been used to characterize EMT. These cells, derived from the ascites of a patient with prostate cancer bony metastasis (Zhau et al., 1996), give rise to epithelial-like ARCaP$_E$ and mesenchymal-like ARCaP$_M$ cells; ARCaP$_M$ cells have increased growth rate, and migratory and invasive

properties relative to ARCaP$_E$ cells. Consistent with this observation, ARCaP$_M$ cells when injected into the left ventricle of immunocompromised SCID mice form shorter latency, more penetrant bone metastases relative to the ARCaP$_E$ cells. Interestingly, tumor cells induced mixed osteoblastic and osteolytic responses in mice (Xu et al., 2006). However, generalization of the findings based on this cell line model is limited by the clonal derivation of the cells from bony metastasis. A non-tumorigenic human prostate cell line BPH1 and its derived tumorigenic sublines have also been used in understanding EMT and invasion (Chu et al., 2009).

The AR negative BPH1 cells form acinus-like spheroids in matrigel with a central lumen surrounded by polarized, growth arrested cells. Ectopic expression of SNAIL in BPH1 cells not only suppressed E-cadherin and induced vimentin expression, but also abolished polarization of the acini and induced invasion into matrigel. This model system provides a 3D-culture system to characterize the mechanism of EMT and invasion (Chu et al., 2009).

Different *in vivo* studies in mice have also been conducted to understand the relationship between EMT and metastasis in prostate cancer. Prostate epithelial cells, with or without stromal cells, have been injected into the kidney capsule of the SCID mice to monitor the regeneration of the prostate gland, tumorigenesis, and invasion into the kidney and distant organs (Karhadkar et al., 2004; Memarzadeh et al., 2007). Direct injection of prostate epithelial cells with luciferase reporter into iliac bone marrow or cardiac ventricle was also used to monitor metastasis by bioluminometer (Buijs et al., 2007a). The genetically engineered mouse prostate model, based on chemical inducers of dimerization for iFGFR1 (inducible FGFR1) in prostate epithelium, demonstrates a progression of pre-invasive hyperplasia to invasive adenocarcinoma and metastasis (Acevedo et al., 2007). This mouse model provides an excellent tool to investigate the iFGFR1-mediated EMT specifically for prostate cancer. A similar concept can be adapted to investigate the role of other receptor kinase pathways.

Another tool to investigate epithelial plasticity *in vivo* was developed in R-3327 Dunning rat prostate adenocarcinoma, utilizing alternative splicing reporters (Oltean et al., 2008). Based on the finding that most epithelial cells express the FGFR2(IIIb) isoform exclusively, and most mesenchymal cells express solely FGFR2(IIIc), a bichromatic fluorescence reporter was developed to evaluate epithelial transitions in culture and *in vivo* and a luciferase reporter to visualize the distribution of mesenchymal epithelial transitions *in vivo*. Some R-3327-derived prostate tumors (DT3) express FGFR2(IIIb), whereas the AT3 tumors, which have lost epithelial markers and display many mesenchymal indicators, express FGFR2(IIIc). Using the reporters in AT3 tumors, MET was detected in the lung micrometastases for the first time in a living animal (Oltean et al., 2006, 2008). Given the transient nature of EMT or MET, the best advantage of this model is the ability to directly visualize the changes of individual cell types both *in vitro* and *in vivo* (Bonano et al., 2007).

In the ensuing sections, we will review the currently available data on EMT in the prostate, generated using the cell culture and animal models described above, and evaluate their validation in annotated primary prostate tumors with the goal of a better understanding of the consequence of EMT in prostate cancer metastasis, which can provide insight into the molecular mechanisms involved in this process.

Aberrant expression of epithelial and mesenchymal markers in prostate cancer

As many different EMT-causing pathways converge on the regulation of E-cadherin, this molecule has been the most widely used proxy of EMT induction in the analysis of clinical specimens. In human prostate adenocarcinoma, Umbas et al. demonstrated that nearly 50% of tumors examined had reduced or absent levels of E-cadherin (Umbas et al., 1992), which in a subsequent study was shown to correlate with tumor grade, tumor stage, Gleason score, and overall survival suggesting that E-cadherin expression can serve as a prognostic indicator for the biological potential of prostate cancer (Richmond et al., 1997; Umbas et al., 1994). miRNA 205, a micro RNA that upregulates E-cadherin expression (Gandellini et al., 2009), is downregulated in prostate tumors (Schaefer et al., 2010); moreover tumors from prostate cancer patients with lymph node dissemination exhibited lower miRNA 205 expression than node-negative patients (Gandellini et al., 2009). Whether suppression of miRNA 205 in prostate cancer correlates with

decreased E-cadherin expression and contributes to tumor cell dissemination remains to be determined.

Conversely, the mesenchymal marker N-cadherin is not expressed in normal prostate tissue but is expressed in poorly differentiated areas of prostate cancer, which also showed aberrant or negative E-cadherin staining. In metastatic lesions, N-cadherin was expressed homogeneously suggesting that cadherin switching during EMT plays an important role in prostate cancer metastasis (Tomita et al., 2000). Consistent with these observations, immmunohisto-chemical staining of 104 radical prostatectomy samples with antibodies against E-cadherin, N-cadherin, and ß-catenin demonstrated that low E-cadherin expression was significantly associated with adverse clinico-pathological features, and switching cadherins (high N-cadherin and low E-cadherin) showed strong and significant associations with multiple end points of progression and cancer-specific death (Gravdal et al., 2007). These studies evaluating the expression of EMT markers in prostate cancer suggest that EMT is likely to be a prognostic factor and may play a functionally significant role in prostate cancer progression.

Deregulation of EMT-inducing transcription factors in prostate cancer
SNAIL genes

Expression analyses of primary prostate adenocarcinoma suggest that deregulated expression of the epithelial and mesenchymal markers, E-cadherin and N-cadherin, respectively, is likely to be functionally involved in prostate tumor progression. Loss or reduction of E-cadherin in human tumors is caused by somatic mutations, chromosomal deletions, and proteolytic cleavage, promoter silencing by DNA hypermethylation, or through the transcriptional repression by EMT inducers, such as TWIST and ZEB1 (Berx et al., 1998; Hirohashi, 1998; Mani et al., 2007; Maretzky et al., 2005; Olmeda et al., 2007; Spaderna et al., 2008; Strathdee, 2002). TWIST, ZEB1, and SNAIL have been shown to directly repress E-cadherin transcription (Batlle et al., 2000; Cano et al., 2000; Eger et al., 2005; Yang et al., 2004). SNAIL has also been shown to activate several mesenchymal genes including ZEB1 (Guaita et al., 2002).

Chronic exposure of prostate cancer cells to EGF induces EMT correlating with the induction of SNAIL (Lu et al., 2003). The expression of SNAIL mRNA in metastatic prostate cancer is elevated compared to normal and benign prostate, and primary prostate adenocarcinoma (Dhanasekaran et al., 2001; Tomlins et al., 2007). SLUG, another member of the SNAIL family of transcription factors, is induced in PC-3 cells by Frzb/secreted Frizzled-related protein 3, a secreted Wnt antagonist, which suppresses PC-3 tumor growth and cellular invasiveness in mice. Thus the anti-tumor activity of Frzb/sFRP3 in a subset of prostate cancers may be caused through the reversal of epithelial–mesenchymal transition (Zi et al., 2005). Dihydro-testosterone and EGF induce SLUG providing a molecular mechanism by which epithelial cell-specific genes are silenced during prostate cancer development and progression (Chen et al., 2006).

Contradictory to these *in vitro* studies, which suggest an inhibitory role for SLUG in prostate cancer metastasis, gene expression profiling studies show that SLUG mRNA expression is decreased in prostatic intraepithelial neoplasia, prostate adenocarcinoma, and in metastatic prostate cancer compared to normal, benign, or non-neoplastic prostate tissue (Dhanase-karan et al., 2001, 2005; Lapointe et al., 2004; LaTulippe et al., 2002; Tomlins et al., 2007). The discrepancy between the *in vitro* experimental data, demonstrating a role for SLUG in EMT and invasion, and the decreased expression of SLUG mRNA observed in prostate tumors could be attributed to the analysis of bulk tumor tissue rather than microdissected individual cell types, and warrants further characterization that includes protein expression and localization analysis of prostate tumors.

TWIST

TWIST, another transcription factor associated with EMT, is highly expressed in about 90% of prostate cancer tissues and its expression levels were positively correlated with Gleason grading and metastasis, suggesting a role for TWIST in the development and progression of prostate cancer. This is consistent with gene expression profiling studies of normal prostate and prostate adenocarcinoma (Lapointe et al., 2004). Knockdown of TWIST in DU145 and PC-3 resulted in increased sensitivity to the anti-cancer drug taxol, leading to activation of the apoptosis pathway. More importantly, inactivation of TWIST suppressed the migration and invasion of androgen-independent prostate cancer cells, and correlated with molecular changes associated with mesenchymal–epithelial transition (Kwok et al., 2005) suggesting that TWIST is a critical regulator of prostate cancer cell growth and that its inactivation could provide a potential therapeutic approach to inhibit the growth and metastasis of androgen-independent prostate cancer.

Analysis of 115 prostate cancers for TWIST and E-cadherin expression demonstrated that high levels of TWIST were significantly associated with aberrant E-cadherin expression, and that the presence of nuclear TWIST was significant in predicting the metastatic potential of the primary prostate cancer (Yuen et al., 2007). Therefore TWIST may be a prognostic marker for high-grade prostatic cancer. Analysis of several EMT inducers suggests that conversion to the mesenchymal state might lead to broader resistance to cell death induced in experimental settings by serum withdrawal or treatment with such genotoxic agents as doxorubicin or pacli-taxel (Catalano et al., 2004; Kajita et al., 2004; Maestro et al., 1999; Vega et al., 2004). Consistent with this observation, TWIST has been implicated in the development of acquired taxol resistance in human cancer cells including LNCaP cells (Kwok et al., 2005; Wang et al., 2004). These fundamental differences in drug responses between epithelial and mesenchymal cells have been particularly well demonstrated in non-small cell lung cancer (NSCLC). In the context of wild type epidermal growth factor receptor (EGFR), epithelial cell lines appear to be sensitive to EGFR inhibition, while mesenchymal cells display significant resistance to EGFR inhibitors (Thomson et al., 2005). Whether molecular markers of EMT can serve as predictors of drug sensitivity remains to be determined.

ZEB genes

The zinc finger E-box binding transcription factor, ZEB1, is emerging as an important regulator of EMT required for development and cancer metastasis. IGF-I induces ZEB1 in prostate cancer cells exhibiting an epithelial phenotype (Graham et al., 2008), and TGFß, a well-documented inducer of EMT, stimulates ZEB1 and ZEB2 through downregulation of the miR-NAs, miRNA 141, miRNA 200a, miRNA 200b, miRNA 200c, miRNA 205, and miRNA 429 (Katoh and Katoh, 2008). ZEB1 is expressed in highly aggressive prostate cancer cells and its expression correlates directly with Gleason grade in human prostate tumors; knockdown of ZEB1 decreases the migratory and invasive potential of prostate cancer cells (Graham et al., 2008). Expression profiling studies show that ZEB1 expression increases in prostate adenocarcinoma but declines again in metastatic lesions (Dhanasekaran et al., 2001; Lapointe et al., 2004; Tomlins et al., 2007). Whether this is reflective of metastatic prostatic lesions undergoing MET remains to be determined.

A subpopulation of PC-3 cells that crossed an endothelial barrier more efficiently exhibited characteristic molecular markers in EMT, including loss of E-cadherin expression and upregulation of ZEB1. Silencing of ZEB1 reduced transendothelial migration. These observations extend the role of EMT in metastasis to transendothelial migration (Drake et al., 2009). In addition to EMT, additional functions have been attributed to ZEB1 expression. In colorectal cancer cells ZEB1 promotes metastasis and loss of cell polarity (Aigner et al., 2007; Spaderna et al., 2008), and mutation of ZEB1 leads to Ink4a-independent premature replicative senescence in mouse embryo fibroblasts (Liu et al., 2008). Thus ZEB1 through linking EMT, loss of polarity, cellular senescence, and metastasis serves as a critical component in malignant tumor progression.

Further characterization of these EMT-inducing transcription factors, elucidation of the signaling pathways that regulate their expression, delineation of the downstream effectors that lead to a metastatic phenotype induced by these transcription factors, and identification of the mechanisms that lead to their aberrant expression in prostate cancers will help validate the usefulness of these biomarkers as prognostic factors that predict recurrence, metastasis, survival, and responsiveness to treatment.

Influence of growth factor-induced signaling on cell fates and prostate cancer progression

In this section, we will focus on the signaling pathways involved in prostate development and prostate cancer progression (Figure 23.1b). We will discuss their role in inducing EMT in the prostate, whether they function by modulating the expression of cadherins and EMT-inducing transcription factors, and evaluate how their ability to stimulate EMT contributes to prostate cancer progression.

Hedgehog (Hh)

The Hh pathway has been implicated in prostate regeneration and prostate cancer metastasis. In mammals, there are three Hh ligands: Sonic Hedgehog (Shh), Indian Hedgehog (Ihh), and Desert Hedgehog (Dhh). Hh binds to the Patched receptor (PTC-PTCH1, PTCH2) to alleviate the suppression of Smoothened (SMO), a putative seven-transmembrane protein that triggers a cascade of intracellular events through GLI-dependent transcription (Jacob and Lum, 2007; Jiang and Hui, 2008; Varjosalo and Taipale, 2008). Activation of Hh signaling is critical for the maintenance of appropriate prostate growth and tissue polarity (Lamm et al., 2002). The Hh pathway is frequently activated in human prostate cancer with Shh-positivity being associated with higher Gleason grade (Sheng et al., 2004). Conversely, Su(Fu), a negative regulator of the Hh pathway, is suppressed in prostate tumors (Fan et al., 2004; Sheng et al., 2004). Genetic analysis of familial prostate cancer suggests that many members of the Hh pathway are localized to chromosomal regions associated with prostate cancer susceptibility (Datta and Datta, 2006).

Hh signaling has been shown to affect EMT (Hay, 1995), and inhibition of this pathway suppresses EMT in pancreatic cancer cell lines (Feldmann et al., 2007). Prostate cell lines with strong metastatic potential have a highly active Hh pathway, and treatment with cyclopamine, an inhibitor for Hh signaling, suppressed the growth and invasiveness of these cells *in vitro* and inhibited their growth and metastasis *in vivo* (Karhadkar et al., 2004). Expression of Gli, the transcriptional target of Hh signaling, in poorly metastatic cells led to enhanced invasion in matrigel *in vitro*, and highly metastatic tumors *in vivo* with an associated increase in SNAIL expression and decreased E-cadherin expression (Karhadkar et al., 2004). Consistent with these observations, Hh signaling and PTCH and GLI1 were preferentially activated in metastatic prostate cancers compared with localized disease (Datta and Datta, 2006; Sheng et al., 2004). Recently, Hh signaling has been implicated in promoting tumor growth and metastasis in patients undergoing androgen deprivation therapy; androgen depletion has been shown to result in a dramatic upregulation of Shh and Ihh in prostate cancer cell lines (Chen et al., 2009).

Members within the miRNA 17~92 cluster, which collaborates with the Shh pathway in medulloblastoma (Uziel et al., 2009), are overexpressed in several solid tumors including the prostate (Mendell, 2008). Interestingly, this miRNA cluster is a direct transcriptional target of c-Myc (Mendell, 2008), which is induced by activation of the Wnt/APC pathway (He et al., 1998) (see below). In a recent study, prevalent upregulation of nuclear c-Myc protein expression was described as an early oncogenic alteration in prostate cancers. In early-stage prostate cancers, a significant correlation was observed between c-Myc overexpression and TMPRSS2–ERG fusion (Hawksworth et al., 2010), which is described to be present in over 50% of prostate tumors (Clark and Cooper, 2009). These results, not so surprisingly, suggest that the complex process of prostate tumor progression is governed by a crosstalk between a myriad of signaling pathways.

Wnt ligands

The Wnt signaling pathway plays important roles in early development and human malignancies, including prostate cancer. Wnt proteins are a large family of cysteine-rich glycoproteins with 19 members identified in humans. Their activity is controlled by a number of soluble extracellular antagonists including secreted frizzled-related proteins (sFRP), Wnt inhibitory factor-1 (WIF1), and Dickkopt (Dkk). Wnts activate intracellular signaling pathways through canonical or non-canonical pathways. In the canonical pathway, Wnts bind to their co-receptors consisting of Frizzled (FZD) and low-density lipoprotein receptor-related protein (LRP), leading to inactivation of glycogen synthase kinase 3β (GSK3β) and reduced targeting of β-catenin degradation by the GSK3β-Axin-APC complex. β-catenin then translocates into the nucleus and acts as a co-activator with the TCF4/LEF1 transcription factor. Non-canonical pathways do not involve β-catenin. While these pathways are still under intense investigation, studies in multiple model organisms suggest that protein kinase C, N-Jun N-terminal kinase (JNK), calcium ions, and Rho family GTPases are involved.

Aberrant expression of different components of the Wnt pathway has been observed in prostate cancers; Wnt-1, Wnt-5A, Wnt-2, and Wnt-11, and β-catenin are elevated in prostate cancers relative to normal prostate tissue (Chen et al., 2004; Verras and Sun, 2006; Zhu et al., 2004). The miRNA cluster, miR-15a and miR-16–1 localized to chromosomal region 13q14 and frequently deleted in cancer (Calin et al., 2002, 2005), targets cyclin D1 and Wnt3A. In advanced prostate tumors, miRNA 15a and miRNA 16 levels were significantly decreased, whereas the expression of cyclin D1 and WNT3A is inversely upregulated (Bonci et al., 2008). Silencing of the ERG transcription factor, deregulated in prostate tumors due to the TMPRSS2–ERG fusion, upregulates Wnt signaling, E-cadherin, and ß1 integrin to modulate EMT and cell adhesion. Frizzled-4 (FZD4) is co-overexpressed with ERG in clinical prostate cancers (Gupta et al., 2010) suggesting a role for Wnt in driving ERG-mediated prostate oncogenesis through modulation of cell fates.

β-catenin has been shown to interact with AR and modulate its transcriptional activity (Truica et al., 2000; Yang et al., 2002). Thus it is speculated that androgen ablation may increase the pool of β-catenin available for activation of the TCF/LEF target genes and contribute to disease relapse (Verras and Sun, 2006). Conditional inactivation of APC in the mouse prostate elevates β-catenin levels and leads to prostate carcinogenesis, and the tumors continue to grow even under conditions of androgen depletion (Bruxvoort et al., 2007).

The Wnt signaling pathway has been implicated in EMT and metastasis of prostate cancers; both canonical and non-canonical Wnt pathways are required for developmental processes involving EMT. Consistent with the role of Wnt signaling in inducing EMT and invasiveness, high levels of Wnt-1 and β-catenin expression were detected in 77% of prostate cancers that metastasized to the lymph node, and in 85% of skeletal metastases with the levels being higher in skeletal metastasis (Chen et al., 2004). Wnt signaling regulates bone mass through multiple mechanisms, including osteoblast differentiation, proliferation, function, and survival coupled to osteoclast formation (Hall et al., 2006). Whether increased Wnt signaling in prostate cancers heralds bone metastasis remains to be determined.

While the expression of positive effectors of the Wnt signaling pathway promote EMT and invasiveness, the expression of components that antagonize Wnt signaling demonstrate an inverse correlation with EMT and tumorigenic characteristics. The expression of the extracellular antagonist WIF1 is reduced in prostate tumors (Wissmann et al., 2003). Expression of Frzb/Secreted Frizzled-Related Protein 3 (sFRP3), a secreted Wnt antagonist, in PC-3 cells reversed EMT and invasion, and dramatically inhibited tumor growth in mice. There was also a decrease in the expression and activities of MMP-2 and MMP-9, and cytosolic β-catenin, SLUG, and TWIST levels (Zi et al., 2005). Similarly, sFRP4 overexpression in LNCaP and PC-3 cell lines resulted in epithelial morphology with increased localization of β-catenin and cadherins to the cell membrane, and decreased invasiveness. In a cohort of human androgen-dependent prostate cancers, increased membranous sFRP4 expression was associated with increased membranous β-catenin expression (Horvath et al., 2007).

Calcitonin (CT), a neuroendocrine peptide, and its receptor (CTR) are highly expressed in advanced prostate cancer and display a positive correlation with the Gleason grade of primary prostate cancers (Sabbisetti et al., 2005; Thomas et al., 2006). Activation of the CT–CTR axis in PC-3 cells led to EMT, loss of cell adhesion, destabilization of tight and adherens junctions, and internalization of ZO-1 and β-catenin. Activation of the CT–CTR axis can increase Wnt1 and Wnt3, which leads to phosphorylation of GSK3β and translocation of β-catenin into the nucleus (Shah et al., 2009). Whether elevated activity of the CT–CTR axis in high Gleason score prostate tumors through activation of the Wnt pathway contributes to tumor progression remains to be determined.

Transforming growth factor-ß family

The TGF-β superfamily, comprising more than 30 ligands, signals through serine/threonine kinase receptors, specifically by binding to a type II receptor, which heterodimerizes with a type I receptor leading to the phosphorylation of Smad proteins. TGF-ß and activin signals phosphorylate Smad2 and Smad3 whereas BMPs induce the phosphorylation of Smad1, Smad5, and Smad8. Once phosphorylated, the Smad proteins form a complex with Smad4, translocate to the nucleus, and regulate gene expression (Massague, 2008). TGF-β signaling plays a crucial role in developmental processes involving EMT, and aberrant TGF-β signaling has been associated with tumor growth and metastasis of several tumors including prostate cancer (Massague, 2008; Yang and Weinberg, 2008).

TGF-β induces EMT in PC-3 and LNCaP cells, and represses the expression of the epithelium-specific Ets transcription factor, PDEF, downregulation of which is required for the induction of vimentin and SNAIL2 (Gu et al., 2007). However, both EGF and TGF-β are required to induce EMT in the epithelial-like clone of ARCap prostate cancer cells.

Interestingly, in this model, Receptor Activator of NF-κB Ligand (RANKL) has been proposed to be a novel EMT marker induced by EGF and TGF-β or by SNAIL. RANKL is normally expressed on osteoblasts and stromal cells and is critically involved in the maturation and activation of osteoclasts (Odero-Marah et al., 2008; Zhau et al., 2008). DU145 and PC-3 cells express the RANK receptor, and RANKL treatment of these cells promotes cell migration (Mori et al., 2007). Although it is proposed that RANKL may function as a link between EMT, bone turnover, and prostate cancer bone metastasis, it is noteworthy that it involves osteolytic lesions, a less common form of prostate tumor bone metastasis.

Consistent with its dual and opposing roles in tumorigenesis, TGF-β inhibits the proliferation of the non-tumorigenic human prostate epithelial cell line, BPH1, while activation of TGF-ß signaling in its derivative tumorigenic subline, BPH1[CAFTD] induces EMT associated with an increase in vimentin and a decrease in cytokeratin and E-cadherin. When xenografted beneath the kidney capsule of male athymic mice, these cells form tumors that invade into the adjacent kidney tissue with vimentin being expressed along the invasive front but not in the body of the tumor (Ao et al., 2006). Consistent with the role of TGF-β in tumor progression, overexpression of Smad3 in human prostate cancer correlates with Gleason score. The functional role for increased expression of Smad3 in prostate cancer was established by overexpressing dominant-negative Smad3 in PC-3MM2 cells, where it reduced tumor growth and metastasis (Lu et al., 2007).

BMP signaling plays a critical role in bone development and thus has been implicated in facilitating bone metastasis of prostate cancer. BMP4 and 6 are expressed in prostatic cancers with known skeletal metastasis (Hamdy et al., 1997); BMP2 and 4 enhance PC-3 cell migration and invasion both *in vitro* and *in vivo* (Feeley et al., 2006).

On the other hand, BMP7 has been previously characterized as a TGF-ß antagonist in kidney epithelial cells where it was able to inhibit TGF-β-induced EMT (Zeisberg et al., 2003). In prostate cancer, BMP7 significantly inhibited bone metastasis of the highly aggressive PC-3M-Pro4 cells injected directly into the tibia or the left cardiac ventricle of immunodeficient mice but had no effect on primary tumor growth. While the exact mechanism of this inhibitory effect by BMP7 is unclear, BMP7 has been shown to suppress TGF-ß-induced vimentin expression in PC-3M-Pro4 cells, and suppress TGF-ß-induced CAGA-luciferase reporters (Buijs et al., 2007b). However, contradictory reports present evidence that BMP7 can itself induce EMT, as treatment of PC-3 with BMP7 leads to a change in morphology, increased motility and invasion, and downregulation of E-cadherin (Yang et al., 2005). It is likely that the functionality of BMP7 is influenced by various signaling modules active within the different cell types.

Fibroblast growth factors

FGFs bind to a family of four distinct tyrosine kinase receptors, FGFR-1 to FGFR-4, which subsequently dimerize and transphosphorylate to activate different downstream effectors that influence diverse cellular functions, including cell proliferation, motility, invasion, and resistance to apoptosis. Extracellular matrix (ECM) and secreted proteins, which mobilize FGFs from the ECM, can also regulate FGF signaling. Stromally produced FGFs through paracrine action regulate normal prostate epithelial growth. In prostate cancer, FGF receptor signaling is enhanced by multiple mechanisms including increased expression of FGF ligands and receptors, increased mobilization of FGFs from the ECM, and loss of negative regulation of FGF signaling (Kwabi-Addo et al., 2004).

Recent studies have indicated that activation of the FGF signaling pathway is involved in EMT and prostate cancer metastasis. FGFRs-1 to -3 undergo an alternative splicing event in which two alternative exons (IIIb and IIIc) can be used to encode the carboxy terminal portion of the third immunoglobulin-like loop. Epithelial cells almost exclusively express the IIIb isoforms of FGFR-2 (Ittman and Mansu-khani, 1997). In the Dunning prostate tumor model, there is an increase in the FGFR-2 IIIc isoform during tumor progression, suggesting a possible role for this protein in EMT (Yan et al., 1993). The differential roles of these splice forms have been demonstrated at the level of ligand binding, as the IIIb form is responsive to FGF7 and the IIIc is not. Conversely, IIIc displays a much higher affinity for FGF2 as compared to the IIIb form (Miki et al., 1992). Interestingly, overexpression of the epithelial IIIb isoform was able to suppress the tumor growth of mesenchymal cells (Matsubara et al., 1998).

FGF8 has been strongly implicated in prostate tumorigenesis and elevated levels are associated with higher Gleason grade. In prostate cell lines, downregulation of the FGF8 inhibitor, hSef, increased FGF8-induced cell migration and invasion, as well as the expression of MMP (Darby et al., 2006). In addition, FGF8 is highly expressed in bone metastasis of prostate cancer and is found to affect both osteoblast and osteoclast differentiation (Valta et al., 2006, 2008).

FGFR1 signaling has been shown to be sufficient to induce EMT and distal metastasis. In some metastatic lesions, FGFR1 signaling upregulates Sox9, an inducer of EMT during neural crest development (Sakai et al., 2006), and a pro-invasive factor overexpressed in recurrent clinical prostate cancers (Wang et al., 2008). Interestingly, this study identified a rare subset of cells, which co-express E-cadherin and vimentin, indicating the transient nature of EMT. In another study, prolonged expression of FGFR1 signaling in the prostate epithelium led to cancer progression with a full spectrum of the disease ranging from prostatic intraepithelial neoplasia, adenocarcinoma, transitional sarcomatoid-carcinoma, to frank sarcoma suggesting an important role for FGF1 in prostate cancer progression (Acevedo et al., 2007). Induction of cancer progression by FGFR1 may result from the induction of Sox9 and subsequent EMT (Abate-Shen and Shen, 2007).

FGF10 plays a critical role in mediating interactions between the epithelium and stroma (Memarzadeh et al., 2007). FGF10 overexpression in the prostate stroma induces hyperproliferation of the epithelium and correlates with upregulation of AR expression. Stromal overexpression of FGF10 can co-operate with epithelial activation of Akt to promote tumorigenesis. The transforming effects of stromal FGF10 can be blocked by epithelial expression of a dominant-negative FGFR1 suggesting that FGF10 exerts its effects through FGFR1 (Memarzadeh et al., 2007). Thus the data suggest a model where the stromal FGF10 functions through epithelial FGFR1 to induce AR expression and activate Akt.

Epidermal growth factor

Epidermal growth factor receptor (EGFR) is an ErbB family receptor tyrosine kinase found on cells of epithelial origin. EGFR can bind to many different ligands including EGF, TGF-α, amphiregulin, epiregulin, and heparin-binding EGF-like factor (HB-EGF). Ligand binding results in homo- or hetero-dimerization of EGFR with other members of ErbB family receptors and activation of downstream signaling leading to a myriad of effects. EGFR activity is frequently dysregulated in many epithelial tumors. In prostate cancer patients, elevated expression of EGFR is associated with increased Gleason score (Di Lorenzo et al., 2002).

Chronic EGF and TGF-α exposure promotes EMT in EGFR overexpressing tumor cells, including DU145 (Lo et al., 2007; Lu et al., 2003). In breast and epidermoid human carcinoma cells, EGFR activation induces TWIST expression and EMT via the janus-activated kinase and activator of transcription 3 (JAK/STAT3) pathway (Lo et al., 2007). Another study demonstrated that downregulation of E-cadherin during EGF-induced EMT is mediated by caveolae-dependent endocytosis. During this process, β-catenin is dissociated from the plasma membrane and translocated to the nucleus resulting in increased transactivation of downstream targets. Prolonged treatment with EGF downregulates caveolin-1 and induces SNAIL leading to further suppression of E-cadherin transcription (Lu et al., 2003).

Conversely, inhibition of EGF activation reverses EMT. Co-culture of DU-145 and PC-3 cells with hepatocytes resulted in decreased EGFR expression with an associated increase in E-cadherin and cytokeratin18 expression and localization of β-catenin to the membrane. Moreover, EGFR kinase inhibitors have been shown to induce E-cadherin expression in DU-145 (Yates et al., 2005) and PC-3 cells (Yates et al., 2007).

Consistent with its ability to influence EMT, EGFR activation increases the motility of many cells, including that of prostate tumor cells (Rajan et al., 1996; Turner et al., 1996; Zolfaghari and Djakiew, 1996). Inhibition of the EGFR pathway with receptor-specific inhibitors, U73122 and Gefitinib, on DU145 and PC-3 cells significantly reduces invasion and metastasis *in vivo* without interfering with cell proliferation (Angelucci et al., 2006; Turner et al., 1996). Consistent with these observations, suppression of protein kinase C delta, a kinase that functions downstream of EGFR, decreased the migration and invasion of DU145 and PC-3 cells *in vitro* (Kharait et al., 2006).

Insulin-like growth factors

The insulin-like growth factor (IGF) axis is activated by two ligands, IGF-I and IGF-II. They bind to two surface tyrosine kinases receptors, IGF-IR and IGF-IIR leading to the activation of several downstream targets including ERK/MAPK, PI3K, and Akt. A group of IGF binding proteins (IGFBP-1 to IGFBP-8) and IGFBP proteinases also participate in the regulation of IGF signaling.

E-cadherin, β-catenin, and IGF-IR form a supramolecular complex in various cell types (Morali et al., 2001), and activation of IGF-IR in mammary epithelial cells (Irie et al., 2005; Morali et al., 2001) and epithelial colorectal cells (Playford et al., 2000) induces EMT. In colorectal cancer cells, IGF-1 induces tyrosine phosphorylation of β-catenin and causes a rapid dissociation of β-catenin from E-cadherin at the cell membrane (Playford et al., 2000). Similarly, IGF-II induces the redistribution of β-catenin from the plasma membrane to the nucleus and an intracellular sequestration and degradation of E-cadherin (Morali et al., 2001). In the human prostate cancer cells, ARCaP$_E$, IGF-I treatment strongly upregulated the ZEB1 mRNA and protein with an associated decrease in E-cadherin, increase in fibronectin and N-cadherin, and enhanced migration and invasion (Graham et al., 2008).

IGF signaling plays an important role in regulating cell growth, survival, and metabolism, and aberrant IGF signaling is detected in several malignancies. A majority of metastatic prostate cancers exhibit increased IGF-IR expression relative to the primary tumor (Krueckl et al., 2004; Reiss et al., 1998). There is also increased expression of IGF-I and IGF-IR in androgen-independent prostate cancers relative to androgen-dependent cells (Krueckl et al., 2004). The 3′ UTR of the insulin receptor substrate-1 (IRS-1) and the IGF-IR (La Rocca et al., 2009) are two of the putative targets of the miRNA 145 (La Rocca et al., 2009). The expression of miRNA 145 is low in all the prostate cell lines due to significant DNA methylation of the promoter region (Zaman et al., 2010). Deep sequencing to define miRNA profiles in prostate tumors demonstrated deregulation of miRNA 145 (Szczyrba et al., 2010), and it is significantly downregulated in aggressive prostate cancer (Wang et al., 2009). A correlation between miRNA 145 and IGF signaling and the functional role for IGF signaling in mediating invasion and metastasis of prostate cancer needs to be established.

Conclusion

Cellular and animal models of prostate cancer progression and expression analysis of normal and tumor tissues strongly suggest that prostate cancer progression is associated with deregulation of a myriad of growth factor-induced pathways, which can alter cell fates and subsequently influence tumor invasion, stem cell characteristics, and drug and radiation resistance. However, the lack of repeated sampling of patients to acquire primary and metastatic tumor tissue has posed significant difficulty in precisely defining the crosstalk between growth factor-induced pathways that promote prostate tumor progression through EMT. This problem is compounded by the fact that mesenchymal cells upon establishing distal metastasis are likely to revert back to the epithelial phenotype via MET (Yates et al., 2007) (Figure 23.1b). Possible characterization of circulating tumor cells as a tool to obtain serial tissue samples from patients with prostate cancer (Nagrath et al., 2007) will be useful in deciphering the role of growth factor-regulated pathways and EMT in prostate metastasis. Elucidation of EMT signaling networks will be critical for the eventual translation of these advances into sensitive molecular diagnostic tools, novel therapeutic interventions, and effective strategies to treat prostate cancer metastasis.

References

Abate-Shen, C., and Shen, M.M. (2007). FGF signaling in prostate tumorigenesis: new insights into epithelial–stromal interactions. *Cancer Cell 12*, 495–497.

Acevedo, V.D., Gangula, R.D., Freeman, K.W., et al. (2007). Inducible FGFR-1 activation leads to irreversible prostate adenocarcinoma and an epithelial-to-mesenchymal transition. *Cancer Cell 12*, 559–571.

Aigner, K., Dampier, B., Descovich, L., et al. (2007). The transcription factor ZEB1 (deltaEF1) promotes tumour cell dedifferentiation by repressing master regulators of epithelial polarity. *Oncogene 26*, 6979–6988.

Allard, S., Adin, P., Gouedard, L., et al. (2000). Molecular mechanisms of hormone-mediated Mullerian duct regression: involvement of beta-catenin. *Development 127*, 3349–3360.

Angelucci, A., Gravina, G.L., Rucci, N., et al. (2006). Suppression of EGF-R signaling reduces the incidence of prostate cancer metastasis in nude mice. *Endocr Relat Cancer 13*, 197–210.

Ao, M., Williams, K., Bhowmick, N.A., and Hayward, S.W. (2006). Transforming growth factor-beta promotes invasion in tumorigenic but not in nontumorigenic human prostatic epithelial cells. *Cancer Res 66*, 8007–8016.

Aumuller, G. (1983). Morphologic and endocrine aspects of prostatic function. *Prostate 4*, 195–214.

Aumuller, G. (1989). Morphologic and regulatory aspects of prostatic function. *Anat Embryol 179*, 519–531.

Batlle, E., Sancho, E., Franci, C., et al. (2000). The transcription factor snail is a repressor of E-cadherin gene expression in epithelial tumour cells. *Nat Cell Biol 2*, 84–89.

Berx, G., Becker, K.F., Hofler, H., and van Roy, F. (1998). Mutations of the human E-cadherin (CDH1) gene. *Hum Mutat 12*, 226–237.

Bonano, V.I., Oltean, S., and Garcia-Blanco, M.A. (2007). A protocol for imaging alternative splicing regulation *in vivo* using fluorescence reporters in transgenic mice. *Nat Protoc 2*, 2166–2181.

Bonci, D., Coppola, V., Musumeci, M., et al. (2008). The miR-15a-miR-16-1 cluster controls prostate cancer by targeting multiple oncogenic activities. *Nat Med 14*, 1271–1277.

Bruxvoort, K.J., Charbonneau, H.M., Giambernardi, T.A., et al. (2007). Inactivation of Apc in the mouse prostate causes prostate carcinoma. *Cancer Res 67*, 2490–2496.

Buijs, J.T., Henriquez, N.V., van Overveld, P.G., et al. (2007a). TGF-beta and BMP7 interactions in tumour progression and bone metastasis. *Clin Exp Metastasis 24*, 609–617.

Buijs, J.T., Rentsch, C.A., van der Horst, G., et al. (2007b). BMP7, a putative regulator of epithelial homeostasis in the human prostate, is a potent inhibitor of prostate cancer bone metastasis *in vivo*. *Am J Pathol 171*, 1047–1057.

Calin, G.A., Dumitru, C.D., Shimizu, M., et al. (2002). Frequent deletions and down-regulation of micro-RNA genes miR15 and miR16 at 13q14 in chronic lymphocytic leukemia. *Proc Natl Acad Sci USA 99*, 15524–15529.

Calin, G.A., Ferracin, M., Cimmino, A., et al. (2005). A microRNA signature associated with prognosis and progression in chronic lymphocytic leukemia. *N Engl J Med 353*, 1793–1801.

Cano, A., Perez-Moreno, M.A., Rodrigo, I., et al. (2000). The transcription factor snail controls epithelial-mesenchymal transitions by repressing E-cadherin expression. *Nat Cell Biol 2*, 76–83.

Catalano, A., Rodilossi, S., Rippo, M.R., Caprari, P., and Procopio, A. (2004). Induction of stem cell factor/c-Kit/slug signal transduction in multidrug-resistant malignant mesothelioma cells. *J Biol Chem 279*, 46706–46714.

Chen, G., Shukeir, N., Potti, A., et al. (2004). Up-regulation of Wnt-1 and beta-catenin production in patients with advanced metastatic prostate carcinoma: potential pathogenetic and prognostic implications. *Cancer 101*, 1345–1356.

Chen, M., Chen, L.M., and Chai, K.X. (2006). Androgen regulation of prostasin gene expression is mediated by sterol-regulatory element-binding proteins and SLUG. *Prostate 66*, 911–920.

Chen, M., Tanner, M., Levine, A.C., et al. (2009). Androgenic regulation of hedgehog signaling pathway components in prostate cancer cells. *Cell Cycle 8*, 149–157.

Chu, J.H., Yu, S., Hayward, S.W., and Chan, F.L. (2009). Development of a three-dimensional culture model of prostatic epithelial cells and its use for the study of epithelial-mesenchymal transition and inhibition of PI3K pathway in prostate cancer. *Prostate 69*, 428–442.

Clark, J.P., and Cooper, C.S. (2009). ETS gene fusions in prostate cancer. *Nat Rev Urol 6*, 429–439.

Cunha, G.R. (1972a). Epithelio-mesenchymal interactions in primordial gland structures which become responsive to androgenic stimulation. *Anat Rec 172*, 179–195.

Cunha, G.R. (1972b). Tissue interactions between epithelium and mesenchyme of urogenital and integumental origin. *Anat Rec 172*, 529–541.

Cunha, G.R. (1994). Role of mesenchymal-epithelial interactions in normal and abnormal development of the mammary gland and prostate. *Cancer 74*, 1030–1044.

Cunha, G.R., Bigsby, R.M., Cooke, P.S., and Sugimura, Y. (1985). Stromal-epithelial interactions in adult organs. *Cell Differ 17*, 137–148.

Cunha, G.R., Donjacour, A.A., and Sugimura, Y. (1986). Stromal-epithelial interactions and heterogeneity of proliferative activity within the prostate. *Biochem Cell Biol 64*, 608–614.

Darby, S., Sahadevan, K., Khan, M.M., et al. (2006). Loss of Sef (similar expression to FGF) expression is associated with high grade and metastatic prostate cancer. *Oncogene 25*, 4122–4127.

Datta, S., and Datta, M.W. (2006). Sonic Hedgehog signaling in advanced prostate cancer. *Cell Mol Life Sci 63*, 435–448.

Dhanasekaran, S.M., Barrette, T.R., Ghosh, D., et al. (2001). Delineation of prognostic biomarkers in prostate cancer. *Nature 412*, 822–826.

Dhanasekaran, S.M., Dash, A., Yu, J., et al. (2005). Molecular profiling of human prostate tissues: insights into gene expression patterns of prostate development during puberty. *FASEB J 19*, 243–245.

Di Lorenzo, G., Tortora, G., D'Armiento, F.P., et al. (2002). Expression of epidermal growth factor receptor correlates with disease relapse and progression to androgen-independence in human prostate cancer. *Clin Cancer Res 8*, 3438–3444.

Drake, J.M., Strohbehn, G., Bair, T.B., Moreland, J.G., and Henry, M.D. (2009). ZEB1 enhances transendothelial migration and represses the epithelial phenotype of prostate cancer cells. *Mol Biol Cell 20*, 2207–2217.

Eger, A., Aigner, K., Sonderegger, S., et al. (2005). DeltaEF1 is a transcriptional repressor of E-cadherin and regulates epithelial plasticity in breast cancer cells. *Oncogene 24*, 2375–2385.

Fan, L., Pepicelli, C.V., Dibble, C.C., et al. (2004). Hedgehog signaling promotes prostate xenograft tumor growth. *Endocrinology 145*, 3961–3970.

Feeley, B.T., Krenek, L., Liu, N., et al. (2006). Overexpression of noggin inhibits BMP-mediated growth of osteolytic prostate cancer lesions. *Bone 38*, 154–166.

Feldmann, G., Dhara, S., Fendrich, V., et al. (2007). Blockade of hedgehog signaling inhibits pancreatic cancer invasion and metastases: a new paradigm for combination therapy in solid cancers. *Cancer Res 67*, 2187–2196.

Gandellini, P., Folini, M., Longoni, N., et al. (2009). miR-205 exerts tumor-suppressive functions in human prostate through down-regulation of protein kinase Cepsilon. *Cancer Res 69*, 2287–2295.

Graham, T.R., Zhau, H.E., Odero-Marah, V.A., et al. (2008). Insulin-like growth factor-I-dependent up-regulation of ZEB1 drives epithelial-to-mesenchymal transition in human prostate cancer cells. *Cancer Res 68*, 2479–2488.

Gravdal, K., Halvorsen, O.J., Haukaas, S.A., and Akslen, L.A. (2007). A switch from E-cadherin to N-cadherin expression indicates epithelial to mesenchymal transition and is of strong and independent importance for the progress of prostate cancer. *Clin Cancer Res 13*, 7003–7011.

Gu, X., Zerbini, L.F., Otu, H.H., et al. (2007). Reduced PDEF expression increases invasion and expression of mesenchymal genes in prostate cancer cells. *Cancer Res 67*, 4219–4226.

Guaita, S., Puig, I., Franci, C., et al. (2002). Snail induction of epithelial to mesenchymal transition in tumor cells is accompanied by MUC1 repression and ZEB1 expression. *J Biol Chem 277*, 39209–39216.

Gupta, S., Iljin, K., Sara, H., et al. (2010). FZD4 as a mediator of ERG oncogene-induced WNT signaling and epithelial-to-mesenchymal transition in human prostate cancer cells. *Cancer Res 70*, 6735–6745.

Hall, C.L., Kang, S., MacDougald, O.A., and Keller, E.T. (2006). Role of Wnts in prostate cancer bone metastases. *J Cell Biochem 97*, 661–672.

Hamdy, F.C., Autzen, P., Robinson, M.C., et al. (1997). Immunolocalization and messenger RNA expression of bone morphogenetic protein-6 in human benign and malignant prostatic tissue. *Cancer Res 57*, 4427–4431.

Hawksworth, D., Ravindranath, L., Chen, Y., et al. (2010). Overexpression of C-MYC oncogene in prostate cancer predicts biochemical recurrence. *Prostate Cancer Prostatic Dis 13*, 311–315.

Hay, E.D. (1995). An overview of epithelio-mesenchymal transformation. *Acta Anat 154*, 8–20.

Hayward, S.W., and Cunha, G.R. (2000). The prostate: development and physiology. *Radiol Clin North Am 38*, 1–14.

He, T.C., Sparks, A.B., Rago, C., et al. (1998). Identification of c-MYC as a target of the APC pathway. *Science 281*, 1509–1512.

Hirohashi, S. (1998). Inactivation of the E-cadherin-mediated cell adhesion system in human cancers. *Am J Pathol 153*, 333–339.

Hogan, B.L. (1999). Morphogenesis. *Cell 96*, 225–233.

Horvath, L.G., Lelliott, J.E., Kench, J.G., et al. (2007). Secreted frizzled-related protein 4 inhibits proliferation and metastatic potential in prostate cancer. *Prostate 67*, 1081–1090.

Irie, H.Y., Pearline, R.V., Grueneberg, D., et al. (2005). Distinct roles of Akt1 and Akt2 in regulating cell migration and epithelial-mesenchymal transition. *J Cell Biol 171*, 1023–1034.

Ittman, M., and Mansukhani, A. (1997). Expression of fibroblast growth factors (FGFs) and FGF receptors in human prostate. *J Urol 157*, 351–356.

Jacob, L., and Lum, L. (2007). Hedgehog signaling pathway. *Sci STKE 2007*, cm6.

Jiang, J., and Hui, C.C. (2008). Hedgehog signaling in development and cancer. *Dev Cell 15*, 801–812.

Kajita, M., McClinic, K.N., and Wade, P.A. (2004). Aberrant expression of the transcription factors snail and slug alters the response to genotoxic stress. *Mol Cell Biol 24*, 7559–7566.

Karhadkar, S.S., Bova, G.S., Abdallah, N., et al. (2004). Hedgehog signalling in prostate regeneration, neoplasia and metastasis. *Nature 431*, 707–712.

Katoh, Y., and Katoh, M. (2008). Hedgehog signaling, epithelial-to-mesenchymal transition and miRNA (review). *Int J Mol Med 22*, 271–275.

Kharait, S., Dhir, R., Lauffenburger, D., and Wells, A. (2006). Protein kinase Cdelta signaling downstream of the EGF receptor mediates migration and invasiveness of prostate cancer cells. *Biochem Biophys Res Commun 343*, 848–856.

Kim, H.G., Kassis, J., Souto, J.C., Turner, T., and Wells, A. (1999). EGF receptor signaling in prostate morphogenesis and tumorigenesis. *Histol Histopathol 14*, 1175–1182.

Klocker, H., Culig, Z., Kaspar, F., et al. (1994). Androgen signal transduction and prostatic carcinoma. *World J Urol 12*, 99–103.

Krueckl, S.L., Sikes, R.A., Edlund, N.M., et al. (2004). Increased insulin-like growth factor I receptor expression and signaling are components of androgen-independent progression in a lineage-derived prostate cancer progression model. *Cancer Res 64*, 8620–8629.

Kumar, V.L., and Majumder, P.K. (1995). Prostate gland: structure, functions and regulation. *Int Urol Nephrol 27*, 231–243.

Kwabi-Addo, B., Ozen, M., and Ittmann, M. (2004). The role of fibroblast growth factors and their receptors in prostate cancer. *Endocr Relat Cancer 11*, 709–724.

Kwok, W.K., Ling, M.T., Lee, T.W., et al. (2005). Up-regulation of TWIST in prostate cancer and its implication as a therapeutic target. *Cancer Res 65*, 5153–5162.

La Rocca, G., Badin, M., Shi, B., et al. (2009). Mechanism of growth inhibition by microRNA 145: the role of the IGF-I receptor signaling pathway. *J Cell Physiol 220*, 485–491.

Lamm, M.L., Catbagan, W.S., Laciak, R.J., et al. (2002). Sonic hedgehog activates mesenchymal Gli1 expression during prostate ductal bud formation. *Dev Biol 249*, 349–366.

Lapointe, J., Li, C., Higgins, J.P., et al. (2004). Gene expression profiling identifies clinically relevant subtypes of prostate cancer. *Proc Natl Acad Sci USA 101*, 811–816.

LaTulippe, E., Satagopan, J., Smith, A., et al. (2002). Comprehensive gene expression analysis of prostate cancer reveals distinct transcriptional programs associated with metastatic disease. *Cancer Res 62*, 4499–4506.

Liu, Y., El-Naggar, S., Darling, D.S., Higashi, Y., and Dean, D.C. (2008). Zeb1 links epithelial-mesenchymal transition and cellular senescence. *Development 135*, 579–588.

Lo, H.W., Hsu, S.C., Xia, W., et al. (2007). Epidermal growth factor receptor cooperates with signal transducer and activator of transcription 3 to induce epithelial-mesenchymal transition in cancer cells via up-regulation of TWIST gene expression. *Cancer Res 67*, 9066–9076.

Lu, S., Lee, J., Revelo, M., Wang, X., and Dong, Z. (2007). Smad3 is overexpressed in advanced human prostate cancer and necessary for progressive growth of prostate cancer cells in nude mice. *Clin Cancer Res 13*, 5692–5702.

Lu, Z., Ghosh, S., Wang, Z., and Hunter, T. (2003). Downregulation of caveolin-1 function by EGF leads to the loss of E-cadherin, increased transcriptional activity of beta-catenin, and enhanced tumor cell invasion. *Cancer Cell 4*, 499–515.

Maestro, R., Dei Tos, A.P., Hamamori, Y., et al. (1999). Twist is a potential oncogene that inhibits apoptosis. *Genes Dev 13*, 2207–2217.

Mani, S.A., Guo, W., Liao, M.J., et al. (2008). The epithelial-mesenchymal transition generates cells with properties of stem cells. *Cell 133*, 704–715.

Mani, S.A., Yang, J., Brooks, M., et al. (2007). Mesenchyme Forkhead 1 (FOXC2) plays a key role in metastasis and is associated with aggressive basal-like breast cancers. *Proc Natl Acad Sci USA 104*, 10069–10074.

Maretzky, T., Reiss, K., Ludwig, A., et al. (2005). ADAM10 mediates E-cadherin shedding and regulates epithelial cell-cell adhesion, migration, and beta-catenin translocation. *Proc Natl Acad Sci USA 102*, 9182–9187.

Massague, J. (2008). TGFbeta in cancer. *Cell 134*, 215–230.

Matsubara, A., Kan, M., Feng, S., and McKeehan, W.L. (1998). Inhibition of growth of malignant rat prostate tumor cells by restoration of fibroblast growth factor receptor 2. *Cancer Res 58*, 1509–1514.

Memarzadeh, S., Xin, L., Mulholland, D.J., et al. (2007). Enhanced paracrine FGF10 expression promotes formation of multifocal prostate adenocarcinoma and an increase in epithelial androgen receptor. *Cancer Cell 12*, 572–585.

Mendell, J.T. (2008). miRiad roles for the miR-17-92 cluster in development and disease. *Cell 133*, 217–222.

Miki, T., Bottaro, D.P., Fleming, T.P., et al. (1992). Determination of ligand-binding specificity by alternative splicing: two distinct growth factor receptors encoded by a single gene. *Proc Natl Acad Sci USA 89*, 246–250.

Morali, O.G., Delmas, V., Moore, R., et al. (2001). IGF-II induces rapid beta-catenin relocation to the nucleus during epithelium to mesenchyme transition. *Oncogene 20*, 4942–4950.

Mori, K., Le Goff, B., Charrier, C., et al. (2007). DU145 human prostate cancer cells express functional receptor activator of NFkappaB: new insights in the prostate cancer bone metastasis process. *Bone 40*, 981–990.

Nagrath, S., Sequist, L.V., Maheswaran, S., et al. (2007). Isolation of rare circulating tumour cells in cancer patients by microchip technology. *Nature 450*, 1235–1239.

Odero-Marah, V.A., Wang, R., Chu, G., et al. (2008). Receptor activator of NF-kappaB Ligand (RANKL) expression is associated with epithelial to mesenchymal transition in human prostate cancer cells. *Cell Res 18*, 858–870.

Olmeda, D., Jorda, M., Peinado, H., Fabra, A., and Cano, A. (2007). Snail silencing effectively suppresses tumour growth and invasiveness. *Oncogene 26*, 1862–1874.

Oltean, S., Febbo, P.G., and Garcia-Blanco, M.A. (2008). Dunning rat prostate adenocarcinomas and alternative splicing reporters: powerful tools to study epithelial plasticity in prostate tumors *in vivo*. *Clin Exp Metastasis 25*, 611–619.

Oltean, S., Sorg, B.S., Albrecht, T., et al. (2006). Alternative inclusion of fibroblast growth factor receptor 2 exon IIIc in Dunning prostate tumors reveals unexpected epithelial mesenchymal plasticity. *Proc Natl Acad Sci USA 103*, 14116–14121.

Playford, M.P., Bicknell, D., Bodmer, W.F., and Macaulay, V.M. (2000). Insulin-like growth factor 1 regulates the location, stability, and transcriptional activity of beta-catenin. *Proc Natl Acad Sci USA 97*, 12103–12108.

Prins, G.S., and Putz, O. (2008). Molecular signaling pathways that regulate prostate gland development. *Differentiation 76*, 641–659.

Rajan, R., Vanderslice, R., Kapur, S., et al. (1996). Epidermal growth factor (EGF) promotes chemomigration of a human prostate tumor cell line, and EGF immunoreactive proteins are present at sites of metastasis in the stroma of lymph nodes and medullary bone. *Prostate 28*, 1–9.

Reiss, K., D'Ambrosio, C., Tu, X., Tu, C., and Baserga, R. (1998). Inhibition of tumor growth by a dominant negative mutant of the insulin-like growth factor I receptor with a bystander effect. *Clin Cancer Res 4*, 2647–2655.

Richmond, P.J., Karayiannakis, A.J., Nagafuchi, A., Kaisary, A.V., and Pignatelli, M. (1997). Aberrant E-cadherin and alpha-catenin expression in prostate cancer: correlation with patient survival. *Cancer Res 57*, 3189–3193.

Sabbisetti, V.S., Chirugupati, S., Thomas, S., et al. (2005). Calcitonin increases invasiveness of prostate cancer cells: role for cyclic AMP-dependent protein kinase A in calcitonin action. *Int J Cancer 117*, 551–560.

Sakai, D., Suzuki, T., Osumi, N., and Wakamatsu, Y. (2006). Cooperative action of Sox9, Snail2 and PKA signaling in early neural crest development. *Development 133*, 1323–1333.

Schaefer, A., Jung, M., Mollenkopf, H.J., et al. (2010). Diagnostic and prognostic implications of microRNA profiling in prostate carcinoma. *Int J Cancer 126*, 1166–1176.

Shah, G.V., Muralidharan, A., Gokulgandhi, M., Soan, K., and Thomas, S. (2009). Cadherin switching and activation of beta-catenin signaling underlie proinvasive actions of calcitonin-calcitonin receptor axis in prostate cancer. *J Biol Chem 284*, 1018–1030.

Sheng, T., Li, C., Zhang, X., et al. (2004). Activation of the hedgehog pathway in advanced prostate cancer. *Mol Cancer 3*, 29.

Spaderna, S., Schmalhofer, O., Wahlbuhl, M., et al. (2008). The transcriptional repressor ZEB1 promotes metastasis and loss of cell polarity in cancer. *Cancer Res 68*, 537–544.

Strathdee, G. (2002). Epigenetic versus genetic alterations in the inactivation of E-cadherin. *Semin Cancer Biol* 12, 373–379.

Szczyrba, J., Loprich, E., Wach, S., et al. (2010). The microRNA profile of prostate carcinoma obtained by deep sequencing. *Mol Cancer Res 8*, 529–538.

Thiery, J.P. (2002). Epithelial-mesenchymal transitions in tumour progression. *Nat Rev Cancer 2*, 442–454.

Thiery, J.P. (2003). Epithelial-mesenchymal transitions in development and pathologies. *Curr Opin Cell Biol 15*, 740–746.

Thomas, S., Chigurupati, S., Anbalagan, M., and Shah, G. (2006). Calcitonin increases tumorigenicity of prostate cancer cells: evidence for the role of protein kinase A and urokinase-type plasminogen receptor. *Mol Endocrinol 20*, 1894–1911.

Thompson, T.C. (1990). Growth factors and oncogenes in prostate cancer. *Cancer Cells 2*, 345–354.

Thomson, S., Buck, E., Petti, F., et al. (2005). Epithelial to mesenchymal transition is a determinant of sensitivity of non-small-cell lung carcinoma cell lines and xenografts to epidermal growth factor receptor inhibition. *Cancer Res 65*, 9455–9462.

Tomita, K., van Bokhoven, A., van Leenders, G.J., et al. (2000). Cadherin switching in human prostate cancer progression. *Cancer Res 60*, 3650–3654.

Tomlins, S.A., Mehra, R., Rhodes, D.R., et al. (2007). Integrative molecular concept modeling of prostate cancer progression. *Nat Genet 39*, 41–51.

Truica, C.I., Byers, S., and Gelmann, E.P. (2000). Beta-catenin affects androgen receptor transcriptional activity and ligand specificity. *Cancer Res 60*, 4709–4713.

Turner, T., Chen, P., Goodly, L.J., and Wells, A. (1996). EGF receptor signaling enhances *in vivo* invasiveness of DU-145 human prostate carcinoma cells. *Clin Exp Metastasis 14*, 409–418.

Umbas, R., Isaacs, W.B., Bringuier, P.P., et al. (1994). Decreased E-cadherin expression is associated with poor prognosis in patients with prostate cancer. *Cancer Res 54*, 3929–3933.

Umbas, R., Schalken, J.A., Aalders, T.W., et al. (1992). Expression of the cellular adhesion molecule E-cadherin is reduced or absent in high-grade prostate cancer. *Cancer Res 52*, 5104–5109.

Uziel, T., Karginov, F.V., Xie, S., et al. (2009). The miR-17~92 cluster collaborates with the Sonic Hedgehog pathway in medulloblastoma. *Proc Natl Acad Sci USA 106*, 2812–2817.

Valta, M.P., Hentunen, T., Qu, Q., et al. (2006). Regulation of osteoblast differentiation: a novel function for fibroblast growth factor 8. *Endocrinology 147*, 2171–2182.

Valta, M.P., Tuomela, J., Bjartell, A., et al. (2008). FGF-8 is involved in bone metastasis of prostate cancer. *Int J Cancer 123*, 22–31.

Varjosalo, M., and Taipale, J. (2008). Hedgehog: functions and mechanisms. *Genes Dev 22*, 2454–2472.

Vega, S., Morales, A.V., Ocana, O.H., et al. (2004). Snail blocks the cell cycle and confers resistance to cell death. *Genes Dev 18*, 1131–1143.

Verras, M., and Sun, Z. (2006). Roles and regulation of Wnt signaling and beta-catenin in prostate cancer. *Cancer Lett 237*, 22–32.

Wang, H., Leav, I., Ibaragi, S., et al. (2008). SOX9 is expressed in human fetal prostate epithelium and enhances prostate cancer invasion. *Cancer Res 68*, 1625–1630.

Wang, L., Tang, H., Thayanithy, V., et al. (2009). Gene networks and microRNAs implicated in aggressive prostate cancer. *Cancer Res 69*, 9490–9497.

Wang, X., Ling, M.T., Guan, X.Y., et al. (2004). Identification of a novel function of TWIST, a bHLH protein, in the development of acquired taxol resistance in human cancer cells. *Oncogene 23*, 474–482.

Wilding, G. (1992). The importance of steroid hormones in prostate cancer. *Cancer Surv 14*, 113–130.

Wissmann, C., Wild, P.J., Kaiser, S., et al. (2003). WIF1, a component of the Wnt pathway, is down-regulated in prostate, breast, lung, and bladder cancer. *J Pathol 201*, 204–212.

Wong, Y.C., and Wang, Y.Z. (2000). Growth factors and epithelial-stromal interactions in prostate cancer development. *Int Rev Cytol 199*, 65–116.

Xu, J., Wang, R., Xie, Z.H., et al. (2006). Prostate cancer metastasis: role of the host microenvironment in promoting epithelial to mesenchymal transition and increased bone and adrenal gland metastasis. *Prostate 66*, 1664–1673.

Yan, G., Fukabori, Y., McBride, G., Nikolaropolous, S., and McKeehan, W.L. (1993). Exon switching and activation of stromal and embryonic fibroblast growth factor (FGF)-FGF receptor genes in prostate epithelial cells accompany stromal independence and malignancy. *Mol Cell Biol 13*, 4513–4522.

Yang, F., Li, X., Sharma, M., et al. (2002). Linking beta-catenin to androgen-signaling pathway. *J Biol Chem 277*, 11336–11344.

Yang, J., Mani, S.A., Donaher, J.L., et al. (2004). Twist, a master regulator of morphogenesis, plays an essential role in tumor metastasis. *Cell 117*, 927–939.

Yang, J., and Weinberg, R.A. (2008). Epithelial-mesenchymal transition: at the crossroads of development and tumor metastasis. *Dev Cell 14*, 818–829.

Yang, S., Zhong, C., Frenkel, B., Reddi, A.H., and Roy-Burman, P. (2005). Diverse biological effect and Smad signaling of bone morphogenetic protein 7 in prostate tumor cells. *Cancer Res 65*, 5769–5777.

Yates, C., Wells, A., and Turner, T. (2005). Luteinising hormone-releasing hormone analogue reverses the cell adhesion profile of EGFR overexpressing DU-145 human prostate carcinoma subline. *Br J Cancer 92*, 366–375.

Yates, C.C., Shepard, C.R., Stolz, D.B., and Wells, A. (2007). Co-culturing human prostate carcinoma cells with hepatocytes leads to increased expression of E-cadherin. *Br J Cancer 96*, 1246–1252.

Yuen, H.F., Chua, C.W., Chan, Y.P., et al. (2007). Significance of TWIST and E-cadherin expression in the metastatic progression of prostatic cancer. *Histopathology 50*, 648–658.

Zaman, M.S., Chen, Y., Deng, G., et al. (2010). The functional significance of microRNA-145 in prostate cancer. *Br J Cancer 103*, 256–264.

Zeisberg, M., Hanai, J., Sugimoto, H., et al. (2003). BMP-7 counteracts TGF-beta1-induced epithelial-to-mesenchymal transition and reverses chronic renal injury. *Nat Med 9*, 964–968.

Zhan, Y., Fujino, A., MacLaughlin, D.T., et al. (2006). Mullerian inhibiting substance regulates its receptor/SMAD signaling and causes mesenchymal transition of the coelomic epithelial cells early in Mullerian duct regression. *Development 133*, 2359–2369.

Zhau, H.E., Odero-Marah, V., Lue, H.W., et al. (2008). Epithelial to mesenchymal transition (EMT) in human prostate cancer: lessons learned from ARCaP model. *Clin Exp Metastasis 25*, 601–610.

Zhau, H.Y., Chang, S.M., Chen, B.Q., et al. (1996). Androgen-repressed phenotype in human prostate cancer. *Proc Natl Acad Sci USA 93*, 15152–15157.

Zhu, H., Mazor, M., Kawano, Y., et al. (2004). Analysis of Wnt gene expression in prostate cancer: mutual inhibition by WNT11 and the androgen receptor. *Cancer Res 64*, 7918–7926.

Zi, X., Guo, Y., Simoneau, A.R., et al. (2005). Expression of Frzb/secreted Frizzled-related protein 3, a secreted Wnt antagonist, in human androgen-independent prostate cancer PC-3 cells suppresses tumor growth and cellular invasiveness. *Cancer Res 65*, 9762–9770.

Zolfaghari, A., and Djakiew, D. (1996). Inhibition of chemomigration of a human prostatic carcinoma cell (TSU-pr1) line by inhibition of epidermal growth factor receptor function. *Prostate 28*, 232–238.

Chapter

24

Colon cancer

Anthony Scott and Zhenghe John Wang

Introduction

The colon and rectum are the final part of the human digestive tract. Cancers derived from these parts are called colorectal cancers. Colorectal cancer is the third most commonly diagnosed cancer with over one million cases reported worldwide annually (Markowitz and Bertagnolli, 2009). Although advances in early detection and chemotherapy have decreased the mortality of this disease, colorectal cancer remains to be the second leading cause of cancer death in the United States with an estimated 57,000 deaths annually (Markowitz and Bertagnolli, 2009). In this chapter, we first give a general introduction to etiology and clinical management of this disease, and then focus on the molecular causes of colorectal cancer, especially recent advances in the colorectal cancer genome, transcriptome, and proteome due to the employment of systems biology approaches.

Etiology, staging, screening, and therapy
Etiology

Most colorectal cancers are derived from predominantly epithelial cells (i.e., adenomas or adenocarcinomas) (Kinzler and Vogelstein, 1996). The causes of colorectal cancer can largely be divided into environmental or molecular factors. Diet has long been hypothesized to affect one's risk of developing the disease, with red meats and specific fats thought to be deleterious while certain vitamins (such as folate and vitamin D) and minerals (calcium and selenium) provide a protective effect. Other important environmental considerations include body habitus (high BMI and sedentary lifestyle conferring risk for disease), ulcerative colitis (which comes with a 20-fold increased risk for colorectal cancer), and tobacco use.

Familial vs. sporadic colorectal cancer

About 15 to 20% of colorectal cancer patients are positive for a family history of the disease, and a sizable proportion of these cases can be classified as having a familial syndrome (Kinzler and Vogelstein, 1996). The two archetypes are familial adenomatous polyposis (FAP) and hereditary non-polyposis colorectal carcinoma (HNPCC, also known as Lynch syndrome). FAP, which accounts for less than 1% of colorectal cancer cases, is caused by a germline mutation of one allele of APC that results in hundreds of adenomas covering the patient's colon by the age of 30; when an acquired mutation occurs in the other allele – which happens almost invariably – that adenoma will progress to carcinoma. FAP can also present with tumors of the nervous system (Turcot syndrome) or bones and soft tissue (Gardner syndrome) (Goss and Groden, 2000). A small percentage of patients with intestinal polyposis may have a mutation in MYH, a member of the base excision repair pathway (Grady and Carethers, 2008). Alternately, 3 to 4% of colorectal cancers are due to HNPCC, in which patients inherit a deficiency in one of a few mismatch repair genes and have an 80% chance of developing a colorectal cancer tumor, usually in the ascending colon and by their 40s, and often have family members with other neoplasms such as endometrial and gastric cancers (also known as Lynch syndrome; Lynch and de la Chapelle, 2003). The molecular phenotype of these familial cases, variously called replication error positive (RER+), mismatch repair deficient, or microsatellite instable (MSI or MIN), is characterized by expansion of mono- and di-nucleotide repeats at various locations throughout the genome (Aaltonen et al., 1993). Most HNPCC cases can be traced to mutations in either hMLH1 (Papadopoulos et al., 1994) or hMSH2 (Fishel et al.,

Systems Biology of Cancer, ed. S. Thiagalingam. Published by Cambridge University Press. © Cambridge University Press 2015.

1993). Like APC, one alteration is transmitted through the germline and the other acquired somatically (Grady and Carethers, 2008). Other inherited forms of colorectal cancer, which collectively make up a small fraction of familial cases, include Cowden syndrome, juvenile polyposis, or Peutz–Jeghers syndrome (Gryfe, 2009). Taken as a whole, these relatively rare syndromes account for 5% of all colorectal cancer cases overall, but only account for a quarter of the patients that have a first-degree relative with cancer. Therefore further studies are currently being done by clinical geneticists to determine the inheritance patterns of the many familial cancers that do not fit into an identified colorectal cancer syndrome.

Staging

Colorectal cancer progression occurs in well-recognized clinical and histopathological stages beginning with dysplastic epithelium and ending with metastatic disease (Fearon and Vogelstein, 1990). Pathologically, colorectal cancers are often divided into four stages. Stage I cancers are confined to the muscular layer. Stage II cancers breach the muscular layer but are limited to the adjacent soft tissues. Cancers metastatic to regional lymph nodes are designated as stage III. Cancers metastatic to distant organ sites such as the liver are designated as stage IV. Surgical resection is a highly effective treatment for early-stage colon cancers, providing cure rates of over 90% in stage I and over 80% in stage II disease. The presence of nodal involvement (stage III) predicts for 60% likelihood for recurrence, with virtually all recurrences being ultimately lethal. Treatment of these high-risk individuals with post-operative chemotherapy reduces the recurrence rate to 40%, increasing overall survival to 60%. On the contrary, the five-year survival rate for the remaining stage IV colorectal cancer patients is only 8% (Houston and O'Connell, 2004).

Screening

Colorectal cancer has become subject to intensive screening efforts, which include colonoscopy, fecal occult blood, and fecal immunochemical tests; these methods are utilized in high-risk patients, specifically those over 50 years of age and individuals with a family history of colon cancer. In addition to screening, there are also clinical findings that can point to the possibility of a colorectal cancer diagnosis. Lesions in the right-sided colon result in patient fatigue and anemia, while signs and symptoms of left-sided cancer include visible red blood in the stool and bowel obstruction and dysfunction.

Therapy

Surgical resection is the standard of care for most cases of disease, and adjuvant therapy is often indicated upon nodal involvement and/or metastasis. 5-Fluorouracil, an anti-metabolite, is used alongside leuvocorin for adjuvant therapy; this modality may be used in combination with oxaliplatin (a platinum-based compound used to disrupt DNA) or irinotecan (an alkaloid that stabilizes topoisomerase I and causes DNA double-stranded breaks). Compounds that target specific molecules, such as bevacizumab or cetuximab (VEGF and EGFR, respectively), are also possible therapies depending on the tumor profile. Radiation is also an option, especially in the case of rectal cancer or in the case of positive margins.

Molecular basis of colorectal cancer

It is now widely accepted that cancer is a genetic disease, mediated by alterations in specific genes (Vogelstein and Kinzler, 2004). Tumor progression from normal epithelia to the final metastatic disease occurs over decades and is accompanied by a series of genetic changes that affect at least one oncogene and several tumor suppressor genes (Figure 24.1). One of the first events in this process is inactivation of the APC tumor suppressor gene or activation of its downstream target β-catenin (Vogelstein and Kinzler, 2004). Mutations that activate KRAS or BRAF oncogenes are thought to follow APC mutations and are observed in at least half of adenomas greater than 1 cm in size as well as in carcinomas (Rajagopalan et al., 2002). As shown in Figure 24.1, subsequent waves of clonal expansion driven by mutations in the PIK3CA/PTEN (Samuels et al., 2004), TP53/BAX (Vogelstein et al., 2000), TGF-βRII/SMAD (Markowitz and Bertagnolli, 2009) pathways and probably other pathways are responsible for the transition of a large adenoma (benign) to an early carcinoma (malignant). In addition to mutation in oncogenes and tumor suppressor genes, genomic instability and aberrant DNA methylation also drive the development of colorectal cancers.

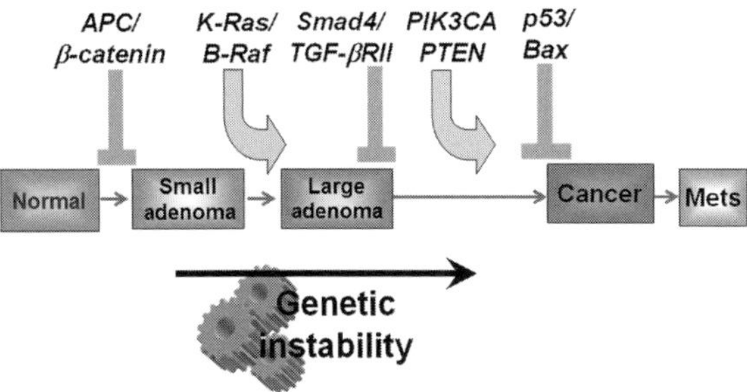

Figure 24.1 A multiple-stage colorectal progression model. Colorectal cancer progression from normal epithelia to the final metastatic disease occurs over decades and is accompanied by a series of genetic changes affecting multiple pathways. Genetic instabilities including microsatellite instability and chromosomal instability are thought to be the driving forces of tumor progression. (Mets: metastasis) (A black and white version of this figure will appear in some formats. For the color version, please refer to the plate section.)

APC/β-catenin pathway

The relationship between APC and colorectal cancer tumorigenesis began when mutations in familial adenomatous polyposis patients were localized to its locus (Groden et al., 1991; Kinzler et al., 1991). APC was later shown to have implications in sporadic forms of the disease too, as it is somatically mutated in over 80% of adenomas and carcinomas (Kinzler and Vogelstein, 1996). The effect these alterations have on colorectal cancer development seems to be mediated by APC's interaction with β-catenin, as APC is required for a complex with axin that eventually leads to β-catenin's phosphorylation by GSK3β and its subsequent degradation. Without this inhibition, or by GSK3β inactivation via Wnt signaling, β-catenin interacts with the transcription factor Tcf-4 and they both transcriptionally activate a variety of genes, especially those involved with cellular proliferation and survival (Goss and Groden, 2000). Accordingly, truncations of APC identified in colorectal cancer patients do not antagonize transcription from β-catenin promoters. Similarly, the reintroduction of wild-type APC into cell lines lacking a functional copy of APC resulted in downregulation of a β-catenin-responsive reporter gene (Korinek et al., 1997). In the odd (10–15%) cases where a colorectal cancer tumor retains wild-type APC, mutations to β-catenin that prevent its phosphorylation by GSK3β lead to constitutive activation of β-catenin's transcriptional activities (Morin et al., 1997). As it relates to the timing of colorectal cancer development, APC most likely plays an early role, as its inactivation leads to the development of adenomas in FAP and because most adenomas harbor APC mutations (Powell et al., 1992).

Ras/Raf pathway

Mutations in K-ras and B-raf are often identified in colorectal cancer patients, in about 50 or 10% of cases, respectively (Rajagopalan et al., 2002). Alterations to these two genes – which are mutually exclusive in a single tumor – are recovered in large adenomas, indicating this mutation may be responsible for transition from a small to large adenoma (Figure 24.1). Relationships have been identified with specific types of colorectal cancer and B-raf mutation, specifically those tumors with microsatellite instability (Rajagopalan et al., 2002) and CpG island methylator phenotype (Weisenberger et al., 2006). K-ras and B-raf proteins are involved in the MAP kinase pathway, which transduces growth signals from receptor tyrosine kinases. Upon stimulation from a growth factor, K-ras becomes GTP bound and activated, interacting most notably with B-raf. B-raf then goes on to phosphorylate its target proteins, especially MEK, a downstream member of the MAP kinase pathway. The cancer-derived mutations in K-ras (Bos et al., 1987) and B-raf (Davies et al., 2002) constitutively activate these proteins' activity, causing tumorigenesis by inappropriate inactivation of growth and pro-survival pathways.

PIK3CA/PTEN pathway

PI3KCA, also known as the p110-alpha subunit of phosphatidylinositol 3-kinase (PI3K), was found to play a role in colorectal cancer tumor progression due to a screen for kinase domain mutations in PI3K genes and related proteins (Samuels et al., 2004). While there are other isoforms of p110 that can constitute the PI3K complex, p110-alpha is most commonly involved in cancer, with alterations often found in specific "hot-spots" centered around its kinase and helical domains (Zhao and Vogt, 2008). These mutations – found in 32% of tumors – were shown to increase its kinase activity, implicating them as oncogenic. PI3K transduces growth signals from activated receptor tyrosine kinases, either by interacting with them directly or by an IRS-1/-2 intermediate. Pro-growth signals release the p110 subunit from inhibition by p85, allowing it to phosphorylate phosphatidylinositol 4,5-bisphosphate to phosphatidylinositol (3,4,5)-trisphosphate (PIP3) (Bader et al., 2005).

PIP3 goes on to activate the growth-promoting and anti-apoptotic Akt and PDK1 pathways. Consequently, PI3K's role as an oncogene has been established in a variety of cancers (Vivanco and Sawyers, 2002). Conversely, PTEN – which reverses the PIP3 phosphorylation catalyzed by PI3K – acts as a tumor suppressor in the colon. In a study of primary colorectal cancer tumor samples, at least one allele of PTEN was either mutated or lost in many of the specimens (in a similar proportion to PI3KCA, about 35%) and its expression decreased in proportion to the progression of the disease (Nassif et al., 2004). These results were consistent with the mouse PTEN knockout model, in which the loss of one allele of PTEN caused hyperplasia in the colon and polyps as well as progression to carcinoma (Vivanco and Sawyers, 2002). Therefore the balance between the oncogenic potential of PI3K and the tumor suppressor activity of PTEN underscores the importance of this pathway in colorectal cancer development.

TP53/BAX pathway

TP53 is mutated in ~50% of all human cancers. TP53's role in colorectal cancer was established when it became mapped to the 17p deletion frequently found in nearly 75% of tumor samples (Baker et al., 1989) and because colorectal cancer cells that either lacked functional TP53 or did not express it sufficiently experienced a significant decrease in proliferation when transfected with a functional copy of the protein (Baker et al., 1990a). TP53 inactivation usually occurs in the carcinoma stage by the allelic deletion of one locus and an inactivating mutation in the other; these mutations were much rarer in adenoma samples, indicating it is responsible for transition to carcinoma (Baker et al., 1990b). In response to a variety of noxious stimuli, such as DNA damage, cytotoxic compounds, and oncogenes, p53 becomes relieved of MDM2-mediated inhibition and activates transcription in certain target genes to prevent a cell from passing on mutations. These include the cell cycle arrest proteins p21 and 14-3-3delta (which act by inhibiting CDKs), members of the DNA repair pathway, and pro-apoptotic factors such as Bax and PUMA (Vogelstein et al., 2000; Yu et al., 2001). Bax, due to its mononucleotide repeat sequence, serves as a target for MMR-induced mutation; the alteration it thus experiences in microsatellite instable tumors results in its inactivation, acting as a surrogate for

TP53 loss in the cases in which TP53 is still functional (Yamamoto et al., 1998). Likewise to the timing of TP53, Bax mutation does not occur as frequently in the adenoma stage of development but instead during carcinoma (Yagi et al., 1998).

TGF-βRII/SMAD pathway

TGF-β is a signaling molecule with anti-neoplastic activity mediated by signaling through TGF-βRII, and inactivating mutations to this receptor often occur in cancer. TGF-β causes TGF-βRII to interact with and phosphorylate TGF-βRI, which activates it to phosphorylate Smad2 and Smad3. Smad2/3 binds Smad4 and together they translocate into the nucleus, affecting the transcription of key proteins: either upregulating cell cycle antagonists such as p15 and p21 or downregulating proliferation-associated genes such as Myc (Massague, 2008). Alterations to TGF-βRII commonly occur in microsatellite instable tumors, as these colorectal cancer samples experienced expansion of mono- or dinucleotide repeats at that locus, which will affect its protein expression (Markowitz et al., 1995). Due to their absence in early adenomas but presence in dysplastic adenomas and carcinomas, loss of TGF-β is a relatively late event in colorectal cancer progression (Grady et al., 1998). Inactivating SMAD4 prevents cells from successfully transducing TGF-β's growth-suppressive signals (Zhou et al., 1998). Accordingly, the chromosome arm containing SMAD4, 18q, is frequently deleted in colorectal cancer (Thiagalingam et al., 1996).

Genomic instability

Genomic instability is largely due to microsatellite instability (MIN) or chromosomal instability (CIN). MIN is caused by inactivation of the mismatch repair (MMR) pathway, which leads to aberrant expansion of mononucleotide or dinucleotide repeats due to faulty repair mechanisms (Grady and Carethers, 2008). Although it is a hallmark of the familial HNPCC tumors, MIN is also evident in 13% of sporadic cancers (Kinzler and Vogelstein, 1996). Many of these sporadic MIN cases are due to hMLH1 becoming transcriptionally inactivated due to hypermethylation at CpG islands (Veigl et al., 1998). Many genes with runs of one- or two-nucleotide repeats are at risk for mutation by MIN, including tumor suppressors like TGF-βRII (Markowitz et al., 1995) and Bax or hMSH3 and hMSH6, binding partners of the crucial hMSH2 and hMLH1 proteins whose abrogation can potentiate the MIN phenotype (Yamamoto et al., 1998).

While MIN is responsible for modest changes to specific genes, gross changes in the genome occur because of CIN. As described above, many of the canonical colorectal cancer tumor suppressors, such as TP53 and SMAD4, were mapped to their physical loci due in part to the large cytological deletions frequently recovered in tumor samples (17p and 18q, respectively). Therefore CIN frequently plays a role in colorectal cancer tumorigenesis by causing deletions of regions that harbor tumor suppressors. Various studies indicate that CIN, since it is evident in both adenomas and carcinomas, is important as a mechanism for tumor initiation (Shih et al., 2001).

Although the specific genes and/or pathways responsible for this phenomenon have not yet been identified, work has been done to elucidate CIN's overall significance. Extreme changes in chromosome number are not required, as karyotypes of the common colorectal cancer cell lines exist on a continuum from the approximately diploid (DLD1) to the decidedly aneuploid (SW480). Merely adding chromosomes to the former lineages was not successful in inducing CIN – the extra chromosomes were stably inherited – but fusing them to cell lines with frank CIN resulted in demonstrable aneuploidy, indicating that CIN is a dominant trait (Lengauer et al., 1997).

Epigenetic abnormality

Aberrant DNA methylation is another driving force of tumorigenesis (Jones and Baylin, 2007). DNA methylation at the CpG sites is an inheritable modification without changes of DNA sequences (epigenetics). Colorectal cancer genomes are globally hypomethylated (Feinberg, 2007), which is suspected to be a cause of genomic instability (Ting et al., 2006). However, focused DNA methylation in the promoter regions, which leads to silence of tumor suppressor genes, is thought to play an important role in tumor initiation and progression. It is well documented that aberrant DNA methylation can result in inactivation of tumor suppressor genes, such as p16, MLH1, E-cadherin, and others, in colorectal cancers (Ting et al., 2006). It appears that some of the aberrant DNA methylation events could happen at the very early stage of cancer development. Interestingly, recent studies have shown that aberrant DNA methylation could be exploited as a diagnostic marker for non-invasive early detection of colorectal cancers.

Systems biology approach

The completion of the Human Genome Project enables systems biology approaches to uncover the molecular bases of colorectal cancer in an unprecedented pace. Recent genomic, epigenomic, transcriptomic, and proteomic analyses of colorectal cancers have yielded an enormous amount of information waiting to be thoroughly deciphered. The emerging field of colorectal cancer metabolomics will have an important impact on our understanding of this disease.

Genomics of colorectal cancer

The initial high-throughput mutational analyses of colorectal cancer focused on gene families that encode protein tyrosine kinases (PTKs), protein tyrosine phosphatases (PTPs), and lipid kinases (Bardelli et al., 2003; Samuels et al., 2004; Wang et al., 2004). Seven PTKs (MLK4, NTRK3, FES, KDR, EPHA3, NTRK2, and GUCY2F) were identified to be mutated somatically in colorectal cancers. These tumor-derived mutations are likely to be activation mutations (Bardelli et al., 2003). In contrast, mutations found in PTPs are inactivation mutations (Wang et al., 2004). Among the six mutated PTPs identified (PTPRT, PTPRG, PTPRF, PTPN13, PTPN14, and PTPN3), PTPRT is the most frequently mutated one. In addition to colorectal cancer, PTPRT is also mutated in lung, stomach, and skin (melanomas) cancers. Several lines of evidence support the premise that PTPRT normally functions as a tumor suppressor gene. First, the spectrum of mutations, which includes nonsense mutations and frameshifts, suggested that these are clearly inactivating mutations (Wang et al., 2004). Second, biochemical analyses demonstrated that missense mutations in the catalytic domains of PTPRT diminished its phosphatase activity (Wang et al., 2004) and that the extracellular domain mutations are defective in cell–cell adhesion (Yu et al., 2008; Zhang et al., 2009), indicating that the missense mutations of PTPRT are also loss-of-function mutations. Third, overexpression of PTPRT inhibits colorectal cancer cell growth (Wang et al., 2004). Fourth, PTPRT knockout mice are highly susceptible to azoxymethane (AOM)-induced colon tumors (Zhao et al., 2010), providing critical *in vivo* evidence that PTPRT normally functions as a tumor suppressor. A proteomic approach identified STAT3 and paxillin as the intracellular substrates of PTPRT. As discussed above, Samuels et al. discovered that PIK3CA was mutated in over 30% of colorectal cancers in the mutational analysis of the PI3K gene family (Samuels et al., 2004). In subsequent studies, PIK3CA was shown to be frequently mutated in hepatocellular carcinomas (Lee et al., 2005), endometrial carcinomas (Oda et al., 2005), breast carcinomas (Bachman et al., 2004), gastric carcinomas (Li et al., 2005), ovarian carcinomas (Campbell et al., 2004), and a smaller fraction of lung carcinomas (Samuels et al., 2004), medullablastomas, and anaplastic astrocytomas (Broderick et al., 2004). Therefore PIK3CA is one of the most frequently mutated oncogenes in human cancer.

Recently, genome-wide mutational analyses of colorectal cancers presented a more complex picture (Sjoblom et al., 2006; Wood et al., 2007). By sequencing the exome, the Vogelstein group at the Johns Hopkins University identified a total of 140 Candidate Cancer genes (CAN genes) in colorectal cancers. On average, individual tumors harbor ~80 non-silent mutations and 15 to 20 of these are likely to drive the cancer development. Besides the few frequently mutated genes (mountains) including APC, p53, RAS, and PIK3CA, most of the genes are mutated in <5% of colorectal cancers (hills). The mutated genes in any given two tumors only overlap to a small extent, indicating a complex genetic heterogeneity of colorectal cancers. Prior to the whole-genome mutational studies, the focus of cancer research had been on the pathways regulated by the gene mountains. The dominance of the gene hills in the cancer genomic landscape presents both challenges and opportunities to cancer research and clinical management of cancer patients.

In addition to point mutations, gene amplifications are another mechanism to activate oncogenes whereas homozygous deletions inactivate tumor suppressor genes. A recent SNP array analysis of 36 colorectal cancers identified several genes that are either amplified or deleted (Leary et al., 2008). Most of these genes encode proteins involved in pathways that are known to drive tumorigenesis (Leary et al., 2008).

Transcriptomics of colorectal cancer

Transcriptomics refers to the global profiling of gene expression patterns. The advent of SAGE and microarray technologies permits such analyses (Schena et al., 1995; Velculescu et al., 1995). Numerous microarray studies have been performed on colorectal cancer specimens with the hope of identifying gene expression signatures that can be utilized as prognostic biomarkers for disease outcomes or predictive biomarkers for drug responses. Although each of the studies found a set of genes that are of either prognostic or predictive values, there were no overlapping gene expression signatures identified from studies undertaken by different groups. Again, these data suggest molecular heterogeneity of colorectal cancers.

Proteomics of colorectal cancer

Proteomics refers to the analysis of the complete set of proteins or proteome. Unlike transcriptomics which is only applicable to cells and tissues, proteomics can be extended to profiling proteins or peptides in body fluids. Thus proteomics holds a great promise of identification of novel diagnostic markers for colorectal cancers. In contrast to the abundant literature on genomics and transcriptomics of colorectal cancers, there were only a few proteomic analyses of colorectal cancer being reported (Nambiar et al., 2010). Whether these findings can be translated into clinical applications it is still too early to tell.

Metabolomics of colorectal cancer

Metabolomics is an emerging field. Metabolomics refers to the analysis of the complete set of small-molecule metabolites (such as metabolic intermediates, hormones and other signaling molecules, and secondary metabolites). It was observed a long time ago that most cancer cells predominantly produce energy by a high rate of glycolysis followed by lactic acid fermentation rather than oxidation of pyruvate in mitochondria, which is also known as the Warburg effect (Warburg, 1956). However, the molecular mechanism of the Warburg effect remained unknown. Recently identification of IDH1 (isocitrate dehydrogenase 1) mutations in colon, brain, and blood cancers rekindled the interest in cancer metabolism (Parsons et al., 2008). A couple of studies of the metabolites of colorectal cancer have been recently reported (Nambiar et al., 2010). With technological improvement, the throughput of metabolite profiling is increasing. More metabolomics analyses of colorectal cancers are on the horizon.

Conclusion and perspective

Recent systems biology approaches, especially genomic, unraveled a trove of information regarding the driving forces of colorectal tumorigenesis, which provides unprecedented opportunities to design novel diagnostic and therapeutic strategies to this disease. How to translate such knowledge into clinic practice should be the future research focus. While developing therapeutic agents is important to manage the late stage disease, breakthrough in non-invasive early detection and screening method will be vital in reducing the mortality of colorectal cancer.

References

Aaltonen, L.A., Peltomaki, P., Leach, F.S., et al. (1993). Clues to the pathogenesis of familial colorectal cancer. *Science 260*, 812–816.

Bachman, K.E., Argani, P., Samuels, Y., et al. (2004). The PIK3CA gene is mutated with high frequency in human breast cancers. *Cancer Biol Ther 3*, 772–775.

Bader, A.G., Kang, S., Zhao, L., and Vogt, P.K. (2005). Oncogenic PI3K deregulates transcription and translation. *Nat Rev Cancer 5*, 921–929.

Baker, S.J., Fearon, E.R., Nigro, J.M., et al. (1989). Chromosome 17 deletions and p53 gene mutations in colorectal carcinomas. *Science 244*, 217–221.

Baker, S.J., Markowitz, S., Fearon, E.R., Willson, J.K., and Vogelstein, B. (1990a). Suppression of human colorectal carcinoma cell growth by wild-type p53. *Science 249*, 912–915.

Baker, S.J., Preisinger, A.C., Jessup, J.M., et al. (1990b). p53 gene mutations occur in combination with 17p allelic deletions as late events in colorectal tumorigenesis. *Cancer Res 50*, 7717–7722.

Bardelli, A., Parsons, D.W., Silliman, N., et al. (2003). Mutational analysis of the tyrosine kinome in colorectal cancers. *Science 300*, 949.

Bos, J.L., Fearon, E.R., Hamilton, S.R., et al. (1987). Prevalence of ras gene mutations in human colorectal cancers. *Nature 327*, 293–297.

Broderick, D.K., Di, C., Parrett, T.J., et al. (2004). Mutations of PIK3CA in anaplastic oligodendrogliomas, high-grade astrocytomas, and medulloblastomas. *Cancer Res 64*, 5048–5050.

Campbell, I.G., Russell, S.E., Choong, D.Y., et al. (2004). Mutation of the PIK3CA gene in ovarian and breast cancer. *Cancer Res 64*, 7678–7681.

Davies, H., Bignell, G.R., Cox, C., et al. (2002). Mutations of the BRAF gene in human cancer. *Nature 417*, 949–954.

Fearon, E.R., and Vogelstein, B. (1990). A genetic model for colorectal tumorigenesis. *Cell 61*, 759–767.

Feinberg, A.P. (2007). Phenotypic plasticity and the epigenetics of human disease. *Nature 447*, 433–440.

Fishel, R., Lescoe, M.K., Rao, M.R., et al. (1993). The human mutator gene homolog MSH2 and its association with hereditary nonpolyposis colon cancer. *Cell 75*, 1027–1038.

Goss, K.H., and Groden, J. (2000). Biology of the adenomatous polyposis coli tumor suppressor. *J Clin Oncol 18*, 1967–1979.

Grady, W.M., and Carethers, J.M. (2008). Genomic and epigenetic instability in colorectal cancer pathogenesis. *Gastroenterology 135*, 1079–1099.

Grady, W.M., Rajput, A., Myeroff, L., et al. (1998). Mutation of the type II transforming growth factor-beta receptor is coincident with the transformation of human colon adenomas to malignant carcinomas. *Cancer Res 58*, 3101–3104.

Groden, J., Thliveris, A., Samowitz, W., et al. (1991). Identification and characterization of the familial adenomatous polyposis coli gene. *Cell 66*, 589–600.

Gryfe, R. (2009). Inherited colorectal cancer syndromes. *Clin Colon Rectal Surg 22*, 198–208.

Houston, A., and O'Connell, J. (2004). The Fas signalling pathway and its role in the pathogenesis of cancer. *Curr Opin Pharmacol 4*, 321–326.

Jones, P.A., and Baylin, S.B. (2007). The epigenomics of cancer. *Cell 128*, 683–692.

Kinzler, K.W., Nilbert, M.C., Su, L.K., et al. (1991). Identification of FAP locus genes from chromosome 5q21. *Science 253*, 661–665.

Kinzler, K.W., and Vogelstein, B. (1996). Lessons from hereditary colorectal cancer. *Cell 87*, 159–170.

Korinek, V., Barker, N., Morin, P.J., et al. (1997). Constitutive transcriptional activation by a beta-catenin-Tcf complex in APC-/- colon carcinoma. *Science 275*, 1784–1787.

Leary, R.J., Lin, J.C., Cummins, J., et al. (2008). Integrated analysis of homozygous deletions, focal amplifications, and sequence alterations in breast and colorectal cancers. *Proc Natl Acad Sci USA 105*, 16224–16229.

Lee, J.W., Soung, Y.H., Kim, S.Y., et al. (2005). PIK3CA gene is frequently mutated in breast carcinomas and hepatocellular carcinomas. *Oncogene 24*, 1477–1480.

Lengauer, C., Kinzler, K.W., and Vogelstein, B. (1997). Genetic instability in colorectal cancers. *Nature 386*, 623–627.

Li, V.S., Wong, C.W., Chan, T.L., et al. (2005). Mutations of PIK3CA in gastric adenocarcinoma. *BMC Cancer 5*, 29.

Lynch, H.T., and de la Chapelle, A. (2003). Hereditary colorectal cancer. *N Engl J Med 348*, 919–932.

Markowitz, S., Wang, J., Myeroff, L., et al. (1995). Inactivation of the type II TGF-beta receptor in colon cancer cells with microsatellite instability. *Science 268*, 1336–1338.

Markowitz, S.D., and Bertagnolli, M.M. (2009). Molecular origins of cancer: molecular basis of colorectal cancer. *N Engl J Med 361*, 2449–2460.

Massague, J. (2008). TGFbeta in cancer. *Cell 134*, 215–230.

Morin, P.J., Sparks, A.B., Korinek, V., et al. (1997). Activation of beta-catenin-Tcf signaling in colon cancer by mutations in beta-catenin or APC. *Science 275*, 1787–1790.

Nambiar, P.R., Gupta, R.R., and Misra, V. (2010). An "omics" based survey of human colon cancer. *Mutat Res 693*, 3–18.

Nassif, N.T., Lobo, G.P., Wu, X., et al. (2004). PTEN mutations are common in sporadic microsatellite stable colorectal cancer. *Oncogene 23*, 617–628.

Oda, K., Stokoe, D., Taketani, Y., and McCormick, F. (2005). High frequency of coexistent mutations of PIK3CA and PTEN genes in endometrial carcinoma. *Cancer Res 65*, 10669–10673.

Papadopoulos, N., Nicolaides, N.C., Wei, Y.F., et al. (1994). Mutation of a mutL homolog in hereditary colon cancer. *Science 263*, 1625–1629.

Parsons, D.W., Jones, S., Zhang, X., et al. (2008). An integrated genomic analysis of human glioblastoma multiforme. *Science 321*, 1807–1812.

Powell, S.M., Zilz, N., Beazer-Barclay, Y., et al. (1992). APC mutations occur early during colorectal tumorigenesis. *Nature 359*, 235–237.

Rajagopalan, H., Bardelli, A., Lengauer, C., et al. (2002). Tumorigenesis: RAF/RAS oncogenes and mismatch-repair status. *Nature 418*, 934.

Samuels, Y., Wang, Z., Bardelli, A., et al. (2004). High frequency of mutations of the PIK3CA gene in human cancers. *Science 304*, 554.

Schena, M., Shalon, D., Davis, R.W., and Brown, P.O. (1995). Quantitative monitoring of gene expression patterns with a complementary DNA microarray. *Science 270*, 467–470.

Shih, I.M., Zhou, W., Goodman, S.N., et al. (2001). Evidence that genetic instability occurs at an early stage of colorectal tumorigenesis. *Cancer Res 61*, 818–822.

Sjoblom, T., Jones, S., Wood, L.D., et al. (2006). The consensus coding sequences of human breast and colorectal cancers. *Science 314*, 268–274.

Thiagalingam, S., Lengauer, C., Leach, F.S., et al. (1996). Evaluation of candidate tumour suppressor genes on chromosome 18 in colorectal cancers. *Nat Genet 13*, 343–346.

Ting, A.H., McGarvey, K.M., and Baylin, S.B. (2006). The cancer epigenome – components and functional correlates. *Genes Dev 20*, 3215–3231.

Veigl, M.L., Kasturi, L., Olechnowicz, J., et al. (1998). Biallelic inactivation of hMLH1 by epigenetic gene silencing, a novel mechanism causing human MSI cancers. *Proc Natl Acad Sci USA 95*, 8698–8702.

Velculescu, V.E., Zhang, L., Vogelstein, B., and Kinzler, K.W. (1995). Serial analysis of gene expression. *Science 270*, 484–487.

Vivanco, I., and Sawyers, C.L. (2002). The phosphatidylinositol 3-kinase AKT pathway in human cancer. *Nat Rev Cancer 2*, 489–501.

Vogelstein, B., and Kinzler, K.W. (2004). Cancer genes and the pathways they control. *Nat Med 10*, 789–799.

Vogelstein, B., Lane, D., and Levine, A.J. (2000). Surfing the p53 network. *Nature 408*, 307–310.

Wang, Z., Shen, D., Parsons, D.W., et al. (2004). Mutational analysis of the tyrosine phosphatome in colorectal cancers. *Science 304*, 1164–1166.

Warburg, O. (1956). On the origin of cancer cells. *Science 123*, 309–314.

Weisenberger, D.J., Siegmund, K.D., Campan, M., et al. (2006). CpG island methylator phenotype underlies sporadic microsatellite instability and is tightly associated with BRAF mutation in colorectal cancer. *Nat Genet 38*, 787–793.

Wood, L.D., Parsons, D.W., Jones, S., et al. (2007). The genomic landscapes of human breast and colorectal cancers. *Science 318*, 1108–1113.

Yagi, O.K., Akiyama, Y., Nomizu, T., et al. (1998). Proapoptotic gene BAX is frequently mutated in hereditary nonpolyposis colorectal cancers but not in adenomas. *Gastroenterology 114*, 268–274.

Yamamoto, H., Sawai, H., Weber, T.K., Rodriguez-Bigas, M.A., and Perucho, M. (1998). Somatic frameshift mutations in DNA mismatch repair and proapoptosis genes in hereditary nonpolyposis colorectal cancer. *Cancer Res 58*, 997–1003.

Yu, J., Becka, S., Zhang, P., et al. (2008). Tumor-derived extracellular mutations of PTPRT/PTPrho are defective in cell adhesion. *Mol Cancer Res 6*, 1106–1113.

Yu, J., Zhang, L., Hwang, P.M., Kinzler, K.W., and Vogelstein, B. (2001). PUMA induces the rapid apoptosis of colorectal cancer cells. *Mol Cell 7*, 673–682.

Zhang, P., Becka, S., Craig, S.E., et al. (2009). Cancer-derived mutations in the fibronectin III repeats of

PTPRT/PTPrho inhibit cell-cell aggregation. *Cell Commun Adhes 16*, 146–153.

Zhao, L., and Vogt, P.K. (2008). Class I PI3K in oncogenic cellular transformation. *Oncogene 27*, 5486–5496.

Zhao, Y., Zhang, X., Guda, K., et al. (2010). Identification and functional characterization of paxillin as a target of protein tyrosine phosphatase receptor T. *Proc Natl Acad Sci USA 107*, 2592–2597.

Zhou, S., Buckhaults, P., Zawel, L., et al. (1998). Targeted deletion of Smad4 shows it is required for transforming growth factor beta and activin signaling in colorectal cancer cells. *Proc Natl Acad Sci USA 95*, 2412–2416.

Chapter

25 Biology of human stomach cancer

Bryan G. Sauer and Steven M. Powell

Introduction

Gastric adenocarcinomas comprise the vast majority of malignant tumors arising from the stomach. While other histopathologic forms of stomach tumors exist, they are rare in occurrence. Gastric cancer is a significant worldwide health burden, second only to lung tumors as a leading cause of cancer deaths.

Gastric cancers exhibit heterogeneity in clinical, biologic, and genetic aspects. Multiple pathological classifications of gastric adenocarcinomas have been proposed including those with morphologic and histologic criteria in efforts to identify subgroups with differing biologic behavior and prognosis. Molecular markers should help facilitate classification of these subgroups. The TNM staging system is generally used as the basis for prognostication in this cancer with depth of infiltration being an important parameter.

Significant geographic and temporal variances are observed in gastric cancer's incidence, predominantly of the intestinal type. Epidemiologic studies indicate a strong environmental component in the acquisition of this cancer. Developing countries, often observed to have a high prevalence of *Helicobacter pylori* infection early in life, have higher rates of gastric adenocarcinomas. The relative importance of bacterial virulence, environmental, and host factors (e.g., age of acquisition, immune response, acid secretion changes) to the clinical outcome of *H. pylori* infections are currently pressing issues.

Most gastric cancers are identified in advanced stages that present in the later decades of adult life, commonly resulting in a lethal outcome shortly thereafter. Thus an improved understanding of the biology of this deadly disease is anxiously awaited. Most cases of stomach cancer are sporadic in nature with genetic alterations being acquired over decades. There are, however, rare inherited gastric cancer predisposition traits such as *E-cadherin* mutations in familial diffuse gastric cancers and hereditary non-polyposis colon cancer associated with gastric cancer as discussed in detail below. The complexity of the genetics involved in gastric adenocarcinoma is reflected in the temporal, regional, and gender variation in gastric cancer incidence rates. A better understanding of these phenomena through molecular and genetic studies of gastric tumor genesis will provide important insights into cancer development in general and is anticipated to lead to earlier diagnosis and better management options.

Epidemiology

Gastric adenocarcinoma is the predominant cancer of the stomach, accounting for over 95% of cases. Lymphomas, leiomyosarcomas, gastrointestinal stromal, and carcinoid lesions account for some of the other stomach tumors. Once the leading cause of cancer deaths in the United States, the incidence of gastric cancer in developed countries has declined dramatically [1]. Adenocarcinoma of the stomach remains a leading cause of cancer death worldwide and continues to be responsible for the majority of cancer deaths in developing countries [2]. The incidence and mortality rates remain high in Japan, Asia, Latin America, and Eastern Europe. Notably, a consistent predominance of gastric cancers in males (approximately 2:1 ratio) is seen across worldwide populations.

Significant geographic variability in incidence both internationally and intranationally is observed [3]. In the early 1990s incidence rates varied from 60.1 per 100,000 in Costa Rica to 7.3 per 100,000 in the United States [4]. Epidemiologic studies, which include migration and temporal analyses, indicate that environmental factors, especially in the first decades of life, are important in the etiology of gastric

Systems Biology of Cancer, ed. S. Thiagalingam. Published by Cambridge University Press. © Cambridge University Press 2015.

cancers [1]. *Helicobacter pylori* infection is now recognized as an etiologic factor in gastric cancer development for both adenocarcinoma and primary non-Hodgkin lymphoma reviewed in [5]. Evidence continues to accumulate that this infection, especially when contracted early in life as commonly occurs in developing countries, leads to chronic gastric inflammation and a five-fold increased risk of gastric cancer [6]. The cofactors and pathophysiologic mechanism of cancer development in *H. pylori* infected subjects remains to be elucidated. Eradication of *H. pylori* at the time of resection for early gastric cancer was associated with reduced metachronous gastric cancers [7]. Similar to *H. pylori*-infection, chronic gastritis and resultant atrophy from pernicious anemia appear to be associated with an increased risk, albeit small, of gastric cancer [8].

Dietary irritants (i.e., salts, or preservatives such as smoked or pickled foods) and potential carcinogens (i.e., nitrates and tobacco) have been suggested as etiologic factors of gastric cancer [9], yet no specific agent has been definitively indicted. Additionally, several protective factors such as fruits, vegetables, ascorbic acid, alpha-tocopherol, onions, and gastric acidity have been suggested; yet the precise agents responsible for the effect remain elusive. Evidence of Epstein–Barr Virus (EBV) infection has been demonstrated in a small proportion (approximately 10%) of gastric carcinomas by *in situ* hybridization with specific RNA probes [10].

Pathology classification

Gastric adenocarcinomas display several distinct morphologic, histologic, and biologic characteristics. Thus multiple tumor classification systems have been created to characterize these lesions pathologically in attempts to confer their natural history. Lauren described histopathologic subtypes of gastric adenocarcinomas as intestinal type (expansive or gland-forming) or diffuse type (infiltrative or scattered neoplastic cells), and this classification is widely applied [11]. Evidence continues to accumulate suggesting that these two subtypes of gastric adenocarcinoma arise in different settings and have distinctive biologic behavior [12]. Intestinal-type gastric adenocarcinomas tend to predominate in high-risk geographic regions, arise in association with precursor lesions (i.e., chronic atrophic gastritis or intestinal metaplasia), and occur more distally and later in life (usually after the sixth decade of life). The risk of subsequent gastric cancer in those with precancerous lesions was followed in a Netherlands cohort and found to be significant once dysplasia, especially high-grade dysplasia, was observed (hazard ratio 40.14; 95% confidence interval 32.2–50.1) [13]. Diffuse types of gastric cancers appear to have a relatively constant incidence, arise without identifiable precursor lesions, and present earlier in life and more diffusely in the stomach. Additionally, intestinal-type gastric cancers tend to spread hematogenously to the liver, while diffuse-type gastric cancers tend to spread more contiguously into the peritoneum. Moreover, although some molecular alterations are shared, distinct genetic abnormalities appear to occur with specific biological phenotypes (see below).

Additional histologic classifications for gastric carcinoma have been developed (i.e., World Health Organization [14] and Ming [15]), which involve tissue architecture and differentiation criteria. The traditional classification of Borman is based on morphologic criteria [16]. Goeski has developed criteria that examine mucin content and degree of cellular atypia as potentially distinctive prognostic features of gastric tumors [17].

The TMN staging classification [18] is primarily used in staging cancers at diagnosis, to assess resectability and prognoses. The primary tumor's depth is the most important parameter in the TMN determination. Individuals with early gastric adenocarcinomas (tumors confined to the mucosa or submucosa) are observed to have much better five-year survival rates (over 90%) versus more advanced lesions (less than 20% for stage III or IV) [19]. Unfortunately most cases of gastric cancer are diagnosed in more advanced stages for which effective systemic therapy is limited and surgery reserved for palliation. Mass screening programs in high-risk regions such as Japan have helped in diagnosing some of these cancers at earlier stages, but significant improvements in diagnosis, prognostication, and therapy are eagerly awaited to make a substantial impact on this cancer's mortality.

Associated host genotypes

Several allelic variants of polymorphisms in proinflammatory cytokines such as IL-1 beta, a potent inhibitor of acid secretion, have been associated with gastric cancers, suggesting the host response may be an important determinant in the clinical outcome of *H. pylori* infection [20, 21]. A polymorphism of TNF-alpha and even the TLR-4 are associated with gastric cancer development as well [22, 23].

A population association study recently demonstrated more frequent occurrence (54%) of the HLA DQB1*0301 allele in Caucasian patients with gastric cancer than in a control non-cancerous group (27%) [24]. If confirmed not to represent ethnic heterogeneity, this association may imply that this locus itself directly influences susceptibility to gastric cancer development or is a marker of linkage disequilibrium with a nearby cancer-predisposing locus. In another study, the frequency of allele DQB1*0401 was significantly higher in those infected with *H. pylori* who developed atrophic gastritis than those who were infected and did not develop atrophic gastritis or those not infected [25]. The potential role of the HLA locus in gastric tumorigenesis has implications regarding the importance of escape from immune surveillance as a causative factor for this disease. The fact that most people with these alleles do not develop gastric cancer illustrates the complexity of this multifactorial disease.

Of note, the blood group A phenotype was associated with gastric cancers in the 1950s [26, 27]. Interestingly, *H. pylori* was shown to adhere to the Lewis[b] blood group antigen, indicating a potentially important host factor that could facilitate this chronic infection and the subsequent risk of gastric cancer [28]. Additionally, small variant alleles of a mucin gene, *Muc1*, were found to be associated with gastric cancer patients compared to a blood donor control population [29]. Data continue to accumulate that reinforce these earlier findings. A recent study demonstrated that among *H. pylori*-infected patients, the risk of patients with a Lewis (a+b−) phenotype having gastric cancer was 3.15-fold higher than the risk of patients with a Lewis (a−b+) phenotype (p = 0.02, 95% confidence interval = 1.26–7.87) [30].

Several studies have noted variant gene and environmental interactions that potentially impact susceptibility to gastric cancer. Park et al. observed that interactions between CYP2E1 gene polymorphism and smoking have the potential to alter susceptibility to cancer development in the stomach [31]. Furthermore, polymorphisms of ADPRT and XRCC1 that reduced their interaction appeared to confer increased host susceptibility to gastric cancer, particularly in smokers [32].

Inherited susceptibility
Familial clustering

Most cases of gastric cancer appear to occur sporadically, without an obvious hereditary component. It is estimated that up to 8 to 10% of gastric cancer cases are familial [33]. Familial clustering has been observed in 12 to 25% of gastric carcinoma cases, with a dominant inheritance pattern observed [34, 35]. Notably, Napoleon Bonaparte apparently suffered from gastric cancer involving most of his stomach and other family members (i.e., his father and sister) may have been afflicted as well [36–38]. In the Swedish Family Cancer Database, the largest published database of familial gastric cancer to date, standardized incidence rate (SIR) was 1.7 (95% confidence interval = 1.08–1.92) when a patient presented with gastric adenocarcinoma. When a parent presented with gastric carcinoma, offspring showed an increased risk of the concordant carcinoma, SIR 1.59, only at ages older than 50 years. The increased risk from sibling gastric carcinoma probands (SIR of 5.75) was noted for those diagnosed before the age of 50 years. Taken together, these findings suggest that some of the familial risk factors are likely to be environmental, siblings being at a higher risk than offspring–parent pairs, consistent with the transmission patterns of *H. pylori* infection [39].

Case-control studies have observed consistent (up to three-fold) increases in risk for gastric cancer among relatives of gastric cancer patients [34, 40]. A population-based control study found an increased risk of developing gastric cancer among first-degree relatives of affected patients (OR = 1.7 with an affected parent, OR = 2.6 with an affected sibling), with the risk increasing (OR up to 8.5) if more than one first-degree relative was affected [41]. Interestingly, a higher risk was noted in individuals with an affected mother versus an affected father. Studies have shown a slight trend toward increased concordance of gastric cancers in monozygotic twins compared to dizygotic twins [42, 43]. A genomic analysis of 170 affected sibling-pairs from 142 Japanese families with gastric cancer yielded several chromosomal regions, with the strongest linkage at 2q33–35, harboring potential susceptibility genes [44].

Several genetic susceptibility traits with an inherited predisposition to gastric cancer development exist. Some are well characterized clinically with their underlying genetic alterations being unveiled and are described below.

Hereditary diffuse gastric cancer (HDGC)

Large families with an obvious autosomal dominant, highly penetrant inherited predisposition to the development of gastric cancer, having sufficient power with which to perform productive linkage studies, are rare. A large Maori kindred manifesting early-onset diffuse gastric cancers was investigated for this type of analysis, revealing linkage to the *E-cadherin/CDH1* locus on 16q22.1 and associated with mutations in this gene [45]. Since then, multiple germline *E-cadherin* mutations have been reported, scattered across the 16 exons this gene encompasses [45–50]. The ages of onset for diffuse gastric cancer in subjects harboring germline *E-cadherin* mutations ranged from 14 to 69 years. The incomplete penetrance of germline *E-cadherin* mutations was seen in several obligate carriers, who remained unaffected even in their eighth and ninth decades of life. Whether this incomplete penetrance is due to the stochastic nature of the second allele alteration or due perhaps to the presence of phenotype altering alleles at other genetic loci remains to be determined.

A study of 38 families diagnosed with HDGC trait having at least two members affected had an *E-cadherin* mutation identified in 40% of cases [50]. If the clinical criteria were more stringent to include one occurring before 50 years of age, then over half the cases had *E-cadherin* mutations detected. A founder (common ancestry) mutation observed in four Newfoundland families with penetrance of symptomatic gastric cancer by 75 years of age was 40% for men and 63% for females. Noteworthy, 12 of 18 asymptomatic gene carriers who underwent prophylactic gastrectomy had occult cancer, often seen as tiny clusters of signet ring cells within and underlying the mucosa. Only one of these individuals had a diffuse gastric cancer detected endoscopically with random biopsies prior to gastrectomy, indicating the ineffectiveness of this type of surveillance for these lesions. Indeed, no lesions were identified endoscopically or radiologically in six CDH1 gene carriers, yet all six had early-stage (T1) multifocal signet ring cell gastric cancers found on pathology exam after prophylactic gastrectomy [51]. A study of 25 "sporadic" diffuse gastric adenocarcinomas identified one case with a germline *E-cadherin* mutation and none in 14 intestinal-type gastric cancers [52]. No germline mutations of this gene were detected in apparent "sporadic"

diffuse gastric cancer cases with a mean age of 62 years in Great Britain [53].

Genetic counseling for a kindred manifesting a strong predisposition toward development of diffuse gastric cancers is imperative to ensure appropriate medical management (Figure 25.1). Once a gene carrier of an *E-cadherin* mutation reaches their mid to late 20s, prophylactic gastrectomy should be considered along with aggressive breast cancer surveillance at the age of 30 [46]. Diffuse gastric cancer families are one kindred subgroup to whom genetic testing can now be offered. It is noteworthy that two thirds of hereditary diffuse gastric cancer families reported to date have proven negative for *E-cadherin* gene mutation [49]. This finding suggests that other molecular alterations yet to be discovered underlie the majority of these cases.

Hereditary non-polyposis colorectal cancer (HNPCC)

A now well-characterized inherited predisposition syndrome that may include gastric cancer development is HNPCC [54]. Germline genetic abnormalities of DNA mismatch repair (MMR) genes underlying this disease entity have been discovered, and HNPCC is characterized by tumor development in a variety of tissue types [55]. Gastric carcinomas occurring in this setting were diagnosed at a mean age of 56 years of age, predominantly intestinal-type, lacking *H. pylori* infection, and exhibiting microsatellite instability in a Finnish HNPCC registry study [56].

Interestingly, the incidence of gastric cancers associated with HNPCC has decreased, similar to the recent general decline in incidence of gastric cancer in developed countries [57]. Renkonen-Sinisalo et al. studied gastric histopathology comparing 73 mutation-positive and 32 mutation-negative families for differences in *H. pylori*, atrophy, inflammation, IM, and dysplasia. They identified only a single case of duodenal cancer among mutation-positive individuals, but there was no evidence of gastric neoplastic lesions in either group [58]. The isolation and characterization of predisposing gene alterations should permit better definition of the proportion of gastric cancers resulting from this trait. Testing for MSH2 and MLH1 gene alterations is currently readily available and can be used to identify gastric cancer cases

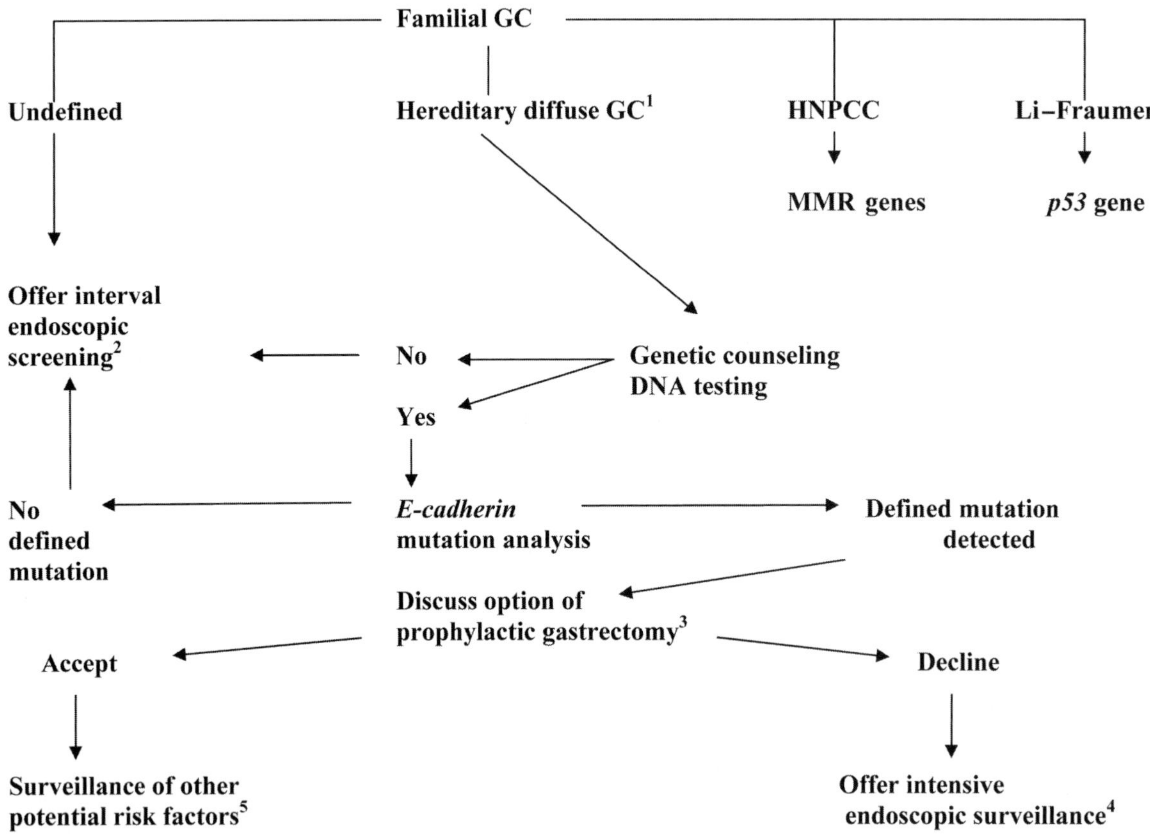

Figure 25.1 Algorithm for guidance in managing familial gastric cancer (GC) kindreds. Those identified to meet clinical criteria for hereditary non-polyposis colon cancer syndrome (HNPCC) and the Li–Fraumeni syndromes are counseled for consideration of genetic testing. MMR = mismatch repair. Microsatellite instability testing can aid in diagnosing HNPCC (1). This algorithm was initially formulated at the consensus symposia where clinical criteria were generated to define hereditary diffuse gastric cancer: two or more cases in first-/second-degree relatives, at least one diagnosed before the age of 50; or three or more cases of documented diffuse gastric cancer in first-/second-degree relatives (2) [48]. Consider individual age-dependent familial expression to determine the initiation, interval, and intensity of screening exams. Prophylactic gastrectomies have been performed in kindreds expressing highly penetrant phenotypes without prior knowledge of causative mutations (3). Consider age-dependent familial expression in determining management (4). Endoscopic ultrasound and chromographic (methylene blue or indigo carmine staining) endoscopy can be applied in attempts to increase the sensitivity of detecting early lesions in the stomach (5). One should have a high index of suspicion for the potential development of other cancers such as that of the breast, colon, and endometrium. (Figure reproduced from Powell, S.M. and Smith, M.F., Jr. (2002). Gastric cancer: molecular biology. In Kelsen, D.P. (ed.) *Gastrointestinal Oncology: Principles and Practice*. Philadelphia: Lippincott, Williams & Wilkins, pp. 325–340)

that are a manifestation of this specific cancer-predisposing genotype.

Li–Fraumeni and Peutz–Jeghers syndromes

Li–Fraumeni syndrome is a rare inherited cancer syndrome defined by a clustering of malignancies including sarcoma, breast cancer, brain tumor, leukemia, and adrenocortical carcinoma due to germline mutation of the tumor suppressor gene p53 [59]. Gastric cancers have also been included in this syndrome. An extended family affected by this syndrome demonstrated strong evidence of linkage to p53, and

three of four gastric tumors analyzed showed loss of the wild-type allele [60]. However, gastric carcinomas account for less than 4% of all neoplasms in this rare syndrome [61]. The Peutz–Jeghers syndrome (PJS) is a rare autosomal-dominant disorder characterized by germline mutations of STK11, hamartomatous polyposis, and pigmentation of the lips, buccal mucosa, and digits that carries an increased risk of many types of cancer, including those of the gastrointestinal tract [62, 63]. Overall, gastric carcinoma is still rare in this setting, and the exact contribution of the polyposis and underlying germline alterations of *LKB1/STK11* (PJS) to gastric cancer development is unclear. Only one missense change of uncertain functional significance in *LKB1/STK11* was noted among 28 sporadic gastric carcinomas [64].

Adenomatous polyposis syndromes

Gastric cancers have been noted to occur in patients with gastrointestinal polyposis disease entities such as familial adenomatous polyposis (FAP). Germline mutations of the APC gene are responsible for this disease [62, 63, 65]. Interestingly, an increased risk of gastric cancer associated with FAP has been reported in high-risk regions such as Asia, whereas no increased risk was exhibited in other populations [66]. Overall, gastric carcinoma is rare in this setting, and the exact contribution of the polyposis and underlying germline alterations of APC is again unclear. In the attenuated adenomatous polyposis syndrome (AFAP), subjects who do not possess germline APC gene mutations at the ends of the gene may have inherited allelic defects of the *MYH* base excision repair gene, which predispose to adenomas and cancers of the colon [67–69]. Recent studies have also linked this gene defect to 2.1% of sporadic cases of gastric cancer, and even to as many as 9.9% of patients with a familial pattern of gastric cancer [68, 69].

Other syndromes

Gastric cancers have been noted to occur in patients with other gastrointestinal polyposis disease entities, such as juvenile polyposis, who harbor germline mutations in SMAD4 or BMPR1A. Up to 24% of those with generalized juvenile gastrointestinal polyposis developed gastric cancer, which was similar to a subset of patients who presented with predominantly gastric polyposis and was found to have neoplastic tissue arising in 25% of resected gastric specimens [70].

Rare kindreds exhibiting site-specific gastric cancer predilection have been reported, occasionally associated with other inherited abnormalities [71, 72]. Familial intestinal gastric cancer not attributed to the above well-characterized inherited cancer susceptibility syndromes has been described as well; though the underlying defect(s) remain uncertain [73]. A constitutional deletion of 18p inherited from her mother, with somatic loss of the remaining long arm of this chromosome, was observed in the gastric carcinoma of a 14-year-old girl with associated mental and cardiac abnormalities, suggesting a predisposing condition [74]. A kindred manifesting the autosomal-dominant inheritance of familial gastric hyperplastic polyposis and gastric cancer development was demonstrated not to have a germline *E-cadherin* mutation, and on linkage analysis was not linked to 16q, the locus of *E-cadherin* [75, 76]. Thus it appears that there exist other loci, which when altered may predispose an individual to gastric cancer development. Indeed, patients with stomach cancer have been rarely found to harbor germline mutations in the ATM (ataxia-telangiectasia mutated) gene and the proto-oncogene MET [77, 78].

Molecular alterations

Molecular analyses of sporadic gastric carcinomas for acquired changes have described many somatic alterations, but the significance of these changes in gastric tumorigenesis remains to be established in most instances. Recently a study by Wright et al. demonstrated that the human gastric mucosa is composed of clonal units with several multipotent stem cells [79]. Analyzing mitochondrial DNA mutations that result in cytochrome c oxidase deficiency, these investigators discovered entirely mutated gastric units revealing monoclonal conversion and clonal derivation from a single stem cell. Moreover, division of these gastric units by fission was illustrated leading to clonal patches of mutated gastric units. Finally, intestinal metaplasia crypts exhibited clonality containing multiple pleuripotent stem cells that can spread by crypt fission. Taken together, these findings provide evidence for how genetic mutations spread clonally in human gastric mucosa.

Chromosomal instability

The majority of gastric cancers exhibit significant gross chromosomal aneuploidy. One study found that 72% of differentiated tumors and 43% of undifferentiated gastric tumors were aneuploid [80]. Variability in the classification of instability or histopathologic subtype and in the number of loci examined account for some of the variation in aneuploidy, with a trend toward more frequent occurrence in intestinal-type cancers at more advanced stages.

Cytogenetic studies of gastric adenocarcinomas are few in number and have failed to identify any consistent chromosomal abnormalities. A variable number of numerical or structural aberrations have been reported in gastric cancer cells, such as those involving chromosomes 3 (rearrangements), 6 (deletion distal to 6q21), 8 (trisomy), 11 (11p13–p15 aberrations), and 13 (monosomy and translocations) [75, 81]. DNA copy number changes in 22 primary gastric cancers and 3 gastric cancer cell lines from patients 45 years or younger were analyzed, revealing frequent DNA copy number increases at chromosomes 17q (52%), 19q (68%), and 20q (64%). Southern blotting probes mapped the chromosomal changes to the 19q12–13.2 region, suggesting cyclin E as one of the candidate target genes [82].

Comparative genomic hybridization (CGH) analyses of xenografted and primary gastric and gastroesophageal junctional adenocarcinomas have revealed several regions of consensus change in DNA copy number indicating the possible location of candidate gastric oncogenes and tumor suppressor genes involved in gastric tumorigenesis [83, 84]. A knowledge of these critical alterations is important, as gastric cancers of different histopathologic features have been shown to be associated with distinct patterns of genetic alterations, supporting the notion that these cancers evolve via distinct genetic pathways. Chromosomal arms 4q, 5q, 9p, 17p, and 18q exhibited frequent decreases in DNA copy number; whereas chromosomes 8q, 17q, and 20q showed frequent increases in DNA copy number [85]. These patterns may harbor prognostic information as well. CGH analysis showed an increased frequency of 20q gains and 18q losses in tumors that metastasized to lymph nodes [86]. As described below, several known or candidate tumor suppressor genes have been isolated within some of these frequently lost regions, but the precise targets of genetic loss providing gastric neoplastic cells with survival or growth advantages for clonal expansion remain to be clarified for many of these loci.

Comprehensive loss of heterozygosity (LOH) analysis in xenografted adenocarcinomas has identified 3p, 4p, 4q, 5p, 8q, 13p, 17p, and 18q to be frequently lost [87]. LOH analysis has identified several arms and regions of chromosomes that contain tumor suppressor genes important in gastric tumorigenesis, such as 17p (over 60% at *p53's* locus) [88], 18q (over 60% at the *SMAD4* and *DCC* loci) [89], and 5q (30 to 40% at or near the *APC* locus) [88, 90]. In LOH analysis of over a hundred archived stomach cancers, allelic loss was most frequently noted on chromosome 3p [91]. Moreover, three distinct regions of chromosome 4q were found to be frequently lost in gastroesophageal junctional adenocarcinomas,

indicating the potential for multiple tumor suppressor genes to be located on this chromosomal arm [92]. Allelic losses of both 17p and 18q in proximal or gastroesophageal junctional tumors were associated with a poorer survival than in those cases without or with only one allelic loss at these loci. The actual targets of genetic loss that provide gastric neoplastic cells with survival or growth advantages for clonal expansion remain to be clarified for many of these loci.

Microsatellite instability

Microsatellite instability (MSI) has been found in 13 to 44% of sporadic gastric carcinomas [93]. The degree of genome-wide instability also varies, with more significant instability (e.g., with microsatellite instability-high (MSI-H) tumors exhibiting instability in >33% of loci tested) occurring in only 10 to 16% of gastric cancers [94].

Alterations responsible for producing the MSI-H phenotype in a subset of sporadic gastric cancers have been elucidated. Abnormal loss of protein expression of either MLH1 or MSH2 was demonstrated in all cases exhibiting MSI-H [95]. Altered expression of MLH1 was associated with increased methylation of the promoter region of MLH1 in MSI-H cases suggesting a silencing role of hypermethylation [96, 97]. Distinct methylation of the promoter of hMLH1 was noted in five of eight MSI-H cases, whereas none of 43 microsatellite-low (MSI-L) or microsatellite stable (MSS) cases exhibited this methylation [98]. Within the MSI phenotype, some gastric cancers have various defects in HLA class I antigen subunits and APM (antigen processing machinery) and is associated with frequent HLA-A inactivation and frameshift mutations of the beta2m and APM genes [99].

MSI-H gastric tumors exhibit distinct clinicopathologic characteristics. Consistent associations of the MSI-H phenotype with intestinal subtype, distal location (e.g., antral), and more favorable prognosis have been observed [94, 100–104]. Additionally, some but not all studies have noted associations between the MSI-H phenotype and less frequent lymph node metastasis [94, 102, 103], greater depth of invasion [103], near-diploid DNA content [94], and tumoral lymphoid infiltration [94, 102, 103]. A possible explanation for the unique clinicopathologic phenotype observed in MSI-H gastric tumors may be the occurrence of mutations in a distinct set of cancer-related genes (see below) differing from those in tumors with no or low level MSI. Because MSI-H gastric carcinomas appear to be clinicopathologically distinct, it may ultimately prove valuable to have markers that identify this subgroup of gastric cancers, such as BAT-26 [95]. Several tumor suppressor genes have been shown to be critical targets of defective mismatch repair (MMR) in MSI-H tumors (see below). Moreover, a proapoptotic gene and additional MMR genes have been demonstrated to be altered in MSI gastric cancer cases. These same genes are observed to be infrequently mutated or altered in MSI-L or MSS tumors [102, 104, 105].

At least one important target of microsatellite instability appears to be the TGF-β type II receptor (*TGFβR2*) at a polyadenine tract within its gene [106]. Altered *TGFβR2* could additionally be found in gastric cancers not displaying microsatellite instability. Several gastric cancer cell lines resistant to the growth inhibitory and apoptotic effects of TGF-β were shown to have abnormal *TGFβR2* genes and/or transcripts [107]. Moreover, some gastric cancer cell lines and five of forty primary gastric cancers (12.5%) exhibited hypermethylation of the promoter region of TGF-beta type 1 receptor gene and decreased mRNA expression [108]. Another gene involved in this signaling pathway, *ACVR2*, was found to be similarly mutated, even in a biallelic fashion, in gastric tumors exhibiting microsatellite instability [109]. Thus alteration of TGF-β receptors and other members of this signaling path appears to be a critical event in the development of at least a subset of gastric cancers, allowing escape from the growth control signal of TGF-β.

Furthermore, somatic non-frameshift mutations have been reported in *BAX* [110]. Moreover, the relatively frequent missense mutations at codon 169 of *BAX* was shown to impair its proapoptotic activity [111]. Additional genes with simple tandem repeat sequences within their coding regions found to be specifically altered in gastric cancers displaying microsatellite instability include *IGFRII*, *hMSH3*, *hMSH6*, and *E2F-4*, which are known to be involved in the regulation of cell cycle progression and apoptotic signaling [110, 112–114].

Acquired somatic alterations

TFF1 loss

Loss of the trefoil peptide TFF1(pS2), a stable three-loop molecule synthesized in mucus-secreting cells, has been described in approximately 50% of gastric

carcinomas [115–118]. The biological significance of this loss was reported in a knockout mouse model of gastric antral neoplasia [119]. Additionally, expression of TFF1 was observed to be lower in some gastric intestinal metaplasia and gastric adenomatous lesions compared to adjacent normal or hyperplastic mucosa [120, 121]. *TFF1* resides on chromosome 21q22, a region noted to be deleted in some gastric cancers in LOH studies [122, 123]. Overexpression of TFF1 in the gastric cancer cell line, AGS, inhibited its growth [124]. Furthermore, the overwhelming majority of gastric cancers studied had absent to minimal transcript levels compared to normal gastric mucosa [125]. Moreover, C/EBP-beta was found to be overexpressed in the majority of gastric cancers in a corresponding manner and binds to the promoter of TFF1, suggesting a regulatory factor role [126, 127]. Cytokine signaling through the co-receptor gp130 with increased STAT-3 expression has also been observed to decrease TFF1 and lead to gastric lesions in mice [128, 129].

E-cadherin alteration

In addition to germline mutations leading to HDGC described above, several sporadic diffuse gastric cancers have displayed altered *E-cadherin*. E-cadherin is a transmembrane, calcium ion-dependent adhesion molecule important in epithelial cell homotypic interactions, which when decreased in expression is associated with invasive properties [130]. Reduced *E-cadherin* expression determined by immunohistochemical analysis was noted in the majority of gastric carcinomas, 92% of 60 cases, when compared to their adjacent normal tissue [131]. Genetic abnormalities of the *E-cadherin* gene and transcripts were demonstrated in a third to one half of diffuse gastric cancers [52, 132]. *E-cadherin* splice site alterations producing exon deletion and skipping, large deletions including allelic loss, and point mutations mostly of the missense nature have been demonstrated in diffuse-type cancers, some even exhibiting alterations in both alleles [133]. Methylation of the *E-cadherin* promoter region was found in 16 (26%) of 61 gastric cancers studied [98]. Additionally, alpha-catenin, which binds to the intracellular domain of *E-cadherin* and links it to actin-based cytoskeletal elements, was noted to have reduced immunohistochemical expression in 70% of 60 gastric carcinomas and correlated with infiltrated growth and poor differentiation [134].

Finally, two gene variants of the potential *E-cadherin* regulator IQGAP1 were found in two diffuse gastric cancers, but not in normal controls [135].

p53 mutations

The *p53* gene has consistently been demonstrated to be altered in gastric adenocarcinomas. Allelic loss occurs in over 60% of cases and mutations are identified in approximately 30 to 50% of cases depending on the mutational screening method employed and variable sample sizes [136]. Some mutations of *p53* have even been identified in early dysplastic and apparent intestinal metaplasia gastric lesions; however, most alterations occurred in the advanced stages of neoplasia. The spectrum of mutations in this gene within gastric tumors is not unusual with a predominance of base transitions, especially at CpG dinucleotides. Immunohistochemical analysis of tumors in an effort to detect excessive expression of p53 as an indirect means to identify mutations of this gene has been studied; but this assay does not appear to have consistent prognostic value in patients with gastric cancers [137, 138].

Kinases/phosphatases

Activation of protein kinases by tyrosine kinases and phosphatases has been shown to affect many cellular activities including cell growth, differentiation, and survival [139]. Phosphatidylinositol 3-kinases (PI3K) are lipid kinases that regulate pathways for neoplastic proliferation, adhesion, survival, and motility [140]. Mutations in *PIC3A*, which encodes a catalytic subunit of PI3K, have been characterized in several cancers including gliobastoma, colon, and up to 25% of gastric cancers analyzed [141]. Further attempts at identifying mutations in *PIC3A* have reported a lower prevalence of 4.3% in gastric cancers; however, also observed was increased expression of this gene product in gastric tumors and an expression profile associated with this upregulation in these tumors [142]. Moreover, increased PI3K/AKT signaling from NF-kappaB has been shown to activate transcription of HuR, which has proliferative and anti-apoptotic effects on gastric cancer cells [143]. Furthermore, a kinase involved in chromosome segregation and stability, STK15 (BTAK, Aurora2), was found to be amplified and overexpressed in gastric cancers [144].

Wang et al. identified 83 somatic mutations in human protein tyrosine phosphatases (PTPs), the majority affecting colon cancer, but also found in 17% of gastric cancers analyzed. The most commonly altered PTP, protine-tyrosinase phospatase-receptor type (PTPRT), was found to decrease its activity, suggesting a role as a tumor suppressor gene [145]. However, an additional PCR-based study suggests a much smaller role for alterations in PTPRT in gastric tumorigenesis, being found in only 1% of cases [146].

FHIT alterations

Evidence of tumor suppressor loci on chromosome 3p has accumulated from a variety of studies including allelic loss in primary gastric tumors (46%) and homozygous deletion in a gastric cancer cell line (KATO III) as well as xenografted tumors [147]. The *FHIT* gene was isolated from the common fragile site FRA3B region at 3p14.2 and found to have abnormal transcripts with deleted exons in five of nine gastric cancers [148]. Furthermore, loss of FHIT protein expression was demonstrated immunohistochemically in the majority of gastric carcinomas [149]. One somatic missense mutation was identified in exon 6 of the *FHIT* gene during a coding region analysis of 40 gastric carcinomas [150]. More studies are needed to identify the critically altered targets on this chromosome, clarify the role *FHIT* plays in gastric tumorigenesis, and determine the role that breakpoints in this region of 3p have in gastric cancer development.

c-MET overexpression

Overexpression of the *MET* gene, which encodes a tyrosine kinase receptor for the hepatocyte growth factor, has been reported to have prognostic value to indicate poorer survival in multivariate analysis [151]. There have been numerous reports in the literature indicating that the *MET* gene is amplified in approximately 15% and its expression elevated in up to 50% of gastric tumors [152–156]. Interestingly, *MET* amplification in colon carcinomas rarely occurs. This evidence, along with the infrequent mutations in the *ras* proto-oncogene found in gastric cancer [157, 158], provides two very good indications that the mechanisms involved in the development of colon and gastric cancers are distinct.

Overexpression of the *MET* gene by itself may not be sufficient for cancer progression as interaction of the precancerous cells with stromal fibroblast-derived growth factors, including HGF, may promote the progression of disease. A hypothesis put forth by Tahara proposes that stromal fibroblasts that have been stimulated by IL-1 or TGF-β secrete HGF, which can then bind to the *MET* protein on cancer cells and promote the morphogenesis and progression of the cancer [159]. In the absence of (or at low levels of) E-cadherin and catenins, HGF may cause the scattering of cancer cells and the development of the poorly differentiated phenotype. Thus interactions between the stromal and epithelial layers of the stomach may have profound effects on the development of disease. Inoue et al. demonstrated that TGF-β and HGF produced by a culture of gastric fibroblasts could stimulate the invasiveness of a human scirrhous-type gastric cancer cell line *in vitro* [160]. It is tempting to speculate that this may be one of the mechanisms through which *H. pylori* could act to promote the development of gastric cancers. Infection with *H. pylori* has in fact been shown to result in an increase in gastric mucosal HGF levels and an increase in HGF gene expression [161–163].

Methylation silencing alterations

No inactivating somatic mutations of $p16^{INK4}$ were detected in over 70 cases of gastric carcinoma screened by PCR-SSCP analysis [164]. On the other hand, $p16^{INK4}$ somatic mutations were noted in several esophageal adenocarcinomas with LOH and others with loss of p16 expression were found to have abnormal hypermethylation of the $p16^{INK4}$ promoter, suggesting this epigenetic gene expression silencing may play a role in esophageal tumorigenesis [165–167]. A significant number (41%) of gastric cancers exhibited CpG island methylation in a study of p16's promoter region [98]. Many of these cases with hypermethylation of promoter regions displayed the MSI-H phenotype and multiple sites of methylation including the *MLH1* promoter region [168].

RUNX3, a tumor suppressor gene, apparently suppresses gastric epithelial cell growth by inducing p21 (WAF1/Cip1) expression in cooperation with transforming growth factor beta-activated SMAD [169]. RUNX3 was found to be altered in 82% of gastric cancers through either gene silencing or protein mislocalization to the cytoplasm [170]. Significant downregulation of RUNX3 through methylation on the promoter region was observed in primary tumors (75%) as well as in all clinical peritoneal metastasis of gastric cancers (100%) compared with normal gastric mucosa. Silencing of RUNX3 affects expression of important genes such as p21 and others involved in aspects of metastasis including cell adhesion, proliferation, and apoptosis. They potentially promote the peritoneal metastasis of gastric cancer [171].

Multiple tumor suppressor genes and candidates including protocadherin 10 (PCDH10), CDKN2A, GSTP1, APC, MGMT, DAKP, XAF1, THBS-1, RUNX1, and CDH1 have been shown to be methylated in gastric cancers and may provide potential molecular markers for these cancers [172–174].

Growth factor alterations

Tumor and stromal cell interaction have been suggested for several other growth factor signaling systems. Approximately 15% of gastric cancers have been observed to express both EGF and EGFR, suggesting the presence of an autocrine mechanism of growth stimulation in these cases [175]. Some studies suggest a worse prognosis for tumors with positive expression of EGF, EGFR, or both. Involvement of several other growth factor signaling paths have been implicated in gastric tumorigenesis including TGF-alpha, interleukin-1, criptor, amphiregulin, platelet-derived growth factor, and K-sam [176]. Expression of bFGF and FGFR have been found in 70 and 60% of gastric cancers, respectively, and noted to be frequently more undifferentiated and in a more advanced invasive stage [177]. Rare amplification of *MYC* and *KSAM* has also been reported in gastric cancers [154]. Overexpression of ERBB2 (HER-2/neu), a transmembrane tyrosine kinase receptor proto-oncogene, has been implicated as a potential marker of poor prognosis [178–180]. For ERBB2, polysomy was a more frequent mechanism than amplification in both esophageal (32.5 vs. 7.5%) and gastric (15 vs. 5%) cancers [181]. ERBB2 has even been detected immunohistochemically in the serum of gastric cancer patients [182]. In 58 gastric adenocarcinomas with lymph node metastasis, one ERBB2 mutation was detected in the lymph node metastasis, but not in the primary tumor of the same patient suggesting a role in metastasis [181]. Other prior studies, however, have not found any prognostic value of ERBB expression in these cancers [183, 184].

Apoptosis signaling alterations

Alterations in the mechanism leading to cell death or apoptosis can allow aberrant cell growth and immortality, playing a crucial role in tumorigenesis. Several mutations in genes usually promoting or regulating apoptosis have been identified. Mutations of the BAX gene, believed to be a promoter of apoptosis, have been identified in up to 33% of gastric cancers as well as colorectal and endometrial cancers [185]. Another study supporting the role of the BAX gene in tumor suppression linked low expression levels of the protein Bax-interacting factor-1 (Bif-1) in gastric cancers, suggesting another breakdown in the pathway of regulating apoptosis [186]. The bcl-2 homolog BAK, another promoter of apoptosis, was found to have missense mutations in 12.5% of gastric cancers analyzed [187]. Defects in cell surface receptors of the apoptosis pathway have also been observed in gastric tumors such as in the death receptors DR4 and DR5. DR4 encodes a tumor necrosis factor apoptosis receptor gene and prior studies had identified a single nucleotide polymorphism present in non-gastric malignancies; however, this polymorphism failed to show increased risk for developing gastric cancer [188]. Mutations in DR5 appear to have a strong role in the development of gastric cancers as up to 7% of analyzed gastric cancers were found to harbor missense mutations of this gene with transfection studies even showing inhibition of apoptotic cell death in all these mutants [189].

Angiogenesis signaling alterations

Thymidine phosphorylase has been noted to correlate with angiogenesis in gastric cancers [190]. Abnormalities of vascular endothelial growth factor (VEGF) have been associated with gastric cancers, and may have a role in promoting lymphangiogenesis and predisposition to metastasis. Reverse transcription PCR was used to demonstrate VEGF expression in 27.8% of gastric cancers and was strongly associated with tumors with high lymphatic vessel density [191]. Several of the VEGF gene polymorphisms were found to be an independent prognostic marker for patients with gastric cancer and the analysis of VEGF gene polymorphisms may help identify patients at greatest risk for poor outcomes [192].

Other alterations

Over expression of DARP32 and a novel truncated isoform were found in the majority of gastric cancers [193]. Moreover, an anti-apoptotic effect of this over-expression of DARP32 and t-DARP was observed *in*

vitro [194]. Only one of 35 gastric cancers contained an intragenic mutation of *SMAD4* along with allelic loss suggesting this gene is infrequently altered in gastric tumorigenesis [169, 195]. Additionally, only one missense change of uncertain functional significance in the *LKB1* (STK11) gene was noted in a study of 28 sporadic gastric carcinomas [64].

Several members of the Wnt signaling pathway, *APC* and *β-Catenin*, have had several somatic alterations noted in a few cases of gastric cancer. Somatic *APC* mutations, mostly missense in nature and of relatively low frequency, have been reported in Japanese patients' gastric adenocarcinomas and adenomas on partial screening [196]. On the other hand, several other reports including Japanese patients have not identified significant *APC* mutations in gastric carcinomas on similar partial screening analysis of the commonly mutated region including direct nucleotide sequencing [197–199]. Significant allelic loss at *APC's* loci suggests the existence of a tumor suppressor gene important in gastric tumorigenesis nearby. Indeed, the interferon regulatory factor-1 gene has been mapped to this commonly deleted region in gastric cancers [200].

Missense somatic mutations in *β-catenin*, which also has connections with the cell adhesion complex involving E-cadherin as well as Lcf/LEF transcription regulation, were identified in a few cases of intestinal-type gastric cancer [201]. However, the functional consequences of these changes remain to be determined as the four commonly mutated serines or threonines, which are known to be involved in regulatory phosphorylation of this signaling pathway, were not affected. No *β-catenin* mutations were observed in diffuse-type gastric cancers [202]. Decreased normal membranous expression of β-catenin was found to be associated with poor differentiation and shorter patient survival [203]. Of note, the gastric cancer cell line HSC-39, established from a patient with a signet ring cell carcinoma, contains a 321-base pair in-frame deletion in the NH_2 terminal region including the GSK3-beta phosphorylation and alpha-catenin binding site [204].

Evidence that cox-2 participates in the proliferation of gastric cancer cells (MKN45) was suggested by Tsuji et al. who also later showed that a role for cox-2 in cancer progression is clearly not limited to tumors of the gastric epithelium [205]. *In vitro* studies suggest that hedgehog signaling contributes to gastric cancer cell growth. Elevated expressions of

Table 25.1 Somatic alterations in human gastric adenocarcinomas.

Frequent	Specifically in MSI-H cases	Infrequent
p53 inactivation	Loss of hMLH1 or hMSH2 protein	APC
TFF1 loss and C/EBP-beta increase	TGF-βIIR	Smad4
FHIT loss	Bax	LKB1
DARP32/t-DARP/c-MET increase	IGFIIR	
HER-2/neu increase	E2F-4	
EGFR/K-sam increase	hMSH3	
E-cadherin loss (diffuse-type)	hMSH6	
p16/p27/RUNX3/PAI-1 loss		
VEGF/MMP-9/Cox-2 increase		

hedgehog target genes, human patched gene 1 (PTCH1), or Glil occurs in 63 of 99 primary gastric cancers. Activation of the hedgehog pathway is associated with poorly differentiated and more aggressive tumors. Treatment of gastric cancer cells with KAAD-cyclopamine, a hedgehog signaling inhibitor, decreases expressions of Glil and PTCH1, resulting in cell growth inhibition and apoptosis [206].

Additionally, increased plasminogin activation has been reported in several gastric tumors [207]. A somatic mitochondrial deletion of 50 bp was even demonstrated in four gastric adenocarcinomas [208]. Telomerase activity, which appears necessary for cell immortality, has been detected by a PCR-based assay frequently in the late stages of gastric tumors and observed to be associated with a poor prognosis [209]. Furthermore, Tahara et al. have identified telomerase activity in the vast majority of primary gastric tumor tissues while no activity was observed in the corresponding normal gastric mucosa [159]. Gastric lavage fluid may provide a window to early diagnosis, as in one study telomerase was detected in 20 patients (80%) with gastric cancer, 7 patients (28%) with peptic ulcer, and none in normal subjects (P < 0.001) [210]. Specific alterations and the true prevalence of significant changes in these genes, gene

products, or phenotypes in human gastric tumors remains to be characterized. Table 25.1 lists alterations that appear consistently in human gastric carcinomas; others will undoubtedly be added as studies accumulate.

Molecular profiling: prognostic and diagnostic markers

High-throughput assays such as microarrays that comprehensively determine gene expression or DNA copy number patterns are being explored in various diseases including gastric cancer. Potential biomarkers such as phospholipase A2, signature profiles of gene expression, or copy number alterations identified with these analyses may provide some prognostic as well as diagnostic information about gastric cancer [211–215]. Several microRNAs have been observed to be expressed aberrantly in stomach cancers as well that affect important signaling pathways such as TGF-beta [216].

Comprehensive serial analysis of gene expression (SAGE) has also identified novel genetic alterations including overexpression of calcium-binding proteins [217]. A proteomic analysis identified 9 proteins with increased expression and 13 proteins with decreased expression in gastric cancers, which included those involved in mitotic checkpoints such as MAD1L1 and EB1 as well as others such as HSP27, CYR61, and CLPP [214].

Elevated tissue levels of certain proteins have been shown to be associated with increased invasiveness of tumor cells, and some even associated with poorer survival including: urokinase-type plasminogen activator (uPA), vascular endothelial growth factor (VEGF), MET, MYC, tie-1 protein tyrosine kinase, CD44v6, PDGF-A, TGF-beta 1, and cyclin D2 [207, 218–223]. On the other hand, decreased levels of other proteins have been noted to be associated with more invasiveness and poorer survival including: loss of p27 (Kip1), p21 (CIP1), plasminogen activator inhibitor type-1 (PAI-1), and tissue-type plasminogen activator (tPA) [207, 224–227]. In a recent study by Zafirellis et al. the expression of p53 and bcl-2 proteins had no significant impact on the outcome of patients with gastric cancer [228]. The combined analysis of p53 and p27 (kip1) was of added prognostic value, but alone these findings were not compelling as a specific marker in gastric carcinomas [229]. Furthermore, patients with a high KLK6 expression had a significantly poorer survival rate than those with a low KLK6 expression (P = 0.03) [230]. Nishigaki et al. identified 9 proteins with increased expression and 13 proteins with decreased expression in gastric carcinomas, notable including mitotic checkpoint (MAD1L1 and EB1) and mitochondrial functions (CLPP, COX5A, and ECH1) [231]. Thus some of these molecules could potentially provide diagnostic and/or prognostic markers that can be evaluated in gastric cancer tissues for clinical utilities.

Chemosensitization markers

Measuring molecular markers in tumor cells may be able to predict how they will respond to specific treatment agents. The feasibility of this kind of testing was demonstrated in a small study of 30 patients. Overexpression of p53 by immunohistochemistry of locally advanced gastric cancers was found to indicate a lower response rate to neoadjuvant cytotoxic therapy [232]. Furthermore, the expression of thymidine phosphorylase or even thymidine synthase in gastric tumors may help predict their response to flourouricil agents [233]. In another small study of 23 patients with gastric cancer, those tumors with positive staining for BAX and Bcl-2 were more chemoresistant and had worse prognosis than those negative for Bcl-2 staining [234, 235]. Moreover, a MTT chemosensitivity assay of gastric tumors predicted responses to certain agents and prolonged survival when used [236].

Serum markers

Additional diagnostic and prognostic markers for gastric cancer patients have been sought in bone marrow and peripheral blood samples. Concentrations of matrix metalloproteinase 9 were higher in the plasma of gastric cancer patients compared to normal controls [237]. High concentrations of tissue inhibitor of metalloproteinase (TIMP)-1, interleukin-10, HGF, soluble receptor for interleukin-2, and soluble fragment of E-cadherin in the serum or plasma

of gastric carcinoma patients have been observed to be associated with more invasiveness and poorer survival [233, 238–241]. Furthermore, detection of cytokeratin 20, CD44 variants, CA-125, CEA, CA 19-9, CA 72-4, and anti-p53 antibodies in the peripheral blood of gastric cancer patients tend to indicate disseminated disease [242–246]. Elevated plasma levels of osteopontin (OPN) were an independent risk factor of poor survival [247]. A proteomic evaluation of serum from gastric cancer patients before and after surgical treatment and compared to controls suggested a panel of five protein markers had diagnostic potential with a sensitivity of 83% and specificity of 95% [248].

Potential targets for treatment and intervention

The advancement in our molecular understanding of tumorigenesis can result in specific targeted chemotherapy with dramatic effects in controlling tumor growth and survival. Indeed, the use of imatinib mesylate to inhibit the overexpressed cKIT tyrosin kinase in a relatively specific fashion in GIST tumors is a prototype example of this kind of specific targeted treatment with significant results [249, 250]. For gastric carcinomas, some potential candidate molecular targets found to alter gastric cellular growth when altered *in vitro* have been identified. A novel KGF-R2 phosphorylation inhibitor, Ki23057, decreased proliferation of two scirrhous gastric cancer cell lines with K-sam amplification [251]. Additionally, cyclopamine, a hedgehog (Hh) inhibitor, decreased the growth of gastric cancer cell lines that expressed high levels of SMO [252]. Also noteworthy, non-steroidal anti-inflammatory drugs have been shown to be associated with decreased incidence of gastric cancers [253, 254]. The mechanism of this effect remains to be elucidated so that more specific targeted agents can be developed. Thus the dismal prognosis of diagnosing disseminated gastric cancer may soon have new avenues of treatment.

References

1. Correa P, Chen V, eds. 1994. *Trends in Cancer Incidence and Mortality: Gastric Cancer*. New York: Cold Spring Harbor Laboratory Press.

2. Crew KD, Neugut AI. 2006. Epidemiology of gastric cancer. *World J Gastroenterol* 12: 354–62.

3. Marugame T, Dongmei Q. 2007. Comparison of time trends in stomach cancer incidence (1973–1997) in East Asia, Europe and USA, from Cancer Incidence in Five Continents Vol. IV-VIII. *Jpn J Clin Oncol* 37:242–3.

4. Parker SL, Tong T, Bolden S, Wingo PA. 1996. Cancer statistics, 1996. *CA Cancer J Clin* 46:5–28.

5. Correa P, Houghton J. 2007. Carcinogenesis of *Helicobacter pylori*. *Gastroenterology* 133: 659–72.

6. Uemura N, Okamoto S, Yamamoto S, et al. 2001. *Helicobacter pylori* infection and the development of gastric cancer. *N Engl J Med* 345:784–9.

7. Fukase K, Kato M, Kikuchi S, et al. 2008. Effect of eradication of *Helicobacter pylori* on incidence of metachronous gastric carcinoma after endoscopic resection of early gastric cancer: an open-label, randomised controlled trial. *Lancet* 372:392–7.

8. Elsborg L, Mosbech J. 1979. Pernicious anaemia as a risk factor in gastric cancer. *Acta Medica Scandinavica* 206:315–8.

9. Joossens JV, Hill MJ, Elliott P, et al. 1996. Dietary salt, nitrate and stomach cancer mortality in 24 countries. European Cancer Prevention (ECP) and the INTERSALT Cooperative Research Group. *Int J Epidemiol* 25:494–504.

10. Rowlands DC, Ito M, Mangham DC, et al. 1993. Epstein–Barr virus and carcinomas: rare association of the virus with gastric adenocarcinomas. *Br J Cancer* 68:1014–19.

11. Lauren P. 1965. The two histological main types of gastric carcinoma: diffuse and so-called intestinal-type carcinoma. *Acta Pathol Microbiol Scand* 64:31–49.

12. Correa P, Shiao YH. 1994. Phenotypic and genotypic events in gastric carcinogenesis. *Cancer Res* 54:1941–3.

13. de Vries AC, van Grieken NC, Looman CW, et al. 2008. Gastric cancer risk in patients with premalignant gastric lesions: a nationwide cohort study in the Netherlands. *Gastroenterology* 134:945–52.

14. Oota K, Sobin LH. 1977. *Histological Typing Gastric and Esophageal Carcinogenesis.* Geneva: World Health Organization.

15. Ming S-C. 1977. Gastric carcinoma: a pathobiological classification. *Cancer* 39:2475–85.

16. Borrmann R. 1926. Gushwulste de Magens and Duodenums. In Henke F, Lubarsh O, eds. *Handbuch der Speziellen Pathologischen Anatomie und Histologie.* Berlin: Springer, p. 865.

17. Dixon MF, Martin JG, Sue-Ling HM, et al. 1994. Goseki grading in gastric cancer: comparison with existing systems of grading and its reproducibility. *Histopathology* 25:309–16.

18. Kennedy BJ. 1987. The unified international gastric cancer staging classification. *Gastroenterology* 1–13.

19. Hirota T, Ming SC, Itabashi M. 1993. Pathology of early gastric cancer. In Nishi M, Ichikawa H, Nakajima T, Maruyama K, Tahara E, eds. *Gastric Cancer.* Tokyo, New York: Springer-Verlag; pp. 66–86.

20. El-Omar EM, Carrington M, Chow WH, et al. 2000. Interleukin-1 polymorphisms associated with increased risk of gastric cancer. *Nature* 404: 398–402.

21. El-Omar EM, Rabkin CS, Gammon MD, et al. 2003. Increased risk of noncardia gastric cancer associated with proinflammatory cytokine gene polymorphisms. *Gastroenterology* 124:1193–201.

22. Machado JC, Figueiredo C, Canedo P, et al. 2003. A proinflammatory genetic profile increases the risk for chronic atrophic gastritis and gastric carcinoma. *Gastroenterology* 125:364–71.

23. Hold GL, Rabkin CS, Chow WH, et al. 2007. A functional polymorphism of toll-like receptor 4 gene increases risk of gastric carcinoma and its precursors. *Gastroenterology* 132:905–12.

24. Lee JE, Lowy AM, Thompson WA, et al. 1996. Association of gastric adenocarcinoma with the HLA class II gene DQB10301. *Gastroenterology* 111:426–32.

25. Sakai T, Aoyama N, Satonaka K, et al. 1999. HLA-DQB1 locus and the development of atrophic gastritis with *Helicobacter pylori* infection. *J Gastroenterol* 34 Suppl 11:24–7.

26. Aird I, Bentall HH, Roberts JA. 1953. A relationship between cancer of stomach and the ABO blood groups. *BMJ* 1:799–801.

27. Buckwalter JA, Wohlwend CB, Colter DC, Tidrick RT, Knowler LA. 1957. The association of the ABO blood groups to gastric carcinoma. *Surg Gynecol Obstet* 104:176–9.

28. Boren T, Falk P, Roth KA, Larson G, Normark S. 1993. Attachment of *Helicobacter pylori* to human gastric epithelium mediated by blood group antigens. *Science* 262:1892–5.

29. Carvalho F, Seruca R, David L, et al. 1997. MUC1 gene polymorphism and gastric cancer – an epidemiological study. *Glycoconj J* 14:107–11.

30. Sheu MJ, Yang HB, Sheu BS, et al. 2006. Erythrocyte Lewis (A+B–) host phenotype is a factor with familial clustering for increased risk of *Helicobacter pylori*-related non-cardiac gastric cancer. *J Gastroenterol Hepatol* 21:1054–8.

31. Park GT, Lee OY, Kwon SJ, et al. 2003. Analysis of CYP2E1 polymorphism for the determination of genetic susceptibility to gastric cancer in Koreans. *J Gastroenterol Hepatol* 18:1257–63.

32. Miao X, Zhang X, Zhang L, et al. 2006. Adenosine diphosphate ribosyl transferase and x-ray repair cross-complementing

1 polymorphisms in gastric cardia cancer. *Gastroenterology* 131: 420–7.

33. La Vecchia C, Negri E, Franceschi S, Gentile A. 1992. Family history and the risk of stomach and colorectal cancer. *Cancer* 70: 50–5.

34. Videbaek A, Mosbech J. 1954. The aetiology of gastric carcinoma elucidated by a study of 302 pedigrees. *Acta Medica Scandinavica* 149:137–59.

35. Goldgar DE, Easton DF, Cannon-Albright LA, Skolnick MH. 1994. Systematic population-based assessment of cancer risk in first-degree relatives of cancer probands. *J Natl Cancer Inst* 86:1600–8.

36. Antommarchi F. 1825. *Les derniers moments de Napoleon, en compement du memorial de Sainte-Helene*, 1st edn. Brusselles, Belgium: H Tarlier.

37. Kubba AK, Young M. 1999. The Napoleonic cancer gene? *J Med Biogr* 7:175–81.

38. Lugli A, Zlobec I, Singer G, et al. 2007. Napoleon Bonaparte's gastric cancer: a clinicopathologic approach to staging, pathogenesis, and etiology. *Nat Clin Pract Gastroenterol Hepatol* 4:52–7.

39. Hemminki K, Jiang Y. 2002. Familial and second gastric carcinomas: a nationwide epidemiologic study from Sweden. *Cancer* 94:1157–65.

40. Zanghieri G, Di Gregorio C, Sacchetti C, et al. 1990. Familial occurrence of gastric cancer in the 2-year experience of a population-based registry. *Cancer* 66: 2047–51.

41. Palli D, Galli M, Caporaso NE, et al. 1994. Family history and risk of stomach cancer in Italy. *Cancer Epidemiol Biomarkers Prev* 3: 15–18.

42. Gorer PA. 1938. Genetic interpretation of studies on cancer in twins. *Annals Eugenics* 8:219.

43. Lee FI. 1971. Carcinoma of the gastric antrum in identical twins. *Postgrad Med J* 47:622–4.

44. Aoki M, Yamamoto K, Noshiro H, et al. 2005. A full genome scan for gastric cancer. *J Med Genet* 42:83–7.

45. Guilford P, Hopkins J, Harraway J, et al. 1998. E-cadherin germline mutations in familial gastric cancer. *Nature* 392:402–5.

46. Blair V, Martin I, Shaw D, et al. 2006. Hereditary diffuse gastric cancer: diagnosis and management. *Clin Gastroenterol Hepatol* 4:262–75.

47. Caldas C, Carneiro F, Lynch HT, et al. 1999. Familial gastric cancer: overview and guidelines for management. *J Med Genet* 36:873–80.

48. Brooks-Wilson AR, Kaurah P, Suriano G, et al. 2004. Germline E-cadherin mutations in hereditary diffuse gastric cancer: assessment of 42 new families and review of genetic screening criteria. *J Med Genet* 41:508–17.

49. Lynch HT, Grady W, Suriano G, Huntsman D. 2005. Gastric cancer: new genetic developments. *J Surg Oncol* 90:114–33.

50. Kaurah P, MacMillan A, Boyd N, et al. 2007. Founder and recurrent CDH1 mutations in families with hereditary diffuse gastric cancer. *JAMA* 297:2360–72.

51. Norton JA, Ham CM, Van Dam J, et al. 2007. CDH1 truncating mutations in the E-cadherin gene: an indication for total gastrectomy to treat hereditary diffuse gastric cancer. *Ann Surg* 245:873–9.

52. Ascano JJ, Frierson H, Jr., Moskaluk CA, et al. 2001. Inactivation of the E-cadherin gene in sporadic diffuse-type gastric cancer. *Mod Pathol* 14:942–9.

53. Stone J, Bevan S, Cunningham D, et al. 1999. Low frequency of germline E-cadherin mutations in familial and nonfamilial gastric cancer. *Br J Cancer* 79:1935–7.

54. Lynch HT, Smyrk TC, Watson P, et al. 1993. Genetics, natural history, tumor spectrum, and pathology of hereditary nonpolyposis colorectal cancer: an updated review. *Gastroenterology* 104:1535–49.

55. Kinzler KW, Vogelstein B. 1996. Lessons from hereditary colorectal cancer. *Cell* 87:159–70.

56. Aarnio M, Salovaara R, Aaltonen LA, Mecklin JP, Jarvinen HJ. 1997. Features of gastric cancer in hereditary non-polyposis colorectal cancer syndrome. *Int J Cancer* 74:551–5.

57. Lynch HT, Krush AJ. 1971. Cancer family "G" revisited: 1895–1970. *Cancer* 27:1505–11.

58. Renkonen-Sinisalo L, Sipponen P, Aarnio M, et al. 2002. No support for endoscopic surveillance for gastric cancer in hereditary non-polyposis colorectal cancer. *Scand J Gastroenterol* 37:574–7.

59. Li FP, Fraumeni JF, Jr. 1969. Soft-tissue sarcomas, breast cancer, and other neoplasms. A familial syndrome? *Ann Intern Med* 71:747–52.

60. Varley JM, McGown G, Thorncroft M, et al. 1995. An extended Li–Fraumeni kindred with gastric carcinoma and a codon 175 mutation in TP53. *J Med Genet* 32:942–5.

61. Kleihues P, Schauble B, zur Hausen A, Esteve J, Ohgaki H. 1997. Tumors associated with p53 germline mutations: a synopsis of 91 families. *Am J Pathol* 150:1–13.

62. Lindor NM, Greene MH. 1998. The concise handbook of family cancer syndromes. Mayo Familial Cancer Program. *J Natl Cancer Inst* 90:1039–71.

63. Hofgartner WT, Thorp M, Ramus MW, et al. 1999. Gastric adenocarcinoma associated with fundic gland polyps in a patient with attenuated familial

adenomatous polyposis. *Am J Gastroenterol* 94:2275–81.

64. Park WS, Moon YW, Yang YM, et al. 1998. Mutations of the STK11 gene in sporadic gastric carcinoma. *Int J Oncol* 13:601–4.

65. Groden J, Thliveris A, Samowitz W, et al. 1991. Identification and characterization of the familial adenomatous polyposis coli gene. *Cell* 66:589–600.

66. Offerhaus GJ, Giardiello FM, Krush AJ, et al. 1992. The risk of upper gastrointestinal cancer in familial adenomatous polyposis. *Gastroenterology* 102:1980–2.

67. Jones S, Emmerson P, Maynard J, et al. 2002. Biallelic germline mutations in MYH predispose to multiple colorectal adenoma and somatic G:C–>T:A mutations. *Hum Mol Genet* 11:2961–7.

68. Kim CJ, Cho YG, Park CH, et al. 2004. Genetic alterations of the MYH gene in gastric cancer. *Oncogene* 23:6820–2.

69. Zhang Y, Liu X, Fan Y, et al. 2006. Germline mutations and polymorphic variants in MMR, E-cadherin and MYH genes associated with familial gastric cancer in Jiangsu of China. *Int J Cancer* 119:2592–6.

70. Hizawa K, Iida M, Yao T, Aoyagi K, Fujishima M. 1997. Juvenile polyposis of the stomach: clinicopathological features and its malignant potential. *J Clin Pathol* 50:771–4.

71. Maimon SN, Zinninger MM. 1953. Familial gastric cancer. *Gastroenterology* 25:139–52; discussion, 53–5.

72. Woolf CM, Isaacson EA. 1961. An analysis of 5 "stomach cancer families" in the state of Utah. *Cancer* 14:1005–16.

73. Oliveira C, Seruca R, Carneiro F. 2006. Genetics, pathology, and clinics of familial gastric cancer. *Int J Surg Pathol* 14:21–33.

74. Dellavecchia C, Guala A, Olivieri C, et al. 1999. Early onset of gastric carcinoma and constitutional deletion of 18p. *Cancer Genet Cytogenet* 113:96–9.

75. Seruca R, Castedo S, Correia C, et al. 1993. Cytogenetic findings in eleven gastric carcinomas. *Cancer Genet Cytogenet* 68:42–8.

76. Gayther SA, Gorringe KL, Ramus SJ, et al. 1998. Identification of germ-line E-cadherin mutations in gastric cancer families of European origin. *Cancer Res* 58:4086–9.

77. Massad M, Uthman S, Obeid S, Majjar F. 1990. Ataxia-telangiectasia and stomach cancer. *Am J Gastroenterol* 85:630–1.

78. Lee JH, Han SU, Cho H, et al. 2000. A novel germ line juxtamembrane Met mutation in human gastric cancer. *Oncogene* 19:4947–53.

79. McDonald SA, Greaves LC, Gutierrez-Gonzalez L, et al. 2008. Mechanisms of field cancerization in the human stomach: the expansion and spread of mutated gastric stem cells. *Gastroenterology* 134:500–10.

80. Sasaki O, Soejima K, Korenaga D, Haraguchi Y. 1999. Comparison of the intratumor DNA ploidy distribution pattern between differentiated and undifferentiated gastric carcinoma. *Anal Quant Cytol Histol* 21:161–5.

81. Panani AD, Ferti A, Malliaros S, Raptis S. 1995. Cytogenetic study of 11 gastric adenocarcinomas. *Cancer Genet Cytogenet* 81:169–72.

82. Varis A, van Rees B, Weterman M, et al. 2003. DNA copy number changes in young gastric cancer patients with special reference to chromosome 19. *Br J Cancer* 88:1914–19.

83. El-Rifai W, Harper JC, Cummings OW, et al. 1998. Consistent genetic alterations in xenografts of proximal stomach and gastro-esophageal junction adenocarcinomas. *Cancer Res* 58:34–7.

84. Moskaluk CA, Hu J, Perlman EJ. 1998. Comparative genomic hybridization of esophageal and gastroesophageal adenocarcinomas shows consensus areas of DNA gain and loss. *Genes Chromosomes Cancer* 22:305–11.

85. Kimura Y, Noguchi T, Kawahara K, et al. 2004. Genetic alterations in 102 primary gastric cancers by comparative genomic hybridization: gain of 20q and loss of 18q are associated with tumor progression. *Mod Pathol* 17:1328–37.

86. Hidaka S, Yasutake T, Kondo M, et al. 2003. Frequent gains of 20q and losses of 18q are associated with lymph node metastasis in intestinal-type gastric cancer. *Anticancer Res* 23:3353–7.

87. Yustein AS, Harper JC, Petroni GR, et al. 1999. Allelotype of gastric adenocarcinoma. *Cancer Res* 59:1437–41.

88. Sano T, Tsujino T, Yoshida K, et al. 1991. Frequent loss of heterozygosity on chromosomes 1q, 5q, and 17p in human gastric carcinomas. *Cancer Res* 51:2926–31.

89. Uchino S, Tsuda H, Noguchi M, et al. 1992. Frequent loss of heterozygosity at the DCC locus in gastric cancer. *Cancer Res* 52:3099–102.

90. Rhyu MG, Park WS, Jung YJ, Choi SW, Meltzer SJ. 1994. Allelic deletions of MCC/APC and p53 are frequent late events in human gastric carcinogenesis. *Gastroenterology* 106:1584–8.

91. Schneider BG, Pulitzer DR, Brown RD, et al. 1995. Allelic imbalance in gastric cancer: an affected site on chromosome arm 3p. *Genes Chromosomes Cancer* 13:263–71.

92. Rumpel CA, Powell SM, Moskaluk CA. 1999. Mapping of genetic deletions on the long arm

of chromosome 4 in human esophageal adenocarcinomas. *Am J Pathol* 154:1329–34.

93. Iacopetta BJ, Soong R, House AK, Hamelin R. 1999. Gastric carcinomas with microsatellite instability: clinical features and mutations to the TGF-beta type II receptor, IGFII receptor, and BAX genes. *J Pathol* 187:428–32.

94. dos Santos NR, Seruca R, Constancia M, Seixas M, Sobrinho-Simoes M. 1996. Microsatellite instability at multiple loci in gastric carcinoma: clinicopathologic implications and prognosis. *Gastroenterology* 110:38–44.

95. Halling KC, Harper J, Moskaluk CA, et al. 1999. Origin of microsatellite instability in gastric cancer. *Am J Pathol* 155:205–11.

96. Fleisher AS, Esteller M, Wang S, et al. 1999. Hypermethylation of the hMLH1 gene promoter in human gastric cancers with microsatellite instability. *Cancer Res* 59:1090–5.

97. Leung SY, Yuen ST, Chung LP, et al. 1999. hMLH1 promoter methylation and lack of hMLH1 expression in sporadic gastric carcinomas with high-frequency microsatellite instability. *Cancer Res* 59:159–64.

98. Suzuki H, Itoh F, Toyota M, et al. 1999. Distinct methylation pattern and microsatellite instability in sporadic gastric cancer. *Int J Cancer* 83:309–13.

99. Hirata T, Yamamoto H, Taniguchi H, et al. 2007. Characterization of the immune escape phenotype of human gastric cancers with and without high-frequency microsatellite instability. *J Pathol* 211:516–23.

100. Strickler JG, Zheng J, Shu Q, et al. 1994. P53 mutations and microsatellite instability in sporadic gastric cancer: when guardians fail. *Cancer Res* 54:4750–5.

101. Ottini L, Palli D, Falchetti M, et al. 1997. Microsatellite instability in gastric cancer is associated with tumor location and family history in a high-risk population from Tuscany. *Cancer Res* 57:4523–9.

102. Oliveira C, Seruca R, Seixas M, Sobrinho-Simoes M. 1998. The clinicopathological features of gastric carcinomas with microsatellite instability may be mediated by mutations of different "target genes": a study of the TGFbeta RII, IGFII R, and BAX genes. *Am J Pathol* 153:1211–19.

103. Wu MS, Lee CW, Shun CT, et al. 1998. Clinicopathological significance of altered loci of replication error and microsatellite instability-associated mutations in gastric cancer. *Cancer Res* 58:1494–7.

104. Yamamoto H, Perez-Piteira J, Yoshida T, et al. 1999. Gastric cancers of the microsatellite mutator phenotype display characteristic genetic and clinical features. *Gastroenterology* 116:1348–57.

105. Ottini L, Falchetti M, D'Amico C, et al. 1998. Mutations at coding mononucleotide repeats in gastric cancer with the microsatellite mutator phenotype. *Oncogene* 16:2767–72.

106. Myeroff LL, Parsons R, Kim SJ, et al. 1995. A transforming growth factor beta receptor type II gene mutation common in colon and gastric but rare in endometrial cancers with microsatellite instability. *Cancer Res* 55:5545–7.

107. Park K, Kim SJ, Bang YJ, et al. 1994. Genetic changes in the transforming growth factor beta (TGF-beta) type II receptor gene in human gastric cancer cells: correlation with sensitivity to growth inhibition by TGF-beta. *Proce Natl Acad Sci USA* 91:8772–6.

108. Kang SH, Bang YJ, Im YH, et al. 1999. Transcriptional repression of the transforming growth factor-beta type I receptor gene by DNA methylation results in the development of TGF-beta resistance in human gastric cancer. *Oncogene* 18:7280–6.

109. Mori Y, Sato F, Selaru FM, et al. 2002. Instabilotyping reveals unique mutational spectra in microsatellite-unstable gastric cancers. *Cancer Res* 62:3641–5.

110. Yamamoto H, Sawai H, Perucho M. 1997. Frameshift somatic mutations in gastrointestinal cancer of the microsatellite mutator phenotype. *Cancer Res* 57:4420–6.

111. Gil J, Yamamoto H, Zapata JM, Reed JC, Perucho M. 1999. Impairment of the proapoptotic activity of Bax by missense mutations found in gastrointestinal cancers. *Cancer Res* 59:2034–7.

112. Souza RF, Appel R, Yin J, et al. 1996. Microsatellite instability in the insulin-like growth factor II receptor gene in gastrointestinal tumours. *Nat Genet* 14:255–7.

113. Souza RF, Yin J, Smolinski KN, et al. 1997. Frequent mutation of the E2F-4 cell cycle gene in primary human gastrointestinal tumors. *Cancer Res* 57:2350–3.

114. Yin J, Kong D, Wang S, et al. 1997. Mutation of hMSH3 and hMSH6 mismatch repair genes in genetically unstable human colorectal and gastric carcinomas. *Hum Mutat* 10:474–8.

115. Luqmani Y, Bennett C, Paterson I, et al. 1989. Expression of the pS2 gene in normal, benign and neoplastic human stomach. *Int J Cancer* 44:806–12.

116. Henry JA, Bennett MK, Piggott NH, et al. 1991. Expression of the pNR-2/pS2 protein in diverse human epithelial tumours. *Br J Cancer* 64:677–82.

117. Muller W, Borchard F. 1993. pS2 protein in gastric carcinoma and normal gastric mucosa:

association with clincopathological parameters and patient survival. *J Pathol* 171:263–9.

118. Wu MS, Shun CT, Wang HP, et al. 1998. Loss of pS2 protein expression is an early event of intestinal-type gastric cancer. *Jpn J Cancer Res* 89:278–82.

119. Lefebvre O, Chenard MP, Masson R, et al. 1996. Gastric mucosa abnormalities and tumorigenesis in mice lacking the pS2 trefoil protein. *Science* 274:259–62.

120. Machado JC, Carneiro F, Blin N, Sobrinho-Simoes M. 1996. Pattern of pS2 protein expression in premalignant and malignant lesions of gastric mucosa. *Eur J Cancer Prev* 5:169–79.

121. Nogueira AM, Machado JC, Carneiro F, et al. 1999. Patterns of expression of trefoil peptides and mucins in gastric polyps with and without malignant transformation. *J Pathol* 187: 541–8.

122. Sakata K, Tamura G, Nishizuka S, et al. 1997. Commonly deleted regions on the long arm of chromosome 21 in differentiated adenocarcinoma of the stomach. *Genes Chromosomes Cancer* 18:318–21.

123. Nishizuka S, Tamura G, Terashima M, Satodate R. 1998. Loss of heterozygosity during the development and progression of differentiated adenocarcinoma of the stomach. *J Pathol* 185:38–43.

124. Calnan DP, Westley BR, May FE, et al. 1999. The trefoil peptide TFF1 inhibits the growth of the human gastric adenocarcinoma cell line AGS. *J Pathol* 188:312–17.

125. Beckler AD, Roche JK, Harper JC, et al. 2003. Decreased abundance of trefoil factor 1 transcript in the majority of gastric carcinomas. *Cancer* 98:2184–91.

126. Sankpal NV, Mayo MW, Powell SM. 2005. Transcriptional repression of TFF1 in gastric epithelial cells by CCAAT/

enhancer binding protein-beta. *Biochim Biophys Acta* 1728:1–10.

127. Sankpal NV, Moskaluk CA, Hampton GM, Powell SM. 2006. Overexpression of CEBPbeta correlates with decreased TFF1 in gastric cancer. *Oncogene* 25:643–9.

128. Tebbutt NC, Giraud AS, Inglese M, et al. 2002. Reciprocal regulation of gastrointestinal homeostasis by SHP2 and STAT-mediated trefoil gene activation in gp130 mutant mice. *Nat Med* 8:1089–97.

129. Howlett M, Judd LM, Jenkins B, et al. 2005. Differential regulation of gastric tumor growth by cytokines that signal exclusively through the coreceptor gp130. *Gastroenterology* 129:1005–18.

130. Birchmeier W, Behrens J. 1994. Cadherin expression in carcinomas: role in the formation of cell junctions and the prevention of invasiveness. *Biochim Biophys Acta* 1198:11–26.

131. Mayer B, Johnson JP, Leitl F, et al. 1993. E-cadherin expression in primary and metastatic gastric cancer: down-regulation correlates with cellular dedifferentiation and glandular disintegration. *Cancer Res* 53:1690–5.

132. Becker KF, Atkinson MJ, Reich U, et al. 1994. E-cadherin gene mutations provide clues to diffuse type gastric carcinomas. *Cancer Res* 54:3845–52.

133. Berx G, Becker KF, Hofler H, van Roy F. 1998. Mutations of the human E-cadherin (CDH1) gene. *Hum Mut* 12:226–37.

134. Matsui S, Shiozaki H, Inoue M, et al. 1994. Immunohistochemical evaluation of alpha-catenin expression in human gastric cancer. *Virchows Arch* 424: 375–81.

135. Morris LE, Bloom GS, Frierson HF, Jr., Powell SM. 2005. Nucleotide variants within the IQGAP1 gene in diffuse-type

gastric cancers. *Genes Chromosomes Cancer* 42:280–6.

136. Hollstein M, Shomer B, Greenblatt M, et al. 1996. Somatic point mutations in the p53 gene of human tumors and cell lines: updated compilation. *Nucleic Acids Res* 24:141–6.

137. Hurlimann J, Saraga EP. 1994. Expression of p53 protein in gastric carcinomas. Association with histologic type and prognosis. *Am J Surg Pathol* 18:1247–53.

138. Gabbert HE, Muller W, Schneiders A, Meier S, Hommel G. 1995. The relationship of p53 expression to the prognosis of 418 patients with gastric carcinoma. *Cancer* 76:720–6.

139. Gschwind A, Fischer OM, Ullrich A. 2004. The discovery of receptor tyrosine kinases: targets for cancer therapy. *Nat Rev Cancer* 4:361–70.

140. Vivanco I, Sawyers CL. 2002. The phosphatidylinositol 3-kinase AKT pathway in human cancer. *Nat Rev Cancer* 2:489–501.

141. Samuels Y, Wang Z, Bardelli A, et al. 2004. High frequency of mutations of the PIK3CA gene in human cancers. *Science* 304:554.

142. Li VS, Wong CW, Chan TL, et al. 2005. Mutations of PIK3CA in gastric adenocarcinoma. *BMC Cancer* 5:29.

143. Kang MJ, Ryu BK, Lee MG, et al. 2008. NF-kappaB activates transcription of the RNA-binding factor HuR, via PI3K-AKT signaling, to promote gastric tumorigenesis. *Gastroenterology* 135:2030–42, 42 e1–3.

144. Sakakura C, Hagiwara A, Yasuoka R, et al. 2001. Tumour-amplified kinase BTAK is amplified and overexpressed in gastric cancers with possible involvement in aneuploid formation. *Br J Cancer* 84:824–31.

145. Wang Z, Shen D, Parsons DW, et al. 2004. Mutational analysis of the tyrosine phosphatome in

colorectal cancers. *Science* 304: 1164–6.

146. Lee JW, Jeong EG, Lee SH, et al. 2007. Mutational analysis of PTPRT phosphatase domains in common human cancers. *APMIS* 115:47–51.

147. Kastury K, Baffa R, Druck T, et al. 1996. Potential gastrointestinal tumor suppressor locus at the 3p14.2 FRA3B site identified by homozygous deletions in tumor cell lines. *Cancer Res* 56:978–83.

148. Ohta M, Inoue H, Cotticelli MG, et al. 1996. The FHIT gene, spanning the chromosome 3p14.2 fragile site and renal carcinoma-associated t(3;8) breakpoint, is abnormal in digestive tract cancers. *Cell* 84:587–97.

149. Baffa R, Veronese ML, Santoro R, et al. 1998. Loss of FHIT expression in gastric carcinoma. *Cancer Res* 58:4708–14.

150. Gemma A, Hagiwara K, Ke Y, et al. 1997. FHIT mutations in human primary gastric cancer. *Cancer Res* 57:1435–7.

151. Nakajima M, Sawada H, Yamada Y, et al. 1999. The prognostic significance of amplification and overexpression of c-met and c-erb B-2 in human gastric carcinomas. *Cancer* 85:1894–902.

152. Kuniyasu H, Yasui W, Kitadai Y, et al. 1992. Frequent amplification of the c-met gene in scirrhous type stomach cancer. *Biochem Biophys Res Comm* 189:227–32.

153. Kuniyasu H, Yasui W, Yokozaki H, Kitadai Y, Tahara E. 1993. Aberrant expression of c-met mRNA in human gastric carcinomas. *Int J Cancer* 55:72–5.

154. Hara T, Ooi A, Kobayashi M, et al. 1998. Amplification of c-myc, K-sam, and c-met in gastric cancers: detection by fluorescence in situ hybridization. *Lab Invest* 78: 1143–53.

155. Taniguchi K, Yonemura Y, Nojima N, et al. 1998. The relation between the growth patterns of

gastric carcinoma and the expression of hepatocyte growth factor receptor (c-met), autocrine motility factor receptor, and urokinase-type plasminogen activator receptor. *Cancer* 82:2112–22.

156. Tsugawa K, Yonemura Y, Hirono Y, et al. 1998. Amplification of the c-met, c-erbB-2 and epidermal growth factor receptor gene in human gastric cancers: correlation to clinical features. *Oncology* 55:475–81.

157. Koshiba M, Ogawa O, Habuchi T, et al. 1993. Infrequent ras mutation in human stomach cancers. *Jpn J Cancer Res* 84: 163–7.

158. Lee KH, Lee JS, Suh C, et al. 1995. Clinicopathologic significance of the K-ras gene codon 12 point mutation in stomach cancer: an analysis of 140 cases. *Cancer* 75:2794–801.

159. Tahara E. 1995. Molecular biology of gastric cancer. *World J Surg* 19:484–8; discussion, 489–90.

160. Inoue T, Chung YS, Yashiro M, et al. 1997. Transforming growth factor-beta and hepatocyte growth factor produced by gastric fibroblasts stimulate the invasiveness of scirrhous gastric cancer cells. *Jpn J Cancer Res* 88:152–9.

161. Kondo S, Shinomura Y, Kanayama S, et al. 1995. *Helicobacter pylori* increases gene expression of hepatocyte growth factor in human gastric mucosa. *Biochem Biophys Res Comm* 210:960–5.

162. Taha AS, Curry GW, Morton R, Park RH, Beattie AD. 1996. Gastric mucosal hepatocyte growth factor in *Helicobacter pylori* gastritis and peptic ulcer disease. *Am J Gastroenterol* 91:1407–9.

163. Yasunaga Y, Shinomura Y, Kanayama S, et al. 1996. Increased production of interleukin 1 beta and hepatocyte growth factor may

contribute to foveolar hyperplasia in enlarged fold gastritis. *Gut* 39:787–94.

164. Igaki H, Sasaki H, Tachimori Y, et al. 1995. Mutation frequency of the p16/CDKN2 gene in primary cancers in the upper digestive tract. *Cancer Res* 55:3421–3.

165. Barrett MT, Sanchez CA, Galipeau PC, et al. 1996. Allelic loss of 9p21 and mutation of the CDKN2/p16 gene develop as early lesions during neoplastic progression in Barrett's esophagus. *Oncogene* 13:1867–73.

166. Wong DJ, Barrett MT, Stoger R, Emond MJ, Reid BJ. 1997. p16INK4a promoter is hypermethylated at a high frequency in esophageal adenocarcinomas. *Cancer Res* 57:2619–22.

167. Klump B, Hsieh CJ, Holzmann K, Gregor M, Porschen R. 1998. Hypermethylation of the CDKN2/p16 promoter during neoplastic progression in Barrett's esophagus. *Gastroenterology* 115:1381–6.

168. Toyota M, Ahuja N, Suzuki H, et al. 1999. Aberrant methylation in gastric cancer associated with the CpG island methylator phenotype. *Cancer Res* 59: 5438–42.

169. Chi XZ, Yang JO, Lee KY, et al. 2005. RUNX3 suppresses gastric epithelial cell growth by inducing p21(WAF1/Cip1) expression in cooperation with transforming growth factor {beta}-activated SMAD. *Mol Cell Biol* 25: 8097–107.

170. Ito K, Liu Q, Salto-Tellez M, et al. 2005. RUNX3, a novel tumor suppressor, is frequently inactivated in gastric cancer by protein mislocalization. *Cancer Res* 65:7743–50.

171. Sakakura C, Hasegawa K, Miyagawa K, et al. 2005. Possible involvement of RUNX3 silencing in the peritoneal metastases of gastric cancers. *Clin Cancer Res* 11:6479–88.

172. Tamura G. 2004. Promoter methylation status of tumor suppressor and tumor-related genes in neoplastic and non-neoplastic gastric epithelia. *Histol Histopathol* 19:221–8.

173. Zou B, Chim CS, Zeng H, et al. 2006. Correlation between the single-site CpG methylation and expression silencing of the XAF1 gene in human gastric and colon cancers. *Gastroenterology* 131: 1835–43.

174. Yu J, Cheng YY, Tao Q, et al. 2009. Methylation of protocadherin 10, a novel tumor suppressor, is associated with poor prognosis in patients with gastric cancer. *Gastroenterology* 136:640–51.

175. Tokunaga A, Onda M, Okuda T, et al. 1995. Clinical significance of epidermal growth factor (EGF), EGF receptor, and c-erbB-2 in human gastric cancer. *Cancer* 75:1418–25.

176. Tahara E, Semba S, Tahara H. 1996. Molecular biological observations in gastric cancer. *Semin Oncol* 23:307–15.

177. Ueki T, Koji T, Tamiya S, Nakane PK, Tsuneyoshi M. 1995. Expression of basic fibroblast growth factor and fibroblast growth factor receptor in advanced gastric carcinoma. *J Pathol* 177:353–61.

178. Yonemura Y, Ninomiya I, Yamaguchi A, et al. 1991. Evaluation of immunoreactivity for erbB-2 protein as a marker of poor short term prognosis in gastric cancer. *Cancer Res* 51:1034–8.

179. Mizutani T, Onda M, Tokunaga A, Yamanaka N, Sugisaki Y. 1993. Relationship of C-erbB-2 protein expression and gene amplification to invasion and metastasis in human gastric cancer. *Cancer* 72:2083–8.

180. Uchino S, Tsuda H, Maruyama K, et al. 1993. Overexpression of c-erbB-2 protein in gastric cancer. Its correlation with long-term survival of patients. *Cancer* 72: 3179–84.

181. Bizari L, Borim AA, Leite KR, et al. 2006. Alterations of the CCND1 and HER-2/neu (ERBB2) proteins in esophageal and gastric cancers. *Cancer Genet Cytogenet* 165:41–50.

182. Chariyalertsak S, Sugano K, Ohkura H, Mori Y. 1994. Comparison of c-erbB-2 oncoprotein expression in tissue and serum of patients with stomach cancer. *Tumour Biol* 15:294–303.

183. Tateishi M, Toda T, Minamisono Y, Nagasaki S. 1992. Clinicopathological significance of c-erbB-2 protein expression in human gastric carcinoma. *J Surg Oncol* 49:209–12.

184. Friess H, Fukuda A, Tang WH, et al. 1999. Concomitant analysis of the epidermal growth factor receptor family in esophageal cancer: overexpression of epidermal growth factor receptor mRNA but not of c-erbB-2 and c-erbB-3. *World J Surg* 23: 1010–18.

185. Ouyang H, Furukawa T, Abe T, Kato Y, Horii A. 1998. The BAX gene, the promoter of apoptosis, is mutated in genetically unstable cancers of the colorectum, stomach, and endometrium. *Clin Cancer Res* 4:1071–4.

186. Lee JW, Jeong EG, Soung YH, et al. 2006. Decreased expression of tumour suppressor Bax-interacting factor-1 (Bif-1), a Bax activator, in gastric carcinomas. *Pathology* 38:312–15.

187. Kondo S, Shinomura Y, Miyazaki Y, et al. 2000. Mutations of the bak gene in human gastric and colorectal cancers. *Cancer Res* 60:4328–30.

188. Kuraoka K, Matsumura S, Sanada Y, et al. 2005. A single nucleotide polymorphism in the extracellular domain of TRAIL receptor DR4 at nucleotide 626 in gastric cancer patients in Japan. *Oncol Rep* 14:465–70.

189. Park WS, Lee JH, Shin MS, et al. 2001. Inactivating mutations of KILLER/DR5 gene in gastric cancers. *Gastroenterology* 121: 1219–25.

190. Saito H, Tsujitani S, Oka S, et al. 1999. The expression of thymidine phosphorylase correlates with angiogenesis and the efficacy of chemotherapy using fluorouracil derivatives in advanced gastric carcinoma. *Br J Cancer* 81:484–9.

191. Hachisuka T, Narikiyo M, Yamada Y, et al. 2005. High lymphatic vessel density correlates with overexpression of VEGF-C in gastric cancer. *Oncol Rep* 13:733–7.

192. Kim J, Sohn S, Chae Y, et al. 2007. Vascular endothelial growth factor gene polymorphisms associated with prognosis for patients with gastric cancer. *Ann Oncol* 18:1030–6.

193. El-Rifai W, Smith MF, Jr., Li G, et al. 2002. Gastric cancers overexpress DARPP-32 and a novel isoform, t-DARPP. *Cancer Res* 62:4061–4.

194. Belkhiri A, Zaika A, Pidkovka N, et al. 2005. Darpp-32: a novel antiapoptotic gene in upper gastrointestinal carcinomas. *Cancer Res* 65:6583–92.

195. Powell SM, Harper JC, Hamilton SR, Robinson CR, Cummings OW. 1997. Inactivation of Smad4 in gastric carcinomas. *Cancer Res* 57:4221–4.

196. Nagase H, Nakamura Y. 1993. Mutations of the APC (adenomatous polyposis coli) gene. *Hum Mut* 2:425–34.

197. Ogasawara S, Maesawa C, Tamura G, Satodate R. 1994. Lack of mutations of the adenomatous polyposis coli gene in oesophageal and gastric carcinomas. *Virchows Arch* 424:607–11.

198. Maesawa C, Tamura G, Suzuki Y, et al. 1995. The sequential

accumulation of genetic alterations characteristic of the colorectal adenoma-carcinoma sequence does not occur between gastric adenoma and adenocarcinoma. *J Pathol* 176:249–58.

199. Powell SM, Cummings OW, Mullen JA, et al. 1996. Characterization of the APC gene in sporadic gastric adenocarcinomas. *Oncogene* 12:1953–9.

200. Tamura G, Ogasawara S, Nishizuka S, et al. 1996. Two distinct regions of deletion on the long arm of chromosome 5 in differentiated adenocarcinomas of the stomach. *Cancer Res 56:* 612–15.

201. Park WS, Oh RR, Park JY, et al. 1999. Frequent somatic mutations of the beta-catenin gene in intestinal-type gastric cancer. *Cancer Res* 59:4257–60.

202. Candidus S, Bischoff P, Becker KF, Hofler H. 1996. No evidence for mutations in the alpha- and beta-catenin genes in human gastric and breast carcinomas. *Cancer Res* 56:49–52.

203. Ramesh S, Nash J, McCulloch PG. 1999. Reduction in membranous expression of beta-catenin and increased cytoplasmic E-cadherin expression predict poor survival in gastric cancer. *Br J Cancer* 81:1392–7.

204. Kawanishi J, Kato J, Sasaki K, et al. 1995. Loss of E-cadherin-dependent cell-cell adhesion due to mutation of the beta-catenin gene in a human cancer cell line, HSC-39. *Mol Cell Biol* 15: 1175–81.

205. Tsuji S, Kawano S, Sawaoka H, et al. 1996. Evidence for involvement of cyclooxygenase-2 in proliferation of two gastrointestinal cancer cell lines. *Prostaglandins Leukot Essent Fatty Acids* 55:179–83.

206. Ma X, Chen K, Huang S, et al. 2005. Frequent activation of the hedgehog pathway in advanced gastric adenocarcinomas. *Carcinogenesis* 26:1698–705.

207. Ito H, Yonemura Y, Fujita H, et al. 1996. Prognostic relevance of urokinase-type plasminogen activator (uPA) and plasminogen activator inhibitors PAI-1 and PAI-2 in gastric cancer. *Virchows Arch* 427:487–96.

208. Burgart LJ, Zheng J, Shu Q, Strickler JG, Shibata D. 1995. Somatic mitochondrial mutation in gastric cancer. *Am J Pathol* 147:1105–11.

209. Hiyama E, Yokoyama T, Tatsumoto N, et al. 1995. Telomerase activity in gastric cancer. *Cancer Res* 55:3258–62.

210. Wong SC, Yu H, So JB. 2006. Detection of telomerase activity in gastric lavage fluid: a novel method to detect gastric cancer. *J Surg Res* 131:252–5.

211. Guan XY, Fu SB, Xia JC, et al. 2000. Recurrent chromosome changes in 62 primary gastric carcinomas detected by comparative genomic hybridization. *Cancer Genet Cytogen* 123:27–34.

212. Hippo Y, Taniguchi H, Tsutsumi S, et al. 2002. Global gene expression analysis of gastric cancer by oligonucleotide microarrays. *Cancer Res* 62: 233–40.

213. Leung SY, Chen X, Chu KM, et al. 2002. Phospholipase A2 group IIA expression in gastric adenocarcinoma is associated with prolonged survival and less frequent metastasis. *Proc Natl Acad Sci USA* 99:16203–8.

214. Boussioutas A, Li H, Liu J, et al. 2003. Distinctive patterns of gene expression in premalignant gastric mucosa and gastric cancer. *Cancer Res* 63:2569–77.

215. Weiss MM, Kuipers EJ, Postma C, et al. 2004. Genomic alterations in primary gastric adenocarcinomas correlate with clinicopathological characteristics and survival. *Cell Oncol* 26:307–17.

216. Petrocca F, Visone R, Onelli MR, et al. 2008. E2F1-regulated microRNAs impair TGFbeta-dependent cell-cycle arrest and apoptosis in gastric cancer. *Cancer Cell* 13:272–86.

217. El-Rifai W, Moskaluk CA, Abdrabbo MK, et al. 2002. Gastric cancers overexpress S100A calcium-binding proteins. *Cancer Res* 62:6823–6.

218. Yamamichi K, Uehara Y, Kitamura N, Nakane Y, Hioki K. 1998. Increased expression of CD44v6 mRNA significantly correlates with distant metastasis and poor prognosis in gastric cancer. *Int J Cancer* 79:256–62.

219. Han S, Kim HY, Park K, et al. 1999. c-Myc expression is related with cell proliferation and associated with poor clinical outcome in human gastric cancer. *J Korean Med Sci* 14:526–30.

220. Kuwahara A, Katano M, Nakamura M, et al. 1999. New therapeutic strategy for gastric carcinoma: a two-step evaluation of malignant potential from its molecular biologic and pathologic characteristics. *J Surgical Oncol* 72:142–9.

221. Lin WC, Li AF, Chi CW, et al. 1999. Tie-1 protein tyrosine kinase: a novel independent prognostic marker for gastric cancer. *Clin Cancer Res* 5: 1745–51.

222. Maeda K, Kang SM, Onoda N, et al. 1999. Vascular endothelial growth factor expression in preoperative biopsy specimens correlates with disease recurrence in patients with early gastric carcinoma. *Cancer* 86:566–71.

223. Takano Y, Kato Y, Masuda M, Ohshima Y, Okayasu I. 1999. Cyclin D2, but not cyclin D1, overexpression closely correlates with gastric cancer progression and prognosis. *J Pathol* 189:194–200.

224. Ganesh S, Sier CF, Heerding MM, et al. 1996. Prognostic value of the plasminogen activation system in patients with gastric carcinoma. *Cancer* 77:1035–43.

225. Jang SJ, Park YW, Park MH, et al. 1999. Expression of cell-cycle regulators, cyclin E and p21WAF1/CIP1, potential prognostic markers for gastric cancer. *Eur J Surg Oncol* 25: 157–63.

226. Kwon OJ, Kang HS, Suh JS, et al. 1999. The loss of p27 protein has an independent prognostic significance in gastric cancer. *Anticancer Res* 19:4215–20.

227. Xiangming C, Hokita S, Natsugoe S, et al. 1999. Cooccurrence of reduced expression of alpha-catenin and overexpression of p53 is a predictor of lymph node metastasis in early gastric cancer. *Oncology* 57:131–7.

228. Zafirellis K, Karameris A, Milingos N, Androulakis G. 2005. Molecular markers in gastric cancer: can p53 and bcl-2 protein expressions be used as prognostic factors? *Anticancer Res* 25: 3629–36.

229. Liu XP, Kawauchi S, Oga A, et al. 2001. Combined examination of p27(Kip1), p21(Waf1/Cip1) and p53 expression allows precise estimation of prognosis in patients with gastric carcinoma. *Histopathology* 39:603–10.

230. Nagahara H, Mimori K, Utsunomiya T, et al. 2005. Clinicopathologic and biological significance of kallikrein 6 overexpression in human gastric cancer. *Clin Cancer Res* 11: 6800–6.

231. Nishigaki R, Osaki M, Hiratsuka M, et al. 2005. Proteomic identification of differentially-expressed genes in human gastric carcinomas. *Proteomics* 5: 3205–13.

232. Cascinu S, Graziano F, Del Ferro E, et al. 1998. Expression of p53 protein and resistance to preoperative chemotherapy in locally advanced gastric carcinoma. *Cancer* 83:1917–22.

233. Saito H, Tsujitani S, Ikeguchi M, Maeta M, Kaibara N. 1999. Serum level of a soluble receptor for interleukin-2 as a prognostic factor in patients with gastric cancer. *Oncology* 56:253–8.

234. Muguruma K, Nakata B, Hirakawa K, et al. 1998. p53 and Bax protein expression as predictor of chemotherapeutic effect in gastric carcinoma. *Cancer & Chemotherapy* 25 Suppl 3: 400–3.

235. Nakata B, Muguruma K, Hirakawa K, et al. 1998. Predictive value of Bcl-2 and Bax protein expression for chemotherapeutic effect in gastric cancer. A pilot study. *Oncology* 55:543–7.

236. Kurihara N, Kubota T, Furukawa T, et al. 1999. Chemosensitivity testing of primary tumor cells from gastric cancer patients with liver metastasis can identify effective antitumor drugs. *Anticancer Res* 19:5155–8.

237. Torii A, Kodera Y, Uesaka K, et al. 1997. Plasma concentration of matrix metalloproteinase 9 in gastric cancer. *Br J Surg* 84: 133–6.

238. Gofuku J, Shiozaki H, Doki Y, et al. 1998. Characterization of soluble E-cadherin as a disease marker in gastric cancer patients. *Br J Cancer* 78:1095–101.

239. De Vita F, Orditura M, Galizia G, et al. 1999. Serum interleukin-10 levels in patients with advanced gastrointestinal malignancies. *Cancer* 86:1936–43.

240. Han SU, Lee JH, Kim WH, Cho YK, Kim MW. 1999. Significant correlation between serum level of hepatocyte growth factor and progression of gastric carcinoma. *World J Surg* 23:1176–80.

241. Yoshikawa T, Saitoh M, Tsuburaya A, et al. 1999. Tissue inhibitor of matrix metalloproteinase-1 in the plasma of patients with gastric carcinoma. A possible marker for serosal invasion and metastasis. *Cancer* 86:1929–35.

242. Nakata B, Hirakawa-Ys Chung K, Kato Y, et al. 1998. Serum CA 125 level as a predictor of peritoneal dissemination in patients with gastric carcinoma. *Cancer* 83:2488–92.

243. Pituch-Noworolska A, Wieckiewicz J, Krzeszowiak A, et al. 1998. Evaluation of circulating tumour cells expressing CD44 variants in the blood of gastric cancer patients by flow cytometry. *Anticancer Res* 18:3747–52.

244. Chausovsky G, Luchansky M, Figer A, et al. 1999. Expression of cytokeratin 20 in the blood of patients with disseminated carcinoma of the pancreas, colon, stomach, and lung. *Cancer* 86:2398–405.

245. Marrelli D, Roviello F, De Stefano A, et al. 1999. Prognostic significance of CEA, CA 19-9 and CA 72-4 preoperative serum levels in gastric carcinoma. *Oncology* 57:55–62.

246. Wu CW, Lin YY, Chen GD, et al. 1999. Serum anti-p53 antibodies in gastric adenocarcinoma patients are associated with poor prognosis, lymph node metastasis and poorly differentiated nuclear grade. *Br J Cancer* 80:483–8.

247. Wu CY, Wu MS, Chiang EP, et al. 2007. Elevated plasma osteopontin associated with gastric cancer development, invasion and survival. *Gut* 56:782–9.

248. Poon TC, Sung JJ, Chow SM, et al. 2006. Diagnosis of gastric cancer by serum proteomic fingerprinting. *Gastroenterology* 130:1858–64.

249. Demetri GD. 2001. Targeting c-kit mutations in solid tumors: scientific rationale and novel therapeutic options. *Semin Oncol* 28:19–26.

250. DeMatteo RP. 2002. The GIST of targeted cancer therapy: a tumor (gastrointestinal stromal tumor), a mutated gene (c-kit), and a molecular inhibitor (STI571). *Ann Surg Oncol* 9:831–9.

251. Nakamura K, Yashiro M, Matsuoka T, et al. 2006. A novel molecular targeting compound as K-samII/FGF-R2 phosphorylation inhibitor, Ki23057, for scirrhous gastric cancer. *Gastroenterology* 131:1530–41.

252. Fukaya M, Isohata N, Ohta H, et al. 2006. Hedgehog signal activation in gastric pit cell and in diffuse-type gastric cancer. *Gastroenterology* 131: 14–29.

253. Abnet CC, Freedman ND, Kamangar F, et al. 2009. Non-steroidal anti-inflammatory drugs and risk of gastric and oesophageal adenocarcinomas: results from a cohort study and a meta-analysis. *Br J Cancer* 100:551–7.

254. Thun MJ, Namboodiri MM, Calle EE, Flanders WD, Heath CW, Jr. 1993. Aspirin use and risk of fatal cancer. *Cancer Res* 53:1322–7.

Chapter 26

Pancreatic cancer

Sergii Ivakhno, Kristopher Frese, Simon Tavaré, Christine
Iacobuzio-Donahue and David Tuveson

Introduction

Pancreatic cancer is highly lethal, with a five-year
survival rate of less than 5% [1]. The most common
form of pancreatic cancer is pancreatic ductal adenocarcinoma (PDA), accounting for over 250,000 deaths
worldwide annually [2] and ranked as the fourth
leading cause of cancer mortality in the United States
[3]. The poor survival of PDA patients has been
attributed to the advanced stage of disease presentation and ineffective therapeutic options. Approximately 20% of PDA patients are diagnosed with
localized disease and are candidates for surgical resection with curative intent; however, the majority of
these patients unfortunately relapse within several
years and carry a five-year post-operative survival of
only 20% despite additional treatment with adjuvant
chemotherapy [4, 5]. The current worldwide standard
treatment for advanced and surgically unresectable
pancreatic cancer is the chemotherapeutic agent gemcitabine; however, this intervention only modestly
improves patients' symptoms and has little measurable effect on overall survival [6]. Despite tremendous
efforts, only the EGFR inhibitor erlotinib has been
approved as an additional agent for PDA patients,
albeit for an incremental increase in median overall
survival of only two weeks [7]. Therefore the development of new therapeutic approaches for PDA patients
that are based upon the underlying biological complexity of this disease is of the utmost importance.
The formulation of new treatment strategies for PDA
patients requires a detailed understanding of the genetic and epigenetic alterations in PDA tumors and the
biochemical signaling networks that are active in this
disease. By collectively integrating such information,
a multimodal molecular network (MMMN) can be
established for PDA and used to probe the biological
vulnerabilities of this nefarious malignancy.

The proposed progression model of pancreatic cancer

The first integrative model of colon cancer progression proposed by Fearon and Vogelstein correlated
the anatomical and histopathological progression
from a pre-invasive intestinal polyp to invasive colon
cancer with the accumulation of genetic and epigenetic mutations thought to be causal events in these
processes [8]. Subsequently, similar models have been
postulated for other cancers, including pancreatic
cancer [9]. PDA histologically arises in the ductal
exocrine compartment of the pancreas due to the
presence of pre-invasive neoplasms in small ducts.
However, the precise nature of the originating cell
type for PDA is unknown and may include either
pancreatic progenitor cells or mature epithelial cells
that dedifferentiate [10]. The cell of origin for PDA is
hypothesized to progress through a series of pre-
invasive histopathological states – termed pancreatic
intraepithelial neoplasms (PanINs) – prior to the
acquisition of invasive and metastatic characteristics.
Additionally, pre-invasive cystic pancreatic neoplasms distinct from PanINs develop in either the
large pancreatic ducts and are termed intraductal
papillary mucinous neoplasms (IPMNs), or peripheral small ducts and are classified as mucinous cystic
neoplams (MCNs). Both IMPNs and MCNs can be
found in close proximity to invasive PDA in resected
specimens, suggesting but not proving that PDA can
also develop from these precursor neoplasms. Since
PanINs appear to be the predominant precursor, we
will focus on these pre-neoplasms for the remainder
of the review. Although PanINs of different grades are
often conceptualized as discrete steps in PDA progression, it is much more plausible that they represent
static pictures of a continuously evolving process.

Systems Biology of Cancer, ed. S. Thiagalingam. Published by Cambridge University Press. © Cambridge University Press 2015.

While normal pancreatic ductal epithelial cells are cuboidal with uniformly small round nuclei, the epithelial cells of grade 1 PanINs are characterized by a transition to a columnar appearance and concomitant accumulation of apical cytoplasmic mucin. More severe grade 2 PanINs harbor cells with less prominent mucin, abnormally shaped nuclei, and loss of polarity. Finally, grade 3 PanINs exhibit all characteristics of grade 2 and in addition have cells budding into the ductal lumen that occasionally form cribriform structures that bridge across the lumen of the duct. Grade 3 PanINs are generally considered to be carcinoma *in situ*, which makes them an attractive therapeutic target.

Molecular genetics of pancreatic cancer

Despite its poor prognosis, the molecular genetics of pancreatic cancer progression are remarkably well defined (Figure 26.1). The most common mutation found in pancreatic cancer is a point mutation of the *KRAS2* gene that produces a constitutively active form of the Kras protein. Active GTP-bound Kras can be self-inactivated through its intrinsic weak GTPase activity resulting in a GDP-bound state. Kras is physiologically activated by a variety of growth factor receptors, most often receptor tyrosine kinases, which promote the membrane recruitment and activation of guanine nucleotide exchange factors (GEFs) that catalyze the exchange of GDP for GTP. Conversely, Kras inactivation is attained through the recruitment of GTPase-activating proteins (GAPs) that dramatically accelerate intrinsic Kras GTPase activity. Oncogenic mutations in Kras invariably occur at amino acids 12, 13, or 61 [11] and produce a constitutively active GTP-bound Kras molecule that is unresponsive to GAP-mediated inactivation. Kras mutations are present in approximately 40% of PanIN-1 and over 90% of invasive pancreatic cancers, suggesting that oncogenic Kras is an initiating factor for pancreatic cancer [12, 13]. Furthermore, *KRAS2* is amplified in approximately 25% and overexpressed in the majority of end-stage cancers. Intriguingly, expression of an oncogenic allele of Kras in the developing murine pancreas promotes the full spectrum of pancreatic cancer, including all grades of PanIN, with complete penetrance [14].

In addition to Kras, mutations in three key tumor suppressor genes (TSG) are frequently found in sporadic pancreatic cancer. Over 90% of all pancreatic cancers harbor $p16^{INK4A}$ mutations including intragenic mutations followed by LOH (40%), homozygous deletions in the 9p21 locus (40%), and promoter hypermethylation (15%) [15–17]. $p16^{INK4A}$ inhibits cdk4/6/cyclin D-mediated phosphorylation of Rb, and thus controls the G1-S transition. Because these mutations are first observed in grade 2 PanINs, this suggests that $p16^{INK4A}$ mutations promote progression rather than initiation of pancreatic cancer [18]. This hypothesis is further supported by the observation that familial atypical multiple mole melanoma (FAMMM) syndrome kindreds with germline mutations in $p16^{INK4A}$ have an increased risk for pancreatic cancer but are not predisposed to early development of the disease [19]. Furthermore, biallelic deletion of the CDKN2A locus in the developing murine pancreas fails to elicit any phenotype; however, in conjunction with a mutant Kras allele results in the development of rapidly invasive PDA [20].

Similar to $p16^{INK4A}$, mutations in *TP53* are found in 55 to 70% of invasive pancreatic cancers [21, 22]. p53 regulates apoptosis and cell cycle arrest through a variety of downstream pathways, and its deletion provides a clear advantage to pre-invasive neoplastic cells. Indeed, p53 mutation may allow cell survival in the presence of genomic instability that will otherwise trigger apoptosis. *TP53* mutations occur relatively late in pancreatic carcinogenesis and are not detected until grade 3 PanINs [23, 24]. The importance of *TP53* mutation can be assessed by studying patterns of Li–Fraumeni syndrome, an autosomal-dominant hereditary disorder that greatly increases the susceptibility to cancer and is linked to germline mutations of *TP53*. Although there is little evidence suggesting that Li–Fraumeni patients with germline *TP53* mutations are predisposed to the development of pancreatic cancer, this is likely because these patients often die of other cancers prior to the development of pancreatic malignancy [25]. Mice lacking p53 are not predisposed to PDA and appear to have normal pancreata; however, in conjunction with mutant Kras this leads to invasive PDA [26].

Deletion of *DPC4/SMAD4* at the 18q21 locus occurs in over 50% of all invasive pancreatic cancers [27], and is found in a higher fraction of tumors from patients with advanced disease [28] and portends a poor overall prognosis [29, 28]. Dpc4/Smad4 is a transcriptional regulator that forms heterodimers with other Smads and is required for a subset of TGFβ

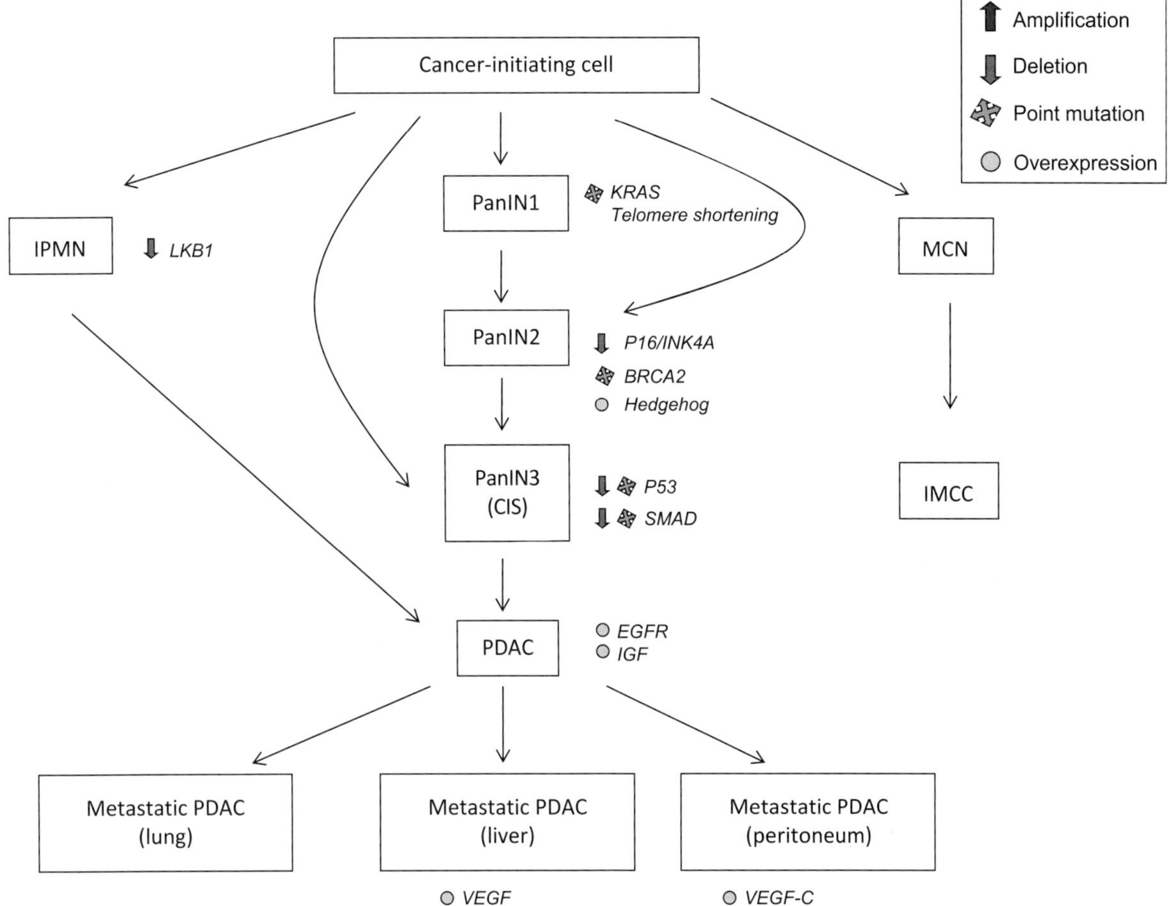

Figure 26.1 Histological and molecular patterns of pancreatic cancer progression. Distinct pathophysiological stages of pancreatic cancer are shown along with the most common genetic alterations prevalent for each stage of disease progression. (A black and white version of this figure will appear in some formats. For the color version, please refer to the plate section.)

signaling. Its loss in high-grade PanINs supports its role in progression rather than initiation of pancreatic cancer [24, 30].

The multimodal molecular network (MMMN) and pancreatic cancer

Because the molecular genetics of pancreatic cancer is relatively well understood, the MMMN for pancreatic cancer is perhaps more defined than for other cancers. Indeed, recent global sequencing projects at the level of genomic copy number alterations and exomic sequencing have revealed that the aforementioned "four peaks" of pancreatic cancer represent the only mutations present in over 20% of all PDA cases [31]. Considering these and other results, the mutations and signaling pathways associated with the

different steps of pancreatic cancer progression are fairly well understood.

The frequency and early occurrence of oncogenic mutations in *KRAS2* makes it the obvious candidate as the gatekeeper of PDA. This is supported by the fact that mutations in Kras occur in practically every pancreatic cancer and that there is no pathological phenotype in the absence of this mutation in mouse models. Therefore a Kras module may be considered a critical event in pancreatic cancer initiation. This module could also encompass upstream receptor tyrosine kinases (RTK) that potentiate Kras signaling or downstream effector molecules such as the Raf/ MEK/ERK, PI3K/Akt, or RalGEF/Ral axes. Genomic amplification or overexpression of the RTKs *EGFR*, *ERBB2, ERBB3* and their ligands *EGF* and *TGFA* can also occur in low-grade PanIN, suggesting that the

EGF pathway may also be important for PDA initiation [32–34]. Indeed, the small molecule inhibitor erlotinib, which inhibits EGFR kinase activity, is the only targeted therapeutic approved for the treatement of PDA [7]. Although mutations in the Raf/MEK/ERK pathway are exceedingly rare, mutations in *PIK3CA* and *AKT2* are found in 11% and 20% of pancreatic cancers, respectively [35–37]. These observations suggest that although the initiator module is predominantly associated with an activating Kras mutation, other mutations in the other pathways may serve to enhance specific aspects of this signal transduction pathway.

As previously mentioned, mutations in the major tumor suppressor genes appear later in pancreatic cancer progression. The earliest of these events is mutation of p16, which first occurs in PanIN-1B/2 and could therefore be considered a part of an intermediate module leading to the loss of cell cycle checkpoint controls. The predominant role of p16 in regulation of cdk4/6-cyclinD complexes suggests that amplification or overexpression of either kinase may be functionally equivalent to p16 inactivation. Mutations in *RB*, which could convey a similar proliferative advantage, are not found in pancreatic cancer. However, there is some evidence that pancreatic tumors express lower levels of pRb [38]. Although the only known function of p16 is regulation of cdk4/6, it is intriguing to note that FAMMM kindreds harboring mutations in CDK4 that render the kinase resistant to inhibition by p16 are not predisposed to pancreatic cancer [39]. This suggests that p16 may have an additional, as yet undiscovered, role in regulating pancreatic homeostasis.

Deletion of *DPC4/SMAD4* is another intermediate-to-late module that promotes PDA progression. TGFβ signaling, of which Smad4 is a critical regulator, has been implicated in immune evasion, epithelial to mesenchymal transition, and invasion [40]. Deficiencies in Smad4 may be mimicked by amplification or overexpression of inhibitory Smads such as Smad7 [41]. Alternatively, deletion or mutation of TGFβ receptors may elicit similar phenotypes. Indeed, incorporating either of these genes in murine pancreatic cancer models promotes accelerated disease [42, 43].

TP53 mutation, as assessed by the nuclear accumulation of the protein product, does not appear until PanIN-3 and therefore can be considered as a late module in the MMMN framework [24].

Pleiotropic effects of p53 mutation make it unlikely that one or even two mutations in downstream effector genes could recapitulate the multitude of p53-dependent activities, although mutations in genes that regulate p53, such as *HDM2* and *ARF*, have been reported. A polymorphism in the promoter of Hdm2, an E3 ligase that directly regulates p53 protein levels, results in elevated Hdm2 protein levels, reduced p53 function, and an increased risk for tumor progression [44]. Mutations in CDKN2A often affect both p16 as well as p14^{ARF}, a negative regulator of Hdm2 (and consequently positive regulator of p53). Interestingly, neither of these mutations result in increased p53 protein levels and therefore would not score positively by immunohistochemistry. Therefore mutations in these genes may significantly increase the number of PDA tumors that are deficient in this critical pathway.

Mutations in the *BRCA2* pathway occur in families predisposed to developing pancreatic cancer [45, 46]. BRCA2 mutations promote genomic instability and may thereby accelerate tumorigenesis through widespread chromosomal aberrations and genomic instability [47–49]. In conjunction with p53 loss of function, BRCA2 mutations abrogate gatekeeper and caretaker functionality of the late MMMN module and subsequently allow cell survival under mutagenic conditions.

Impact of high-throughput genomic approaches on systems biology of PDA

The construction of an MMMN implicated in pancreatic cancer progression was enabled by applying two separate experimental strategies to study the cancer genome. The first strategy involved now standard molecular biological approaches to identify oncogenes and tumor suppressor genes in PDA specimens. The second approach involved recent high-throughput genomic (HTG) techniques that allow the analysis of thousands of genes simultaneously. It is this second set of HTG approaches that holds the most promise for the discovery of new genetic aberrations and epistatic interactions between mutations in PDA. The ultimate aim of oncogenomic research is to identify and catalog both genetic and epigenetic changes in cancer at the genome-wide level in both temporal (e.g., oncogenic progression) and spatial (e.g., metastasis-specific mutations) dimensions. The construction of such a catalog will be pivotal for selecting patient-tailored treatment programs [50]. From the MMMN network's perspective, the comprehensive oncogenomic information can be utilized to build exploratory and predictive mathematical models for detecting interactions between different genetic and epigenetic alterations and elucidating their impact on temporal and spatial progression of pancreatic cancer. Here we provide some examples of how new genomic approaches gave novel insights into the systems biology of pancreatic cancer.

By application of HTG techniques, current research efforts can be grouped into two categories: those that use only one HTG approach to profile molecular aberrations and those that apply several HTG techniques simultaneously to comprehensively characterize genomic, epigenomic, and physiological changes in cancer. The first category usually includes proof-of-concept applications, new technologies, and studies where an answer to a specific question is the primary concern. It includes early gene sequencing efforts to identify point mutations in either a specific protein family, such as protein kinases, or all protein-coding sequences in several different cancers [51, 52]. Since the number of PDA samples in most previous studies was usually limited, most of the conclusions concerned broad mutation patterns in different cancers. For instance, Greenman et al. sequenced coding exons from 518 protein kinase genes in 210 diverse human cancers [51] and later (see below) Jones and colleagues [31] sequenced protein-coding genes of 24 pancreatic cancers. When analyzing such datasets the most challenging computational task lies in selecting driver point mutations among a large number of passenger aberrations. Each somatic mutation in a cancer genome, whatever its structural nature, may be classified according to its consequences for cancer development. "Driver" mutations confer a growth advantage on the cells carrying them and have been positively selected during the evolution of the cancer. They reside, by definition, in the subset of genes known as "cancer genes." The remainder of mutations are "passengers" that do not confer growth advantage, but happened to be present in an ancestor of the cancer cell when it acquired one of its drivers. Most algorithms developed to identify relevant mutations utilize the proportion of synonymous to non-synonymous changes for particular nucleotide position to define the baseline for driver mutations.

Use of single nucleotide polymorphism (SNP) and oligonucleotide microarrays for copy number analyses is another widely used approach that can detect both copy number changes (CNCs) and copy neutral allelic losses, such as uniparental disomy. In one of the earliest applications to PDA, cDNA microarrays containing 14,160 cDNA clones were used to define copy number alterations in a panel of 24 pancreatic adenocarcinoma cell lines and 13 primary tumor specimens [53]. The high frequency of genomic instability in PDA, in part due to loss of p53, can make it difficult to identify driver CNCs. The authors of the paper developed a prioritization scheme for identifying minimal common regions, which used cross-sample information to find genes most frequently altered in many cancer samples. Sixty-four regions of recurrent copy number changes were identified by this approach that harbored genes known to play important roles in the pathogenesis of pancreatic adenocarcinoma, including the tumor suppressors $p16^{INK4A}$ and *TP53* and the oncogenes *MYC*, *KRAS2*, and *AKT2*.

Since the most informative copy number changes for functional validation of candidate genes are focal high-level amplifications and homozygous deletions, the prominent trend in array comparative genomic hybridization (aCGH) manufacturing was to increase the number and density of features on microarrays. For example, in 2003 Affymetrix, one of the largest SNP array producers, released the Mapping 10 K array, a high-density oligonucleotide array suitable for genotyping and estimation of copy numbers of 10,000 SNPs. Following that, Affymetrix has released platforms that interrogate 100 K and 500 K SNPs. More recently, Affymetrix made available Genome-wide SNP 5.0 and Genome-wide SNP 6.0 arrays, which in addition to SNPs interrogate a large number of non-polymorphic (NP) loci. The 6.0 chip interrogates 900 K SNPs and 900 K NP loci. Following the trend of applying denser arrays to study copy number changes in cancer, Harada et al. used Affymetrix 100 K SNPs arrays to profile 27 microdissected PDA samples [54]. With increased resolution, these arrays allowed more precise delineation of the physical boundaries of chromosomal breakpoints in PDA. For example, homozygous deletions at 9p21.3 (45 kb) and high-level amplifications in three regions of 8q: 8q24.13–q24.21 (2.2 Mb), 8q24.22 (177 kb), and 8q24.23–q24.3 (2.7 Mb) (19% of cases) were identified. *SCAP2* (*SKAP2*, 7p15.2) was the most frequently amplified gene (63% of cases), which has not been described in any type of cancer. Increased copy number was also identified for *MYC* (8q24.21, 48%), *NCOA3/AIB1* (20q13.12, 44%), *KRAS* (12p12.1, 44%), *ERBB2* (17q12, 41%), and *EGFR* (7p11.2, 33%) genes. On the other hand, two tumor suppressor genes, *CDKN2A* and *CDKN2B*, were included in the locus of 9p21.3 that was deleted at the highest frequency (63% of cases). Genetic losses were also found in genes such as *DCC* (18q21.1, 48%), *SMAD4* (18q21.1, 33%), *MAP2K4* (17p12, 30%), *TP53* (17p13.1, 26%), and *RUNX3* (1p36.11, 22%). As can be seen from this study, high-density arrays allowed finding both common (i.e., *MYC*) and PDA-specific (i.e., *SKAP2*) genes implicated in pancreatic cancer progression.

Jones and colleagues [31] undertook a large-scale HTG survey of mutations in 24 pancreatic cancers, providing the largest genome-wide search for new candidate genes reported to date. This study determined the sequences of more than 23,000 transcripts representing 20,661 protein-coding genes and also searched for homozygous deletions and amplifications in the tumor DNA using Illumina Human1M-Duo SNP arrays. PDA tumors were found to contain on average 63 genetic alterations, the majority of which were point mutations. A core set of 12 cellular signaling pathways and processes were each found to be genetically altered in 67 to 100% of the tumors. Different modes of genetic alterations in these pathways were identified. First were those pathways in which a single, frequently altered gene predominated, such as KRAS signaling, and the regulation of the G1/S cell cycle transition. Some pathways had a few predominant altered genes, such as TGFβ signaling. A large number of pathways had many different altered genes such as integrin signaling, regulation of invasion, homophilic cell adhesion, and small guanine triphosphatase (GTPase)-dependent signaling. Unfortunately, it is still unclear how pair-wise interactions within and between different pathways influence cancer progression. Larger sample size will be needed to identify and explain the functional consequences of interactions between different genetic mutations.

The use of DNA microarrays for studying gene expression among different phenotypic categories in PDA was the first oncogenomic technique to be applied to this type of cancer [55]. Buchholz et al. [56] conducted a large-scale expression profiling of microdissected cells from normal pancreatic ducts, PanINs of different grades, and PDA from a total of 51 patients with pancreatic cancer using whole-genome oligonucleotide microarrays representing

21,329 genes in the form of optimized oligonucleotide probes. Differentially expressed genes between different stages were organized into functional categories such as development, structure, and signal transduction according to the Gene Ontology annotations with further divisions into subtypes based on tumor invasiveness. For example, within the "structure" category, clusters of benign/hyperplastic tissue-specific genes and dysplastic/neoplastic tissue-specific genes were readily distinguishable: while the former included the matrix metalloproteinases 3 and 17 (MMP3, MMP17) as well as laminin gamma 3 (LAMC3), the latter encompassed fibronectin 1 (FN1), keratin 16 (KRT16), plastin 3 (PLS3), matrix metalloproteinase 7 (MMP7), and collagen type III alpha-1 (COL3A1).

An alternative to the CNCs detection application of SNP arrays is their use in genome-wide association studies (GWAS) that aim to identify cancer susceptibility alleles [7]. Such studies provide an ample ground for the development of various statistical genetics methods for identifying association of single alleles or a subset/combination of alleles with cancer phenotype. Indeed, a GWAS study has confirmed the linkage of ABO blood types to the risk of developing PDA [57].

The second category represents oncogenomic studies that use multiple sources of HTG information. The most prominent example in this category is a recently published paper from the Cancer Genome Atlas pilot project applied to glioblastoma [58]. The large-scale multidimensional analysis in this study assessed DNA sequence changes, copy number aberrations, chromosomal rearrangements, and DNA methylation. Much smaller studies attempted to use gene expression data from DNA microarrays to pinpoint functional consequences of copy number changes in PDA. For example, Heidenblad et al. carried out copy number analysis of 29 pancreatic carcinoma cell lines using aCGH arrays and compared the results with matching transcriptomic profiling data [59]. They showed that a strong association between DNA copy numbers and mRNA expression levels is present in pancreatic cancer, and demonstrated that as many as 60% of the genes within highly amplified genomic regions display associated overexpression. Another study by Fu and colleagues exemplified the advantage of using several HTG technologies for finding candidate cancer genes. These authors used representational oligonucleotide microarray analysis (ROMA) to identify copy number changes in pancreatic cancer xenografts, and validated these findings using FISH, quantitative PCR, Western blotting, and immunohistochemical labelling. With this approach, they identified a 0.36-Mb amplification at 18q11.2 containing two known genes, GATA-6 and cTAGE1. However, combined genetic and transcriptional analyses showed consistent overexpression of GATA-6 in all carcinomas with 18q11.2 gain, as well as in the majority of pancreatic cancers examined (17 of 30 cancers, 56.7%) that did not have gain of this region. By contrast, overexpression of cTAGE1 was rare in these same samples suggesting GATA-6 is the true target of this copy number increase.

For ultimate understanding of oncogenic pathways in PDA and for development of targeted therapeutics, not only the use of different genomic technologies, but also utilization of appropriate model systems will be required.

Cancer cell lines in general are invaluable for studying the genetics of PDA, and some recent studies showed that the cancer genomes of a panel of human cancer cell lines reflect the genomic diversity of human cancers [60]. However, cell line models are limited by their inability to recapitulate tumor–stroma interactions. The mouse has become a model system of choice to study PDA and many other types of cancer [61]. Substantial functional genomic evidence corroborates extensive use of this model organism. For example, Maser et al. compared genomes of mouse tumor cells with genetically engineered chromosomal instability to the genomes of various human cancers and showed that there is a significant non-random number of syntenic events [62]. The observations suggest that mouse and human cells can experience common biological processes driven by orthologous genetic events during transformation. In our laboratory, mouse models of pancreatic cancer are used to find syntenic copy number changes with human PDA tissue samples (as assessed by several aCGH/SNP array platforms), followed by functional validation of potential candidate genes. Similar approaches have proven invaluable in prioritizing genes functionally relevant to melanoma, prostate cancer, and hepatocellular carcinoma [63–65].

Informatics support for systems biology of PDA

The vast and diverse information on various pathogenic alterations in the PDA MMMN requires elaborate IT support for data storage, retrieval, processing, and analysis. Both PDA-specific and general cancer databases exist for storage and curation of HTG data. Cancer-specific databases store specialist knowledge of alterations in one particular cancer by combining diverse data types into one database schema, while general cancer databases allow inter-cancer comparisons. Pancreatic Expression database is an example of the earlier approach [66]. It is a data management system based on the BioMart technology, which stores pancreatic gene expression data alongside the human genome, gene and protein annotations, sequence, gene homolog, SNP and antibody data. Interrogation of the database can be achieved through both a web-based query interface and through web services using combined criteria from pancreatic cancer-specific (disease stages, regulation, differential expression, expression, platform technology, publication) and/or public data (antibodies, genomic region, gene-related accessions, ontology, expression patterns, multi-species comparisons, protein data, SNPs).

The Oncomine database, which stores more than 28,000 gene expression and aCGH microarrays from 41 cancer types (with more than 200 microarrays representing PDA) is the most comprehensive general cancer microarray database [67]. Oncomine creates a set of differentially regulated genes in a particular set of experiments, commonly referred to as gene signatures, and then compares this set with sets from other experiments using several different strategies including differential expression, correlation, meta-analysis, and COPA score. For instance, the COPA score is calculated by searching for gene expression profiles that display the most profound overexpression in a subset of tumors using rank-ordered statistics. A separate module, Molecular Concept Data, compares gene signatures with independently derived gene sets representing molecular pathways and other biological concepts.

Development of new mathematical tools for comparing and analyzing modules and other entities in the PDA MMMN network is another active area of research spawned by the rapid accumulation of HTG data. There are four different approaches that can be considered for mining MMMN of PDA:

1. Making prognostic and predictive estimates of pancreatic cancer progression based on the HTG data.
2. Finding whether genes altered in some way between different pancreatic cancer phenotypes

have overrepresentation of a particular set of features, i.e., genes in the same signaling pathway.

3. Defining the topological properties of the network, such as identity of hubs or connectivity patterns.

4. Integrating different HTG data types to gain comprehensive insight into pancreatic cancer progression.

Again, oncogenomic databases provide both data and programmatic access that make such approaches possible. For instance, data generated by the TCGA consortium can be retrieved and analyzed using extensive sets of database queries [68].

Development of appropriate visualization schemes is also a highly relevant activity for identifying patterns and relationships between genome-wide distribution of genetic aberrations in PDA. Pathway diagrams, networks, and gene signatures are frequently used for identification and visualization of various biological relationships and a number of customizable software tools are available for this purpose [69, 70]. Often it is necessary to create a custom data representation scheme. For example, we use copy number aberration image maps to find regions of recurrent copy number changes in different tumor samples and associate them with cancer phenotype (Figure 26.2).

Future directions in PDA systems biology research

Recent technological improvements in next-generation sequencing provide a great opportunity for rapid advancement in our understanding of genome-wide patterns of somatic aberrations in PDA. The following example illustrates the impact of sequencing technologies on oncogenomic research. Researchers reported approximately 100,000 somatic mutations from cancer genomes in the quarter of a century since the first somatic mutation was found in *HRAS*. With projected advancements in next-generation sequencing technologies, over the next few years several hundred million more will be revealed by large-scale, complete sequencing of cancer genomes [71]. With sufficient genome coverage,

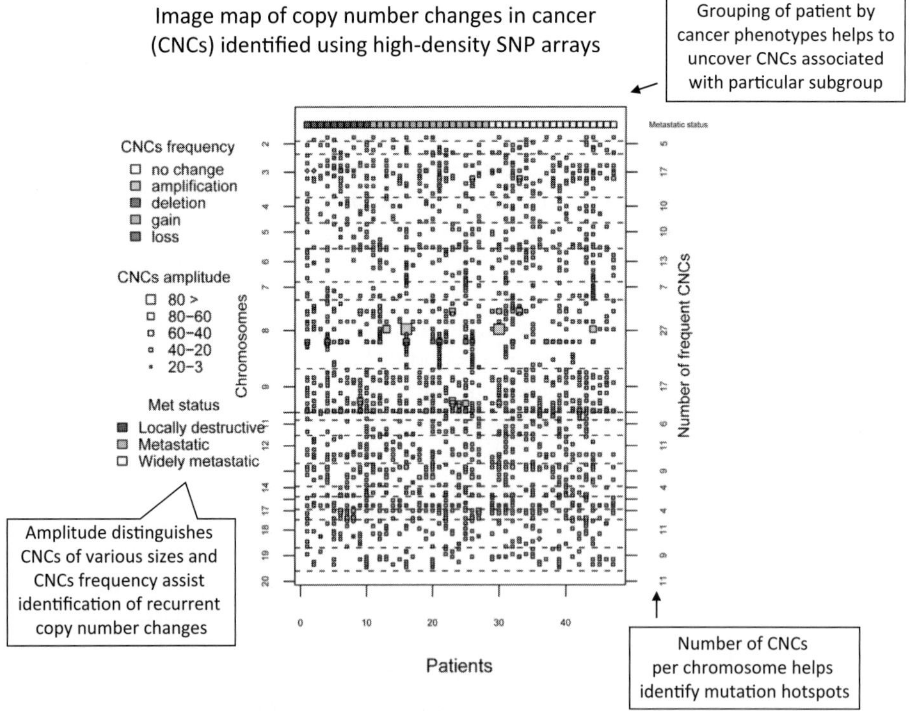

Figure 26.2 Example of an image map of recurrent copy number changes (CNCs) in cancer. Each box represents a CNC event in a particular sample. The size and color of boxes represents amplitude of copy number changes (i.e., gains versus high-level amplifications) and their size. The samples can also be associated with different cancer phenotypes (metastatic status in this example). (A black and white version of this figure will appear in some formats. For the color version, please refer to the plate section.)

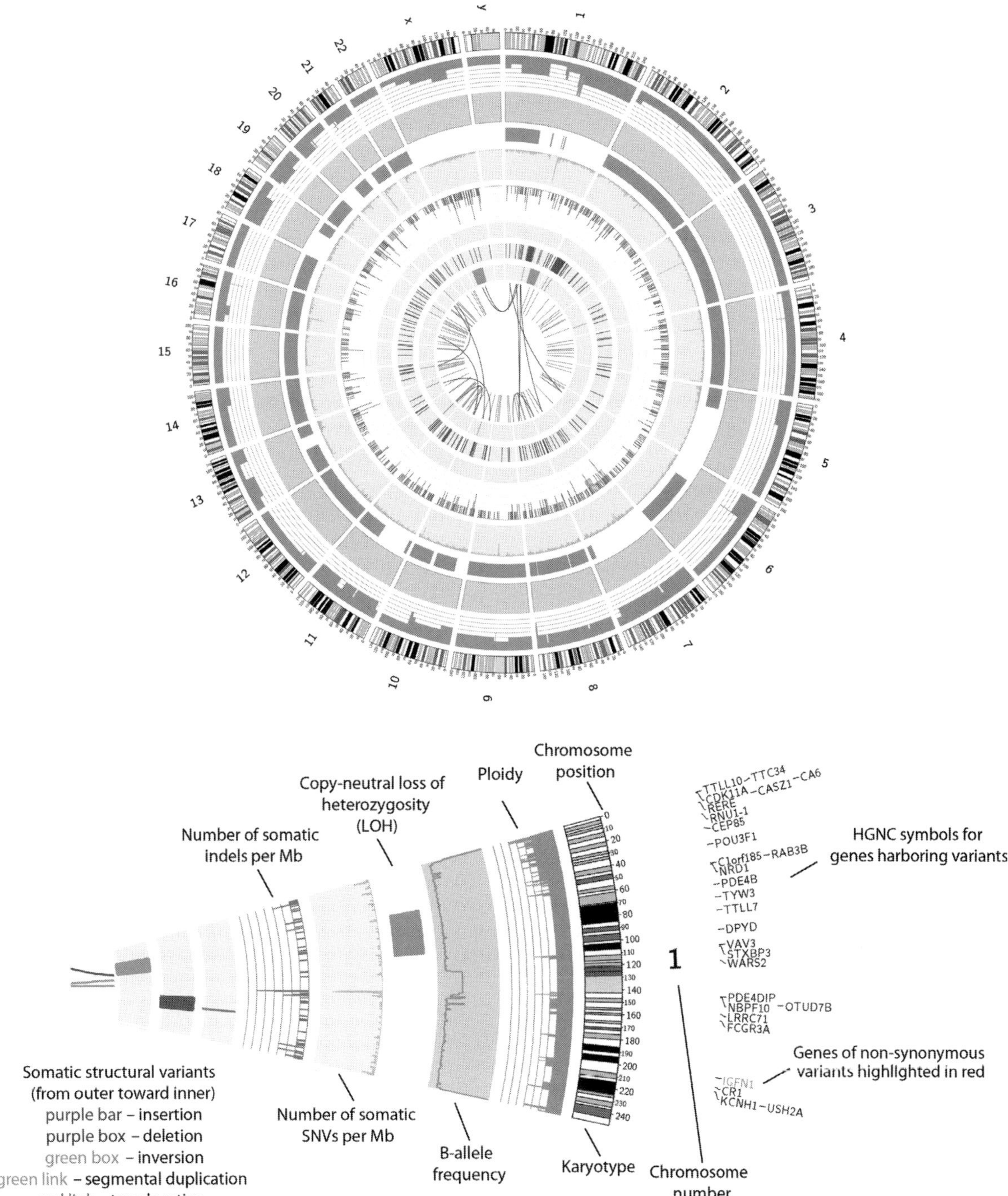

Figure 26.3 Comprehensive categorization of genetic abnormalities in cancers using next-generation sequencing technologies. The example shows part of a catalog of somatic mutations. Individual chromosomes are depicted on the outer circle followed by concentric tracks for point mutation, copy number, and rearrangement data relative to mapping position in the genome. Arrows indicate examples of the various types of somatic mutation present in this cancer genome. (A black and white version of this figure will appear in some formats. For the color version, please refer to the plate section.)

sequencing allows comprehensive detection of major mutation types, including copy number changes, point mutations, and genomic rearrangements, which could not have been possible with microarray technologies (Figure 26.3). For example, recent sequencing of an AML genome identified ten non-synonymous somatic mutations, of which only two were previously known AML-associated mutations. The other eight somatic mutations detected were all single base changes, and none had previously been detected in an AML genome. Moreover, four among these eight somatic mutations occurred in genes not previously implicated in cancer pathogenesis, but whose potential functions in metabolic pathways suggest mechanisms by which they could act to promote cancer, thereby providing new avenues for exploration of diagnostic and therapeutic approaches for AML treatment [72].

Sequencing of the lung cancer genome, albeit at low coverage to robustly detect point mutations, in addition to copy number changes identified 103 somatic rearrangements to the base-pair level of resolution. The other advantage of sequencing is that it, on the contrary to microarray technologies, provides an unlimited dynamic range for detecting copy number changes in cancer.

Finally, characterization of the cancer cell transcriptome and active set of transcription factors, combined with genomic aberration profile, will build a comprehensive picture of cancer cell physiology. Further single-cell sequencing will uncover subclones carrying drug-resistance mutations and allow reconstruction of cancer cells' lineage [71].

We therefore expect in the next few years the appearance of a number of reports that categorize PDA genomes using next-generation sequencing. Analyses of genomic information will be only the first step in our understanding of the molecular genetics of PDA progression. The vast amount of new data will create new and exciting opportunities for systems biology research. Mining DNA sequence datasets will require application of statistical methodologies at two distinct levels. First, there will be a growing need for methods to facilitate discovery of recurrent "driver" mutations. Once the driver mutations are identified, the second-level analysis will focus on the interactions between driver mutations present in different individuals. For example, structurally different mutations inactivating protein phosphatases or activating protein kinases could lead to functionally identical cellular responses (i.e., constitutive kinase activity). The use of pathway-based information derived from literature curation and other data sources will be pivotal for reconstructing MMMN PDA networks from sequencing data and proposing novel therapeutic approaches.

References

1. Jemal, A. et al. 2008. Cancer statistics, 2008. *CA Cancer J Clin*, **58**(2): 71–96.

2. Cancer Research UK 2009. Pancreatic cancer mortality statistics. http://info. cancerresearchuk.org/cancerstats/ types/pancreas/mortality/.

3. Jemal, A. et al. 2009. Cancer statistics, 2009. *CA Cancer J Clin*, **59**(4): 225–49.

4. Wang, M.L. and Foo, K.F. 2009. Adjuvant chemoradiotherapy for high-risk pancreatic cancer. *Singapore Med J*, **50**(1): 43–8.

5. Russo, S. et al. 2007. Locally advanced pancreatic cancer: a review. *Semin Oncol*, **34**(4): 327–34.

6. Lowy, A.M. 2008. Neoadjuvant therapy for pancreatic cancer. *J Gastrointest Surg*, **12**(9): 1600–8.

7. Moore, M.J. et al. 2007. Erlotinib plus gemcitabine compared with gemcitabine alone in patients with advanced pancreatic cancer: a phase III trial of the National Cancer Institute of Canada Clinical Trials Group. *J Clin Oncol*, **25**(15): 1960–6.

8. Fearon, E.R. and Vogelstein, B. 1990. A genetic model for colorectal tumorigenesis. *Cell*, **61**(5): 759–67.

9. Hruban, R.H., Wilentz, R.E. and Kern, S.E. 2000. Genetic progression in the pancreatic ducts. *Am J Pathol*, **156**(6): 1821–5.

10. Maitra, A. and Hruban, R.H. 2008. Pancreatic cancer. *Annu Rev Pathol*, **3**: 157–88.

11. Caldas, C. and Kern, S.E. 1995. K-ras mutation and pancreatic adenocarcinoma. *Int J Pancreatol*, **18**(1): 1–6.

12. Lohr, M. et al. 2005. Frequency of K-ras mutations in pancreatic intraductal neoplasias associated with pancreatic ductal adenocarcinoma and chronic pancreatitis: a meta-analysis. *Neoplasia*, **7**(1): 17–23.

13. Smit, V.T. et al. 1988. KRAS codon 12 mutations occur very frequently in pancreatic adenocarcinomas. *Nucleic Acids Res*, **16**(16): 7773–82.

14. Hingorani, S.R. et al. 2003. Preinvasive and invasive ductal pancreatic cancer and its early detection in the mouse. *Cancer Cell*, **4**(6): 437–50.

15. Caldas, C. et al. 1994. Frequent somatic mutations and homozygous deletions of the p16 (MTS1) gene in pancreatic adenocarcinoma. *Nat Genet*, **8**(1): 27–32.

16. Schutte, M. et al. 1997. Abrogation of the Rb/p16 tumor-suppressive pathway in virtually all pancreatic carcinomas. *Cancer Res*, **57**(15): 3126–30.

17. Ueki, T. et al. 2000. Hypermethylation of multiple genes in pancreatic adenocarcinoma. *Cancer Res*, **60**(7): 1835–9.

18. Wilentz, R.E. et al. 1998. Inactivation of the p16 (INK4A) tumor-suppressor gene in pancreatic duct lesions: loss of intranuclear expression. *Cancer Res*, **58**(20): 4740–4.

19. Landi, S. 2009. Genetic predisposition and environmental risk factors to pancreatic cancer: a review of the literature. *Mutat Res*, **681**(2–3): 299–307.

20. Aguirre, A.J. et al. 2003. Activated Kras and Ink4a/Arf deficiency cooperate to produce metastatic pancreatic ductal adenocarcinoma. *Genes Dev*, **17**(24): 3112–26.

21. Redston, M.S. et al. 1994. p53 mutations in pancreatic carcinoma and evidence of common involvement of homocopolymer tracts in DNA microdeletions. *Cancer Res*, **54**(11): 3025–33.

22. Rozenblum, E. et al. 1997. Tumor-suppressive pathways in pancreatic carcinoma. *Cancer Res*, **57**(9): 1731–4.

23. Boschman, C.R. et al. 1994. Expression of p53 protein in precursor lesions and adenocarcinoma of human pancreas. *Am J Pathol*, **145**(6): 1291–5.

24. Luttges, J. et al. 2001. Allelic loss is often the first hit in the biallelic inactivation of the p53 and DPC4 genes during pancreatic carcinogenesis. *Am J Pathol*, **158**(5): 1677–83.

25. Birch, J.M. et al. 2001. Relative frequency and morphology of cancers in carriers of germline TP53 mutations. *Oncogene*, **20**(34): 4621–8.

26. Bardeesy, N. et al. 2002. Obligate roles for p16(Ink4a) and p19 (Arf)-p53 in the suppression of murine pancreatic neoplasia. *Mol Cell Biol*, **22**(2): 635–43.

27. Wilentz, R.E. et al. 2000. Loss of expression of Dpc4 in pancreatic intraepithelial neoplasia: evidence that DPC4 inactivation occurs late in neoplastic progression. *Cancer Res*, **60**(7): 2002–6.

28. Embuscado, E.E. et al. 2005. Immortalizing the complexity of cancer metastasis: genetic features of lethal metastatic pancreatic cancer obtained from rapid autopsy. *Cancer Biol Ther*, **4**(5): 548–54.

29. Blackford, A. et al. 2009. SMAD4 gene mutations are associated with poor prognosis in pancreatic cancer. *Clin Cancer Res*, **15**(14): 4674–9.

30. Maitra, A. et al. 2003. Multicomponent analysis of the pancreatic adenocarcinoma progression model using a pancreatic intraepithelial neoplasia tissue microarray. *Mod Pathol*, **16**(9): 902–12.

31. Jones, S. et al. 2008. Core signaling pathways in human pancreatic cancers revealed by global genomic analyses. *Science*, **321**(5897): 1801–6.

32. Dancer, J. et al. 2007. Coexpression of EGFR and HER-2 in pancreatic ductal adenocarcinoma: a comparative study using immunohistochemistry correlated with gene amplification by fluorescencent in situ hybridization. *Oncol Rep*, **18**(1): 151–5.

33. Hermanova, M. et al. 2004. Amplification and overexpression of HER-2/neu in invasive ductal carcinomas of the pancreas and pancreatic intraepithelial neoplasms and the relationship to the expression of p21(WAF1/CIP1). *Neoplasma*, **51**(2): 77–83.

34. Yamanaka, Y. et al. 1993. Coexpression of epidermal growth factor receptor and ligands in human pancreatic cancer is associated with enhanced tumor aggressiveness. *Anticancer Res*, **13**(3): 565–9.

35. Miwa, W. et al. 1996. Isolation of DNA sequences amplified at chromosome 19q13.1-q13.2 including the AKT2 locus in human pancreatic cancer. *Biochem Biophys Res Commun*, **225**(3): 968–74.

36. Ruggeri, B.A. et al. 1998. Amplification and overexpression of the AKT2 oncogene in a subset of human pancreatic ductal adenocarcinomas. *Mol Carcinog*, **21**(2): 81–6.

37. Schonleben, F. et al. 2006. PIK3CA mutations in intraductal papillary mucinous neoplasm/carcinoma of the pancreas. *Clin Cancer Res*, **12**(12): 3851–5.

38. Huang, L. et al. 1996. Molecular and immunochemical analyses of RB1 and cyclin D1 in human ductal pancreatic carcinomas and cell lines. *Mol Carcinog*, **15**(2): 85–95.

39. Borg, A. et al. 2000. High frequency of multiple melanomas and breast and pancreas carcinomas in CDKN2A mutation-positive melanoma families. *J Natl Cancer Inst*, **92**(15): 1260–6.

40. Bierie, B. and Moses, H.L. 2006. Tumour microenvironment: TGFbeta: the molecular Jekyll and Hyde of cancer. *Nat Rev Cancer*, **6**(7): 506–20.

41. Kleeff, J. et al. 1999. The TGF-beta signaling inhibitor Smad7 enhances tumorigenicity in pancreatic cancer. *Oncogene*, **18**(39): 5363–72.

42. Ijichi, H. et al. 2006. Aggressive pancreatic ductal adenocarcinoma in mice caused by pancreas-specific blockade of transforming growth factor-beta signaling in cooperation with active Kras expression. *Genes Dev*, **20**(22): 3147–60.

43. Kuang, C. et al. 2006. In vivo disruption of TGF-beta signaling by Smad7 leads to premalignant ductal lesions in the pancreas. *Proc Natl Acad Sci USA*, **103**(6): 1858–63.

44. Asomaning, K. et al. 2008. MDM2 promoter polymorphism and pancreatic cancer risk and prognosis. *Clin Cancer Res*, **14**(12): 4010–15.

45. Maitra, A., Kern, S.E. and Hruban, R.H. 2006. Molecular pathogenesis of pancreatic cancer. *Best Pract Res Clin Gastroenterol*, **20**(2): 211–26.

46. Jones, S. et al. 2009. Exomic sequencing identifies PALB2 as a pancreatic cancer susceptibility gene. *Science*, **324**(5924): 217.

47. Shi, C., Hruban, R.H. and Klein, A.P. 2009. Familial pancreatic cancer. *Arch Pathol Lab Med*, **133**(3): 365–74.

48. Ozcelik, H. et al. 1997. Germline BRCA2 6174delT mutations in Ashkenazi Jewish pancreatic cancer patients. *Nat Genet*, **16**(1): 17–18.

49. Goggins, M., Hruban, R.H. and Kern, S.E. 2000. BRCA2 is inactivated late in the development of pancreatic intraepithelial neoplasia: evidence and implications. *Am J Pathol*, **156**(5): 1767–71.

50. Chin, L. and Gray, J.W. 2008. Translating insights from the cancer genome into clinical practice. *Nature*, **452**(7187): 553–63.

51. Greenman, C. et al. 2007. Patterns of somatic mutation in human cancer genomes. *Nature*, **446**: 153–8.

52. Wood, L.D. et al. 2007. The genomic landscapes of human breast and colorectal cancers. *Science*, **318**: 1108–13.

53. Aguirre, A.J. et al. 2004. High-resolution characterization of the pancreatic adenocarcinoma genome. *Proc Natl Acad Sci USA*, **101**(24): 9067–72.

54. Harada, T. et al. 2008. Genome-wide DNA copy number analysis in pancreatic cancer using high-density single nucleotide polymorphism arrays. *Oncogene*, **27**(13): 1951–60.

55. Han, H. et al. 2002. Identification of differentially expressed genes in pancreatic cancer cells using cDNA microarray. *Cancer Res*, **62**(10): 2890–6.

56. Buchholz, M. et al. 2005. Transcriptome analysis of microdissected pancreatic intraepithelial neoplastic lesions. *Oncogene*, **24**(44): 6626–36.

57. Amundadottir, L. et al. 2009. Genome-wide association study identifies variants in the ABO locus associated with susceptibility to pancreatic cancer. *Nat Genet*, **41**(9): 986–90.

58. Cancer Genome Atlas Research Network. 2008. Comprehensive genomic characterization defines human glioblastoma genes and core pathways. *Nature*, **455**(7216): 1061–8.

59. Heidenblad, M. et al. 2005. Microarray analyses reveal strong influence of DNA copy number alterations on the transcriptional patterns in pancreatic cancer: implications for the interpretation of genomic amplifications. *Oncogene*, **24**(10): 1794–801.

60. Neve, R.M. et al. 2006. A collection of breast cancer cell lines for the study of functionally distinct cancer subtypes. *Cancer Cell*, **10**(6): 515–27.

61. Tuveson, D.A. and Jacks, T. 2002. Technologically advanced cancer modeling in mice. *Curr Opin Genet Dev*, **12**(1): 105–10.

62. Maser, R.S. et al. 2007. Chromosomally unstable mouse tumours have genomic alterations similar to diverse human cancers. *Nature*, **447**(7147): 966–71.

63. Ellwood-Yen, K. et al. 2003. Myc-driven murine prostate cancer shares molecular features with human prostate tumors. *Cancer Cell*, **4**(3): 223–38.

64. Kim, M. et al. 2006. Comparative oncogenomics identifies NEDD9 as a melanoma metastasis gene. *Cell*, **125**(7): 1269–81.

65. Lee, J.S. et al. 2004. Application of comparative functional genomics to identify best-fit mouse models to study human cancer. *Nat Genet*, **36**(12): 1306–11.

66. Chelala, C. et al. 2007. Pancreatic Expression database: a generic model for the organization, integration and mining of complex cancer datasets. *BMC Genomics*, **8**: 16.

67. Rhodes, D.R. et al. 2007. Oncomine 3.0: genes, pathways, and networks in a collection of 18,000 cancer gene expression profiles. *Neoplasia*, **9**(2): 166–80.

68. *The Cancer Genome Atlas (TCGA) Data Portal Use Case Workshop.* 2008, National Cancer Institute and National Human Genome Research Institute, National Institutes of Health, p. 5.

69. Freeman, T.C. et al. 2007. Construction, visualisation, and clustering of transcription networks from microarray expression data. *PLoS Comput Biol*, **3**(10): e206.

70. Shannon, P. et al. 2003. Cytoscape: a software environment for integrated models of biomolecular interaction networks. *Genome Res*, **13**(11): 2498–504.

71. Stratton, M.R., Campbell, P.J. and Futreal, P.A. 2009. The cancer genome. *Nature*, **458**(7239): 719–24.

72. Ley, T.J. et al. 2008. DNA sequencing of a cytogenetically normal acute myeloid leukaemia genome. *Nature*, **456**(7218): 66–72.

Chapter

27

Deregulated signaling networks in lung cancer

Anurag Singh

Introduction

The lung is a complex organ composed of various cell types, performing very specialized functions. The overall purpose is to facilitate efficient oxygen and carbon dioxide exchange between inhaled air and blood in the systemic circulation (Warburton et al., 2000). Furthermore, the lung performs essential filter and barrier functions to limit the deleterious effects of inhaled toxicants and microbes. Since the lung is a primary contact point with air-borne agents in the environment, it is a site of extensive cellular damage. Tissue damage can lead to many lung-associated diseases including chronic obstructive pulmonary disorder (COPD), bronchitis, and, in cases of DNA mutagenesis, cancer may develop.

Lung cancer is one of the most deadly of all human diseases, leading to over 900,000 annual deaths worldwide (Pao and Chmielecki, 2010). It is the leading cause of cancer death in the United States, accounting for 30 and 25% of all cancer-related deaths in males and females, respectively. Overall five-year survival rates have not changed significantly in the last three decades, remaining at around 10 to 15% for most forms of the disease. Lung tumors are diverse and can be subdivided based on histology into small cell (SCLC) and non-small cell subtypes (NSCLC). SCLCs have neuroendocrine-like properties whereas NSCLCs are cancers of bronchiolar epithelial cells. NSCLCs can be further subdivided into three histological subtypes: squamous cell cancer, adenocarcinoma, or large cell carcinoma. These subtypes have contrasting molecular profiles and respond differently to anti-cancer therapeutics.

Lung cancer incidences have steadily risen and plateaued over a 40-year period beginning in the 1940s (Travis et al., 1996). These trends are largely attributable to the rise in popularity of cigarette smoking in the latter half of the twentieth century. Cigarette smoke-associated carcinogens function as potent mutagenic agents that directly promote the mutational disruption of cancer genes such as *KRAS* and *TP53* that lead to lung tumor initiation and progression (Hecht, 2003; Sun et al., 2007). Therefore lung cancer is seen as a largely preventable disease with smoking cessation being a major clinical and public health goal. However, a large fraction of lung cancers arise in so-called "never smokers," accounting for around 25% of all lung cancer cases worldwide (Sun et al., 2007). The etiology and pathophysiology of lung cancer in "never smokers" is complex and somewhat distinct from the disease in smokers (Sun et al., 2007). Of particular note, mutational activation of *EGFR* and *ALK* receptor tyrosine kinases (RTKs) have recently been identified specifically in "never smokers," and EGFR/ALK selective tyrosine kinase inhibitors (TKIs) have shown dramatic effects in subsets of patients who harbor mutations in these genes. The success of these targeted agents provides a strong rationale for "personalized or precision cancer medicine." However, resistance to these therapies is now a critical clinical problem, which must be tackled to prolong disease-free survival.

Many oncogenic driver mutations in lung cancer, such as *EGFR* and *KRAS*, regulate signaling networks associated with cell fate specification during lung morphogenesis. Cancer can arise from improper regulation of developmental cues that lead to uncontrolled tissue growth and proliferation. This may occur in sites prone to tissue injury such as the epithelial lining of the lung, where timely repair of damaged tissue is critical. Indeed, many tumors have properties of "wounds that never heal," with microenvironments that are composed of complex cell types including endothelial cells, and immune cells both of myeloid and lymphoid origin (Dvorak, 1986; Karnoub et al., 2007). To add to this complexity, rare subsets of multipotent cancer stem cells (CSCs) with properties of self-renewal have now been identified (Singh et al., 2004). An appreciation for this cellular heterogeneity in tumors is critical if effective anti-cancer therapeutics for lung cancer are to be designed and used in the clinic. The advent of next-generation massive parallel sequencing has provided unprecedented insight into lung cancer genomics and oncogenic driver events, revealing novel therapeutic opportunities. This chapter will provide an overview of lung cancer molecular etiology and pathology with an emphasis on aberrant regulation of signaling pathways and networks that normally maintain lung tissue homeostasis. The use of functional genomics and systems biology to identify lung cancer-associated oncogenic signaling and gene regulatory networks will be highlighted.

Lung morphogenesis and histopathology

The lung is a ductal epithelial structure derived from the primitive foregut of the endodermal germ layer during embryonic development (Warburton et al., 2000). Highly conserved morphogenetic signaling pathways act on lung progenitor cells resulting in branching morphogenesis and the establishment of a bi-lobular structure in the adult. The adult lung is composed of a number of differentiated cell types with highly specialized functions. Unlike hematopoietic stem cells, lung cells do not fit a classical hierarchical stem cell model. However, upon tissue damage and injury a number of resident progenitor cells with stem-like properties can self-renew and repopulate the lung with differentiated epithelial cell types (Kotton and Fine, 2008). Indeed, multipotent human lung stem cells have been identified that can differentiate and form highly specialized structures in damaged mouse lungs (Kajstura et al., 2011). Furthermore, a common bronchoalveolar stem cell (BASC) has been proposed as the progenitor for all the differentiated cell types in the lung, and a potential cell of origin for NSCLCs (Kim et al., 2005).

Lung histology

The rudimentary lung bronchus buds ventrally from a common esophageal–tracheal tube during early embryonic development. The trachea branches into two proximal bronchi, which progressively branch into smaller distal or peripheral bronchioles that eventually terminate in alveolar ducts where the majority of gas exchange occurs (Figure 27.1) (Kotton and Fine, 2008). The proximal bronchi of the lung are lined by cartilage and smooth muscle, which are essential for mechanical integrity and the control of airflow during breathing. Proximal bronchial epithelial cells are columnar cuboidal and exhibit classical apical versus basolateral polarity defined by adherens and tight junctions. The large bronchus is composed of ciliated and mucous-secreting cells that line the ducts for protection (Figure 27.1). In addition there are secretory neuroendocrine cells as well as basal cells that are believed to have stem cell-like properties (Rawlins and Hogan, 2006). These can be identified by the histological markers cytokeratins 5 and 14 (CK5/CK14). These cells are also positive for the p53 family transcription factor p63, which is a key regulator of squamous epithelial differentiation (Kajstura et al., 2011; Yang et al., 1999).

Distal epithelial structures lack smooth muscle and contractility but are well vascularized by blood vessels composed of endothelial cells. The distal alveolar cells have squamous epithelial morphology and are less polarized than bronchial cells. Distal structures also have an increased concentration of secretory lung progenitor cells known as Clara cells. These can be identified histologically by the marker CC10/SCGB1A1 (Rawlins and Hogan, 2006). In addition, two other cell types with stem- or progenitor-like functions have been described. Within the alveolar duct are secretory cells known as type II pneumocytes, which are derived from type II alveolar ductal epithelial cells. Type II pneumocytes express the surfactant protein C (SPC). Recently an additional stem cell has been identified that can give rise to all the mature differentiated cellular lineages of the lung, referred to as the bronchoalveolar stem cell (BASC) (Kim et al., 2005). BASCs reside at the bronchoalveolar ductal junction (BADJ) and are CC10 and SPC double positive (CC10+/SPC+). Importantly, BASCs are believed to be the cell of origin for oncogenic *KRAS*-driven lung tumorigenesis, although other cell types such as type II pneumocytes can serve this purpose (Xu et al., 2012).

In addition to polarized epithelia, the bronchi play host to pulmonary neuroendocrine (PNE) cells, also known as neuroendocrine bodies (NEBs). PNE cells contain dense granular structures with exocytic products that facilitate communication with nerve fibers

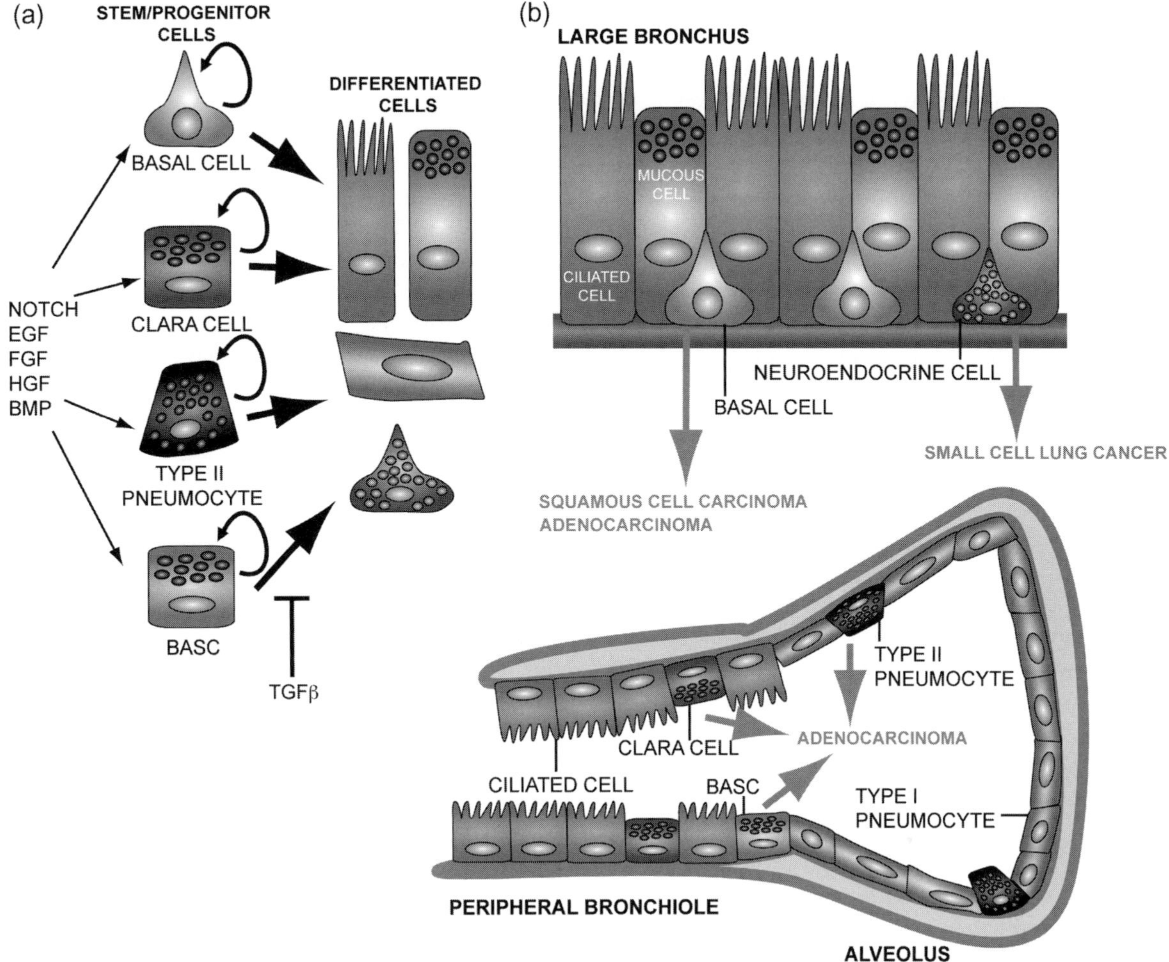

Figure 27.1 Lung morphogenesis and tissue structure.
(a) Different tissue progenitor cells that are resident in the lung parenchyma have stem cell-like properties and can give rise to the various differentiated cell types of the lung. This can occur in response to tissue damage. When these progenitor cells acquire oncogenic mutations, lung cancer can develop. Stimuli such as NOTCH, FGF, and BMPs, sometimes secreted by stromal cells, bind to cognate receptors in target progenitor cells, eliciting a differentiation response. On the other hand, the renewal and expansion of the progenitor cells can be promoted by TGFβ.
(b) Lung cancers can arise in various subcompartments of the lung. In the large bronchus, basal cells can serve as the cells of origin for NSCLCs and neuroendocrine (PNE) cells can give rise to SCLCs. In distal epithelial cells of the peripheral bronchiole and alveolus, Clara cells, type II pneumocytes, and BASCs all have the potential to initiate tumor formation. (A black and white version of this figure will appear in some formats. For the color version, please refer to the plate section.)

and capillaries in the interstitium, as well as peribronchial smooth muscle cells that relax or contract to control airflow into the lungs. PNE cells can be identified histologically by expression of calcitonin gene-related protein (CGRP) (Kotton and Fine, 2008). PNE cells are derived from primitive multipotent bronchial epithelial progenitor cells. PNE cell differentiation is under the control of the basic helix-loop-helix transcription factor hASH1, an ortholog of the *Drosophila* achaete-scute complex 1 gene, which has conserved roles in neuronal development (Borges et al., 1997).

Regulation of lung development

The branched ductal structure of the lung is strikingly similar to structures such as mammary ducts or ductal epithelia of the exocrine pancreas (Affolter et al., 2009). Indeed, molecular and pathophysiological similarities have been noted in cancers of these tissues, particularly adenocarcinomas (Su et al., 2001). Branching morphogenesis of ductal structures is an evolutionarily conserved developmental process in higher metazoans (Affolter et al., 2009; Metzger et al., 2008). The branching of the lung is highly stereotypical and reproducible in human lungs suggesting very tight regulatory control of the process. As detailed below, lung developmental signaling networks are regulated by a number of extracellular factors that are deregulated in human cancers. Indeed, gene expression signatures of lung cancers relate to misregulated lung development (Borczuk et al., 2003).

In the fruit fly *Drosophila melanogaster*, morphogenesis of ducts in respiratory and salivary glands has been extensively studied using classical forward and reverse genetics. Early formation of the common tracheal tube is controlled by the *tracheless/trh* gene, an ortholog of mammalian hypoxia-inducible factor 1α (HIF1α) (Wilk et al., 1996). Primary and secondary branching are controlled by the *branchless/bnl* and *breathless/btl* genes, which are orthologous to fibroblast growth factor and its receptor (FGF/FGFR) (Glazer and Shilo, 1991). The role of FGFR RTK signaling in mammalian lung morphogenesis is conserved, since conditional expression of dominant negative FGFR2 in lung progenitor cells results in branching defects. In addition, FGFR3 and FGFR4 function coordinately to regulate alveolar differentiation (Weinstein et al., 1998). FGF10 is a direct homolog of the *Drosophila bnl* gene. Genetic studies in mice demonstrate a critical role for FGF10 in the initial stages of cell budding during embryonic lung morphogenesis (Min et al., 1998; Sekine et al., 1999). FGF10 is secreted by the primordial lung mesenchyme where it elicits an inductive signal in epithelial cells that express the cognate FGFR2 receptor. Such stromal interactions are crucial for the correct spatio-temporal control of branching morphogenesis.

Activation of FGFR2 leads to the expression of a number of signaling molecules that modulate cell fate specification. Sprouty2 is one the earliest genes to be upregulated whose expression is critical for proper lung morphogenesis (Kramer et al., 1999). Sprouty proteins function as negative regulators of the RAS-mitogen-activated protein kinase (MAPK) signaling cascade to control the spatial amplitude of the FGF10 signal. FGFR signaling also activates and coordinates with a number of evolutionarily conserved signaling

modules to control appropriate lung developmental dynamics. For instance, bone morphogenetic protein 4 (BMP-4) is activated by FGFR, which acts through the type I BMP receptor BMPR1A to spatially constrict FGFR signaling. Activation of these signaling cascades then leads to specific changes in gene regulatory networks mediated by transcription factors such as TTF-1/NKX2.1 and GATA6, which control alveolar cell fate specification via expression of lung-specific differentiation markers (Cheung et al., 2013; Yuan et al., 2000).

A number of factors also control proximal to distal polarity of the lung bronchi and bronchioles. These include components of the Wnt, sonic hedgehog (shh), and retinoic acid (RA) signaling pathways, which facilitate epithelial–mesenchymal cellular communication to coordinate spatio-temporal control of branching morphogenesis (Mucenski et al., 2003; Pepicelli et al., 1998). Of note, RA signaling modulates lung morphogenesis through control of EGFR/RAS signaling (Schuger et al., 1993). A major pathway that counteracts lung differentiation and maintains lung progenitor cell identity is the TGFβ pathway. TGFβ blocks terminal differentiation and promotes stem cell self-renewal (Saito et al., 2009). As discussed below, many of these developmental pathways are severely deregulated in lung cancer.

Lung cancer pathophysiology

Lung cancers can be histologically categorized into two major groups, small cell (SCLC) and non-small cell (NSCLC). SCLCs, also called lung carcinoid tumors, are less frequent but faster growing and often more aggressive. NSCLCs are further comprised of three histological subtypes, squamous cell carcinoma (SCC), adenocarcinoma, and large cell carcinoma. An interesting feature of lung cancers is that the major histological subtypes arise in different subcompartments of the lung. For example, the proximal compartment including the bronchi and PNE cells are common sites for SCC and SCLC (Sun et al., 2007). In contrast, adenocarcinomas are primarily derived from bronchoalveolar cells in the peripheral or distal compartments (Figure 27.1).

SCLCs are believed to derive from PNE cells due to their neuroendocrine-like features. PNE cells secrete a number of neurotransmitters that communicate with underlying neuronal cells and smooth muscle. In addition, PNE cells secrete peptide morphogens such as gastrin-releasing peptide (GRP) also known as bombesin, which plays a key inductive role during fetal lung development and has mitogenic properties on cultured lung epithelial cells (Aguayo et al., 1994). Lending credence to the theory that PNE cells give rise to SCLCs, these cancers express very high levels of GRP/bombesin (Moody et al., 1981). Also of note, the PNE differentiation inducing transcription factor, hASH1, drives proliferation and survival of lung cancer cells with neuroendocrine features and could be a therapeutic target for SCLCs (Osada et al., 2005).

SCLCs and NSCLCs are diagnosed and staged based on histopathological determinations. SCLCs are graded into two stages, limited and extensive, which relate to the growth and spread of the disease within the lungs and to distal tissues. NSCLCs are categorized into four stages relating to lymph node infiltration. Lung cancers can be diverse in nature when analyzed for histopathological markers. IHC analysis of lung tumors can differentiate between adenocarcinoma and SCC subtypes. Specific markers such as TTF-1, CK5/6, and CK7 can be used to that end. Adenocarcinomas are TTF-1 and CK7 positive in contrast to SCCs, which are CK5/6 positive, CK7 negative (Tan et al., 2003). Finally, adenocarcinomas can be subdivided by histological marker expression and gene expression clustering into bronchoid, squamoid, and magnoid (large cell) subtypes (Hayes et al., 2006). These tumor subtypes are either poorly, moderately, or well differentiated. Differentiation is judged by the presence of bronchoalveolar ductal structures and mucin expression using conventional histochemical staining methods. Mucinous adenocarcinomas can be clearly distinguished from non-mucinous counterparts (Travis et al., 2011). Of note, bronchoid tumors diagnosed at stages I and II are usually well differentiated, found predominantly in female non-smokers and are associated with a much better prognosis than squamoid and magnoid adenocarcinomas (Hayes et al., 2006).

All the major subtypes of lung cancer are significantly associated with cigarette smoking. This is the most important single risk factor for lung cancer development. However, smoking-related cancers are not necessarily associated with worse prognoses than those in non-smoking individuals (Hecht, 1999). Since public awareness of this relationship has improved, lung cancer incidences have steadily declined in males, although it is still rising in females (Travis et al., 1996). Smoking-associated carcinogens such as nitrosamine, nitrosamine 4-(methylnitrosamine)-1-(3-pyridyl)-1-butanone (NNK), forms DNA adducts and causes accumulation of oncogenic *KRAS* mutations (Rivenson et al., 1988). In experimental rodent models, treatment with N-nitrosamine is sufficient to induce lung adenocarcinoma (Rivenson et al., 1988). NNK is also an agonist of the β-adrenergic receptor, which contributes to its tumor-promoting properties (Hecht, 2003). Interestingly, the histopathological spectrum of lung cancers has shifted in the last 30 years, with fewer cases of SCC and an increasing rate in adenocarcinomas, which now account for over 40% of all lung cancer cases. This trend has been attributed to the introduction of filtered cigarettes that alter the composition of inhaled carcinogens and subsequently the sites of lung carcinogenesis (Stellman et al., 1997).

Lung cancer genomics

Lung cancers harbor common oncogenic driver mutations and gene copy number variations (CNVs) that promote uncontrolled cell growth and cell survival. Activating mutations in the prototypical oncogene *KRAS* were first discovered in lung cancers in the early 1980s, contributing to the growing field of cancer genetics and the central dogma of cancer as a genetic disorder driven by gene mutations (Santos et al., 1984). Subsequent advances in cancer genetics have provided unprecedented insight into the mechanistic underpinnings of disease etiology and pathology. Numerous large-scale and high-throughput next-generation sequencing (NGS), cDNA and single nucleotide polymorphism/SNP-based microarray studies have amassed detailed cancer genomics information. Lung cancer has featured prominently in these efforts (Ding et al., 2008; Imielinski et al., 2012; Weir et al., 2007). Cancer genomics datasets can be used to identify oncogenic driver mutations, to classify lung tumors into discrete molecular subtypes, and to relate the data to clinical parameters such as prognosis. Furthermore, functional genomics can be used to generate detailed signaling network models that can be exploited for therapeutic target identification. Finally, a recent appreciation for alterations in epigenetic and non-coding RNA expression profiles in lung cancer have added another layer of complexity that must be appreciated for a comprehensive understanding of disease etiology and progression (Lockwood et al., 2012; Yanaihara et al., 2006).

Genetic driver mutations and alterations

Frequently mutated cancer genes have been identified in the past through a number of approaches. Early studies identified oncogenic *RAS* mutations in human cancers by homology to transforming oncogenes of retroviruses (Der et al., 1982; Parada et al., 1982). These mutations were gain of function and highly potent at transforming normal fibroblast cells. Subsequent larger scale sequencing studies identified mutations in the three major RAS isoforms, H, K, and N. *KRAS* mutations in lung cancer can be detected by automated sequencing methods and are found in approximately 30% of all lung cancers. The most frequently mutated gene in lung cancer is *TP53*, encoding the p53 tumor suppressor gene (TSG) gene product. These mutations were identified following the discovery of p53 as a transforming oncogene (Hollstein et al., 1991). However, the mutations in *TP53* were mostly loss-of-function mutations resulting in reduced transcriptional activity. Thus p53 has traditionally been classified as a TSG although recent reports have shown oncogenic and pro-metastatic activities for this enigmatic gene (Muller and Vousden, 2013).

Another frequently mutated gene in lung cancer is *EGFR*, which encodes the epidermal growth factor receptor tyrosine kinase (RTK). The oncogenic potential of EGFR had been recognized for some time and its overexpression or amplification had been noted in many cancers including lung. However, *EGFR* mutations were not identified until 2004 in two seminal studies that reported robust responsiveness to the EGFR tyrosine kinase inhibitor (TKI) gefitinib in patients with activating mutations in the kinase domain of the receptor (Lynch et al., 2004; Paez et al., 2004). Common mutations are the L858R missense mutation and in-frame deletions of exon 19.

Following the elucidation and release of the human genome sequence, efforts began to identify somatic mutations in known protein-coding genes through "exome" sequencing. Initial studies were limited to subsets of genes, for example kinases, which are tractable therapeutic targets (Davies et al., 2005). Subsequent studies were much larger in scale and were directed at mapping entire "Cancer Genomes" (Sjoblom et al., 2006). As a result, both well-established and novel mutated genes have been identified. For well-established oncogenes such as the RAS family, sequencing efforts in large patient cohorts have provided valuable statistical information on mutation frequencies and associated clinical correlates such as prognosis. The new era of cancer genomics has also led to the identification of a number of novel, previously unappreciated gene mutations, such as those in *PIK3CA* and *BRAF*. Furthermore, known cancer gene mutations have been identified in tissues not traditionally associated with those mutations. An example in lung cancer is bi-allelic inactivating mutations in the RAS GTPase activating protein (GAP) *NF1* (Ding et al., 2008).

On the other hand, exome sequencing studies have also revealed a large degree of "mutational noise," which may occur in aneuploid cancer cells exhibiting genomic instability. The challenge, therefore, has been to differentiate significant mutations that promote cancer progression from those that occur stochastically. This has been referred to as the "driver" versus "passenger" mutational dichotomy (Greenman et al., 2006). Therefore exome sequencing of very large patient cohorts is required to provide enough statistical power to differentiate between driver and passenger mutations. Driver mutations are typically non-synonymous (i.e., they cause a missense mutation) and cluster in hotspot exons/codons in multiple individual samples. For example, *KRAS* gene mutations frequently cluster in exons 1 and 2 with a large number encoding for glycine 12 (G12), G13, or glutamine 61 (Q61) position variants. In addition to missense mutations, cancer gene function may be altered by virtue of frameshift mutations, premature stop codons, or in-frame deletions. These latter classes of mutations typically occur in TSGs reflecting loss-of-function activities.

In addition to somatic mutations, genomic instability in cancer cells can lead to gene copy number variations (CNVs), namely amplifications and deletions. Commonly amplified oncogenes include *EGFR*, *HER2*, *MYC*, and *KRAS*. Mutations in TSGs are often followed by deletion of the wild-type TSG allele resulting in loss of heterozygosity (LOH) (Knudson, 1971). Another class of genetic alterations is chromosomal rearrangement that results in chimeric fusion proteins with altered and often oncogenic activities. With these types of alterations in mind, a logical progression from exome sequencing has been the recent focus on "integrated cancer genomics." This entails cataloging detailed information incorporating somatic exome mutations, CNVs, and chromosomal rearrangements, which are all prevalent in virtually all tumor types. This information is then overlapped to accurately identify driver mutations or lesions based on "mutational hotspots" in the genome. All these alterations can be detected by high resolution NGS-based approaches, which have made cancer genomics affordable and cost-effective. The Wellcome Trust Sanger Institute has magnanimously established a curated and searchable database of cancer genomics data from patients (Forbes et al., 2010). As part of the Cancer Genome Project (CGP), the Catalog of Somatic Mutations in Cancer (COSMIC) website provides an excellent user-friendly resource to identify genetic alterations in various cancer types and their relative prevalence. This is often accompanied by clinical information including tumor histopathology and some patient history.

The lung cancer genome

Since lung cancer is a highly prevalent disease, many cancer genomics studies have focused on defining the lung cancer genome. The first large-scale cancer exome sequencing project from the Sanger Institute provided data on 518 protein kinases in 210 human cancers of various origins (Greenman et al., 2007). Lung cancers were found to have a disproportionately high rate of base-pair substitutions, at 4.21 mutations per megabase (Mb) of DNA. This was almost two times higher than for breast cancer and four times higher than colorectal cancer. The high mutational rate could be attributed to DNA damage induced as a result of alkylating agents in cigarette smoke. In all cancers, the C:G>T:A transitions are the most common form of mutation. Lung cancers, in contrast to other cancers, also exhibit very high rates of C:G>A:T transversions, which are typically associated with mutagenic agents in cigarette smoke (Hecht, 2003).

Many non-synonymous somatic mutations have been identified in kinase genes. However, the likelihood that all these are driver mutations may be called into question. It has been hypothesized that many of the kinase mutations identified are loss-of-function mutations that create "dominant negative" kinase variants. Indeed, in lung cancers loss-of-function *STK11* and *ATM* mutations are highly prevalent and cluster in mutational hotspots (Greenman et al., 2007; Sanchez-Cespedes et al., 2002). Other large-scale sequencing studies have identified frequent driver mutations in RTK genes and RTK signaling pathway components, including *EGFR* and *KRAS* (Figure 27.2) (Ding et al., 2008). Of note, many of these gene mutations exhibit mutual exclusivity. For example, *EGFR* and *KRAS* mutations are not found in the same tumors (Ding et al., 2008). These genes form components of the same canonical signaling pathway, which may explain their mutual exclusivity. Alternatively, there may be a negative selective pressure against acquiring mutations in the same pathway.

Lung cancers, in particular adenocarcinomas, exhibit large-scale CNVs. The identification of activating *EGFR* mutations in lung cancer led to the discovery that both mutant and wild-type *EGFR* alleles are amplified, with as many as 20 gene copies per cell (Bell et al., 2005). A landmark study from the Broad Institute subsequently identified 57 recurrent CNVs in 371 lung adenocarcinomas using high-density SNP arrays (Weir et al., 2007). Some CNVs span large regions of the genome but the majority are focal. Of 31 focal CNVs, 24 were amplifications and 7 were homozygous deletions. Interestingly, only 6 of the 24 recurrent events were associated with known driver mutations. Recurrent gene amplifications included well-characterized oncogenes, such as *MDM2, MYC, EGFR, CDK4*, and *KRAS*. Strikingly, the most frequent copy number gain was *NKX2-1*, which encodes the lineage transcription factor TTF-1 that controls lung cell fate commitment. Common deletions include *CDKN2A/CDKN2B*, which encode the p16/p14 cell cycle inhibitors and *PTEN*, a negative regulator of PI3-kinase signaling (discussed below).

Structural genomic rearrangements are common in hematopoietic malignancies leading to chromosomal translocations that generate activated fusion variants of oncoproteins. A classic example is the *BCR-ABL* fusion oncogene product in chronic myelogenous leukemia (CML). Chromosomal translocations are less frequent in solid tumors but have recently come into the spotlight following a number of provocative findings. In lung adenocarcinomas, *EGFR* and *KRAS* mutations account for almost 50% of all tumors and are found in a mutually exclusive manner. This observation led many to speculate on the existence of other oncogenic drivers in the remaining 50% of cases. A seminal study then identified a translocation event involving the anaplastic lymphoma kinase (ALK) through expression cloning of a transforming oncogene from lung adenocarcinomas that were *KRAS* and *EGFR* wild-type. The fusion

(a)

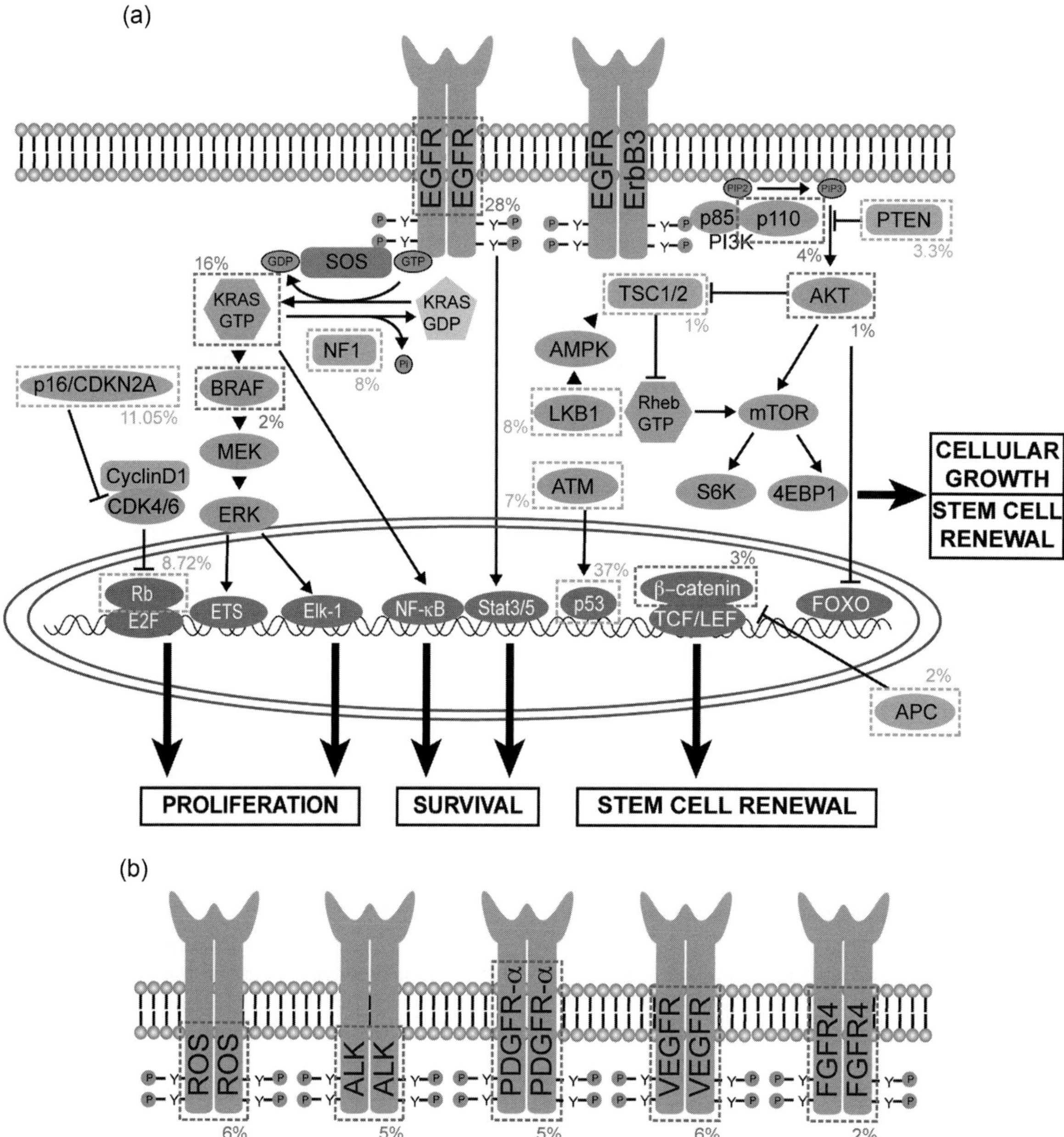

Figure 27.2 Mutationally activated signaling networks in lung cancer.
(a) Components of the EGFR RTK signaling network are frequently altered in lung cancer. Gain-of-function oncogenic mutations are highlighted by a dark dashed line, tumor suppressor genes are highlighted by a lighter dashed line. The mutational frequencies of these genes are also depicted based on data from COSMIC. EGFR activates RAS-MAPK and PI3K signaling, the major arms of the network that drive tumor progression. Activation of these modules leads to changes in gene expression through control of transcription factor function. The net effects in tumors are proliferation, cellular growth, and survival as well as stem cell renewal and altered differentiation patterns.
(b) RTKs are frequently deregulated in lung cancer. Many of these activated RTKs serve complementary function roles to EGFR, and have the capacity to activate RAS and PI3K signaling. Importantly, many of these RTKs that are mutationally activated play roles in normal lung physiology and tissue hoemestasis. (A black and white version of this figure will appear in some formats. For the color version, please refer to the plate section.)

product *EML4-ALK* was shown to have potent oncogenic activities and is now a validated therapeutic target for subsets of lung cancers. Subsequent studies have identified additional rearrangements in lung cancer, including fusions of the ROS1 RTK with CD74 and SLC34A2 (Rikova et al., 2007).

The majority of high-profile genomics studies have focused on lung adenocarcinomas due to their high prevalence compared to other lung cancer subtypes. However, a number of studies have identified alterations in other subtypes such as SCCs. The *SOX2* gene, which encodes a stem cell transcription factor, has been found amplified in lung as well as esophageal SCCs (Bass et al., 2009). Also of note, *FGFR1* is amplified in lung SCCs implicating this RTK as an oncogenic driver in the disease (Heist et al., 2012; Zhao et al., 2005). *FGFR1* is also amplified in a subset of SCLCs, which have been studied in a large-scale integrated cancer genomics investigation (Peifer et al., 2012). In a prior report from the Sanger Institute, next-generation massive parallel sequencing was carried out on an SCLC cell line and showed mutational signatures associated with tobacco carcinogen exposure (Pleasance et al., 2010). The study identified copy number alterations and in-frame fusions involving the DNA helicase and chromatin modifier *CHD7*. Thus aberrant chromatin modification and epigenetic inheritance is integral to SCLC pathology.

Epigenetic alterations

In addition to oncogenic driver mutations in exomes, recent attention has been paid to aberrant cancer epigenetics. A central dogma in genetics that has persisted for many years is the idea that eukaryotic genomes contain extremely large proportions of "junk DNA" or non-coding DNA (ncDNA). In the human genome ncDNA comprises approximately 98% of total DNA. These segments of the genome were believed to serve little or no regulatory purpose in gene expression and played a merely structural role. Large-scale genomics studies as part of the ENCODE (Encyclopedia of DNA Elements) project have now identified novel DNA regulatory elements that control gene transcription and the dogma of "junk DNA" has been challenged (Bernstein et al., 2012). These systems level approaches have allowed for the detailed mapping of genome structure and how information is transmitted both genetically and epigenetically. Epigenetics is defined as genetic information, other than that encoded by DNA sequences, inherited by a daughter cell after cell division (Feinberg and Tycko, 2004). Cancer genomics studies have identified mutations or structural alterations in key epigenetic regulators. Thus altered epigenetic state is now considered a central hallmark of cancer. Importantly, the existence of cancer-associated mutations in ncDNA that alter epigenetic regulation remains a tantalizing possibility.

In cancers, two main types of epigenetic inheritance have been noted, gene regulation through promoter methylation and histone modification, including methylation, acetylation, and phosphorylation. A third type, genomic imprinting, occurs during normal development, in which paternal or maternal alleles of certain genes are epigenetically altered such that the paternal allele predominates. Loss of imprinting (LOI) of the IGF2 locus on chromosome 11p15 results in two copies being aberrantly expressed and has been linked to the pathology of colon cancers and pediatric Wilms' tumors, as well as lung cancers (Kondo et al., 1995; Ogawa et al., 1993).

Gene promoters may be methylated by DNA methyl transferases (DNMTs) at cytosine residues in so-called CpG islands. Gene silencing through promoter hypermethylation prevents the recruitment of active transcriptional complexes and shuts off target gene expression. In contrast, CpG hypomethylation can result in gene activation, for example RAS oncogenes in lung and colon cancers (Feinberg and Vogelstein, 1983). Although gene hypomethylation is a well-established trait in cancer, the majority of attention has been paid to inactivation of TSGs through promoter hypermethylation. A common example of this in lung as well as other cancers is methylation and silencing of the *p16/CDKN2A/INK4A* tumor suppressor locus, resulting in deregulated cell cycle progression (Belinsky et al., 1998; Merlo et al., 1995). Of clinical importance, aberrant promoter hypermethylation is associated with a risk of recurrence in surgically resected stage I NSCLCs (Brock et al., 2008). The study identified a four-gene methylation signature resulting in inactivation of *CDKN2A/p16*, *CDH13*/H-cadherin, the Ras effector *RASSF1A*, and the Wnt pathway negative regulator *APC*.

The second type of epigenetic inheritance is reversible histone modification. Modified histones have the ability to recruit transcriptional regulators to either repress or activate DNA expression in a dynamic manner. In heterochromatic areas such as pericentromeric regions, DNA methylation at CpG islands is required for chromatin compaction and gene repression (Keshet et al., 1986). This is mediated by protein complexes that recognize CpG methylated DNA and methylated histone H4. Histone deacetylases (HDACs) can concomitantly promote chromatin compaction by allowing histones to wrap DNA more tightly. Evidence indicates that histone deacetylation at specific lysine residues in histone H4, K16, and K20 is a universal hallmark in many cancers, causing widespread gene repression (Fraga et al., 2005).

In regions where genes are being actively transcribed, activating histone marks can recruit nucleosome remodelling factors such as the SWI/SNF complex and histone acetyl transferases (HATs). Of note, an integrated genomics report from the Broad Institute has identified lung adenocarcinoma driver mutations in a number of epigenetic regulators, such as *ARID1A* (Imielinski et al., 2012). *ARID1A* encodes an SWI/SNF family member and is a strong candidate TSG. Loss-of-function mutations in other SWI/SNF complex components including *PBRM1* have been identified through exome sequencing studies (Varela et al., 2011). *PBRM1* is mutated in approximately 40% of renal cell carcinomas and 4% of NSCLCs, predominantly of squamous cell histology. Another well-characterized epigenetic regulator with established roles in oncogenesis is the *EZH2* gene product, a component of the polycomb repressor complex 2 (PRC2). PRC2 mediates gene silencing through histone H3 lysine 27 trimethylation. Recurrent EZH2 Y641 mutations were originally identified in lymphomas (Nikoloski et al., 2010). Furthermore, solid tumors such as SCCs of the lung carry loss-of-function EZH2 mutations consistent with the role of epigenetic gene regulation as a lung cancer driver mechanism. Finally, exome sequencing of SCLC has revealed driver mutations in a number of epigenetic regulators, namely *CREBBP*, *EP300*, and *MLL* (Peifer et al., 2012). Taken together, epigenetic regulation is clearly altered in lung cancer with global patterns of gene repression being the most conspicuous net effect.

Non-coding RNAs

The discovery that double-stranded RNA can cause gene silencing via interference with target mRNA has led to a complete paradigm shift in molecular genetics (Fire et al., 1998). This led to subsequent discoveries of endogenous small non-coding microRNAs (miRNAs) *lin-4 and let-7* that regulate gene expression (Reinhart et al., 2000), as well as long non-coding RNAs (lncRNAs) (Mercer et al., 2009). The latter class has received much less attention in the context of cancer biology, although the imprinted H19 transcript, which is downregulated in tandem with *IGF2* upregulation in a number of cancers, represents a famous example. Studies of miRNA function and regulation now constitute one of the most active areas of biomedical research since the turn of the millennium, with over 24,000 publications in a 12-year period since 2001. The processing and maturation of miRNAs has been very well characterized genetically and biochemically. Long pri-miRNA transcripts are expressed under constitutive or regulated conditions followed by processing to shorter hairpin-like precursors by the DROSHA complex. Finally, mature short miRNAs, which are approximately 22 nucleotides in length, are generated by the DICER enzyme. Mature miRNAs subsequently repress gene expression by forming duplexes with target mRNA transcripts, usually in 3′ untranslated regions (UTRs) causing mRNA degradation or blocking translation (Bartel, 2004).

Deregulated expression and function of miRNAs in cancer progression have been documented and studied extensively. Following from the discovery of miRNAs and their integral role in gene regulatory networks, efforts were made to globally profile miRNA expression in cancers. An early study found that miRNA expression profiles could be used as cancer subtype classifiers. Furthermore, global repression of miRNA expression was noted in all cancer types suggesting that miRNAs may play a tumor-suppressive role (Lu et al., 2005). In support of this, genetic depletion of the *DICER* gene, which promotes miRNA maturation causes increased lung adenocarcinoma formation in cooperation with oncogenic *KRAS* in a transgenic mouse model (Kumar et al., 2007). However, a subsequent profiling study in lung cancers identified both up- and downregulated miRNA expression profiles that related to prognosis and SCC versus adenocarcinoma histology (Yanaihara et al., 2006).

A number of studies indicate that aberrant miRNA regulation in lung cancer contributes to disease progression. Since the prototypical miRNA *let-7* was found to target RAS genes in an evolutionarily conserved manner, many reports emerged of let-7 downregulation in human lung cancers, associated with poorer clinical prognosis (Johnson et al., 2005; Takamizawa et al., 2004). Later reports illustrated the potent tumor-suppressive properties of *let-7* by virtue of its ability to reduce KRAS-dependent lung tumor growth in genetically engineered mouse (GEM) models (Esquela-Kerscher et al., 2008; Kumar et al., 2008). Further supporting evidence for let-7 in tumorigenesis was documented through the discovery of a variant SNP in the 3′ UTR of *KRAS* in a higher proportion of NSCLC patients compared to the general population, suggesting an increased risk for tumor development (Chin et al., 2008).

In contrast to the TSG-like properties attributed to *let-7*, pro-oncogenic microRNAs have been identified. For example, the *miR-17–92* cluster is frequently overexpressed in lung cancer and contributes to lung cancer progression, particularly SCLC (Hayashita et al., 2005). Similarly, *mir-21* is overexpressed in lung cancers, which predicts poor patient outcome and forced *mir-21* overexpression promotes KRAS-induced lung tumorigenesis in GEM models (Hatley et al., 2010). Finally, in an effort to identify SNPs in pre- or pri-miRNA segments, a SNP was identified in the *mir-196a2* locus (Hu et al., 2008). This SNP is inherited in a recessive manner and tumors in individuals homozygous for the SNP are associated with poorer survival. Biochemically, the SNP in the *mir-196a2* locus causes increased processing of the pre-miRNA to its mature active form, resulting in higher steady-state levels of the miRNA, suggesting a pro-oncogenic role. Taken together, aberrant expression and function of specific microRNAs can contribute to lung cancer pathology. This may be due to gain or loss of function, which appears to be context dependent.

Molecular subtype classification of lung cancer

The ability to rapidly profile gene expression levels on a whole-genome scale has revolutionized molecular biology research. The use of cDNA microarrays to perform multivariate analyses and to identify significant gene expression patterns in tumor subtypes has broad clinical implications. Early studies elegantly demonstrated that differential expression patterns in breast cancers can be used to accurately classify clinically relevant subtypes that relate to histology and/or oncogenic backgrounds, such as HER2 and estrogen receptor (ER) positivity (Perou et al., 2000; Sorlie

et al., 2001). Shortly thereafter, similar "hierarchical clustering" approaches were applied to lung adenocarcinomas to demonstrate the existence of distinct molecular subtypes (Bhattacharjee et al., 2001). Of note, an adenocarcinoma subtype with high expression of neuroendocrine genes was identified. This subtype was associated with the worst clinical outcome. Numerous studies have now been published on signatures and classifiers that can be used to identify clinically relevant lung cancer subtypes. However, a number of discrepancies have emerged and many of the published signatures are not reproducible. A recent study has attempted to reconcile these discrepancies by employing consistent statistical methodologies to identify reproducible lung adenocarcinoma subtypes (Hayes et al., 2006). Three major histological subtypes were identified that could be molecularly classified, squamoid, bronchoid, and magnoid, and confirmed by standard tissue microarray (TMA)-based histology. Furthermore, the derived gene signatures were capable of predicting stage-specific survival in each of these subtypes. In summary, molecular profiling and the derivation of robust, reproducible subtype classifiers offer promise for the use of cDNA microarray technology as a clinical diagnostic and prognostic platform.

Deregulated signaling networks in lung cancer

As detailed above, key signal transduction networks are central to both normal lung morphogenesis and lung cancer pathology. Many of the modulators that control cell fate and branching morphogenesis are either mutated or misexpressed in human lung cancers. Growth factors and hormones can elicit profound changes in gene transcription leading to cellular reprogramming leading to alterations in cell fate and differentiation. Cooperation between these factors fine-tune lung tissue development leading to appropriate differentiation of progenitor cells into the varied cell types seen in the lung. However, these very same factors when deregulated can promote uncontrolled cell growth and resistance to apoptotic cues, which are the major requisites for lung tumor development. Although lung cancers exhibit varied mutational profiles, many of the deregulated networks have overlapping nodes and common transcriptional outputs. For example, many oncogenic signaling pathways result in activation of the c-Myc transcription factor, which is frequently amplified in subsets of lung cancers (Weir et al., 2007). The most frequent oncogenic lesions are seen in components of the canonical EGFR-RAS signaling pathway, which is evolutionarily conserved as a mediator of cell fate and lineage commitment. Other RTKs, such as ALK, ROS, VEGFR, PDGFR, FGFR, and MET, are also frequently altered and deregulated in lung cancer, highlighting the importance of aberrant RTK function in lung cancer progression (Figure 27.2).

RTK signaling

EGFR and related members of the ErbB/HER family (e.g., HER2) represent prototypical examples of the RTK signaling paradigm. In general, RTKs form homo- or heterodimers when ligated by cognate extracellular peptide growth factors. In the case of EGFR, EGF or TGFα are well-known ligands. Receptor dimerization brings the intracellular kinase domains of the RTK into close proximity culminating in receptor transphosphorylation at key tyrosine residues. RTK mutations in lung cancer lead to constitutive and ligand-independent activation of kinase function (Pao and Chmielecki, 2010). Indeed, *EGFR* mutations in lung cancer cluster in the kinase domain. The L858R, G719S, and L861Q missense mutations, as well as exon 19 in-frame deletions are the most common lesions. Thus lung cancer-derived EGFR mutants display high phosphorylation levels of tyrosine residues in the intracellular region, such as Y1068. These phosphotyrosine residues then serve as docking sites for the recruitment of effector molecules that function as adaptors or have specific enzymatic activities. These docking proteins have conserved SH2 (Src-homology 2) domains that form high-affinity protein–protein interactions with specific phosphotyrosine residues in activated RTKs.

Activation of SH2-domain containing proteins results in downstream signaling events that lead to changes in gene transcription and diverse cell biological outcomes. Proteins recruited and activated by EGFR and other RTKs include phospholipase Cγ (PLCγ), Janus kinases (JAK), which activate the STAT family of transcription factors, tyrosine phosphatases such as SHP1 and PTP1, as well as the E3 ubiquitin ligase c-Cbl. Cbl functions to promote EGFR/RTK internalization and degradation following activation to attenuate signaling during normal cellular function. In addition to these pathways, those mediated by RAS and the phosphoinositide 3-kinase (PI3K) have been most well studied. Importantly, many components and modules of the RAS and PI3K signaling networks are frequently mutated in lung cancer strongly indicative of a causal role for these networks in disease progression (Figure 27.2).

Oncogenic RAS signaling

A very well-characterized RTK docking protein is the Grb2-SOS complex, which serves as an activator of RAS family GTPases. The *KRAS* isoform is mutationally activated in approximately 30% of cases, predominantly in NSCLCs. In contrast, closely related RAS isoforms *HRAS* and *NRAS* are mutated much less frequently in lung cancer, but are found more frequently mutated in other settings such as melanoma. This may suggest context specificity in RAS isoform function in these various cancers. The pro-oncogenic role of mutant KRAS in lung cancer has been firmly established through the use of GEM models in which oncogenic KRAS alleles are conditionally activated in specific tissues. KRAS mutational activation in concert with other genetic lesions, such as *CDKN2A/INK4A, TP53*, or *APC* deletions, lead to the formation of lung, pancreatic, and colorectal adenocarcinomas, which recapitulate the full histopathological spectrum of their human counterparts (Hingorani et al., 2003). Moreover, in human cancers, KRAS mutations arise in early pre-neoplastic lesions that can progress to metastatic disease. Thus KRAS mutational activation plays a direct causal role in the initiation of epithelial tumorigenesis.

Tumor-associated oncogenic mutations in KRAS cluster in hotspot codons 12, 13, and 61, which are sites of GTP and Mg^{2+} coordination. Mutations render the GTP hydrolytic activity of the protein dysfunctional, such that the protein remains in a constitutively active GTP-bound "on state." GTP hydrolysis is accelerated by GTPase activating proteins (GAPs). RAS GAPs function as tumor suppressors, most notably the neurofibromatosis type-1 (*NF1*) gene product, which is mutated in a subset of lung cancers and is often found mutually exclusive to *KRAS* mutations (Ding et al., 2008). RAS GDP–GTP exchange is accelerated by guanine nucleotide exchange factors (GEFs), such as SOS, which can be activated by RTKs. Although *SOS1* and other GEF-encoding gene mutations are rare in lung cancer, *EGFR* and *ALK* are frequently altered in lung cancers and these can potently activate GEFs. As such, cancers with activating RTK aberrations are likely to display hyperactive RAS signaling (Figure 27.2). Therefore in addition to mutational activation of RAS itself, loss of appropriate RAS regulation either by RAS GEF hyperactivation or by loss of RAS GAP function can promote tumor development.

RAS proteins can associate with cellular membranes by virtue of lipid modifications. Palmitoylation and isoprenylation in the C-terminus of RAS allows for membrane anchorage. Active GTP-bound RAS undergoes a conformational change that brings together two "switch domains," which form the effector binding region. Effectors bind to membrane localized RAS and are recruited to membrane microdomains that serve as sites of signal integration to facilitate the initiation of signal transduction cascades (Downward, 2003). Well-studied RAS effectors include the RAF family of serine/threonine (S/T) kinases and the p110 subunit of the PI3K complex. Activating mutations in *BRAF* and *PIK3CA*, which encodes p110, are common in some cancer types but infrequent in lung cancer. RAF activates the canonical MEK-ERK MAP kinase pathway culminating in the phosphorylation, nuclear translocation, and transactivation of transcription factors such as ETS and Elk. The results are gene expression changes that lead to context-dependent regulation of cell fate specification, cell cycle progression, and modulation of inflammatory signaling that can lead to protection against apoptotic cell death. Adverse manifestations of these cell biological outcomes all contribute to RAS-dependent tumorigenesis.

PI3K/mTOR signaling

The PI3K complex, composed of the p85 regulatory subunit and the p110 catalytic subunit, is activated by a number of stimuli and is integral to stress-induced signaling networks that result from energy deprivation. Once recruited to the plasma membrane by RAS and RTKs, PI3K mediates phosphorylation of the 3′ position of the membrane phospholipid phosphatidylinositol 4,5 bisphosphate (PIP2) converting it to PIP3 (Cantley, 2002). PIP3 serves as a membrane recruitment site for proteins that contain a canonical pleckstrin-homology (PH) domain. PIP3 is hydrolyzed to PIP2 catalyzed by PTEN protein and results in PI3K inactivation. *PTEN* is frequently deleted or epigenetically silenced in lung cancer. PH domain-

containing proteins include GEFs for the Rho family of small GTPases and, most notably, the AKT S/T kinase, in which activating mutations can play a role in tumorigenesis (Carpten et al., 2007). AKT has a number of cellular substrates including MDM2, which promotes p53 degradation, Foxo transcription factors, and BAD, an anti-apoptotic mediator. These coordinated processes promote cell survival and proliferation in the face of stress-induced apoptosis. Full activation of AKT is dependent on a phosphorylation event at S473, which is mediated by the mTOR S/T kinase, as part of the mTORC2 complex.

The mammalian target of rapamycin (mTOR) is an evolutionarily conserved kinase that integrates energy sensing with control of cell fate specification through pronounced effects on protein translation and cellular growth. It can reside in two very distinct multiprotein complexes, mTORC1 and mTORC2, which exhibit different modes of regulation and signaling outputs (Zoncu et al., 2011). The regulation of mTORC2 is largely through growth factor signaling and PI3K activation although specific details remain to be elucidated. Energy deprivation and other cellular stressors such as DNA damage can modulate mTORC1 activity via activation of the AMP-activated kinase (AMPK). This phosphorylates and activates the tuberous sclerosis complex (TSC) gene products *TSC1/2*, which function as GAPs for the RHEB small GTPase. When active, under nutrient-rich conditions, GTP-bound RHEB is capable of directly activating mTORC1. Under energy stress, AMPK is activated via LKB1 activation and promotes the TSC1/2-mediated downregulation of RHEB and subsequent inhibition of mTORC1 activity. The results of mTORC1 activation are ribosomal protein S6 phosphorylation through the S6 kinase and phosphorylation of 4EBP resulting in activation of the translation/elongation factor eIF4E. This results in general increased protein translation and effects on autophagy. LKB1, encoded by the *STK11* TSG and *TSC1* are frequently mutated and deleted in lung cancers, resulting in constitutive mTORC1 activation. Thus this pathway has teleological significance in lung cancer pathogenesis. It is important to note that much attention is paid to the effects of mTOR on regulation of translation but a major cell biological regulatory role is in the process of stem cell renewal and proliferation. When mTOR is deleted in mice, embryonic lethality results due to a failure of embryonic stem (ES) cells to proliferate (Murakami et al., 2004).

Furthermore, mTOR activity promotes ES cell pluripotency and suppresses the induction of endoderm and mesoderm cell fates (Zhou et al., 2009). Indeed, control of stemness may be central to the pro-oncogenic properties of PI3K and mTOR in lung cancer as detailed below.

Developmental gene regulatory networks

The predominant signaling networks that are mutationally disrupted in lung cancers affect lung morphogenesis and development. Indeed, gene expression profiling of lung cancers have revealed signatures that relate to lung developmental signaling pathways (Borczuk et al., 2003). Evidence for this relationship has emerged in other cancers such as melanoma, where the lineage transcription factor MITF is amplified and can drive tumor growth and survival (Garraway and Sellers, 2006). Similar examples have been noted in various lung cancer subtypes. In SCLCs, the hASH1 factor that controls neuroendocrine cell fate plays a causal role in the tumor progression (Chen et al., 1997; Osada et al., 2005). In lung SCCs, the stem cell factor SOX2 is frequently amplified and SOX2 depletion causes growth arrest in SCC cell lines that overexpress SOX2 (Bass et al., 2009). In lung adenocarcinomas, a number of RTK signaling networks are frequently disrupted, which control cell fate and differentiation (Figure 27.2). Many of these RTKs, such as FGFR, control branching morphogenesis through regulation of specific transcription factors. Of note, a transcriptional program regulated by lineage factors GATA6 and HOPX limit metastatic progression of lung adenocarcinomas that originate in distal alveolar epithelial cells (Cheung et al., 2013). Finally, genomic amplification of the obligate lung lineage factor TTF1/NKX2.1 has been noted in a significant fraction of lung adenocarcinomas, which are dependent on NKX2.1 for growth and proliferation (Weir et al., 2007). In contrast to human disease, studies in GEM models of KRAS-driven lung adenocarcinomas show that NKX2.1 functions to suppress tumor metastasis (Winslow et al., 2011). The mechanisms by which lineage factors such as NKX2.1 modulate tumorigenesis remain to be fully elucidated. A possibility is that they can maintain cancer cells in a stem-like state or promote differentiation and survival depending on context. This regulation of cancer stem cell renewal and differentiation may occur through control of epithelial–mesenchymal transitions (EMT), which has been noted for the NKX2.1 factor in lung cancer cells, *in vitro* (Saito et al., 2009).

EMT is a developmental program that promotes transdifferentiation of polarized epithelial cells to more migratory mesenchymal cells (Thiery and Sleeman, 2006). Pathophysiological EMT is observed during tissue injury such as bronchial epithelial damage in the lung leading to protective tissue fibrosis. Furthermore, metastatic cancer cells have EMT-like properties that allow for invasion into the vasculature. The EMT program is characterized by key stereotyped gene expression changes resulting in the repression of epithelial genes such as E-cadherin by zinc-finger transcription factors such as Snail and Zeb1. This is accompanied by upregulation of mesenchymal genes such as vimentin and N-cadherin. The EMT program is often subverted during tumorigenesis resulting in the aggressive growth of epithelial tumors ultimately resulting in the dissemination of metastatic cells from the primary tumor during malignant progression. Of note, a distinct miRNA expression profile has been identified in cells that have undergone EMT (Gibbons et al., 2009; Wellner et al., 2009). Specifically, members of the mir-200 family of related miRNAs play a functional role in the EMT process. The mir-200 family targets the E-cadherin transcriptional repressors Zeb1 and Zeb2/SIP1. Furthermore, reciprocal feedback loops result in Zeb repression of mir-205 and the mir-200 family. Thus Zeb factors are critical mediators of epithelial plasticity. This is supported by the observation that manipulation of Zeb expression levels either via RNAi or forced expression of the mir-200 family result in mesenchymal–epithelial (MET) metaplasia (Burk et al., 2008). Interestingly, mir-200 and mir-205 expression levels are significantly deregulated in aggressive lung cancer subtypes, which are causal in tumorigenesis (Wellner et al., 2009; Yu et al., 2008). Taken together, altered regulation of cell fate specification is intrinsic to lung cancer pathogenesis. An appreciation of this will lead to new opportunities for the development of therapeutic paradigms.

Therapeutic strategies in lung cancer treatment

Lung cancer treatment has evolved dramatically since combinations of DNA-damaging agents and antimetabolites were routinely administered for both

SCLC and NSCLC in the late 1970s (Hansen et al., 1976). This was followed almost two decades later by trials showing significant benefit with platinum and paclitaxel combination therapies specifically for NSCLC (Schiller et al., 2002). Overall survival was limited to five or six months with these non-specific agents. The turn of the millennium has seen a growing interest in targeted therapeutics and personalized medicine, reflected by the mantra "finding the right drug for the right patient." In landmark studies, the anti-VEGFR monoclonal antibody bevacizumab (Avastin®) (Genentech/Roche Inc.) demonstrated marked efficacy in NSCLC patients when combined with platinum-based therapeutic regimens, almost doubling overall survival (Sandler et al., 2006). Blockade of kinase function has now become a bellwether paradigm for anti-cancer drug discovery efforts since imatinib (Gleevec®), the BCR-ABL inhibitor, showed almost curative efficacy in Philadelphia chromosome positive CML. Such vulnerabilities in clinical and experimental models are indicative of oncogene "addiction" or dependency, which has been exploited with some success in lung cancer therapeutics. However, acquired drug resistance emerges and methods to tackle this problem are now being explored.

Exploiting oncogene "addiction"

Oncogene dependency, a phenomenon originally described for the MYC oncogene, occurs when a single oncogenic event can profoundly reprogram a cell. During the evolution of the resulting tumor, the initiating oncogene drives a persistent cell survival program despite the accumulation of other oncogenic events. Blockade of the initiating/addicting oncogene results in dramatic tumor regression either through apoptotic cell death or a differentiation response (Weinstein, 2002). In lung cancer, the discovery that EGFR TKIs exhibit a narrow pharmacogenomic efficacy profile in NSCLC patients demonstrated oncogene dependency in solid tumors for the first time (Lynch et al., 2004). Since then, inhibitors of BRAF in melanoma and ALK in EML4-ALK positive lung cancers provide further supporting evidence for the validity of the oncogene dependency model (Kwak et al., 2010). Erlotinib (Tarciva®), which targets EGFR and crizotinib (Xalkori®), which targets ALK, are now first-line therapies in patients harboring activating lesions in these RTKs (Shaw and Engelman, 2013).

EGFR and *ALK* are mutated or disrupted in 20% of NSCLCs, mostly in "never smokers." In another 30% of cases, *KRAS* mutations are found, mostly in patients with a history of smoking. EGFR and ALK TKIs show little or no efficacy in KRAS mutant NSCLCs (Pao and Chmielecki, 2010). Efforts to develop anti-KRAS therapies have been unsuccessful thus far due to the lack of amenability of the KRAS protein to pharmacological inhibition. Previous attempts have focused on inhibitors of KRAS post-translational farnesylation – farnesyl transferase inhibitors or FTIs. These have failed in the clinic due to alternative isoprenylation of KRAS in the presence of these inhibitors (Downward, 2003). An alternative strategy to block KRAS function could be to inhibit its interaction with PDEδ, which has been shown to effectively block KRAS-dependent pancreatic cancer growth (Zimmermann et al., 2013).

Although oncogenic KRAS seems like an attractive therapeutic target, it is not clear whether it represents a universal "Achilles' heel" akin to BCR-ABL or mutant EGFR. In other words, do *KRAS* mutant cancers remain "addicted" to the mutated KRAS protein? Our studies have shown that a subset of KRAS mutant lung NSCLC cell lines can be classified as KRAS independent (Singh et al., 2009). For lung cancers that are KRAS-dependent, KRAS oncogene dependency is not as robust as that seen with EGFR and ALK. Many alternative signaling pathways become activated if KRAS activity is blocked through feedback loops, including activation of the two closely related HRAS and NRAS isoforms. A number of alternative strategies to treat KRAS-dependent cancers have been posited. An intuitive approach is to block the immediate downstream effectors of RAS, namely PI3K, MEK, and downstream kinases such as mTOR. As single agents, MEK and PI3K inhibitors have not shown clear anti tumor efficacy in KRAS-dependent lung cancers. This could be due to alternative PI3K activation mechanisms other than those mediated by KRAS.

The role of KRAS-mediated activation of PI3K in lung tumorigenesis is controversial. Biochemical studies suggest that PI3K activation kinetics by the three different RAS isoforms vary considerably. HRAS and NRAS activate PI3K directly whereas KRAS may form a ternary complex with activated RTKs to fully elicit PI3K activation. However, oncogenic KRAS-driven lung tumor formation in a GEM model requires intact PI3K signaling providing evidence for KRAS-PI3K interactions as a requisite for oncogenesis. A more effective approach would be to block both the MEK and PI3K arms of KRAS signal transduction. Indeed, combination anti-MEK/PI3K therapeutics show dramatic effects in mouse models of KRAS-driven lung tumorigenesis (Engelman et al., 2008). Clinical trials are now under way based on this therapeutic paradigm.

Finally, a novel approach that has received attention but remains controversial is to exploit the phenomenon of synthetic lethality. A number of RNAi screens have been conducted to identify synthetic lethal interactors with KRAS, yielding a variety of tractable genes including the kinases *TBK1, STK33, CDK4* as well as the transcription factor *GATA2* (Barbie et al., 2009; Kumar et al., 2012; Puyol et al., 2010; Scholl et al., 2009). However, some results from these screens, most notably the role of STK33 as a survival mediator, have been called into question due to lack of reproducibility (Babij et al., 2011).

An emerging concept in cancer genetics is the idea of "lineage addiction," which has been reported in a number of contexts, most notably amplification of the melanocyte lineage factor *MITF* in melanoma (Garraway and Sellers, 2006). In NSCLCs, amplification of NKX2.1 suggests lineage addiction in this context and could open avenues to develop inhibitors of NKX2.1 function. Inhibitors of transcription factors are now receiving attention from the pharmaceutical industry following the discovery of a protein–protein interaction inhibitor, JQ1 that blocks MYC-dependent transcriptional activity (Delmore et al., 2011). Alternatively, cancers exhibiting lineage addiction to transcription factors such as NKX2.1 may be addicted to NKX2.1 regulated gene products that drive tumor cell survival. Identification of these genes through functional genomics and RNAi-based screening may prove useful in developing therapies for NKX2.1 amplified lung cancers.

Cancer genomics and personalized medicine

The growth in our understanding of the lung cancer genomic landscape has laid a solid foundation for personalized medicine in this arena. A large array of genome-scale datasets have been generated from lung cancer patient tumors, which highlight the promise of being able to identify underlying tumorigenic processes on an individual basis. Most notably, NSCLCs are routinely genotyped for *EGFR* mutations and *EML4-ALK* translocations to inform clinical treatment decisions on the use of erlotinib and crizotinib, respectively. In other cancers, underlying oncogenic driver lesions may be more difficult to identify. In these cases, gene expression profiling may prove useful. A number of studies have identified lung cancer molecular subtypes though genome-wide expression profiling. These studies have also revealed "pathway signatures" of oncogene activation that may help to guide therapeutic decisions in the future (Bild et al., 2006). For example, *KRAS* is mutated in 30% of lung cancers but a subset of *KRAS* wild-type cancers display a RAS gene expression signature. Thus RAS-directed therapies may show efficacy beyond *KRAS* mutant cancers. Furthermore, these pathway signatures could be used to derive functional information and to develop signaling network models that could be exploited for the identification of synthetic lethal genes and for therapeutic target discovery (Kaelin, 2005).

Drug resistance

The use of cancer therapeutics invariably leads to resistance mechanisms and tumor relapse. Drug resistance can exist *de novo*, in that some agents will not be effective for a particular subset of tumors that are not dependent on a particular drug target. For example, EGFR inhibitors fail to show efficacy in lung cancer patients with *KRAS* mutant tumors, most likely because these tumors rely on KRAS and not EGFR for survival (Pao et al., 2005b). More troubling for clinicians is the emergence of acquired resistance. This is particularly problematic for the use of targeted TKIs in lung cancer. Following the dramatic effects seen in *EGFR* mutant NSCLCs treated with erlotinib, resistance to the drug emerged rapidly. A number of drug resistance mechanisms have been described that could be tackled to treat relapse in these patients.

Drug resistance occurs either through genetic or epigenetic mechanisms. In the case of erlotinib resistance, secondary mutations in EGFR that prevent drug binding are selected for, namely T790M, which is referred to as the gatekeeper residue (Pao et al., 2005a). This occurs in approximately 50% of cases with acquired drug resistance. In the remaining cases, a number of alternative mechanisms have been described. These include amplification of the MET RTK, loss of PTEN, and emergence of CD74-ROS1 translocations (Awad et al., 2013; Engelman et al., 2007; Sos et al., 2009). In addition to these genetic alterations, reversible epigenetic changes that are less tangible to detect may be responsible for drug resistance (Sharma et al., 2010). In such cases, alterations in the differentiation of the tumor elicited as a result of drug treatment may lead to inherent alterations in drug sensitivity. These differentiation changes or cell fate switches may be related to the process of EMT.

A number of studies provide evidence for a link between EMT during malignant growth and loss of oncogene dependency (Singh and Settleman, 2010). For example, inducible expression of oncogenic FGFR-1 in a mouse model of prostate cancer results in EMT during late stages of tumor progression, which is associated with loss of FGFR-1 addiction (Acevedo et al., 2007). Also of note, EMT is associated with a general lack of sensitivity to EGFR TKIs in lung cancers (Fuchs et al., 2008). Interestingly, a recent clinical study indicates that a subset of lung cancer patients with *EGFR* mutant tumors display either tumor transdifferentiation from adeno to squamous histology, or expression of EMT markers when resistance develops to EGFR TKIs (Sequist et al., 2011). Thus the EMT program in cancer cells appears to play a general role in modifying oncogene dependency and drug sensitivity. However, the underlying basis and teleological significance of this relationship has yet to be determined. A possible mechanism may be provided by a report of upregulation of the AXL RTK in erlotinib-resistant lung tumors, which is associated with development of EMT (Zhang et al., 2012).

Epigenetic alterations in drug-resistant tumors may be difficult to manage in the clinic. DNA methylation patterns that result in gene silencing can be reversed through the use of methylation inhibitors such as 5-azaCdR. A possible caveat for this treatment strategy is that DNA hypomethylation has been linked to increased genomic instability, which could accelerate tumor growth in some cases (Feinberg and Tycko, 2004). Since histone deacetylation is required to silence genes such as E-cadherin during EMT, HDAC inhibitors could be useful in reversing EMT-related drug resistance. HDAC inhibitors such as vorinostat are in the clinic for the treatment of some cancers such as cutaneous T-cell lymphoma and could be leveraged for the treatment of drug-resistant lung cancers when used in combination with targeted TKI-based therapies (Lane and Chabner, 2009).

Conclusions

Lung cancers are diverse yet invariably aggressive and deadly. Progress in basic cancer biology and cancer genomics has provided detailed mechanistic insight into the processes that shape lung tumor initiation and malignant progression. Molecular mechanisms of tumorigenesis can vary significantly in the various lung cancer subtypes. This diversity must be appreciated in the design and use of novel anti-cancer therapeutics. Lung cancer is a disease that likely arises when lung tissue is damaged due to environmental exposures, such as cigarette smoke, and repair of the damage is not carried out correctly. Hence the disease can be the result of defects in lung morphogenetic pathways. Indeed, mutations in RTKs such as FGFR that normally promote lung development are indicative of this abnormal relationship. The idea of lineage dependency is further supported by genomic amplification of the transcription factor NKX2.1 that is required for lung morphogenesis and tissue regeneration. Next-generation sequencing efforts have revealed the staggering complexity of lung cancer genomes. These efforts have also identified common genetic alterations that drive lung cancer progression, most notably deregulated RTK signaling networks and aberrations in epigenetic regulation. Thus a path has been forged for the discovery of novel therapeutic paradigms for the treatment of lung cancer. Lung cancer genomics can also open avenues for the development of personalized medicine, firstly to identify driver mutations and subsequently to stratify patient populations based on lung tumor molecular subtypes and to subsequently predict responsiveness to targeted anti-cancer agents. Finally, resistance to anti-cancer therapies such as erlotinib and crizotinib is a serious clinical problem. Many resistance mechanisms may exist, complicating efforts to tackle this problem successfully. In an era of cancer genomics, where a wealth of information is at the fingertips of cancer researchers, these clinical problems may soon be rectified through a deeper understanding of the disease at the systems level.

References

Acevedo, V.D., Gangula, R.D., Freeman, K.W., et al. (2007). Inducible FGFR-1 activation leads to irreversible prostate adenocarcinoma and an epithelial-to-mesenchymal transition. *Cancer Cell 12*, 559–571.

Affolter, M., Zeller, R., and Caussinus, E. (2009). Tissue remodelling through branching morphogenesis. *Nat Rev Mol Cell Biol 10*, 831–842.

Aguayo, S.M., Schuyler, W.E., Murtagh, J.J., Jr., and Roman, J. (1994). Regulation of lung branching morphogenesis by bombesin-like peptides and neutral endopeptidase. *Am J Respir Cell Mol Biol 10*, 635–642.

Awad, M.M., Katayama, R., McTigue, M., et al. (2013). Acquired resistance to crizotinib from a mutation in CD74-ROS1. *N Engl J Med 368*, 2395–2401.

Babij, C., Zhang, Y., Kurzeja, R.J., et al. (2011). STK33 kinase activity is nonessential in KRAS-dependent cancer cells. *Cancer Res 71*, 5818–5826.

Barbie, D.A., Tamayo, P., Boehm, J.S., et al. (2009). Systematic RNA interference reveals that oncogenic KRAS-driven cancers require TBK1. *Nature 462*, 108–112.

Bartel, D.P. (2004). MicroRNAs: genomics, biogenesis, mechanism, and function. *Cell 116*, 281–297.

Bass, A.J., Watanabe, H., Mermel, C.H., et al. (2009). SOX2 is an amplified lineage-survival oncogene in lung and esophageal squamous cell carcinomas. *Nat Genet 41*, 1238–1242.

Belinsky, S.A., Nikula, K.J., Palmisano, W.A., et al. (1998). Aberrant methylation of p16(INK4a) is an early event in lung cancer and a potential biomarker for early diagnosis. *Proc Natl Acad Sci USA 95*, 11891–11896.

Bell, D.W., Lynch, T.J., Haserlat, S.M., et al. (2005). Epidermal growth factor receptor mutations and gene amplification in non-small-cell lung cancer: molecular analysis of the IDEAL/INTACT gefitinib trials. *J Clin Oncol 23*, 8081–8092.

Bernstein, B.E., Birney, E., Dunham, I., et al. (2012). An integrated encyclopedia of DNA elements in the human genome. *Nature 489*, 57–74.

Bhattacharjee, A., Richards, W.G., Staunton, J., et al. (2001). Classification of human lung carcinomas by mRNA expression profiling reveals distinct adenocarcinoma subclasses. *Proc Natl Acad Sci USA 98*, 13790–13795.

Bild, A.H., Yao, G., Chang, J.T., et al. (2006). Oncogenic pathway signatures in human cancers as a guide to targeted therapies. *Nature 439*, 353–357.

Borczuk, A.C., Gorenstein, L., Walter, K.L., et al. (2003). Non-small-cell lung cancer molecular signatures recapitulate lung developmental pathways. *Am J Pathol 163*, 1949–1960.

Borges, M., Linnoila, R.I., van de Velde, H.J., et al. (1997). An achaete-scute homologue essential for neuroendocrine differentiation in the lung. *Nature 386*, 852–855.

Brock, M.V., Hooker, C.M., Ota-Machida, E., et al. (2008). DNA methylation markers and early recurrence in stage I lung cancer. *N Engl J Med 358*, 1118–1128.

Burk, U., Schubert, J., Wellner, U., et al. (2008). A reciprocal repression between ZEB1 and members of the miR-200 family promotes EMT and invasion in cancer cells. *EMBO Rep 9*, 582–589.

Cantley, L.C. (2002). The phosphoinositide 3-kinase pathway. *Science 296*, 1655–1657.

Carpten, J.D., Faber, A.L., Horn, C., et al. (2007). A transforming mutation in the pleckstrin homology domain of AKT1 in cancer. *Nature 448*, 439–444.

Chen, H., Biel, M.A., Borges, M.W., et al. (1997). Tissue-specific expression of human achaete-scute homologue-1 in neuroendocrine tumors: transcriptional regulation by dual inhibitory regions. *Cell Growth Differ 8*, 677–686.

Cheung, W.K., Zhao, M., Liu, Z., et al. (2013). Control of alveolar differentiation by the lineage transcription factors GATA6 and HOPX inhibits lung adenocarcinoma metastasis. *Cancer Cell 23*, 725–738.

Chin, L.J., Ratner, E., Leng, S., et al. (2008). A SNP in a let-7 microRNA complementary site in the KRAS 3′ untranslated region increases non-small cell lung cancer risk. *Cancer Res 68*, 8535–8540.

Davies, H., Hunter, C., Smith, R., et al. (2005). Somatic mutations of the protein kinase gene family in human lung cancer. *Cancer Res 65*, 7591–7595.

Delmore, J.E., Issa, G.C., Lemieux, M.E., et al. (2011). BET bromodomain inhibition as a therapeutic strategy to target c-Myc. *Cell 146*, 904–917.

Der, C.J., Krontiris, T.G., and Cooper, G.M. (1982). Transforming genes of human bladder and lung carcinoma cell lines are homologous to the ras genes of Harvey and Kirsten sarcoma viruses. *Proc Natl Acad Sci USA 79*, 3637–3640.

Ding, L., Getz, G., Wheeler, D.A., et al. (2008). Somatic mutations affect key pathways in lung adenocarcinoma. *Nature 455*, 1069–1075.

Downward, J. (2003). Targeting RAS signalling pathways in cancer therapy. *Nat Rev Cancer 3*, 11–22.

Dvorak, H.F. (1986). Tumors: wounds that do not heal. Similarities between tumor stroma generation and wound healing. *N Engl J Med 315*, 1650–1659.

Engelman, J.A., Chen, L., Tan, X., et al. (2008). Effective use of PI3K and

MEK inhibitors to treat mutant Kras G12D and PIK3CA H1047R murine lung cancers. *Nat Med 14*, 1351–1356.

Engelman, J.A., Zejnullahu, K., Mitsudomi, T., et al. (2007). MET amplification leads to gefitinib resistance in lung cancer by activating ERBB3 signaling. *Science 316*, 1039–1043.

Esquela-Kerscher, A., Trang, P., Wiggins, J.F., et al. (2008). The let-7 microRNA reduces tumor growth in mouse models of lung cancer. *Cell Cycle 7*, 759–764.

Feinberg, A.P., and Tycko, B. (2004). The history of cancer epigenetics. *Nat Rev Cancer 4*, 143–153.

Feinberg, A.P., and Vogelstein, B. (1983). Hypomethylation of ras oncogenes in primary human cancers. *Biochem Biophys Res Commun 111*, 47–54.

Fire, A., Xu, S., Montgomery, M.K., et al. (1998). Potent and specific genetic interference by double-stranded RNA in *Caenorhabditis elegans. Nature 391*, 806–811.

Forbes, S.A., Tang, G., Bindal, N., et al. (2010). COSMIC (the Catalogue of Somatic Mutations in Cancer): a resource to investigate acquired mutations in human cancer. *Nucleic Acids Res 38*, D652–657.

Fraga, M.F., Ballestar, E., Villar-Garea, A., et al. (2005). Loss of acetylation at Lys16 and trimethylation at Lys20 of histone H4 is a common hallmark of human cancer. *Nat Genet 37*, 391–400.

Fuchs, B.C., Fujii, T., Dorfman, J.D., et al. (2008). Epithelial-to-mesenchymal transition and integrin-linked kinase mediate sensitivity to epidermal growth factor receptor inhibition in human hepatoma cells. *Cancer Res 68*, 2391–2399.

Garraway, L.A., and Sellers, W.R. (2006). Lineage dependency and lineage-survival oncogenes in human cancer. *Nat Rev Cancer 6*, 593–602.

Gibbons, D.L., Lin, W., Creighton, C.J., et al. (2009). Contextual extracellular cues promote tumor cell EMT and metastasis by regulating miR-200 family expression. *Genes Dev 23*, 2140–2151.

Glazer, L., and Shilo, B.Z. (1991). The *Drosophila* FGF-R homolog is expressed in the embryonic tracheal system and appears to be required for directed tracheal cell extension. *Genes Dev 5*, 697–705.

Greenman, C., Stephens, P., Smith, R., et al. (2007). Patterns of somatic mutation in human cancer genomes. *Nature 446*, 153–158.

Greenman, C., Wooster, R., Futreal, P.A., Stratton, M.R., and Easton, D.F. (2006). Statistical analysis of pathogenicity of somatic mutations in cancer. *Genetics 173*, 2187–2198.

Hansen, H.H., Selawry, O.S., Simon, R., et al. (1976). Combination chemotherapy of advanced lung cancer: a randomized trial. *Cancer 38*, 2201–2207.

Hatley, M.E., Patrick, D.M., Garcia, M.R., et al. (2010). Modulation of K-Ras-dependent lung tumorigenesis by MicroRNA-21. *Cancer Cell 18*, 282–293.

Hayashita, Y., Osada, H., Tatematsu, Y., et al. (2005). A polycistronic microRNA cluster, miR-17-92, is overexpressed in human lung cancers and enhances cell proliferation. *Cancer Res 65*, 9628–9632.

Hayes, D.N., Monti, S., Parmigiani, G., et al. (2006). Gene expression profiling reveals reproducible human lung adenocarcinoma subtypes in multiple independent patient cohorts. *J Clin Oncol 24*, 5079–5090.

Hecht, S.S. (1999). Tobacco smoke carcinogens and lung cancer. *J Natl Cancer Inst 91*, 1194–1210.

Hecht, S.S. (2003). Tobacco carcinogens, their biomarkers and tobacco-induced cancer. *Nat Rev Cancer 3*, 733–744.

Heist, R.S., Mino-Kenudson, M., Sequist, L.V., et al. (2012). FGFR1 amplification in squamous cell carcinoma of the lung. *J Thorac Oncol 7*, 1775–1780.

Hingorani, S.R., Petricoin, E.F., Maitra, A., et al. (2003). Preinvasive and invasive ductal pancreatic cancer and its early detection in the mouse. *Cancer Cell 4*, 437–450.

Hollstein, M., Sidransky, D., Vogelstein, B., and Harris, C.C. (1991). p53 mutations in human cancers. *Science 253*, 49–53.

Hu, Z., Chen, J., Tian, T., et al. (2008). Genetic variants of miRNA sequences and non-small cell lung cancer survival. *J Clin Invest 118*, 2600–2608.

Imielinski, M., Berger, A.H., Hammerman, P.S., et al. (2012). Mapping the hallmarks of lung adenocarcinoma with massively parallel sequencing. *Cell 150*, 1107–1120.

Johnson, S.M., Grosshans, H., Shingara, J., et al. (2005). RAS is regulated by the let-7 microRNA family. *Cell 120*, 635–647.

Kaelin, W.G., Jr. (2005). The concept of synthetic lethality in the context of anticancer therapy. *Nat Rev Cancer 5*, 689–698.

Kajstura, J., Rota, M., Hall, S.R., et al. (2011). Evidence for human lung stem cells. *N Engl J Med 364*, 1795–1806.

Karnoub, A.E., Dash, A.B., Vo, A.P., et al. (2007). Mesenchymal stem cells within tumour stroma promote breast cancer metastasis. *Nature 449*, 557–563.

Keshet, I., Lieman-Hurwitz, J., and Cedar, H. (1986). DNA methylation affects the formation of active chromatin. *Cell 44*, 535–543.

Kim, C.F., Jackson, E.L., Woolfenden, A.E., et al. (2005). Identification of bronchioalveolar stem cells in normal lung and lung cancer. *Cell 121*, 823–835.

Knudson, A.G., Jr. (1971). Mutation and cancer: statistical study of retinoblastoma. *Proc Natl Acad Sci USA 68*, 820–823.

Kondo, M., Suzuki, H., Ueda, R., et al. (1995). Frequent loss of imprinting of the H19 gene is often associated with its overexpression in human lung cancers. *Oncogene 10*, 1193–1198.

Kotton, D.N., and Fine, A. (2008). Lung stem cells. *Cell Tissue Res 331*, 145–156.

Kramer, S., Okabe, M., Hacohen, N., Krasnow, M.A., and Hiromi, Y. (1999). Sprouty: a common antagonist of FGF and EGF signaling pathways in *Drosophila*. *Development 126*, 2515–2525.

Kumar, M.S., Erkeland, S.J., Pester, R.E., et al. (2008). Suppression of non-small cell lung tumor development by the let-7 microRNA family. *Proc Natl Acad Sci USA 105*, 3903–3908.

Kumar, M.S., Hancock, D.C., Molina-Arcas, M., et al. (2012). The GATA2 transcriptional network is requisite for RAS oncogene-driven non-small cell lung cancer. *Cell 149*, 642–655.

Kumar, M.S., Lu, J., Mercer, K.L., Golub, T.R., and Jacks, T. (2007). Impaired microRNA processing enhances cellular transformation and tumorigenesis. *Nat Genet 39*, 673–677.

Kwak, E.L., Bang, Y.J., Camidge, D.R., et al. (2010). Anaplastic lymphoma kinase inhibition in non-small-cell lung cancer. *N Engl J Med 363*, 1693–1703.

Lane, A.A., and Chabner, B.A. (2009). Histone deacetylase inhibitors in cancer therapy. *J Clin Oncol 27*, 5459–5468.

Lockwood, W.W., Wilson, I.M., Coe, B.P., et al. (2012). Divergent genomic and epigenomic landscapes of lung cancer subtypes underscore the selection of different oncogenic pathways during tumor development. *PLoS One 7*, e37775.

Lu, J., Getz, G., Miska, E.A., et al. (2005). MicroRNA expression profiles classify human cancers. *Nature 435*, 834–838.

Lynch, T.J., Bell, D.W., Sordella, R., et al. (2004). Activating mutations in the epidermal growth factor receptor underlying responsiveness of non-small-cell lung cancer to gefitinib. *N Engl J Med 350*, 2129–2139.

Mercer, T.R., Dinger, M.E., and Mattick, J.S. (2009). Long non-coding RNAs: insights into functions. *Nat Rev Genet 10*, 155–159.

Merlo, A., Herman, J.G., Mao, L., (1995). 5′ CpG island methylation is associated with transcriptional silencing of the tumour suppressor p16/CDKN2/MTS1 in human cancers. *Nat Med 1*, 686–692.

Metzger, R.J., Klein, O.D., Martin, G.R., and Krasnow, M.A. (2008). The branching programme of mouse lung development. *Nature 453*, 745–750.

Min, H., Danilenko, D.M., Scully, S.A., et al. (1998). Fgf-10 is required for both limb and lung development and exhibits striking functional similarity to *Drosophila* branchless. *Genes Dev 12*, 3156–3161.

Moody, T.W., Pert, C.B., Gazdar, A.F., Carney, D.N., and Minna, J.D. (1981). High levels of intracellular bombesin characterize human small-cell lung carcinoma. *Science 214*, 1246–1248.

Mucenski, M.L., Wert, S.E., Nation, J.M., et al. (2003). Beta-catenin is required for specification of proximal/distal cell fate during lung morphogenesis. *J Biol Chem 278*, 40231–40238.

Muller, P.A., and Vousden, K.H. (2013). p53 mutations in cancer. *Nat Cell Biol 15*, 2–8.

Murakami, M., Ichisaka, T., Maeda, M., et al. (2004). mTOR is essential for growth and proliferation in early mouse embryos and embryonic stem cells. *Mol Cell Biol 24*, 6710–6718.

Nikoloski, G., Langemeijer, S.M., Kuiper, R.P., et al. (2010). Somatic mutations of the histone methyltransferase gene EZH2 in myelodysplastic syndromes. *Nat Genet 42*, 665–667.

Ogawa, O., Eccles, M.R., Szeto, J., et al. (1993). Relaxation of insulin-like growth factor II gene imprinting implicated in Wilms' tumour. *Nature 362*, 749–751.

Osada, H., Tatematsu, Y., Yatabe, Y., Horio, Y., and Takahashi, T. (2005). ASH1 gene is a specific therapeutic target for lung cancers with neuroendocrine features. *Cancer Res 65*, 10680–10685.

Paez, J.G., Janne, P.A., Lee, J.C., et al. (2004). EGFR mutations in lung cancer: correlation with clinical response to gefitinib therapy. *Science 304*, 1497–1500.

Pao, W., and Chmielecki, J. (2010). Rational, biologically based treatment of EGFR-mutant non-small-cell lung cancer. *Nat Rev Cancer 10*, 760–774.

Pao, W., Miller, V.A., Politi, K.A., et al. (2005a). Acquired resistance of lung adenocarcinomas to gefitinib or erlotinib is associated with a second mutation in the EGFR kinase domain. *PLoS Med 2*, e73.

Pao, W., Wang, T.Y., Riely, G.J., et al. (2005b). KRAS mutations and primary resistance of lung adenocarcinomas to gefitinib or erlotinib. *PLoS Med 2*, e17.

Parada, L.F., Tabin, C.J., Shih, C., and Weinberg, R.A. (1982). Human EJ bladder carcinoma oncogene is homologue of Harvey sarcoma virus ras gene. *Nature 297*, 474–478.

Peifer, M., Fernandez-Cuesta, L., Sos, M.L., et al. (2012). Integrative genome analyses identify key somatic driver mutations of small-cell lung cancer. *Nat Genet 44*, 1104–1110.

Pepicelli, C.V., Lewis, P.M., and McMahon, A.P. (1998). Sonic hedgehog regulates branching morphogenesis in the mammalian lung. *Curr Biol 8*, 1083–1086.

Perou, C.M., Sorlie, T., Eisen, M.B., et al. (2000). Molecular portraits of human breast tumours. *Nature 406*, 747–752.

Pleasance, E.D., Stephens, P.J., O'Meara, S., et al. (2010). A small-cell lung cancer genome with complex signatures of tobacco exposure. *Nature 463*, 184–190.

Puyol, M., Martin, A., Dubus, P., et al. (2010). A synthetic lethal interaction between K-Ras oncogenes and Cdk4 unveils a therapeutic strategy for non-small cell lung carcinoma. *Cancer Cell 18*, 63–73.

Rawlins, E.L., and Hogan, B.L. (2006). Epithelial stem cells of the lung: privileged few or opportunities for many? *Development 133*, 2455–2465.

Reinhart, B.J., Slack, F.J., Basson, M., et al. (2000). The 21-nucleotide let-7 RNA regulates developmental timing in *Caenorhabditis elegans*. *Nature 403*, 901–906.

Rikova, K., Guo, A., Zeng, Q., et al. (2007). Global survey of phosphotyrosine signaling identifies oncogenic kinases in lung cancer. *Cell 131*, 1190–1203.

Rivenson, A., Hoffmann, D., Prokopczyk, B., Amin, S., and Hecht, S.S. (1988). Induction of lung and exocrine pancreas tumors in F344 rats by tobacco-specific and Areca-derived N-nitrosamines. *Cancer Res 48*, 6912–6917.

Saito, R.A., Watabe, T., Horiguchi, K., et al. (2009). Thyroid transcription factor-1 inhibits transforming growth factor-beta-mediated epithelial-to-mesenchymal transition in lung adenocarcinoma cells. *Cancer Res 69*, 2783–2791.

Sanchez-Cespedes, M., Parrella, P., Esteller, M., et al. (2002). Inactivation of LKB1/STK11 is a common event in adenocarcinomas of the lung. *Cancer Res 62*, 3659–3662.

Sandler, A., Gray, R., Perry, M.C., et al. (2006). Paclitaxel-carboplatin alone or with bevacizumab for non-small-cell lung cancer. *N Engl J Med 355*, 2542–2550.

Santos, E., Martin-Zanca, D., Reddy, E.P., et al. (1984). Malignant activation of a K-ras oncogene in lung carcinoma but not in normal tissue of the same patient. *Science 223*, 661–664.

Schiller, J.H., Harrington, D., Belani, C.P., et al. (2002). Comparison of four chemotherapy regimens for advanced non-small-cell lung cancer. *N Engl J Med 346*, 92–98.

Scholl, C., Frohling, S., Dunn, I.F., et al. (2009). Synthetic lethal interaction between oncogenic KRAS dependency and STK33 suppression in human cancer cells. *Cell 137*, 821–834.

Schuger, L., Varani, J., Mitra, R., Jr., and Gilbride, K. (1993). Retinoic acid stimulates mouse lung development by a mechanism involving epithelial-mesenchymal interaction and regulation of epidermal growth factor receptors. *Dev Biol 159*, 462–473.

Sekine, K., Ohuchi, H., Fujiwara, M., et al. (1999). Fgf10 is essential for limb and lung formation. *Nat Genet 21*, 138–141.

Sequist, L.V., Waltman, B.A., Dias-Santagata, D., et al. (2011). Genotypic and histological evolution of lung cancers acquiring resistance to EGFR inhibitors. *Sci Transl Med 3*, 75ra26.

Sharma, S.V., Lee, D.Y., Li, B., et al. (2010). A chromatin-mediated reversible drug-tolerant state in cancer cell subpopulations. *Cell 141*, 69–80.

Shaw, A.T., and Engelman, J.A. (2013). ALK in lung cancer: past, present, and future. *J Clin Oncol 31*, 1105–1111.

Singh, A., Greninger, P., Rhodes, D., et al. (2009). A gene expression signature associated with "K-Ras addiction" reveals regulators of EMT and tumor cell survival. *Cancer Cell 15*, 489–500.

Singh, A., and Settleman, J. (2010). EMT, cancer stem cells and drug resistance: an emerging axis of evil in the war on cancer. *Oncogene 29*, 4741–4751.

Singh, S.K., Hawkins, C., Clarke, I.D., et al. (2004). Identification of human brain tumour initiating cells. *Nature 432*, 396–401.

Sjoblom, T., Jones, S., Wood, L.D., et al. (2006). The consensus coding sequences of human breast and colorectal cancers. *Science 314*, 268–274.

Sorlie, T., Perou, C.M., Tibshirani, R., et al. (2001). Gene expression patterns of breast carcinomas distinguish tumor subclasses with clinical implications. *Proc Natl Acad Sci USA 98*, 10869–10874.

Sos, M.L., Koker, M., Weir, B.A., et al. (2009). PTEN loss contributes to erlotinib resistance in EGFR-mutant lung cancer by activation of Akt and EGFR. *Cancer Res 69*, 3256–3261.

Stellman, S.D., Muscat, J.E., Hoffmann, D., and Wynder, E.L. (1997). Impact of filter cigarette smoking on lung cancer histology. *Prev Med 26*, 451–456.

Su, A.I., Welsh, J.B., Sapinoso, L.M., et al. (2001). Molecular classification of human carcinomas by use of gene expression signatures. *Cancer Res 61*, 7388–7393.

Sun, S., Schiller, J.H., and Gazdar, A.F. (2007). Lung cancer in never smokers – a different disease. *Nat Rev Cancer 7*, 778–790.

Takamizawa, J., Konishi, H., Yanagisawa, K., et al. (2004). Reduced expression of the let-7 microRNAs in human lung cancers in association with shortened postoperative survival. *Cancer Res 64*, 3753–3756.

Tan, D., Li, Q., Deeb, G., et al. (2003). Thyroid transcription factor-1 expression prevalence and its clinical implications in non-small cell lung cancer: a high-throughput tissue microarray and immunohistochemistry study. *Hum Pathol 34*, 597–604.

Thiery, J.P., and Sleeman, J.P. (2006). Complex networks orchestrate epithelial-mesenchymal transitions. *Nat Rev Mol Cell Biol 7*, 131–142.

Travis, W.D., Brambilla, E., Noguchi, M., et al. (2011). International Association for the Study of Lung Cancer/American Thoracic Society/European Respiratory Society international multidisciplinary classification of lung adenocarcinoma. *J Thorac Oncol 6*, 244–285.

Travis, W.D., Lubin, J., Ries, L., and Devesa, S. (1996). United States lung carcinoma incidence trends: declining for most histologic types among males, increasing among females. *Cancer 77*, 2464–2470.

Varela, I., Tarpey, P., Raine, K., et al. (2011). Exome sequencing identifies frequent mutation of the SWI/SNF complex gene PBRM1 in renal carcinoma. *Nature 469*, 539–542.

Warburton, D., Schwarz, M., Tefft, D., et al. (2000). The molecular basis of lung morphogenesis. *Mech Dev 92*, 55–81.

Weinstein, I.B. (2002). Cancer. Addiction to oncogenes – the Achilles heel of cancer. *Science 297*, 63–64.

Weinstein, M., Xu, X., Ohyama, K., and Deng, C.X. (1998). FGFR-3 and FGFR-4 function cooperatively to direct alveogenesis in the murine lung. *Development 125*, 3615–3623.

Weir, B.A., Woo, M.S., Getz, G., et al. (2007). Characterizing the cancer genome in lung adenocarcinoma. *Nature 450*, 893–898.

Wellner, U., Schubert, J., Burk, U.C., et al. (2009). The EMT-activator ZEB1 promotes tumorigenicity by repressing stemness-inhibiting microRNAs. *Nat Cell Biol 11*, 1487–1495.

Wilk, R., Weizman, I., and Shilo, B.Z. (1996). tracheless encodes a bHLH-PAS protein that is an inducer of tracheal cell fates in *Drosophila*. *Genes Dev 10*, 93–102.

Winslow, M.M., Dayton, T.L., Verhaak, R.G., et al. (2011). Suppression of lung adenocarcinoma progression by Nkx2-1. *Nature 473*, 101–104.

Xu, X., Rock, J.R., Lu, Y., et al. (2012). Evidence for type II cells as cells of origin of K-Ras-induced distal lung adenocarcinoma. *Proc Natl Acad Sci USA 109*, 4910–4915.

Yanaihara, N., Caplen, N., Bowman, E., et al. (2006). Unique microRNA molecular profiles in lung cancer diagnosis and prognosis. *Cancer Cell 9*, 189–198.

Yang, A., Schweitzer, R., Sun, D., et al. (1999). p63 is essential for regenerative proliferation in limb, craniofacial and epithelial development. *Nature 398*, 714–718.

Yu, S.L., Chen, H.Y., Chang, G.C., et al. (2008). MicroRNA signature predicts survival and relapse in lung cancer. *Cancer Cell 13*, 48–57.

Yuan, B., Li, C., Kimura, S., et al. (2000). Inhibition of distal lung morphogenesis in Nkx2.1(–/–) embryos. *Dev Dyn 217*, 180–190.

Zhang, Z., Lee, J.C., Lin, L., et al. (2012). Activation of the AXL kinase causes resistance to EGFR-targeted therapy in lung cancer. *Nat Genet 44*, 852–860.

Zhao, X., Weir, B.A., LaFramboise, T., et al. (2005). Homozygous deletions and chromosome amplifications in human lung carcinomas revealed by single nucleotide polymorphism array analysis. *Cancer Res 65*, 5561–5570.

Zhou, J., Su, P., Wang, L., et al. (2009). mTOR supports long-term self-renewal and suppresses mesoderm and endoderm activities of human embryonic stem cells. *Proc Natl Acad Sci USA 106*, 7840–7845.

Zimmermann, G., Papke, B., Ismail, S., et al. (2013). Small molecule inhibition of the KRAS-PDEdelta interaction impairs oncogenic KRAS signalling. *Nature 497*, 638–642.

Zoncu, R., Efeyan, A., and Sabatini, D.M. (2011). mTOR: from growth signal integration to cancer, diabetes and ageing. *Nat Rev Mol Cell Biol 12*, 21–35.

Chapter 28

Modular signaling in hematopoietic malignancies

Adam Lerner

Introduction

Among the most influential concepts in cancer biology over the last decade is that cancer cells, regardless of origin, acquire a common set of "hallmarks" during their progression to a fully developed malignant phenotype [1]. These hallmarks include unlimited proliferation, self-sufficiency in growth signals, resistance to anti-proliferative and anti-apoptotic signals, sustained angiogenesis, and the ability to invade tissues and metastasize. Such changes are generally the result of gain-of-function mutation, amplification, or overexpression of oncogenes, or loss-of-function mutation, deletion, or epigenetic silencing of tumor suppressors.

Given the morphologic and clinical heterogeneity found in hematologic malignances, it is not surprising that there is comparable molecular heterogeneity in the genetic events that the neoplastic hematopoietic cells acquire that are responsible for their malignant phenotype. Within each subtype of hematopoietic malignancy, however, it appears that identified genetic abnormalities in subsets of patients tend to cluster in common molecular modules [2]. Such observations certainly emphasize the importance of the specific deranged signaling pathway in these malignancies and serve as evidence that therapies targeting this pathway have a higher likelihood of therapeutic success than in patients without such genetic abnormalities. There is therefore quite appropriate interest in the development of biomarkers that allow identification of such deranged molecular modular signaling regardless of the specific genetic lesion underlying the disease. An example of this would be the presence of nuclear p65 in malignant myeloma cells containing genetic abnormalities that result in constitutive signaling through the classic NF-κB signaling pathway.

An extension of the "hallmarks" concept has been the observation that the achievement of such characteristics of malignancy as self-sufficiency in growth signals results in concurrent "stress phenotypes" common to a wide variety of malignancies [3]. Survival in the setting of constitutive DNA damage, metabolic, oxidative, mitotic, or proteotoxic stressors results in unusual reliance of malignant cells on signaling pathways whose function may not be as critical under basal conditions in non-malignant cells.

The usefulness of this malignancy associated stress pathway concept comes in its ability to explain both "oncogene addiction" and the more recent concept of "non-oncogene addiction" [4, 5]. Non-oncogene addiction posits that while a specific component of a signaling pathway may not itself be the target of acquired genetic changes in a particular malignancy, the stress pathways activated in that malignancy may render the function of that component critical to cell survival. Again, a relevant example would be the function of the proteosome in multiple myeloma. While components of the proteosome have not themselves been identified as targets of recurrent genetic changes in this disease, the results of clinical trials with the proteosomal inhibitor bortezomib would suggest that "proteosomal addiction" might be a manifestation of stress pathways induced in the malignant plasma cell. However, from a practical standpoint, these models suggest that it will be important in the years ahead to identify not only the full range of molecular modules activated in each hematopoietic malignancy but also the means by which to determine reliably and inexpensively whether the function of a particular module is required for cell survival in a clinical sample.

Primary malignant hematopoietic cells can often be purified directly from patient samples and in some

---

instances, utilized for *in vitro* experiments, both of which are generally quite difficult for samples derived from solid tumors. In those cases where primary malignant cells have been compared with cell lines derived from such primary cells, many aspects of the behavior of the primary cells are found to be altered in the cell lines. The crisis required for establishment of cell lines results in additional genetic changes that appear to have a substantial effect on proliferation, differentiation, and sensitivity to apoptosis. Nonetheless, some important signaling features of malignant hematopoietic cell lines remain intact and there are many instances in which experiments carried out on cell lines have allowed insights into the disease from which the cell lines were derived. On the other hand, the ease with which experiments can be carried out in malignant cell lines is sometimes beguiling and should be no substitute for the admittedly difficult task of carrying out a careful examination in primary malignant cells of the hypotheses generated by initial experiments in cell lines.

This chapter will summarize some recent highlights of researchers' efforts to identify examples of recurrent dysregulated modular signaling in myeloid and lymphoid malignancies.

Acute myeloid leukemia: a malignancy that requires the combined effect of at least two classes of mutations

Acute myeloid leukemia (AML) is characterized by the accumulation of morphologically immature myeloid cells in the bone marrow. Historically, AML was classified into eight subtypes (M0 to M7) defined by characteristic morphology and immunophenotype. As a series of recurrent translocations were identified in AML cells that resulted in aberrant transcription factor fusion products, their presence was determined to frequently correlate with particular morphologic characteristics. AML cells containing the t(8;21)(q22; q22) AML1/ETO translocation typically have the morphology of differentiated myeloblasts (M2); those with the t(15;17)(q22;q12) PML/RARα translocation have a promyelocytic (M3) morphology; while those with the inv(16)(p13;q22) CBFβ/MYH11 translocation have a predominantly myelomonocytic morphology with abnormal eosinophils (M4a). AML1 and CBFβ normally heterodimerize as components of corebinding factor (CBF), a key regulator of definitive

hematopoiesis, and both the AML1/ETO and CBFβ/MYH11 fusion proteins drive aberrant recruitment of the transcriptional repressor NCoR/HDAC complex to promoters required for normal myeloid development.

The correlation between translocation and characteristic morphology suggest that the altered transcription factor products of such translocations are responsible for arresting AML cells at a specific stage of differentiation and might be sufficient to induce the disease. However, experiments in which such translocation products are constitutively expressed in murine hematopoietic cells demonstrate that expression of these mutant proteins is not in itself sufficient to induce an AML-like malignancy. Mice expressing a conditional "knockin" AML1-ETO fusion protein have hematopoietic progenitor cells with increased self-renewal capability but such mice have no increased frequency of AML [6]. However, treatment of such mice with the mutagen N-ethyl-N-nitrosourea (ENU) led to a 31% incidence of granulocytic sarcomas while no such tumors were identified in comparably treated wild-type mice. Mice transgenic for PML/RARα under the control of the cathepsin G gene have an increased percentage of immature myeloid cells in bone marrow, blood, and spleen [7]. Thirty percent of such mice develop an AML-like disease, but only after a long latency period, suggesting the acquisition of further genetic events is required for such a transition. Finally, bone marrow from chimeric mice generated with heterozygous "knockin" Cbfβ+/Cbfβ–MYH11 ES cells demonstrate a selective defect in myeloid and lymphoid development [8]. Leukemic transformation in such chimeric mice is again, however, only observed following treatment with ENU, whereupon 84% of the Cbfβ+/Cbfβ–MYH11 chimeras but none of the wild-type chimeras develop leukemia.

In addition to transcription factor translocations that dysregulate myeloid differentiation, AML is also characterized by mutations in genes such as FLT3, Kit, and NRAS that are known to play important roles in signal transduction pathways that regulate myeloid proliferation. While it is uncommon to identify AML patients with mutation of more than one of these myeloid proliferation genes, it is common to observe concurrent expression of fusion transcriptional factors with Flt3, Kit, or NRas mutations. In one well-characterized set of 135 AML patients with the AML1-ETO rearrangement, 10.3% of patients had FLT3 mutations, 8.1% the KITD816 mutation, and

9.6% had NRAS codon 12/13/61 mutations [9]. Such observations led to the proposal that these two classes of genetic aberrations collaborate to induce leukemogenesis [10]. To address whether concurrent expression of the AML1-ETO fusion protein and a FLT3 length mutation (FLT3-LM) is sufficient to induce an AML-like disease, Schessl and colleagues carried out long-term bone marrow transplantation studies using marrow singly or doubly transduced with vectors driving expression of these mutant proteins [9]. Whereas none of the mice transplanted with marrow expressing AML1-ETO or FLT3-LM alone developed leukemia over a 20-month period, all recipients of doubly transduced bone marrow rapidly succumbed to an aggressive leukemia.

AML fusion proteins: more than just arrest of differentiation

All-trans retinoic acid (ATRA) was reported to induce remissions in patients with acute promyelocytic leukemia (APL) in 1987 [11]. Treatment of APL patients with ATRA was often accompanied by a rapid rise in the peripheral blood neutrophil count, suggesting ATRA-induced differentiation of bone marrow leukemic promyelocytes. Three years later, the t(15;17) translocation characteristic of this form of AML was shown to result in fusion of the retinoic acid receptor gene to the promyelocytic leukemia (PML) gene on chromosome 15, resulting in a PML-RARα fusion protein [12]. *In vitro* studies demonstrated that pharmacologic doses of ATRA induced the release of transcriptional co-repressor complexes from PML-RARα, with subsequent activation of transcription of genes required for myeloid differentiation [13, 14]. This confluence of clinical and molecular data on APL generated considerable enthusiasm for the concept that other cancers characterized by morphology consistent with oncogene-induced developmental arrest might prove to respond to comparable "differentiation therapy."

When used alone, however, ATRA is insufficient to induce lasting remissions in APL. In contrast, addition of ATRA to standard AML chemotherapy reduces deaths from the coagulopathy that results from release of promyelocyte granules during induction therapy. Addition of ATRA to standard anthracycline-based chemotherapy has also improved the long-term remission rate in APL. Arsenic trioxide, another strikingly effective treatment regimen for this leukemia, induces SUMOylation of the PML moiety of the PML-RARα fusion protein, with subsequent proteosome-mediated destruction of the protein [15]. Recent work by Nasr et al. suggests the molecular basis for the ineffectiveness of ATRA-induced APL differentiation to cure this disease. These authors observed that while ATRA induced differentiation in a murine model of APL, it was insufficient to cause complete depletion of transplantable cells capable of re-initiating APL in a second animal, termed "leukemia initiating cells" (LIC) [16]. In contrast, addition of ATRA to arsenic augmented proteosomal destruction of PML/RARα and led to a dramatic reduction in LIC, and substantially improved survival in this murine APL model. Co-treatment of mice with the proteosomal inhibitor bortezomib reduced ATRA/ arsenic trioxide-induced PML/RARα degradation and largely eliminated the protective effect of these two agents on leukemic progression. Importantly, studies with a second set of mice expressing another APL fusion oncoprotein, PLZF/RARα, suggested that the differentiating activity of ATRA could be separated from its ability to reduce LIC.

These studies suggest that the translocation-induced fusion proteins present in AML not only arrest myeloid differentiation but also directly contribute to activation of signaling pathways necessary for the survival of the APL "stem cell." Consistent with this model, gene chip expression studies in hematopoietic cells expressing the PML-RARα and AML1/ETO oncoproteins have revealed constitutive activation of the Jagged1/NOTCH pathway, a pathway known to be associated with stem cell renewal [17]. Functional studies confirmed upregulation of Jagged1 mRNA and protein following overexpression of PML/RARα or AML1/ETO in U937 cells as well as activation of a known luciferase construct containing the Notch-responsive HES1 promoter. These studies suggest that fusion proteins expressed in AML may be selected to induce both arrest of differentiation and acquisition of "stemness." Pharmacologic agents such as ATRA that counteract only one aspect of such signaling while leaving the protein intact are understandably less effective than those such as arsenic that effectuate the destruction of the fusion oncoprotein.

Role for HOXA homeobox proteins in MLL-translocation leukemias

Myeloid and lymphoid leukemias carrying translocations at the mixed lineage leukemia locus (MLL) at chromosome 11q23 carry a poor prognosis. More than 30 distinct translocations involving this locus have been identified that generate novel chimeric proteins containing the amino terminal 1,400 residues of the transcription factor MLL fused to a partner protein [18]. While the DNA-binding domain of MLL is conserved, the PHD, transactivation, and SET domains of MLL are consistently replaced by partner protein sequences. Analyses of the multiple fusion partner proteins of MLL suggest that these sequences contribute either transactivation domains or the ability to homodimerize [18]. Loss of MLL results in embryonic lethality and myeloid progenitor defects, with a reduction, but not absence, of CFU-M and CFU-GEMM colonies and mature myeloid cells [19].

MLL is required during development for maintaining expression of pattern-forming HOX genes, suggesting that MLL may be the mammalian counterpart of *Drosophila* trithorax [20]. Leukemic samples containing MLL translocations have an identifiable gene expression pattern, characterized in part by high-level expression of 5′-HOXA genes (HOXA5-11) as well as MEIS1, a DNA-binding cofactor of HOX proteins [21]. Both MEIS1 and HOXA genes are expressed in myeloid progenitor cells but their expression is downregulated upon differentiation. Retroviral transduction of an MLL-AF9 construct into granulocyte macrophage progenitors induced expression of HOXA5, 10, and particularly 9 in the subsequently induced leukemic stem cells [22]. siRNA-mediated downregulation of HOXA9 induced apoptosis to a greater degree in MLL translocation-bearing cell lines and primary AML cells than in those that do not express this translocation [23]. Thus HOXA9 is an example of non-oncogene addiction in MLL leukemias. Of note, knockdown of HOXA9 reduced expression of MEIS1. While the specific mechanism by which MLL translocations contribute to leukemogenesis remains unknown, these studies suggest that MLL fusion protein-mediated alterations in a MLL/HOXA/MEIS1 transcriptional cascade may arrest differentiation and be required for leukemic cell survival.

Assessing response of AML blasts to cytokines by phosphoflow cytometry predicts response to therapy

An important novel approach to identifying signaling modules with prognostic significance is that of examining phospho-proteins by flow cytometry following stimulation of primary leukemic cells with cytokines. In one such study, leukemic blasts from 30 AML patients previously assessed for Flt3 mutations and cytogenetic abnormalities were exposed for 15 minutes to five cytokines (Flt3 ligand, GM-CSF, G-CSF, IL3, and IFNγ) followed by intracellular phospho-FACS analysis for five phosphoproteins (Stat1, Stat3, Stat5, p38, and ERK1/2) [24]. Unsupervised clustering of phospho-protein biosignatures resulted in separation of the AML samples into four groups that were then shown to correlate significantly with presence of FLT3 mutations, response to chemotherapy, and cytogenetic abnormalities. One striking result of this analysis was the observation that leukemic cells with potentiated Stat3 and Stat5 phosphorylation responses to the myeloid cytokines G-CSF and GM-CSF were associated with Flt3 mutation and failure to respond to initial chemotherapy, while two Flt3 mutation-positive leukemic samples without this biosignature proved responsive to chemotherapy. This methodologic approach suggests that in the future, interrogation of the signaling response of leukemic blasts may prove to be a particularly powerful prognostic tool and a means by which to identify patients likely to respond to specific targeted therapies.

Dyregulated NF-κB signaling in multiple myeloma

Upon surveying the landscape of genetic abnormalities observed in lymphoid malignancies, one is immediately struck by the frequency with which dysregulated NF-κB signaling plays a role in the molecular pathogenesis of such diseases. This has been particularly well demonstrated in primary samples of malignant plasma cells from patients with multiple myeloma.

Annunziata et al. observed that an inhibitor of IκB kinase beta (IKKβ) that was notably an ineffective inhibitor for IKKα, induced apoptosis in 15 of 21 multiple myeloma cell lines. Following Affymetrix analysis of the cell lines before and after IKKβ inhibitor treatment, 11 genes were identified as significantly altered in expression and such altered expression correlated with apoptosis induction. When cDNA was generated from 451 CD138-sorted primary multiple myeloma cell samples, expression of the same 11-gene signature was elevated in 82% of samples relative to MM-M1, an NF-κB-dependent cell line, and such gene expression correlated with nuclear p65 expression.

Examination of primary multiple myeloma samples for "outlier" gene expression identified four NF-κB regulators in which at least five patient samples demonstrated markedly elevated or depressed expression. Two of these genes, NIK and CD40, were upregulated. CD40 is a TNFR family member that plays a critical role in B cell regulation and which is known to activate NFkB upon overexpression even in the absence of ligand. Six CD40 outlier cases of myeloma demonstrated greater than a 32-fold increase in CD40 transcript levels: this process was not due to CD40 locus amplification. NIK1 is a kinase capable of activating both the classical and alternative NF-κB pathways. Among five NIK1 outlier cases, three had chromosomal translocations involving the IgH or IgL loci, while the two others had monoallelic expression and were likely to have alternate translocations.

Two other genes, TRAF3 and the tumor suppressor cyclindromatosis protein CYLD, were downregulated. TRAF2, a CARMA1 signalosome component, which activates NF-κB by binding MALT1 and Bcl10 and recruiting RIP1, is inhibited by TRAF3 by forming TRAF2/TRAF3 heterotrimers [25]. Among six TRAF3 myeloma sample outliers examined, two had TRAF3 locus biallelic deletion, three had monoallelic deletion, and one had a frameshift mutation. The ubiquitylase CYLD inhibits NF-κB signaling downstream of innate immune receptors by removing polyubiquitin chains from components of the CARMA1 signalosome [26, 27]. Among five myeloma sample CYLD outliers examined, all had CYLD locus biallelic deletions.

In summary, from this and other studies, there is striking genetic evidence that activation of the classical NF-κB signaling module by a wide variety of mechanisms is a recurrent although not invariable event in multiple myeloma. While this observation is consistent with the demonstrated efficacy of the proteosomal inhibitor bortezomib in multiple myeloma, it also suggests that IKKβ is a logical therapeutic target in this disease.

Dysregulated NF-κB signaling in DLBCL

NF-κB signaling also appears to be a relevant molecular module in non-Hodgkin's lymphoma (NHL). Following B lymphocyte contact with antigen, the BCR signalosome induces recruitment, phosphorylation, and activation of phospholipase C gamma, with resultant cleavage of PIP2 and generation of IP3 and DAG. These two products activate protein kinase C beta (PKCβ). One of the critical substrates for this enzyme is the PKC-regulated domain of the adapter protein CARMA1 (also known as CARD11), the initial component of the CARMA1 signalosome responsible for antigen receptor-mediated IKKβ phosphorylation and activation [27]. Of note, inducible knockout murine studies have demonstrated that normal B lymphocytes require both surface immunoglobulin and IKKβ function for survival [28, 29]. Thus a clinically relevant question in this field will be whether partial but not complete ablation of the CARMA1/IKKβ signaling module shows a significant "therapeutic window" and effectively targets subsets of NHL while leaving normal B cell function intact.

Staudt and colleagues carried out gene expression studies of diffuse large B cell lymphoma (DLBCL), the most common form of NHL, and then sought to categorize the patient samples into groups that resembled specific types of normal B cells. This effort yielded three groups: germinal center B cell-like (GCB), activated B cell-like (ABC), and primary mediastinal B cell lymphoma [30, 31]. Five-year survival after anthracycline-based chemotherapy is significantly lower (35%) in patients with ABC-type than in those with GCB-type DLBCL (60%) [31]. c-rel amplification was found in 17 of 115 GCB cases but in none of 73 ABC cases [31]. While the selective presence of c-rel amplification might have suggested that the NF-κB signaling module would prove to be of particular importance in GCB DLBCL, gene expression profiling of ABC and GCB-derived DLBCL cell lines revealed surprisingly that target genes of the NF-κB pathway were expressed at high levels in ABC lines while they were suppressed in GCB lines [32]. Consistent with this finding, retroviral transduction of a "super-repressor" form of IκBα induced apoptosis in ABC but not GCB-derived DLBCL cell lines [32]. Transduction of two ABC and two GCB-derived DLBCL cell lines with a doxycycline-regulated retroviral vector expressing a library of "bar-coded" shRNAs, targeting 2,500 human genes demonstrated that after three weeks of culture, all of the shRNAs selectively depleted from the ABC but not the GCB cell lines targeted the NF-κB pathway, suggesting that loss of such gene products resulted in death of the

ABC cells [33]. Among the shRNAs depleted from the ABC DLBCL cell lines were those targeting CARMA1, BCL10, and MALT1, all components of the CARMA1 signalosome (see discussion of the role of BCL10 and MALT1 in MALT lymphomas below).

Although these studies demonstrate an unusual dependence of ABC DLBCL on the CARMA1/IKKβ signaling module and resulted in ongoing clinical trials examining the utility of proteosome or IKKβ inhibitors in this lymphoma subtype, they did not elucidate the molecular mechanisms responsible for such tonic NF-κB signaling. Staudt and colleagues therefore turned to direct sequencing of all coding exons in CARMA1/CARD11 in DLBCL biopsies. Missense base substitutions in the coiled-coil domain of CARMA1/CARD11 were identified in 7/73 ABC and 3/79 GCB primary DLBCL biopsies as well as in DLBCL cell lines [34]. *In vitro* studies with ABC cell lines demonstrated that the missense mutations identified in CARMA1/CARD11 led to constitutively active NF-κB signaling as a result of CARMA1 oligomerization. While these studies clearly demonstrate the importance of CARMA1/CARD11 mutations in the pathophysiology of some ABC-type DLBCL cases, they leave open the question as to whether all ABC lymphomas will show comparable NF-κB dependence and, if so, what molecular defects account for the 90% of ABC lymphomas that do not have CARMA1 mutations.

Dysregulated NF-κB signaling in MALT lymphomas

MALT lymphomas are indolent marginal zone B cell lymphomas of the gastrointestinal tract, salivary gland, lung, ocular adnexa, skin, thyroid, and breast. These lymphomas often arise out of a background of chronic inflammation and in some cases have been linked to specific infectious organisms. Gastric MALTs are associated with *Helicobacter pylori* infections, and ocular adnexal MALTs to *Chlamydia psittica* infections. Remarkably, patients with early-stage cases of these MALT lymphomas may achieve lasting remissions with appropriate antibiotic therapy alone, suggesting that the MALT lymphomas are initiated as an antigen-driven and antigen-dependent process [35].

A subset of MALT lymphomas have been found to harbor translocations of either of two genes encoding components of the CARMA1 signalosome, Bcl10 and MALT1 [36]. Three types of translocations have been identified: t(11;18)/API2-MALT1 resulting in a MALT1 fusion protein, t(14;18)/IGH-MALT in which MALT1 expression is driven by the IgH promoter, and t(1;14)/IGH-BCL10 in which BCL10 expression is driven by the IgH promoter. Presence of any of these translocations is associated with more advanced clinical stage and, in the case of gastric lymphomas, with resistance to cure by *H. pylori* eradication. The association of these initially antigen-driven B cell lymphomas with translocations involving components of the CARMA1 signalosome suggests that at later stages, acquisition of genetic alterations in this signaling module allow MALT lymphoma B cell proliferation to become independent of antigen and thus unresponsive to antimicrobial therapy. Despite the fact that MALT1 is the immediate downstream target of CARMA1/CARD11 in the CARMA1 signalosome, and that CARD11 is mutated in a subset of DLBCL patients, the t(11;18) translocation appears to be specifically associated with MALT lymphomas and is not found in other types of lymphomas [36]. It would thus appear that the signaling consequences of overexpression of each constituent of this signaling complex differ and are consequently selected variably among different forms of lymphoma.

Bcl6-mediated ATR and BLIMP repression critical for BCR-type DLBCL

The transcriptional repressor Bcl6 is constitutively expressed in normal germinal center centroblasts as well as in roughly 50% of diffuse large B cell lymphomas (DLBCL). Constitutive expression of Bcl6 in DLBCL results either from promoter substitution as a result of translocations involving the Bcl6 gene at band 3q27 or somatic Bcl6 promoter mutations [37]. Bcl6 promoter mutations in DLBCL appear to occur as a consequence of aberrant somatic hypermutation, as is true of a variety of other transcriptionally active loci including PIM1, Myc, RhoH/TTF, and PAX5 [38]. In DLBCLs, a fraction of these Bcl6 promoter mutations abrogate a normal negative feedback loop that reduces Bcl6 expression [39]. Thus it is clear that tonic expression of this repressor protein confers an advantage to a significant fraction of cases of DLBCL.

During the germinal center reaction, B cells that have been recruited to follicles by T cell stimulation undergo somatic hypermutation and class switch recombination of immunoglobulin genes. Knockout of the Bcl6 gene results in loss of germinal centers as well as loss of affinity maturation following vaccination with NP-KLH [40]. Recent studies focused on identifying the critical genes whose expression is regulated by Bcl6 have begun to shed light on why this transcriptional repressor is required for germinal center formation and why aberrant Bcl6 expression is selected during the development of DLBCL. While a wide array of candidate genes regulated by Bcl6 has been identified including CCL3, CCND2, CD80, CD69, and FCER2, attention has focused on repression of genes involved in responding to DNA damage (p53, ATR) and plasma cell differentiation (PRDM1/BLIMP1) [41–44].

Melnick and colleagues have developed a remarkable reagent for the study of Bcl6 function that has also served as a starting point for a promising novel class of therapeutic agents to treat Bcl6-overexpressing DLBCLs [45, 46]. This group took advantage of the fact that the co-repressors SMRT, NCoR, and BCoR bind to Bcl6's N-terminal BTB domain by a conserved 17-residue BTB binding domain (BBD). In order to disrupt the association of these co-repressors with Bcl6, a 21-residue SMRT BBD motif was combined with a pTAT protein transduction domain allowing the BBD peptide inhibitor (BPI) to enter cells efficiently and localize to the nucleus. Treatment of Bcl6-expressing DLBCL lines with BPI led to loss of SMRT and NCoR from the promoter of the Bcl6-regulated genes CCL3 (with maintenance of Bcl6 association), a transition from promoter H3K9 methylation and H4 deacetylation to H3K9 demethylation and H4 acetylation, and CCL3 re-expression. BPI treatment also led to growth suppression and apoptosis of almost all Bcl6-expressing DLBCL cell lines, suggesting that Bcl6 is required for the survival of DLBCL cells that constitutively express this protein. In contrast to the prior work cited by Staudt et al., a transcriptional profiling effort by Shipp and colleagues has categorized DLBCL samples into three groups: B cell receptor/proliferation (BCR), oxidative phosphorylation (OxPhos), and host response (HR) [47]. The BCR subgroup was substantially enriched for alterations in Bcl6 target genes and BCR DLBCL cell lines had a significantly lower BPI IC_{50} than OxPhos lines (12.7 vs. 50.1 μM) [48].

ATR is known to play a critical role in replication and DNA damage-sensing checkpoints. When BPI was used in microarray assays to identify Bcl6-dependent gene repression in the Ly1 DLBCL cell line, the ATR gene was shown to undergo 25-fold upregulation after five hours of BPI treatment [44]. Both this observation and prior studies demonstrating Bcl6-mediated regulation of p53 suggested the model that Bcl6 lowers the sensitivity of germinal center centroblasts to DNA damage signals that result from class-switch recombination and somatic hypermutation processes. Such DNA damage signals, it is argued, would under normal circumstances induce cell cycle arrest and potentially apoptosis. The obvious corollary of this hypothesis is that constitutive expression of Bcl6 in DLBCL may contribute to genomic instability as well as resistance to DNA damage-inducing therapeutic agents in this disease. Consistent with this model, downregulation of Bcl6 in Bcl6-expressing DLBCL lines results in restored expression of ATR, augmented H2AX phosphorylation, and cell death. H2AX is a variant histone whose phosphorylation plays a role in both chromatin rearrangement and recruitment of repair factors around the site of a double-stranded DNA break [49]. Remarkably, concurrent downregulation of ATR and Bcl6 rescued DLBCL cell lines from both H2AX phosphorylation and cell death, suggesting that, at least in this model system, ATR downregulation is sufficient to account for DLBCL "addiction" to Bcl6 signaling [44].

In addition to its effects on cell proliferation and survival, Bcl6 represses differentiation of B cells into plasma cells through recruitment of the MTA3 subunit of the Mi-2/NuRD (nucleosome remodelling and deacetylase) complex to a central RD2 region of Bcl6 that is distinct from the NH2-terminal BTB domain [50]. Among the genes known to be repressed by Bcl2 through this inhibitory complex is prdm1, the gene encoding B lymphocyte-inducing maturation protein 1 (BLIMP-1), a transcriptional repressor that plays a central role in plasma cell differentiation [43]. Formal studies contrasting the effects of SMRT/NCor recruitment and MTA3 recruitment have demonstrated that the functional consequences of recruiting these two Bcl6 effector proteins is largely distinct [42]. Inhibition of SMRT/NCor recruitment with BPI reduced cell viability but had no effect on prdm1 expression, while siRNA-mediated MTA3 depletion restored prdm1 expression as well as the plasma cell marker syndecan, induced phenotypic plasma cell differentiation but failed to alter cell viability. While MTA3 siRNA was inactive alone, combined treatment with BPI and MTA3 siRNA enhanced the induction of apoptosis relative to BPI alone. Given the ample evidence that Bcl6 plays a critical role in the biology of a large subset of cases of this common lymphoid malignancy, it will be of great interest to determine whether Bcl6-targeted therapies prove to enhance cure rate in Bcl6-expressing BCR-type DLBCLs.

Burkitt's lymphoma: escape from Myc-driven apoptosis

In Burkitt's lymphoma, a high-grade non-Hodgkin's lymphoma, expression of the proto-oncogene c-Myc on chromosome 8 is dysregulated due to translocation to the IgH locus promoter on chromosome 14, or less commonly to light chain loci on chromosomes 2 or 22 [51]. Although myc is now known, among other properties, for its ability to drive proliferation, early studies demonstrated that constitutive overexpression of c-myc in cell lines induced apoptosis by a p53-dependent pathway triggered by inappropriate bypass of the G1-S checkpoint [52]. Following the cloning of Bcl2 at the t(14;18) translocation characteristic of low-grade follicular lymphomas, it was initially unclear what function this unknown protein carried out in such malignant cells. However, subsequent studies demonstrated that Bcl2 rescued myc-transfected or transduced cell lines or primary hematopoietic cells from apoptosis, suggesting that Bcl2 was the founding member of a novel class of anti-apoptotic oncogenes [53, 54]. Equally important, these experiments were the first to confirm the concept that overcoming oncogene-induced apoptosis by acquisition of anti-apoptotic oncogenes or deletion of components of the apoptotic signal cascade were likely to be critical events in the evolution of cancer cells.

Given the absence of t(14;18) translocations in Burkitt's lymphoma, researchers next focused on the mechanisms by which such primary lymphoma cells escaped myc-induced apoptosis. While p53 mutations are generally unusual in other forms of non-Hodgkin's lymphomas, they were relatively frequently identified in Burkitt's lymphoma, albeit in a minority of samples tested (9/27 biopsies, 17/27 cell lines) [55]. In studies of murine cells, p53-mediated apoptosis was shown to be in part mediated by upregulation of p19 ARF message and protein [56]. Arf stabilizes p53 both by sequestration of MDM2 in nucleoli and inhibition of MDM2 ubiquitin ligase activity [57]. In mice in which a c-Myc transgene was overexpressed from the Eu immunoglobulin heavy chain enhancer-promoter, Burkitt's-like lymphomas developed within one year [58]. When examined at early ages, Eu-*Myc*

mice had B cell proliferation that appeared to be balanced by cell death. At later stages, when proliferation was not balanced by apoptosis, at least half of the lymphomas had lost either p53 or the CDKN2A p16 INK4A/p14 Arf locus [59]. As might be predicted, lymphomagenesis in Eu-Myc mice was significantly accelerated by crossing such mice with strains lacking the CDKN2A locus [60]. Similarly, deletion of the CDKN2A locus is common in Burkitt's cell lines that are p53 wild-type. Although this locus encodes not only ARF but also the cyclin-dependent kinase inhibitor INK4A, which acts upstream of Rb, experiments with mice hemizygous for ARF or INK4A have established that ARF appeared to be the relevant product of the CDKN2A locus whose absence accelerates lymphomagenesis in Eu-Myc mice [57, 59].

Given this background, it is somewhat surprising that while four instances of p53 deletion and ten instances of overexpression of Mdm2 mRNA were identified in 24 children with sporadic Burkitt's/B-ALL, no instances of deletion of the CDKN2A locus were observed in this study [61]. In contrast, loss of p16 INK4A expression, often by gene methylation, is common in primary Burkitt's tumors [62–64]. Of note, ARF methylation has not been observed in these studies and ARF protein expression appears to be conserved. The discrepancy between the animal studies and the observations in clinical materials remains unexplained.

The studies described above demonstrate that acquired alterations in the p53/Mdm2 axis are capable of bypassing myc-induced apoptosis in primary Burkitt's lymphoma. However, it is also apparent that such abnormalities are generally observed in only half of the Burkitt's specimens analyzed, suggesting that other anti-apoptotic mechanisms are likely at play. Egle et al. noted that the pro-apoptotic Bcl2 family member Bim was upregulated in Eu-*Myc* lymphomas and that loss of Bim accelerated lymphoma development in such mice [65]. Hemann and colleagues then observed that in lymphomas expressing two fairly common *Myc* point mutations, P57S and T58A, *Myc* signaling induced p53 and p19 ARF but failed to induce Bim [66]. In lymphomas derived from p53$^{+/-}$ stem cells, both wild-type and mutant *Myc* rapidly produced lymphomas but the remaining p53 allele was lost only in the tumors induced by wild-type myc. Consistent with these observations, in primary Burkitt's lymphoma samples that contained wild-type *Myc*, immunohistochemistry showed Bim expression, while Bim expression was lost in six of seven samples with *Myc* box 1 mutations, including P57S and T58A. Conversely, despite the relatively common occurrence of *Myc* box 1 mutations (12 of 71 samples), none of the 17 Burkitt's cases with p53 mutations contained *Myc* mutations.

The picture that emerges from these studies is a provocative one, in that it suggests that proliferation-inducing oncogenes such as Myc may activate more than one "circuit-breaker" apoptotic pathway. Remarkably, while the sum of these apoptotic signals appears to be sufficient to successfully impede tumor progression, loss of either apoptotic pathway allows Burkitt's lymphoma cells to survive and proliferate.

CLL tumor suppressors: the del13q14 controversy

Whereas Burkitt's lymphoma is among the most high grade of hematopoietic malignancies, chronic lymphocytic leukemia (CLL) often has an indolent clinical course. Despite focused research for many years, there is as of yet no unifying explanation for the genetic basis of CLL. A variety of cytogenetic abnormalities have been identified in CLL samples, several of which, such as 11q23 and 17p deletions, have significant prognostic importance. Deletions in the 13q14 region are the single most common cytogenetic abnormality in CLL, occur in roughly 50% of CLL patients, and are associated with a better prognosis. Deletion of this region is observed in multiple other cancers as well.

Identification of the target gene or genes deleted in del13q14 has proven difficult. Although the Rb gene is located in this region, 60% of del13q14 lesions fail to delete this gene [67]. Efforts by multiple groups to implicate a specific gene in the roughly 1 MB region of 13q14 that is commonly deleted have been unsuccessful in identifying mutations in the remaining allele of the candidate gene targets. There was therefore considerable interest when Croce and colleagues reported that two microRNAs, *miR15a/miR16*, were deleted in cases of del13q14 and in 68% of cases of CLL overall [68]. In subsequent work, this group suggested that these microRNAs target the anti-apoptotic gene Bcl2 and that levels of *miR15a/miR16* correlate inversely with Bcl2 levels [69]. However, subsequently other groups reported that while very low levels of *miR15a/miR16* are found in 10 to 15% of patients with CLL, often corresponding to bi-allelic deletion of the genomic region encoding these miRNAs, *miR15a/miR16* expression more generally fails to correlate with del13q14 lesions or with Bcl2 expression [67, 70]. Interest in the role of miRNAs in CLL pathophysiology remains high but the relevant targets of these and other miRNAs remain to be established in this disease.

CLL: poised for death, rescued by BCR signaling?

As noted previously, B lineage cells require tonic signaling through the BCR for survival [71]. Prognosis in CLL correlates significantly with several aspects of BCR signaling. CLL with unmutated immunoglobulin variable regions, presumably derived from a B cell precursor that did not undergo variable region somatic hypermutation in the germinal center, have a markedly worse prognosis than those with mutated variable regions. The traditional cut-off defining "mutated" and "unmutated" variable regions has been generally accepted as a 2% difference from germline sequence. The tyrosine kinase ZAP-70, initially thought to be limited to T and NK cells, is expressed in CLL. ZAP-70 expression correlates with expression of unmutated immunoglobulin variable regions and may be an even better predictor of poor clinical outcome [72, 73]. Expression of ZAP-70 also correlates with stronger BCR signaling, as judged by higher levels of phosphorylated p72^{Syk}, BLNK, and phospholipase-Cγ as well as intracellular calcium levels following IgM ligation [74]. ZAP-70 appears to augment BCR signaling independent of its kinase activity but rather perhaps through its role as an adapter protein [75, 76].

Two of the strongest predictors of prognosis in CLL, variable region mutation and ZAP-70 expression, are related to BCR function. Consistent with this, work from a wide variety of investigators has demonstrated that the clonal BCR present in CLL cells are genetically non-random, suggesting that the BCRs of leukemic cells from at least a subset of CLL patients may be specific for a common set of antigens [77]. Provocatively, Chiorazzi and colleagues have determined that a significant portion of CLL BCRs react with antigens expressed on the surface of apoptotic cells such as non-muscle myosin heavy chain IIA [78]. As a result of such studies, recent studies have sought to establish whether selective inhibitors of BCR function might be therapeutically effective in CLL. The protein tyrosine kinase Syk is structurally related to ZAP-70, plays a critical role in normal BCR signaling, and has been implicated in defective BCR signaling in a subset of CLL patients. Gobessi et al. detected constitutive Syk Y352 phosphorylation in CLL samples and reported that the Syk inhibitor R406 induced apoptosis in the majority of leukemic samples examined [79]. R406 also blocked the enhanced CLL viability, Akt activation, and Mcl-1 upregulation induced following BCR cross-linking, suggesting that this drug may disrupt both antigen-independent and antigen-dependent survival signals. In an early clinical trial examining the activity of the R406 prodrug fostamatinib disodium in relapsed non-Hodgkin's lymphoma, objective responses were observed in 55% of patients with relapsed CLL/SLL [80]. Overall, these studies suggest that the BCR signaling module is perhaps another example of "non-oncogene addiction" for CLL cells and that therapies directed toward this module may prove to be a powerful new class of therapeutics in the treatment of B lymphoid malignancies.

Conclusion

As is apparent from many of the hematopoietic malignancies examined above, the molecular mechanisms that drive dependence ("non-oncogene dependence") on a particular signaling pathway have in many or most instances not yet been clearly established for specific malignancies. Similarly, a clear relationship between specific pro-apoptotic oncogenes and compensatory anti-apoptotic pathways has been elucidated for only a minority of such malignancies. Myc-induced apoptosis in Burkitt's lymphoma is perhaps one of the best studied examples, yet even here, the mechanisms by which lymphoma cells escape Myc-induced apoptosis remain unknown for many lymphoma samples.

A second phenomenon that is apparent from this brief survey is the remarkable heterogeneity of molecular defects that underlie hematopoietic malignancies. Nonetheless, certain signaling modules are of recurrent importance. Dysregulation of both differentiation and proliferation pathways is recurrently observed in AML. The BCR signaling module plays a critical role in the biology of CLL. NF-κB signaling is of fundamental importance in a variety of B lymphoid malignancies. Yet despite such commonalities, the specifics of the mechanisms by which such critical nodal signaling pathways are dysregulated varies markedly among lymphoid neoplasms, as in the examples cited of tonic NF-κB signaling in myeloma, MALT lymphoma, and DLBCL. Despite all the complexity of the oncogenic mechanisms underlying hematopoietic malignancies, there seems to be considerable reason for optimism that the identification of these recurrent dysregulated signaling pathways will ultimately lead to increasingly effective targeted therapies for these deadly illnesses.

References

1. Hanahan D, Weinberg RA. 2000. The hallmarks of cancer. *Cell* 100:57–70.

2. Thiagalingam S. 2006. A cascade of modules of a network defines cancer progression. *Cancer Res* 66:7379–7385.

3. Luo J, Solimini NL, Elledge SJ. 2009. Principles of cancer therapy: oncogene and non-oncogene addiction. *Cell* 136:823–837.

4. Weinstein IB. 2002. Cancer. addiction to oncogenes – the Achilles heel of cancer. *Science* 297:63–64.

5. Solimini NL, Luo J, Elledge SJ. 2007. Non-oncogene addiction and the stress phenotype of cancer cells. *Cell* 130:986–988.

6. Higuchi M, O'Brien D, Kumaravelu P, et al. 2002. Expression of a conditional AML1-ETO oncogene bypasses embryonic lethality and establishes a murine model of human t(8;21) acute myeloid leukemia. *Cancer Cell* 1:63–74.

7. Grisolano JL, Wesselschmidt RL, Pelicci PG, Ley TJ. 1997. Altered myeloid development and acute leukemia in transgenic mice expressing PML-RAR alpha under control of cathepsin G regulatory sequences. *Blood* 89:376–387.

8. Castilla LH, Garrett L, Adya N, et al. 1999. The fusion gene Cbfb-MYH11 blocks myeloid differentiation and predisposes mice to acute myelomonocytic leukaemia. *Nat Genet* 23: 144–146.

9. Schessl C, Rawat VP, Cusan M, et al. 2005. The AML1-ETO fusion gene and the FLT3 length mutation collaborate in inducing acute leukemia in mice. *J Clin Invest* 115:2159–2168.

10. Gilliland DG, Tallman MS. 2002. Focus on acute leukemias. *Cancer Cell* 1:417–420.

11. Huang ME, Ye YC, Chen SR, et al. 1987. All-trans retinoic acid with or without low dose cytosine arabinoside in acute promyelocytic leukemia. Report

of 6 cases. *Chin Med J* 100: 949–953.

12. de The H, Chomienne C, Lanotte M, Degos L, Dejean A. 1990. The t(15;17) translocation of acute promyelocytic leukaemia fuses the retinoic acid receptor alpha gene to a novel transcribed locus. *Nature* 347:558–561.

13. Lin RJ, Nagy L, Inoue S, et al. 1998. Role of the histone deacetylase complex in acute promyelocytic leukaemia. *Nature* 391:811–814.

14. Grignani F, De Matteis S, Nervi C, et al. 1998. Fusion proteins of the retinoic acid receptor-alpha recruit histone deacetylase in promyelocytic leukaemia. *Nature* 391:815–818.

15. Zhu J, Zhou J, Peres L, et al. 2005. A sumoylation site in PML/RARA is essential for leukemic transformation. *Cancer Cell* 7:143–153.

16. Nasr R, Guillemin MC, Ferhi O, et al. 2008. Eradication of acute promyelocytic leukemia-initiating cells through PML-RARA degradation. *Nat Med* 14: 1333–1342.

17. Alcalay M, Meani N, Gelmetti V, et al. 2003. Acute myeloid leukemia fusion proteins deregulate genes involved in stem cell maintenance and DNA repair. *J Clin Invest* 112:1751–1761.

18. Ayton PM, Cleary ML. 2001. Molecular mechanisms of leukemogenesis mediated by MLL fusion proteins. *Oncogene.* 20:5695–5707.

19. Hess JL, Yu BD, Li B, Hanson R, Korsmeyer SJ. 1997. Defects in yolk sac hematopoiesis in Mll-null embryos. *Blood* 90:1799–1806.

20. Yu BD, Hanson RD, Hess JL, Horning SE, Korsmeyer SJ. 1998. MLL, a mammalian trithorax-group gene, functions as a transcriptional maintenance factor in morphogenesis. *Proc Natl Acad Sci USA* 95: 10632–10636.

21. Armstrong SA, Staunton JE, Silverman LB, et al. 2002. MLL translocations specify a distinct gene expression profile that distinguishes a unique leukemia. *Nat Genet* 30:41–47.

22. Krivtsov AV, Twomey D, Feng Z, et al. 2006. Transformation from committed progenitor to leukaemia stem cell initiated by MLL-AF9. *Nature* 442:818–822.

23. Faber J, Krivtsov AV, Stubbs MC, et al. 2009. HOXA9 is required for survival in human MLL-rearranged acute leukemias. *Blood* 113:2375–2385.

24. Irish JM, Hovland R, Krutzik PO, et al. 2004. Single cell profiling of potentiated phospho-protein networks in cancer cells. *Cell* 118:217–228.

25. He L, Grammer AC, Wu X, Lipsky PE. 2004. TRAF3 forms heterotrimers with TRAF2 and modulates its ability to mediate NF-{kappa}B activation. *J Biol Chem* 279:55855–55865.

26. Liu YC, Penninger J, Karin M. 2005. Immunity by ubiquitylation: a reversible process of modification. *Nat Rev Immunol* 5:941–952.

27. Rawlings DJ, Sommer K, Moreno-Garcia ME. 2006. The CARMA1 signalosome links the signalling machinery of adaptive and innate immunity in lymphocytes. *Nat Rev Immunol* 6:799–812.

28. Lam KP, Kuhn R, Rajewsky K. 1997. *In vivo* ablation of surface immunoglobulin on mature B cells by inducible gene targeting results in rapid cell death. *Cell* 90:1073–1083.

29. Pasparakis M, Schmidt-Supprian M, Rajewsky K. 2002. IkappaB kinase signaling is essential for maintenance of mature B cells. *J Exp Med* 196:743–752.

30. Alizadeh AA, Eisen MB, Davis RE, et al. 2000. Distinct types of diffuse large B-cell lymphoma identified by gene expression profiling. *Nature* 403:503–511.

31. Rosenwald A, Wright G, Chan WC, et al. 2002. The use of molecular profiling to predict survival after chemotherapy for diffuse large-B-cell lymphoma. *N Engl J Med* 346:1937–1947.

32. Davis RE, Brown KD, Siebenlist U, Staudt LM. 2001. Constitutive nuclear factor kappaB activity is required for survival of activated B cell-like diffuse large B cell lymphoma cells. *J Exp Med* 194:1861–1874.

33. Ngo VN, Davis RE, Lamy L, et al. 2006. A loss-of-function RNA interference screen for molecular targets in cancer. *Nature* 441: 106–110.

34. Lenz G, Davis RE, Ngo VN, et al. 2008. Oncogenic CARD11 mutations in human diffuse large B cell lymphoma. *Science* 319:1676–1679.

35. Ferreri AJ, Ponzoni M, Guidoboni M, et al. 2005. Regression of ocular adnexal lymphoma after *Chlamydia psittaci*-eradicating antibiotic therapy. *J Clin Oncol* 23:5067–5073.

36. Du MQ. 2007. MALT lymphoma: recent advances in aetiology and molecular genetics. *J Clin Exp Hematop* 47:31–42.

37. Ye BH, Lista F, Lo Coco F, et al. 1993. Alterations of a zinc finger-encoding gene, BCL-6, in diffuse large-cell lymphoma. *Science* 262:747–750.

38. Pasqualucci L, Neumeister P, Goossens T, et al. 2001. Hypermutation of multiple proto-oncogenes in B-cell diffuse large-cell lymphomas. *Nature* 412:341–346.

39. Pasqualucci L, Migliazza A, Basso K, et al. 2003. Mutations of the BCL6 proto-oncogene disrupt its negative autoregulation in diffuse large B-cell lymphoma. *Blood* 101:2914–2923.

40. Ye BH, Cattoretti G, Shen Q, et al. 1997. The BCL-6 proto-oncogene controls germinal-centre formation and Th2-type

inflammation. *Nat Genet* 16: 161–170.

41. Phan RT, Dalla-Favera R. 2004. The BCL6 proto-oncogene suppresses p53 expression in germinal-centre B cells. *Nature* 432:635–639.

42. Parekh S, Polo JM, Shaknovich R, et al. 2007. BCL6 programs lymphoma cells for survival and differentiation through distinct biochemical mechanisms. *Blood* 110:2067–2074.

43. Tunyaplin C, Shaffer AL, Angelin-Duclos CD, et al. 2004. Direct repression of prdm1 by Bcl-6 inhibits plasmacytic differentiation. *J Immunol* 173:1158–1165.

44. Ranuncolo SM, Polo JM, Dierov J, et al. 2007. Bcl-6 mediates the germinal center B cell phenotype and lymphomagenesis through transcriptional repression of the DNA-damage sensor ATR. *Nat Immunol* 8:705–714.

45. Polo JM, Dell'Oso T, Ranuncolo SM, et al. 2004. Specific peptide interference reveals BCL6 transcriptional and oncogenic mechanisms in B-cell lymphoma cells. *Nat Med* 10:1329–1335.

46. Cerchietti LC, Yang SN, Shaknovich R, et al. 2009. A peptomimetic inhibitor of BCL6 with potent antilymphoma effects *in vitro* and *in vivo*. *Blood* 113:3397–3405.

47. Monti S, Savage KJ, Kutok JL, et al. 2005. Molecular profiling of diffuse large B-cell lymphoma identifies robust subtypes including one characterized by host inflammatory response. *Blood* 105:1851–1861.

48. Polo JM, Juszczynski P, Monti S, et al. 2007. Transcriptional signature with differential expression of BCL6 target genes accurately identifies BCL6-dependent diffuse large B cell lymphomas. *Proc Natl Acad Sci USA* 104:3207–3212.

49. Fernandez-Capetillo O, Lee A, Nussenzweig M, Nussenzweig A. 2004. H2AX: the histone guardian of the genome. *DNA Repair* 3:959–967.

50. Fujita N, Jaye DL, Geigerman C, et al. 2004. MTA3 and the Mi-2/NuRD complex regulate cell fate during B lymphocyte differentiation. *Cell* 119:75–86.

51. Dalla-Favera R, Bregni M, Erikson J, et al. 1982. Human c-myc onc gene is located on the region of chromosome 8 that is translocated in Burkitt lymphoma cells. *Proc Natl Acad Sci USA* 79:7824–7827.

52. Hermeking H, Eick D. 1994. Mediation of c-Myc-induced apoptosis by p53. *Science* 265:2091–2093.

53. Vaux DL, Cory S, Adams JM. 1988. Bcl-2 gene promotes haemopoietic cell survival and cooperates with c-myc to immortalize pre-B cells. *Nature* 335:440–442.

54. Bissonnette RP, Echeverri F, Mahboubi A, Green DR. 1992. Apoptotic cell death induced by c-myc is inhibited by bcl-2. *Nature* 359:552–554.

55. Gaidano G, Ballerini P, Gong JZ, et al. 1991. p53 mutations in human lymphoid malignancies: association with Burkitt lymphoma and chronic lymphocytic leukemia. *Proc Natl Acad Sci USA* 88:5413–5417.

56. Zindy F, Eischen CM, Randle DH, et al. 1998. Myc signaling via the ARF tumor suppressor regulates p53-dependent apoptosis and immortalization. *Genes Dev* 12:2424–2433.

57. Lowe SW, Sherr CJ. 2003. Tumor suppression by Ink4a-Arf: progress and puzzles. *Curr Opin Genet Dev* 13:77–83.

58. Adams JM, Harris AW, Pinkert CA, et al. 1985. The c-myc oncogene driven by immunoglobulin enhancers induces lymphoid malignancy in transgenic mice. *Nature* 318:533–538.

59. Eischen CM, Weber JD, Roussel MF, Sherr CJ, Cleveland JL. 1999. Disruption of the ARF-Mdm2-p53 tumor suppressor pathway in Myc-induced lymphomagenesis. *Genes Dev* 13:2658–2669.

60. Schmitt CA, McCurrach ME, de Stanchina E, Wallace-Brodeur RR, Lowe SW. 1999. INK4a/ARF mutations accelerate lymphomagenesis and promote chemoresistance by disabling p53. *Genes Dev* 13:2670–2677.

61. Wilda M, Bruch J, Harder L, et al. 2004. Inactivation of the ARF-MDM-2-p53 pathway in sporadic Burkitt's lymphoma in children. *Leukemia* 18:584–588.

62. Villuendas R, Sanchez-Beato M, Martinez JC, et al. 1998. Loss of p16/INK4A protein expression in non-Hodgkin's lymphomas is a frequent finding associated with tumor progression. *Am J Pathol* 153:887–897.

63. Klangby U, Okan I, Magnusson KP, et al. 1998. p16/INK4a and p15/INK4b gene methylation and absence of p16/INK4a mRNA and protein expression in Burkitt's lymphoma. *Blood* 91:1680–1687.

64. Baur AS, Shaw P, Burri N, et al. 1999. Frequent methylation silencing of p15(INK4b) (MTS2) and p16(INK4a) (MTS1) in B-cell and T-cell lymphomas. *Blood* 94:1773–1781.

65. Egle A, Harris AW, Bouillet P, Cory S. 2004. Bim is a suppressor of Myc-induced mouse B cell leukemia. *Proc Natl Acad Sci USA* 101:6164–6169.

66. Hemann MT, Bric A, Teruya-Feldstein J, et al. 2005. Evasion of the p53 tumour surveillance network by tumour-derived MYC mutants. *Nature* 436: 807–811.

67. Ouillette P, Erba H, Kujawski L, et al. 2008. Integrated genomic profiling of chronic lymphocytic leukemia identifies subtypes of deletion 13q14. *Cancer Res* 68:1012–1021.

68. Calin GA, Dumitru CD, Shimizu M, et al. 2002. Frequent deletions and down-regulation of micro-RNA genes miR15 and miR16 at 13q14 in chronic lymphocytic leukemia. *Proc Natl Acad Sci USA* 99:15524–15529.

69. Cimmino A, Calin GA, Fabbri M, et al. 2005. miR-15 and miR-16 induce apoptosis by targeting BCL2. *Proc Natl Acad Sci USA* 102:13944–13949.

70. Fulci V, Chiaretti S, Goldoni M, et al. 2007. Quantitative technologies establish a novel microRNA profile of chronic lymphocytic leukemia. *Blood* 109:4944–4951.

71. Kraus M, Alimzhanov MB, Rajewsky N, Rajewsky K. 2004. Survival of resting mature B lymphocytes depends on BCR signaling via the Igalpha/beta heterodimer. *Cell* 117:787–800.

72. Rassenti LZ, Huynh L, Toy TL, et al. 2004. ZAP-70 compared with immunoglobulin heavy-chain gene mutation status as a predictor of disease progression in chronic lymphocytic leukemia. *N Engl J Med* 351:893–901.

73. Rassenti LZ, Jain S, Keating MJ, et al. 2008. Relative value of ZAP-70, CD38, and immunoglobulin mutation status in predicting aggressive disease in chronic lymphocytic leukemia. *Blood* 112:1923–1930.

74. Chen L, Apgar J, Huynh L, et al. 2005. ZAP-70 directly enhances IgM signaling in chronic lymphocytic leukemia. *Blood* 105:2036–2041.

75. Gobessi S, Laurenti L, Longo PG, et al. 2007. ZAP-70 enhances B-cell-receptor signaling despite absent or inefficient tyrosine kinase activation in chronic lymphocytic leukemia and lymphoma B cells. *Blood* 109:2032–2039.

76. Chen L, Huynh L, Apgar J, et al. 2008. ZAP-70 enhances IgM signaling independent of its kinase activity in chronic lymphocytic leukemia. *Blood* 111:2685–2692.

77. Ghia EM, Jain S, Widhopf GF, 2nd, et al. 2008. Use of IGHV3-21 in chronic lymphocytic leukemia is associated with high-risk disease and reflects antigen-driven, post-germinal center leukemogenic selection. *Blood* 111:5101–5108.

78. Chu CC, Catera R, Zhang L, et al. 2010. Many chronic lymphocytic leukemia antibodies recognize apoptotic cells with exposed nonmuscle myosin heavy chain IIA: implications for patient outcome and cell of origin. *Blood* 115:3907–3915.

79. Gobessi S, Laurenti L, Longo PG, et al. 2009. Inhibition of constitutive and BCR-induced Syk activation downregulates Mcl-1 and induces apoptosis in chronic lymphocytic leukemia B cells. *Leukemia* 23:686–697.

80. Friedberg JW, Sharman J, Sweetenham J, et al. 2010. Inhibition of Syk with fostamatinib disodium has significant clinical activity in non-Hodgkin lymphoma and chronic lymphocytic leukemia. *Blood* 115:2578–2585.

Chapter

29

Role of network biology and network medicine in early detection of cancer

Asad Umar and Simon Rosenfeld

Introduction

Despite recent lowering trends in some cancers, it remains as a major cause of morbidity and mortality in the United States and the world (Jemal et al., 2010). Over 1.5 million new cancer cases and over half a million deaths from cancer, and nearly a quarter of all deaths were projected to occur in the United States in 2010 (Jemal et al., 2010). Although the use of many newer technologies such as mass spectrometry, deep sequencing, and protein and DNA arrays, combined with our understanding of the human and, more recently, the cancer genome, has enabled simultaneous examination of multiple genes and proteins and, more interestingly, profiles has fueled early detection of cancers. These enormous advances in technologies should have led to a better success rate in the discovery and validation of cancer biomarkers but it has proven difficult due to multiple reasons. In this chapter we will attempt to identify current progress, challenges, and some alternative strategies to improve future success.

Cancer is a complex disease characterized by the transformation of normal cells into malignant and invasive cells, taking a relatively long period of time and accumulating several molecular alterations on their way to achieve control over one or more of the hallmarks of cancers (Hanahan and Weinberg, 2000). Carcinogenesis is a multistage, evolutionary process within the organism and is influenced by environmental, genetic, social, and behavioral factors. The multistage nature of carcinogenesis also provides opportunities to detect early and intermediate lesions in any given organ site before they reach a fully invasive stage. However, not enough effort has been made at identifying the molecular changes in the temporal and spatial entities leading up to the advanced stage. Another challenge is that the identification of a single marker or a set of markers that will be sufficient in predicting disease risk, onset, or outcome has not been well established. A great body of research has been performed into the discovery and development of single biomarkers and validation of these markers remains challenging mainly due to the above-mentioned caveats. As cancer is a representation of a complex set of diseases, multiple biomarkers rather than a single marker is needed for successful development and validation. These two challenges are related and a pre-cancer genomic atlas effort is badly needed before such multiple biomarker panels can be delineated for early detection of cancers.

Regardless of the development and validation process of biomarkers it is clear that in order to effectively diagnose individuals that are at risk of carcinogenesis and to assign screening and intervention strategies, effective and clinically validated biomarkers are essential. Furthermore, it is becoming clear that the response to interventions needs to be tailored, especially in cancer prevention, where special emphasis is needed to minimize toxicity by knowing an individual's ability to metabolize a particular drug.

One of the major problems in the war on cancer is the difficulty to detect cancers early enough. As a result of either difficulty of an early detection modality or penetration of screening tests into the population, cancer is detected in its later stages, when it has compromised the function of one or more vital organ system and is widespread throughout the body. Hence, it is not surprising that methods for early detection of cancer are of paramount significance in cancer prevention and treatment and are an active area of cancer research. There will likely be increase in research activity as the significance of early detection becomes more important in the war on cancer.

Systems Biology of Cancer, ed. S. Thiagalingam. Published by Cambridge University Press. © Cambridge University Press 2015.

In addition to the importance of early detection for preventing progress of frank disease after the initial detection, accurate diagnosis and staging of the disease are also essential for the design of a treatment plan.

Types of biomarkers for cancer research

A biomarker can be any property that is objectively measured and evaluated as an indicator of normal biological processes, pathogenic processes, or pharmacologic responses to a therapeutic intervention (Biodefinitions Working Group, 2001). The cancer biomarker field has been one of the most prolific in recent years for publications as well as discovery and development. Biomarkers can be classified into several categories, as they require special consideration while being developed and validated. Biomarkers can be image-based or physiologic/pathologic indicators. The recent use and advent of the targeted-therapy modalities, cellular and molecular, that include genetic, epigenetic, proteomic, and metabolomic biomarkers is becoming increasingly important. Biological considerations have a key role in the initial identification of early biomarkers for cancer, and remain important – alongside statistical analysis of clinical trials – during their evaluation and adoption into clinical practice. Definitions and specific classifications of cancer biomarkers are shown in Figure 29.1.

Screening biomarkers (diagnostic)

These are the biomarkers that can be used either singly or in combination with other markers to reliably detect and identify a given type of cancer. Hence, it is generally expected that these markers have a high degree of specificity (the proportion of normal individuals who test negative) and sensitivity (the proportion of individuals with confirmed disease who test positive).

Prognostic biomarkers

Markers used after the onset of disease, which predict the course of disease, such as recurrence, etc., to help decide treatment options.

Predictive (stratification) biomarkers

These markers serve to predict response to an intervention usually before a treatment is started; hence providing information for responders and non-responders to a particular intervention.

Surrogate end point biomarkers

These markers provide early and accurate prediction of both a clinical end point, and the effects of treatment on this end point and may or may not be directly associated with intervention. Arguably, surrogate end-point biomarkers can be considered as the Achilles' heel of predictive oncology; however, there are only a few accepted surrogate end points that are in use in clinical oncology (Buyse et al., 2000). It is unrealistic for a surrogacy relationship to be established; the surrogate should predict the clinical end point, treatment should have a significant effect on both the candidate surrogate and the clinical end point, and the treatment effect on the surrogate should capture the full effect of treatment on the clinical end point. The current thinking in the field is that validation of surrogate end point biomarkers can be on a correlation-based approach, where the demonstration can preferably be in a randomized clinical trial and meta-analysis can be used for hypothesis generation such that the effect on the surrogate is sufficiently correlated with the effect on the actual end point (Prentice, 2009). For example, advanced adenoma can be a surrogate end point biomarker for colorectal cancer.

Clinical validation of biomarkers

Clinical validation involves confirmation by robust statistical methods that a candidate prognostic biomarker, predictive biomarker, or surrogate end point biomarker fulfills a set of conditions that are necessary and sufficient for its use in the clinic. The receiver operating characteristic (ROC) curve represents graphically a relationship between sensitivity and specificity of a particular biomarker. This curve is used to evaluate the efficacy of a tumor marker at various cut-off points; hence an ideal graph gives maximum area under the curve (AUC) to distinguish between an effective (useful) and ineffective (useless) test.

New era of molecular markers of early detection

With the advent of the molecular advances we have come to a point where we can start to think of precision treatment and preventative approaches. What strategy might be good for one individual may not work, or may even harm, another set of individuals. Furthermore, comparative effectiveness has to come to the reality where randomized clinical trials are designed with the efficacy and effectiveness instituted at the population

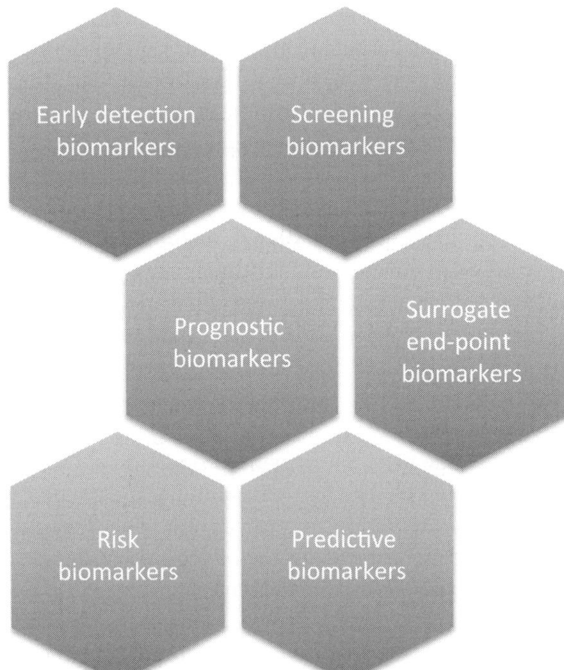

Figure 29.1 Biomarkers in the continuum of carcinogenesis. Biomarkers can serve a number of purposes including early detection of cancer, screening biomarkers, surrogate end point biomarkers (where the biomarker is not directly related to the outcome, hence they substitute for key clinical outcomes, predicting desired response of the clinical end point), prognostic biomarkers, biomarkers of risk of developing cancer, and predictive biomarkers that may predict likelihood of response to treatment, specifically the likelihood that the tumor will respond to a mechanism-based intervention. (A black and white version of this figure will appear in some formats. For the color version, please refer to the plate section.)

level. For a case example, beta-carotene – a vitamin A derivative – when given for lung cancer prevention did not help most of the people taking it but, worse than that, individuals who smoked even showed an increase in lung cancer incidence when on this dietary component (Goodman et al., 1996). Similarly, adenoma prevention with celecoxib – a popular selective inhibitor of cyclooxygenase 2 and an anti-arthritis anti-inflammatory – showed that individuals with underlying cardiovascular disease risk although receiving the cancer preventative effect on their advanced adenoma also had an increased risk for cardiovascular events; no such cardiovascular risk was seen in individuals that did not have underlying disease (Bertagnolli et al., 2009; Solomon et al., 2005). These clinical trials are not only key in defining the role of cancer prevention for a specific population, but also toward a need for personalized approaches

that we must take into account with our new knowledge of molecular biology.

Importance of early detection

It can be argued that early detection and prevention are probably the only successful strategies in the war on cancer, while the only successful treatment options are surgical in nature. Despite this fact, cancer treatment has gotten a lot more attention offering only dismal effectiveness of many approved therapeutics. What is the reason that both the fields of early detection, i.e., the discovery and validation of cancer and cancer prevention have suffered from a lackluster performance?

The major reasons are more likely due to lack of incentives rendered due to their socioeconomic factors rather than lack of inventiveness on the part of the research community (Bernards, 2010). Early detection research is not hypothesis generating and is less exciting to many of the young research fellows that have the highest creativity and productivity. Furthermore there is no incentive for any industrial ventures to delve into early detection and prevention research and development due to several competing economic realities. First, preventing a disease that may have albeit less effective treatment product cuts into the business model of certain self-obliteration that does not make sense. Second, early detection and prevention research takes long-term commitment and serious investment in time, money, and resources that in today's tough economic environment are even more difficult to assign. Finally, early detection and prevention clinical trials have not gotten a fair share of resources, even through the non-profit and academic institutes, as fewer lead compounds are available due to above-mentioned factors and a low tolerance for risks and side effects and higher standard for prevention research than associated with treatment.

Overdiagnosis and risk stratification

Despite all the advances in biomarkers for early detection of cancer, there are only a handful that have been validated and fully vetted as useful with a meaningful outcome. A different but more significant problem is the overdiagnosis of precancerous lesions of undetermined significance and the underdiagnosis of lesions that are truly dangerous and meaningful. Recent advances in omics allow for a deeper understanding of the meaning of these screen-positive lesions and individuals to further clarify how selection bias, lead-time biased sampling, and overdiagnosis be minimized and eventually make clinical decisions based on the real risk of developing cancer. It is well established that randomized clinical trials are the best method to ensure the efficacy of cancer screening tests.

Examples of these overdiagnoses are rampant throughout, but to highlight an extreme example Barrett's esophagus (BE) is a well-established pre-cancerous lesion for esophageal adenocarcinoma (EA). Esophageal adenocarcinoma cases and deaths continue to increase in number each year as it is a highly lethal cancer, although still relatively rare, but has the fastest growing incidence rate of all cancers in the United States and other Western countries. Notwithstanding the progress in multimodal therapies, the five-year survival rate is still less than 15%. Combined with an overdiagnosis of BE, operative resection, with its significant attendant morbidities, offers a grim picture. Unfortunately, the completion of therapy is most often followed by recurrence of tumor growth or distant metastasis that causes severe morbidities and eventual death. In addition, existing non-surgical therapies are relatively non-specific and highly toxic, often killing normally proliferating cells as well as tumor cells. Furthermore, the dismal natural course of EA tends to become clinically apparent only in advanced stages that are refractory to treatment. Effective measures to prevent or delay the onset of EA would have a substantial effect on the cancer burden in the United States (and other Western countries). The crux of the matter is that greater than 90% of patients who screen positive for BE through endoscopy will not develop esophageal adenocarcinoma, yet many individuals that develop cancer are never identified with BE. It highlights this challenge of underdiagnosis of early EA and overdiagnosis of BE without knowing the meaning of lesions in each individual person.

The situation is similar for other cancers where prostate, breast, and colorectal pre-malignant lesions are notoriously unpredictive. Clearly, a better approach with molecular identification of these lesions with validated biomarkers for early detection needs to be incorporated to move the field of cancer screening forward.

Network biology, network medicine, and biomarker discovery

Sir Ronald A. Fisher, one of the founders of modern biomedical statistics wrote (Fisher, 1948): "The rise of biometry in this 20th century, like that of geometry in the 3rd century BC, seems to mark out one of the great ages or critical periods in the advance of the human understanding. It is a tidal movement of our time, which has already begun to refresh and reinforce the means of research in all the biological sciences."

Throughout the twentieth century, the biomedical field has been a cradle and testing bed for countless statistical ideas, computational methodologies, and measurement paradigms. This tendency continues in accelerated pace into the twenty-first century. Long gone is the time when a researcher could comprehend an experimental outcome just by inspecting a one-page scatter-plot or a simple two-way table. Now, at the age of terabyte-size datasets and high-speed internet access, the need for formalized statistical summaries, supported by efficient data mining and mentally digestible presentation, is a rapidly growing demand. Just during the last decade, the biological community has witnessed the birth and fast development of a number of new statistical fields, such as microarray data analysis, computational proteomics, genome-wide association studies, and many others. A prominent role among these innovative developments belongs to the discipline of systems biology (SB), a far-reaching synthesis of biology, biophysics, biochemistry, mathematics, statistics, computer science, and systems theory (Ideker et al., 2001). In a recent review (Rosenfeld and Kapetanovic, 2008), we have already touched upon some aspects of SB that show promise in cancer prevention. In this section, we narrow down the general topic of SB to a newly emerging field of *network biology and network medicine*. More specifically, we focus here on the aspects of these fields that pertain to early cancer detection.

The concept of *network* is a centerpiece of systems theory. Graphically, a network is a set of *nodes* (a.k.a. *vertices*) connected by lines or arrows called *edges*. The nodes may represent various entities of a biological system; whereas the edges reflect the interrelations between these entities. Thus in protein–protein interaction networks (PPI), the nodes are the protein molecules or their aggregates, and the edges are their pair-wise affinities or potentially feasible chemical interactions (Bornke, 2008). In genetic regulatory networks (GRN), the nodes are genes and the edges are the *symbolic interactions* between them, which occur due to mutually interdependent sets of transcription factors (Potapov, 2008). Mathematically, a network is a well-defined mathematical object called a *graph*, which is completely characterized by the so-called *adjacency matrix.* An element a_{ij} of the adjacency matrix is a positive integer indicating how many edges connect node i to node j. Formal mathematical manipulations with the adjacency matrix allow inferring rich information regarding the network's topological structure and dynamical properties (Bornholdt and Shuster, 2003).

Cancer: a systems biology disease paradigm

Numerous research unambiguously confirms the key system biology premise that disease onset and proliferation are expressed in concerted reactions of the intracellular networks as a whole, rather than in changing the expression levels of individual molecular species such as mRNAs, proteins, and enzymes. This premise is fully applicable to cancer; due to this reason, cancer is sometimes called "a systems biology disease" (Hornberg et al., 2006). Among all the cancer biomarkers, those intended for early detection are expected to be sensitive to subtle shifts in intracellular organization, which may appear in the organism prior to the symptomatic stage. Taken individually, these biomarkers may not necessarily be biologically significant, and even less so should they necessarily be the culprits of the disease and the targets for therapeutic intervention; only a system-wide view of a disease may tell the whole story. These considerations may seem to be trivial and fairly old; however, it should be noted that the technical capabilities for fulfilling the promise of systemic view are just emerging. High-throughput techniques for genome-wide expression profiling combined with powerful computational capabilities make systemic genome-wide exploration practically feasible. The newly emerging disciplines of network biology and network medicine (Barabasi, 2007; Loscalzo et al., 2007; Zanzoni et al., 2009) are the embodiments of such systemic approaches. In what follows, we give a brief overview of the current status of this innovative field.

Role of modularity in biologic network connectevity

Modularity is an important characteristic of the network connectivity and its most visible graphical feature. It is typical for biological networks to have a set of highly linked nodes (usually called *hubs*) with the rest of the nodes having much smaller link densities. In PPIs, the hubs are the proteins that participate in many interactions; like big transportation hubs they link many remote areas of the network, thus having a substantial impact on its overall "health." Thus it has been demonstrated (Taylor et al., 2009) that the modular structure of human PPI may serve as an efficient predictor of breast cancer outcome. It has been observed in this work that alterations in gene expression (not necessarily significant on their own) may nevertheless noticeably disturb the higher level PPI organization and change its modular structure. Analysis of two breast cancer patient cohorts "revealed that altered modularity of the human interactome may be useful as an indicator of breast cancer prognosis."

Sub-network approach to gene expression profiles

A different network-based approach to the classification of breast cancer outcomes and prediction of likelihood of metastasis in unknown samples has been undertaken (Chuang et al., 2007). In this approach, the panels of gene expression profiles are not used as multiple predictors directly; rather, they are used as sub-networks of interacting proteins in a larger human PPI network. The authors observe that the genes with known breast cancer mutations (e.g., P53, KRAS, HRAS, HER-2/neu, PIK3CA) may not be always detectable through differential expression profiling. Nevertheless, they may play a central role in protein networks by interconnecting many expression-responsive genes. Also, the authors (Chuang et al., 2007) have come to two important methodological conclusions. First, they claim that the identified sub-networks are *significantly* more reproducible than the individual marker genes selected without network information. Second, the network-based classification is capable of achieving a higher accuracy in prediction. These two conclusions indicate that network-based analysis may be the basis for more statistically robust estimates as compared to differential expression profiling or protein mass-spectrometry taken on their own.

Mode of action network identification

A network-based approach to prostate cancer has been undertaken (as described in Ergun et al., 2007). Using a special technique called *mode of action network identification* (MNI) these authors studied non-

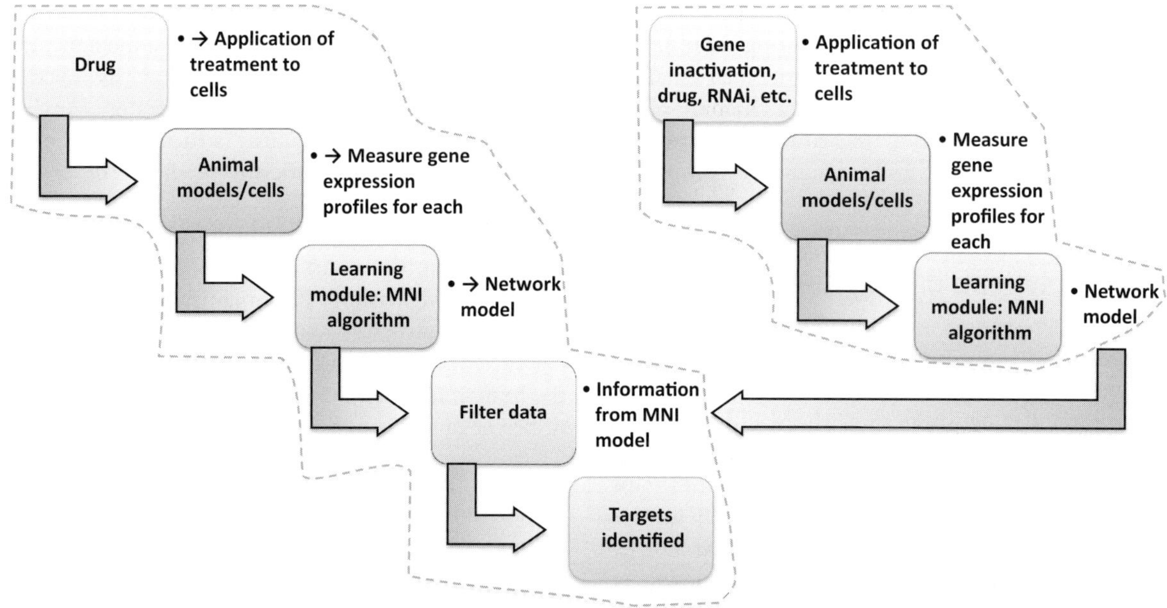

Figure 29.2 A set of treatments, including gene inactivation, interventions, and RNAi, may be used in the initial screen. Gene expression data from the samples are processed as shown. The MNI model filters data to identify the molecular targets of the test treatment. (A black and white version of this figure will appear in some formats. For the color version, please refer to the plate section.)

recurrent primary and metastatic prostate cancer data. They identified the androgen receptor (AR) gene among the top genetic mediators, and the AR pathway as a highly enriched pathway for metastatic prostate cancer. Importantly, these results have not been obtained on the basis of expression profiling alone. Rather, it has been shown that the AR gene can be used as a marker of agressiveness of primary prostate cancer more effectively if taken not alone but in the context of a network. The authors reach an overall conclusion that "a network biology approach can be used advantageously to identify the genetic mediators of the pathways associated with disease."

A systemic view of a disease through the prism of network biology allows for higher-level generalizations. Thus not only the individual biomarkers may be seen as players in a larger game of network-level disease, but also the diseases themselves may be seen as different facets of the same process encompassing the biomolecular network (Goh et al., 2007). A basis for such a logical leap is the observation that oftentimes the mutations in the same gene or group of genes led to quite different manifestations, thus giving rise to the phenotypes classified as different disorders. For example, mutations in p53 have been linked to many clinically distinguishable cancers. Considering diseases as nodes in the "human disease network"

(HDN), and common biomolecular markers as links between them, it would be possible to identify tightly interrelated disorders. In the network terminology, such "small worlds" of disorders could be classified as separate diseases. This classification, however, would be based not on the phenotypic manifestations but rather on their common biomolecular origins. The authors (Goh et al., 2007) indicate: "Given the highly interlinked internal organization of the cell, it should be possible to improve the *single gene–single disorder* approach by developing a conceptual framework to link systematically all genetic disorders with complete list of disease genes." That would be a step toward a more holistic view of human health.

Summary

It is estimated that in the United States approximately 30% of total health care costs for an individual happen during the last years of life – usually when cancer strikes (Earle and Schrag, 2008). More and more effort is being invested into precision medicine – especially in oncology where targeted interventions are being developed without much effort in developing and validating early biomarkers of cancer that can prevent and hence delay the onset of cancer by several more years. In this chapter we have attempted to show the importance of early detection by taking into consideration the aspects of network biology and network medicine, which although currently in their infancy, might be promising avenues in systems biology of cancer. In the presence of a barrage of molecular discovery data, further development and application of these approaches to decide patient benefit may bring some hope for the future (Figure 29.2).

References

Barabasi, A.L. (2007). Network medicine: from obesity to the "diseasome". *N Engl J Med 357*, 404–407.

Bernards, R. (2010). It's diagnostics, stupid. *Cell 141*, 13–17.

Bertagnolli, M.M., Eagle, C.J., Zauber, A.G., et al. (2009). Five-year efficacy and safety analysis of the Adenoma Prevention with Celecoxib Trial. *Cancer Prev Res 2*, 310–321.

Biomarkers Definitions Working Group (2001). Biomarkers and surrogate endpoints: preferred definitions and conceptual framework. *Clin Pharmacol Ther 69*, 89–95.

Bornholdt, S., and Shuster, H.G. (2003). *Handbook of Graphs and Networks: From the Genome to the Internet*. Wiley-VCH.

Bornke, F. (2008). Protein interaction networks. In B.H. Junker, and F. Schreiber (eds) *Analysis of Biological Networks*. John Wiley & Sons, pp. 207–231.

Buyse, M., Molenberghs, G., Burzykowski, T., Renard, D., and Geys, H. (2000). The validation of surrogate endpoints in meta-analyses of randomized experiments. *Biostatistics 1*, 49–67.

Chuang, H.Y., Lee, E., Liu, Y.T., Lee, D., and Ideker, T. (2007). Network-based classification of breast cancer metastasis. *Mol Syst Biol 3*, 140.

Earle, C.C., and Schrag, D. (2008). Principles of health services research. In V. DeVita, T. Lawrence, and S. Rosenberg, (eds) *DeVita, Hellman and Rosenberg's Cancer: Principles & Practice of Oncology*. Philadelphia, PA, Lippincott Williams & Wilkins, pp. 369–384.

Ergun, A., Lawrence, C.A., Kohanski, M.A., Brennan, T.A., and Collins, J.J. (2007). A network biology approach to prostate cancer. *Mol Syst Biol 3*, 82.

Fisher, R.A. (1948). Biometry. *Biometrics 4*, 217–219.

Goh, K.I., Cusick, M.E., Valle, D., et al. (2007). The human disease network. *Proc Natl Acad Sci USA 104*, 8685–8690.

Goodman, G.E., Thornquist, M., Kestin, M., et al. (1996). The association between participant characteristics and serum concentrations of beta-carotene, retinol, retinyl palmitate, and alpha-tocopherol among participants in the Carotene and Retinol Efficacy Trial (CARET) for prevention of lung cancer. *Cancer Epidemiol Biomarkers Prev 5*, 815–821.

Hanahan, D., and Weinberg, R.A. (2000). The hallmarks of cancer. *Cell 100*, 57–70.

Hornberg, J.J., Bruggeman, F.J., Westerhoff, H.V., and Lankelma, J. (2006). Cancer: a systems biology disease. *Biosystems 83*, 81–90.

Ideker, T., Galitski, T., and Hood, L. (2001). A new approach to decoding life: systems biology. *Annu Rev Genomics Hum Genet 2*, 343–372.

Jemal, A., Siegel, R., Xu, J., and Ward, E. 2010. Cancer statistics 2010. *CA Cancer J Clin 60*, 277–300.

Loscalzo, J., Kohane, I., and Barabasi, A.L. (2007). Human disease classification in the postgenomic era: a complex systems approach to human pathobiology. *Mol Syst Biol 3*, 124.

Potapov, A.P. (2008). Signal transduction and gene regulatory networks. In B.H. Junker, and F. Schreiber (eds) *Analysis of Biological Networks*. John Wiley & Sons, pp. 184–205.

Prentice, R.L. (2009). Surrogate and mediating endpoints: current status and future directions. *J Natl Cancer Inst 101*, 216–217.

Rosenfeld, S., and Kapetanovic, I. (2008). Systems biology and cancer prevention: all options on the table. *Gene Regul Syst Bio 2*, 307–319.

Solomon, S.D., McMurray, J.J., Pfeffer, M.A., et al. (2005). Cardiovascular risk associated with celecoxib in a clinical trial for colorectal adenoma prevention. *N Engl J Med 352*, 1071–1080.

Taylor, I.W., Linding, R., Warde-Farley, D., et al. (2009). Dynamic modularity in protein interaction networks predicts breast cancer outcome. *Nat Biotechnol 27*, 199–204.

Zanzoni, A., Soler-Lopez, M., and Aloy, P. (2009). A network medicine approach to human disease. *FEBS Lett 583*, 1759–1765.

Chapter

30

Systems biology in cancer biomarkers for early detection, diagnosis, and prognosis

Sudhir Srivastava and Karl Krueger

Introduction

Detecting cancer in its initial stages presents the opportunity to treat the disease before it spreads. Further, the ability to reduce a person's risk of developing cancer opens the way for optimum prevention strategies. The National Cancer Institute (NCI) is committed to progress in cancer detection and risk assessment that allows interventions to focus on the earliest stages of disease. The burden of cancer has been significantly reduced since the enactment in the United States of the 1971 National Cancer Act, but sustained research and clinical efforts are necessary to continue the trend. While cancer incidence and mortality rates declined overall by about 1.1% per year from 1992 to 1998, cancer remains a leading cause of death in the United States (Weir et al., 2003).

Throughout the history of medicine, man has developed effective ways to treat many diseases that he has been presented with. The treatments range from prevention in the form of vaccination to the use of medicines and surgical intervention as cures or even treatments to control the diseases' effects on patients' quality of life. For many diseases, such as bacterial infections, heart disease, and cancer, early detection plays a pivotal role in the survival of the patient. When detected early, many such lethal diseases can be effectively treated with existing remedies. The difficulty remains, however, how to effectively detect such conditions at the earliest possible stage with a high enough positive predictive value so that they can be treated effectively without arriving at false positive cases or overdiagnoses.

Definition of systems approach

A number of definitions have been proposed by researchers focusing on their areas of research and interplay between disciplines aiding their focus of research. Systems biology is a biology-based interdisciplinary field of study that focuses on complex interactions in biological systems, claiming that it uses a new perspective (holism instead of reductionism). However, the definition seems to vary across a wide spectrum of approaches. In a generalized term, a systems approach is an understanding of the fundamental biological processes underlying disease progression, detection, prevention, and treatment leading to interdisciplinary research: systems biology, synthetic biology, and artificial cell technologies. In relation to cancer, systems approaches focus on the analysis of cancer as a complex biological system relevant to cancer prevention, diagnostics, and therapeutics. The integration of experimental biology with mathematical modelling will result in new insights in the biology and new approaches to the management of cancer encompassing clinical and basic science, mathematics, physics, bioinfomatics, imaging sciences, and computer science to work on key questions in cancer biology.

What is systems biology now?

The potential of molecular techniques to significantly improve the detection of localized cancer provides an unprecedented opportunity to understand the biology, improve diagnosis, enhance treatment, and reduce mortality. These strategies have just begun to explore the utility of imaging, proteomics, and genomic analysis of tumors or other specimens. These approaches have the potential to identify small and early lesions that have not been readily accessible in clinical practice through more conventional detection methods. While there are effective markers for diseases such as prostate and ovarian cancer (i.e., prostate-specific antigen and cancer-antigen 125, respectively), these markers do not possess the sensitivity and specificity necessary to be used as a general screening tool in the diagnosis of early-stage cancer. What is required is the identification of highly predictive biomarkers, as well as other types of technologies, that can aid in the diagnosis of early-stage diseases (Kellof et al., 2007; Moncada and Srivastava, 2008; Srivastava, 2006, 2007).

The challenge is how to identify more effective biomarkers or technologies that can provide an earlier indication of a disease with a higher positive predictive value than presently utilized methods. Fortunately, the past decade has been witness to a new revolution in biological science. In this revolution, investigators now have the opportunity to study thousands of different biological molecules in a single experiment or in a series of high-throughput studies. This rate of characterization allows for the rapid comparison of samples from many different patients with the hope of finding species that are directly related to the disease state.

Currently only a handful of biomarker tests are presently in use despite the hype and hope. In a systematic review of scientific literature in the last ten years, more than 4,000 primary research articles published each year were identified that reported positive findings on diagnostic performance of biomarkers (Ludwig and Weinstein, 2005). This has fueled expectations in the field generating hopes for biomarker-based diagnostics in the near future. However, the reality is far from this expectation. Only a handful of biomarkers have been approved by the US Food and Drug Administration (FDA) over the last ten years. Many of the reported findings had only one study with a small number of convenience samples, which lacked protocol-driven sample collection, processing, and storage. The predominance of nominally significant P values suggested the presence of reporting bias. These findings may not be applicable to individuals of differing ethnicity because of sample collection bias.

Systems approach to biomarker research?

Only a few biomarkers are in clinical practice despite the fact that hundreds of thousands of research papers claim to have discovered biomarkers for potential utility in clinical application. Why is only marginal progress being made in this field? Let us examine this

issue by understanding the requirements that a potential biomarker must meet before being considered for clinical application. It is unlikely that a single molecular biomarker with sufficient performance would be found able to distinguish disease from non-disease states because of the high degrees of heterogeneity across larger populations. Also, many of the assays for markers are either not reproducible or not sensitive and specific enough to warrant further consideration. Sensitivity is a measure of the proportion of individuals with a disease who test positive for the disease. Specificity is the proportion of individuals without the disease who test negative.

To illustrate how these parameters influence performance first consider a biomarker test that may have high sensitivity but low specificity. In this case, most individuals with disease will test positive, but the low specificity rating indicates that many people without the disease will also test positive. Biomarkers with this performance profile certainly would not be desirable since unnecessary and often costly or invasive treatment might be administered to patients who are free of disease. Conversely, a biomarker with low sensitivity and high specificity will help ensure that an individual is disease free, but will often fail to detect the genuine cases of disease. At times this type of performance might have limited clinical value, particularly in helping to rule out negative cases where additional tests were performed on a patient yielding uncertainties in the prediction of disease. Clearly, a biomarker possessing the qualities of both high sensitivity and specificity is desired for accurate diagnosis of disease. It is likely that many diagnostic tests of the future will encompass analysis of panels of such discriminatory molecules.

Therefore there is an urgent need to refine the biomarker discovery paradigm based on sound basic knowledge about the marker in question and its relation to carcinogenesis at a systems level understanding of biological processes, where a group of biological networks and pathways are connected to one another and work together. It is time for a systems approach to the discovery and validation of disease biomarkers, one that takes full advantage of the revolution in genomic technologies and in the development of computational tools for the analysis of large complex datasets. Unlike traditional approaches, a systems approach relies on the types of data used, and the tools used for data analysis, both of which reflect revolution in high-throughput analytical methods and high-throughput computing. But what is a "systems biology" approach? According to Wang et al. (2010) a systems biology approach comprises five features: (1) measuring and quantifying biological information on a global scale; (2) integrating distinct modalities of biological information such as DNA, RNA, and proteins; (3) capturing the dynamics of biological systems and networks; (4) using these sources of information to model the system; and then (5) iteratively testing and refining the model. Although the description of systems biology by Wang et al. (2010) prominently features the generation of large quantitative datasets using a variety of "omics" technologies (features 1 through 3), numbers alone do not constitute a systems biology study. Systems biology also incorporates the flow of information in the system that may emanate from global changes in gene expression or protein expression. The building of this information would require "expert knowledge" of the relevant biology. This can be illustrated by the following example.

Virtually all proteins on the cell surface and secreted proteins are glycosylated and thus changes in this post-translational modification will influence how the cell interacts with its environment and other cells. To understand how systems biology plays into the scheme of protein glycosylation it is first important to explain how these complex structures are synthesized.

Glycoproteins are glycosylated at asparagine residues (N-linked) and/or serine/threonine (O-linked). The mechanisms of building N-linked and O-linked glycans differ but these events occur in the lumen of the endoplasmic reticulum and Golgi apparatus as the polypeptide is being synthesized and matures in its transit through this system. Synthesis of both types of glycans involves sequential addition of monosaccharides to the growing glycan by a class of enzymes called glycosyltransferases. Every glycosyltransferase is specific for transferring only one monosaccharide from a sugar-nucleotide donor and they have very discrete specificity for the substrate to which the monosaccharide is being added. Furthermore, the location of attachment to the accepting site (normally four available hydroxyl groups are available at the growing end) and the anomeric configuration of the resulting glycosidic bond (alpha or beta) is also dictated by the glycosyltransferase.

Although N-linked and O-linked glycans are built by glycosyltransferases, they are initiated by very different processes (Rini et al., 2009). O-linked glycans are initiated in the cis-Golgi by any of a series of over 20 glycosyltransferases that covalently attach N-acetylgalactoseamine to serine/threonine. These glycosyltransferases differ in their polypeptide substrate specificity. N-linked oligosaccharides are initiated in the endoplasmic reticulum through a lipid-linked carrier dolichol pyrophosphate, which through the action of many glycosyltransferases results in a glycan bearing a core of two N-acetylglucoseamine, nine mannose, and three glucose units. This larger glycan is then transferred to appropriate asparagine residues of the growing polypeptide chain within the lumen of the endoplasmic reticulum. The N-glycan is further processed through the action of several glycosidases to trim terminal sugar units off the glycan core, ultimately to serve as a substrate for other glycosyltransferases in the development of a series of branched structures that can have from two to four antennae. This latter process of trimming and extension of glycan structures plays a major role in the proper folding of the growing polypeptide as a number of chaperone proteins recognize intermediates of this N-linked glycan pathway and assist with the proper folding and processing of the glycoprotein as it is being synthesized and transported to the Golgi. After completion of a glycan structure, it can be further modified through site-specific sulfation, phosphorylation, acetylation, or methylation.

Based on these schemes it should be apparent that the synthesis of glycans is highly dependent on the repertoire of glycosyltransferases, glycosidases, and other enzymes expressed by the cell. In particular, loss of or elevated expression of a glycosyltransferase participating in an early step of a synthetic pathway can have a dramatic effect on the downstream steps and substantially alter the structure of the glycan synthesized.

Over 180 confirmed human glycosyltransferase genes have been identified with still other genes potentially belonging to this family remaining to be determined placing their number well over 200. This accounts for nearly 2% of the entire human genome. Because these genes are not clustered but distributed throughout most chromosomes, genetic amplifications and deletions, almost always associated with cancers, are likely to involve changes in copy number and hence expression of glycosyltransferases. Cancer is also typified by profound changes in transcriptional regulation where the abundance of glycosyltransferases implies some of these genes would be affected. Similar principles could apply for sulfotransferases, acetylases, or methylases. On a broader scale many other factors impact the expression of the cellular glycome including availability and transport of sugar-nucleotide donors as substrates for glycosyltransferases, accessibility of appropriate acceptor sites for glycosylation throughout the entire synthetic process within the Golgi, proper localization of glycosyltransferases to enable sequential addition of monosaccharides, and adequate recruitment of accessory proteins participating in these pathways (Nairn et al., 2008).

In essence, systems biology approaches to develop models of interdependent networks and pathways correlating glycomic profiles with the expression of all genes participating in glycan synthesis have only recently been started (Liu et al., 2008; Marathe et al., 2008). To have a comprehensive understanding in order to predict the profile of carbohydrate structures produced by a cell would require a detailed systems biology approach to characterize all of these glycomics participants as well as the signaling pathways that regulate their expression.

In addition, a systems approach can also accelerate the validation of biomarkers for clinical use. The introduction of new biomarkers in clinical settings will require directly measuring the impact of the biomarker on mortality in a prospective, randomized trial; however, the costs of conducting large trials combined with regulatory and health-care market constraints make such assessments expensive and time consuming. Simulation modelling to predict how new biomarker technologies affect overall outcomes could be an alternative, inexpensive approach as opposed to conducting a large, expensive trial for every new biomarker developed. A simulation approach can help optimize the sensitivity, specificity, and cost trade-offs at each step as well as identify situations where more definitive biomarkers are needed. For example, mathematical models can be built to simulate tumor growth over time, either at the cellular level or spatially (that is, representing tumor size). The models are often calibrated to observed data, typically longitudinal population studies of tumor growth and clinical disease. Real or hypothetical screening or diagnostic tests are then introduced into the model, with sensitivity and specificity based on their ability to detect either markers that increase in the blood in proportion to tumor burden, resolution for detecting tumors of a particular size, or ability to detect metabolic changes. Based on known or hypothesized distributions representing ranges in the rates of tumor onset and growth, the models are then run for a population either with or without disease to generate life histories both in the absence and presence of testing. The outcomes (stage at diagnosis, tumor response rate, cancer-specific or overall survival, medical care costs) are then compared under the test and no-test scenarios. This approach was used to estimate the cost-effectiveness of flexible sigmoidoscopy screening for colorectal cancer in the United States, and will likely become more common as new biomarkers are introduced.

Systematic approach to the development of diagnostic biomarkers

The systems approach for diagnostic biomarkers could occur in the following phases (Figure 30.1). (1) Defining clinical needs: in this phase the clinical need for diagnostic biomarkers is analyzed and active search and creation of new technologies or biomarkers that might detect cancer or pre-cancer events at earlier points in disease progression are conceptualized. The concept may be based on the fact the biomarkers may have been found in other model systems, such as cell lines, animal models, or yeast, in which biomarkers might have been discovered and found to be associated with the carcinogenic process. (2) Data gathering: such biomarkers then undergo partial development in humans yielding preliminary promising results to justify their further evaluation.

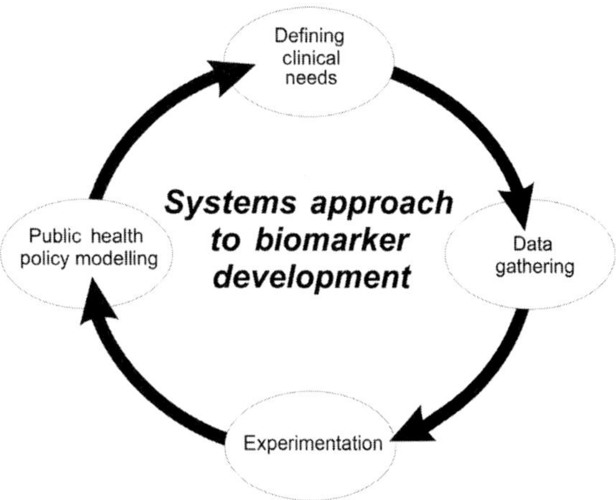

Figure 30.1 Four-phase approach to developing biomarkers for diagnostic use and measuring their impact in clinical care.

As we may expect only a few biomarkers produce sufficiently promising findings derived from model systems. At this point a systems approach may be applied to see the worthiness of the biomarker in the natural history of the disease, interdependence with other disease networks or pathways, and the feasibility of its application in cancer diagnosis. This approach, hopefully, will eliminate unproductive experimentation and allow us to focus on only those that are likely to succeed. (3) Experimentation: this phase will focus on candidate biomarkers that have sufficient evidence, both biological and mechanistic, coupled with data modelling with the aid of a systems approach. In case of high-throughput, high-dimensional data, statistical modelling may be employed to identify candidate biomarkers as explained later. Once the biomarkers are identified and verified, these will be rigorously tested and validated with appropriate biologic samples to determine their association with the disease. Another associated task in this phase is the development of an appropriate technology or an assay technology to measure such an association. (4) Public heath phase: depending on the intended clinical use of the biomarkers, a large clinical trial, preferably prospective and randomized, may be required to measure the efficacy of the biomarker. However, it is not feasible and cost-effective to conduct a large, randomized trial for every biomarker developed. Therefore it may be beneficial to use a data modelling approach to compare the cost-effectiveness of the proposed tests in the context of population screening, considering the natural history of disease, risk factors that can modify the onset and progression of the disease, cost saving in comparison to existing tests, person-years life saved as the result of intervention (diagnosis). Data modelling along with a large case-control study based on the PRoBE design (Pepe et al., 2008) may suffice the needs for validation to meet the public health application requirements.

Wet lab discovery

In general, biomarker research occurs in phases: initial discovery of biological markers, evaluation of the most promising biomarkers, and validation to determine that they can work in a clinical setting. There is a need to speed translation of basic scientific discoveries in genomics and proteomics into medical benefits in a systematic and concerted fashion. The National Cancer Institute's Early Detection Research Network (EDRN) has adopted a five-phase approach to discovering, developing, and validating biomarkers. This five-phase approach establishes both a standard and a road map for successfully translating research on biomarker applications from the laboratory to the bedside (Pepe et al., 2001).

Phase 1 includes exploratory study to identify potentially useful biomarkers. In Phase 2, biomarkers are studied to determine their capacity for distinguishing between cases with cancer and those without. Phase 2 is called the validation phase. Repositories of longitudinally collected clinical specimens from research cohorts are used in Phase 3 to determine the capacity of a biomarker to detect preclinical disease.

Prospective screening studies are considered part of the next phase, Phase 4. Finally, large-scale population studies that evaluate not just the role of the biomarker for detection of cancer, but the overall impact of screening on the population comprise Phase 5. Researchers have welcomed the five-phase structure that EDRN proposes for biomarker development because it provides an orderly series of studies that build upon each other to yield an efficient and thorough approach to biomarker development.

The key aspects of the study design for each of the phases has been discussed and published. The challenges to the discovery and validation of biomarkers are being met by following the principles and theories of systems biology in which disciplines such as biology, chemistry, and computational science along with the clinical sciences are integrated in a seamless way to

- yield clinically useful assays for cancer detection and diagnosis
- create novel approaches to validation studies
- improve informatics and information flow
- standardize data reporting
- create general statistical and computational tools and standardize reagents and assays.

From a biomarkers viewpoint, the systems approach can benefit the discovery and development of cancer biomarkers that have the chance to succeed in becoming clinically useful by generating strong scientific evidence in support of their ability to represent the human condition and have immediate diagnostic potential through the development of scientific research involving model systems, e.g., yeast, cell, animal, or other genetically engineered models of cancer-relevant diseases, such as:

- diseases or conditions wherein models exist that have strong scientific evidence for mimicking key aspects of the human condition
- comparative analyses of existing models to rigorously establish through comparison with human clinical data that the model(s) is a valid surrogate for the human disease or condition
- cancer biomarker research based on screening of animal models leading to the development and validation of biomarkers in humans. This will require clearly defined, clinically relevant screening criteria, and documented agreements with ongoing studies with human materials
- approaches for validating the relevance of the model to the human clinical condition.

In silico discovery from data mining

Presently, biomarker discovery efforts are plagued with problems related to reproducibility, generalizabilty to low sample size, tumor heterogeneity, and over-fitting of the data due to wired-in bias with sample collection. A number of models have been developed elsewhere utilizing the *a priori* knowledge of the biology and natural history of the disease; however, these models do not provide much insight into the discovery and development of biomarkers. Recently, a number of models have appeared using genomic and proteomic data to develop classifiers to distinguish cancer from normal and in so doing developed a set of features as biomarkers. It is beyond the scope of this article to describe the mathematics behind the models. Qiu et al. (2007) used a statistical modelling approach for cancer biomarker discovery and provide new insights into early cancer detection. They proposed the concept of a dependence network and applied it to identifying cancer biomarkers, and study the difference between the protein or gene samples from cancer and non-cancer subjects based on mass spectrometry (MS) and microarray data from three MS and two gene microarray datasets. They observed clear differences in the dependence networks for cancer and non-cancer samples. Protein/gene features are examined three at a time through an exhaustive search. Dependence networks are constructed by binding triples identified by the eigenvalue pattern of the dependence model, and are further compared to identify cancer biomarkers. According to Wikipedia, eigenvalues are properties of a matrix. (Note: they have applications in areas of applied mathematics as diverse as economics and quantum mechanics.) In dependence models the dependence matrix and noise statistics are used to model the expression data. However, it is difficult to tell the dependence relationship from the model parameters. Therefore Qiu et al. (2007) used the term, the eigenvalues, to describe the dependence relationship. Such dependence network-based biomarkers show much greater consistency under ten-fold cross-validation than the classification performance-based biomarkers. They applied this model to build a supervised classifier for individual protein features and used their classification performance as a biomarker identification criterion.

Chromosomal translocations have long been known to occur in various types of leukemias, lymphomas, and sarcomas, but few have been documented in most types of solid cancers. Recently a chromosomal rearrangement has been implicated as a potential cause of some prostate cancers, a finding that could lead to new diagnostic and therapeutic strategies (Tomlins et al., 2005).

What is interesting about this finding is the fact the discovery of the gene fusion was made through analyzing a database called Oncomine, a collection of data from numerous cancer studies across the globe, studies that use devices called microarrays to simultaneously screen the activity of thousands of genes (Hampton, 2004). The Oncomine database integrates 132 gene expression datasets representing 10,486 microarray experiments.

By surveying Oncomine for genes that are over-expressed in prostate tumors, Tomlins et al. (2005) found two candidate genes – ERG and ETV1 – that encode transcription factors, which are proteins that regulate the expression of other genes. In addition, they reported that either gene was frequently fused to the regulatory region of a third gene (TMPRSS2), which encodes an enzyme activated by the male hormone androgen. The translocation of ERG and ETV1 may be redundant because only one or the other was found fused to the TMPRSS2 regulatory region in any given tumor. While the actions of these gene fusions are not known, they may play an important role in prostate cancer development or progression because one or the other fusion was present in cultured lines of prostate cancer cells and in 23 of the 29 primary prostate cancers examined in the study.

Encouraged by these findings, Chinnaiyan and his colleagues have begun studying the downstream effect of these gene fusions on phenotypic expression. A recent study from Dr. Chinnaiyan's group revealed that sarcosine, an N-methyl derivative of glycine, is highly increased during prostate cancer progression to metastasis and can be detected non-invasively in urine (Sreekumar et al., 2009). Functional genomic studies have indicated that the androgen receptor and ERG gene fusion product associated with prostate cancer development and progression (Tomlins et al., 2005) coordinately regulates the sarcosine synthesis pathway via transcriptional control of its regulatory enzymes. Hence by profiling metabolomic alterations in prostate cancer, new insights were revealed in prostate cancer progression, invasion, and aggressiveness. Furthermore, components of the sarcosine synthesis pathway may have potential, not only as diagnostic/ prognostic biomarkers of prostate cancer, but may also serve as targets for developing new therapeutic modalities. This example nicely illustrates a systems approach to discovering biomarkers from high-volume expression data to a metabolite discovery whose expression was also regulated by the gene fusion products.

Data are rapidly accumulating from genomic and proteomic analyses. A key concern is the development of algorithms to combine information from multiple biomarkers to identify preclinical cancer. The problem is huge when gene expression profiles or protein mass spectrometer profiles are the basis for identification since the number of potential markers (genes or proteins) is enormous. Innovative analytic tools are needed to provide us with the best metric for evaluation of biomarkers.

The design and evaluation of exploratory studies is perhaps the most promising and difficult statistical issue in biomarker development. Consider the analysis of gene expression data and the problem of ranking genes in regards to their differential expression in cases and controls. Classical statistical measures, such as the t-statistic do not necessarily identify the most useful genes.

System screening

One of the important potential uses of markers for early cancer detection is to implement their monitoring via screening programs in target populations. Hence it is crucial to explore and understand factors that influence the accuracy (sensitivity and specificity) of such programs.

While the issue of evaluating the accuracy of a single diagnostic test has been the focus of much attention in the statistical and applied literature, little attention has been paid to factors affecting accuracy characteristics of a sequence of tests, as would be used in routine screening. With repeated marker measurements over time in a target population, true cases and false positives will be identified cumulatively, at each time of testing. In practice, when a single marker is used repeatedly in routine screening, the same screening threshold is typically used to define positivity at each screening visit.

By application of a systems biology approach, particularly predictive modelling, one can develop an algorithm to guide choices in designing a screening program. For example, if the modelling suggests that adaptive thresholds may perform better than constant thresholds and that screening biannually has accuracy characteristics almost as good as annual screens, with very little difference between annual and biannual screening designs in the mean time prior to clinical diagnosis at which cases are identified, biannual screening may be preferred. For instance, when PSA is used to screen for prostate cancer, a common threshold is 4 ng/ml (Parekh et al., 2007). One possible alternative is to adjust the threshold at successive visits according to individual-specific characteristics. It has been shown that adaptive thresholds may perform better than constant thresholds (such as 4 ng/ml) and that screening biannually has accuracy characteristics almost as good as annual screens (Feng et al., 2009).

Future roles of systems biology in cancer biomarker discovery and validation

In the history of biomarker discovery, serendipity has played its role in biomarker research. Many presently used biomarkers, such as CA 125 or PSA, were not the outcome of systematic discovery approach, but were the byproducts of another method of discovery. However, these biomarkers though useful, do not provide sufficient sensitivity and specificity for the early detection of cancer. With the advent of new molecular tools and technologies, focus has now shifted toward biomarker discovery based on carcinogenic networks and pathways. Mathematical modelling has proven extremely useful in deciphering molecular and biochemical mechanisms, such as the control of the cell cycle (Novak and Tyson, 2004). In some instances, the development of mathematical modelling inspired by the desire to analyze the difficult, complex, non-linear gene-metabolic networks has stimulated the development of novel mathematical methodologies (Ball et al., 2006; Craciun et al., 2006).

Mathematical modelling of cancer-relevant biological pathways may provide technology developers and biomarker discoverers with a new powerful tool to search for new biomarkers that may result from the analysis of complex datasets, e.g., proteomic MS spectra, or genomic microarray data. One should be aware that these models have strengths and limitations and must be well understood to benefit from their use and avoid misinterpretation. A model will never be perfect, in part because it is usually based on incomplete and simplified biological information. Models can be used as a means to test or generate hypotheses that may be experimentally tested or verified. Modelling will allow us to link experimental data and derived biomarkers to systems biology and will help drive our understanding of cancer biomarkers in relation to established pathways and networks (Hartwell et al., 2006) underlying the disease process.

References

Ball, K., Kurtz, T.G., Popovic, L., and Rempala, G. (2006). Asymptomatic analysis of multiscale approximations to reaction networks. *Ann Appl Prob 16*, 1925–1961.

Craciun, G., Tang, Y., and Feinberg, M. (2006). Understanding bistability in complex enzyme-driven reaction networks. *Proc Natl Acad Sci USA 103*, 8697–8702.

Feng, Z., Kagan, J., and Srivastava, S. (2009). *Toward a Robust System for Biomarker Triage and Validations: EDRN Experience*, 2nd edn. Totowa, NJ, Humana Press.

Hampton, T. (2004). Tool helps cancer scientists mine genes. *JAMA 292*, 2073.

Hartwell, L., Mankoff, D., Paulovich, A., Ramsey, S., and Swisher, E. (2006). Cancer biomarkers: a systems approach. *Nat Biotechnol 24*, 905–908.

Kellof, G.J., Sullivan, D.C., Baker, H., Clarke, L.P., and MacAulay, C. (2007). Workshop on imaging science development for cancer prevention and preemption. *Cancer Biomarkers 3*, 1–33.

Liu, G., Marathe, D.D., Matta, K.L., and Neelamegham, S. (2008). Systems-level modeling of cellular glycosylation reaction networks: O-linked glycan formation on natural selectin ligands. *Bioinformatics 24*, 2740–2747.

Ludwig, J.A., and Weinstein, J.N. (2005). Biomarkers in cancer staging, prognosis and treatment selection. *Nat Rev Cancer 5*, 845–856.

Marathe, D.D., Chandrasekaran, E.V., Lau, J.T., Matta, K.L., and Neelamegham, S. (2008). Systems-level studies of glycosyltransferase gene expression and enzyme activity that are associated with the selectin binding function of human leukocytes. *FASEB J 22*, 4154–4167.

Moncada, V., and Srivastava, S. (2008). Biomarkers in oncology research and treatment: early detection research network: a collaborative approach. *Biomark Med 2*, 181–195.

Nairn, A.V., York, W.S., Harris, K., et al. (2008). Regulation of glycan structures in animal tissues: transcript profiling of glycan-related genes. *J Biol Chem 283*, 17298–17313.

Novak, B., and Tyson, J.J. (2004). A model for restriction point control of the mammalian cell cycle. *J Theor Biol 230*, 563–579.

Parekh, D.J., Ankerst, D.P., Troyer, D., Srivastava, S., and Thompson, I.M. (2007). Biomarkers for prostate cancer detection. *J Urol 178*, 2252–2259.

Pepe, M.S., Etzioni, R., Feng, Z., et al. (2001). Phases of biomarker development for early detection of cancer. *J Natl Cancer Inst 93*, 1054–1061.

Pepe, M.S., Feng, Z., Janes, H., Bossuyt, P.M., and Potter, J.D. (2008). Pivotal evaluation of the accuracy of a biomarker used for classification or prediction: standards for study design. *J Natl Cancer Inst 100*, 1432–1438.

Qiu, P., Wang, Z.J., Liu, K.J., Hu, Z.Z., and Wu, C.H. (2007). Dependence network modeling for biomarker identification. *Bioinformatics 23*, 198–206.

Rini, J., Esko J., and Varki, A. (2009). Glycosyltransferases and glycan-processing enzymes. In Varki, A., Cummings, R.D., Esko, J.D., et al. (eds) *Essentials of Glycobiology*. Cold Spring Harbor, Cold Spring Harbor Laboratory Press.

Sreekumar, A., Poisson, L.M., Rajendiran, T.M., et al. (2009). Metabolomic profiles delineate potential role for sarcosine in prostate cancer progression. *Nature 457*, 910–914.

Srivastava, S. (2006). Molecular screening of cancer: the future is here. *Mol Diagn Ther 10*, 221–230.

Srivastava, S., and. Wagner, P.D (2007). Risk-based and diagnostic-lined personal medicine. *Person Med 4*, 33–34.

Tomlins, S.A., Rhodes, D.R., Perner, S., et al. (2005). Recurrent fusion of TMPRSS2 and ETS transcription factor genes in prostate cancer. *Science 310*, 644–648.

Wang, K., Lee, I., Carlson, G., Hood, L., and Galas, D. (2010). Systems biology and the discovery of diagnostic biomarkers. *Dis Markers 28*, 199–207.

Weir, H.K., Thun, M.J., Hankey, B.F., et al. (2003). Annual report to the nation on the status of cancer, 1975–2000, featuring the uses of surveillance data for cancer prevention and control. *J Natl Cancer Inst 95*, 1276–1299.

Chapter

Prognosis of cancer

31

Sharyn Katz and Wafik S. El-Deiry

Introduction

In this chapter we will explore the prognostic and predictive patient and tumor factors that influence cancer outcomes and impact overall survival and progression-free survival. Prognostic factors are defined as patient and tumor characteristics that influence the natural history of a disease. For example, triple negative breast tumors carry a very poor prognosis with low overall five-year survival rate. Predictive factors are patient and tumor factors that predict the likelihood of a tumor response to specific clinical interventions. For example, the presence of elevated expression of DNA damage repair enzyme, *ERCC1*, predicts a poor response to DNA damaging platinum-based chemotherapies [1]. In the last several decades, many cancer biomarkers representing a spectrum of molecules have been identified to have prognostic and predictive value and have been incorporated into conventional oncologic management and many more are in the process of being validated in clinical trials. A biomarker is a molecule found in the body or body products that can be objectively measured as an indicator of a normal physiological or abnormal pathophysiological state. Cancer biomarkers are biomarkers that have prognostic or predictive value in the setting of malignancy and include genetic mutations, gene amplifications, gene polymorphisms, epigenetic alterations, oncofetal markers, markers of epithelial–mesenchymal transition, and altered cellular metabolism. As tumor biology becomes better understood, these cancer biomarkers will become increasingly refined in their use improving individualized, personalized prognoses and tailored therapeutics to a particular tumor phenotype with the goal of providing overall survival and minimizing morbidities. In addition to refining cancer prognosis and treatment selection, some of these cancer biomarkers are being explored for medical imaging of cancer, allowing for functional tumor imaging to characterize tumors *in situ* at diagnosis and over the course of treatment.

Historical perspectives

It is important to realize that the prognosis of cancer at any point in time throughout medicine has been predicated on the knowledge of the factors that can significantly impact on cancer prognosis, the ability to accurately detect those factors and the treatment options available. In the early 1900s prognosis was generally rendered based on the patient's general medical condition and estimated size of the tumor. Over the decades, assessment of the primary tumor became more sophisticated as factors such as degree of local invasion, presence of metastasis, and regional lymph node involvement became recognized to impact on patient prognosis. An example of the evolution of cancer prognostication was summarized in a review on melanoma staging by Thompson et al. [2]. In 1953, it was reported in the literature that deeper melanomas had a worse prognosis than more superficial melanomas [3]. Eventually, similar observations led to the proposal of the first melanoma staging system in 1962, which incorporated histological parameters of invasion, dividing melanoma prognosis into three stages: stage I, no dermal invasion; stage II, invasion of the superficial dermis; and stage III, tumor formation with or without a pigmented flare [4]. By 1988, the staging system had evolved to include four stages with parameters of microscopic tumor depth, extent of dermal invasion, presence of regional lymph node metastasis, and presence of distant metastasis [2]. By 2002, the staging system demonstrated refinements on the categories of depth of tumor invasion and included

---

Systems Biology of Cancer, ed. S. Thiagalingam. Published by Cambridge University Press. © Cambridge University Press 2015.

presence of tumor ulceration, microscopic nodules, and in-transit metastasis.

In recent years, there has been a movement to advance traditional staging systems, like that of melanoma, beyond histological and morphological parameters to also include biomarkers of subsets of tumor phenotype. For example, in 2000 it was reported by Perou et al. that there are at least five distinct molecular subtypes of breast cancer identified by gene expression microarray and these are: basal-like, luminal A (lumA), luminal B (lumB), human epidermal growth factor receptor 2 (HER2)-enriched, and normal-like [5]. Of these subtypes, the basal-type and HER2 receptor-enriched subtypes were demonstrated to have the shortest survival times [6]. In the decade that followed, there has been further refinement of these molecular subtypes and advancement of therapeutics to specifically target subtypes such as the Her2 overexpressors and basal-type that carry the most dismal prognoses.

Similar progress has been made in a wide range of malignancies leading to the identification of an ever-expanding pool of cancer biomarkers, which include genetic mutations, gene amplifications, genetic polymorphisms, oncofetoproteins, inappropriately expressed normal gene products, and markers of altered cellular metabolism. These cancer biomarkers can be utilized to predict overall survival and specific tumor characteristics, such as local invasiveness, glycolytic activity, tumor vascularity, proclivity to metastasize, and likelihood of sensitivity to a specific therapeutic regimen. Some cancer biomarkers expressed in the serum, usually oncofetal biomarkers, have been used in following disease response to therapy as an approximation of tumor burden. Specific biomarkers of therapeutic success are of great value since the early detection of treatment failure avoids lost time to an ineffective therapy and exposure to the toxicities of an ineffective therapy. With the rapid proliferation of targeted therapies over the last decade, patients are often now limited by the pace of tumor progression and accumulation of toxicities to therapy rather than the number of treatment options. They simply run out of time to find an effective therapy. Beyond the individual patient management, specific markers of therapeutic response are also invaluable for clinical trial development of a chemotherapeutic or specific intervention to allow for triage of proposed treatment regimens. There is also much interest in utilizing tumor biomarkers for medical imaging of cancer to allow *in situ* definition of tumor characteristics, such as glycolytic activity, vascularity, oxygen tension, and fluid restriction during a given therapy to provide multidimensional feedback on therapeutic performance.

General prognostic factors

General prognostic patient factors include age at diagnosis, comorbidities such as emphysema, pulmonary interstitial diseases, chronic infectious diseases, allergies, liver or renal disease, cardiac insufficiency, diabetes, and immunosuppressive conditions (Figure 31.1). These comorbidities can significantly impact prognosis through unwanted drug–drug interactions, sensitivities to therapeutic toxicities, and restriction of access to available effective therapies due to impaired drug clearance. In addition to significant comorbidities, patients who have had prior cytotoxic therapies, such as radiation therapy, cardiotoxic or neurotoxic therapies may be limited by available therapeutic options, potentially excluding access to a more effective regimen, and thus negatively impacting overall prognosis. Finally, non-medical patient factors can also impact the success of therapy, and thus prognosis, including socioeconomic status, access to oncologic treatment centers, patient compliance, use of recreational drugs or alcohol, and degree of supportive family infrastructure.

At diagnosis, these patient characteristics are weighed together with the tumor characteristics to guide selection of therapeutic options and estimates of overall patient survival. Tumor characteristics included in this assessment include tumor stage, which typically incorporates tumor size, location, locoregional spread and presence/absence of metastasis, and tumor histological phenotype. Assessment of tumor histological phenotype is complex and subject to the inclusion of a dynamic range of cancer biomarkers that have implications on overall survival and can be predictive of therapeutic success. Example of these include mutations of known tumor suppressors, such as *BRCA* and *p53*, oncogene amplifications, such as c-*MYC* and *EGFR*, and markers of dysregulated cellular metabolism, such as polymorphisms of thymidylate synthase, a rate-limiting enzyme in thymidine metabolism, and overexpression of Ki-67, an antigen expressed during DNA synthesis. When expressed, serum biomarkers, usually oncofetal proteins, are extremely valuable for assessment of

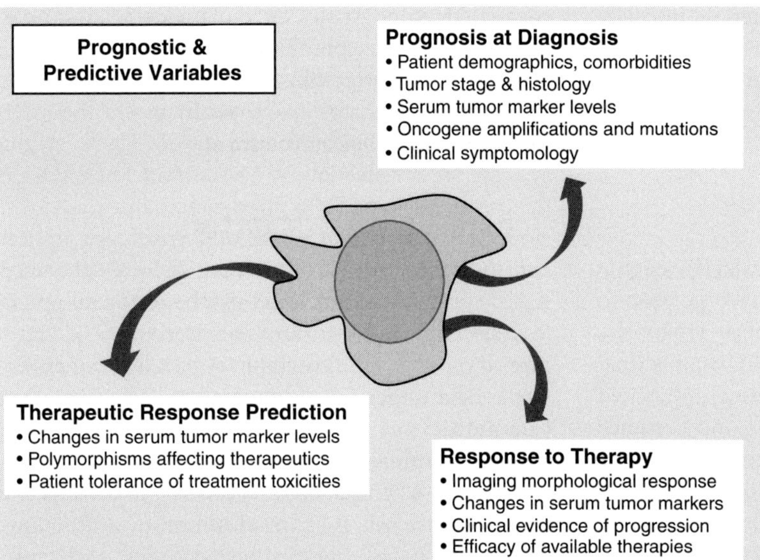

Figure 31.1 General prognostic factors.

Prognostic and Predictive Variables in Cancer. Generation of a cancer prognosis in modern oncology includes a number of variables including tumor characteristics, patient factors, and access to therapeutics. Cancer biomarkers of individual tumor phenotype can be exploited to predict therapeutic success and improve sensitivity and specificity of indicators of therapeutic response.

therapeutic success. The sooner therapy failure can be determined, the sooner a switch to a more effective therapy can be made saving time in constraining tumor growth and patient exposure to therapy-limiting drug toxicities. Specific early markers of cancer therapy success or failure are greatly needed and represent an area of great future potential improvement in cancer prognosis.

General prognostic histological biomarkers

Tumor grade

Histological tumor grade is an assessment of the degree of differentiation of the tumor tissue and is a powerful tool in cancer prognostication. Histological grading systems have been developed for a range of human malignancies and remain a major cornerstone for generating a cancer prognosis. A "low grade" tumor refers to a less aggressive, highly differentiated tumor that closely resembles the normal tissue of origin and generally carries a more favorable prognosis. A "high grade" tumor refers to a highly aggressive, poorly differentiated tumor characterized by disruption and distortion of the normal tissue architecture,

often with areas of necrosis and invasion of adjacent tissue structures such as nerves, lymphatics, and capillaries, and generally carries a corresponding less favorable prognosis. Included in the assessment of tumor grade is nuclear grade, which involves measurements of tumor cell nuclear morphometry.

The course of transition to malignancy requires a number of profound genetic changes including changes in DNA ploidy, pattern of DNA methylation, usually toward a generally demethylated phenotype, DNA mutations in genes and chromosomal translocations. These changes give rise to measurable nuclear morphological changes, which include distortion of the normally smoothly rounded nuclear contour and increased, inhomogeneous nuclear DNA content. As a result, nuclear morphometry has become incorporated into the tissue analysis of potential cancers and included as a marker of risk assessment. Generally the analysis of tumor cell nuclear morphometry in a specimen is semi-automated using computer-assisted image analysis via an artificial neural network (ANN). These computer-generated nuclear assessments have given rise to a new oncologic biomarker termed quantitative nuclear grade (QNG) [7]. This biomarker indicates the degree of cellular genetic instability manifested by alterations in nuclear size, shape,

DNA content, and DNA complexity, and its integrity relies on the quality of the tissue staining and imaging equipment, especially spatial resolution and ability to handle cytologic specimens as well as histological specimens.

Tumor stage

There is a TNM (tumor–node–metastasis) staging system available for each human cancer type used to triage a diagnosed malignancy into a prognostic cancer stage based on tumor size and locoregional extent (T0 to T4), presence and number of infiltrated or "positive" lymph nodes (N1 to N3), and presence of distant metastasis (M0 or M1), in order to generate a prognosis and to guide therapeutic intervention. In addition to a letter designation (T, N, or M), there is a number (i.e., 0 to 4), to assign degree of severity, and often further subcategories (i.e., a to c) to further refine the severity of disease. The resulting TNM formula for a particular cancer is then compared to the established current TNM cut-offs accepted for each of the four stages (stage 1 to 4) of malignancies, each of which carry an established prognosis, with stage 4 always carrying the worst prognosis. For example, a hypothetical patient with a 4 cm non-small cell lung primary malignancy that does not invade the mediastinum or chest wall and is found to have three positive hilar lymph nodes on the same side of the cancer but no metastasis would, based on the current staging system, have a TMN stage of $T_{2a}N_1M_0$ corresponding to stage 2b disease. This tumor would likely be surgically resectable, provided that the patient is a good surgical candidate, and this stage would predict an overall five-year survival rate of approximately 30%. By comparison, a different patient with 1.5 cm non-small cell lung cancer, which does not invade the mediastinum or chest wall, has no positive lymph nodes but does have one metastatic nodule detected in the left adrenal gland would have a TMN stage of $T_{1a}N_0M_{1b}$ corresponding to stage 4 disease, which carries a predicted overall five-year survival of 1%.

Decisions on whether to surgically resect a tumor, give radiation therapy, or chemotherapy are largely informed by the tumor stage, along with other factors such as tumor grade, previous treatment therapies, available therapeutic interventions, and patient factors. The determination of tumor stage is often accomplished by multimodality imaging including computed tomography (CT), magnetic resonance imaging (MRI), positron emission tomography (PET), bone scintigraphy, and usually surgical tissue sampling in order to unequivocally assess the extent of disease and exclude metastasis. These staging systems are a reflection of our current knowledge of the pathophysiology of a given cancer integrated with clinically proven diagnostic and treatment options and their track record of success. As a result, cancer staging remains in flux and must be routinely revised to reflect the most current literature, a task that is carried out by a multidisciplinary task force of experts in the field under the auspices of the American Joint Committee on Cancer (AJCC). Issues addressed include dilemmas such as how to define a "micrometastasis." Is a lymph node "positive" if there is one single positive cell, 0.1 mm of tumor, or if the tumor wasn't seen on routine hematoxylin and eosin staining but was seen on specialized immunohistochemical (IHC) staining? Do these vanishingly small metastases matter in terms of prognosis or treatment? How questions such as these are answered not only impacts individual patient care but also clinical trial development and the questions that can be answered through clinical trials.

Intra-tumoral lymphocytic infiltration

It has been increasingly recognized that the tumor micro-environment plays a critical supportive role in malignancy and that chronic inflammation promotes the malignant state. However, while intra-tumoral lymphocytic infiltration is observed in a range of malignancies, it appears to have different prognostic implications depending on the nature of the subset of infiltrating lymphocytes. For example, in a large study evaluating lymphocytic infiltration in colorectal cancer, signs of an immune response within colorectal cancers were associated with the absence of pathological evidence of early metastatic invasion and with prolonged survival [8]. Similarly, CD8+ lymphocytic infiltration is an independent favorable prognostic indicator in basal-like breast cancer [9] and the intensity of the lymphocyte infiltration was associated with tumor differentiation in ovarian carcinomas [10]. Conversely, a study in breast cancer demonstrated that lymph node metastases, immunopositivity of Ki-67, and non-triple negative tumors were associated with high regulatory T-cell infiltration [11], and the presence of intra-tumoral neutrophils is an independent factor for poor prognosis in colorectal cancer [12]. Recent evidence suggests that the prognostic significance of intra-tumoral infiltration relies on the subset of infiltrating lymphocytes. For example, it appears that dendritic cells, M1 macrophages, cytotoxic CD8+ T-cells, T-helper 1 (TH1) CD4+ T-cells, natural killer cells, and TH17 CD4+ T-cells present in the tumor bed assist in reducing cancer growth, whereas tumor infiltration with neutrophils, M2 macrophages, myeloid-derived suppressor cells, T-helper 2 (TH2) CD4+ T-cells, Foxp3+CD4+ T-regulatory (TREG) cells, and B lymphocytes is suspected to stimulate cancer growth [13]. As such, intra-tumoral infiltration is not yet a reliable prognostic marker in cancer management.

Epithelial–mesenchymal transition tumor phenotype (EMT)

The epithelial–mesenchymal transition (EMT) is a process during which epithelial cells undergo a programmatic genetic switch to a more dedifferentiated state and transform into the phenotype of a mesenchymal cell. EMT plays a key role during embryogenesis and its dysregulation in the malignancy is now understood to play a profound role in the development of malignant invasion. The switch from epithelial phenotype to mesenchymal phenotype is manifested by a reprogramming of the transcriptional machinery where epithelial-type genes, such as intercellular adhesion gene E-cadherin, are downregulated and expression of mesenchymal gene products, such as vimentin, fibronectin, and N-cadherin, are upregulated. The process of EMT frees the cell of epithelial restraints in cell–cell and cell–matrix interactions, and generates cytoskeletal reorganization and increased cellular mobility. It appears that the expression of genetic regulators of EMT, such as TWIST and SNAIL, and expression of mesenchymal gene products in malignancies may serve a prognostic and potentially predictive purpose in the future [14–18]; however, they are still early in development as cancer biomarkers.

Oncogenes

An oncogene is the mutated form or amplified state of a normal cellular gene that controls or maintains the pathophysiological state of malignancy. These altered genetic states can be inherited or can occur spontaneously either through inherent genetic instability or acquired through infection with an oncogenic virus or exposure to carcinogens in the environment. A number of oncogenes have been validated as cancer biomarkers predictive of overall prognosis and, in some cases, predictive of the likelihood of therapeutic success for specific interventions. Figure 31.2 provides a few examples of common oncogenic biomarkers; it is by no means an exhaustive list.

KRAS (KI-RAS2 Kirsten rat sarcoma viral oncogene homolog)

KRAS is a small plasma cell membrane-bound GTPase and part of the "ras superfamily" of over 150 mammalian ras genes two of which, KRAS1 and KRAS2, are present in the human genome. KRAS2 has two variants based on whether coding exon 6 is or is not transcribed, termed KRASB and KRASA, respectively. The product of KRASB is the predominant form of KRAS expressed in humans. KRAS functions as a molecular switch transducing signals to the nucleus through an amplifying cascade of effectors. KRAS signaling begins with a signaling trigger with resulting binding of GTP to KRAS causing the KRAS protein to undergo an activating conformational change. The activated KRAS then transduces this signal to the nucleus through downstream effectors, also known as GTPase-activating proteins (GAPs), which amplify the KRAS signal up to 100,000-fold [19] generating changes in cellular proliferation, differentiation, cytoskeletal organization, and hence cellular shape and motility. Due to the magnitude of consequence to KRAS activation, there are several negative regulation loops that maintain safety in the setting of wild-type KRAS. However, with oncogenic KRAS, these safety valves have been removed such that KRAS is insensitive to negative regulation by GAPs and negative regulation through cellular stress mechanisms and is thus permanently switched to the activated conformation. In addition, mutation of the codon 12 results in altered affinity of the GTP binding site, which affects the rate of GDP/GTP exchange allowing the KRAS to remain in the activated conformation for longer periods of time.

The presence of activating KRAS mutations is strongly predictive of a poor overall survival and is present in greater than 80% of pancreatic carcinomas (usually mutated in codon 12), 40 to 50% of colon carcinomas (usually codon 12), and 30 to 50% of lung

Biomarker	Alteration	Predictive/Prognostic Implication
Microsatellite Instability	Presence	Favorable overall prognosis
IDH1	Mutation	Favorable overall prognosis
BRAF	Activating mutation	Poor overall prognosis
BRCA	Presence of mutation	Poor overall prognosis
Thymidylate Synthase (TS)	Overexpression	Resistance to 5-FU therapy
EGFR	Activating mutation	Sensitivity to targeted therapy to EGFR
KRAS	Activating mutation	Resistance to EGFR targeted therapy
ERCC1	Overexpression	Resistance to platin-based therapy
PTEN	Loss of gene expression	Resistance to HER2 and EGFR targeted therapy

Figure 31.2 Oncologic biomarkers.

Predictive and Prognostic Oncologic Biomarkers in Clinical Use. Above is a representative list of selected prognostic and predictive biomarkers used in the management of cancer. IDH1 (isocitrate dehydrogenase 1 (NADP+)); BRAF (v-raf murine sarcoma viral oncogene homolog B1); BRCA (breast cancer 1/breast cancer 2, early onset); EGFR (epidermal growth factor receptor); KRAS (v-Ki-ras2 Kristen rat sarcoma viral oncogene homolog); ERCC1 (excision repair cross-complementing rodent repair deficiency, complementation group 1); PTEN (phosphatase and tensin homolog)

carcinomas (usually codons 12 and 13) and, in lower frequencies, in a number of other malignancies including breast cancers, which have been observed to have an estimated KRAS prevalence of about 5 to 12% [20]. In addition to being implicated as a causative factor in these cancers and conferring a poor overall prognosis, the presence of an activating KRAS mutation can be predictive of treatment failure with radiation therapy and a number of chemotherapies including targeted inhibitors of the epidermal growth factor receptor (EGFR), a receptor often overexpressed or overactive in malignancy. In fact tumor KRAS expression is one of the most important determinants of the likelihood of success of EGFR (epidermal growth factor receptor) inhibitors such as gefitinib (Iressa®, ZD1839) and erlotinib (Tarceva®) and monoclonal antibody antagonists of EGFR including cetuximab (Erbitux®) and panitumumab (Vectibix®). Since the KRAS pathway links EGFR to cell proliferation and survival, in the presence of activating mutations in KRAS, EGFR inhibitors are generally not effective regardless of EGFR activational status. In addition to EGFR inhibitors, activational mutations in KRAS have been associated with resistance to

platelet-derived growth factor receptor (PDGFR), tyrosine kinase inhibitor imatinib mesylate (Glivec®), and conventional DNA damaging agent, paclitaxel.

BRAF (v-raf-1 murine leukemia viral oncogene homolog B1)

The BRAF gene encodes a serine-threonine kinase that functions as a downstream effector of the KRAS oncogene and potent activator of the MAP kinase pathway, which is frequently overactive in cancers. Activating mutations in BRAF have been implicated in the pathogenesis of a number of malignancies including melanoma, papillary thyroid cancer, and colon cancer. Activating BRAF mutations in cancer, the majority of which are characterized by an amino acid substitution of valine to glutamic acid (V600E BRAF), are associated with a poor outcome and proclivity for metastasis and invasion, and are found in approximately 50 to 70% of malignant melanomas, 30 to 70% of papillary thyroid carcinomas, 10 to 25% of colorectal adenocarcinomas, and a small percentage of ovarian carcinomas. In addition, the presence of an activating mutation in BRAF has been demonstrated to have predictive value for resistance to EGFR-antagonist therapy [21]. The V600E BRAF mutation has been associated with sensitivity to MEK inhibitors, since MEK 1 and MEK 2 are the only known substrates of V600E BRAF.

MYC (myelocytomatosis viral oncogene homolog)

MYC belongs to a family of oncogenic transcription factors, which includes c-myc, l-myc, n-myc, and v-myc that are frequently aberrantly expressed, usually by gene amplification, but also by translocation in approximately 70% of human malignancies. Aberrant *MYC* expression is implicated in tumor initiation and is associated with an unfavorable prognosis [22] and less well differentiated tumors [23]. MYC has been observed to play a role in a number of pathways important to cell growth and proliferation, and hence cancer, including ribosome biogenesis, repression of genes involved in cell–cell and cell–substratum contacts, tissue differentiation, and angiogenesis. In addition to providing prognostic value, the presence of MYC gene amplifications appears predictive of resistance to therapy with inhibitors of the PI3K-mTOR pathway [24], cisplatin [25], and anti-estrogen therapy [26], and may ultimately serve as a biomarker of drug resistance in these chemotherapeutic settings.

PI3KCA/PTEN

PTEN and PI3KCA play a regulatory role in the PI3K pathway, which is one of the most highly mutated pathways in human malignancy. The PI3K pathway serves as an intracellular conduit for cell growth, and survival signals transmitted from transmembrane receptor tyrosine kinases including EGFR family members and the insulin-like growth factor receptor (IGFR) through the downstream central mediators AKT/mTOR to the nucleus. PI3K pathway activation has been observed to result in alterations in cellular metabolism, including enhancement of glucose uptake and glucose synthase activity in muscles and fat, increased proliferation through repression of FOXO-mediated transcription of cell cycle regulators $p27^{Kip1}$ and p21, cell growth, and subversion of programmed cell death through inhibition of pro-apoptotic genes such as Fas ligand, Bim, and BAD. The PI3KCA gene encodes one of three iso-forms (PI3KCA, PI3KCB, PI3KCD) of the p110

catalytic subunit that encodes the catalytic subunits for the class 1A PI3K family enzymes. Of the three p110 isoforms, PI3KCA is the most frequently ampli-fied or activationally mutated, which is observed in approximately 60 to 70% of mantle cell lymphomas and oral cell squamous cell carcinomas, 30 to 40% of ovarian cancers, 30 to 35% of breast cancers, 25 to 30% of bladder cancers, 10 to 20% of colon and gastric cancers, and 7 to 10% of synovial sarcomas. PTEN functions as a negative regulator of the PI3K pathway by antagonizing the conversion of phospha-tidylinositol 4,5-diphosphate (PIP2) to the active phosphatidylinositol 3,4,5-triphosphate (PIP3) and its loss of function is commonly observed in malignancy either through loss of heterozygosity or reduced functional activity. PTEN loss or hypofunc-tion is observed in approximately 60% of malignant pleural mesotheliomas, 50 to 60% of bladder cancers, 50% of colon cancers, 40% of cervical carcinomas, 40% of osteosarcomas, 20% of salivary gland cancers, 20% of glioblastoma multiformes, 20% of malignant melanomas, 15 to 75% of non-small cell lung cancers, and 15 to 20% of prostate cancers.

Mutations and amplifications in PI3KCA and PTEN loss are generally associated with a poor prognosis although there is some evidence to suggest that PI3KCA hyperfunction in breast cancer can be associated with a more favorable prognosis and is associated with post-menopausal, hormone receptor positive malignancies [27] although this remains controversial. Activation of the PI3K pathway, either through mutations or amplifications in PIK3CA or loss of PTEN function are associated with drug resistance to EGFR antagonists, including cetuxi-mab, lapatinib, and HER2/neu receptor antagonists, such as trastuzumab [28, 29] and suggest possible sensitivity to inhibitors of AKT and mTOR since they are central downstream mediators of the PI3K pathway signal.

p53

Dysfunction of p53, often due to genetic mutation, is observed in up to 80% of human malignancies. p53 is a transcriptional factor and critical component of the genome protective DNA damage mechanism acti-vated in the setting of cellular stress, such as hypoxia, DNA damage, and nutritional deficit (Figure 31.3). p53 activation induces cell cycle arrest at the G_1/S cell cycle checkpoint to allow for DNA repair, induction

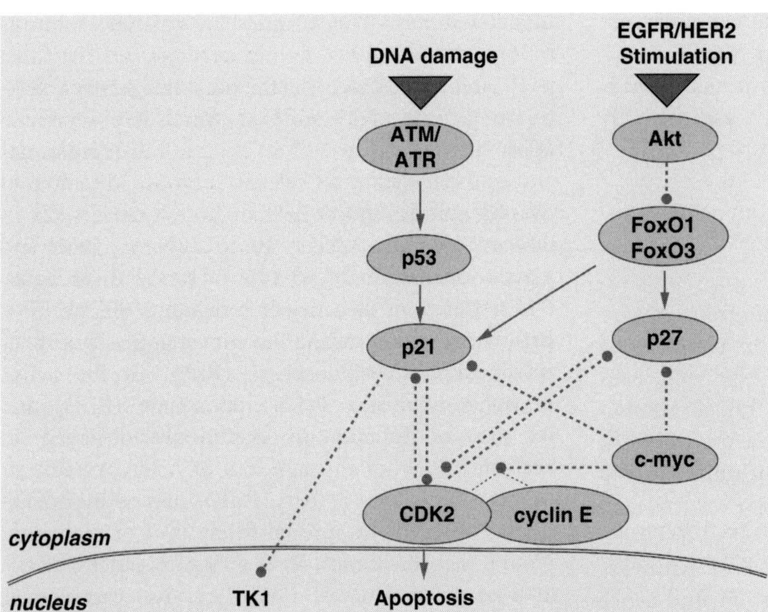

Figure 31.3 DNA damage pathway.

The DNA damage pathway is frequently disabled in malignancy allowing cells otherwise destined for cell death to survive and proliferate. p53 is very commonly hypofunctional or absent in cancer. In addition, a number of other commonly amplified pathways participate in hampering the effectiveness of the DNA damage pathway to initiate apoptosis.

of cellular senescence, or initiation of programmed cell death. In addition to a genome protective role, during cellular stress, p53 also plays a role in differentiation and meiosis. The wide range of biological activities observed for p53 is likely due in part to the potential expression of nine alternately spliced p53 isoforms, as well as the interplay between the isoforms of p53 and the highly homologous p53 family members, p63 and p73, which exist in 6 and 35 alternately spliced isoforms, respectively [30].

The importance of intact p53 function is demonstrated with the heritable Li–Fraumeni syndrome, which occurs as a result of germline missense, non-transcriptionally active mutations in the p53 gene. Li–Fraumeni syndrome, originally characterized in a study of children with rhabomyosarcoma and their families, is characterized by an increased likelihood of developing a wide spectrum of cancers, often with multiple cancers over a lifetime, most frequently bone and soft tissue sarcoma, brain cancer, and adrenocortical carcinoma. Other cancers frequent in this population include germ cell tumors, acute leukemia, pre-menopausal breast cancer, pancreatic cancer, gastric cancer, Wilms' tumors, and melanoma, all of

which typically develop at an age younger than characteristic for that malignancy [31]. In addition to inherited missense mutations, there are common polymorphisms of the p53 gene characterized by either a proline (p53pro) or arginine (p53arg) at codon 72, located in the proline-rich domain. The p53arg polymorphism is notable as a target for degradation by the human papilloma virus (HPV) E6 protein [32]. The p53pro polymorphism is a weaker inducer of apoptosis than p53arg and homozygosity for the p53pro has been associated with an elevated risk of adenocarcinoma of the lung and decreased survival both in lung and colon cancers [33].

While these inherited p53 conditions shed light on the importance of intact p53 function and provide prognostic significance in those settings, the majority of malignancies exhibit somatic missense mutations in the p53 gene, which typically occur non-randomly at six hotspots in the DNA-binding domain of the p53 gene: codons 175, 245, 248, 249, 273, and 282 [31]. The presence of a somatic mutation at a particular codon may suggest an etiology. For example, p53 mutation at codon 249 is observed in hepatocytes exposed to the hepatocellular carcinoma-inducing aflatoxin [34], and p53 mutation at codon 157 is observed in lung cancer of smokers [35].

While the presence of mutated p53 in malignancy has been well established, the prognostic and predictive significance of the presence and type of p53 mutation is not fully understood for many cancers. There are numerous clinical studies supporting an association between p53 mutational status and poor overall survival for chronic lymphocytic leukemia, diffuse large B-cell lymphoma, breast cancer, hepatocellular carcinoma, and squamous cell carcinomas of the head and neck [31]. In fact, p53 mutational status is considered an independent risk factor for breast cancer especially in the presence of a concomitant progesterone receptor negativity, or basal-type and ERBB2+ histology [36]. Furthermore, in basal cell type breast cancer, p53 mutation has been demonstrated to predict therapeutic response with combination epi-rubicin and cyclophosphamide therapy [37]. Other malignancies, including colorectal, pancreatic, stomach, esophageal, bladder, prostate, ovarian, brain, and lung cancers, have less clear associations between prognosis and p53 mutational status either due to conflicting clinical trial results or insufficient data. The lack of clarity in correlation of p53 status with prognosis in cancer is also clouded by the complexity of interactions between the isoforms of p53, and the many isoforms of p53 family members, p63 and p73. Further investigation is needed to fully understand the intricacies of the interplay of these family members in cancer.

Breast cancer susceptibility gene (BRCA) mutation

The exact mechanism by which breast cancer susceptibility gene (BRCA) inactivation results in cancer is not completely understood. The *BRCA1* gene plays an important role in the control of the DNA damage mechanism and its absence results in sensitivity to DNA mutations and deficiency in the repair of DNA double-stranded breaks resulting in reliance on poorly conservative mechanisms of repair such as non-homologous end joining and single-strand annealing. Breast cancer susceptibility gene 1 (BRCA1) appears to play a role in DNA repair, transcriptional regulation, and chromatic remodelling; while breast cancer susceptibility gene 2 (BRCA2) appears to be primarily restricted to DNA recombination and repair processes. A subset of phenotypic characteristics have been associated with the presence of BRCA mutation in breast cancer, termed "BRCA-ness," all of which carry poor prognostic implications including high-grade histology, estrogen receptor (ER) and HER2 (erbb2; a member of the epidermal growth factor receptor family) receptor negativity, *c-MYC* amplification, medullary and basal histologic phenotype, p53 mutations, X-chromosome inactivation, and sensitivity to DNA damage [38]. In addition to breast cancer, BRCA hypofunction is observed in approximately 10 to 15% of ovarian cancers and 40 to 50% of lung cancers.

BRCA mutational status is used to assess lifetime risk of breast and ovarian cancer in healthy patients carrying the germline mutation and as a prognostic indicator for those with newly diagnosed breast cancer. Women carrying germline mutations in the *BRCA1* or *BRCA2* have a 60 to 85% lifetime risk of developing breast cancer, and a 15 to 40% risk of developing ovarian cancer [39]. Approximately 10% of women with a first diagnosis of breast cancer under the age of 40 years are positive for BRCA mutations [40], either *BRCA1* or *BRCA2*, and have an approximately 20 to 40% ten-year risk of developing breast cancer in the non-affected breast. Those carrying the *BRCA1* germline mutation have an extremely high likelihood of developing breast cancer (80 to 100% lifetime risk) usually an extremely aggressive "basal-type" triple negative, ER negative, progesterone receptor (PR) negative, and HER2 receptor negative histology, which carries a very poor prognosis. Often known *BRCA1* germline carriers will have prophylatic mastectomies. In addition to germline *BRCA* mutation, it is now well established that sporadic breast cancers can suppress BRCA1 functionality through epigenetic inactivation [41–43]. *BRCA1* methylation is seen in breast and ovarian cancer but not the normal tissue counterparts and is rare in cancers of other tissue origin.

Ataxia-Telangiectasia-Mutated (ATM)

ATM is a serine-threonine kinase that serves as a critical step in the DNA damage pathway and mediates double-stranded DNA damage repair and cell cycle checkpoint activation. ATM function was first characterized through study of the rare heritable ataxia-telangiectasia, which results from germline biallelic loss of function mutations of *ATM*. Ataxia-telangiectasia syndrome includes cerebellar ataxia, immunosuppression, occulo-cutaneous telangiectasia, premature aging, susceptibility to radiation-induced DNA damage, and marked increased frequency of malignancy, most often leukemia and lymphoma [44]. In addition to the heritable loss of ATM function, there is also ATM hypofunction observed in a range of sporadic malignancies including head and neck squamous cell carcinoma (approximately 25%), chronic lymphocytic leukemia (approximately 10 to 20%), lung cancer (approximately 40 to 60%), advanced gastric cancer (approximately 65 to 75%), breast (approximately 50 to 65%), cervical, and colon cancers, and appears to be associated with an unfavorable prognosis and can be predictive of treatment failure. For example, ATM loss of function is associated with radiation treatment failure and poor prognosis of head and neck squamous cell carcinoma [45], shorter progression-free survival in chronic lymphocytic leukemia [46], invasiveness in cervical cancer [47], and poor differentiation, lymph node metastasis, and poor five-year survival in gastric cancer [48], and resistance to genotoxic anti-tumor therapeutics in glioblastoma multiforme [49]. Conversely, high levels of ATM gene expression were found to be associated with a favorable prognosis in breast cancer [50].

Excision repair cross-complementing rodent repair deficiency, complementation group 1 (ERCC1)

ERCC1 is a critical component of the nucleotide excision repair pathway (NER), responsible for repair of DNA damage, and is an excisional nuclease. As with many other components of the DNA damage pathway, ercc1 is often hypofunctional in malignancy. For example, ERCC1 hypofunction has been observed in approximately 30 to 70% of lung cancers, 40% of transitional cell carcinoma of the bladder, 70 to 75% of ovarian cancers, and 60% of pancreatic adenocarcinoma. The NER capacity for DNA repair has been associated with tumor resistance to DNA alkylating platinum agents, such as cisplatin, that rely on DNA damage to induce tumor apoptosis [51]. Following treatment with cisplatin, there is upregulation of ERCC1 mRNA via increased activity of the AP-1 binding ERCC1 promoter. Polymorphisms in ERCC1 and ERCC1 expression levels have been demonstrated to impact platin-based chemotherapy resistance, and

as such ERCC1 expression has become a predictive biomarker for platin-based chemotherapy response and tolerance of toxicity [51]. For example, in completely resected primary transitional cell carcinoma of the bladder and non-small cell lung cancer of the lung, those with ERCC1-negative tumors seemed to benefit more from cisplatin-based adjuvant therapy than those with ERCC1-positive tumors [52, 53]. High levels of ERCC1 mRNA have also been shown to be associated with response of colon cancer to irinotecan.

Altered cellular metabolism
Thymidylate synthase (TS)

The rate of tumor cell proliferation impacts overall survival, propensity for metastasis, and often effectiveness of therapy, and is considered a reliable biomarker of prognosis often assessed by surrogate markers of cell cycle progression such as Ki-67 and PCNA (proliferating cell nuclear antigen), both antigens expressed during cell cycle progression. Sustained proliferation requires a steady intracellular pool of deoxynucleotide-triphosphates (dNTPs) of which thymidine is specific to DNA synthesis. Thymidine metabolism (Figure 31.4) is tightly regulated by cell cycle control mechanisms, which source thymidine through two pathways, de novo synthesis, regulated by thymidylate synthase (TS), and the salvage pathway, regulated by thymidine kinase (TK1). TS catalyzes the conversion of deoxyuridine monophosphate (dUMP) to deoxythymidine monophosphate (dTMP) via methylation and is elevated in expression in a number of malignancies including approximately 25 to 30% of pancreatic cancers, 20 to 25% of non-small cell lung cancers, and approximately 70 to 75% of colon cancers. It is common for TS to either be elevated in expression or exist in a more active polymorphic form, typically double (2R) and triple (3R) tandem gene repeats, in the setting of cancer, and overexpression/overactivity of TS is generally associated with an unfavorable prognosis. In colon cancer, elevated expression of TS is associated with a poor prognosis and low expression is associated with an increased response rate to anti-folate chemotherapeutic 5-FU [54]. In one study of 71 patients with colon cancer, those with an elevated TS expression had a five-year survival of 40% compared to 69% for low expression TS [55]. In addition, the 3R/3R homozygous polymorphism

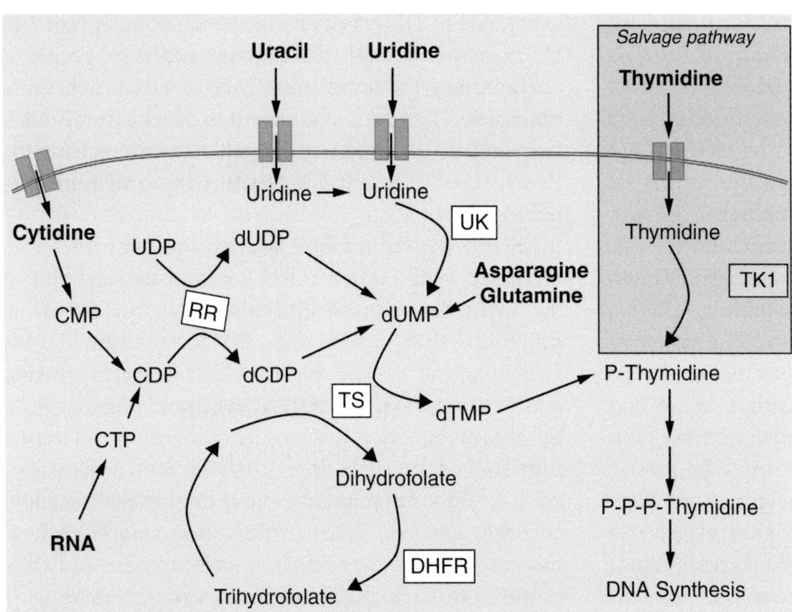

Figure 31.4 Thymidine metabolism.

CMP (Cytidine monophosphate); CDP (Cytidine diphosphate); CTP (Cytidine triphosphate); dCDP (deoxy-Cytidine diphosphate); UDP (Uridine diphosphate); UMP (Uridine monophosphate); dUDP (deoxy-Uridine diphosphate); dUMP (deoxy-Uridine monophosphate); dTMP (deoxy-Thymidine monophosphate); TS (Thymidylate synthase); DHFR (Dihydrofolate reductase); UK (Urokinase); and TK1 (Thymidine kinase 1)

has been demonstrated to be the most refractory to treatment in colon cancer metastasis among the possible combinations of tandem repeat polymorphisms [54]. Low TS expression has been associated with improved progression-free survival, and overall survival in patients with malignant pleural mesothelioma treated with combination carboplatin-pemetrexed [56], and has been demonstrated to be predictive of successful pemetrexed, an anti-folate, therapy in a number of malignancies including non-small cell lung cancer [57] and colon cancer.

Des-γ-carboxy prothrombin (DCP)

DCP is an abnormal prothrombin, which is produced in the absence of vitamin K2 and is increased in the serum of patients with hepatocellular carcinoma. There is accumulating evidence that DCP may serve as a biomarker of disease to prognosticate HCC recurrence in patients who have had liver transplantation [58]. Interestingly, DCP may function as a growth factor and promote angiogenesis, proliferation, and migration, and might play significant roles in cancer progression [59]. DCP also has been demonstrated to induce epidermal growth factor

receptor (EGFR) and vascular endothelial growth factor (VEGFR) overexpression [60].

Cell surface receptor expression
CXCR4

CXCR4 is a member of a family of G-protein coupled cell surface receptors that function as chemokine receptors performing a homing function in stem cells and select adult tissues including the pituitary, adrenal glands, lymph nodes, bone marrow, blood, spleen, and thymus. Pathologically, CXCR4 serves as a co-receptor for cell entry by HIV and appears to play a central role in the cancer stem cell physiology, epithelial–mesenchymal transition, and metastatic homing of several cancers including non-small cell lung cancer (NSCLC), breast, prostate, ovarian, thyroid, liver, hematologic malignancies, rhabdomyosarcoma, and neuroblastoma. The homing function of CXCR4 is mediated through binding of stromal cell-derived factor 1 (SDF-1, CXCL12), which is produced at high levels within the lymph nodes, lung, liver, and bone, producing downstream activation of the MAPK/ERK1/2 and PI3K/AKT pathways. In addition to functioning as a homing mechanism, CXCR4 promotes tumor progression by activating the production of matrix metalloproteases and induction of tumor-associated integrin activation and signaling. CXCR4 appears to be regulated by epigenetic mechanisms including DNA methylation negative regulation, and has been observed to be elevated in the setting of hypoxia, exposure to vascular endothelial growth factor (VEGF), and activation of nuclear factor NF-κB.

Tumor expression of CXCR4 has also been shown to have promise as a prognostic biomarker. Clinical trials evaluating CXCR4 as a prognostic biomarker in hepatoma and pancreatic cancer demonstrated that patients with hepatic tumors expressing CXCR4 had shorter time to progression and overall survival than those that did not express CXCR4 [61]. In gastric cancer, CXCR4 expression has been associated with lymphatic metastasis, higher pathological stage, and reduced five-year survival rate [62]. Overexpression of CXCR4 has been shown to be an independent adverse prognostic factor and associated with VEGF expression in soft tissue sarcoma and nasopharyngeal carcinoma [63, 64], associated with decreased disease-free survival in melanoma [65], prognostic of a poor five-year overall and disease-free survival in HER2 negative breast cancer, and associated with worse prognosis in triple negative (progesterone, estrogen, and HER2 receptor negative) breast cancer [66–69]. In early breast cancer, CXCR4 expression was associated with increased likelihood of bone metastasis [70]. Data such as these have led to interest in CXCR4 as a prognostic biomarker and drug therapy target.

Epidermal growth factor receptor (EGFR)

The epidermal growth factor receptor (EGFR) is a transmembrane receptor tyrosine kinase of the ErbB family. Four members of the ErbB (avian erythroblastic leukemia viral (v-erb-b) oncogene homolog) family have been identified, EGFR (ErbB1, HER1), ErbB2 (HER2), ErbB3 (HER3), and ErbB4 (HER4), and each have modifications that allow for unique signaling characteristics and differential growth factor signaling. EGFR family signaling initiates with ligand receptor binding which results in homo- and hetero-dimerization and resulting cytosolic tyrosine kinase activation. For EGFR, this tyrosine kinase activation initiates a growth factor signaling cascade that culminates in the nucleus promoting proliferation, cell growth, and survival. EGFR signaling also plays a significant role in the setting of DNA damage. EGFR

can mediate DNA repair via the catalytic subunit of DNA protein kinase (DNA-PKcs) and is suspected of contributing to tumor resistance in DNA damaging therapies. Therefore in addition to exerting an inhibitory effect on DNA damage pathway activation, EGFR speeds the rate of DNA repair and improves cell viability.

EGFR hyperfunction typically results from either increased gene copy number or increased activity of the cytosolic tyrosine kinase through an activating mutation that leaves the EGFR tyrosine kinase switched into the "on" position. EGFR overexpression and/or overactivity is a very common phenotype in human malignancy with overexpression or activating mutation observed in approximately 80% of osteosarcomas, 75% of malignant pleural mesotheliomas, 50% of colon cancers, 50 to 70% of non-small cell lung cancers, 30 to 40% of thymic carcinomas, and 20% of breast cancers. EGFR overexpression/overactivity is a strong prognosticator of poor prognosis in ovarian, head and neck, bladder, and esophageal cancers, and a modest poor prognostic indicator in colorectal, breast, gastric, and endometrial cancers [72]. The presence of activating mutations in EGFR is a favorable predictive indicator for patients who can be treated with targeted drugs against EGFR activity. For example, in a study of non-small cell lung cancer patients receiving chemotherapy plus cetuximab, activating EGFR mutations were identified as indicators of improved response, with patients in both treatment groups whose tumors carried such mutations having improved survival compared with those whose tumors did not [73]. EGFR mutation testing is now recommended as part of routine care of non-small cell lung cancer (NSCLC) patients to guide decisions about first-line treatment. At present, EGFR gene copy number testing is not recommended in the selection of first- or second-line treatment of advanced NSCLC patients. No data exist about the role of EGFR polymorphisms as predictors for treatment outcome with anti-EGFR monoclonal antibodies [74].

HER2 (Human epidermal growth factor receptor type 2)

Human epidermal growth factor receptor type 2 (HER2), also known as v-erb-b2 erythroblastic leukemia viral oncogene homolog 2 (ERBB2), is an EGFR family member and frequent heterodimer partner for EGFR (HER1) serving as a finely tuned amplifier for

EGFR signaling. The extracellular domain of the HER2 molecule is configured as if it is ligand bound, thus obviating the need for ligand binding, and existing in a primed configuration for receptor dimerization. The HER2 receptor is thought to serve as a signal amplifier by creating high affinity binding with ligands along with its heterodimer partner. Other HER family member HER3, also known as ERBB3, has a defective tyrosine kinase domain, and also forms heterodimers to exert its function [75], which includes recruitment of PI3K. HER1 and HER4 are similar in their binding and signaling properties both able to activate the RAS and MAPK signaling cascades, although HER4 is not able to bind the adapter protein Grb2.

As a group, overexpression and/or overactivity of the HER family is associated with a poor prognosis in cancer. Overfunction of HER2 has been observed in approximately 25 to 75% of lung cancers, 50 to 60% of thymic carcinomas, 20% of metastatic gastric cancers, 10 to 50% of all gastric cancers, 15% of colon cancers, and 15 to 20% of breast cancers. HER2 overexpression is a negative prognostic indicator in ovarian, cervical, bladder, head and neck, endometrial, colorectal, gastric, NSCLC, esophageal, and breast cancers [72]. However, while HER2 overexpression in breast malignancy has an overall unfavorable prognosis with an increased risk of relapse, breast cancers expressing HER2 predict a favorable response to tratuzumab (Herceptin®) [79, 80].

Oncofetal serum biomarkers

Oncofetal serum biomarkers are gene products that are normally expressed in pregnancy and fetal development but can become inappropriately expressed in pathophysiological conditions such as malignancy (Table 31.1). These serum biomarkers, when present, often carry a prognostic significance and can also be exploited as indicators of therapeutic response, since the serum concentration of oncofetal biomarkers is generally a reflection of overall tumor burden. While oncofetal biomarkers share a common abnormal context of gene expression, they represent a wide range of molecular structures including mucins, carbohydrates, glycolipids, glycoproteins, proteoglycans, and hormones. Common examples of clinically relevant oncofetal biomarkers include alpha-fetoprotein (AFP), human chorionic gonadotropin (hCG), carcinoembryonic antigen (CEA), cancer antigen 125

Table 31.1 Oncofetal biomarkers. Oncofetal biomarkers are typically differentially expressed in a range of malignancies. When present, some of these biomarkers are typically used by oncologists to follow therapeutic success. Abbreviations for some of the more commonly used oncofetal proteins are as follows: AFP (alpha-fetoprotein); hCG (human chorionic gonadotropin); CA-125 (cancer antigen 125); CA-19-9 (cancer antigen 19-9); CEA (carcinoembryonic antigen).

Pancreatic	Ovarian	Germ cell & GTN
CA19-9	CA-125	hCG
Colorectal	**Breast**	**Renal/bladder**
CA19-9	CA15-3	hCG
CEA	CEA	CEA
	CA-125	
Gastric	**Cervical**	**Lung**
CA19-9	CA-125	CEA
		hCG
Endometrial	**Hodgkin's lymphoma**	**Hepatocellular**
hCG	CA-125	AFP
CA-125		

(CA 125), squamous cell carcinoma antigen (SCC), mucin-like carcinoma-associated antigen (MCA), tissue polypeptide-specific antigen (TPS), carbohydrate antigen 19-9 (CA 19-9), and prostate-specific antigen (PSA).

As a group, oncofetal biomarkers are thought to primarily serve an immune modulatory function, inducing tolerance in the maternal immune system to the developing embryo and fetus. Levels of these oncofetal antigens can be influenced by pathological conditions such as inflammation and malignancy, and normal physiological conditions such as pregnancy and menstruation [81]. In malignancy, serum levels of these antigens may be influenced not only by tumor bulk and expression level, but also the degree of basement membrane disruption and renal function. And while a malignancy may not express an oncofetal biomarker initially, tumor evolution may lead to the expression of one or more oncofetal biomarkers over time. These variables are important to keep in mind when using these molecules as prognostic tools in cancer management. Also, in order to fully realize the prognostic potential of oncofetal proteins in malignancy, an understanding of the natural biologic functions of these proteins and the mechanisms underlying their expression is necessary. As a group, these cancer biomarkers have great potential to serve as early and relatively specific markers of therapeutic response, which is highly desirable not only for individual patient management but also clinical trial development.

Cancer antigen 125 (CA-125)

CA-125, also known as mucin-16 (MUC 16), is a large transmembrane glycoprotein of the mucin family, a heavy glycosylated family expressed in the apical epithelia of all wet surfaces in the body where they serve a barrier function, with a dominant, heavily glycosylated extracellular domain [82], and short cytoplasmic tail. The synthesis of the CA-125 glycoprotein appears to be stimulated by interferons and mediated by EGFR pathway activation and by progesterone, and is found expressed to varying degrees in seminal fluid, the fallopian tube and endometrium, folliculo–luteal menstrual transition of spontaneous human cycle, early pregnancy, and the amnion and decidua into the amnionic cavity. The expression of CA-125 during pregnancy is thought to protect the fetus from maternal complement activation at the maternal–fetal interface. Low levels of CA-125 are also detectable in the lung and conjunctiva. CA-125 has also been reported to inhibit natural killer (NK) cell activity by preventing the synapse between the ovarian cancer cell and the NK cell [83]. In fact a receptor for CA-125, Siglec-9, is expressed on NK cells, B cells, and monocytes [84] indicating additional roles in immune system evasion. It has been suggested that CA-125 plays a pathologic role in malignancy [85], with amplifications and mutations in 5% of high-grade ovarian cancers [86], and overexpression in a range of malignancies including 80% of epithelial, non-mucinous ovarian carcinomas, 60% of Hodgkin's lymphomas, 60 to 70% of endometrial carcinomas, 10 to 15% of breast cancers, and 10% of colon cancers. Interestingly, CA-125 binds mesothelin [87], which is expressed on the surface of normal mesothelial cells and may mediate peritoneal adhesion of ovarian and pancreatic cancer cells.

CA-125 is a sensitive biomarker of tumor bulk in epithelial, non-mucinous ovarian carcinomas, approximately 80% of which express CA-125. For this subset of ovarian carcinoma, CA-125 is used to evaluate the possibility of complete resection during surgery, to estimate sensibility for adjuvant or neo-adjuvant chemotherapy, and for diagnosis of disease recurrences. Pre-treatment CA-125 serum level in advanced ovarian cancer following tumor debulking has been demonstrated as a prognostic indicator of progression-free survival [90]. The rate of change in serum levels of CA-125 post-chemotherapy has also been demonstrated as a useful prognostic indicator of tumor response to therapy in advanced ovarian cancer for serous subtypes [91–94] and, recently, mucinous and clear cell epithelial subtypes as well [95]. In addition to ovarian cancers, CA-125 can be elevated in the serum of patients with endometrial, cervical, fallopian tube, pancreatic, breast, colon, and lung cancers [96], and non-Hodgkin's lymphoma [97, 98], and may be used as an adjunct biomarker to follow response to therapy when expressed. Overexpression of CA-125 at the time of diagnosis has been associated with an unfavorable prognosis in Hodgkin's lymphoma and endometrial, non-small cell lung, cervical, and breast cancers [97–104].

Carcinoembryonic antigen (CEA)

Carcinoembryonic antigen (CEA) is a member of the immunoglobulin superfamily, a co-receptor for CEACAM6, and member of the glycosyl-phosphatidylinositol (GPI)–anchored CEA family, all of which function as homotypic intercellular adhesion molecules. CEA plays a physiological role during development when it is produced in the embryonic then fetal gastrointestinal tract and pancreatic tissues with subsequent suppression after birth. While the exact role that CEA plays during development is not known, CEA is noted to be highly homologous to lymphoid tissue marker CD66, which is known to induce immune tolerance through several mechanisms [117]. It is known that TGF-β induces CEA secretion in a dose-dependent manner. CEA has also been observed to cooperate with c-myc and Bcl-2 to promote cellular transformation [118] and blocks DR5-mediated apoptosis [119, 120].

CEA is suspected to play a pathophysiological role in malignancy and, although predominantly expressed in the gastrointestinal tract, is overexpressed in as many as 70% of all human cancers. It is highly expressed by the majority of colorectal cancers, as well as most other gastrointestinal tract neoplasms and pancreatic carcinomas, 50% of breast cancers, 70% of non-small cell lung carcinomas, and 50 to 60% of endometrial carcinomas. High levels of CEA at diagnosis are associated with a poor overall survival in non-small cell lung [99, 121], colorectal [122, 123], and breast cancers. Serum CEA levels have also been demonstrated to be associated with response to therapy in subsets of tumors expressing CEA [124–127], although interpretation of serum levels may be complicated by "flare" responses, transient elevations in serum CEA levels, to therapy. Post-treatment "flare" responses have been associated with improved prognosis [128].

Cancer antigen 19-9 (CA 19-9)

CA 19-9 is a member of the blood-group related sialyl Lewis (sLe) antigen carbohydrates. It has been hypothesized that sialyl groups on the cell surface identify cells as "self" to the Siglec-9 receptors on neutrophils and dampen local neutrophil activity [130]. CA 19-9 is also expressed in normal tissues, including the pancreas, gall bladder, stomach, colon, bronchial tree, endometrium, salivary glands, prostate [131], renal pelvis [132], and liver, and in cirrhosis, cholecystitis, pancreatitis, and hepatitis [133]. As a result, CA 19-9 can be elevated in a number of benign diseases including hydronephrosis, urinary tract infection, nephrolithiasis, proteinuria [132], bronchiectasis, emphysema, and pulmonary interstital lung disease [134]. Interestingly, Lewis antigens have also been identified in the extravillous trophoblast [135], which invades the myometrial wall and remodels the spiral arteries in order to create lower vascular resistance and increase in vessel diameter to allow increased blood flow to the rapidly growing fetus [136]. In addition, CA 19-9 is the cognate receptor for E-selectin, expressed in the vascular endothelium, and is suspected of promoting blood-borne metastasis and invasion [137].

Serum CA 19-9 level is a key biomarker used in the management of pancreatic cancer and is elevated in approximately 70 to 90% of patients with pancreatic cancer, 90% of endometrial carcinomas, and 60% of lung cancers serving as an indication of tumor burden. High CA 19-9 serum levels are associated with poor prognosis in pancreatic cancer [138] and a post-therapeutic fall in serum CA 19-9 levels is predictive of a favorable therapeutic response and improved overall survival [139–141]. In addition to pancreatic cancer, elevated serum levels of CA 19-9 have been observed to correlate with an unfavorable overall survival in malignant intraductal papillary mucinous neoplasms [142], gastric [144] and colon [145] cancers, and squamous cell carcinomas of the larynx [146]. Patients with a Lewis-negative (a–/b–) status (approximately 10% of the population) are not able to synthesize CA 19-9 from the precursor carbohydrate antigen

DUPAN-2. In such cases, DUPAN-2 is a better parameter in monitoring than CA 19-9.

Cancer antigen (CA 15-3)

CA 15-3, also known as MUC-1, is a mucin-like glycoprotein expressed in the apical surface of secretory epithelia of the mammary gland, and the gastrointestinal, respiratory, urinary, and reproductive tracts, and likely serves a barrier function like other mucin family members. In addition, CA 15-3 appears to play a role during implantation of the embryo via binding of CA 15-3 to galectin-1, a member of the mammalian beta-galactoside-binding proteins, resulting in downregulation of CA 15-3 and induction of the syncytial formation during implantation [148]. Although downregulated during trophoblast invasion, MUC1 in placental villi is expressed by syncytiotrophoblasts throughout pregnancy increasing with gestational age and present in the term placenta [149]. MUC1 is also known to bind to intercellular adhesion molecule-1 (ICAM-1), which may facilitate transendothelial trophoblast migration during implantation.

CA 15-3 may also play a role in the pathophysiology of cancer and is overexpressed in a number of malignancies including approximately 90% of breast cancers, 40 to 45% of gastric carcinomas, 25 to 30% of prostate cancers, 75 to 80% of papillary and metastatic clear cell renal cell carcinomas, 50% of squamous cell lung cancers, 80% of adenocarcinomas of the lung, and 60% of mucinous adenocarcinomas of the cervix. CA 15-3 overexpression activates IGF-1R and phosphatidylinositol 3-kinase/AKT pathways, a crucial event for the regulation of VEGF, one of the pivotal genes implicated in angiogenesis [151]. Although CA 15-3 overexpression has been found equally in poorly and highly vascularized tumors, a significant co-expression with multiple angiogenic factors and their receptors, thymidine phosphorylase, VEGF, KDR, bFGF, and FGFR-2, has been noted [152]. In general, overexpression of CA 15-3 in cancer is associated with an unfavorable prognosis. For example, CA 15-3 overexpression has been demonstrated to correlate with high metastatis c potential and poor prognosis in gastric carcinoma. Increased serum CA 15-3 levels are prognostic for poor clinical response and reduced overall survival in platinum-resistant or platinum-refractory ovarian cancer [153], and were an independent prognostic factor for disease-free survival in mucinous adenocarcinoma of the cervix [100].

Alpha-feto-protein (AFP)

Alpha-feto-protein (AFP) has been classified as a member the albuminoid gene family that currently consists of four members: albumin (ALB), vitamin D-binding protein, AFP, and α-ALB [157]. Similar to albumin, serum AFP functions as a ligand carrier/transporter, able to bind a variety of ligands including bilirubin, fatty acids, retinoids, steroids, heavy metals, dyes, flavonoids, phytoestrogens, dioxin, and various drugs. AFP may also act as a modulator of the immune system, including impairment of dendritic cells, suppression of B- and T-cell blast cell expansion and apoptosis of antigen-processing cells allowing escape from immune surveillance, and growth during embryonic and fetal development. Growth effects are likely exerted through the cyclic AMP–protein kinase A activation pathway [158] and include pro-angiogenic properties able to upregulate expression of the c-FOS, c-JUN, and n-RAS oncogenes leading to cellular proliferation. There is evidence that AFP plays a role in gonadal development and access of sex hormones to the developing brain. Finally, AFP is present in the placenta throughout gestation and may play a role in a receptor-mediated transport mechanism for AFP placental transfer to maternal tissues [161, 162].

AFP may also play a role in the pathophysiology of cancer; however, this has yet to be fully defined. AFP is expressed in germ cell tumors, especially endodermal sinus tumors and embryonal cell carcinomas, and is overexpressed in the majority of hepatocellular carcinomas (HCC). In HCC, serum AFP levels play an important role in prognosticating the likelihood of tumor recurrence, which is used not only to assess prognosis but also in the decision to perform liver transplantation [155]. AFP levels are also followed closely during treatment [156] and following liver transplantation in order to detect tumor recurrence.

Human chorionic gonadotropin (hCG)

The glycoprotein human chorionic gonadotropin (hCG) is a placental hormone and marker for the differentiation process of cytotrophoblast cells to syncytial trophoblasts during normal pregnancy. hCG is composed of two subunits that are structurally dissimilar, an α-subunit that closely resembles luteinizing hormone (LH) and follicular stimulating hormone (FSH) and a β-subunit that has a distinct composition. Sustained corpus luteal stimulation of the shared hCG/LH receptor leads to hCG secretion by the villous syncytiotrophoblasts of the placenta, which continues to escalate until the syncytiotrophoblast is abundant enough to take over progesterone production from the corpus luteum. hCG appears to play a number of other roles including the prevention of fetoplacental rejection by the maternal immune system, uterine growth and angiogenesis, LH-like functions, and umbilical cord growth and differentiation [173]. Recent literature has also demonstrated hCG/LH receptors in the fetal lung, liver, kidney, spleen, and intestines, indicating that hCG may play a role in fetal growth and differentiation [174, 175]. A hyperglycosylated form of hCG is produced by the extravillous cytotrophoblast cells, which acts as an autocrine stimulation to promote cell growth and invasion. These additional functions of hCG suggest possible mechanisms for hCG-mediated pathogenesis in malignancies that express this oncofetoprotein.

hCG is expressed in germ cell tumors and gestational trophoblastic neoplasia (GTN) and serves as an important biomarker in these cancers. For GTN, a decline in serum hCG levels is prognostic for response to chemotherapy and tumor burden in hCG-expressing germ cell tumors [165]. Neoplasms associated with hCG production may only express the α- or β- subunit. The expression of the free β-subunit is associated with a poor prognosis in a subset of malignancies including cancers of the naso-pharynx, lung, neuroendocrine, germ cell, renal, bladder, cervix, and ovaries [166–171], and likely plays a role in the promotion of malignant transformation [172] and metastasis.

Role of imaging in cancer prognosis

If pathologists can detect the tiniest of tumors and subtype them with a wide range of molecular tools, why is there a need for radiological staging? The answer is that both fields yield oncologic data that is vital and complementary. The advantages of radiological staging are as follows. First, pathologists can only analyze tissue extracted by surgical excision or biopsy, which is often not inclusive of the entire tumor. Second, pathologists often analyze representative pieces of the tumor and may or may not get a comprehensive look at tumor heterogeneity. Radiologists image the tumor *in situ*, still whole and relatively undisturbed. Finally, pathologists look at *ex vivo* tissue, while radiologists have the benefit of *in vivo* imaging allowing accurate 3D tumor measurements and live assessments of tumor physiology. Together, the pathologist's histological diagnosis and molecular phenotypic data and the radiologist's imaging assessments of gross tumor measurements, and the local and distant extent of tumor, yield a cancer stage with associated prognosis and therapeutic options.

CT (computed tomography)

In order to understand the strengths and limitations of the radiological contribution to cancer staging, a basic understanding of how the imaging equipment is utilized and operates is essential. Imaging of gross tumor morphology is usually assessed with CT (computed tomography), which now generates isometric voxels allowing high resolution 3D reconstruction with spatial and temporal resolution as high as 0.3 mm and 42 ms (useful for freezing and reconstructing cardiac motion). Therefore among the available cross-sectional imaging options, CT is by far the highest in resolution, fastest, most readily available, and least expensive modality. For all these reasons, CT has become the workhorse of oncologic imaging.

CT images are performed using a high-energy X-ray beam source (sometimes two sources) mounted on an imaging detector ring, which rotates around the patient at very high velocities irradiating the region of interest as the patient slides through the imaging ring. The CT image acquisition is achieved using crystal detectors imbedded in the imaging ring. These detectors absorb radiation after passing through the patient's body, and determine the attenuation of the radiation beam by the patient's body. The signature of beam attenuation transmitted to the detector imparts information regarding the type and amounts of tissues in the thin cross-sectional plane through the body. Through a mathematical "backprojection" technique, the acquired volumetric data is then reconstructed into a series of 2D CT images that provide the anatomical data used for medical interpretation. Thus CT image tissue contrast, the ability to discriminate between different tissues, relies on differences in tissue densities and the resulting differential strength in attenuation of the CT X-ray beam.

The radiologist relies on differential tissue density to detect the sometimes subtle mass effect of the tumor on regional tissue and abnormal tissue contours. For certain areas of the body, such as the lung, this system of differential tissue density measurement is an advantage. In the normal lung, the complex sponge-like architecture of the parenchyma is fully etched by air, providing excellent tissue contrast. In other areas of the body, such as the bowel and in solid organs, the inherent tissue contrast is limited, necessitating the use of radiodense tissue contrast enhancers such as intravenous or oral CT contrast. For some tumors, multiple acquisitions are performed after the administration of IV contrast in order to elevate tumor conspicuity. For tumors in the bowel, oral contrast is useful for intraluminal distention and definition of bowel wall contours and thickness. CT imaging strengths and limitations impact significantly on the accuracy of cancer staging and prognostication, and must be taken into consideration in the construction of staging criteria, clinical trial development, and individual patient management.

Beyond the staging criteria, there are general morphological CT characteristics that are considered to be indicative of a more aggressive malignancy. Rapid tumor enlargement and poorly marginated permeative margins are considered a poor prognostic sign as they indicate a more metabolically active tumor. Tumors that cross natural anatomic boundaries such as the pericardium, pleura, chest wall, bone, peritoneum, and arterial walls are more likely to be aggressive, probably due to elevated tumor levels of destructive proteases. Tumors that exhibit central necrosis tend to be more aggressive, which in the absence of therapy is usually an indication of an accelerated tumor growth that outstrips the vascular supply either because it is growing too quickly or forming less well developed tumor vasculature. This is also important since hypoxic tumors are generally more refractory to radiation therapy. In addition to the characteristics of the tumor itself, tumors are also evaluated on CT for their locoregional extent with note of tumor proximity to, or invasion of, vital structures. For example, lung cancers that invade the aorta and brain tumors located in the brain stem. Tumors that are difficult to access or invade vital structures can have a strong negative impact on prognosis as they may cripple the range of therapeutic options available to the patient and are reflected in the staging criteria.

Finally, there are secondary findings on CT that are not specific for spread of malignancy but are highly suggestive and may indicate a poorer prognosis. Included in this category are prominent lymph nodes that do not meet pathologic size criteria, pleural effusion (may indicate tiny pleural metastases), ascites (can be an indication of tiny serosal implants), multiple subcentimeter pulmonary nodules, and low-density lesions in the liver that are too small to characterize by density measurements both of which may be manifestations of metastasis. Often additional imaging, tissue/fluid sampling or surgical exploration is undertaken to further characterize these findings to obtain a more accurate prognosis and staging.

Role of adjunct imaging with MRI

Magnetic resonance imaging (MRI) is often used as an adjunct to CT in order to more fully characterize tissue characteristics. The MRI signal is generated by placing the patient in a magnetic field and pulsing a gradient encoded orthogonal field of magnetism through the patient, forcing hydrogen proton spins to realign into a higher energy state, in the direction dictated by the secondary magnetic field. As the spins relax to the resting state, an electrical current is generated and this current is converted into a cross-sectional image using Fourier transformation. MRI tissue contrast is generated by the subtle differences within the molecular environment of a tissue that results in subtle changes in the speed, direction, and magnitude of magnetic relaxation. Tissues that contain high amounts of fluid, fat, or iron behave very differently in a magnetic field allowing for a tissue contrast that does not rely on inherent tissue density. Stepwise acquisition of a multidirectional spectrum of magnetic gradients allows full cross-sectional coverage of anatomical regions. MRI "pulse" sequences, can be "weighted" to detect subtle differences in water, fat, or soft tissue density. Pulses can also be timed in acquisition to obtain arterial or venous phases, gated by ECG to obtain real-time 3D cardiac contractions or timed with respiration to eliminate respiratory motion. Imaging of the lung is suboptimal due to magnetic susceptibility artifacts generated by air–tissue interfaces.

For the staging of most malignancies, MRI is implemented to further characterize indeterminate lesions detected on CT. Since CT relies on differential tissue densities for signal contrast, tissues of similar tissue density, such as a metastasis in a solid organ, can be a challenge for characterization on CT. Often, suspicious lesions in solid organs such as the liver, kidney, adrenals, brain, cardiac, and spine are further characterized with MRI prior to an attempt at biopsy. This is especially helpful in characterizing lesions that are difficult to access by biopsy, such as those located in

proximity to vascular structures. MRI is sometimes utilized as a primary staging modality in malignancies of the pancreas, liver, prostate, uterus, cervix, vagina, and ovaries since these are areas that are less well visualized on CT. In the pelvis, depth of extension into pelvic organs, extension into surrounding fat, and detection of small abnormal lymph nodes is best seen on MRI where there is a stark contrast between tumor, muscle, and fat. Aggressive malignancies of the pancreas are often amorphous and infiltrative into the adjacent fat, bowel, and along the vasculature. The extent of this tumor is often best depicted on MRI where the contrast between pancreas and surrounding fat, bowel, lymph nodes, and vasculature is more easily discerned.

In addition to defining the extent of tumor invasion and characterizing suspicious lesions, MRI can be useful in detecting the influence of tumor masses on their adjacent structures. Brain masses often exhibit peri-tumoral edema, which, along with the tumor bulk itself, can exert pressure on adjacent brain structures with possible herniation of brain tissue through skull apertures resulting in hydrocephalus, brain ischemia, venous clotting and hemorrhage, and possibly death. MRI is very sensitive to the detection of fluid and has excellent tissue resolution in the brain making it very useful for characterization of impending herniation and extent of mass extension. MRI can also detect direction and speed of blood flow through vessels and can be helpful in cases of suspected vascular steal, where the blood flow to the tumor competes with other vital structures for blood flow.

Functional tumor imaging with PET (positron emission tomography)

Positron emission tomography is currently extensively utilized in the practice of clinical oncology. This imaging modality operates through the intravenous administration of a physiologic radio-labelled tracer, currently almost exclusively FDG (^{18}F-fluoro-2-deoxy-D-glucose), allowance of time for that tracer to be taken into target tissues, followed by image acquisition. Patients lie on an imaging platform surrounded by rotating crystal detectors that receive photons emitted by the patient's body from the FDG retained within the body. Photon pairs are ejected from the body at 180 degrees from one another and impact the crystal detectors, each pair impacting crystal detectors at exactly the same time. This coincident bombardment of the detector generates an electrical signal, which is encoded spatially using backprojection techniques allowing for reconstruction of data into a 2D image.

As mentioned above, the functional tracer that is currently routinely used in cancer imaging is FDG (^{18}F-fluoro-2-deoxy-D-glucose), which is an analog of glucose. FDG is rapidly taken up by metabolically active cells, then trapped within these cells by phosphorylation, allowing for accumulation within metabolically active tissue. The amount of glucose taken up by a specific tissue during the specified uptake time is a function of their glucose metabolism, which is typically much accelerated in malignancy. In fact, FDG–PET is exquisitely sensitive for malignancy in lesions greater than 1 cm in size and is routinely employed in oncologic staging to detect occult metastasis. Furthermore, the pre-therapy maximal glycolytic activity of a tumor expressed as the standard uptake value (SUV_{MAX}) has been demonstrated to be an independent prognostic indicator in a number of malignancies. In general, the higher the tumor SUV_{MAX} prior to therapy, the more aggressive the malignancy and worse the prognosis.

More recently, novel PET tracers that are functional markers of malignancies have been developed and are beginning to be tested in clinical trials. FLT ($3'$-deoxy-$3'$-^{18}F-fluorothymidine) is an analog of thymidine, which is not incorporated into the cellular DNA and has been validated as a surrogate marker of proliferation. This drug has been the subject of intense interest since early clinical trials have demonstrated that a drop/lack of drop in the tumor accumulation of FLT shortly after beginning a chemotherapy can prognosticate ultimate success/ failure of the chemotherapy regimen. Thus FLT could potentially yield information on usefulness of a chemotherapeutic as early as 24 to 48 hours after therapy initiation, rather than after months of therapy as currently performed with CT. There are other candidate imaging markers of proliferation including ^{11}C-FMAU (1-($2'$-deoxy-$2'$-fluoro-ß-D-arabinofuranosyl)-[methyl-^{11}C]) thymine, which have met with varying success in preclinical and pilot clinical trials.

In addition to ^{18}F labelled markers of proliferation, there are also evolving markers of tumor hypoxia [^{18}F-fluoromisonidazole (FMISO) and ^{18}F-2-(2-nitro-1H-imidazol-1-yl)-N-(2,2,3,3,3-pentafluoropropyl)-acetamide (EF5)], angiogenesis [vascular endothelial integrin ligand RGD (^{18}F-galacto-RGD)],

and cell death (^{18}F-annexin V). These radiotracers are currently being explored for implantation into the staging and management algorithms. The ability to characterize tumor metabolism and heterogeneity *in vivo* may lead to a more effective choice of chemotherapeutic and hence improved prognosis. Also, many of these evolving PET radiotracers have potential as specific metabolic markers of therapeutic response to targeted agents, for example anti-angiogenic agents such as sorafenib. With the expanding choices of targeted therapies, multifactorial physiologic characterization of tumors has gained of increasing importance.

Future directions

In the twenty-first century, the process of generating a cancer prognosis has evolved from gross estimations of tumor size and origin to sophisticated staging systems that take into account highly accurate measurements of tumor size and invasion, histologic subtype, histologic grade, locoregional invasion, and metastasis. With the wealth of knowledge that has accumulated on molecular phenotypes in cancer, it has become clear that within a cancer type, there are distinct molecular subtypes that carry prognostic and predictive significance. This awareness has led to an increasingly individualized portrait of tumor phenotype that includes assessment of tumor biomarker expression and altered cellular metabolism. These advances are a direct reflection of the insights made in cancer biology. A major goal of modern oncology is to use these cancer biomarkers not only to generate a cancer prognosis and direct patient therapy, but also to assess the effectiveness of therapy in order to more quickly triage therapeutic benefit both for the individual patient and for expeditious clinical trial development of novel therapeutics. Not only are such biomarkers of therapeutic response invaluable to individual patient management, they also expedite the process of clinical trial development both in animal models and in humans.

Now, as our understanding of tumor genetics, signaling, and receptor expression contribution to prognosis becomes better understood, prognosis will not only be based on anatomical assessments, as traditionally done, but also will include functional parameters such as proliferative rate, glycolytic activity, degree of vascularity, molecular expression profile, and expression of predictive cancer biomarkers.

References

1. Saijo N (2012) Critical comments for roles of biomarkers in the diagnosis and treatment of cancer. *Cancer Treat Rev* 38: 63–67.

2. Thompson JF, Shaw HM, Hersey P, Scolyer RA (2004) The history and future of melanoma staging. *J Surg Oncol* 86: 224–235.

3. Allen AC, Spitz S (1953) Malignant melanoma: a clinicopathological analysis of the criteria for diagnosis and prognosis. *Cancer* 6: 1–45.

4. Petersen NC, Bodenham DC, Lloyd OC (1962) Malignant melanomas of the skin. A study of the origin, development, aetiology, spread, treatment, and prognosis. I. *Br J Plast Surg* 15: 49–94.

5. Perou CM, Sorlie T, Eisen MB, et al. (2000) Molecular portraits of human breast tumours. *Nature* 406: 747–752.

6. Sorlie T, Perou CM, Tibshirani R, et al. (2001) Gene expression patterns of breast carcinomas distinguish tumor subclasses with clinical implications. *Proc Natl Acad Sci USA* 98: 10869–10874.

7. Shim HJ, Yun JY, Hwang JE, et al. (2010) BRCA1 and XRCC1 polymorphisms associated with survival in advanced gastric cancer treated with taxane and cisplatin. *Cancer Sci* 101: 1247–1254.

8. Pages F, Berger A, Camus M, et al. (2005) Effector memory T cells, early metastasis, and survival in colorectal cancer. *N Engl J Med* 353: 2654–2666.

9. Liu S, Lachapelle J, Leung S, et al. (2012) CD8+ lymphocyte infiltration is an independent favorable prognostic indicator in basal-like breast cancer. *Breast Cancer Res* 14: R48.

10. Yigit S, Demir L, Tarhan MO, et al. (2012) The clinicopathological significance of Bax and Bcl-2 protein expression with tumor infiltrating lymphocytes in ovarian carcinoma. *Neoplasma* 59: 475–485.

11. Kim ST, Jeong H, Woo OH, et al. (2013) Tumor-infiltrating lymphocytes, tumor characteristics, and recurrence in patients with early breast cancer. *Am J Clin Oncol* 36: 224–231.

12. Rao HL, Chen JW, Li M, et al. (2012) Increased intratumoral neutrophil in colorectal carcinomas correlates closely with malignant phenotype and predicts patients' adverse prognosis. *PLoS One* 7: e30806.

13. Zitvogel L, Kepp O, Kroemer G (2011) Immune parameters affecting the efficacy of chemotherapeutic regimens. *Nat Rev Clin Oncol* 8: 151–160.

14. Gasparotto D, Polesel J, Marzotto A, et al. (2011) Overexpression of TWIST2 correlates with poor prognosis in head and neck squamous cell carcinomas. *Oncotarget* 2: 1165–1175.

15. Kim H, Choi GH, Na DC, et al. (2011) Human hepatocellular carcinomas with "Stemness"-related marker expression: keratin 19 expression and a poor prognosis. *Hepatology* 54: 1707–1717.

16. Li Y, Wang W, Yang R, et al. (2012) Correlation of TWIST2 up-regulation and epithelial–mesenchymal transition during tumorigenesis and progression of cervical carcinoma. *Gynecol Oncol* 124: 112–118.

17. Ryu HS, Park do J, Kim HH, Kim WH, Lee HS (2012) Combination of epithelial-mesenchymal transition and cancer stem cell-like phenotypes has independent prognostic value in gastric cancer. *Hum Pathol* 43: 520–528.

18. van Nes JG, de Kruijf EM, Putter H, et al. (2012) Co-expression of SNAIL and TWIST determines prognosis in estrogen receptor-positive early breast cancer patients. *Breast Cancer Res Treat* 133: 49–59.

19. Gideon P, John J, Frech M, et al. (1992) Mutational and kinetic analyses of the GTPase-activating protein (GAP)-p21 interaction: the C-terminal domain of GAP is not sufficient for full activity. *Mol Cell Biol* 12: 2050–2056.

20. Jancik S, Drabek J, Radzioch D, Hajduch M (2010) Clinical relevance of KRAS in human cancers. *J Biomed Biotechnol* 2010: 150960.

21. Tol J, Nagtegaal ID, Punt CJ (2009) BRAF mutation in metastatic colorectal cancer. *N Engl J Med* 361: 98–99.

22. Wolfer A, Wittner BS, Irimia D, et al. (2010) MYC regulation of a "poor-prognosis" metastatic cancer cell state. *Proc Natl Acad Sci USA* 107: 3698–3703.

23. Ben-Porath I, Thomson MW, Carey VJ, et al. (2008) An embryonic stem cell-like gene expression signature in poorly differentiated aggressive human tumors. *Nat Genet* 40: 499–507.

24. Ilic N, Utermark T, Widlund HR, Roberts TM (2011) PI3K-targeted therapy can be evaded by gene amplification along the MYC-eukaryotic translation initiation factor 4E (eIF4E) axis. *Proc Natl Acad Sci USA* 108: E699–708.

25. Pyndiah S, Tanida S, Ahmed KM, et al. (2011) c-MYC suppresses BIN1 to release poly(ADP-ribose) polymerase 1: a mechanism by which cancer cells acquire cisplatin resistance. *Sci Signal* 4: ra19.

26. Miller TW, Balko JM, Ghazoui Z, et al. (2011) A gene expression signature from human breast cancer cells with acquired hormone independence identifies MYC as a mediator of antiestrogen resistance. *Clin Cancer Res* 17: 2024–2034.

27. Di Cosimo S, Baselga J (2009) Phosphoinositide 3-kinase

mutations in breast cancer: a "good" activating mutation? *Clin Cancer Res* 15: 5017–5019.

28. Wang L, Zhang Q, Zhang J, et al. (2011) PI3K pathway activation results in low efficacy of both trastuzumab and lapatinib. *BMC Cancer* 11: 248.

29. Berns K, Horlings HM, Hennessy BT, et al. (2007) A functional genetic approach identifies the PI3K pathway as a major determinant of trastuzumab resistance in breast cancer. *Cancer Cell* 12: 395–402.

30. Machado-Silva A, Perrier S, Bourdon JC (2010) p53 family members in cancer diagnosis and treatment. *Semin Cancer Biol* 20: 57–62.

31. Robles AI, Harris CC (2010) Clinical outcomes and correlates of TP53 mutations and cancer. *Cold Spring Harb Perspect Biol* 2: a001016.

32. Storey A, Thomas M, Kalita A, et al. (1998) Role of a p53 polymorphism in the development of human papillomavirus-associated cancer. *Nature* 393: 229–234.

33. Mechanic LE, Bowman ED, Welsh JA, et al. (2007) Common genetic variation in TP53 is associated with lung cancer risk and prognosis in African Americans and somatic mutations in lung tumors. *Cancer Epidemiol Biomarkers Prev* 16: 214–222.

34. Aguilar F, Hussain SP, Cerutti P (1993) Aflatoxin B1 induces the transversion of G–>T in codon 249 of the p53 tumor suppressor gene in human hepatocytes. *Proc Natl Acad Sci USA* 90: 8586–8590.

35. Hainaut P, Pfeifer GP (2001) Patterns of p53 G–>T transversions in lung cancers reflect the primary mutagenic signature of DNA-damage by tobacco smoke. *Carcinogenesis* 22: 367–374.

36. Langerod A, Zhao H, Borgan O, et al. (2007) TP53 mutation status and gene expression profiles are powerful prognostic markers of breast cancer. *Breast Cancer Res* 9: R30.

37. Bertheau P, Turpin E, Rickman DS, et al. (2007) Exquisite sensitivity of TP53 mutant and basal breast cancers to a dose-dense epirubicin-cyclophosphamide regimen. *PLoS Med* 4: e90.

38. Turner N, Tutt A, Ashworth A (2004) Hallmarks of "BRCAness" in sporadic cancers. *Nat Rev Cancer* 4: 814–819.

39. Wooster R, Weber BL (2003) Breast and ovarian cancer. *N Engl J Med* 348: 2339–2347.

40. Papelard H, de Bock GH, van Eijk R, et al. (2000) Prevalence of BRCA1 in a hospital-based population of Dutch breast cancer patients. *Br J Cancer* 83: 719–724.

41. Catteau A, Harris WH, Xu CF, Solomon E (1999) Methylation of the BRCA1 promoter region in sporadic breast and ovarian cancer: correlation with disease characteristics. *Oncogene* 18: 1957–1965.

42. Turner NC, Reis-Filho JS, Russell AM, et al. (2007) BRCA1 dysfunction in sporadic basal-like breast cancer. *Oncogene* 26: 2126–2132.

43. Birgisdottir V, Stefansson OA, Bodvarsdottir SK, et al. (2006) Epigenetic silencing and deletion of the BRCA1 gene in sporadic breast cancer. *Breast Cancer Res* 8: R38.

44. Micol R, Ben Slama L, Suarez F, et al. (2011) Morbidity and mortality from ataxia-telangiectasia are associated with ATM genotype. *J Allergy Clin Immunol* 128: 382–389.

45. Lim AM, Young RJ, Collins M, et al. (2012) Correlation of Ataxia-Telangiectasia-Mutated (ATM) gene loss with outcome in head and neck squamous cell carcinoma. *Oral Oncol* 48: 698–702.

46. Guarini A, Marinelli M, Tavolaro S, et al. (2012) ATM gene alterations in chronic lymphocytic leukemia patients induce a distinct gene expression profile and predict disease progression. *Haematologica* 97: 47–55.

47. Mazumder Indra D, Mitra S, Roy A, et al. (2011) Alterations of ATM and CADM1 in chromosomal 11q22.3–23.2 region are associated with the development of invasive cervical carcinoma. *Hum Genet* 130: 735–748.

48. Kang B, Guo RF, Tan XH, et al. (2008) Expression status of ataxia-telangiectasia-mutated gene correlated with prognosis in advanced gastric cancer. *Mutat Res* 638: 17–25.

49. Seol HJ, Yoo HY, Jin J, et al. (2011) Prognostic implications of the DNA damage response pathway in glioblastoma. *Oncol Rep* 26: 423–430.

50. Ye C, Cai Q, Dai Q, et al. (2007) Expression patterns of the ATM gene in mammary tissues and their associations with breast cancer survival. *Cancer* 109: 1729–1735.

51. Gossage L, Madhusudan S (2007) Current status of excision repair cross complementing-group 1 (ERCC1) in cancer. *Cancer Treat Rev* 33: 565–577.

52. Sun JM, Sung JY, Park SH, et al. (2012) ERCC1 as a biomarker for bladder cancer patients likely to benefit from adjuvant chemotherapy. *BMC Cancer* 12: 187.

53. Olaussen KA, Dunant A, Fouret P, et al. (2006) DNA repair by ERCC1 in non-small-cell lung cancer and cisplatin-based adjuvant chemotherapy. *N Engl J Med* 355: 983–991.

54. Winder T, Lenz HJ (2010) Molecular predictive and

prognostic markers in colon cancer. *Cancer Treat Rev* 36: 550–556.

55. Tsourouflis G, Theocharis SE, Sampani A, et al. (2008) Prognostic and predictive value of thymidylate synthase expression in colon cancer. *Dig Dis Sci* 53: 1289–1296.

56. Zucali PA, Giovannetti E, Destro A, et al. (2011) Thymidylate synthase and excision repair cross-complementing group-1 as predictors of responsiveness in mesothelioma patients treated with pemetrexed/carboplatin. *Clin Cancer Res* 17: 2581–2590.

57. Rossi A, Galetta D, Gridelli C (2009) Biological prognostic and predictive factors in lung cancer. *Oncology* 77 Suppl 1: 90–96.

58. Bertino G, Ardiri AM, Calvagno GS, Bertino N, Boemi PM (2010) Prognostic and diagnostic value of des-gamma-carboxy prothrombin in liver cancer. *Drug News Perspect* 23: 498–508.

59. Inagaki Y, Tang W, Makuuchi M, et al. (2011) Clinical and molecular insights into the hepatocellular carcinoma tumour marker des-gamma-carboxyprothrombin. *Liver Int* 31: 22–35.

60. Wang SB, Cheng YN, Cui SX, et al. (2009) Des-gamma-carboxy prothrombin stimulates human vascular endothelial cell growth and migration. *Clin Exp Metastasis* 26: 469–477.

61. Marechal R, Demetter P, Nagy N, et al. (2009) High expression of CXCR4 may predict poor survival in resected pancreatic adenocarcinoma. *Br J Cancer* 100: 1444–1451.

62. Lee HJ, Kim SW, Kim HY, et al. (2009) Chemokine receptor CXCR4 expression, function, and clinical implications in gastric cancer. *Int J Oncol* 34: 473–480.

63. Oda Y, Tateishi N, Matono H, et al. (2009) Chemokine receptor CXCR4 expression is correlated with VEGF expression and poor survival in soft-tissue sarcoma. *Int J Cancer* 124: 1852–1859.

64. Segawa Y, Oda Y, Yamamoto H, et al. (2009) Close correlation between CXCR4 and VEGF expression and their prognostic implications in nasopharyngeal carcinoma. *Oncol Rep* 21: 1197–1202.

65. Franco R, Cantile M, Scala S, et al. (2010) Histomorphologic parameters and CXCR4 mRNA and protein expression in sentinel node melanoma metastasis are correlated to clinical outcome. *Cancer Biol Ther* 9: 423–429.

66. Mizell J, Smith M, Li BD, Ampil F, Chu QD (2009) Overexpression of CXCR4 in primary tumor of patients with HER-2 negative breast cancer was predictive of a poor disease-free survival: a validation study. *Ann Surg Oncol* 16: 2711–2716.

67. Chu QD, Holm NT, Madumere P, et al. (2011) Chemokine receptor CXCR4 overexpression predicts recurrence for hormone receptor-positive, node-negative breast cancer patients. *Surgery* 149: 193–199.

68. Chu QD, Panu L, Holm NT, et al. (2010) High chemokine receptor CXCR4 level in triple negative breast cancer specimens predicts poor clinical outcome. *J Surg Res* 159: 689–695.

69. Hiller DJ, Li BD, Chu QD (2011) CXCR4 as a predictive marker for locally advanced breast cancer post-neoadjuvant therapy. *J Surg Res* 166: 14–18.

70. Andre F, Xia W, Conforti R, et al. (2009) CXCR4 expression in early breast cancer and risk of distant recurrence. *Oncologist* 14: 1182–1188.

71. Minamiya Y, Saito H, Takahashi N, et al. (2010) Expression of the chemokine receptor CXCR4 correlates with a favorable prognosis in patients with adenocarcinoma of the lung. *Lung Cancer* 68: 466–471.

72. Nicholson RI, Gee JM, Harper ME (2001) EGFR and cancer prognosis. *Eur J Cancer* 37 Suppl 4: S9–15.

73. O'Byrne KJ, Gatzemeier U, Bondarenko I, et al. (2011) Molecular biomarkers in non-small-cell lung cancer: a retrospective analysis of data from the phase 3 FLEX study. *Lancet Oncol* 12: 795–805.

74. Pallis AG, Fennell DA, Szutowicz E, et al. (2011) Biomarkers of clinical benefit for anti-epidermal growth factor receptor agents in patients with non-small-cell lung cancer. *Br J Cancer* 105: 1–8.

75. Citri A, Yarden Y (2006) EGF-ERBB signalling: towards the systems level. *Nat Rev Mol Cell Biol* 7: 505–516.

76. Ford AC, Grandis JR (2003) Targeting epidermal growth factor receptor in head and neck cancer. *Head Neck* 25: 67–73.

77. Ross JS, Fletcher JA, Linette GP, et al. (2003) The Her-2/neu gene and protein in breast cancer 2003: biomarker and target of therapy. *Oncologist* 8: 307–325.

78. Slamon DJ, Godolphin W, Jones LA, et al. (1989) Studies of the HER-2/neu proto-oncogene in human breast and ovarian cancer. *Science* 244: 707–712.

79. Slamon DJ, Leyland-Jones B, Shak S, et al. (2001) Use of chemotherapy plus a monoclonal antibody against HER2 for metastatic breast cancer that overexpresses HER2. *N Engl J Med* 344: 783–792.

80. Wang Y, Yin Q, Yu Q, et al. (2011) A retrospective study of breast cancer subtypes: the risk of relapse and the relations with treatments. *Breast Cancer Res Treat* 130: 489–498.

81. Sarandakou A, Protonotariou E, Rizos D (2007) Tumor markers in biological fluids associated with

pregnancy. *Crit Rev Clin Lab Sci* 44: 151–178.

82. O'Brien TJ, Beard JB, Underwood LJ, et al. (2001) The CA 125 gene: an extracellular superstructure dominated by repeat sequences. *Tumour Biol* 22: 348–366.

83. Patankar MS, Jing Y, Morrison JC, et al. (2005) Potent suppression of natural killer cell response mediated by the ovarian tumor marker CA125. *Gynecol Oncol* 99: 704–713.

84. Belisle JA, Horibata S, Jennifer GA, et al. (2010) Identification of Siglec-9 as the receptor for MUC16 on human NK cells, B cells, and monocytes. *Mol Cancer* 9: 118.

85. Bast RC, Jr., Spriggs DR (2011) More than a biomarker: CA125 may contribute to ovarian cancer pathogenesis. *Gynecol Oncol* 121: 429–430.

86. Rosen DG, Wang L, Atkinson JN, et al. (2005) Potential markers that complement expression of CA125 in epithelial ovarian cancer. *Gynecol Oncol* 99: 267–277.

87. Rump A, Morikawa Y, Tanaka M, et al. (2004) Binding of ovarian cancer antigen CA125/MUC16 to mesothelin mediates cell adhesion. *J Biol Chem* 279: 9190–9198.

88. Hogdall E (2008) Cancer antigen 125 and prognosis. *Curr Opin Obstet Gynecol* 20: 4–8.

89. Coussy F, Chereau E, Darai E, et al. (2011) Interest of CA 125 level in management of ovarian cancer. *Gynecol Obstet Fertil* 39: 296–301.

90. Zorn KK, Tian C, McGuire WP, et al. (2009) The prognostic value of pretreatment CA 125 in patients with advanced ovarian carcinoma: a Gynecologic Oncology Group study. *Cancer* 115: 1028–1035.

91. Rustin GJ, Marples M, Nelstrop AE, Mahmoudi M, Meyer T (2001) Use of CA-125 to define progression of ovarian cancer in

patients with persistently elevated levels. *J Clin Oncol* 19: 4054–4057.

92. Rustin GJ, Timmers P, Nelstrop A, et al. (2006) Comparison of CA-125 and standard definitions of progression of ovarian cancer in the intergroup trial of cisplatin and paclitaxel versus cisplatin and cyclophosphamide. *J Clin Oncol* 24: 45–51.

93. Bridgewater JA, Nelstrop AE, Rustin GJ, et al. (1999) Comparison of standard and CA-125 response criteria in patients with epithelial ovarian cancer treated with platinum or paclitaxel. *J Clin Oncol* 17: 501–508.

94. Vergote I, Rustin GJ, Eisenhauer EA, et al. (2000) Re: new guidelines to evaluate the response to treatment in solid tumors (ovarian cancer). Gynecologic Cancer Intergroup. *J Natl Cancer Inst* 92: 1534–1535.

95. Tian C, Markman M, Zaino R, et al. (2009) CA-125 change after chemotherapy in prediction of treatment outcome among advanced mucinous and clear cell epithelial ovarian cancers: a Gynecologic Oncology Group study. *Cancer* 115: 1395–1403.

96. Bast RC, Jr., Xu FJ, Yu YH, et al. (1998) CA 125: the past and the future. *Int J Biol Markers* 13: 179–187.

97. Batlle M, Ribera JM, Oriol A, et al. (2005) Usefulness of tumor markers CA 125 and CA 15.3 at diagnosis and during follow-up in non-Hodgkin's lymphoma: study of 200 patients. *Leuk Lymphoma* 46: 1471–1476.

98. Bairey O, Blickstein D, Stark P, et al. (2003) Serum CA 125 as a prognostic factor in non-Hodgkin's lymphoma. *Leuk Lymphoma* 44: 1733–1738.

99. Cedres S, Nunez I, Longo M, et al. (2011) Serum tumor markers CEA, CYFRA21-1, and CA-125 are associated with worse prognosis in advanced non-small-

cell lung cancer (NSCLC). *Clin Lung Cancer* 12: 172–179.

100. Togami S, Nomoto M, Higashi M, et al. (2010) Expression of mucin antigens (MUC1 and MUC16) as a prognostic factor for mucinous adenocarcinoma of the uterine cervix. *J Obstet Gynaecol Res* 36: 588–597.

101. Berruti A, Tampellini M, Torta M, et al. (1994) Prognostic value in predicting overall survival of two mucinous markers: CA 15–3 and CA 125 in breast cancer patients at first relapse of disease. *Eur J Cancer* 30A: 2082–2084.

102. Pollan M, Varela G, Torres A, et al. (2003) Clinical value of p53, c-erbB-2, CEA and CA125 regarding relapse, metastasis and death in resectable non-small cell lung cancer. *Int J Cancer* 107: 781–790.

103. Kotowicz B, Kaminska J, Fuksiewicz M, et al. (2010) Clinical significance of serum CA-125 and soluble tumor necrosis factor receptor type I in cervical adenocarcinoma patients. *Int J Gynecol Cancer* 20: 588–592.

104. Denschlag D, Tan L, Patel S, et al. (2007) Stage III endometrial cancer: preoperative predictability, prognostic factors, and treatment outcome. *Am J Obstet Gynecol* 196: 546, e541–547.

105. Chiu CH, Shih YN, Tsai CM, et al. (2007) Serum tumor markers as predictors for survival in advanced non-small cell lung cancer patients treated with gefitinib. *Lung Cancer* 57: 213–221.

106. Hoskins PJ, Le N, Correa R (2011) CA 125 normalization with chemotherapy is independently predictive of survival in advanced endometrial cancer. *Gynecol Oncol* 120: 52–55.

107. Hollingsworth MA, Swanson BJ (2004) Mucins in cancer: protection and control of the cell surface. *Nat Rev Cancer* 4: 45–60.

108. Fendrick JL, Konishi I, Geary SM, et al. (1997) CA125 phosphorylation is associated with its secretion from the WISH human amnion cell line. *Tumour Biol* 18: 278–289.

109. Blalock TD, Spurr-Michaud SJ, Tisdale AS, Gipson IK (2008) Release of membrane-associated mucins from ocular surface epithelia. *Invest Ophthalmol Vis Sci* 49: 1864–1871.

110. Govindarajan B, Gipson IK (2010) Membrane-tethered mucins have multiple functions on the ocular surface. *Exp Eye Res* 90: 655–663.

111. Bon GG, Kenemans P, Dekker JJ, et al. (1999) Fluctuations in CA 125 and CA 15-3 serum concentrations during spontaneous ovulatory cycles. *Hum Reprod* 14: 566–570.

112. Gipson IK, Blalock T, Tisdale A, et al. (2008) MUC16 is lost from the uterodome (pinopode) surface of the receptive human endometrium: in vitro evidence that MUC16 is a barrier to trophoblast adherence. *Biol Reprod* 78: 134–142.

113. Baalbergen A, Janssen JW, van der Weiden RM (2000) CA-125 levels are related to the likelihood of pregnancy after in vitro fertilization and embryo transfer. *Am J Reprod Immunol* 43: 21–24.

114. Paoletti AM, Serra GG, Mais V, et al. (1995) Involvement of ovarian factors magnified by pharmacological induction of multiple follicular development (MFD) in the increase in Ca125 occurring during the luteal phase and the first 12 weeks of induced pregnancies. *J Assist Reprod Genet* 12: 263–268.

115. Xiong L, Woodward AM, Argueso P (2011) Notch signaling modulates MUC16 biosynthesis in an in vitro model of human corneal and conjunctival epithelial cell differentiation. *Invest Ophthalmol Vis Sci* 52: 5641–5646.

116. Blalock TD, Spurr-Michaud SJ, Tisdale AS, et al. (2007) Functions of MUC16 in corneal epithelial cells. *Invest Ophthalmol Vis Sci* 48: 4509–4518.

117. Pickford WJ, Watson AJ, Barker RN (2007) Different forms of helper tolerance to carcinoembryonic antigen: ignorance and regulation. *Clin Cancer Res* 13: 4528–4537.

118. Screaton RA, Penn LZ, Stanners CP (1997) Carcinoembryonic antigen, a human tumor marker, cooperates with Myc and Bcl-2 in cellular transformation. *J Cell Biol* 137: 939–952.

119. Samara RN, Laguinge LM, Jessup JM (2007) Carcinoembryonic antigen inhibits anoikis in colorectal carcinoma cells by interfering with TRAIL-R2 (DR5) signaling. *Cancer Res* 67: 4774–4782.

120. Camacho-Leal P, Stanners CP (2008) The human carcinoembryonic antigen (CEA) GPI anchor mediates anoikis inhibition by inactivation of the intrinsic death pathway. *Oncogene* 27: 1545–1553.

121. Takahashi N, Suzuki K, Takamochi K, Oh S (2011) Prognosis of surgically resected lung cancer with extremely high preoperative serum carcinoembryonic antigen level. *Gen Thorac Cardiovasc Surg* 59: 699–704.

122. Yeh CY, Hsieh PS, Chiang JM, et al. (2011) Preoperative carcinoembryonic antigen (CEA) elevation in colorectal cancer. *Hepatogastroenterology* 58: 1171–1176.

123. Basbug M, Arikanoglu Z, Bulbuller N, et al. (2011) Prognostic value of preoperative CEA and CA 19-9 levels in patients with colorectal cancer. *Hepatogastroenterology* 58: 400–405.

124. Di Gioia D, Heinemann V, Nagel D, et al. (2011) Kinetics of CEA and CA15–3 correlate with treatment response in patients undergoing chemotherapy for metastatic breast cancer (MBC). *Tumour Biol* 32: 777–785.

125. Ishihara S, Watanabe T, Kiyomatsu T, Yasuda K, Nagawa H (2010) Prognostic significance of response to preoperative radiotherapy, lymph node metastasis, and CEA level in patients undergoing total mesorectal excision of rectal cancer. *Int J Colorectal Dis* 25: 1417–1425.

126. Perez RO, Sao Juliao GP, Habr-Gama A, et al. (2009) The role of carcinoembriogenic antigen in predicting response and survival to neoadjuvant chemoradiotherapy for distal rectal cancer. *Dis Colon Rectum* 52: 1137–1143.

127. Moreno Garcia V, Cejas P, Blanco Codesido M, et al. (2009) Prognostic value of carcinoembryonic antigen level in rectal cancer treated with neoadjuvant chemoradiotherapy. *Int J Colorectal Dis* 24: 741–748.

128. Strimpakos AS, Cunningham D, Mikropoulos C, et al. (2010) The impact of carcinoembryonic antigen flare in patients with advanced colorectal cancer receiving first-line chemotherapy. *Ann Oncol* 21: 1013–1019.

129. Li Y, Cao H, Jiao Z, et al. (2010) Carcinoembryonic antigen interacts with TGF-{beta} receptor and inhibits TGF-{beta} signaling in colorectal cancers. *Cancer Res* 70: 8159–8168.

130. Carlin AF, Uchiyama S, Chang YC, et al. (2009) Molecular mimicry of host sialylated glycans allows a bacterial pathogen to engage neutrophil Siglec-9 and dampen the innate immune response. *Blood* 113: 3333–3336.

131. Ohshio G, Ogawa K, Kudo H, et al. (1990) Immunohistochemical distribution of CA19-9 in normal and tumor tissues of the kidney. *Urol Int* 45: 1–3.

132. Suzuki K, Muraishi O, Tokue A (2002) The correlation of serum carbohydrate antigen 19-9 with benign hydronephrosis. *J Urol* 167: 16–20.

133. Osswald BR, Klee FE, Wysocki S (1993) The reliability of highly elevated CA 19-9 levels. *Dis Markers* 11: 275–278.

134. Kim HR, Lee CH, Kim YW, et al. (2009) Increased CA 19-9 level in patients without malignant disease. *Clin Chem Lab Med* 47: 750–754.

135. King A, Loke YW (1988) Differential expression of blood-group-related carbohydrate antigens by trophoblast subpopulations. *Placenta* 9: 513–521.

136. Minas V, Mylonas I, Schiessl B, et al. (2007) Expression of the blood-group-related antigens Sialyl Lewis a, Sialyl Lewis x and Lewis y in term placentas of normal, preeclampsia, IUGR- and HELLP-complicated pregnancies. *Histochem Cell Biol* 128: 55–63.

137. Sawada R, Sun SM, Wu X, et al. (2011) Human monoclonal antibodies to sialyl-Lewis (CA19.9) with potent CDC, ADCC, and antitumor activity. *Clin Cancer Res* 17: 1024–1032.

138. Boeck S, Stieber P, Holdenrieder S, Wilkowski R, Heinemann V (2006) Prognostic and therapeutic significance of carbohydrate antigen 19-9 as tumor marker in patients with pancreatic cancer. *Oncology* 70: 255–264.

139. Maisey NR, Norman AR, Hill A, et al. (2005) CA19-9 as a prognostic factor in inoperable pancreatic cancer: the implication for clinical trials. *Br J Cancer* 93: 740–743.

140. Ko AH, Hwang J, Venook AP, et al. (2005) Serum CA19-9 response as a surrogate for clinical outcome in patients receiving fixed-dose rate gemcitabine for advanced pancreatic cancer. *Br J Cancer* 93: 195–199.

141. Ziske C, Schlie C, Gorschluter M, et al. (2003) Prognostic value of CA 19-9 levels in patients with inoperable adenocarcinoma of the pancreas treated with gemcitabine. *Br J Cancer* 89: 1413–1417.

142. Xu B, Zheng WY, Jin DY, et al. (2011) Predictive value of serum carbohydrate antigen 19-9 in malignant intraductal papillary mucinous neoplasms. *World J Surg* 35: 1103–1109.

143. Tsavaris N, Kosmas C, Papadoniou N, et al. (2009) CEA and CA-19.9 serum tumor markers as prognostic factors in patients with locally advanced (unresectable) or metastatic pancreatic adenocarcinoma: a retrospective analysis. *J Chemother* 21: 673–680.

144. Fan B, Xiong B (2011) Investigation of serum tumor markers in the diagnosis of gastric cancer. *Hepatogastroenterology* 58: 239–245.

145. Sato H, Maeda K, Sugihara K, et al. (2011) High-risk stage II colon cancer after curative resection. *J Surg Oncol* 104: 45–52.

146. Wiest I, Alexiou C, Mayr D, et al. (2010) Expression of the carbohydrate tumor marker Sialyl Lewis a (Ca19-9) in squamous cell carcinoma of the larynx. *Anticancer Res* 30: 1849–1853.

147. Heimburg-Molinaro J, Lum M, Vijay G, et al. (2011) Cancer vaccines and carbohydrate epitopes. *Vaccine* 29: 8802–8826.

148. Fischer I, Jeschke U, Friese K, Daher S, Betz AG (2011) The role of galectin-1 in trophoblast differentiation and signal transduction. *J Reprod Immunol* 90: 35–40.

149. Paszkiewicz-Gadek A, Porowska H, Sredzinska K (2008) Expression of MUC1 mucin in full-term pregnancy human placenta. *Adv Med Sci* 53: 54–58.

150. Saeland E, de Jong MA, Nabatov AA, et al. (2009) MUC1 in human milk blocks transmission of human immunodeficiency virus from dendritic cells to T cells. *Mol Immunol* 46: 2309–2316.

151. Woo JK, Choi Y, Oh SH, et al. (2012) Mucin 1 enhances the tumor angiogenic response by activation of the AKT signaling pathway. *Oncogene* 31: 2187–2198.

152. Giatromanolaki A, Koukourakis MI, Sivridis E, et al. (2000) Coexpression of MUC1 glycoprotein with multiple angiogenic factors in non-small cell lung cancer suggests coactivation of angiogenic and migration pathways. *Clin Cancer Res* 6: 1917–1921.

153. Budiu RA, Mantia-Smaldone G, Elishaev E, et al. (2011) Soluble MUC1 and serum MUC1-specific antibodies are potential prognostic biomarkers for platinum-resistant ovarian cancer. *Cancer Immunol Immunother* 60: 975–984.

154. Ziecik AJ, Waclawik A, Kaczmarek MM, et al. (2011) Mechanisms for the establishment of pregnancy in the pig. *Reprod Domest Anim* 46 Suppl 3: 31–41.

155. Burak KW (2011) Prognosis in the early stages of hepatocellular carcinoma: predicting outcomes and properly selecting patients for curative options. *Can J Gastroenterol* 25: 482–484.

156. Kim BK, Ahn SH, Seong JS, et al. (2011) Early alpha-fetoprotein response as a predictor for clinical outcome after localized concurrent chemoradiotherapy for advanced hepatocellular carcinoma. *Liver Int* 31: 369–376.

157. Mizejewski GJ (1995) The phylogeny of alpha-fetoprotein in vertebrates: survey of biochemical and physiological data. *Crit Rev Eukaryot Gene Expr* 5: 281–316.

158. Li MS, Li PF, Yang FY, et al. (2002) The intracellular mechanism of alpha-fetoprotein promoting the proliferation of

497

NIH 3T3 cells. *Cell Res* 12: 151–156.

159. Pekkanen-Mattila M, Pelto-Huikko M, Kujala V, et al. (2010) Spatial and temporal expression pattern of germ layer markers during human embryonic stem cell differentiation in embryoid bodies. *Histochem Cell Biol* 133: 595–606.

160. Setiyono A, Budiyati AD, Purwantomo S, et al. (2011) Immunoregulatory effects of AFP domains on monocyte-derived dendritic cell function. *BMC Immunol* 12: 4.

161. Newby D, Dalgliesh G, Lyall F, Aitken DA (2005) Alphafetoprotein and alphafetoprotein receptor expression in the normal human placenta at term. *Placenta* 26: 190–200.

162. Brownbill P, Edwards D, Jones C, et al. (1995) Mechanisms of alphafetoprotein transfer in the perfused human placental cotyledon from uncomplicated pregnancy. *J Clin Invest* 96: 2220–2226.

163. Alazzam M, Young T, Coleman R, et al. (2011) Predicting gestational trophoblastic neoplasia (GTN): is urine hCG the answer? *Gynecol Oncol* 122: 595–599.

164. Kang WD, Kim CH, Cho MK, et al. (2010) Serum hCG level and rising World Health Organization score at second-line chemotherapy (pulse dactinomycin): poor prognostic factors for methotrexate-failed low-risk gestational trophoblastic neoplasia. *Int J Gynecol Cancer* 20: 1424–1428.

165. Olofsson SE, Tandstad T, Jerkeman M, et al. (2011) Population-based study of treatment guided by tumor marker decline in patients with metastatic nonseminomatous germ cell tumor: a report from the Swedish-Norwegian Testicular Cancer Group. *J Clin Oncol* 29: 2032–2039.

166. Andrusiewicz M, Szczerba A, Wolun-Cholewa M, et al. (2011) CGB and GNRH1 expression analysis as a method of tumor cells metastatic spread detection in patients with gynecological malignances. *J Transl Med* 9: 130.

167. Shah T, Srirajaskanthan R, Bhogal M, et al. (2008) Alpha-fetoprotein and human chorionic gonadotrophin-beta as prognostic markers in neuroendocrine tumour patients. *Br J Cancer* 99: 72–77.

168. Vartiainen J, Lassus H, Lehtovirta P, et al. (2008) Combination of serum hCG beta and p53 tissue expression defines distinct subgroups of serous ovarian carcinoma. *Int J Cancer* 122: 2125–2129.

169. Hotakainen K, Lintula S, Jarvinen R, et al. (2007) Overexpression of human chorionic gonadotropin beta genes 3, 5 and 8 in tumor tissue and urinary cells of bladder cancer patients. *Tumour Biol* 28: 52–56.

170. Iles RK (2007) Ectopic hCGbeta expression by epithelial cancer: malignant behaviour, metastasis and inhibition of tumor cell apoptosis. *Mol Cell Endocrinol* 260–262: 264–270.

171. Hotakainen K, Lintula S, Ljungberg B, et al. (2006) Expression of human chorionic gonadotropin beta-subunit type I genes predicts adverse outcome in renal cell carcinoma. *J Mol Diagn* 8: 598–603.

172. Bellet D, Lazar V, Bieche I, et al. (1997) Malignant transformation of nontrophoblastic cells is associated with the expression of chorionic gonadotropin beta genes normally transcribed in trophoblastic cells. *Cancer Res* 57: 516–523.

173. Cole LA (2010) Biological functions of hCG and hCG-related molecules. *Reprod Biol Endocrinol* 8: 102.

174. Goldsmith PC, McGregor WG, Raymoure WJ, Kuhn RW, Jaffe RB (1983) Cellular localization of chorionic gonadotropin in human fetal kidney and liver. *J Clin Endocrinol Metab* 57: 654–661.

175. Abdallah MA, Lei ZM, Li X, et al. (2004) Human fetal nongonadal tissues contain human chorionic gonadotropin/luteinizing hormone receptors. *J Clin Endocrinol Metab* 89: 952–956.

Chapter

32

Cancer pharmacogenomics: challenges, promises, and its application to cancer drug discovery

Lihua Yu and Kevin Webster

Introduction

Future success in oncology drug discovery and development (R&D) is reliant on strategies to identify novel disease targets, appropriate preclinical disease models, and biomarkers to drive patient enrichment and clinical trial design. The discipline of cancer pharmacogenomics[1] underpins each of these future success factors. Today the field of cancer pharmacogenomics is at a tipping point, where the technologies have sufficiently advanced and are now being applied to significantly sized sample sets enabling acquisition of quality data. However, the analytical strategies are lagging and are largely inaccessible to the broad oncology scientific community. The future potential of cancer pharmacogenomics hinges on analytical solutions that enable the "experimentalist" to build and test hypotheses in an iterative manner and thus build a deeper knowledge base.

Cancer pharmacogenomics is a term frequently used to describe the study of molecular variations (genetics, genomics, heritable, or somatic) of patient tumors and/or preclinical disease model systems to variability in cancer drug response, in terms of either

efficacy or adverse drug reactions. Today the field of cancer pharmacogenomics is uniquely active due to both the promises it holds in personalized medicine (Parkinson and Ziegler, 2009; Raponi et al., 2008) and the recent advances in the area of large-scale cancer cell line profiling (McDermott, et al., 2007; Sharma et al., 2010), integrative genomics analysis (Rhodes et al., 2007; Subramanian, et al., 2005), and in particular the depth of data coming out of cancer genomics initiatives. Such as the Cancer Genome Atlas (TCGA, http://cancergenome.nih.gov), the Cancer Genome Project (CGP, http://www.sanger.ac.uk/genetics/CGP), and International Cancer Genome Consortium (ICGC, https://icgc.org).

For our purpose of discussion, we would like to highlight the potential of cancer pharmacogenomics in drug discovery and development. Drug discovery and development is an integrated process (Figure 32.1) where application of cancer pharmacogenomics can directly impact target selection and validation, preclinical model development, biomarker discovery, disease hypothesis generation/testing, and ultimately clinical trial design including patient selection strategies and development of companion diagnostics. Therefore it is important to view it in the context of whole drug discovery/development process. Successful development of new cancer therapies will rely heavily on joining the progress in our understanding of disease at the *molecular level*, in terms of disease heterogeneity and molecular aberrations, at a *system level*, in terms of key molecular pathway drivers, concurrent or complementary roles of different pathways in tumor cells and their interaction with the microenvironment, and the iterative learning from these scientific advances in the discovery phase.

---

[1] Even though earlier distinctions were made between pharmacogenomics and pharmacogenetics, where pharmacogenetics tends to be slightly more narrowly defined as the study of the relationship between inheritabe variations, such as SNPs, to drug metabolism and response, the distinctions are arbitrary and the two terms are used interchangeably in practice (NCBI Science primer: Pharamcogenomics Factsheet, 2004). Therefore throughout this chapter we will use the term pharmacogenomics exclusively and make no attempt to differentiate between the two terms.

Systems Biology of Cancer, ed. S. Thiagalingam. Published by Cambridge University Press. © Cambridge University Press 2015.

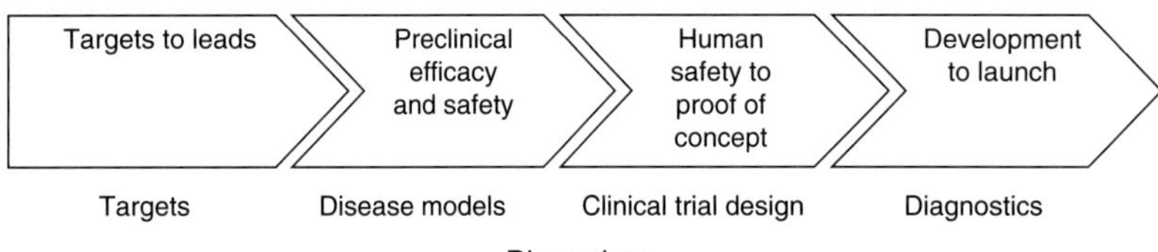

Figure 32.1 Impact of cancer pharmacogenomics across drug discovery and development.

Recent advances in cancer genomics and pharmacogenomics

Cancer genomics has been associated primarily with gene expression profiling data until a few years ago. Some of the early challenges and criticisms of lack of translational success in applying genomics to drug R&D can be attributed to two main limitations: the availability of data and the ability to analyze the data in the context of disease biology. There are two aspects that limited data availability in the past. One aspect is the availability of large-scale, high-quality genomic profiling data with well-characterized sample information. Even though there were commercial data providers, the data have not been broadly accessible to the scientific community and the tumor representations tend to concentrate on a few large tumor types, such as breast, lung, and colon cancers. The other aspect is the availability of multidimensional data beyond gene expression. We now know that gene expression alone is a one-dimensional view of complex biological systems. The second significant limitation is our ability to analyze and interpret data in the broader context of cancer biology. Early data analysis efforts focused largely on statistical methodology development addressing a major challenge posed by making a large number of observations on an intrinsically complex and noisy biological system. While necessary, this did not in and of itself enable more complex problems to be engaged. With these limitations, most early applications of cancer pharmacogenomics in cancer drug discovery/development relied heavily on small datasets due to sample availability, technical feasibility, and financial resources to carry out large experiments and statistical data analysis approaches. Not surprisingly, many pharmaceutical/biotechnology companies' opinions on pharmacogenomics swung from extremely hopeful and excited to deeply disappointed, after they failed to see the benefit in proportion to their large investment in this area.

However, the situation has changed significantly over the last few years. With scientific and technological advancement, commitment of the public and scientific community to conquer cancer, and financial support from government agencies, we have seen the accelerated increase of data available in the public domain. Today data range from gene expression, gene copy number, mutation data from conventional sequencing approaches, to microRNA, epigenetic profiling, and most recently data generated by deep sequencing. Often all of these measures are taken on a single sample and linked to clinical information. Equally important is the recent advancement in our ability to analyze and mine multidimension, large datasets across studies, increasingly under biological context. Most importantly, the broad adoption and appreciation by the oncology community of molecular profiling data, in particular, genetic data, and their significant impact in advancing our understanding of cancer disease at the molecular level, and applying the most recent scientific learning into clinical practice (Molecular Fingerprint at Massachusetts General Hospital Cancer Center, www.massgeneral.org/cancer/news/molecularprofiling.axpx; Lung Cancer Mutation Consortium Protocol, http://clinicaltrials.gov/ct2/show/NCT0104286), i.e., personalized medicine, has challenged cancer drug discovery into adopting cancer pharmacogenomics more systematically across the R&D process.

Advances in genomics data generation

Both scientific and technological advances have defined a new era of productivity, yielding vast quantities of pharmacogenomic data. For example, it is now routine to profile biological samples for not only mRNA expressions, but also DNA copy number alterations, gene mutations, microRNA expressions (Barbarotto et al., 2008), DNA methylations (Gargiulo and Minucci, 2009), and protein expressions (Speer et al., 2007). Next-generation sequencing technologies have been applied broadly to a variety of applications, e.g., whole-genome sequencing, exome sequencing, RNA sequencing, and the analysis of epigenetic modifications of histones (ChIP-Seq) and DNA (bisulfite sequencing). For a detailed review of next-generation sequencing technologies and their applications in functional genomics, please refer to Morozova and Marra (2008). A detailed review of next-generation sequencing technology's application in cancer genomics was provided by Mardis and Wilson (2009).

Recently, the Cancer Genome Atlas (TCGA) program, a comprehensive US-based project, designed to accelerate our understanding of the molecular basis of cancer through the application of genome analysis technologies, especially large-scale genome sequencing, has finished its three-year, $100 million pilot project. The pilot effort focused on the evaluation of brain and ovarian cancers. The data generated and analysis that followed have provided new genetic mutations and genomic alterations, key deregulated pathways, and new molecular insights into the heterogeneities of diseases with potential implications for disease diagnosis and treatment (Cancer Genome Atlas Research Network, 2008; Verhaak et al., 2010). As a result an additional $275 million will be invested in TCGA to cover 20 more cancer types. Complementary to the US-led TCGA, the International Cancer Genome Consortium (ICGC), a confederation of funding and research members from different countries, will generate comprehensive data covering genomic, transcriptomic, and epigenomic changes across 50 tumor types or subtypes. Both projects build on the learnings from early gene expression surveys by ensuring: large sample set for each tumor type, comprehensive coverage of human genome, high-resolution sequencing and profiling technologies, and high quality. The large number, high quality, and multidimensional data, plus their open data access policy (data available as rapidly as possible, with minimal restrictions), give the scientific community both the statistical power for data analysis and a holistic view of the tumors.

The Cancer Genome Project (CGP), undertaken by the Wellcome Trust Sanger Institute, dedicates its effort in sequencing and cataloging somatically acquired sequence variants or mutations. Complementary to other public data sources its rich collection of cancer cell line data covers Affymetrix SNP 6.0 Array, Affymetrix Human Genome U133A gene expression, systematic sequencing of 64 known cancer genes[2], and a catalog of somatic mutations extracted

from the literature. Very recently, the cancer translation project has initiated under a five-year collaboration with the Massachusetts General Hospital Cancer Center focusing on drug sensitivity biomarker identification. This project promises to provide compound pharmacology data alongside the molecular data profile of cell lines spanning multiple cancer types. Since preclinical models are at the core of cancer drug discovery and development, these datasets are important information bridges linking preclinical hypothesis generation, efficacy in disease model systems, to tumor genomics derived from patient samples.

Advances in data mining and exploitation

Compared to merely five to ten years ago when genomics and pharmacogenomics data were scarce, we are now working in an environment that is data rich but knowledge poor. The challenge has shifted from acquiring, housing, and accessing data, to the mining and exploitation of data, and the generation of new knowledge. The importance and recognition of this challenge is highlighted by TCGA's investment in genomics data analysis centers. The complexity of the data and cancer biology will surely require different approaches. Several noticeable advances in the area of data mining and exploitation, namely meta-analysis or comparison across a large number of genomic datasets, analytical methods reflecting cancer biology, and by pathway-based approaches have contributed to recent improvement in data interpretation and hypothesis generation.

Meta-analysis across multiple datasets covering the same disease or similar experiments is increasingly adopted in cancer pharmacogenomics research (Desmedt et al., 2008; Fan et al., 2006; Haibe-Kains et al., 2008; Wirapati et al., 2008; Yu et al., 2007). The value of meta-analysis lies in its ability to increase the confidence of a particular observation or hypothesis. Earlier genomic data analysis was heavily focused on statistical analytical method development, and the interpretability of the data was often limited by the practicality and financial constraints of not being able to run a sufficiently large number of experiments, or to profile a large number of samples to gain enough statistical power. Even in rare cases when that was done, data interpretation was in general limited to a single study due to lack of a proper method/tool to compare across experiments and the inability to access multiple datasets from outside sources.

---

[2] The Sanger Institute has recently made available mutations from exome sequencing of 1,015 cell lines.

Widely accessible genomics data through systems such as Gene Expression Omnibus of NCBI (GEO, www.ncbi.nlm.nih.gov/geo) (Barrett and Edgar, 2006) and ArrayExpress of EBI (www.ebi.ac.uk/microarray-as/ae) (Parkinson et al., 2009) enabled scientists to conduct meta-analysis across multiple datasets. The publication and broad adoption of gene set enrichment analysis (GSEA) (Subramanian et al., 2005) offers an analytical solution to this challenge and is significant in several aspects. First, it allows the comparison of a gene set with another ranked gene set typically generated from a gene expression study to focus on the overall trend, hence enrichment, rather than the overlap of the absolute top-ranked genes. This is liberating and allows even weak signals to be detected and distinguished from the noise that is intrinsic to any biological experiment. Second, the comparison of a gene set vs. another ranked gene set by GSEA is independent of any technology platform, the nature of the samples, or statistical method of how the ranked genes were generated, all of which were limiting factors in the past for meta-analysis across experiments. Thus GSEA allows gene set-based meta-analysis to be done flexibly across multiple similar or related experiments to further increase the confidence of a weak signal, i.e., if a gene set is consistently enriched across multiple similar and independent experiments it is more likely to be a real signal. The Connectivity Map (Lamb et al., 2006) exemplifies the impact of this approach, adopting GSEA as its core query engine to identify small molecules with similar cellular effects on the basis of gene expression perturbation.

Oncomine (Rhodes et al., 2007), a large compendium of published cancer microarray data that has been annotated, standardized, and analyzed, delivers both broad data accessibility and meta-analysis capability through its biologist-friendly web application. It enables biologists and bioinformaticians alike to focus their effort on data mining/hypothesis generation instead of data collection and statistical analysis.

Another trend of data analysis is the development of analytical methods better reflecting cancer biology. The two group comparison-based methods (t-test, SAM, signal/noise ratio) for analyzing gene expression data was widely used for target hypothesis generation, where the hypothesis is that cancer relevant genes are likely overexpressed in one group of samples – usually a specific tumor type compared to normal samples of the same tissue origin. This approach frequently identified genes involved in cell cycle regulation perhaps reflecting the high proliferative state of cancer cells in general. Its ability to identify other oncogenes has been limited by the heterogeneous nature of tumor samples. Two recent genomic studies of pancreatic cancers and glioblastoma multiforme revealed that there are a few hundreds to over a thousand somatic mutations detected in each study and dozens of alterations in each patient (Jones et al., 2008; Parsons et al., 2008). Another approach, the cancer outlier profile analysis (COPA) (Tomlins et al., 2005) instead seeks to identify outlier profiles, i.e., a subset of samples with marked overexpression that are the likely result of genetic rearrangement or amplification[3]. It has been successfully applied to identify an important family of rearrangements in prostate cancer, the recurrent fusion of TMPRSS2 and ETS transcription factor genes including ERG, ETV1, and ETV4 (Rubin and Chinnaiyan, 2006) and later on AGTR1 (Rhodes et al., 2009), a gene selectively overexpressed in 10 to 20% of ER positive breast cancers and functionally promoting cancer cell invasion *in vitro*.

Analyzing and interpreting genomic data at the level of individual genes will likely fall short of expectations. This is perhaps best exemplified by the reactions to the publications of large-scale cancer genome studies of pancreatic cancer and glioblastoma (Check Hayden, 2008; Nature Editorial, 2008). The complexity of genetic changes from either individual patient or across tumor types revealed by these studies suggests that no single driver gene for these tumors exists. Instead, it highlights the importance of understanding complex genetic aberrations in the context of cancer pathways, and the richness of the data offers the opportunity to do just that. Almost all large-scale cancer genomics studies include a component of functional gene group or pathway enrichment analysis. The basic concept is to identify statistically significant enrichment of certain gene groups or pathways among commonly altered genes. The assumption is that if a pathway or gene network is frequently altered

---

[3] At the time of writing of this chapter, whole-genome sequencing and RNA sequencing have not been adopted broadly yet. Gene expression profiling was still the largest data source, therefore the need of inferring genetic rearrangement from gene expression profiling data. Since then, whole-genome sequencing and RNA sequencing have made it possible to directly map genetic rearrangements from tumor DNAs or RNAs.

in a certain tumor type, it is likely functionally relevant. The typical enrichment analysis tends to treat functional groups or pathways as a collection of genes not utilizing the information of protein–protein interactions. The network-based approach goes one level up to identify enriched networks where not only the number of genes but also the known gene connections were assessed to identify highly connected network modules (Cerami et al., 2010). This approach has the potential to identify deregulated gene network modules beyond known functional groups or canonical pathways, or draw connections among multiple related pathways, which may not be obvious at first sight. Such a network-based approach still relies on known information of protein interactions. It, however, does not have the capability to infer new relationships among genes; therefore does not take full advantage of multidimensional data within the same tumor sample set. A new proposal has brought forward a Bayesian probabilistic approach to build statistical models of gene networks as a new way to mine the TCGA ovarian cancer dataset (Khalil et al., 2010). The hope is that such network models would have the potential to both learn the "key drivers" of diseases and provide simulation tools for network intervention (Schadt, 2009). Although computer-aided simulation is routinely used in many other scientific disciplines or engineering fields, its benefit to basic cancer biology and cancer drug discovery is still yet to be convincingly demonstrated.

Cancer genomics' impact on disease understanding

Cancer as a disease is intrinsically heterogeneous as demonstrated by varied prognosis and response to therapies. Analysis of cancer genomics data not only can readily identify a known cancer type or subtype, but more importantly, can resolve these into novel molecular subtypes.

Unsupervised methods are a common approach used to define disease subtypes reflecting intrinsic diversity of underlying cancer biology. The pioneering work of Perou et al. (2000) that redefined breast cancer subtypes based on the transcriptome demonstrated the feasibility of characterizing disease diversity using genomics. A series of follow-up studies since then have reached a consensus definition of "intrinsic" molecular subtypes of breast cancer: luminal A, luminal B, Her2-enriched, basal-like, and normal-like. Compared to many other molecular classification studies, the wider acceptance of this particular subtype definition is largely due to two factors. First, similar analysis was carried out using independent datasets representing different patient cohorts from different laboratories generated by different technological platforms, and the same subtype classification was uncovered (Hu et al., 2006; Sørlie et al., 2001). Second, and the most important factor, is its biological and clinical relevance. Even though the approach was "unsupervised," the discovered classification scheme reflects the best known molecular markers of breast cancer, i.e., estrogen receptor (ER), progesterone receptor (PR), and HER2. Luminal subtypes contain mostly ER positive patients. The HER2-enriched subtype contains mostly HER2 amplification positive patients. Disease samples representing the basal-like subtype contain mostly triple negative (ER, PR, HER2 negative) patients. However, ER and HER2 status alone are not equivalent for "intrinsic subtype" status (Parker et al., 2009). In addition, these tumor subtypes are associated with significant difference in clinical outcome. "Her2 positive" and "basal" subtypes are associated with much shorter disease-free survival. This is perhaps not surprising and consistent with existing clinical knowledge. However, even within ER positive patients, two subclasses were identified. The "luminal A" subgroup has the best clinical outcome and is associated with the longest disease-free survival, while the "luminal B" subgroup is associated with shorter disease-free survival. Similar unsupervised approaches have been adopted to reveal molecular heterogeneity of several other tumor types. Table 32.1 lists the representative studies in different tumor types, where classification schemes were derived from relatively large sample sets and the molecular subtypes were related back both to clinical outcomes and key molecular markers of diseases.

Until recently, disease molecular reclassification was defined almost exclusively by transcriptome profiling. In hematological tumors, it is recognized that the main subtypes defined by transcriptome profiling tend to reflect major known genetic aberrations (Alizadeh et al., 2000; Fine et al., 2004; Zhan et al., 2006). This association of key genetic abnormalities with gene expression profiles has proven difficult in solid tumor types due to the lack of data or understanding of major genetic aberrations in solid tumors (Futreal et al., 2004). Data linking molecular subtypes to driver genetic abnormalities were at best anecdotal (Lapointe et al., 2007; Sørlie et al., 2001). However, the

Table 32.1 A list of representative studies classified tumors into molecular subtypes based on transcription profiling of patient samples.

Disease	References	Molecular subtypes	Association with clinopathological features	Association with clinical outcomes	Association with molecular marker
Acute myeloid leukemia	Bullinger et al., 2004	Two groups: I and II	Group I: FAB M1 and M2 Group II: FAB M4 and M5	Group I: poor overall survival	Group I: FLT3
Pediatric acute lymphoblastic leukemia	Fine et al., 2004				Sample clusters were associated primarily with selected chromosomal abnormalities –TEL/AML1, BCR/ABL, or MLL
Breast cancer	Parker et al., 2009; Perou et al., 2000; Sørlie et al., 2001	Five subtypes: luminal A, luminal B, Her2-enriched, basal-like, normal-like	Basal-like: high grade	Luminal A: long overall survival and relapse-free survival Luminal B: medium relapse-free survival Basal-like, HER2-enriched: shortest overall survival and relapse-free survival	Luminal A, luminal B: estrogen receptor positive Her2-enriched: HER2 amplification positive Basal-like: TP53 mutant, BRCA1 mutant
Diffuse large B cell lymphoma	Alizadeh et al., 2000; Rosenwald et al., 2003; Staudt and Dave, 2005	Three subtypes: GCB, ABC and PMBL	PBML: molecular subtype is consistent with clinical subgroup of primary mediastinal B cell lymphoma	Five-year survival: 31% ABC, 59% GCB, 64% PMBL	GCB: BCL6, t(14;18) BCL2 PMBL: 9p24 amplification
Diffuse large B cell lymphoma	Monti et al., 2005	Three subtypes: OxPhos, BCR, HR	HR: younger age, higher incidence of splenic and bone marrow involvement	Similar five-year survival	OxPhos: t(14;18) BCL2 BCR: BCL6
Glioblastoma	Verhaak et al., 2010	Four subtypes: proneural, neural, classical, and mesenchymal	Proneural: younger age	Proneural: longer survival Classical and mesenchymal: better response to aggressive treatment measured by reduced mortality rate	Proneural: TP53 mutation, IDH1 mutation, PDGFRA amplification Classical: EGFR mutation Mesenchymal: NF1 mutation
Lung adenocarcinoma	Bhattacharjee et al., 2001	Four subtypes: C1 to C4	C1: poorly differentiated tumor C2, C3: well-differentiated tumors C4: bronchioloalveolar carcinoma, lower smoking history	C2: least favorable overall survival C4: most favorable overall survival	

Table 32.1 (*cont.*)

Disease	References	Molecular subtypes	Association with clinopathological features	Association with clinical outcomes	Association with molecular marker
Multiple myeloma	Zhan et al., 2006	Seven subtypes: PR, LB, MS, HY, CD-1, CD-2, MF	PR: high serum B2M and LDH, low albumin LB: low incidence of focal bone lesion	HY, CD-1, CD-2, LB: longer event-free and overall survival	MS: t(4;14)(p16;q32) FGFR3/MMSET CD-1 and CD-2: t (11;14)(q13;q32) CCND1 and t(6;14) (p21;q32) CCND3 MF: t(14;16)(q32;q23) cMAF and t(14;20) (q32;q11) MAFB
Ovarian cancer	Tothill et al., 2008	Six subtypes: C1 to C6	C1, C2, C4, and C5: high-grade C3 and C6: serous limited malignant potential, or low-grade, early-stage endometrioid tumors	C1: poor survival C3 and C6: long progression-free survival	
Prostate cancer	Lapointe et al., 2004, 2007	Three subtypes: I, II, and III	Subgroups II and III: high tumor grade, advanced stage, and lymph node metastasis	Subgroups II and III: early recurrence	Subgroup II: TMPRSS2-ERG fusion Subgroup III: PTEN loss, Myc gain

integrated approach adopted by TCGA, i.e., profiling the same sample set for chromosomal aberrations (gene sequencing, DNA copy number profiling), transcriptional profile (mRNA and microRNA), and epigenetic modifications, is going to change this as demonstrated by the publication identifying four clinical subtypes of glioblastoma (Verhaak et al., 2010). The four clinical subtypes, proneural, neural, classical, and mesenchymal, were defined by analysis of gene expression data from 200 glioblastoma and two normal brain samples. Significant amplifications and deletions were identified by the GISTIC (Beroukhim et al., 2007) method using copy number data of a subset of 170 core samples. Each disease subtype was further examined for association with significant amplification/deletion and gene mutations (see Table 32.1).

Classical clinical and tumor characteristics have provided valuable information for managing patient care including informing prognosis and guiding treatment choices. The molecular classification of diseases provides further insights into deregulated genes and pathways. Such information is valuable to inform choices of therapeutic targets, patient selection to guide clinical trials, and ultimately matching targeted therapy with patients who are most likely to benefit in clinical practice.

Impact of cancer pharmacogenomics in drug discovery and developments
Personalized medicine

Personalized medicine is essential to the future success of cancer drug discovery and development, from the discovery of more effective targeted therapies to securing regulatory approval and reimbursement by health-care providers. The challenges, successes, and lessons learned through the discovery and development of receptor tyrosine kinase inhibitors (RTKis), in particular therapeutic agents targeting epidermal growth factor receptor (EGFR), perhaps best illustrate this point.

Gefitinib (Iressa®), an EGFR inhibitor, first won FDA conditional approval in May 2003 for the treatment of patients with locally advanced or metastatic non-small cell lung cancer (NSCLC) after failure of both platinum-based and docetaxel chemotherapies. The conditional approval was based on two randomized phase II trials of gefitinib monotherapy at 250p mg/day or 500 mg/day (IDEAL 1&2) where gefitinib demonstrated anti-tumor activity in about 10% of unselected patient populations (Fukuoka et al., 2003; Kris et al., 2003). The setback came in December 2004, when gefitinib failed to demonstrate survival benefit in an unselected population of NSCLC patients in a placebo-controlled phase III study (Iressa Survival Evaluation in Lung Cancer, ISEL; Thatcher et al., 2005). However, pronounced heterogeneity in both clinical responses and survival outcomes was observed among patient groups, with evidence of benefits among patients of Asian ethnicity, female, never or former light smoker, and adenocarcinoma histology. This population of patients also have a relatively high incidence of specific EGFR mutations that correlated with tumor response to gefitinib (Lynch et al., 2004). On the basis of these observations from multiple studies, a large-scale randomized phase III study in a selected patient population of non-smokers or former light smokers in East Asia who have lung adenocarcinoma was launched to compare the efficacy, safety, and toxicity of gefitinib vs. carboplatin-paclitaxel as first-line treatments (Iressa Pan-Asia Study, IPASS) (Mok et al., 2009). The study met its primary objective of demonstrating the non-inferiority of gefitinib and also showed its superiority vs. carboplatin-paclitaxel with a 12-month progression-free survival rate of 24.9% and 6.7%, respectively. It also convincingly demonstrated that EGFR mutation status is an important predictive marker for EGFR inhibitors. The IPASS trial result is significant with multiple implications. First, it represents a paradigm change of first-line lung cancer treatment in Asia and for patients with specific EGFR mutations. Second, it demonstrated that personalized medicine for NSCLC is feasible. Third, it was one of the pivotal trials together with the INTEREST trial (Kim et al., 2008) to support gefitinib's license for use in the EU for adults with locally advanced or metastatic NSCLC with activating mutations of EGFR across all lines of therapy in July 2009.

In advanced breast cancer, trastuzumab (Herceptin®), a monoclonal humanized antibody targeting another member of the EGFR family, HER-2, is widely used in both the adjuvant and metastatic settings for HER-2 positive patients. The patient selection strategy originated from the observation that HER-2 negative tumors by IHC rarely responded to the antibody, diluting the response rate seen in HER-2 positive patients. Trastuzumab induced tumor regression in 12 to 34% of HER2 positive patients (Demonty et al., 2007) as compared to less than 10% in an unselected patient population.

Gefitinib and trastuzumab exemplify the significance of selecting patients who are most likely to benefit from a given therapy as personalized medicine strategy.

A different but equally important patient selection strategy would be to identify and exclude patients who are either unlikely to benefit or potentially harmed by a given therapy. It is noticed in the IPASS trial that EGFR activating mutation negative patients had worse progression-free survival with treatment of gefitinib as compared to patients treated with carboplatin-paclitaxel. In colorectal cancer, patients with Kras mutation rarely respond to treatment with panitumumab (Vectibix®) and cetuximab (Erbitux®), two antibody therapies targeting EGFR (Freeman et al., 2008; Khambata-Ford et al., 2007). Similar associations of Kras mutation with non-responders of anti-EGFR treatments of gefitinib and erlotinib in NSCLC were also reported (Pao et al., 2005). In July 2009, the FDA granted class labelling changes to cetuximab and panitumumab not recommending these antibodies for the treatment of colorectal cancer with Kras mutations (FDA, 2009). One important lesson of this class labelling change is the rapid adoption of personalized medicine by oncology health-care providers and the influence of their change of practice on both regulatory agencies, such as the US FDA and pharmaceutical companies, in terms of how clinical trials can be conducted and what information is sufficient to warrant FDA approval of labelling related to patient selection. In the example of anti-EGFR antibodies, the clinical adoption of Kras mutation as a biomarker to exclude patients unlikely to benefit from cetuximab or panitumumab treatment by practicing oncologists made it both potentially unethical and unpractical to run prospective clinical trials to validate Kras mutation as a patient selection marker (Jimeno et al., 2009; Lancet Editorial, 2009). Instead, the FDA advisory board recommended that retrospective analysis of biomarkers from completed trials to support labelling claims needs to be hypothesis driven, well controlled, adequately powered, and with pre-specified tissue

Table 32.2 Timeline of EML4-ALK discovery and PF-02341066 clinical development in non-small cell lung cancer (NSCLC) patients.

Time	Event
August 2007	EML4-ALK activating fusion in patients of NSCLC identified and reported (Soda et al., 2007).
December 2007	Phase I (NCT00585195 clinicaltrials.gov) study of PF-02341066 on an expanded cohort including EML4-ALK positive NSCLC patients initiated. PF-02341066 was developed as a c-met inhibitor with known inhibition against ALK.
May 2009	Positive clinical activity from a phase I study in NSCLC patients carrying EML4-ALK activating rearrangements reported at ASCO 2009 (Kwak, 2009): 1/10 partial response, 2/10 unconfirmed partial response, 4/10 stable disease.
September 2009	Updated study results of a phase I study in NSCLC patients carrying EML4-ALK activating rearrangements reported at ECCO15/ESMO34: overall response rate 65% (20/31), disease control rate 84% (26/31).
September 2009	Phase III clinical trial (NCT00932893 clinicaltrials.gov) of PF-02341066 vs. Standard of Care in EML4-ALK positive patients of advanced NSCLC initiated.
August 2011	Crizotinib was approved by the US FDA to treat late-stage (locally advanced or metastatic) NSCLCs that express the abnormal ALK gene. The approval required a companion molecular diagnostic test for the EML4-ALK fusion.

collection and data analysis (Mack, 2009). This recommendation's impact on future clinical trials is clear: that future clinical trials should consider large and organized tissue collection to enable retrospective biomarker analysis.

Other personalized medicine initiatives launched by leading cancer medical centers, such as the genotyping effort at Massachusetts General Hospital Cancer Center (2009) and the Lung Cancer Mutation Consortium Protocol (2009) project, mean, on the one hand, that it is important to have a personalized medicine strategy for a new drug to be partner of choice for clinical trials in these medical centers. On the other hand, it also means that conducting clinical trials with a patient selection strategy is much more feasible, i.e., instead of prospectively screening for patients suitable for the trial, patients can be easily identified through existing databases with information of patients' genetic and genomic profiles in the future.

The recent rapid advancement of PF-02341066 (crizotinib or Xalkori®), an ALK inhibitor developed by Pfizer, through clinical trials provides another proof-of-concept example of what personalized medicine means for oncology drug development, one being the paradigm shift of clinical trial design from testing a large number of patients to detect small benefit toward testing a smaller number of well-characterized patients to detect bigger effect. The most significant aspect of PF-02341066 clinical development is the fast translation of basic research to the clinic as illustrated in Table 32.2. Another potential

impact is on the debate of first-in-class vs. best-in-class medication. The small population of patients with certain molecular markers, e.g., gene fusion like EML4-ALK in NSCLC (Shaw et al., 2009; Soda et al., 2007), and their remarkable response to the right drug, in this case PF-02341066 (Kwak et al., 2009), would argue that it is important to be first-in-class to access this small population of patients. The major differentiation factor of later candidates coming into clinical trials is likely to be their effectiveness at treating patients who are resistant to the first drug either *de novo* or acquired.

Cancer cell line panels as important discovery platform for personalized medicine strategy development

The realization of personalized medicine is dependent on the integration of cancer pharmacogenomics into drug discovery and development. A small percentage response rate of any targeted therapy in an unselected patient population is likely the norm rather than the exception. Cancer genomic studies will continue to reveal the molecular diversity at both disease and individual-patient level, essential to the delivering of personalized treatment for patients and the identification of promising new therapeutic targets. Cancer pharmacogenomic studies using cell line-based platforms have emerged as important preclinical tools for the development of personalized medicine strategy.

A major hurdle for cancer drug discovery and development is the lack of translation of preclinical model efficacy into clinical patient response. The poor predictability of preclinical models has led to a general sentiment that preclinical cancer models (mostly cell lines and cell line derived animal models) are not disease relevant. There are likely multiple factors contributing to the poor performance of preclinical models. For example, preclinical efficacy is often measured as growth inhibition or delay compared to control. This is not consistent with the definition of clinical response, which is typically defined as percentage tumor shrinkage (partial response) or disappearance of all signs of cancer (complete response). Another important factor that perhaps has not been given sufficient attention in the past is the molecular makeup of preclinical models. Given the heterogeneity of any tumor type, each preclinical model, typically derived from a single patient, is unlikely to represent a cancer type in general. Model availability, experimental feasibility, ease of manipulation, and the tendency to follow published precedence have all contributed to the overreliance on relatively few cancer cell line models, as shown in Figure 32.2. Not surprisingly,

we often fail to predict clinical efficacy in an unselected large patient cohort by generalizing preclinical observations based on a few preclinical models.

The molecular redefinition of disease coupled with molecular profiling of preclinical models has started to provide more accurate classification and identification of cancer cell line collections that better represent disease subtypes. Instead of generally asking the question of whether preclinical models are disease relevant or not, people have started to ask the question of what disease subtype or patient segment a particular disease model is most relevant to. Virtanen et al. identified systematic differences between lung tumor samples and cell lines with proliferation-associated genes more highly expressed in cell lines (Virtanen et al., 2002). Once the cell line-specific gene expression features were removed, multiple studies have demonstrated that cancer cell lines share common molecular features with either the histopathology subtypes (Virtanen et al., 2002) or genomic aberrations (Fine et al., 2004; Neve et al., 2006; Rücker

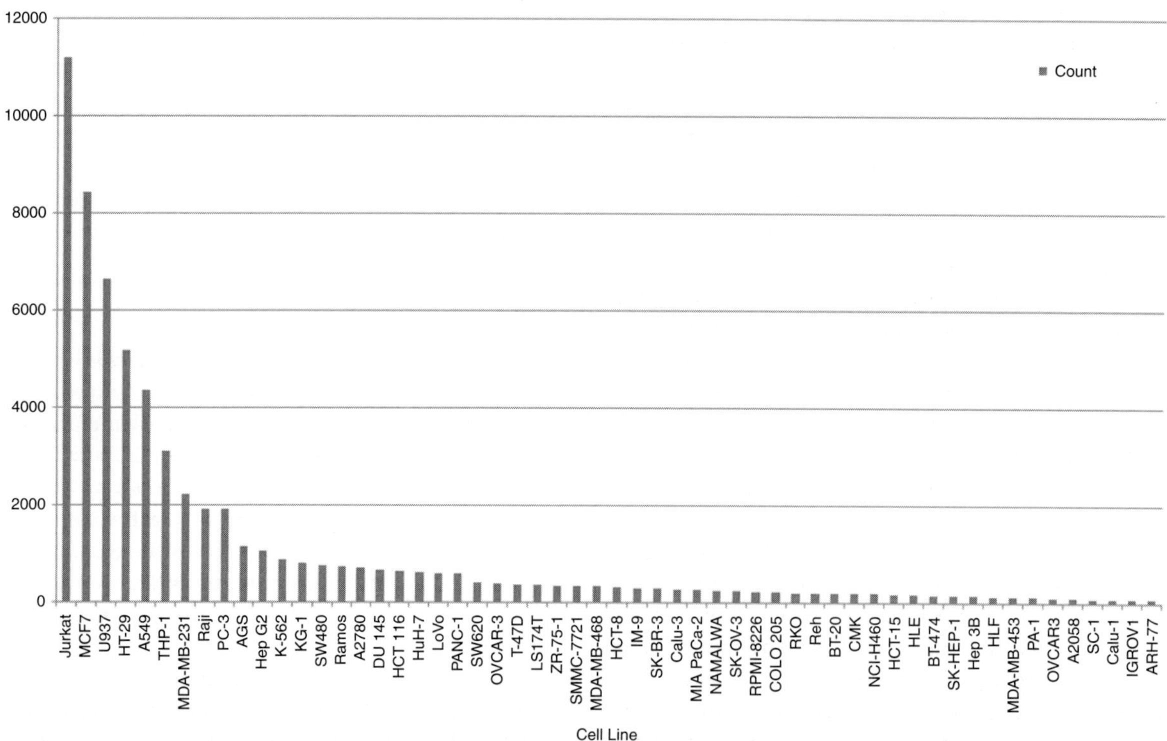

Figure 32.2 Number of published papers citing cancer cell lines. Total of 56 commonly used and public accessible cancer cell lines covering diverse tumor types were included in the search. The numbers in this figure were obtained by using search terms "cell line" and cell line name as it is. (A black and white version of this figure will appear in some formats. For the color version, please refer to the plate section.)

et al., 2006). Andersson et al. provided a powerful example of this approach in which they demonstrated the similarity of an alleged acute monoblastic leukemia (AML) cell line CTV-1 to T cell acute lymphoblastic leukemia (T-ALL) (Andersson et al., 2005). The observation was later followed up and confirmed by FISH analysis of t(1;7) rearrangement and immune profile, which resulted in its reclassification as an T-ALL cell line by the German Collection of Microorganisms and Cell Cultures (DSMZ, n.d.). If a particular drug shows exclusive sensitivity in preclinical models representing a particular disease segment, it is predicted to be likely efficacious in that patient segment (Finn et al., 2007; Kuo et al., 2009; Mirzoeva et al., 2009).

Not all cancer types have been stratified into molecular subclasses or have a generally agreed classification scheme. Even in cancer types such as breast cancer, where molecular classification is widely accepted and frequently applied, there is still heterogeneity within each molecular subtype. For example, BRCA1 mutations are almost exclusively associated with basal-like subtype of breast cancer (Foulkes et al., 2003), but only a fraction of basal subtype patients harbor BRCA1 mutations (Foulkes et al., 2003; Sørlie et al., 2001; Young et al., 2009). In addition, there are not always large numbers of cancer cell lines representing disease diversity in a particular cancer type (Lacroix and Leclercq, 2004; Neve et al., 2006). Therefore it is not always possible to make statistical association of preclinical model response with disease segment by profiling large numbers of cell lines from a single disease type. The emerging picture from recent cancer genomics studies across multiple diseases (Cancer Genome Atlas Research Network, 2008; Ding et al., 2008; Parsons et al., 2008; Jones et al., 2008; Wood et al., 2007) suggest there are common cancer modules/pathways deregulated across all cancer types. The detail of genomic aberrations are patient and disease dependent, but they all tend to converge on a relatively small number of cancer pathways or biological processes such as receptor tyrosine kinases/Ras/PI3K, P53 signaling pathways, and "apoptosis." In a pooled analysis of 3,131 cancer specimens covering 26 histological types of cancer, common somatic copy number alterations (SCNAs) covering functionally validated oncogenes and tumor suppressors are found across tumor types (Beroukhim et al., 2010). In fact, it is observed that the most focal SCNAs identified in any one of 17 cancer types

are also found in the pooled analysis excluding that cancer type. These observations suggest the possibility of identifying drug response/resistance biomarkers by associating drug response/resistance with common molecular markers or pathways deregulated in large cell line panels of either mixed tumor types or a single tumor type.

The National Cancer Institute's (NCI) NCI60 human tumor cell line panel was developed in the late 1980s as an *in vitro* anti-cancer drug screening tool (Shoemaker, 2006). The NCI60 cell line panel provides valuable information of anti-cancer activities of compounds with its broad tumor coverage including leukemia (5), non-small cell lung cancer (8), colon cancer (7), CNS cancer (6), melanoma (8), ovarian cancer (6), renal cancer (8), prostate cancer (2), and breast cancer (7). It was primarily used to screen compounds for indication of anti-tumor activity as measured by growth inhibition or cell kill. With the rapid growth of pharmacology data, an algorithm (COMPARE) was developed and made available as a tool to infer a potential mechanism of action by comparing one compound's pharmacology profile against the large collection of compound pharmacology profiles with known mechanism of action. Due to its broad accessibility, extensive pharmacology characterization, and open data access, NCI60 is frequently used by the cancer community as a testing platform for new molecular profiling technologies and analytical methods. Many of the pharmacogenomic approaches/concepts were first developed and tested using NCI60 data. As a proof-of-concept example for drug response biomarker identification, asparagine synthetase was first identified as a predictive biomarker for L-asparaginase activity in ovarian cancer cells by observing an inverse correlation between asparagine synthetase gene expression levels and sensitivity to L-asparaginase (Bussey et al., 2006; Lorenzi et al., 2006, 2008). A set of gene expression signatures predictive of cell line sensitivity to chemotherapeutic drugs was developed using drug sensitivity data and Affymetrix microarray data of NCI60 cell lines (Potti et al., 2006).

The mixed tumor type coverage of NCI60 limits the number of cell lines representing any particular disease. This may not have been a limiting factor for identifying response/resistance biomarkers to chemotherapeutics as their mechanisms are conserved across cancer types. However, this is proving to be a significant challenge when attempting to define disease and patient subsets most or least likely to respond to molecularly targeted agents., The small number of cell lines representing each tumor type within the NCI60 panel is not sufficient to detect statistically significant correlation of response to molecular markers. One solution is to develop disease-specific cell line panels including a large number of tumor cell lines from a single disease that is richly annotated and characterized at the molecular level, such as for bladder (Smith et al., 2010), breast (Finn et al., 2007), lung (Choi et al., 2007; Sos et al., 2009), or colorectal (Liu and Bodmer, 2006) cell line panels. Another complementary approach is to develop an even larger panel of cell lines covering both more tumor types and more cell lines in each tumor type if possible (Greshock, 2010; OncoPanel™ 240; Sharma et al., 2010). Both approaches coupled with integrated pharmacogenomic data mining have generated patient selection biomarkers (Dry et al., 2010; Finn et al., 2007; Huang et al., 2007; Kuo et al., 2009; McDermott et al., 2007; Mirzoeva et al., 2009; Pratilas et al., 2008; Sos et al., 2009), and some are being tested in clinical trials, such as dasatinib (Sprycel®) in triple-negative breast cancer. For a review of dasatinib's preclinical activity in basal subtype or triple-negative breast cancer and its clinical development in triple-negative breast cancer, the reader is referred to Araujo and Logothetis (2010).

These well-characterized cancer cell line panels are now being used across the drug discovery process from target identification/validation (Scholl et al., 2009), mechanistic understanding of compound pharmacology profile (Shoemaker, 2006), to response biomarker identification (Dry et al., 2010; Finn et al., 2007; Kuo et al., 2009), and even potential rational combination strategy identification (Chresta et al., 2008; Mirzoeva et al., 2009). The broad profiling of candidate drugs against diverse diseases of preclinical models also allows additional disease indications to be identified.

The recent publication of two gene amplifications LAPTM4B and YWHAZ in chromosome 8q22 as contributing factors to chemoresistance and disease recurrence in breast cancer provided a nice example demonstrating the power of the integrated approach of genomics, functional-genomics, and pharmacogenomics (Li et al., 2010). In this example, 75 genes were identified as differentially expressed between distance metastasis recurrent patients and patients without recurrence in one dataset. An enrichment

analysis leads to the identification of chromosome 8q22 as overpopulated with 12 out of the 75 genes. This observation suggests the possibility of 8q22 amplification, which was confirmed by both SNP array and fluorescent *in situ* hybridization (FISH) analysis. The result was then further validated by a meta-analysis of six independent breast cancer gene expression datasets. Breast cancer cell line BT549 carrying 8q22 amplification was identified as a preclinical model for functional genomics siRNA screening experiment. Depletion of two genes LAPTM4B and YWHAZ sensitize BT549 to anthracyclines. Further examination of a panel of 16 breast cancer cell lines revealed the correlation of LAPTM4B and YWHAZ mRNA expression levels and low sensitivity to anthracyclines. These data taken together with several other lines of evidence provided in the study support 8q22 amplification, in particular two genes LAPTM4B and YWHAZ, as a contributing factor to chemotherapy resistance in breast cancer.

Cancer biomarker identification is increasingly focused on genetic aberrations, such as somatic mutations, gene fusions, and translocations, or DNA copy number alterations. The publication of the cancer gene census (Futreal et al., 2004) and launch of the Catalogue of Somatic Mutations in Cancer (Forbes et al., 2010) catalyzed this paradigm shift by re-emphasizing the causal roles many known somatic mutations play in cancer and providing access to high-quality somatic mutation data. Pharmacogenomics studies have shown that cell lines extremely sensitive to targeted inhibitors normally carry genetic aberrations of target genes, which is remarkably consistent with clinical findings (Sharma et al., 2010), supporting cell line-based platforms as important tools for personalized medicine strategy development.

One of the exciting new developments in the area of cancer pharmacogenomics is the launch of the Cancer Translation Project, a five-year collaboration between the Cancer Genome Project (CGP) at the Wellcome Trust Sanger Institute and the Center for Molecular Therapeutics Massachusetts General Hospital (MGH) Cancer Center. Even though the details are not fully disclosed to the public yet, it will most likely combine the expertise and experience of MGH in profiling compounds in large cancer cell line panels and CGP's expertise in molecular characterization of cancer cell lines, in particular sequence analysis of gene mutations, and their combined expertise in correlating cancer cell line genomics and pharmacology profiles of compounds to identify drug response biomarkers. It is possible, that this platform and data generated from this platform will become the new test ground for emerging new technologies and analytical methods, leading to the identification and clinical validation/adoption of drug response and patient selection biomarkers.

Discussion

Within the short period of a few years, cancer pharmacogenomics are making an impact from the early phase of cancer drug discovery to the clinical practice of personalized health care. Looking ahead, there are both opportunities and challenges.

Technology will continue to evolve at a fast pace. For example, the growing power and reduced cost of next-generation sequencing potentially at a single-cell level not only will challenge us with an unprecedented amount of data, but also will provide us with opportunities to have a much more detailed view of cancer.

Another area of both challenge and opportunity is advancement in cancer biology. This could mean an area of biology that is relatively new to cancer or areas where relevance to cancer is newly appreciated, such as cancer stem cell, tumor micro-environment, or cell metabolism. New biology areas will likely require and lead to new types of high-throughput data generation. Their integration into the existing multidimensional data of cancer genomics will again provide us with a more holistic view of cancer.

One clear challenge is how to make sense of the large and increasing amount of data, i.e., how to translate data into knowledge. There are several levels to this problem.

At the most basic level is how to manage and process the data. Although a clear challenge, it is probably the most manageable one since it has been the core function of the genomics and informatics community. It is worth pointing out, however, that biological data management including biomaterials, study-related information, and sometimes molecular measurements is an underserved area, perhaps due to historical reasons, such as relatively small data volume and in many cases, the qualitative or semi-quantitative nature of the data. In sharp contrast is how we manage compound-related data, where all chemical materials are registered, uniquely identified by their structure, and tracked, and almost all the test results are managed in some database system.

Now biological measurements are increasingly high-throughput and quantitative, it demands a culture change toward biological data management, i.e., biomaterials should also be uniquely identified, tracked, and specifically linked to any data generated using that material. In the area of cancer pharmacogenomics data management, one difficulty we encounter is related to cancer cell line naming and identification. The multi-plural spelling of cell line names makes linking data to each cell line through disparate sources a constant struggle for both scientists and computers. This is similar to the challenge the biology community faced in the mid to late 1990s when dealing with a large amount of genomics data. The lack of unique identifiers of genes, the ambiguity of gene names, and the lack of standardization of gene function description were viewed as major obstacles to unifying gene-related information at a genomic scale using computational approaches. It took a major effort, the Gene Ontology Consortium, to rectify the problem (Ashburner et al., 2000). It is perhaps time to call for similar public effort to standardize cell line names. Another even more serious problem related to cancer cell lines is the cross-contamination and mis-identification of cancer cell lines as discussed in these references (Nardone, 2007; Potash and Anderson, 2009). This issue highlights the importance of uniquely identifying and tracking biological materials such as cancer cell lines through effective data management.

Managing data simply gives scientists means to retrieve data. The disparate nature of data and information related to cancer genomics and pharmacogenomics still presents a major hurdle and requires significant initial effort for non-informaticians and informaticians alike to retrieve all relevant data and information from multiple sources for analysis. Therefore the next level of challenge and priority for turning data into knowledge is data integration. Compared to data storage and access, this is an area perhaps even less emphasized. Taking the GEO (Barrett and Edgar, 2006), for example, after years of creation it still remains largely a genomics data repository with data being deposited as individual datasets. Another example of cancer pharmacogenomics would be the cancer cell line-based platforms as discussed previously. Although cancer cell line-based platforms as a major tool for cancer pharmacogenomics have been adopted broadly across industry and academic institutions, the cell line-related data from genomics profiling data of cell lines to pharmacology data generated by testing hundreds to thousands of compounds in those cell line panels have largely being managed separately. Without proper data integration, data analysis is both a time-consuming endeavor, and a specialty limited to a small group of bioinformaticians or statisticians.

Toward this goal, AstraZeneca Cancer Drug Discovery has developed a data integration, analysis, and visualization system integrating cell line panel-related information including cell line, gene, compound annotations, cell line molecular profiling data from gene expression, CGH, to mutations, and pharmacology data of all compounds tested in these cell line panels. Figure 32.3 shows a high-level view of data integration, which captures the *biological relationships* of different data types. As simple as it may seem, it has guided us over the last few years to evolve the system as more data or data types become available. Most importantly, it allows easy query and visualization of molecular data alongside of drug pharmacology profile in the same panel of cancer cell lines for any scientist accessing this system, which affords them the time to concentrate on data interpretation (e.g., does this profile look like what I expected based on molecular target or disease indication?) or hypothesis generation (e.g., what

molecular marker seems to be associated with my compound's sensitivity or resistance?). Today, almost all cancer drug discovery projects within AstraZeneca use this system. An application example is provided by Chresta et al. (2008).

The real impact of data integration is to significantly lower the hurdle of data access; therefore to increase the ratio of knowledge obtained vs. the effort involved as illustrated in Figure 32.4a, b. The one untapped area to unleash the power of cancer pharmacogenomics is information intelligence (Community cleverness required, 2008), i.e., how we can move toward a system/platform where not only data are seamlessly merged and integrated, but community knowledge is also captured and integrated with digital data, and the system has the capability to co-evolve with the community as we build more knowledge around cancer biology. The broader acceptance and more usage such a system has, the more powerful and intelligent it becomes. The best analogy here is Google Flu Trends (www.google.org/flutrends), which demonstrated the power of this concept (Ginsberg et al., 2009). Instead of the typical approach of providing a database and requiring health-care professionals to enter the data for flu tracking, Google captures the user search terms and mines the most relevant terms as an indicator of flu activity. With the amount and the complexity of multidimensional data we are faced with, a system relying solely on the data underneath could become increasingly complex to render it hard to use and even harder to interpret. It is conceivable that a system that captures both the relationship of actual data and the information of how users are using them, and use that user information to draw relationships of different data and connect different users, would greatly increase its utility by leading scientists to insights or opportunities they may not set out to ask. The hope is to increase the ratio of knowledge obtained vs. effort involved even higher (Figure 32.4c).

Parallel to the need of developing more intelligent systems for knowledge transformation is the development of better analytical approaches at mining cancer pharmacogenomics data. As illustrated by multiple examples cited in this review, different approaches were taken at linking compounds' pharmacology profiles with molecular markers to derive potential sensitivity or resistance markers, or proposing novel combinations, as well as linking drug response with disease subtypes. This is an evolving field as we are just starting to pull all the data together to enable

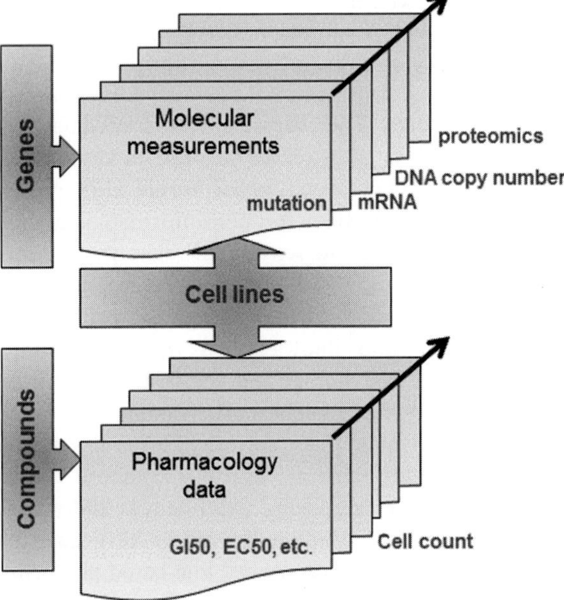

Figure 32.3 Schematic view of a cancer pharmacogenomics data integration system developed by AstraZeneca. (A black and white version of this figure will appear in some formats. For the color version, please refer to the plate section.)

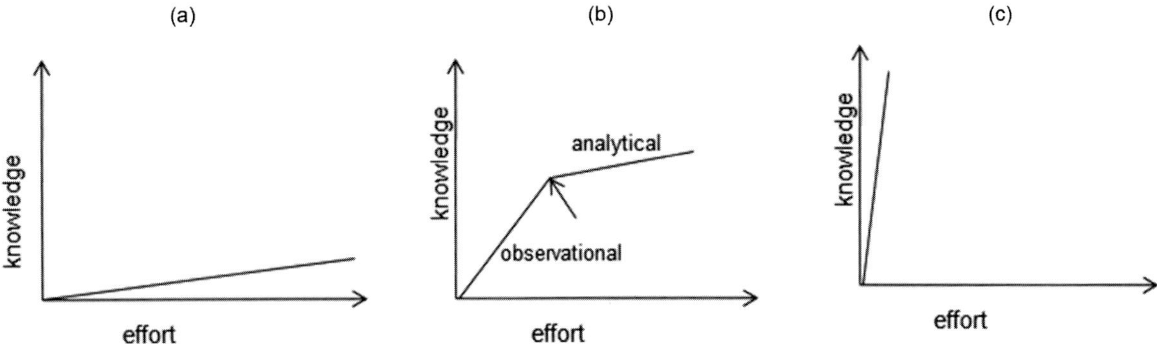

Figure 32.4 A progressive view of how data management, integration, information intelligence, and data mining will impact the translation of data into knowledge for cancer pharmacogenomics. (a) Without a data integration system, a large amount of initial effort is needed to retrieve multiple dimensional data from disparate sources. The ratio of knowledge generated vs. effort involved is low. (b) With a proper data integration system, effort needed to retrieve data is greatly reduced. It is possible to generate knowledge from data by observations from all. However, to reach to the next level insight of complex data, significant analytical exercise carried out mostly by informaticians or statisticians is still required. (c) An information system combining data integration and user community intelligence is likely to increase the knowledge vs. effort ratio even higher. Under such a system, most effort is spent on acquiring new knowledge.

more holistic analysis of all measurements. A better analytical approach is necessary to help us move from observational to analytical to derive the next level value from all data (Figure 32.4b), and is likely to fuel the engine of the next-generation information intelligence system together with community intelligence (Figure 32.4c).

In summary, cancer pharmacogenomics powered by the advancement in cancer genomics, molecular disease redefinition, large-scale cell line-based drug sensitivity screening, and data integration and analysis has demonstrated its broader impact across cancer drug discovery, development, and patient care. This area will continue to evolve quickly propelled by the scientific and technological advancements and pulled by the urgent need of better care. To realize its full potential, i.e., more efficacious drugs in the right patients, will no doubt require "community cleverness."

Acknowledgment

The authors thank David Brandon for his assistance in generating Figure 32.2.

References

Alizadeh, A., Eisen, M., Davis, R., et al. (2000). Distinct types of diffuse large B-cell lymphoma identified by gene expression profiling. *Nature, 403* (6769), 503–11.

Andersson, A., Edén, P., Lindgren, D., et al. (2005). Gene expression profiling of leukemic cell lines reveals conserved molecular signatures among subtypes with specific genetic aberrations. *Leukemia, 19* (6), 1042–50.

Araujo, J., and Logothetis, C. (2010). Dasatinib: a potent SRC inhibitor in clinical development for the treatment of solid tumors. *Cancer Treat Rev, 36* (6), 492–500.

Ashburner, M., Ball, C., Blake, J., et al. (2000). Gene ontology: tool for the unification of biology. The Gene Ontology Consortium. *Nat Genet, 25* (1), 25–9.

Barbarotto, E., Schmittgen, T., and Calin, G. (2008). MicroRNAs and cancer: profile, profile, profile. *Int J Cancer, 122* (5), 969–77.

Barrett, T., and Edgar, R. (2006). Gene expression omnibus: microarray data storage, submission, retrieval, and analysis. *Methods Enzymol, 411,* 352–69.

Beroukhim, R., Getz, G., Nghiemphu, L., et al. (2007). Assessing the significance of chromosomal aberrations in cancer: methodology and application to glioma. *Proc Natl Acad Sci USA, 104* (50), 20007–12.

Beroukhim, R., Mermel, C., Porter, D., et al. (2010). The landscape of somatic copy-number alteration across human cancers. *Nature, 463* (7283), 899–905.

Bhattacharjee, A., Richards, W., Staunton, J., et al. (2001). Classification of human lung carcinomas by mRNA expression profiling reveals distinct adenocarcinoma subclasses. *Proc Natl Acad Sci USA, 98* (24), 13790–5.

Bullinger, L., Döhner, K., Bair, E., et al. (2004). Use of gene-expression profiling to identify prognostic subclasses in adult acute myeloid leukemia. *N Engl J Med*, 350 (16), 1605–16.

Bussey, K., Chin, K., Lababidi, S., et al. (2006). Integrating data on DNA copy number with gene expression levels and drug sensitivities in the NCI-60 cell line panel. *Mol Cancer Ther*, 5 (4), 853–67.

Cancer Genome Atlas Research Network. (2008). Comprehensive genomic characterization defines human glioblastoma genes and core pathways. *Nature*, 455 (7216), 1061–8.

Cerami, E., Demir, E., Schultz, N., Taylor, B., and Sander, C. (2010). Automated network analysis identifies core pathways in glioblastoma. *PLoS One*, 5 (2), e8918.

Check Hayden, E. (2008). Cancer complexity slows quest for cure. *Nature*, 455 (7210), 148.

Choi, K., Creighton, C., Stivers, D., Fujimoto, N., and Kurie, J. (2007). Transcriptional profiling of non-small cell lung cancer cells with activating EGFR somatic mutations. *PLoS One*, 2 (11), e1226.

Chresta, C., Butterworth, S., Cosulich, S., et al. (2008). Cellular differentiation of rapamycin, PI3 kinase and TOR kinase inhibitors. *99th AACR Annual Meeting*, (abstr 3564). San Diego, CA.

Community cleverness required. (2008). *Nature*, 455 (7209), 1.

Demonty, G., Bernard-Marty, C., Puglisi, F., Mancini, I., and Piccart, M. (2007). Progress and new standards of care in the management of HER-2 positive breast cancer. *Eur J Cancer*, 43 (3), 497–509.

Desmedt, C., Haibe-Kains, B., Wirapati, P., et al. (2008). Biological processes associated with breast cancer clinical outcome depend on the molecular subtypes. *Clin Cancer Res*, 14 (16), 5158–65.

Ding, L., Getz, G., Wheeler, D., et al. (2008). Somatic mutations affect key pathways in lung

adenocarcinoma. *Nature*, 455 (7216), 1069–75.

Dry, J., Pavey, S., Pratilas, C., et al. (2010, Mar). Transcriptional pathway signatures predict MEK addiction and response to selumetinib (AZD6244). *Cancer Res*, 70 (6), 2264–73.

DSMZ (n.d.). http:/www.dsmz.de

Fan, C., Oh, D., Wessels, L., et al. (2006). Concordance among gene-expression-based predictors for breast cancer. *N Engl J Med*, 355 (6), 560–9.

FDA (2009). Class labeling changes to anti-EGFR monoclonal antibodies, cetuximab (Erbitux) and panitumumab (Vectibix): KRAS mutations. Retrieved from www.fda.gov/aboutfda/centersoffices/officeofmedical productsandtobacco/cder/ucm172905.htm

Fine, B., Stanulla, M., Schrappe, M., et al. (2004). Gene expression patterns associated with recurrent chromosomal translocations in acute lymphoblastic leukemia. *Blood*, 103 (3), 1043–9.

Finn, R., Dering, J., Ginther, C., et al. (2007). Dasatinib, an orally active small molecule inhibitor of both the src and abl kinases, selectively inhibits growth of basal-type/"triple-negative" breast cancer cell lines growing in vitro. *Breast Cancer Res Treat*, 105 (3), 319–26.

Forbes, S., Tang, G., Bindal, N., et al. (2010). COSMIC (the Catalogue of Somatic Mutations in Cancer): a resource to investigate acquired mutations in human cancer. *Nucleic Acids Res*, 38 (Database issue), D652–7.

Foulkes, W., Stefansson, I., Chappuis, P., et al. (2003). Germline BRCA1 mutations and a basal epithelial phenotype in breast cancer. *J Natl Cancer Inst*, 95 (19), 1482–5.

Freeman, D., Juan, T., Reiner, M., et al. (2008). Association of K-ras mutational status and clinical outcomes in patients with metastatic colorectal cancer

receiving panitumumab alone. *Clin Colorectal Cancer*, 7 (3), 184–90.

Fukuoka, M., Yano, S., Giaccone, G., et al. (2003). Multi-institutional randomized phase II trial of gefitinib for previously treated patients with advanced non-small-cell lung cancer (The IDEAL 1 Trial) (corrected). *J Clin Oncol*, 21 (12), 2237–46.

Futreal, P., Coin, L., Marshall, M., et al. (2004). A census of human cancer genes. *Nat Rev Cancer*, 4 (3), 177–83.

Gargiulo, G., and Minucci, S. (2009). Epigenomic profiling of cancer cells. *Int J Biochem Cell Biol*, 41 (1), 127–35.

Ginsberg, J., Mohebbi, M., Patel, R., et al. (2009). Detecting influenza epidemics using search engine query data. *Nature*, 457 (7232), 1012–14.

Greshock, J., Bachman, K.E., Degenhardt, Y.Y., et al. (2010). Molecular target class is predictive of *in vitro* response profile. *Cancer Res*, 70 (9), 3677–86.

Haibe-Kains, B., Desmedt, C., Piette, F., et al. (2008). Comparison of prognostic gene expression signatures for breast cancer. *BMC Genomics*, 9 (394).

Hu, Z., Fan, C., Oh, D., et al. (2006). The molecular portraits of breast tumors are conserved across microarray platforms. *BMC Genomics*, 7 (96).

Huang, F., Reeves, K., Han, X., et al. (2007). Identification of candidate molecular markers predicting sensitivity in solid tumors to dasatinib: rationale for patient selection. *Cancer Res*, 67 (5), 2226–38.

Jimeno, A., Messersmith, W., Hirsch, F., Franklin, W., and Eckhardt, S. (2009). KRAS mutations and susceptibility to cetuximab and panitumumab in colorectal cancer. *Cancer J*, 15 (2), 110–13.

Jones, S., Zhang, X., Parsons, D., et al. (2008). Core signaling pathways in human pancreatic cancers revealed by global genomic analyses. *Science, 321* (5897), 1801–6.

Khalil, I., Brewer, M., Neyarapally, T., and Runowicz, C. (2010). The potential of biologic network models in understanding the etiopathogenesis of ovarian cancer. *Gynecol Oncol, 116* (2), 282–5.

Khambata-Ford, S., Garrett, C., Meropol, N., et al. (2007). Expression of epiregulin and amphiregulin and K-ras mutation status predict disease control in metastatic colorectal cancer patients treated with cetuximab. *J Clin Oncol, 25* (22), 3230–7.

Kim, E., Hirsh, V., Mok, T., et al. (2008). Gefitinib versus docetaxel in previously treated non-small-cell lung cancer (INTEREST): a randomised phase III trial. *Lancet, 372* (9652), 1809–18.

Kris, M., Natale, R., Herbst, R., et al. (2003). Efficacy of gefitinib, an inhibitor of the epidermal growth factor receptor tyrosine kinase, in symptomatic patients with non-small cell lung cancer: a randomized trial. *JAMA, 290* (16), 2149–58.

Kuo, W., Das, D., Ziyad, S., et al. (2009). A systems analysis of the chemosensitivity of breast cancer cells to the polyamine analogue PG-11047. *BMC Med, 7* (77).

Kwak, E., Camidge, D., Clark, J., et al. (2009). Clinical activity observed in a phase I dose escalation trial of an oral c-met and ALK inhibitor, PF-02341066. *J Clin Oncol, 27* (15s), abstr 3509.

Lacroix, M., and Leclercq, G. (2004). Relevance of breast cancer cell lines as models for breast tumours: an update. *Breast Cancer Res Treat, 83* (3), 249–89.

Lamb, J., Crawford, E., Peck, D., et al. (2006). The Connectivity Map: using gene-expression signatures to connect small molecules, genes, and disease. *Science, 313* (5795), 1929–35.

Lancet Editorial. (2009). A favourable (molecular) signal for personalised medicine. *Lancet, 373* (9657), 2.

Lapointe, J., Li, C., Giacomini, C., et al. (2007). Genomic profiling reveals alternative genetic pathways of prostate tumorigenesis. *Cancer Res, 67* (18), 8504–10.

Lapointe, J., Li, C., van de Rijn, M., et al. (2004). Gene expression profiling identifies clinically relevant subtypes of prostate cancer. *Proc Natl Acad Sci USA, 101* (3), 811–16.

Li, Y., Zou, L., Li, Q., et al. (2010). Amplification of LAPTM4B and YWHAZ contributes to chemotherapy resistance and recurrence of breast cancer. *Nat Med, 16* (2), 214–18.

Liu, Y., and Bodmer, W. (2006). Analysis of P53 mutations and their expression in 56 colorectal cancer cell lines. *Proc Natl Acad Sci USA, 103* (4), 976–81.

Lorenzi, P., Llamas, J., Gunsior, M., et al. (2008). Asparagine synthetase is a predictive biomarker of L-asparaginase activity in ovarian cancer cell lines. *Mol Cancer Ther, 7* (10), 3123–8.

Lorenzi, P., Reinhold, W., Rudelius, M., et al. (2006). Asparagine synthetase as a causal, predictive biomarker for L-asparaginase activity in ovarian cancer cells. *Mol Cancer Ther, 5* (11), 2613–23.

Lynch, T., Bel, D., Sordella, R., et al. (2004). Activating mutations in the epidermal growth factor receptor underlying responsiveness of non-small-cell lung cancer to gefitinib. *N Engl J Med, 350* (21), 2129–39.

Mack, G. (2009). FDA holds court on post hoc data linking KRAS status to drug response. *Nat Biotechnol, 27* (2), 110–12.

Mardis, E., and Wilson, R. (2009). Cancer genome sequencing: a review. *Hum Mol Genet, 18* (R2), R163–8.

McDermott, U., Sharma, S., Dowell, L., et al. (2007). Identification of genotype-correlated sensitivity to selective kinase inhibitors by using high-throughput tumor cell line profiling. *Proc Natl Acad Sci USA, 104* (50), 19936–41.

Mirzoeva, O., Das, D., Heiser, L., et al. (2009). Basal subtype and MAPK/ERK kinase (MEK)-phosphoinositide 3-kinase feedback signaling determine susceptibility of breast cancer cells to MEK inhibition. *Cancer Res, 69* (2), 565–72.

Mok, T., Wu, Y., Thongprasert, S., et al. (2009). Gefitinib or carboplatin-paclitaxel in pulmonary adenocarcinoma. *N Engl J Med, 361* (10), 947–57.

Monti, S., Savage, K., Kutok, J., et al. (2005). Molecular profiling of diffuse large B-cell lymphoma identifies robust subtypes including one characterized by host inflammatory response. *Blood, 105* (5), 1851–61.

Morozova, O., and Marra, M. (2008). Applications of next-generation sequencing technologies in functional genomics. *Genomics, 92* (5), 255–64.

Nardone, R. (2007). Eradication of cross-contaminated cell lines: a call for action. *Cell Biol Toxicol, 23* (6), 367–72.

Nature Editorial. (2008). A bigger picture. *Nature, 455* (7210), 138.

NCBI Science primer: Pharamcogenomics Factsheet. (2004, March 31). Retrieved from NCBI Science Primer: www.ncbi.nil.nih.gov/About/primer/pharm.html

Neve, R., Chin, K., Fridlyand, J., et al. (2006). A collection of breast cancer cell lines for the study of functionally distinct cancer subtypes. *Cancer Cell, 10* (6), 515–27.

OncoPanel™ 240. https://www.eurofinspanlabs.com/Marketing/Microsite/oncopanel%E2%84%A2-240.html.

Pao, W., Wang, T., Riely, G., et al. (2005). KRAS mutations and primary resistance of lung adenocarcinomas to gefitinib or erlotinib. *PLoS Med, 2* (1), e17.

Parker, J., Mullins, M., Cheang, M., et al. (2009). Supervised risk predictor of breast cancer based on intrinsic subtypes. *J Clin Oncol, 27* (8), 1160–7.

Parkinson, D., and Ziegler, J. (2009). Educating for personalized medicine: a perspective from oncology. *Clin Pharmacol Ther, 86* (1), 23–5.

Parkinson, H., Kapushesky, M., Kolesnikov, N., et al. (2009). ArrayExpress update: from an archive of functional genomics experiments to the atlas of gene expression. *Nucleic Acids Res, 37* (Database issue), D868–72.

Parsons, D., Jones, S., Zhang, X., et al. (2008). An integrated genomic analysis of human glioblastoma multiforme. *Science, 321* (5897), 1807–12.

Perou, C., Sørlie, T., Eisen, M., et al. (2000). Molecular portraits of human breast tumours. *Nature, 406* (6797), 747–52.

Potash, J., and Anderson, K. (2009). What's your line? *Clin Cancer Res, 15* (13), 4251.

Potti, A., Dressman, H., Bild, A., et al. (2006). Genomic signatures to guide the use of chemotherapeutics. *Nat Med, 12* (11), 1294–300.

Pratilas, C., Hanrahan, A., Halilovic, E., et al. (2008). Genetic predictors of MEK dependence in non-small cell lung cancer. *Cancer Res, 68* (22), 9375–83.

Raponi, M., Winkler, H., and Dracopoli, N. (2008). KRAS mutations predict response to EGFR inhibitors. *Curr Opin Pharmacol, 8* (4), 413–18.

Rhodes, D., Ateeq, B., Cao, Q., et al. (2009). AGTR1 overexpression defines a subset of breast cancer and confers sensitivity to losartan, an AGTR1 antagonist. *Proc Natl Acad Sci USA, 106* (25), 10284–9.

Rhodes, D., Kalyana-Sundaram, S., Mahavisno, V., et al. (2007). Oncomine 3.0: genes, pathways, and networks in a collection of 18,000 cancer gene expression profiles. *Neoplasia, 9* (2), 166–80.

Rosenwald, A., Wright, G., Leroy, K., et al. (2003). Molecular diagnosis of primary mediastinal B cell lymphoma identifies a clinically favorable subgroup of diffuse large B cell lymphoma related to Hodgkin lymphoma. *J Exp Med, 198* (6), 851–62.

Rubin, M., and Chinnaiyan, A. (2006). Bioinformatics approach leads to the discovery of the TMPRSS2:ETS gene fusion in prostate cancer. *Lab Invest, 86* (11), 1099–102.

Rücker, F., Sander, S., Döhner, K., et al. (2006). Molecular profiling reveals myeloid leukemia cell lines to be faithful model systems characterized by distinct genomic aberrations. *Leukemia, 20* (6), 994–1001.

Schadt, E. (2009). Molecular networks as sensors and drivers of common human diseases. *Nature, 461* (7261), 218–23.

Scholl, C., Fröhling, S., Dunn, I., et al. (2009). Synthetic lethal interaction between oncogenic KRAS dependency and STK33 suppression in human cancer cells. *Cell, 137* (5), 821–34.

Sharma, S., Haber, D., and Settleman, J. (2010). Cell line-based platforms to evaluate the therapeutic efficacy of candidate anticancer agents. *Nat Rev Cancer, 10* (4), 241–52.

Shaw, A., Yeap, B., Mino-Kenudson, M., et al. (2009). Clinical features and outcome of patients with non-small-cell lung cancer who harbor EML4-ALK. *J Clin Oncol, 27* (26), 4247–53.

Shoemaker, R. (2006). The NCI60 human tumour cell line anticancer drug screen. *Nat Rev Cancer, 6* (10), 813–23.

Smith, S., Baras, A., Lee, J., and Theodorescu, D. (2010). The COXEN principle: translating signatures of in vitro chemosensitivity into tools for clinical outcome prediction and drug discovery in cancer. *Cancer Res, 70* (5), 1753–8.

Soda, M., Choi, Y., Enomoto, M., et al. (2007). Identification of the transforming EML4-ALK fusion gene in non-small-cell lung cancer. *Nature, 448* (7153), 561–6.

Sørlie, T., Perou, C., Tibshirani, R., et al. (2001). Gene expression patterns of breast carcinomas distinguish tumor subclasses with clinical implications. *Proc Natl Acad Sci USA, 98* (19), 10869–74.

Sos, M., Michel, K., Zander, T., et al. (2009). Predicting drug susceptibility of non-small cell lung cancers based on genetic lesions. *J Clin Invest, 119* (6), 1727–40.

Speer, R., Wulfkuhle, J., Espina, V., et al. (2007). Development of reverse phase protein microarrays for clinical applications and patient-tailored therapy. *Cancer Genomics Proteomics, 4* (3), 157–64.

Staudt, L., and Dave, S. (2005). The biology of human lymphoid malignancies revealed by gene expression profiling. *Adv Immunol, 87*, 163–208.

Subramanian, A., Tamayo, P., Mootha, V., et al. (2005). Gene set enrichment analysis: a knowledge-based approach for interpreting genome-wide expression profiles. *Proc Natl Acad Sci USA, 102* (43), 15545–50.

Thatcher, N., Chang, A., Parikh, P., et al. (2005). Gefitinib plus best supportive care in previously treated patients with refractory advanced non-small-cell lung cancer: results from a randomised, placebo-controlled, multicentre study (Iressa Survival Evaluation in Lung Cancer). *Lancet, 366* (9496), 1527–37.

Tomlins, S., Rhodes, D., Perner, S., et al. (2005). Recurrent fusion of TMPRSS2 and ETS transcription factor genes in prostate cancer. *Science, 310* (5748), 644–8.

Tothill, R., Tinker, A., George, J., et al. (2008). Novel molecular subtypes of serous and endometrioid ovarian cancer linked to clinical outcome. *Clin Cancer Res, 14* (16), 5198–208.

Verhaak, R., Hoadley, K., Purdom, E., et al. (2010). Integrated genomic analysis identifies clinically relevant subtypes of glioblastoma characterized by abnormalities in PDGFRA, IDH1, EGFR, and NF1. *Cancer Cell, 17* (1), 98–110.

Virtanen, C., Ishikawa, Y., Honjoh, D., et al. (2002). Integrated classification of lung tumors and cell lines by expression profiling. *Proc Natl Acad Sci USA, 99* (19), 12357–62.

Wirapati, P., Sotiriou, C., Kunkel, S., et al. (2008). Meta-analysis of gene expression profiles in breast cancer: toward a unified understanding of breast cancer subtyping and prognosis signatures. *Breast Cancer Res, 10* (4), R65.

Wood, L., Parsons, D., Jones, S., et al. (2007). The genomic landscapes of human breast and colorectal cancers. *Science, 318* (5853), 1108–13.

Young, S., Pilarski, R., Donenberg, T., et al. (2009). The prevalence of BRCA1 mutations among young women with triple-negative breast cancer. *BMC Cancer, 9* (86).

Yu, J., Sieuwerts, A., Zhang, Y., et al. (2007). Pathway analysis of gene signatures predicting metastasis of node-negative primary breast cancer. *BMC Cancer, 7*, 182.

Zhan, F., Huang, Y., Colla, S., et al. (2006). The molecular classification of multiple myeloma. *Blood, 108* (6), 2020–8.

Index